100 Years of Gravity and Accelerated Frames

The Deepest Insights of Einstein and Yang–Mills

ADVANCED SERIES ON THEORETICAL PHYSICAL SCIENCE
A Collaboration between World Scientific and Institute of Theoretical Physics

---

Vol. 1: Yang-Baxter Equation and Quantum Enveloping Algebras
(Zhong-Qi Ma)

Vol. 2: Geometric Methods in the Elastic Theory of Membrane in Liquid Crystal Phases
(Ouyang Zhong-Can, Xie Yu-Zhang & Liu Ji-Xing)

Vol. 4: Special Relativity and Its Experimental Foundation
(Yuan Zhong Zhang)

Vol. 6: Differential Geometry for Physicists
(Bo-Yu Hou & Bo-Yuan Hou)

Vol. 7: Einstein's Relativity and Beyond
(Jong-Ping Hsu)

Vol. 8: Lorentz and Poincaré Invariance: 100 Years of Relativity
(J.-P. Hsu & Y.-Z. Zhang)

Vol. 9: 100 Years of Gravity and Accelerated Frames: The Deepest Insights of Einstein and Yang–Mills
(J.-P. Hsu & D. Fine)

Advanced Series on Theoretical Physical Science Volume 9

100 Years of Gravity and Accelerated Frames

The Deepest Insights of Einstein and Yang–Mills

Editors

Jong-Ping Hsu

Dana Fine

University of Massachusetts Dartmouth, USA

World Scientific

NEW JERSEY • LONDON • SINGAPORE • BEIJING • SHANGHAI • HONG KONG • TAIPEI • CHENNAI

Published by

World Scientific Publishing Co. Pte. Ltd.
5 Toh Tuck Link, Singapore 596224
USA office: 27 Warren Street, Suite 401-402, Hackensack, NJ 07601
UK office: 57 Shelton Street, Covent Garden, London WC2H 9HE

Library of Congress Cataloging-in-Publication Data
100 years of gravity and accelerated frames : the deepest insights of Einstein and Yang–Mills / editor, Jong-Ping Hsu; advisory editor, Dana Fine.
p. cm. -- (Advanced series on theoretical physical science; v. 9)
Includes bibliographical references.
ISBN 981-256-335-0 (alk. paper)
1. Gravitation. 2. Relativity (Physics). 3. Einstein field equations. 4. Yang–Mills theory. I. Title: One hundred years of gravity and accelerated frames. II. Hsu, J. P. (Jong-Ping). III. Fine, Dana. IV. Series.

QC178.A15 2005
530.11--dc22

2005050077

British Library Cataloguing-in-Publication Data
A catalogue record for this book is available from the British Library.

Printed in Singapore by B & JO Enterprise

Deepest Insights

The Deepest Insights of Einstein and Yang-Mills

The Deepest Insights of Einstein and Yang-Mills

The Deepest Insights of Einstein and Yang-Mills

The Deepest Insights of Einstein and Yang-Mills

To

Arthur Fine

Hu/Shih and Tsien/Shue-Sen

The Deepest Insights of Einstein and Yang-Mills

The Deepest Insights of Einstein and Yang-Mills

The Deepest Insights of Einstein and Yang-Mills

The Deepest Insights of Einstein and Yang-Mills

Deepest Insights

"In 1953-1954, I was visiting Brookhaven and Bob was my office mate. We discussed many things in physics, from the experimental results pouring out of the new Cosmotron, to theoretical topics like renormalization and the Ward identity. It was in that year that we found the very elegant and unique generalization of Maxwell's equation. We were pleased by the beauty of the generalization, but neither of us had anticipated its great impact on physics 20 years later."

C. N. Yang,
in "Remembering Robert L. Mills" by Samuel L. Marateck,
Physics Today, p. 14, October 2003.

Robert Mills and Kitty,
the oldest of his 5 kids.

A Hair perhaps divides the False and True;
Yes, and a single Alif were the Clue-
Could you but find it - to the Treasure-house,
and peradventure to The Master too.

"The Rubaiyat"
Omar Khayyam
(transl. by Edward Fitzgerald)

Preface

This book is a collection of papers and writings from the past 100 years on ideas and problems related to gravity, gauge fields and accelerated frames. The grand triumphs of Einstein's theory of gravity and Yang-Mills' theory in physics are well known. It is believed that both theories are based on the principle of 'gauge invariance,' although not on the same kind of action. Einstein's theory is linear in spacetime curvature, while Yang-Mills' theory is quadratic in gauge curvature. Now, at the dawn of the 21st century, invariance principles in physics have transcended the kinematical and dynamical contexts from which they originated to become the foundation of our understanding of the physical world. Using this framework of invariance principles, this book surveys the development of gravitational and Yang-Mills fields, as well as spacetime transformations of accelerated frames. It also attempts to reveal the problems and limitations of various formulations of gravitational and Yang-Mills fields. The intent is to enlarge and broaden the reader's views on the subjects.

As TIME magazine's person of the 20th century (cf. TIME magazine), Einstein's contributions to physics are arguably incomparable, aside from Newton's. The gravitational force and accelerated frames were two ingredients in the young Einstein's 'happiest thoughts' in 1907. The simple thought that 'If a person falls freely he will not feel his own weight,' made a deep impression on him and impelled him toward a successful theory of gravitation. Unfortunately, accelerated spacetime transformations for non-inertial frames have still not been well developed. However, they are important because one cannot claim to have a complete understanding of the physical world, especially the basic gravitational and Yang-Mills fields, if one understands physics only from the viewpoint of the special and limited class of inertial frames. Strictly speaking, all real frames of reference in the physical world are non-inertial because of the long range of the gravitational force. In particular, when one talks about an inherent property of nature (e.g., values of fundamental constants such as the fine structure constant and the speed of light), a reasonable criterion is that the property must be present in both inertial and non-inertial frames. In this sense, the book suggests that the present understanding of gravitational and Yang-Mills fields is far from complete.

The formulations of the gravitational and Yang-Mills theories are both an effect and a cause of scientific development in experiment and theory. Progress in physics is made through the collective effort of many physicists. The community of physicists is like the thousand-hand Guan-Yin: Each hand accomplishes only a partial or small task, yet the overall accomplishment is enormous. As we shall see in this volume, in the pursuit of physical laws, the right track has often been discovered only after many failures by well-known and not-so-well-known pathfinders.

Nowadays, the spectacular success of Einstein and Yang-Mills' profound thoughts is often emphasized while the lessons of the struggle in their birth and development is lost. Furthermore, there is little chance for making progress simply by going over their success repeatedly.

The aim here is to present some of the leading ideas and problems discussed by physicists and mathematicians, highlighting three aspects:

(1) the idea of gravity as a Yang-Mills field, first discussed by Utiyama;
(2) the problems of quantum gravity, discussed by Feynman, Dyson and others;
(3) spacetime properties and the physics of particles and fields in accelerated frames.

It is hoped that the present volume will bring some of the unfulfilled aspects of the profound thoughts to the attention of physicists and mathematicians of the 21st century. For various reasons, research in the areas of spacetime symmetry and special relativity are sometimes discouraged by general editorial policy. Fortunately, so far this is not true in the cases of general relativity and Yang-Mills theory. If the deepest insights of Einstein and Yang-Mills can inspire its readers to pursue these subjects further, the chief purpose of the book will have been achieved.

We are grateful that Chairman M. Ninomiya of the Progress of Theoretical Physics, Acta Physica Polonica B, Elsevier Ltd, and Editorial Administrator R. W. Brown of the Annals of the New York Academy of Sciences freely granted us permission to reprint their papers. We would like to thank L. Hsu, H. L. Chen, N. Cleffi and E. M. Winiarz for their help. This book was supported in part by the Prof. George Leung Memorial Fund of the University of Massachusetts Dartmouth Foundation, the Potz Science Fund, and the World Scientific Publishing Company.

Hsu/Jong-Ping

Contents

Acknowledgements

'The Mathematical Principles of Natural Philosophy' (1687) (Extract)
From The Principia by Isaac Newton, translated by Andrew Motte (1729) (Amherst, NY; Prometheus Books). Published 1995.

'Hypotheses Concerning Gravitation' (Extract)
Extracted from 'The Dynamics of the Electron' by H. Poincaré, Rend. Circ. Mat. Palermo **21**, 129 (1906), with comments by A. A. Logunov, translated by G. Pontecorvo. Reprinted with permission from A. A. Logunov, Dubna Joint Institute for Nuclear Research, published 2001.

'On the Relativity Principle and the Conclusions Drawn from It' (Extract)
EINSTEIN, ALBERT; THE COLLECTED PAPERS OF ALBEERT EINSTEIN © 1987–2004 Hebrew University and Princeton University Press. Reprinted by permission of Princeton University Press.

'Outline of a Generalized Theory of Relativity and of a Theory of Gravitation'
EINSTEIN, ALBERT; THE COLLECTED PAPERS OF ALBEERT EINSTEIN © 1987–2004 Hebrew University and Princeton University Press. Reprinted by permission of Princeton University Press.

'The Foundation of the General Theory of Relativity'
Reprinted from A. Einstein, in *The Principles of Relativity* (Translated by W. Perrett and G. B. Jeffery, Methuen and Company, 1923)

'The Foundation of Physics'
Translated from D. Hilbert, "Die grundlagen der Physik. (Erste Mitteilung)." Goett. Nachr. 395 (1915), by Dana Fine.

'On a Generalization of the Concept of Riemann Curvature and Spaces with Torsion'
Translated from E. Cartan, C. R. Acad. Sci. (Paris) **174**, 597 (1922) by Arthur Fine.

'Formation of the Stars and Development of the Universe'
Reprinted with permission from P. Jordan, Nature (London) **164**, 637 (1949). © 1949 Macmillan Magazines Ltd.

'On the Physical Interpretation of P. Jordan's Extended Theory of Gravitation' (Extract)
Translated from M. Fierz, Helv. Phys. Acta **29**, 128 (1956) by Dana Fine.

'Mach's Principle and a Relativistic Theory of Gravitation'
Reprinted with permission from C. Brans and R. H. Dicke, Phys. Rev. **124**, 925 (1961). © 1961 The American Physical Society.

'Conservation of Isotopic Spin and Isotopic Gauge Invariance'
Reprinted with permission from C. N. Yang and R. L. Mills, Phys. Rev. **96**, 191 (1954). © 1954 The American Physical Society.

'Conservation of Heavy Particles and Generalized Gauge Transformations'
Reprinted with permission from T. D. Lee and C. N. Yang, Phys. Rev. **98**, 1501 (1955). © 1955 The American Physical Society.

'Invariant Theoretical Interpretation of Interaction'
Reprinted with permission from R. Utiyama, Phys. Rev. **101**, 1597 (1956).

'Lorentz Invariance and the Gravitational Field'
Reprinted with permission from T. W. B. Kibble, J. Math. Phys. **2**, 212 (1961).

'On Homogeneous Gravitational Fields in the General Theory of Relativity and the Clock Paradox'
Reprinted with permission from C. Møller, Danske Vid. Sel. Mat-Fys. **20**, No. 19 (1943).

'Physical Consequences of a Co-ordinate Transformation to a Uniformly Accelerating Frame'
Reprinted with permission from T. Fulton, R. Rohrlich and L. Witten, Nuovo Cimento **xxvi**, 652 (1962).

'The Clock Paradox in the Relativity Theory'
Reprinted with permission from T. Y. Wu and Y. C. Lee, Int. J. Theor. Phys. **5**, 307 (1972).

'Four-dimensional Symmetry of Taiji Relativity and Coordinate Transformations Based on a Weaker Postulate for the Speed of Light' (Extract)
Reprinted with permission from J. P. Hsu and L. Hsu, Nuovo Cimento, **112**, 575 (1997).

'Generalized Lorentz Transformations for Linearly Accelerated Frames with Limiting Four-Dimensional Symmetry'
Reprinted with permission from J. P. Hsu and L. Hsu, Chinese Journal of Physics **35**, 407 (1997).

'Generalizing Lorentz Transformations for Accelerated Frames and Their Physical Implications'
Contributed by D. T. Schmitt and Tobias Kleinschmidt. Based on a paper submitted to the Int. J. Modern Phys. (to be published).

'Quantum Theory of Gravitation'
Reprinted with permission from R. P. Feynman, Acta Physica Polonica **24**, 697 (1963).

'Quantum Theory of Gravity. II. The Manifestly Covariant Theory' (Extract)
Reprinted with permission from B. S. DeWitt, Phys. Rev. **162**, 1195 (1967).

'Quantum Theory of Gravity, III. Applications of the Covariant Theory'
Reprinted with permission from B. S. DeWitt, Phys. Rev. **162**, 1239 (1967).

'Feynman Diagram for the Yang–Mills Field'
Reprinted from L. D. Faddeev and V. N. Popov, Phys. Lett. **25B**, 29 (1967).

'Feynman Rules for Electromagnetic and Yang–Mills Fields from the Gauge-Independent Field-Theoretic Formalism' (Extract)
Reprinted with permission from S. Mendelstam, Phys. Rev. **175**, 1580 (1968).
© 1968 The American Physical Society.

'*S* Matrix for Yang–Mills and Gravitational Fields' (Extract)
Reprinted with permission from E. S. Fradkin and I. V. Tyutin, Phys. Rev. **D2**, 2841 (1970).
© 1970 The American Physical Society.

'Missed Opportunities' (Extract, with a brief comment of the author)
Introduction, General Coordinate Invariance
Reprinted with permission from F. J. Dyson, Bull. Am. Math. Soc. **78**, 635 (1972).
© 1972 American Mathematical Society.

'Extended Translation Invariance and Associated Gauge Fields'
Reprinted with permission from K. Hayashi and T. Nakano, Prog. Theor. Phys. **38**, 491 (1967).
© 1967 Prog. Theor. Phys.

'Gravitational Field as a Generalized Gauge Field'
Reprinted with permission from R. Utiyama and T. Fukuyama, Prog. Theor. Phys. **45**, 612 (1971). © 1971 Prog. Theor. Phys.

'Integral Formalism for Gauge Fields'
Reprinted with permission from C. N. Yang, Phys. Rev. Letters **33**, 445 (1974).
© 1974 The American Physical Society.

'Yang's Gravitational Field Equations'
Reprinted with permission from Wei-Tou Ni, Phys. Rev. Letters **35**, 319 (1975).
© 1975 The American Physical Society.

'Einstein Lagrangian as the Translational Yang–Mills Lagrangian'
Reprinted with permission from Y. M. Cho, Phys. Rev. **14**, 2515 (1976).
© 1976 The American Physical Society.

'De Sitter and Poincaré Gauge-Invariant Fermion Lagrangians and Gravity'
Reprinted from J. P. Hsu, Phys. Letters, **119B**, 328 (1982).
© 1982, with permission from Elsevier.

'Fixation of Coordinates in the Hamiltonian Theory of Gravitation'
Reprinted with permission from P. A. M. Dirac, Phys. Rev. **114**, 924 (1959).
© 1959 The American Physical Society.

'New General Relativity' (with Addendum)
Reprinted with permission from K. Hayashi and T. Shirafuji, Phys. Rev. **19**, 3524 (1979).
© 1979 The American Physical Society.

'Relativistic Theory of Gravitation'
Reprinted with permission from A. A. Logunov and M. A. Mestvirishvili, Prog. Theor. Phys. **74**, 31 (1985). © 1985 Prog. Theor. Phys.

'Yang–Mills Gravity: A Union of Einstein–Grossmann Metric and Yang–Mills Tensor Fields in Flat Spacetime with Translation Symmetry'
Contributed by J. P. Hsu. Based on a paper in Annual Report of the National Center for Theoretical Sciences (2005).

'Empirical Foundations of the Relativistic Gravity'
Reprinted with permission from Wei-Tou Ni, Int. J. Mod. Phys. D (to be published) © 2005 World Scientific.

'Binary Pulsars and Relativistic Gravity'
Reprinted with permission from J. H. Taylor, Jr., Rev. Mod. Phys. **66**, 711 (1994). © 1994 The American Physical Society.

'Concept of Nonintegrable Phase Factors and Global Formulation of Gauge Fields' (Extract)
Reprinted with permission from T. T. Wu and C. N. Yang, Phys. Rev. **D12**, 3845 (1975). © 1975 The American Physical Society.

'Gauge Fields'
Reprinted with permission from Robert Mills, Am. J. Phys. **57**, 493 (1989), © 1989 American Institute of Physics.

'Magnetic Monopoles, Fiber Bundles, and Gauge Fields'
Reprinted with permission from C. N. Yang , Ann. of the New York Academy of Sciences, **294**, 86 (1977). © 1977 New York Academy of Sciences.

'Gauge Theory: Historical Origins and Some Modern Developments'
Reprinted with permission from H. L. O'Raifeartaigh and N. Straumann, Rev. Mod. Phys. **72**, 1 (2000). © 2000 The American Physical Society.

'String Theory as a Generalization of Gauge Symmetry'
Contributed by Pei-Ming Ho, National Taiwan University.

'The Cosmological Constant Problem'
Reprinted with permission from S. Weinberg, Rev. Mod. Phys. **61**, 1 (1989). © 1989 The American Physical Society.

'The Cosmological Constant and Dark Energy' (Extract)
Reprinted with permission from P. J. E. Peebles and B. Ratra, Rev. Mod. Phys. **75**, 559 (2003). © 2003 The American Physical Society.

'Marcel Grossmann (1878–1936)'
Contributed by Jong-Ping Hsu and Dana Fine, University of Massachusetts Dartmouth.

'Remembering Robert L. Mills'
Reprinted with permission from S. L. Marateck, Phys. Today, Oct. 2003. © 2003 American Institute of Physics.

Marcel Grossmann (Photo courtesy of Anna and Carlo Révay-Grossmann)

Bryce DeWitt (Photo courtesy of Cecile DeWitt-Morette)

Robert Mills (Photo courtesy of Mrs. Mills)

Chen Ning Yang and Tai Tsun Wu in Leiden (1984) (Photo courtesy of Judy Wong)

T. D. Lee and C. N. Yang at the Institute for Advanced Study (Courtesy of the Archives of the Institute for Advanced Study, Princeton, New Jersey, USA)

Remarks on the Development of the Gravitational and Yang-Mills Fields, and Accelerated Frames

Jong-Ping Hsu and Dana Fine

1 Introduction

In the past 100 years, the ideas of general coordinate invariance and of gauge invariance have played leading roles in the investigation of the fundamental interactions of nature (gravitational, weak, electromagnetic and strong interactions). Physics Today calls 2005 the 'World Year of Physics' "to celebrate physics in its broadest context as part of the human experience" and "to raise awareness of physics within the broad population."[1] It also happens to be roughly the 50th birth-year of Yang-Mills theory, the 100th birth-year of Einstein's happiest thought, as well as the 100th anniversary of the publication of the celebrated theory of special relativity. [2] It is thus a fitting time to review Einstein and Yang-Mills' ideas, their impacts on both physics and mathematics, and some open problems in related areas.

In 1907, a simple thought flashed through young Einstein's mind:

"I was sitting in a chair in the patent office at Bern when all of a sudden a thought occurred to me: 'If a person falls freely he will not feel his own weight.' I was startled. This simple thought made a deep impression on me. It impelled me toward a theory of gravitation."

He told this story in his Kyoto lecture. [3] This 'happiest thought,' as Einstein called it, involves two ingredients:

i. gravitational force, and

ii. accelerated frames.

These two related physical subjects were first discussed by Einstein in his 1907 review paper entitled "On the Relativity Principle and the Conclusions Drawn From it." [4] Apparently, these problems had engrossed Einstein's thought shortly after he published his landmark paper on special relativity 100 years ago. Einstein realized that his knowledge of the relevant mathematics was inadequate, so he turned to his mathematician friend and university colleague Grossmann. Grossmann's help proved significant for Einstein in realizing his dream theory.

Around 1947, a graduate student at the University of Chicago, C. N. Yang, also had a simple thought: [5]

"Since Maxwell's equations and the conservation of electric charge are intimately related, and the conservation of isotopic spin has been established by experiments, should it imply another kind of gauge field?

Yang tried and tried, but he was just unable to overcome a key difficulty. [6] (See sec. 3) Such efforts, and failures, are typical, everyday occurrences in graduate students' offices. It seems this thought "made a deep impression" on him. Seven years later, it also "impelled" Yang and Mills toward a non-Abelian gauge theory, when a spark from their discussions grew to illuminate the key difficulty.

These two instances, and many others, raise awareness of a remarkable fact that new thoughts and ideas often emerge from refreshingly young minds. As Nobert Wiener said it eloquently to young mathematicians: "You must devote this brief springtime of top creative ability to the discovery of new fields and new problems, of such richness and compelling character that you can scarcely exhaust them in your life." This goal appears to apply equally well to all young researchers.

2 Gravity

Historically, Newton created the basic framework for understanding the static property of gravitational force in his grand PRINCIPIA (1686): the inverse-square law of the universal gravity and the laws of motion. Newton achieved an extraordinary unification, perhaps the first in physics: he demonstrated that the same force and laws of motion apply in the celestrial and the sub-lunar spheres. The style of Newton's PRINCIPIA, follows closely that of ELEMENTS by a pioneer and profound mathematical thinker Euclid. For example, book III of PRINCIPIA started from 'rules of reasoning in philosophy', introduced 'phenomena', and followed these with 'proposition.' [7] Newton proposed the first scientific understanding of the solar system based on his universal law for (static) gravitational force and his powerful general methods of mathematics.

About 220 years later, Poincaré investigated the Lorentz invariant and kinematic properties of gravity in his comprehensive paper on relativity finished in 1905. After he discussed and derived all essential results of special relativity for mechanics and electrodynamics, he reached an insightful conclusion that the gravitational action propagates with the speed of light, based on the invariants of the Lorentz group. [8] Poincaré is known for his broad and universal mind and for his contributions to mathematics and physics, comparable to those of Gauss.

With his persistence and ingenuity, Einstein's 'happiest thought' of 1907 eventually lead to arguably the greatest leap forward in the human endeavor to understand the universe, "The Foundation of the General Theory of Relativity." This followed 8 years of hard work and about 30 "not-completely-correct" papers on gravity and/or general relativity. [9] The result is Einstein's splendid equation for gravitation which has been extensively discussed and tested by experiments.

Einstein conceived and worked on the idea of general coordinate invariance during 1911-12. His knowledge of mathematics was inadequate to express his radically new ideas, so he sought help, pleading to his mathematician friend: "Grossmann, you must help me, or else I'll go crazy!"; thus started one of the most beautiful collaborations between two scientists in different disciplines. [10] Einstein's insight was that the law of gravity must be invariant under arbitrary spacetime coordinate transformations (one-to-one and twice-differentiable). This idea implies that the physics of gravity should be formulated and understood in any frame of reference, and it eventually led to a theory of gravity which revolutionized our concepts of the physical universe, endowing the universe with pseudo-Riemannian geometry. The impact on mathematics can be seen from the participation of development of Einstein's idea by such brilliant mathematicians as Hilbert, Cartan, Levi-Civita, Weyl and others. Indeed the impact went well beyond the sciences to include literature and art, which is part of how Einstein became an international celebrity and TIME magazine's "person of the century."

In 1915, after attending Einstein's talks and many discussions and extensive correspondence, Hilbert was able to grasp the essence of Einstein's idea and express it through an invariant action involving the linear scalar curvature. He finished the paper "The Foundation of Physics" containing this invariant action about a week before Einstein completed his landmark paper. [11] Hilbert was a newcomer in this field, and one can feel a fierce competition. Indeed, their relation appears to have been strained for a short period of time. Presumably this was due to their difference in philosophy (action principle versus detailed dynamical analysis) rather than a dispute over priority. In fact, Hilbert clearly credits Einstein with the idea for the theory. Apparently, Einstein liked Hilbert's method of deriving the gravitational equation 'from one single principle of variation.' He also published a paper "Hamilton's Principle and the General Theory of Relativity" in 1916. The recent public dissemination of a hand-edited galley proof of Hilbert's paper has shed new light on the development of Hilbert's formulation and its relation to Einstein's, as discussed in detail in [12]. Hilbert's treatment of gravity (and electromagnetism) was based on an elegant invariant formalism which he argued mathematically was unique given, as axioms, the requirement of general covariance and some reasonable additions assumptions. [11] He even believed what we would now call the "theory of everything" could be constructed on the basis of an axiomatically-determined geometrically-invariant action. This situation resembles the way the mathematician Poincaré grasped the essence of relativity principle in 1905 through his complete understanding of the symmetry group of the Lorentz transformations. He derived, for the first time, the invariant law for the motion of a charged particle by using a Lorentz invariant action. [8] The young Einstein was unable to do so; in his 1905 landmark paper he only obtained an approximate and non-invariant equation for the motion of charged particle with small accelerations. The principle of least action is now a powerful and standard method for the formulations of physical theories.

Furthermore, Cartan recognized in 1922 an "unsatisfactory feature" in Einstein's equation of gravity, namely, the energy-momentum tensor does not have geometrical meaning. [13] He showed how, in an Einstein universe with a given ds^2, the energy tensor attached to each volume element of that universe can be defined geometrically. Following this line of research, he published a paper 'On a generalization of the concept of Riemann curvature and spaces with torsion.' [14] Cartan's work created the only satisfactory mathematical framework for physicists to be able to introduce fermions or spinors into Einstein's theory. This is the only known way for fermions to be coupled to gravitational field according to the requirement of general coordinate invariance. All these examples clearly show the needs of collaborations between physicists and mathematicians to develop a physical theory.

In some sense, Einstein's idea of general coordinate invariance suggests a drastic new approach in physics: namely, the geometrization of all physical fields. If one follows this approach to treat the electromagnetic force, which is velocity-dependent, one might choose to employ the Finsler geometry rather than the Riemann geometry. The reason is that the fundamental metric tensors of the Finsler geometry depend on both position and "velocity" (i.e., the differentials of coordinates). [15] It appears that, so far, all attempts to geometrize the classical electromagnetic field has not been successful, let alone quantum electrodynamics and Einstein's unified theory based on Riemannian geometry.

Einstein's tremendously successful theory of gravity also presents a huge problem in physics. When Dyson gave the Gibbs lectures on "Missed Opportunities" under the auspices of the American Mathematical Society in 1972, he stressed that the most glaring incompatibility of concepts in contemporary physics is that between the principle of general coordinate invariance and all quantum-mechanical and quantum-field-theoretic descriptions of nature. [16] Such an incompatibility is intimately related to the difficulty of quantization in curved spacetime. As a perturbative theory in flat spacetime Einstein's theory is very complicated and is not renormalizable. [17] This is of course a challenge to mathematicians and physicists, as Bohr was fond of saying: "How wonderful that we have met with a paradox. Now we have some hope of making progress."

3 Yang-Mills Fields and Gauge Theory of Gravity

In the decades from the 1930's to the 1950's, most physicists were excited by new discoveries in quantum mechanics for atomic structure, nuclear physics, elementary particles and quantum fields. Gravitational theory was no longer in the main stream of physics research. Around 1940, experiments of strong interaction indicated the existence of a new conservation law of isospin (or isotopic spin). The concept of isospin was originally introduced by Heisenberg in 1932 as a convenient mathematical representation to characterize the two states of the nucleon–'isospin up' as proton and 'isospin down' as neutron, similar to the usual electron spin states. Nevertheless, it proves to be a useful and good physical quantum number for strong interactions, but not for all interactions.

The story of the celebrated Yang-Mills collaboration is briefly as follows: Robert Mills was a research associate from 1953 to 1955 at Brookhaven National Laboratory, while C. N. Yang joined the Institute for Advanced Study in Princeton as a postdoc in 1949 and became a professor in 1955. Mills shared an office with Yang who visited the Laboratory from 1953 to 1954. They were both young and dreamed about physics. Both were at the right place and at the right time. They earnestly discussed many things in physics, including the experimental data flowing out of the new Cosmotron (then, at 3GeV, the world's largest proton accelerator) in the National Laboratory, and possible implications of isospin conservation. During their discussions, they had the idea of adding a new term of the form $\epsilon^{abc} b_\mu^b b_\nu^c$ to the field strength $f_{\mu\nu}^a$. The gauge-covariant derivative D_μ and field strength $f_{\mu\nu}^a$ of the isovector field b_μ^a are given by

$$D_\mu = \partial_\mu - i g b_\mu^a \tau^a, \text{ and} \tag{1}$$

$$f_{\mu\nu}^a = \partial_\nu b_\mu^a - \partial_\mu b_\nu^a - g\epsilon^{abc} b_\mu^b b_\nu^c, \tag{2}$$

where $a, b, c, = 1, 2, 3$, and τ^a's are the constant matrix representations of the isospin SU(2) group. This addition solved Yang's original difficulty, and the rest is history. This type of new term in the field strength is essential for all non-Abelian gauge fields. Adding this term to the gauge field strength is comparable in its significance to adding the displacement current term to the original Ampere law by Maxwell, which made the electromagnetic theory consistent and complete.

Nevertheless, this non-trivial and unique generalization of Maxwell's equations did not attract much attention initially. When Dyson wrote the article "Innovation in Physics" in 1958, he failed to mention the discovery of non-Abelian gauge fields by Yang and Mills. [16] Even when Yang himself gave a talk on "The Future of Physics" at MIT in 1961, he just asked a related question : "What are the basis of the invariance under charge conjugation, and the invariance under isotopic spin rotation, both of which, unlike space-time symmetries, are known to be violated?" Yang did not mention the non-Abelian gauge fields.

There were in fact serious problems associated with the original Yang-Mills theory. One of the problems is that Yang-Mills field is massless, while all observed particles with strong interactions have mass. This incompatibility between Yang-Mills field theory and experiment became the central issue when Yang was invited by Oppenheimer to give a talk at the Institute for Advanced Study, Princeton, on this work. Pauli persistently and repeatedly asked Yang the question regarding the mass of the new field. Yang could not give him a satisfactory answer. Pauli made a strong criticism, and the talk was almost stopped by him. [19] It was well known that Pauli was super-critical on every new physical idea, including Einstein's ideas. In fact, Pauli himself had a similar idea and investigated the same problem before. He also obtained the expression for the field strength of the new gauge field. According to Pauli's colleague Gulmanelli, [20] Pauli gave up the whole investigation because the new field quantum was a massless vector particle and, hence, contradicted experiments. To all practical physicists at that time, it was obvious that the Yang-Mills theory with zero mass field did not exist in nature, because a zero mass field would have been easily detected in strong-interaction experiments.

Here, one sees clearly the difference between Yang and Pauli regarding their tastes in physics research. Yang said later: "We did not know how to make the theory fit experiment. It was our judgment, however, the beauty of the idea alone merited attention."[21] Even so, when Mills passed away in 2003, Yang, recalling their creation of the gauge theory, acknowledged "We were pleased by the beauty of the generalization, but neither of us had anticipated its great impact on physics 20 years later."[22]

Although most physicists ignored Yang-Mills' work at that time, T. D. Lee and R. Utiyama were immediately attracted to their idea. Lee and Yang proposed 'generalized gauge transformations' to understand the conservation of heavy particle (or "baryon") number in 1955. They argued that such a new conservation law implies the existence of a new long- range repulsive force between baryons (e.g., protons and neutrons). The corresponding force would be attractive between baryons and anti-baryons. They used Eötvös experiment to estimate the strength of such a new force. They found that it to be about 10^{-5} times smaller than that of the gravitational force. Similar consideration of the conservation law for lepton number lead to a new and very weak long-range repulsive force between electrons. It is quite possible that such a 'Lee-Yang force' between baryons in galaxies played a role similar to the so-called 'dark energy' and has direct relevance to the observed accelerating expansion of the universe. [23]

One year later in 1956, Utiyama generalized Yang-Mills discussion of the SU(2) gauge group to a general group with n gauge functions, and grasped the essential idea expressed in his paper "Invariant Theoretical Interpretation of Interaction." He stressed that the form of the interactions between some well known fields can be uniquely determined by postulating invariance under a certain group of transformations.[24] Furthermore, he applied his generalized viewpoint of Yang-Mills fields to discuss gravity. Thus Utiyama opened a new approach to gravity, which was later termed 'Gauge Theory of Gravity.' Over the years since, many people, including Kibble, Hayashi, Nakano, Utiyama, Fukuyama, Yang, Thompson, Ni, Hsu, Cho, Ashtekar and others continue to explore such gauge theories of gravity. These discussions are based on the Lorentz group, the Poincaré group, the de Sitter group, spacetime translation group and other groups. [25] We shall suggest below that, from the viewpoint of symmetry and its related conservation law, the spacetime translation group appears to be the simplest and the most natural candidate for the gauge group of gravity. The problem of gravity as a Yang-Mills field is still open.

A few years later in 1960, J. J. Sakurai [26] took advantage of Yang-Mills' gauge invariant principle to propose the strongly interacting vector mesons for discussing universality and conserved current. But the presence of masses of vector mesons in strong interaction physics violates gauge invariance. The mass problem in the Yang-Mills theory was later solved by Higgs' mechanism of spontaneous symmetry breaking, which enabled Weinberg to make a wonderful prediction of gauge boson masses in 1967. [27] Thus Yang-Mills fields took center stage in physics only gradually.

The first breakthrough in the development of Yang-Mills theory did not address the mass problem. It came from Feynman's 1962 lecture at the "Conference on Relativistic Theories of Gravitation" in Poland. This is a rare opportunity to see the detailed process of discovery in a genius's mind: Feynman's discovery of a serious problem through direct calculations and his original idea of using a 'fictitious particle' to fix it. Presumably, such a crazy and not-even-half-baked idea may have hard time to be published in "the damned Physical Review," (Feynman's term) if the paper were submitted. A conference was no doubt the ideal forum for him to present such an original and crazy idea and to challenge directly the experts in the audience. Feynman considered Einstein's gravitational theory with a scalar field as the source. This grand master of calculations with rules and diagrams evaluated amplitudes at the one-loop level and encountered a serious problem: namely, the resultant amplitudes do not satisfy unitarity (a necessary condition for their interpretation as determining probabilities). He was worried and puzzled. At the suggestion of Gell-Mann, he looked into the Yang-Mills theory which, like Einstein's theory of gravity, is non-linear, but is otherwise much simpler. Feynman was happy to discover that Yang-Mills theory has the same problem! He remarked that the unitarity difficulty "should have been noticed by meson physicists who had been fooling around the Yang-Mills theory. They had not noticed it because they are practical, and the Yang-Mills theory with zero mass obviously does not exist, ..." Based on the simpler Yang-Mills theory, Feynman was able to find a suitable and crazy way to cure the difficulty: "... You must subtract from the answer, the result that you get by imagining that in the ring which involves only a graviton going around, instead you calculate with a different particle going around, an artificial, dopey particle is coupled to it. It is a vector particle, artificially coupled to the external field, so designed as to correct the error in this one." Nevertheless, Feynman was unable to solve the problem beyond one-loop at that time. B. S. DeWitt was attracted to Feynman's crazy idea of "ghost particle" and asked Feynman about its structure and nature.

After the conference, DeWitt made an heroic effort to work out all the details of Feynman rules beyond one-loop to show that Feynman's idea of a 'ghost particle' works for both the Yang-Mills theory and the Einstein theory. He published three long and detailed papers in 1967: "Quantum Theory of Gravity I, II, and III." [28] The whole thing is very complicated and the physical reason for the presence of the ghost particle only in the intermediate states of a physical process is not completely clear. However, the results of Feynman and DeWitt inspired two Russian physicists Faddeev and Popov to write a most elegant paper "Feynman Diagrams for the Yang-Mills Field" (with only 2 pages!) in the same year. [29] This paper completely clarified and solved the problem of unitarity and gauge invariance to all orders, on the basis of the Feynman path integral.

Basically, Faddeev and Popov solve Feynman's problem of unitarity by considering the gauge invariance of physical amplitudes. They showed that the gauge condition such as $\partial_\mu B_a^\mu = 0$ in the Yang-Mills theory cannot be consistently imposed for all time, in contrast to that in quantum electrodynamics. Faddeev and Popov proposed a new method to enforce the gauge condition for all time and, hence, maintain the gauge invariance of physical amplitudes. They showed that it leads to the closed loop with a (scalar) 'ghost particle' propagating along it with a specific interaction, which restores the unitarity of the physical amplitudes. The same method based on Feynman's path integral can also be applied to Einstein's theory of gravity.[30]

The path integral is another not-completely-baked but (probably) profound idea of Feynman's. It is useful in physics but still lacks a mathematical foundation. The status of Dirac's delta function around 1930's was similar. Now we have a foundation for the delta function in the mathematical theory of distributions, but we still lack a mathematical basis for Feynman's path integral. This problem was also discussed by Dyson in his lecture "Missed Opportunities." [31] In this sense, the results of Faddeev and Popov may not be considered as rigorously established within the framework of quantum field theory. Mandelstam's obtained the same results on the basis of quantum field theory.[32]

Thus, the Yang-Mills theory (with or without masses) can be established as a gauge theory which satisfies both unitarity and renormalizability. So far, this is the best theory of strong, weak and electromagnetic interactions that physicists can construct within the framework of local field theory. [33] As noted above Einstein's theory of gravity can also be regarded as a gauge theory. However, there are important differences: Einstein's theory is based on curved spacetime and its action is linear in the spacetime curvature. In contrast, the conventional Yang-Mills theory is based on flat spacetime and its action is quadratic in gauge curvature. As a result, Einstein's theory is gauge invariant and unitary, but it is not renormalizable. Only when one can construct a unitary, renormalizable and consistent theory of gravity, can one claim to have solve the problem of quantum gravity.

4 Transformations of Spacetime for Accelerated Frames

The first ingredient, 'gravitational force', in Einstein's happiest thought has been well developed and understood through his theory of gravity. Indeed, his idea of general coordinate invariance implies that the physics of gravity should be formulated and understood in any frame of reference. However, the concept of specific accelerated frames, the second ingredient in his happiest thought, turns out to be very difficult to define precisely and, hence, has not been well developed theoretically or experimentally. So far, we even do not have a well-established definition of constant linear accelerations as transformations of spacetime. In one simple, smooth generalization of Lorentz transformations to reference frames with linear accelerations, spacetime as seen from these accelerated frames is non-uniform, and the group properties of the spacetime transformations, with the associated Lie algebra appears to be non-trivial. For instance, the Lie algebra has infinite rank.

In his 1907 paper [4], Einstein asked:

'Is it conceivable that the principle of relativity also holds for systems which accelerate relative to each other?

Apparently, he has in mind the physical properties of space and time and exact physical laws in non-inertial frames of reference. He was able to obtain some physical results with the help of the principle of equivalence. Nevertheless, he could only write down approximate spacetime transformations involving constant linear accelerations, which do not reduce to the Lorentz transformations in the limit of zero acceleration. It appears that Møller is the first to attempt to derive the spacetime transformations for accelerated frames. [35] In 1943, he used Einstein's field equation in vacuum to derive a spacetime transformation between an inertial frame $F_I(w_I, x_I, y_I, z_I)$ and a frame $F(w^*, x, y, z)$ which moves with a constant acceleration α_o^* along the x-direction:

$$
\begin{aligned}
w_I &= (x + \frac{1}{\alpha_o^*}) sinh(\alpha_o^* w^*), \\
x_I &= (x + \frac{1}{\alpha_o^*}) cosh(\alpha_o^* w^*) - \frac{1}{\alpha_o^*}, \\
y_I &= y, \\
z_I &= z;
\end{aligned}
\tag{3}
$$

where $w_I = ct_I$ and $w^* = ct$. In 1972, T.Y. Wu and Y.C. Lee derived the same accelerated transformations from kinematical considerations based on Lorentz transformations for short spacetime intervals. [35] In the limit of zero acceleration, they reduce to the identity transformation, and although there are reasons to term these 'constant accelerations' the velocity β of the $F(w^*, x, y, z)$ frame is not a linear function of time w^*, $\beta = tanh(\alpha^* w^*)$.

J.P. Hsu and L. Hsu reparametrized the time coordinate and generalized further to arrive at transformations with two properties: [36]

i. minimal departure from the Lorentz transformation: the velocity β is a linear function of the (accelerated-frame) time w, $\beta = \beta_o + \alpha_o w$, and

ii. limiting 4-dimensional symmetry: the transformation to an allowed accelerated frame reduces to a (generally non-trivial) Lorentz boost in the limit of zero acceleration.

The resulting transformation between an inertial frame F_I and an allowed accelerated frame F, moving with a constant acceleration α_o in the x-direction, is

$$
\begin{aligned}
w_I &= \gamma\beta(x + \frac{1}{\alpha_o \gamma_0^2}) - \frac{\beta_o}{\alpha_o \gamma_o} \\
x_I &= \gamma(x + \frac{1}{\alpha_o \gamma_o^2}) - \frac{1}{\alpha_o \gamma_o} \\
y_I &= y, \\
z_I &= z;
\end{aligned}
\tag{4}
$$

where $\beta = \alpha_o w + \beta_o$, $\gamma_o = 1/\sqrt{1-\beta_o^2}$, $\gamma = 1/\sqrt{1-\beta^2}$. This is called the Wu transformation. [36] It implies that a particle at rest in the F frame will gain constant energy per unit length, as measured in an inertial frame F_I. This property, which defines the constant linear acceleration, means the frame F may serve as the rest frame for a particle in a high-energy linear accelerator.

The accelerated transformations of spacetime (4) reveal the following property for physics in accelerated frames of reference: The speed of light is described by the equations, $ds^2 = dw_I^2 - dr_I^2 = W^2 dw^2 - dr^2 = 0$, where $W = \gamma^2(\gamma_o^{-2} + \alpha_o x)$ can be obtained from (4). Thus, in the inertial frames F_I, the speed of a light signal in vacuum is constant, $dr_I/dw_I = 1$. On the contrary, the speed of this light signal as measured in the accelerated frame F is given by the function W, i.e., $dr/dw = W = \gamma^2(\gamma_o^{-2} + \alpha_o x)$. This implies there is no relativity or equivalence between an inertial frame F_I and an accelerated frame F in the sense of the invariance of physical laws familiar in special relativity theory.

It is interesting to note that these accelerated transformations of spacetime are non-linear, in contrast to their zero acceleration limit, the linear Lorentz transformations. As a result, the finite point transformations cannot be derived from infinitesimal transformations through the usual process of exponentiation of the transformation matrices.

The group operations and the Lie algebra for the group generators for the MWL and Wu transformations involving two parameters are not trivial. For example, the Lie algebra is not finitely generated, in contrast to the Lorentz and Poincaré groups. [37]

The corresponding transformations of the differentials dx_I^μ and dx^μ (for fixed values of the spacetime coordinates) are necessarily linear. They can be described by a pair of group operations:

i. multiplying by a synchronization factor W to dw:

$$(dw, dx, dy, dz) \to (Wdw, dx, dy, dz), \qquad W = \gamma^2(\gamma_o^{-2} + \alpha_o x), \tag{5}$$

and then

ii. making the usual 4-dimensional Lorentz boost for (Wdw,dx,dy,dz) involving the (spacetime-dependent) velocity $\beta = \beta_0 + \alpha_o w$,

$$dw_I = \gamma(Wdw + \beta dx), \quad dx_I = \gamma(dx + \beta W dw), \quad dy_I = dy, \quad dz_I = dz. \tag{6}$$

These equations can be obtained from (4) by differentiation. The synchronization factor W in the first operation guarantees that the differential equations in (6) are integrable and that the time in the accelerated frame F automatically becomes the synchronized time in the Lorentz transformation when the acceleration α_o approaches zero. [36]

5 Yang-Mills and the Interplay between Mathematics and Physics

Einstein's general relativity involved some of the period's mathematics in its original formulation. Yang and Mills' theory did not, yet it had a swift and lasting impact in mathematics, particularly in the areas of geometry and topology. Mathematicians, who had developed fiber bundles as tools for understanding the global geometry of manifolds, soon understood the vector potential and the Yang-Mills action as a connection on a principal fiber bundle and the square norm of this connection's curvature, respectively. Indeed, Wu and Yang published a dictionary [38] allowing the physicist and the mathematician each to translate the terms the other had developed for the same concepts. The recognition of this overlap of interests quickly led to a renewal of interaction between what had become two distinct communities. As outlined below, basic questions in the new physics suggested new realms of mathematical exploration; conversely, existing mathematical constructs, notably index theory, provided insights in the new physics.

The formulation of an action immediately raises the question of its critical points; that is, the solutions of the corresponding Euler-Lagrange equations. For the Yang-Mills action these solutions can be non-trivial field configurations known as instantons. Hitchin, Atiyah, Singer, Drinfeld and Manin [39] began studying the spaces of instantons. In the case of self-dual Yang-Mills instantons Donaldson [40] obtained results which firmly established the Yang-Mills equations as a useful tool in the study of manifolds. He developed what is now known as the Donaldson invariant, which proves sensitive not only to the topology but to the differentiable structure of a manifold. This served as the key to resolve a long-standing question of whether there could be inequivalent differentiable structures on a given topological four-manifold. (The answer is "yes".)

The chiral and non-Abelian anomaly, first discovered using the technique of current algebras, provided another avenue from Yang-Mills physics to mathematics. This avenue proved to be very much a two-way street. Bringing techniques from global analysis of manifolds, including index theory, to bear on a puzzle arising from their study of the Yang-Mills path integral, Singer and Atiyah [41] developed a new, global formulation of the anomaly. In so doing, they resolved some outstanding issues in quantum physics, and inspired a generation of physicists to study new techniques of differential geometry. On the mathematical side, this inaugurated the study of the topology and geometry of certain infinite-dimensional spaces. Fine and Fine [42] present the history of the anomaly and the attendant interplay between mathematics and physics in detail.

The profound influence of Yang-Mills on mathematics is ongoing. Attempts to rigorously define the quantum Yang-Mills field theory continue. Moreover, there is a direct connection between the mathematical formulation of the anomaly in Yang-Mills theory and the contemporary formulation of string theory, which, in turn, is a continuing source of mathematical studies. String theory is also a descendant of general relativity: it is supposed to reduce to general relativity in a classical limit, and is conjectured to reduce to higher-dimensional supergravity in an appropriate parameter regime.

Topological quantum field theory is an area of active research in which trying to follow ideas as they move between mathematics and physics is like watching a tennis match from center court. After Donaldson developed his invariants by mathematical analyzing the classical Yang-Mills equations, Atiyah conjectured there should be a quantum field theory in which they arise naturally. In response, Witten wrote down a Lagrangian for a topological Yang-Mills theory, followed shortly by a topological gravity theory. Atiyah and Jeffrey promptly re-interpreted the topological Yang-Mills theory in terms of equivariant cohomology. Since then topological quantum field theories have led to new mathematical conjectures, most of which have then been rigorously proven, and provided both mathematicians and physicists with examples of path integrals which can be treated non-perturbatively.

Arguably, the most important influence of Yang and Mills work has been bringing significant portions of the mathematics and physics communities back into their historically close contact, once again sharing a language and working on aspects of the same problems, after a long period of divergence.

There is an interesting story in which Yang described his joy of comprehension of the relation of physics and mathematics: [43]

In 1975, impressed with the fact that gauge fields are connections on fiber bundles, I drove to the house of Shiing Shen Chern in El Cerrito, near Berkeley. ..... I told him that I had finally learned from Jim Simons the beauty of fiber-bundle theory and the profound Chern-Weil theorem. I said I found it amazing that gauge fields are exactly connections on fiber bundles, which the mathematicians developed without reference to the physical world. I added, "this is both thrilling and puzzling, since you mathematicians dreamed up these concepts out of nowhere." He immediately protested, "No, no. These concepts were not dreamed up. They were natural and real."

6 Perspectives on General Relativity and Gauge Theory

The terms 'general relativity' and 'gauge fields' are widely used and have their conventional meanings. But these meanings are not fixed in terms of underlying physical ideas. Rather, they show the evolution of ideas in physics research and also show the difficulty for physicists to change the accepted names. The modern understanding of 'gauge fields' turn out to have nothing to do with 'scale invariance', as originally proposed by H. Weyl, but have everything to do with the 'phase' of a wave function, as pointed out by V. Fock and F. London. [43]

Fock's Comments on General Relativity

However, when Einstein created his theory of gravitation, he put forward the term 'general relativity' which confused everything. This term was adopted in the sense of 'general covariance', i.e. in the sense of the covariance of equations with respect to arbitrary transformations of coordinates accompanied by transformations of the $g_{\mu\nu}$. But we have seen that this kind of covariance has nothing to do with the uniformity of space, while in one way or another relativity is connected with uniformity. This means that 'general relativity' has nothing to do with "relativity as such." At the same time the latter received the name "special" relativity, which purports to indicate that it is a special case of "general" relativity. [44]

Fock's criticism is constructive because it helps to clarify the relation or the lack of relation between 'special relativity' and 'general relativity.' In special relativity, as discussed by Lorentz, Poincaré, Einstein and Minkowski, the spacetime is flat. However, in Einstein's general relativity, the spacetime is curved, so that he could introduce the physical effects of gravity into Riemannian curvature tensor. Perhaps, the idea of "generalization of special relativity" first came to mind when he was thinking about generalizing physical laws from inertial frames to non-inertial frames with an arbitrary velocities. [4] This does not justify the term 'general relativity' in the sense of frame-independence, because the accelerated transformations in (3) and (4) show that there is no relativity between inertial and accelerated frames. Note that the curvature of spacetime of in these accelerated frame must still be zero. Nevertheless, the terms reveal the continuity of Einstein's reasoning after 1905.

Synge's Comments on the Principle of Equivalence

V. L. Synge made a constructive critism concerning the Principle of Equivalence: [45]

..... Perhaps they speak of the Principle of Equivalence. If so, it is my turn to have a blank mind, I have never been able to understand this principle. Does it mean that the signature of the space-time metric is +2 (or -2 if you prefer the other convention)? If so, it is important, but hardly a Principle. Does it mean that the effects of a gravitational field are indistinguishable from the effects of an observer's acceleration? If so, it is false. In Einstein's theory, either there is a gravitational field or there is none, according as the Riemann tensor does not or does vanish. This is an absolute property; it has nothing to do with any observer's world-line. Space-time is either flat or curved, and in several places in the book I have been at considerable pains to separate truly gravitational effects due to curvature of space-time from those due to curvature of the observer's world-line (in most ordinary cases the latter predominate). The Principle of Equivalence performed the essential office of midwife at the birth of general relativity, but, as Einstein remarked, the infant would never have got beyond its long-clothes had it not been for Minkowski's concept. I suggest that the midwife be now buried with appropriate honours and the facts of absolute space-time faced.

As noted above the Riemann curvature tensor of the spacetime of accelerated frames characterized by the transformations (4) vanishes, just as in the inertial frames. Thus, the physical effects related to these accelerated transformations have nothing to do with the gravitational field.

The physics in non-inertial frames deserves more theoretical and experimental investigations. Clearly, one cannot be contented with the understanding of physics only in the usual inertial frames, which is the basic framework for the standard models and particle physics. The Lorentz and Poincaré transformations are linear and carry the whole arithmetic spacetime into itself. The accelerated transformations such as those of Møller and the Wu transformations, are nonlinear, and they carry only portion of spacetime in accelerated frame F to the whole spacetime in an inertial frame F_I. The notion of pseudo-group was discussed by Veblen and Whitehead in order to deal with the transformations which carry the whole space into portions of space. A set of transformations is called a pseudo-group if it satisfies the conditions: (i) If the resultant of two transformations in the set exists it is also in the set. (ii) The set contains the inverse of each transformation in the set. Thus, it is possible that the concept of 'pseudo-group' could become relevant in both differential geometry and physics of general accelerated transformations of spacetime. [46]

Gauge Theory of Gravity, Yang-Mills Gravity and Translation Symmetry

In early 1950, Einstein said that he was not sure whether differential geometry was to be the framework for further progress, but if it was then he believed he was on the right track. [47] Indeed, in the past 50 years, most researchers in this area followed Einstein's approach to investigate gauge theory of gravity, based on Riemannian geometry and a gauge symmetry group in curved spacetime. [48] However, the difficulties related to the ultraviolet divergence and the energy-momentum tensor in Einstein's theory have not been overcome. These problems are probably related to the fact that Einstein's approach to gravitational field is characterized by a drastic departure from the grand tradition of all classical and quantum fields, as stressed by Dyson [31] It appears that the framework of Riemannian geometry is too general for field theory in the sense that once a gauge symmetry is introduced in it, the gauge symmetry, however powerful it may be, cannot harness the ultraviolet divergence. In contrast, we know that gauge symmetry in Yang-Mills theory can exercise its full power to harness ultraviolet divergence within the usual framework of local field theory based on flat spacetime. This appears to be the 'secret' essence for the success of unified electroweak theory and quantum chromodynamics.

Thus, a burning question is: Is it possible to realize a union of Einstein's theory and Yang-Mills theory to overcome the divergence difficulties and to understand gravitational experiments? Two key features for connecting Einstein's theory to experiments and observations are the field equation (involving a spacetime curvature tensor) and the 'Einstein-Grossmann metric' $g_{\mu\nu}dx^\mu dx^\nu$ which is essential for the motion of classical objects and light rays. Similarly, there are also two basic features of Yang-Mills theory; namely, an action involving quadratic gauge curvature with a symmetry group and an underlying flat spacetime. In view of the divergence difficulty associated with curved spacetime, it is worthwhile to consider a 'Yang-Mills gravity' [49] which is characterized by: (a) an action with quadratic gauge curvature with translational gauge symmetry, (b) flat spacetime, and (c) an effective 'Einstein-Grossmann metric'. This is interesting because the external spacetime translation gauge symmetry naturally leads to an effective 'Einstein-Grossmann metric', provided the external symmetry group is implemented in the action according to the Yang-Mills approach for internal gauge groups. Furthermore, the Yang-Mills theory has the advantage of connecting translation gauge symmetry to the conserved energy-momentum tensor which is the source of the gravitational field. Table 1 shows a comparison of key features of Yang-Mills theory, Yang-Mills gravity, gauge theory of gravity and Einstein's theory. [50]

In the bundle language, Einstein's gravitational potential is a metric which determines the Levi-Civita connection on the tangent (vector) bundle, while Yang-Mills potentials are connections on principal fiber bundles. Thus, Einstein's gravitational field is not exactly the same as the Yang-Mills field. In this context, it is worth noting Ashtekar's approach to loop quantum gravity actually does formulate Einstein's gravitational theory as an SU(2) Yang-Mills theory defined on the principal fiber bundle associated to the frame bundle. The dynamical variable is no longer the metric but the vierbein and a spin connection. [51]

In the past 100 years, the track record of researches in the fundamental physics of particles and fields could perhaps be briefly summarized as follows:

Local quantum fields with gauge symmetry appear to be the most effective key at hand to unlock the secret of the basic interactions of quantum particles and, hence, the last mystery of quantum gravity. The concepts of Regge poles, current algebra and superconvergence, once very hot, turn out to be not viable. Geometrization of all physical fields based on Riemann-Cartan or Finsler geometry is still a dream, while supersymmetric string theory and loop quantum gravity are only visions.

Table 1. A comparison of key features of gauge fields with internal and various external spacetime symmetries.

$Yang-Mills\ \ Theory$	$Yang-Mills\ \ Gravity$	$Gauge\ Theory\ of\ Gravity$	$Einstein's\ Theory\ of\ Gravity$
$(Minkowski\ spacetime)^*$	$(Flat\ spacetime)^*$	$(Curved\ spacetime)$	$(Curved\ spacetime)$
$A^\mu_a, \quad \Phi_i$	$\phi_{\mu\nu}, \quad \phi,$	$B^\mu_i, \quad \phi,$	$\Gamma^\lambda_{\mu\nu}, \quad \phi,$
$[Lie\ \ groups\ (internal)$	$[translation\ and\ arbitrary$		
$\delta\Phi(x)_i = -i\omega_a(x)(\tau_a)_{ik}\Phi_k(x)]^{**}$	$coordinate\ transformations:$	$[Spacetime\ Tanslation:$	$[Arbitrary\ coord.\ transformations$
$group\ generators: \quad \tau_a$	$x'^\mu = x^\mu + \Lambda^\mu(x),]$	$x'^\mu = x^\mu + \Lambda^\mu(x),]$	$x'^\mu = x^\mu + \Lambda^\mu(x)$
$A'_a{}^\mu = A^\mu_a + \frac{1}{f}\partial^\mu\omega_a + K_{abc}\omega_b A^\mu_c,$	$group\ generators: \ p_\mu = -iD_\mu,^{**}$	$group\ \ generators: \ \xi_\mu,$	$\Lambda^\mu(x) = infinitesimal\ \ func.]$
$\Delta^\mu\Phi = (\partial^\mu + ifA^\mu_a\tau_a)\Phi,$	$\phi'_{\mu\nu} = \phi_{\mu\nu} - \Lambda^\lambda\partial_\lambda\phi_{\mu\nu} - \phi_{\mu\alpha}\partial_\nu\Lambda^\alpha - \phi_{\alpha\nu}\partial_\mu\Lambda^\alpha,$	$B'^\alpha_i = B^\alpha_i + \partial_i\Lambda^\alpha/f,$	$\Delta_\mu V_\nu = \partial_\mu V_\nu + \Gamma^\lambda_{\mu\nu}V_\lambda,$
$[\Delta^\mu, \Delta^\nu] = ifF^{\mu\nu}_a\tau_a,$	$(D_\mu + f\phi_{\mu\nu}D^\nu)\phi \equiv J_{\mu\alpha}D^\alpha\phi, \quad J_{\mu\alpha} = P_{\mu\alpha} + f\phi_{\mu\alpha},$	$(\partial_i + B^\mu_i\xi_\mu)\phi \equiv \Delta_i\phi,^*$	$[\Delta_\mu, \Delta_\nu]V_\lambda = R^\alpha_{\mu\nu\lambda}V_\alpha,$
$F^{\mu\nu}_a = \partial^\mu A^\nu_a - \partial^\nu A^\mu_a - fK_{abc}A^\mu_b A^\nu_c,$	$[J_{\mu\alpha}D^\alpha, J_{\nu\beta}D^\beta] = C_{\mu\nu\lambda}D^\lambda,$	$[\Delta_i, \Delta_j] = T^k_{ij}\Delta_k + G^\mu_{ij}\xi_\mu,$	$R^\alpha_{\mu\nu\lambda} = \partial_\mu\Gamma^\alpha_{\nu\lambda} - \partial_\nu\Gamma^\alpha_{\mu\lambda}$
	$C_{\mu\nu\lambda} = J_{\mu\alpha}D^\alpha J_{\nu\lambda} - J_{\nu\beta}D^\beta J_{\mu\lambda},$	$T^k_{ij} = -h^\mu_i h^\nu_j(\partial_\mu h^k_\nu - \partial_\nu h^k_\mu),$	$-\Gamma^\alpha_{\mu\beta}\Gamma^\beta_{\nu\lambda} + \Gamma^\alpha_{\nu\beta}\Gamma^\beta_{\mu\lambda},$
$L = \left[-\frac{1}{4}F_{\mu\nu}\cdot F^{\mu\nu}\right.$	$L = \left[-(C_{\mu\alpha\beta}C^{\mu\beta\alpha} - C^\alpha_{\mu\alpha}C^{\mu\beta}_\beta)/(2f^2)\right.$	$L = \sqrt{-g}\left\{\frac{1}{2f^2}[\frac{1}{4}T^{ijk}T_{ijk} + \frac{1}{2}T^{ijk}T_{ikj}\right.$	$L = \sqrt{-g}\left\{\frac{R}{16\pi G}\right.$
$\left.+\eta_{\mu\nu}(\Delta^\mu\Phi)^\dagger\cdot(\Delta^\nu\Phi) - m\Phi^\dagger\cdot\Phi\right].$	$\left.+\frac{1}{2}(G_{\alpha\beta}\partial^\alpha\phi\partial^\beta\phi - m^2\phi^2)\right]\sqrt{-P}.$	$\left.-T_{ij}{}^jT^{ik}{}_k] + \frac{1}{2}(g_{\mu\nu}\partial^\mu\phi\partial^\nu\phi - m^2\phi^2)\right\}.^{**}$	$\left.+\frac{1}{2}(g_{\mu\nu}\partial^\mu\phi\partial^\nu\phi - m^2\phi^2)\right\}.$
	$G_{\alpha\beta} = P^{\mu\nu}J_{\mu\alpha}J_{\nu\beta} \equiv\ 'effective\ metrtic\ tensor,'$		
$maximum\ coupling: 4-vertex$	$maximum\ coupling: 4-vertex$	$maximum\ coupling: \infty-vertex$	$maximum\ coupling: \infty-vertex^*$
[* inertial frames,] [** $\lvert\omega^b(x)\rvert$ is small.]	[* inertial and non-inertial frames with the metric tensor $P_{\mu\nu}$,] [** D_μ is covariant derivative w.r.t the Levi-Civita connection in flat spacetime.]	[*$assume\ \xi_\mu\phi = 0, \ \ and\ \ \partial_i = h^\mu_i\partial_\mu,$ $where\ \ h^\mu_i = \delta^\mu_i + fB^\mu_i.$] [** $h^\mu_i h^\nu_k\eta^{ik} = g^{\mu\nu}.$]	[* in the $\sqrt{-g}$ term.]

References

[1] S. G. Benka, Physics Today, Jan. 2005, p.10.

[2] It is more reasonable to view the creation of the celebrated theory of special relativity as both a cause and an effect of scientific progress, through the collective effort of Lorentz (1904), Poincaré (1905), Einstein (1905) and Minkowski (1908).

[3] J. Ishiwara, "Einstein Koen-Roku" Tokyo-Tosho, Tokyo, 1977. See also A. Pais, "Subtle is the Lord..." (Oxford Univ. Press, 1982) p. 179.

[4] A. Einstein, paper C in Chapter 1 of this volume.

[5] C. N. Yang. "Talking about Physics Research and Teaching," (The 5th talk at the Graduate School of the Chinese University of Science and Technology, Beijing, China), (1986/5/27 - 1986/6/12). "This kind of repeated failure at some seemingly good idea is, of course, a common experience for all research workers," he said later, see Chen Ning Yang, *Selected Papers 1945-1980 With Commentary* (W. H. Freeman and Comp. 1983) p. 19. Robert Mills, Am. J. Phys. **57**, 493 (1989).

[6] It was related to the existence of the third term $\epsilon^{abc} b_\mu^b b_\nu^c$ in the field strength $f_{\mu\nu}^a$.

[7] I. Newton, paper A in Chapter 1 of this volume. Euclid's ELEMENTS started with the definitions of point, line, etc. and followed by 5 postulates; while Newton's PRINCIPIA started with the definitions of mass, kinetic energy,... and followed by the 3 laws of motion.

[8] H. Poincaré, Rend. Circ. Mat. Palermo 21, 129 (1906), "On the Dynamics of the Electron" Section 9 "Hypothesis Concerning Gravitational Force." Its English translation in paper B in Chapter 1 is an extract from A. A. Logunov's book *On the Articles by Henry Poincaré "ON THE DYNAMICS OF THE ELECTRON"*

[9] The Collected Papers of Albert Einstein (Ed. J. Stachel, Princeton University Press, 1993.)

[10] A. Pais, *Subtle is the Lord..., The Science and the Life of Albert Einstein* (Oxford Univ. Press, 1982), p. 212. Pais gave a lucid and detailed discussion about Einstein-Grossmann collaboration in Chapter 12 in his book. As far as final result is concerned, the Einstein-Grossmann collaboration resembles, perhaps, the discovery of the structure of DNA in 1953 by a collaboration of Physicist Francis Crick and biologist James Watson. Grossmann's contribution and help to Einstein's theory deserves to be more widely known. There exists a first page of Einstein's landmark paper "The Foundation of the General Theory of Relativity", in which he said: "Finally, I want to acknowledge gratefully my friend, the mathematician Grossmann, whose help not only saved me the effort of studying the pertinent mathematical literature, but who also helped me in my search for the field equations of gravitation." See ref. 4. Unfortunately, this page was not published together with Einstein's landmark paper. It is fitting that since 1975 there has been a regular International Grossmann Conference on Gravity to commemorate his contribution.

[11] D. Hilbert, paper C in Chapter 2 of this volume. In this paper, Hilbert already attempted to unify electromagnetic and gravitational forces in his theory. Also, we note that the addition of a cosmological term in the Einstein-Hilbert gravitational equation will upset the elegant and unique invariant action for Einstein's theory of gravity and, hence, is untenable from the viewpoint of symmetry.

[12] L. Cory, J. Renn and J. Stachel, *Science* **278** 1270-73 (1997); T. Sauer, *Archive for History of Exact Sciences* **53** # 6 529-575 (1999); echo.mpiwg-berlin.mpg.de/content/relativityrevolution/hilbert

[13] So far, the energy-momentum tensor in Einstein's equation is still not well understood.

[14] E. Cartan, paper D in Chapter 2 of this volume. For a more detailed discussion of torsion in gauge theory, see J. M. Nester, in *An Introduction to Kaluza-Klein Theories* (Ed. H. C. Lee, World Scientific, 1984) pp. 83-115.

[15] J. P. Hsu, Nuovo Cimento **109B**, 645 (1994).

[16] F. Dyson, paper G in Chapter 6 of this volume, and Bull. Am. Math. Soc. 78, 635 (1972).

[17] See, for example, B. DeWitt, papers B and C in Chapter 6 of this volume. For a review of quantum gravity, see E. Alvarez, Rev. Mod. Phys. **61** 561 (1989).

[18] F. Dyson, Sci. Am. **199**, 74 (1958).

[19] The "rascal", as Einstein once called Pauli, was so extraordinary that Weyl said: "Pauli combines in an exemplary way physical insight and mathematical skill." But one also knew that, in 1925, Pauli did not believe that Kronig's new idea of the electron spin had any connection with reality, and discouraged Kronig from publishing it. In the same year, Pauli refused to collaborate with Born to develop the matrix mechanics.

[20] C. P. Enz, *No Time to be Brief, A scientific biography of Wolfgang Pauli* (Oxford Univ. Press, 2002) pp. 481-482. Aside from Pauli, Ronald Shaw also discussed possible generalization of gauge invariance in his Ph.D. thesis (1954, unpublished); see R. Mills, ref. 5.

[21] This belief resembles Voigt's belief when he published his paper on Doppler effects in 1887. Without having any experimental evidence, Voigt postulated (i) the invariance of the laws for the propagation of the light wave (in aether) and (ii) the universal constant of the speed of light to derive a (conformal) 4-dimensional space and time transformations and the Doppler effects. Such a Voigt transformation differs from the Lorentz transformation by an overall constant. Based on Doppler effects, Voigt obtained an approximate relativistic time and presented the first challenge to Newton's absolute time about 20 years before the discovery of special relativity. See, W. Voigt, Nachr. Ges. Wiss. Goettingen. 41 (1887). For an English translation, see A. Ernst and J. P. Hsu, Chin. J. Phys. **39**, 211, (2001).

[22] Article B of Appendices in this volume.

[23] Jong-Ping Hsu and Leonardo Hsu, UMassD preprint (2005).

[24] R. Utiyama, paper C in Chapter 4 of this volume.

[25] See, for example, papers in Chapter 7 of this volume. For the gauge theory with the Poincaré group, see J. M. Nester, in *An Introduction to Kaluza-Klein Theories* (Ed. H. C. Lee, World Scientific, 1984) pp. 83-115; F. W. Hehl, P. von der Heyde, G. D. Kerlick and J. M. Nester, Rev. Mod. Phys. **48** 393 (1976).

[26] J. J. Sakurai, Ann. of Phys. **11**, 1 (1960).

[27] P. W. Higgs, Phys. Rev. Lett. **12**, 132 (1964). The Higgs' mechanism for introducing masses to gauge bosons is related to 'spontaneous symmetry breaking,' which is only an "apparent breaking" of symmetry because the essence of gauge symmetry is still preserved. This mechanism is crucial for Weinberg to construct explicitly a celebrated lepton model to "unify" the electromagnetic and weak interactions, to show the power of Yang-Mills fields through the astonishing predictions of gauge-boson masses and others. All Weinberg's predictions were confirmed by experiments later. This is one of the greatest achievements of the 20th century science. Glashow and Salem also had similar ideas.

[28] B DeWitt, papers B and C in Chapter 6 of this volume.

[29] L. D. Faddeev and V.. N. Popov, paper D in Chapter 6 of this volume.

[30] E. S. Fradkin and I. V. Tyutin, paper F in Chapter 6 of this volume.

[31] F. Dyson, Bull. Am. Math. Soc. 78, 635 (1972). See also paper G in Chapter 6.

[32] S. Mandelstam, paper E in Chapter 6 of this volume.

[33] The ultraviolet divergence in local field theory may be the price to pay for the idealization of physical particles as point particles. And renormalization with counter terms may be considered as a way to compensate the deficiency of the mathematics of local functions.

[34] V. Fock, *The Theory of Space, Time, and Gravitation* (Pergamon Press, 1964, 2nd Revised Edition, transl. by N. Kemmer), p. 207.

[35] C. Møller, paper A in Chapter 5 of this volume. T. Y. Wu and Y. C. Lee, paper C in Chapter 5 of this volume. The Møller transformations can be generalized to include a constant velocity so that they reduces to the Lorentz transformation in the limit of zero acceleration. See Jong-Ping Hsu, *Einstein's Relativity and Beyond-New Symmetry Approaches* (World Scientific, 2000) Ch. 21.

[36] J. P. Hsu and L. Hsu, papers D and E in Chapter 5 of this volume. The Wu transformation is named to honor Ta-You Wu's idea of a kinematic approach to finding an accelerated transformation within the framework of spacetime with a vanishing Riemann curvature tensor. The time w in the Wu transformation (4) can be synchronized by a set of computerized clocks. See, J. P. Hsu, ref. 35, pp. 289-290.

[37] D. Fine and J. P. Hsu, preprint (UMassD, 2005).

[38] T. T. Wu and C. N. Yang, Phys. Rev. **D12**, 3845 (1975); paper A in Chapter 10 of this volume. In order to understand the geometrical meaning of gauge theory, Yang asked his colleague J. Simons. Simons told Yang that gauge theory is related to the connections on fiber bundles. Yang tried to read books on fiber bundles and felt that modern mathematical language is too dull and too abstract to physicists. In early 1975, Yang invited Simons to give lunch talks on differential forms and fiber bundles. These talks helped Yang and others to understand the mathematical meaning of Dirac's quantization rules, magnetic monopoles and the Chern-Weil theorem. Simons' talk stimulated Wu and Yang to write this paper in ref. 38. See, D. Z. Zhang, *C. N. Yang and Contemporary Mathematics* in Mathematical Intelligencer, **15**, no. 4 (1993).

[39] M. F. Atiyah, N. J. Hitchin and I. M. Singer, proc. Roy. Soc. London Ser. A. **362**, 425-461 (1978); Atiyah, M. F., Drinfeld, V. , Hitchin N. J., and Manin, Yu. I., Phys. Lett. A **65** 185-187 (1978); Atiyah, M. F. , Hitchin N., and Singer, I. M. Proc. Nat. Acad. Sci. USA, **7**, 2662 (1977).

[40] S. K. Donaldson, Topology **29** 257-315 (1990); S. K. Donaldson, J. Differential Geom. **18**, 279-315 (1983).

[41] M. F. Atiyah and I. M. Singer, Proc. Nat. Acad. Sci. U.S.A. **81** 2597-2600 (1984)

[42] A. Fine and D. Fine, Studies in History and Philosophy of Modern Physics, **28B** # 2 307-323 (1997)

[43] V. Fock, Z. Phys. **39**, 226 (1927); F. London, Z. Phys. **42** 375 (1927). For a review of the history of the idea of gauge symmetry, see Chen-Ning Yang, Phys. Today, June 1980, pp.42-49; or in *JingShin Theoretical Physics Symposium in Honor of Prof. Ta-You Wu* (Ed. J. P. Hsu and L. Hsu, World Scientific, 1998) pp. 61-71.

[44] V. Fock, *The Theory of Space, Time, and Gravitation* (Pergamon Press, 1964, 2nd Revised Edition, transl. by N. Kemmer), p. xvii.

[45] J. L. Synge, *Relativity: The General Theory* (North-Holland, 1966), Preface.

[46] O. Veblen and J.H.C. Whitehead, *The Foundations of Differential Geometry* (Cambridge Univ. Press, 1953) pp. 37-38.

[47] A. Pais, *Subtle is the Lord..., The Science and the Life of Albert Einstein* (Oxford Univ. Press, 1982), p. 467.

[48] See papers C and D in Chapter 4 and those in Chapter 7 of this volume. Y. M. Cho, Phys. Rev. **D14**, 3341 (1976).

[49] See paper D in Chapter 8 of this volume.

[50] See papers A in Chapter 4, E in Chapter 7, D in Chapter 8 and B in Chapter 2 of this volume.

[51] A. Ashtekar, Phys. Rev. D **36**, 1587 (1987); A. Ashtekar, Phys. Rev. Lett. **57**, 2244 (1986); Rovelli, C. and Smolin, L. Nucl. Phys. B **331**, 80 (1990).

Chapter 1

The Dawn of Gravitation*

*I. Newton, H. Poincaré, A. Einstein

THE

MATHEMATICAL PRINCIPLES

OF

NATURAL PHILOSOPHY.

ISAAC
NEWTON

Translated by Andrew Motte

BOOK III.

In the preceding Books I have laid down the principles of philosophy, principles not philosophical, but mathematical; such, to wit, as we may build our reasonings upon in philosophical inquiries. These principles are the laws and conditions of certain motions, and powers or forces, which chiefly have respect to philosophy; but, lest they should have appeared of themselves dry and barren, I have illustrated them here and there with some philosophical scholiums, giving an account of such things as are of more general nature, and which philosophy seems chiefly to be founded on; such as the density and the resistance of bodies, spaces void of all bodies, and the motion of light and sounds. It remains that, from the same principles, I now demonstrate the frame of the System of the World. Upon this subject I had, indeed, composed the third Book in a popular method, that it might be read by many; but afterward, considering that such as had not sufficiently entered into the principles could not easily discern the strength of the consequences, nor lay aside the prejudices to which they had been many years accustomed, therefore, to prevent the disputes which might be raised upon such accounts, I chose to reduce the substance of this Book into the form of Propositions (in the mathematical way), which should be read by those only who had first made themselves masters of the principles established in the preceding Books: not that I would advise any one to the previous study of every Proposition of those Books; for they abound with such as might cost too much time, even to readers of good mathematical learning. It is enough if one carefully reads the Definitions, the Laws of Motion, and the first three Sections of the first Book. He may then pass on to this Book, and consult such of the remaining Propositions of the first two Books, as the references in this, and his occasions, shall require.

RULES OF REASONING IN PHILOSOPHY.

RULE I.

We are to admit no more causes of natural things than such as are both true and sufficient to explain their appearances.

To this purpose the philosophers say that Nature does nothing in vain, and more is in vain when less will serve; for Nature is pleased with simplicity, and affects not the pomp of superfluous causes.

RULE II.

Therefore to the same natural effects we must, as far as possible, assign the same causes.

As to respiration in a man and in a beast; the descent of stones in *Europe* and in *America;* the light of our culinary fire and of the sun; the reflection of light in the earth, and in the planets.

RULE III.

The qualities of bodies, which admit neither intension nor remission of degrees, and which are found to belong to all bodies within the reach of our experiments, are to be esteemed the universal qualities of all bodies whatsoever.

For since the qualities of bodies are only known to us by experiments, we are to hold for universal all such as universally agree with experiments; and such as are not liable to diminution can never be quite taken away. We are certainly not to relinquish the evidence of experiments for the sake of dreams and vain fictions of our own devising; nor are we to recede from the analogy of Nature, which uses to be simple, and always consonant to itself. We no other way know the extension of bodies than by our senses, nor do these reach it in all bodies; but because we perceive extension in all that are sensible, therefore we ascribe it universally to all others also. That abundance of bodies are hard, we learn by experience; and because the hardness of the whole arises from the hardness of the parts, we therefore justly infer the hardness of the undivided particles not only of the bodies we feel but of all others. That all bodies are impenetrable, we gather not from reason, but from sensation. The bodies which we handle we find impenetrable, and thence conclude impenetrability to be an universal property of all bodies whatsoever. That all bodies are moveable, and endowed with certain powers (which we call the *vires inertiæ*) of persevering in their motion, or in their rest, we only infer from the like properties observed in the bodies which we have seen. The extension, hardness, impenetrability, mobility, and *vis inertiæ* of the whole, result from the extension, hardness, impenetrability, mobility, and *vires inertiæ* of the parts; and thence we conclude the least particles of all bodies to be also all extended, and hard and impenetrable, and moveable, and endowed with their proper *vires inertia*. And this is the foundation of all philosophy. Moreover, that the divided but contiguous particles of bodies may be separated from one another, is matter of observation; and, in the particles that remain undivided, our minds are able to distinguish yet lesser parts, as is mathematically demonstrated. But whether the parts so distinguished, and not yet divided, may, by the powers of Nature, be actually divided and separated from one another, we cannot certainly determine. Yet, had we the proof of but one experiment that any undivided particle, in breaking a hard and solid body, suffered a division, we might by virtue of this rule conclude that the undivided as well as the divided particles may be divided and actually separated to infinity.

Lastly, if it universally appears, by experiments and astronomical observations, that all bodies about the earth gravitate towards the earth, and that in proportion to the quantity of matter which they severally contain; that the moon likewise, according to the quantity of its matter, gravitates towards the earth; that, on the other hand, our sea gravitates towards the moon; and all the planets mutually one towards another; and the comets in like manner towards the sun; we must, in consequence of this rule, universally allow that all bodies whatsoever are endowed with a principle of mutual gravitation. For the argument from the appearances concludes with more force for the universal gravitation of all bodies than for their impenetrability; of which, among those in the celestial regions, we have no experiments, nor any manner of observation. Not that I affirm gravity to be essential to bodies: by their *vis insita* I mean nothing but their *vis inertiæ*. This is immutable. Their gravity is diminished as they recede from the earth.

RULE IV.

In experimental philosophy we are to look upon propositions collected by general induction from phænomena as accurately or very nearly true, notwithstanding any contrary hypotheses that may be imagined, till such time as other phænomena occur, by which they may either be made more accurate, or liable to exceptions.

This rule we must follow, that the argument of induction may not be evaded by hypotheses.

PHÆNOMENA, OR APPEARANCES.

PHÆNOMENON I.

That the circumjovial planets, by radii drawn to Jupiter's centre, describe areas proportional to the times of description; and that their periodic times, the fixed stars being at rest, are in the sesquiplicate proportion of their distances from its centre.

This we know from astronomical observations. For the orbits of these planets differ but insensibly from circles concentric to Jupiter; and their motions in those circles are found to be uniform. And all astronomers agree that their periodic times are in the sesquiplicate proportion of the semi-diameters of their orbits; and so it manifestly appears from the following table.

The periodic times of the satellites of Jupiter.

$1^d. 18^h. 27'. 34''$. $3^d. 13^h. 13' 42''$. $7^d. 3^h. 42' 36''$. $16^d. 16^h. 32' 9''$.

The distances of the satellites from Jupiter's centre.

From the observations of	1	2	3	4	
Borelli	$5\frac{2}{3}$	$8\frac{2}{3}$	14	$24\frac{2}{3}$	*semi-diameter of Jupiter.*
Townly *by the Microm.* . . .	5,52	8,78	13,47	24,72	
Cassini *by the Telescope* . . .	5	8	13	23	
Cassini *by the eclip. of the satel.* . .	$5\frac{2}{3}$	9	$14\frac{23}{60}$	$25\frac{3}{10}$	
From the periodic times	5,667	9,017	14,384	25,299	

Mr. *Pound* has determined, by the help of excellent micrometers, the diameters of Jupiter and the elongation of its satellites after the following manner. The greatest heliocentric elongation of the fourth satellite from Jupiter's centre was taken with a micrometer in a 15 feet telescope, and at the mean distance of Jupiter from the earth was found about 8′ 16″. The elongation of the third satellite was taken with a micrometer in a telescope of 123 feet, and at the same distance of Jupiter from the earth was found 4′ 42″. The greatest elongations of the other satellites, at the same distance of Jupiter from the earth, are found from the periodic times to be 2′ 56″ 47‴, and 1′ 51″ 6‴.

The diameter of Jupiter taken with the micrometer in a 123 feet telescope several times, and reduced to Jupiter's mean distance from the earth, proved always less than 40″, never less than 38″, generally 39″. This diameter in shorter telescopes is 40″, or 41″; for Jupiter's light is a little dilated by the unequal refrangibility of the rays, and this dilatation bears a less ratio to the diameter of Jupiter in the longer and more perfect telescopes than in those which are shorter and less perfect. The times in which two satellites, the first and the third, passed over Jupiter's body, were observed, from the beginning of the ingress to the beginning of the egress, and from the complete ingress to the complete egress, with the long telescope. And from the transit of the first satellite, the diameter of Jupiter at its mean distance from the earth came forth $37\frac{1}{8}''$. and from the transit of the third $37\frac{3}{8}''$. There was observed also the time in which the shadow of the first satellite passed over Jupiter's body, and thence the diameter of Jupiter at its mean distance from the earth came out about $37''$. Let us suppose its diameter to be $37\frac{1}{4}''$ very nearly, and then the greatest elongations of the first, second, third, and fourth satellite will be respectively equal to 5,965, 9,494, 15,141, and 26,63 semi-diameters of Jupiter.

PHÆNOMENON II.

That the circumsaturnal planets, by radii drawn to Saturn's centre, describe areas proportional to the times of description; and that their periodic times, the fixed stars being at rest, are in the sesquiplicate proportion of their distances from its centre.

For, as *Cassini* from his own observations has determined; their distances from Saturn's centre and their periodic times are as follow.

The periodic times of the satellites of Saturn.

$1^d. 21^h. 18' 27''$. $2^d. 17^h. 41' 22''$. $4^d. 12^h. 25' 12''$. $15^d. 22^h. 41' 14''$. $79^d. 7^h. 48' 00''$.

The distances of the satellites from Saturn's centre, in semi-diameters of its ring.

From observations	$1\frac{19}{20}$.	$2\frac{1}{2}$.	$3\frac{1}{2}$.	8.	24.
From the periodic times . . .	1,93.	2,47.	3,45.	8.	23,35.

The greatest elongation of the fourth satellite from Saturn's centre is commonly determined from the observations to be eight of those semi-diameters very nearly. But the greatest elongation of this satellite from Saturn's centre, when taken with an excellent micrometer in Mr. *Huygens'* telescope of 123 feet, appeared to be eight semi-diameters and $\frac{7}{10}$ of a semi-diameter. And from this observation and the periodic times the distances of the satellites from Saturn's centre in semi-diameters of the ring are 2,1. 2,69. 3,75. 8,7. and 25,35. The diameter of Saturn observed in the same telescope was found to be to the diameter of the ring as 3 to 7; and the diameter of the ring, *May* 28–29, 1719, was found to be $43''$; and thence the diameter of the ring when Saturn is at its mean distance from the earth is $42''$, and the diameter of Saturn $18''$. These things appear so in very long and excellent telescopes, because in such telescopes the apparent magnitudes of the heavenly bodies bear a greater proportion to the dilatation of light in the extremities of those bodies than in shorter telescopes.

If we, then, reject all the spurious light, the diameter of Saturn will not amount to more than 16″.

PHÆNOMENON III.

That the five primary planets, Mercury, Venus, Mars, Jupiter, and Saturn, with their several orbits, encompass the sun.

That Mercury and Venus revolve about the sun, is evident from their moon-like appearances. When they shine out with a full face, they are, in respect of us, beyond or above the sun; when they appear half full, they are about the same height on one side or other of the sun; when horned, they are below or between us and the sun; and they are sometimes, *when directly under*, seen like spots traversing the sun's disk. That Mars surrounds the sun, is as plain from its full face when near its conjunction with the sun, and from the gibbous figure which it shews in its quadratures. And the same thing is demonstrable of Jupiter and Saturn, from their appearing full in all situations; for the shadows of their satellites that appear sometimes upon their disks make it plain that the light they shine with is not their own, but borrowed from the sun.

PHÆNOMENON IV.

That the fixed stars being at rest, the periodic times of the five primary planets, and (whether of the sun about the earth, or) of the earth about the sun, are in the sesquiplicate proportion of their mean distances from the sun.

This proportion, first observed by *Kepler*, is now received by all astronomers; for the periodic times are the same, and the dimensions of the orbits are the same, whether the sun revolves about the earth, or the earth about the sun. And as to the measures of the periodic times, all astronomers are agreed about them. But for the dimensions of the orbits, *Kepler* and *Bullialdus*, above all others, have determined them from observations with the greatest accuracy; and the mean distances corresponding to the periodic times differ but insensibly from those which they have assigned, and for the most part fall in between them; as we may see from the following table.

The periodic times with respect to the fixed stars, of the planets and earth revolving about the sun, in days and decimal parts of a day.

♄	♃	♂	♁	♀	☿
10759,275.	4332,514.	686,9785.	365,2565.	224,6176.	87,9692.

The mean distances of the planets and of the earth from the sun.

	♄	♃	♂
According to *Kepler*	951000.	519650.	152350.
" to *Bullialdus*	954198.	522520.	152350.
" to the periodic times	954006.	520096.	152369

	♁	♀	☿
According to *Kepler*	100000.	72400.	38806
" to *Bullialdus*	100000.	72398.	38585
" to the periodic times	100000.	72333.	38710.

As to Mercury and Venus, there can be no doubt about their distances from the sun; for they are determined by the elongations of those planets from the sun; and for the distances of the superior planets, all dispute is cut off by the eclipses of the satellites of Jupiter. For by those eclipses the position of the shadow which Jupiter projects is determined; whence we have the heliocentric longitude of Jupiter. And from its heliocentric and geocentric longitudes compared together, we determine its distance.

PHÆNOMENON V.

Then the primary planets, by radii drawn to the earth, describe areas no wise proportional to the times; but that the areas which they describe by radii drawn to the sun are proportional to the times of description.

For to the earth they appear sometimes direct, sometimes stationary, nay, and sometimes retrograde. But from the sun they are always seen direct, and to proceed with a motion nearly uniform, that is to say, a little swifter in the perihelion and a little slower in the aphelion distances, so as to maintain an equality in the description of the areas. This a noted proposition among astronomers, and particularly demonstrable in Jupiter, from the eclipses of his satellites; by the help of which eclipses, as we have said, the heliocentric longitudes of that planet, and its distances from the sun, are determined.

PHÆNOMENON VI.

That the moon, by a radius drawn to the earth's centre, describes an area proportional to the time of description.

This we gather from the apparent motion of the moon, compared with its apparent diameter. It is true that the motion of the moon is a little disturbed by the action of the sun: but in laying down these Phænomena, I neglect those small and inconsiderable errors.

PROPOSITIONS.

PROPOSITION I. THEOREM I.

That the forces by which the circumjovial planets are continually drawn off from rectilinear motions, and retained in their proper orbits, tend to Jupiter's centre; and are reciprocally as the squares of the distances of the places of those planets *from that centre.*

The former part of this Proposition appears from Phæn. I, and Prop. II or III, Book I; the latter from Phæn. I, and Cor. 6, Prop. IV, of the same Book.

The same thing we are to understand of the planets which encompass Saturn, by Phæn. II.

PROPOSITION II. THEOREM II.

That the forces by which the primary planets are continually drawn off from rectilinear motions, and retained in their proper orbits, tend to the sun; and are reciprocally as the squares of the distances of the places of those planets *from the sun's centre.*

The former part of the Proposition is manifest from Phæn. V, and Prop. II, Book I; the latter from Phæn. IV, and *Cor.* 6, Prop. IV, of the same Book. But this part of the Proposition is, with great accuracy, demonstrable from the quiescence of the aphelion points; for a very small aberration from the *reciprocal* duplicate proportion would (by Cor. 1, Prop. XLV, Book I) produce a motion of the apsides sensible enough in every single revolution, and in many of them enormously great.

PROPOSITION III. THEOREM III.

That the force by which the moon is retained in its orbit tends to the earth; and is reciprocally as the square of the distance of its place from the earth's centre.

The former part of the Proposition is evident from Phæn. VI, and Prop. II or III, Book I; the latter from the very slow motion of the moon's apogee; which in every single revolution amounting but to 3° 3′ *in consequentia*, may be neglected. For (by Cor. 1, Prop. XLV, Book I) it appears, that, if the distance of the moon from the earth's centre is to the semi-diameter of the earth as D to 1, the force, from which such a motion will result, is reciprocally as $D^{2\frac{4}{243}}$, i. e., reciprocally as the power of D, whose exponent is $2\frac{4}{243}$; that is to say, in the proportion of the distance something greater than reciprocally duplicate, but which comes $59\frac{3}{4}$ times nearer to the duplicate than to the triplicate proportion. But in regard that this motion is owing to the action of the sun (as we shall afterwards shew), it is here to be neglected. The action of the sun, attracting the moon from the earth, is nearly as the moon's distance from the earth; and therefore (by what we have shewed in Cor. 2, Prop. XLV, Book I) is to the centripetal force of the moon as 2 to 357,45, or nearly so; that is, as 1 to $178\frac{29}{40}$. And if we neglect so inconsiderable a force of the sun, the remaining force, by which the moon is retained in its orb, will be reciprocally as D^2. This will yet more fully appear from comparing this force with the force of gravity, as is done in the next Proposition.

Cor. If we augment the mean centripetal force by which the moon is retained in its orb, first in the proportion of $177\frac{29}{40}$ to $178\frac{29}{40}$, and then in the duplicate proportion of the semi-diameter of the earth to the mean distance of the centres of the moon and earth, we shall have the centripetal force of the moon at the surface of the earth; supposing this force, in descending to the earth's surface, continually to increase in the reciprocal duplicate proportion of the height.

PROPOSITION IV. THEOREM IV.

That the moon gravitates towards the earth, and by the force of gravity is continually drawn off from a rectilinear motion, and retained in its orbit.

The mean distance of the moon from the earth in the syzygies in semi-diameters of the earth, is, according to *Ptolemy* and most astronomers, 59; according to *Vendelin* and *Huygens*, 60; to *Copernicus*, $60\frac{1}{3}$; to *Street*, $60\frac{2}{5}$; and to *Tycho*, $56\frac{1}{2}$. But *Tycho*, and all that follow his tables of refraction, making the refractions of the sun and moon (altogether against the nature of light) to exceed the refractions of the fixed stars, and that by four or five minutes *near the horizon*, did thereby increase the moon's *horizontal* parallax by a like number of minutes, that is, by a twelfth or fifteenth part of the whole parallax. Correct this error, and the distance will become about $60\frac{1}{2}$ semi-diameters of the earth, near to what others have assigned. Let us assume the mean distance of 60 diameters in the syzygies; and suppose one revolution of the moon, in respect of the fixed stars, to be completed in $27^d. 7^h. 43'$, as astronomers have determined; and the circumference of the earth to amount to 123249600 *Paris* feet, as the French have found by mensuration. And now if we imagine the moon, deprived of all motion, to be let go, so as to descend towards the earth with the impulse of all that force by which (by Cor. Prop. III) it is retained in its orb, it will in the space of one minute of time, describe in its fall $15\frac{1}{12}$ *Paris* feet. This we gather by a calculus, founded either upon Prop. XXXVI, Book I, or (which comes to the same thing) upon Cor. 9, Prop. IV, of the same Book. For the versed sine of that arc, which the moon, in the space of one minute of time, would by its mean motion describe at the distance of 60 semi-diameters of the earth, is nearly $15\frac{1}{12}$ *Paris* feet, or more accurately 15 feet, 1 inch, and 1 line $\frac{4}{9}$. Wherefore, since that force, in approaching to the earth, increases in the reciprocal duplicate proportion of the distance, and, upon that account, at the surface of the earth, is 60×60 times greater than at the moon, a body in our regions, falling with that force, ought in the space of one minute of time, to describe $60 \times 60 \times 15\frac{1}{12}$ *Paris* feet; and, in the space of one second of time, to describe $15\frac{1}{12}$ of those feet; or more accurately 15 feet, 1 inch, and 1 line $\frac{4}{9}$. And with this very force we actually find that bodies here upon earth do really descend; for a pendulum oscillating seconds in the latitude of *Paris* will be 3 *Paris* feet, and 8 lines $\frac{1}{2}$ in length, as Mr. *Huygens* has observed. And the space which a heavy body describes by falling in one second of time is to half the length of this pendulum in the duplicate ratio of the circumference of a circle to its diameter (as Mr. *Huygens* has also shewn), and is therefore 15 *Paris* feet, 1 inch, 1 line $\frac{7}{9}$. And therefore the force by which the moon is retained in its orbit becomes, at the very surface of the earth, equal to the force of gravity which we observe in heavy bodies there. And therefore (by Rule I and II) the force by which the moon is retained in its orbit is that very same force which we commonly call gravity; for, were gravity another force different from that, then bodies descending to the earth with the joint impulse of both forces would fall with a double velocity, and in the space of one second of time would describe $30\frac{1}{6}$ *Paris* feet; altogether against experience.

This calçulus is founded on the hypothesis of the earth's standing still; for if both earth and moon move about the sun, and at the same time about their common centre of gravity, the distance of the centres of the moon and earth from one another will be $60\frac{1}{2}$ semi-diameters of the earth; as may be found by a computation from Prop. LX, Book I.

A.A.Logunov

ON THE ARTICLES BY HENRI POINCARE

«ON THE DYNAMICS OF THE ELECTRON»

Translated by G.Pontecorvo

H.Poincare

THE DYNAMICS OF THE ELECTRON[1]

(23 July 1905)

INTRODUCTION

It would seem at first sight that the aberration of light and the optical and electrical effects related thereto should afford a means of determining the absolute motion of the Earth, or rather its motion relative to the ether instead of relative to the other celestial bodies. An attempt at this was made, indeed, by Fresnel, but he soon perceived that the Earth's motion does not affect the laws of refraction and reflection. Similar experiments, such as that using a waterfilled telescope, or any in which only the first-order terms relative to the aberration were considered, likewise yielded only negative results. The explanation of this was soon found; but Michelson, who devised an experiment wherein the terms involving the square of the aberration should be detectable, was equally unsuccessful.

This impossibility of experimentally demonstrating the absolute motion of the Earth appears to be a general law of Nature; it is reasonable to assume the existence of this law, which we shall call **the relativity postulate**, and to assume that it is universally valid. Whether this postulate, which so far is in agreement with experiment, be later confirmed or disproved by more accurate tests, it is, in any case, of interest to see what consequences follow from it.

* *We note that* **the relativity postulate**, *the postulate of total impossibility of revealing absolute motion, as formulated in his first short note «On the dynamics of the electron», was first mentioned by Poincare in his report to the Congress of art and science held in Saint Louis in 1904*[2].

In this report Poincare lists the main principles of theoretical physics and formulates the «relativity principle, in accordance with

[1]Poincare H.O. On the dynamics of the electron // The relativity principle: Collection of works of the classics of relativity. Leningrad, 1935. P.51–129. Rend. del Circ. Mat. di Palermo 21 (1906) P.129–175.

[2]Poincare H. The present and the future of mathematical physics // The relativity principle: Coll. of works on special relativity theory. — Moscow, 1973. — P.27–44; Poincare H. L'etat actuel et l'avenir de la Physique mathematique. — Janvier 1904. — V.28. Ser. 2. — P.302–324 // The Monist. — 1905. — V.XV, N1.

which the laws of physical phenomena must be the same for a motionless observer and for an observer experiencing uniform motion along a straight line, so the laws governing physical phenomena should be the same for a motionless observer and for an observer experiencing uniform motion, so there is no way and cannot be any way of determining whether one experiences such motion or not».

One explanation, suggested by Lorentz and Fitzgerald, involves the hypothesis that all bodies undergo a contraction in the direction of the Earth's motion, of an amount proportional to the square of the aberration; such a contraction, which we shall call **the Lorentz contraction**, would explain the result of Michelson's experiment and of all others conducted heretofore. The hypothesis would nevertheless be inadequate if the relativity postulate were valid in its most general form.

Lorentz has sought to extend and modify the hypothesis so as to make it fully compatible with the relativity postulate. This he has succeeded in doing, in his paper «Electromagnetic phenomena in a system moving with any velocity smaller than that of light» (Proceedings of the Section of Sciences, Koninklijke Akademie van Wetenschappen te Amsterdam 6, 809—831, 1904)[3].

In view of the importance of this problem, I resolved to examine it further. The results which I have obtained agree with those of Lorentz in all the principal points, and I have needed only to modify and augment them in certain details. These differences, which are of but minor importance, will be shown in later sections.

* *These lines can be properly understood, if one remembers that already in his first work Poincare fully attributes his own understanding of the problem to Lorentz and speaks of it as of the «idea of Lorentz». Below he again formulates the «idea of Lorentz», although Lorentz nowhere and never wrote any such thing before Poincare.*

When Poincare writes about an agreement with Lorentz, he attributes his own understanding to Lorentz. This circumstance was not understood hitherto by many people, although it was formulated quite definitely by Poincare, if one reads his works carefully.

Lorentz's concept may be summarized thus: if a common translatory motion may be imparted to the entire system without any alternation of the observable phenomena, then the equations of an electromagnetic medium are unaltered by certain transformations, which we shall call **Lorentz transformations**. In this way two systems, of which one is fixed and the other is in translatory motion, become exact images of each other.

[3]Lorentz H.A. Electromagnetic phenomena in a system moving at any velocity smaller than the velocity of light // The relativity principle: Collection of works of the classics of relativity. Leningrad, 1935. P.16–48.

* *The last words show how profoundly Poincare understood and formulated clearly the equality, contained in the transformations of the Lorentz group, of all reference systems moving along a straight line uniformly with respect to each other, which leads to the total impossibility of determining absolute motion.*

We may note, here, the essential difference, from the point of view of Lorentz, concerning the interaction of two reference systems related by relativistic transformations. Lorentz, in his own words, «considered there being an essential difference between the systems x, y, z, t *and* x', y', z', t'*». In one reference system used are — such was my reasoning — coordinate axes having a definite location in ether and what can be called «true» time; in the other system, contrariwise, we are simply dealing with auxiliary quantities introduced only with the aid of a mathematical trick. Thus, for instance, the variable* t' *should not be called «time» in the same sense as the variable* t.

Given such reasoning, I had no intention of describing phenomena in system x', y', z', t' *in precisely the same way as in system* x, y, z, t*,...»*[4].

Lorentz further, in the same article, writes: «I tried to choose the transformation formulae so as to obtain the simplest possible equations in the new system. Later I saw from Poincare's article that, had my approach been more methodical I could have achieved greater simplicity. Having not noticed this circumstance I didn't achieve complete invariance in the equations».

Langevin[5] sought to derive a modification of Lorentz's concept. Both authors consider that an electron in motion assumes the form of an oblate spheroid; but Lorentz considers that two of the axes of this spheroid remain constant, whereas Langevin supposes that its volume remains constant. These two authors have shown that the two hypotheses are in agreement with the experiments of Kaufmann, as is Abraham's original hypothesis of a rigid spherical electron.

The advantage of Langevin's theory is that it involves only the electromagnetic forces and the constraints; but it is not compatible with the relativity postulate. This was shown by Lorentz, and I have likewise proved it by a different method, based upon the use of group theory.

[4]Lorentz H.A. Two articles by Henri Poincare about mathematical physics // The relativity principle: Coll. of works on special relativity theory. — Moscow, 1973. — P.189–196.

[5]Bucherer expressed the same idea before Langevin in Bonn. See: Bucherer A.H. Mathematische Einführung in die Elektronentheorie. — Leipzig: Druck und Verlag von B.G.Teubner, 1904. — P.148.

We must return therefore to Lorentz's theory, but, in order to maintain this free from unacceptable contradictions, a special force must be invoked to account both for the contraction and for the constancy of two of the axes. I have attempted to determine this force, and have found that **it can be regarded as a constant external pressure acting upon an electron capable of deformation and compression, the work done being proportional to the change in the volume of the electron**.

Then, if the inertia of matter is exclusively of electromagnetic origin, as has been customarily supposed since Kaufmann's experiment, and if all forces (other than the constant pressure to which I have just alluded) are of electromagnetic origin, the relativity postulate can be accepted as strictly valid. I show this by means of a very simple calculation based upon the principle of least action.

But this is not all. Lorentz, in his paper already mentioned, has deemed it necessary to extend his hypothesis in such a manner that the postulate remains valid when there exist forces other than the electromagnetic forces. In Lorentz's view, all forces, no matter how originating, are affected by the Lorentz transformation (and therefore by a translatory motion) in the same manner as the electromagnetic forces.

* *And in spite of the assertion made by Poincare being extremely clear and unambiguous, P.A.M.Dirac, even in 1980, wrote*[6]*: «In one aspect Einstein went much farther than Lorentz, Poincare and others, namely in assuming that the Lorentz transformations should be applied in all physics and not only in the case of phenomena related to electrodynamics. Any physical forces, that may be introduced in the future, must be consistent with the Lorentz transformations». Similar things are written by physicists of a somewhat lower rank, not to speak of those authors who have quite an unclear idea of physics, but also write that Poincare didn't make the decisive step towards the creation of relativity theory.*

All the above is actually not true, and the relevant answer is found in the two articles of Poincare, and if someone has not understood or has not made any attempt to do so, it is certainly not the fault of Poincare.

By comparing what was written by Poincare and the assertion made by Dirac one can readily verify that everything considered by Dirac to be the merit of Einstein is actually already all found in the first paper of Poincare. So Dirac's claim, quoted above, that «in one aspect Einstein went much farther...» simply has no grounds. It was Poincare who extended the Lorentz transformations to all the forces of Nature.

[6]Collection dedicated to Einstein. 1982–1983, Moscow: Nauka, 1986. P.218.

Surprisingly, many people exert great efforts in attributing their own incomprehension to Poincare. In formulating the relativity postulate, or the postulate of total impossibility of determining absolute motion as he terms it in his first paper, Poincare stresses its universality in applications to all natural forces. It must be noted that although here, also, Poincare, as usually, modestly attributes precedence to Lorentz, it is precisely to Poincare himself, who first established the group nature of the Lorentz transformations as well as the correct transformation laws of forces and other physical quantities, that we owe the discovery of the overall general nature of relativistic laws, independently of the nature of the forces.

It was necessary to consider this hypothesis more closely, and in particular to ascertain the changes which it would compel us to apply to the laws of gravitation.

First of all, we find that gravitational action would be propagated with the velocity of light, and not instantaneously. This might in itself appear to be sufficient reason to reject the hypothesis, for Laplace has shown that such propagation cannot occur. But, in fact, the effects of this are largely counterbalanced by another phenomenon, and there is, therefore, no contradiction between the proposed law and astronomical observations.

The question arises whether it is possible to discover a law which satisfies Lorentz's condition and which yet reduces to Newton's whenever the velocities of the bodies are so small that the squares of these velocities (and the products of the accelerations and the distances) may be neglected in comparison with the square of the velocity of light.

It will be seen later that the answer must be affirmative.

Is the law, thus modified, compatible with astronomical observations?

At first sight it appears to be so, but a more detailed discussion is necessary to settle the question.

Even assuming, however, that the new hypothesis survives this test, what conclusion is to be drawn? If the gravitational attraction is propagated with the velocity of light, this cannot occur by mere chance, but must be dependent on the ether; we should then have to investigate the nature of this dependence, and attempt to relate it to other such dependences.

* *It is interesting to note that although the relativity postulate put forward by Poincare implies total impossibility of determining the motion of matter relative to ether, the concept itself of ether is not discarded, since it is difficult to imagine any greater absurdity than empty space. Many people consider the removal of ether to be a most important revolutionary achievement of relativity theory. This is not true. It is quite evident that relativity theory did not simply «destroy» ether, but only assigned it a certain state of motion.*

This was subsequently noted by Einstein, who wrote: «... a closer examination reveals that special relativity theory does not require unconditional negation of ether. One can accept the existence of ether, but one needs not care about assigning it any definite state of motion; in other words, it must, in abstract terms, be deprived of its last mechanical sign still left it by Lorentz. We shall further see that general relativity theory justifies such an approach...»[7].

In modern theoretical physics the notion of ether gave up its place to the concept of physical vacuum, in which there are inevitably present quantum fluctuations — zero oscillations of quantum fields.

We cannot be satisfied with formulae that are merely placed side by and agree only by a lucky chance; these formulae must, as it were, interlock. The mind will consent only when it sees the reason for the agreement, and when this agreement even seems to have been predictable.

But the matter may be viewed in a different light, as an analogy will show. Let us imagine some astronomer before Copernicus, pondering upon the Plotemainc system. He would notice that, for every planet, either the epicycle or the deferent is traversed in the same time. This cannot be due to chance, and there must be some mysterious bond between all the planets of the system.

Then Copernicus, by a simple change of the coordinate axes which were supposed fixed, did away with this seeming relationship: every planet described one circular orbit only, and the periods of revolution became independent of one another — until Kepler once more established the relationship that had apparently been destroyed.

Now, there may be an analogy with our problem. If we assume the relativity postulate, we find a quantity common to the law of gravitation and the laws of electromagnetism, and this quantity is the velocity of light; and this same quantity appears in every other force, of whatever origin. There can be only two explanations.

Either, everything in the universe is of electromagnetic origin; or, this constituent which appears common to all the phenomena of physics has no real existence, but arises from our methods of measurement. What are these methods? One might first reply, the bringing into juxtaposition of objects regarded as invariable solid things; but this is no longer so in our present theory, if the Lorentz contraction is assumed. In this theory, two lengths are by definition equal if they are traversed by light in the same time.

Perhaps the abandonment of this definition would suffice to overthrow Lorentz's theory as decisively as the system of Ptolemy was by the work of Copernicus. Should this ever happen, it would by no means argue the futility of

[7]Einstein A. Collection of scientific papers: in 4 volumes. Ed. I.Ye.Tamm, Ya.A.Smorodinsky, B.G.Kuznetsov. — Moscow: Nauka, 1965, V.1. P.685–686.

Lorentz's analysis: whatever the faults of the Ptolemainc theory, it was the necessary foundation for Copernicus to build upon.

I have therefore not hesitated to publish these incomplete results, even though at the present time the entire theory may seem to be threatened by the discovery of cathode rays.

SECTION 9.
HYPOTHESES CONCERNING GRAVITATION

> * *«Mass has two aspects: these are both inertia and the gravitational mass, which is involved in Newtonian gravity as a multiplication factor. If the coefficient of inertia is constant, can the gravitational mass also be constant? That is the question».*
>
> *H.Poincare*
> *The present and future of mathematical physics (1904)*[18]

Thus, Lorentz' theory would entirely account for the impossibility of demonstrating absolute motion, provided that all forces were of electromagnetic origin.

But there exist forces, such as gravitation, which cannot be regarded as being of electromagnetic origin. It may happen that two systems of bodies create equivalent electromagnetic fields, in the sense of exerting the same action upon electrified bodies and currents, while at the same time these two systems do not exert the same gravitational action upon Newtonian masses.

The gravitational field is therefore not identical with the electromagnetic field. Lorentz was thus compelled to augment his hypothesis by assuming that **forces, of whatever origin, and in particular gravitation, are affected by translation** (or, if one prefers, by the Lorentz transformation) **in the same way as the electromagnetic forces.**

We must now examine this hypothesis in detail. If the Newtonian force is to behave in such a way under the Lorentz transformation, we can no longer

[18]Poincare H. L'avenir de la Physique mathematique // Bulletin des Sciences Mathematiques. Janvier 1904 — V.28. Ser.2. — P.302–324.

suppose that this force depends only on the relative position of the attracting and the attracted body at the instant concerned; it must depend also on the velocities of the two bodies. Moreover, we may reasonably assume that the force acting upon the attracted body, at an instant t, depends on the position and velocity of the body at that instant; but it will also depend on the position and velocity of the attracting body, not at the instant but at some previous instant, as if gravitation required a certain time for its propagation.

Let us consider therefore the position of the attracted body at the instant t_0, and let its coordinates at that instant be $\vec{r}_0 = (x_0, y_0, z_0)$ and the components of its velocity be $\vec{v}$, and let us consider the attracting body at the corresponding instant $t_0 + t$, its coordinates at that instant being $\vec{r}_0 + \vec{r}$ and its velocity components $\vec{v}_1$.

First of all, we must have a relationship

$$\varphi(t, \vec{r}, \vec{v}, \vec{v}_1) = 0 \tag{1}$$

to determine the time t. This relationship express the law of propagation of gravitational action; I shall by no means impose the condition that propagation occurs with the same velocity in every direction.

Next, let $\vec{F} = (F_z, F_y, F_z)$ be the three components of the action exerted upon the attracted body at the instant t. We have to express $\vec{F} = (F_x, F_y, F_z)$ as functions of

$$t, \vec{r}, \vec{v}, \vec{v}_1. \tag{2}$$

The conditions to be satisfied are as follows:

1. The relationship (1) must not be affected by the transformations of the Lorentz group.

2. The components $\vec{F} = (F_x, F_y, F_z)$ must behave, under the Lorentz transformations, in the same manner as the electromagnetic forces denoted by the same letters, that is, as shown by equations (11′) of section 1.

3. When both bodies are at rest, the usual law of attraction must apply.

In the latter case, however, is should be noted that the relationship (1) plays no part, since the time t is of no significance if both bodies are at rest.

The problem thus stated is clearly indeterminate. We shall therefore seek to satisfy as many further conditions as possible.

4. Astronomical observations do not appear to reveal any perceptible deviation from Newton's law, and we shall therefore choose the solution which differs least from this law when the velocities of the two bodies are small.

5. We shall attempt to ensure that t is always negative; for, whereas it is reasonable that the effect of gravitation should require a certain time for its propagation, we should find it more difficult to understand how this effect could depend on a position of the attracting body which the latter has not yet reached.

There is one case where the problem is no longer indeterminate, namely if the two bodies are at relative rest, i.e. if

$$\vec{v} = \vec{v}_1,$$

we shall therefore first investigate this case, assuming that these velocities are constant, and therefore that the two bodies are executing a common uniform motion of translation in a straight line.

We may assume that the x-axis has been taken to be parallel to this motion of translation, so that $\vec{v}_\perp = 0$, and we shall take $\beta = v_x = v_\parallel$.

If, under these conditions, we apply the Lorentz transformation, the two bodies will be at rest after the transformation, with

$$\vec{v}' = 0.$$

The components $\vec{F'}$ must then be in accordance with Newton's law and we have, apart from a constant factor,

$$\vec{F}' = -\vec{r}'/r'^3; \quad r'^2 = \vec{r}^2 = x'^2 + y'^2 + z'^2. \tag{3}$$

But, from section 1,

$$x' = \gamma(x - \beta t), \quad y' = y, \quad z' = z, \quad t' = \gamma(t - \beta x).$$

$$\frac{\rho'}{\rho} = \gamma(1 - \beta v_x) = \gamma(1 - \beta^2) = \gamma^{-1}, \quad \vec{F}\vec{v} = F_\parallel \beta.$$

$$F'_\parallel = \gamma \frac{\rho}{\rho'}(F_\parallel - \beta \vec{F} \cdot \vec{v}) = \gamma^2 F_\parallel (1 - \beta^2) = F_\parallel.$$

$$\vec{F}'_\perp = \frac{\rho}{\rho'}\vec{F}_\perp = \gamma \vec{F}_\perp.$$

Moreover,

$$x - \beta t = x - v_\parallel t, \quad r'^2 = \gamma^2 (x - vt)^2 + y^2 + z^2$$

and

$$F_\parallel = -\gamma(x - v_\parallel t)/t'^3, \quad \vec{F}_\perp = -\vec{r}_\perp/\gamma r'^3, \tag{4}$$

which may also be written

$$F_x = \frac{\partial V}{\partial x}, \quad F_y = \frac{\partial V}{\partial y}, \quad F_z = \frac{\partial V}{\partial z}, \quad V = 1/\gamma r'. \tag{4'}$$

It seems at first sight that the indeterminacy remains, since no hypotheses have been made concerning the value of t, that is, concerning the velocity of propagation. Moreover, x is a function of t. But it is easily seen that the quantities $x - v_{\|}t, y, z$ and which appear in the formulae do not depend on t.

Thus, if two bodies have a common translatory motion, the force acting upon the attracted body is normal to an ellipsoid having the attracting body at its centre.

In order to proceed further, it is necessary to ascertain **the invariants of the Lorentz group.**

> * *The further exposition represents a general formulation of the mathematical principles of relativity theory, including tensor calculus and investigation of invariants of the Lorentz group corresponding to relativistically invariant physical quantities and relations.*
>
> *Here Poincare for the first time considers the transformations of the Lorentz group as elements of transformations of the four-dimensional manifold, the coordinates of points of which serve as sets of the space coordinates x, y, z and of the «imaginary» time $t\sqrt{-1}$, conserving the quadratic form $x^2 + y^2 + z^2 - t^2$.*
>
> *Following Poincare, Minkowski subsequently developed these arguments concerning the unity of space and time, the geometry of which is pseudo-Euclidean. In his famous talk at a meeting of German scientists and doctors in Köln, which turned out to be decisive for relativity theory acquiring wide-spread recognition, G.Minkowski expressed these ideas as follows:*
>
> *«Dear gentlemen! The notions of space and time, which I intend to develop here, are based on physical experiment. That is why they are powerful. Their tendency is radical. From now on, space by itself and time by itself will have to become fictions, while only a certain form joining the two together will still retain independence»*[19].

It is known that the substitutions forming this group (if $l = 1$) are linear and such that the quadratic form

$$x^2 + y^2 + z^2 - t^2$$

is invariant. Putting

$$\vec{v} = \delta\vec{r}/\delta t, \quad \delta\vec{r} = (\delta x,\, \delta y,\, \delta z),$$

$$\vec{v}_1 = \delta_1\vec{r}/\delta_1 t, \quad \delta_1\vec{r} = (\delta_1 x,\, \delta_1 y,\, \delta_1 z).$$

We see that the Lorentz transformation causes δx, δy, δz, δt and $\delta_1 x$, $\delta_1 y$, $\delta_1 z$, $\delta_1 t$ to undergo the same linear substitutions as x, y, z, t.

[19]Minkowski H. Raum und Zeit: Vortäge von der 80. Naturforchersammlung zu Köln // Physikalische Zeitschrift. — 1909. — B.10. N.3. — P.104–111.

If

$$x,\ y,\ z,\ t\sqrt{-1},$$

$$\delta x,\ \delta y,\ \delta z,\ \delta t\sqrt{-1},$$

$$\delta_1 x,\ \delta_1 y,\ \delta_1 z,\ \delta_1 t\sqrt{-1},$$

are regarded as the coordinates of three points P, P', P'' in four-dimensional space, we see that the Lorentz transformation is simply a rotation of this space about a fixed origin.

The only distinct invariants are therefore the six distances of the points P, P', P'' from one another and from the origin, or alternatively the two expressions

$$x^2 + y^2 + z^2 - t^2, \quad x\delta x + y\delta y + z\delta z - t\delta t,$$

and the four expressions of the same form obtained by permuting the three points P, P', P'' in any manner.

What we are seeking, however, is invariant functions of the ten variables (2); we must therefore find, among combinations of the six invariants, those which depend only on these ten variables, i.e. those which are homogeneous and of degree zero with respect to δx, δy, δz, δt and with respect to $\delta_1 x$, $\delta_1 y$, $\delta_1 z$, $\delta_1 t$. This leaves four distinct invariants, namely

$$\vec{r}^{\,2} - t^2, \quad \frac{t - \vec{r}\,\vec{v}}{\sqrt{1-\vec{v}^{\,2}}}, \quad \frac{t - \vec{r}\,\vec{v}_1}{\sqrt{1-\vec{v}_1^{\,2}}}, \quad \frac{1 - \vec{v}\,\vec{v}_1}{\sqrt{1-\vec{v}^{\,2}}\sqrt{1-\vec{v}_1^{\,2}}}. \tag{5}$$

Let us now consider how the components of the force are transformed. We return to equations (11) of section 1, which refer not to the force $\vec{F}$, discussed here but to the force per unit volume. Putting

$$f_t = \vec{f}\cdot\vec{v},$$

we see that these equations (11) may be written (with $l = 1$)

$$F_x' = \gamma(f_x - \beta f_t), \quad f_x' = \gamma(f_t - \beta f_x),$$

$$f_y' = f_y, \quad f_z' = f_z. \tag{6}$$

Thus, f_x, f_y, f_z, f_t are transformed in the same manner as x, y, z, t. The invariants of the group will therefore be

$$\vec{f}^{\,2} - f_t^2, \quad \vec{f}\vec{r} - f_t t, \quad \vec{f}\delta\vec{r} - f_t\delta t, \quad \vec{f}\delta_1\vec{r} - f_t\delta_1 t.$$

The quantities in which we are interested are not $\vec{f}$, but $\vec{F}$, with

$$T = \vec{F}\cdot\vec{v}.$$

Evidently

$$\vec{F}/\vec{f} = T/f_t = 1\rho.$$

Thus the Lorentz transformation will act upon $\vec{F}$, T, in the same way as upon $\vec{f}$, f_t, except that these expressions will in addition be multiplied by

$$\frac{\rho}{\rho'} = \frac{1}{\gamma(1-\beta v_x)} = \frac{\delta t}{\delta t'}.$$

Likewise, the transformation will act upon $\vec{v}$ in the same way as upon δx, δy, δz, δt, except that these expressions will in addition be multiplied by the same factor,

$$\frac{\delta t}{\delta t'} = \frac{1}{\gamma(1-\beta v_x)}.$$

Let us now regard f_x, f_y, f_z, f_t as being the coordinates of a fourth point Q; The invariants will then be functions of the distances between the five points O, P, P', P'', Q and these functions must be homogeneous of degree zero, firstly with respect to f_x, f_y, f_z, f_t, δx, δy, δz, δt (which variables can subsequently be replaced by F_x, F_y, F_z, T, $\vec{v}$, 1), and secondly with respect to $\delta_1 x$, $\delta_1 y$, $\delta_1 z$, 1 (which variables can subsequently be replaced by $\vec{v}_1$, 1).

In this way we find, in addition to the four invariants (5), four further and distinct invariants, namely

$$\frac{\vec{F}^2 - T^2}{1-\vec{v}^2,} \quad \frac{\vec{F}\vec{r} - Tt}{\sqrt{1-v^2}}, \quad \frac{\vec{F}\vec{v}_1 - T}{\sqrt{1-\vec{v}^2}\sqrt{1-\vec{v}_1^2}}, \quad \frac{\vec{F}\vec{v} - T}{1-\vec{v}^2}. \tag{7}$$

The last of these is always zero, according to the definition of T.

* *Having discovered the Lorentz group and the fundamental invariant $x^2 + y^2 + z^2 - t^2$, Poincare established that a series of physical quantities must be individual components of unique four-dimensional quantities varying under the Lorentz transformations like time and coordinates. Below we present some such quantities constructed by Poincare:*

time and space coordinates — $[t, \vec{x}]$;

work per unit time and force reduced to unit volume — $[\vec{f}\vec{v},\ \vec{f}]$;

work per unit time and force reduced to unit charge — $[\gamma(\vec{F}\vec{v}),\ \gamma\vec{F}]$;

four-component velocity, or four-momentum — $[\gamma,\ \gamma\vec{v}]$;

charge and current — $[\rho,\ \rho\vec{v}]$;

scalar and vector potential — $[\varphi,\ \vec{A}]$ *with* $\gamma = 1/\sqrt{1-v^2}$.

Which are the conditions that must now be satisfied?

1. The left-hand side of equation (1), which defines the velocity of propagation, must be a function of the four invariants (5).

It is obvious that a large number of hypotheses could be constructed. We shall consider only two of these.

(A) It may be that

$$\vec{r}^2 - t^2 = r^2 - t^2 = 0,$$

whence $t = \pm r$, and, since t must be negative, $t = -r$.

This means that the velocity of propagation is equal to that of light.

At first, it seems that this hypothesis should be rejected immediately; for Laplace has shown that the propagation is either instantaneous or much more rapid than that of light. But Laplace was discussing the hypothesis of a finite velocity of propagation alone, whereas here it is compounded with many others, and there may happen to be some more or less complete mutual compensation between them, a situation of which many examples have already appeared in the applications of the Lorentz transformation.

(B) It may be that

$$\frac{t - \vec{r}\vec{v}_1}{\sqrt{1 - \vec{v}_1^2}} = 0, \quad t = \vec{r}\vec{v}_1.$$

The velocity of propagation is then much more rapid than that of light, but in certain cases might be negative, which, as we have said, seems hardly acceptable. We shall therefore abide by hypothesis (A).

2. The four invariants (7) must be functions of the invariants (5).

3. When both bodies are at absolute rest, $\vec{F}$, must have the values given by Newton's law; when the bodies are at relative rest, the values must be those given by equations (4).

In the case of absolute rest, the first two invariants (7) must reduce to

$$\vec{F}^2, \quad \vec{F}\vec{r},$$

or, by Newton's law, to

$$\frac{1}{r^4}, \quad -\frac{1}{r}.$$

According to hypothesis (A), the second and third of the invariants (5) become

$$-\frac{r + \vec{r}\vec{v}}{\sqrt{1 - \vec{v}^2}}, \quad -\frac{r + \vec{r}\vec{v}_1}{\sqrt{1 - \vec{v}_1^2}},$$

that is, for absolute rest,

$$-r, \; -r.$$

We may therefore assume, for example, that the first two invariants (5) reduce to

$$(1 - \vec{v}_1^2)^2/(r + \vec{r}\vec{v}_1)^4, \quad -\sqrt{1 - \vec{v}_1^2}/(r + \vec{r}\vec{v}_1),$$

but other combinations are possible.

It is necessary to choose some combination, and a third equation is also needed to determine $\vec{F}$. In making the choice, we shall attempt to remain as close as possible to Newton's law. Let us then examine the result when the squares of the velocities $\vec{v}, \vec{v}_1$, etc., are neglected (and $t = -r$).

The four invariants (5) then become

$$0, -r - \vec{r}\vec{v}, \quad -r - \vec{r}\vec{v}_1, 1,$$

and the four invariants (7) become

$$\vec{F}^2, \quad \vec{F}(\vec{r} + \vec{v}r), \quad \vec{F}(\vec{v}_1 - \vec{v}), \quad 0.$$

In order to compare this with Newton's law, however, a further transformation is necessary. In these equations, $x_0 + x$, $y_0 + y$, $z_0 + z$ represent the coordinates of the attracting body at the instant $t_0 + r$, and $r = |\vec{r}|$; in Newton's law, we have to consider the coordinates $x_0 + x_1$, $y_0 + y_1$, $z_0 + z_1$ of the attracting body at the instant t_0, and the distance $r_1 = |\vec{r}_1|$.

We may neglect the square of the time t occupied by the propagation, and therefore regard the motion as uniform; then

$$\vec{r} = \vec{r}_1 + \vec{v}_1 t, \quad r(r - r_1) = t(\vec{r}v_1)$$

or, since,

$$t = -r, \quad \vec{r} = \vec{r}_1 + \vec{v}_1 r, \quad r = r_1 - \vec{r}\vec{v}_1,$$

and the four invariants (5) become

$$0, -r_1 + \vec{r}(\vec{v}_1 - \vec{v}), \quad -r_1, \quad 1,$$

and the four invariants (7) become

$$\vec{F}^2, \quad \vec{F}[\vec{r}_1 + (\vec{v} - \vec{v}_1)r_1], \quad \vec{F}(\vec{v}_1 - v), \quad 0.$$

In the second of these expressions I have written r_1, in place of r, since r is multiplied by $v - v_1$ and the square of v is neglected.

Newton's law gives, for these four invariants (7),

$$\frac{1}{r_1^4}, \quad -\frac{1}{r_1} - \frac{\vec{r}_1(\vec{v} - \vec{v}_1)}{r_1^2}, \quad \frac{\vec{r}_1(v - \vec{v}_1)}{r_1^3}, \quad 0.$$

If, therefore, we denote the second and third invariants (5) by A and B, and the first three invariants (7) by M, N and P, Newton's law will be obeyed, to within terms of the order of the squares of the velocities, by putting

$$M = \frac{1}{B^4}, \quad N = \pm\frac{A}{B^2}, \quad P = \frac{A-B}{B^3}. \tag{8}$$

This solution is not unique: if the fourth invariant (5) is denoted by C, then $C - 1$ is of the order of v, as is $(A - B)^2$.

We may therefore add to the right-hand side of each of the equations (8) a term consisting of $C - 1$ multiplied by any function of A, B and C, and a term consisting of $(A - B)^2$ also multiplied by any function of A, B, and C.

The solution (8) appears the simplest at first sight, but it cannot be accepted. Since M, N and P are functions of $\vec{F}$, and $T = \vec{F}\vec{v}$, these equations yield values of $\vec{F}$; but the resulting values may in some cases be imaginary.

In order to avoid this difficulty, we proceed differently, putting

$$\gamma_0 = \frac{1}{\sqrt{1-\vec{v}^{\,2}}}, \quad \gamma_1 = \frac{1}{\sqrt{1-\vec{v}_1^{\,2}}},$$

by analogy with

$$\gamma = \frac{1}{\sqrt{1-\beta^2}},$$

as in the Lorentz substitution.

Then, with the condition $-r = t$, the invariants (5) become

$$0, \quad A = -\gamma_0(r + \vec{r}\cdot\vec{v}), \quad B = -\gamma_1(r + r\cdot\vec{v}_1),$$

$$C = \gamma_0\gamma_1(1 - \vec{v}\cdot\vec{v}_1).$$

Moreover, the following systems of quantities:

$$\begin{array}{cccc} x, & y, & z, & t = -r \\ \gamma_0 F_x, & \gamma_0 F_y, & \gamma_0 F_z, & \gamma_0 T \\ \gamma_0 v_x, & \gamma_0 v_y, & \gamma_0 v_z, & \gamma_0 \\ \gamma_1 v_{1x}, & \gamma_1 v_{1y}, & \gamma_1 v_{1z}, & \gamma_1 \end{array}$$

are seen to undergo the same linear substitutions when the transformations of the Lorentz group are applied to them.

* *Here Poincare introduces, for the first time, the sets of quantities $F_\mu = (\gamma_0 T, \gamma_0\vec{F})$ and $u_\mu = (\gamma_0, \gamma_0\vec{v})$ transforming with respect to the same linear irreducible (tensor) law like the set of space-time coordinates $x_\mu = (t, \vec{x})$ and presently termed the four-vectors of force and velocity. In this case the relations $u_\mu^2 = 1$ and $u_\mu F_\mu = 0$ are identities.*

We therefore put

$$\vec{F} = a\vec{r}/\gamma_0 + b\vec{v} + c\vec{v}_1\gamma_1/\gamma_0,$$
$$T = -ar/\gamma_0 + b + c\gamma_1/\gamma_0. \tag{9}$$

It is evident that, if a, b, c are invariants $\vec{F}$, T will satisfy the fundamental condition, i.e. will undergo an appropriate linear substitution when the Lorentz transformations are applied to them.

If the equations (9) are compatible, we must have

$$\vec{F}\vec{v} - T = 0.$$

When $\vec{F}$, T are replaced by their values (9), the result is, after multiplication by γ_0^2,

$$-Aa - b - Cc = 0. \tag{10}$$

The desired conclusion is that the values of $\vec{F}$ should remain in accordance with Newton's law when the square of the velocities $\vec{v}$, $\vec{v}_1$, etc., and the products of the accelerations and the distances are neglected in comparison with the square of the velocity of light.

We can take

$$b = 0, \quad c = -aA/C.$$

To the approximation used,

$$\gamma_0 = \gamma_1 = 1, \quad C = 1, \quad A = -r_1 + \vec{r}(\vec{v}_1 - \vec{v}), \quad B = -r_1,$$
$$\vec{r} = \vec{r}_1 + \vec{v}_1 t = \vec{r}_1 - \vec{v}_1 r.$$

Then the first equation (9) becomes

$$\vec{F} = a(\vec{r} - A\vec{v}_1).$$

But, if v^2 is neglected, Av_1 may be replaced by $-r_1v$, or by $-rv_1$, whence

$$\vec{F} = a(\vec{r} + \vec{v}_1 r) = a\vec{r}_1.$$

Newton's law would give

$$\vec{F} = -\vec{r}_1/r_1^3.$$

We must therefore take as the invariant a one which reduces to $-1/r_1^3$ within the approximation adopted, that is, $1/B^3$. The equations (9) then become

$$\vec{F} = \vec{r}/\gamma_0 B^3 - \gamma A\vec{v}_1/\gamma_0 B^3 C,$$
$$T = -r/\gamma_0 B^3 - \gamma_1 A/\gamma_0 B^3 C. \tag{11}$$

It is seen, first of all, that the corrected attraction consists of two components, one parallel to the vector joining the position of the two bodies, and the other parallel to the velocity of the attracting body.

When we speak of the position or the velocity of the attracting body, we mean its position or velocity at the instant when the gravitational wave leaves it; but the position or the velocity of the attracted body means its position or velocity at the instant when the gravitational wave reaches it, this wave being assumed to be propagated with the velocity of light.

I believe that it would be premature to attempt to continue the discussion of these formulae, and I shall therefore confine myself to making a few comments.

1. The solutions (11) are not unique; for the common factor may be replaced by

$$\frac{1}{B^3} + (C-1)f_1(A.\ B.\ C) + (A-B)^2 f_2(A.\ B.\ C).$$

where f_1 and f_2 are any functions of A, B and C. Moreover b, need not be taken as zero; any additional terms may be added to a, b and c which satisfy the condition (10) and are of the second order in $\vec{v}$ for a, and of the first order in $\vec{v}$ for b and c.

2. The first equation (11) may be written

$$\vec{F} = -\frac{\gamma_1}{B^3 C}[\vec{r}(1-\vec{v}\vec{v}_1) + \vec{v}_1(r+\vec{r}\vec{v})], \tag{11'}$$

and the quantity in the brackets may turn be written

$$(\vec{r} + \vec{v}_1 r) + [\vec{v}[\vec{v}_1 \times \vec{r}]] \tag{12}$$

so that the total force is divisible into three components corresponding to the three parentheses in equation (12). The first component is somewhat similar to the mechanical force due to the electric field, the other two to the mechanical force due to the magnetic field. By virtue of comment 1, I may replace $1/B^3$ in equations (11) by C/B^3, so that $\vec{F}$ are linear functions of the velocity $\vec{v}$ of the attracted body, having been eliminated from the denominator of (11′). This completes the analogy.

Putting then

$$\vec{e} = \gamma_1(\vec{r} + r\vec{v}_1), \quad \vec{h} = \gamma_1[\vec{v}_1 \times \vec{r}], \tag{13}$$

with eliminated C from the denominator of (11′) we obtain

$$\vec{F} = \frac{1}{B^3}\vec{e} + \frac{1}{B^3}[\vec{v} \times \vec{h}], \tag{14}$$

and also

$$B^2 = \vec{e}^2 - \vec{h}^2. \tag{15}$$

Thus $\vec{e}$ or $\vec{e}/B^3$ is a kind of electric field, while $\vec{h}$ or $\vec{h}/B^3$ is a kind of magnetic field.

3. The relativity postulate would compel us to use either the solution (11) or the solution (14) or any one of the solutions obtained therefrom by using comment 1. But the prime question is whether these are compatible with astronomical observations. The deviation from Newton's law is of the order of v^2, that is 10000 times less than if it had been of the order of v, as it would have been with the velocity of propagation equal to that of light and the other conditions unchanged. We may therefore hope that the deviation will not be very great; but only a more extended investigation will furnish the answer to this question.

* *Let us now briefly list some of the main results obtained by Poincare in this work.*

A formulation is presented of the relativity principle, formulated by Poincare in 1904 (see footnote 2 on page 7) for all physical phenomena.

It has been shown, for the first time, that the Lorentz transformations form, together with space rotations, a group termed by Poincare the Lorentz group. Infinitesimal operators of the Lorentz group have been constructed. Invariance has been revealed of the quadratic form $t^2 - x^2 - y^2 - z^2$ *under transformations of the Lorentz group.*

Poincare's discovery of the Lorentz group and of the fundamental invariant $t^2 - x^2 - y^2 - z^2$ *permitted him to construct a series of four-dimmensional quantities, which under Lorentz transformations vary like the time and the space coordinates.*

Here follow some of these quantities:
work per unit time and force reduced to unit volume — $[\vec{f}\vec{v}, \vec{f}]$;
work per unit time and force reduced to unit charge — $[\gamma(\vec{F}\vec{v}), \gamma\vec{F}]$;
four-component velocity, or four-momentum — $[\gamma, \gamma\vec{v}]$;
charge and current — $[\rho, \rho\vec{v}]$;
scalar and vector potential — $[\varphi\vec{A}]$.

With the aid of these transformation laws Poincare established, for the first time, that the Maxwell–Lorentz equations and, also, which is extremely important, the Lorentz force acting on a elementary charge in unit volume do not alter their form under transformations of the Lorentz group. Thus, it was shown that no phenomena can be used for establishing, whether one is in the state of rest or of linear and uniform motion.

Thus, Poincare demonstrated that the relativity principle for electromagnetic phenomena follows from the Maxwell–Lorentz equations as a rigorous mathematical truth.

The relativistic law for adding velocities was first established by Poincare.

Poincare was the first to demonstrate the invariance of the action integral for an electromagnetic field under transformations of the Lorentz group and to discover the fundamental invariants of the electromagnetic field,

$$\vec{E}^{\,2} - \vec{H}^{\,2}, \quad \vec{E} \cdot \vec{H}.$$

Poincare was the first to discover the equations of relativistic mechanics (in units $m = c = 1$)

$$\frac{d}{dt}\left(\frac{\vec{v}}{\sqrt{1-\vec{v}^{\,2}}}\right) = \vec{F},$$

$$\frac{d}{dt}\left(\frac{1}{\sqrt{1-\vec{v}^{\,2}}}\right) = \vec{F}\vec{v},$$

and to write the corresponding expression for the Lagrangian functions of a moving material point.

Poincare introduced four-dimensional space with the coordinates (x, y, z, $t\sqrt{-1}$) and showed that transformations of the Lorentz group correspond to various rotations in this space about the origin. He constructed various invariants of the Lorentz group.

He put forward the hypothesis that all natural forces, including gravitational forces, must transform in the same manner under Lorentz transformations.

Poincare introduced the concept of gravitational waves travelling with the speed of light. He demonstrated that the hypothesis asserting propagation of the forces of gravity with the speed of light does not contradict observational data.

Even this short list of results reveals that Poincare discovered, in an extremely precise and general form, almost all the essential constituents of relativity theory.

Doc. 47

ON THE RELATIVITY PRINCIPLE AND THE CONCLUSIONS DRAWN FROM IT

by A. Einstein

[*Jahrbuch der Radioaktivität und Elektronik* 4 (1907): 411-462]

Newton's equations of motion retain their form when one transforms to a new system of coordinates that is in uniform translational motion relative to the system used originally according to the equations

$$x' = x - vt$$
$$x' = y$$
$$z' = z \ .$$

As long as one believed that all of physics can be founded on Newton's equations of motion, one therefore could not doubt that the laws of nature are the same without regard to which of the coordinate systems moving uniformly (without acceleration) relative to each other they are referred. However, this independence from the state of motion of the system of coordinates used, which we will call "the principle of relativity," seemed to have been suddenly called into question by the brilliant confirmations of H. A. Lorentz's electrodynamics of moving bodies.[1] That theory is built on the presupposition of a resting, immovable, luminiferous ether; its basic equations are not such that they transform to equations of the same form when the above transformation equations are applied.

After the acceptance of that theory, one had to expect that one would succeed in demonstrating an effect of the terrestrial motion relative to the luminiferous ether on optical phenomena. It is true that in the study cited
[2] Lorentz proved that in optical experiments, as a consequence of his basic assumptions, an effect of that relative motion on the ray path is not to be expected as long as the calculation is limited to terms in which the ratio

---

[1] [1]H. A. Lorentz, *Versuch einer Theorie der elektrischen und optischen Erscheinungen in bewegten Körpern.* [Attempt at a theory of electric and optical phenomena in moving bodies] Leiden, 1895. Reprinted Leipzig, 1906.

v/c of the relative velocity to the velocity of light in vacuum appears in the first power. But the negative result of Michelson and Morley's experiment[1] showed that in a particular case an effect of the second order (proportional to v^2/c^2) was not present either, even though it should have shown up in the experiment according to the fundamentals of the Lorentz theory.

It is well known that this contradiction between theory and experiment was formally removed by the postulate of H. A. Lorentz and FitzGerald, [4]
according to which moving bodies experience a certain contraction in the direction of their motion. However, this ad hoc postulate seemed to be only an artificial means of saving the theory: Michelson and Morley's experiment had actually shown that phenomena agree with the principle of relativity even where this was not to be expected from the Lorentz theory. It seemed therefore as if Lorentz's theory should be abandoned and replaced by a theory whose foundations correspond to the principle of relativity, because such a theory would readily predict the negative result of the Michelson and Morley experiment. [5]

Surprisingly, however, it turned out that a sufficiently sharpened conception of time was all that was needed to overcome the difficulty discussed. One had only to realize that an auxiliary quantity introduced by H. A. Lorentz and named by him "local time" could be defined as "time" in [6]
general. If one adheres to this definition of time, the basic equations of Lorentz's theory correspond to the principle of relativity, provided that the above transformation equations are replaced by ones that correspond to the new conception of time. H. A. Lorentz's and FitzGerald's hypothesis appears then as a compelling consequence of the theory. Only the conception of a luminiferous ether as the carrier of the electric and magnetic forces does not fit into the theory described here; for electromagnetic forces appear here not as states of some substance, but rather as independently existing things that are similar to ponderable matter and share with it the feature of inertia. [7]

The following is an attempt to summarize the studies that have resulted to date from the merger of the H. A. Lorentz theory and the principle of relativity.

---

[1]A. A. Michelson and E. W. Morley, *Amer. J. of Science* 34, (1887): 333. [3]

The first two parts of the paper deal with the kinematic foundations as well as with their application to the fundamental equations of the Maxwell-Lorentz theory, and are based on the studies[1] by H. A. Lorentz (*Versl. Kon.*
[9] *Akad. v. Wet., Amsterdam* (1904)) and A. Einstein (*Ann. d. Phys.* 16 (1905)).

In the first section, in which only the kinematic foundations of the theory are applied, I also discuss some optical problems (Doppler's principle, aberration, dragging of light by moving bodies); I was made aware of the possibility of such a mode of treatment by an oral communication and a paper
[10] by Mr. M. Laue (*Ann. d. Phys.* 23 (1907): 989), as well as a paper (though in
[11] need of correction) by Mr. J. Laub (*Ann. d. Phys.* 32 (1907)).

In the third part I develop the dynamics of the material point (electron). In the derivation of the equations of motion I used the same method as
[12] [13] in my paper cited earlier. Force is defined as in Planck's study. The reformulations of the equations of motion of material points, which so clearly demonstrate the analogy between these equations of motion and those of classical mechanics, are also taken from that study.

The fourth part deals with the general inferences regarding the energy and momentum of physical systems to which one is led by the theory of relativity. These have been developed in the original studies,

[14] A. Einstein, *Ann. d. Phys.* 18 (1905): 639 and *Ann. d. Phys.* 23 (1907): 371, as well as M. Planck, *Sitzungsber. d. Kgl. Preuss. Akad. d. Wissensch.* XXIX (1907),

but are here derived in a new way, which, it seems to me, shows especially clearly the relationship between the above application and the foundations of the theory. I also discuss here the dependence of entropy and temperature on the state of motion; as far as entropy is concerned, I kept completely to the Planck study cited, and the temperature of moving bodies I defined as did Mr. Mosengeil in his study on moving black-body radiation.[2]

The most important result of the fourth part is that concerning the inertial mass of the energy. This result suggests the question whether energy also possesses *heavy* (gravitational) mass. A further question suggesting itself is whether the principle of relativity is limited to *nonaccelerated* moving systems. In order not to leave this question totally undiscussed, I added to the present paper a fifth part that contains a novel consideration, based on the principle of relativity, on acceleration and gravitation.

---

[8] [1]E. Cohn's studies on the subject are also pertinent, but I did not make use of them here.

[15] [2]Kurd von Mosengeil, *Ann. d. Phys.* 22 (1907): 867.

V. PRINCIPLE OF RELATIVITY AND GRAVITATION

§17. *Accelerated reference system and gravitational field*

So far we have applied the principle of relativity, i.e., the assumption that the physical laws are independent of the state of motion of the reference system, only to *nonaccelerated* reference systems. Is it conceivable that the principle of relativity also applies to systems that are accelerated relative to each other?

While this is not the place for a detailed discussion of this question, it will occur to anybody who has been following the applications of the principle of relativity. Therefore I will not refrain from taking a stand on this question here.

We consider two systems Σ_1 and Σ_2 in motion. Let Σ_1 be accelerated
[93] in the direction of its X-axis, and let γ be the (temporally constant) magnitude of that acceleration. Σ_2 shall be at rest, but it shall be located in a homogeneous gravitational field that imparts to all objects an acceleration $-\gamma$ in the direction of the X-axis.

[94] As far as we know, the physical laws with respect to Σ_1 do not differ from those with respect to Σ_2; this is based on the fact that all bodies are equally accelerated in the gravitational field. At our present state of experience we have thus no reason to assume that the systems Σ_1 and Σ_2 differ from each other in any respect, and in the discussion that follows, we shall therefore assume the complete physical equivalence of a gravitational field and a corresponding acceleration of the reference system.

This assumption extends the principle of relativity to the uniformly accelerated translational motion of the reference system. The heuristic value of this assumption rests on the fact that it permits the replacement of a homogeneous gravitational field by a uniformly accelerated reference system, the latter case being to some extent accessible to theoretical treatment.

§18. *Space and time in a uniformly accelerated reference system*

We first consider a body whose individual material points, at a given time t of the nonaccelerated reference system S, possess no velocity relative to S, but a certain acceleration. What is the influence of this acceleration γ on the shape of the body with respect to S?

If such an influence is present, it will consist of a constant-ratio dilatation in the direction of acceleration and possibly in the two directions perpendicular to it, since an effect of another kind is impossible for reasons of symmetry. The acceleration-caused dilatations (if such exist at all) must be even functions of γ; hence they can be neglected if one restricts oneself to the case in which γ is so small that terms of the second or higher power in γ may be neglected. Since we are going to restrict ourselves to that case, we do not have to assume that the acceleration has any influence on the shape of the body.

We now consider a reference system Σ that is uniformly accelerated relative to the nonaccelerated system S in the direction of the latter's X-axis. The clocks and measuring rods of Σ, examined at rest, shall be identical with the clocks and measuring rods of S. The coordinate origin of Σ shall move along the X-axis of S, and the axes of Σ shall be perpetually parallel to those of S. At any moment there exists a nonaccelerated reference system S' whose coordinate axes coincide with the coordinate axes of Σ at the moment in question (at a given time t' of S'). If the coordinates of a point event occurring at this time t' are ξ, η, ζ with respect to Σ, we will have

$$\left.\begin{aligned} x' &= \xi \\ y' &= \eta \\ z' &= \zeta \end{aligned}\right\} ,$$

because in accordance with what we said above, we are not to assume that acceleration affects the shape of the measuring instruments used for measuring ξ, η, ζ. We shall also imagine that the clocks of Σ are set at time t' of S' such that their readings at that moment equal t'. What about the rate of the clocks in the next time element τ?

First of all, we have to bear in mind that a specific effect of *acceleration* on the rate of the clocks of Σ need not be taken into account, since it would have to be of the order γ^2. Furthermore, since the effect of the velocity attained during τ on the rate of the clocks is negligible, and the distances traveled by the clocks during the time τ relative to those traveled by S' are also of the order τ^2, i.e., negligible, the readings of the clocks of Σ may be fully replaced by readings of the clocks of S' for the time element τ. [95]

From the foregoing it follows that, relative to Σ, light in vacuum is propagated during the time element τ with the universal velocity c if we define simultaneity in the system S' which is momentarily at rest relative to Σ, and if the clocks and measuring rods we use for measuring the time and length are identical with those used for the measurement of time and space in nonaccelerated systems. Thus the principle of constancy of the velocity of light can be used here too to define simultaneity if one restricts oneself to very short light paths.

We now imagine that the clocks of Σ are adjusted, in the way described, at that time $t = 0$ of S at which Σ is instantaneously at rest relative to S. The totality of readings of the clocks of Σ adjusted in
[96] this way is called the "local time" σ of the system Σ. It is immediately evident that the physical meaning of the local time σ is as follows. If one uses the local time σ for the temporal evaluation of processes occurring in the individual space elements of Σ, then the laws obeyed by these processes cannot depend on the position of these space elements, i.e., on their coordinates, if not only the clocks, but also the other measuring tools used in the
[97] various space elements are identical.

However, we must not simply refer to the local time σ as the "time" of Σ, because according to the definition given above, two point events occurring at different points of Σ are not simultaneous when their local times σ are equal. For if at time $t = 0$ two clocks of Σ are synchronous with respect to S and are subjected to the same motions, then they remain forever synchronous with respect to S. However, for this reason, in accordance with §4, they do not run synchronously with respect to a reference system S' instantaneously at rest relative to Σ but in motion relative to S, and hence according to our definition they do not run synchronously with respect to Σ either.

We now define the "time" τ of the system Σ as the totality of those readings of the clock situated at the coordinate origin of Σ which are, according to the above definition, simultaneous with the events which are to be temporally evaluated.[1]

We shall now determines the relation between the time τ and the local time σ of a point event. It follows from the first of equations (1) that

[1]Thus the symbol "τ" is used here in a different sense than above.

two events are simultaneous with respect to S', and thus also with respect to Σ, if

$$t_1 - \frac{v}{c^2} x_1 = t_2 - \frac{v}{c^2} x_2 ,$$

where the subscripts refer to the one or to the other point event, respectively. We shall first confine ourselves to the consideration of times that are so short[1] that all terms containing the second or higher power of τ or v can be omitted; taking (1) and (29) into account, we then have to put [98]

$$x_2 - x_1 = x_2' - x_2' = \xi_2 - \xi_1$$
$$t_1 = \sigma_1 \qquad t_2 = \sigma_2$$
$$v = \gamma t = \gamma \tau ,$$ [99]

so that we obtain from the above equation

$$\sigma_2 - \sigma_1 = \frac{\gamma \tau}{c^2} (\xi_2 - \xi_1) .$$

If we move the first point event to the coordinate origin, so that $\sigma_1 = \tau$ and $\xi_1 = 0$, we obtain, omitting the subscript for the second point event,

$$\sigma = \tau \left[1 + \frac{\gamma \xi}{c^2} \right] . \tag{30}$$

This equation holds first of all if τ and ξ lie below certain limits. It is obvious that it holds for arbitrarily large τ if the acceleration γ is constant with respect to Σ, because the relation between σ and τ must then be linear. Equation (30) does not hold for arbitrarily large ξ. From the fact that the choice of the coordinate origin must not affect the relation, one must conclude that, strictly speaking, equation (30) should be replaced by the equation

$$\sigma = \tau e^{\frac{\gamma \xi}{c^2}} .$$

Nevertheless, we shall maintain formula (30).

[1]In accordance with (1), we thereby also assume a certain restriction with respect to the values of $\xi = x'$.

According to §17, equation (30) is also applicable to a coordinate system in which a homogeneous gravitational field is acting. In that case we have to put $\Phi = \gamma\xi$, where Φ is the gravitational potential, so that we obtain

$$\sigma = \tau\left[1 + \frac{\Phi}{c^2}\right] . \tag{30a}$$

We have defined two kinds of times for Σ. Which of the two definitions do we have to use in the various cases? Let us assume that at two locations of different gravitational potentials $(\gamma\xi)$ there exists one physical system each, and we want to compare their physical quantities. To do this, the most natural procedure might be as follows: First we take our measuring tools to the first physical system and carry out our measurements there; then we take our measuring tools to the second system to carry out the same measurement here. If the two sets of measurements give the same results, we shall denote the two physical systems as "equal." The measuring tools include a clock with which we measure local times σ. From this it follows that to define the physical quantities at some position of the gravitational field, it is natural to use the time σ.

However, if we deal with a phenomenon in which objects situated at positions with different gravitational potentials must be considered simultaneously, we have to use the time τ in those terms in which time occurs explicitly (i.e., not only in the definition of physical quantities), because otherwise the simultaneity of the events would not be expressed by the equality of the time values of the two events. Since in the definition of the time τ a clock situated in an arbitrarily chosen position is used, but not an arbitrarily chosen instant, when using time τ the laws of nature can vary with position but not with time.

§19. *The effect of the gravitational field on clocks*

If a clock showing local time is located in a point P of gravitational potential Φ, then, according to (30a), its reading will be $(1 + \frac{\Phi}{c^2})$ times greater than the time τ, i.e., it runs $(1 + \frac{\Phi}{c^2})$ times faster than an identical clock located at the coordinate origin. Suppose an observer located somewhere in space perceives the indications of the two clocks in a certain way, e.g., optically. As the time $\Delta\tau$ that elapses between the instants at which a clock indication occurs and at which this indication is perceived by the observer is independent of τ, for an observer situated somewhere in space the clock in point P runs $(1 + \frac{\Phi}{c^2})$ times faster than the clock at the coordinate origin. In this sense we may say that the process occurring in the clock, and, more generally, any physical process, proceeds faster the greater the gravitational potential at the position of the process taking place.

There exist "clocks" that are present at locations of different gravitational potentials and whose rates can be controlled with great precision; these are the producers of spectral lines. It can be concluded from the aforesaid[1] that the wave length of light coming from the sun's surface, which originates from such a producer, is larger by about one part in two millionth than that of light produced by the same substance on earth. [100]

§20. *The effect of gravitation on electromagnetic phenomena*

If we refer an electromagnetic process at some point of time to a non-accelerated reference system S' that is instantaneously at rest relative to the reference system Σ accelerated as above, then the following equations will hold according to (5) and (6):

$$\frac{1}{c}\left[\rho' u'_x + \frac{\partial X'}{\partial t'}\right] = \frac{\partial N'}{\partial y'} - \frac{\partial M'}{\partial z'}\ , \text{ etc.}$$

and

$$\frac{1}{c}\frac{\partial L'}{\partial t'} = \frac{\partial Y'}{\partial z'} - \frac{\partial Z'}{\partial y'}\ , \text{ etc.}$$

In accordance with the above, we may readily equate the S'-referred quantities ρ', u', X', L', x', etc., with the corresponding Σ-referred

[1]While assuming that equation (30a) holds for an inhomogeneous gravitational field as well.

quantities ρ, u, X, L, ξ, etc., if we limit ourselves to an infinitesimally short period[1] that is infinitesimally close to the time of relative rest of S' and Σ. Further, we have to replace t' by the local time σ. However, we must not simply put

$$\frac{\partial}{\partial t'} = \frac{\partial}{\partial \sigma} ,$$

because a point which is at rest relative to Σ, and to which equations transformed to Σ should refer, changes its velocity relative to S' during the time element $dt' = d\sigma$, to which change, according to equations (7a) and (7b), there corresponds a temporal change of the Σ-related field component. Hence we have to put

$$\frac{\partial X'}{\partial t'} = \frac{\partial X}{\partial \sigma} \qquad \frac{\partial L'}{\partial t'} = \frac{\partial L}{\partial \sigma}$$

$$\frac{\partial Y'}{\partial t'} = \frac{\partial Y}{\partial \sigma} + \frac{\gamma}{c} N \qquad \frac{\partial M'}{\partial t'} = \frac{\partial M}{\partial \sigma} - \frac{\gamma}{c} Z$$

$$\frac{\partial Z'}{\partial t'} = \frac{\partial Z}{\partial \sigma} - \frac{\gamma}{c} M \qquad \frac{\partial N'}{\partial t'} = \frac{\partial N}{\partial \sigma} + \frac{\gamma}{c} Y .$$

Hence the Σ-referred electromagnetic equations are

$$\frac{1}{c}\left[\rho u_\xi + \frac{\partial X}{\partial \sigma}\right] = \frac{\partial N}{\partial \eta} - \frac{\partial M}{\partial \zeta}$$

$$\frac{1}{c}\left[\rho u_\eta + \frac{\partial Y}{\partial \sigma} + \frac{\gamma}{c} N\right] = \frac{\partial L}{\partial \zeta} - \frac{\partial N}{\partial \xi}$$

$$\frac{1}{c}\left[\rho u_\xi + \frac{\partial Z}{\partial \sigma} - \frac{\gamma}{c} M\right] = \frac{\partial M}{\partial \xi} - \frac{\partial L}{\partial \eta}$$

$$\frac{1}{c}\frac{\partial L}{\partial \sigma} = \frac{\partial Y}{\partial \zeta} - \frac{\partial Z}{\partial \eta}$$

$$\frac{1}{c}\left[\frac{\partial M}{\partial \sigma} - \frac{\gamma}{c} Z\right] = \frac{\partial Z}{\partial \xi} - \frac{\partial X}{\partial \zeta}$$

$$\frac{1}{c}\left[\frac{\partial N}{\partial \sigma} + \frac{\gamma}{c} Y\right] = \frac{\partial X}{\partial \eta} - \frac{\partial Y}{\partial \xi} .$$

[1]This restriction does not affect the range of validity of our results because inherently the laws to be derived cannot depend on the time.

We multiply these equations by $\left[1 + \frac{\gamma\xi}{c^2}\right]$ and put for the sake of brevity

$$X^* = X\left[1 + \frac{\gamma\xi}{c^2}\right], \qquad Y^* = Y\left[1 + \frac{\gamma\xi}{c^2}\right], \text{ etc.}$$

$$\rho^* = \rho\left[1 + \frac{\gamma\xi}{c^2}\right].$$

Neglecting terms of the second power in γ, we obtain the equations

$$\left.\begin{aligned} \frac{1}{c}\left[\rho^* u_\xi + \frac{\partial X^*}{\partial \sigma}\right] &= \frac{\partial N^*}{\partial \eta} - \frac{\partial M^*}{\partial \zeta} \\ \frac{1}{c}\left[\rho^* u_\eta + \frac{\partial Y^*}{\partial \sigma}\right] &= \frac{\partial L^*}{\partial \zeta} - \frac{\partial N^*}{\partial \xi} \\ \frac{1}{c}\left[\rho^* u_\zeta + \frac{\partial Z^*}{\partial \sigma}\right] &= \frac{\partial M^*}{\partial \xi} - \frac{\partial L^*}{\partial \eta} \end{aligned}\right\} \tag{31a}$$

$$\left.\begin{aligned} \frac{1}{c}\frac{\partial L^*}{\partial \sigma} &= \frac{\partial Y^*}{\partial \zeta} - \frac{\partial Z^*}{\partial \eta} \\ \frac{1}{c}\frac{\partial M^*}{\partial \sigma} &= \frac{\partial Z^*}{\partial \xi} - \frac{\partial X^*}{\partial \zeta} \\ \frac{1}{c}\frac{\partial N^*}{\partial \sigma} &= \frac{\partial X^*}{\partial \eta} - \frac{\partial Y^*}{\partial \xi}. \end{aligned}\right\} \tag{32a}$$

These equations show first of all how the gravitational field affects the static and stationary phenomena. The same laws hold as in the gravitation-free field, except that the field components X, etc. are replaced by $X\left[1 + \frac{\gamma\xi}{c^2}\right]$, etc., and ρ is replaced by $\rho\left[1 + \frac{\gamma\xi}{c^2}\right]$.

Furthermore, to follow the development of nonstationary states, we make use of the time τ in the terms differentiated with respect to time as well as in the definition of the velocity of electricity, i.e., we put according to (30)

$$\frac{\partial}{\partial \tau} = \left[1 + \frac{\gamma\xi}{c^2}\right]\frac{\partial}{\partial \tau}$$ [101]

and

$$w_\xi = \left[1 + \frac{\gamma\xi}{c^2}\right].$$ [102]

We thus obtain

$$\frac{1}{c\left[1+\frac{\gamma\xi}{c^2}\right]}\left[\rho^* w_\xi + \frac{\partial X^*}{\partial\tau}\right] = \frac{\partial N^*}{\partial\eta} - \frac{\partial M^*}{\partial\zeta} \quad \text{etc.} \tag{31b}$$

and

[103]

$$\frac{1}{c\left[1+\frac{\gamma\xi}{c^2}\right]}\frac{\partial L^*}{\partial\tau} = \frac{\partial Y^*}{\partial\zeta} = \frac{\partial Z^*}{\partial\eta} \quad \text{etc.} \tag{32b}$$

These equations too have the same form as the corresponding equations of the nonaccelerated or gravitation-free space; however, c is here replaced by the value

$$c\left[1+\frac{\gamma\xi}{c^2}\right] = c\left[1+\frac{\Phi}{c^2}\right] .$$

From this it follows that those light rays that do not propagate along the ξ-axis are bent by the gravitational field; it can easily be seen that the change of direction amounts to $\frac{\gamma}{c^2}\sin\varphi$ per cm light path, where φ
[104] denotes the angle between the direction of gravity and that of the light ray.

With the help of these equations and the equations relating the field strength and the electric current of one point, which are known from the optics of bodies at rest, we can calculate the effect of the gravitational field on optical phenomena in bodies at rest. One has to bear in mind, however, that the above-mentioned equations from the optics of bodies at rest hold for the local time σ. Unfortunately, the effect of the terrestrial gravitational field is so small according to our theory (because of the smallness of $\frac{\gamma x}{c^2}$) that there is no prospect of a comparison of the results of
[105] the theory with experience.

If we successively multiply equations (31a) and (32a) by $\frac{X^*}{4\pi} \cdots\cdots \frac{N^*}{4\pi}$ and integrate over infinite space, we obtain, using our earlier notation,

[106]

$$\int\left[1+\frac{\gamma\xi}{c^2}\right]^2\frac{\rho}{4\pi}(uX+u_\eta Y+uZ)\,d\omega+\int\left[1+\frac{\gamma\xi}{c^2}\right]^2\cdot\frac{1}{8\pi}\frac{\partial}{\partial\sigma}(X^2+Y^2+\cdots+N^2)\,d\omega = 0 .$$

[107] $\frac{\rho}{4\pi}(uX + u_\eta Y + u_\zeta Z)$ is the energy η_σ supplied to the matter per unit volume and unit local time σ if this energy is measured by measuring tools situated at the corresponding location. Hence, according to (30),

$\eta_\tau = \eta^\sigma\left[1 + \frac{\gamma\xi}{c^2}\right]$ is the (similarly measured) energy supplied to the matter per unit volume and unit local time τ; $\frac{1}{8\pi}(X^2 + Y^2 \cdots + N^2)$ is the electromagnetic energy ϵ per unit volume, measured the same way. If we take into account that according to (30) we have to set $\frac{\partial}{\partial\sigma} = \left[1 - \frac{\gamma\xi}{c^2}\right]\frac{\partial}{\partial\tau}$, we obtain [108]

$$\int \left[1 + \frac{\gamma\xi}{c^2}\right]\eta_\tau d\omega + \frac{d}{d\tau}\left\{\int \left[1 + \frac{\gamma\xi}{c^2}\right]\epsilon \; d\omega\right\} = 0 \; .$$

This equation expresses the principle of conservation of energy and contains a very remarkable result. An energy, or energy input, that, measured locally, has the value $E = \epsilon d\omega$ or $E = \eta \; d\omega d\tau$, respectively, contributes to the energy integral, in addition to the value E that corresponds to its magnitude, also a value $\frac{E}{c^2}\,\gamma\xi = \frac{E}{c^2}\,\Phi$ that corresponds to its *position*. Thus, to each energy E in the gravitational field there corresponds an energy of position that equals the potential energy of a "ponderable" mass of magnitude $\frac{E}{c^2}$.

Thus the proposition derived in §11, that to an amount of energy E there corresponds a mass of magnitude $\frac{E}{c^2}$, holds not only for the *inertial* but also for the *gravitational* mass, if the assumption introduced in §17 is correct.

(Received on 4 December 1907)

Chapter 2

Einstein's Deepest Insight and Its Early Impacts*

*A. Einstein, M. Grossmann, D. Hilbert, E. Cartan

Doc. 13

Outline of a Generalized Theory of Relativity and of a Theory of Gravitation

I. Physical Part

by Albert Einstein

II. Mathematical Part

by Marcel Grossmann

[Teubner, Leipzig, 1913]

I
Physical Part

The theory expounded in what follows derives from the conviction that the proportionality between the inertial and the gravitational mass of bodies is an exactly valid law of nature that must already find expression in the very foundation of theoretical physics. I already sought to give expression to this conviction in several earlier papers by seeking to reduce the *gravitational* mass to the *inertial* mass;[1] this endeavor led me to the hypothesis that, from a physical point of view, an (infinitesimally extended, homogeneous) gravitational field can be completely replaced by a state of acceleration of the reference system. This hypothesis can be expressed pictorially in the following way: An observer enclosed in a box can in no way decide whether the box is at rest in a static gravitational field, or whether it is in accelerated motion, maintained by forces acting on the box, in a space that is free of gravitational fields (equivalence hypothesis). [2]

We know the fact that the law of proportionality of inertial and gravitational mass is satisfied to an extraordinary degree of accuracy from the fundamentally important investigation by Eötvös,[2] which is based on the following argument. A body at rest on the surface of the Earth is acted upon by gravity as well as by the centrifugal force resulting from Earth's rotation. The first of these forces is

[1]A. Einstein, *Ann. d. Phys.* 35 (1911): 898; 38 (1912):355; 38 (1912): 443. [1]

[2]B. Eötvös, *Mathematische und naturwissenschaftliche Berichte aus Ungarn* 8 (1890); Wiedemann's *Beiblätter* 15 (1891): 688. [3]

proportional to the gravitational mass, and the second to the inertial mass. Thus, the direction of the resultant of these two forces , i.e., the direction of the apparent gravitational force (direction of the plumb) would have to depend on the physical nature of the body under consideration if the proportionality of the inertial and gravitational mass were not satisfied. In that case the apparent gravitational forces acting on parts of a heterogeneous rigid system would, in general, not merge into a resultant; instead, in general, there would still be a torque associated with the apparent gravitational forces that would have to make itself noticeable if the system were suspended from a torsion-free thread. By having established the absence of such torques with great care, Eötvös proved that, for the bodies that he investigated, the relationship of the two masses was independent of the nature of the body to such a degree of exactness that the relative difference in this relationship that might still exist from one substance to another must be smaller than one twenty-millionth.

The decomposition of radioactive substances occurs with a release of such significant quantities of energy that the change in the inertial mass of the system that corresponds to that energy decrease according to the theory of relativity is not very small relative to the total mass.[3] In the case of the decay of radium, for example, this decrease amounts to one ten-thousandth of the total mass. If these changes of the inertial mass did not correspond to changes in the *gravitational* mass, then there would have to be deviations of the inertial mass from the gravitational mass much
[5] greater than those allowed by Eötvös's experiments. Hence it must be considered very probable that the identity of the inertial and gravitational mass is exactly satisfied. For these reasons it seems to me that the equivalence hypothesis, which asserts the essential physical identity of the gravitational with the inertial mass, possesses a high degree of probability.[4]

§3. The Significance of the Fundamental Tensor of the $g_{\mu\nu}$ for the Measurment of Space and Time

From the foregoing, one can already infer that there cannot exist relationships between the space-time coordinates x_1, x_2, x_3, x_4 and the results of measurements obtainable by means of measuring rods and clocks that would be as simple as those in the old relativity theory. With regard to time, this has already found to be true in the case of the static gravitational field.[8] The question therefore arises, what is the

---

[6]Cf. Part II, §1.

[7]Cf. Part II, §1.

[16] [8]Cf., e.g., A. Einstein, *Ann. d. Phys.* 35 (1911): 903 ff.

physical meaning (measurability in principle) of the coordinates x_1, x_2, x_3, x_4.

We note in this connection that *ds* is to be conceived as the invariant measure of the distance between two infinitely close space-time points. For that reason, *ds* must also possess a physical meaning that is independent of the chosen reference system. We will assume that *ds* is the "naturally measured" distance between the two space-time points, and by this we will understand the following. [17]

The immediate vicinity of the point (x_1, x_2, x_3, x_4) with respect to the coordinate system is determined by the infinitesimal variables dx_1, dx_2, dx_3, dx_4. We assume that, in their place, new variables $d\xi_1$, $d\xi_2$, $d\xi_3$, $d\xi_4$ are introduced by means of a linear transformation in such a way that

$$ds^2 = d\xi_1^2 + s\xi_2^2 + d\xi_3^2 - d\xi_4^2.$$

In this transformation the $g_{\mu\nu}$ are to be viewed as constants; the real cone $ds^2 = 0$ appears referred to its principal axes. Then the ordinary theory of relativity holds in this elementary $d\xi$ system, and the physical meaning of lengths and times shall be the same in this sytem as in the ordinary theory of relativity, i.e., *ds* is the square of the four-dimensional distance between two infinitely close space-time points, measured by means of a rigid body that is not accelerated in the $d\xi$-system, and by means of unit measuring rods and clocks at rest relative to it. [18]

From this one sees that, for given dx_1, dx_2, dx_3, dx_4, the natural distance that corresponds to these differentials can be determined only if one knows the quantities $g_{\mu\nu}$ that determine the gravitational field. This can also be expressed in the following way: the gravitational field influences the measuring bodies and clocks in a determinate manner. [19]

From the fundamental equation

$$ds^2 = \sum_{\mu\nu} g_{\mu\nu} dx_\mu dx_\nu$$

one sees that, in order to fix the physical dimensions of the quantities $g_{\mu\nu}$ and x_ν, yet another stipulation is required. The quantity *ds* has the dimension of a length. Likewise, we wish to view the x_ν (x_4 too) as lengths, and thus we do not ascribe any physical dimension to the quantities $g_{\mu\nu}$.

§5. The Differential Equations of the Gravitational Field

Having established the momentum-energy equation for material processes (mechanical, electrical, and other processes) in relation to the gravitational field, there remains for us only the following task. Let the tensor $\Theta_{\mu\nu}$ for the material process be given.

What differential equations permit us to determine the quantities g_{ik}, i.e., the gravitational field? In other words, we seek the generalization of Poisson's equation

$$\Delta\varphi = 4\pi k\rho .$$

We have not found a method for the solution of this problem as thoroughly compelling as that for the solution of the problem discussed previously. It would be necessary to introduce several assumptions whose correctness seems plausible but not evident.

The generalization that we seek would likely have the form

$$\kappa\cdot\Theta_{\mu\nu} = \Gamma_{\mu\nu}, \tag{11}$$

where κ is a constant and $\Gamma_{\mu\nu}$ a second-rank contravariant tensor derived from the fundamental tensor $g_{\mu\nu}$ by differentiatial operations. In line with the Newton-Poisson law one would be inclined to require that these equations (11) be *second order*. But it must be stressed that, given this assumption, it proves impossible to find a differential expression $\Gamma_{\mu\nu}$ that is a generalization of $\Delta\varphi$ and that proves to be a *tensor* with respect to *arbitrary* transformations.[10] To be sure, it cannot be negated a priori that the final, exact equations of gravitation could be of higher than second order. Therefore there still exists the possibility that the perfectly exact differential equations of gravitation could be covariant with respect to *arbitrary* substitutions. But given the present state of our knowledge of the physical properties of the gravitational field, the attempt to discuss such possibilities would be premature. For that reason we have to confine ourselves to the second order, and we must therefore forgo setting up gravitational equations that are covariant with respect to arbitrary transformations. Besides, it should be emphasized that we have no basis whatsoever
[24] for assuming a general covariance of the gravitational equations.[11]

The Laplacian scalar $\Delta\varphi$ is obtained from the scalar φ if one forms the expansion (the gradient) of the latter and then the inner operator (the divergence) of this. Both operations can be generalized in such a way that one can carry them out on every tensor of arbitrarily high rank, namely while permitting arbitrary substitutions of the basic variables.[12] But these operations degenerate if they are carried out on the fundamental tensor $g_{\mu\nu}$.[13] From this it seems to follow that the equations sought will be covariant only with respect to a particular group of transformations,
[25] which group however, is as yet unknown to us.

---

[10]Cf. Part II, §4, No. 2.

[11]Cf. also the arguments given at the beginning of §6.

[12]Part II, §2.

[13]Cf. the remark on p. 28 in Part II, §2.

Given this state of affairs, and in view of the old theory of relativity, it seems natural to assume that *the transformation group we are seeking also includes the linear transformations.* Hence we require that $\Gamma_{\mu\nu}$ be a tensor with respect to arbitrary linear transformations.

Now it is easy to prove (by carrying out the transformation) the following theorems: [26]

1. If $\Theta_{\alpha\beta\ldots\lambda}$ is a contravariant tensor of rank n with respect to linear transformations, then

$$\sum_{\mu} \gamma_{\mu\nu} \frac{\partial \Theta_{\alpha\beta\ldots\lambda}}{\partial x_{\mu}}$$

is a contravariant tensor of rank $n + 1$ with respect to linear transformations (expansion).[14]

2. If $\Theta_{\alpha\beta\ldots\lambda}$ is a contravariant tensor of rank n with respect to linear transformations, then

$$\sum_{\lambda} \frac{\partial \Theta_{\alpha\beta\ldots\lambda}}{\partial x_{\lambda}}$$

is a contravariant tensor of rank $n - 1$ with respect to linear transformations (divergence).

If one carries out these two operations on a tensor in succession, one obtains a tensor of the same rank as the original one (operation Δ, carried out on a tensor). For the fundamental tensor $\gamma_{\mu\nu}$ one obtains

$$\text{(a)} \qquad \sum_{\alpha\beta} \frac{\partial}{\partial x_{\alpha}} \left(\gamma_{\alpha\beta} \frac{\partial \gamma_{\mu\nu}}{\partial x_{\beta}} \right). \qquad [27]$$

One can also see from the following argument that this operator is related to the Laplacian operator. In the theory of relativity (absence of gravitational field) one would have to set

$$g_{11} = g_{22} = g_{33} = -1, \quad g_{44} = c^2, \quad g_{\mu\nu} = 0, \quad \text{for } \mu \neq \nu;$$

hence $\quad \gamma_{11} = \gamma_{22} = \gamma_{33} = -1, \quad \gamma_{44} = \dfrac{1}{c^2}, \quad \gamma_{\mu\nu} = 0, \text{ for } \mu \neq \nu.$

If a gravitational field is present that is sufficiently weak, i.e., if the $g_{\mu\nu}$ and $\gamma_{\mu\nu}$ differ only infinitesimally from the values just given, then one obtains instead of the expression (a), neglecting the second-order terms,

---

[14] $\gamma_{\mu\nu}$ is the contravariant tensor reciprocal to $g_{\mu\nu}$ (Part II, §1).

$$-\left(\frac{\partial^2\gamma_{\mu\nu}}{\partial x_1^2}+\frac{\partial^2\gamma_{\mu\nu}}{\partial x_2^2}+\frac{\partial^2\gamma_{\mu\nu}}{\partial x_3^2}-\frac{1}{c^2}\frac{\partial^2\gamma_{\mu\nu}}{\partial x_4^2}\right).$$

If the field is static and only $g_{\mu\nu}$ is variable, we thus arrive at the case of the Newtonian theory of gravitation if we take the expression obtained for the quantity
[28] $\Gamma_{\mu\nu}$ up to a constant.

Hence one might think that, up to a constant factor, the expression (a) must already be the generalization of $\Delta\varphi$ that we are seeking. But this would be a mistake; for alongside this expression, in a generalization of this kind there could also appear terms that are themselves tensors and that vanish when we neglect the kinds of terms just indicated. This always occurs when two first derivatives of the $g_{\mu\nu}$ or $\gamma_{\mu\nu}$ are multiplied by each other. Thus, for example,

$$\sum_{\alpha\beta}\frac{\partial g_{\alpha\beta}}{\partial x_\mu}\cdot\frac{\partial\gamma_{\alpha\beta}}{\partial x_\nu}$$

is a covariant tensor of the second rank (with respect to linear transformations); it becomes infinitesimally small to the second order if the quantities $g_{\alpha\beta}$ and $\gamma_{\alpha\beta}$ deviate from constant values only infinitesimally to the first order. We must therefore allow still other terms in $\Gamma_{\mu\nu}$, in addition to (a), which terms, for now, must satisfy only the condition that, taken together, they must possess the character of a tensor with respect to linear transformations.

We make use of the momentum-energy law to find these terms. To make myself
[29] clear about the method used, I will first apply it to a generally known example.

In *electrostatics* $-\frac{\partial\varphi}{\partial x_\nu}\rho$ is the νth component of the momentum transferred to the matter per unit volume, if φ denotes the electrostatic potential and ρ the electric density. We seek a differential equation for φ of such kind that the law of the conservation of momentum is always satisfied. It is well known that the equation

$$\sum_\nu\frac{\partial^2\varphi}{\partial x_\nu^2}=\rho$$

solves the problem. The fact that the momentum law is satisfied follows from the identity

$$\sum_\mu\frac{\partial}{\partial x_\mu}\left(\frac{\partial\varphi}{\partial x_\nu}\frac{\partial\varphi}{\partial x_\mu}\right)-\frac{\partial}{\partial x_\nu}\left(\frac{1}{2}\sum_\mu\left(\frac{\partial\varphi}{\partial x_\mu}\right)^2\right)=\frac{\partial\varphi}{\partial x_\nu}\sum_\mu\frac{\partial^2\varphi}{\partial x_\mu^2}\left(=-\frac{\partial\varphi}{\partial x_\nu}\cdot\rho\right).$$

Thus, if the momentum law is satisfied, then an identity of the following construction must exist for every ν: On the right side, $-\frac{\partial\varphi}{\partial x_\nu}$ is multiplied by the left side of the differential equation; on the left side of the identity there is a sum of the differential quotients.

If the differential equation for φ were not yet known, the problem of finding it would be reduced to that of finding this identity. What is essential for us to realize is that this identity can be derived *if one of the terms occurring in it is known.* All one has to do is to apply repeatedly the product differentiation rule in the forms

$$\frac{\partial}{\partial x_\nu}(uv) = \frac{\partial u}{\partial x_\nu}v + \frac{\partial v}{\partial x_\nu}u$$

and

$$u\frac{\partial v}{\partial x_\nu} = \frac{\partial}{\partial x_\nu}(uv) - \frac{\partial u}{\partial x_\nu}v,$$

and then finally to put the terms that are differential quotients on the left side and the rest of the terms on the right side. For example, if one starts with the first term of the above identity, one obtains, one after another,

$$\sum_\mu \frac{\partial}{\partial x_\mu}\left(\frac{\partial\varphi}{\partial x_\nu}\frac{\partial\varphi}{\partial x_\mu}\right) = \sum_\mu \frac{\partial\varphi}{\partial x_\nu}\cdot\frac{\partial^2\varphi}{\partial x_\mu^2} + \sum_\mu \frac{\partial\varphi}{\partial x_\mu}\cdot\frac{\partial^2\varphi}{\partial x_\nu\partial x_\mu}$$
$$= \frac{\partial\varphi}{\partial x_\nu}\cdot\sum_\mu \frac{\partial^2\varphi}{\partial x_\mu^2} + \frac{\partial}{\partial x_\nu}\left\{\frac{1}{2}\sum_\mu\left(\frac{\partial\varphi}{\partial x_\mu}\right)^2\right\},$$

from which we obtain the above identity upon rearrangement.

Now we turn again to our problem. It follows from equation (10) that

$$\frac{1}{2}\sum_{\mu\nu}\sqrt{-g}\cdot\frac{\partial g_{\mu\nu}}{\partial x_\sigma}\Theta_{\mu\nu} \qquad (\sigma = 1,2,3,4)$$

is the momentum (or energy) imparted by the gravitational field to the matter per unit volume. For the energy-momentum law to be satisfied, the differential expressions $\Gamma_{\mu\nu}$ of the fundamental quantities $\gamma_{\mu\nu}$ that enter the gravitational equations

$$\kappa\cdot\Theta_{\mu\nu} = \Gamma_{\mu\nu}$$

must be chosen such that

$$\frac{1}{2\kappa}\sum_{\mu\nu}\sqrt{-g}\cdot\frac{\partial g_{\mu\nu}}{\partial x_\sigma}\Gamma_{\mu\nu}$$

can be rewritten in such a way that it appears as the sum of differential quotients. On the other hand, we know that the term (a) appears in the expression sought for $\Gamma_{\mu\nu}$. Hence the identity that is being sought has the following form:

Sum of differential quotients

$$= \frac{1}{2}\sum_{\mu\nu}\sqrt{-g}\cdot\frac{\partial g_{\mu\nu}}{\partial x_\sigma}\left\{\sum_{\alpha\beta}\frac{\partial}{\partial x_\alpha}\left(\gamma_{\alpha b}\frac{\partial\gamma_{\mu\nu}}{\partial x_\beta}\right)\right.$$
$$\left.+ \text{ the other terms, which vanish with the first approximation.}\right\}$$

The identity that is being sought is thereby uniquely determined; if one [30] constructs it according to the procedure indicated,[15] one obtains

$$(12)\quad \begin{cases} \displaystyle\sum_{\alpha\beta\tau\rho} \frac{\partial}{\partial x_\alpha}\left(\sqrt{-g}\cdot\gamma_{\alpha\beta}\frac{\partial\gamma_{\tau\rho}}{\partial x_\beta}\cdot\frac{\partial g_{\tau\rho}}{\partial x_\sigma}\right) - \frac{1}{2}\cdot\sum_{\alpha\beta\tau\rho}\frac{\partial}{\partial x_\sigma}\left(\sqrt{-g}\cdot\gamma_{\alpha\beta}\frac{\partial\gamma_{\tau\rho}}{\partial x_\alpha}\frac{\partial g_{\tau\rho}}{\partial x_\beta}\right) \\ \displaystyle = -\sum_{\mu\nu}\sqrt{-g}\cdot\frac{\partial g_{\mu\nu}}{\partial x_\sigma}\left\{\sum_{\alpha\beta}\frac{1}{\sqrt{-g}}\cdot\frac{\partial}{\partial x_\alpha}\left(\gamma_{\alpha\beta}\sqrt{-g}\cdot\frac{\partial\gamma_{\mu\nu}}{\partial x_\beta}\right) - \sum_{\alpha\beta\tau\rho}\gamma_{\alpha\beta}g_{\tau\rho}\frac{\partial\gamma_{\mu\tau}}{\partial x_\alpha}\frac{\partial\gamma_{\nu\rho}}{\partial x_\beta}\right. \\ \displaystyle \left. + \frac{1}{2}\sum_{\alpha\beta\tau\rho}\gamma_{\alpha\mu}\gamma_{\beta\nu}\frac{\partial g_{\tau\rho}}{\partial x_\alpha}\frac{\partial\gamma_{\tau\rho}}{\partial x_\beta} - \frac{1}{4}\sum_{\alpha\beta\tau\rho}\gamma_{\mu\nu}\gamma_{\alpha\beta}\frac{\partial g_{\tau\rho}}{\partial x_\alpha}\frac{\partial\gamma_{\tau\rho}}{\partial x_\beta}\right\}. \end{cases}$$

Thus, the expression for $\Gamma_{\mu\nu}$ that is enclosed between the curly brackets on the right-hand side is the tensor that is being sought that enters into the gravitational equations

$$\kappa\Theta_{\mu\nu} = \Gamma_{\mu\nu}.$$

To make these equations more comprehensible, we introduce the following abbreviations:

$$(13)\qquad -2\kappa\cdot\vartheta_{\mu\nu} = \sum_{\alpha\beta\tau\rho}\left(\gamma_{\alpha\mu}\gamma_{\beta\nu}\frac{\partial g_{\tau\rho}}{\partial x_\alpha}\cdot\frac{\partial\gamma_{\tau\rho}}{\partial x_\beta} - \frac{1}{2}\gamma_{\mu\nu}\gamma_{\alpha\beta}\frac{\partial g_{\tau\rho}}{\partial x_\alpha}\frac{\partial\gamma_{\tau\rho}}{\partial x_\beta}\right).$$

We will designate $\vartheta_{\mu\nu}$ as the "*contravariant stress-energy tensor of the* [31] *gravitational field*." The covariant tensor reciprocal to it will be denoted by $t_{\mu\nu}$; then we have

$$(14)\qquad -2\kappa\cdot t_{\mu\nu} = \sum_{\alpha\beta\tau\rho}\left(\frac{\partial g_{\tau\rho}}{\partial x_\mu}\cdot\frac{\partial\gamma_{\tau\rho}}{\partial x_\nu} - \frac{1}{2}g_{\mu\nu}\gamma_{\alpha\beta}\frac{\partial g_{\tau\rho}}{\partial x_\alpha}\frac{\partial\gamma_{\tau\rho}}{\partial x_\beta}\right).$$

Likewise, for the sake of brevity, we introduce the following notations for differential operations carried out on the fundamental tensors γ and g:

$$(15)\qquad \Delta_{\mu\nu}(\gamma) = \sum_{\alpha\beta}\frac{1}{\sqrt{-g}}\cdot\frac{\partial}{\partial x_\alpha}\left(\gamma_{\alpha\beta}\sqrt{-g}\cdot\frac{\partial\gamma_{\mu\nu}}{\partial x_\beta}\right) - \sum_{\alpha\beta\tau\rho}\gamma_{\alpha\beta}g_{\tau\rho}\frac{\partial\gamma_{\mu\tau}}{\partial x_\alpha}\frac{\partial\gamma_{\nu\rho}}{\partial x_\beta},$$

and

$$(16)\qquad D_{\mu\nu}(\gamma) = \sum_{\alpha\beta}\frac{1}{\sqrt{-g}}\cdot\frac{\partial}{\partial x_\alpha}\left(\gamma_{\alpha\beta}\sqrt{-g}\cdot\frac{\partial g_{\mu\nu}}{\partial x_\beta}\right) - \sum_{\alpha\beta\tau\rho}\gamma_{\alpha\beta}\gamma_{\tau\rho}\frac{\partial g_{\mu\tau}}{\partial x_\alpha}\frac{\partial g_{\nu\rho}}{\partial x_\beta}.$$

[15]Cf. Part II, §4, No. 8.

Each of these operators yields again a tensor of the same kind (w. resp. to linear transformations).

With the application of these abbreviations the identity (12) assumes the form

$$\sum_{\mu\nu} \frac{\partial}{\partial x_\nu}\{\sqrt{-g}\cdot g_{\sigma\mu}\cdot\kappa\vartheta_{\mu\nu}\} = \frac{1}{2}\sum_{\mu\nu}\sqrt{-g}\cdot\frac{\partial g_{\mu\nu}}{\partial x_\sigma}\{-\Delta_{\mu\nu}(\gamma) + \kappa\vartheta_{\mu\nu}\}, \tag{12a}$$

or also

$$\sum_{\mu\nu} \frac{\partial}{\partial x_\nu}\{\sqrt{-g}\cdot\gamma_{\mu\nu}\cdot\kappa t_{\mu\sigma}\} = \frac{1}{2}\sum_{\mu\nu}\sqrt{-g}\cdot\frac{\partial\gamma_{\mu\nu}}{\partial x_\sigma}\{-D_{\mu\nu}(g) + \kappa\cdot t_{\mu\nu}\}. \tag{12b}$$

If we write the conservation law (10) for matter and the conservation law (12a) for the gravitational field in the form

$$\sum_{\mu\nu} \frac{\partial}{\partial x_\nu}\{\sqrt{-g}\cdot g_{\sigma\mu}\cdot\Theta_{\mu\nu}\} - \frac{1}{2}\sum_{\mu\nu}\sqrt{-g}\cdot\frac{\partial g_{\mu\nu}}{\partial x_\sigma}\cdot\Theta_{\mu\nu} = 0 \tag{10}$$

$$\begin{aligned}&\sum_{\mu\nu} \frac{\partial}{\partial x_\nu}(\sqrt{-g}\cdot g_{\sigma\mu}\cdot\vartheta_{\mu\nu}) - \frac{1}{2}\sum_{\mu\nu}\sqrt{-g}\cdot\frac{\partial g_{\mu\nu}}{\partial x_\mu}\cdot\vartheta_{\mu\nu}\\ &= -\frac{1}{2\kappa}\cdot\sum_{\mu\nu}\sqrt{-g}\cdot\frac{\partial g_{\mu\nu}}{\partial x_\sigma}\cdot\Delta_{\mu\nu}(\gamma),\end{aligned} \tag{12c}$$

then one recognizes that the stress-energy tensor $\vartheta_{\mu\nu}$ of the gravitational field enters the conservation law for the gravitational field in exactly the same way as the tensor $\Theta_{\mu\nu}$ of the material process enters the conservation law for this process; this is a noteworthy circumstance considering the difference in the derivation of the two laws.

From equation (12a) follows the expression for the differential tensor entering into the gravitational equations

$$\Gamma_{\mu\nu} = \Delta_{\mu\nu}(\gamma) - \kappa\cdot\vartheta_{\mu\nu}. \tag{17}$$

Thus, the gravitational equations (11) are of the form

$$\Delta_{\mu\nu}(\gamma) = \kappa(\Theta_{\mu\nu} + \vartheta_{\mu\nu}). \tag{18}$$ [32]

These equations satisfy a requirement that, in our opinion, must be imposed on a relativity theory of gravitation; that is to say, they show that the tensor $\vartheta_{\mu\nu}$ of the gravitational field acts as a field generator in the same way as the tensor $\Theta_{\mu\nu}$ of the material processes. An exceptional position of gravitational energy in comparison [33] with all other kinds of energies would lead to untenable consequences.

Adding equations (10) and (12a) while taking into account equation (18), one finds

[34] (19) $$\sum_{\mu\nu} \frac{\partial}{\partial x_\nu}\left\{\sqrt{-g}\cdot g_{\sigma\mu}(\Theta_{\mu\nu} + \vartheta_{\mu\nu})\right\} = 0 \qquad (\sigma = 1,2,3,4)$$

This shows that the conservation laws hold for the matter and the gravitational field taken together.

In the foregoing we have given preference to the contravariant tensors, because the contravariant stress-energy tensor of the flow of incoherent masses can be expressed in an especially simple manner. However, we can express the fundamental relations that we have obtained just as simply by using covariant tensors. Instead of $\Theta_{\mu\nu}$, we must then take $T_{\mu\nu} = \sum_{\alpha\beta} g_{\mu\alpha} g_{\nu\beta} \Theta_{\alpha\beta}$ as the stress-energy tensor of the material process. Instead of equation (10), we obtain through term-by-term reformulation

(20) $$\sum_{\mu\nu} \frac{\partial}{\partial x_\nu}(\sqrt{-g}\cdot\gamma_{\mu\nu} T_{\mu\sigma}) + \frac{1}{2}\sum_{\mu\nu}\sqrt{-g}\cdot\frac{\partial\gamma_{\mu\nu}}{\partial x_\sigma}\cdot T_{\mu\nu} = 0.$$

It follows from this equation and equation (16) that the equations of the gravitational field can also be written in the form

(21) $$-D_{\mu\nu}(g) = \kappa(t_{\mu\nu} + T_{\mu\nu});$$

these equations can also be derived directly from (18). The equation that corresponds to (19) reads

(22) $$\sum_{\nu} \frac{\partial}{\partial x_\nu}\left\{\sqrt{-g}\cdot\gamma_{\sigma\mu}(T_{\mu\nu} + t_{\mu\nu})\right\} = 0.$$

II
Mathematical Part

by Marcel Grossman

The mathematical tools for developing the vector analysis of a gravitational field, which is characterized by the invariance of the line element

$$ds^2 = \sum_{\mu\nu} g_{\mu\nu} dx_\mu dx_\nu ,$$

derive from Christoffel's fundamental paper on the transformation of quadratic differential forms.[1] Taking Christoffel's results as their starting point, Ricci and Levi-Cività[2] developed their methods of the absolute differential calculus—i.e., a differential calculus that is independent of the coordinate system—which permit our
[47] giving an invariant form to the differential equations of mathematical physics. But since the vector analysis of a Euclidean space referred to arbitrary curvilinear coordinates is formally identical with the vector analysis of an arbitrary manifold specified by its line element, the extension of the vector-analytical conceptions that
[48] Minkowski, Sommerfeld, Laue, et al. worked out for the theory of relativity in recent years to the general theory of Einstein's expounded above does not present any difficulty.

With some practice, *the general vector analysis* obtained in this way is as simple to handle as the special vector analysis of three- or four-dimensional Euclidean space; in fact, the greater generality of its conceptions lends it a clarity that is lacking often enough in the special case.

The theory of special tensors (§3) has been treated to the full in a paper by Kottler,[3] published while this work was in progress; the treatment is based on the theory of integral forms, something that is not possible in the general case.

Since more detailed mathematical investigations will have to be done in
connection with Einstein's theory of gravitation, and especially in connection with the [50]
problem of the differential equations of the gravitational field, a systematic presentation of the general vector analysis might be in order. I have purposely not employed geometrical aids because, in my opinion, they contribute very little to an intuitive understanding of the conceptions of vector analysis.

---

[1]Christoffel, "Über die Transformation der homogenen Differentialausdrücke zweiten
[45] Grades," *J. f. Math.* 70 (1869): 46.

[2]Ricci et Levi-Cività, "Méthodes de calcul différentiel absolu et leurs application."
[46] *Math. Ann.* 54 (1901): 125.

[49] [3]Kottler, "Über die Raumzeitlinien der Minkowskischen Welt." *Wien. Ber.* 121 (1912).

§4. Mathematical Supplements to the Physical Part

1. Proof of the Covariance of the Momentum-Energy Equations

It has to be proved that the equations (10) of Part I, page 10, which, neglecting the factor $\sqrt{-1}$, read

$$\sum_{\mu\nu} \frac{\partial}{\partial x_\nu}(\sqrt{g}\cdot g_{\sigma\mu}\cdot \Theta_{\mu\nu}) - \frac{1}{2}\sqrt{g}\cdot \sum_{\mu\nu} \frac{\partial g_{\mu\nu}}{\partial x_\sigma}\cdot \Theta_{\mu\nu} = 0, \qquad (\sigma = 1,2,3,4)$$

are covariant with respect to arbitrary transformations.

According to formula (35), the divergence of the contravariant tensor $\Theta_{\mu\nu}$ is

$$\Theta_\mu = \sum_\nu \frac{1}{\sqrt{g}} \frac{\partial}{\partial x_\nu}(\sqrt{g}\cdot \Theta_{\mu\nu}) + \sum_{\nu k} \begin{Bmatrix} \nu k \\ \mu \end{Bmatrix} \Theta_{\nu k}.$$

The covariant vector T_σ reciprocal to this contravariant vector Θ_μ is thus

---

[11]The factor ½ serves to simplify the result but is inconsequential from the point of view of the theory of invariants.

$$T_\sigma = \sum_\mu g_{\sigma\mu}\Theta_\mu = \sum_{\mu\nu k}\left(\frac{1}{\sqrt{g}}\frac{\partial}{\partial x_\nu}(\sqrt{g}\cdot g_{\sigma\mu}\cdot\Theta_{\mu\nu}) - \frac{\partial g_{\sigma\mu}}{\partial x_\nu}\cdot\Theta_{\mu\nu} + g_{\sigma\mu}\begin{Bmatrix}\nu k\\ \mu\end{Bmatrix}\cdot\Theta_\nu k\right).$$

But the last term of this sum is equal to

$$\sum_{\nu k}\begin{bmatrix}\nu k\\ \sigma\end{bmatrix}\Theta_{\nu k} = \sum_{\mu\nu}\frac{1}{2}\left(\frac{\partial g_{\mu\sigma}}{\partial x_\nu} + \frac{\partial g_{\nu\sigma}}{\partial x_\mu} - \frac{\partial g_{\mu\nu}}{\partial x_\sigma}\right)\cdot\Theta_{\mu\nu}.$$

Hence, we end up with

$$T_\sigma = \sum_{\mu\nu}\frac{1}{\sqrt{g}}\frac{\partial}{\partial x_\nu}(\sqrt{g}\cdot g_{\sigma\mu}\Theta_{\mu\nu}) - \frac{1}{2}\sum_{\mu\nu}\frac{\partial g_{\mu\nu}}{\partial x_\sigma}\cdot\Theta_{\mu\nu},$$

i.e., the left side of the investigated equation, up to the factor $\frac{1}{\sqrt{g}}$. Thus, if that equation is divided by $\sqrt{g}$, then its left side represents the σ-component of a covariant vector, and is, therefore, in fact, covariant. For that reason, the content of those four equations can also be expressed thus:

The divergence of the (contravariant) stress-energy tensor of the material flow or of the physical process vanishes.

2. *Differential Tensors of a Manifold Given by Its Line Element*

The problem of constructing the differential equations of a gravitational field (Part I, §5) draws one's attention to the *differential invariants* and *differential covariants* of the quadratic differential form

$$ds^2 = \sum_{\mu\nu} g_{\mu\nu}\,dx_\mu dx_\nu.$$

In the sense of our general vector analysis, the theory of these differential covariants leads to the *differential tensors* that are given with a gravitational field. The complete system of these differential tensors (with respect to arbitrary transformations) goes back to a covariant differential tensor of fourth rank found by Riemann[12] and, independently of him, by Christoffel,[13] which we shall call the *Riemann differential tensor*, and which reads as follows:

$$(43)\qquad R_{iklm} = (ik,lm) = \frac{1}{2}\left(\frac{\partial^2 g_{im}}{\partial x_k\partial x_l} + \frac{\partial^2 g_{kl}}{\partial x_i\partial x_m} - \frac{\partial^2 g_{il}}{\partial x_k\partial x_m} - \frac{\partial^2 g_{mk}}{\partial x_l\partial x_i}\right)$$

---

[66] [12]Riemann, *Ges. Werke,* p. 270.

[67] [13]Christoffel, l.c., p. 54.

$$+\sum_{\rho\sigma}\gamma_{\rho\sigma}\left(\begin{bmatrix}im\\ \rho\end{bmatrix}\begin{bmatrix}kl\\ \sigma\end{bmatrix}-\begin{bmatrix}il\\ \rho\end{bmatrix}\begin{bmatrix}km\\ \sigma\end{bmatrix}\right).$$ [68]

By means of covariant algebraic and differential operations we obtain the complete system of differential tensors (thus also the differential invariants) of the manifold from the Riemann differential tensor and the discriminant tensor (§3, formula 38). [69]

The *(ik, lm)* are also called the *Christoffel four-index symbols of the first kind.* In addition to these, of importance are also the *four-index symbols of the second kind*

(44) $$\{ik,\ lm\} = \frac{\partial\begin{Bmatrix}il\\ k\end{Bmatrix}}{\partial x_m} - \frac{\partial\begin{Bmatrix}im\\ k\end{Bmatrix}}{\partial x_l} + \sum_{\rho}\left(\begin{Bmatrix}il\\ \rho\end{Bmatrix}\begin{Bmatrix}\rho m\\ k\end{Bmatrix} - \begin{Bmatrix}im\\ \rho\end{Bmatrix}\begin{Bmatrix}\rho l\\ k\end{Bmatrix}\right),$$

which are related to the former in the following way:

(45) $$\begin{cases}\{i\rho,\ lm\} = \sum_k \gamma_{\rho k}(ik,\ lm), \text{ or, when solved,}\\[2ex] (ik,\ lm) = \sum_\rho g_{k\rho}\ \{i\rho,\ lm\}.\end{cases}$$

In general vector analysis, the four-index symbols of the second kind take on the meaning of the components of a *mixed tensor* that is covariant of third rank and contravariant of first rank.[14]

The extraordinary importance of these conceptions for the *differential geometry*[15] of a manifold that is given by its line element makes it a priori probable that these general differential tensors may also be of importance for the problem of the differential equations of a gravitational field. To begin with, it is, in fact, possible [71]
to specify a covariant differential tensor of second rank and second order G_{im} that could enter into those equations, namely,

(46) $$G_{im} = \sum_{kl}\gamma_{kl}\,(ik,\ lm) = \sum_k \{ik,\ km\}.$$

It turns out, however, that in the special case of the infinitely weak, static gravitational field this tensor does *not* reduce to the expression $\Delta\varphi$. We must [72]
therefore leave open the question to what extent the general theory of the differential tensors associated with a gravitational field is connected with the problem of the

---

[14]This follows from the first of equations 45.

[15]The identical vanishing of the tensor R_{iklm} constitutes a necessary and sufficient condition for the differential form's being transformable to the form $\sum_i dx_i^2$. [70]

gravitational equations. Such a connection would have to exist insofar as the gravitational equations are to permit *arbitrary* substitutions; but in that case, it seems that it would be impossible to find *second-order* differential equations. On the other hand, if it were established that the gravitational equations permit only a particular group of transformations, then it would be understandable if one could not manage with the differential tensors yielded by the general theory. As has been explained in the physical part, we are not able to take a stand on these questions.—

3. On the Derivation of the Gravitational Equations

The derivation of the gravitational equations described by Einstein (Part I, §5) is carried out step by step in the following way:

We start out from the term that is definitely to be expected in the energy balance,

$$U = \sum_{\alpha\beta\mu\nu} \frac{\partial g_{\mu\nu}}{\partial x_\sigma} \frac{\partial}{\partial x_\alpha}\left(\sqrt{g}\gamma_{\alpha\beta}\frac{\partial \gamma_{\mu\nu}}{\partial x_\beta}\right), \tag{47}$$

and reformulate it by integrating by parts.[16] In this way we obtain

$$U = \sum_{\alpha\beta\mu\nu} \frac{\partial}{\partial x_\alpha}\left(\sqrt{g}\gamma_{\alpha\beta}\frac{\partial \gamma_{\mu\nu}}{\partial x_\beta}\frac{\partial g_{\mu\nu}}{\partial x_\sigma}\right) - \sum_{\alpha\beta\mu\nu} \sqrt{g}\gamma_{\alpha\beta}\frac{\partial \gamma_{\mu\nu}}{\partial x_\beta}\cdot\frac{\partial^2 g_{\mu\nu}}{\partial x_\sigma \partial x_\alpha}.$$

The first sum on the right-hand side has the desired form of a sum of differential quotients and shall be denoted by A, so that we have

$$A = \sum_{\alpha\beta\mu\nu} \frac{\partial}{\partial x_\alpha}\left(\sqrt{g}\gamma_{\alpha\beta}\frac{\partial \gamma_{\mu\nu}}{\partial x_\beta}\frac{\partial g_{\mu\nu}}{\partial x_\sigma}\right). \tag{48}$$

We once again integrate by parts in the second sum on the right-hand side. The identity will then take the form

$$U = A - \sum_{\alpha\beta\mu\nu} \frac{\partial}{\partial x_\sigma}\left(\sqrt{g}\cdot\gamma_{\alpha\beta}\frac{\partial \gamma_{\mu\nu}}{\partial x_\beta}\cdot\frac{\partial g_{\mu\nu}}{\partial x_\alpha}\right) + \sum_{\alpha\beta\mu\nu} \frac{\partial g_{\mu\nu}}{\partial x_\alpha}\cdot\frac{\partial}{\partial x_\sigma}\left(\sqrt{g}\cdot\gamma_{\alpha\beta}\frac{\partial \gamma_{\mu\nu}}{\partial x_\beta}\right).$$

The first of the sums obtained on the right-hand side can be written as a sum of differentials and shall be denoted by

$$B = \sum_{\alpha\beta\mu\nu} \frac{\partial}{\partial x_\sigma}\left(\sqrt{g}\gamma_{\alpha\beta}\frac{\partial \gamma_{\mu\nu}}{\partial x_\beta}\frac{\partial g_{\mu\nu}}{\partial x_\alpha}\right). \tag{49}$$

We differentiate in the second sum. Then we get

---

[16] The derivation of the identity we are seeking becomes simpler, without affecting the result, if we put the factor $\sqrt{g}$ inside the differentiation sign.

$$U = A - B + \sum_{\alpha\beta\mu\nu} \frac{\partial g_{\mu\nu}}{\partial x_\alpha}\left(\gamma_{\alpha\beta}\frac{\partial\gamma_{\mu\nu}}{\partial x_\beta}\frac{\partial\sqrt{g}}{\partial x_\sigma} + \sqrt{g}\cdot\frac{\partial\gamma_{\mu\nu}}{\partial x_\beta}\cdot\frac{\partial\gamma_{\alpha\beta}}{\partial x_\sigma} + \sqrt{g}\cdot\gamma_{\alpha\beta}\frac{\partial^2\gamma_{\mu\nu}}{\partial x_\beta\partial x_\sigma}\right),$$

or if we apply formula (29) of §2 in the second summand and integrate by parts in the third summand

$$U = A - B + \sum_{\alpha\beta\mu\nu ik}\gamma_{\alpha\beta}\frac{\partial g_{\mu\nu}}{\partial x_\alpha}\frac{\partial\gamma_{\mu\nu}}{\partial x_\beta}\cdot\frac{\sqrt{g}}{2}\gamma_{ik}\frac{\partial g_{ik}}{\partial x_\sigma} - \sum_{\alpha\beta\mu\nu ik}\sqrt{g}\cdot\frac{\partial g_{\mu\nu}}{\partial x_\alpha}\cdot\frac{\partial\gamma_{\mu\nu}}{\partial x_\beta}\cdot\gamma_{\alpha i}\gamma_{\beta k}\frac{\partial g_{ik}}{\partial x_\sigma}$$

$$+\sum_{\alpha\beta\mu\nu}\frac{\partial}{\partial x_\beta}\left(\sqrt{g}\gamma_{\alpha\beta}\frac{\partial g_{\mu\nu}}{\partial x_\alpha}\cdot\frac{\partial\gamma_{\mu\nu}}{\partial x_\sigma}\right) - \sum_{\alpha\beta\mu\nu}\frac{\partial\gamma_{\mu\nu}}{\partial x_\sigma}\frac{\partial}{\partial x_\beta}\left(\sqrt{g}\gamma_{\alpha\beta}\frac{\partial g_{\mu\nu}}{\partial x_\alpha}\right).$$

The first two sums have the form of terms such as we place on the left side of our identity. We denote them by

$$V = \frac{1}{2}\sum_{\alpha\beta\mu\nu ik}\frac{\partial g_{ik}}{\partial x_\sigma}\cdot\sqrt{g}\cdot\gamma_{\alpha\beta}\gamma_{ik}\frac{\partial g_{\mu\nu}}{\partial x_\alpha}\cdot\frac{\partial\gamma_{\mu\nu}}{\partial x_\beta} \tag{50}$$

$$W = \sum_{\alpha\beta\mu\nu ik}\frac{\partial g_{ik}}{\partial x_\sigma}\cdot\sqrt{g}\cdot\gamma_{\alpha i}\gamma_{\beta k}\frac{\partial g_{\mu\nu}}{\partial x_\alpha}\cdot\frac{\partial\gamma_{\mu\nu}}{\partial x_\beta}. \tag{51}$$

The third of the sums appearing on the right has the form of a sum of differential quotients; if we eliminate $\dfrac{\partial\gamma_{\mu\nu}}{\partial x_\sigma}$ from it with the help of the above formula (29), this sum proves to be the quantity A that has already been introduced. Finally, we replace $\dfrac{\partial\gamma_{\mu\nu}}{\partial x_\sigma}$ in the last sum in accord with the same formula. In this way we find

$$U - V + W = 2A - B + \sum_{\alpha\beta\mu\nu ik}\gamma_{\mu i}\gamma_{\nu k}\frac{\partial g_{ik}}{\partial x_\sigma}\frac{\partial}{\partial x_\beta}\left(\sqrt{g}\gamma_{\alpha\beta}\frac{\partial g_{\mu\nu}}{\partial x_\alpha}\right),$$

or

$$U - V + W = 2A - B + \sum_{\alpha\beta\mu\nu ik}\frac{\partial g_{ik}}{\partial x_\sigma}\cdot\frac{\partial}{\partial x_\beta}\left(\sqrt{g}\cdot\gamma_{\alpha\beta}\gamma_{\mu i}\gamma_{\nu k}\frac{\partial g_{\mu\nu}}{\partial x_\alpha}\right)$$

$$- \sum_{\alpha\beta\mu\nu ik}\frac{\partial g_{ik}}{\partial x_\sigma}\frac{\partial g_{\mu\nu}}{\partial x_\alpha}\sqrt{g}\cdot\gamma_{\alpha\beta}\frac{\partial}{\partial x_\beta}(\gamma_{\mu i}\gamma_{\nu k}).$$

By virtue of (29), i.e., by virtue of

$$\sum_{\mu\nu}\gamma_{i\mu}\gamma_{\nu k}\frac{\partial g_{\mu\nu}}{\partial x_\alpha} = -\frac{\partial\gamma_{ik}}{\partial x_\alpha},$$

the first of these sums becomes

$$-\sum_{\alpha\beta ik}\frac{\partial g_{ik}}{\partial x_\sigma}\frac{\partial}{\partial x_\beta}\left(\sqrt{g}\gamma_{\alpha\beta}\frac{\partial\gamma_{ik}}{\partial x_\alpha}\right) = -U.$$

Since i is interchangeable with k, and μ with ν, we can write the second sum as

$$2X = 2 \cdot \sum_{\alpha\beta\mu\nu ik} \frac{\partial g_{ik}}{\partial x_\sigma} \cdot \sqrt{g} \cdot \gamma_{\alpha\beta} \gamma_{\mu i} \frac{\partial g_{\mu\nu}}{\partial x_\alpha} \cdot \frac{\partial \gamma_{\nu k}}{\partial x_\beta}$$

$$= -2 \cdot \sum_{\alpha\beta\mu\nu ik} \frac{\partial g_{ik}}{\partial x_\sigma} \cdot \sqrt{g} \gamma_{\alpha\beta} g_{\mu\nu} \frac{\partial \gamma_{i\mu}}{\partial x_\alpha} \frac{\partial \gamma_{k\nu}}{\partial x_\beta} .$$

Hence, the identity sought reads

$$2U - V + W + 2X = 2A - B,$$

and is thus identical with the one given in Part I, §5.

Einstein and Marcel Grossmann in the garden of the Grossmann home in Thalwil, May 28, 1899.

Doc. 30

[p. 769] **The Foundation of the General Theory of Relativity**

by A. Einstein

[This first page was missing in the existing translation.]

The theory which is presented in the following pages conceivably constitutes the
farthest-reaching generalization of a theory which, today, is generally called the
[1] "theory of relativity"; I will call the latter one—in order to distinguish it from the
[2] first named—the "special theory of relativity," which I assume to be known. The
generalization of the theory of relativity has been facilitated considerably by
Minkowski, a mathematician who was the first one to recognize the formal
[3] equivalence of space coordinates and the time coordinate, and utilized this in the
construction of the theory. The mathematical tools that are necessary for general
relativity were readily available in the "absolute differential calculus," which is based
upon the research on non-Euclidean manifolds by Gauss, Riemann, and Christoffel,
and which has been systematized by Ricci and Levi-Civita and has already been
[4] applied to problems of theoretical physics. In section B of the present paper I
developed all the necessary mathematical tools—which cannot be assumed to be
known to every physicist—and I tried to do it in as simple and transparent a manner
as possible, so that a special study of the mathematical literature is not required for
[5] the understanding of the present paper. Finally, I want to acknowledge gratefully my
friend, the mathematician Grossmann, whose help not only saved me the effort of
[6] studying the pertinent mathematical literature, but who also helped me in my search
for the field equations of gravitation.

[The balance of this translation is reprinted from H. A. Lorentz et al., *The Principle of Relativity*, trans. W. Perrett and G. B. Jeffery (Methuen, 1923; Dover rpt., 1952).]

THE FOUNDATION OF THE GENERAL THEORY OF RELATIVITY

By A. EINSTEIN

A. Fundamental Considerations on the Postulate of Relativity

§ 1. Observations on the Special Theory of Relativity

THE special theory of relativity is based on the following postulate, which is also satisfied by the mechanics of Galileo and Newton.

If a system of co-ordinates K is chosen so that, in relation to it, physical laws hold good in their simplest form, the *same* laws also hold good in relation to any other system of co-ordinates K′ moving in uniform translation relatively to K. This postulate we call the "special principle of relativity." The word "special" is meant to intimate that the principle is restricted to the case when K′ has a motion of uniform translation relatively to K, but that the equivalence of K′ and K does not extend to the case of non-uniform motion of K′ relatively to K.

Thus the special theory of relativity does not depart from classical mechanics through the postulate of relativity, but through the postulate of the constancy of the velocity of light *in vacuo*, from which, in combination with the special principle of relativity, there follow, in the well-known way, the relativity of simultaneity, the Lorentzian transformation, and the related laws for the behaviour of moving bodies and clocks.

The modification to which the special theory of relativity has subjected the theory of space and time is indeed far-reaching, but one important point has remained unaffected.

For the laws of geometry, even according to the special theory of relativity, are to be interpreted directly as laws relating to the possible relative positions of solid bodies at rest; and, in a more general way, the laws of kinematics are to be interpreted as laws which describe the relations of measuring bodies and clocks. To two selected material points of a stationary rigid body there always corresponds a distance of quite definite length, which is independent of the locality and orientation of the body, and is also independent of the time. To two selected positions of the hands of a clock at rest relatively to the privileged system of reference there always corresponds an interval of time of a definite length, which is independent of place and time. We shall soon see that the general theory of relativity cannot adhere to this simple physical interpretation of space and time.

§ 2. The Need for an Extension of the Postulate of Relativity

In classical mechanics, and no less in the special theory of relativity, there is an inherent epistemological defect which
[7] was, perhaps for the first time, clearly pointed out by Ernst Mach. We will elucidate it by the following example:—Two fluid bodies of the same size and nature hover freely in space at so great a distance from each other and from all other masses that only those gravitational forces need be taken into account which arise from the interaction of different parts of the same body. Let the distance between the two bodies be invariable, and in neither of the bodies let there be any relative movements of the parts with respect to one another. But let either mass, as judged by an observer at rest relatively to the other mass, rotate with constant angular velocity about the line joining the masses. This is a verifiable relative motion of the two bodies. Now let us imagine that each of the bodies has been surveyed by means of measuring instruments at rest relatively to itself, and let the surface of S_1 prove to be a sphere, and that of S_2 an ellipsoid of revolution. Thereupon we put the question—What is the reason for this difference in the two bodies? No answer can

be admitted as epistemologically satisfactory,* unless the reason given is an *observable fact of experience.* The law of causality has not the significance of a statement as to the world of experience, except when *observable facts* ultimately appear as causes and effects.

Newtonian mechanics does not give a satisfactory answer to this question. It pronounces as follows:—The laws of mechanics apply to the space R_1, in respect to which the body S_1 is at rest, but not to the space R_2, in respect to which the body S_2 is at rest. But the privileged space R_1 of Galileo, thus introduced, is a merely *factitious* cause, and not a thing that can be observed. It is therefore clear that Newton's mechanics does not really satisfy the requirement of causality in the case under consideration, but only apparently does so, since it makes the factitious cause R_1 responsible for the observable difference in the bodies S_1 and S_2.

The only satisfactory answer must be that the physical system consisting of S_1 and S_2 reveals within itself no imaginable cause to which the differing behaviour of S_1 and S_2 can be referred. The cause must therefore lie *outside* this system. We have to take it that the general laws of motion, which in particular determine the shapes of S_1 and S_2, must be such that the mechanical behaviour of S_1 and S_2 is partly conditioned, in quite essential respects, by distant masses which we have not included in the system under consideration. These distant masses and their motions relative to S_1 and S_2 must then be regarded as the seat of the causes (which must be susceptible to observation) of the different behaviour of our two bodies S_1 and S_2. They take over the rôle of the factitious cause R_1. Of all imaginable spaces R_1, R_2, etc., in any kind of motion relatively to one another, there is none which we may look upon as privileged *a priori* without reviving the above-mentioned epistemological objection. *The laws of physics must be of such a nature that they apply to systems of reference in any kind of motion.* Along this road we arrive at an extension of the postulate of relativity.

In addition to this weighty argument from the theory of

* Of course an answer may be satisfactory from the point of view of epistemology, and yet be unsound physically, if it is in conflict with other experiences.

knowledge, there is a well-known physical fact which favours an extension of the theory of relativity. Let K be a Galilean system of reference, i.e. a system relatively to which (at least in the four-dimensional region under consideration) a mass, sufficiently distant from other masses, is moving with uniform motion in a straight line. Let K′ be a second system of reference which is moving relatively to K in *uniformly accelerated* translation. Then, relatively to K′, a mass sufficiently distant from other masses would have an accelerated motion such that its acceleration and direction of acceleration are independent of the material composition and physical state of the mass.

Does this permit an observer at rest relatively to K′ to infer that he is on a "really" accelerated system of reference? The answer is in the negative; for the above-mentioned relation of freely movable masses to K′ may be interpreted equally well in the following way. The system of reference K′ is unaccelerated, but the space-time territory in question is under the sway of a gravitational field, which generates the
[8] accelerated motion of the bodies relatively to K′.

This view is made possible for us by the teaching of experience as to the existence of a field of force, namely, the gravitational field, which possesses the remarkable property of imparting the same acceleration to all bodies.* The mechanical behaviour of bodies relatively to K′ is the same as presents itself to experience in the case of systems which we are wont to regard as "stationary" or as "privileged." Therefore, from the physical standpoint, the assumption readily suggests itself that the systems K and K′ may both with equal right be looked upon as "stationary," that is to say, they have an equal title as systems of reference for the physical description of phenomena.

It will be seen from these reflexions that in pursuing the general theory of relativity we shall be led to a theory of gravitation, since we are able to "produce" a gravitational field merely by changing the system of co-ordinates. It will also be obvious that the principle of the constancy of the velocity of light *in vacuo* must be modified, since we easily

[9] * Eötvös has proved experimentally that the gravitational field has this property in great accuracy.

recognize that the path of a ray of light with respect to K′ must in general be curvilinear, if with respect to K light is propagated in a straight line with a definite constant velocity.

§ 3. The Space-Time Continuum. Requirement of General Co-Variance for the Equations Expressing General Laws of Nature

In classical mechanics, as well as in the special theory of relativity, the co-ordinates of space and time have a direct physical meaning. To say that a point-event has the X_1 co-ordinate x_1 means that the projection of the point-event on the axis of X_1, determined by rigid rods and in accordance with the. rules of Euclidean geometry, is obtained by measuring off a given rod (the unit of length) x_1 times from the origin of co-ordinates along the axis of X_1. To say that a point-event has the X_4 co-ordinate $x_4 = t$, means that a standard clock, made to measure time in a definite unit period, and which is stationary relatively to the system of co-ordinates and practically coincident in space with the point-event,* will have measured off $x_4 = t$ periods at the occurrence of the event.

This view of space and time has always been in the minds of physicists, even if, as a rule, they have been unconscious of it. This is clear from the part which these concepts play in physical measurements; it must also have underlain the reader's reflexions on the preceding paragraph (§ 2) for him to connect any meaning with what he there read. But we shall now show that we must put it aside and replace it by a more general view, in order to be able to carry through the postulate of general relativity, if the special theory of relativity applies to the special case of the absence of a gravitational field.

In a space which is free of gravitational fields we introduce a Galilean system of reference K (x, y, z, t), and also a system of co-ordinates K′ (x', y', z', t') in uniform rotation relatively to K. Let the origins of both systems, as well as their axes

* We assume the possibility of verifying "simultaneity" for events immediately proximate in space, or—to speak more precisely—for immediate proximity or coincidence in space-time, without giving a definition of this fundamental concept.

of Z, permanently coincide. We shall show that for a space-time measurement in the system K′ the above definition of the physical meaning of lengths and times cannot be maintained. For reasons of symmetry it is clear that a circle around the origin in the X, Y plane of K may at the same time be regarded as a circle in the X′, Y′ plane of K′. We suppose that the circumference and diameter of this circle have been measured with a unit measure infinitely small compared with the radius, and that we have the quotient of the two results. If this experiment were performed with a measuring-rod at rest relatively to the Galilean system K, the quotient would be π. With a measuring-rod at rest relatively to K′, the quotient would be greater than π. This is readily understood if we envisage the whole process of measuring from the "stationary" system K, and take into consideration that the measuring-rod applied to the periphery undergoes a Lorentzian contraction, while the one applied along the radius does not. Hence Euclidean geometry does not apply to K′. The notion of co-ordinates defined above, which presupposes the validity of Euclidean geometry, therefore breaks
[10] down in relation to the system K′. So, too, we are unable to introduce a time corresponding to physical requirements in K′, indicated by clocks at rest relatively to K′. To convince ourselves of this impossibility, let us imagine two clocks of identical constitution placed, one at the origin of co-ordinates, and the other at the circumference of the circle, and both envisaged from the "stationary" system K. By a familiar result of the special theory of relativity, the clock at the circumference—judged from K—goes more slowly than the other, because the former is in motion and the latter at rest. An observer at the common origin of co-ordinates, capable of observing the clock at the circumference by means of light, would therefore see it lagging behind the clock beside him. As he will not make up his mind to let the velocity of light along the path in question depend explicitly on the time, he will interpret his observations as showing that the clock at the circumference "really" goes more slowly than the clock at the origin. So he will be obliged to define time in such a way that the rate of a clock depends upon where the clock may be.

We therefore reach this result:—In the general theory of relativity, space and time cannot be defined in such a way that differences of the spatial co-ordinates can be directly measured by the unit measuring-rod, or differences in the time co-ordinate by a standard clock.

The method hitherto employed for laying co-ordinates into the space-time continuum in a definite manner thus breaks down, and there seems to be no other way which would allow us to adapt systems of co-ordinates to the four-dimensional universe so that we might expect from their application a particularly simple formulation of the laws of nature. So there is nothing for it but to regard all imaginable systems of co-ordinates, on principle, as equally suitable for the description of nature. This comes to requiring that:—

The general laws of nature are to be expressed by equations which hold good for all systems of co-ordinates, that is, are co-variant with respect to any substitutions whatever (generally co-variant).

It is clear that a physical theory which satisfies this postulate will also be suitable for the general postulate of relativity. For the sum of *all* substitutions in any case includes those which correspond to all relative motions of three-dimensional systems of co-ordinates. That this requirement of general co-variance, which takes away from space and time the last remnant of physical objectivity, is a natural one, will be seen from the following reflexion. All our space-time verifications invariably amount to a determination
of space-time coincidences. If, for example, events consisted [11]
merely in the motion of material points, then ultimately nothing would be observable but the meetings of two or more of these points. Moreover, the results of our measurings are nothing but verifications of such meetings of the material points of our measuring instruments with other material points, coincidences between the hands of a clock and points on the clock dial, and observed point-events happening at the same place at the same time.

The introduction of a system of reference serves no other purpose than to facilitate the description of the totality of such coincidences. We allot to the universe four space-time variables x_1, x_2, x_3, x_4 in such a way that for every point-event there is a corresponding system of values of the variables $x_1 \ldots x_4$. To two coincident point-events there corresponds one system of values of the variables $x_1 \ldots x_4$, i.e. coincidence is characterized by the identity of the co-ordinates. If, in place of the variables $x_1 \ldots x_4$, we introduce functions of them, x'_1, x'_2, x'_3, x'_4, as a new system of co-ordinates, so that the systems of values are made to correspond to one another without ambiguity, the equality of all four co-ordinates in the new system will also serve as an expression for the space-time coincidence of the two point-events. As all our physical experience can be ultimately reduced to such coincidences, there is no immediate reason for preferring certain systems of co-ordinates to others, that is to say, we arrive at the requirement of general co-variance.

§ 4. The Relation of the Four Co-ordinates to Measurement in Space and Time

It is not my purpose in this discussion to represent the general theory of relativity as a system that is as simple and logical as possible, and with the minimum number of axioms; but my main object is to develop this theory in such a way that the reader will feel that the path we have entered upon is psychologically the natural one, and that the underlying assumptions will seem to have the highest possible degree of security. With this aim in view let it now be granted that:—

For infinitely small four-dimensional regions the theory of relativity in the restricted sense is appropriate, if the co-ordinates are suitably chosen.

For this purpose we must choose the acceleration of the infinitely small ("local") system of co-ordinates so that no gravitational field occurs; this is possible for an infinitely small region. Let X_1, X_2, X_3, be the co-ordinates of space, and X_4 the appertaining co-ordinate of time measured in the appropriate unit.* If a rigid rod is imagined to be given as the unit measure, the co-ordinates, with a given orientation of the system of co-ordinates, have a direct physical meaning

* The unit of time is to be chosen so that the velocity of light *in vacuo* as measured in the "local" system of co-ordinates is to be equal to unity.

in the sense of the special theory of relativity. By the special theory of relativity the expression

$$ds^2 = -dX_1^2 - dX_2^2 - dX_3^2 + dX_4^2 \quad . \quad . \quad (1)$$

then has a value which is independent of the orientation of the local system of co-ordinates, and is ascertainable by measurements of space and time. The magnitude of the linear element pertaining to points of the four-dimensional continuum in infinite proximity, we call ds. If the ds belonging to the element $dX_1 \ldots dX_4$ is positive, we follow Minkowski in calling it time-like; if it is negative, we call it space-like.

To the "linear element" in question, or to the two infinitely proximate point-events, there will also correspond definite differentials $dx_1 \ldots dx_4$ of the four-dimensional co-ordinates of any chosen system of reference. If this system, as well as the "local" system, is given for the region under consideration, the dX_ν will allow themselves to be represented here by definite linear homogeneous expressions of the dx_σ:—

$$dX_\nu = \sum_\sigma a_{\nu\sigma} dx_\sigma \quad . \quad . \quad . \quad . \quad . \quad (2)$$

Inserting these expressions in (1), we obtain

$$ds^2 = \sum_{\tau\sigma} g_{\sigma\tau} dx_\sigma dx_\tau, \quad . \quad . \quad . \quad . \quad (3)$$

where the $g_{\sigma\tau}$ will be functions of the x_σ. These can no longer be dependent on the orientation and the state of motion of the "local" system of co-ordinates, for ds^2 is a quantity ascertainable by rod-clock measurement of point-events infinitely proximate in space-time, and defined independently of any particular choice of co-ordinates. The $g_{\sigma\tau}$ are to be chosen here so that $g_{\sigma\tau} = g_{\tau\sigma}$; the summation is to extend over all values of σ and τ, so that the sum consists of 4×4 terms, of which twelve are equal in pairs.

The case of the ordinary theory of relativity arises out of the case here considered, if it is possible, by reason of the particular relations of the $g_{\sigma\tau}$ in a finite region, to choose the system of reference in the finite region in such a way that the $g_{\sigma\tau}$ assume the constant values

$$\begin{matrix} -1 & 0 & 0 & 0 \\ 0 & -1 & 0 & 0 \\ 0 & 0 & -1 & 0 \\ 0 & 0 & 0 & +1 \end{matrix} \Bigg\} \quad . \quad . \quad . \quad (4)$$

We shall find hereafter that the choice of such co-ordinates is, in general, not possible for a finite region.

From the considerations of § 2 and § 3 it follows that the quantities $g_{\tau\sigma}$ are to be regarded from the physical standpoint as the quantities which describe the gravitational field in relation to the chosen system of reference. For, if we now assume the special theory of relativity to apply to a certain four-dimensional region with the co-ordinates properly chosen, then the $g_{\sigma\tau}$ have the values given in (4). A free material point then moves, relatively to this system, with uniform motion in a straight line. Then if we introduce new space-time co-ordinates x_1, x_2, x_3, x_4, by means of any substitution we choose, the $g^{\sigma\tau}$ in this new system will no longer be constants, but functions of space and time. At the same time the motion of the free material point will present itself in the new co-ordinates as a curvilinear non-uniform motion, and the law of this motion will be independent of the nature of the moving particle. We shall therefore interpret this motion as a motion under the influence of a gravitational field. We thus find the occurrence of a gravitational field connected with a space-time variability of the g_{σ} . So, too, in the general case, when we are no longer able by a suitable choice of co-ordinates to apply the special theory of relativity to a finite region, we shall hold fast to the view that the $g_{\sigma\tau}$ describe the gravitational field.

Thus, according to the general theory of relativity, gravitation occupies an exceptional position with regard to other forces, particularly the electromagnetic forces, since the ten functions representing the gravitational field at the same time define the metrical properties of the space measured.

B. Mathematical Aids to the Formulation of Generally Covariant Equations

Having seen in the foregoing that the general postulate of relativity leads to the requirement that the equations of physics shall be covariant in the face of any substitution of the co-ordinates $x_1 \ldots x_4$, we have to consider how such generally covariant equations can be found. We now turn to this purely mathematical task, and we shall find that in its solution a fundamental rôle is played by the invariant ds given in equation (3), which, borrowing from Gauss's theory of surfaces, we have called the "linear element."

The fundamental idea of this general theory of covariants is the following:—Let certain things ("tensors") be defined with respect to any system of co-ordinates by a number of functions of the co-ordinates, called the "components" of the tensor. There are then certain rules by which these components can be calculated for a new system of co-ordinates, if they are known for the original system of co-ordinates, and if the transformation connecting the two systems is known. The things hereafter called tensors are further characterized by the fact that the equations of transformation for their components are linear and homogeneous. Accordingly, all the components in the new system vanish, if they all vanish in the original system. If, therefore, a law of nature is expressed by equating all the components of a tensor to zero, it is generally covariant. By examining the laws of the formation of tensors, we acquire the means of formulating generally covariant laws.

§ 5. Contravariant and Covariant Four-vectors

Contravariant Four-vectors.—The linear element is defined by the four "components" dx_ν, for which the law of transformation is expressed by the equation

$$dx'_\sigma = \sum_\nu \frac{\partial x'_\sigma}{\partial x_\nu} dx_\nu \quad . \quad . \quad . \quad . \quad (5)$$

The dx'_σ are expressed as linear and homogeneous functions of the dx_ν. Hence we may look upon these co-ordinate differentials as the components of a "tensor" of the particular kind which we call a contravariant four-vector. Any thing which is defined relatively to the system of co-ordinates by four quantities A^ν, and which is transformed by the same law

$$A'^\sigma = \sum_\nu \frac{\partial x'_\sigma}{\partial x_\nu} A^\nu, \quad . \quad . \quad . \quad . \quad (5a)$$

we also call a contravariant four-vector. From (5a) it follows at once that the sums $A^\sigma \pm B^\sigma$ are also components of a four-vector, if A^σ and B^σ are such. Corresponding relations hold for all "tensors" subsequently to be introduced. (Rule for the addition and subtraction of tensors.)

Covariant Four-vectors.—We call four quantities A_ν the components of a covariant four-vector, if for any arbitrary choice of the contravariant four-vector B^ν

$$\sum_\nu A_\nu B^\nu = \text{Invariant} \quad . \quad . \quad . \quad (6)$$

The law of transformation of a covariant four-vector follows from this definition. For if we replace B^ν on the right-hand side of the equation

$$\sum_\sigma A'_\sigma B'^\sigma = \sum_\nu A_\nu B^\nu$$

by the expression resulting from the inversion of (5a),

$$\sum_\sigma \frac{\partial x_\nu}{\partial x'_\sigma} B'^\sigma,$$

we obtain

$$\sum_\sigma B'^\sigma \sum_\nu \frac{\partial x_\nu}{\partial x'_\sigma} A_\nu = \sum_\sigma B'^\sigma A'_\sigma.$$

Since this equation is true for arbitrary values of the B'^σ, it follows that the law of transformation is

$$A'_\sigma = \sum_\nu \frac{\partial x_\nu}{\partial x'_\sigma} A_\nu \quad . \quad . \quad . \quad . \quad (7)$$

Note on a Simplified Way of Writing the Expressions.—A glance at the equations of this paragraph shows that there is always a summation with respect to the indices which occur twice under a sign of summation (e.g. the index ν in (5)), and only with respect to indices which occur twice. It is therefore possible, without loss of clearness, to omit the sign of summation. In its place we introduce the convention:—If an index occurs twice in one term of an expression, it is
[12] always to be summed unless the contrary is expressly stated.

The difference between covariant and contravariant four-vectors lies in the law of transformation ((7) or (5) respectively). Both forms are tensors in the sense of the general remark above. Therein lies their importance. Following Ricci and

Levi-Civita, we denote the contravariant character by placing the index above, the covariant by placing it below. [13]

§ 6. Tensors of the Second and Higher Ranks

Contravariant Tensors.—If we form all the sixteen products $A^{\mu\nu}$ of the components A^{μ} and B^{ν} of two contravariant four-vectors

$$A^{\mu\nu} = A^{\mu}B^{\nu} \quad . \quad . \quad . \quad . \quad . \quad (8)$$

then by (8) and (5a) $A^{\mu\nu}$ satisfies the law of transformation

$$A'^{\sigma\tau} = \frac{\partial x'_{\sigma}}{\partial x_{\mu}} \frac{\partial x'_{\tau}}{\partial x_{\nu}} A^{\mu\nu} \quad . \quad . \quad . \quad . \quad (9)$$

We call a thing which is described relatively to any system of reference by sixteen quantities, satisfying the law of transformation (9), a contravariant tensor of the second rank. Not every such tensor allows itself to be formed in accordance with (8) from two four-vectors, but it is easily shown that any given sixteen $A^{\mu\nu}$ can be represented as the sums of the $A^{\mu}B^{\nu}$ of four appropriately selected pairs of four-vectors. Hence we can prove nearly all the laws which apply to the tensor of the second rank defined by (9) in the simplest manner by demonstrating them for the special tensors of the type (8).

Contravariant Tensors of Any Rank.—It is clear that, on the lines of (8) and (9), contravariant tensors of the third and higher ranks may also be defined with 4^3 components, and so on. In the same way it follows from (8) and (9) that the contravariant four-vector may be taken in this sense as a contravariant tensor of the first rank.

Covariant Tensors.—On the other hand, if we take the sixteen products $A_{\mu\nu}$ of two covariant four-vectors A_{μ} and B_{ν},

$$A_{\mu\nu} = A_{\mu}B_{\nu}, \quad . \quad . \quad . \quad . \quad . \quad (10)$$

the law of transformation for these is

$$A'_{\sigma\tau} = \frac{\partial x_{\mu}}{\partial x'_{\sigma}} \frac{\partial x_{\nu}}{\partial x'_{\tau}} A_{\mu\nu} \quad . \quad . \quad . \quad . \quad (11)$$

This law of transformation defines the covariant tensor of the second rank. All our previous remarks on contravariant tensors apply equally to covariant tensors.

NOTE.—It is convenient to treat the scalar (or invariant) both as a contravariant and a covariant tensor of zero rank.

Mixed Tensors.—We may also define a tensor of the second rank of the type

$$A_\mu^\nu = A_\mu B^\nu \quad . \quad . \quad . \quad . \quad (12)$$

which is covariant with respect to the index μ, and contravariant with respect to the index ν. Its law of transformation is

$$A'^\tau_\sigma = \frac{\partial x'_\tau}{\partial x_\nu}\frac{\partial x_\mu}{\partial x'_\sigma}A^\nu_\mu \quad . \quad . \quad . \quad (13)$$

Naturally there are mixed tensors with any number of indices of covariant character, and any number of indices of contravariant character. Covariant and contravariant tensors may be looked upon as special cases of mixed tensors.

Symmetrical Tensors.—A contravariant, or a covariant tensor, of the second or higher rank is said to be symmetrical if two components, which are obtained the one from the other by the interchange of two indices, are equal. The tensor $A^{\mu\nu}$, or the tensor $A_{\mu\nu}$, is thus symmetrical if for any combination of the indices μ, ν,

$$A^{\mu\nu} = A^{\nu\mu}, \quad . \quad . \quad . \quad . \quad (14)$$

or respectively,

$$A_{\mu\nu} = A_{\nu\mu}. \quad . \quad . \quad . \quad . \quad (14a)$$

It has to be proved that the symmetry thus defined is a property which is independent of the system of reference. It follows in fact from (9), when (14) is taken into consideration, that

$$A'^{\sigma\tau} = \frac{\partial x'_\sigma}{\partial x_\mu}\frac{\partial x'_\tau}{\partial x_\nu}A^{\mu\nu} = \frac{\partial x'_\sigma}{\partial x_\mu}\frac{\partial x'_\tau}{\partial x_\nu}A^{\nu\mu} = \frac{\partial x'_\sigma}{\partial x_\nu}\frac{\partial x'_\tau}{\partial x_\mu}A^{\mu\nu} = A'^{\tau\sigma}.$$

The last equation but one depends upon the interchange of the summation indices μ and ν, i.e. merely on a change of notation.

Antisymmetrical Tensors.—A contravariant or a covariant tensor of the second, third, or fourth rank is said to be antisymmetrical if two components, which are obtained the one from the other by the interchange of two indices, are equal and of opposite sign. The tensor $A^{\mu\nu}$, or the tensor $A_{\mu\nu}$, is therefore antisymmetrical, if always

$$A^{\mu\nu} = - A^{\nu\mu}, \quad . \quad . \quad . \quad . \quad (15)$$

or respectively,

$$A_{\mu\nu} = - A_{\nu\mu} \quad . \quad . \quad . \quad . \quad (15a)$$

Of the sixteen components $A^{\mu\nu}$, the four components $A^{\mu\mu}$ vanish; the rest are equal and of opposite sign in pairs, so that there are only six components numerically different (a six-vector). Similarly we see that the antisymmetrical tensor of the third rank $A^{\mu\nu\sigma}$ has only four numerically different components, while the antisymmetrical tensor $A^{\mu\nu\sigma\tau}$ has only one. There are no antisymmetrical tensors of higher rank than the fourth in a continuum of four dimensions.

§ 7. Multiplication of Tensors

Outer Multiplication of Tensors.—We obtain from the components of a tensor of rank n and of a tensor of rank m the components of a tensor of rank $n + m$ by multiplying each component of the one tensor by each component of the other. Thus, for example, the tensors T arise out of the tensors A and B of different kinds,

$$\begin{aligned} T_{\mu\nu\sigma} &= A_{\mu\nu}B_{\sigma}, \\ T^{\mu\nu\sigma\tau} &= A^{\mu\nu}B^{\sigma\tau}, \\ T^{\sigma\tau}_{\mu\nu} &= A_{\mu\nu}B^{\sigma\tau}. \end{aligned}$$

The proof of the tensor character of T is given directly by the representations (8), (10), (12), or by the laws of transformation (9), (11), (13). The equations (8), (10), (12) are themselves examples of outer multiplication of tensors of the first rank.

"Contraction" of a Mixed Tensor.—From any mixed tensor we may form a tensor whose rank is less by two, by equating an index of covariant with one of contravariant character, and summing with respect to this index ("contraction"). Thus, for example, from the mixed tensor of the fourth rank $A^{\sigma\tau}_{\mu\nu}$, we obtain the mixed tensor of the second rank,

$$A^{\tau}_{\nu} = A^{\mu\tau}_{\mu\nu} \quad \left(= \sum_{\mu} A^{\mu\tau}_{\mu\nu} \right),$$

and from this, by a second contraction, the tensor of zero rank,

$$A = A^{\nu}_{\nu} = A^{\mu\nu}_{\mu\nu}.$$

The proof that the result of contraction really possesses the tensor character is given either by the representation of a tensor according to the generalization of (12) in combination with (6), or by the generalization of (13).

Inner and Mixed Multiplication of Tensors.—These consist in a combination of outer multiplication with contraction.

Examples.—From the covariant tensor of the second rank $A_{\mu\nu}$ and the contravariant tensor of the first rank B^{σ} we form by outer multiplication the mixed tensor

$$D^{\sigma}_{\mu\nu} = A_{\mu\nu}B^{\sigma}.$$

On contraction with respect to the indices ν and σ, we obtain the covariant four-vector

$$D_{\mu} = D^{\nu}_{\mu\nu} = A_{\mu\nu}B^{\nu}.$$

This we call the inner product of the tensors $A_{\mu\nu}$ and B^{σ}. Analogously we form from the tensors $A_{\mu\nu}$ and $B^{\sigma\tau}$, by outer multiplication and double contraction, the inner product $A_{\mu\nu}B^{\mu\nu}$. By outer multiplication and one contraction, we obtain from $A_{\mu\nu}$ and $B^{\sigma\tau}$ the mixed tensor of the second rank $D^{\tau}_{\mu} = A_{\mu\nu}B^{\nu\tau}$. This operation may be aptly characterized as a mixed one, being "outer" with respect to the indices μ and τ, and "inner" with respect to the indices ν and σ.

We now prove a proposition which is often useful as evidence of tensor character. From what has just been explained, $A_{\mu\nu}B^{\mu\nu}$ is a scalar if $A_{\mu\nu}$ and $B^{\sigma\tau}$ are tensors. But we may also make the following assertion: If $A_{\mu\nu}B^{\mu\nu}$ is a scalar *for any choice of the tensor* $B^{\mu\nu}$, then $A_{\mu\nu}$ has tensor character. For, by hypothesis, for any substitution,

$$A'_{\sigma\tau}B'^{\sigma\tau} = A_{\mu\nu}B^{\mu\nu}.$$

But by an inversion of (9)

$$B^{\mu\nu} = \frac{\partial x_{\mu}}{\partial x'_{\sigma}}\frac{\partial x_{\nu}}{\partial x'_{\tau}}B'^{\sigma\tau}.$$

This, inserted in the above equation, gives

$$\left(A'_{\sigma\tau} - \frac{\partial x_{\mu}}{\partial x'_{\sigma}}\frac{\partial x_{\nu}}{\partial x'_{\tau}}A_{\mu\nu}\right)B'^{\sigma\tau} = 0.$$

This can only be satisfied for arbitrary values of $B'^{\sigma\tau}$ if the bracket vanishes. The result then follows by equation (11). This rule applies correspondingly to tensors of any rank and character, and the proof is analogous in all cases.

The rule may also be demonstrated in this form: If B^μ and C^ν are any vectors, and if, for all values of these, the inner product $A_{\mu\nu}B^\mu C^\nu$ is a scalar, then $A_{\mu\nu}$ is a covariant tensor. This latter proposition also holds good even if only the more special assertion is correct, that with any choice of the four-vector B^μ the inner product $A_{\mu\nu}B^\mu B^\nu$ is a scalar, if in addition it is known that $A_{\mu\nu}$ satisfies the condition of symmetry $A_{\mu\nu} = A_{\nu\mu}$. For by the method given above we prove the tensor character of $(A_{\mu\nu} + A_{\nu\mu})$, and from this the tensor character of $A_{\mu\nu}$ follows on account of symmetry. This also can be easily generalized to the case of covariant and contravariant tensors of any rank.

Finally, there follows from what has been proved, this law, which may also be generalized for any tensors: If for any choice of the four-vector B^ν the quantities $A_{\mu\nu}B^\nu$ form a tensor of the first rank, then $A_{\mu\nu}$ is a tensor of the second rank. For, if C^μ is any four-vector, then on account of the tensor character of $A_{\mu\nu}B^\nu$, the inner product $A_{\mu\nu}B^\nu C^\mu$ is a scalar for any choice of the two four-vectors B^ν and C^μ. From which the proposition follows.

§ 8. Some Aspects of the Fundamental Tensor $g_{\mu\nu}$

The Covariant Fundamental Tensor.—In the invariant expression for the square of the linear element,

$$ds^2 = g_{\mu\nu}dx_\mu dx_\nu,$$

the part played by the dx_μ is that of a contravariant vector which may be chosen at will. Since further, $g_{\mu\nu} = g_{\nu\mu}$, it follows from the considerations of the preceding paragraph that $g_{\mu\nu}$ is a covariant tensor of the second rank. We call it the "fundamental tensor." In what follows we deduce some properties of this tensor which, it is true, apply to any tensor of the second rank. But as the fundamental tensor plays a special part in our theory, which has its physical basis in the peculiar effects of gravitation, it so happens that the relations to be developed are of importance to us only in the case of the fundamental tensor.

The Contravariant Fundamental Tensor.—If in the determinant formed by the elements $g_{\mu\nu}$, we take the co-factor of each of the $g_{\mu\nu}$ and divide it by the determinant $g = | g_{\mu\nu} |$, we obtain certain quantities $g^{\mu\nu} (= g^{\nu\mu})$ which, as we shall demonstrate, form a contravariant tensor.

By a known property of determinants

$$g_{\mu\sigma} g^{\nu\sigma} = \delta_{\mu}^{\nu} \quad . \quad . \quad . \quad . \quad (16)$$

where the symbol δ_{μ}^{ν} denotes 1 or 0, according as $\mu = \nu$ or $\mu \neq \nu$.

Instead of the above expression for ds^2 we may thus write

$$g_{\mu\sigma} \delta_{\nu}^{\sigma} dx_{\mu} dx_{\nu}$$

or, by (16)

$$g_{\mu\sigma} g_{\nu\tau} g^{\sigma\tau} dx_{\mu} dx_{\nu}.$$

But, by the multiplication rules of the preceding paragraphs, the quantities

$$d\xi_{\sigma} = g_{\mu\sigma} dx_{\mu}$$

form a covariant four-vector, and in fact an arbitrary vector, since the dx_{μ} are arbitrary. By introducing this into our expression we obtain

$$ds^2 = g^{\sigma\tau} d\xi_{\sigma} d\xi_{\tau}.$$

Since this, with the arbitrary choice of the vector $d\xi_{\sigma}$, is a scalar, and $g^{\sigma\tau}$ by its definition is symmetrical in the indices σ and τ, it follows from the results of the preceding paragraph that $g^{\sigma\tau}$ is a contravariant tensor.

It further follows from (16) that δ_{μ} is also a tensor, which we may call the mixed fundamental tensor.

The Determinant of the Fundamental Tensor.—By the rule for the multiplication of determinants

$$| g_{\mu\alpha} g^{\alpha\nu} | = | g_{\mu\alpha} | \times | g^{\alpha\nu} |.$$

On the other hand

$$| g_{\mu\alpha} g^{\alpha\nu} | = | \delta_{\mu}^{\nu} | = 1.$$

It therefore follows that

$$| g_{\mu\nu} | \times | g^{\mu\nu} | = 1 \quad . \quad . \quad . \quad (17)$$

The Volume Scalar.—We seek first the law of transformation of the determinant $g = | g_{\mu\nu} |$. In accordance with (11)

$$g' = \left| \frac{\partial x_\mu}{\partial x'_\sigma} \frac{\partial x}{\partial x'_\tau} g_{\mu\nu} \right| .$$

Hence, by a double application of the rule for the multiplication of determinants, it follows that

$$g' = \left| \frac{\partial x_\mu}{\partial x'_\sigma} \right| \cdot \left| \frac{\partial x_\nu}{\partial x'_\tau} \right| \cdot | g_{\mu\nu} | = \left| \frac{\partial x_\mu}{\partial x'_\sigma} \right|^2 g,$$

or

$$\sqrt{g'} = \left| \frac{\partial x_\mu}{\partial x'_\sigma} \right| \sqrt{g}.$$

On the other hand, the law of transformation of the element of volume

$$d\tau = \int dx_1 dx_2 dx_3 dx_4$$

is, in accordance with the theorem of Jacobi,

$$d\tau' = \left| \frac{\partial x'_\sigma}{\partial x_\mu} \right| d\tau.$$

By multiplication of the last two equations, we obtain

$$\sqrt{g'} d\tau' = \sqrt{g} d\tau \quad . \quad . \quad . \quad (18).$$

Instead of $\sqrt{g}$, we introduce in what follows the quantity $\sqrt{-g}$, which is always real on account of the hyperbolic character of the space-time continuum. The invariant $\sqrt{-g} d\tau$ is equal to the magnitude of the four-dimensional element of volume in the "local" system of reference, as measured with rigid rods and clocks in the sense of the special theory of relativity.

Note on the Character of the Space-time Continuum.—Our assumption that the special theory of relativity can always be applied to an infinitely small region, implies that ds^2 can always be expressed in accordance with (1) by means of real quantities $dX_1 \ldots dX_4$. If we denote by $d\tau_0$ the "natural" element of volume dX_1, dX_2, dX_3, dX_4, then

$$d\tau_0 = \sqrt{-g} d\tau \quad . \quad . \quad . \quad (18a)$$

If $\sqrt{-g}$ were to vanish at a point of the four-dimensional continuum, it would mean that at this point an infinitely small "natural" volume would correspond to a finite volume in the co-ordinates. Let us assume that this is never the case. Then g cannot change sign. We will assume that, in the sense of the special theory of relativity, g always has a finite negative value. This is a hypothesis as to the physical nature of the continuum under consideration, and at the same time a convention as to the choice of co-ordinates.

But if $-g$ is always finite and positive, it is natural to settle the choice of co-ordinates *a posteriori* in such a way that this quantity is always equal to unity. We shall see later that by such a restriction of the choice of co-ordinates it is possible to achieve an important simplification of the laws of nature.

In place of (18), we then have simply $d\tau' = d\tau$, from which, in view of Jacobi's theorem, it follows that

$$\left| \frac{\partial x'_\sigma}{\partial x_\mu} \right| = 1 \quad . \quad . \quad . \quad . \quad (19)$$

Thus, with this choice of co-ordinates, only substitutions for which the determinant is unity are permissible.

But it would be erroneous to believe that this step indicates a partial abandonment of the general postulate of relativity. We do not ask "What are the laws of nature which are co-variant in face of all substitutions for which the determinant is unity?" but our question is "What are the generally co-variant laws of nature?" It is not until we have formulated these that we simplify their expression by a particular choice of the system of reference.

The Formation of New Tensors by Means of the Fundamental Tensor.—Inner, outer, and mixed multiplication of a tensor by the fundamental tensor give tensors of different character and rank. For example,

$$A^\mu = g^{\mu\sigma} A_\sigma,$$
$$A = g_{\mu\nu} A^{\mu\nu}.$$

The following forms may be specially noted:—

$$A^{\mu\nu} = g^{\mu\alpha} g^{\nu\beta} A_{\alpha\beta},$$
$$A_{\mu\nu} = g_{\mu\alpha} g_{\nu\beta} A^{\alpha\beta}$$

(the " complements " of covariant and contravariant tensors respectively), and

$$B_{\mu\nu} = g_{\mu\nu}g^{\alpha\beta}A_{\alpha\beta}.$$

We call $B_{\mu\nu}$ the reduced tensor associated with $A_{\mu\nu}$. Similarly,

$$B^{\mu\nu} = g^{\mu\nu}g_{\alpha\beta}A^{\alpha\beta}.$$

It may be noted that $g^{\mu\nu}$ is nothing more than the complement of $g_{\mu\nu}$, since

$$g^{\mu\alpha}g^{\nu\beta}g_{\alpha\beta} = g^{\mu\alpha}\delta_{\alpha}^{\nu} = g^{\mu\nu}.$$

§ 9. The Equation of the Geodetic Line. The Motion of a Particle

As the linear element ds is defined independently of the system of co-ordinates, the line drawn between two points P and P′ of the four-dimensional continuum in such a way that $\int ds$ is stationary—a geodetic line—has a meaning which also is independent of the choice of co-ordinates. Its equation is

$$\delta\int_{P}^{P'} ds = 0 \quad . \quad . \quad . \quad . \quad (20)$$

Carrying out the variation in the usual way, we obtain from this equation four differential equations which define the geodetic line ; this operation will be inserted here for the sake of completeness. Let λ be a function of the co-ordinates x_ν, and let this define a family of surfaces which intersect the required geodetic line as well as all the lines in immediate proximity to it which are drawn through the points P and P′. Any such line may then be supposed to be given by expressing its co-ordinates x_ν as functions of λ. Let the symbol δ indicate the transition from a point of the required geodetic to the point corresponding to the same λ on a neighbouring line. Then for (20) we may substitute

$$\left.\begin{aligned} \int_{\lambda_1}^{\lambda_2}\delta w d\lambda &= 0 \\ w^2 &= g_{\mu\nu}\frac{dx_\mu}{d\lambda}\frac{dx_\nu}{d\lambda} \end{aligned}\right\} \quad . \quad . \quad . \quad (20a)$$

But since

$$\delta w = \frac{1}{w}\left\{\frac{1}{2}\frac{\partial g_{\mu\nu}}{\partial x_\sigma}\frac{dx_\mu}{d\lambda}\frac{dx_\nu}{d\lambda}\delta x_\sigma + g_{\mu\nu}\frac{dx_\mu}{d\lambda}\delta\left(\frac{dx_\nu}{d\lambda}\right)\right\},$$

and

$$\delta\left(\frac{dx_\nu}{d\lambda}\right) = \frac{d}{d\lambda}(\delta x_\nu),$$

we obtain from (20a), after a partial integration,

$$\int_{\lambda_1}^{\lambda_2} \kappa_\sigma \delta x_\sigma d\lambda = 0,$$

where

[14] $$\kappa_\sigma = \frac{d}{d\lambda}\left\{\frac{g_{\mu\nu}}{w}\frac{dx_\mu}{d\lambda}\right\} - \frac{1}{2w}\frac{\partial g_{\mu\nu}}{\partial x_\sigma}\frac{dx_\mu}{d\lambda}\frac{dx_\nu}{d\lambda} \quad . \quad (20b)$$

Since the values of δx_σ are arbitrary, it follows from this that

$$\kappa_\sigma = 0 \quad . \quad . \quad . \quad . \quad . \quad (20c)$$

are the equations of the geodetic line.

If ds does not vanish along the geodetic line we may choose the "length of the arc" s, measured along the geodetic line, for the parameter λ. Then $w = 1$, and in place of (20c) we obtain

[15] $$g_{\mu\nu}\frac{d^2x_\mu}{ds^2} + \frac{\partial g_{\mu\nu}}{\partial x_\sigma}\frac{dx_\sigma}{ds}\frac{dx_\mu}{ds} - \frac{1}{2}\frac{\partial g_{\mu\nu}}{\partial x_\sigma}\frac{dx_\mu}{ds}\frac{dx_\nu}{ds} = 0$$

or, by a mere change of notation,

$$g_{\alpha\sigma}\frac{d^2x_\alpha}{ds^2} + [\mu\nu, \sigma]\frac{dx_\mu}{ds}\frac{dx_\nu}{ds} = 0 \quad . \quad . \quad (20d)$$

where, following Christoffel, we have written

$$[\mu\nu, \sigma] = \frac{1}{2}\left(\frac{\partial g_{\mu\sigma}}{\partial x_\nu} + \frac{\partial g_{\nu\sigma}}{\partial x_\mu} - \frac{\partial g_{\mu\nu}}{\partial x_\sigma}\right) \quad . \quad . \quad (21)$$

Finally, if we multiply (20d) by $g^{\sigma\tau}$ (outer multiplication with respect to τ, inner with respect to σ), we obtain the equations of the geodetic line in the form

$$\frac{d^2x_\tau}{ds^2} + \{\mu\nu, \tau\}\frac{dx_\mu}{ds}\frac{dx_\nu}{ds} = 0 \quad . \quad . \quad . \quad (22)$$

where, following Christoffel, we have set

$$\{\mu\nu, \tau\} = g^{\tau\alpha}[\mu\nu, \alpha] \quad . \quad . \quad . \quad (23)$$

§ 10. The Formation of Tensors by Differentiation

With the help of the equation of the geodetic line we can now easily deduce the laws by which new tensors can be formed from old by differentiation. By this means we are able for the first time to formulate generally covariant differential equations. We reach this goal by repeated application of the following simple law :—

If in our continuum a curve is given, the points of which are specified by the arcual distance s measured from a fixed point on the curve, and if, further, ϕ is an invariant function of space, then $d\phi/ds$ is also an invariant. The proof lies in this, that ds is an invariant as well as $d\phi$.

As

$$\frac{d\phi}{ds} = \frac{\partial\phi}{\partial x_\mu}\frac{dx_\mu}{ds}$$

therefore

$$\psi = \frac{\partial\phi}{dx_\mu}\frac{dx_\mu}{ds}$$

is also an invariant, and an invariant for all curves starting from a point of the continuum, that is, for any choice of the vector dx_μ. Hence it immediately follows that

$$A_\mu = \frac{\partial\phi}{\partial x_\mu} \quad . \quad . \quad . \quad . \quad . \quad (24)$$

is a covariant four-vector—the "gradient" of ϕ.

According to our rule, the differential quotient

$$\chi = \frac{d\psi}{ds}$$

taken on a curve, is similarly an invariant. Inserting the value of ψ, we obtain in the first place

$$\chi = \frac{\partial^2\phi}{\partial x_\mu \partial x_\nu}\frac{dx_\mu}{ds}\frac{dx_\nu}{ds} + \frac{\partial\phi}{\partial x_\mu}\frac{d^2x_\mu}{ds^2}.$$

The existence of a tensor cannot be deduced from this forthwith. But if we may take the curve along which we have differentiated to be a geodetic, we obtain on substitution for d^2x_ν/ds^2 from (22),

$$\chi = \left(\frac{\partial^2\phi}{\partial x_\mu \partial x_\nu} - \{\mu\nu, \tau\}\frac{\partial\phi}{\partial x_\tau}\right)\frac{dx_\mu}{ds}\frac{dx_\nu}{ds}.$$

Since we may interchange the order of the differentiations,

and since by (23) and (21) $\{\mu\nu, \tau\}$ is symmetrical in μ and ν, it follows that the expression in brackets is symmetrical in μ and ν. Since a geodetic line can be drawn in any direction from a point of the continuum, and therefore dx_μ/ds is a four-vector with the ratio of its components arbitrary, it follows from the results of § 7 that

$$A_{\mu\nu} = \frac{\partial^2 \phi}{\partial x_\mu \partial x_\nu} - \{\mu\nu, \tau\}\frac{\partial \phi}{\partial x_\tau} \quad . \quad . \quad . \quad (25)$$

is a covariant tensor of the second rank. We have therefore come to this result: from the covariant tensor of the first rank

$$A_\mu = \frac{\partial \phi}{\partial x_\mu}$$

we can, by differentiation, form a covariant tensor of the second rank

$$A_{\mu\nu} = \frac{\partial A_\mu}{\partial x_\nu} - \{\mu\nu, \tau\}A_\tau \quad . \quad . \quad . \quad (26)$$

We call the tensor $A_{\mu\nu}$ the " extension " (covariant derivative) of the tensor A_μ. In the first place we can readily show that the operation leads to a tensor, even if the vector A_μ cannot be represented as a gradient. To see this, we first observe that

$$\psi \frac{\partial \phi}{\partial x_\mu}$$

is a covariant vector, if ψ and ϕ are scalars. The sum of four such terms

$$S_\mu = \psi^{(1)}\frac{\phi\partial^{(1)}}{\partial x_\mu} + . + . + \psi^{(4)}\frac{\partial \phi^{(4)}}{\partial x_\mu},$$

is also a covariant vector, if $\psi^{(1)}, \phi^{(1)} \ldots \psi^{(4)}, \phi^{(4)}$ are scalars. But it is clear that any covariant vector can be represented in the form S_μ. For, if A_μ is a vector whose components are any given functions of the x_ν, we have only to put (in terms of the selected system of co-ordinates)

$$\begin{aligned} \psi^{(1)} &= A_1, & \phi^{(1)} &= x_1, \\ \psi^{(2)} &= A_2, & \phi^{(2)} &= x_2, \\ \psi^{(3)} &= A_3, & \phi^{(3)} &= x_3, \\ \psi^{(4)} &= A_4, & \phi^{(4)} &= x_4, \end{aligned}$$

in order to ensure that S_μ shall be equal to A_μ.

Therefore, in order to demonstrate that $A_{\mu\nu}$ is a tensor if *any* covariant vector is inserted on the right-hand side for A_μ, we only need show that this is so for the vector S_μ. But for this latter purpose it is sufficient, as a glance at the right-hand side of (26) teaches us, to furnish the proof for the case

$$A_\mu = \psi \frac{\partial \phi}{\partial x_\mu}.$$

Now the right-hand side of (25) multiplied by ψ,

$$\psi \frac{\partial^2 \phi}{\partial x_\mu \partial x_\nu} - \{\mu\nu, \tau\} \psi \frac{\partial \phi}{\partial x_\tau}$$

is a tensor. Similarly

$$\frac{\partial \psi}{\partial x_\mu} \frac{\partial \phi}{\partial x_\nu}$$

being the outer product of two vectors, is a tensor. By addition, there follows the tensor character of

$$\frac{\partial}{\partial x_\nu}\left(\psi \frac{\partial \phi}{\partial x_\mu}\right) - \{\mu\nu, \tau\}\left(\psi \frac{\partial \phi}{\partial x_\tau}\right).$$

As a glance at (26) will show, this completes the demonstration for the vector

$$\psi \frac{\partial \phi}{\partial x_\mu}$$

and consequently, from what has already been proved, for any vector A_μ.

By means of the extension of the vector, we may easily define the "extension" of a covariant tensor of any rank. This operation is a generalization of the extension of a vector. We restrict ourselves to the case of a tensor of the second rank, since this suffices to give a clear idea of the law of formation.

As has already been observed, any covariant tensor of the second rank can be represented * as the sum of tensors of the

* By outer multiplication of the vector with arbitrary components A_{11}, A_{12}, A_{13}, A_{14} by the vector with components 1, 0, 0, 0, we produce a tensor with components

$$\begin{matrix} A_{11} & A_{12} & A_{13} & A_{14} \\ 0 & 0 & 0 & 0 \\ 0 & 0 & 0 & 0 \\ 0 & 0 & 0 & 0. \end{matrix}$$

By the addition of four tensors of this type, we obtain the tensor $A_{\mu\nu}$ with any assigned components.

type $A_\mu B_\nu$. It will therefore be sufficient to deduce the expression for the extension of a tensor of this special type. By (26) the expressions

$$\frac{\partial A_\mu}{\partial x_\sigma} - \{\sigma\mu, \tau\}A_\tau,$$

$$\frac{\partial B_\nu}{\partial x_\sigma} - \{\sigma\nu, \tau\}B_\tau,$$

are tensors. On outer multiplication of the first by B_ν, and of the second by A_μ, we obtain in each case a tensor of the third rank. By adding these, we have the tensor of the third rank

$$A_{\mu\nu\sigma} = \frac{\partial A_{\mu\nu}}{\partial x_\sigma} - \{\sigma\mu, \tau\}A_{\tau\nu} - \{\sigma\nu, \tau\}A_{\mu\tau} \quad . \quad . \quad (27)$$

where we have put $A_{\mu\nu} = A_\mu B_\nu$. As the right-hand side of (27) is linear and homogeneous in the $A_{\mu\nu}$ and their first derivatives, this law of formation leads to a tensor, not only in the case of a tensor of the type $A_\mu B_\nu$, but also in the case of a sum of such tensors, i.e. in the case of any covariant tensor of the second rank. We call $A_{\mu\nu\sigma}$ the extension of the tensor $A_{\mu\nu}$.

It is clear that (26) and (24) concern only special cases of extension (the extension of the tensors of rank one and zero respectively).

In general, all special laws of formation of tensors are included in (27) in combination with the multiplication of tensors.

§ 11. Some Cases of Special Importance

The Fundamental Tensor.—We will first prove some lemmas which will be useful hereafter. By the rule for the differentiation of determinants

$$dg = g^{\mu\nu} g dg_{\mu\nu} = -g_{\mu\nu} g dg^{\mu\nu} \quad . \quad . \quad . \quad (28)$$

The last member is obtained from the last but one, if we bear in mind that $g_{\mu\nu} g^{\mu'\nu} = \delta_\mu^{\mu'}$, so that $g_{\mu\nu} g^{\mu\nu} = 4$, and consequently

$$g_{\mu\nu} dg^{\mu\nu} + g^{\mu\nu} dg_{\mu\nu} = 0.$$

From (28), it follows that

$$\frac{1}{\sqrt{-g}}\frac{\partial\sqrt{-g}}{\partial x_\sigma} = \tfrac{1}{2}\frac{\partial \log(-g)}{\partial x_\sigma} = \tfrac{1}{2}g^{\mu\nu}\frac{\partial g_{\mu\nu}}{\partial x_\sigma} = \tfrac{1}{2}g_{\mu\nu}\frac{\partial g^{\mu\nu}}{\partial x_\sigma}. \quad (29)$$

Further, from $g_{\mu\sigma}g^{\nu\sigma} = \delta^\nu_\mu$, it follows on differentiation that

$$\left.\begin{aligned} g_{\mu\sigma}dg^{\nu\sigma} &= -g^{\nu\sigma}dg_{\mu\sigma} \\ g_{\mu\sigma}\frac{\partial g^{\nu\sigma}}{\partial x_\lambda} &= -g^{\nu\sigma}\frac{\partial g_{\mu\sigma}}{\partial x_\lambda} \end{aligned}\right\} \quad . \quad . \quad . \quad (30)$$

From these, by mixed multiplication by $g^{\sigma\tau}$ and $g_{\nu\lambda}$ respectively, and a change of notation for the indices, we have

$$\left.\begin{aligned} dg^{\mu\nu} &= -g^{\mu\alpha}g^{\nu\beta}\,dg_{\alpha\beta} \\ \frac{\partial g^{\mu\nu}}{\partial x_\sigma} &= -g^{\mu\alpha}g^{\nu\beta}\frac{\partial g_{\alpha\beta}}{\partial x_\sigma} \end{aligned}\right\} \quad . \quad . \quad . \quad (31)$$

and

$$\left.\begin{aligned} dg_{\mu\nu} &= -g_{\mu\alpha}g_{\nu\beta}\,dg^{\alpha\beta} \\ \frac{\partial g_{\mu\nu}}{\partial x_\sigma} &= -g_{\mu\alpha}g_{\nu\beta}\frac{\partial g^{\alpha\beta}}{\partial x_\sigma} \end{aligned}\right\} \quad . \quad . \quad . \quad (32)$$

The relation (31) admits of a transformation, of which we also have frequently to make use. From (21)

$$\frac{\partial g_{\alpha\beta}}{\partial x_\sigma} = [\alpha\sigma, \beta] + [\beta\sigma, \alpha] \quad . \quad . \quad . \quad (33)$$

Inserting this in the second formula of (31), we obtain, in view of (23)

$$\frac{\partial g^{\mu\nu}}{\partial x_\sigma} = -g^{\mu\tau}\{\tau\sigma, \nu\} - g^{\nu\tau}\{\tau\sigma, \mu\} \quad . \quad . \quad (34)$$ [16]

Substituting the right-hand side of (34) in (29), we have

$$\frac{1}{\sqrt{-g}}\frac{\partial\sqrt{-g}}{\partial x_\sigma} = \{\mu\sigma, \mu\} \quad . \quad . \quad (29a)$$ [17]

The "Divergence" of a Contravariant Vector.—If we take the inner product of (26) by the contravariant fundamental tensor $g^{\mu\nu}$, the right-hand side, after a transformation of the first term, assumes the form

$$\frac{\partial}{\partial x_\nu}(g^{\mu\nu}A_\mu) - A_\mu\frac{\partial g^{\mu\nu}}{\partial x_\nu} - \tfrac{1}{2}g^{\tau\alpha}\left(\frac{\partial g_{\mu\alpha}}{\partial x_\nu} + \frac{\partial g_{\nu\alpha}}{\partial x_\mu} - \frac{\partial g_{\mu\nu}}{\partial x_\alpha}\right)g^{\mu\nu}A_\tau.$$ [18]

[19] In accordance with (31) and (29), the last term of this expression may be written

[20]
$$\frac{1}{2}\frac{\partial g^{\tau\nu}}{\partial x_\nu}\mathrm{A}_\tau + \frac{1}{2}\frac{\partial g^{\tau\mu}}{\partial x_\mu}\mathrm{A}_\tau + \frac{1}{\sqrt{-g}}\frac{\partial\sqrt{-g}}{\partial x_\alpha}g^{\mu\nu}\mathrm{A}_\tau.$$

As the symbols of the indices of summation are immaterial, the first two terms of this expression cancel the second of the one above. If we then write $g^{\mu\nu}\mathrm{A}_\mu = \mathrm{A}^\nu$, so that A^ν like A_μ is an arbitrary vector, we finally obtain

$$\Phi = \frac{1}{\sqrt{-g}}\frac{\partial}{\partial x_\nu}(\sqrt{-g}\mathrm{A}^\nu). \quad . \quad . \quad (35)$$

This scalar is the *divergence* of the contravariant vector A^ν.

The "Curl" of a Covariant Vector.—The second term in (26) is symmetrical in the indices μ and ν. Therefore $\mathrm{A}_{\mu\nu} - \mathrm{A}_{\nu\mu}$ is a particularly simply constructed antisymmetrical tensor. We obtain

$$\mathrm{B}_{\mu\nu} = \frac{\partial\mathrm{A}_\mu}{\partial x_\nu} - \frac{\partial\mathrm{A}_\nu}{\partial x_\mu} \quad . \quad . \quad . \quad (36)$$

Antisymmetrical Extension of a Six-vector.—Applying (27) to an antisymmetrical tensor of the second rank $\mathrm{A}_{\mu\nu}$, forming in addition the two equations which arise through cyclic permutations of the indices, and adding these three equations, we obtain the tensor of the third rank

$$\mathrm{B}_{\mu\nu\sigma} = \mathrm{A}_{\mu\nu\sigma} + \mathrm{A}_{\nu\sigma\mu} + \mathrm{A}_{\sigma\mu\nu} = \frac{\partial\mathrm{A}_{\mu\nu}}{\partial x_\sigma} + \frac{\partial\mathrm{A}_{\nu\sigma}}{\partial x_\mu} + \frac{\partial\mathrm{A}_{\sigma\mu}}{\partial x_\nu} \quad (37)$$

which it is easy to prove is antisymmetrical.

The Divergence of a Six-vector.—Taking the mixed product of (27) by $g^{\mu\alpha}g^{\nu\beta}$, we also obtain a tensor. The first term on the right-hand side of (27) may be written in the form

$$\frac{\partial}{\partial x_\sigma}(g^{\mu\alpha}g^{\nu\beta}\mathrm{A}_{\mu\nu}) - g^{\mu\alpha}\frac{\partial g^{\nu\beta}}{\partial x_\sigma}\mathrm{A}_{\mu\nu} - g^{\nu\beta}\frac{\partial g^{\mu\alpha}}{\partial x_\sigma}\mathrm{A}_{\mu\nu}.$$

If we write $\mathrm{A}^{\alpha\beta}_\sigma$ for $g^{\mu\alpha}g^{\nu\beta}\mathrm{A}_{\mu\nu\sigma}$ and $\mathrm{A}^{\alpha\beta}$ for $g^{\mu\alpha}g^{\nu\beta}\mathrm{A}_{\mu\nu}$, and in the transformed first term replace

$$\frac{\partial g^{\nu\beta}}{\partial x_\sigma} \text{ and } \frac{\partial g^{\mu\alpha}}{\partial x_\sigma}$$

by their values as given by (34), there results from the right-hand side of (27) an expression consisting of seven terms, of which four cancel, and there remains

$$A^{\alpha\beta}_{\sigma} = \frac{\partial A^{\alpha\beta}}{\partial x_\sigma} + \{\sigma\gamma, \alpha\}A^{\gamma\beta} + \{\sigma\gamma, \beta\}A^{\alpha\gamma}. \quad . \quad (38)$$

This is the expression for the extension of a contravariant tensor of the second rank, and corresponding expressions for the extension of contravariant tensors of higher and lower rank may also be formed.

We note that in an analogous way we may also form the extension of a mixed tensor :—

$$A^{\alpha}_{\mu\sigma} = \frac{\partial A^{\alpha}_{\mu}}{\partial x_\sigma} - \{\sigma\mu, \tau\}A^{\alpha}_{\tau} + \{\sigma\tau, \alpha\}A^{\tau}_{\mu} \quad . \quad . \quad (39)$$

On contracting (38) with respect to the indices β and σ (inner multiplication by δ^{σ}_{β}), we obtain the vector

$$A^{\alpha} = \frac{\partial A^{\alpha\beta}}{\partial x_\beta} + \{\beta\gamma, \beta\}A^{\alpha\gamma} + \{\beta\gamma, \alpha\}A^{\gamma\beta}.$$

On account of the symmetry of $\{\beta\gamma, \alpha\}$ with respect to the indices β and γ, the third term on the right-hand side vanishes, if $A^{\alpha\beta}$ is, as we will assume, an antisymmetrical tensor. The second term allows itself to be transformed in accordance with (29a). Thus we obtain

$$A^{\alpha} = \frac{1}{\sqrt{-g}} \frac{\partial(\sqrt{-g}A^{\alpha\beta})}{\partial x_\beta} \quad . \quad . \quad . \quad (40)$$

This is the expression for the divergence of a contravariant six-vector.

The Divergence of a Mixed Tensor of the Second Rank.—Contracting (39) with respect to the indices α and σ, and taking (29a) into consideration, we obtain

$$\sqrt{-g}A_{\mu} = \frac{\partial(\sqrt{-g}A^{\sigma}_{\mu})}{\partial x_\sigma} - \{\sigma\mu, \tau\}\sqrt{-g}A^{\sigma}_{\tau} \quad . \quad (41)$$

If we introduce the contravariant tensor $A^{\rho\sigma} = g^{\rho\tau}A^{\sigma}_{\tau}$ in the last term, it assumes the form

$$-[\sigma\mu, \rho]\sqrt{-g}A^{\rho\sigma}.$$

If, further, the tensor $A^{\rho\sigma}$ is symmetrical, this reduces to

$$-\tfrac{1}{2}\sqrt{-g}\frac{\partial g_{\rho\sigma}}{\partial x_\mu}A^{\rho\sigma}.$$

Had we introduced, instead of $A^{\rho\sigma}$, the covariant tensor $A_{\rho\sigma} = g_{\rho\alpha}g_{\sigma\beta}A^{\alpha\beta}$, which is also symmetrical, the last term, by virtue of (31), would assume the form

$$\tfrac{1}{2}\sqrt{-g}\frac{\partial g^{\rho\sigma}}{\partial x_\mu}A_{\rho\sigma}.$$

In the case of symmetry in question, (41) may therefore be replaced by the two forms

$$\sqrt{-g}A_\mu = \frac{\partial(\sqrt{-g}A_\mu^\sigma)}{\partial x_\sigma} - \tfrac{1}{2}\frac{\partial g_{\rho\sigma}}{\partial x_\mu}\sqrt{-g}A^{\rho\sigma} \quad . \quad (41a)$$

$$\sqrt{-g}A_\mu = \frac{\partial(\sqrt{-g}A_\mu^\sigma)}{\partial x_\sigma} + \tfrac{1}{2}\frac{\partial g^{\rho\sigma}}{\partial x_\mu}\sqrt{-g}A_{\rho\sigma} \quad . \quad (41b)$$

which we have to employ later on.

§ 12. The Riemann-Christoffel Tensor

We now seek the tensor which can be obtained from the fundamental tensor *alone*, by differentiation. At first sight the solution seems obvious. We place the fundamental tensor of the $g_{\mu\nu}$ in (27) instead of any given tensor $A_{\mu\nu}$, and thus have a new tensor, namely, the extension of the fundamental tensor. But we easily convince ourselves that this extension vanishes identically. We reach our goal, however, in the following way. In (27) place

$$A_{\mu\nu} = \frac{\partial A_\mu}{\partial x_\nu} - \{\mu\nu, \rho\}A_\rho,$$

i.e. the extension of the four-vector A_μ. Then (with a somewhat different naming of the indices) we get the tensor of the third rank

$$A_{\mu\sigma\tau} = \frac{\partial^2 A_\mu}{\partial x_\sigma \partial x_\tau} - \{\mu\sigma, \rho\}\frac{\partial A_\rho}{\partial x_\tau} - \{\mu\tau, \rho\}\frac{\partial A_\rho}{\partial x_\sigma} - \{\sigma\tau, \rho\}\frac{\partial A_\mu}{\partial x_\rho}$$

$$+ \left[-\frac{\partial}{\partial x_\tau}\{\mu\sigma, \rho\} + \{\mu\tau, \alpha\}\{\alpha\sigma, \rho\} + \{\sigma\tau, \alpha\}\{\alpha\mu, \rho\}\right]A_\rho.$$

This expression suggests forming the tensor $A_{\mu\sigma\tau} - A_{\mu\tau\sigma}$. For, if we do so, the following terms of the expression for $A_{\mu\sigma\tau}$ cancel those of $A_{\mu\tau\sigma}$, the first, the fourth, and the member corresponding to the last term in square brackets; because all these are symmetrical in σ and τ. The same holds good for the sum of the second and third terms. Thus we obtain

$$A_{\mu\sigma\tau} - A_{\mu\tau\sigma} = B^{\rho}_{\mu\sigma\tau}A_{\rho} \quad . \quad . \quad . \quad (42)$$

where

$$B^{\rho}_{\mu\sigma\tau} = -\frac{\partial}{\partial x_{\tau}}\{\mu\sigma, \rho\} + \frac{\partial}{\partial x_{\sigma}}\{\mu\tau, \rho\} - \{\mu\sigma, \alpha\}\{\alpha\tau, \rho\}$$
$$+ \{\mu\tau, \alpha\}\{\alpha\sigma, \rho\} \quad (43)$$

The essential feature of the result is that on the right side of (42) the A_{ρ} occur alone, without their derivatives. From the tensor character of $A_{\mu\sigma\tau} - A_{\mu\tau\sigma}$ in conjunction with the fact that A_{ρ} is an arbitrary vector, it follows, by reason of § 7, that $B^{\rho}_{\mu\sigma\tau}$ is a tensor (the Riemann-Christoffel tensor).

The mathematical importance of this tensor is as follows: If the continuum is of such a nature that there is a co-ordinate system with reference to which the $g_{\mu\nu}$ are constants, then all the $B^{\rho}_{\mu\sigma\tau}$ vanish. If we choose any new system of co- [21]
ordinates in place of the original ones, the $g_{\mu\nu}$ referred thereto will not be constants, but in consequence of its tensor nature, the transformed components of $B^{\rho}_{\mu\sigma\tau}$ will still vanish in the new system. Thus the vanishing of the Riemann tensor is a necessary condition that, by an appropriate choice of the system of reference, the $g_{\mu\nu}$ may be constants. In our problem this corresponds to the case in which,* with a suitable choice of the system of reference, the special theory of relativity holds good for a *finite* region of the continuum.

Contracting (43) with respect to the indices τ and ρ we obtain the covariant tensor of second rank

* The mathematicians have proved that this is also a *sufficient* condition.

$$
\left.\begin{aligned}
G_{\mu\nu} &= B^{\rho}_{\mu\nu\rho} = R_{\mu\nu} + S_{\mu\nu} \\
\text{where} \qquad & \\
R_{\mu\nu} &= -\frac{\partial}{\partial x_\alpha}\{\mu\nu, \alpha\} + \{\mu\alpha, \beta\}\{\nu\beta, \alpha\} \\
S_{\mu\nu} &= \frac{\partial^2 \log\sqrt{-g}}{\partial x_\mu \partial x_\nu} - \{\mu\nu, \alpha\}\frac{\partial \log\sqrt{-g}}{\partial x_\alpha}
\end{aligned}\right\} \qquad (44)
$$

Note on the Choice of Co-ordinates.—It has already been observed in § 8, in connexion with equation (18a), that the choice of co-ordinates may with advantage be made so that $\sqrt{-g} = 1$. A glance at the equations obtained in the last two sections shows that by such a choice the laws of formation of tensors undergo an important simplification. This applies particularly to $G_{\mu\nu}$, the tensor just developed, which plays a fundamental part in the theory to be set forth. For this specialization of the choice of co-ordinates brings about the vanishing of $S_{\mu\nu}$, so that the tensor $G_{\mu\nu}$ reduces to $R_{\mu\nu}$.

On this account I shall hereafter give all relations in the simplified form which this specialization of the choice of co-ordinates brings with it. It will then be an easy matter to revert to the *generally* covariant equations, if this seems desirable in a special case.

C. Theory of the Gravitational Field

§ 13. Equations of Motion of a Material Point in the Gravitational Field. Expression for the Field-components of Gravitation

A freely movable body not subjected to external forces moves, according to the special theory of relativity, in a straight line and uniformly. This is also the case, according to the general theory of relativity, for a part of four-dimensional space in which the system of co-ordinates K_0, may be, and is, so chosen that they have the special constant values given in (4).

If we consider precisely this movement from any chosen system of co-ordinates K_1, the body, observed from K_1, moves, according to the considerations in § 2, in a gravitational field. The law of motion with respect to K_1 results without difficulty from the following consideration. With respect to K_0 the law of motion corresponds to a four-dimensional straight line, i.e. to a geodetic line. Now since the geodetic line is defined independently of the system of reference, its equations will also be the equation of motion of the material point with respect to K_1. If we set

$$\Gamma^{\tau}_{\mu\nu} = - \{\mu\nu, \tau\} \qquad . \quad . \quad . \quad (45)$$

the equation of the motion of the point with respect to K_1, becomes

$$\frac{d^2x_\tau}{ds^2} = \Gamma^{\tau}_{\mu\nu} \frac{dx_\mu}{ds} \frac{dx_\nu}{ds} \qquad . \quad . \quad . \quad (46)$$

We now make the assumption, which readily suggests itself, that this covariant system of equations also defines the motion of the point in the gravitational field in the case when there is no system of reference K_0, with respect to which the special theory of relativity holds good in a finite region. We have all the more justification for this assumption as (46) contains only *first* derivatives of the $g_{\mu\nu}$, between which even in the special case of the existence of K_0, no relations subsist.*

If the $\Gamma^{\tau}_{\mu\nu}$ vanish, then the point moves uniformly in a straight line. These quantities therefore condition the deviation of the motion from uniformity. They are the components of the gravitational field.

§ 14. The Field Equations of Gravitation in the Absence of Matter

We make a distinction hereafter between "gravitational field" and "matter" in this way, that we denote everything but the gravitational field as "matter." Our use of the word therefore includes not only matter in the ordinary sense, but the electromagnetic field as well.

Our next task is to find the field equations of gravitation in the absence of matter. Here we again apply the method

* It is only between the second (and first) derivatives that, by § 12, the relations $B^{\rho}_{\mu\sigma\tau} = 0$ subsist.

employed in the preceding paragraph in formulating the equations of motion of the material point. A special case in which the required equations must in any case be satisfied is that of the special theory of relativity, in which the $g_{\mu\nu}$ have certain constant values. Let this be the case in a certain finite space in relation to a definite system of co-ordinates K_0. Relatively to this system all the components of the Riemann tensor $B^{\rho}_{\mu\sigma\tau}$, defined in (43), vanish. For the space under consideration they then vanish, also in any other system of co-ordinates.

Thus the required equations of the matter-free gravitational field must in any case be satisfied if all $B^{\rho}_{\mu\sigma\tau}$ vanish. But this condition goes too far. For it is clear that, e.g., the gravitational field generated by a material point in its environment certainly cannot be "transformed away" by any choice of the system of co-ordinates, i.e. it cannot be transformed to the case of constant $g_{\mu\nu}$.

This prompts us to require for the matter-free gravitational field that the symmetrical tensor $G_{\mu\nu}$, derived from the tensor $B^{\rho}_{\mu\nu\tau}$, shall vanish. Thus we obtain ten equations for the ten quantities $g_{\mu\nu}$, which are satisfied in the special case of the vanishing of all $B^{\rho}_{\mu\nu\tau}$. With the choice which we have made of a system of co-ordinates, and taking (44) into consideration, the equations for the matter-free field are

$$\left.\begin{aligned} \frac{\partial \Gamma^{\alpha}_{\mu\nu}}{\partial x_{\alpha}} + \Gamma^{\alpha}_{\mu\beta}\Gamma^{\beta}_{\nu\alpha} &= 0 \\ \sqrt{-g} &= 1 \end{aligned}\right\} \quad . \quad . \quad . \quad (47)$$

It must be pointed out that there is only a minimum of arbitrariness in the choice of these equations. For besides $G_{\mu\nu}$ there is no tensor of second rank which is formed from the $g_{\mu\nu}$ and its derivatives, contains no derivations higher than second, and is linear in these derivatives.*

These equations, which proceed, by the method of pure

* Properly speaking, this can be affirmed only of the tensor

$$G_{\mu\nu} + \lambda g_{\mu\nu} g^{\alpha\beta} G_{\alpha\beta},$$

where λ is a constant. If, however, we set this tensor = 0, we come back again to the equations $G_{\mu\nu} = 0$.

mathematics, from the requirement of the general theory of relativity, give us, in combination with the equations of motion (46), to a first approximation Newton's law of attraction, and to a second approximation the explanation of the motion of the perihelion of the planet Mercury discovered by Leverrier (as it remains after corrections for perturbation have been made). These facts must, in my opinion, be taken as a convincing proof of the correctness of the theory.

§ 15. The Hamiltonian Function for the Gravitational Field. Laws of Momentum and Energy

To show that the field equations correspond to the laws of momentum and energy, it is most convenient to write them in the following Hamiltonian form :—

$$\left.\begin{aligned} \delta\int \mathrm{H}d\tau &= 0 \\ \mathrm{H} &= g^{\mu\nu}\Gamma^{\alpha}_{\mu\beta}\Gamma^{\beta}_{\nu\alpha} \\ \sqrt{-g} &= 1 \end{aligned}\right\} \quad . \quad . \quad . \quad (47\mathrm{a})$$

where, on the boundary of the finite four-dimensional region of integration which we have in view, the variations vanish.

We first have to show that the form (47a) is equivalent to the equations (47). For this purpose we regard H as a function of the $g^{\mu\nu}$ and the $g^{\mu\nu}_{\sigma}$ $(= \partial g^{\mu\nu}/\partial x_{\sigma})$.
Then in the first place

$$\begin{aligned} \delta\mathrm{H} &= \Gamma^{\alpha}_{\mu\beta}\Gamma^{\beta}_{\nu\alpha}\,\delta g^{\mu\nu} + 2g^{\mu\nu}\Gamma^{\alpha}_{\mu\beta}\,\delta\Gamma^{\beta}_{\nu\alpha} \\ &= -\,\Gamma^{\alpha}_{\mu\beta}\Gamma^{\beta}_{\nu\alpha}\,\delta g^{\mu\nu} + 2\Gamma^{\alpha}_{\mu\beta}\,\delta(g^{\mu\nu}\Gamma^{\beta}_{\nu\alpha}). \end{aligned}$$

But

$$\delta\left(g^{\mu\nu}\Gamma^{\beta}_{\nu\alpha}\right) = -\tfrac{1}{2}\delta\left[g^{\mu\nu}g^{\beta\lambda}\left(\frac{\partial g_{\nu\lambda}}{\partial x_{\alpha}} + \frac{\partial g_{\alpha\lambda}}{\partial x_{\nu}} - \frac{\partial g_{\alpha\nu}}{\partial x_{\lambda}}\right)\right]. \qquad [22]$$

The terms arising from the last two terms in round brackets are of different sign, and result from each other (since the denomination of the summation indices is immaterial) through interchange of the indices μ and β. They cancel each other in the expression for δH, because they are multiplied by the quantity $\Gamma^{\alpha}_{\mu\beta}$, which is symmetrical with respect to the indices μ and β. Thus there remains only the first term in round brackets to be considered, so that, taking (31) into account, we obtain

$$\delta \mathrm{H} = -\Gamma^{\alpha}_{\mu\beta}\Gamma^{\beta}_{\nu\alpha}\delta g^{\mu\nu} + \Gamma^{\alpha}_{\mu\beta}\delta g^{\mu\beta}_{\alpha}.$$

Thus

[23]

$$\left.\begin{aligned} \frac{\partial \mathrm{H}}{\partial g^{\mu\nu}} &= -\Gamma^{\alpha}_{\mu\beta}\Gamma^{\beta}_{\nu\alpha} \\ \frac{\partial \mathrm{H}}{\partial g^{\mu\nu}_{\sigma}} &= \Gamma^{\sigma}_{\mu\nu} \end{aligned}\right\} \quad . \quad . \quad . \quad (48)$$

Carrying out the variation in (47a), we get in the first place

$$\frac{\partial}{\partial x_{\alpha}}\left(\frac{\partial \mathrm{H}}{\partial g^{\mu\nu}_{\alpha}}\right) - \frac{\partial \mathrm{H}}{\partial g^{\mu\nu}} = 0, \quad . \quad . \quad . \quad (47\mathrm{b})$$

which, on account of (48), agrees with (47), as was to be proved.

If we multiply (47b) by $g^{\mu\nu}_{\sigma}$, then because

$$\frac{\partial g^{\mu\nu}_{\sigma}}{\partial x_{\alpha}} = \frac{\partial g^{\mu\nu}_{\alpha}}{\partial x_{\sigma}}$$

and, consequently,

$$g^{\mu\nu}_{\sigma}\frac{\partial}{\partial x_{\alpha}}\left(\frac{\partial \mathrm{H}}{\partial g^{\mu\nu}_{\alpha}}\right) = \frac{\partial}{\partial x_{\alpha}}\left(g^{\mu\nu}_{\sigma}\frac{\partial \mathrm{H}}{\partial g^{\mu\nu}_{\alpha}}\right) - \frac{\partial \mathrm{H}}{\partial g^{\mu\nu}_{\alpha}}\frac{\partial g^{\mu\nu}_{\alpha}}{\partial x_{\sigma}},$$

we obtain the equation

$$\frac{\partial}{\partial x_{\alpha}}\left(g^{\mu\nu}_{\sigma}\frac{\partial \mathrm{H}}{\partial g^{\mu\nu}_{\alpha}}\right) - \frac{\partial \mathrm{H}}{\partial x_{\sigma}} = 0$$

or *

$$\left.\begin{aligned} \frac{\partial t^{\alpha}_{\sigma}}{\partial x_{\alpha}} &= 0 \\ -2\kappa t^{\alpha}_{\sigma} &= g^{\mu\nu}_{\sigma}\frac{\partial \mathrm{H}}{\partial g^{\mu\nu}_{\alpha}} - \delta^{\alpha}_{\sigma}\mathrm{H} \end{aligned}\right\} \quad . \quad . \quad . \quad (49)$$

where, on account of (48), the second equation of (47), and
[24] (34)

$$\kappa t^{\alpha}_{\sigma} = \tfrac{1}{2}\delta^{\alpha}_{\sigma}g^{\mu\nu}\Gamma^{\lambda}_{\mu\beta}\Gamma^{\beta}_{\nu\lambda} - g^{\mu\nu}\Gamma^{\alpha}_{\mu\beta}\Gamma^{\beta}_{\nu\sigma} \quad . \quad . \quad (50)$$

* The reason for the introduction of the factor -2κ will be apparent later.

It is to be noticed that t_σ^α is not a tensor; on the other hand (49) applies to all systems of co-ordinates for which $\sqrt{-g} = 1$. This equation expresses the law of conservation of momentum and of energy for the gravitational field. Actually the integration of this equation over a three-dimensional volume V yields the four equations

$$\frac{d}{dx_4}\int t_\sigma^4 dV = \int (lt_\sigma^1 + mt_\sigma^2 + nt_\sigma^3) dS \,. \qquad (49a)$$

where l, m, n denote the direction-cosines of direction of the inward drawn normal at the element dS of the bounding surface (in the sense of Euclidean geometry). We recognize in this the expression of the laws of conservation in their usual form. The quantities t_σ^α we call the "energy components" of the gravitational field.

I will now give equations (47) in a third form, which is particularly useful for a vivid grasp of our subject. By multiplication of the field equations (47) by $g^{\nu\sigma}$ these are obtained in the "mixed" form. Note that

$$g^{\nu\sigma}\frac{\partial \Gamma_{\mu\nu}^\alpha}{\partial x_\alpha} = \frac{\partial}{\partial x_\alpha}\left(g^{\nu\sigma}\Gamma_{\mu\nu}^\alpha\right) - \frac{\partial g^{\nu\sigma}}{\partial x_\alpha}\Gamma_{\mu\nu}^\alpha,$$

which quantity, by reason of (34), is equal to

$$\frac{\partial}{\partial x_\alpha}\left(g^{\nu\sigma}\Gamma_{\mu\nu}^\alpha\right) - g^{\nu\beta}\Gamma_{\alpha\beta}^\sigma\Gamma_{\mu\nu}^\alpha - g^{\sigma\beta}\Gamma_{\beta\alpha}^\nu\Gamma_{\mu\nu}^\alpha,$$

or (with different symbols for the summation indices)

$$\frac{\partial}{\partial x_\alpha}\left(g^{\sigma\beta}\Gamma_{\mu\beta}^\alpha\right) - g^{\gamma\delta}\Gamma_{\gamma\beta}^\sigma\Gamma_{\delta\mu}^\beta - g^{\nu\sigma}\Gamma_{\mu\beta}^\alpha\Gamma_{\nu\alpha}^\beta.$$

The third term of this expression cancels with the one arising from the second term of the field equations (47); using relation (50), the second term may be written

$$\kappa(t_\mu^\sigma - \tfrac{1}{2}\delta_\mu^\sigma t),$$

where $t = t_\alpha^\alpha$. Thus instead of equations (47) we obtain

$$\left.\begin{aligned} \frac{\partial}{\partial x_\alpha}\left(g^{\sigma\beta}\Gamma_{\mu\beta}^\alpha\right) &= -\kappa(t_\mu^\sigma - \tfrac{1}{2}\delta_\mu^\sigma t) \\ \sqrt{-g} &= 1 \end{aligned}\right\} \qquad (51)$$

§ 16. The General Form of the Field Equations of Gravitation

The field equations for matter-free space formulated in § 15 are to be compared with the field equation

$$\nabla^2\phi = 0$$

of Newton's theory. We require the equation corresponding to Poisson's equation

$$\nabla^2\phi = 4\pi\kappa\rho,$$

where ρ denotes the density of matter.

The special theory of relativity has led to the conclusion that inert mass is nothing more or less than energy, which finds its complete mathematical expression in a symmetrical tensor of second rank, the energy-tensor. Thus in the general theory of relativity we must introduce a corresponding energy-tensor of matter T_σ^α, which, like the energy-components t_σ [equations (49) and (50)] of the gravitational field, will have mixed character, but will pertain to a symmetrical covariant tensor.*

The system of equation (51) shows how this energy-tensor (corresponding to the density ρ in Poisson's equation) is to be introduced into the field equations of gravitation. For if we consider a complete system (e.g. the solar system), the total mass of the system, and therefore its total gravitating action as well, will depend on the total energy of the system, and therefore on the ponderable energy together with the gravitational energy. This will allow itself to be expressed by introducing into (51), in place of the energy-components of the gravitational field alone, the sums $t_\mu^\sigma + T_\mu^\sigma$ of the energy-components of matter and of gravitational field. Thus instead of (51) we obtain the tensor equation

$$\left.\begin{aligned} \frac{\partial}{\partial x_\alpha}(g^{\sigma\beta}T_{\mu\beta}^\alpha) &= -\kappa\left[(t_\mu^\sigma + T_\mu^\sigma) - \tfrac{1}{2}\delta_\mu^\sigma(t + T)\right], \\ \sqrt{-g} &= 1 \end{aligned}\right\} . \quad (52)$$

where we have set $T = T_\mu^\mu$ (Laue's scalar). These are the

[25] * $g_{\alpha\tau}T_\sigma^\alpha = T_{\sigma\tau}$ and $g^{\sigma\beta}T_\sigma^\alpha = T^{\alpha\beta}$ are to be symmetrical tensors.

required general field equations of gravitation in mixed form. Working back from these, we have in place of (47)

$$\left.\begin{aligned}\frac{\partial}{\partial x_a}\Gamma^a_{\mu\nu} + \Gamma^a_{\mu\beta}\Gamma^\beta_{\nu a} &= -\kappa(T_{\mu\nu} - \tfrac{1}{2}g_{\mu\nu}T),\\ \sqrt{-g} = 1\end{aligned}\right\} \quad . \quad (53)$$

It must be admitted that this introduction of the energy-tensor of matter is not justified by the relativity postulate alone. For this reason we have here deduced it from the requirement that the energy of the gravitational field shall act gravitatively in the same way as any other kind of energy. But the strongest reason for the choice of these equations lies in their consequence, that the equations of conservation of momentum and energy, corresponding exactly to equations (49) and (49a), hold good for the components of the total energy. This will be shown in § 17.

§ 17. The Laws of Conservation in the General Case

Equation (52) may readily be transformed so that the second term on the right-hand side vanishes. Contract (52) with respect to the indices μ and σ, and after multiplying the resulting equation by $\frac{1}{2}\delta^\sigma_\mu$, subtract it from equation (52). This gives

$$\frac{\partial}{\partial x_a}(g^{\sigma\beta}\Gamma^a_{\mu\beta} - \tfrac{1}{2}\delta^\sigma_\mu g^{\lambda\beta}\Gamma^a_{\lambda\beta}) = -\kappa(t^\sigma_\mu + T^\sigma_\mu). \qquad (52a)$$

On this equation we perform the operation $\partial/\partial x_\sigma$. We have

$$\frac{\partial^2}{\partial x_a \partial x_\sigma}\left(g^\sigma\Gamma^{\ a}_{\beta\mu}\right) = -\tfrac{1}{2}\frac{\partial^2}{\partial x_a \partial x_\sigma}\left[g^{\sigma\beta}g^{a\lambda}\left(\frac{\partial g_{\mu\lambda}}{\partial x_\beta} + \frac{\partial g_{\beta\lambda}}{\partial x_\mu} - \frac{\partial g_{\mu\beta}}{\partial x_\lambda}\right)\right].$$

The first and third terms of the round brackets yield contributions which cancel one another, as may be seen by interchanging, in the contribution of the third term, the summation indices a and σ on the one hand, and β and λ on the other. The second term may be re-modelled by (31), so that we have

$$\frac{\partial^2}{\partial x_a \partial x_\sigma}\left(g^{\sigma\beta}\Gamma^a_{\mu\beta}\right) = \tfrac{1}{2}\frac{\partial^3 g^{a\beta}}{\partial x_a \partial x_\beta \partial x_\mu} \quad . \quad . \quad (54)$$

The second term on the left-hand side of (52a) yields in the first place

$$-\tfrac{1}{2}\frac{\partial^2}{\partial x_\alpha \partial x_\mu}\left(g^{\lambda\beta}\Gamma^\alpha_{\lambda\beta}\right)$$

or

$$\tfrac{1}{4}\frac{\partial^2}{\partial x_\alpha \partial x_\mu}\left[g^{\lambda\beta}g^{\alpha\delta}\left(\frac{\partial g_{\delta\lambda}}{\partial x_\beta}+\frac{\partial g_{\delta\beta}}{\partial x_\lambda}-\frac{\partial g_{\lambda\beta}}{\partial x_\delta}\right)\right].$$

With the choice of co-ordinates which we have made, the term deriving from the last term in round brackets disappears by reason of (29). The other two may be combined, and together, by (31), they give

$$-\tfrac{1}{2}\frac{\partial^3 g^{\alpha\beta}}{\partial x_\alpha \partial x_\beta \partial x_\mu},$$

so that in consideration of (54), we have the identity

$$\frac{\partial^2}{\partial x_\alpha \partial x_\sigma}\left(g^{\rho\beta}\Gamma_{\mu\beta}-\tfrac{1}{2}\delta^\sigma_\mu g^{\lambda\beta}\Gamma^\alpha_{\lambda\beta}\right)\equiv 0 \quad . \quad . \quad (55)$$

From (55) and (52a), it follows that

$$\frac{\partial(t^\sigma_\mu + \mathrm{T}^\sigma_\mu)}{\partial x_\sigma} = 0. \quad . \quad . \quad . \quad (56)$$

Thus it results from our field equations of gravitation that the laws of conservation of momentum and energy are satisfied. This may be seen most easily from the consideration which leads to equation (49a); except that here, instead of the energy components t^σ of the gravitational field, we have to introduce the totality of the energy components of matter and gravitational field.

§ 18. The Laws of Momentum and Energy for Matter, as a Consequence of the Field Equations

Multiplying (53) by $\partial g^{\mu\nu}/\partial x_\sigma$, we obtain, by the method adopted in § 15, in view of the vanishing of

$$g_{\mu\nu}\frac{\partial g^{\mu\nu}}{\partial x_\sigma},$$

the equation

[26]
$$\frac{\partial t^\alpha_\sigma}{\partial x_\alpha} + \tfrac{1}{2}\frac{\partial g^{\mu\nu}}{\partial x_\sigma}\mathrm{T}_{\mu\nu} = 0,$$

or, in view of (56),

$$\frac{\partial T_\sigma^\alpha}{\partial x_\alpha} + \tfrac{1}{2}\frac{\partial g^{\mu\nu}}{\partial x_\sigma}T_{\mu\nu} = 0 \quad . \quad . \quad . \quad (57)$$

Comparison with (41b) shows that with the choice of system of co-ordinates which we have made, this equation predicates nothing more or less than the vanishing of divergence of the material energy-tensor. Physically, the occurrence of the second term on the left-hand side shows that laws of conservation of momentum and energy do not apply in the strict sense for matter alone, or else that they apply only when the $g^{\mu\nu}$ are constant, i.e. when the field intensities of gravitation vanish. This second term is an expression for momentum, and for energy, as transferred per unit of volume and time from the gravitational field to matter. This is brought out still more clearly by re-writing (57) in the sense of (41) as

$$\frac{\partial T_\sigma^\alpha}{\partial x_\alpha} = -\Gamma_{\alpha\sigma}^\beta T_\beta^\alpha \quad . \quad . \quad . \quad (57a)$$

The right side expresses the energetic effect of the gravitational field on matter.

Thus the field equations of gravitation contain four conditions which govern the course of material phenomena. They give the equations of material phenomena completely, if the latter is capable of being characterized by four differential equations independent of one another.*

D. Material Phenomena

The mathematical aids developed in part B enable us forthwith to generalize the physical laws of matter (hydrodynamics, Maxwell's electrodynamics), as they are formulated in the special theory of relativity, so that they will fit in with the general theory of relativity. When this is done, the general principle of relativity does not indeed afford us a further limitation of possibilities ; but it makes us acquainted with the influence of the gravitational field on all processes,

* On this question cf. H. Hilbert, Nachr. d. K. Gesellsch. d. Wiss. zu Göttingen, Math.-phys. Klasse, 1915, p. 3. [27]

without our having to introduce any new hypothesis whatever.

Hence it comes about that it is not necessary to introduce definite assumptions as to the physical nature of matter (in the narrower sense). In particular it may remain an open question whether the theory of the electromagnetic field in conjunction with that of the gravitational field furnishes a sufficient basis for the theory of matter or not. The general postulate of relativity is unable on principle to tell us anything about this. It must remain to be seen, during the working out of the theory, whether electromagnetics and the doctrine of gravitation are able in collaboration to perform what the former by itself is unable to do.

§ 19. Euler's Equations for a Frictionless Adiabatic Fluid

Let p and ρ be two scalars, the former of which we call the "pressure," the latter the "density" of a fluid; and let an equation subsist between them. Let the contravariant symmetrical tensor

$$\mathrm{T}^{\alpha\beta} = -g^{\alpha\beta}p + \rho\frac{dx_\alpha}{ds}\frac{dx_\beta}{ds} \quad . \quad . \quad . \quad (58)$$

be the contravariant energy-tensor of the fluid. To it belongs the covariant tensor

$$\mathrm{T}_{\mu\nu} = -g_{\mu\nu}p + g_{\mu\alpha}g_{\mu\beta}\frac{dx_\alpha}{ds}\frac{dx_\beta}{ds}\rho, \quad . \quad . \quad (58a)$$

as well as the mixed tensor *

$$\mathrm{T}_\sigma^\alpha = -\delta_\sigma^\alpha p + g_{\sigma\beta}\frac{dx_\beta}{ds}\frac{dx_\alpha}{ds}\rho \quad . \quad . \quad (58b)$$

Inserting the right-hand side of (58b) in (57a), we obtain the Eulerian hydrodynamical equations of the general theory of relativity. They give, in theory, a complete solution of the problem of motion, since the four equations (57a), together

* For an observer using a system of reference in the sense of the special theory of relativity for an infinitely small region, and moving with it, the density of energy T_4^4 equals $\rho - p$. This gives the definition of ρ. Thus ρ is not constant for an incompressible fluid.

with the given equation between p and ρ, and the equation

$$g_{\alpha\beta}\frac{dx_\alpha}{ds}\frac{dx_\beta}{ds} = 1,$$

are sufficient, $g_{\alpha\beta}$ being given, to define the six unknowns

$$p, \rho, \frac{dx_1}{ds}, \frac{dx_2}{ds}, \frac{dx_3}{ds}, \frac{dx_4}{ds}.$$

If the $g_{\mu\nu}$ are also unknown, the equations (53) are brought in. These are eleven equations for defining the ten functions $g_{\mu\nu}$, so that these functions appear over-defined. We must remember, however, that the equations (57a) are already contained in the equations (53), so that the latter represent only seven independent equations. There is good reason for this lack of definition, in that the wide freedom of the choice of co-ordinates causes the problem to remain mathematically undefined to such a degree that three of the functions of space may be chosen at will.*

§ 20. Maxwell's Electromagnetic Field Equations for Free Space [28]

Let ϕ_ν be the components of a covariant vector—the electromagnetic potential vector. From them we form, in accordance with (36), the components $F_{\rho\sigma}$ of the covariant six-vector of the electromagnetic field, in accordance with the system of equations

$$F_{\rho\sigma} = \frac{\partial\phi_\rho}{\partial x_\sigma} - \frac{\partial\phi_\sigma}{\partial x_\rho} \quad . \quad . \quad . \quad (59)$$

It follows from (59) that the system of equations

$$\frac{\partial F_{\rho\sigma}}{\partial x_\tau} + \frac{\partial F_{\sigma\tau}}{\partial x_\rho} + \frac{\partial F_{\tau\rho}}{\partial x_\sigma} = 0 \quad . \quad . \quad . \quad (60)$$ [29]

is satisfied, its left side being, by (37), an antisymmetrical tensor of the third rank. System (60) thus contains essentially four equations which are written out as follows:—

* On the abandonment of the choice of co-ordinates with $g = -1$, there remain *four* functions of space with liberty of choice, corresponding to the four arbitrary functions at our disposal in the choice of co-ordinates.

$$\left.\begin{aligned}\frac{\partial F_{23}}{\partial x_4}+\frac{\partial F_{34}}{\partial x_2}+\frac{\partial F_{42}}{\partial x_3}&=0\\\frac{\partial F_{34}}{\partial x_1}+\frac{\partial F_{41}}{\partial x_3}+\frac{\partial F_{13}}{\partial x_4}&=0\\\frac{\partial F_{41}}{\partial x_2}+\frac{\partial F_{12}}{\partial x_4}+\frac{\partial F_{24}}{\partial x_1}&=0\\\frac{\partial F_{12}}{\partial x_3}+\frac{\partial F_{23}}{\partial x_1}+\frac{\partial F_{31}}{\partial x_2}&=0\end{aligned}\right\} \quad . \quad . \quad (60a)$$

This system corresponds to the second of Maxwell's systems of equations. We recognize this at once by setting

$$\left.\begin{aligned}F_{23}&=H_x, & F_{14}&=E_x\\F_{31}&=H_y, & F_{24}&=E_y\\F_{12}&=H_z, & F_{34}&=E_z\end{aligned}\right\} \quad . \quad . \quad . \quad (61)$$

Then in place of (60a) we may set, in the usual notation of three-dimensional vector analysis,

$$\left.\begin{aligned}-\frac{\partial H}{\partial t}&=\text{curl } E\\\text{div } H&=0\end{aligned}\right\} \quad . \quad . \quad . \quad (60b)$$

We obtain Maxwell's first system by generalizing the form given by Minkowski. We introduce the contravariant six-vector associated with $F^{\alpha\beta}$

[30] $$F^{\mu\nu}=g^{\mu\alpha}g^{\nu\beta}F_{\alpha\beta} \quad . \quad . \quad . \quad (62)$$

and also the contravariant vector J^μ of the density of the electric current. Then, taking (40) into consideration, the following equations will be invariant for any substitution whose invariant is unity (in agreement with the chosen co-ordinates) :—

$$\frac{\partial}{\partial x_\nu}F^{\mu\nu}=J^\mu \quad . \quad . \quad . \quad . \quad . \quad (63)$$

Let

$$\left.\begin{aligned}F^{23}&=H'_x, & F^{14}&=-E'_x\\F^{31}&=H'_y, & F^{24}&=-E'_y\\F^{12}&=H'_z, & F^{34}&=-E'_z\end{aligned}\right\} \quad . \quad . \quad (64)$$

which quantities are equal to the quantities H_x . . . E^z in the special case of the restricted theory of relativity ; and in addition

$$J^1 = j_x,\ J^2 = j_y,\ J^3 = j_z,\ J^4 = \rho,$$

we obtain in place of (63)

$$\left.\begin{aligned} \frac{\partial E'}{\partial t} + j &= \operatorname{curl} H' \\ \operatorname{div} E' &= \rho \end{aligned}\right\} \quad . \quad . \quad . \quad (63a)$$

The equations (60), (62), and (63) thus form the generalization of Maxwell's field equations for free space, with the convention which we have established with respect to the choice of co-ordinates.

The Energy-components of the Electromagnetic Field.— We form the inner product

$$\kappa_\sigma = F_{\sigma\mu} J^\mu \quad . \quad . \quad . \quad . \quad (65)$$

By (61) its components, written in the three-dimensional manner, are

$$\left.\begin{aligned} \kappa_1 &= \rho E_x + [j \,.\, H]^x \\ &\cdot \quad \cdot \quad \cdot \quad \cdot \\ &\cdot \quad \cdot \quad \cdot \quad \cdot \\ \kappa_4 &= -(jE) \end{aligned}\right\} \quad . \quad . \quad . \quad (65a)$$

κ_σ is a covariant vector the components of which are equal to the negative momentum, or, respectively, the energy, which is transferred from the electric masses to the electromagnetic field per unit of time and volume. If the electric masses are free, that is, under the sole influence of the electromagnetic field, the covariant vector κ_σ will vanish.

To obtain the energy-components T_σ^ν of the electromagnetic field, we need only give to equation $\kappa_\sigma = 0$ the form of equation (57). From (63) and (65) we have in the first place

$$\kappa_\sigma = F_{\sigma\mu}\frac{\partial F^{\mu\nu}}{\partial x_\nu} = \frac{\partial}{\partial x_\nu}(F_{\sigma\mu}F^{\mu\nu}) - F^{\mu\nu}\frac{\partial F_{\sigma\mu}}{\partial x_\nu}.$$

The second term of the right-hand side, by reason of (60), permits the transformation

$$F^{\mu\nu}\frac{\partial F_{\sigma\mu}}{\partial x_\nu} = -\tfrac{1}{2}F^{\mu\nu}\frac{\partial F_{\mu\nu}}{\partial x_\sigma} = -\tfrac{1}{2}g^{\mu\alpha}g^{\nu\beta}F_{\alpha\beta}\frac{\partial F_{\mu\nu}}{\partial x_\sigma},$$

which latter expression may, for reasons of symmetry, also be written in the form

$$- \tfrac{1}{4}\left[g^{\mu\alpha}g^{\nu\beta}\mathrm{F}_{\alpha\beta}\frac{\partial \mathrm{F}_{\mu\nu}}{\partial x_\sigma} + g^{\mu\alpha}g^{\nu\beta}\frac{\partial \mathrm{F}_{\alpha\beta}}{\partial x_\sigma}\mathrm{F}_{\mu\nu}\right].$$

But for this we may set

$$- \tfrac{1}{4}\frac{\partial}{\partial x_\sigma}(g^{\mu\alpha}g^{\nu\beta}\mathrm{F}_{\alpha\beta}\mathrm{F}_{\mu\nu}) + \tfrac{1}{4}\mathrm{F}_{\alpha\beta}\mathrm{F}_{\mu\nu}\frac{\partial}{\partial x_\sigma}(g^{\mu\alpha}g^{\nu\beta}).$$

The first of these terms is written more briefly

$$- \tfrac{1}{4}\frac{\partial}{\partial x_\sigma}(\mathrm{F}^{\mu\nu}\mathrm{F}_{\mu\nu});$$

the second, after the differentiation is carried out, and after some reduction, results in

[31] $$- \tfrac{1}{2}\mathrm{F}^{\mu\tau}\mathrm{F}_{\mu\nu}g^{\nu\rho}\frac{\partial g_{\sigma\tau}}{\partial x_\sigma}.$$

Taking all three terms together we obtain the relation

$$\kappa_\sigma = \frac{\partial \mathrm{T}_\sigma^\nu}{\partial x_\nu} - \tfrac{1}{2}g^{\tau\mu}\frac{\partial g_{\mu\nu}}{\partial x_\sigma}\mathrm{T}_\tau^\nu \quad . \quad . \quad . \quad (66)$$

where

$$\mathrm{T}_\sigma^\nu = -\mathrm{F}_{\sigma\alpha}\mathrm{F}^{\nu\alpha} + \tfrac{1}{4}\delta_\sigma^\nu\mathrm{F}_{\alpha\beta}\mathrm{F}^{\alpha\beta}.$$

Equation (66), if κ_σ vanishes, is, on account of (30), equivalent to (57) or (57*a*) respectively. Therefore the T_σ^ν are the energy-components of the electromagnetic field. With the help of (61) and (64), it is easy to show that these energy-components of the electromagnetic field in the case of the special theory of relativity give the well-known Maxwell-Poynting expressions.

We have now deduced the general laws which are satisfied by the gravitational field and matter, by consistently using a system of co-ordinates for which $\sqrt{-g} = 1$. We have thereby achieved a considerable simplification of formulæ and calculations, without failing to comply with the requirement of general covariance; for we have drawn our equations from generally covariant equations by specializing the system of co-ordinates.

Still the question is not without a formal interest, whether with a correspondingly generalized definition of the energy-components of gravitational field and matter, even without specializing the system of co-ordinates, it is possible to formulate laws of conservation in the form of equation (56), and field equations of gravitation of the same nature as (52) or (52a), in such a manner that on the left we have a divergence (in the ordinary sense), and on the right the sum of the energy-components of matter and gravitation. I have found that in both cases this is actually so. But I do not think that the communication of my somewhat extensive reflexions on this subject would be worth while, because after all they do not give us anything that is materially new.

E

§ 21. Newton's Theory as a First Approximation [32]

As has already been mentioned more than once, the special theory of relativity as a special case of the general theory is characterized by the $g_{\mu\nu}$ having the constant values (4). From what has already been said, this means complete neglect of the effects of gravitation. We arrive at a closer approximation to reality by considering the case where the $g_{\mu\nu}$ differ from the values of (4) by quantities which are small compared with 1, and neglecting small quantities of second and higher order. (First point of view of approximation.)

It is further to be assumed that in the space-time territory under consideration the $g_{\mu\nu}$ at spatial infinity, with a suitable choice of co-ordinates, tend toward the values (4); i.e. we are considering gravitational fields which may be regarded as generated exclusively by matter in the finite region.

It might be thought that these approximations must lead us to Newton's theory. But to that end we still need to approximate the fundamental equations from a second point of view. We give our attention to the motion of a material point in accordance with the equations (16). In the case of the special theory of relativity the components

$$\frac{dx_1}{ds}, \frac{dx_2}{ds}, \frac{dx_3}{ds}$$

may take on any values. This signifies that any velocity

$$v = \sqrt{\left(\frac{dx_1}{dx_4}\right)^2 + \left(\frac{dx_2}{dx_4}\right)^2 + \left(\frac{dx_3}{dx_4}\right)^2}$$

may occur, which is less than the velocity of light *in vacuo*. If we restrict ourselves to the case which almost exclusively offers itself to our experience, of v being small as compared with the velocity of light, this denotes that the components

$$\frac{dx_1}{ds}, \frac{dx_2}{ds}, \frac{dx_3}{ds}$$

are to be treated as small quantities, while dx_4/ds, to the second order of small quantities, is equal to one. (Second point of view of approximation.)

Now we remark that from the first point of view of approximation the magnitudes $\Gamma^\tau_{\mu\nu}$ are all small magnitudes of at least the first order. A glance at (46) thus shows that in this equation, from the second point of view of approximation, we have to consider only terms for which $\mu = \nu = 4$. Restricting ourselves to terms of lowest order we first obtain in place of (46) the equations

$$\frac{d^2x_\tau}{dt^2} = \Gamma^\tau_{44}$$

where we have set $ds = dx_4 = dt$; or with restriction to terms which from the first point of view of approximation are of first order :—

$$\frac{d^2x_\tau}{dt^2} = [44, \tau] \quad (\tau = 1, 2, 3)$$

$$\frac{d^2x_4}{dt^2} = -[44, 4].$$

If in addition we suppose the gravitational field to be a quasi-static field, by confining ourselves to the case where the motion of the matter generating the gravitational field is but slow (in comparison with the velocity of the propagation of light), we may neglect on the right-hand side differentiations with respect to the time in comparison with those with respect to the space co-ordinates, so that we have

$$\frac{d^2x_\tau}{dt^2} = -\frac{1}{2}\frac{\partial g_{44}}{\partial x_\tau} \quad (\tau = 1, 2, 3) \quad . \quad . \quad (67)$$

This is the equation of motion of the material point according to Newton's theory, in which $\frac{1}{2}g_{44}$ plays the part of the gravitational potential. What is remarkable in this result is that the component g_{44} of the fundamental tensor alone defines, to a first approximation, the motion of the material point.

We now turn to the field equations (53). Here we have to take into consideration that the energy-tensor of "matter" is almost exclusively defined by the density of matter in the narrower sense, i.e. by the second term of the right-hand side of (58) [or, respectively, (58a) or (58b)]. If we form the approximation in question, all the components vanish with the one exception of $T_{44} = \rho = T$. On the left-hand side of (53) the second term is a small quantity of second order; the first yields, to the approximation in question,

$$\frac{\partial}{\partial x_1}[\mu\nu, 1] + \frac{\partial}{\partial x_2}[\mu\nu, 2] + \frac{\partial}{\partial x_3}[\mu\nu, 3] - \frac{\partial}{\partial x_4}[\mu\nu, 4].$$

For $\mu = \nu = 4$, this gives, with the omission of terms differentiated with respect to time,

$$-\frac{1}{2}\left(\frac{\partial^2 g_{44}}{\partial x_1^2} + \frac{\partial^2 g_{44}}{\partial x_2^2} + \frac{\partial^2 g_{44}}{\partial x_3^2}\right) = -\frac{1}{2}\nabla^2 g_{44}.$$

The last of equations (53) thus yields

$$\nabla^2 g_{44} = \kappa\rho \quad . \quad . \quad . \quad . \quad (68)$$

The equations (67) and (68) together are equivalent to Newton's law of gravitation.

By (67) and (68) the expression for the gravitational potential becomes

$$-\frac{\kappa}{8\pi}\int\frac{\rho d\tau}{r} \quad . \quad . \quad . \quad . \quad (68a)$$

while Newton's theory, with the unit of time which we have chosen, gives

$$-\frac{K}{c^2}\int\frac{\rho d\tau}{r}$$

in which K denotes the constant $6{\cdot}7 \times 10^{-8}$, usually called the constant of gravitation. By comparison we obtain

$$\kappa = \frac{8\pi K}{c^2} = 1{\cdot}87 \times 10^{-27} \quad . \quad . \quad (69)$$

§ 22. Behaviour of Rods and Clocks in the Static Gravitational Field. Bending of Light-rays. Motion of the Perihelion of a Planetary Orbit

To arrive at Newton's theory as a first approximation we had to calculate only one component, g_{44}, of the ten $g_{\mu\nu}$ of the gravitational field, since this component alone enters into the first approximation, (67), of the equation for the motion of the material point in the gravitational field. From this, however, it is already apparent that other components of the $g_{\mu\nu}$ must differ from the values given in (4) by small quantities of the first order. This is required by the condition $g = -1$.

For a field-producing point mass at the origin of co-ordinates, we obtain, to the first approximation, the radially symmetrical solution

$$\left.\begin{aligned} g_{\rho\sigma} &= -\delta_{\rho\sigma} - a\frac{x_\rho x_\sigma}{r^3} \; (\rho, \sigma = 1, 2, 3) \\ g_{\rho 4} &= g_{4\rho} = 0 \qquad\quad (\rho = 1, 2, 3) \\ g_{44} &= 1 - \frac{a}{r} \end{aligned}\right\} \quad . \quad (70)$$

[33] where $\delta_{\rho\sigma}$ is 1 or 0, respectively, accordingly as $\rho = \sigma$ or $\rho \neq \sigma$, and r is the quantity $+\sqrt{x_1^2 + x_2^2 + x_3^2}$. On account of (68a)

$$a = \frac{\kappa M}{4\pi}, \quad . \quad . \quad . \quad . \quad (70a)$$

if M denotes the field-producing mass. It is easy to verify that the field equations (outside the mass) are satisfied to the first order of small quantities.

We now examine the influence exerted by the field of the mass M upon the metrical properties of space. The relation

$$ds^2 = g_{\mu\nu} dx_\mu dx_\nu.$$

always holds between the "locally" (§ 4) measured lengths and times ds on the one hand, and the differences of co-ordinates dx_ν on the other hand.

For a unit-measure of length laid " parallel " to the axis of x, for example, we should have to set $ds^2 = -1$; $dx_2 = dx_3 = dx_4 = 0$. Therefore $-1 = g_{11}dx_1^2$. If, in addition, the unit-measure lies on the axis of x, the first of equations (70) gives

$$g_{11} = -\left(1 + \frac{a}{r}\right).$$

From these two relations it follows that, correct to a first order of small quantities,

$$dx = 1 - \frac{a}{2r} \quad . \quad . \quad . \quad . \quad (71)$$

The unit measuring-rod thus appears a little shortened in relation to the system of co-ordinates by the presence of the gravitational field, if the rod is laid along a radius.

In an analogous manner we obtain the length of co-ordinates in tangential direction if, for example, we set

$ds^2 = -1$; $dx_1 = dx_3 = dx_4 = 0$; $x_1 = r$, $x_2 = x_3 = 0$.

The result is

$$-1 = g_{22}dx_2^2 = -dx_2^2 \quad . \quad . \quad . \quad (71a)$$

With the tangential position, therefore, the gravitational field of the point of mass has no influence on the length of a rod.

Thus Euclidean geometry does not hold even to a first approximation in the gravitational field, if we wish to take one and the same rod, independently of its place and orientation, as a realization of the same interval; although, to be sure, a glance at (70a) and (69) shows that the deviations to be expected are much too slight to be noticeable in measurements of the earth's surface.

Further, let us examine the rate of a unit clock, which is arranged to be at rest in a static gravitational field. Here we have for a clock period $ds = 1$; $dx_1 = dx_2 = dx_3 = 0$

Therefore

$$1 = g_{44}dx_4^2;$$

$$dx_4 = \frac{1}{\sqrt{g_{44}}} = \frac{1}{\sqrt{(1 + (g_{44} - 1))}} = 1 - \tfrac{1}{2}(g_{44} - 1)$$

or

$$dx_4 = 1 + \frac{\kappa}{8\pi}\int \rho\frac{d\tau}{r} \quad . \quad . \quad . \quad (72)$$

Thus the clock goes more slowly if set up in the neighbourhood of ponderable masses. From this it follows that the spectral lines of light reaching us from the surface of large stars must appear displaced towards the red end of the spectrum.*

We now examine the course of light-rays in the static gravitational field. By the special theory of relativity the velocity of light is given by the equation

$$- dx_1^2 - dx_2 - dx_3^2 + dx_4^2 = 0$$

and therefore by the general theory of relativity by the equation

$$ds^2 = g_{\mu\nu}dx_\mu dx_\nu = 0 \quad . \quad . \quad . \quad (73)$$

If the direction, i.e. the ratio $dx_1 : dx_2 : dx_3$ is given, equation (73) gives the quantities

$$\frac{dx_1}{dx_4}, \frac{dx_2}{dx_4}, \frac{dx_3}{dx_4}$$

and accordingly the velocity

$$\sqrt{\left(\frac{dx_1}{dx_4}\right)^2 + \left(\frac{dx_2}{dx_4}\right)^2 + \left(\frac{dx_3}{dx_4}\right)^2} = \gamma$$

defined in the sense of Euclidean geometry. We easily recognize that the course of the light-rays must be bent with regard to the system of co-ordinates, if the $g_{\mu\nu}$ are not constant. If n is a direction perpendicular to the propagation of
[35] light, the Huyghens principle shows that the light-ray, envisaged in the plane (γ, n), has the curvature $-\partial\gamma/\partial n$.

We examine the curvature undergone by a ray of light passing by a mass M at the distance Δ. If we choose the system of co-ordinates in agreement with the accompanying diagram, the total bending of the ray (calculated positively if

* According to E. Freundlich, spectroscopical observations on fixed stars of
[34] certain types indicate the existence of an effect of this kind, but a crucial test of this consequence has not yet been made.

concave towards the origin) is given in sufficient approximation by

$$B = \int_{-\infty}^{+\infty} \frac{\partial \gamma}{\partial x_1} dx_2,$$

while (73) and (70) give

$$\gamma = \sqrt{\left(-\frac{g_{44}}{g_{22}}\right)} = 1 - \frac{a}{2r}\left(1 + \frac{x_2^2}{r^2}\right).$$

Carrying out the calculation, this gives

$$B = \frac{2a}{\Delta} = \frac{\kappa M}{2\pi\Delta}. \quad . \quad . \quad . \quad (74)$$ [36]

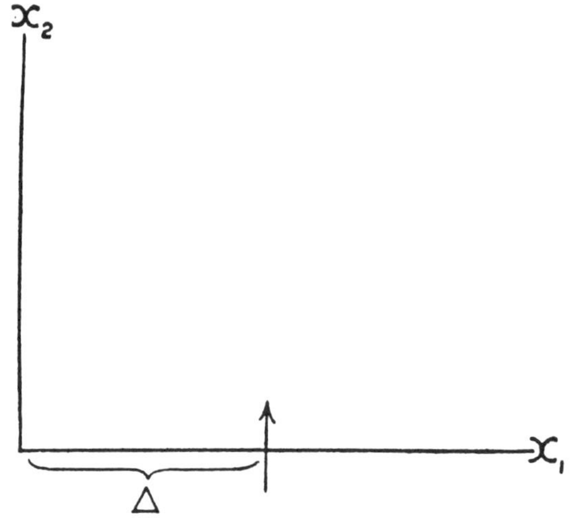

Fig. 8.

According to this, a ray of light going past the sun undergoes a deflexion of 1·7″; and a ray going past the planet Jupiter a deflexion of about ·02″.

If we calculate the gravitational field to a higher degree of approximation, and likewise with corresponding accuracy the orbital motion of a material point of relatively infinitely small mass, we find a deviation of the following kind from the Kepler-Newton laws of planetary motion. The orbital ellipse of a planet undergoes a slow rotation, in the direction of motion, of amount

$$\epsilon = 24\pi^3 \frac{a^2}{T^2 c^2 (1 - e^2)} \quad . \quad . \quad . \quad (75)$$

per revolution. In this formula a denotes the major semi-axis, c the velocity of light in the usual measurement, e the eccentricity, T the time of revolution in seconds.*

Calculation gives for the planet Mercury a rotation of the orbit of 43″ per century, corresponding exactly to astronomical observation (Leverrier); for the astronomers have discovered in the motion of the perihelion of this planet, after allowing for disturbances by other planets, an inexplicable remainder of this magnitude.

*For the calculation I refer to the original papers: A. Einstein, Sitzungsber. d. Preuss. Akad. d. Wiss., 1915, p. 831; K. Schwarzschild,
[37] *ibid.*, 1916, p. 189.

The Foundations of Physics.

(First communication.)

by

David Hilbert.

Presented in the session of November 20th, 1915.

(Translated by D. Fine.)

The powerful posing of problems by Einstein,[1] as well as his penetrating methods devised for their solution, and the deep thoughts and original conceptualizations by means of which Mie[2] builds his electrodynamics, have opened new paths for the investigation of the foundations of physics.

I would like in the following – in the sense of the axiomatic method – to establish, essentially from two simple axioms, a new system of basic equations of physics, that are of ideal beauty, and in which, so I believe, is included **simultaneously** the solutions of the problems of Einstein and Mie. I reserve for later communications the more precise exposition as well as, above all, the specific application of my basic equations to the fundamental questions of the science of electricity.

Let $w_s (s = 1, 2, 3, 4)$ be some coordinates naming the world-points essentially unambiguously (most general spacetime coordinates). The quantities characterizing what happens in w_s are

1. the ten gravitational potentials $g_{\mu\nu} (\mu, \nu = 1, 2, 3, 4)$, first introduced by Einstein, with symmetric tensor character under an arbitrary transformation of the world-parameters w_s;

2. the four electrodynamical potentials q_s with vector character in the same sense.

What happens physically is not haphazard; on the contrary, the following two axioms hold true:

Axiom I (Mie's Axiom of the world-function[3]): *The law of what happens physically is determined by a world-function H that contains the following as arguments:*

$$g_{\mu\nu}, \quad g_{\mu\nu l} = \frac{\partial g_{\mu\nu}}{\partial w_l}, \quad g_{\mu\nu lk} = \frac{\partial^2 g_{\mu\nu}}{\partial w_l \partial w_k}, \tag{1}$$

$$q_s, \quad q_{sl} = \frac{\partial q_s}{\partial w_l}, \qquad (l, k = 1, 2, 3, 4) \tag{2}$$

and indeed the variation of the integral

$$\int H\sqrt{g}dw$$

$$(g = |g_{\mu\nu}|, \, dw = dw_1 dw_2 dw_3 dw_4)$$

must vanish for each of the 14 potentials $g_{\mu\nu}, q_s$.

[1] Sitzungsber. d. Berliner Akad. 1914 S.1030, 115 S.778, 799, 831, 844.

[2] Ann. d. Phys. 1912, Bd. 37 S. 511, Bd. 39 S. 1, 1913, Bd. 40 S. 1.

[3] Mie's world-functions do not contain exactly these arguments; in particular, the use of the arguments (2) goes back to Born; nevertheless it is precisely the introduction and use of such a world-function in the Hamiltonian principle which is characteristic of Mie's electrodynamics.

In place of the arguments (1) the arguments

$$g^{\mu\nu},\ g_l^{\mu\nu} = \frac{\partial g^{\mu\nu}}{\partial w_l},\ g_{lk}^{\mu\nu} = \frac{\partial^2 g^{\mu\nu}}{\partial w_l \partial w_k} \tag{3}$$

obviously can appear, where $g^{\mu\nu}$ means the minor determinant of the determinant of g corresponding to $g_{\mu\nu}$, divided by g.

Axiom II (Axiom of general invariance [4]): *The world-function H is an invariant with respect to an arbitrary transformation of the world-parameter w_s.*

Axiom II is the simplest mathematical expression for the requirement that the coupling of the potentials $g_{\mu\nu}, q_s$ is in and of itself completely independent of the way one chooses to name world-points using world-parameters.

The following mathematical theorem, whose proof I will lay out elsewhere, provides the "Leitmotiv" for the construction of my theory.

Theorem I. If J is an invariant with respect to arbitrary transformations of the four world-parameters which contains n quantities and their derivatives, and if one forms from

$$\delta \int J\sqrt{g}dw = 0$$

in reference to those n quantities the n Lagrangian variational equations, then, in this system of n differential equations for the n quantities, four are always a consequence of the $n-4$ others – in the sense that among the n differential equations and their total derivatives four mutually-independent linear combinations are identically fulfilled.

In regard to the derivatives with respect to $g^{\mu\nu}, g_k^{\mu\nu}, g_{kl}^{\mu\nu}$ as they appear in (4) and following formulas, let it be noted once and for all, that due to the symmetry in μ, ν on the one hand, and k, l on the other hand the derivatives with respect to $g^{\mu\nu}, g_k^{\mu\nu}$ are to be taken as multiplied by 1, respectively $\frac{1}{2}$, according to whether $\mu = \nu$, respectively $\mu \neq \nu$ occurs; further, the derivatives with respect to $g_{kl}^{\mu\nu}$ are to be taken as multiplied by 1, respectively $\frac{1}{2}$, respectively $\frac{1}{4}$, according to whether $\mu = \nu$ and $k = l$, respectively $\mu = \nu$ and $k \neq l$ or $\mu \neq \nu$ and $k = l$, respectively $\mu \neq \nu$ and $k \neq l$ occurs.

From Axiom I follow first in regard to the ten gravitational potentials $g^{\mu\nu}$ the ten Lagrangian differential equations

$$\frac{\partial\sqrt{g}H}{\partial g^{\mu\nu}} - \sum_k \frac{\partial}{\partial w_k}\frac{\partial\sqrt{g}H}{\partial g_k^{\mu\nu}} + \sum_{k,l}\frac{\partial^2}{\partial w_k \partial w_l}\frac{\partial\sqrt{g}H}{\partial g_{kl}^{\mu\nu}} = 0, \quad (\mu,\nu = 1,2,3,4). \tag{4}$$

and then in regard to the four electrodynamic potentials q_s the four Lagrangian differential equations

$$\frac{\partial\sqrt{g}H}{\partial q_h} - \sum_k \frac{\partial}{\partial w_k}\frac{\partial\sqrt{g}H}{\partial q_{hk}} = 0, \quad (h = 1,2,3,4). \tag{5}$$

[4]Mies has already imposed the requirement of orthogonal invariance. In Axiom II above, the **Einsteinian** fundamental basic idea of general invariance is given the simplest expression, albeit in Einstein the Hamiltonian principle plays only a secondary role and his functions H are certainly not general invariants nor do they contain the electric potentials.

For brevity we denote the left-hand sides of the equations (4), (5) respectively as

$$[\sqrt{g}H]_{\mu\nu}, [\sqrt{g}H]_h .$$

The equations (4) might be called the basic equations of gravitation, the equations (5) the basic electrodynamics equations or the generalized Maxwell's equations. Due to the theorem given above, the equations (5) may be viewed as a consequence of the equations (4); that is, based on that mathematical proposition, we can immediately state the claim *that in the indicated sense the electrodynamic phenomena are effects of gravitation.* In this realization I discern the simple and quite surprising solution of the problem of Riemann who was the first to seek the theoretical relationship between gravitation and light.

In the following, we use the easily proven fact that if p^j $(j = 1, 2, 3, 4)$ denotes an *arbitrary* contravariant vector, the expression

$$p^{\mu\nu} = \sum_s (g_s^{\mu\nu} p^s - g^{\mu s} p_s^\nu - g^{\nu s} p_s^\mu), \quad \left(p_s^j = \frac{\partial p^j}{\partial w_s}\right)$$

represents a symmetric contravariant tensor and the expression

$$p_l = \sum_s (q_{ls} p^s + q_s p_l^s)$$

a covariant vector.

Furthermore, we present two mathematical theorems, that read as follows: **Theorem II.** If J is an invariant depending on $g^{\mu\nu}, g_l^{\mu\nu}, g_{lk}^{\mu\nu}, q_s, q_{sk}$ then identically in all arguments and for every arbitrary contravariant vector p^s

$$\sum_{\mu,\nu,l,k} \left(\frac{\partial J}{\partial g^{\mu\nu}} \Delta g^{\mu\nu} + \frac{\partial J}{\partial g_l^{\mu\nu}} \Delta g_l^{\mu\nu} + \frac{\partial J}{\partial g_{lk}^{\mu\nu}} \Delta g_{lk}^{\mu\nu}\right) + \sum_{s,k} \left(\frac{\partial J}{\partial q_s} \Delta q_s + \frac{\partial J}{\partial q} \Delta q_{sk}\right) = 0,$$

wherein†

$$\begin{aligned}
\Delta g^{\mu\nu} &= \sum_m (g^{\mu m} p_m^\nu + g^{\nu m} p_m^\mu), \\
\Delta g_l^{\mu\nu} &= -\sum_m g_m^{\mu\nu} p_l^m + \frac{\partial \Delta g^{\mu\nu}}{\partial w_l}, \\
\Delta g_{lk}^{\mu\nu} &= -\sum_m (g_m^{\mu\nu} p_{lk}^m + g_{lm}^{\mu\nu} p_k^m + g_{km}^{\mu\nu} p_l^m) + \frac{\partial^2 \Delta g^{\mu\nu}}{\partial w_l \partial w_k}, \\
\Delta q_s &= -\sum_m q_m p_s^m, \\
\Delta q_{sk} &= -\sum_m q_{sm} p_k^m + \frac{\partial \Delta q_s}{\partial w_k}.
\end{aligned}$$

This Theorem II may also be stated in the following form:

† *The last partial derivative above is $\frac{\partial J}{\partial q}$, not $\frac{\partial J}{\partial q_{sk}}$ in the original. (Translator)*

If J is an invariant and p^s an arbitrary vector as before, then the identity

$$\sum_s \frac{\partial J}{\partial w_s} p^s = PJ \tag{6}$$

is valid; here we set

$$\begin{aligned} P &= P_g + P_q, \\ P_q &= \sum_{\mu,\nu,l,k} \left(p^{\mu\nu} \frac{\partial}{\partial g^{\mu\nu}} + p_l^{\mu\nu} \frac{\partial}{\partial g_l^{\mu\nu}} + p_{lk}^{\mu\nu} \frac{\partial}{\partial g_{lk}^{\mu\nu}} \right), \\ P_q &= \sum_{l,k} \left(p_l \frac{\partial}{\partial q_l} + p_{lk} \frac{\partial}{\partial q_{lk}} \right), \end{aligned}$$

and abbreviate

$$p_k^{\mu\nu} = \frac{\partial p^{\mu\nu}}{\partial w_k}, \quad p_{kl}^{\mu\nu} = \frac{\partial^2 p^{\mu\nu}}{\partial w_k \partial w_l}, \quad p_{lk} = \frac{\partial p_l}{\partial w_k}.$$

The proof of (6) falls out easily, because this identity is obviously correct when p^s is a constant vector from which it follows in general due to its invariance.

Theorem III If J is an invariant depending *only* on the $g^{\mu\nu}$ and their derivatives, and, as above, the variational derivative of $\sqrt{g}J$ with respect to $g^{\mu\nu}$ is denoted $[\sqrt{g}J]_{\mu\nu}$, then the expression – with $h^{\mu\nu}$ understood to be some contravariant tensor

$$\frac{1}{\sqrt{g}} \sum_{\mu,\nu} [\sqrt{g}J]_{\mu\nu} h^{\mu\nu}$$

represents an invariant; if in this sum we replace $h^{\mu\nu}$ by the particular tensor $p^{\mu\nu}$ and write

$$\sum_{\mu,\nu} [\sqrt{g}J]_{\mu\nu} p^{\mu\nu} = \sum_{s,l} (i_s p^s + i_s^l p_l^s),$$

where the expressions

$$\begin{aligned} i_s &= \sum_{\mu,\nu} [\sqrt{g}J]_{\mu\nu} g_s^{\mu\nu}, \\ i_s^l &= -2 \sum_{\mu} [\sqrt{g}J]_{\mu s} g^{\mu l}, \end{aligned}$$

depend only on the $g^{\mu\nu}$ and their derivatives, then

$$i_s = \sum_l \frac{\partial i_s^l}{\partial w_l} \tag{7}$$

in the manner that this equation is fulfilled for all arguments, namely the $g^{\mu\nu}$ and their derivatives.

For the proof we consider the integral

$$\int J\sqrt{g}dw, \; dw = dw_1 dw_2 dw_3 dw_4$$

extending over a finite piece of the four-dimensional world. Further p^s should be a vector that along with its derivatives vanishes on the three-dimensional boundary of this piece of the world. As $P = P_g$, it follows from the formula (12a) appearing below that

$$P_g(\sqrt{g}J) = \sum_s \frac{\partial \sqrt{g} J p^s}{\partial w_s},$$

which gives

$$\int P_g(J\sqrt{g})dw = 0,$$

and, due to the manner of construction of the Lagrangian derivative, is also accordingly

$$\int \sum_{\mu\nu} [\sqrt{g}J]_{\mu\nu} p^{\mu\nu} dw = 0.$$

The introduction of i_s, i_s^l in this identity shows finally that

$$\int \left(\sum_l \frac{\partial i_s^l}{\partial w_l} - i_s \right) p^s dw = 0$$

and therefore also that the claim of our theorem is correct.

The most important goal at this point is the establishment of the concept of energy and the derivation of the energy proposition on the basis of the axioms I and II alone.

Towards that we construct first:

$$P_g(\sqrt{g}H) = \sum_{\mu,\nu,k,l} \left(\frac{\partial \sqrt{g} H}{\partial g^{\mu\nu}} p^{\mu\nu} + \frac{\partial \sqrt{g} H}{\partial g_k^{\mu\nu}} p_k^{\mu\nu} + \frac{\partial \sqrt{g} H}{\partial g_{kl}^{\mu\nu}} p_{kl}^{\mu\nu} \right).$$

Now, $\frac{\partial H}{\partial g_{kl}^{\mu\nu}}$ is a mixed tensor of fourth order and therefore if one sets

$$\begin{aligned} A_k^{\mu\nu} &= p_k^{\mu\nu} + \sum_\rho \left(\begin{Bmatrix} k\rho \\ \mu \end{Bmatrix} p^{\rho\nu} + \begin{Bmatrix} k\rho \\ \nu \end{Bmatrix} p^{\rho\mu} \right), \\ \begin{Bmatrix} k\rho \\ \mu \end{Bmatrix} &= \frac{1}{2} \sum_\sigma g^{\mu\sigma} (g_{k\sigma\rho} + g_{\rho\sigma k} - g_{k\rho\sigma}) \end{aligned}$$

the expression

$$a^l = \sum_{\mu,\nu,k} \frac{\partial H}{\partial g_{kl}^{\mu\nu}} A_k^{\mu\nu} \tag{8}$$

is a contragradient vector.

If we therefore form the expression

$$P_g(\sqrt{g}H) - \sum_l \frac{\partial \sqrt{g} a^l}{\partial w_l},$$

this no longer contains the second derivatives $p^{\mu\nu}_{kl}$ and is therefore of the form

$$\sqrt{g}\sum_{\mu,\nu,k}(B_{\mu\nu}p^{\mu\nu}+B^k_{\mu\nu}p^{\mu\nu}_k),$$

wherein

$$B^k_{\mu\nu}=\sum_{\rho,l}\left(\frac{\partial H}{\partial g^{\mu\nu}_k}-\frac{\partial}{\partial w_l}\frac{\partial H}{\partial g^{\mu\nu}_{kl}}-\frac{\partial H}{\partial g^{\rho\nu}_{kl}}\left\{{l\mu \atop \rho}\right\}-\frac{\partial H}{\partial g^{\mu\rho}_{kl}}\left\{{l\nu \atop \rho}\right\}\right)$$

is again a mixed tensor.

At this point we form the vector

$$b^l=\sum_{\mu,\nu}B^l_{\mu\nu}p^{\mu\nu} \tag{9}$$

and then obtain

$$P_g(\sqrt{g}H)-\sum_l\frac{\partial\sqrt{g}(a^l+b^l)}{\partial w_l}=\sum_{\mu,\nu}[\sqrt{g}H]_{\mu\nu}p^{\mu\nu}. \tag{10}$$

On the other hand, we form

$$P_g(\sqrt{g}H)=\sum_{k,l}\left(\frac{\partial\sqrt{g}H}{\partial q_k}p_k+\frac{\partial\sqrt{g}H}{\partial q_{kl}}p_{kl}\right)$$

then $\frac{\partial H}{\partial q_{kl}}$ is a tensor, and the expression

$$c^l=\sum_k\frac{\partial H}{\partial q_{kl}}p_k \tag{11}$$

therefore represents a contragradient vector. Correspondingly, as above,

$$P_g(\sqrt{g}H)-\sum_l\frac{\partial\sqrt{g}c^l}{\partial w_l}=\sum_k[\sqrt{g}H]_kp_k. \tag{12}$$

Recalling now the basic equations (4) and (5), it follows from addition of (10) and (12) that

$$P_g(\sqrt{g}H)=\sum_l\frac{\partial\sqrt{g}(a^l+b^l+c^l)}{\partial w_l}.$$

Now,

$$\begin{aligned}P_g(\sqrt{g}H)&=\sqrt{g}PH+H\sum_{\mu,\nu}\frac{\partial\sqrt{g}}{\partial g^{\mu\nu}}p^{\mu\nu}\\&=\sqrt{g}PH+H\sum_s\left(\frac{\partial\sqrt{g}}{\partial w_s}p^s+\sqrt{g}p^s_s\right),\end{aligned}$$

and therefore by dint of the identity (6)

$$(12a)\qquad P_g(\sqrt{g}H)=\sqrt{g}\sum_s\frac{\partial H}{\partial w_s}p^s+H\sum_s\left(\frac{\partial\sqrt{g}}{\partial w_s}p^s+\sqrt{g}p^s_s\right)=\sum_s\frac{\partial\sqrt{g}Hp^s}{\partial w_s}.$$

With this we finally obtain the equation

$$\sum_l\frac{\partial}{\partial w_l}\sqrt{g}(Hp^l-a^l-b^l-c^l)=0.$$

Now we recall that

$$\frac{\partial H}{\partial q_{lk}}-\frac{\partial H}{\partial q_{kl}}$$

is a skew-symmetric tensor; as a consequence

$$(13)\qquad d^l=\frac{1}{2\sqrt{g}}\sum_{k,s}\frac{\partial}{\partial w_k}\left\{\left(\frac{\partial\sqrt{g}H}{\partial q_{lk}}-\frac{\partial\sqrt{g}H}{\partial q_{kl}}\right)p^s q_s\right\}$$

will be a contravariant vector and indeed the latter obviously satifies the identity

$$\sum_l\frac{\partial\sqrt{g}d^l}{\partial w_l}=0.$$

If at this point we define

$$(14)\qquad e^l=Hp^l-a^l-b^l-c^l-d^l$$

as the **energy-vector**, *then the energy-vector is a contravariant vector that still depends linearly on the arbitrary vector p^s and for every choice of this vector p^s satisfies identically the* **invariant energy-equation**

$$\sum_l\frac{\partial\sqrt{g}e^l}{\partial w_l}=0.$$

As concerns the world-function H, further axioms are necessary to make its choice unambiguous. If the gravitational equations are to contain only second derivatives of the potentials $g^{\mu\nu}$, then H must have the form

$$H=K+L,$$

where K denotes the invariant arising from the Riemann tensor (curvature of the four-dimensional manifold)

$$K=\sum_{\mu,\nu}g^{\mu\nu}K_{\mu\nu},$$

$$K_{\mu\nu}=\sum_\chi\left(\frac{\partial}{\partial w_\nu}\begin{Bmatrix}\mu\chi\\ \chi\end{Bmatrix}-\frac{\partial}{\partial w_\chi}\begin{Bmatrix}\mu\nu\\ \chi\end{Bmatrix}\right)+\sum_{\chi,\lambda}\left(\begin{Bmatrix}\mu\chi\\ \lambda\end{Bmatrix}\begin{Bmatrix}\lambda\nu\\ \chi\end{Bmatrix}-\begin{Bmatrix}\mu\nu\\ \lambda\end{Bmatrix}\begin{Bmatrix}\lambda\chi\\ \chi\end{Bmatrix}\right)$$

and L depends only on $g^{\mu\nu}, g_l^{\mu\nu}, q_s, q_{sk}$. Finally we make in the following the simplifying assumption that L does not contain the $g_l^{\mu\nu}$.

We then apply Theorem II to the invariant L and obtain

$$(15) \quad \sum_{\mu,\nu,m} \frac{\partial L}{\partial g^{\mu\nu}}(g^{\mu m}p_m^\nu + g^{\nu m}p_m^\mu) - \sum_{s,m} \frac{\partial L}{\partial q_s} q_m p_s^m - \sum_{s,k,m} \frac{\partial L}{\partial q_{sk}}(q_{sm}p_k^m + q_{mk}p_s^m + q_m p_{sk}^m) = 0.$$

Equating to zero the coefficients of p_{sk}^m on the left-hand side yields the equation

$$\left(\frac{\partial L}{\partial q_{sk}} + \frac{\partial L}{\partial q_{ks}}\right) q_m = 0,$$

or

$$(16) \qquad \frac{\partial L}{\partial q_{sk}} + \frac{\partial L}{\partial q_{ks}} = 0;$$

that is, the derivatives of the electrodynamic potentials q_s appear only in the combination

$$M_{ks} = q_{sk} - q_{ks}.$$

We thereby recognize that with our assumptions, the invariant L beyond the potentials $g^{\mu\nu}, q_s$ depends solely on the components of the skew-symmetric invariant tensor

$$M = (M_{ks}) = \mathrm{Rot}(q_s);$$

that is, the so-called electromagnetic six-vector. *This result, by which the character of Maxwell's equations is really conditioned, ensues here essentially as a consequence of general invariance, thus on the basis of Axiom II.*

If we set the coefficient of p_m^ν on the left-hand side of the identity (15) to zero, then we obtain, using (16),

$$(17) \qquad 2\sum_\mu \frac{\partial L}{\partial g^{\mu\nu}} g^{\mu m} - \frac{\partial L}{\partial q_m} q_\nu - \sum_s \frac{\partial L}{\partial M_{ms}} M_{\nu s} = 0, \quad (\mu = 1, 2, 3, 4).$$

This equation permits an important reformulation of the electromagnetic energy, that is the part of the energy vector coming from L. This part is given in fact from (11), (13) (14) as follows:

$$Lp^l - \sum_k \frac{\partial L}{\partial q_{kl}} p_k - \frac{1}{2\sqrt{g}} \sum_{k,s} \frac{\partial}{\partial w_k}\left\{\left(\frac{\partial\sqrt{g}L}{\partial q_{lk}} - \frac{\partial\sqrt{g}L}{\partial q_{kl}}\right) p^s q_s\right\}.$$

Due to (16) and recalling (5) this expression becomes

$$(18) \qquad \sum_{s,k}\left(L\delta_s^l - \frac{\partial L}{\partial M_{lk}} M_{sk} - \frac{\partial L}{\partial q_l} q_s\right) p^s$$

$$(\delta_s^l = 0, l \neq s; \delta_s^s = 1);$$

that is, due to (17)

$$(19) \qquad -\frac{2}{\sqrt{g}}\sum_{\mu,s}\frac{\partial\sqrt{g}L}{\partial g^{\mu s}}g^{\mu l}p^{s}.$$

Due to the formula (21) developed in the following, we see from this in particular that the electromagnetic energy and with it also the total energy-vector e^l may be expression in terms of K alone, so that only the $g^{\mu\nu}$ and their derivatives, but not the q_s and their derivatives appear therein. If in the expression (18) one goes to the boundary case

$$\begin{aligned} g_{\mu\nu} &= 0, \ (\mu \neq \nu) \\ g_{\mu\mu} &= 1 \end{aligned}$$

then the same agrees exactly with that which Mie set out in his electrodymics: **the Mie electromagnetic energy-tensor is thus nothing but the general invariant tensor arising through differentiation of the invariant L with respect to the gravitational potential $g^{\mu\nu}$** – a circumstance which first directed me to the necessarily close relationship between the Einsteinian general theory of relativity and the Miesian electrodymics and gave me the conviction of the correctness of the theory developed here.

It remains yet to show directly from the assumption

$$(20) \qquad H = K + L$$

how the generalized Maxwell's equations (5) presented above are a result of the gravitational equations (4) in the sense given above. Using the notation introduced above for the variational derivatives with respect to the $g^{\mu\nu}$ the gravitational equations due to (20) take the form

$$(21) \qquad [\sqrt{g}K]_{\mu\nu} + \frac{\partial\sqrt{g}L}{\partial g^{\mu\nu}} = 0.$$

The first term on the left becomes

$$[\sqrt{g}K]_{\mu\nu} = \sqrt{g}(K_{\mu\nu} - \frac{1}{2}Kg_{\mu\nu}),$$

as follows easily without calculation from the fact $K_{\mu\nu}$ is, other than $g_{\mu\nu}$, the only tensor of second rank and K is the only invariant, that can be constructed with only the $g^{\mu\nu}$ and their first and second derivatives $g^{\mu\nu}_k, g^{\mu\nu}_{kl}$.

The differential equations of gravity coming about thusly appear to me to be in agreement with the ambitious theory of general relativity presented by Einstein in his later treatments. [5]

If we continue in general as above to denote the variational derivatives of $\sqrt{g}J$ with respect to the electrodynamic potential q_h by

$$[\sqrt{g}J]_h = \frac{\partial\sqrt{g}J}{\partial q_h} - \sum_k \frac{\partial}{\partial w_k}\frac{\partial\sqrt{g}J}{\partial q_{hk}}$$

[5] *loc. cit.* Berliner Sitzungsber. 1915.

then due to (20) the electrodynamic basic equation takes the form

$$[\sqrt{g}J]_h = 0. \tag{22}$$

Now as K is an invariant depending soley on the $g^{\mu\nu}$ and their derivatives, according to Theorem III the equality (7) obtains identically, wherein

$$i_s = \sum_{\mu,\nu}[\sqrt{g}K]_{\mu\nu}g_s^{\mu\nu} \tag{23}$$

and

$$i_s^l = -2\sum_{\mu}[\sqrt{g}K]_{\mu s}g^{\mu l}, \quad (\mu = 1,2,3,4) \tag{24}$$

Due to (21) and (24), (19) equals $-\frac{1}{\sqrt{g}}i_\nu^m$. Through differentiation by w_m and summation over m we obtain due to (7)

$$\begin{aligned}
i_\nu &= \sum_m \frac{\partial}{\partial w_m}\left(-\sqrt{g}L\delta_\nu^m + \frac{\partial\sqrt{g}L}{\partial q_m}q_\nu + \sum_s \frac{\partial\sqrt{g}L}{\partial M_{sm}}M_{s\nu}\right) \\
&= -\frac{\partial\sqrt{g}L}{\partial w_\nu} + \sum_m\left\{q_\nu\frac{\partial}{\partial w_m}\left([\sqrt{g}L]_m + \sum_s\frac{\partial}{\partial w_s}\frac{\partial\sqrt{g}L}{\partial q_{ms}}\right)\right. \\
&\quad \left. + q_{\nu m}\left([\sqrt{g}L]_m + \sum_s\frac{\partial}{\partial w_s}\frac{\partial\sqrt{g}L}{\partial q_{ms}}\right)\right\} \\
&\quad + \sum_s\left([\sqrt{g}L]_s - \frac{\partial\sqrt{g}L}{\partial q_s}\right)M_{s\nu} + \sum_{s,m}\frac{\partial\sqrt{g}L}{\partial M_{sm}}\frac{\partial M_{s\nu}}{\partial w_m},
\end{aligned}$$

as

$$\frac{\partial\sqrt{g}L}{\partial q_m} = [\sqrt{g}L]_m + \sum_s\frac{\partial}{\partial w_s}\frac{\partial\sqrt{g}L}{\partial q_{ms}},$$

and

$$-\sum_m\frac{\partial}{\partial w_m}\frac{\partial\sqrt{g}L}{\partial q_{sm}} = [\sqrt{g}L]_s - \frac{\partial\sqrt{g}L}{\partial q_s}.$$

We now recall that due to (16)

$$\sum_{m,s}\frac{\partial^2}{\partial w_m\partial w_s}\frac{\partial\sqrt{g}L}{\partial q_{ms}} = 0$$

and obtain then through appropriate rearrangement:

$$\begin{aligned}
i_\nu &= -\frac{\partial\sqrt{g}L}{\partial w_\nu} + \sum_m\left(q_\nu\frac{\partial}{\partial w_m}[\sqrt{g}L]_m + M_{m\nu}[\sqrt{g}L]_m\right) \\
&\quad + \sum_m\frac{\partial\sqrt{g}L}{\partial q_m}q_{m\nu} + \sum_{s,m}\frac{\partial\sqrt{g}L}{\partial M_{sm}}\frac{\partial M_{s\nu}}{\partial w_m}.
\end{aligned} \tag{25}$$

On the other hand,

$$-\frac{\partial\sqrt{g}L}{\partial w_\nu} = -\sum_{s,m}\frac{\partial\sqrt{g}L}{\partial g^{sm}}g_\nu^{sm} - \sum_m \frac{\partial\sqrt{g}L}{\partial q_m}q_{m\nu} - \sum_{m,s}\frac{\partial\sqrt{g}L}{\partial q_{ms}}\frac{\partial q_{ms}}{\partial w_\nu}.$$

The first term on the right-hand side is, due to (21) and (23), nothing other than i_ν. The last term on the right proves to be cancelled by the last term on right in (25); in fact,

$$\sum_{s,m}\frac{\partial\sqrt{g}L}{\partial M_{sm}}\left(\frac{\partial M_{s\nu}}{\partial w_m} - \frac{\partial q_{ms}}{\partial w_\nu}\right) = 0, \tag{26}$$

as the expression

$$\frac{\partial M_{s\nu}}{\partial w_m} - \frac{\partial q_{ms}}{\partial w_\nu} = \frac{\partial^2 q_\nu}{\partial w_s \partial w_m} - \frac{\partial^2 q_s}{\partial w_\nu \partial w_m} - \frac{\partial^2 q_m}{\partial w_\nu \partial w_s}$$

is symmetric in s, m and the first factor in the summation in (26) comes out skew-symmetric in s, m.

The equation

$$\sum_m \left(M_{m\nu}[\sqrt{g}L]_m + q_\nu \frac{\partial}{\partial w_m}[\sqrt{g}L]_m\right) = 0 \tag{27}$$

follows immediately from (25); that is, from the gravitational equations (4) follow indeed the four mutually linearly independent combinations (27) of the electrodynamic basic equations (5) and their first derivatives. *This is the precise mathematical expression of the above claim expressed generally about the character of electrodynamics as a consequence of gravity.*

As L in consequence of our assumptions should not depend on the derivatives of the $g^{\mu\nu}$, L must be a function of four certain general invariants which correspond to the special orthogonal invariants given by Mie and of which the two most simple are these:

$$Q = \sum_{k,l,m,n} M_{mn}M_{lk}g^{mk}g^{nl}$$

and

$$q = \sum_{k,l} q_k q_l g^{kl}.$$

The most simple, and looking at the construction of K, most obvious "Ansatz" for L is likewise the one which corresponds to Mie's electrodynamics; namely,

$$L = \alpha Q + f(q)$$

or, more specifically following Mie:

$$L = \alpha Q + \beta q^s,$$

where $f(q)$ denotes any function of q and α, β denote constants.‡

‡ *The last term on the right-hand side is q^s not q in the original. (Translator)*

As one sees, the few simple assumptions expressed in the Axioms I and II suffice with appropriate interpretation for the construction of the theory: thereby not only are our conceptions of space, time and motion transformed from the ground up in the sense laid forth by Einstein, but I am also of the conviction that through the basic equations presented here the most intimate, hitherto hidden, processes within the atom will receive clarification and most particularly it must be generally possible to trace all physical constants back to mathematical constants – as then thereby the possibility first edges nearer that in principle a science of the style of geometry will evolve from physics: certain is the most excellent reputation of the axiomatic method which, as we see here, takes into its service the powerful instruments of analysis; namely, variational calculus and invariant theory.

Translator's notes:

1. Hilbert writes in the erudite manner of a well-educated mathematician of the early 20th century. I have endeavored to translate his German into an English preserving the feel of his style. Insofar as I have succeeded, credit is due to my wife, Heidemarie Floerke, whose help in sorting through the nuances of Hilbert's German was invaluable. The extent to which the reader may mistake Hilbert here for an elderly German academic writing in broken English is my fault entirely.
2. This paper does not appear in Hilbert's collected works.[(i)] As Pais notes,[(ii)] Hilbert's claim that electrodynamics can be viewed as a consequence of gravity rests on a misunderstanding. The identities among the gravitational potentials $g_{\mu\nu}$ and the electrodynamical potentials q_μ on which Hilbert bases his claim express the Bianchi identities, which are automatically true, and of which Hilbert was apparently unaware. In a subsequent version,[(iii)] Hilbert corrects this and other errors. This version, which presents verbatim much of the present paper, is included in his collected works. Presumably, Hilbert was not anxious to have the erroneous early version more widely read. From a historical perspective, however, it provides a fascinating glimpse of Hilbert poised, in 1915, to give an axiomatic, geometric foundation to all of theoretical physics.

(i) D. Hilbert *Gesammelte Abhandlungen* Chelsea, New York, 1981

(ii) A. Pais, *Subtle is the Lord: The Science and the Life of Albert Einstein*, p. 358 Oxford U. Press, Oxford,1982

(iii) D. Hilbert "Die Grundlagen der Physik" *Math. Ann.* **92** p. 1 (1924)

Meeting of February 27, 1922

On a generalization of the concept of Riemann curvature and spaces with torsion.

Note by Mr. E. Cartan, presented by Mr. E. Borel.

(Translated by Arthur Fine)

In a recent note([1])I showed how, in an Einstein universe with a given ds^2, the energy tensor attached to each volume element of that universe can be defined geometrically; this is the tensor which, set equal to zero, gives the laws of gravitation in any region devoid of matter. The definition that I gave makes the curvature of the universe depend on a certain rotation associated with every closed, infinitesimal contour, and this rotation was introduced on the basis of the concept of parallel transport of Levi-Civita. This last concept itself, although it was originally presented using geometrical considerations, is rather difficult to define in a precise way without calculation. But it is possible, it seems to me, to show the major significance of it by generalizing the concept even of space; at the same time that will lead us to geometrical images of material universes physically richer than our universe, at least as it is usually considered; that will also show us the true rational of the fundamental laws governing the energy tensor (law of symmetry, law of conservation).

Let us restrict ourselves to the case of three dimensions, the generalization to four dimensions being easy. Imagine a space which, in the immediate neighborhood of each point, has all the characteristics of Euclidean space. The inhabitants of this space will know, for example, how to locate points infinitely close to a point **A** by means of an orthogonal triple having this point **A** as origin; but we will suppose further that they have a law enabling them to orient, in relation to the triple at origin **A**, every coordinate triple having its origin **A'** close to **A**; in particular that will give a sense for them to say that two directions, one coming from **A** and another from **A'**, are parallel. Ultimately, *such a space will be defined by the law of mutual orientation* (of a Euclidean nature) *of two triples with origins infinitely close.*

A space of the preceding kind *is not completely defined by its ds^2*. The ds^2, indeed, determines only one part of the operation that allows the passage from a triple with origin **A** to an infinitely close triple with origin **A'**, namely a *translation* $\mathbf{A} \rightarrow \mathbf{A'}$; in addition, as one knows, ds^2 being fixed, a rotation can still be defined according to an arbitrary law.

That granted, when one describes a closed, infinitesimal contour starting from point **A** and returning there, the divergence between the space considered and Euclidean space

---

([1]) *Comptes rendus*, vol. 174, 1922, p. 437.

will show itself in the following way. Let us attach a coordinate triple to each point **M** of the contour; to pass from the triple attached to **M** to the triple attached to the infinitely close point **M'**, one needs to make an infinitesimal translation and rotation whose components one knows with respect to the moving triple with origin **M**.

Imagine that this collection of infinitesimal displacements is carried out *in a Euclidean space* starting from an initial triple chosen arbitrarily. When the point **M** of non-Euclidean space that starts from **A** returns there after having described the closed contour, in Euclidean space *one will not recover the initial triple*, but for that to obtain it will be necessary to carry out a complementary displacement whose components will be well defined with respect to the initial triple. This complementary displacement is otherwise independent of the law whereby one attached a triple to each point **M** of the contour.

In sum, associated with any infinitesimal closed contour of the given space are an infinitesimal translation and rotation (on the order of magnitude of the surface area bounded by the contour) and which express the divergence between this space and Euclidean space. The rotation can be represented by a vector with origin **A** and the translation by a couple. One can then prove the following *conservation law*: *If one considers an infinitesimal volume, the vectors and the couples associated with different elements of the surface bounding the volume are in equilibrium.*

Thus one has a geometrical image of a continuous material medium in equilibrium under the sole action of its elastic forces, but in a situation where these forces would be expressed on each surface element, not only by one single force (tension or pressure), but by a couple (torsion).

Return now to the case where we are given simply ds^2. An easy calculation shows that, among all the laws of mutual orientation of two triples of infinitely close origin compatible with the given ds^2, *there is only one for which the translation associated with an arbitrary, infinitesimal closed contour is null.* It is this law which leads to the concept of parallel displacement of Levi-Civita. The couple in question above disappears, *and this is why the elastic tensor satisfied the law of symmetry.*

In the general case where there is a translation associated with any infinitesimal closed contour, one can say that the given space is different from Euclidean space in two respects: 1) by a curvature in the sense of Riemann, which results in rotation; 2) by a torsion, which results in the translation.

In a space with curvature and torsion, the method of moving triples, as in Euclidean space, allows one to build a theory of the curvature of curves (and even of surfaces). A straight line will be characterized by the property of having null (relative) curvature at all of its points; i.e., of preserving the same direction locally. A straight line is no longer necessarily the shortest path from one point to another; it is in spaces devoid of torsion; exceptionally, it can also be in certain special torsion spaces.

A very simple example of this last case is the following. Imagine a space $\mathcal{E}$ that corresponds point by point with a Euclidean space **E**, the correspondence preserving distances. The difference between the two spaces will be as follows: two orthogonal triples originating from two infinitely close points **A** and **A'** of **E** will be parallel when the corresponding triples of **E** can result one from the other by a helicoidal displacement at a given rate in a given sense (righthanded, for example), with the line that connects their origins as axis. Lines of $\mathcal{E}$ then correspond to lines of **E**: they are still geodesics. Space $\mathcal{E}$ thus defined admits a 6-parameter group of transformations; it would be our ordinary space seen by observers all of whose perceptions would be twisted. Mechanically it would correspond to a medium with constant pressure and constant torsion.

I will add that the preceding considerations which, from the point of view of mechanics, are connected with the beautiful work of Mssrs E. and F. Cosserat on the Euclidean action, are also connected with the theory of generalized spaces of H. Weyl and can themselves be extended.

Chapter 3

The Scalar-Tensor Theory of Gravity*

*P. Jordan, von M. Fierz, C. Brans, R. H. Dicke

No. 4172 October 15, 1949 NATURE 637

FORMATION OF THE STARS AND DEVELOPMENT OF THE UNIVERSE

By Prof. PASCUAL JORDAN
Hamburg

Introduction

In a fascinating article which appeared in *Nature* of February 6, under the title "Stellar Evolution and the Expanding Universe", Mr. F. Hoyle has brought forward convincing arguments for the permanent creation of matter in space. Many physicists will find it difficult to accept this hypothesis. For if there is any law which has withstood all changes and revolutions in physics, it is the law of conservation of energy, which according to Einstein's formula $E = mc^2$ is equivalent to the conservation of mass. The same strange conclusion has, during recent years, been formulated by Prof. Pascual Jordan, but with an important modification, whereby the conservation law is not violated. This is achieved by taking account of the loss of gravitational energy connected with the creation of particles. As Jordan's papers do not seem to be known to many English-speaking physicists, I have asked him to write a short report of his work, and the following article is a translation of his article made by my collaborator, Dr. H. S. Green. Max Born

---

Since the formulation by Einstein of the general theory of relativity, many theoretical considerations have been advanced in which models of the universe are derived *deductively* from certain physical principles (expressed in the form of field equations), which are adopted as hypotheses. The treatment here explained is an attempt to approach the problem more cautiously. It is necessary to reckon with the possibility that processes and natural laws are involved which lie outside our past experience, and which must be learnt by the examination of the universe: an essentially *inductive* approach to the cosmological problem must be sought. The dimensional analysis of the empirical facts presents itself as the method appropriate to this task. This method proceeds by the combination of the experimental data to form dimensionless quantities, and the comparison of these dimensionless numbers with regard to their order of magnitude.

Dimensionless constants, the significance of which remains to be explained, are, as is well known, already apparent in atomic physics. Many attempts have been made to elucidate the significance of Sommerfeld's fine-structure constant and the ratio of the masses of the proton and electron. Though Eddington's well-known considerations have not led to the solution of these problems, he has very usefully made apparent their great urgency and importance. The discovery of different kinds of mesons has intensified these problems: a theory which could explain the mass ratios of the different kinds of mesons could probably account also for the two constants already mentioned. The reciprocity theory of Born and Green promises some interesting progress in this direction.

There is, however, an even more significant dimensionless constant in microphysics. If the ratio of the electrostatic and gravitational attractions of an electron and a proton is formed, the number 2×10^{39} results.

Can it reasonably be hoped that this value can ever be established by theoretical considerations? Eddington cherished the optimistic view that it could; but his speculations really only served to make one more acutely conscious of how improbable it is, on the basis of such natural laws as are known, that an explanation could be given of why a dimensionless constant should have just this enormous value, and not some other. It is not so much a question of the particular natural laws known to us to-day, as of the *mathematical type* common to all the known laws of physics. The mathematical type which characterizes physical laws allows the explanation of such natural constants as π, or 4π, or $\sqrt{2}$; but a number of the order of magnitude 10^{40} can never arise from the mathematically simple, fundamental laws of Nature. But should it now be believed that there are, in the laws of Nature, numerical constants which by a meaningless chance have their actual values—which could equally well assume other arbitrary values, without producing a fundamental discord in the harmony of the natural laws? It is an outstanding merit of Eddington that he emphasized that we should never believe in meaningless coincidences.

There are six distinct quantities which comprise the sum of our knowledge of the structure of the universe on a large scale. After the velocity of light, c, and the gravitational constant, $k = 8\pi f/c^2$, where f is the Newtonian gravitational constant, comes thirdly the maximum age of the oldest celestial bodies known to us. From a variety of data and considerations, one arrives at the conclusion that none of all the known stars and systems of stars can have existed for longer than a certain maximum time, which is approximately 4×10^9 years. Thus one has a quantity of the order of magnitude $A = 10^{17}$ sec.

Three constants additional to those already mentioned are furnished by the statistics of the spiral nebulæ. From the enumeration of the nebulæ, and the determination of the masses of a cross-section of typical nebulæ, one obtains the value of the mean mass-density μ throughout the universe; this comes to about 10^{-28} gm. cm.$^{-3}$. The second of the three 'constants of the spiral nebulæ' is the constant α of the Hubble effect. The spectral lines of the very distant nebulæ show a displacement to the red, which may be described simply by the empirical fact that if a spectral line has the wave-length λ, then the displacement $\Delta\lambda$ is proportional to λ, and also to the distance r of the nebula from us, so that one obtains always the same value for $\alpha = c\Delta\lambda/\lambda r$. The third constant of the spiral nebulæ is empirically the worst determined. It seems, however, that the number of those spiral nebulæ the separation of which from us is in the interval $r, r + \Delta r$, is rather less than proportional to $4\pi r^2 \Delta r$, that is, approximately proportional to $4\pi r^2(1 - \text{const.} r^2)\Delta r$. Here the constant defines the radius of the universe R, to the square of which it is inversely proportional.

The assembly of cosmological data is now complete; and it may be inquired what dimensionless constants can be constructed from c, k, A, μ, α and R. Such are αA; R/cA; $k\mu c^2 A^2$; and it is an extraordinary thing that all three are of the order 1. This is the justification for an attempt to interpret these constants by means of the following simple and illuminating picture. The 'radius of curvature' R is interpreted—and is for that reason given such a name—in the sense of Riemannian geometry: instead of the infinite Euclidean space, one has thus a closed, finite, Riemannian space, the volume of which is of the order of magnitude R^3. In the same intuitive way the Hubble effect is interpreted as a Doppler effect. Different interpretations of the Hubble effect have, indeed, often been attempted, but here the intuitive concept of an expanding space is retained without modification. The empirical relation $R = cA$ then means that the radius of the universe increases at a rate which is just of the magnitude of the velocity of light—an attractive and revealing result. Also $\alpha A = 1$ means that this space, which has been expanding with the velocity of light ever since that time which we recognize to be most remote in the history of the universe, must once have been very small.

But what is meant by the fact that the last of the three dimensionless constants mentioned is of the order of magnitude unity? This relation appears already in the well-known model of the universe which was Einstein's bold attempt, in framing the general theory of relativity, to realize the idea of a closed, Riemannian, physical space. Indeed, if M is understood to be the total mass of the universe—which is then of the order of magnitude μR^3—and cA is replaced as previously by R, the equation in question can be written in the form $kM \simeq R$; and Einstein had previously obtained the relation $kM = 4\pi^2 R$, for a closed space with a *time-independent* radius R, on the basis of the relativistic theory of gravitation.

From a consideration of the Hubble effect, however, the concept of a growing universe has been reached; and then the empirical relation $kM \simeq R$ gives a disturbing conclusion. If R is not fixed, but is always growing, then neither can kM be regarded any longer as invariable; and, of the two factors, the gravitational constant k, and the mass M of the universe, at least one must vary likewise with time. An important contribution to the elucidation of this situation is contained in an observation by A. Haas: the relation $kM \simeq R$ can also be written in the form $fM^2/R \simeq Mc^2$, which means that the negative potential energy of gravitation for the whole universe is equal to the sum of the rest-energies of the masses of the stars. This provides a surprising solution of the problem of the universe: it is possible for the total energy of the universe to have the value zero exactly—through the cancelling out of the positive and negative contributions to the energy. The relation $kM \simeq R$ would then appear as a direct consequence of the conservation of energy, which requires that the evolving universe should continue in a sequence of states the total energy of which always has the value zero.

To proceed further, another dimensionless constant is required; those which can be produced from the six cosmological quantities c, k, A, μ, α and R are exhausted by the three dimensionless constants already mentioned. More are obtained if the cosmological quantities are compared with those derived from microphysics. Hitherto the radius of the universe R and the mass of the universe $M \simeq \mu R^3$ have been expressed in centimetres and grams; now they will be expressed in terms of the fundamental units of microphysics. The choice of these units is, of course, made uncertain by the appearance of the unexplained dimensionless constants of microphysics: it is questionable whether one should take, as the natural unit of mass, the mass of the proton, or that of the meson or electron, and whether the natural unit of length should be Bohr's radius of the hydrogen atom, or the electronic radius e^2/m_0c^2, the ratio of which is essentially the square of the fine-structure constant. However, the dimensional ratios with which one has to do are so enormously large that these distinctions are of little consequence; moreover, there is sufficient reason to give preference to the mass m_M of the meson, and to the electronic radius or 'elementary length' $l = 2 \times 10^{-13}$ cm. If the ratio $M : m_M$, which is then of the order of magnitude of the number of elementary particles in the whole universe, is formed, then a number of the colossal magnitude 10^{80} is obtained. Eddington had the boldness to claim that this number should not be simply accepted as something incapable or without need of explanation; he sought, by means of some curious considerations, to give theoretical grounds for supposing that the number of protons in the universe must have the value 2^{256}. The true solution would appear to lie in another direction: an idea which Dirac has put forward in quite a different connexion will now be introduced.

The ratio R/l is of the same order of magnitude, 10^{40}, already encountered in the comparison of the gravitational and electromagnetic attractions between two elementary particles; if these two dimensionless numbers are divided, therefore, a new dimensionless quantity of the gratifying order of magnitude 1 is obtained. The ratio of the two attractive forces is thereby compared, however, with a number which, as is already known, is not constant: the ratio R/l increases proportionally to the age of the universe; it is, in fact, equal to the age of the universe, expressed in terms of the 'elementary time' $\tau = l/c \simeq 10^{-23}$ sec. Also, from the significant discovery of a quotient of the order of magnitude unity, it must be concluded, with Dirac, that the constant of gravitation is in reality not a constant, but inversely proportional to the age A of the universe.

What is now the obstacle to applying this idea also to the ratio M/m_M, and asserting that the number of elementary particles in the universe obviously increases as the square of the age of the universe? Dirac, who was led away from this application by a special train of thought (concerning the formulation of a cosmology with an infinite Euclidean space and an infinite total mass of the universe), was probably influenced by a fear of contradicting the principle of conservation of energy. However, the foregoing consideration has cleared the way in this respect: with the perception that $kM \simeq R$ guarantees the conservation of energy, the complete harmony of all the statements concerning the proportionality of k to A^{-1}, M to A^2, and R to A is attained.

It is, therefore, accepted that there is a continual creation of matter in the space of the universe, and the question arises, how and where this generation of matter, connected with the growth of the universe, occurs. To answer this, the individual stars, instead of the universe as a whole, are now taken as the subject for careful consideration in the light of dimensional analysis. The mass of the sun amounts to 2×10^{33} gm.; there are certainly many stars with still smaller masses, and an established lower limit scarcely exists. There are, however, on the other hand, few stars with very large masses: 10^{35} gm. = 50 suns may be taken as an approximate value for an upper limit to stellar masses which is very seldom exceeded. It may be asked how many protons such a star would contain: the answer is of the order of magnitude 10^{60}.

Here is a fresh opportunity to apply Dirac's principle, from which it may be concluded that this upper limit, too, must be a function of A; indeed, it must then be interpreted as $A^{3/2}$. According to this hypothesis, the mass of a star (on the average) depends on the age attained by the universe at its formation; this formation being settled, the mass of the star (the number of its nucleons) will remain unchanged (apart from secondary processes, as occasional gathering of dust). Therefore, the normal mass of a star which was generated at an age A_0 of the world must be proportional to $A_0^{3/2}$. (In consequence of this, to-day not only the average mass of recently created stars, but also the average mass of now existing stars has an order of magnitude 10^{60}, proportional to $A^{3/2}$.) Naturally, this dependence of the average of stellar masses on the age of the universe necessarily implies a corresponding dependence of the *gravitational constant*, which is then itself a function of the age of the universe; and, in fact, the proportionality $M_{st} = A_0^{3/2}$, inferred in the first instance from a comparison of the empirical numbers 10^{40} (present age of the world) and 10^{60} (present mass of the normal star), is in agreement with and corroborated by the fact that the present theories of the internal constitution of the stars continually presuppose stellar quantities which are proportional to $k^{-3/2}$.

It has been made clear above that the requirement of conservation of energy on a large scale is satisfied, if a creation of matter is assumed which leads to an increase of M proportional to A^2. Nothing is said there, however, of the way in which this creation of matter can occur. Though the possibility of a localization of energy, like that encountered in Maxwell's theory, can no longer be maintained in Einstein's theory of gravitation, it is certain that in some way a rigorous formulation of the principle of conservation of energy must hold such that, not only for the universe as a whole, but also for a finite region, where the creation takes place, a balance of energy is maintained. Consider now the rough model of a homogeneous sphere with mass M_0 and radius R_0 in an approximately flat, Euclidean space. If it satisfies the relation $kM_0 = R_0$ (apart from a certain numerical factor of the order of magnitude unity) then this space has zero total energy: if it were required to effect a transformation to a cloud of gas of virtually infinite radius, then it would be necessary, in order to overcome the gravitational force between the atoms, to expend just as much energy as would be present in the final state as the sum of the rest-energies. The conjecture suggests itself that the cosmic creation of matter does not take place as a diffuse creation of protons, but by the sudden appearance of whole *drops* of matter, for which M_0 and R_0 are chosen so that each drop, even taken by itself, remains true to the principle of conservation of energy from the instant of its formation—in such a way that the total energy has the value zero.

The critical question at once arises, What density must be ascribed to the drop of newly created matter? It can certainly take only the density of the atomic nucleus, which is of the order of the proton mass, contained in a space of the dimensions of the elementary length. Hence the size of the drop is determined as a function of the gravitational constant at any given time: drops are obtained which contain a number of elementary particles to-day of the order of magnitude 10^{60}, and in general proportional to $k^{-3/2}$ or $A^{3/2}$. Each drop which appears in the space of our universe is therefore a star in its own right.

This picture can be worked out in rather more detail. The universe, regarded as a four-dimensional space-time manifold, is, according to the above, a cone: its apex, $A = 0$, marks the starting-point in time, at which also R and M are zero. It may be imagined, however, that this cone stage does not necessarily possess only one apex, but may have a large number of subsidiary apices. A 'cut' t = const. may then in certain circumstances cut the space-time manifold into several unconnected three-dimensional parts: another isolated, smaller space may exist apart from the large universe, which, by the gradual unfolding of time, and also by its own expansion, may begin to coalesce with the large universe. Supposing that in this small 'world' the mass-density remains constant, and the gravitational constant decreases as the inverse square of its age, then the conservation of energy holds there also. If it is supposed further that when, by such a process, the little world reaches the same value of k as attained in the large universe, it can coalesce with the universe, then it will assume a mass of exactly the same value, at the instant of coalescence, as has been recognized to be characteristic of a newly formed star. Such a world may therefore be regarded as the embryonic state of a star.

A careful discussion of the astrophysical facts seems to indicate that the ideas suggested above are well adapted to give a better understanding of many known empirical facts. Instances will not be developed here; the last two of my papers cited below contain a full discussion. I mention only that I suspect that the supernovæ I (which appear to belong to the star population II, and which, one may venture to say, probably account also for the planetary nebulæ) are stars created in this way.

Finally, it may be noted that the theory, sketched here in a purely inductive manner, has received also a quantitative mathematical treatment. For this, of course, a certain generalization of Einstein's theory of gravitation was required, as this theory treats the constant of gravitation essentially as a genuine constant; and the substitution of a variable k (in Dirac's sense), which must then be treated as a scalar field quantity, requires a fundamental generalization of the theory. For this purpose I found, to all appearances, a very natural starting-point in the five-dimensional (or projective) theory of relativity. The generalization in question has also been studied by Einstein and Bergmann, and by Lichnerowicz and Thiry. In a paper by Ludwig and Mueller, it is shown that, with this new and more general form of the theory of relativity, a deductive foundation and quantitative precision can be given to the model sketched above. This applies to the model of the universe on a large scale as well as to the model of a star in its embryonic state. In this way also the correctness and practicability of the present interpretation of the problem of cosmological energy are confirmed.

After having written this article, I had the opportunity of studying the paper by Mr. F. Hoyle mentioned above by Prof. Max Born, and which will be sure to give a new great impetus to cosmological discussion. Several decisive ideas of Hoyle's are in full harmony with my own theory; especially his very convincing arguments against oscillatory models, and concerning the irreversible conversion of hydrogen. But by far the greatest encouragement which I gain from his very interesting discussion is given by his hypothesis of creation of matter. When I put forward in 1939, following Dirac, the idea of an increasing mass M of the universe, I myself believed this to be a rather astonishing hypothesis. Surely this hypothesis will now be discussed earnestly. But there are also considerable differences between Hoyle's theory and my own. The principal point has already been emphasized by Prof. Born. Will Hoyle's thesis (F), namely, the existence of internebular matter, really be accepted by empirical astronomers? I learn from Dr. Baade that he does not believe it.

Bibliography

Jordan, P., *Ann. d. Physik*, **36**, 64 (1939); *Phys. Z.*, **45**, 183 (1944); "Die Herkunft der Sterne" (Stuttgart, 1947); *Ann. d. Physik*, **1**, 219 (1947); *Astro. Nachr.*, **276**, 193 (1948); *Acta Physica Austriaca* (in the press).

Ludwig, G., and Mueller, C., *Ann. d. Physik*, **2**, 76 (1948).

Bergmann, P. G., *Ann. Math.*, **49**, 255 (1948).

Lichnerowicz, A., *C.R. Acad. Sci., Paris*, **222**, 432 (1946).

Thiry, J., *C.R. Acad. Sci., Paris*, **226**, 216 (1948).

Born, M., and Green, H. S., *Nature*, **163**, 201 and 208 (1949).

On the physical interpretation of P. Jordans extended theory of gravitation

by **M. Fierz.**

(11/24/56)

(Translated by D.Fine)

Abstract

The metric of the Jordan theory can be defined through the postulate that point-masses move along geodesics. This postulate is equivalent to another: the Compton wavelength of the elementary particles provides a natural length-scale. The gravitational constant is defined as the ratio of gravitational to inertial mass. Further, there generally follows from the theory a vacuum dielectric constant $\epsilon_0 = 1/\mu_0$, whose dependence on χ depends on the choice of the exponent η introduced by Jordan: $\epsilon_0 = \chi^{1+1/\eta}.(\eta \neq 0)$.

Introduction

It is well-known to be possible to formally combine the equations of the gravitational field and the electromagnetic field by interpreting them as describing a five-dimensional projective space[1]. So that the correct number of field equations still obtain one must normalize the metric components through the side condition

$$J = g_{\mu\nu} X^\mu X^\nu = 1, \tag{1}$$

where X^ν denotes the 5 homgeneous coordinates.

P. JORDAN[2] has suggested extending the theory by dropping the side condition (1) and introducing J as a variable scalar field in the theory. JORDAN assumes that the field equations follow from the 5-dimensional variational principle

$$\delta \int J^\alpha \left(R - \lambda (J_{|\mu} J^{|\mu} / J^2) \right) \sqrt{-g^5}\, d^5X. \tag{2}$$

(P.J. §26 (22)). Here J is the invariant (1), R is the contracted 5-dimensional curvature tensor. α and λ are arbitrary constants. This corresponds in 4 dimensions to the variational principle:

$$\delta \int J^\eta \left(G - (\lambda + 1/2) J_{|j} J^{|j} / J^2 - \frac{J}{2c^2} F_{kl} F^{kl} \right) \sqrt{-g}\, d^4X = 0 \tag{3}$$

(*c.f.* P.J. §27 (22)). Here $\eta = \alpha + 1/2$. JORDAN chooses for η the value 1. As we do not wish to do that it is necessary, in the case $\eta \neq 0$, to set

$$J^\eta = \chi.$$

[1]See for example W. PAULI, Annalen d. Ph. **18**, 305 (1933).

[2]P. JORDAN, Schwerkraft und Weltall, 2. erw. Auflage (Braunschweig 1955), S. 128 ff. We cite in the following P.J. followed by §and the equation number.

Then (3) takes the following form:

$$\delta \int \left\{ \chi \left(G - \zeta \frac{\chi_{|j}\chi^{|j}}{\chi^2} \right) - \frac{\epsilon_0}{2c^2} F_{kl} F^{kl} \right\} \sqrt{-g}\, d^4X = 0 \tag{3'}$$

with

$$\zeta = \frac{\lambda + 1/2}{\eta^2}, \; \epsilon_0 = \chi^{1-1/\eta}.$$

JORDAN takes χ to be the "Gravitational constant" which in this theory becomes variable. Further, he interprets the g_{ik} appearing in G as the metric components of the four-space.

W. PAULI (P.J. §28) has pointed out that through the principle (3), or through the differential equations which follow from it, the metric is far from unambiguously determined. Indeed, one can, through a conformal transformation

$$g^*_{kl} = w(\chi) g_{kl}$$

introduce a new metric which could serve in place of g_{ik} as the "correct" metric. In this, $w(\chi)$ is largely arbitrary. This has to do with the fact that in the variational principle (3) only the electomagnetic field F_{ik} represents matter. The light rays are however in this theory, too, always null geodesics. This property is preserved by conformal transformations. Therefore light does not define a unit of length.

JORDAN's interpretation of g_{ik} as the metric and of χ as gravitational constant is thereby unfounded. He admits this himself (P.J. S. 170). Solely the interpretation of F_{ik} as electromagnetic field $(\mathcal{E}, \mathcal{B})$ may not be doubted, as these quantities are rotations of the potentials.

A decision on what the metric and gravitational constant are to be first becomes possible when real matter, for example point-masses, are introduced into the theory.

The physical interpretation of the quantities appearing in the theory depends upon the form of the additional terms which are to describe matter in the variational principle. When one has decided on the form of these terms, the interpretation is, under very general assumptions, unambiguous. The goal of our work is to demonstrate this.

Whether the Jordan theory describes reality better than the usual relativity theory of Einstein is another question entirely, into which we will not go in more detail here. Admittedly it appears to us that no convincing physical reasons exist to prefer this sort of extension.

PHYSICAL REVIEW VOLUME 124, NUMBER 3 NOVEMBER 1, 1961

Mach's Principle and a Relativistic Theory of Gravitation*

C. BRANS† AND R. H. DICKE
Palmer Physical Laboratory, Princeton University, Princeton, New Jersey
(Received June 23, 1961)

The role of Mach's principle in physics is discussed in relation to the equivalence principle. The difficulties encountered in attempting to incorporate Mach's principle into general relativity are discussed. A modified relativistic theory of gravitation, apparently compatible with Mach's principle, is developed.

INTRODUCTION

IT is interesting that only two ideas concerning the nature of space have dominated our thinking since the time of Descartes. According to one of these pictures, space is an absolute physical structure with properties of its own. This picture can be traced from Descartes vortices[1] through the absolute space of Newton,[2] to the ether theories of the 19th century. The contrary view that the geometrical and inertial properties of space are meaningless for an empty space, that the physical properties of space have their origin in the matter contained therein, and that the only meaningful motion of a particle is motion relative to other matter in the universe has never found its complete expression in a physical theory. This picture is also old and can be traced from the writings of Bishop Berkeley[3] to those of Ernst Mach.[4] These ideas have found a limited expression in general relativity, but it must be admitted that, although in general relativity spatial geometries are affected by mass distributions, the geometry is not uniquely specified by the distribution. It has not yet been possible to specify boundary conditions on the field equations of general relativity which would bring the theory into accord with Mach's principle. Such boundary conditions would, among other things, eliminate all solutions without mass present.

It is necessary to remark that, according to the ideas of Mach, the inertial forces observed locally in an accelerated laboratory may be interpreted as gravitational effects having their origin in distant matter accelerated relative to the laboratory. The imperfect expression of this idea in general relativity can be seen by considering the case of a space empty except for a lone experimenter in his laboratory. Using the traditional, asymptotically Minkowskian coordinate system fixed relative to the laboratory, and assuming a normal laboratory of small mass, its effect on the metric is minor and can be considered in the weak-field approximation. The observer would, according to general relativity, observe normal behavior of his apparatus in accordance with the usual laws of physics. However, also according to general relativity, the experimenter could set his laboratory rotating by leaning out a window and firing his 22-caliber rifle tangentially. Thereafter the delicate gyroscope in the laboratory would continue to point in a direction nearly fixed relative to the direction of motion of the rapidly receding bullet. The gyroscope would rotate relative to the walls of the laboratory. Thus, from the point of view of Mach, the tiny, almost massless, very distant bullet seems to be more important that the massive, nearby walls of the laboratory in determining inertial coordinate frames and the orientation of the gyroscope.[5] It is clear that what is being described here is more nearly an absolute space in the sense of Newton rather than a physical space in the sense of Berkeley and Mach.

The above example poses a problem for us. Apparently, we may assume one of at least three things:

1. that physical space has intrinsic geometrical and inertial properties beyond those derived from the matter contained therein;
2. that the above example may be excluded as nonphysical by some presently unknown boundary condition on the equations of general relativity.
3. that the above physical situation is not correctly described by the equations of general relativity.

These various alternatives have been discussed previously. Objections to the first possibility are mainly philosophical and, as stated previously, go back to the time of Bishop Berkeley. A common inheritance of all present-day physicists from Einstein is an appreciation for the concept of relativity of motion.

As the universe is observed to be nonuniform, it would appear to be difficult to specify boundary conditions which would have the effect of prohibiting unsuitable mass distributions relative to the laboratory *arbitrarily placed*; for could not a laboratory be built near a massive star? Should not the presence of this massive star contribute to the inertial reaction?

The difficulty is brought into sharper focus by con-

* Supported in part by research contracts with the U. S. Atomic Energy Commission and the Office of Naval Research.
† National Science Foundation Fellow; now at Loyola University, New Orleans, Louisiana.
[1] E. T. Whittaker, *History of the Theories of Aether and Electricity* (Thomas Nelson and Sons, New York, 1951).
[2] I. Newton, *Principia Mathematica Philosophiae Naturalis* (1686) (reprinted by University of California Press, Berkeley, California, 1934).
[3] G. Berkeley, *The Principles of Human Knowledge*, paragraphs 111–117, 1710-*De Motu* (1726).
[4] E. Mach, *Conservation of Energy*, note No. 1, 1872 (reprinted by Open Court Publishing Company, LaSalle, Illinois, 1911), and *The Science of Mechanics*, 1883 (reprinted by Open Court Publishing Company, LaSalle, Illinois, 1902), Chap. II, Sec. VI.
[5] Because of the Thirring-Lense effect, [H. Thirring and J. Lense, Phys. Zeits. **19**, 156 (1918)], the rotating laboratory would have a weak effect on the axis of the gyroscope.

sidering the laws of physics, including their quantitative aspects, inside a static massive spherical shell. It is well known that the interior Schwarzschild solution is flat and can be expressed in a coordinate system Minkowskian in the interior. Also, according to general relativity all Minkowskian coordinate systems are equivalent and the mass and radius of the spherical shell have no discernible effects upon the laws of physics as they are observed in the interior. Apparently the spherical shell does not contribute in any discernible way to inertial effects in the interior. What would happen if the mass of the shell were decreased, or its radius increased without limit? It might be remarked also that Komar[6] has attempted, without success, to find suitable boundary- and initial-value conditions for general relativity which would bring into evidence Mach's principle.

The third alternative is the subject of this paper. Actually the objectives of this paper are more limited than the formulation of a theory in complete accord with Mach's principle. Such a program would consist of two parts, the formulation of a suitable field theory and the formulation of suitable boundary- and initial-value conditions for the theory which would make the space geometry depend uniquely upon the matter distribution. This latter part of the problem is treated only partially.

At the end of the last section we shall briefly return again to the problem of the rotating laboratory.

A principle as sweeping as that of Mach, having its origins in matters of philosophy, can be described in the absence of a theory in a qualitative way only. A model of a theory incorporating elements of Mach's principle has been given by Sciama.[7] From simple dimensional arguments[8,9] as well as the discussion of Sciama, it has appeared that, with the assumption of validity of Mach's principle, the gravitational constant G is related to the mass distribution in a uniform expanding universe in the following way:

$$GM/Rc^2 \sim 1. \tag{1}$$

Here M stands for the finite mass of the visible (i.e., causally related) universe, and R stands for the radius of the boundary of the visible universe.

The physical ideas behind Eq. (1) have been given in references 7–9 and can be summarized easily. As stated before, according to Mach's principle the only meaningful motion is that relative to the rest of the matter in the universe, and the inertial reaction experienced in a laboratory accelerated relative to the distant matter of the universe may be interpreted equivalently as a gravitational force acting on a fixed laboratory due to the presence of distant accelerated matter.[7] This interpretation of the inertial reaction carries with it an interesting implication. Consider a test body falling toward the sun. In a coordinate system so chosen that the object is not accelerating, the gravitational pull of the sun may be considered as balanced by another gravitational pull, the inertial reaction.[8] Note that the balance is not disturbed by a doubling of all gravitational forces. Thus the acceleration is determined by the mass distribution in the universe, but is independent of the strength of gravitational interactions. Designating the mass of the sun by m_s and its distance by r enables the acceleration to be expressed according to Newton as $a=Gm_s/r^2$ or, from dimensional arguments, in terms of the mass distribution as $a\sim mRc^2/Mr^2$. Combining the two expressions gives Eq. (1).

This relation has significance in a rough order-of-magnitude manner only, but it suggests that either the ratio of M to R should be fixed by the theory, or alternatively that the gravitational constant observed locally should be variable and determined by the mass distribution about the point in question. The first of these two alternatives is of course, in part, simply the limitation of mass distribution which it might be hoped would result from some boundary condition on the field equations of general relativity. The second alternative is not compatible with the "strong principle of equivalence"[10] and general relativity. The reasons for this will be discussed below.

If the inertial reaction may be interpreted as a gravitational force due to distant accelerated matter, it might be expected that the locally observed values of the inertial masses of particles would depend upon the distribution of matter about the point in question. It should be noted, however, that there is a fundamental ambiguity in a statement of this type, for there is no direct way in which the mass of a particle such as an electron can be compared with that of another at a different space-time point. Mass ratios can be compared at different points, but not masses. On the other hand, gravitation provides another characteristic mass

$$(\hbar c/G)^{\frac{1}{2}}=2.16\times10^{-5}\text{ g}, \tag{2}$$

and the mass ratio, the dimensionless number

$$m(G/\hbar c)^{\frac{1}{2}}\cong 5\times10^{-23}, \tag{3}$$

provides an unambiguous measure of the mass of an electron which can be compared at different space-time points.

It should also be remarked that statements such as "$\hbar$ and c are the same at all space-time points" are in the same way meaningless within the same context until a method of measurement is prescribed. In fact, it should be noted that $\hbar$ and c may be defined to be constant. A set of physical "constants" may be defined as constant if they cannot be combined to form one or

[6] A. Komar, Ph.D. thesis, Princeton University, 1956 (unpublished).
[7] D. W. Sciama, Monthly Notices Roy. Astron. Soc. **113**, 34 (1953); *The Unity of the Universe* (Doubleday & Company, Inc., New York, 1959), Chaps. 7–9.
[8] R. H. Dicke, Am. Scientist **47**, 25 (1959).
[9] R. H. Dicke, Science **129**, 621 (1959).
[10] R. H. Dicke, Am. J. Phys. **29**, 344 (1960).

more dimensionless numbers. The necessity for this limitation is obvious, for a dimensionless number is invariant under a transformation of units and the question of the constancy of such dimensionless numbers is to be settled, not by definition, but by measurements. A set of such independent physical constants which are constant by definition is "complete" if it is impossible to include another without generating dimensionless numbers.

It should be noted that if the number, Eq. (3), should vary with position and $\hbar$ and c are defined as constant, then either m or G, or both, could vary with position. There is no fundamental difference between the alternatives of constant mass or constant G. However, one or the other may be more convenient, for the formal structure of the theory would, in a superficial way, be quite different for the two cases.

To return to Eq. (3), the odd size of this dimensionless number has often been noticed as well as its apparent relation to the large dimensionless numbers of astrophysics. The apparent relation of the square of the reciprocal of this number [Eq. (3)] to the age of the universe expressed as a dimensionless number in atomic time units and the square root of the mass of the visible portion of the universe expressed in proton mass units suggested to Dirac[11] a causal connection that would lead to the value of Eq. (3) changing with time. The significance of Dirac's hypothesis from the standpoint of Mach's principle has been discussed.[8]

Dirac postulated a detailed cosmological model based on these numerical coincidences. This has been criticized on the grounds that it goes well beyond the empirical data upon which it is based.[8] Also in another publication by one of us (R. H. D.), it will be shown that it gives results not in accord with astrophysical observations examined in the light of modern stellar evolutionary theory.

On the other hand, it should be noted that a large dimensionless physical constant such as the reciprocal of Eq. (3) must be regarded as either determined by nature in a completely capricious fashion or else as related to some other large number derived from nature. In any case, it seems unreasonable to attempt to derive a number like 10^{23} from theory as a purely mathematical number involving factors such as $4\pi/3$.

It is concluded therefore, that although the detailed structure of Dirac's cosmology cannot be justified by the weak empirical evidence on which it is based, the more general conclusion that the number [Eq. (3)] varies with time has a more solid basis.

If, in line with the interpretation of Mach's principle being developed, the dimensionless mass ratio given by Eq. (3) should depend upon the matter distribution in the universe, with $\hbar$ and c constant by definition, either the mass m or the gravitational constant, or both, must vary. Although these are alternative descriptions of the same physical situation, the formal structure of the theory would be very different for the two cases. Thus, for example, it can be easily shown that uncharged spinless particles whose masses are position dependent no longer move on geodesics of the metric. (See Appendix I.) Thus, the definition of the metric tensor is different for the two cases. The two metric tensors are connected by a conformal transformation.

The arbitrariness in the metric tensor which results from the indefiniteness in the choice of units of measure raises questions about the physical significance of Riemannian geometry in relativity.[12] In particular the 14 invariants which characterize the space are generally not invariant under a conformal transformation interpreted as a redefinition of the metric tensor in the *same* space.[13] Matters are even worse, for a more general redefinition of the units of measure can be used to reduce all 14 invariants to zero. It should be said that these remarks should not be interpreted as casting doubt on the correctness or usefulness of Riemannian geometry in relativity, but rather that each such geometry is but a particular representation of the theory. It would be expected that the physical content of the theory should be contained in the invariants of the group of position-dependent transformations of units and coordinate transformations. The usual invariants of Riemannian geometry are not invariants under this wider group.

In general relativity the representation is one in which units are chosen so that atoms are described as having physical properties independent of location. It is assumed that this choice is possible!

In accordance with the above, a particular choice of units is made with the realization that the choice is arbitrary and without an invariant significance. The theoretical structure appears to be simpler if one defines the inertial masses of elementary particles to be constant and permits the gravitational constant to vary. It should be noted that this is possible only if the mass ratios of elementary particles are constant. There may be reasonable doubt about this.[9,10] On the other hand, it would be expected that such quantities as particle mass ratios or the fine-structure constant, if they depend upon mass distributions in the universe, would be much less sensitive in their dependence[9] rather than the number given by Eq. (3) and their variation could be neglected in a first crude theory. Also it should be remarked that the requirements of the approximate constancy of the ratio of inertial to passive gravitational mass,[14] and the extremely stringent requirement of spatial isotropy,[15] impose conditions so severe that it has been found to be difficult, if not impossible, to

[11] P. A. M. Dirac, Proc. Roy. Soc. (London) **A165**, 199 (1938).

[12] E. P. Wigner has questioned the physical significance of Riemannian geometry on other grounds [Relativity Seminar, Stevens Institute, May 9, 1961 (unpublished)].

[13] B. Hoffman, Phys. Rev. **89**, 49 (1953).

[14] R. Eötvös, Ann. Physik **68**, 11 (1922).

[15] V. W. Hughes, H. G. Robinson, and V. Beltran-Lopez, Phys. Rev. Letters **4**, 342 (1960).

construct a satisfactory theory with a variable fine-structure constant.

It should be emphasized that the above argument involving the large dimensionless numbers, Eq. (3), does not concern Mach's principle directly, but that Mach's principle and the assumption of a gravitational "constant" dependent upon mass distributions gives a reasonable explanation for varying "constants."

It would be expected that both nearby and distant matter should contribute to the inertial reaction experienced locally. If the theory were linear, which one does not expect, Eq. (1) would suggest that it is the reciprocal of the gravitational constant which is determined locally as a linear superposition of contributions from the matter in the universe which is causally connected to the point in question. This can be expressed in a somewhat symbolic equation:

$$G^{-1}\sim\sum_i(m_i/r_ic^2), \tag{4}$$

where the sum is over all the matter which can contribute to the inertial reaction. This equation can be given an exact meaning only after a theory has been constructed. Equation (4) is also a relation from Sciama's theory.

It is necessary to say a few words about the equivalence principle as it is used in general relativity and as it relates to Mach's principle. As it enters general relativity, the equivalence principle is more than the assumption of the local equivalence of a gravitational force and an acceleration. Actually, in general relativity it is assumed that the laws of physics, including numerical content (i.e., dimensionless physical constants), as observed locally in a freely falling laboratory, are independent of the location in time or space of the laboratory. This is a statement of the "strong equivalence principle."[9,10] The interpretation of Mach's principle being developed here is obviously incompatible with strong equivalence. The local equality of all gravitational accelerations (to the accuracy of present experiments) is the "weak equivalence principle." It should be noted that it is the "weak equivalence principle" that receives strong experimental support from the Eötvös experiment.

Before attempting to formulate a theory of gravitation which is more satisfactory from the standpoint of Mach's principle than general relativity, the physical ideas outlined above, and the assumptions being made, will be summarized:

1. An approach to Mach's principle which attempts, with boundary conditions, to allow only those mass distributions which produce the "correct" inertial reaction seems foredoomed, for there do exist large localized masses in the universe (e.g., white dwarf stars) and a laboratory could, in principle, be constructed near such a mass. Also it appears to be possible to modify the mass distribution. For example, a massive concrete spherical shell could be constructed with the laboratory in its interior.

2. The contrary view is that locally observed inertial reactions depend upon the mass distribution of the universe about the point of observation and consequently the quantitative aspects of locally observed physical laws (as expressed in the physical "constants") are position dependent.

3. It is possible to reduce the variation of physical "constants" required by this interpretation of Mach's principle to that of a single parameter, the gravitational "constant."

4. The separate but related problem posed by the existence of very large dimensionless numbers representing quantitative aspects of physical laws is clarified by noting that these large numbers involve G and that they are of the same order of magnitude as the large numbers characterizing the size and mass distribution of the universe.

5. The "strong principle of equivalence" upon which general relativity rests is incompatible with these ideas. However, it is only the "weak principle" which is directly supported by the very precise experiments of Eötvös.

A THEORY OF GRAVITATION BASED ON A SCALAR FIELD IN A RIEMANNIAN GEOMETRY

The theory to be developed represents a generalization of general relativity. It is not a completely geometrical theory of gravitation, as gravitational effects are described by a scalar field in a Riemannian manifold. Thus, the gravitational effects are in part geometrical and in part due to a scalar interaction. There is a formal connection between this theory and that of Jordan,[16] but there are differences and the physical interpretation is quite different. For example, the aspect of mass creation[17] in Jordan's theory is absent from this theory.

In developing this theory we start with the "weak principle of equivalence." The great accuracy of the Eötvös experiment suggests that the motion of uncharged test particles in this theory should be, as in general relativity, a geodesic in the four-dimensional manifold.

With the assumption that only the gravitational "constant" (or active gravitational masses) vary with position, the laws of physics (exclusive of gravitation) observed in a freely falling laboratory should be unaffected by the rest of the universe as long as self-gravitational fields are negligible. The theory should be constructed in such a way as to exhibit this effect.

If the gravitational "constant" is to vary, it should be

[16] P. Jordon, *Schwerkraft and Weltall* (Friedrich Vieweg and Sohn, Braunschweig, 1955); Z. Physik **157**, 112 (1959). In this second reference, Jordan has taken cognizance of the objections of Fierz (see reference 19) and has written his variational principle in a form which differs in only two respects from that expressed in Eq. (16). See also reference 20.

[17] For a discussion of this, see H. Bondi, *Cosmology*, 2nd edition, 1960.

a function of some scalar field variable. The contracted metric tensor is a constant and devoid of interest. The scalar curvature and the other scalars formed from the curvature tensor are also devoid of interest as they contain gradients of the metric tensor components, and fall off more rapidly than r^{-1} from a mass source. Thus such scalars are determined primarily by nearby mass distributions rather than by distant matter.

As the scalars of general relativity are not suitable, a new scalar field is introduced. The primary function of this field is the determination of the local value of the gravitational constant.

In order to generalize general relativity, we start with the usual variational principle of general relativity from which the equations of motion of matter and nongravitational fields are obtained as well as the Einstein field equation, namely,[18]

$$0=\delta\int[R+(16\pi G/c^4)L](-g)^{\frac{1}{2}}d^4x. \tag{5}$$

Here, R is the scalar curvature and L is the Lagrangian density of matter including all nongravitational fields.

In order to generalize Eq. (5) it is first divided by G, and a Lagrangian density of a scalar field ϕ is added inside the bracket. G is assumed to be a function of ϕ. Remembering the discussion in connection with Eq. (4), it would be reasonable to assume that G^{-1} varies as ϕ, for then a simple wave equation for ϕ with a scalar matter density as source would give an equation roughly the same as (4).

The required generalization of Eq. (6) is clearly

$$0=\delta\int[\phi R+(16\pi/c^4)L-\omega(\phi_{,i}\phi^{,i}/\phi)](-g)^{\frac{1}{2}}d^4x. \tag{6}$$

Here ϕ plays a role analogous to G^{-1} and will have the dimensions $ML^{-3}T^2$. The third term is the usual Lagrangian density of a scalar field, and the scalar in the denominator has been introduced to permit the constant ω to be dimensionless. In any sensible theory ω must be of the general order of magnitude of unity.

It should be noted that the term involving the Lagrangian density of matter in Eq. (6) is identical with that in Eq. (5). Thus the equations of motion of matter in a given externally determined metric field are the same as in general relativity. The difference between the two theories lies in the gravitational field equations which determine g_{ij}, rather than in the equations of motion in a given metric field.

It is evident, therefore, that, as in general relativity, the energy-momentum tensor of matter must have a vanishing covariant divergence,

$$T^{ij}{}_{;j}=0, \tag{7}$$

where

$$T^{ij}=[2/(-g)^{\frac{1}{2}}](\partial/\partial g_{ij})[(-g)^{\frac{1}{2}}L]. \tag{8}$$

It is assumed that L does not depend explicitly upon derivatives of g_{ij}.

Jordan's theory has been criticized by Fierz[19] on the grounds that the introduction of matter into the theory required further assumptions concerning the standards of length and time. Further, the mass creation aspects of this theory and the nonconservation of the energy-momentum tensor raise serious questions about the significance of the energy-momentum tensor. To make it clear that this objection cannot be raised against this version of the theory, we hasten to point out that L is assumed to be the normal Lagrangian density of matter, a function of matter variables and of g_{ij} only, *not* a function of ϕ. It is a well-known result that for *any* reasonable metric field distribution g_{ij} (a distribution which need not be a solution of the field equations of g_{ij}), the matter equations of motion, obtained by varying matter variables in Eq. (6), are such that Eq. (7) is satisfied with T^{ij} defined by Eq. (8). Thus Eq. (7) is satisfied and this theory does not contain a mass creation principle.

The wave equation for ϕ is obtained in the usual way by varying ϕ and $\phi_{,i}$ in Eq. (6). This gives

$$2\omega\phi^{-1}\Box\phi-(\omega/\phi^2)\phi^{,i}\phi_{,i}+R=0. \tag{9}$$

Here the generally covariant d'Alembertian $\Box$ is defined to be the covariant divergence of $\phi^{,i}$:

$$\Box\phi=\phi^{,i}{}_{;i}=(-g)^{-\frac{1}{2}}[(-g)^{\frac{1}{2}}\phi^{,i}]_{,i}. \tag{10}$$

From the form of Eq. (9), it is evident that ϕR and the Lagrangian density of ϕ serves as the source term for the generation of ϕ waves. Remarkably enough, as will be shown below, this equation can be transformed so as to make the source term appear as the contracted energy-momentum tensor of matter alone. Thus, in accordance with the requirements of Mach's principle, ϕ has as its sources the matter distribution in space.

By varying the components of the metric tensor and their first derivatives in Eq. (6), the field equations for the metric field are obtained. This is the analog of the Einstein field equation and is

$$R_{ij}-\tfrac{1}{2}g_{ij}R=(8\pi\phi^{-1}/c^4)T_{ij}+(\omega/\phi^2)(\phi_{,i}\phi_{,j}-\tfrac{1}{2}g_{ij}\phi_{,k}\phi^{,k})+\phi^{-1}(\phi_{,i;j}-g_{ij}\Box\phi). \tag{11}$$

The left side of Eq. (11) is completely familiar and needs no comment. Note that the first term on the right is the usual source term of general relativity, but with the variable gravitational coupling parameter ϕ^{-1}. Note also that the second term is the energy-momentum tensor of the scalar field, also coupled with the gravitational coupling ϕ^{-1}. The third term is foreign and results from the presence of second derivatives of the metric

[18] L. Landau and E. Liftschitz, *Classical Theory of Fields* (Addison-Wesley Publishing Company, Reading, Massachusetts, 1951).

[19] M. Fierz, Helv. Phys. Acta. **29**, 128 (1956).

tensor in R in Eq. (6). These second derivatives are eliminated by integration by parts to give a divergence and the extra terms. It should be noted that when the first term dominates the right side of Eq. (11), the equation differs from Einstein's field equation by the presence of a variable gravitational constant only.

While the "extra" terms in Eq. (12) may at first seem strange, their role is essential. They are needed if Eq. (7) is to be consistent with Eqs. (9) and (11). This can be seen by multiplying Eq. (11) by ϕ and then taking the covariant divergence of the resulting equation. The divergence of these two terms cancels the term $\phi_{,i}R_j{}^i=\phi^{;i}R_{ji}$. To show this, use is made of the well-known property of the full curvature tensor that it serves as a commutator for two successive gradient operations applied to an arbitrary vector.

If Eq. (11) is contracted there results

$$-R=(8\pi\phi^{-1}/c^4)T-(\omega/\phi^2)\phi_{,k}\phi^{,k}-3\phi^{-1}\Box\phi. \quad (12)$$

Equation (12) can be combined with Eq. (9) to give a new wave equation for ϕ[20]:

$$\Box\phi=[8\pi/(3+2\omega)c^4]T. \quad (13)$$

With the sign convention

$$ds^2=g_{ij}dx^idx^j \quad \text{and} \quad g_{00}<0,$$

for a fluid

$$T_{ij}=-(p+\epsilon)u_iu_j+pg_{ij}, \quad (14)$$

so that

$$T=-\epsilon+3p, \quad (15)$$

where ϵ is the energy density of the matter in comoving coordinates and p is the pressure in the fluid. With this sign convention and ω positive, the contribution to ϕ from a local mass is positive. Note, however, that there is no direct electromagnetic contribution to T, as the contracted energy-momentum tensor of an electromagnetic field is identically zero. However, bound electromagnetic energy does contribute indirectly through the stress terms in other fields, the stresses being necessary to confine the electromagnetic field.[21] In conclusion, ω must be positive if the contribution to the inertial reaction from nearby matter is to be positive.

[20] There are but two formal differences between the field equations of this theory and those of the particular form of Jordan's theory given in Z. Physik **157**, 112 (1959). First, Jordan has defined his scalar field variable reciprocal to ϕ. Thus, the simple wave character of the scalar field equation [Eq. (13)] is not so clear and the physical arguments based on Mach's principle and leading to Eq. (4) have not been satisfied. Second, as a result of its outgrowth from his five-dimensional theory, Jordan has limited his matter variables to those of the electromagnetic field.

[21] C. Misner and P. Putnam, Phys. Rev. **116**, 1045 (1959).

Chapter 4

Yang-Mills' Deepest Insight and Its Relation to Gravity*

C. N. Yang, R. L. Mills, T. D. Lee, R. Utiyama, T. W. B. Kibble

PHYSICAL REVIEW VOLUME 96, NUMBER 1 OCTOBER 1, 1954

Conservation of Isotopic Spin and Isotopic Gauge Invariance*

C. N. YANG † AND R. L. MILLS
Brookhaven National Laboratory, Upton, New York

(Received June 28, 1954)

It is pointed out that the usual principle of invariance under isotopic spin rotation is not consistant with the concept of localized fields. The possibility is explored of having invariance under local isotopic spin rotations. This leads to formulating a principle of isotopic gauge invariance and the existence of a **b** field which has the same relation to the isotopic spin that the electromagnetic field has to the electric charge. The **b** field satisfies nonlinear differential equations. The quanta of the **b** field are particles with spin unity, isotopic spin unity, and electric charge $\pm e$ or zero.

INTRODUCTION

THE conservation of isotopic spin is a much discussed concept in recent years. Historically an isotopic spin parameter was first introduced by Heisenberg[1] in 1932 to describe the two charge states (namely neutron and proton) of a nucleon. The idea that the neutron and proton correspond to two states of the same particle was suggested at that time by the fact that their masses are nearly equal, and that the light stable even nuclei contain equal numbers of them. Then in 1937 Breit, Condon, and Present pointed out the approximate equality of $p-p$ and $n-p$ interactions in the ^1S state.[2] It seemed natural to assume that this equality holds also in the other states available to both the $n-p$ and $p-p$ systems. Under such an assumption one arrives at the concept of a total isotopic spin[3] which is conserved in nucleon-nucleon interactions. Experi-

* Work performed under the auspices of the U. S. Atomic Energy Commission.

† On leave of absence from the Institute for Advanced Study, Princeton, New Jersey.

[1] W. Heisenberg, Z. Physik **77**, 1 (1932).

[2] Breit, Condon, and Present, Phys. Rev. **50**, 825 (1936). J. Schwinger pointed out that the small difference may be attributed to magnetic interactions [Phys. Rev. **78**, 135 (1950)].

[3] The total isotopic spin **T** was first introduced by E. Wigner, Phys. Rev. **51**, 106 (1937); B. Cassen and E. U. Condon, Phys. Rev. **50**, 846 (1936).

ments in recent years[4] on the energy levels of light nuclei strongly suggest that this assumption is indeed correct, An implication of this is that all strong interactions such as the pion-nucleon interaction, must also satisfy the same conservation law. This and the knowledge that there are three charge states of the pion, and that pions can be coupled to the nucleon field *singly*, lead to the conclusion that pions have isotopic spin unity. A direct verification of this conclusion was found in the experiment of Hildebrand[5] which compares the differential cross section of the process $n+p\rightarrow\pi^0+d$ with that of the previously measured process $p+p\rightarrow\pi^++d$.

The conservation of isotopic spin is identical with the requirement of invariance of all interactions under isotopic spin rotation. This means that when electromagnetic interactions can be neglected, as we shall hereafter assume to be the case, the orientation of the isotopic spin is of no physical significance. The differentiation between a neutron and a proton is then a purely arbitrary process. As usually conceived, however, this arbitrariness is subject to the following limitation: once one chooses what to call a proton, what a neutron, at one space-time point, one is then not free to make any choices at other space-time points.

It seems that this is not consistent with the localized field concept that underlies the usual physical theories. In the present paper we wish to explore the possibility of requiring all interactions to be invariant under *independent* rotations of the isotopic spin at all space-time points, so that the relative orientation of the isotopic spin at two space-time points becomes a physically meaningless quantity (the electromagnetic field being neglected).

We wish to point out that an entirely similar situation arises with respect to the ordinary gauge invariance of a charged field which is described by a complex wave function ψ. A change of gauge[6] means a change of phase factor $\psi\rightarrow\psi'$, $\psi'=(\exp i\alpha)\psi$, a change that is devoid of any physical consequences. Since ψ may depend on x, y, z, and t, the relative phase factor of ψ at two different space-time points is therefore completely arbitrary. In other words, the arbitrariness in choosing the phase factor is local in character.

We define *isotopic gauge* as an arbitrary way of choosing the orientation of the isotopic spin axes at all space-time points, in analogy with the electromagnetic gauge which represents an arbitrary way of choosing the complex phase factor of a charged field at all space-time points. We then propose that all physical processes (not involving the electromagnetic field) be invariant under an isotopic gauge transformation, $\psi\rightarrow\psi'$, $\psi'=S^{-1}\psi$, where S represents a space-time dependent isotopic spin rotation.

To preserve invariance one notices that in electrodynamics it is necessary to counteract the variation of α with x, y, z, and t by introducing the electromagnetic field A_μ which changes under a gauge transformation as

$$A_\mu'=A_\mu+\frac{1}{e}\frac{\partial\alpha}{\partial x_\mu}.$$

In an entirely similar manner we introduce a B field in the case of the isotopic gauge transformation to counteract the dependence of S on x, y, z, and t. It will be seen that this natural generalization allows for very little arbitrariness. The field equations satisfied by the twelve independent components of the B field, which we shall call the **b** field, and their interaction with any field having an isotopic spin are essentially fixed, in much the same way that the free electromagnetic field and its interaction with charged fields are essentially determined by the requirement of gauge invariance.

In the following two sections we put down the mathematical formulation of the idea of isotopic gauge invariance discussed above. We then proceed to the quantization of the field equations for the **b** field. In the last section the properties of the quanta of the **b** field are discussed.

ISOTOPIC GAUGE TRANSFORMATION

Let ψ be a two-component wave function describing a field with isotopic spin $\frac{1}{2}$. Under an isotopic gauge transformation it transforms by

$$\psi=S\psi', \tag{1}$$

where S is a 2×2 unitary matrix with determinant unity. In accordance with the discussion in the previous section, we require, in analogy with the electromagnetic case, that all derivatives of ψ appear in the following combination:

$$(\partial_\mu-i\epsilon B_\mu)\psi.$$

B_μ are 2×2 matrices such that[7] for $\mu=1$, 2, and 3, B_μ is Hermitian and B_4 is anti-Hermitian. Invariance requires that

$$S(\partial_\mu-i\epsilon B_\mu')\psi'=(\partial_\mu-i\epsilon B_\mu)\psi. \tag{2}$$

Combining (1) and (2), we obtain the isotopic gauge transformation on B_μ:

$$B_\mu'=S^{-1}B_\mu S+\frac{i}{\epsilon}S^{-1}\frac{\partial S}{\partial x_\mu}. \tag{3}$$

The last term is similar to the gradiant term in the gauge transformation of electromagnetic potentials. In analogy to the procedure of obtaining gauge invariant field strengths in the electromagnetic case, we

[4] T. Lauritsen, Ann. Rev. Nuclear Sci. 1, 67 (1952); D. R. Inglis, Revs. Modern Phys. 25, 390 (1953).
[5] R. H. Hildebrand, Phys. Rev. 89, 1090 (1953).
[6] W. Pauli, Revs. Modern Phys. 13, 203 (1941).

[7] We use the conventions $\hbar=c=1$, and $x_4=it$. Bold-face type refers to vectors in isotopic space, not in space-time.

define now

$$F_{\mu\nu}=\frac{\partial B_\mu}{\partial x_\nu}-\frac{\partial B_\nu}{\partial x_\mu}+i\epsilon(B_\mu B_\nu-B_\nu B_\mu). \quad (4)$$

One easily shows from (3) that

$$F_{\mu\nu}'=S^{-1}F_{\mu\nu}S \quad (5)$$

under an isotopic gauge transformation.‡ Other simple functions of B than (4) do not lead to such a simple transformation property.

The above lines of thought can be applied to any field ψ with arbitrary isotopic spin. One need only use other representations S of rotations in three-dimensional space. It is reasonable to assume that different fields with the same total isotopic spin, hence belonging to the same representation S, interact with the same matrix field B_μ. (This is analogous to the fact that the electromagnetic field interacts in the same way with any charged particle, regardless of the nature of the particle. If different fields interact with different and independent B fields, there would be more conservation laws than simply the conservation of total isotopic spin.) To find a more explicit form for the B fields and to relate the B_μ's corresponding to different representations S, we proceed as follows.

Equation (3) is valid for any S and its corresponding B_μ. Now the matrix $S^{-1}\partial S/\partial x_\mu$ appearing in (3) is a linear combination of the isotopic spin "angular momentum" matrices T^i ($i=1, 2, 3$) corresponding to the isotopic spin of the ψ field we are considering. So B_μ itself must also contain a linear combination of the matrices T^i. But any part of B_μ in addition to this, $\bar{B}_\mu$, say, is a scalar or tensor combination of the T's, and must transform by the homogeneous part of (3), $\bar{B}_\mu'=S^{-1}\bar{B}_\mu S$. Such a field is extraneous; it was allowed by the very general form we assumed for the B field, but is irrelevant to the question of isotopic gauge. Thus the relevant part of the B field is of the form

$$B_\mu=2\mathbf{b}_\mu\cdot\mathbf{T}. \quad (6)$$

(Bold-face letters denote three-component vectors in isotopic space.) To relate the $\mathbf{b}_\mu$'s corresponding to different representations S we now consider the product representation $S=S^{(a)}S^{(b)}$. The B field for the combination transforms, according to (3), by

$$B_\mu'=[S^{(b)}]^{-1}[S^{(a)}]^{-1}BS^{(a)}S^{(b)}$$
$$+\frac{i}{\epsilon}[S^{(a)}]^{-1}\frac{\partial S^{(a)}}{\partial x_\mu}+\frac{i}{\epsilon}[S^{(b)}]^{-1}\frac{\partial S^{(b)}}{\partial x_\mu}.$$

But the sum of $B_\mu^{(a)}$ and $B_\mu^{(b)}$, the B fields corresponding to $S^{(a)}$ and $S^{(b)}$, transforms in exactly the same way, so that

$$B_\mu=B_\mu^{(a)}+B_\mu^{(b)}$$

(plus possible terms which transform homogeneously, and hence are irrelevant and will not be included). Decomposing $S^{(a)}S^{(b)}$ into irreducible representations, we see that the twelve-component field $\mathbf{b}_\mu$ in Eq. (6) is the same for all representations.

To obtain the interaction between any field ψ of arbitrary isotopic spin with the $\mathbf{b}$ field one therefore simply replaces the gradiant of ψ by

$$(\partial_\mu-2i\epsilon\mathbf{b}_\mu\cdot\mathbf{T})\psi, \quad (7)$$

where T^i ($i=1, 2, 3$), as defined above, are the isotopic spin "angular momentum" matrices for the field ψ.

We remark that the nine components of $\mathbf{b}_\mu$, $\mu=1, 2, 3$ are real and the three of $\mathbf{b}_4$ are pure imaginary. The isotopic-gauge covariant field quantities $F_{\mu\nu}$ are expressible in terms of $\mathbf{b}_\mu$:

$$F_{\mu\nu}=2\mathbf{f}_{\mu\nu}\cdot\mathbf{T}, \quad (8)$$

where

$$\mathbf{f}_{\mu\nu}=\frac{\partial \mathbf{b}_\mu}{\partial x_\nu}-\frac{\partial \mathbf{b}_\nu}{\partial x_\mu}-2\epsilon\mathbf{b}_\mu\times\mathbf{b}_\nu. \quad (9)$$

$\mathbf{f}_{\mu\nu}$ transforms like a vector under an isotopic gauge transformation. Obviously the same $\mathbf{f}_{\mu\nu}$ interact with all fields ψ irrespective of the representation S that ψ belongs to.

The corresponding transformation of $\mathbf{b}_\mu$ is cumbersome. One need, however, study only the infinitesimal isotopic gauge transformations,

$$S=1-2i\mathbf{T}\cdot\delta\boldsymbol{\omega}.$$

Then

$$\mathbf{b}_\mu'=\mathbf{b}_\mu+2\mathbf{b}_\mu\times\delta\boldsymbol{\omega}+\frac{1}{\epsilon}\frac{\partial}{\partial x_\mu}\delta\boldsymbol{\omega}. \quad (10)$$

FIELD EQUATIONS

To write down the field equations for the $\mathbf{b}$ field we clearly only want to use isotopic gauge invariant quantities. In analogy with the electromagnetic case we therefore write down the following Lagrangian density:[8]

$$-\tfrac{1}{4}\mathbf{f}_{\mu\nu}\cdot\mathbf{f}_{\mu\nu}.$$

Since the inclusion of a field with isotopic spin $\frac{1}{2}$ is illustrative, and does not complicate matters very much, we shall use the following total Lagrangian density:

$$\mathcal{L}=-\tfrac{1}{4}\mathbf{f}_{\mu\nu}\cdot\mathbf{f}_{\mu\nu}-\bar{\psi}\gamma_\mu(\partial_\mu-i\epsilon\boldsymbol{\tau}\cdot\mathbf{b}_\mu)\psi-m\bar{\psi}\psi. \quad (11)$$

One obtains from this the following equations of motion:

$$\begin{aligned}&\partial\mathbf{f}_{\mu\nu}/\partial x_\nu+2\epsilon(\mathbf{b}_\nu\times\mathbf{f}_{\mu\nu})+\mathbf{J}_\mu=0,\\&\gamma_\mu(\partial_\mu-i\epsilon\boldsymbol{\tau}\cdot\mathbf{b}_\mu)\psi+m\psi=0,\end{aligned} \quad (12)$$

‡ *Note added in proof.*—It may appear that B_μ could be introduced as an auxiliary quantity to accomplish invariance, but need not be regarded as a field variable by itself. It is to be emphasized that such a procedure violates the principle of invariance. Every quantity that is not a pure numeral (like 2, or M, or any definite representation of the γ matrices) should be regarded as a dynamical variable, and should be varied in the Lagrangian to yield an equation of motion. Thus the quantities B_μ must be regarded as independent fields.

[8] Repeated indices are summed over, except where explicitly stated otherwise. Latin indices are summed from 1 to 3, Greek ones from 1 to 4.

where

$$\mathbf{J}_\mu = i\epsilon\bar{\psi}\gamma_\mu \boldsymbol{\tau}\psi. \quad (13)$$

The divergence of $\mathbf{J}_\mu$ does not vanish. Instead it can easily be shown from (13) that

$$\partial \mathbf{J}_\mu/\partial x_\mu = -2\epsilon \mathbf{b}_\mu \times \mathbf{J}_\mu. \quad (14)$$

If we define, however,

$$\mathfrak{J}_\mu = \mathbf{J}_\mu + 2\epsilon \mathbf{b}_\nu \times \mathbf{f}_{\mu\nu}, \quad (15)$$

then (12) leads to the equation of continuity,

$$\partial \mathfrak{J}_\mu/\partial x_\mu = 0. \quad (16)$$

$\mathfrak{J}_{1,2,3}$ and $\mathfrak{J}_4$ are respectively the isotopic spin current density and isotopic spin density of the system. The equation of continuity guarantees that the total isotopic spin

$$\mathbf{T} = \int \mathfrak{J}_4 d^3x$$

is independent of time and independent of a Lorentz transformation. It is important to notice that $\mathfrak{J}_\mu$, like $\mathbf{b}_\mu$, does not transform exactly like vectors under isotopic space rotations. But the total isotopic spin,

$$\mathbf{T} = -\int \frac{\partial \mathbf{f}_{4i}}{\partial x_i} d^3x,$$

is the integral of the divergence of $\mathbf{f}_{4i}$, which transforms like a true vector under isotopic spin space rotations. Hence, under a general isotopic gauge transformation, if $S \to S_0$ on an infinitely large sphere, $\mathbf{T}$ would transform like an isotopic spin vector.

Equation (15) shows that the isotopic spin arises both from the spin-$\frac{1}{2}$ field ($\mathbf{J}_\mu$) and from the $\mathbf{b}_\mu$ field itself. Inasmuch as the isotopic spin is the source of the $\mathbf{b}$ field, this fact makes the field equations for the $\mathbf{b}$ field nonlinear, even in the absence of the spin-$\frac{1}{2}$ field. This is different from the case of the electromagnetic field, which is itself chargeless, and consequently satisfies linear equations in the absence of a charged field.

The Hamiltonian derived from (11) is easily demonstrated to be positive definite in the absence of the field of isotopic spin $\frac{1}{2}$. The demonstration is completely identical with the similar one in electrodynamics.

We must complete the set of equations of motion (12) and (13) by the supplementary condition,

$$\partial \mathbf{b}_\mu/\partial x_\mu = 0, \quad (17)$$

which serves to eliminate the scalar part of the field in $\mathbf{b}_\mu$. This clearly imposes a condition on the possible isotopic gauge transformations. That is, the infinitesimal isotopic gauge transformation $S = 1 - i\boldsymbol{\tau}\cdot\delta\boldsymbol{\omega}$ must satisfy the following condition:

$$2\mathbf{b}_\mu \times \frac{\partial}{\partial x_\mu}\delta\boldsymbol{\omega} + \frac{1}{\epsilon}\frac{\partial^2}{\partial x_\mu^2}\delta\boldsymbol{\omega} = 0. \quad (18)$$

This is the analog of the equation $\partial^2\alpha/\partial x_\mu^2 = 0$ that must be satisfied by the gauge transformation $A_\mu' = A_\mu + e^{-1}(\partial\alpha/\partial x_\mu)$ of the electromagnetic field.

QUANTIZATION

To quantize, it is not convenient to use the isotopic gauge invariant Lagrangian density (11). This is quite similar to the corresponding situation in electrodynamics and we adopt the customary procedure of using a Lagrangian density which is not obviously gauge invariant:

$$\mathfrak{L} = -\frac{1}{2}\frac{\partial \mathbf{b}_\mu}{\partial x_\nu}\cdot\frac{\partial \mathbf{b}_\mu}{\partial x_\nu} + 2\epsilon(\mathbf{b}_\mu \times \mathbf{b}_\nu)\frac{\partial \mathbf{b}_\mu}{\partial x_\nu} - \epsilon^2(\mathbf{b}_\mu \times \mathbf{b}_\nu)^2 + \mathbf{J}_\mu\cdot\mathbf{b}_\mu - \bar{\psi}(\gamma_\mu\partial_\mu + m)\psi. \quad (19)$$

The equations of motion that result from this Lagrangian density can be easily shown to imply that

$$\frac{\partial^2}{\partial x_\nu^2}\mathbf{a} + 2\epsilon \mathbf{b}_\nu \times \frac{\partial}{\partial x_\nu}\mathbf{a} = 0,$$

where

$$\mathbf{a} = \partial \mathbf{b}_\mu/\partial x_\mu.$$

Thus if, consistent with (17), we put on one space-like surface $\mathbf{a} = 0$ together with $\partial\mathbf{a}/\partial t = 0$, it follows that $\mathbf{a} = 0$ at all times. Using this supplementary condition one can easily prove that the field equations resulting from the Lagrangian densities (19) and (11) are identical.

One can follow the canonical method of quantization with the Lagrangian density (19). Defining

$$\boldsymbol{\Pi}_\mu = -\partial \mathbf{b}_\mu/\partial x_4 + 2\epsilon(\mathbf{b}_\mu \times \mathbf{b}_4),$$

one obtains the equal-time commutation rule

$$[b_\mu^i(x), \Pi_\nu^j(x')]_{t=t'} = -\delta_{ij}\delta_{\mu\nu}\delta^3(x - x'), \quad (20)$$

where b_μ^i, $i = 1, 2, 3$, are the three components of $\mathbf{b}_\mu$. The relativistic invariance of these commutation rules follows from the general proof for canonical methods of quantization given by Heisenberg and Pauli.[9]

The Hamiltonian derived from (19) is identical with the one from (11), in virtue of the supplementary condition. Its density is

$$\begin{aligned} H &= H_0 + H_{\text{int}}, \\ H_0 &= -\tfrac{1}{2}\boldsymbol{\Pi}_\mu\cdot\boldsymbol{\Pi}_\mu + \frac{1}{2}\frac{\partial \mathbf{b}_\mu}{\partial x_j}\cdot\frac{\partial \mathbf{b}_\mu}{\partial x_j} + \bar{\psi}(\gamma_j\partial_j + m)\psi, \\ H_{\text{int}} &= 2\epsilon(\mathbf{b}_i \times \mathbf{b}_4)\cdot\boldsymbol{\Pi}_i - 2\epsilon(\mathbf{b}_\mu \times \mathbf{b}_j)\cdot(\partial \mathbf{b}_\mu/\partial x_j) \\ &\quad + \epsilon^2(\mathbf{b}_i \times \mathbf{b}_j)^2 - \mathbf{J}_\mu\cdot\mathbf{b}_\mu. \end{aligned} \quad (21)$$

The quantized form of the supplementary condition is the same as in quantum electrodynamics.

[9] W. Heisenberg and W. Pauli, Z. Physik **56**, 1 (1929).

PROPERTIES OF THE *b* QUANTA

The quanta of the **b** field clearly have spin unity and isotopic spin unity. We know their electric charge too because all the interactions that we proposed must satisfy the law of conservation of electric charge, which is exact. The two states of the nucleon, namely proton and neutron, differ by charge unity. Since they can transform into each other through the emission or absorption of a **b** quantum, the latter must have three charge states with charges $\pm e$ and 0. Any measurement of electric charges of course involves the electromagnetic field, which necessarily introduces a preferential direction in isotopic space at all space-time points. Choosing the isotopic gauge such that this preferential direction is along the z axis in isotopic space, one sees that for the nucleons

$$Q=\text{electric charge}=e(\tfrac{1}{2}+\epsilon^{-1}T^z),$$

and for the **b** quanta

$$Q=(e/\epsilon)T^z.$$

The interaction (7) then fixes the electric charge up to an additive constant for all fields with any isotopic spin:

$$Q=e(\epsilon^{-1}T^z+R). \tag{22}$$

The constants R for two charge conjugate fields must be equal but have opposite signs.[10]

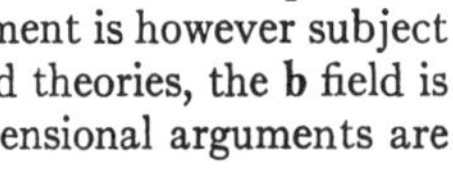

FIG. 1. Elementary vertices for **b** fields and nucleon fields. Dotted lines refer to **b** field, solid lines with arrow refer to nucleon field.

We next come to the question of the mass of the **b** quantum, to which we do not have a satisfactory answer. One may argue that without a nucleon field the Lagrangian would contain no quantity of the dimension of a mass, and that therefore the mass of the **b** quantum in such a case is zero. This argument is however subject to the criticism that, like all field theories, the **b** field is beset with divergences, and dimensional arguments are not satisfactory.

One may of course try to apply to the **b** field the methods for handling infinities developed for quantum electrodynamics. Dyson's approach[11] is best suited for the present case. One first transforms into the interaction representation in which the state vector Ψ

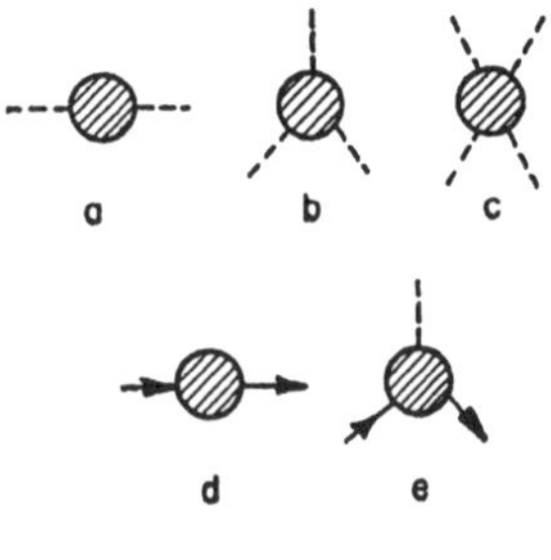

FIG. 2. Primitive divergences.

satisfies

$$i\partial\Psi/\partial t=H_{\text{int}}\Psi,$$

where H_{int} was defined in Eq. (21). The matrix elements of the scattering matrix are then formulated in terms of contributions from Feynman diagrams. These diagrams have three elementary types of vertices illustrated in Fig. 1, instead of only one type as in quantum electrodynamics. The "primitive divergences" are still finite in number and are listed in Fig. 2. Of these, the one labeled a is the one that effects the propagation function of the **b** quantum, and whose singularity determines the mass of the **b** quantum. In electrodynamics, by the requirement of electric charge conservation,[12] it is argued that the mass of the photon vanishes. Corresponding arguments in the **b** field case do not exist[13] even though the conservation of isotopic spin still holds. We have therefore not been able to conclude anything about the mass of the **b** quantum.

A conclusion about the mass of the **b** quantum is of course very important in deciding whether the proposal of the existence of the **b** field is consistent with experimental information. For example, it is inconsistent with present experiments to have their mass less than that of the pions, because among other reasons they would then be created abundantly at high energies and the charged ones should live long enough to be seen. If they have a mass greater than that of the pions, on the other hand, they would have a short lifetime (say, less than 10^{-20} sec) for decay into pions and photons and would so far have escaped detection.

[10] See M. Gell-Mann, Phys. Rev. **92**, 833 (1953).
[11] F. J. Dyson, Phys. Rev. **75**, 486, 1736 (1949).
[12] J. Schwinger, Phys. Rev. **76**, 790 (1949).
[13] In electrodynamics one can formally prove that $G_{\mu\nu}k_\nu=0$, where $G_{\mu\nu}$ is defined by Schwinger's Eq. (A12). ($G_{\mu\nu}A_\nu$ is the current generated through virtual processes by the arbitrary external field A_ν.) No corresponding proof has been found for the present case. This is due to the fact that in electrodynamics the conservation of charge is a consequence of the equation of motion of the electron field alone, quite independently of the electromagnetic field itself. In the present case the **b** field carries an isotopic spin and destroys such general conservation laws.

PHYSICAL REVIEW VOLUME 98, NUMBER 5 JUNE 1, 1955

Conservation of Heavy Particles and Generalized Gauge Transformations

T. D. LEE, *Columbia University, New York, New York*

AND

C. N. YANG, *Institute for Advanced Study, Princeton, New Jersey*

(Received March 2, 1955)

The possibility of a heavy-particle gauge transformation is discussed.

THE conservation laws of nature fall into two distinct categories: those that are related to invariance under space-time displacements and rotations, and those that are not. In the former category there are the conservation laws of momentum, energy, and angular momentum. In the latter category we find the conservation laws of electric charge, of heavy particles, and the approximate conservation laws of isotopic spin, and perhaps others.[1] We notice that the best known within this second category, the conservation of electric charge, is related to invariance under gauge transformations,[2] which expresses the nonmeasurability of the phase of the complex wave function of a charged particle.

We want to ask here whether similar gauge invariances should be related to all conservation laws of the second category. This question has been discussed in connection with the conservation of isotopic spin by Yang and Mills.[3] We wish here to discuss the problem in connection with the conservation of heavy particles.

If we take the conservation of heavy particles to mean invariance under the transformation

$$\psi_N \rightarrow e^{i\alpha}\psi_N, \quad \psi_P \rightarrow e^{i\alpha}\psi_P, \tag{1}$$

for the wave function of the heavy particles (neutrons and protons), a general gauge transformation (heavy-particle gauge transformation) is a transformation like (1) with the phase α an arbitrary function of space-time. Invariance under such a transformation means that the relative phase of the wave function of a heavy particle at two different space-time points is not measurable.

Such a gauge transformation is formally completely identical with the electromagnetic gauge transformation. Invariance under such a transformation therefore necessitates the existence of a neutral vector massless field coupled to all heavy particles. A nucleon would have a "heavy-particle charge" of $+\eta$ in such a field and an antinucleon would have a "heavy-particle charge" of $-\eta$. The force between two massive bodies therefore would contain a contribution from the Coulomb-like repulsion between such "heavy-particle charges." The total force including the gravitational attraction is:

$$\text{Force} = -G(M_1M_2/R^2) + \eta^2(A_1A_2/R^2). \tag{2}$$

Here M_1, M_2, A_1, and A_2 are the inertia masses and mass numbers of the two bodies. There should also be a magnetic-dipole-like interaction between individual nuclei because the nucleons are in constant motion in a nucleus. But in a macroscopic object the nuclear spins average out so that (2) is correct unless the two bodies are spinning at high speeds.

Now the packing fraction of various atoms differ so that M/A varies fractional-wise from substance to substance by $\sim 10^{-3}$. This means that the observed gravitation mass [which contains a contribution from the η^2 term in (2)] divided by the inertia mass would vary fractional-wise from substance to substance by $10^{-3}\eta^2/G(M_P)^2$, where M_P is the mass of the proton[4] Very careful measurements by Eötvös and co-workers. have shown this variation to be $<10^{-8}$. Therefore

$$\eta^2/G(M_P)^2 < 10^{-5}.$$

It may be remarked that since the packing fraction differs most between hydrogen and, say, carbon, Eötvös' experiment could yield a more sensitive detection of η^2 by a factor of 10 if repeated with a comparison of hydrogen and carbon.

The assumption that leads to the above line of reasoning and the force expression (2) is that the phase factor α in (1) should be space-time-dependent. It should be noticed that in addition the assumption has also been made that the transformation that generates the conservation of heavy particles is of the specific form (1).

We wish to thank Dr. J. Robert Oppenheimer for an interesting discussion.

[1] See M. Gell-Mann and A. Pais, Proceedings of the Glasgow Conference, July, 1954 (to be published).
[2] W. Pauli, Revs. Modern Phys. **13**, 203 (1941).
[3] C. N. Yang and R. L. Mills, Phys. Rev. **96**, 191 (1954).
[4] Eötvös, Pekár, and Fekete, Ann. Physik **68**, 11 (1922).

T. D. Lee and C. N. Yang at the Institute for Advanced Study
(Courtesy of the Archives of the Institute for Advanced Study,
Princeton, New Jersey, U.S.A.)

PHYSICAL REVIEW VOLUME 101, NUMBER 5 MARCH 1, 1956

Invariant Theoretical Interpretation of Interaction

RYOYU UTIYAMA*
Institute for Advanced Study, Princeton, New Jersey

(Received July 7, 1955)

Some systems of fields have been considered which are invariant under a certain group of transformations depending on n parameters. A general rule is obtained for introducing a new field in a definite way with a definite type of interaction with the original fields by postulating the invariance of these systems under a wider group derived by replacing the parameters of the original group with a set of arbitrary functions. The transformation character of this new field under the wider group is determined from the invariance postulate. The possible types of the equations of the new fields can be also derived, giving rise to a certain conservation law owing to the invariance. As examples, the electromagnetic, the gravitational and the Yang-Mills fields are reconsidered following this line of approach.

INTRODUCTION

THE form of the interactions between some well known fields can be determined by postulating invariance under a certain group of transformations. For example, let us consider the electromagnetic interaction of a charged field $Q(x)$, $Q^*(x)$. The electromagnetic interaction appears in the Lagrangian through the expressions

$$\frac{\partial Q}{\partial x^\mu}-ieA_\mu Q \quad \text{or} \quad \frac{\partial Q^*}{\partial x^\mu}+ieA_\mu Q^*. \tag{1}$$

The gauge invariance of this system is easily verified in virtue of the combinations of Q, Q^*, and A_μ in (1), if this system is invariant under the phase transformation

$$Q\rightarrow e^{i\alpha}Q, \quad Q^*\rightarrow Q^*e^{-i\alpha}, \quad \alpha=\text{const.} \tag{2}$$

Reversing the argument, the combination (1) can be uniquely introduced by the following line of reasoning. In the first place, let us suppose that the Lagrangian $L(Q,Q_{,\mu})$ is invariant under the constant phase transformation (2). Let us replace this phase transformation with the wider one (gauge transformation) having the phase factor $\alpha(x)$ instead of the constant α. In order to make the Lagrangian still invariant under this wider transformation it is necessary to introduce the electromagnetic field through the combination (1). This combination and the transformation character of A_μ under the gauge transformation can be uniquely determined from the gauge invariance postulate of the Lagrangian $L(Q,Q_{,\mu},A_\mu)$.

This approach was taken by Yang and Mills[1] to introduce their new field $\mathbf{B}_\mu$ which interacts with fields having nonvanishing isotopic spins. The gravitational interaction also can be introduced in this fashion.

It may be worthwhile to investigate this approach for a more general case, for if there is a system of fields $Q^A(x)$ which is invariant under some transformation group depending on parameters $\epsilon_1, \epsilon_2, \cdots\epsilon_n$, then according to the aforementioned viewpoint we may have the possibility of introducing a new field, say $A(x)$, in a definite way. In addition, the transformation character of this new field and the interaction form with the Q's can be determined uniquely.

Let us tentatively call a family of the interactions derived in this way "the interactions of the first class," while other types of interactions are denoted as "the interactions of the second class." The electromagnetic, gravitational and $\mathbf{B}_\mu$-field interactions belong to the first class and the meson-nucleon interaction to the second class, at least at the present stage.

The main purpose of the present paper is to investigate the following problem. Let us consider a system of fields $Q^A(x)$, which is invariant under some transformation group G depending on parameters $\epsilon_1, \epsilon_2, \cdots\epsilon_n$. Suppose that the aforementioned parameter-group G is replaced by a wider group G', derived by replacing the parameters ϵ's by a set of arbitrary functions $\epsilon(x)$'s, and that the system considered is invariant under this wider group G'. Then, can we answer the following questions by using only the postulate of invariance stated above? (1) What kind of field, $A(x)$, is introduced on account of the invariance? (2) How is this new field A transformed under G'? (3) What form does the interaction between the field A and the original field Q take? (4) How can we determine the new Lagrangian $L'(Q,A)$ from the original one $L(Q)$? (5) What type of field equations for A are allowable?

The solution of these problems will be stated in Sec. 1. In Secs. 2, 3, and 4 the well-known examples of the interactions of the first class will be reconsidered following the line of reasoning of Sec. 1. We shall find an analogy between the transformation characters of the electromagnetic field A_μ, the Yang-Mills field $\mathbf{B}_\mu$, and Christoffel's affinity $\Gamma_{\mu\nu}{}^\lambda$ in the theory of the general relativity. Furthermore we shall understand the reason why in the Yang-Mills field strength the quadratic term, $\mathbf{B}_\mu\times\mathbf{B}_\nu$, appears which is quite similar to that occurring in the Riemann-Christoffel tensor $R^\lambda{}_{\mu\nu\rho}$, namely, to the term $\Gamma\Gamma-\Gamma\Gamma$ in R.

In the usual textbooks of general relativity the covariant derivative of any tensor is introduced by

* On leave of absence from the University of Osaka, Osaka, Japan.

[1] C. N. Yang and R. L. Mills, Phys. Rev. 96, 191 (1954).

using the concept of parallel displacement. On the other hand, we shall see in Sec. 4 that the covariant derivative of any tensor or spinor can be derived from the postulate of invariance under the "generalized Lorentz transformations" derived by replacing the six parameters of the usual Lorentz group with a set of six arbitrary functions of x. In deriving such covariant derivatives it is unnecessary to use explicitly the notion of parallel displacement.

Now the above stated classification of the interactions has only a tentative meaning. Some of the interactions of the second class might be translated to the first class if we could find a transformation group by means of which we can derive that interaction following the general scheme in Sec. 1. For example, if the interaction between mesons and nucleons could be reinterpreted in a fashion analogous to those of the first class, then one might presumably be able to get a wider viewpoint for interpreting the interactions between the new unstable particles and the nucleons.

1. GENERAL THEORY

Let us consider a set of fields $Q^A(x)$, $(A=1, 2, \cdots N)$, with the Lagrangian density

$$L(Q^A,Q^A{}_{,\mu}), \quad Q^A{}_{,\mu}=\partial Q^A/\partial x^\mu.$$

Now let us postulate that the action integral referred to some arbitrary four-dimensional domain Ω,

$$I=\int_\Omega Ld^4x,$$

is invariant under the following infinitesimal transformation:

$$\begin{aligned} &Q^A\rightarrow Q^A+\delta Q^A,\\ &\delta Q^A=T_{(a),}{}^A{}_B\,\epsilon^a\,Q_B,\\ &\quad \epsilon^a=\text{infinitesimal parameter}\ (a=1,2,\cdots n),\\ &T_{(a),}{}^A{}_B=\text{constant coefficient.}\end{aligned} \tag{1.1}$$

In addition, the transformation (1.1) is assumed to be a Lie group G depending on the n parameters ϵ^a.

Thus there must be a set of constants $f_b{}^a{}_c$ called the "structure constants," which are defined by

$$\begin{aligned}[T_{(a)},T_{(b)}]^A{}_B&=T_{(a),}{}^A{}_C\,T_{(b)}{}^C{}_B-T_{(b),}{}^A{}_C\,T_{(a),}{}^C{}_B\\ &=f_a{}^c{}_b\,T_{(c),}{}^A{}_B.\end{aligned} \tag{1.2}$$

These constants, $f_b{}^a{}_c$, have the following important properties:

$$\begin{aligned}&f_a{}^m{}_b\,f_m{}^l{}_c+f_b{}^m{}_c\,f_m{}^l{}_a+f_c{}^m{}_a\,f_m{}^l{}_b=0,\\ &f_a{}^c{}_b=-f_b{}^c{}_a.\end{aligned} \tag{1.3}$$

The relations (1.3) can be easily obtained from Jacobi's identity and the definition (1.2).

Now from the invariant character of I under the transformation (1.1) and from the fact that this invariance is always preserved for an arbitrary domain Ω, we have the invariance of the Lagrangian density itself. Namely we have

$$\delta L\equiv\frac{\partial L}{\partial Q^A}\delta Q^A+\frac{\partial L}{\partial Q^A{}_{,\mu}}\delta Q^A{}_{,\mu}\equiv 0. \tag{1.4}$$

The symbol $\equiv$ means that δL must vanish at any world point and further that this relation does not depend on the behavior of Q^A and $Q^A{}_{,\mu}$. Substituting (1.1) into (1.4) we get

$$\frac{\partial L}{\partial Q^A}T_{(a),}{}^A{}_B\,Q^B+\frac{\partial L}{\partial Q^A{}_{,\mu}}T_{(a),}{}^A{}_B\,Q^B{}_{,\mu}\equiv 0, \qquad (a=1,2,\cdots n) \tag{1.5}$$

since the ϵ's are independent of each other. These n identities are the necessary and sufficient conditions for the invariance of I under G.

If we take into account the field equation for Q^A, we obtain from (1.5) the following n conservation laws:

$$\partial J^\mu{}_a/\partial x^\mu=0, \quad J^\mu{}_a=\frac{\partial L}{\partial Q^A{}_{,\mu}}T_{(a),}{}^A{}_B Q^B. \tag{1.6}$$

This is so because (1.5) can be rewritten as follows:

$$\left\{\frac{\partial L}{\partial Q^A}-\frac{\partial}{\partial x^\mu}\left(\frac{\partial L}{\partial Q^A{}_{,\mu}}\right)\right\}\delta Q^A+\frac{\partial}{\partial x^\mu}\left[\frac{\partial L}{\partial Q^A{}_{,\mu}}\delta Q^A\right]\equiv 0.$$

The first term, $\{\quad\}\delta Q^A$, vanishes on account of the field equation.

Now let us consider the following transformation:

$$\begin{aligned}&\delta Q^A(x)=T_{(a),}{}^A{}_B\,\epsilon^a(x)\,Q^B,\\ &T_{(a),}{}^A{}_B=\text{constant},\\ &\quad\epsilon^a(x)=\text{infinitesimal arbitrary function,}\end{aligned} \tag{1.1$'$}$$

instead of (1.1). In this case δL does not vanish but becomes

$$\delta L\equiv\{(1.5)\}_a\,\epsilon^a(x)+\frac{\partial L}{\partial Q^A{}_{,\mu}}T_{(a),}{}^A{}_B\,Q^B\frac{\partial\epsilon^a}{\partial x^\mu},$$

or

$$\delta L\equiv\frac{\partial L}{\partial Q^A{}_{,\mu}}T_{(a),}{}^A{}_B\,Q^B\frac{\partial\epsilon^a}{\partial x^\mu}, \tag{1.5$'$}$$

by virtue of the identity (1.5).

In order to preserve the invariance of the Lagrangian under (1.1)$'$, it is necessary to introduce a new field

$$A'^J(x), \quad J=1,2,\cdots M,$$

in such a way that the right-hand side of (1.5)$'$ can be cancelled with the contribution from this new field A'^J.

Now let us denote the new Lagrangian by

$$L'(Q^A, Q^A{}_{,\mu}, A'^J),$$

and consider the following transformation:

$$\delta Q^A = T_{(a),}{}^A{}_B\, Q^B\, \epsilon^a(x),$$

$$\delta A'^J = U_{(a)}{}^J{}_K\, A'^K\, \epsilon^a(x) + C^J{}_{,}{}^{\mu}{}_a \frac{\partial \epsilon^a}{\partial x^\mu}, \qquad (1.7)$$

where the coefficients U and C are unknown constants which will be determined later. In addition, let us propose that the new action integral I' is invariant under the transformation (1.7).

Our problem is to answer the five questions listed in the Introduction.

From the invariance postulate we get the following identity:

$$\delta L'(x) \equiv \frac{\partial L'}{\partial Q^A}\, \delta Q^A + \frac{\partial L'}{\partial Q^A{}_{,\mu}}\, \delta Q^A{}_{,\mu} + \frac{\partial L'}{\partial A'^J}\, \delta A'^J \equiv 0.$$

Inserting (1.7) into the above and taking account of the arbitrariness of choosing ϵ^a and $\partial\epsilon^a/\partial x^\mu$, we see that each coefficient of ϵ and $\partial\epsilon/\partial x$ must vanish independently. Namely, we have the identities

$$\frac{\partial L'}{\partial Q^A} T_{(a),}{}^A{}_B\, Q^B + \frac{\partial L'}{\partial Q^A{}_{,\mu}} T_{(a),}{}^A{}_B\, Q^B{}_{,\mu} + \frac{\partial L'}{\partial A'^J} U_{(a)}{}^J{}_{K,}\, A'^K \equiv 0, \qquad (1.8)$$

$$\frac{\partial L'}{\partial Q^A{}_{,\mu}} T_{(a),}{}^A{}_B\, Q^B + \frac{\partial L'}{\partial A'^J} C^J{}_{,}{}^{\mu}{}_a \equiv 0. \qquad (1.9)$$

Now, in order to be able to determine uniquely the A'^J-dependence of L', the number of the components of A'^J should be equal to the number of Eqs. (1.9), namely,

$$M = 4n$$

should hold. In addition, the matrix $C^J{}_{,}{}^{\mu}{}_a$ must be nonsingular. Thus there is the inverse of C defined by

$$C^J{}_{,}{}^{\mu}{}_a\, C^{-1a}{}_{\mu,K} = \delta^J{}_K, \quad C^{-1a}{}_{\mu,J}\, C^J{}_{,}{}^{\nu}{}_b = \delta^a{}_b\, \delta^\nu{}_\mu.$$

Then (1.9) can be rewritten as

$$\frac{\partial L'}{\partial A^a{}_\mu} + \frac{\partial L'}{\partial Q^A{}_{,\mu}} T_{(a),}{}^A{}_B\, Q^B \equiv 0,$$

where we have put

$$A^a{}_\mu = C^{-1a}{}_{\mu,J}\, A'^J.$$

Thus A'^J should be contained in L' only through the combination

$$\nabla_\mu Q^A \equiv \frac{\partial Q^A}{\partial x^\mu} - T_{(a),}{}^A{}_B\, Q^B\, C^{-1a}{}_{\mu,J}\, A'^J,$$

or

$$\nabla_\mu Q^A \equiv \frac{\partial Q^A}{\partial x^\mu} - T_{(a),}{}^A{}_B\, Q^B\, A^a{}_\mu. \qquad (1.10)$$

By using $A^a{}_\mu$ in place of A'^J, the transformation character of A turns into

$$\delta A^a{}_\mu = S_{(c)}{}^a{}_{,\mu,}{}^\nu{}_b\, A^b{}_\nu\, \epsilon^c(x) + \partial\epsilon^a/\partial x^\mu, \qquad (1.11)$$

where

$$S_{(c)}{}^a{}_{,\mu,}{}^\nu{}_b = C^{-1a}{}_{\mu,J}\, U_{(c)}{}^J{}_K\, C^K{}_{,}{}^\nu{}_b.$$

Now the new Lagrangian must have the form

$$L'(Q^A, Q^A{}_{,\mu}, A^a{}_\mu) = L''(Q^A, \nabla_\mu Q^A).$$

Therefore we have the relations

$$\frac{\partial L'}{\partial Q^A} = \left.\frac{\partial L''}{\partial Q^A}\right|_{\nabla Q=\text{const}} - \left.\frac{\partial L''}{\partial \nabla_\mu Q^B}\right|_{Q=\text{const}} T_{(a),}{}^B{}_A\, A^a{}_\mu,$$

$$\frac{\partial L'}{\partial Q^A{}_{,\mu}} = \left.\frac{\partial L''}{\partial \nabla_\mu Q^A}\right|_{Q=\text{const}},$$

$$\frac{\partial L'}{\partial A'^J} = -\left.\frac{\partial L''}{\partial \nabla_\nu Q^A}\right|_{Q=\text{const}} T_{(b),}{}^A{}_B\, Q^B\, C^{-1b}{}_{\nu,J}.$$

By using these relations, (1.8) becomes

$$\left.\frac{\partial L''}{\partial Q^A}\right|_{\nabla Q=\text{const}} T_{(a),}{}^A{}_B\, Q^B + \left.\frac{\partial L''}{\partial \nabla_\mu Q^A}\right|_{Q=\text{const}} T_{(a),}{}^A{}_B\, \nabla_\mu Q^B + \left.\frac{\partial L''}{\partial \nabla_\mu Q^A}\right|_{Q=\text{const}} Q^B A^b{}_\nu \{[T_{(a)}\, T_{(b)}]^A{}_B\, \delta^\nu{}_\mu - S_{(a)}{}^d{}_{,\mu,}{}^\nu{}_b\, T_{(d),}{}^A{}_B\} \equiv 0. \qquad (1.12)$$

If we put[2]

$$L''(Q^A, \nabla_\mu Q^A) = L(Q^A, \nabla_\mu Q^A),$$

namely, put L'' to be what is obtained by replacing $\partial Q^A/\partial x^\mu$ in the original Lagrangian L with the "covariant derivative," $\nabla_\mu Q^A$, then on account of the identity (1.5), the first and the second terms in (1.12) cancel each other. The remaining terms of (1.12) can be rewritten in the following way owing to the group character (1.2):

$$\left.\frac{\partial L}{\partial \nabla_\mu Q^A}\right|_{Q=\text{const}} Q^B A^b{}_\nu \{f_a{}^d{}_b\, \delta^\nu{}_\mu - S_{(a)}{}^d{}_{,\mu,}{}^\nu{}_b\} T_{(d),}{}^A{}_B \equiv 0.$$

Therefore we can determine the unknown coefficient S as follows:

$$S_{(a)}{}^c{}_{,\mu,}{}^\nu{}_b = \delta^\nu{}_\mu\, f_a{}^c{}_b. \qquad (1.13)$$

Using this expression for S, we can easily show the covariant character of the derivative $\nabla_\mu Q^A$, i.e.,

$$\delta \nabla_\mu Q^A = T_{(a),}{}^A{}_B\, \epsilon^a(x)\, \nabla_\mu Q^B. \qquad (1.14)$$

[2] This particular choice of L'' is due to the requirement that when the field A is assumed to vanish, we must have the original Lagrangian L.

Now let us investigate the possible type of the Lagrangian for the free A-field. Let it be denoted by

$$L_0(A^a{}_\mu, A^a{}_{\mu,\nu}), \quad A^a{}_{\mu,\nu}=\partial A^a{}_\mu/\partial x^\nu.$$

The invariance postulate for L_0 under the transformation (1.11) leads to

$$\frac{\partial L_0}{\partial A^a{}_\mu} f_b{}^a{}_c\, A^c{}_\mu + \frac{\partial L_0}{\partial A^a{}_{\mu,\nu}} f_b{}^a{}_c\, A^c{}_{\mu,\nu} \equiv 0, \tag{1.15}$$

$$\frac{\partial L_0}{\partial A^a{}_\mu} + \frac{\partial L_0}{\partial A^b{}_{\nu,\mu}} f_a{}^c{}_c\, A^c{}_\nu \equiv 0, \tag{1.16}$$

$$\frac{\partial L_0}{\partial A^a{}_{\mu,\nu}} + \frac{\partial L_0}{\partial A^a{}_{\nu,\mu}} \equiv 0. \tag{1.17}$$

From (1.17) we see that the derivative of A should be contained in L_0 through the combination

$$A^a{}_{[\mu,\nu]} \equiv \frac{\partial}{\partial x^\mu} A^a{}_\nu - \frac{\partial}{\partial x^\nu} A^a{}_\mu.$$

Thus (1.16) can be written as

$$\frac{\partial L_0}{\partial A^a{}_\mu} \equiv \frac{\partial L_0}{\partial A^c{}_{[\nu,\mu]}} f_a{}^c{}_b\, A^b{}_\nu. \tag{1.16$'$}$$

(1.16)′ means that the derivative of A appears in L_0, only through the particular combination

$$F^a{}_{\mu\nu} = \frac{\partial A^a{}_\nu}{\partial x^\mu} - \frac{\partial A^a{}_\mu}{\partial x^\nu} - \tfrac{1}{2} f_b{}^a{}_c (A^b{}_\mu A^c{}_\nu - A^b{}_\nu A^c{}_\mu). \tag{1.18}$$

Finally, substituting (1.16) into the first term of (1.15), we get

$$\frac{1}{2}\frac{\partial L_0}{\partial F^a{}_{\nu\mu}} \{ f_c{}^a{}_b\, A^b{}_{[\nu,\mu]} + \tfrac{1}{2}(f_d{}^a{}_b\, f_c{}^d{}_e - f_d{}^a{}_e\, f_c{}^d{}_b)(A^b{}_\nu A^e{}_\mu - A^b{}_\mu A^e{}_\nu)\} \equiv 0,$$

or by virtue of (1.3) we have

$$\frac{1}{2}\frac{\partial L_0}{\partial F^a{}_{\mu\nu}} f_c{}^a{}_b\, F^b{}_{\mu\nu} \equiv 0. \tag{1.19}$$

(See Appendix I.) Since L_0 must have the form

$$L_0(A, \partial A/\partial x) \equiv L_0'(A^a{}_\mu, F^a{}_{\mu\nu}),$$

we have the relations

$$\left.\frac{\partial L_0}{\partial A^a{}_{\nu,\mu}}\right|_{A=\text{const}} = \left.\frac{\partial L_0'}{\partial F^a{}_{\mu\nu}}\right|_{A=\text{const}},$$

$$\left.\frac{\partial L_0}{\partial A^a{}_\mu}\right|_{\partial A/\partial x=\text{const}} = \left.\frac{\partial L_0'}{\partial A^a{}_\mu}\right|_{F=\text{const}} - \left.\frac{\partial L_0'}{\partial F^b{}_{\mu\nu}}\right|_{A=\text{const}} f_a{}^b{}_c\, A^c{}_\nu.$$

From these relations and (1.16), we have

$$\left.\frac{\partial L_0'}{\partial A^a{}_\mu}\right|_{F=\text{const}} \equiv 0.$$

Namely L_0 must be a function of F alone and must satisfy the identity (1.19).

As may easily be seen, the transformation character of F is given by

$$\delta F^a{}_{\mu\nu} = \epsilon^b(x)\, f_b{}^a{}_c\, F^c{}_{\mu\nu}. \tag{1.20}$$

Equation (1.20) can be verified by using the relation (1.3).

Now let us define a set of matrices, $M_{(1)}$, $M_{(2)}$, $\cdots M_{(n)}$, in the following way:

$$(a,b)\text{-element of } M_{(c)} \equiv M_{(c)}{}^a{}_b = f_c{}^a{}_b, \qquad (a, b, c=1, 2, \cdots n).$$

Then these matrices are a representation of degree n for the generators of the Lie group G, since the relation (1.3) can be written as

$$[M_{(a)}, M_{(b)}]^l{}_c = f_a{}^m{}_b\, M_{(m),}{}^l{}_c.$$

Therefore (1.20) shows that n quantities, $F^1{}_{\mu\nu}$, $F^2{}_{\mu\nu}$, $\cdots F^n{}_{\mu\nu}$, are transformed cogradiently to the transformation of Q.

So far we have not used the field equations of A and Q. The variation of the total Lagrangian density

$$L_T = L_0(F) + L(Q, \nabla Q)$$

can be rewritten as

$$\frac{\delta L_T}{\delta Q^A}\delta Q^A + \frac{\delta L_T}{\delta A^a{}_\mu} f_b{}^a{}_c\, \epsilon^b A^c{}_\mu - \frac{\partial}{\partial x^\mu}\left(\frac{\delta L_T}{\delta A^a{}_\mu}\right)\epsilon^a + \frac{\partial}{\partial x^\mu}\left\{\frac{\partial L}{\partial \nabla_\mu Q^A}\delta Q^A + \frac{\partial L_0}{\partial F^a{}_{\mu\nu}}\delta A^a{}_\nu + \frac{\delta L_T}{\delta A^a{}_\mu}\epsilon^a\right\} \equiv 0, \tag{1.21}$$

where the following abbreviations have been used:

$$\frac{\delta L_T}{\delta Q^A} = \frac{\partial L_T}{\partial Q^A} - \frac{\partial}{\partial x^\mu}\left(\frac{\partial L_T}{\partial Q^A{}_{,\mu}}\right), \quad \frac{\delta L_T}{\delta A^a{}_\mu} = \frac{\partial L_T}{\partial A^a{}_\mu} - \frac{\partial}{\partial x^\nu}\left(\frac{\partial L_T}{\partial A^a{}_{\mu,\nu}}\right).$$

Now let us choose the arbitrary function $\epsilon^a(x)$ in such a way that the values of all the ϵ's and $\partial\epsilon/\partial x$'s vanish on the boundary surface of the integration domain Ω. Then the integration of (1.21) over the domain Ω becomes

$$\int_\Omega K d^4x \equiv 0, \tag{1.22}$$

with the abbreviation

$$K = \frac{\delta L_T}{\delta Q^A}\delta Q^A + \frac{\delta L_T}{\delta A^a{}_\mu} f_b{}^a{}_c\, \epsilon^b A^c{}_\mu - \frac{\partial}{\partial x^\mu}\left(\frac{\delta L_T}{\delta A^a{}_\mu}\right)\epsilon^a,$$

because the integration of the divergence term in (1.21) vanishes on account of our special choice of the ϵ's. Since the ϵ's can be chosen arbitrarily within Ω, K must vanish at every point in Ω, as is easily seen from (1.22).

Consequently the identity (1.21) are separated into the following two relations:

$$K \equiv 0, \tag{1.23}$$

and

$$\frac{\partial}{\partial x^\mu}\left\{\frac{\partial L}{\partial \nabla_\mu Q^A}\delta Q^A + \frac{\partial L_0}{\partial F^a{}_{\mu\nu}}\delta A^a{}_\nu + \frac{\delta L_T}{\delta A^a{}_\mu}\epsilon^a\right\} \equiv 0. \tag{1.24}$$

From (1.24), we have

$$\frac{\partial}{\partial x^\mu}\left\{\frac{\partial L}{\partial \nabla_\mu Q^A}T_{(a)},{}^A{}_B\, Q^B + \frac{\partial L_0}{\partial F^b{}_{\mu\nu}}f_a{}^b{}_c A^c{}_\nu + \frac{\delta L_T}{\delta A^a{}_\mu}\right\} \equiv 0, \tag{1.25}$$

$$\frac{\partial L}{\partial \nabla_\mu Q^A}T_{(a)},{}^A{}_B\, Q^B + \frac{\partial L_0}{\partial F^b{}_{\mu\nu}}f_a{}^b{}_c A^c{}_\nu + \frac{\partial L_T}{\partial A^a{}_\mu} \equiv 0, \tag{1.26}$$

and

$$\frac{\partial L_0}{\partial F^a{}_{\mu\nu}} + \frac{\partial L_0}{\partial F^a{}_{\nu\mu}} \equiv 0.$$

Put

$$J^\mu{}_a = \partial L_T / \partial A^a{}_\mu. \tag{1.27}$$

Then (1.26) leads to

$$J^\mu{}_a \equiv -\left(\frac{\partial L}{\partial \nabla_\mu Q^A}T_{(a)},{}^A{}_B\, Q^B + \frac{\partial L_0}{\partial F^b{}_{\mu\nu}}f_a{}^b{}_c A^c{}_\nu\right), \tag{1.28}$$

and (1.25) becomes

$$\frac{\partial J^\mu{}_a}{\partial x^\mu} \equiv \frac{\partial}{\partial x^\mu}\left\{\frac{\delta L_T}{\delta A^a{}_\mu}\right\}. \tag{1.29}$$

If we use the field equation

$$\delta L_T / \delta A^a{}_\mu = 0,$$

then we have the conservation of the "current," i.e.,

$$\partial J^\mu{}_a / \partial x^\mu = 0 \quad (a = 1, 2, \cdots n). \tag{1.30}$$

Thus we have obtained a general rule for introducing a new field A in a definite way when there exists some conservation law such as (1.6) or there is a Lie group depending upon some parameters under which the system is invariant.

In the following sections we shall consider the following groups as examples of the original Lie group: (1) the phase transformation of a charged field, (2) the rotation group in the isotopic spin space, and (3) the Lorentz group.

2. PHASE TRANSFORMATION GROUP AND THE ELECTROMAGNETIC FIELD

Let us consider a charged field Q and Q^*. The Lagrangian of this system is assumed to be invariant under the phase transformation

$$\delta Q^A = i\alpha Q^A, \quad \delta Q^{A*} = -i\alpha Q^{A*}, \quad \alpha = \text{a real constant}.$$

Since this one-parameter group is commutative, the structure constant, of course, vanishes. By replacing the constant α with a scalar function $\lambda(x)$, a vector field $A_\mu(x)$ is introduced. The transformation character of $A_\mu(x)$ is given by

$$\delta A_\mu = \partial\lambda / \partial x^\mu,$$

following the general formula (1.11). The new Lagrangian L' has the form

$$L' = L(Q, Q^*, \nabla_\mu Q, \nabla_\mu Q^*),$$

where $\nabla_\mu Q$ and $\nabla_\mu Q^*$ are given by

$$\nabla_\mu Q^A = \frac{\partial Q^A}{\partial x^\mu} - iA_\mu Q^A, \quad \nabla_\mu Q^{A*} = \frac{\partial Q^{A*}}{\partial x^\mu} + iA_\mu Q^{A*},$$

because in the present case

$$T^A{}_B = i\delta^A{}_B \quad \text{for} \quad Q^A, \quad T^A{}_B = -i\delta^A{}_B \quad \text{for} \quad Q^{A*}.$$

The Lagrangian L_0 for the free A_μ-field is

$$L_0 = L_0(F_{\mu\nu}),$$

where

$$F_{\mu\nu} = \frac{\partial A_\nu}{\partial x^\mu} - \frac{\partial A_\mu}{\partial x^\nu}.$$

The current J^μ can be obtained from the two different expressions

$$J^\mu = \frac{\partial L_T}{\partial A_\mu} = -i\left(\frac{\partial L}{\partial \nabla_\mu Q^A}Q^A - \frac{\partial L}{\partial \nabla_\mu Q^{A*}}Q^{A*}\right).$$

3. ROTATION GROUP IN ISOTOPIC SPIN SPACE AND THE YANG-MILLS FIELD

As an example let us consider a system of proton and neutron fields;

$$\psi^\alpha = \begin{pmatrix}\psi^1\\ \psi^2\end{pmatrix} = \begin{pmatrix}\text{proton}\\ \text{neutron}\end{pmatrix}.$$

The Lagrangian in the charge-independent theory is invariant under the rotation in the three-dimensional isotopic spin space:

$$\delta\psi^\alpha = i\sum_{c=1}^{3}\epsilon^c\,\tau_{(c)}{}^\alpha{}_\beta\,\psi^\beta, \quad \delta\bar\psi_\alpha = -i\sum_{c=1}^{3}\epsilon^c\,\bar\psi_\beta\,\tau_{(c)}{}^\beta{}_\alpha, \tag{3.1}$$

where $\tau_{(1)}$, $\tau_{(2)}$, and $\tau_{(3)}$ are the usual isotopic spin matrices.

In this case the general notation T in Sec. 2 corresponds to τ as follows

$$T_{(c),}{}^{A}{}_{B}\to i\tau_{(c),}{}^{\alpha}{}_{\beta},\quad \begin{pmatrix}\alpha,\beta=1,2\\ c=1,2,3\end{pmatrix}.$$

By replacing the parameters, ϵ^a, with a set of functions, $\epsilon^a(x)$, the Yang-Mills field

$$B^c{}_\mu(x)\quad(c=1,2,3)$$

is introduced, and this appears in the Lagrangian through the combination [see (1.10)]

$$\nabla_\mu\psi^\alpha=\partial\psi^\alpha/\partial x^\mu-i\tau_{(c),}{}^\alpha{}_\beta\,\psi^\beta\,B^c{}_\mu. \tag{3.2}$$

The variation of $B^c{}_\mu$ is given by [see (1.10) and (1.13)]

$$\delta B^c{}_\mu=f_a{}^c{}_b\,\epsilon^a(x)\,B^b{}_\mu+\frac{\partial\epsilon^c}{\partial x^\mu}, \tag{3.3}$$

where $f_a{}^c{}_b$ is defined by

$$[i\tau_{(a)},i\tau_{(b)}]=f_a{}^c{}_b\cdot i\tau_{(c)}. \tag{3.4}$$

The derivative of $B^a{}_\mu$ can appear only through the combination [see (1.18)]

$$F^a{}_{\mu\nu}=\frac{\partial B^a{}_\nu}{\partial x^\mu}-\frac{\partial B^a{}_\mu}{\partial x^\nu}-\frac{1}{2}f_b{}^a{}_c(B^b{}_\mu\,B^c{}_\nu-B^b{}_\nu\,B^c{}_\mu). \tag{3.5}$$

The variations of $\nabla_\mu\psi$ and $F^a{}_{\mu\nu}$ are as follows:

$$\delta\nabla_\mu\psi^\alpha=i\epsilon^c\,\tau_{(c),}{}^\alpha{}_\beta\,\nabla_\mu\psi^\beta$$

and

$$\delta F^a{}_{\mu\nu}=\epsilon^c\,f_c{}^a{}_b\,F^b{}_{\mu\nu}. \tag{3.6}$$

As was stated in Sec. 1, $F^a{}_{\mu\nu}$ is transformed under the rotation group as a vector, namely, the isotopic spin of this B-field is unity. The expression for the "current" has the form [see (1.25) and (1.24)]:

$$J^\mu{}_c=\frac{\partial L_T}{\partial B^c{}_\mu}=-i\frac{\partial L}{\partial\nabla_\mu\psi^\alpha}\tau_{(c),}{}^\alpha{}_\beta\,\psi^\beta-\frac{\partial L_0}{F^a{}_{\mu\nu}}f_c{}^a{}_b\,B^b{}_\nu.$$

4. LORENTZ GROUP AND THE GRAVITATIONAL FIELD

Let us consider a system of fields $Q^A(x)$ being defined with respect to some Lorentz frame. In addition, let us assume that the action integral

$$I=\int L(Q^A,Q^A{}_{,k})d^4x$$

is invariant under any Lorentz transformation.

Now besides the x-system, let us introduce an arbitrary system of the curvilinear coordinates u^μ ($\mu=1,2,3,4$). In what follows, the Latin and Greek indices represent quantities defined with respect to the x-system (or the local Lorentz frame) and to the u-system respectively.

The square of the invariant length of the infinitesimal line element is given by

$$ds^2=g^*{}_{ik}\,dx^i\,dx^k=g_{\mu\nu}\,du^\mu\,du^\nu,$$

where

$$g^*{}_{11}=g^*{}_{22}=g^*{}_{33}=-g^*{}_{44}=1,\quad g^*{}_{ik}=0\quad\text{for}\quad i\neq k,$$

and

$$g_{\mu\nu}(u)=\frac{\partial x^i}{\partial u^\mu}\frac{\partial x^k}{\partial u^\nu}g^*{}_{ik}.$$

Let us introduce two sets of functions defined by

$$h^k{}_\mu(u)=\partial x^k/\partial u^\mu,$$

and

$$h_k{}^\mu(u)=\partial u^\mu/\partial x^k. \tag{4.1}$$

Then we have the following relations:

$$g^*{}_{kl}\,h^k{}_\mu\,h^l{}_\nu=g_{\mu\nu}(u),\quad g_{\mu\nu}\,h_k{}^\mu\,h_l{}^\nu=g^*{}_{kl},\quad h_k{}^\mu\,h^l{}_\mu=\delta^l{}_k,$$

$$g^{kl*}\,h_k{}^\mu\,h_l{}^\nu=g^{\mu\nu}(u),\quad g^{\mu\nu}\,h^k{}_\mu\,h^l{}_\nu=g^{kl*},\quad h_k{}^\mu\,h^k{}_\nu=\delta^\mu{}_\nu,$$

$$\det(g_{\mu\nu})=g=-h^2\equiv-[\det(h^k{}_\mu)]^2.$$

Raising or lowering of both kinds of suffices can be done by means of $g_{\mu\nu}$, $g^{\mu\nu}$ or $g^*{}_{kl}$ and g^{kl*}. The geometrical meaning of the sets of $h^k{}_\mu$ and $h_k{}^\mu$, is obvious. The introduction of the four-world vector[3] $h_1{}^\mu$, $h_2{}^\mu$, $h_3{}^\mu$, and $h_4{}^\mu$ assigns respectively a local Lorentz frame to every world point. Of course, the local frames at every world point are transformed in the same way under any Lorentz transformation, i.e.,

$$x^k\to x^k+\epsilon^k{}_l x^l,$$

$$h_k{}^\mu\to h_k{}^\mu+\delta h_k{}^\mu,\quad \epsilon^{kl}=-\epsilon^{lk}$$

$$\delta h_k{}^\mu=-\epsilon^l{}_k\,h_l{}^\mu.$$

On account of this geometrical meaning of the h's, we can transform the world tensor into the corresponding local tensor defined with respect to the local frame, or vice versa, using $h_k{}^\mu$ or $h^k{}_\mu$. For example,

$$Q^k(u)=h^k{}_\mu(u)\,Q^\mu(u),\quad Q^\mu(u)=h_k{}^\mu(u)\,Q^k(u),$$

where the abbreviation

$$Q^k(u)=Q^k\{x(u)\}$$

has been used.

In this way we can rewrite the action integral as follows:

$$I=\int\mathfrak{L}(Q^A(u),Q^A{}_{,\mu}(u),h^k{}_\mu(u))d^4u,$$

where $\mathfrak{L}$ is defined by

$$\mathfrak{L}=L(Q^A(u),h_k{}^\mu(u)\,Q^A{}_{,\mu}(u))\,h, \tag{4.2}$$

and $Q^A{}_{,\mu}$ stands for

$$\partial Q^A(u)/\partial u^\mu.$$

---

[3] The world vector means a vector which is defined with respect to the u-system.

The reason for the fact that the Q^A in (4.2) is not transformed into the corresponding world quantity is that if Q^A is a spinor this rewriting is not possible, because the spinor can be well defined only with respect to a Lorentz frame.

Now I is invariant under the following two kinds of transformations[4,5]:

(1) The Lorentz transformation

$$\delta h^k{}_\mu = \epsilon^k{}_l\, h^l{}_\mu,\quad \delta Q^A = \tfrac{1}{2} T_{(kl),}{}^A{}_B\, Q^B\, \epsilon^{kl},\quad u^\mu = \text{unchanged}, \tag{4.3}$$

where $T_{(kl),}{}^A{}_B$ is the (A,B)-element of the $N\times N$ matrix $T_{(kl)}$ which is the representation of the generator of the Lorentz group. The matrix $T_{(kl)}$ satisfies the relation

$$[T_{(kl)}, T_{(m,n)}] = \tfrac{1}{2} f_{kl,}{}^{ab}{}_{mn}\, T_{(ab)},\quad T_{(kl)} = -T_{(lk)}.$$

(2) The general point transformation

$$u^\mu \to u^\mu + \lambda^\mu(u) = u'^\mu,$$

$$\lambda^\mu(u) = \text{an arbitrary function of } u,$$

$$\delta h^k{}_\mu = -\frac{\partial\lambda^\nu}{\partial u^\mu} h^k{}_\nu,\quad \delta Q^A(u) \equiv Q^{A\prime}(u') - Q^A(u) = 0,\quad \delta Q^A{}_{,\mu} = -\frac{\partial\lambda^\nu}{\partial u^\mu} Q^A{}_{,\nu}. \tag{4.4}$$

Now our Lagrangian (4.2) has the suitable form for the application of the general method stated in I, if the given functions, $h^k{}_\mu$, are regarded as a set of field quantities satisfying the condition:

$$\partial h^k{}_\mu / \partial u^\nu = \partial h^k{}_\nu / \partial u^\mu, \tag{4.5}$$

and having the transformation character (4.3) under the Lorentz group. Though we will omit the condition (4.5), the invariance of I under the transformations (4.3) and (4.4) still holds. The only role of (4.5) is to guarantee the possibility of finding the simplest and most convenient system of coordinates $(x^1, \cdots, x^4)$. In fact if we replace the parameters, ϵ^{ik}, with a set of arbitrary functions, $\epsilon^{ik}(u)$, after the Lorentz transformation depending on such $\epsilon(u)$'s, the relation (4.5) is destroyed.

The condition (4.5) is inconsistent with the application of the general procedure of Sec. 1 to the present problem. Accordingly we shall consider the h's as a set of *16 independent given functions.*

Now following the prescription of Sec. 1, let us consider the "generalized Lorentz transformation" depending upon a set of arbitrary functions $\epsilon^{ik}(u)$ instead of the parameters ϵ^{ik}. Under this transformation, Q^A and $h^k{}_\mu$ are assumed to be transformed as

$$\delta Q^A = \tfrac{1}{2}\epsilon^{kl}(u)\, T_{(kl),}{}^A{}_B\, Q^B,\quad \delta h^k{}_\mu = \epsilon^k{}_l(u)\, h^l{}_\mu. \tag{4.6}$$

Then, in order to retain the invariance of I under the transformation (4.6), it is necessary to introduce a new field

$$A^{kl}{}_\mu(u) = -A^{lk}{}_\mu(u),$$

which has the following transformation character according to (1.11):

$$\delta A^{kl}{}_\mu = \tfrac{1}{4} f_{ab,}{}^{kl}{}_{h g}\, \epsilon^{ab}(u)\, A^{hg}{}_\mu + \frac{\partial\epsilon^{kl}}{\partial u^\mu} = \epsilon^k{}_m\, A^{ml}{}_\mu + \epsilon^l{}_m\, A^{km}{}_\mu + \frac{\partial\epsilon^{kl}}{\partial u^\mu}. \tag{4.7}$$

Furthermore the new Lagrangian is given by

$$\mathfrak{L}(Q^A, \nabla_\mu Q^A, h^k{}_\mu) = hL\{Q^A, (h_k{}^\mu \nabla_\mu Q^A)\} \tag{4.8}$$

where[6]

$$\nabla_\mu Q^A = \frac{\partial Q^A}{\partial u^\mu} - \frac{1}{2} A^{kl}{}_\mu\, T_{(kl),}{}^A{}_B\, Q^B. \tag{4.9}$$

[see (1.10)].

The factors $\frac{1}{2}$ in (4.9) and $\frac{1}{4}$ in (4.7) are necessary because in summing up the terms in these expressions with respect to the dummy suffices the same contributions are counted twice or four times.

Because of the "general Lorentz transformation," under which each local frame at each world point is transformed differently, the relation (4.5) was abandoned. Since this relation is satisfied only when the basic world is flat, we are forced to take as our basic space-time some Riemannian space with the metric

$$g_{\mu\nu}(u) = h^k{}_\mu\, h_{k\nu},$$

and the affine connection

$$\Gamma_{\mu\nu}{}^\lambda = \tfrac{1}{2} g^{\lambda\sigma}\left(\frac{\partial g_{\sigma\mu}}{\partial u^\nu} + \frac{\partial g_{\nu\sigma}}{\partial u^\mu} - \frac{\partial g_{\mu\nu}}{\partial u^\sigma}\right).$$

Accordingly we would expect that there exists some relationship between $A^{kl}{}_\mu$ and $h^k{}_\mu$.

In order to obtain this relationship let us consider, as an example, the local tensor

$$Q^{kl}(u)\,(=Q^A).$$

Then from (4.9) we have

$$\nabla_\mu Q^{kl} = \frac{\partial Q^{kl}}{\partial u^\mu} - A^{km}{}_\mu\, Q_m{}^l - A^{lm}{}_\mu\, Q^k{}_m.$$

---

[4] R. Utiyama, Progr. Theoret. Phys. (Japan) **2**, 38 (1947).

[5] L. Rosenfeld, Ann. Physik **5**, 113 (1930).

[6] F. J. Belinfante, Physica **7**, 305 (1940); K. Husimi, Proc. National Research Council of Japan **4**, 81 (1943).

By using h, this can be rewritten as

$$\nabla_\mu Q^{k\nu}=\frac{\partial Q^{k\nu}}{\partial u^\mu}-A^{km}{}_\mu\, Q_m{}^\nu+\Gamma'_{\rho\mu}{}^\nu\, Q^{k\rho}, \tag{4.10}$$

where $Q^{m\nu}$ and Γ' are defined by

$$Q^{k\nu}=h_m{}^\nu\, Q^{km},$$

and

$$\Gamma'_{\nu\mu}{}^\rho=h_l{}^\rho\frac{\partial h^l{}_\nu}{\partial u^\mu}-h_k{}^\rho\, h_{l\nu}\, A^{kl}{}_\mu. \tag{4.11}$$

In general, the following relation is easily derived from (4.9):

$$\begin{aligned}\nabla_\mu Q^{kl\cdots,\rho\sigma\cdots}{}_{ab\cdots,\alpha\beta\cdots}&=\frac{\partial}{\partial u^\mu}Q^{kl\cdots,\rho\sigma\cdots}{}_{ab\cdots,\alpha\beta\cdots}\\ &-A^k{}_{i\mu}\, Q^{il\cdots,\rho\sigma\cdots}{}_{ab\cdots,\alpha\beta\cdots}-A^l{}_{i\mu}\, Q^{ki\cdots,\rho\sigma\cdots}{}_{ab\cdots,\alpha\beta\cdots}-\cdots\\ &+A^i{}_{a\mu}\, Q^{kl\cdots,\rho\sigma\cdots}{}_{ib\cdots,\alpha\beta\cdots}+A^i{}_{b\mu}\, Q^{kl\cdots,\rho\sigma\cdots}{}_{ai\cdots,\alpha\beta\cdots}+\cdots\\ &+\Gamma'_{\lambda\mu}{}^\rho\, Q^{kl\cdots,\lambda\sigma\cdots}{}_{ab\cdots,\alpha\beta\cdots}+\Gamma'_{\lambda\mu}{}^\sigma\, Q^{kl\cdots,\rho\lambda\cdots}{}_{ab\cdots,\alpha\beta\cdots}\\ &+\cdots-\Gamma'_{\alpha\mu}{}^\lambda\, Q^{kl\cdots,\rho\sigma\cdots}{}_{ab\cdots,\lambda\beta\cdots}\\ &-\Gamma'_{\beta\mu}{}^\lambda\, Q^{kl\cdots,\rho\sigma\cdots}{}_{ab\cdots,\alpha\lambda\cdots}-\cdots.\end{aligned} \tag{4.12}$$

This relation is nothing but the usual covariant derivative with the exception that for the Latin indices the $A^{kl}{}_{,\mu}$ must be inserted in place of the usual affinity Γ, and for the Greek indices our Γ' must be used instead of the Γ. Therefore for the world tensor $Q^{\mu\nu\cdots}$ our "covariant derivative" agrees with the usual covariant derivative with the affinity Γ'. Namely, if we use the symbol δ_μ for the usual covariant derivative, we get

$$\nabla_\mu Q^{\rho\nu}=\frac{\partial Q^{\rho\nu}}{\partial u^\mu}+\Gamma'_{\sigma\mu}{}^\rho\, Q^{\sigma\nu}+\Gamma'_{\sigma\mu}{}^\nu\, Q^{\rho\sigma}\equiv\delta_\mu\, Q^{\rho\nu}.$$

The relationship between A and h can be derived from the following consideration:

$$\begin{aligned}\nabla_\mu g^{kl*}&=-A^{kl}{}_\mu-A^{lk}{}_\mu=0\\ &=h^k{}_\rho\, h^l{}_\nu\, \nabla_\mu g^{\rho\nu}=h^k{}_\rho\, h^l{}_\nu\, \delta_\mu g^{\rho\nu}.\end{aligned}$$

From this expression we have

$$\delta_\mu g^{\rho\nu}=\frac{\partial g^{\rho\nu}}{\partial u^\mu}+\Gamma'_{\sigma\mu}{}^\rho\, g^{\sigma\nu}+\Gamma'_{\sigma\mu}{}^\nu\, g^{\rho\sigma}=0. \tag{4.13}$$

If we assume that[7]

$$\Gamma'_{\mu\nu}{}^\rho=\Gamma'_{\nu\mu}{}^\rho,$$

then we can solve (4.13) for Γ'. The solution is

$$\Gamma'_{\mu\nu}{}^\rho=\tfrac{1}{2}g^{\rho\sigma}\left(\frac{\partial g_{\sigma\mu}}{\partial u^\nu}+\frac{\partial g_{\nu\sigma}}{\partial u^\mu}-\frac{\partial g_{\mu\nu}}{\partial u^\sigma}\right)\equiv\Gamma_{\mu\nu}{}^\rho,$$

or

$$h_l{}^\rho\frac{\partial h^l{}_\nu}{\partial u^\mu}-A^\rho{}_{\nu,\mu}=\Gamma_{\nu\mu}{}^\rho, \tag{4.14}$$

where

$$A^\rho{}_{\nu,\mu}=h_k{}^\rho\, h_{l\nu}\, A^{kl}{}_{,\mu}.$$

(4.14) is just the relation desired.

Now from (4.14) we see that $A^\rho{}_{\nu,\mu}$ is a world tensor of the third rank under the general point transformation (4.4) because the inhomogeneous term

$$-\frac{\partial^2\lambda^\rho}{\partial u^\nu\partial u^\mu}$$

arising from $\delta\Gamma$ is cancelled out in virtue of the term $\delta(h\partial h/\partial u)$. Consequently the $A^{kl}{}_\mu$ is a covariant world vector under the u-transformation (4.4). Then it is easily seen that our "covariant derivative" ∇_μ is in fact covariant under both kinds of transformations, namely, under any "general Lorentz transformation" and any u-transformation (4.4).

Thus we have obtained the general expression for the covariant derivative without using the concept of parallel displacement. For example, if the Q^A were the spinor field ψ, we have

$$\nabla_\mu\psi=\frac{\partial\psi}{\partial x^\mu}-\frac{i}{4}A^{kl}{}_\mu\,[\gamma_k,\gamma_l]\psi,$$

where the γ_k's are the usual Dirac γ matrices.

Now let us consider the Lagragian $\mathfrak{L}_0$ of the free A-field:

$$\mathfrak{L}_0(h^k{}_\mu,A^{kl}{}_\mu,\partial A^{kl}{}_\mu/\partial u^\nu),$$

where the $h^k{}_\mu$ is necessary to raise or lower both kinds of tensor suffices.

From the invariance postulate for $\mathfrak{L}_0$ under the "general Lorentz transformation," we see that $\mathfrak{L}_0$ must have the form

$$\mathfrak{L}_0(h^k{}_\mu,F^{kl}{}_{\mu\nu}),$$

---

[7] In general (4.13) gives the following expression for Γ', if the symmetry $\Gamma'_{\mu\nu}{}^\lambda=\Gamma'_{\nu\mu}{}^\lambda$ is not assumed:

$$\Gamma'_{\mu\nu}{}^\rho=\Gamma_{\mu\nu}{}^\rho-(B^\rho{}_{\mu,\nu}+B^\rho{}_{\nu,\mu})+B_{\mu\nu,}{}^\rho,$$

where Γ is the Christoffel affinity, while B is an arbitrary tensor with the symmetry character $B_{\mu\nu,\rho}=-B_{\nu\mu,\rho}$. [See E. Schrödinger, *Space-time Structure* (Cambridge University Press, Cambridge, 1950), p. 66.] Therefore our new field $A^{kl}{}_\mu$ (or $A_{\rho\nu,\mu}=h_{k\rho}\, h_{l\nu}\, A^{kl}{}_\mu$) can be represented in the following way in terms of h, $\partial h/\partial u$ and B:

$$A_{\rho\nu,\mu}=h_{l\rho}\frac{\partial h^l{}_\nu}{\partial u^\mu}-\Gamma_{\nu\mu,\rho}+B_{\rho\nu,\mu}+B_{\rho\mu,\nu}-B_{\nu\mu,\rho}.$$

The right-hand side of the above expression is antisymmetric in ρ and ν because of the fact that

$$\left(h_{l\rho}\frac{\partial h^l{}_\nu}{\partial u^\mu}-\Gamma_{\nu\mu,\rho}\right)+(\nu\text{ and }\rho\text{ interchanged})=\frac{\partial g_{\rho\nu}}{\partial u^\mu}-\Gamma_{\nu\mu,\rho}-\Gamma_{\rho\mu,\nu}=0.$$

Therefore the antisymmetry of $A_{\rho\nu,\mu}$ in ρ and ν does not give any restriction to the symmetry character of B. Now if B is assumed to vanish, we obtain the relation (4.14). On the other hand, if the basic space-time is flat, then A takes the form $A_{\rho\nu,\mu}=B_{\rho\nu,\mu}+B_{\rho\mu,\nu}-B_{\nu\mu,\rho}$, on account of the relation $\partial h^l{}_\nu/\partial u^\mu=\partial h^l{}_\mu/\partial u^\nu$, or, equivalently, $h_{l\rho}\,\partial h^l{}_\nu/\partial u^\mu=\Gamma_{\nu\mu,\rho}$.

where F is defined by

$$F^{kl}{}_{\mu\nu}=\frac{\partial A^{kl}{}_{\nu}}{\partial u^{\mu}}-\frac{\partial A^{kl}{}_{\mu}}{\partial u^{\nu}}-\frac{1}{4}f_{ab,}{}^{kl}{}_{mn}(A^{ab}{}_{\mu}A^{mn}{}_{\nu}-A^{ab}{}_{\nu}A^{mn}{}_{\mu})$$

$$=\frac{\partial A^{kl}{}_{\nu}}{\partial u^{\mu}}-\frac{\partial A^{kl}{}_{\mu}}{\partial u^{\nu}}+A^{kb}{}_{\mu}A^{l}{}_{b\nu}-A^{kb}{}_{\nu}A^{l}{}_{b\mu}. \quad (4.15)$$

(4.15) can be rewritten formally as follows:

$$F^{kl}{}_{\mu\nu}=\nabla_{\mu}A^{kl}{}_{\nu}-\nabla_{\nu}A^{kl}{}_{\mu}-A^{kb}{}_{\mu}A^{l}{}_{b\nu}+A^{kb}{}_{\nu}A^{l}{}_{b\mu}, \quad (4.15)'$$

where $\nabla_{\mu}A^{kl}{}_{\nu}$ does not behave like a local tensor of the second rank, but is a covariant world tensor as the suffices μ and ν show. Using the expression (4.15)′ we can prove the following relation (see Appendix II)

$$F^{kl}{}_{\mu\nu}=h^{l\lambda}h^{k}{}_{\alpha}R^{\alpha}{}_{\lambda\mu\nu}, \quad (4.16)$$

where R is the Riemann-Christoffel curvature tensor:

$$R^{\alpha}{}_{\lambda\mu\nu}=\frac{\partial\Gamma_{\lambda\mu}{}^{\alpha}}{\partial u^{\nu}}-\frac{\partial\Gamma_{\lambda\nu}{}^{\alpha}}{\partial u^{\mu}}+\Gamma_{\lambda\mu}{}^{\beta}\Gamma_{\beta\nu}{}^{\alpha}-\Gamma_{\lambda\nu}{}^{\beta}\Gamma_{\beta\mu}{}^{\alpha}. \quad (4.17)$$

Though $h^{k}{}_{\mu}$ is contained in our Lagrangian as well as A and $\partial A/\partial u$, we can still prove that $A^{kl}{}_{\mu}$ appears in $\mathfrak{L}_0$ only through the combination F.

So far we have assumed that $h^{k}{}_{\mu}$ is a given function. The behavior of $h^{k}{}_{\mu}$ in general relativity is defined by the field equation derived by the variational principle.

The total Lagrangian density is now given by

$$\mathfrak{L}_T(Q^A,\nabla_{\mu}Q^A,h^{k}{}_{\mu},\partial h^{k}{}_{\mu}/\partial u^{\nu},\partial^2h^{k}{}_{\mu}/\partial u^{\nu}\partial u^{\lambda})$$
$$=\mathfrak{L}(Q^A,\nabla_{\mu}Q^A,h^{k}{}_{\mu})+\mathfrak{L}_0(h^{k}{}_{\mu},F^{kl}{}_{\mu\nu}).$$

The field equations for Q and h are[8]

$$\delta\mathfrak{L}/\delta Q^A=0$$

and

$$\frac{\delta\mathfrak{L}_T}{\delta h^{i}{}_{\mu}}=\frac{\partial\mathfrak{L}_T}{\partial h^{i}{}_{\mu}}-\frac{\partial}{\partial u^{\nu}}\left(\frac{\partial\mathfrak{L}_T}{\partial h^{i}{}_{\mu,\nu}}\right)+\frac{\partial^2}{\partial u^{\nu}\partial u^{\lambda}}\left(\frac{\partial\mathfrak{L}_T}{\partial h^{i}{}_{\mu,\nu\lambda}}\right)=0,$$

with the abbreviations

$$h^{i}{}_{\mu,\nu}=\partial h^{i}{}_{\mu}/\partial u^{\nu},\quad h^{i}{}_{\mu,\nu\lambda}=\partial^2h^{i}{}_{\mu}/\partial u^{\nu}\partial u^{\lambda}.$$

[8] If the $B_{\mu\nu,\rho}$ is taken into account, the Riemann-Christoffel tensor $R^{\rho}{}_{\lambda\mu\nu}$ is the same function as (4.17) but Γ' must be inserted in place of Γ in (4.17). Here Γ' is the affinity given in reference 7. In this case, in addition to the equations for the Q and h fields, we have the following equation:

$$\frac{\delta\mathfrak{L}_T}{\delta B_{\mu\nu,\lambda}}=\frac{1}{2}\frac{\partial\mathfrak{L}}{\partial A^{kl}{}_{\rho}}\frac{\partial A^{kl}{}_{\rho}}{\partial B_{\mu\nu,\lambda}}+\frac{1}{4}\frac{\partial\mathfrak{L}_0}{\partial F^{kl}{}_{\alpha\beta}}\frac{\partial F^{kl}{}_{\alpha\beta}}{\partial B_{\mu\nu,\lambda}}-\frac{1}{4}\frac{\partial}{\partial u^{\rho}}\left\{\frac{\partial\mathfrak{L}_0}{\partial F^{kl}{}_{\alpha\beta}}\frac{\partial F^{kl}{}_{\alpha\beta}}{\partial\left(\frac{\partial B_{\mu\nu,\lambda}}{\partial u^{\rho}}\right)}\right\}=0.$$

Of course, in all these equations for the Q, h, and B fields, the affinity Γ' must be used instead of Γ. (4.18) and (4.19) also hold in this case, since B is invariant under the "general Lorentz transformations." Consequently this B field is of no use in avoiding the trivial result: $\mathfrak{M}^{\mu}{}_{ik}\equiv0$.

Now corresponding to (1.23) and (1.24) we have the following identities:

$$\frac{\delta\mathfrak{L}_T}{\delta Q^A}\delta Q^A+\frac{\delta\mathfrak{L}_T}{\delta h^{i}{}_{\mu}}\delta h^{i}{}_{\mu}\equiv0, \quad (4.18)$$

and

$$(\partial/\partial u^{\mu})\mathfrak{M}^{\mu}\equiv0, \quad (4.19)$$

where $\mathfrak{M}^{\mu}$ is

$$\mathfrak{M}^{\mu}=\frac{\partial\mathfrak{L}}{\partial\nabla_{\mu}Q^A}\delta Q^A+\frac{\partial\mathfrak{L}_T}{\partial h^{i}{}_{\rho,\mu}}\delta h^{i}{}_{\rho}+\frac{\partial\mathfrak{L}_T}{\partial h^{i}{}_{\rho,\mu\nu}}\delta h^{i}{}_{\rho,\nu}-\frac{\partial}{\partial u^{\nu}}\left(\frac{\partial\mathfrak{L}_T}{\partial h^{i}{}_{\rho,\mu\nu}}\right)\cdot\delta h^{i}{}_{\rho}. \quad (4.20)$$

Accordingly, the coefficient of $\partial^2\epsilon/\partial u^2$ in (4.19) identically vanishes:

$$\frac{\partial\mathfrak{L}_T}{\partial h^{i}{}_{\rho,\mu\nu}}h_{k\rho}-\frac{\partial\mathfrak{L}_T}{\partial h^{k}{}_{\rho,\mu\nu}}h_{i\rho}\equiv0.$$

Thus $\mathfrak{M}^{\mu}$ becomes

$$\mathfrak{M}^{\mu}=\tfrac{1}{2}\epsilon^{ik}\mathfrak{M}^{\mu}{}_{ik},$$

$$\mathfrak{M}^{\mu}{}_{ik}=\frac{\partial\mathfrak{L}}{\partial\nabla_{\mu}Q^A}T_{(ik),}{}^{A}{}_{B}Q^B+\left[\left\{\frac{\partial\mathfrak{L}_T}{\partial h^{i}{}_{\rho,\mu}}h_{k\rho}+\frac{\partial\mathfrak{L}_T}{\partial h^{i}{}_{\rho,\mu\nu}}h_{k\rho,\nu}-\frac{\partial}{\partial u^{\nu}}\left(\frac{\partial\mathfrak{L}_T}{\partial h^{i}{}_{\rho,\mu\nu}}\right)h_{k\rho}\right\}-\{k\text{ and }i\text{ interchanged}\}\right].$$

Inserting this expression into (4.19), we have the trivial result

$$\partial\mathfrak{M}^{\mu}{}_{ik}/\partial u^{\mu}\equiv0, \quad (4.21)$$

as

$$\mathfrak{M}^{\mu}{}_{ik}\equiv0.$$

Since the Eulerian derivative $\delta L/\delta A$ appeared in (1.24), the nonvanishing "current" could be derived in the general theory in Sec. 1. In the present case, however, we have no Eulerian derivatives in $\mathfrak{M}$. Thus the field equations do not play any role in deriving the "current."

Now the usual equation for the gravitational field is derived by taking a particular Lagrangian

$$\mathfrak{L}_0=hR,$$

where R is defined as follows:

$$R=g^{\mu\nu}R_{\mu\nu}=h_{l}{}^{\mu}h_{k}{}^{\nu}F^{kl}{}_{\mu\nu},\quad R_{\mu\nu}=R^{\lambda}{}_{\mu\nu\lambda}.$$

Taking the variation with regard to h, we get

$$\frac{\delta\mathfrak{L}_0}{\delta h^{i}{}_{\mu}}+\frac{\delta\mathfrak{L}}{\delta h^{i}{}_{\mu}}=0.$$

The following relation is now easily verified:

$$\frac{\delta\mathfrak{L}_0}{\delta h^i{}_\mu}\delta h^i{}_\mu=\frac{\delta\mathfrak{L}_0}{\delta g_{\rho\mu}}h_{i\rho}\,\delta h^i{}_\mu+\frac{\partial}{\partial u^\mu}\left\{\frac{\partial\mathfrak{L}_0}{\partial g_{\rho\sigma,\mu}}h_{i\rho}\,\delta h^i{}_\sigma\right\},$$

where[9]

$$\delta\mathfrak{L}_0/\delta g_{\rho\sigma}=-h(R^{\rho\sigma}-\tfrac{1}{2}g^{\rho\sigma}R)=-\mathfrak{G}^{\rho\sigma}.$$

Thus we have

$$h_{i\sigma}\,\mathfrak{G}^{\rho\sigma}=-\mathfrak{T}^\rho{}_i,$$

with

$$\mathfrak{T}^\rho{}_i=-\delta\mathfrak{L}/\delta h^i{}_\rho,$$

or we have

$$\mathfrak{G}^{\rho\sigma}=-\mathfrak{T}^{\rho\sigma}.$$

Here $\mathfrak{T}$ is the symmetric energy-momentum-tensor density of the original field Q. The symmetry character of $\mathfrak{T}$ can be proved in the following way.

Since the Lagrangian $\mathfrak{L}$ for the Q-field is also invariant under the "general Lorentz transformation," we have an identity similar to (4.18):

$$\frac{\delta\mathfrak{L}}{\delta Q^A}\delta Q^A+\frac{\delta\mathfrak{L}}{\delta h^i{}_\mu}\delta h^i{}_\mu\equiv 0.$$

Inserting the field equation for the Q-field into this identity, we get

$$\mathfrak{T}_{ik}=\frac{\delta\mathfrak{L}}{\delta h^i{}_\mu}h_{k\mu}=\frac{\delta\mathfrak{L}}{\delta h^k{}_\mu}h_{i\mu}=\mathfrak{T}_{ki}.$$

From this relation,

$$\mathfrak{T}^{\mu\rho}=h^{i\mu}\,h^{k\rho}\,\mathfrak{T}_{ik}=\mathfrak{T}^{\rho\mu}$$

can be easily derived.

ACKNOWLEDGMENTS

The author is most grateful to the Institute for Advanced Study for a grant-in-aid and to Professor Robert Oppenheimer for the kind hospitality extended him there. He is also indebted to members of the Institute, especially to Dr. R. Arnowitt, for helpful conversations.

APPENDIX I. CONDITION (1.19)

Here we shall show how to construct an invariant in terms of $F^a{}_{\mu\nu}$.

Consider a quantity G_a, the transformation character of which is contragradient to that of $F^a{}_{\mu\nu}$ under the transformation (1.20).

Since

$$G_aF^a{}_{\mu\nu}$$

is invariant by definition, δG_a is given by

$$\delta G_a=-\epsilon^c\,f_c{}^b{}_a\,G_b.$$

For example,

$$K_{a,\mu\nu}=f_a{}^l{}_m\,f_l{}^m{}_b\,F^b{}_{\mu\nu}$$

is transformed contragradiently to $F^a{}_{\mu\nu}$, for $\delta K_{a,\mu\nu}$ is

$$\delta K_{a,\mu\nu}=f_a{}^l{}_m\,f_l{}^m{}_b\,\delta F^b{}_{\mu\nu}$$
$$=f_a{}^l{}_m\,f_l{}^m{}_b\,f_j{}^b{}_k\,F^k{}_{\mu\nu}\,\epsilon^j.$$

By taking into account the relation (1.3), the above expression becomes

$$\delta K_{a,\mu\nu}=f_a{}^l{}_m(f_k{}^b{}_l\,f_b{}^m{}_j+f_l{}^b{}_j\,f_b{}^m{}_k)\,F^k{}_{\mu\nu}\,\epsilon^j.$$

Using (1.3) again, we can rewrite the first term of this expression as follows:

$$\delta K_{a,\mu\nu}=f_k{}^b{}_l(f_j{}^m{}_a\,f_m{}^l{}_b+f_a{}^m{}_b\,f_m{}^l{}_j)\,F^k{}_{\mu\nu}\epsilon^j-f_k{}^m{}_b\,f_a{}^l{}_m\,f_l{}^b{}_j\,F^k{}_{\mu\nu}\,\epsilon^j.$$

Since the second term is cancelled with the last term, we have

$$\delta K_{a,\mu\nu}=-\epsilon^j\,f_j{}^m{}_a\,f_m{}^l{}_b\,f_l{}^b{}_k\,F^k{}_{\mu\nu}=-\epsilon^j\,f_j{}^m{}_a\,K_{m,\mu\nu}.$$

Now let us call $F^a{}_{\mu\nu}$ a contravariant vector, and $K_{a,\mu\nu}$ a covariant vector with regard to the transformation (1.20). In addition, let us propose that $f_b{}^a{}_c$ is contravariant with respect to the suffix a and covariant with respect to b and c. Then we see that $f_b{}^a{}_c$ is a constant and invariant tensor owing to the following fact that

$$\delta f_b{}^a{}_c=\epsilon^j(f_j{}^a{}_k\,f_b{}^k{}_c-f_j{}^k{}_b\,f_k{}^a{}_c-f_j{}^k{}_c\,f_b{}^a{}_k),$$

which vanishes in virtue of the relation (1.3). Hence this proposal concerning the transformation character of $f_b{}^a{}_c$ is compatible with the covariant character of $K_{a,\mu\nu}$.

Using the quantity

$$g_{ab}=f_a{}^l{}_m\,f_l{}^m{}_b=g_{ba}$$

and its inverse g^{ab}, we can easily construct a tensor algebra similar to that used in the theory of relativity. For example, we have invariants

$$H_{\mu\nu,\rho\sigma}=g_{ab}\,F^a{}_{\mu\nu}\,F^b{}_{\rho\sigma}=H_{\rho\sigma,\mu\nu}.$$

In the case of the rotation group in three-dimensional isotopic spin space (see Sec. 3), $f_b{}^a{}_c$ has the following values:

$$\begin{cases} f_1{}^3{}_2=f_3{}^2{}_1=f_2{}^1{}_3=-1,\\ f_i{}^l{}_k=-f_k{}^l{}_i,\\ \text{otherwise } f=0.\end{cases}$$

Therefore, we have

$$g_{ab}=2\delta_{ab},$$

and

$$H_{\mu\nu,\rho\sigma}=2\delta_{ab}\,F^a{}_{\mu\nu}\,F^b{}_{\rho\sigma}.$$

Another familiar example is the case of the Lorentz group. Here we have

$$\tfrac{1}{4}f_{jk,}{}^{ab}{}_{cd}\,f_{ab,}{}^{cd}{}_{lm}=g^*{}_{jl}\,g^*{}_{km}-g^*{}_{jm}\,g^*{}_{kl},$$

[9] W. Pauli, Encyklopaedie der Mathematischen Wissenschaften (B. G. Teubner, Leipzig, 1904–1922), Vol. 5, Chap. 19, p. 621.

and

$$H_{\mu\nu,\rho\sigma}=\frac{1}{2^4}F^{jk}{}_{\mu\nu}\,f_{jk,}{}^{ab}{}_{cd}\,f_{ab,}{}^{cd}{}_{lm}\,F^{lm}{}_{\rho\sigma}$$

$$=\tfrac{1}{2}F^{jk}{}_{\mu\nu}\,F^{lm}{}_{\rho\sigma}\,g^*{}_{jl}\,g^*{}_{km}=\tfrac{1}{2}F^{jk}{}_{\mu\nu}\,F_{jk,\rho\sigma}.$$

If L_0 is a function of the invariant $H_{\mu\nu,\beta\sigma}$ alone, we can easily prove the identity

$$\frac{\partial L_0}{\partial F^a{}_{\mu\nu}}f_c{}^a{}_b\,F^b{}_{\mu\nu}\equiv 0. \tag{1.19}$$

Namely, the left-hand side can be written as

$$\frac{1}{4}\frac{\partial L_0}{\partial H_{\rho\sigma,\alpha\beta}}\frac{\partial H_{\rho\sigma,\alpha\beta}}{\partial F^a{}_{\mu\nu}}f_c{}^a{}_b\,F^b{}_{\mu\nu}$$

$$=\frac{1}{4}\frac{\partial L_0}{\partial H_{\rho\sigma,\mu\nu}}F^d{}_{\rho\sigma}\,F^b{}_{\mu\nu}(g_{ad}\,f_c{}^a{}_b+g_{ab}\,f_c{}^a{}_d).$$

The factor in the bracket vanishes on account of the relation (1.3). Consequently there exists, in fact, a family of invariant Lagrangians, L_0, which are functions of $F^a{}_{\mu\nu}$ alone and satisfy the condition (1.19).

APPENDIX II. PROOF OF THE RELATION $F^{kl}{}_{\mu\nu}=h^{l\lambda}h^k{}_\alpha R^\alpha{}_{\lambda\mu\nu}$

$F^{kl}{}_{\mu\nu}$ is given by

$$F^{kl}{}_{\mu\nu}=\nabla_\mu A^{kl}{}_\nu-\nabla_\nu A^{kl}{}_\mu-A^{kb}{}_\mu\,A^l{}_{b\nu}+A^{kb}{}_\nu\,A^l{}_{b\mu}. \tag{A.1}$$

Now according to the general rule (4.12)

$$\nabla_\mu A^{kl}{}_\nu=h^k{}_\rho\,h^l{}_\sigma\,\delta_\mu A^{\rho\sigma}{}_\nu, \tag{A.2}$$

where δ_μ is the usual covariant derivative with the Christoffel affinity.

$A^{\rho\sigma}{}_{,\nu}$ is by virtue of (4.14) rewritten as

$$A^{\rho\sigma}{}_{,\nu}=g^{\sigma\lambda}\,A^\rho{}_{\lambda,\nu}$$

$$=g^{\sigma\lambda}\,h_l{}^\rho\left(\frac{\partial h^l{}_\lambda}{\partial u^\nu}-\Gamma_{\nu\lambda}{}^\alpha\,h^l{}_\alpha\right).$$

If we suppose $h^l{}_\lambda$ as a simple covariant world vector and ignore the local suffix l, then the factor in the parenthesis in the above expression is just the usual covariant derivative of $h^l{}_\lambda$. Therefore we get

$$A^{\rho\sigma}{}_{,\nu}=g^{\sigma\lambda}\,h_l{}^\rho\,\delta_\nu h^l{}_\lambda.$$

On the other hand from (4.14) we have the relation

$$\delta_\nu h^l{}_\lambda=A^l{}_{m,\nu}\,h^m{}_\lambda. \tag{A.3}$$

Thus we have

$$\delta_\mu A^{\rho\sigma}{}_{,\nu}=g^{\sigma\lambda}\,\delta_\mu(h_l{}^\rho\,\delta_\nu h^l{}_\lambda).$$

By using (A.3) this becomes

$$\delta_\mu A^{\rho\sigma}{}_{,\nu}=g^{\sigma\lambda}\,h_l{}^\rho\,\delta_\mu\delta_\nu h^l{}_\lambda-A^m{}_{l,\mu}\,A^l{}_{k,\nu}\,h_m{}^\rho\,h^{k\sigma}.$$

Inserting this expression into (A.1) with (A.2), we have

$$F^{kl}{}_{\mu\nu}=h^{l\lambda}(\delta_\mu\delta_\nu-\delta_\nu\delta_\mu)h^k{}_\lambda.$$

As is well known, the Riemann-Christoffel tensor is defined by

$$(\delta_\mu\delta_\nu-\delta_\nu\delta_\mu)V^\lambda=R^\rho{}_{\lambda\mu\nu}\,V_\rho$$

for an arbitrary covariant vector V_ρ.

Thus we get

$$F^{kl}{}_{\mu\nu}=h^{l\lambda}\,h^k{}_\alpha\,R^\alpha{}_{\lambda\mu\nu}.$$

JOURNAL OF MATHEMATICAL PHYSICS VOLUME 2, NUMBER 2 MARCH-APRIL, 1961

Lorentz Invariance and the Gravitational Field

T. W. B. KIBBLE*
Department of Mathematics, Imperial College, London, England
(Received August 19, 1960)

An argument leading from the Lorentz invariance of the Lagrangian to the introduction of the gravitational field is presented. Utiyama's discussion is extended by considering the 10-parameter group of inhomogeneous Lorentz transformations, involving variation of the coordinates as well as the field variables. It is then unnecessary to introduce *a priori* curvilinear coordinates or a Riemannian metric, and the new field variables introduced as a consequence of the argument include the vierbein components $h_k{}^\mu$ as well as the "local affine connection" $A^i{}_{j\mu}$. The extended transformations for which the 10 parameters become arbitrary functions of position may be interpreted as general coordinate transformations and rotations of the vierbein system. The free Lagrangian for the new fields is shown to be a function of two covariant quantities analogous to $F_{\mu\nu}$ for the electromagnetic field, and the simplest possible form is just the usual curvature scalar density expressed in terms of $h_k{}^\mu$ and $A^i{}_{j\mu}$. This Lagrangian is of first order in the derivatives, and is the analog for the vierbein formalism of Palatini's Lagrangian. In the absence of matter, it yields the familiar equations $R_{\mu\nu}=0$ for empty space, but when matter is present there is a difference from the usual theory (first pointed out by Weyl) which arises from the fact that $A^i{}_{j\mu}$ appears in the matter field Lagrangian, so that the equation of motion relating $A^i{}_{j\mu}$ to $h_k{}^\mu$ is changed. In particular, this means that, although the covariant derivative of the metric vanishes, the affine connection $\Gamma^\lambda{}_{\mu\nu}$ is nonsymmetric. The theory may be reexpressed in terms of the Christoffel connection, and in that case additional terms quadratic in the "spin density" $S^k{}_{ij}$ appear in the Lagrangian. These terms are almost certainly too small to make any experimentally detectable difference to the predictions of the usual metric theory.

1. INTRODUCTION

IT has long been realized that the existence of certain fields, notably the electromagnetic field, can be related to invariance properties of the Lagrangian.[1] Thus, if the Lagrangian is invariant under phase transformations $\psi\rightarrow e^{ie\lambda}\psi$, and if we wish to make it invariant under the general gauge transformations for which λ is a function of x, then it is necessary to introduce a new field A_μ which transforms according to $A_\mu\rightarrow A_\mu-\partial_\mu\lambda$, and to replace $\partial_\mu\psi$ in the Lagrangian by a "covariant derivative" $(\partial_\mu+ieA_\mu)\psi$. A similar argument has been applied by Yang and Mills[2] to isotopic spin rotations, and in that case yields a triplet of vector fields. It is thus an attractive idea to relate the existence of the gravitational field to the Lorentz invariance of the Lagrangian. Utiyama[3] has proposed a method which leads to the introduction of 24 new field variables $A^{ij}{}_\mu$ by considering the homogeneous Lorentz transformations specified by six parameters ϵ^{ij}. However, in order to do this it was necessary to introduce *a priori* curvilinear coordinates and a set of 16 parameters $h_k{}^\mu$. Initially, the $h_k{}^\mu$ were treated as given functions of x, but at a later stage they were regarded as field variables and interpreted as the components of a vierbein system in a Riemannian space. This is a rather unsatisfactory procedure since it is the purpose of the discussion to supply an argument for introducing the gravitational field variables, which include the metric as well as the affine connection. The new field variables $A^{ij}{}_\mu$ were subsequently related to the Christoffel connection $\Gamma^\lambda{}_{\mu\nu}$ in the Riemannian space, but this could only be done uniquely by making the *ad hoc* assumption that the quantity $\Gamma'^\lambda{}_{\mu\nu}$ calculated from $A^{ij}{}_\mu$ was symmetric.

It is the purpose of this paper to show that the vierbein components $h_k{}^\mu$, as well as the "local affine connection" $A^{ij}{}_\mu$, can be introduced as new field variables analogous to A_μ if one considers the full 10-parameter group of inhomogeneous Lorentz transformations in place of the restricted six-parameter group. This implies that one must consider transformations of the coordinates as well as the field variables, which will necessitate some changes in the argument, but it also means that only one system of coordinates is required, and that a Riemannian metric need not be introduced *a priori*. The interpretation of the theory in terms of a Riemannian space may be made later if desired. The starting point of the discussion is the ordinary formulation of Lorentz invariance (including translational invariance) in terms of rectangular coordinates in flat space. We shall follow the analogy with gauge transformations as far as possible, and for purposes of comparison we give in Sec. 2 a brief discussion of linear transformations of the field variables. This is essentially a summary of Utiyama's argument, though the emphasis is rather different, particularly with regard to the covariant and noncovariant conservation laws.

In Sec. 3 we discuss the invariance under Lorentz transformations, and in Sec. 4 we extend the discussion to the corresponding group in which the ten parameters become arbitrary functions of position. We show that to maintain invariance of the Lagrangian, it is necessary to introduce 40 new variables so that a suitable covariant derivative may be constructed. To make the action integral invariant, one actually requires the Lagrangian to be an invariant density rather than an invariant, and one must, therefore, multiply the invariant by a suitable (and uniquely determined) function of the new fields. In Sec. 5 we consider the possible forms of the free Lagrangian for the new fields. As in the case of the

* NATO Research Fellow.

[1] See, for example, H. Weyl, *Gruppentheorie und Quantenmechanik* (S. Hirzel, Leipzig, 1931), 2nd ed., Chap. 2, p. 89; and earlier references cited there.

[2] C. N. Yang and R. L. Mills, Phys. Rev. **96**, 191 (1954).

[3] Ryoyu Utiyama, Phys. Rev. **101**, 1597 (1956).

electromagnetic field, we choose the Lagrangian of lowest degree which satisfies the invariance requirements.

The geometrical interpretation in terms of a Riemannian space is discussed in Sec. 6, where we show that the free Lagrangian we have obtained is just the usual curvature scalar density, though expressed in terms of an affine connection $\Gamma^{\lambda}{}_{\mu\nu}$ which is not necessarily symmetric. In fact, when no matter is present it is symmetric as a consequence of the equations of motion, but otherwise it has an antisymmetric part expressible in terms of the "spin density" $\mathfrak{S}^{\mu}{}_{ij}$. Thus there is a difference between this theory and the usual metric theory of gravitation. This difference was first pointed out by Weyl,[4] and has more recently been discussed by Sciama.[5] It arises from the fact that our free Lagrangian is of first order in the derivatives, with the $h_k{}^{\mu}$ and $A^{ij}{}_{\mu}$ as independent variables. It is possible to re-express the theory in terms of the Christoffel connection ${}^0\Gamma^{\lambda}{}_{\mu\nu}$ or its local analog ${}^0A^{ij}{}_{\mu}$, and this is done in Sec. 7. In that case, additional terms quadratic in $\mathfrak{S}^{\mu}{}_{ij}$, and multiplied by the gravitational constant, appear in the Lagrangian.

2. LINEAR TRANSFORMATIONS

We consider a set of field variables $\chi_A(x)$, which we regard as the elements of a column matrix $\chi(x)$, with the Lagrangian

$$L(x)\equiv L\{\chi(x),\chi_{,\mu}(x)\},$$

where $\chi_{,\mu}=\partial_\mu\chi$. We also consider linear transformations of the form

$$\delta\chi=\epsilon^a T_a\chi, \tag{2.1}$$

where the ϵ^a are n constant infinitesimal parameters, and the T_a are n given matrices satisfying commutation rules appropriate to the generators of a Lie group,

$$[T_a,T_b]=f_a{}^c{}_bT_c.$$

The Lagrangian is invariant under these transformations if the n identities

$$(\partial L/\partial\chi)T_a\chi+(\partial L/\partial\chi_{,\mu})T_a\chi_{,\mu}\equiv 0, \tag{2.2}$$

are satisfied, and we shall assume that this is so. Note that $\partial/\partial\chi$ must be regarded as a row matrix. The equations of motion imply n conservation laws

$$J^{\mu}{}_{a,\mu}=0,$$

where the "currents" are defined by[6]

$$J^{\mu}{}_a\equiv-(\partial L/\partial\chi_{,\mu})T_a\chi. \tag{2.3}$$

Now, under the more general transformations of the form (2.1), but in which the parameters ϵ^a become arbitrary functions of position, the Lagrangian is no longer invariant, because the derivatives transform according to

$$\delta\chi_{,\mu}=\epsilon^aT_a\chi_{,\mu}+\epsilon^a{}_{,\mu}T_a\chi, \tag{2.4}$$

and the terms in $\epsilon^a{}_{,\mu}$ do not cancel. In fact, one finds

$$\delta L\equiv-\epsilon^a{}_{,\mu}J^{\mu}{}_a.$$

However, one can obtain a modified Lagrangian which is invariant by replacing $\chi_{,\mu}$ in L by a quantity $\chi_{;\mu}$ which transforms according to

$$\delta\chi_{;\mu}=\epsilon^aT_a\chi_{;\mu}. \tag{2.5}$$

To do this[7] it is necessary to introduce $4n$ new field variables $A^a{}_{\mu}$ whose transformation properties involve $\epsilon^a{}_{,\mu}$. In fact, if one takes

$$\chi_{;\mu}\equiv\chi_{,\mu}+A^a{}_{\mu}T_a\chi, \tag{2.6}$$

then the condition (2.5) determines the transformation properties of the new fields uniquely. They are

$$\delta A^a{}_{\mu}=\epsilon^bf_b{}^a{}_cA^c{}_{\mu}-\epsilon^a{}_{,\mu}. \tag{2.7}$$

In this way one obtains the invariant Lagrangian

$$L'\{\chi,\chi_{,\mu},A^a{}_{\mu}\}\equiv L\{\chi,\chi_{;\mu}\}.$$

The expression $\chi_{;\mu}$ may be called the covariant derivative of χ with respect to the transformations (2.1). One may define covariant currents by

$$J'^{\mu}{}_a\equiv-(\partial L'/\partial A^a{}_{\mu})\equiv-(\partial L/\partial\chi_{;\mu})T_a\chi, \tag{2.8}$$

where L is regarded as a function of χ and $\chi_{;\mu}$. They transform linearly according to

$$\delta J'^{\mu}{}_a=-\epsilon^bf_b{}^c{}_aJ'^{\mu}{}_c,$$

and their covariant divergences vanish in virtue of the equations of motion and the identities (2.2):

$$J'^{\mu}{}_{a;\mu}\equiv J'^{\mu}{}_{a,\mu}-A^b{}_{\mu}f_b{}^c{}_aJ'^{\mu}{}_c$$
$$=0.$$

Two covariant differentiations do not in general commute. From (2.6) one finds

$$\chi_{;\mu\nu}-\chi_{;\nu\mu}=F^a{}_{\mu\nu}T_a\chi,$$

where

$$F^a{}_{\mu\nu}\equiv A^a{}_{\mu,\nu}-A^a{}_{\nu,\mu}-f_b{}^a{}_cA^b{}_{\mu}A^c{}_{\nu}. \tag{2.9}$$

Unlike $A^a{}_{\mu}$, the expression $F^a{}_{\mu\nu}$ is a covariant quantity transforming according to

$$\delta F^a{}_{\mu\nu}=\epsilon^bf_b{}^a{}_cF^c{}_{\mu\nu},$$

and one may, therefore, define its covariant derivative in an obvious manner. It satisfies the cyclic identity

$$F^a{}_{\mu\nu;\rho}+F^a{}_{\nu\rho;\mu}+F^a{}_{\rho\mu;\nu}\equiv 0.$$

---

[4] H. Weyl, Phys. Rev. 77, 699 (1950).

[5] D. W. Sciama, *Festschrift for Infeld* (Pergamon Press, New York), to be published.

[6] We have defined $J^{\mu}{}_a$ with the opposite sign to that used by Utiyama.[3] This is because with this choice of sign the analogous quantity for translations is $T^{\mu}{}_{\nu}$ rather than $-T^{\mu}{}_{\nu}$. The change may be considered as a change of sign of ϵ^a and T_a, and there is a corresponding change of sign in (2.6). This convention has the additional advantage that the "local affine connection" $A^i{}_{j\mu}$ defined in Sec. 4 specifies covariant derivatives according to the same rule as $\Gamma^{\lambda}{}_{\mu\nu}$.

[7] For a full discussion, see footnote 3.

It remains to find a free Lagrangian L_0 for the new fields. Clearly L_0 must be separately invariant, and it is easy to see[8] that this implies that it must contain $A^a{}_\mu$ only through the covariant combination $F^a{}_{\mu\nu}$. The simplest such Lagrangian is[8]

$$L_0=-\tfrac{1}{4}F^a{}_{\mu\nu}F_a{}^{\mu\nu}, \tag{2.10}$$

where the tensor indices are raised with the flat-space metric $\eta^{\mu\nu}$ with diagonal elements $(1, -1, -1, -1)$, and the index a is lowered with the metric[8a]

$$g_{ab}\equiv f_a{}^c{}_d f_c{}^d{}_b$$

associated with the Lie group (except of course for a one-parameter group). It is clear that this Lagrangian is not unique. All that is required is that it should be a scalar both in coordinate space and in the Lie-group space, and one could add to it terms of higher degree in $F^a{}_{\mu\nu}$. However, it seems reasonable to choose the Lagrangian of lowest degree which satisfies the invariance requirements.

With the choice (2.10) of L_0, the equations of motion for the new fields are

$$F_a{}^{\mu\nu}{}_{;\nu}=J'^\mu{}_a.$$

Because of the antisymmetry of $F_a{}^{\mu\nu}$ one can define another current which is conserved in the strict sense:

$$(J'^\mu{}_a+j^\mu{}_a)_{,\mu}=0, \tag{2.11}$$

where

$$j^\mu{}_a\equiv A^b{}_\nu f_b{}^c{}_a F_c{}^{\mu\nu}.$$

This extra current $j^\mu{}_a$ may be regarded as the current of the new field $A^a{}_\mu$ itself, since it is expressible in the form

$$j^\mu{}_a\equiv-(\partial L_0/\partial A^a{}_\mu)\equiv-(\partial L_0/\partial A^b{}_{\nu,\mu})f_a{}^b{}_c A^c{}_\nu, \tag{2.12}$$

which should be compared with (2.8). Note, however, that it is not a covariant quantity. To obtain a strict conservation law one must sacrifice the covariance of the current.

3. LORENTZ TRANSFORMATIONS

We now wish to consider infinitesimal variations of both the coordinates and the field variables,

$$\begin{aligned} x^\mu &\rightarrow x'^\mu=x^\mu+\delta x^\mu, \\ \chi(x) &\rightarrow \chi'(x')=\chi(x)+\delta\chi(x). \end{aligned} \tag{3.1}$$

It will be convenient to allow for the possibility that the Lagrangian may depend on x explicitly. Then, under a variation (3.1), the change in L is

$$\delta L\equiv(\partial L/\partial\chi)\delta\chi+(\partial L/\partial\chi_{,\mu})\delta\chi_{,\mu}+(\partial L/\partial x^\mu)\delta x^\mu,$$

where $\partial L/\partial x^\mu$ denotes the partial derivative with fixed χ. It is sometimes useful to consider also the variation at a fixed value of x,

$$\delta_0\chi=\chi'(x)-\chi(x)=\delta\chi-\delta x^\mu\chi_{,\mu}. \tag{3.2}$$

In particular, it is obvious that δ_0 commutes with ∂_μ, whence

$$\delta\chi_{,\mu}=(\delta\chi)_{,\mu}-(\delta x^\nu)_{,\mu}\chi_{,\nu}. \tag{3.3}$$

The action integral

$$I(\Omega)\equiv\int_\Omega L(x)d_4x$$

over a space-time region Ω is transformed under (3.1) into

$$I'(\Omega)\equiv\int_\Omega L'(x')\|\partial_\nu x'^\mu\|d_4x.$$

Thus the action integral over an arbitrary region is invariant if[9]

$$\delta L+L(\delta x^\mu)_{,\mu}\equiv\delta_0 L+(L\delta x^\mu)_{,\mu}\equiv 0. \tag{3.4}$$

This is of course the typical transformation law of an invariant density.

We now consider the specific case of Lorentz transformations,

$$\delta x^\mu=\epsilon^\mu{}_\nu x^\nu+\epsilon^\mu, \quad \delta\chi=\tfrac{1}{2}\epsilon^{\mu\nu}S_{\mu\nu}\chi, \tag{3.5}$$

where ϵ^μ and $\epsilon^{\mu\nu}=-\epsilon^{\nu\mu}$ are 10 real infinitesimal parameters, and the $S_{\mu\nu}$ are matrices satisfying

$$S_{\mu\nu}+S_{\nu\mu}=0,$$

$$[S_{\mu\nu},S_{\rho\sigma}]=\eta_{\nu\rho}S_{\mu\sigma}+\eta_{\mu\sigma}S_{\nu\rho}-\eta_{\nu\sigma}S_{\mu\rho}-\eta_{\mu\rho}S_{\nu\sigma}\equiv\tfrac{1}{2}f_{\mu\nu}{}^{\kappa\lambda}{}_{\rho\sigma}S_{\kappa\lambda}.$$

From (3.3) one has

$$\delta\chi_{,\mu}=\tfrac{1}{2}\epsilon^{\rho\sigma}S_{\rho\sigma}\chi_{,\mu}-\epsilon^\rho{}_\mu\chi_{,\rho}. \tag{3.6}$$

Moreover, since $(\delta x^\mu)_{,\mu}=\epsilon^\mu{}_\mu=0$, the condition (3.4) for invariance of the action integral again reduces to $\delta L\equiv 0$, and yields the 10 identities[10]

$$\partial L/\partial x^\rho\equiv L_{,\rho}-(\partial L/\partial\chi)\chi_{,\rho}-(\partial L/\partial\chi_{,\mu})\chi_{,\mu\rho}\equiv 0, \tag{3.7}$$

$$\begin{aligned}(\partial L/\partial\chi)S_{\rho\sigma}\chi+(\partial L/\partial\chi_{,\mu})(S_{\rho\sigma}\chi_{,\mu} \\ +\eta_{\mu\rho}\chi_{,\sigma}-\eta_{\mu\sigma}\chi_{,\rho})\equiv 0.\end{aligned} \tag{3.8}$$

These are evidently the analogs of the identities (2.2), and we shall assume that they are satisfied. Note that (3.7), which express the conditions for translational invariance, are equivalent to the requirement that L be explicitly independent of x, as might be expected.

As before, the equations of motion may be used to obtain 10 conservation laws which follow from these identities, namely,

$$T^\mu{}_{\rho,\mu}=0, \quad (S^\mu{}_{\rho\sigma}-x_\rho T^\mu{}_\sigma+x_\sigma T^\mu{}_\rho)_{,\mu}=0,$$

---

[8] There could of course be a constant factor multiplying (2.10), but this can be absorbed by a trivial change of definition of $A^a{}_\mu$ and T_a.

[8a] The discussion here applies only to semisimple groups since otherwise g_{ab} is singular. (I am indebted to the referee for this remark.)

[9] See L. Rosenfeld, Ann. Physik 5, 113 (1930).

[10] Compare L. Rosenfeld, Ann. inst. Henri Poincaré 2, 25 (1931).

where

$$T^\mu{}_\rho \equiv (\partial L/\partial \chi_{,\mu})\chi_{,\rho} - \delta^\mu{}_\rho L, \quad S^\mu{}_{\rho\sigma} \equiv -(\partial L/\partial \chi_{,\mu}) S_{\rho\sigma}\chi.$$

These are the conservation laws of energy, momentum, and angular momentum.

It is instructive to examine these transformations in terms of the variation $\delta_0\chi$ also, which in this case is

$$\delta_0\chi = -\epsilon^\rho \partial_\rho \chi + \tfrac{1}{2}\epsilon^{\rho\sigma}(S_{\rho\sigma} + x_\rho \partial_\sigma - x_\sigma \partial_\rho)\chi.$$

On comparing this with (2.1), one sees that the role of the matrices T_a is played by the differential operators $-\partial_\mu$ and $S_{\rho\sigma} + x_\rho\partial_\sigma - x_\sigma\partial_\rho$. Thus, by analogy with the definition (2.3) of the currents $J^\mu{}_a$, one might expect the currents in this case to be

$$J^\mu{}_\rho \equiv (\partial L/\partial \chi_{,\mu})\chi_{,\rho}, \quad J^\mu{}_{\rho\sigma} \equiv S^\mu{}_{\rho\sigma} - x_\rho J^\mu{}_\sigma + x_\sigma J^\mu{}_\rho,$$

corresponding to the parameters ϵ^ρ, $\epsilon^{\rho\sigma}$, respectively. However, in terms of δ_0, the condition for invariance (3.4) is not simply $\delta_0 L \equiv 0$, and the additional term $\delta x^\rho L_{,\rho}$ is responsible for the appearance of the term $L_{,\rho}$ in the identities (3.7), and hence for the term $\delta^\mu{}_\rho L$ in $T^\mu{}_\rho$.

4. GENERALIZED LORENTZ TRANSFORMATIONS

We now turn to a consideration of the generalized transformations (3.5) in which the parameters ϵ^μ and $\epsilon^{\mu\nu}$ become arbitrary functions of position. It is more convenient, and clearly equivalent, to regard as independent functions $\epsilon^{\mu\nu}$ and

$$\xi^\mu \equiv \epsilon^\mu{}_\nu x^\nu + \epsilon^\nu,$$

since this avoids the explicit appearance of x. Moreover, one could consider generalized transformations with $\xi^\mu = 0$ but nonzero $\epsilon^{\mu\nu}$, so that the coordinate and field transformations can be completely separated. In view of this fact, it is convenient to use Latin indices for ϵ^{ij} (and for the matrices S_{ij}), retaining the Greek ones for ξ^μ and x^μ. Thus the transformations under consideration are

$$\delta x^\mu = \xi^\mu, \quad \delta\chi = \tfrac{1}{2}\epsilon^{ij} S_{ij}\chi \tag{4.1}$$

or

$$\delta_0\chi = -\xi^\mu \chi_{,\mu} + \tfrac{1}{2}\epsilon^{ij} S_{ij}\chi. \tag{4.2}$$

This notation emphasizes the similarity of the ϵ^{ij} transformations to the linear transformations discussed in Sec. 2. These transformations alone were considered by Utiyama.[3] Evidently, the four functions ξ^μ specify a general coordinate transformation. The geometrical significance of the ϵ^{ij} will be discussed in Sec. 6.

According to our convention, the differential operator ∂_μ must have a Greek index. However, in the Lagrangian function L it would be inconvenient to have two kinds of indices, and we shall, therefore, regard L as a given function of χ and χ_k (no comma),[11] satisfying the identities (3.7) and (3.8). The original Lagrangian is then obtained by setting

$$\chi_k = \delta_k{}^\mu \chi_{,\mu}.$$

It is of course not invariant under the generalized transformations (4.1), but we shall later obtain an invariant expression by replacing χ_k by a suitable quantity $\chi_{;k}$.

The transformation of $\chi_{,\mu}$ is given by

$$\delta\chi_{,\mu} = \tfrac{1}{2}\epsilon^{ij} S_{ij}\chi_{,\mu} + \tfrac{1}{2}\epsilon^{ij}{}_{,\mu} S_{ij}\chi - \xi^\nu{}_{,\mu}\chi_{,\nu}, \tag{4.3}$$

and so the original Lagrangian transforms according to

$$\delta L \equiv -\xi^\rho{}_{,\mu} J^\mu{}_\rho - \tfrac{1}{2}\epsilon^{ij}{}_{,\mu} S^\mu{}_{ij}.$$

Note that it is $J^\mu{}_\rho$ rather than $T^\mu{}_\rho$ which appears here. The reason for this is that we have not included the extra term $L(\delta x^\mu)_{,\mu}$ in (3.4). The left-hand side of (3.4) actually has the value

$$\delta L + L(\delta x^\mu)_{,\mu} \equiv -\xi^\rho{}_{,\mu} T^\mu{}_\rho - \tfrac{1}{2}\epsilon^{ij}{}_{,\mu} S^\mu{}_{ij}.$$

We now look for a modified Lagrangian which makes the action integral invariant. The additional term just mentioned is of a different kind to those previously encountered, in that it involves L and not $\partial L/\partial\chi_k$. In particular, it includes contributions from terms in L which do not contain derivatives. Thus it is clear that we cannot remove it by replacing the derivative by a suitable covariant derivative. For this reason, we shall consider the problem in two stages. We first eliminate the noninvariance arising from the fact that $\chi_{,\mu}$ is not a covariant quantity, and thus obtain an expression L' satisfying

$$\delta L' \equiv 0. \tag{4.4}$$

Then, because the condition (3.4) for invariance of the action integral requires the Lagrangian to be an invariant density rather than an invariant, we make a further modification, replacing L' by $\mathfrak{L}'$, which satisfies

$$\delta\mathfrak{L}' + \xi^\mu{}_{,\mu}\mathfrak{L}' \equiv 0. \tag{4.5}$$

The first part of this program can be accomplished by replacing χ_k in L by a "covariant derivative" $\chi_{;k}$ which transforms according to

$$\delta\chi_{;k} = \tfrac{1}{2}\epsilon^{ij} S_{ij}\chi_{;k} - \epsilon^i{}_k \chi_{;i}. \tag{4.6}$$

The condition (4.4) then follows from the identities (3.8). To do this it is necessary to introduce forty new field variables. We consider first the ϵ^{ij} transformations, and eliminate the $\epsilon^{ij}{}_{,\mu}$ term in (4.3) by setting[12]

$$\chi_{|\mu} \equiv \chi_{,\mu} + \tfrac{1}{2}A^{ij}{}_\mu S_{ij}\chi, \tag{4.7}$$

where the $A^{ij}{}_\mu = -A^{ji}{}_\mu$ are 24 new field variables. We can then impose the condition

$$\delta\chi_{|\mu} = \tfrac{1}{2}\epsilon^{ij} S_{ij}\chi_{|\mu} - \xi^\nu{}_{,\mu}\chi_{|\nu} \tag{4.8}$$

which determines the transformation properties of $A^{ij}{}_\mu$

[11] Note that since we are using Latin indices for S_{ij} the various tensor components of χ must also have Latin indices, and for spinor components the Dirac matrices must be γ^k.

[12] Our $A^{ij}{}_\mu$ differs in sign from that of Utiyama.[3] Compare footnote 6.

uniquely. They are

$$\delta A^{ij}{}_{\mu}=\epsilon^{i}{}_{k}A^{kj}{}_{\mu}+\epsilon^{j}{}_{k}A^{ik}{}_{\mu}-\xi^{\nu}{}_{,\mu}A^{ij}{}_{\nu}-\epsilon^{ij}{}_{,\mu}. \quad (4.9)$$

The position with regard to the last term in (4.3) is rather different. The term involving $\epsilon^{ij}{}_{,\mu}$ is inhomogeneous in the sense that it contains χ rather than $\chi_{,\mu}$, just like the second term of (2.4), but this is not true of the last term.[13] Correspondingly, the transformation law (4.8) of $\chi_{|\mu}$ is already homogeneous. This means that to obtain an expression $\chi_{;k}$ transforming according to (4.6) we should add to $\chi_{|\mu}$ not a term in χ but rather a term in $\chi_{|\mu}$ itself. In other words, we can merely multiply by a new field:

$$\chi_{;k}\equiv h_{k}{}^{\mu}\chi_{|\mu}. \quad (4.10)$$

Here the $h_k{}^\mu$ are 16 new field variables with transformation properties determined by (4.6) to be

$$\delta h_{k}{}^{\mu}=\xi^{\mu}{}_{,\nu}h_{k}{}^{\nu}-\epsilon^{i}{}_{k}h_{i}{}^{\mu}. \quad (4.11)$$

It should be noted that the fields $h_k{}^\mu$ and $A^{ij}{}_\mu$ are quite independent and unrelated at this stage, though of course they will be related by equations of motion.

We have now found an invariant L'. We can easily obtain an invariant density $\mathfrak{L}'$ by multiplying by a suitable function of the fields already introduced:

$$\mathfrak{L}'\equiv\mathfrak{H}L'.$$

Then (4.5) is satisfied provided that $\mathfrak{H}$ is itself an invariant density,

$$\delta\mathfrak{H}+\xi^{\mu}{}_{,\mu}\mathfrak{H}\equiv 0.$$

It is easy to see that the only function of the new fields which obeys this transformation law, and does not involve derivatives, is

$$\mathfrak{H}=[\det(h_{k}{}^{\mu})]^{-1},$$

where the arbitrary constant factor has been chosen so that $\mathfrak{H}$ reduces to 1 when $h_k{}^\mu$ is set equal to $\delta_k{}^\mu$.[14]

The final form of our modified Lagrangian is

$$\mathfrak{L}\{\chi,\chi_{,\mu},h_{k}{}^{\mu},A^{ij}{}_{\mu}\}\equiv\mathfrak{H}L\{\chi,\chi_{;k}\}.$$

(We can drop the prime without risk of confusion.) It may be asked whether this Lagrangian is unique in the same sense as the modified Lagrangian L' of Sec. 2, and in fact it is easy to see that it is not. The reason for this is that if one starts with two Lagrangians L_1 and L_2 which differ by an explicit divergence, and are therefore equivalent, then the modified Lagrangians $\mathfrak{L}_1$ and $\mathfrak{L}_2$ are not necessarily equivalent. Consider for example the Lagrangian for a real scalar field written in its first-order form

$$L_{1}=\pi^{k}\varphi_{,k}-\tfrac{1}{2}\pi^{k}\pi_{k}-\tfrac{1}{2}m^{2}\varphi^{2}. \quad (4.12)$$

This is equivalent to

$$L_{2}=-\pi^{k}{}_{,k}\varphi-\tfrac{1}{2}\pi^{k}\pi_{k}-\tfrac{1}{2}m^{2}\varphi^{2}, \quad (4.13)$$

but the corresponding modified Lagrangians differ by

$$\begin{aligned}\mathfrak{L}_{1}-\mathfrak{L}_{2}&\equiv\mathfrak{H}(\pi^{k}\varphi)_{;k}\\&\equiv\mathfrak{H}h_{k}{}^{\mu}[(\pi^{k}\varphi)_{,\mu}+A^{k}{}_{i\mu}\pi^{i}\varphi]\end{aligned} \quad (4.14)$$

which is not an explicit divergence. Thus in order to define the modified Lagrangian $\mathfrak{L}$ completely it would be necessary to specify which of the possible equivalent forms of the original Lagrangian is to be chosen. The reasons for this situation and the problem of choosing the correct form are discussed in the Appendix.

As in Sec. 2, one may define modified "currents" in terms of $L\equiv L\{\chi,\chi_{;k}\}$ by

$$\mathfrak{T}^{k}{}_{\mu}\equiv\partial\mathfrak{L}/\partial h_{k}{}^{\mu}\equiv\mathfrak{H}b^{i}{}_{\mu}\{(\partial L/\partial\chi_{;k})\chi_{;i}-\delta^{k}{}_{i}L\}, \quad (4.15)$$

$$\mathfrak{S}^{\mu}{}_{ij}\equiv-2(\partial\mathfrak{L}/\partial A^{ij}{}_{\mu})\equiv-\mathfrak{H}h_{k}{}^{\mu}(\partial L/\partial\chi_{;k})S_{ij}\chi, \quad (4.16)$$

where $b^i{}_\mu$ is the inverse of $h_i{}^\mu$, satisfying

$$b^{i}{}_{\mu}h_{i}{}^{\nu}=\delta_{\mu}{}^{\nu},\quad b^{i}{}_{\mu}h_{j}{}^{\mu}=\delta^{i}{}_{j}.$$

To express the "conservation laws" which these currents satisfy in a simple form, it is convenient to extend the definition of the covariant derivative $\chi_{|\mu}$ (not $\chi_{;k}$). Originally, it is defined for χ and, therefore, by a trivial extension for any other quantity which is invariant under ξ^μ transformations, and transforms linearly under ϵ^{ij} transformations. We wish to extend it to any quantity which transforms linearly under ϵ^{ij} transformations, by simply ignoring the ξ^μ transformation properties altogether. Thus, for example, we would have

$$h_{i}{}^{\mu}{}_{|\nu}\equiv h_{i}{}^{\mu}{}_{,\nu}-A^{k}{}_{i\nu}h_{k}{}^{\mu}, \quad (4.17)$$

according to the ϵ^{ij} transformation law of $h_i{}^\mu$. We shall call this the ϵ covariant derivative. Later we shall define another covariant derivative which takes account of ξ^μ transformations also.

One can easily calculate the commutator of two ϵ covariant differentiations.[15] This gives

$$\chi_{|\mu\nu}-\chi_{|\nu\mu}=\tfrac{1}{2}R^{ij}{}_{\mu\nu}S_{ij}\chi, \quad (4.18)$$

where

$$R^{i}{}_{j\mu\nu}\equiv A^{i}{}_{j\mu,\nu}-A^{i}{}_{j\nu,\mu}-A^{i}{}_{k\mu}A^{k}{}_{j\nu}+A^{i}{}_{k\nu}A^{k}{}_{j\mu}.$$

This quantity is covariant under ϵ^{ij} transformations, and satisfies the cyclic identity

$$R^{i}{}_{j\mu\nu|\rho}+R^{i}{}_{j\nu\rho|\mu}+R^{i}{}_{j\rho\mu|\nu}\equiv 0.$$

[13] The reason for this may be seen in terms of the variation $\delta_0\chi$ given by (4.2). The analogs of the matrices T_a are clearly $-\partial_\mu$ and S_{ij}, so that the presence of the derivative $\chi_{,\mu}$ in the last term of (4.3) is to be expected. By analogy with (2.6) we should expect the covariant derivative to have the form

$$\chi_{;k}=\delta_{k}{}^{\mu}\chi_{,\mu}+\tfrac{1}{2}A^{ij}{}_{k}S_{ij}\chi-A^{\mu}{}_{k}\partial_{\mu}\chi.$$

Because of the appearance of derivatives, the first and last terms can be combined in the form $h_k{}^\mu\chi_{,\mu}$, where $h_k{}^\mu=\delta_k{}^\mu-A^\mu{}_k$. If we then set $A^{ij}{}_k=h_k{}^\mu A^{ij}{}_\mu$, we arrive at the same form for $\chi_{;k}$ as that obtained in the text.

[14] Multiplication of the entire Lagrangian by a constant factor is of course unimportant.

[15] Note that this could not be done without extending the definition, since one must know how to treat the index on $\chi_{|\mu}$. Here, as in Sec. 2, we simply ignore it.

It is thus closely analogous to $F^a{}_{\mu\nu}$. Note that $R^{ij}{}_{\mu\nu}$ is antisymmetric in both pairs of indices.

In terms of the ϵ covariant derivative, the "conservation laws" can be expressed in the form[5]

$$(\mathfrak{T}^k{}_\nu h_k{}^\mu)_{|\mu}+\mathfrak{T}^k{}_\mu h_k{}^\mu{}_{|\nu}=\tfrac{1}{2}\mathfrak{S}^\mu{}_{ij}R^{ij}{}_{\mu\nu}, \tag{4.19}$$

$$\mathfrak{S}^\mu{}_{ij|\mu}=\mathfrak{T}_{i\mu}h_j{}^\mu-\mathfrak{T}_{j\mu}h_i{}^\mu. \tag{4.20}$$

5. FREE GRAVITATIONAL LAGRANGIAN

We now wish to examine the quantity $\chi_{;k}$, rather than $\chi_{|\mu}$. As before, the covariant derivative of any quantity which transforms in a similar way to χ may be defined analogously. Now in particular $\chi_{;k}$ itself (unlike $\chi_{|\mu}$) is such a quantity, and therefore without extending the definition of covariant derivative one can evaluate the commutator $\chi_{;kl}-\chi_{;lk}$. However, this quantity is not simply obtained by multiplying $\chi_{|\mu\nu}-\chi_{|\nu\mu}$ by $h_k{}^\mu h_l{}^\nu$, as one might expect. The reason for this is that in evaluating $\chi_{;kl}$ one differentiates the $h_k{}^\mu$ in $\chi_{;k}$, and moreover adds an extra $A^i{}_{k\mu}$ term on account of the index k. Thus one finds

$$\chi_{;kl}-\chi_{;lk}=\tfrac{1}{2}R^{ij}{}_{kl}S_{ij}\chi-C^i{}_{kl}\chi_{;i}, \tag{5.1}$$

where

$$R^{ij}{}_{kl}\equiv h_k{}^\mu h_l{}^\nu R^{ij}{}_{\mu\nu}, \tag{5.2}$$

$$C^i{}_{kl}\equiv(h_k{}^\mu h_l{}^\nu-h_l{}^\mu h_k{}^\nu)b^i{}_{\mu|\nu}. \tag{5.3}$$

Note that (5.1) is not simply proportional to χ, but involves $\chi_{;i}$ also.[16]

We now look for a free Lagrangian $\mathfrak{L}_0$ for the new fields. Clearly $\mathfrak{L}_0$ must be an invariant density, and if we set

$$\mathfrak{L}_0\equiv\mathfrak{H}L_0,$$

then it is easy to see, as in the case of linear transformations, that the invariant L_0 must be a function only of the covariant quantities $R^{ij}{}_{kl}$ and $C^i{}_{kl}$. As before, there are many possible forms for $\mathfrak{L}_0$, but there is a difference between this case and the previous one in that all the indices on these expressions are of the same type (unlike $F^a{}_{\mu\nu}$), and one can, therefore, contract the upper indices with the lower. In fact, the condition that L_0 be a scalar in two separate spaces is now reduced to the condition that it be a scalar in one space. In particular, this means that there exists a linear invariant which has no analog in the previous case, namely,

$$R\equiv R^{ij}{}_{ij}.$$

There are in addition several quadratic invariants. However, if we again choose for L_0 the form of lowest possible degree, then we are led to the free Lagrangian[17]

$$\mathfrak{L}_0=\tfrac{1}{2}\mathfrak{H}R \tag{5.4}$$

which differs from (2.10) in being only linear in the derivatives.

With this choice of Lagrangian, the equations of motion for the new fields are

$$\mathfrak{H}(R^{ik}{}_{jk}-\tfrac{1}{2}\delta^i{}_jR)=-\mathfrak{T}^i{}_\mu h_j{}^\mu, \tag{5.5}$$

$$-[\mathfrak{H}(h_i{}^\mu h_j{}^\nu-h_j{}^\mu h_i{}^\nu)]_{|\nu} \equiv\mathfrak{H}(h_k{}^\mu C^k{}_{ij}-h_j{}^\mu C^k{}_{ik}-h_i{}^\mu C^k{}_{kj})=\mathfrak{S}^\mu{}_{ij}. \tag{5.6}$$

From Eq. (5.6) one can immediately obtain a strict conservation law

$$(\mathfrak{S}^\mu{}_{ij}+\mathfrak{s}^\mu{}_{ij})_{,\mu}=0, \tag{5.7}$$

where

$$\mathfrak{s}^\mu{}_{ij}\equiv\mathfrak{H}A^k{}_{i\nu}(h_j{}^\mu h_k{}^\nu-h_k{}^\mu h_j{}^\nu)-\mathfrak{H}A^k{}_{j\nu}(h_i{}^\mu h_k{}^\nu-h_k{}^\mu h_i{}^\nu).$$

This quantity is expressible in the form

$$\mathfrak{s}^\mu{}_{ij}\equiv-2(\partial\mathfrak{L}_0/\partial A^{ij}{}_\mu)\equiv-\tfrac{1}{2}(\partial\mathfrak{L}_0/\partial A^{mn}{}_{\nu,\mu})f_{ij}{}^{mn}{}_{kl}A^{kl}{}_\nu,$$

which is closely analogous to (2.12), and should be compared with (4.16). Equation (5.7) is a rather surprising result, since $\mathfrak{S}^\mu{}_{ij}$ may very reasonably be interpreted as the spin density of the matter field,[18] so that it appears to be a law of conservation of spin with no reference to the orbital angular momentum. In fact, however, the orbital angular momentum appears in the corresponding "covariant conservation law" (4.20), and therefore part of the "spin" of the gravitational field, $\mathfrak{s}^\mu{}_{ij}$, may be regarded as arising from this source. Nevertheless, Eq. (5.7) differs from other statements of angular momentum conservation in that the coordinates do not appear explicitly.

It would also be possible to deduce from Eq. (5.5) a strict conservation law

$$[h_k{}^\nu(\mathfrak{T}^k{}_\mu+\mathfrak{t}^k{}_\mu)]_{,\nu}=0, \tag{5.8}$$

but there is a considerable amount of freedom in choosing $\mathfrak{t}^k{}_\mu$. The most natural definition, by analogy with (4.15) would be

$$\mathfrak{t}^k{}_\mu\equiv\partial\mathfrak{L}_0/\partial h_k{}^\mu,$$

and this quantity does indeed satisfy (5.8). However, in this case the expression within the parentheses itself vanishes, so that (5.8) is rather trivial. We shall not discuss the question of the correct choice of $\mathfrak{t}^k{}_\mu$ further, as this lies beyond the scope of the present paper.[19]

It should be noted that Eq. (5.6) can be solved, at least in principle, for $A^{ij}{}_\mu$. In the simple case when $\mathfrak{S}^\mu{}_{ij}$ vanishes, one finds[20]

$$A_{ij\mu}={}^0A_{ij\mu}\equiv\tfrac{1}{2}b^k{}_\mu(c_{kij}-c_{ijk}-c_{jki}), \quad c^k{}_{ij}\equiv(h_i{}^\mu h_j{}^\nu-h_j{}^\mu h_i{}^\nu)b^k{}_{\mu,\nu}. \tag{5.9}$$

[16] This is another example of the fact that for ξ^μ transformations derivatives play the role of the matrices T_a. Compare footnote 13.

[17] We choose units in which $\kappa=1$ (as well as $c=\hbar=1$).

[18] See H. J. Belinfante, Physica 6, 887 (1939), and footnote 5.

[19] It is well known in the case of the ordinary metric theory of gravitation that many definitions of the energy pseudotensor are possible. See, for example, P. G. Bergmann, Phys. Rev. 112, 287 (1958).

[20] The ${}^0A^{ij}{}_\mu$ are Ricci's coefficients of rotation. See for instance V. Fock, Z. Physik 57, 261 (1929).

In general, if we write

$$\mathfrak{S}^{\mu}{}_{ij} \equiv \mathfrak{H} h_k{}^{\mu} S^k{}_{ij},$$

then

$$A_{ij\mu} = {}^0A_{ij\mu} - \tfrac{1}{2} b^k{}_{\mu}(S_{kij} - S_{ijk} - S_{jki} - \eta_{ki} S^l{}_{lj} - \eta_{kj} S^l{}_{il}). \quad (5.10)$$

If the original Lagrangian L is of first order in the derivatives, then $S^k{}_{ij}$ is independent of $A^{ij}{}_{\mu}$ so that (5.10) is an explicit solution. Otherwise, however, $A^{ij}{}_{\mu}$ also appears on the right-hand side of this equation.

We conclude this section with a discussion of the Lagrangian for the fields $A^a{}_{\mu}$ introduced in Sec. 2 when the "gravitational" fields $h_k{}^{\mu}$ and $A^{ij}{}_{\mu}$ are also introduced. The fields $A^a{}_{\mu}$ should not be regarded merely as components of χ when dealing with Lorentz transformations, since one must preserve the invariance under the linear transformations. To find the correct form of the Lagrangian, one should consider simultaneously Lorentz transformations and these linear transformations. This can be done provided that the matrices T^a commute with the S_{ij}, a condition which is always fulfilled in practice. Then one finds that χ_k in L should be replaced by a derivative which is covariant under both (2.1) and (4.1), namely,

$$\chi_{;k} = h_k{}^{\mu}(\chi_{,\mu} + \tfrac{1}{2} A^{ij}{}_{\mu} S_{ij}\chi + A^a{}_{\mu} T_a \chi).$$

The commutator $\chi_{;kl} - \chi_{;lk}$ then contains the extra term

$$F^a{}_{kl} T^a \chi,$$

where

$$F^a{}_{kl} \equiv h_k{}^{\mu} h_l{}^{\nu} F^a{}_{\mu\nu},$$

with $F^a{}_{\mu\nu}$ given by (2.9). It is important to notice that the derivatives of $A^a{}_{\mu}$ in $F^a{}_{\mu\nu}$ are ordinary derivatives, not covariant ones. (We shall see in the next section that the ordinary and covariant curls are not equal, because the affine connection is in general nonsymmetric.) As before, one can see that any invariant function of $A^a{}_{\mu}$ must be a function of $F^a{}_{kl}$ only, and the simplest free Lagrangian for $A^a{}_{\mu}$ is, therefore,

$$-\tfrac{1}{4}\mathfrak{H} F^a{}_{kl} F_a{}^{kl}. \quad (5.11)$$

6. GEOMETRICAL INTERPRETATION

Up to this point, we have not given any geometrical significance to the transformations (4.1), or to the new fields $h_k{}^{\mu}$ and $A^{ij}{}_{\mu}$, but it is useful to do so in order to be able to compare the theory with the more familiar metric theory of gravitation.

Now the ξ^{μ} transformations are general coordinate transformations, and according to (4.11) $h_k{}^{\mu}$ transforms like a contravariant vector under these transformations, while $b^k{}_{\mu}$ and $A^{ij}{}_{\mu}$ transform like covariant vectors. Thus the quantity

$$g_{\mu\nu} \equiv b^k{}_{\mu} b_{k\nu} \quad (6.1)$$

is a symmetric covariant tensor, and may therefore be interpreted as the metric tensor of a Riemannian space. It is moreover invariant under the ϵ^{ij} transformations. Evidently, the Greek indices may be regarded as world tensor indices, and we must of course abandon for them the convention that all indices are to be raised or lowered with the flat-space metric $\eta_{\mu\nu}$, and use $g_{\mu\nu}$ instead. It is easy to see that the scalar density $\mathfrak{H}$ is equal to $(-g)^{\frac{1}{2}}$, where $g = \det(g_{\mu\nu})$.

Now, in view of the relation (6.1), $h_k{}^{\mu}$ and $b^k{}_{\mu}$ are the contravariant and covariant components, respectively, of a vierbein system in the Riemannian space.[21] Thus the ϵ^{ij} transformations should be interpreted as vierbein rotations, and the Latin indices as local tensor indices with respect to this vierbein system. The original field χ may be decomposed into local tensors and spinors,[22] and from the tensors one can form corresponding world tensors by multiplying by $h_k{}^{\mu}$ or $b^k{}_{\mu}$. For example, from a local vector v^i one can form

$$v^{\mu} = h_i{}^{\mu} v^i, \quad v_{\mu} = b^i{}_{\mu} v_i. \quad (6.2)$$

No confusion can be caused by using the same symbol v for the local and world vectors, since they are distinguished by the type of index, and indeed we have already used this convention in (5.2). Note that $v_{\mu} = g_{\mu\nu} v^{\nu}$, so that (6.2) is consistent with the choice of metric (6.1). We shall frequently use this convention of associating world tensors with given local tensors without explicit mention on each occasion.

The field $A^i{}_{j\mu}$ may reasonably be called a "local affine connection" with respect to the vierbein system, since it specifies the covariant derivatives of local tensors or spinors.[23] For a local vector, this takes the form

$$v^i{}_{|\nu} = v^i{}_{,\nu} + A^i{}_{j\nu} v^j, \quad v_{j|\nu} = v_{j,\nu} - A^i{}_{j\nu} v_i. \quad (6.3)$$

It may be noticed that the relation (4.10) between $\chi_{|\mu}$ and $\chi_{;k}$ is of the same type as (6.2) and could be written simply as

$$\chi_{;\mu} = \chi_{|\mu} \quad (6.4)$$

according to our convention. However, we shall retain the use of two separate symbols because we wish to extend the definition of covariant derivative in a different way to that of Sec. 4. It seems natural to define the covariant derivative of a world tensor in terms of the covariant derivative of the associated local tensor. Thus, for instance, to define the covariant derivatives of the world vectors (6.2) one would form the world tensors corresponding to (6.3). This gives

$$v^{\lambda}{}_{;\nu} \equiv h_i{}^{\lambda} v^i{}_{|\nu} = v^{\lambda}{}_{,\nu} + \Gamma^{\lambda}{}_{\mu\nu} v^{\mu},$$

$$v_{\mu;\nu} \equiv b^i{}_{\mu} v_{i|\nu} = v_{\mu,\nu} - \Gamma^{\lambda}{}_{\mu\nu} v_{\lambda},$$

where

$$\Gamma^{\lambda}{}_{\mu\nu} \equiv h_i{}^{\lambda} b^i{}_{\mu|\nu} \equiv -b^i{}_{\mu} h_i{}^{\lambda}{}_{|\nu}. \quad (6.5)$$

Note that this definition of $\Gamma^{\lambda}{}_{\mu\nu}$ is equivalent to the

[21] See for instance H. Weyl, Z. Physik **56**, 330 (1929).
[22] H. J. Belinfante, Physica **7**, 305 (1940).
[23] Compare J. A. Schouten, J. Math. and Phys. **10**, 239 (1931).

requirement that the covariant derivatives of the vierbein components should vanish,

$$h_i{}^\lambda{}_{;\nu}\equiv 0, \quad b^i{}_{\mu;\nu}\equiv 0. \tag{6.6}$$

For a generic quantity α transforming according to

$$\delta\alpha=\tfrac{1}{2}\epsilon^{ij}S_{ij}\alpha+\xi^\lambda{}_{,\mu}\Sigma_\lambda{}^\mu\alpha, \tag{6.7}$$

the covariant derivative is defined by[21]

$$\alpha_{;\nu}\equiv\alpha_{,\nu}+\tfrac{1}{2}A^{ij}{}_\nu S_{ij}\alpha+\Gamma^\lambda{}_{\mu\nu}\Sigma_\lambda{}^\mu\alpha, \tag{6.8}$$

whereas the ϵ covariant derivative defined in Sec. 4 is obtained by simply omitting the last term of (6.8). Note that the two derivatives are equal for purely local tensors or spinors, but not otherwise. One easily finds that the commutator of two covariant differentiations is given by

$$\alpha_{;\mu\nu}-\alpha_{;\nu\mu}=\tfrac{1}{2}R^{ij}{}_{\mu\nu}S_{ij}\alpha+R^\rho{}_{\sigma\mu\nu}\Sigma_\rho{}^\sigma\alpha-C^\lambda{}_{\mu\nu}\alpha_{;\lambda},$$

where $R^\rho{}_{\sigma\mu\nu}$ and $C^\lambda{}_{\mu\nu}$ are defined in the usual way in terms of $R^i{}_{j\mu\nu}$ and $C^i{}_{kl}$. They are both world tensors, and can easily be expressed in terms of $\Gamma^\lambda{}_{\mu\nu}$, in the form[24]

$$R^\rho{}_{\sigma\mu\nu}=\Gamma^\rho{}_{\sigma\mu,\nu}-\Gamma^\rho{}_{\sigma\nu,\mu}-\Gamma^\rho{}_{\lambda\mu}\Gamma^\lambda{}_{\sigma\nu}+\Gamma^\rho{}_{\lambda\nu}\Gamma^\lambda{}_{\sigma\nu}, \tag{6.9}$$

$$C^\lambda{}_{\mu\nu}=\Gamma^\lambda{}_{\mu\nu}-\Gamma^\lambda{}_{\nu\mu}. \tag{6.10}$$

Thus one sees that $R^\rho{}_{\sigma\mu\nu}$ is just the Riemann tensor formed from the affine connection $\Gamma^\lambda{}_{\mu\nu}$.

From (6.6) it follows that

$$g_{\mu\nu;\rho}\equiv 0, \tag{6.11}$$

so that it is consistent to interpret $\Gamma^\lambda{}_{\mu\nu}$ as an affine connection in the Riemannian space. However, the definition (6.5) evidently does not guarantee that it is symmetric, so that in general it is not the Christoffel connection. The curvature scalar R has the usual form

$$R\equiv R^\mu{}_\mu, \quad R_{\mu\nu}\equiv R^\lambda{}_{\mu\lambda\nu},$$

so that the free gravitational Lagrangian is just the usual one except for the nonsymmetry of $\Gamma^\lambda{}_{\mu\nu}$. It should be remarked that it would be incorrect to treat the 64 components of $\Gamma^\lambda{}_{\mu\nu}$ as independent variables, since there are only 24 components of $A^{ij}{}_\mu$. In fact the $\Gamma^\lambda{}_{\mu\nu}$ are restricted by the 40 identities (6.11). Thus there is no contradiction with the well-known fact that the first-order Palatini Lagrangian with nonsymmetric $\Gamma^\lambda{}_{\mu\nu}$ does not yield (6.11) as equations of motion.[25]

The equations of motion (5.5) and (5.6) can be rewritten in the form

$$\mathfrak{H}(R_{\mu\nu}-\tfrac{1}{2}g_{\mu\nu}R)=-\mathfrak{T}_{\mu\nu}, \tag{6.12}$$

$$\mathfrak{H}C^\lambda{}_{\mu\nu}=\mathfrak{S}^\lambda{}_{\mu\nu}-\tfrac{1}{2}\delta^\lambda{}_\mu\mathfrak{S}^\rho{}_{\rho\nu}-\tfrac{1}{2}\delta^\lambda{}_\nu\mathfrak{S}^\rho{}_{\mu\rho}. \tag{6.13}$$

From Eqs. (6.10) and (6.13) one sees that in the absence of matter the affine connection $\Gamma^\lambda{}_{\mu\nu}$ is symmetric, and therefore equal to the Christoffel connection ${}^0\Gamma^\lambda{}_{\mu\nu}$. (This is the analog for world tensors of ${}^0A^i{}_{j\mu}$.) Then $R_{\mu\nu}$ is symmetric, and Eq. (6.12) yields Einstein's familiar equations for empty space,

$$R_{\mu\nu}=0.$$

However, when matter is present, $\Gamma^\lambda{}_{\mu\nu}$ is no longer symmetric, and its antisymmetric part is given by (6.13). Then the tensor $R_{\mu\nu}$ is also nonsymmetric, and correspondingly the energy tensor density $\mathfrak{T}_{\mu\nu}$ is in general nonsymmetric, because $h_k{}^\mu$ does not appear in $\mathfrak{L}$ only through the symmetric combination $g^{\mu\nu}$. Thus the theory differs slightly from the usual one, in a way first noted by Weyl.[4] In the following section, we shall investigate this difference in more detail.[5]

Finally, we can rewrite the covariant conservation laws in terms of world tensors. It is convenient to define the contraction

$$C_\mu\equiv C^\lambda{}_{\mu\lambda},$$

since the covariant divergence of a vector density $\mathfrak{f}^\mu$ is then

$$\mathfrak{f}^\mu{}_{;\mu}=\mathfrak{f}^\mu{}_{,\mu}+C_\mu\mathfrak{f}^\mu. \tag{6.14}$$

The conservation laws become

$$\mathfrak{T}^\nu{}_{\mu;\nu}-C_\nu\mathfrak{T}^\nu{}_\mu+C^\lambda{}_{\mu\nu}\mathfrak{T}^\nu{}_\lambda=\tfrac{1}{2}R^{\rho\sigma}{}_{\mu\nu}\mathfrak{S}^\nu{}_{\rho\sigma},$$

$$\mathfrak{S}^\mu{}_{\rho\sigma;\mu}-C_\mu\mathfrak{S}^\mu{}_{\rho\sigma}=\mathfrak{T}_{\rho\sigma}-\mathfrak{T}_{\sigma\rho}.$$

It may be noticed that these are slightly more complicated than the expressions in terms of the ϵ covariant derivative.

7. COMPARISON WITH METRIC THEORY

For simplicity, we shall assume in this section that L is only of first order in the derivatives, so that (5.10) is an explicit solution for $A^{ij}{}_\mu$. The difference between the theory presented here and the usual one arises because we are using a Lagrangian $\mathfrak{L}_0$ of first order, in which $h_k{}^\mu$ and $A^{ij}{}_\mu$ are independent variables. The situation is entirely analogous to that which obtains for any theory with "derivative" interaction. In first-order form, the "momenta" $A^{ij}{}_\mu$ are not just equal to derivatives of the "coordinates" $h_k{}^\mu$, or in other words to ${}^0A^{ij}{}_\mu$. Thus an interaction which appears simple in first-order form will be more complicated if a second-order Lagrangian is used, and vice versa.

The second-order form of the Lagrangian may be obtained by substituting for $A^{ij}{}_\mu$ the expression (5.10). This gives

$$\mathfrak{L}'={}^0\mathfrak{L}+{}^0\mathfrak{L}_0+{}^1\mathfrak{L},$$

where ${}^0\mathfrak{L}$ and ${}^0\mathfrak{L}_0$ are obtained from $\mathfrak{L}$ and $\mathfrak{L}_0$ by replacing $A^{ij}{}_\mu$ by ${}^0A^{ij}{}_\mu$ (or equivalently $\Gamma^\lambda{}_{\mu\nu}$ by ${}^0\Gamma^\lambda{}_{\mu\nu}$), and ${}^1\mathfrak{L}$ is an additional term quadratic in $S^k{}_{ij}$, namely,

$${}^1\mathfrak{L}=\tfrac{1}{8}\mathfrak{H}(2S_{ijk}S^{jki}-S_{ijk}S^{ijk}+2S^i{}_{ik}S_j{}^{jk}). \tag{7.1}$$

In this Lagrangian, only $h_k{}^\mu$ and χ are treated as inde-

[24] This is a generalization to nonsymmetric affinities of the result proved in the appendix to footnote 3. See also footnotes 4 and 5.

[25] See for instance E. Schrödinger, *Space-time Structure* (Cambridge University Press, New York, 1950).

pendent variables. The equations of motion are equivalent to those previously obtained if the variables $A^{ij}{}_\mu$ are eliminated from the latter by using (5.10).

The usual metric theory, on the other hand, is given by the Lagrangian

$$\mathfrak{L}'' = {}^0\mathfrak{L} + {}^0\mathfrak{L}_0,$$

without the extra terms (7.1). If this Lagrangian were written in a first-order form by introducing additional independent variables $A^{ij}{}_\mu$, then one would arrive at a form identical to the one given here except for the appearance of extra terms equal to (7.1) with a negative sign.

Thus we see that the only difference between the two theories is the presence or absence of these "direct-interaction" terms. Now if we had not set $\kappa=1$, then $\mathfrak{L}_0$ would have a factor κ^{-1}, whereas the terms (7.1) would appear with the factor κ. They are, therefore, extremely small in comparison to other interaction terms. In particular, for a Dirac field, they would be proportional to (see Appendix)

$$\kappa\bar{\psi}\gamma_k\gamma_5\psi\bar{\psi}\gamma^k\gamma_5\psi.$$

Thus they are similar in form to the Fermi interaction terms, but much smaller in magnitude, so that it seems impossible that they would lead to any observable difference between the predictions of the two theories. Hence we must conclude that for all practical purposes the theory presented here is equivalent to the usual one.

ACKNOWLEDGMENTS

The author is indebted to Drs. J. L. Anderson, P. W. Higgs, and D. W. Sciama for helpful discussions and comments.

APPENDIX

In this appendix we shall discuss the remaining ambiguity in the modified Lagrangian. It was pointed out in Sec. 4 that the generally covariant Lagrangians obtained from two equivalent Lagrangians L_1 and L_2 are in general inequivalent. One can now see that in fact they differ by a covariant divergence. Thus (4.14) can be written in the form

$$\mathfrak{L}_1 - \mathfrak{L}_2 = (\mathfrak{H} h_k{}^\mu \pi^k \varphi)_{;\mu},$$

but in view of (6.14) this is not equal to the ordinary divergence. It is clear that quite generally changing L by a divergence must change $\mathfrak{L}$ by the covariant divergence of a quantity which is a vector density under coordinate transformations, and invariant under all other transformations. This is the reason for the difference between this case and that of the linear transformations of Sec. 2.

We now wish to investigate the possibility of choosing a criterion which will select a particular form of L, and thus specify $\mathfrak{L}$ completely. There does not seem to be any really compelling reason for one choice rather than another, but there are plausible arguments for a particular choice.

The most obvious criterion would be to require that the Lagrangian should be written in the symmetrized first-order form suggested by Schwinger,[26] which in the case of the scalar field discussed in Sec. 4 is

$$L = \tfrac{1}{2}(L_1 + L_2).$$

This corresponds to treating φ and π^k on a symmetrical footing. However, this may not in fact be the correct choice, because for some purposes φ and π^k should not be treated in this way. In fact, the two Lagrangians differ in one important respect: $\mathfrak{L}_1$ is independent of $A^{ij}{}_\mu$, whereas $\mathfrak{L}_2$ is not. Correspondingly, for L_1 the quantity $S^k{}_{ij}$ vanishes, whereas for L_2 one finds

$$S^k{}_{ij} = (\delta^k{}_i\pi_j - \delta^k{}_j\pi_i)\varphi.$$

The conservation laws in the two cases are of course the same, because the quantities $T^k{}_i$ also differ. Now the tensor $S^k{}_{ij}$ has often been interpreted as the spin density,[18] so that the two cases differ with regard to the separation of the total angular momentum into orbital and spin terms. The scalar field is normally regarded as a field of spinless particles, so that one would naturally expect $S^k{}_{ij}$ to vanish. This, therefore, furnishes a possible criterion, which would select L_1 rather than L_2. With this choice, a preferred position is assigned to the "wave function" φ rather than the "momenta" π^k, and the derivatives are written on φ only. In this way one achieves a vanishing spin tensor, because the matrices S_{ij} are zero for the scalar field φ, but not for the vector π^k. It may be noticed that L_1 is automatically selected if one writes the Lagrangian in its second-order form in terms of φ only:

$$L_1' = \tfrac{1}{2}\varphi_{,k}\varphi^{,k} - \tfrac{1}{2}m^2\varphi^2,$$

which yields the modified Lagrangian

$$\mathfrak{L}_1' = \tfrac{1}{2}\mathfrak{H}(g^{\mu\nu}\varphi_{,\mu}\varphi_{,\nu} - m^2\varphi^2),$$

equivalent to $\mathfrak{L}_1$.[27] This should be contrasted with the second-order form of $\mathfrak{L}_2$, which is

$$\mathfrak{L}_2' = \tfrac{1}{2}\mathfrak{H}^{-1}(\mathfrak{H} h_i{}^\mu\varphi)_{;\mu}(\mathfrak{H} h^{i\nu}\varphi)_{;\nu} - \tfrac{1}{2}\mathfrak{H} m^2\varphi^2,$$

and clearly differs from $\mathfrak{L}_1'$ by a covariant divergence.

This seems to be a resonable criterion, but the arguments for it cannot be regarded as conclusive. For, although it is true that the spin tensor obtained from L_2 is nonzero, it is still true that the three space-space components of the total spin

$$\mathcal{S}_{ij} = \int d_3x\, S^0{}_{ij}$$

are zero. Thus L_1 and L_2 differ only in the values of the

[26] J. Schwinger, Phys. Rev. 91, 713 (1953).

[27] Here $\mathfrak{L}_1$ is a "linearization" of $\mathfrak{L}_1'$ in the sense of T. W. B. Kibble and J. C. Polkinghorne, Nuovo cimento 8, 74 (1958).

spin part of the $(0i)$ components of angular momentum. Indeed, one easily sees that it is true in general that adding a divergence to L will change only the $(0i)$ components of $\mathcal{S}_{ij}$. Since it is not at all clear what significance should be attached to the separation of these components into "orbital" and "spin" terms, it might be questioned whether one should expect the spin terms to vanish even for a spinless particle. Even so, the choice of L_1 seems in this case to be the most reasonable.

For a field of spin 1, the corresponding choice would be

$$L_1 = -\tfrac{1}{2} f^{ij}(a_{i,j} - a_{j,i}) + \tfrac{1}{4} f^{ij} f_{ij} + \tfrac{1}{2} m^2 a_i a^i,$$

which is again equivalent to the choice of the second-order Lagrangian in terms of a_i only. It yields

$$S^k{}_{ij} = a_i f_j{}^k - a_j f_i{}^k,$$

which is a reasonable definition of the spin density.[28] The modified Lagrangian may be expressed in terms of the world vector a_μ as

$$\mathfrak{L} = -\tfrac{1}{4}\mathfrak{H} g^{\mu\rho} g^{\nu\sigma}(a_{\mu;\nu} - a_{\nu;\mu})(a_{\rho;\sigma} - a_{\sigma;\rho}) + \tfrac{1}{2}\mathfrak{H} m^2 g^{\mu\nu} a_\mu a_\nu. \quad \text{(A.1)}$$

It should be noticed that the electromagnetic Lagrangian is not obtained simply by putting $m=0$ in (A.1). The difference is that the derivatives in (A.1) are covariant derivatives, and since $\Gamma^\lambda{}_{\mu\nu}$ is nonsymmetric the covariant curl is not equal to the ordinary curl (though both are of course tensors). In fact, (A.1) with $m=0$ would not be gauge invariant. The reason for the difference is that a_i is here treated simply as a component of χ, whereas A_μ is introduced along with the gravitational variables to ensure gauge invariance.[29]

For a spinor field ψ, symmetry between ψ and $\bar\psi$ appears to demand that one should choose the symmetrized Lagrangian

$$L = \tfrac{1}{2}(\bar\psi i\gamma^k \psi_{,k} - \bar\psi_{,k} i\gamma^k \psi) - m\bar\psi\psi,$$

which yields the spin density

$$S_{kij} = \tfrac{1}{2}\epsilon_{kijl}\bar\psi i\gamma^l \gamma_5 \psi.$$

Since the Lagrangian $\mathfrak{L}$ must be Hermitian, one could not write the derivative on ψ alone. There remains, however, another possible choice: We could introduce a distinction between the left- and right-handed components, $\psi_\pm = \tfrac{1}{2}(1 \pm i\gamma_5)\psi$, treating one of them line φ and the other like π^k. This gives the Lagrangian

$$L = \tfrac{1}{2}\bar\psi i\gamma^k(1+i\gamma_5)\psi_{,k} - \tfrac{1}{2}\bar\psi_{,k} i\gamma^k(1-i\gamma_5)\psi - m\bar\psi\psi.$$

This form of Lagrangian may seem rather unnatural, but it should be mentioned because there are other grounds for treating ψ_+ and ψ_- on a nonsymmetrical footing.[30]

---

[28] Compare footnote 18.

[29] This has the rather strange consequence that for the electromagnetic field the "spin" tensor $S^k{}_{ij}$ vanishes, since the Lagrangian is independent of $A^{ij}{}_\mu$.

[30] See R. P. Feynman and M. Gell-Mann, Phys. Rev. **109**, 193 (1958).

Chapter 5

Accelerated Frames: Generalizing the Lorentz Transformations*

---

*C. Møller, T. Fulton, R. Rohrlich, L. Witten, T. Y. Wu, Y. C. Lee, J. P. Hsu, L. Hsu, D. T. Schmitt, T. Kleinschmidt

DET KGL. DANSKE VIDENSKABERNES SELSKAB
MATEMATISK-FYSISKE MEDDELELSER, BIND XX, NR. 19

ON HOMOGENEOUS GRAVITATIONAL FIELDS IN THE GENERAL THEORY OF RELATIVITY AND THE CLOCK PARADOX

BY

C. MØLLER

KØBENHAVN
I KOMMISSION HOS EJNAR MUNKSGAARD
1943

1. Introduction and statement of the problem.

When the behaviour of clocks is treated according to the principles of the special theory of relativity, without making due allowance for the principles of the general theory, a well-known paradox can arise, which was already mentioned in EINSTEIN's original paper[1]) and later was discussed in detail by LANGEVIN[2]), LAUE[3]), and LORENTZ[4]). With a slight simplification of the usual representation*, the problem may be stated as follows. Consider two identically constructed clocks, C_1 and C_2, one of which, say C_1, is permanently situated at rest at a point A on the positive X-axis of a definite Lorentz frame of reference K, while C_2 is moving with constant velocity $-v$ in the direction of the X-axis (see Fig. 1). At the moment of coincidence between C_1 and C_2, the readings of the two clocks are compared. After having travelled with constant velocity for a long time, C_2 for a short time is attacked by a constant force F which brings it to rest at the origin 0 of K and starts it back to A with reversed velocity v. At the moment of the second encounter, the clocks C_1 and C_2 are compared again. Let Δt_1 and Δt_2 denote the measurements on the two clocks of the time elapsed between the two encounters. Now, assuming that the force F is so large that the time during which C_2 is accelerated is negligible compared with the time of travel at the constant velocity v, we have, according to the special theory of relativity, the formula**

* Usually, the two clocks are assumed to be initially and finally at rest, which necessitates the further introduction of a force at the beginning and at the end of the experiment.

** Throughout this paper, we shall use a time unit which makes the velocity of light equal to unity. The transition to ordinary units is then performed by replacing in our formulae all time variables t, velocities v, accelerations g, and gravitational potentials Φ by ct, $\frac{v}{c}$, $\frac{g}{c^2}$, $\frac{\Phi}{c^2}$, respectively, where c is the velocity of light in ordinary units.

$$\Delta t_2 = \Delta t_1 \sqrt{1-v^2} \tag{1}$$

which shows that C_2 will register a smaller number of divisions than C_1 at the end of the indicated experiment.

The paradox in question now arises, if we introduce a frame of reference k moving together with C_2 in such a way that C_2 is permanently situated at the origin of k. Since the motion of C_1 with respect to k then is similar to the motion of C_2 with respect to K, it seems that an observer in k should arrive at the conclusion that Δt_1 must be smaller than Δt_2 and must be given by the formula

$$\Delta t_1 = \Delta t_2 \sqrt{1-v^2} \tag{2}$$

in contradiction to (1). In the papers quoted above, it was pointed out, however, that the equation

$$d\tau = dt \sqrt{1-u^2} \tag{3}$$

connecting the proper time $d\tau$ of a clock moving with the velocity u in a given system of reference with the time dt of this system is valid only if the frame of reference is a system of inertia like K. The application of (3) in K thus leads to the correct formula (1), while the application of (3) in k which leads to formula (2) is not justified, since k is accelerated in the middle of the experiment and, therefore, does not constitute a simple system of inertia during this interval.

In the space-time continuum introduced by MINKOWSKI, the two events marked by the first and second encounters of the clocks are represented by two points connected by the world lines of C_1 and C_2, of which the first mentioned is a straight line. Since the lengths of these world lines, on account of (3), are proportional to the proper times Δt_1 and Δt_2 of the two clocks, the statement expressed by (1) may be considered a special case of the general statement that a straight line connecting two points in Minkowski space is of greater length than any other curve (of everywhere time-like character) connecting the two points.

Thus, it was clear that the discussion of the indicated experiment could not lead to any difficulties for the special theory of relativity, since this theory does not make any statement at all regarding the behaviour of clocks in accelerated systems like k. The paradox arose again, however, in the general theory of relativity, according to which a treatment of the behaviour of C_1, from the point of view of an observer in k, must be possible. Neglecting the short interval during which k is no system of inertia, we then find again the formula (2) for the time increase of C_1 measured with the time scale of k and, at first sight, it is difficult to understand how it is possible to account for the difference between (2) and (1) by consideration of the short interval in which k is accelerated. The whole question was clarified by EINSTEIN[5]) who pointed out that, during this interval, the distant masses of the universe are accelerated relative to k, and thus temporarily create a gravitational field which influences the time rates of the clocks in such a way that the total time increase of C_1 measured in the time scale of k is again given by (1).

In his paper just quoted, EINSTEIN did not give any explicit calculations, but it is clear beforehand that the result of a calculation must be as stated above. In fact, since Δt_1 and Δt_2 are proportional to the lengths of the world lines of C_1 and C_2 and these lengths, according to the basic assumptions of the general theory of relativity, are independent of the space-time coordinates used in their evaluation, it is obvious that we shall get the same value for $\frac{\Delta t_1}{\Delta t_2}$ whether the calculation is performed in K or in k. Nevertheless, it is instructive to calculate directly the time increase of C_1 during the existence of the gravitational field in k. For small values of v, this has been done by TOLMAN[6]) who assumed that terms in v higher than the second can be neglected. In order to account for the lack of symmetry between the treatment given to the clock C_1, which was at no time subjected to any force, and that given to the clock C_2, which was subjected to the force F in the middle of the experiment, TOLMAN introduces a temporary homogeneous gravitational field in the description where C_1 is taken as the moving clock and C_2 as the one which remains at rest. This gravitational field is allowed to act on C_1 and C_2 in such a way as to produce the desired change in velocity of C_1, while C_2 remains at rest on account of the force F. By means of the well-known formula for the relative rates of two clocks situated at points of different potential in a weak static gravitational field, TOLMAN then finds for the total increase in time of C_1 and C_2 during the considered experiment the relation

$$\Delta t_1 = \Delta t_2 \left(1 + \frac{1}{2} v^2\right) \tag{4}$$

which, for small v, is in accordance with (1).

Apart from the restriction to the case of small v, this treatment does not seem to us to be complete, since it remains to be shown that the transformation from K to the accelerated system k leads to a system of space-time coordinates in which the components of the metrical tensor are constant in time and are of the form corresponding to the gravitational field explicitly introduced by TOLMAN. In the present paper, we shall investigate this point more closely and without making the assumption of a small velocity v. It is shown that the accelerated frame of reference k may be defined in such a way that the gravitational field in k is static in the sense of the general theory of relativity. The equations by which the space-time coordinates of k are expressed as functions of the coordinates of the system K during the whole experiment are explicitly written down. By means of these equations, the behaviour of the clocks C_1 and C_2 may easily be treated from the alternative standpoints of the observers in the two systems K and k, thus leading to a complete solution of the clock paradox.

2. Uniformly accelerated frames of reference and homogeneous gravitational fields.

In a general discussion of the clock paradox, we need a formula connecting the space-time coordinates X, Y, Z, and T of a Lorentz frame K with the coordinates x, y, z, and t of a "uniformly accelerated" frame of reference k. If the direction of acceleration is chosen as x-axis, the desired transformation must have the form

$$\left.\begin{aligned} x &= f(X, T), \quad y = Y, \quad z = Z \\ t &= h(X, T), \end{aligned}\right\} \tag{5}$$

f and h being functions of X and T, only.

Taking for f and h the expressions

$$f = X - \frac{1}{2} g T^2, \quad h = T, \tag{6}$$

where g is a constant, (5) represents the ordinary transformation to accelerated axes which, at least for small velocities, might be regarded as a reasonable change of coordinates. A free particle in k has then a constant acceleration $-g$, just like a particle in a constant Newtonian field of gravitation. The gravitational field in k, however, is not static in the sense of the general theory of relativity, since the components of the metrical tensor are varying with t.

In fact, introducing (5) and (6) into the expression

$$ds^2 = dX^2 + dY^2 + dZ^2 - dT^2 \tag{7}$$

for the line element in Minkowski space, we get

$$ds^2 = dx^2 + dy^2 + dz^2 + 2gtdxdt - dt^2(1 - g^2 t^2), \tag{8}$$

i. e. the non-vanishing components of the metrical tensor defined by the general expression*

$$ds^2 = g_{ik}\, dx^i\, dx^k, \qquad (x^i) = (x, y, z, t) \tag{9}$$

are

$$\left.\begin{array}{l} g_{11} = g_{22} = g_{33} = 1, \quad g_{44} = -(1 - g^2 t^2) \\ g_{14} = g_{41} = gt. \end{array}\right\} \tag{10}$$

Even the geometry in physical space defined by the three-dimensional line element

$$\left.\begin{array}{l} d\sigma^2 = \sum\limits_{i,k=1}^{3} \gamma_{ik}\, dx^i\, dx^k \\ \gamma_{ik} = g_{ik} - \dfrac{g_{i4}\, g_{k4}}{g_{44}} \end{array}\right\} \tag{11}$$

is seen to vary with t.

From (10) and (11), we get

$$d\sigma^2 = \frac{dx^2}{1 - g^2 t^2} + dy^2 + dz^2 \tag{12}$$

* Here, the usual convention is made regarding the summation over dummy indices from 1 to 4.

in accordance with the fact that the measuring rods in k are subjected to a Lorentz contraction.

The gravitational field in the frame of reference defined by (6) has, therefore, not much resemblance with the gravitational fields assumed in the previous discussions of the clock paradox. Our first task will be, if possible, to choose the functions f and h in (5) in such a way that the gravitational field in k is static. The expression for the element of interval in the new coordinates will then be of the form

$$ds^2 = A \cdot dx^2 + dy^2 + dz^2 - D \cdot dt^2, \tag{13}$$

where A and D are functions of x, only. This expression may be further simplified by taking as coordinate $\int \sqrt{A}\, dx$ instead of x so that the line element takes the form

$$ds^2 = dx^2 + dy^2 + dz^2 - D \cdot dt^2. \tag{13'}$$

If the desired transformation is at all possible, the functions g_{ik} defined by (9) and (13′) must satisfy EINSTEIN's field equations for an empty space

$$G_i^k \equiv R_i^k - \frac{1}{2}\delta_i^k R = 0, \tag{14}$$

where R_i^k is the contracted Riemann-Christoffel tensor, and $R \equiv R_i^i$ is obtained from R_i^k by further contraction. The components of G_i^k have been calculated by DINGLE[7]) for a general line element of the form

$$ds^2 = A(dx^1)^2 + B(dx^2)^2 + C(dx^3)^2 - D(dx^4)^2 \tag{15}$$

with A, B, C, and D being any functions of the coordinates.

Using DINGLE's formula in the special case of (13′), we get simply

$$G_2^2 = G_3^3 = -\frac{1}{2D}\left[D'' - \frac{(D')^2}{2D}\right] = -\frac{1}{D^{1/2}}(D^{1/2})'',$$

where the accents indicate differentiation with respect to x, and all other components G_i^k vanish identically.

The equations (14) thus reduce to the single equation

$$(D^{1/2})'' = 0 \tag{16}$$

with the general solution

$$D = a(1+gx)^2 \tag{16'}$$

containing two arbitrary constants, a and g.

By adequate choice of the time variable, the constant a may be made equal to one, giving for the line element (13') the expression

$$ds^2 = dx^2 + dy^2 + dz^2 - (1+gx)^2\, dt^2. \tag{17}$$

The functions g_{ik}, defined by (9) and (17), which were found as solutions of the equations (14), may now by a simple calculation be shown also to satisfy the more strict conditions

$$R^i_{klm} = 0, \tag{18}$$

where R^i_{klm} is the uncontracted Riemann-Christoffel tensor. This means that the geometry in the space-time continuum corresponding to (17) is pseudo-Euclidean and that the line element (17) may be brought into the simple form (7) by a suitable transformation of the type (5). Apart from an arbitrary Lorentz transformation, which does not change the form (7), this transformation is uniquely determined.

Before we write down explicitly this transformation, which inversely gives the transition from an inertial system K to the desired frame of reference k, we note that the gravitational field in k, according to (17), is uniform in that part of the space for which gx is a small quantity. In fact, neglecting all terms of higher order in gx than the first, we have

$$g_{44} = -1 - 2gx, \tag{19}$$

and the Newtonian gravitational potential Φ_w, which, in the case of "weak" fields, is defined by the equation[8])

$$g_{44} = -1 - 2\Phi_w, \tag{20}$$

has therefore the simple form

$$\Phi_w = gx. \tag{21}$$

The line element (17) has, however, well defined physical consequences for large values of gx also, so that the gravitational field defined by (17) is a generalization of the "weak" uniform field postulated in previous discussions of the clock paradox. The only necessary restriction regarding the values of x is the condition $x > -\frac{1}{g}$.

The geometry of physical space in k is Euclidean, x, y, and z being Cartesian coordinates. The time variable t is the time measured by a standard clock situated at rest at the origin $x = 0$. The increase of time $d\tau$ of a standard clock situated at any other place is given by the formula

$$d\tau = \sqrt{-g_{44}}\, dt = (1 + gx)\, dt, \tag{22}$$

$d\tau$ thus being zero in the singular plane $x = -\frac{1}{g}$.

Turning now to the explicit derivation of the transformation connecting the space-time variables of the two systems K and k, we start with the system k and try to find a transformation by which the gravitational field of k is "transformed away". This may be effected by introduction of a frame of reference consisting of material points which are allowed to fall freely in the gravitational field of k. The world line of a free particle is a geodesic given by the equations

$$\frac{d^2 x^i}{d\tau^2} + \Gamma^i_{kl} \frac{dx^k}{d\tau} \frac{dx^l}{d\tau} = 0, \tag{23}$$

where $d\tau = \frac{1}{i} ds$ is the proper time of the particle and Γ^i_{kl} denote the ordinary Christoffel three index symbols. The values of Γ^i_{kl} in the case of (17) may also be taken from DINGLE's paper[7]), and we get

$$\Gamma^1_{44} = g(1 + gx), \quad \Gamma^4_{14} = \Gamma^4_{41} = \frac{g}{1 + gx},$$

all other components being zero.

The equations (23) with $i = 1, 2, 3$ are then simply

$$\left.\begin{aligned} &\frac{d^2 x}{d\tau^2} + g(1 + gx)\left(\frac{dt}{d\tau}\right)^2 = 0 \\ &\frac{d^2 y}{d\tau^2} = \frac{d^2 z}{d\tau^2} = 0 \end{aligned}\right\} \tag{24}$$

and from (17) we get, as a first integral of (23),

$$\left(\frac{dt}{d\tau}\right)^2 = \frac{1+\left(\frac{dx}{d\tau}\right)^2+\left(\frac{dy}{d\tau}\right)^2+\left(\frac{dz}{d\tau}\right)^2}{(1+gx)^2}. \tag{25}$$

For a particle at rest (or small velocity), (24) reduces to

$$\frac{d^2x}{dt^2} = -g(1+gx)$$

$$\frac{d^2y}{dt^2} = \frac{d^2z}{dt^2} = 0.$$

If the gravitational potential Φ is defined by the equation

$$\frac{d^2\vec{x}}{dt^2} = -\text{grad}\,\Phi, \quad \vec{x} = (x, y, z),$$

we thus get

$$\Phi = gx + \frac{1}{2}g^2x^2, \tag{26}$$

an expression which may be regarded as the generalization of (21) to the case of strong fields.

The equation (20) is seen to hold also in this case, since we get from (17) and (26)

$$g_{44} = -(1+2\Phi). \tag{27}$$

Finally, (22) may be written

$$d\tau = \sqrt{1+2\Phi}\,dt \tag{28}$$

which, for small Φ, reduces to the well-known formula[8]) for weak fields

$$d\tau = (1+\Phi_w)\,dt. \tag{29}$$

Returning now to the general equations (24) and (25), we see that the motion of the particle in the directions of the y- and z-axes is uniform if the proper time τ is used as time scale. We are here only interested in the case where the velocities are zero at $t = \tau = 0$, so that we have the solutions

$$y = y_0, \quad z = z_0, \tag{30}$$

y_0 and z_0 being the initial values of y and z.

From (25) and (24) we then get

$$\frac{dt}{d\tau} = \frac{\sqrt{1+\left(\frac{dx}{d\tau}\right)^2}}{1+gx} \tag{31}$$

and

$$\frac{d^2 x}{d\tau^2} + \frac{g}{1+gx}\left(\frac{dx}{d\tau}\right)^2 = -\frac{g}{1+gx}$$

which may also be written

$$\frac{d^2}{d\tau^2}(1+gx)^2 = -2g^2. \tag{32}$$

When the initial velocity is zero and x_0 denotes the initial value of x, we get by integration of (32)

$$x = \frac{1}{g}\left\{\sqrt{(1+gx_0)^2 - g^2\tau^2} - 1\right\}. \tag{33}$$

Introduction of (33) into (31) gives

$$\frac{dt}{d\tau} = \frac{1+gx_0}{(1+gx_0)^2 - g^2\tau^2}$$

which by integration yields

$$t = \int_0^{\tau} \frac{dt}{d\tau} d\tau = \frac{1}{2g} \ln \frac{1+gx_0+g\tau}{1+gx_0-g\tau}. \tag{34}$$

From (33) and (34) it follows that a free particle initially at rest at some point in k will move with increasing velocity in the direction of the negative x-axis, later the velocity will decrease and, finally, the particle will come to rest again in the singular plane $x = -\frac{1}{g}$ at the time $t = \infty$ or $\tau = x_0 + \frac{1}{g}$.

We now get the desired transformation, if we put $x_0 = X$, $y_0 = Y$, $z_0 = Z$, and $\tau = T$ in the equations (30), (33), and (34), X, Y, Z, T then being the space-time coordinates of a freely falling frame of reference K which, at the time $t = T = 0$, coincides with the system k. In this way, we get

$$\left.\begin{aligned} x &= \frac{1}{g}\{\sqrt{(1+gX)^2 - g^2 T^2} - 1\} \\ y &= Y,\ z = Z \\ t &= \frac{1}{2g} ln \frac{1+gX+gT}{1+gX-gT}. \end{aligned}\right\} \tag{35}$$

By a simple calculation, it may be verified that the line element (17) is really brought into the form (7) by the transformation (35), showing that the system K is actually a system of inertia.

Any fixed point in k with constant coordinates x, y, and z is moving relative to K in accordance with the equations

$$\left.\begin{aligned} X &= \frac{1}{g}\{\sqrt{(1+gx)^2 + g^2 T^2} - 1\} \\ Y &= y,\ Z = z \end{aligned}\right\} \tag{36}$$

obtained by solving (35) with respect to X, Y, Z.

This motion is, according to the laws of the special theory of relativity, identical with the motion of a particle of rest mass m subjected to a constant force $\frac{mg}{1+gx}$ in the direction of the X-axis in a system of inertia, *i. e.* (36) represents the "hyperbolic motion"[9]) of a "uniformly accelerated" particle with acceleration

$$\gamma = \frac{g}{1+gx}. \tag{37}$$

On account of the dependence of γ on x, the distance, measured by an observer in K, between two fixed points in the frame k will not, in general, be constant in time. Since, however, the same distance is constant when measured by a comoving meter stick, the system k deserves the name of a u n i f o r m l y a c c e l e r a t e d r i g i d frame of reference, and the transformation (35) plays a similar part as does the Lorentz transformation in the case of a rigid frame moving with c o n s t a n t velocity.

Since the variables x and t, defined by (35), must be real, we shall have to confine ourselves to the consideration of events satisfying the condition

$$-(1+gX) < gT < 1+gX. \tag{38}$$

For later use, we also write down the Lorentz transformation connecting the space-time coordinates of two systems of inertia with the relative velocity v

$$\left.\begin{aligned} x &= \frac{X - X_0 - v(T - T_0)}{\sqrt{1 - v^2}} \\ y &= Y, \; z = Z \\ t - t_0 &= \frac{T - T_0 - v(X - X_0)}{\sqrt{1 - v^2}}. \end{aligned}\right\} \quad (39)$$

In (39), the space and time variables have been chosen in such a way that the origin $x = 0$ of the system k at the time $t = t_0$ corresponds to the coordinate $X = X_0$ and the time $T = T_0$ in K.

3. The clock paradox.

a. In the first part of this section, we shall treat the problem from the point of view of an observer in K. While the clock C_1 is permanently situated at rest at the point A on the positive X-axis, C_2 at the beginning is travelling with constant velocity $-v$ in the direction of the X-axis. At the point B, the clock C_2 is subjected to a constant force F, which brings it to rest at the origin O and starts it back to B with reversed velocity. At the time of arrival in B, C_2 will have regained the velocity v which it retains during the travel from B to A. Let us assume for simplicity that the coincidence of C_2 with O takes place at the time $T = 0$ and that the proper time τ of C_2 is also zero at this moment. Since the problem is then completely symmetrical with respect to this event, we only need explicitly to consider the behaviour of C_2 during its travel from O to B and onwards to A.

Let T′ and T″ be the times, measured in the time scale of the system K, during which C_2 travels from O to B and from B to A, respectively, and let τ' and τ'' be the corresponding proper times measured by the clock C_2 itself. The motion of C_2 from B to O and back to B will be a hyperbolic motion given by the equation

$$X = \frac{1}{g}\left\{\sqrt{1 + g^2 T^2} - 1\right\}, \quad (40)$$

where the constant g is connected with the force F and the rest mass m of C_2 by the relation

$$F = mg. \tag{41}$$

According to (40), the velocity $u = \frac{dX}{dT}$ is given by

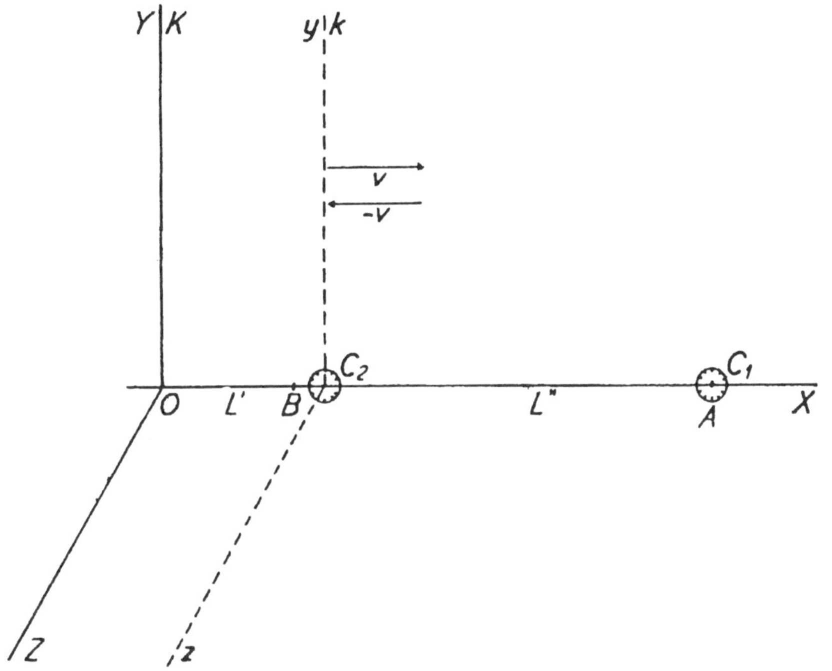

Fig. 1.

$$u = \frac{gT}{\sqrt{1+g^2T^2}} \tag{42}$$

and, since $u = v$ for $T = \mathrm{T}'$, we have

$$g\mathrm{T}' = \frac{v}{\sqrt{1-v^2}}. \tag{43}$$

Introducing (42) into (3), we get by integration

$$g\mathrm{T}' = \sinh g\tau'. \tag{44}$$

The corresponding relation between T'' and τ'' is, according to the well-known formula from the special theory of relativity,

$$T'' = \frac{\tau''}{\sqrt{1-v^2}}. \tag{45}$$

From (43) and (44) we further obtain

$$\left.\begin{aligned} v &= tgh\, g\tau' \\ \frac{1}{\sqrt{1-v^2}} &= \cosh g\tau'. \end{aligned}\right\} \tag{46}$$

Now, let $\varDelta'_K t_1$ denote the number of divisions registered by C_1 during the travel of C_2 from O to B, as judged by an observer in K, and let $\varDelta''_K t_1$ be the corresponding number during the period of uniform motion of C_2 from B to A. We then have

$$\varDelta'_K t_1 = T' \quad \text{and} \quad \varDelta''_K t_1 = T'' \tag{47}$$

and, for the total time elapsed between the two encounters of C_1 and C_2, measured by C_1 and C_2, respectively, we get

$$\left.\begin{aligned} \varDelta t_1 &= 2(\varDelta'_K t_1 + \varDelta''_K t_1) = 2(T' + T'') \\ \varDelta t_2 &= 2(\tau' + \tau''). \end{aligned}\right\} \tag{48}$$

When the applied force F is chosen so large that T' and τ', given by (43) and (44), become negligible, the connection between $\varDelta t_1$ and $\varDelta t_2$, according to (48) and (45), is again given by the simple formula (1).

If L' and L'' denote the distances OB and BA, measured with the measuring rods of the system K, we get from (40) and (43)

$$L' = \frac{1}{g}\left(\sqrt{1+g^2T'^2}-1\right) = \frac{1}{g}\left(\frac{1}{\sqrt{1-v^2}}-1\right) \tag{49}$$

while, obviously,

$$L'' = vT''. \tag{50}$$

We shall now introduce a frame of reference k moving together with C_2 and we may take C_2 as the origin of k. While the motion of the origin is, thus, completely determined, the motion of any other fixed point of k may, beforehand, be chosen arbitrarily. In the previous discussions of the clock paradox, it has, however, tacitly been assumed that k should be a rigid frame of reference. According to the considerations in Section 2, it is then clear that the transformation connecting the space-time variables of K and k must be given by (35) during the accelerated motion of C_2 from B to O and back. The motion of the origin $x = 0$ relative to K is then, on account of (36), identical with the motion of C_2 given by (40), and the time variable t is simply the proper time of the clock C_2.

For all events satisfying the conditions $-\tau' < t < \tau'$, the connection between the coordinates of K and k is, thus, given by (35). For $t > \tau'$, the system k is a simple system of inertia, and the corresponding space-time transformation is obtained from (39) by putting

$$t_0 = \tau', \quad T_0 = \mathrm{T}', \text{ and } X_0 = L'. \tag{51}$$

Similarly, we have for $t < -\tau'$ the transformations (39) with reversed signs of v, τ' and T'.

In the following, we shall use the equations (35) and (39) in a somewhat different form. Solving the last equation (35) with respect to gT and introducing into the first equation, we get, if we omit the trivial transformations of the y and z variables,

$$\left.\begin{aligned} gT &= (1+gX)\,\mathrm{tgh}\,gt \\ 1+gX &= (1+gx)\cosh gt \end{aligned}\right\} \tag{52}$$

for

$$-\tau' < t < \tau'.$$

By a similar procedure, we get from (39) and (51) the transformation

$$\left.\begin{aligned} T - \mathrm{T}' &= (t-\tau')\sqrt{1-v^2} + v(X - L') \\ X - L' &= \frac{x + v(t-\tau')}{\sqrt{1-v^2}} \end{aligned}\right\} \tag{53}$$

for

$$t > \tau'.$$

For $t < -\tau'$, the corresponding transformation (53') is obtained from (53) by reversing the signs of v, T', and τ'.

In spite of the great difference in form between the equations (52) and (53), they are easily seen to be identical for $t = \tau'$. For this particular value of t, the equations (52) reduce to

$$gT = (1+gX)\,\mathrm{tgh}\,g\tau'$$
$$1+gX = (1+gx)\cosh g\tau'$$

which, by means of (43), (46), and (49) may be written

$$T = v(X-L') + T'$$
$$X-L' = \frac{x}{\sqrt{1-v^2}},$$

in accordance with (53) for $t = \tau'$.

On account of the symmetry inherent in our problem, a similar result would be obtained for $t = -\tau'$, so that the correlation of the coordinates x, y, z, t and the physical events is performed in a continuous way by the equations (52), (53), and (53′). Also the velocity of any fixed point in k relative to K varies continuously at $t = \tau'$ (and $-\tau'$). From (36) and (52) we get for constant values of x, y, z

$$\left(\frac{dX}{dT}\right)_x = \frac{gT}{\sqrt{(1+gx)^2+g^2T^2}} = \mathrm{tgh}\,gt. \tag{54}$$

On the other hand, $\left(\frac{dX}{dT}\right)_x$ is equal to v and $-v$ for $t>\tau'$ and $t<-\tau'$, respectively, which, on account of (46), is seen to be in accordance with (54) for t equal to τ' and $-\tau'$.

While, thus, the velocities of the different points of k vary continuously, it is clear that the accelerations must be discontinuous for $t = \tau'$ and $-\tau'$, since the force F is assumed to set in abruptly. This is also the reason for the sudden change in the gravitational potential from the value zero to the value given by (26) at these moments.

The system k defined by (52), (53), and (53′) thus seems to be the most natural frame of reference to be used in the discussion of the clock paradox. The applicability of this system of coordinates is only restricted by the condition that (38) must be satisfied for $-\tau'<t<\tau'$, *i. e.* for

$$-v(1+gX) < gT < v(1+gX), \tag{55}$$

on account of (52) and (46). Since v is smaller than one, a comparison of (38) and (55) shows that this condition is satisfied for all events which take place at points $X > -\frac{1}{g}$.

b. We shall now treat the problem from the point of view of an observer in k, according to which C_2 is permanently situated at rest at the origin o of k, while C_1 at the beginning is travelling with constant velocity v. The first encounter between C_1 and C_2 takes place at the time $t = -\tau' - \tau''$. At $t = -\tau'$, C_1 has arrived at a point b on the positive x-axis with the coordinate $x = l''$. During the time $-\tau' < t < \tau'$, C_1 is subjected to the gravitational field which brings it to rest at the time $t = o$ at a point a in the distance l' from b, and starts it back to o with reversed motion. In spite of this gravitational field everywhere present during this period, C_2 remains at rest on account of the force F which just counterbalances the gravitational force.

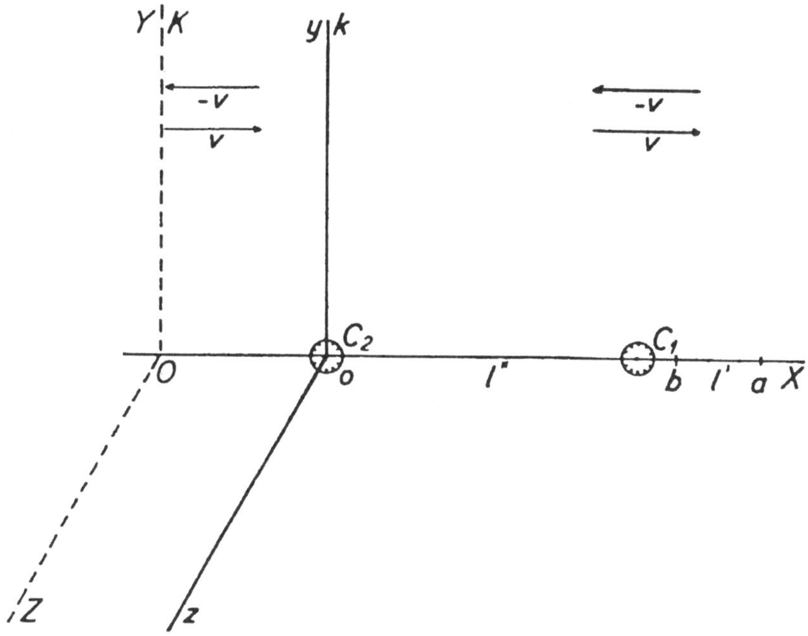

Fig. 2.

The behaviour of the clock C_1 is now simply obtained from (52) and (53) if we remember that the X-coordinate of C_1 has the constant value $X = L' + L''$.

From the second equation (53) we then get

$$l'' = L'' \sqrt{1 - v^2}, \tag{56}$$

since l'' is the value of x for $t = \tau'$.

Further, since the x-value of C_1 at $t = o$ is $l' + l''$, we get from the second equation (52)

$$l' + l'' = L' + L'' \tag{57}$$

and, therefore,

$$l' = L' + L''\left(1 - \sqrt{1 - v^2}\right). \tag{58}$$

On account of the Lorentz contraction factor in (56), the distance travelled by C_1 with constant velocity v relative to k will, thus, be shorter than the distance which C_2 travels with constant velocity in K. Nevertheless, the total distances travelled by the two clocks along the x-axes will be equal. In the extreme case of $v \to c$, we have simply $l'' \to 0$ and $l' \to L' + L''$.

If $\Delta'_k t_1$ denotes the number of divisions registered by C_1 during the travel from a to b (or from b to a), we get from the first equation (52), by putting $X = L' + L''$, $t = \tau'$, and $T = \Delta'_k t_1$,

$$\Delta'_k t_1 = \frac{1}{g}\left[1 + g(L' + L'')\right] \operatorname{tgh} g\tau' = \mathrm{T}' + v^2 \mathrm{T}'' \tag{59}$$

by means of (44), (46), (49), and (50).

For the corresponding number of divisions $\Delta''_k t_1$ registered by C_1 during the period of travel with constant velocity, we have, according to the special theory of relativity,

$$\Delta''_k t_1 = \tau'' \sqrt{1 - v^2}. \tag{60}$$

This formula is also easily obtained from the first equation (53) if we remember that $\Delta''_k t_1$ is the increase in T for $X = L' + L''$ during the interval $\tau' < t < \tau' + \tau''$. On account of (45), we may also write

$$\Delta''_k t_1 = \mathrm{T}''(1 - v^2). \tag{61}$$

Although, thus, $\Delta''_k t_1$ is smaller than $\Delta''_K t_1$ in (47), it follows from (59), and (61) that the total time elapsed between the two encounters of C_1 and C_2 measured by C_1 and C_2, respectively, is again given by

$$\left.\begin{aligned} \Delta t_1 &= 2(\Delta'_k t_1 + \Delta''_k t_1) = 2(\mathrm{T}' + \mathrm{T}'') \\ \Delta t_2 &= 2(\tau' + \tau'') \end{aligned}\right\} \tag{48'}$$

in accordance with the expressions (48) derived from the standpoint of an observer in K.

It is interesting to note that $\varDelta'_k t_1$ remains finite in the limiting case of very large forces F, where τ', T', and $\varDelta_K t_1$ vanish, since $\varDelta'_k t_1$ in (59) contains a term which only depends upon v and T''. It is just this term which is essential for the solution of the clock paradox.

Since $\varDelta t_2$ in any case is smaller than $\varDelta t_1$, and $\varDelta''_k t_1$, according to (60), is smaller than τ'', $\varDelta'_k t_1$ must be greater than τ', *i. e.* the clock C_1 goes faster than C_2 during this period. From the point of view of an observer in k, the reason for this difference in rate is to be sought mainly in the difference in gravitational potential Φ at the places of the two clocks. The behaviour of C_1, however, will in general not be like that of a clock at rest at the point $x = l' + l'' = L' + L''$, even if T' and τ' are made small by use of a large force F. In fact, the number of divisions registered by a clock at rest during the time $\varDelta t = \tau'$ is, according to (26), (28), or (22), given by

$$(\varDelta'_k t_1)_0 = \left[1 + g\,(L' + L'')\right]\cdot\tau', \tag{62}$$

a number which is greater than $\varDelta'_k t_1$ in (59), since we have

$$\frac{\operatorname{tgh} g\tau'}{g} < \tau'.$$

From (17) and (26), we get the expression

$$\left.\begin{aligned} d\tau &= dt\sqrt{(1+gx)^2 - \left(\frac{dx}{dt}\right)^2 - \left(\frac{dy}{dt}\right)^2 - \left(\frac{dz}{dt}\right)^2} \\ &= dt\sqrt{1 + 2\,\Phi - u^2} \end{aligned}\right\} \tag{63}$$

for the proper time of a particle moving with velocity u in the gravitational potential Φ. This general formula, which comprises the special formulae (3) and (28), clearly shows that $\varDelta'_k t_1$ in general must be smaller than $(\varDelta'_k t_1)_0$, since C_1 during the time in question falls freely with increasing velocity from the place $x = L' + L''$ towards smaller values of x, *i. e.* smaller values of the potential Φ.

Only in the case $v \ll 1$ considered by Tolman, where $\operatorname{tgh} g\tau'$ is equal to $g\tau'$, apart from terms of the third order in v (cf.

(46)), it is allowed to treat C_1 as a clock at rest during the period of acceleration, since the difference between $\Delta'_k t_1$ and $(\Delta'_k t_1)_0$ is then of higher order in v. Even in this case, where gt may be treated as a small quantity, the equations (52), however, do not reduce to the transformations given by (6). If we neglect terms of higher order in gt, we obtain instead

$$T = (1+gX)\,t = (1+gx)\,t$$
$$X = x + \frac{1}{2}\,gt^2\,(1+gx)\,.$$

To get the transformation (6), we should, thus, have to replace the factor $1+gx$ by 1 and this would mean neglect of just those terms which, in the preceding discussion, have been seen to be essential for the treatment of the clock paradox.

4. Rigid frames of reference in arbitrary motion.

In Section 2, it was shown that the transformation (35) is essentially determined by the condition that the gravitational field of the accelerated system k should be static, and the line element and the gravitational potential in the transformed system are given by (17) and (26), respectively. Since the motion of the origin of k in this case is a hyperbolic motion, the applicability of the transformation (35) in the preceding discussion is confined to the case where the clock C_2 is subjected to a constant force during the period of acceleration. For any other motion the gravitational field in the comoving system will not be static. Anyway, it is always possible to choose the time variable t in the transformations (5) in such a way that the line element takes the form (13), where A and D in general are functions of both variables x and t. If we want the system k to be a rigid frame of reference, A must, however, be independent of t, so that the line element may be brought into the simple form (13′) by a suitable choice of the variable x. Then, the spacial geometry is again Euclidean, x, y, and z being Cartesian coordinates.

Using DINGLE's general formulae[7]), one finds that EINSTEIN's field equations (14) in this case reduce to the single equation

$$\frac{\partial^2}{\partial x^2}(D^{1/2}) = 0$$

which is obtained from (16) by replacing the ordinary differentiations with respect to x by partial differentiations. The general solution is again of the form (16′), a and g here being arbitrary functions of t. Finally, the time variable t may be chosen such that the line element takes the same simple form (17) as in the special case treated in Section 2, g in the general case being an arbitrary function of t. The equations (18)—(29) and (63) are seen, therefore, to hold also in the general case.

In order to find the transformation (5) by which the expression (7) for interval is transformed into (17) and by which, conversely, the gravitational field in k is transformed away, we may proceed exactly as in Section 2. First, we solve the equations (24) and (25) for the motion of a free particle initially at rest. After that, the proper time τ of the particle and the initial values x_0, y_0, z_0 of the space coordinates in k are identified with the time and space coordinates T, X, Y, Z in K. The solution of the equations (24) and (25) is only somewhat more complicated than in the case of constant g considered in Section 2. Since g may be regarded as a known function of t, it is convenient to use t as parameter in (24) instead of τ. The elimination of τ is easily performed by means of (25) and, finally, applying elementary methods, a complete solution of the problem is possible.

We shall here give the results, only. For the transformations connecting the space-time variables of the systems K and k, we get

$$\left.\begin{aligned} X &= x\cosh\theta + \int_0^t \sinh\theta\, dt, \quad Y = y, \quad Z = z \\ T &= x\sinh\theta + \int_0^t \cosh\theta\, dt \\ \text{with}\quad \Theta(t) &= \int_0^t g(t)\, dt. \end{aligned}\right\} \qquad (64)$$

It is easily verified by means of direct calculation that the line element (7) is really brought into the form (17) by the transformation (64). Further, we see that the equations (64) in the case of constant g reduce to the equations (52) which are equivalent to (35). On the other hand, if g is assumed to be finite and constant for $-\tau' < t < \tau'$ and zero for all other times, (64) leads to the transformations (52) and (53) used in the discussions of Section 3.

When g is given as a function of t, the transformation (64) and, consequently, also the motion of the origin of k with respect to K is completely determined. Conversely, the function g and the transformation (64) are uniquely determined by the motion of the origin of k. Differentiating (64) by constant x, we get

$$\left.\begin{aligned} dX &= \sinh\theta\cdot(1+gx)\,dt \\ dT &= \cosh\theta\cdot(1+gx)\,dt. \end{aligned}\right\} \quad (65)$$

The velocity $U = \left(\frac{dX}{dT}\right)_x$ of a fixed point in k with respect to K is, thus,

$$U = \mathrm{tgh}\,\theta, \quad (66)$$

an equation which may be regarded as a generalization of (54). Moreover, we get from (66), (65), and from the definition of θ in (64)

$$\frac{d}{dT}\left(\frac{U}{\sqrt{1-U^2}}\right) = \frac{g}{1+gx} \quad (67)$$

which shows that the motion of a point x = constant is the same as that of a particle of mass m attacked by a force $\frac{mg}{1+gx}$, just like in the case of constant g (cf. p. 13).

When the motion of the origin $x = 0$ is given by the equation

$$X = \Psi(T), \quad (68)$$

the corresponding g is obtained as a function of t by elimination of the variable T from the equations

$$\left.\begin{aligned} g &= \frac{d}{dT}\left(\frac{\Psi'}{\sqrt{1-(\Psi')^2}}\right) \\ t &= \int_0^T \sqrt{1-(\Psi')^2}\,dT \end{aligned}\right\} \quad (69)$$

which are easily derived from (67), (68), (66), and (65).

By means of the general equations (64), it is now easy to treat the clock paradox for an arbitrary motion of the clock C_2 during the interval of acceleration. Since, however, the treatment of this general case does not exhibit any essentially new features as compared with the treatment of the special case discussed in Section 3, we shall confine ourselves to the general remarks already made in this section.

IL NUOVO CIMENTO Vol. XXVI, N. 4 16 Novembre 1962

Physical Consequences of a Co-ordinate Transformation to a Uniformly Accelerating Frame (*).

T. Fulton (**)
The Johns Hopkins University - Baltimore, Md.

F. Rohrlich (**)
University of Iowa - Iowa City, Ia.

L. Witten
RIAS - Baltimore, Md.

(ricevuto il 16 Febbraio 1962)

Summary. — The classical equations of particle motion and of classical electrodynamics are known to be covariant with respect to conformal co-ordinate transformations. This paper deals with a simple conformal co-ordinate transformation from an inertial frame to one which accelerates uniformly; it examines the physical consequences predicted by the covariant theory and the transformation. Expressions are derived for various effects of the conformal invariance on mass, length, and time measurements; the predictions of the theory are shown to be the same as the corresponding predictions derived from the general theory of relativity for the transformation being studied. The predictions of the conformal theory depend on the particular co-ordinate system chosen (with an explicit dependence on the origin). The corresponding situation prevails in general relativity with the components of the metric tensor being a function of space and time. An example of the « twin paradox » is studied. Finally, after considering the behavior of four-momentum conservation under conformal transformations, the frequency shift of radiation emitted by an excited atom falling freely in a uniform field is calculated to first order in acceleration. Identical results are obtained from the conformal and general relativistic points of view. The frequency shifts are what one would expect to obtain by properly combining the Doppler shifts and the shift arising as a consequence of the equivalence of inertial mass to gravitational mass.

(*) Supported in part by the National Science Foundation and in part by the Aeronautical Research Laboratory.

1. – Introduction.

The study of conformal transformations and conformal invariance has a rather extensive history. The transformations are well understood mathematically, but their consequences for physics have been somewhat less extensively examined, particularly regarding specific experimental predictions. In fact, most physicists probably believe that conformal invariance should not play any important role in the description of nature while some want to consider it as a logical generalization of Lorentz invariance.

WEYL (1), EINSTEIN (2), and PAULI (3), among others, have considered in a general way the consequences of conformal relativity as they may affect physical measurements. These questions are of interest in view of the conformal invariance of the basic equations of classical physics (4,6). To the best of our knowledge, no one has discussed the specific physical effects of conformal relativity for a simple case, though very likely a number of physicists have thought about and have been aware of the matters to be discussed in this paper. These matters are very restricted in scope. We shall not consider all conformal co-ordinate transformations but only those which take us from flat space to flat space. In fact, we shall discuss in detail only one specific example of such a transformation, the one responsible for transforming from an inertial co-ordinate system to one which is accelerated uniformly with respect to it. Though this transformation is a simple one, it does have physical significance. In the one-dimensional case and for weak fields, its inverse represents a transformation from a frame in which a particle falls freely in a static and uniform force field to the rest frame of the particle (7).

In the next section, we discuss in detail the properties of our specific transformation, its singularities and other peculiar features. In Section **3** we treat various thought experiments—relating to the variation of rest mass, length, and time under a transformation to a uniformly accelerating co-ordinate frame. We also discuss the so-called « twin paradox. » Finally, in Section **4**, we deal with a specific experiment: we calculate the frequency shift of the spectral line of a freely falling emitter.

---

(1) H. WEYL: *Sitzungsber. Preuss. Akad. Wiss.* (1918), p. 465; *Math. Zeits.*, **2**, 384 (1918); *Ann. Phys.*, **59**, 101 (1919); *Space, Time, and Matter* (New York, 1950).

(2) A. EINSTEIN: *Sitzungsber. Preuss. Akad. Wiss.* (1921), p. 261.

(3) W. PAULI: *Theory of Relativity* (New York, 1958).

(4) J. A. SCHOUTEN and J. HAANTJES: *Physica*, **1**, 869 (1934).

(5) J. A. SCHOUTEN and J. HAANTJES: *Proc. Ned. Akad. Wet.*, **39**, 1063 (1936).

(6) F. ROHRLICH, T. FULTON and L. WITTEN: *Bull. Am. Phys. Soc.*, **6**, 346 (1961).

(7) F. ROHRLICH: *The Static Homogeneous Gravitational Field* (in preparation).

2. – The acceleration transformation.

We consider the co-ordinate transformation (8) from a reference frame S to a reference frame S'. The co-ordinates of a point in S are given by x^λ; the co-ordinates, x'^μ, of the same point in S' are given by

$$x'^\mu = \Psi^\mu(x^\lambda) \,. \tag{2.1}$$

The metric associated with S is $g_{\alpha\beta}(x)$. If the frame S' is a Riemann space (*i.e.* we deal with a general relativistic co-ordinate transformation) the metric in S' is given by

$$g'_{\mu\nu}(x') = \frac{\partial x^\alpha}{\partial x'^\mu}\frac{\partial x^\beta}{\partial x'^\nu}\, g_{\alpha\beta}(x) \,. \tag{2.2}$$

The metric transforms as a second rank tensor, and as a consequence, in general relativity, we have

$$\mathrm{d}\tau^2(x) = \mathrm{d}\tau'^2(x') \,, \tag{2.3}$$

where

$$\mathrm{d}\tau^2(x) = -\, g_{\alpha\beta}(x)\, \mathrm{d}x^\alpha\, \mathrm{d}x^\beta \,, \tag{2.4}$$

and

$$\mathrm{d}\tau'^2(x') = -\, g'_{\mu\nu}(x')\, \mathrm{d}x'^\mu\, \mathrm{d}x'^\nu \,. \tag{2.5}$$

A conformal co-ordinate transformation differs from a general relativistic co-ordinate transformation in that the metric no longer transforms as a tensor. Instead we have (9)

$$\overset{c}{g}_{\mu\nu}(x') = \sigma(x)\,\frac{\partial x^\alpha}{\partial x'^\mu}\frac{\partial x^\beta}{\partial x'^\nu}\, g_{\alpha\beta}(x) = \sigma(x)\, g'_{\mu\nu}(x') \,. \tag{2.6}$$

The function $\sigma(x)$ represents a scale transformation, and, as a consequence, the proper time is no longer an invariant under conformal transformation.

---

(8) This type of transformation is sometimes referred to as a passive co-ordinate transformation. For a discussion of the relationship between conformal point transformations and conformal co-ordinate transformation see the note by T. FULTON, F. ROHRLICH and L. WITTEN: in preparation.

(9) J. HAANTJES: *Proc. Ned. Akad. Wet.*, **43**, 1288 (1940).

By definition, the proper time is

(2.7) $$\mathrm{d}\tau^{c^2} = -\, g^c_{\mu\nu}(x')\,\mathrm{d}x'^\mu\,\mathrm{d}x'^\nu\,.$$

It follows that

(2.8) $$\mathrm{d}\tau^{c^2} = \sigma(x)\,\mathrm{d}\tau^2\,.$$

The light cones are still invariant under conformal transformations:

(2.9) $$\mathrm{d}\tau^{c^2} = \mathrm{d}\tau^2 = 0\,.$$

Since, in dealing with equations of motion of charged particles (Lorentz equation, or Lorentz-Dirac equation), we have to deal with derivatives of vectors, the definition of conformal co-ordinate transformations as given by (2.1), (2.2), and (2.6) is not complete. It must be augmented by discussing covariant differentiation; in particular the behavior of the affine connection under the transformation must be given (see Appendix). A flat space is one in which the curvature tensor, $\boldsymbol{R}$, computed from the affine connection, $\boldsymbol{\Gamma}$, vanishes. A conformal co-ordinate transformation from flat space to flat space is one in which both $\boldsymbol{R}$ and $\boldsymbol{R}^c$ vanish. These co-ordinate transformations are given by the 15-parameter Lie group often referred to as the « conformal group ». In case derivatives enter the problem being considered, it is important to realize that although both S and S' are flat under the conformal transformations, not both $\boldsymbol{\Gamma}$ and $\boldsymbol{\Gamma}^c$ vanish.

In this paper we restrict ourselves to conformal co-ordinate transformations from flat space to flat space and emphasize the details of one such transformation. We deal only with frames S which are Lorentz frames. That is, we set

(2.10) $$g_{\mu\nu}(x) = \eta_{\mu\nu}\,,$$

where $\eta_{\mu\nu}$ is the Minkowski metric with signature $+2$. We shall consider transformations to uniformly accelerating frames of reference. The $\Psi - s$ of (2.1) will represent an inversion in the origin, followed by translation by a constant vector and then by a second inversion about the new origin (10). In other words, we let (11)

(2.11) $$x'^\mu = \frac{f^\mu}{f^2},$$

(10) Such a transformation is discussed among other authors by T. Fulton and F. Rohrlich: *Ann. Phys.*, **9**, 499 (1960); S. A. Bludman: *Phys. Rev.*, **107**, 1163 (1957).

(11) For any vector A^μ, we define $A^2 = A^\mu A_\mu$:

where

$$f^\mu = \frac{x^\mu}{x^2} - a^\mu \, . \tag{2.12}$$

and a^μ is a constant vector ([12]).

We will also demand that the conformal transformation take us to a flat space S', *i.e.*

$$g^c_{\mu\nu}(x') = \eta_{\mu\nu} \, . \tag{2.13}$$

We can imagine, for definiteness, that the affine connection in S vanishes, but in S' it does not. However, the affine connections will not explicitly enter into the sequel.

Since a general co-ordinate transformation for (2.11) and (2.12) gives, upon explicit computation,

$$g'_{\mu\nu}(x') = \lambda^2(x)\eta_{\mu\nu} \, , \tag{2.14}$$

where

$$\lambda(x) = f^2x^2 = 1 - 2a^\mu x_\mu + x^2a^2 \, , \tag{2.15}$$

we identify the scale function $\sigma(x)$, using (2.13), as

$$\sigma(x) = \lambda^{-2}(x) \, . \tag{2.16}$$

The transformation inverse to (2.11) and (2.12) is simple to exhibit. It too must be an inversion in the origin, followed by a translation, this time in the opposite sense, by the same constant vector a^μ, and a second inversion:

$$x^\mu = \frac{h^\mu}{h^2} \, , \tag{2.17}$$

where

$$h^\mu = \frac{x'^\mu}{x'^2} + a^\mu \, , \tag{2.18}$$

and

$$\lambda'(x') = 1 + 2a^\mu x'_\mu x'^2 a^2 \equiv \frac{1}{\lambda(x)} \, . \tag{2,.19}$$

In what follows, we shall limit ourselves further to a special choice of a^μ,

([12]) We put $c = 1$ throughout this paper, unless explicitly stated otherwise.

namely

$$a^\mu = \left(0;\, 0,\, 0,\, -\frac{g}{2}\right). \tag{2.20}$$

We observe that the transformation (2.11), (2.12), with a^μ defined as in (2.20) does describe a uniformly accelerating frame of reference. This may be made more obvious, if we do not set $c=1$ and instead take the limit $c \to \infty$. The conformal transformation in this limit becomes

$$\begin{cases} \lim\limits_{c\to\infty} x'^\mu = x^\mu; \qquad \mu = 0, 1, 2\,. \\ \lim\limits_{c\to\infty} z' = z - \tfrac{1}{2} g t^2\,; \\ \lim\limits_{c\to\infty} \lambda(x) = 1\,. \end{cases} \tag{2.21}$$

In the limit of vanishing g, the transformation reduces to the identity transformation, as it should.

Next, we examine how a particle, at rest at the origin for all time in the S frame, will move as seen by an observer at rest in the S' frame. The observer will determine the co-ordinates at the accelerating particle to be

$$\begin{cases} x' = y' = 0\,, \\ z' = -\dfrac{gt^2}{2\lambda}\,, \\ t' = \dfrac{t}{\lambda}\,, \end{cases} \tag{2.22}$$

with

$$\lambda = 1 - \tfrac{1}{4} g^2 t^2\,. \tag{2.23}$$

The orbit of the particle, described by the observer at rest in the S' frame, lies in the z'-t' plane and is hyperbolic, *i.e.* represents uniform acceleration:

$$\left(z' - \frac{1}{g}\right)^2 - t'^2 = \frac{1}{g^2}\,. \tag{2.24}$$

In the nonrelativistic limit, or alternatively the small g limit, (2.24) reduces to

$$z' = -\tfrac{1}{2} g t'^2\,. \tag{2.25}$$

Note that we have used only the transformation eqs. (2.11), (2.12) to obtain (2.24), so it holds both for general relativistic co-ordinate transformations and for conformal co-ordinate transformations. Not until we specify our metric or, alternatively, the transformation of proper time, do we restrict ourselves to one or the other. Hyperbolic motion thus appears as a solution of either a conformally [13,14] or a general relativistically covariant equation of motion which reduces to the equation of motion for a free particle in the frame S. The hyperbolic motion as seen by the observer at rest in S' would equally well result if we considered S' to be an inertial frame in which the Lorentz invariant equation of motion with constant rest mass and constant force held. Thus we have three alternative and equivalent descriptions of the motion as seen by the observer in S':

A) S' is a frame in which only a constant force is acting.

B_1) S' is a frame in a Weyl space, uniformly accelerating with respect to an inertial frame. This considers the transformation to be a conformal co-ordinate transformation.

B_2) S' is a frame in a Riemannian space uniformly accelerating with respect to an inertial frame. This considers the transformation to be a general relativistic co-ordinate transformation.

The principle of equivalence as usually considered regards the descriptions A and B_2 to be equivalent. We shall shortly discuss the equivalence of the descriptions B_1 and B_2 for the simple class of transformations (2.11) and (2.12) that we are considering. Hence, the principle of equivalence can be regarded as an equivalence between A and B_2 or between A and B_1. For the case of uniformly accelerating motion, the description used is a matter of computational convenience.

For the sake of completeness, since we will use them later, we give here the expression for the instantaneous velocity

$$v \equiv \frac{\mathrm{d}z'}{\mathrm{d}t'} = -\frac{gt'}{\gamma'}\,, \tag{2.26}$$

where

$$\gamma' \equiv \frac{1}{(1-v'^2)^{\frac{1}{2}}} = 1 - gz'\,. \tag{2.27}$$

Next we study the mapping of the z-t plane into the z'-t' plane for

---

(13) F. GÜRSEY: *Nuovo Cimento*, **3**, 988 (1956); J. A. MC LENNAN: *Nuovo Cimento* **5**, 640 (1957); H. A. BUCHDAHL: *Nuovo Cimento*, **11**, 496 (1959). References to earlier work can be found in these papers.

(14) L. INFELD and A. SCHILD: *Phys. Rev.*, **70**, 410 (1946).

the transformation (2.11), (2.12). (This mapping is not sufficient to characterize the transformation completely.) The salient features of the mapping can be most easily represented graphically.

The transformation is singular in the z-t plane for the lines

$$t = \pm\left(z + \frac{2}{g}\right). \tag{2.28}$$

The fact that the transformation is singular is in no way disturbing. One must merely obey the injunction to stay away from the singularities in discussing any physical process. Using this transformation, we discuss physical processes only for regions in space-time for which the transformation is non-singular.

In order to determine the mapping, we consider the most general straight line in the z-t plane, parallel to the t-axis,

$$z = \frac{2}{g}(\alpha - 1) \quad \text{for all } t\,,(\alpha \gtrless 0,\ \alpha \neq 0)\,. \tag{2.29}$$

The transformed equations in the S' frame for these lines are

$$\left[z' + \frac{2}{g}\left(\frac{1}{2\alpha} - 1\right)\right]^2 - t'^2 = \frac{1}{g^2\alpha^2}\,. \tag{2.30}$$

Thus, straight lines parallel to the t-axis map into a family of hyperbolas witl parallel asymptotes of slopes ± 1. The vertices of the two arms of the hyper

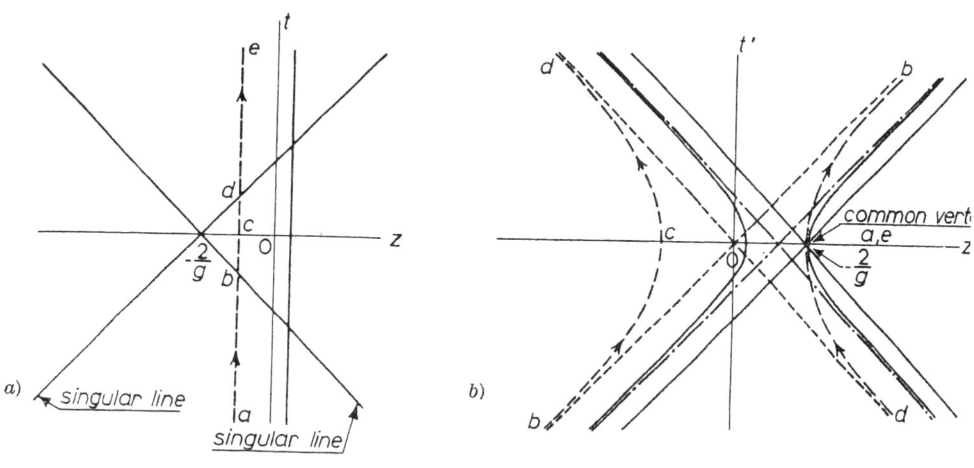

Fig. 1. – The mapping of particles at rest in the S frame (z-t plane) at two typic positions into their corresponding hyperbolic motions in the S' plane (z'-t' plan by means of the acceleration transformation.

bolae are located at the points

$$z' = \frac{2}{g} \quad \text{and} \quad z' = \frac{2}{g}\left(1 - \frac{1}{\alpha}\right). \tag{2.31}$$

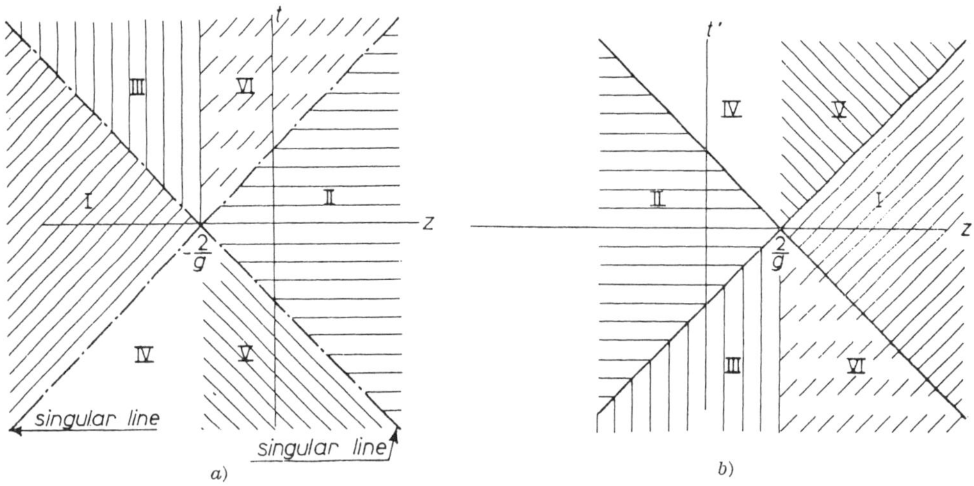

Fig. 2. – The mapping of the z-t plane into the z'-t' plane for the transformation of uniform acceleration. The singular lines in the z-t plane map into infinity in the z'-t' plane.

Figure 1 illustrates the way typical lines in the z-t plane map into the z'-t' plane. Figure 2 indicates the mapping of corresponding domains in the two planes. The mapping is one to one for all regions of the plane except along the singular lines.

3. – Kinematical consequences; transformation of mass.

In this section, we shall be concerned mostly with certain kinematical properties of the transformation as they relate to effects on the measurement of length and time (15). We shall also discuss an example of the « twin paradox » and the transformation of mass under a conformal co-ordinate transformation.

3.1. *Transformation of rods.* – Consider first the effect of the transformation on length. Suppose we have a very small measuring rod, at rest in frame S and lying along the z-axis. The length, dl, measured by S is given by $dl = dz$.

(15) Some of these questions were recently discussed in a different context by S. N. Gupta: *Science*, **134**, 1360 (1961).

An observer in S' will have to make measurements at a given instant t_0 to measure the length of the rod. In order to compare the results of his measurement with the results of the observer S he will have to realize that he has achieved a Minkowski metric by the introduction of a scale factor. To make comparisons he must account for the scale factor by saying that $dl' = (dz'_0/\lambda')_{t'_0}$. In order to predict the result of his measurement, we must calculate $(\lambda' \partial z/\partial z')_{t'_0, x'=y'=0}$ by using the inverse transformation. If S and S' were inertial frames connected by a Lorentz transformation ($\lambda = 1$), the resulting partial derivative would correctly give the Lorentz contraction. For our case, we have, from eqs. (2.17)–(2.20)

$$\left[\lambda' \left(\frac{\partial z}{\partial z'}\right)\right]_{t'_0,\, x'=y'=0} = \left[\frac{1}{\lambda'}\left(\lambda' + \frac{1}{2} g\, t'^2\right)\right]_{t'_0,\, x'=y'=0} \tag{3.1}$$

In order to obtain the length effects, we now have to substitute the motion satisfied in S' by a particle at rest in S. We used infinitesimal lengths in (3.1) in order to enable us to substitute such particle motion. No such need arises for Lorentz transformations so that finite lengths may be used when comparing measurements in two inertial frames. The distinction is one between a transformation for which the new variables are non-linear functions of the old and one for which they are linear.

Substitution of the motion (2.24) in the expression (2.19) for λ' yields

$$\lambda'|_{\text{orbit}} = 1 - \tfrac{1}{2} g z'_0 , \tag{3.2}$$

and insertion in (3.1) results in

$$\frac{dl}{dl'} = \left(\lambda' \frac{\partial z'}{\partial z}\right)_{\text{orbit}} = 1 - g z'_0 = (1 - v'^2_0)^{-\frac{1}{2}} = \gamma'_0 , \tag{3.3}$$

where z'_0 is the z' co-ordinate at t'_0. This agrees with the predictions of the special theory of relativity. Observer S' was only able to make this prediction if he knew explicitly the value $1/\lambda'$ by which he expanded space.

A general relativistic observer would make exactly the same calculation but would introduce λ' not as a scale factor but with the appropriate components of the metric tensor. The distinction for this transformation is purely formal.

3·2. *Transformation of clocks.* – We can proceed in a similar way to study effects on time. Assume a clock at rest in S (*i.e.*, at a fixed co-ordinate in S). In order to compare the rate of this clock with clocks in S', we require the expressions $(\lambda'^{-1}\,\partial t'/\partial t)_{r=r_0}$. In particular, take $r_0 = 0$. Then we have

$$\left(\frac{1}{\lambda'}\frac{\partial t'}{\partial t}\right)_{r=0} = \frac{(\lambda + \frac{1}{2}g^2t^2)}{\lambda} = 1 + \frac{g^2t'^2}{2\lambda'}\,. \tag{3.4}$$

Evaluation of (3.4) along the orbit gives, with the use of (3.2)

$$\left(\frac{1}{\lambda'}\frac{\partial t'}{\partial t}\right)_{\text{orbit}} = (1 - gz_0') = \gamma_0'\,. \tag{3.5}$$

Thus we get a time dilation which is the same as the one which would be obtained from the use of instantaneous Lorentz frames. Again the general relativist would make the same calculation using λ' as the appropriate components of the metric tensor.

3·3. *Twin paradox.* – It is convenient to use the transformations (2.11) and (2.17) to make explicit calculations regarding the twin paradox. One « twin » is at rest in a constant force field and sees the other one start from the same point with a given velocity opposite to the force, decelerate, stop, and eventually return to the starting position with his initial speed.

Consider an observer A, at rest in S at the origin, and his twin, B, at rest in S' at the position z_B', $x' = y' = 0$. Assume that both twins are at the same space-time point $z_A = z_B = 0$, at the time $-t_0$ and again at the time t_0. The proper time that has elapsed between coincidences of A and B is, of course, a function of the specific path taken (proper time is not an exact differential). The proper time that has elapsed on the path of A between coincidences is greater than on the path of B.

The situation of each of the two observers, A and B, seems on the face of it to be similar. Observer A is stationary in his frame of reference and sees observer B traverse a hyperbolic orbit and return. Similarly observer B is stationary in his frame of reference and sees A travel a hyperbolic orbit. It might be reasoned that if observer A concludes that his proper elapsed time is greater than that of B, observer B will conclude similarly that his own proper elapsed time is greater than that of A. These two conclusions disagree with each other and hence the paradox.

The resolution of the paradox is of course well known. Observer A is in an inertial frame; observer B is not. His frame appears inertial because of his having scaled space everywhere by the factor λ'. Observer B realizes that he cannot talk about proper time at all in a physically meaningful way (*i.e.*, a way that corresponds with measurements on a clock) unless he scales space again in such a way that at each space-time point all proper intervals or proper times are invariant. Having accomplished this, he will realize that his estimate of the reading of a proper time interval between two world points on a given world line will necessarily agree with the estimate of observer A for the same world line going between the same two points.

Should observer B fail to keep track of his scaling factor, he will be powerless to make any remarks about the ratio of times or distances at different parts of his space. The interval $d\tau^c$ has no physical significance because there is no clock which directly measures it. However, he can compare ratios of infinitesimal lengths or distances at the same world-point (λ' drops out of the ratio); equivalently he can compare angles between world lines or any other physically meaningful quantity that can be expressed independently of λ' (a conformal invariant).

Realizing now that there is no paradox and that calculations made in the frame, S, in which A is stationary must agree with those made in the frame, S', in which B is stationary, we proceed to calculate a special case as an exercise.

In S, the observer is stationary at $z_A = 0$. His path in S' is

$$\left(z'_A - \frac{1}{g}\right)^2 - t'^2_A = \frac{1}{g^2}\,. \tag{3.6}$$

For definiteness let observer B be located in S' at $z'_B = 1/g$. His path in in S is

$$\left(z_B + \frac{4}{3g}\right)^2 - t_B^2 = \frac{4}{9g^2}\,. \tag{3.7}$$

The two paths will intersect at the times $t = \pm\, t_0 = \pm\, 2/(\sqrt{3}g)$, $t' = \pm\, t'_0 = \pm\, \sqrt{3}/g$. The elapsed time for A as calculated by the observer A is

$$\tau_A = \int\limits_{-t_0}^{t_0} (dt_A^2 - dz_A^2)^{\frac{1}{2}} = \int\limits_{-2/\sqrt{3}g}^{2/\sqrt{3}g} dt_A = \frac{4}{\sqrt{3}g}\,, \tag{3.8}$$

where the integral is calculated over the path $z_A = 0$. Observer B would find for this elapsed time

$$\tau_A = \int\limits_{-t'_0}^{t'_0} \frac{(dt'^2_A - dz'^2_A)^{\frac{1}{2}}}{\lambda'}\,, \tag{3.9}$$

where the integral is calculated over the path given by (3.6). Of course, the transformation has been constructed so that $(dt'^2 - dz'^2)/\lambda'^2 = dt^2 - dz^2$ and observer B will find that (3.9) gives the same result as (3.8). This can be shown explicitly. By use of (3.6) and (3.2), the integral (3.9) reduces after some manipulation to

$$\tau_A = 2\int_{-\sqrt{3}/g}^{\sqrt{3}/g} \frac{dt'_A(1+g^2t'^2_A)^{-\frac{1}{2}}}{1+(1+g^2t'^2_A)} . \tag{3.10}$$

This reduces to (3.8) by use of the transformation

$$t'_A = \frac{t_A}{1-g^2t_A^2/4} . \tag{3.11}$$

The elapsed time for B calculated by the observer A is

$$\tau_B = \int_{-t_0}^{t_0} (dt_B^2 - dz_B^2)^{\frac{1}{2}} , \tag{3.12}$$

calculated along the path (3.7). This can be rewritten as

$$\tau_B = \int_{-2/\sqrt{3}g}^{2/\sqrt{3}g} \frac{dt_B}{(1+9g^2t_B^2/4)^{\frac{1}{2}}} = \frac{4}{3g}\left(\sinh^{-1}\sqrt{3}\right) . \tag{3.13}$$

The fact that τ_B is less than τ_A is in accordance with expectations. Observer B would calculate this time by doing the following integration along the path $z' = -1/g$:

$$\tau_B = \int_{-\sqrt{3}/g}^{\sqrt{3}/g} \left(\frac{dt'^2_B - dz'^2_B}{\lambda'^2}\right)^{\frac{1}{2}} = 4\int_{-\sqrt{3}/g}^{\sqrt{3}/g} \frac{dt'_B}{9-g^2t'^2_B} . \tag{3.14}$$

He knows without calculation that (3.13) must produce the same answer as (3.14) and can also readily verify that the integrals give the same results by using the transformation

$$t_B = \frac{4t'_B}{9-g^2t'^2_B} , \tag{3.15}$$

to transform the integral (3.13) into (3.14).

Again the general relativist would make exactly the same calculation as the conformal relativist. He would use the metric involving λ'.

3·4. *Transformation of mass.* – SCHOUTEN and HAANTJES ([5]) have shown that the Lorentz equation of motion of charged particles is covariant under conformal transformations, providing the rest mass is not invariant but transforms appropriately. The charge of the particle and the velocity of light are invariant. The proper mass must transform in the following way

$$m^c = \lambda(x)m = \frac{m}{\lambda'(x')} . \tag{3.16}$$

The quantity $m\,\mathrm{d}\tau$ is invariant under conformal transformations:

$$m^c\,\mathrm{d}\tau^c = m\,\mathrm{d}\tau . \tag{3.17}$$

For the transformation (2.11), (2.12) by virtue of (2.19), (2.20), in the plane $x' = y' = 0$,

$$m^c = \frac{m}{1 - gz' + (g^2/4)(z'^2 - t'^2)} . \tag{3.18}$$

In the region of space-time for which $gz' \gg (g^2/4)(z'^2 - t'^2)$,

$$m^c \approx m(1 + gz') . \tag{3.19}$$

Equation (3.17) suggests how the point of view of general relativity differs from that of conformal transformations. For general relativity, the Lorentz equation would be kept invariant, as would (3.17), by keeping each factor m and $\mathrm{d}\tau$ individually invariant; conformal relativity varies both. The value of the transformed mass (3.18) is origin-dependent; this corresponds to the analogous situation in general relativity which has an origin-dependent metric tensor, $g'_{\mu\nu}(x') = \lambda'(x')^{-2}\eta_{\mu\nu}$. The origin-dependence is not particularly surprising, and occurs quite often in metrics commonly considered in general relativity.

Equation (3.19) which is an appropriate approximation for certain regions of space-time suggests an interpretation in classical terms. m^c is the total energy of the particle which contains contributions due to the rest mass, m, and to the gravitational potential, mgz'. The full eq. (3.18) apparently adds a correction due to velocity and its effects to the total energy. The origin-dependence of (3.19) can now be thought of as corresponding to the arbitrary addition of a constant to the gravitational potential; the mass difference between two points will according to (3.19) be origin-independent. However the exact expression (3.18) yields an origin-dependence even to the mass difference between two points. This is probably related to the actual physical model that must be used to produce the acceleration field with which we are dealing, much as the gravitational potential from a point mass will yield an origin-dependence for the potential difference of two equally spaced points along the same radial line.

The discussion of mass in general relativity is not obvious and straightforward and a simple comparison of results is difficult. General considerations would lead one to believe that not mass but a notion like total energy content of a particle (whatever that may mean in general relativity) is important and that the results of a calculation in general relativity would agree with one in conformal relativity. As an example of this we make a calculation in the next section explicitly using the mass transformation.

4. – Frequency shift of a freely falling emitter.

Let S again be the frame which is freely falling in a uniform gravitational field of intensity g. Let S' be the frame which is at rest in this field. Thus, S is an inertial frame, while S' is a frame in which only a constant force is acting.

Let an atom of rest mass m_* (the mass measured by S while the atom is at rest in S) emit a photon. Let the final atom have rest mass m.

We want to investigate the frequency of the emitted photon as seen by an observer in S', using the methods of special relativity, conformal relativity, and general relativity, respectively. These three ways of looking at the situation correspond exactly to the three points of view, A, B_1, and B_2, listed in Section 2.

4·1. *Special relativity.* – Although there is a gravitational field present in S', this frame can nevertheless be regarded as an inertial frame in a certain sense. If the atom decays at the space-time point, z'_0, at which the observer is located, the presence of a uniform gravitational field can be ignored. One applies the conservation laws as in special relativity

$$p^\mu_* = p^\mu + k^\mu \,, \tag{4.1}$$

where p^μ_* and p^μ are the momentum four-vectors of the atom immediately before and after the emission, and k^μ is the photon four-momentum. We have ignored the irrelevant dimensions x' and y', for motion along the z'-axis (we also put $\hbar = 1$).

$$\left\{\begin{aligned} p^\mu_* &= m_*(1+gz'_0)\frac{\mathrm{d}x'^\mu}{\mathrm{d}\tau_*} &&= m_*(1+gz'_0)\,\gamma_*\,(1;\,0,\,0,\,v_*)\,; \\ p^\mu &= m(1+gz'_0)\,\frac{\mathrm{d}x'^\mu}{\mathrm{d}\tau} &&= m(1+gz'_0)\,\gamma\,(1;\,0,\,0,\,v)\,; \\ k^\mu &= (1+gz')\varkappa^\mu(z') &&\equiv (1+gz')(\omega;\,0,\,0,\,k)\,; \qquad \omega = -k > 0\,. \end{aligned}\right. \tag{4.2}$$

Here v_* and v are the velocities of m_* and m as seen by the observer in S'. The symbols τ_* and τ denote the fact that each particle has its proper time. The factor $1+gz'$ takes into account the presence of the gravitational field to first order in g. The principle of equivalence demands that it be applied to k^μ as well.

Elimination of v from these equations yields for the frequency seen at z_0' and produced at z_0' from an atom of velocity v_*,

$$\omega(z_0'; z_0', v_*) = (m_* - m)\frac{m_* + m}{2m_*}\gamma_*(1 - v_*). \tag{4.3}$$

The factor $m_* - m$ would yield ω if recoil were neglected and v_* were zero; the term $(m_* + m)/2m_*$ is a recoil correction; $\gamma_*(1 - v_*)$ is the Doppler effect.

If the observer is located at a point $z_1' \neq z_0'$, energy conservation will give

$$\omega(z_1')(1 + gz_1') = \omega(z_0')(1 + gz_0') \tag{4.4}$$

with $\omega(z'; z', 0) \equiv \omega(z')$. Thus, the frequency seen at z_1' of a photon produced at z_0' by an atom of velocity v_* is

$$\omega(z_1'; z_0', v_*) = \omega(z_0'; z_0', v_*)[1 + g(z_0' - z_1')]. \tag{4.5}$$

B_1) *Conformal relativity.* – The observer in S' realizes that he is not in an inertial frame. The momentum four-vector is defined by

$$p^{\mu^c} = m^c \frac{dx'^\mu}{d\tau^c}, \tag{4.6}$$

where m^c is related to m by eq. (3.16). The conservation law (assured by the invariance of the theory under the translation group) is

$$p_*^{\mu^c} = p^{\mu^c} + k^{\mu^c}. \tag{4.7}$$

This involves, when both observer and atom are at z_0', according to (3.16) and (3.18)

$$\begin{cases} p_*^{\mu^c} = \dfrac{m_*}{\lambda'(x')}\dfrac{dx'^\mu}{d\tau_*^c} = \dfrac{m_*}{1 - gz_0'}\dfrac{dx'^\mu}{d\tau_*^c} = \dfrac{m_*\gamma_*}{1 - gz_0'}(1; 0, 0, v_*); \\[2ex] p^{\mu^c} = \dfrac{m}{\lambda'(x')}\dfrac{dx'^\mu}{d\tau^c} = \dfrac{m\gamma}{1 - gz_0'}(1; 0, 0, v); \\[2ex] k^{\mu^c} = \dfrac{1}{\lambda'(x')}\varkappa^\mu(x') = \dfrac{1}{1 - gz_0'}(\omega; 0, 0, k). \end{cases} \tag{4.8}$$

It is evident that k^{μ^c} must be related to $\varkappa^\mu$ in the same way that m^c is related to m, (3.16), in order to preserve the covariance of (4.7).

Equations (4.7) and (4.8) lead to the same results as in case A. In particular, the observer S' will again find the result (4.3) for $\omega(z_0'; z_0', v_*)$.

If the conformal observer is located at z_1', he will apply a different scale factor at z_1' than at z_0'. Since the conformal observer considers himself to be in field free space, k^{μ^c} is independent of position. Hence the ratio $\omega(z')/\lambda'(z')$ is a constant, and he finds

$$\omega(z_1'; z_0', v_*) = \frac{\lambda'(z_1')}{\lambda'(z_0')}\,\omega_0'(z_0'; z_0', v_*) \approx \omega(z_0'; z_0', v_*)[1 + g(z_0' - z_1')]\,, \tag{4.9}$$

which is identical with (4.5).

It may be thought that, since ionized atoms can obtain large accelerations in readily available electric fields, a large frequency shift of a spectral line emitted by an ion in such a field should be predicted by the conformal calculation. However, the photon is not accelerated by the electric field in moving from z_0 to z_1 and hence its behavior in an electric field is quite different from that in a gravitational field. It acquires no additional potential energy and hence there is no frequency shift.

B_2) *General relativity*. – The general relativist might attack the problem in a way which partially resembles both of the above points of view. When both he and the emitter are at z_0', he would say that he can apply the results of special relativity and write (4.3) directly.

Now suppose the general relativist is at z_1' and the emitter is at z_0'. He will apply an invariance argument which will tell him that [16]

$$\frac{\mathrm{d}\tau(z_1')}{\mathrm{d}\tau(z_0')} = \frac{[g_{\mu\nu}'(x_1')\,(\mathrm{d}x_1'^\mu/\mathrm{d}t_1')(\mathrm{d}x_1'^\nu/\mathrm{d}t_1')]^{\frac12}\,\delta t_1'}{[g_{\mu\nu}'(x_0')\,(\mathrm{d}x_0'^\mu/\mathrm{d}t_0')(\mathrm{d}x_0'^\nu/\mathrm{d}t_0')]^{\frac12}\,\delta t_0'} = \frac{\lambda'(z_0')}{\lambda'(z_1')}\,\frac{1}{\gamma_*(1-v_*)}\,. \tag{4.10}$$

It should be pointed out that the equivalence of the conformal relativistic result, eq. (4.9) and its corresponding general relativistic counterpart, eq. (4.10) holds for arbitrary values of gz. Equation (4.5) is necessarily valid only for small gz, because gravitational fields can be taken into account in the framework of special relativity only in the Newtonian sense.

5. – Discussion.

We have considered how predictions of the results of some specific measurements can be made from the point of view of conformal invariance. In

[16] See, for example, R. C. TOLMAN: *Relativity, Thermodynamics and Cosmology* (Oxford, 1934), p. 288.

special relativity referring to field-free space, the theory makes contact with experience because the proper time is an invariant and is measured by a physical clock. The comparison of clocks at two different space-time points offers no difficulty. In general relativity, proper time is also an invariant measured by a physical clock. A co-ordinate system can always be chosen sot hat the description of a physical situation at a point is identical with that of special relativity. (Note that all the effects we have discussed are independent of curvature.) To compare measurements at two different space-time points, the general relativist must know the metric tensor at these points. He can determine it in either of two ways. He can calculate it from the theory if he knows the physical situation in detail, or he can measure it.

The case of the general relativist is entirely analogous to that of the conformal relativist. The latter knows that before making contact with experience he must determine a scale factor. By an appropriate choice of co-ordinates, he can choose this factor to be equal to unity at his space-time point of observation; hence again he can reduce the situation at a single point to that of special relativity. To make comparisons at different space-time points, he must determine the scale factor at the different points. This he can do either by making calculations from the physical situation that prevails or by making measurements.

The situation of the conformal relativist and of the general relativist are thus completely parallel. Neither can make measurements without determining λ'; one will consider it to be a scale factor, the other a factor that multiplies $\eta_{\mu\nu}$ to give the metric tensor. Obviously, general relativity deals with a much broader class of transformations than does conformal relativity. Where conformal relativity applies, general relativity agrees with it in detail. Since general relativity is the broader theory, there seems to be no advantage at all in considering conformal relativity as presenting a physically significant point of view.

In this paper, we have treated only conformal transformations from flat space to flat space. If more general conformal transformations are considered, the above conclusions may no longer hold.

Appendix

The affine connection.

In Riemannian geometry, the affine connection is given by the Christoffe symbol

$$\left\{ \begin{matrix} \mu \\ \alpha \ \beta \end{matrix} \right\} = \frac{1}{2} g^{\mu\lambda} \left(\frac{\partial g_{\lambda\beta}}{\partial x^\alpha} + \frac{\partial g_{\lambda\alpha}}{\partial x^\beta} - \frac{\partial g_{\alpha\beta}}{\partial x^\lambda} \right), \tag{A.1}$$

which transforms under a general relativistic co-ordinate transformation from the frame S to the grame S' as

$$\left\{ \begin{matrix} \mu \\ \alpha\ \beta \end{matrix} \right\}' = \left\{ \begin{matrix} \nu \\ \varrho\ \sigma \end{matrix} \right\} \frac{\partial x^\varrho}{\partial x'^\alpha} \frac{\partial x^\sigma}{\partial x'^\beta} \frac{\partial x'^\mu}{\partial x^\nu} + \frac{\partial x'^\mu}{\partial x^\nu} \frac{\partial^2 x^\nu}{\partial x'^\alpha \partial x'^\beta} . \tag{A.2}$$

If the transformation from S to S' were a conformal co-ordinate transformation so that in place of $\left\{ \begin{matrix} \mu \\ \alpha\ \beta \end{matrix} \right\}'$ we would deal with $\left\{ \begin{matrix} \mu \\ \alpha\ \beta \end{matrix} \right\}^c$,

$$\left\{ \begin{matrix} \mu \\ \alpha\ \beta \end{matrix} \right\}^c = \left\{ \begin{matrix} \mu \\ \alpha\ \beta \end{matrix} \right\}' + \frac{1}{2} \{\delta^\mu_\alpha s_\beta + \delta^\mu_\beta s_\alpha - s^\mu g_{\alpha\beta}\} , \tag{A 3}$$

where

$$s_\mu = \frac{\partial}{\partial x'^\mu} \ln \sigma . \tag{A.4}$$

It is more convenient in dealing with conformal transformations to define the affine connection, $\Gamma^\mu_{\alpha\beta}$, in such a way that under a conformal co-ordinate transformation

$$\Gamma^{\mu c}_{\alpha\beta} = \Gamma'^\mu_{\alpha\beta} , \tag{A.5}$$

where $\Gamma'^\mu_{\alpha\beta}$ is related to $\Gamma^\mu_{\alpha\beta}$ in the same way as are the Cristoffel symbols, (A.2). This requires the definition

$$\Gamma^\mu_{\alpha\beta} \equiv \left\{ \begin{matrix} \mu \\ \alpha\ \beta \end{matrix} \right\} - \frac{1}{2} (\delta^\mu_\alpha \varkappa_\beta + \delta^\mu_\beta \varkappa_\alpha - \varkappa^\mu g_{\alpha\beta}) , \tag{A.6}$$

provided $\varkappa_\mu$ transforms according to

$$\varkappa^c_\mu = \varkappa_\alpha \frac{\partial x^\alpha}{\partial x'^\mu} + s_\mu . \tag{A.7}$$

With the affine connection defined in this way, and considering conformal co-ordinate transformations, we are dealing with an example of a Weyl space; in showing the invariance of the equation of motion of charged particles, this affine connection is particularly helpful.

The Clock Paradox in the Relativity Theory

TA-YOU WU and Y. C. LEE

Department of Physics, State University of New York, Buffalo, New York 14214

Received: 11 *May* 1971

Abstract

A system S' (rocket) starts from rest in an inertial system S, and after a series of accelerated, uniform and decelerated motions, comes back to rest at its initial position in S. An exact calculation is carried out, from the standpoint of S, of the time intervals for the arrivals at S of light signals sent back by S'. From the standpoint of S', S has made a round trip after undergoing a series of free falls in gravitational fields and coasting motions. An exact calculation is carried out for the 'proper time' intervals in S from the standpoint of S'. It is shown that there is exact agreement between S and S' in their reckonings of the total time intervals for the two frames, namely, both S and S' agree quantitatively, to them, the time interval is longer for S than for S'.

The accelerated motion of S' relative to S explicitly used in the treatment of the problem in the present work is that under time-independent field and subject to the condition of *local* Lorentz contraction and dilation; the resulting motion turns out to be that obtained earlier by Møller on entirely different considerations. The result of the present treatment is, however, more general than this particular motion seems to imply, since by an arbitrary coordinate transformation, it can be made to include an infinite number of accelerated frames including time-dependent fields, all within the framework of flat space-time. General remarks are given for the clock problem in the general theory of relativity in the sense of Einstein's curved space.

1. *The Clock Paradox*

The question concerned is the following: Imagine a pair of clocks, one of which remains at rest in an inertial frame, and the other sets out on a trip (on a rocket, say), and after a time returns to rest in the inertial frame. Will the travelling clock be slower than the one at home? Will they both agree exactly by how much one is slower than the other?

This problem is sixty years old. In a paper in 1911, Einstein (1911) gave a simple theory in which (1) he employed the Doppler effect formula of the special theory of relativity and obtained the effect of uniform acceleration of a reference frame on the Doppler shift, and (2) he introduced the equivalence principle for the acceleration of a frame and a gravitational field. Einstein concluded that a clock that has travelled, say in a circular path, will 'lose time', because the rate of the clock is slower in the accelerated motion.

That a returned clock should have lost time compared with the one at home is so strange a conclusion that Einstein specifically wrote an article (Einstein, 1918) in 1918, in the form of a dialogue between a critic and himself, to show (1) how the trip will be viewed from the standpoints of both frames, (2) how the reciprocal symmetry (in the sense of the special theory of relativity) will be destroyed in this case by the accelerated motion of the rocket, and that both frames will agree that the returned one will be slow, and (3) that this is due to the loss of time, or the 'slowing down of the clock', of the rocket *during* the accelerated portion of the rocket's trip when it *turns back*.

For definiteness, let us pose the following situation. From the standpoint of the inertial system S, the rocket (or the travelling twin) S' goes through the following sequence of events:

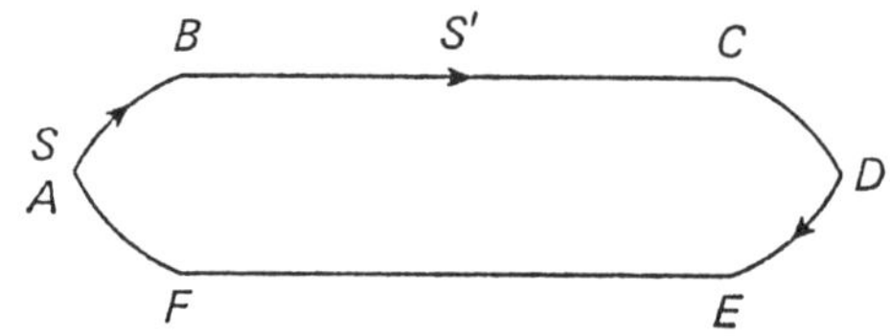

(1.1)

A–B: S' starts off, with acceleration a (in the positive x direction), reaching the velocity v at B.

B–C: S' shuts off its engine and moves with a uniform velocity v relative to S.

C–D: S' starts its engine and decelerates, reducing its velocity relative to S to zero.

D–E: S' keeps its engine and starts accelerating toward S, reaching the velocity $-v$ at E.

E–F: S' shuts off its engine and moves with constant velocity $(-v)$ toward S.

F–A: S' starts its braking engine, and moves with acceleration $-a$, coming to rest at A.

From the standpoint of the rocket S' who regards itself as at rest, S will go through the following events:

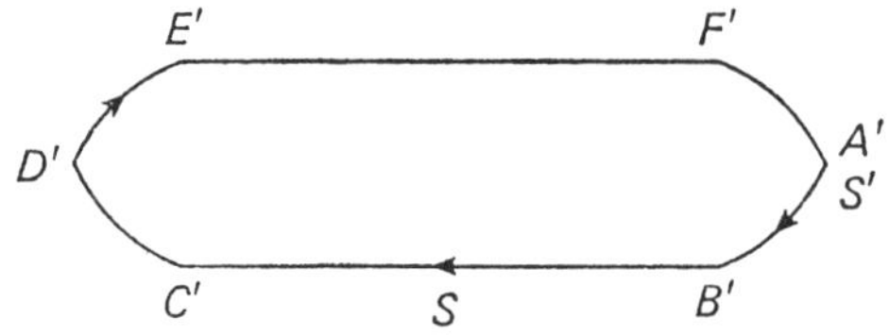

(1.2)

A'–B': S starts 'falling' in a (universal) gravitation field $-g$ (in the negative x direction), attaining a velocity $-v$ (relative to S') at B'.

B'–C': The gravitational field is removed, and S keeps on moving with the (constant) velocity $-v$.

C'–D': A (universal) gravitational field g in the positive x-direction is turned on. S comes to stop (relative to S') at D'.

D'–E': The same field g continues to act, and S 'falls' from D' to E', attaining the velocity v (relative to S').

E'–F': The field g is removed and S moves with the constant velocity v.

F'–A': A gravitational field $-g$ is turned on, and S is brought to rest at A'.

During A'–B', C'–D', D'–E' and F'–A', S' itself is in the same universal gravitational field as S, but S' is held fixed by some external agency.

During the uniform relative motion parts B'–C', E'–F', S will say that the clock of S' is slow according to the time dilatation relation. If $\Delta\tau_{S'}$ is the proper time interval recorded by S' for each of these parts, and ΔT_S is the time interval recorded by synchronised clocks attached to the frame S, then

$$2\Delta T_S = \frac{2\Delta\tau_{S'}}{\sqrt{(1-\beta^2)}} \tag{1.3}$$

But with equal right, S' will say that the clock of S is slow compared with that of S', and if $\Delta\tau_S$ is the (proper) time interval recorded by S (for each of the uniform relative motion parts) and $\Delta T_{S'}$ the time interval recorded by synchronised clocks attached to the frame S', then

$$2\Delta T_{S'} = \frac{2\Delta\tau_S}{\sqrt{(1-\beta^2)}} \tag{1.4}$$

Einstein pointed out, however, that during the 'turning around' parts C'–D', D'–E' in (1.2), S is at a higher gravitational potential than S', and the clock of S is faster than that of S'. The clock in S will during C'–D'' and D'–E' 'gain' time and more than compensate the 'loss' as given by (1.4). If the time intervals during C'–D', D'–E' are very short compared with those for the uniform motion parts, the 'gain' during C'–D', D'–E' by the clock of S will be such as to bring the total time ΔT recorded by S (for the trip B'–C', C'–D', D'–E', E'–F') to be longer than that $\Delta T'$ recorded by S', in accordance with (1.3), i.e.,

$$\Delta T = \frac{\Delta T'}{\sqrt{(1-\beta^2)}} \tag{1.5}$$

The above statement of Einstein has been expressed in explicit form by Tolman (1934). Let us view the trip from the standpoint of S' as in (1.2). Let τ_{AB}, τ_{BC}, τ_{CDE}, τ_{EF} $(=\tau_{BC})$, τ_{FA} $(=\tau_{AB})$ be the proper time intervals as recorded by S, and let t_{BC}, t_{CDE}, t_{EF} $(=t_{BC})$, t_{FA} $(=t_{AB})$ be the time intervals as recorded by synchronised clocks at various points in (1.2) attached to S'.† Then, by (1.4),

$$\tau_{BC}\,(=\tau_{EF}) = \sqrt{(1-\beta^2)}\,t_{BC}\,(=\sqrt{(1-\beta^2)}\,t_{EF}) \tag{1.6}$$

† We shall, without causing confusion, drop the prime for B', C', etc. in the subscripts for the τ's and t's.

Let the average distance between S and S' during the turning around portion C'–D'–E' be approximately taken to be $x = vt_{BC}$. Since $v = gt_{CD}$, the Doppler effect separation gives

$$\tau_{CDE} \cong \left(1 + \frac{gx}{c^2}\right) t_{CDE} \tag{1.7}$$

$$= t_{CDE} + 2\beta^2 t_{BC} \tag{1.7a}$$

The total time for the whole trip is, for S, from (1.6) and (1.7a),

$$\begin{aligned}\tau &= \tau_{AB} + 2\tau_{BC} + \tau_{CDE} + \tau_{FA} \\ &= 2[\surd(1-\beta^2) + \beta^2]\, t_{BC} + 2\tau_{AB} + t_{CDE} \\ &= 2(1 + \tfrac{1}{2}\beta^2 + \cdots)\, t_{BC} + 2\tau_{AB} + t_{CDE} \end{aligned} \tag{1.8}$$

If we make the time intervals τ_{AB}, T_{CDE} very short compared with τ_{BC} and t_{BC}, then (1.8) becomes

$$\tau \cong \frac{1}{\surd(1-\beta^2)} \times \text{time for the trip recorded in } S' \tag{1.9}$$

This is in approximate agreement with (1.3). It is important to note that the $\cong$ sign in (1.9) arises not so much because of the neglect of τ_{AB} and t_{CDE} in (1.8) as because of the approximations made in obtaining (1.7).

In 1956, Dingle (1956) in a series of articles renewed the question of whether the returned twin from a rocket trip is younger than his brother who has stayed home. He believed that there should be no difference in their aging, that all earlier conclusions, including Einstein's, are erroneous. His questioning of these earlier works by many physicists has led to a great flux of discussions. Most authors (Arzelies, 1966) maintain the conclusion of Einstein. In most cases, the arguments amount to the simple statement that since the rocket S' has undergone accelerated and decelerated motions, it is not on equal footing with S which is an inertial system, and hence the reciprocal symmetry in the sense of the special theory of relativity has been removed. This part of the argument is of course correct. But then, because of attempts to simplify the problem for the non-specialist, the following argument is usually put forward: One can make the time intervals for the accelerated and decelerated parts very short compared with the time intervals for the uniform relative motion parts [see (1.1) or (1.2)] and in the limit negligible. Then, since only S is a 'preferred' (in the sense that it is an inertial) frame, one must only employ the relation (1.3). This part of the argument is unfortunately misleading. We have seen in the preceding section from the approximate treatment by Tolman that it is precisely the acceleration (or, an equivalent gravitation field) during the 'turning around' of the rocket that 'slows down its clock' (relative to the inertial frame S), and that one obtains the result (1.9) in an *approximation* only, which is *not* exactly the relation (1.3). The point that seems to have been forgotten in many 'elementary' discussions of the clock paradox is that while the 'compensations' in (1.8) and (1.9) must come from the accelerated motions, the correct result [(3.20) and (3.22) in the following] should really be independent of the strength of the accelerating and decelerating field which determine the length of time for these accelerated and decelerated parts. The undue prominence given the uniform relative motion parts (coasting of the rocket) and the consequent appearance of the Lorentz relation (1.3) are unfortunate, for they tend to divert the attention from the accelerated motion, which is essential in the clock paradox problem, to the expression (1.3) for uniform relative motion. In actual fact, the uniform relative motion parts [B–C, E–F, B'–C', E'–F' in (1.1) and (1.2)] are non-essential, and one would have essentially the same 'paradox' if one does away with these (coasting) parts entirely. In the following section, an analysis of the clock problem with, and also without, the uniform relative motion parts in (1.1) and (1.2) will be carried out to illustrate this point.

In the literature, attempts have been made to convince one of the immediate applicability of the expression (1.3) for the time intervals for the whole trip as recorded by S and S' (assuming negligible times for the accelerated parts of the motion) by the following argument. Let there be a set of triplets instead of a pair of twins. Let C stay home (in an inertial frame); let A be moving away in a rocket with velocity v (relative to C). At a certain point in space, A meets B who is travelling toward C with velocity $-v$ (relative to C). A and B do not stop; B just sets his clock according to that of A. B finally passes by C. It is then claimed that the time recorded by C (for the interval between the passing by of A and that of B) is longer than that recorded by B, in accordance with (1.3). This argument is a special case of a general theorem in the special theory of relativity, namely, that in a Minkowski diagram, the time interval measured on a straight line AB (which is the time axis) is longer than the sum of time intervals measured along a series of straight lines AC, CD, . . . DB (each being a time axis in another Lorentz frame) which together with AB form a polygon. The implication of this argument is that we make use of the awareness of the accelerations to remove the reciprocal symmetry of the Lorentz frames, but have ignored the effects of the accelerations on the time measures of the systems. On this argument, one might as well contend with two Lorentz frames since the use of a third frame does not add to the resolution of the problem.

2. *Arbitrary Motion Relative to an Inertial Frame*

We shall study an accelerated motion that can be treated exactly in the clock problem.

Let (X, T) be the space and time coordinates in an inertial frame S and let x, t be those in a frame S' which may be accelerated under the time-independent field. Let $v(x, T)$ be the velocity of a fixed point x in S' relative to S at time T and let X be the space coordinate of the point p so that $X = X(x, T)$ and

$$v(x,T) = v(x(X,T),T) \tag{2.1}$$

The velocity of the point p in S is

$$v = \left(\frac{\partial X}{\partial T}\right)_x \tag{2.2}$$

If we assume that in S' the unit of length is the same as that in S, then the condition of local Lorentz contraction is expressed by

$$\left(\frac{\partial X}{\partial x}\right)_T = \sqrt{\left(1 - \frac{v^2}{c^2}(x, T)\right)} \tag{2.3}$$

From this, one obtains the equation

$$\left(\frac{\partial v(x, T)}{\partial x}\right)_T = \left[\frac{\partial}{\partial T}\sqrt{\left(1 - \frac{v^2}{c^2}(x, T)\right)}\right]_x \tag{2.4}$$

This is equivalent to

$$\left(\frac{\partial v(X, T)}{\partial X}\right)_T = -\frac{1}{c^2}v(X, T)\frac{\partial v}{\partial T}(X, T) \tag{2.5}$$

in which v is regarded as a function of X, T through the transformation (2.1).

To obtain the relation between the time t and the coordinates X and T, we shall introduce the conditions of local Lorentz time dilatation and relativity of motion of S and S', namely,

$$\sqrt{(g_{44})}\, d\tau = \sqrt{\left(1 - \frac{v^2}{c^2}(x, T)\right)}\, dT \tag{2.6}$$

and

$$v(X, T) = \left(\frac{\partial X}{\partial T}\right)_x = -\frac{1}{\sqrt{(g_{44})}}\left(\frac{\partial x}{\partial t}\right)_X = v(x, t) \tag{2.7}$$

where τ is the proper time in S' in the sense that if the metric in S' is $(dy = dz = 0)$

$$ds^2 = -(dx^2 + dy^2 + dz^2) + g_{44}\, dt^2 \tag{2.8}$$

then

$$ds^2 = g_{44}\, d\tau^2 \tag{2.9}$$

so that

$$d\tau = \sqrt{(1 - v^2)}\, dt \tag{2.10}$$

where

$$v^2 = \frac{1}{g_{44}}\left(\frac{\partial x}{\partial t}\right)_X^2 \tag{2.7$'$}$$

Note that the τ called the proper time above and defined by (2.9) is not the normal proper time τ_0 defined by $d\tau_0 = ds$ which will be related to τ here by $d\tau_0 = \sqrt{(g_{44})}d\tau$. In the present work, we make use of τ in the calculation of τ_0.

The conditions (2.3) and (2.6) are valid at a point (X,T) or (x,t). Our object is to find a class of accelerated frames S' (with respect to S) with the transformation

$$x = x(X,T), \qquad t = t(X,T) \tag{2.11}$$

and satisfying (2.3) and (2.6). The hypothesis that a transformation (2.11) exists between the (X,T) in an inertial frame and the (x,t) coordinates implies that the space is Euclidean. In this case, we can integrate the differential equations (2.5) and (2.6) since no curvature of space is involved.

Equation (2.5) can be solved by the method of separation of variables, namely, by setting

$$v = \xi(X)\,\zeta(T)$$

which leads to

$$v = \frac{\lambda c^2 T + C_1}{\lambda X + C_2}, \qquad \lambda, C_1, C_2 \text{ being constants.}$$

If the initial condition is

$$v \to aT \qquad \text{at } X = 0 \qquad \text{as } T \to 0$$

then

$$v = \frac{aT}{1 + (aX/c^2)} \tag{2.12}$$

Putting this into (2.2), one obtains the equation of motion of the point p ($x = 0$ in S'),

$$\frac{dX}{dT} = \frac{aT}{1 + (aX/c^2)} \tag{2.13}$$

To obtain the relationship between X and x, we may first integrate (2.13) to obtain

$$\left(1 + \frac{aX}{c^2}\right)^2 - \frac{a^2 T^2}{c^2} = f(x) \tag{2.14}$$

where $f(x)$ is an arbitrary function of x. On the other hand we can also integrate the Lorentz contraction equation (2.3) in which v is given by (2.12). The integration yields

$$\left(1 + \frac{aX}{c^2}\right)^2 - \frac{a^2 T^2}{c^2} = \left(\frac{ax}{c^2} + b(T)\right)^2 \tag{2.14'}$$

where $b(T)$ is an arbitrary function of T. However, by comparing the above two equations (2.14) and (2.14′) we see that $b(T)$ must actually be a pure constant b, independent of T, and $f(x)$ is just $[(ax/c^2) + b]^2$. Our initial conditions then require $b = 1$, so that we have

$$\left(1 + \frac{aX}{c^2}\right)^2 = \left(1 + \frac{ax}{c^2}\right)^2 + \left(\frac{aT}{c}\right)^2 \tag{2.15}$$

From (2.7), (2.13) and (2.15) we obtain

$$\frac{(aT/c)}{1+(aX/c^2)} = v = -\frac{1}{\sqrt{(g_{44})}}\left(\frac{dx}{dt}\right)_X = \frac{aT/c}{\sqrt{(g_{44})}\,[1+(ax/c^2)]}\frac{dT}{dt}$$

From Equations (2.6), (2.7), (2.8) and the assumption that g_{44} does not depend on t, one can show that

$$g_{44} = \left(1+\frac{ax}{c^2}\right)^2 \tag{2.16}$$

The above equation in dT/dt can be integrated, and with the initial condition

$$t = 0 \qquad \text{when } T = 0$$

we obtain

$$\frac{2at}{c} = \ln\left(1+\frac{aX}{c^2}+\frac{aT}{c}\right) - \ln\left(1+\frac{aX}{c^2}-\frac{aT}{c}\right) \tag{2.17}$$

Using (2.15) in (2.17), we obtain

$$+\frac{1}{[1+(ax/c^2)]}\left(\frac{dx}{dt}\right)_X = -\frac{aT/c}{\sqrt{\{[1+(ax/c^2)]^2+(aT/c)^2\}}} = -\tanh\frac{at}{c} \tag{2.18}$$

and (2.7) becomes

$$\left(\frac{dX}{dT}\right)_x = \frac{aT/c}{1+(aX/c^2)} = \tanh\frac{at}{c} \tag{2.18a}$$

In (2.18a), since x is held fixed, t is also the proper time τ in S'.

Equations (2.15) and (2.17) now transform the metric

$$ds^2 = -dX^2 + dT^2 \tag{2.19}$$

into

$$ds^2 = -dx^2 + \left(1+\frac{ax}{c^2}\right)^2 dt^2 \tag{2.20}$$

In (2.20), for a given constant a, x is restricted to the region $x > -(c^2/a)$. It is seen from (2.13) and (2.15) that the limiting value $x = -(c^2/a)$ corresponds to $v = c$ beyond which v should not pass.

For convenience, we write three consequences of equations (2.15) and (2.17), namely,

$$\frac{aT}{c} = \left(1+\frac{aX}{c^2}\right)\tanh\left(\frac{at}{c}\right) \tag{2.21}$$

$$\frac{aT}{c} = \left(1+\frac{ax}{c^2}\right)\sinh\left(\frac{at}{c}\right) \tag{2.22}$$

$$1+\frac{aX}{c^2} = \left(1+\frac{ax}{c^2}\right)\cosh\frac{at}{c} \tag{2.23}$$

Equation (2.15) describes the motion of a fixed point p in S' from the standpoint of S. It is the so-called hyperbolic motion.

Equation (2.23) describes the motion of a fixed point p in S from the standpoint of S'.

If S' is accelerated along the $-X$ direction, we have only to replace a in all the equations (2.12)–(2.23) by $-a$, and obtain

$$\left(1-\frac{a(x-x_0)}{c^2}\right)^2=\left(1-\frac{a(X-X_0)}{c^2}\right)^2-\frac{a^2}{c^2}(T-T_0)^2 \tag{2.24}$$

$$\frac{a(T-T_0)}{c}=\left(1-\frac{a(X-X_0)}{c^2}\right)\tanh\frac{a(t-t_0)}{c} \tag{2.25}$$

$$\frac{a(T-T_0)}{c}=\left(1+\frac{a(X-X_0)}{c^2}\right)\sinh\frac{a(t-t_0)}{c} \tag{2.26}$$

$$1-\frac{a(X-X_0)}{c^2}=\left(1-\frac{a(x-x_0)}{c^2}\right)\cosh\frac{a(t-t_0)}{c} \tag{2.27}$$

where x_0, X_0, T_0, t_0 are constants to be determined by the appropriate initial conditions. In this case, equations (2.7) and (2.13) become

$$v=\left(\frac{dX}{dT}\right)_x=-\frac{a(T-T_0)}{1-(a/c^2)(X-X_0)} \tag{2.28}$$

$$v=-\frac{1}{1-[a(x-x_0)/c^2]}\left(\frac{dx}{dt}\right)_X=\left(\frac{dX}{dT}\right)_x=-\tanh a(t-t_0) \tag{2.28a}$$

At this point, it is of considerable interest to note that the transformation (2.15) and (2.17) are precisely that derived by Møller (1943) on completely different considerations. Møller starts from a static metric assumed to be

$$ds^2=-(dx^2+dy^2+dz^2)+g_{44}\,dt^2 \tag{2.8}$$

where $g_{44}=g_{44}(x)$. g_{44} is determined by the Einstein equations $R_{\mu\nu}=0$, which lead to $g_{44}=[1+(gx/c^2)]^2$, g being a constant. The curvature tensor $R^{\lambda}_{\mu\nu\sigma}$ for this metric vanishes, showing the space to be Euclidean. With the help of the equations of the geodesic, the transformation (2.15) and (2.17) is found. We have arrived at (2.15) and (2.17) on the basis of the local Lorentz transformation properties (2.3), (2.6), (2.7) with the assumption that in (2.7), g_{44} is a function of x but not of t.

3. *Resolution of the Clock Paradox*

We shall now study the clock problem as stated in (1.1) and (1.2) of Section 1, by treating the accelerated parts of the trip A–B, C–D–E, F–A, and A'–B', C'–D'–E', F'–A' by means of the accelerated motion described by equations (2.9)–(2.24) and (2.25)–(2.28) of the preceding section.

(A) *From the Standpoint of S*

Referring to the figure in (1.1), let the origins of the coordinate systems $S(X)$ and $S'(x)$ be coincident at $T = t = 0$.

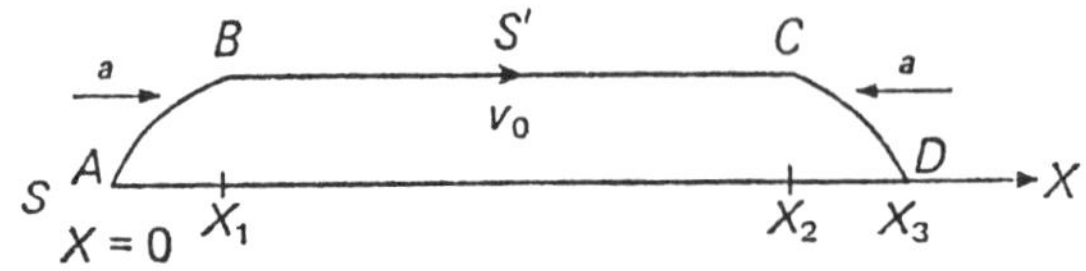

(3.1)

The rocket $S'(x=0)$ moves according to (2.9), (2.7), (2.18)–(2.18a), (2.21)–(2.23). Part *A–B*: For the motion of the point $x = 0$ (fixed in S'), the t in (2.18a), (2.22), (2.23) becomes the proper time τ_0 in S' (i.e., the time registered by one and the same clock at $x = 0$ fixed in S'). When $x = 0$ reaches the velocity v_0 (relative to S frame) we have†

$$\tanh a\Delta\tau_1 = v_0 \tag{3.2}$$

$$aT_1 = \sinh(a\tau_{01}) = \frac{v_0}{\sqrt{(1 - v_0^2)}} \tag{3.3}$$

$$1 + aX_1 = \cosh(a\tau_{01}) = \frac{1}{\sqrt{(1 - v_0^2)}} \tag{3.4}$$

where T_1 is time, measured by the synchronised clocks attached to S, X_1 is the distance traversed by S' $(x = 0)$ when it has reached the velocity v_0 (i.e., the part *A–B*). In the following, the subscript 1, 2, 3, . . . refer to the parts *A–B*, *B–C*, *C–D*, etc. respectively of the trip. The same subscripts 1, 2, 3, . . . are also used for the *A′–B′*, *B′–C′*, *C′–D′*, etc. in the following section from the standpoint of S'.

To obtain the time interval ΔT_1 recorded by one clock fixed at $X = 0$ in S, let S' send light signals back to S. Let τ be the proper time in S'. Then‡

$$\Delta T_1 = \int_0^{\Delta\tau_1} \sqrt{\left(\frac{1+v}{1-v}\right)}\, d\tau \tag{3.5}$$

where by (3.2) $\Delta\tau_1 = \tanh^{-1} v_0$. Thus

$$\Delta T_1 = \frac{1}{a}\left[\frac{v_0}{\sqrt{(1 - v_0^2)}} + \frac{1}{\sqrt{(1 - v_0^2)}} - 1\right]$$

† In the following, we simplify writing by choosing the unit of time such that $c = 1$. All time, velocity, acceleration T, t, v, v_0, a are to be replaced by cT, ct, v/c, v_0/c, a/c^2, to convert to c.g.s. units.

‡ ΔT_1 only represents the time interval for the clock at $X = 0$ to intercept all the light signals sent to it by the clock attached to $x = 0$ within $\Delta\tau_1$. It does *not* really represent the time of travel of the rocket $(x = 0)$ from A to B as recorded by the clock at $X = 0$. The sum in (3.10) is, however, the total time interval for the whole trip of S', as recorded by one and the same clock at $X = 0$ in S.

From the symmetry of the situation, it is clear that for the part C–D,

$$\Delta T_3 = \Delta T_1 \tag{3.6}$$

and for the parts D–E, F–A,

$$\Delta T_4 = \int_0^{\Delta\tau_1} \sqrt{\left(\frac{1-v}{1+v}\right)}\, d\tau \tag{3.7}$$

$$= \frac{1}{a}\left[\frac{v_0}{\sqrt{(1-v_0^2)}} - \frac{1}{\sqrt{(1-v_0^2)}} + 1\right]$$

$$\Delta T_6 = \Delta T_4 \tag{3.8}$$

For the parts B–C, E–F, if $\Delta\tau_2 = \Delta\tau_5$ is the proper time intervals in S', the sum of the intervals for the arrivals of the signals sent back by S' during these intervals is

$$\Delta T_2 + \Delta T_5 = \Delta\tau_2 \sqrt{\left(\frac{1+v_0}{1-v_0}\right)} + \Delta\tau_5 \sqrt{\left(\frac{1-v_0}{1+v_0}\right)}$$

$$= \frac{2\Delta\tau_2}{\sqrt{(1-v_0^2)}} \tag{3.9}$$

Thus the total interval recorded by one single clock in S (at rest at $X = 0$) from all signals sent back by S' during its round trip is

$$\sum_{j=1}^{6} \Delta T_j = \frac{4}{a} \frac{v_0}{\sqrt{(1-v_0^2)}} + \frac{2\Delta\tau_2}{\sqrt{(1-v_0^2)}} \tag{3.10}$$

The proper time (recorded by one clock) in S' for the whole round trip is

$$\Delta\tau_0 = 4\Delta\tau_{01} + 2\Delta\tau_2 = \frac{4}{a}\tanh^{-1} v_0 + 2\Delta\tau_2 \tag{3.11}$$

(B) *From the Standpoint of S'*

The fixed point $X = 0$ in S moves in the negative x direction.

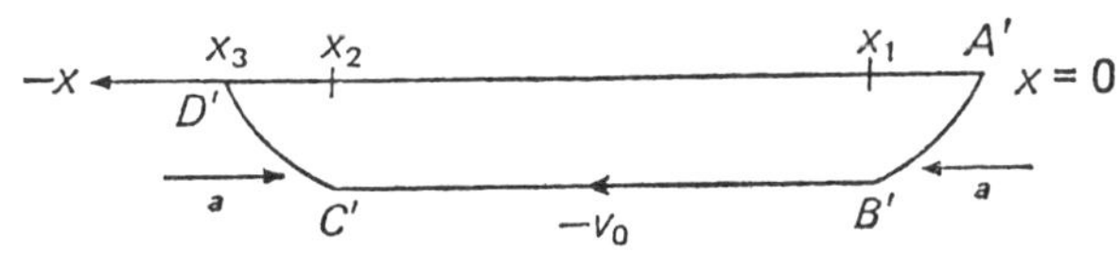

(3.12)

Part A'–B'. Let ΔT_1 be the proper time interval (registered by a clock at $X = 0$ in S) for $X = 0$ to reach the velocity $-v_0$ (relative to S'). From (2.18a),

$$a\Delta T_1 = \tanh\frac{a\tau_1}{c} = v_0 \tag{3.13}$$

and the distance traversed by $X=0$ during this interval is given by (2.23) (with $X=0$) and (2.18) with $\tanh at_1 = v_0$, i.e.,

$$1+ax_1 = \frac{1}{\cosh at_1} = \sqrt{(1-v_0^2)} \tag{3.14}$$

Parts B'–C' and E'–F'. Since the time interval for B'–C' or E'–F' in S' is $\Delta\tau_2$ in (3.8), S' would have deduced from the special theory of relativity that the combined intervals would have added to the proper time intervals of S the value $2\Delta T_2$ which can be calculated as follows. Let dT be an element of proper time in S. For the combined B'–C' and E'–F', light signals sent by S back to S' will reach S' in interval

$$\left[\Delta T_2 \sqrt{\left(\frac{1+v_0}{1-v_0}\right)} + \Delta T_5 \sqrt{\left(\frac{1-v_0}{1+v_0}\right)}\right] = \frac{2\Delta T_2}{\sqrt{(1-v_0^2)}}$$

This is now the time interval recorded by the clock in S', and is hence the proper time interval $2\Delta\tau_2$, i.e.,

$$2\Delta T_2 = \sqrt{(1-v_0^2)}\, 2\Delta\tau_2 \tag{3.15}$$

Part C'–D'. During C'–D' and D'–E', S' would describe S as being acted on by a gravitational field a in the positive x-direction, the motion being described by equations (2.24)–(2.27). Equation (2.24) is

$$[1-a(x-x_0)]^2 = [1-a(X-X_0)]^2 - a^2(T-T_0)^2 \tag{2.24}$$

The constants x_0, X_0, T_0 are determined as follows. From the standpoint of S_1 [see (3.1), (3.2) and (3.8)], at C,

$$aT = \frac{v_0}{\sqrt{(1-v_0^2)}} + \frac{a\Delta\tau_2}{\sqrt{(1-v_0^2)}}, \qquad v = v_0 \tag{3.16}$$

and (2.28) leads to

$$-\frac{a\left[\frac{1}{\sqrt{(1-v_0^2)}}(v_0 + a\Delta\tau_2) - T_0\right]}{1+aX_0} = v_0$$

At D,

$$T = \frac{2v_0}{\sqrt{(1-v_0^2)}} + \frac{a\Delta\tau_2}{\sqrt{(1-v_0^2)}},$$

$v=0$ and

$$X = 2X_1 + \frac{v_0\Delta\tau_2}{\sqrt{(1-v_0^2)}}$$

[see (3.4)]. These lead to

$$X_0 = 2X_1 + \frac{v_0\Delta\tau_2}{\sqrt{(1-v_0^2)}}$$

$$aT_0 = \frac{2v_0}{\sqrt{(1-v_0^2)}} + \frac{a\Delta\tau_2}{\sqrt{(1-v_0^2)}} \tag{3.17}$$

$$x_0 = 0$$

From (2.28a),

$$\tanh a(t-t_0)=-v$$

the condition $v=v_0$ when $t=\Delta\tau_1+\Delta\tau_2$ leads to

$$t_0=2\Delta\tau_1+\Delta\tau_2 \tag{3.18}$$

Thus, equation (2.25) corresponding to the case (2.24) is now

$$a\left(T-\frac{2v_0}{\sqrt{(1-v_0^2)}}-\frac{\Delta\tau_2}{\sqrt{(1-v_0^2)}}\right)=\left[1-a\left(X-2X_1-\frac{v_0\Delta\tau_2}{\sqrt{(1-v_0^2)}}\right)\right] \times\tanh(t-2\Delta\tau_1-\Delta\tau_2) \tag{3.19}$$

At D', $v=\tanh(t-2\Delta\tau-\Delta\tau_2)=0$ and the point $X=0$ of S has the time $T=\Delta T_{A'-B'-C'-D'}$

$$\Delta T_{A'-B'-C'-D'}=\frac{2v_0}{\sqrt{(1-v_0^2)}}+\frac{\Delta\tau_2}{\sqrt{(1-v_0^2)}} \tag{3.20}$$

From the symmetry of the situation, it is clear that the time in S for the trip A'–B'–C'–D'–E'–F'–A' as seen (or, calculated) from the standpoint of S' is twice the $\Delta T_{A'-D'}$ in (3.20), i.e.,

Total time ΔT in $S=$

$$2\left(\frac{2v_0}{a\sqrt{(1-v_0^2)}}+\frac{\Delta\tau_2}{\sqrt{(1-v_0^2)}}\right) \tag{3.21}$$

which can be written

$$=2(\Delta T_{A-B}+\Delta T_{B-C}+\Delta T_{C-D})$$

or,

$$a\Delta T=2\left[v_0+a\sqrt{(1-v_0^2)}\,\Delta\tau_2+\frac{2v_0}{\sqrt{(1-v_0^2)}}+\frac{a\Delta\tau_2}{\sqrt{(1-v_0^2)}} -(v_0+a\sqrt{(1-v_0^2)}\,\Delta\tau_2)\right] \tag{3.22}$$

the proper time in

$$S'=4/a\tanh^{-1}v_0+2\Delta\tau_2 \tag{3.23}$$

Equation (3.21) shows complete agreement with the value given in (3.10) obtained from the standpoint of S. Equation (3.22) shows that the 'loss of time' during B'–C' and E'–F' by S in the view of S' [i.e., $2\sqrt{(1-v_0^2)}\Delta\tau_2$ in (3.15) compared with $2\Delta\tau_2/\sqrt{(1-v_0^2)}$ in (3.9)] is more than made up by the 'gain in time' by the S clock during C'–D', D'–E' when S is at a higher equivalent gravitational potential than S' [see (3.12)], and the clock of S is 'faster' on account of the factor $g_{44}=(1-gx)^2$ in

$$ds^2=dx^2+(1-gx)^2\,dt^2$$

(with $x=0$ at A' and $ax=2[\sqrt{(1-v_0^2)}-1]-v_0\Delta\tau_2$ at D'). The smaller 'loss of time' of the clock of S during A'–B' (when S is at a lower gravitational potential, $ds^2=-dx^2+(1+gx)^2\,dt^2$) is also more than made up by the 'gain' during C'–D'. The total result is to bring the two reckonings of the proper time intervals, by S and S', of the round trip into *exact* agreement with each other.†

It is seen from the foregoing results that all the calculations are *exact*, and no approximations involving the assumption of making the accelerated parts A–B, C–D–E, F–A (or A'–B', C'–D'–E', F'–A') very short compared with the uniform relative motion part B–C, E–F (or B'–C', E'–F') have been made. In fact, as emphasised by Einstein as early as in the 1918 paper and brought out approximately by Tolman (1934) and exactly in (3.22) above that is precisely the accelerated parts that resolve the 'paradox'. Had one literally 'neglected' the accelerated parts, (3.10) and (3.22) would have become

$$\text{Total time in } S \text{ (as reckoned by } S) = \frac{2\Delta\tau_2}{\sqrt{(1-v_0^2)}}$$

$$\text{Total time in } S \text{ (as reckoned by } S') = 2\sqrt{(1-v_0^2)}\,\Delta\tau_2$$

On the other hand, had one done away entirely with the uniform relative motion (coasting of rocket) parts B–C, E–F (B'–C', E'–F'), the results (3.10) and (3.22) would have become:

	Standpoint of S	Standpoint of S'
Total proper time in S'	$\frac{4}{a}\tanh^{-1} v_0$	$\frac{4}{a}\tanh^{-1} v_0$
Total proper time in S	$\frac{4}{a}\frac{v_0}{\sqrt{(1-v_0^2)}}$	$\frac{2}{a}\left[v_0 + \frac{2v_0}{\sqrt{(1-v_0^2)}} - v_0\right]$

† The results (3.10), (3.11), (3.22), (3.23) above are a little more complete than those of Møller (1943) in that here S' starts out from rest and comes back at rest to S. There are differences in details between this and Møller's work. For example, we calculate the time intervals ΔT_1, ΔT_3, ΔT_4, ΔT_6 in (3.5)–(3.8) as recorded by one clock in S, and not Møller's times T', T'' which are not the proper times of one clock. Also, as remarked in Section 2 above, the starting points in the two works are different. In an application of Møller's work, Fock (1959) has obtained an erroneous conclusion.

Fock (1959) states that the time intervals recorded by the clocks A, B in S, S' are given by

$$\tau_A - \tau_B = \frac{v^2}{c^2}\left(\tfrac{1}{2}T - \tfrac{2}{3}t\right)$$

where $t = 2v/g$ is the time for the turning around part (C–D–E in (1.8) in the present article), and T = uniformly moving part $(B–C) + (E–F) + t$. Thus $\tau_A - \tau_B$ can be $\gtreqless 0$, in disagreement with the results of everyone else. This strange result arises from the error of the + sign in (62.09), which should have read $U = U_0 - g(x_1 - x)$. When this correction is made, one would have

$$\tau_A - \tau_B = \frac{v^2}{c^2}\frac{T}{4}$$

which is in agreement with the approximate result of (1.8) of Tolman and others.

Here is, of course, exact agreement between the reckonings of the total proper time intervals from the standpoints of both frames.

In the calculations above, we had employed the accelerated motion represented by equations (2.15) and (2.17) [or, (2.24) and (2.25)] which correspond to motion under a time-independent field. The result, however, is in fact quite general since from $S'(x,t)$, one can carry out any arbitrary coordinate transformation to a frame S'',

$$x'' = x''(x,t), \qquad t'' = t''(x,t)$$

which will lead in general to

$$ds^2 = \sum g_{ij}\, dx_i''\, dx_j''$$

where the g_{ij} are functions of x'' and t'' and hence no longer static. The space is, however, Euclidean. The motions of S'' relative to S can be quite arbitrary and very complicated, but the description can be reduced to that of S' by the transformation above so that in a sense the treatment of the clock problem by means of S' has covered a whole (infinite number) class of accelerated motions relative to S. This class of accelerated motions has not brought in any curved space properties in the sense of Einstein's general theory of relativity.

The present work has thus treated and resolved the clock problem without having really made recourse to Einstein's theory of gravitation involving curved space. This is worth noting in view of the usual statement in the literature that an exact treatment of the clock problem (i.e., to all orders of v_0/c) calls for the general theory of relativity.

4. *General Remarks on the 'Clock Paradox' Problem*

We are now in a position to summarise what we believe is relevant in the clock problem in the relativity theory.

(1) Invariance of proper time under coordinate transformations. In the theory of relativity (special and general),

$$\begin{aligned} ds^2 &= g_{\mu\nu}\, d\chi_\mu\, d\chi_\nu \\ &= \text{invariant} \end{aligned} \tag{4.1}$$

in each group of coordinate transformations (the group in flat space-time which includes the Lorentz group, and the general group in curved space-time). Thus for a given world line C between two world points P_1 and P_2, the proper time interval

$$\Delta\tau \int_{P_1}^{P_2} ds \text{ along } C = \text{invariant} \tag{4.2}$$

i.e., has the same value in all frames satisfying (4.1)

(2) Proper time intervals between two world-points along different world lines.

Consider a given field $g_{\mu\nu} = g_{\mu\nu}(x_1, x_2, x_3, x_4)$. The motion of a particle from one world point P_1 to another P_2 is uniquely given by the geodesic C

$$\delta \int_1^2 ds = 0$$

Other paths C_1, C_2 joining P_1 and P_2 will not correspond to the 'free' motion in the field $g_{\mu\nu}$, but will correspond to motions under agencies other than the field representative by $g_{\mu\nu}$, and

$$\int_{1\,C_1}^{2} ds \neq \int_{1\,C}^{2} ds \tag{4.3}$$

which follows from the definition of the geodesic.

(3) For a given field $g_{\mu\nu}$, between two arbitrarily given points P_1 and P_2, there is one and only one geodesic.

From a point P_1, there are ∞^3 geodesics issuing in all directions. Of these, one, say C, goes through P_2. Suppose another, say C_1, makes an angle θ with C at P_1. Since a geodesic is a line generated by an infinitesimal vector in a continuous series of infinitesimal parallel displacements, and since the angle between two vectors is invariant under parallel displacements, it follows that in general two geodesics from a given point P_1 cannot intersect at another arbitrarily chosen point P_2. (In the case of a spherical surface, geodesics from a point P intersect at the antipode of P only.)

(4) The clock paradox. Let S and S' be two (material) frames whose coordinates transform according to (4.1). Let us follow Einstein's argument in Section 1, namely, from the standpoint of S' (the rocket), S undergoes a series of free falls in certain universal gravitation fields during $A'B'$, $C'D'$, $D'E'$, $F'A'$ and coasting $B'C'$, $E'E'$ in (1.2). Suppose these fields are represented by a field $g_{\mu\nu}$. From the standpoint S', the frame S passes from the initial point P_1 to P_2 (in 4-space) along the geodesic of $g_{\mu\nu}$, but S' itself passes from P_1 to P_2 along a pure-time trajectory since S' has been held fixed (at rest) by means of some *external* agency. Thus the world line of S' is not a geodesic of $g_{\mu\nu}$. In general, the proper time intervals along the two world lines between P_1 and P_2 are different, according to (4.3).

The above result is general, holding for curved space as well as for flat space. It is possible to make an explicit and exact calculation of the proper time intervals in the flat space case, using the spirit of the general theory of relativity (acceleration represented by a $g_{\mu\nu}$ field). This has been done by Møller (1943), and in the present work (Sections 2 and 3).

(5) Let $S(X, Y, Z, T)$ be a strictly inertial frame, i.e., frame in flat space-time, and $S'(x, y, z, t)$ be a frame in a curved space-time (i.e., in a gravitational field in Einstein's theory). Then there is no coordinate transformation which transforms $ds^2 = -(dx^2 + dY^2 + dZ^2) + dT^2$ into $ds^2 = \sum g_{\mu\nu}\, dx_\mu\, dx_\nu$ with the curvature tensor $R^{\lambda}_{\mu\nu\sigma} \neq 0$. In this case there is no invariant ds^2 and there is no exact (only approximate) connection between the space-time

description in S and that in S'. One can no longer compare $d\tau_0$ in S and the $d\tau_0$ in S', and the 'clock paradox' does not have any clear and exact meaning.

References

Arzelies, H. (1966). *Relativistic Kinematics*. Pergamon Press. Contains an extensive bibliography on the clock problem.

Dingle, H. (1956). *Nature, London*, **177**, 782.

Dingle, H. (1957a). *Nature, London*, **179**, 1242.

Dingle, H. (1957b). *Nature, London*, **180**, 499, 1275.

Darwin, C. G. (1957). *Nature, London*, **180**, 976.

Einstein, A. (1911). *Annalen der Physik*, **35**, 898. Translated and contained in Einstein *et al.*, *The Principle of Relativity*. Dover Publ. Inc., New York.

Einstein, A. (1918). *Naturwissenschaften*, **6**, 697. An exposition of the relativity theory, and of the clock paradox, in the form of a dialogue.

Fock, V. (1959). *The Theory of Space, Time and Gravitation*, Section 62, p. 214, eq. (62.16). Pergamon Press.

McCrea, W. H. (1956). *Nature, London*, **177**, 783.

McMillan, E. M. (1957). *Science, New York*, **126**, 381.

Møller, C. (1943). *Danske Vid. Sel. Mat-Fys. Med. XX*, No. 19.

Tolman, R. T. (1934). *Relativity, Thermodynamics and Cosmology*. Oxford University Press.

IL NUOVO CIMENTO VOL. 112 B, N. 4 Aprile 1997

Four-dimensional symmetry of taiji relativity and coordinate transformations based on a weaker postulate for the speed of light. – II

JONG PING HSU (1) and LEONARDO HSU (2)

(1) *Physics Department, University of Massachusetts Dartmouth North Dartmouth, MA 02747, USA*

(2) *Physics Department, University of California at Berkeley - Berkeley, CA 94720, USA*

(ricevuto il 15 Maggio 1996; approvato il 9 Luglio 1996)

Summary. — Extended relativity is a theory of four-dimensional symmetry with Reichenbach's time, in which only the 2-way speed of light is a universal constant. It includes special relativity as a special case. The theory is shown to be consistent with experiments such as Fizeau's experiment, aberration of light and precision Doppler shifts. The formulations of classical and quantum electrodynamics are discussed. They are shown to be dependent on the four-dimensional symmetry rather than on the usual constant one-way speed of light. The four-dimensional symmetry also dictates a new coordinate transformation, called the "Wu transformation", for constant-linear-acceleration frames.

PACS 03.30 – Special relativity.
PACS 11.30.Cp – Lorentz and Poincaré invariance.

We continue to demonstrate that the four-dimensional symmetry is necessary and essential [1] for discussing physical laws from Reichenbach's viewpoint of time [2] or Edwards' weaker postulate for the speed of light [3]. Edwards attempted in 1963 to formulate a relativity theory based on a weaker postulate that the 2-way speed of light in a vacuum is a universal constant. He derived space and time transformations which involve Reichenbach's time but which do not form a four-dimensional Lorentz group in general. As a result, it leads to an incorrect expression for the relativistic energy-momentum of a particle in the Lagrangian formalism of mechanics and electrodynamics, as shown in paper I [1]. Furthermore, it appears to be impossible to obtain invariant forms of Maxwell equations and the Dirac equation if Reichenbach's time is used as an evolution variable, as we shall see later. The reason is that Reichenbach's time does not transform covariantly as the zeroth component of the coordinate 4-vector in general. Therefore, the lack of four-dimensional symmetry makes Edwards' original transformations untenable.

Recently, we have formulated and discussed taiji relativity [4] based solely on the first postulate of relativity, *i.e.* the invariance of physical laws. A four-dimensional transformation between any two inertial frames, $F(w, x, y, z)$ and $F'(w', x', y', z')$, is derived. *Since taiji relativity does not make any assumption regarding the speed of light, the zeroth components w and w' cannot be factored into a well-defined speed of light and time.* In fact, the speed of light and usual time (measured in seconds) are unspecified and undefined in the theory. However, the theory of taiji relativity possesses the four-dimensional symmetry which is shown to be the only essential ingredient for the theory to be consistent with previous experiments. This sheds light on the difficulty encountered by Edwards' transformations.

In paper I, we show that, guided by the four-dimensional symmetry of taiji relativity, Reichenbach's general convention of time (or, equivalently, the universal 2-way speed of light) can be used as the "second postulate" for the construction of a new four-dimensional formalism of coordinate transformation which is termed extended relativity. The second postulate is necessary to factorize, say, w' into a well-defined velocity function b' (called "ligh") and Reichenbach's time t' in the F' frame, *i.e.* $w' = b' t'$ which is called "lightime". (See eqs. (2.1)-(2.3) in sect. **2**.) It turns out that the lightime w', rather than Reichenbach's time, plays the role of evolution variable in physical laws and makes extended relativity consistent with established energy-momentum of a particle, the Lorentz group, etc. Furthermore, the covariant lightime embedded in the four-dimensional symmetry is also crucial for the formulation of a covariant quantum electrodynamics (QED) based on extended relativity, as we shall see in sect. **6**.

7. – Limiting four-dimensional symmetry and accelerated Wu transformation

One may wonder whether the power of four-dimensional symmetry is strong enough to say something about constant-linear-acceleration (CLA) frames. The attempt to generalize the coordinate transformation for inertial frames to that for CLA frames through a symmetry consideration is very natural because the transformation for a CLA frame must reduce to that for an inertial frame in the limit of zero acceleration. So far, no satisfactory transformation for such non-inertial frames has been obtained in the literature, even though one has general relativity and the correspondence principle [6]. All those accelerated transformations discussed previously are not based on a symmetry principle and do not naturally reduce to the four-dimensional transformation for an intertial frame when the acceleration approaches zero.

By a stroke of luck, we have found a transformation for CLA frames which does reduce to the correct four-dimensional transformation in the limit of zero acceleration and reduces to the Galilean accelerated transformation when the velocity is small. For simplicity of notation and calculations in an accelerated frame, let us denote a CLA frame by $F(w, x, y, z)$ and an inertial frame by $F_I(w_I, x_I, y_I, z_I)$. Suppose a CLA frame $F(w, x, y, z)$ is moving with a constant acceleration α, so that its velocity is $\beta = \alpha w + \beta_0$, along the $+x$ axis. We find that the accelerated transformation between F_I and F is given by

$$
(7.1)\qquad \begin{cases} w_{\mathrm{I}} = \gamma\beta(x + 1/\alpha\gamma_0^2) - \beta_0/\alpha\gamma_0, \\ x_{\mathrm{I}} = \gamma(x + 1/\alpha\gamma_0^2) - 1/\alpha\gamma_0, \qquad y_{\mathrm{I}} = y, \qquad z_{\mathrm{I}} = z\,; \\ \beta = \alpha w + \beta_0, \qquad \gamma = 1/(1-\beta^2)^{1/2}, \qquad \gamma_0 = 1/(1-\beta_0^2)^{1/2}, \end{cases}
$$

which will be called the "Wu transformation". If one wishes, one may define $w_{\mathrm{I}} = ct_{\mathrm{I}}$, where t_{I} is the usual Einstein time, in (7.1) for easy comparson with special relativity. (But this definition is not necessary for deriving experimental results.) The inverse Wu transformation of (7.1) is

$$
(7.2)\qquad \begin{cases} w = \dfrac{w_{\mathrm{I}} + \beta_0/\alpha\gamma_0}{\alpha(x_{\mathrm{I}} + 1/\alpha\gamma_0)} - \dfrac{\beta_0}{\alpha}, \\ x = [(x_{\mathrm{I}} + 1/\alpha\gamma_0)^2 - (w_{\mathrm{I}} + \beta_0/\alpha\gamma_0)^2]^{1/2} - 1/\alpha\gamma_0^2, \\ y = y_{\mathrm{I}}, \qquad z = z_{\mathrm{I}}. \end{cases}
$$

One can verify that (7.1) and (7.2) reduce to four-dimensional transformations of the form (2.2) in the limit of zero acceleration α. (See appendix.) We may remark that the coordinate transformation between two CLA frames can be derived on the basis of (7.1) or (7.2).

From the viewpoint of limiting four-dimensional symmetry, the CLA transformation must be expressed in terms of the Cartesian coordinates rather than other coordinates, just like the Lorentz transformation. Furthermore, the coordinates of CLA frames should play the same role and have a similar physical meaning as those of inertial frames. This appears to be different from the usual viewpoint that coordinates for accelerated frames have no physical meaning. The Wu transformation (7.1), based on the four-dimensional symmetry, differs from that obtained by Møller [6] based on the approximate principle of equivalence in general relativity because they give different spatial measurements by meter sticks or the Bohr radius of hydrogen atoms. We believe that such a difference should be tested by, say, measuring a Doppler shift of wavelength emitted from a source with a constant linear acceleration. We may remark that the constant acceleration α in (7.1) can be shown to be related to constant change of "energy" (or "moving mass") per unit length measured in an inertial frame. This differs from the usual definition of acceleration in (2.4). It is interesting to note that such a constant acceleration α dictated by the limiting four-dimensional symmetry is precisely what has been actually realized in linear accelerators in laboratories. Physical implications of the Wu transformation and their experimental tests will be discussed in a separate paper.

8. – Remarks and discussions

In the formulation of QED in sect. **6**, the electron is, as usual, assumed to be a point particle. However, if the physical electron is really a fuzzy point (in the sense of fuzzy set theory with a bell-shape membership function having a width L_0) rather than a geometric point, then there will be a departure from the four-dimensional symmetry at short distances or large momentum [7]. A fuzzy-point model of a particle has been interpretated as follows: a particle by itself is a structureless-point particle, but it can simultaneously exist at different places with a different probabilities. As a result, the position uncertainty of such a quantum particle has a minimum width $\Delta x \sim L_0$. The Coulomb potential will be modified when $r < L_0$, and the photon propagator in (6.19) will be modified when momentum becomes larger than $\hbar/L_0$. For a detailed discussion of the fuzzy-point model of particles, we refer to ref. [7].

Let us compare and summarize basic differences in various relativity theories:

a) Taiji relativity: it is based solely on the first postulate of relativity, namely, the invariance of physical laws. A four-dimensional transformation between two inertial frames, $F(w, x, y, z)$ and $F'(w', x', y', z')$, can be derived. The usual concept of time and speed of light are undefined and completely unknown; nevertheless, the theory agrees with all experiments. Taiji-times w and w' play the roles of evolution variables, and the dimensionless "taiji-velocities" dx/dw, dx'/dw', etc. are well-defined. It is interesting that there are only two universal and fundamental constants in QED based on taiji relativity.

b) Extended relativity: it is based on two postulates. In addition to the invariance of physical laws, its second postulate is the universality of the 2-way speed of light. Reichenbach's time and one-way speeds of light (non-isotropic in general) are well-defined. However, lightime w plays the role of the evolution variable in four-dimensional physical laws. There are three universal and fundamental constants in QED based on extended relativity. Special relativity is a special case ($q = q' = 0$ in (2.2)).

c) Common relativity: it is based on two postulates. The additional second postulate is a common time $t' = t$ for all observers [8]. The speed of light is, roughly speaking, relative. Lightimes w and w' are evolution variables in four-dimensional laws. There are only two universal and fundamental constants in QED based on common relativity, precisely the same as those in taiji relativity. Common relativity has the unique advantage for dealing with many-particle systems where canonical evolution of the system is essential and for obtaining covariant thermodynamics and invariant Planck's law of black-body radiations [9].

One may ask: how can one realize the evolution variable w' in the extended coordinate transformation (2.2) by physical means? Since the invariant phase of an electromagnetic wave in the F' frame is given by $k_0' w' - \mathbf{k}' \cdot \mathbf{r}'$, where $k_0' = |\mathbf{k}'|$, we can define the lightime w' in terms of k_0', just as the length can be defined by the wavelength λ' or $|\mathbf{k}'|$. We note that the "clocks", which show lightime in this theory, are the same as those in taiji relativity [4] because they have exactly the same four-dimensional transformation property. However, the taiji-time w' in F' cannot be factored into two well-defined b' and t' because of the absence of a second postulate while the lightime w' in extended relativity and common relativity can be factored into two well-defined functions b' and t', as shown in (2.1), (2.2) and ref. [8].

Our discussions show that it is extremely important to be aware of what quantities are actually measured in the experiments and what effects the assumption of a universal speed of light may have had on the interpretation of the results. For example, we have seen in paper I that the lifetime dilatation of unstable particle decay in flight has little to do with the property of Reichenbach's time with a general parameter q or q', because the lifetime τ is basically defined as the decay length divided by the universal 2-way speed of light c. The basic reason is that the four-dimensional symmetry dictates that the decay rates in, say, QED based on extended relativity can only be defined in terms of the covariant lightime w or w' which has the dimension of length.

The constant 2-way speed of light in extended relativity is in general not the maximum speed of physical objects in the universe. Rather, it is the one-way speed of light in a given direction, that is the maximum speed of any object in that direction, as shown in (2.3). This holds for any inertial frame. It is worthwhile to note that this property of light, being the "maximum speed" of all physical objects in any given direction, is a logical consequence of the first postulate of relativity, as shown in taiji relativity [4].

We have examined a number of experimental tests of special relativity and the formulations of classical electrodynamics and QED. All of them are consistent with extended relativity. These discussions can be generalized to other field theories such as unified electroweak theory and quantum chromodynamics. As we have seen, only the four-dimensional symmetry of physical laws is absolutely essential for explanations of experimental results and for the formulation of classical electrodynamics and QED; the universality of the one-way speed of light is irrelevant. In this connection, we stress that according to the theory of taiji relativity [4], the universality of the one-way or the two-way speed of light is a convention rather than an inherent part of the physical world [10].

Note added in proofs

Suppose one writes $dw_I = \gamma(T\,dw + U\beta\,dx)$, $dx_I = \gamma(V\,dx + W\beta\,dw)$, $dy_I = dy$, $dz_I = dz$; $\gamma = 1/(1-\beta^2)^{1/2}$, where T, U, V and W are four unknown functions of x and w. The new Wu transformation (7.1) for a constant-linear-acceleration (CLA) frame can be derived from the postulate of the limiting four-dimensional symmetry of taiji relativity and the initial condition that a CLA transformation reduces to the spatial identity $\mathbf{r}_I = \mathbf{r}$ when the taiji-time $w = 0$ and the initial "velocity" $\beta_0 = 0$. This initial condition holds also for the Lorentz transformation. Thus, once the principle (or the first postulate) of relativity is rigorously stated to include the limiting cases, the concept of acceleration is determined in the physical theory based on "extended four-dimensional framework".

Within the present conceptual framework, the taiji-time w in the Wu transormation (7.1) or (A.2) is a *primary concept* and has the dimension of length. The motion of physical objects, including light signals, is a derived concept and described by dimensionless "taiji-velocities" $d\mathbf{r}/dw$. The taiji-time w can be realized by computerized "Leonardo clocks" [4]: We could program any Leonardo clock in a CLA frame F to obtain a reading w_I from the nearest clock in an inertial frame F_I and, based on its F_I frame position x_I and given parameters α and β_0, compute the taiji-time w it should display, $w = (w_I + \beta_0/\alpha\gamma_0)/[\alpha(x_I + 1/\alpha\gamma_0)] - \beta_0/\alpha$. (See (7.2).) In the limit of zero acceleration w shown on a Leonardo clock will automatically reduce to the taiji-time in the four-dimensional transformation, $w = \gamma_0(w_I + \beta_0 x_I)$. It will not reduce to relativistic time, unless the second postulate of universal constant for the speed of light ($w \equiv ct$, $w_I \equiv ct_I$) is made in this limit [4].

* * *

This paper is dedicated to Prof. TA-YOU WU for his wonderful and tireless teaching of physics and his ninetieth birthday. The work was supported in part by The Jing Shin Research Fund of the UMass Dartmouth and by a grant from the Potz Science Fund.

APPENDIX

Limiting four-dimensional symmetry and constant-linear-acceleration frames

For simplicity, let us denote a CLA frame by $F(w, x, y, z)$ and an inertial frame by $F_I(w_I, x_I, y_I, z_I)$. Suppose a CLA frame $F(w, x, y, z)$ is moving with a constant acceleration α, so that its velocity is

$$\beta = \alpha w + \beta_0, \tag{A.1}$$

along the $+x$ axis. Guided by the limiting four-dimensional symmetry, we find that the linearly accelerated transformation between F_I and F should be

$$\text{(A.2)} \qquad \begin{cases} w_I = \gamma\beta(x + 1/a\gamma_0^2) - \beta_0/a\gamma_0 , \\ x_I = \gamma(x + 1/a\gamma_0^2) - 1/a\gamma_0 , \qquad y_I = y , \qquad z_I = z , \\ \beta = aw + \beta_0 , \qquad \gamma = 1/(1-\beta^2)^{1/2} , \qquad \gamma_0 = 1/(1-\beta_0^2)^{1/2} . \end{cases}$$

This is a generalization of the accelerated transformation obtained by Wu and Lee [7] based on a kinematic approach to satisfy limiting four-dimensional symmetry. It will be called the "Wu transformation". When β_0 approaches zero, the accelerated transformation (A.2) can reduce to the well-known transformation obtained in ref. [7], provided one uses a new time $\tau : \beta = aw + \beta_0 = \mathrm{tgh}(a\tau)$. Furthermore, one can verify that the Wu transformation (A.2) indeed reduces to four-dimensional transformations in the limit of zero acceleration $a \to 0$: $w_I = \gamma_0(w + \beta_0 x)$, $x_I = \gamma_0(x + \beta_0 w)$, $y_I = y$, $z_I = z$; where $\gamma_0 = 1/(1-\beta_0^2)^{1/2}$.

With the definitions $w_I = ct_I$ and $w = ct$, the "extented relativistic time" t in the CLA frame F is completely determined by (A.2). In other words, if the time t_I and the position x_I of an event as observed in the inertial frame F_I is known, then the corresponding time t in the CLA frame can be calculated, provided the constants c, a and β_0 are given. Such a time t in a CLA frame can be physically realized by computerized "Leonardo clocks" [4]: evidently, the time t in the CLA frame is not the relativistic time in general and, hence, the constant c by itself is not physically meaningful. Only when the acceleration a vanishes, the time t in (A.2) for the frame F reduces to the relativistic time and c becomes physically meaningful. We may remark that the definitions $w_I = ct_I$ and $w = ct$ are not necessary for deriving observable results because we may directly use w as evolution variables.

When $\beta_0 \to 0$, the inverse of the Wu transformation (A.2) leads to

$$\text{(A.3)} \qquad \begin{cases} w = ct_I/(1 + ax_I) \approx ct_I(1 - ax_I) , \qquad ct_I = w_I , \\ x = (1/a)[1 + 2ax_I + a^2(x_I^2 - c^2t_I^2)]^{1/2} - 1/a \approx x_I - c^2at_I^2/2 . \end{cases}$$

Thus, c^2a is related to a constant acceleration g in Newtonian mechanics by the relation

$$\text{(A.4)} \qquad a = g/c^2$$

when velocities are small. In this sense, the Wu transformation (A.2) is a four-dimensional generalization of the Galilean transformation for accelerated frames in classical mechanics.

From (A.1) we obtain

$$\text{(A.5)} \qquad ds^2 = c^2\,dt_I^2 - dr_I^2 = g_{00}\,dw^2 - dr^2 , \qquad g_{00} = \gamma^4(\gamma_0^{-2} + ax)^2 ,$$

where x in a CLA frame F is restricted to the region $x > x_s \equiv -1/(a\gamma_0^2)$ which may be pictured as a "wall singularity" at x_s. We may remark that finite Wu transformation (A.2) implies that the space-time of the CLA frame $F(W, x, y, z)$ is flat, *i.e.* the Riemann curvature tensor vanishes, $R^i{}_{jkm} = 0$, which can also be directly calculated by using the metric tensors in (A.5).

The velocity of a fixed point x_I in F_I as measured by F-observers using evolution variable [4] w is dx/dw with x_I fixed. From (A.2), we find

$$\text{(A.6)} \qquad (dx/dw)_{x_I} = \beta(g_{00})^{1/2} ; \qquad x > -1/(a\gamma_0^2) , \qquad 1 > \beta^2 .$$

We see that only in the approximation $g_{00} \approx 1$ do we have $(dx/dw)_{x_I} \approx \beta$ and

(A.7) $$(\mathrm{d}^2 x/\mathrm{d}w^2)_{X_\mathrm{I}} \approx \alpha = \text{constant}\,.$$

We note that the Wu transformation (A.2) holds for general w_I and w. In the limit of zero acceleration, it reduces to the four-dimensional taiji transformation [4]. If one wishes, one may define

(A.8) $$w_\mathrm{I} = ct_\mathrm{I} \text{ and } w = bt\,, \quad b \equiv (ct_\mathrm{I} - \beta x_\mathrm{I})/[t_\mathrm{I}(1-\beta q') - (\beta - q')\,x_\mathrm{I}/c]\,,$$

where t_I and t are, respectively, Einstein's time and "extended" Reichenbach's time (and b is the corresponding "ligh function"), then the limit of zero acceleration of (A.2) is the extended transformation (2.2) (where the inertial frame F' corresponds to the CLA frame F of (A.2) in the limit of zero acceleration). One can formulate, say, classical electrodynamics in a CLA frame. According to taiji relativity, physical results in the CLA frame F should be independent of the definition in (A.8).

REFERENCES

[1] Hsu L., Hsu J. P. and Schneble D., *Nuovo Cimento B*, **111** (1996) 1299. This is referred as paper I in the text.

[2] Reichenbach H., *The Philosophy of Space and Time* (Dover, New York) 1958.

[3] Edwards W.F., *Am. J. Phys.*, **31** (1963) 482.

[4] Hsu J. P. and Hsu L., *Phys. Lett. A*, **196** (1994) 1; **217** (1996) 359; Hsu L. and Hsu J. P., *Nuovo Cimento B*, **111** (1996) 1283.

[5] See, for example, Bjorken J. D. and Drell S. D., *Relativistic Quantum Mechanics* (McGraw-Hill, New York) 1964, pp. 261-268 and pp. 285-286; Sakurai J. J., *Advanced Quantum Mechanics* (Addison-Wesley, Reading, Mass.), 1967, pp. 171-172 and pp.181-188; Weinberg S., *The Quantum Theory of Fields* (Cambridge University Press, New York) 1995, pp. 134-147.

[6] Møller C., *Danske Vid, Sel. Mat.-Fyz.*, xx, No. 19 (1943); Fock V., *The Theory of Space Time and Gravitation* (Pergamon, New York) 1958, pp. 206-211; Wu T. Y. and Lee Y. C., *Int. J. Theor. Phys.*, **5** (1972) 307; Ta-You Wu, *Theoretical Physics*, Vol. 4, *Theory of Relativity* (Lian Jing Publishing Co., Taipei) 1978, pp. 172-175.

[7] Hsu J. P., *Nuovo Cimento B*, **80** (1984) 183; **88** (1985) 140; Hsu J. P. and Pei S. Y., *Phys. Rev. A*, **37** (1988) 1406.

[8] For a detailed discussion of common time in four-dimensional framework and its implications, see Hsu J. P., *Nuovo Cimento B*, **74** (1983) 67; **88** (1985) 140; **89** (1985) 30; *Phys. Lett. A*, **97** (1983) 137; Hsu J. P. and Whan C., *Phys. Rev. A*, **38** (1988) 2248, appendix.

[9] Hsu J. P., *Nuovo Cimento B*, **93** (1986) 178.

[10] In other words, all physical results in taiji relativity or extended relativity can be derived by simply using the quantities (w, x, y, z) and (w', x', y', z') without ever mention time t or t' (measured in seconds) and speeds of light or other physical objects.

CHINESE JOURNAL OF PHYSICS VOL. 35, NO. 4 AUGUST 1997

Generalized Lorentz Transformations for Linearly Accelerated Frames with Limiting Four-Dimensional Symmetry

Jong-Ping Hsu[1] and Leonardo Hsu[2]

[1]*Department of Physics, University of Massachusetts Dartmouth, North Dartmouth, MA 02747-2300, U.S.A.*

[2]*Department of Physics, University of California at Berkeley, Berkeley, CA 94720-7300, U.S.A.*

(Received March 24, 1997)

Based on the principle of limiting four-dimensional symmetry, we discuss coordinate transformations for constant-linear-acceleration (CLA) frames. We derive a new "Wu transformation" which reduces to the Lorentz transformation in the limit of zero acceleration. The time for an accelerated frame can be realized by "computerized clocks". A 'CLA coordinate' (w, x, y, z) is preferred for the accelerated transformation and has as much physical meaning for an accelerated frame F as (w_I, x_I, y_I, z_I) for an inertial frame F_I. Furthermore, constant-linear-acceleration α must be constant increase of a particle's "energy" per unit length, in consistent with what has been realized in high energy linear accelerators. Some experimental implications are discussed.

PACS. 04.20.−q – Classical general relativity.
PACS. 11.30.Cp – Lorentz and Poincare invariance.
PACS. 03.30.+p – Special relativity.

I. Introduction

Classical physics using Galilean transformations has been satisfactorily improved in modern physics by the use of Lorentz transformations. Yet, the corresponding classical physics in constant-linear-acceleration (CLA) frames has not been equally improved in modern physics. This is due to the lack of a generalized Lorentz transformation for CLA frames. In 1943, Moeller obtained a transformation for an inertial frame F_I and a uniformly accelerated frame F, moving along the x-axis. This was based on (a) Einstein's vacuum equation $R_{ij} = 0$ and (b) time-independent metric tensors, $ds^2 = g_{00}(x)c^2dt^2 - dx^2 - dy^2 - dz^2$. He obtained $g_{00}(x) = (1 + gx)^2$ and a transformation for uniformly accelerated frames [1]. The reason for using $R_{ik} = 0$ is suggested by a heuristic view that the 'inertial force' of accelerated frames and the 'gravitational force' may be considered as being "unified" by Einstein's equation. Nevertheless, these two forces satisfy quite different "boundary conditions": Namely, in contrast to the gravitational force, the inertial force does not vanish at spatial infinity and the transformations for accelerated frames should reduce to the Lorentz transformation when inertial forces vanish. In 1972, Wu and Lee derived the same transformation based on a "*kinematic approach*" without using Einstein's equation for gravity [1]. But, Moeller's transformations cannot be smoothly connected to the Lorentz transformations in the limit of zero acceleration. This is due to the stringent assumption that the metric tensor g_{00} is time-independent.

In this paper, we follow the kinematic approach and obtain a satisfactory CLA transformation by postulating a new and natural principle of "limiting four-dimensional symmetry" [2]: Any accelerated transformation of coordinates must reduce to the form with 4-dimensional symmetry in the limit of zero acceleration. We show that the set of transformations for the CLA frames forms a new group, which is termed the "Wu group". The "Wu transformation" is a natural and simple generalization of the Lorentz transformation and the Galilean transformation with constant acceleration. The limiting four-dimensional symmetry principle contains more definite and satisfactory physical results than the equation $R_{ik} = 0$, as far as CLA transformations are concerned. In the gravitational approach, Einstein's covariant equation holds for any coordinate. However, the Lorentz transformation prefers the Cartesian coordinate. Therefore, the natural assumption of the smooth connection between a linearly accelerated frame and an inertial frame dictates that a CLA coordinate is preferred for CLA transformations. This is an important difference between kinematic and gravitational approaches.

In our discussions, a CLA frame $F(w,x,y,z)$ with the usual definition $w = ct$ is introduced. But we know that the constant speed of light c has no operational definition in any CLA frame. Fortunately, the physical results in the paper are actually independent of the definition $w = ct$. In previous papers, we have shown that the logically simplest theory of relativity, called "taiji relativity," can be formulated *solely on the basis of the first postulate of relativity, without making any second postulate concerning the speed of light* [3]. The first postulate of relativity states that the laws of physics have the same form in all inertial frames. We are able to formulate a 4-dimensional physical theory with the coordinate $x_I^\mu = (w_I, x_I, y_I, z_I)$ for an inertial frame F_I, where w_I is the 'taiji-time' with the dimension of length. The absence of the second postulate forbids one to express the taiji-time w_I in terms of the usual time t_I (measured in seconds) and velocity because they cannot be defined for all inertial frames in taiji relativity. Nevertheless, the taiji-time w_I can be directly used as the evolution variable. Furthermore, the taiji-time with the unit of, say, centimeters, can be physically realized by computerized clocks. Also, the invariant law for the propagation of light, $ds^2 = dw_I^2 - dr_I^2 = 0$, implies the 'taiji-speed' of a light signal to be *dimensionless* and has the universal value, $\beta_L = |dr_I/dw_I| = 1$, for all inertial frames. A careful examination shows that taiji relativity is consistent with all previous experiments [3]. Indeed, one can simply consider w in a CLA frame as the evolution variable for a physical system. One can have a grid of computerized clocks in a CLA frame. These clocks can be synchronized without relying on the constant speed of light signals and will automatically read taiji-time w_I in the limit of zero acceleration.

II. Coordinate transformations for linearly accelerated frames

Suppose we have an inertial frame $F_I(w_I, x_I, y_I, z_I)$, $w_I = ct_I$, and a CLA frame $F(w,x,y,z)$ moving with a constant acceleration a along the x-axis. Based on the preceding discussions, it is natural to assume that ds^2 takes the form

$$\begin{aligned} ds^2 &= c^2dt_I^2 - dx_I^2 - dy_I^2 - dz_I^2 \\ &= g_{00}(x,w)dw^2 + g_{11}(x)dx^2 + g_{22}(x)dy^2 + g_{33}(x)dz^2, \end{aligned} \tag{2.1}$$

where t_I is shown by the Einstein clocks in the inertial frame F_I. As usual, one may define $w = ct$, where the realization of t through a grid of computerized clocks will be discussed later in sec. 6. Although $R_{ik} = 0$ holds for arbitrary coordinates, we postulate the metric (2.1) so that ds^2 and the resultant transformations are compatible with both Einstein's vacuum equation $R_{ik} = 0$ and the new "boundary conditions" of limiting four-dimensional symmetry. Since the CLA frame F moves along the x-axis, we look for axial symmetric solutions with

$$g_{22} = g_{33} = -Y^2(x), \tag{2.2}$$

and all metric tensors are functions of x, except that g_{00} may be a function of x and w. This property of $g_{00}(x,w)$ is crucial for the new CLA transformation.

II-1. Gravitational approach

Let us first consider the conventional "gravitational approach" based on Einstein's equation to obtain uniformly accelerated transformations. Based on (2.1) and (2.2). We can calculate Christoffel symbols $G^i_{jk} = g^{im}(\partial_k g_{mj} + \partial_j g_{mk} - \partial_m g_{jk})$ and the Ricci tensor $R_{ik} = \partial_m G^m_{ik} + \partial_k G^m_{im} + G^n_{ik}G^m_{nm} - G^m_{in}G^n_{km}$. The equations $R_{ii} = 0,\ i = 0,1,2$, lead to

$$\partial_x^2 W - \partial_x W(\partial_x X/X - 2\partial_x Y/Y) = 0, \tag{2.3}$$

$$\partial_x^2 W/W + 2\partial_x^2 Y/Y - (\partial_x X/X)[2\partial_x Y/Y + \partial_x W/W] = 0, \tag{2.4}$$

$$\partial_x^2 Y/Y - \partial_x Y \partial_x X/(YX) + (\partial_x Y/Y)^2 + \partial_x Y \partial_x W/(YW) = 0, \tag{2.5}$$

respectively, where Y is given by (2.2) and

$$W^2 = g_{00}(x,w), \quad X^2 = -g_{11}(x); \quad W = W_1(x)W_2(w). \tag{2.6}$$

We note that $R_{33} = 0$ gives the same equation as (2.5) and other components of R_{ik} vanish identically.

If $\partial_x Y \neq 0$, equations (2.3)–(2.5) leads to an exact solution

$$W_1 = f_3/(f_1 x + f_o)^{1/2}, \quad X = f_2(f_1 x + fo)^{1/2}, \ \ Y = f_1 x + f_o, \tag{2.7}$$

where $f's$ are constants. We stress that $W_2(w)$ in (2.6) is arbitrary because it cannot be determined by Einstein's equation. Physically, one expects that the metric tensor g_{22} should satisfies $-g_{22} = Y^2 = 1$ rather than $Y^2 = (f_1 x + f_o)^2$ since there is no motion along the y-axis. Furthermore, the accelerated transformation based on this solution cannot be smoothly connected to the Lorentz transformation (i.e., it does not satisfy the "limiting four-dimensional symmetry" or the integrability conditions in (2.15) below.) Therefore, the solution (2.7) is not physically meaningful.

Let us concentrate on the case $\partial_x Y = 0$ and $\partial_x W \neq 0$. We have the solution $Y = 1$, which satisfies the boundary conditions $g_{22}(0) = g_{33}(0) = -1$ at the origin. From Eqs. (2.3) and (2.8), we deduce a general relation between $W_1(x)$ and $X(x)$:

$$dW_1(x)/dx = fX(x), \tag{2.8}$$

where f is a constant of integration. Furthermore, the time-dependent part of g_{00}, i.e., $W_2(w)$, still cannot be determined by Einstein's equation, just like the previous case $\partial_x Y \neq 0$. Thus we have seen that Einstein's covariant equation by itself does not lead to a specific form for $X(x)$, $W_1(x)$ and $W_2(w)$. (If one further postulates the limiting 4-dimensional symmetry, then $W_2(w)$ can be determined and one can obtain a CLA transformation based on $W_2(w)$, Eq. (2.8) and a suitable initial condition. But if one postulate the limiting 4-dimensional symmetry, then one can obtain the same CLA transformation without using Einstein's equation at all, as we shall see below in sec. II-2.)

Moeller made two additional postulates, $X(x) = W_2(w) = 1$, and obtained [1]

$$ds^2 = g_{00}(x)c^2dt^2 - dx^2 - dy^2 - dz^2, \quad g_{00}(x) = (1+gx)^2. \tag{2.9}$$

This leads to Moeller's transformation involving only one parameter g. When the acceleration g approaches zero, it does not reduce to a Lorentz transformation with a constant velocity. This is due to the lack of a velocity parameter in (2.9) which is intimately related to his stringent assumption that the metric tensor g_{00} is time-independent, i.e., $W_2(w) = 1$, as we shall see below.

II-2. Kinematic approach

Next, let us consider a new "kinematic approach" based on the principle of limiting 4-dimensional symmetry. Since F accelerates along the x-axis, the perpendicular coordinates y and z should not appear in the metric tensors g_{ik} and $g_{22} = g_{33} = -1$ in (2.1). The length of a measuring rod along the x-axis or the component g_{11} should not depend on time or w because the acceleration is characterized by a constant.

Based on the limiting four-dimensional symmetry, the invariant interval (2.1) with $g_{22} = g_{33} = -1$ leads to the following differential form of transformation between F_I and F,

$$d(ct_I) = \gamma(Wdw + \beta X dx), \quad dx_I = \gamma(Xdx + \beta W dw), \quad dy_I = dy, \quad dz_I = dz; \tag{2.10}$$

where

$$\beta = \alpha w + \beta_o, \quad \gamma(w) = (1-\beta^2)^{-1/2} \equiv \gamma; \tag{2.11}$$

$$[g_{00}(w,x)]^{1/2} = G(w)Z(x) \equiv W(w,x) > 0; \tag{2.12}$$

$$(-g_{11})^{1/2} = X(x) > 0, \tag{2.13}$$

The CLA transformation (2.10) is characterized by two parameters: acceleration α and initial velocity β_0. In order to satisfy the limiting four-dimensional symmetry, we must have

$$g_{00} \to 1 \quad \text{and} \quad g_{11} \to -1 \quad \text{as } \alpha \to 0. \tag{2.14}$$

Also, the coefficients of dx and dw in (2.10) must satisfy the integrability conditions

$$\partial(\gamma W)/\partial x = \partial(\gamma\beta X)/\partial w, \qquad \partial(\gamma X)/\partial w = \partial(\gamma\beta W)/\partial x, \tag{2.15}$$

so that we have a finite coordinate transformation. It follows from (2.12) and (2.15) that

$$G(w) = \alpha\gamma^2(w)/f, \tag{2.16}$$

$$dZ(x)/dx = fX(x), \tag{2.17}$$

where the constant f is coming from separation of variables w and x. Using (2.16) and (2.17), we can integrate (2.10) to obtain the finite transformation between F_I and F,

$$ct_I = \gamma\beta Z(x)/\alpha - \beta_0/\alpha\gamma_0, \quad x_I = \gamma Z(x)/\alpha - 1/\alpha\gamma_0, \quad y_I = y, \quad z_I = z; \tag{2.18}$$

$$Z(x) \sim (\gamma_0^{-2} + \alpha x + \cdots) \quad \text{for small } \alpha; \quad f = \alpha; \quad \gamma_0 = (1-\beta_0^2)^{-1/2}. \tag{2.19}$$

Note that the constants of integration in (2.18) and the relations in (2.19) are all determined by the limiting four-dimensional symmetry and a boundary condition at the origin, $Z(0) < \infty$. In order to determined the precise form for the function $Z(x)$, we observe that the Lorentz transformation reduces to the identity transformation, $\mathbf{r}'=\mathbf{r}$, when time and velocity vanish, $t = 0$ and $V = 0$. Thus, it is natural to impose the same initial condition to the accelerated transformation (2.18): Namely, when time $t_I = 0$ and velocity $\beta_0 = 0$, transformation (2.18) reduces to the identity transformation,

$$\mathbf{r} = \mathbf{r}_I, \tag{2.20}$$

for all values of acceleration α. It follows from (2.18)–(2.20) that $Z(x)$ must be

$$Z(x) = (\gamma_0^{-2} + \alpha x), \quad \text{or} \quad X = 1 = (-g_{11})^{1/2}, \tag{2.21}$$

Thus, *the metric tensor in* (2.1) *is now completely determined by the limiting four-dimensional symmetry and the initial condition* (2.20).

From Eqs. (2.18) and (2.21), we obtain a definite coordinate transformation between the inertial frame F_I and the accelerated frame F:

$$\begin{aligned} w_I &= \gamma\beta(x + 1/\alpha\gamma_0^2) - \beta_0/\alpha\gamma_0, \\ x_I &= \gamma(x + 1/\alpha\gamma_0^2) - 1/\alpha\gamma_0, \quad y_I = y, \quad z_I = z; \quad w_I = ct_I, \end{aligned} \tag{2.22}$$

where the time $t_I = w_I/c$ is shown by the conventional Einstein clocks in the inertial frame F_I. We shall call the result (2.22) the "Wu transformation" [4]. One can verify that the transformation (2.22) with $w = ct$ includes the Lorentz transformation, $w_I = \gamma_0(x + \beta_0 w)$, $x_I = \gamma_0(x + \beta w)$, as a special case, $\alpha \to 0$. In this sense, it satisfies the limiting four-dimensional symmetry, $s^2 = c^2t_I^2 - \mathbf{r}_I^2 = w^2 - \mathbf{r}^2$ as $\alpha \to 0$. The inverse Wu transformation of (2.22) can be deduced:

$$\begin{aligned} w &= (ct_I + \beta_0/\alpha\gamma_0)/[\alpha(x_I + 1/\alpha\gamma_0)] - \beta_0/\alpha, \\ x &= [(x_I + 1/\alpha\gamma_0)^2 - (ct_I + \beta_0/\alpha\gamma_0)^2]^{1/2} - 1/\alpha\gamma_o^2. \quad y = y_I, \quad z = z_I. \end{aligned} \tag{2.23}$$

We may remark that when $\beta_0 \to 0$, one can verify that (2.23) leads to the accelerated Galilean transformation, $x \approx x_I - c^2\alpha t_I^2/2$. Thus, the acceleration α can be approximately related to a constant acceleration g in Newtonian mechanics, $\alpha \approx g/c^2$. The transformation for the covariant differential operators $(\partial/\partial w,\ \partial/\partial \mathbf{r})$ can be deduced from (2.23):

$$\begin{aligned} \partial/\partial w_I &= \gamma(w^{-1}\partial/\partial w - \beta\partial/\partial x), \\ \partial/\partial x_I &= \gamma(\partial/\partial x - \beta W^{-1}\partial/\partial w), \quad \partial/\partial y_I = \partial/\partial y, \quad \partial/\partial z_I = \partial/\partial z. \end{aligned} \tag{2.24}$$

These relations will be useful for wave equations in quantum mechanics.

III. The Wu group of constant-linear-acceleration transformations

To show the group properties of the Wu transformation (2.22) for CLA frames, we must first find the coordinate transformations between two CLA frames F and F'. Let us consider another frame F' accelerating with a velocity $\beta' = \alpha' w'$ with respect to F_I. For simplicity, we shall ignore trivial y' and z' axes and set all initial velocities to zero in the following discussions. *However, these group properties can be shown to be true also for non-zero initial velocities.* Similar to (2.22), we can write down the Wu transformation between the inertial frame F_I and the accelerated frame F',

$$\begin{aligned} w_I &= \gamma' Z' w', \quad x_I = \gamma' Z'/\alpha' - 1/\alpha'; \\ Z' &= 1 + \alpha' x', \quad \beta' = \alpha' w', \quad \gamma' = (1 - \beta'^2)^{-1/2}. \end{aligned} \tag{3.1}$$

Using (2.14) with $\beta_0 = y = z = 0$ and (3.1), we deduce the Wu transformation between F' and F:

$$\begin{aligned} w &= \gamma' Z' w'/\{\alpha[Z'\gamma'/\alpha' + 1/\alpha - /\alpha']\}, \\ x &= \{[\gamma' Z'/\alpha' + 1/\alpha - 1/\alpha']^2 - \gamma'^2 Z'^2 w'^2\}^{1/2} - 1/\alpha. \end{aligned} \tag{3.2}$$

In order to see the group property of the Wu transformation (3.2) for two accelerated frames, we need to consider a third accelerated frame F'' with a velocity $\beta'' = \alpha'' w''$. In analogy to (3.2), we can write down the transformation between F'' and F,

$$\begin{aligned} w &= \gamma'' Z'' w''/\{\alpha[Z''\gamma''/\alpha'' + 1/\alpha - /\alpha'']\}, \\ x &= \{[\gamma'' Z''/\alpha'' + 1/\alpha - 1/\alpha'']^2 - \gamma''^2 Z''^2 w''^2\}^{1/2} - 1/\alpha; \\ Z'' &= 1 + \alpha'' x'', \quad \gamma'' = (1 - \beta''^2)^{-1/2}. \end{aligned} \tag{3.3}$$

Based on (3.3) and the inverse of (3.2), the transformation between F' and F'' is

$$\begin{aligned} w' &= \gamma'' Z'' w''/\{\alpha'[Z''\gamma''/\alpha'' + 1/\alpha' - /\alpha'']\}, \\ x' &= \{[\gamma'' Z''/\alpha'' + 1/\alpha' - 1/\alpha'']^2 - \gamma''^2 Z''^2 w''^2\}^{1/2} - 1/\alpha'. \end{aligned} \tag{3.4}$$

which has the same form as that of (3.2). Using (3.1) and (3.4), $s^2 = (ct_I)^2 - x_I^2$ can be expressed in terms of the coordinate variables in F' and F'' as follows:

$$s^2 = (ct_I)^2 - x_I^2 = [2Z'\gamma' - Z'^2 - 1]/\alpha'^2 = [2Z''\gamma'' - Z''^2 - 1]/\alpha''^2. \tag{3.5}$$

The existence of identity transformation and the associate law can also be verified. Thus we conclude that the set of Wu transformations for accelerated frames form a group which is termed the Wu group. Since a Wu transformation (with $w = ct$ or $w' = ct'$, etc.) reduces to a Lorentz transformation in the limit of zero acceleration, the Wu group includes the Lorentz group as a special case when all accelerations vanish.

One can generalize the Wu transformation in such a way that the velocity $\mathbf{u} \equiv \vec{u}$ is in arbitrary and fixed direction. This implies that both the acceleration $\vec{\alpha}$ and the initial velocity $\vec{\beta}_0$ must be in the same direction:

$$\mathbf{u} = \vec{\alpha} w + \vec{\beta}_0. \tag{3.6}$$

By differentiating (2.14), we obtain

$$d(ct_I) = \gamma(W dw + \beta dx), \quad dx_I = \gamma(dx + \beta W dw), \quad dy_I = dy, \quad dz_I = dz; \tag{3.7}$$

where

$$W = \gamma^2(\gamma_0^{-2} + \alpha x); \qquad \gamma_0 = (1 - \beta_0^2)^{-1/2}.$$

In analogy with the Lorentz transformation in an arbitrary direction, (3.7) can be generalized to the following form:

$$\begin{aligned} d\mathbf{r}_I &= d\mathbf{r} + (\gamma - 1)(\mathbf{u} \cdot d\mathbf{r})\mathbf{u}/u^2 + [\gamma^3\gamma_0^{-2}\mathbf{u} + \alpha\gamma^3(\mathbf{u} \cdot \mathbf{r})\mathbf{u}/u]dw, \\ &= d\mathbf{r} + (\gamma - 1)(\mathbf{u}/u)d(\mathbf{u} \cdot \mathbf{r}/u) + (1/\alpha\gamma_0^2)(\mathbf{u}/u)d\gamma + (\mathbf{u} \cdot \mathbf{r})(\mathbf{u}/u^2)d(\gamma - 1), \\ dw_I &= \gamma^3[\gamma_0^{-2} + \alpha \mathbf{u} \cdot \mathbf{r}/u]dw + \gamma \mathbf{u} \cdot d\mathbf{r}. \\ &= \gamma^3\gamma_0^{-2} dw + \gamma(\mathbf{u} \cdot \mathbf{r}/u)du(\gamma^2 - 1) - \gamma^3(\mathbf{u} \cdot \mathbf{r})u du + d(\gamma \mathbf{u} \cdot \mathbf{r}), \end{aligned} \tag{3.8}$$

where $u = |\mathbf{u}|$. It is straightforward to carried out the integrations, we have

$$\begin{aligned} \mathbf{r}_I &= \mathbf{r} + (\gamma - 1)(\mathbf{u} \cdot \mathbf{r})\mathbf{u}/u^2 + \mathbf{u}[\gamma\gamma_0^{-2} - \gamma_0^{-1}]/(\alpha u), \\ w_I &= (1/\alpha\gamma_0^2)(\gamma u - \gamma_0\beta_0) + \gamma \mathbf{u} \cdot \mathbf{r}. \end{aligned} \tag{3.9}$$

It can be verified that the general transformation (3.9) reduces to (2.22) if $\mathbf{u}$ is in the x-direction, $\mathbf{u} = (\beta, 0, 0)$. In the zero acceleration limit, (3.9) reduces to

$$\begin{aligned} \mathbf{r}_I &= \mathbf{r} + (\gamma_o - 1)(\vec{\beta}_o \cdot \mathbf{r})\vec{\beta}_o/\beta_o^2 + \vec{\beta}_o\gamma_o w, \\ w_I &= w\gamma_o + \gamma_o\vec{\beta}_o \cdot \mathbf{r}, \end{aligned} \tag{3.10}$$

which is the well-known form of the general Lorentz transformation.

IV. Physics in linearly accelerated frames

Within the four-dimensional symmetry framework of taiji relativity based solely on the first postulate of relativity [3], it was shown that there are only two universal and fundamental constants: $J = 3.5177293 \times 10^{-38}$ g·cm and $\bar{e} = -1.6021891 \times 10^{-20}(4\pi)^{1/2}(\text{g} \cdot \text{cm})^{1/2}$, instead of the usual three, h, e (in esu) [5] and c. These results can be applied to the present formalism of physics in CLA frames.

Since the speed of light in a CLA frame F is not a universal constant, we shall write $w \equiv bt$ and $u \equiv dw/dt$, where b is a variable in this section, so that one can see that the physics in F does not depend on $w = ct$. The invariant action for a charged particle and the electromagnetic potential $a_\mu(x)$ in F is assumed to be

$$S = \int(-mds - \bar{e}a_\mu dx^\mu) - (1/4)\int f_{\mu\nu}f^{\mu\nu}Wd^4x = \int Ldt - (1/4)\int f_{\mu\nu}f^{\mu\nu}Wd^4x; \tag{4.1}$$

$$ds^2 = W^2dw^2 - dx^2 - dy^2 - dz^2, \quad (-det\ g_{\mu\nu})^{1/2} = W; \quad w = bt; \tag{4.2}$$

$$L = -m[W^2u^2 - v_x^2 - v_y^2 - v_z^2]^{1/2} - \bar{e}(a_0u + a_iv^i), \quad u = dw/dt, \quad \mathbf{v} = d\mathbf{r}/dt; \tag{4.3}$$

$$f_{\mu\nu} = \partial_\mu a_\nu - \partial_\nu a_\mu. \quad g_{\mu\nu} = (W^2, -1, -1, -1). \tag{4.4}$$

In the limit $\alpha \to 0$, F becomes an inertial frame and, hence, $\bar{e}$ and a_μ correspond to the usual charge e (in esu) and the electromagnetic potential A_μ by $\bar{e} = e/c$ and $a_\mu = A_\mu/c$ respectively. The canonical momentum P_i of a particle in F is given by

$$P_i = -\partial L/\partial v^i = p_i + \bar{e}a_i; \quad i = 1, 2, 3; \tag{4.5}$$

$$p_i = (-m\gamma v_x/C, -m\gamma v_y/C, -m\gamma v_z/C) = g_{ik}p^k, \quad p^i \equiv mdx^i/ds; \tag{4.6}$$

$$\gamma = (1 - v^2/C^2)^{-1/2}; \ C \equiv uW, \quad v^2 = v_x^2 + v_y^2 + v_z^2 \equiv -v_iv^i. \tag{4.7}$$

The "Hamiltonian" $H = P_0$. with the same dimension as that of P_i, is defined by

$$P_0 \equiv [(\partial L/\partial v^i)v^i - L]/u = p_0 + \bar{e}a_0 = W[(P_i - \bar{e}a_i)^2 + m^2]^{1/2} + \bar{e}a_0 \equiv H; \tag{4.8}$$

$$p_0 = m\gamma W = g_{00}p^0, \quad p^0 \equiv mdx^0/ds = mdw/ds. \tag{4.9}$$

The transformation of the covariant momenta p_i and p_0 is given by

$$p_{I0} = \gamma(p_0/W - \beta p_1), \quad p_{I1} = \gamma(p_1 - \beta p_0/W), \quad p_{I2} = p_2, \quad p_{I3} = p_3; \tag{4.10}$$

where

$$p_\mu = (p_0, -\mathbf{p}), \quad \mathbf{p} = (p_x, p_y, p_z) = (p^1, p^2, p^3) = (-p_1, -p_2, -p_3).$$

Note that (4.10) is consistent with (2.24) because p_μ and $\partial/\partial x^\mu$ should have the same transformation property. Also, the CLA transformation of the covariant vector dx_μ is the same as (4.10) because $dx_\mu = g_{\mu\nu}dx^\nu$ and $dx_{I\mu} = \eta_{\mu\nu}dx_I^\nu$, where $\eta_{\mu\nu} = (1, -1, -1, -1)$. The invariant relation $g^{\mu\nu}p^\nu p^\nu = m^2$ implies

$$\begin{aligned} &g^{\mu\nu}(P_\mu - \bar{e}a_\mu)(P_\nu - \bar{e}a_\nu) \\ &= W^{-2}(P_0 - \bar{e}a_0)^2 - (P_1 - \bar{e}a_1)^2 - (P_2 - \bar{e}a_2)^2 - (P_3 - \bar{e}a_3)^2 = m^2, \end{aligned} \tag{4.11}$$

where we have used (4.5), (4.6), (4.8), (4.9), $g^{\mu\nu} = (W^{-2}, -1, -1, -1)$ and $P^\mu = g^{\mu\nu} P_\nu$. This equation suggests that the generalized Dirac equation for the accelerated frame F should have the form

$$[\gamma^{*\mu}(x)(P_\mu - \bar{e}a_\mu) - m]\Psi = 0, \qquad P_\mu = i\mathbf{J}\partial/\partial x^\mu. \tag{4.12}$$

If one wishes, one can relate $\gamma^*(x)$ in (4.12) to constant Dirac matrices $\overline{\gamma}^\mu$ by the relation $\gamma^{*\mu}(x) = e^\mu_{\nu|}(x)\overline{\gamma}^\nu$, where $e^\mu_{\nu|}(x)$ is a tetrad.

V. Experimental Implications and Discussions

The result (4.10) and the new transformation (2.22) can be experimentally tested by measuring Doppler shift of wavelength of light emitted from a CLA source. From Eq. (4.10) one obtains the transformation of the covariant wave 4-vector $k_\mu = p_\mu/J$ between an inertial frame F_I and a CLA frame F. Note that Jk_{I0} and Jk_0 are moving masses of the same photon measured from F_I and F respectively. Suppose the radiation source is at rest at the origin of the F frame, $\mathbf{r} = 0$, and $k_\mu = (k_0, -k_1, 0, 0)$, where $k_0 = k_0$ (rest) and $k_1 = k_1$ (rest). Experimentally, it is difficult to measure k_0 (rest) and k_1 (rest) in the CLA frame. Thus we have to express them in terms of quantities measured in the inertial frame (or laboratory) F_I. Using (4.10), the relation [6] $k_0(\text{rest}) = k_{I0}(\text{rest})$ and $Z(0) = \gamma_o^{-2}$, we obtain the shifts of k_{I0} (related to photon's "moving mass" or atomic mass level [3]) for waves emitted from a CLA source,

$$k_{I0} = k_{I0}(\text{rest})[\gamma_o^2(1+\beta)/\gamma], \quad k_1 = -k_x; \quad \beta = \alpha w + \beta_o, \tag{5.1}$$

where (rest) denotes the source being at rest in F_I. A similar relation can be obtained for the wavelength. Such new effects predicted by the Wu transformation for waves emitted from a CLA source may be termed Wu-Doppler effect. Note that Moeller's transformation will lead to a result different from (5.1) because t in (2.9) and (2.22) with $w = ct$ [or (5.2) below] must have the *same* physical interpretation. Such a difference can be tested by measuring the Wu-Doppler effect (5.1) in the laboratory frame F_I by using the method of Ives-Stilwell [7].

We stress that the Wu transformation (2.22) does not depend on a specific relation between w and t. Suppose one assumes $w = ct$ in (2.22). The time t,

$$t = [t_I + \beta_o/(c\alpha\gamma_o)]/[\alpha(x_I + 1/\alpha\gamma_o)] - \beta_o/(c\alpha), \tag{5.2}$$

in the accelerated frame F can be physically realized by the "computerized clocks" [3]: By basing "computerized clocks" on a computer chip, one could program any clock in F to obtain a time reading t_I from the nearest Einstein clock in F_I and, based on c, β_o, α and its F_I frame position x_I, compute the time t it should display according to (5.2). This is a general method for synchronization of computerized clocks in a reference frame, without relying on the constant speed of light. If one compares the *rate of ticking* of the Einstein and the computerized clocks at a fixed position x_I, one has $(\partial t/\partial t_I)_{x_I} = 1/[\alpha x_I + 1/\gamma_0]$. *This can be physically realized because the reading and the rate of ticking of a clock are adjustable. The "computerized clocks" in F will automatically becomes Einstein clocks as* *the acceleration approaches zero,* $\alpha \to 0$. In this sense, these sophisticated computerized clocks are generalized Einstein clocks for both inertial and non-inertial frames. We note that the choice of $w = ct$ in (2.22) to synchronize computerized clocks in F does not imply that the speed of light is a constant c in the accelerated frame. For the general case $w = bt$, where t is defined by an arbitrarily preassigned function $t(x_I, t_I)$, the time t can also be physically realized by the "computerized clocks". It appears that all these different times $t(x_I, t_I)$ are equally physical, in principle, for describing physical phenomena, as discussed in [3]. However, from fundamental laws of physics such as (4.11) and (4.12), we can see that the real evolution variable is w rather than t. Of course, we can also make these computerized clocks to read w directly.

The boundary condition $Z(0) < \infty$, which leads to $f = \alpha$ in (2.19), is imposed for simplicity. It is not necessary: If $Z(0) < \infty$ is not imposed, then one has $f \neq \alpha$ in general. However, $G(w)$ and $Z(x)$ involve the factors $1/f$ and f respectively, so that $W(w,x) = G(w)Z(x)$ and the resultant physics do not depend on f.

The equivalence of the effects of a gravitational field and those of an observer's acceleration played an essential role at the birth of general relativity. Nevertheless, some authors suggested that it be buried with appropriate honors because it is false [9]. It is the equivalence of gravitational and inertial mass which is precise and necessary for general relativity.

What is the operational meaning of the constant acceleration α? We show that the constant acceleration α of a particle is directly and uniquely related the change of its energy per unit length as measured in an inertial frame F_I [10]:

$$(dp_{I0}/dx_I)_x = m\alpha/Z(x), \qquad x = \text{fixed}, \tag{5.3}$$

where we have used the differential transformation (2.10) and the momentum transformation (4.10) with $p_\mu = Jk_\mu$. We stress that, within the framework of the four-dimensional symmetry, the concept of uniform acceleration of a particle can only be defined in the sense of (5.3), i.e., constant change of a particle's "energy" p_{I0} per unit length, as measured in an inertial frame F_I. It is gratifying to see that this is precisely what has been used in high energy laboratory. Other definition of acceleration such as the change of velocity per unit time is only an approximation for small velocities and is, strictly speaking, incompatible with the 4-dimensional symmetry.

Our results suggest that the kinematic approach first discussed by Wu and Lee [1] is more fruitful than the conventional gravitational approach, provided the limiting 4-dimensional symmetry is postulated.

The work is supported in part by the Potz Science Fund. This paper is written as an affectionate jubilee greeting to Taida's Physics Department. Appropriately, it deals with 4-dimensional symmetry and time which engrossed JP's thoughts for many years as a student at Taida.

References

[1] C. Møller, Danske Vid. Sel. Mat-Fys. xx, No.19 (1943) and *The Theorg of Relativity* (Clarendon, Oxford 1969) pp.255-258. Ta-You Wu and Y. C. Lee, Intern. J. Theor. Phys. **5**, 307 (1972). Wu and Lee assumed (i) local FitzGerald-Lorentz contraction of length, (ii) local

time-dilatation and (iii) a time-independent g_{00} for CLA coordinates, and derived the same transformation.

[2] Jong-Ping Hsu and Leonardo Hsu, Nuovo Cim. **B112** (to be published in April, 1997).

[3] J. P. Hsu and L. Hsu, Phys. Letters **A196**, 1-6 (1994); (Erratum) *ibid*, **217**, 359 (1996); Leonardo Hsu and Jong-Ping Hsu, Nuovo Cimento **B111**, 1283 (1996).

[4] Ta-You Wu, *Theoretical Physics*, vol. 4: *Theory of Relativity* (in Chinese, Lian Jing Publishing Co., 1978) pp. 172-175; see also ref. 1. Roughly speaking, Wu explored local relation between "accelerated transformation" and Lorentz transformation; while we consider their global relations based on the limiting four-dimensional symmetry principle.

[5] Note that the universal constant $\bar{e}$ is the charge measured in the electromagnetic unit (emu) rather than in the electrostatic unit (esu).

[6] This equality is only approximate because there is no relativity or equivalence between F and F_I. Nevertheless, it turns out to be an extremely good approximation because atomic structure is very stable against constant-linear-acceleration. This is basically related to the metric tensor g_{00} and the smallness of atomic sizes, $\sim 10^{-8}$ cm.

[7] When we define $w = ct$, the speed of light measured in the CLA frame F is $C \equiv dr/dt = cW(t, x)$ because the propagation of light is described by equation (2.1), $ds = 0$. Note that C is anisotropic and depends on space and time in general. We stress that one can also set $w = bt_I$ in (2.22) and (2.23) *without upsetting its Wu group property*. Thus, we have a common time, $t = t_I$, for all frames and, hence, the speed of light measured in F by using such a common time will be $C = c\gamma(1-\beta)$, if $dx_I/dt_I = +c$. For discussions of common time within the 4-dimensional framework, see J. P. Hsu, Phys. Lett. **A97**, 137 (1983); Nuovo Cimento **B74**, 67 (1983); J. P. Hsu and C. Whan, Phys. Rev. **A38**, 2248 (1988), Appendix.

[8] H. E. Ives and G. R. Stilwell, J. Opt. Soc. Am. **28**, 215 (1938); **31**, 369 (1941).

[9] J. L. Synge, *Relativity: The General Thory* (North-Holland, Amsterdam, 1966) pp. ix-x. See also V. A. Fock, *The theory of space, time and gravitation* (Pergamon, London, 1959).

[10] For an object at rest in F_I, we have $(dp_0/dx)_{x_I} = m\alpha(\gamma^{-1}-1)/(\gamma Z^2) \neq (dp_{I0}/dx_I)_x$. It also shows the lack of symmetry between a CLA frame F and an inertial frame F_I.

Generalizing Lorentz Transformations for Accelerated Frames and Their Physical Implications

Daniel T. Schmitt*and Tobias Kleinschmidt †

April 6, 2005

We discuss reference frames with constant-linear-accelerations and their generalized spacetime transformations with minimal departure from the Lorentz transformations. The requirement of limiting 4-dimensional symmetry of the Lorentz and Poincaré groups assures that the generalized transformations reduce to the Lorentz transformations in the limit of zero acceleration. The changes in the geometry of spacetime in accelerated frames are shown graphically, including singularities and horizons. Physical implications and a feasible experimental test are discussed.

*daniel.schmitt@physik.uni-ulm.de
Department of Theoretical Physics, University of Ulm, Germany
†tobias.kleinschmidt@gmx.de
Department of Physics, University of Massachusetts Dartmouth
North Dartmouth, MA 02747-2300, USA

1 Introduction

In the physics of the future, it is desirable that particle physics and quantum field theory, including gravity, can be understood in both inertial and non-inertial frames. The reason is that all physical frames of reference in the universe are, strictly speaking, non-inertial because of the existence of the long range gravitational force. The inertial frame is only an approximation or idealization of physically realizable reference frames. To understand physics in inertial frames, we have the Lorentz transformations or Lorentz and Poincaré invariance, so that physical theories can be formulated covariantly and tested experimentally. In contrast, to understand physics in non-inertial frames, there is a group of all point transformations (one-to-one and twice differentiable) of spacetime in which the differential form $ds^2 = g_{\mu\nu}dx^\mu dx^\nu$ is invariant. Its invariant theory is the tensor calculus of the general theory of relativity. However, this group for non-inertial frames is too general for quantum field theory and have little specific predictions. For example, there is no Quantum Field Theory (QED) for accelerated reference frames which would allow us to calculate particle lifetimes in accelerated frames.

Some physical phenomena involving accelerations of one particle can be treated, by introducing a co-moving frame[1] , on the basis of the invariant theory of the Lorentz group. But such an invariant theory is inadequate for treating many other physical phenomena. In particular, quantizations of fields in non-inertial frames cannot be treated on the basis of the invariant theory of the Lorentz group.

This motivates us to investigate a simple subset of reference frames which has constant-linear-accelerations. [1] Our discussions are based on the principle of limiting 4-dimensional symmetry which requires that all accelerated transformations be reduced to the Lorentz and Poincaré transformations in the limit of zero acceleration. [2] We discuss specific spacetime transformations and their geometry and physical properties. Their geometry and light propagation is illustrated in several graphs. We also discuss a specific experiment of decay-length dilation to test the generalized Lorentz transformations.

2 Simple constant-acceleration transformations with limiting 4-dimensional symmetry

The concept of linearly accelerated frames was not defined with complete satisfaction because most accelerated transformations of spacetime are not smoothly connected to the Lorentz transformations in the limit of zero acceleration. [3] We shall adopt the following definition for constant-linear-acceleration (CLA) frame: (i) A particle at rest at any point in a CLA frame must have a constant change of energy per unit length (or momentum per unit time) as measured in high energy laboratory or in an inertial frame, whereas the values of accelerations, measured from an inertial frame, may be different for particles located at different points in the CLA frame. (ii) A finite spacetime transformation between inertial and accelerated frames exists and it satisfies the limiting 4-dimensional symmetry in the limit of zero acceleration.

A generalized Lorentz transformation must be consistent with the fact that any accelerated frame reduces to an inertial frame in the limit of zero acceleration. This is a natural, basic and necessary requirement from the physical viewpoint. But such a generalization turns out to be not unique and, hence, theoretically there are infinitely many generalizations. So far, no established principle can lead to a simple and unique generalization. Of

[1] The co-moving frame is usually treated to be an inertial frame which, of course, is an approximation.

course, experiment has the final say regarding the correct accelerated spacetime transformation. Indeed, the Wu and the Møller transformations for constant accelerations are just two of simple generalizations, and only the Wu transformation includes the Lorentz transformation as a limiting case of zero acceleration. [1] There are other simple generalizations of the Lorentz transformation. In all these minimal generalizations, the spacetime of CLA frames is characterized by a metric tensor of the form $(W^2, -1, -1, -1) \equiv P_{\mu\nu}$, which may be called the Poincaré metric tensor. Furthermore, there exist finite transformations for inertial and CLA transformations and, therefore, the spacetime of these CLA frames is flat, i.e., having a vanishing Riemann curvature tensor. This property is useful for formulations of field theory.

Let us consider the transformations between an inertial frame $F_I(w_I, x_I, y_I, z_I)$ and a new constant-linear-acceleration (CLA) frame $F(w, x, y, z)$ which moves with a time-dependent velocity $\beta(w)$ along the x-axis. Suppose the metric tensors in the inertial and the CLA frames are:

$$\eta_{\mu\nu} = diag.(1, -1, -1, -1) \tag{1}$$

$$P_{\mu\nu} = diag.(W^2, -1, -1, -1) \tag{2}$$

where $W = W(w, \mathbf{r})$ is any real-valued function of spacetime in $F(w, x, y, z)$ with the property $W \to 1$ for vanishing acceleration. Thus, the Poincaré metric tensor $P_{\mu\nu}$ reduces to the Minkowski metric tensor $\eta_{\mu\nu}$ in the limit of zero acceleration.

To derive the spacetime transformation for CLA frames, one may start with a linear relation for the differentials (dw_I, dx_I) and (dw, dx) with some unknown coefficients which are functions of spacetime. Other components, dy and dz, are unchanged. Based on physical considerations, a finite transformation of spacetime must exist between F_I and F. Thus, these coefficients must satisfy integrability condition, and the finite transformation must reduce to the Lorentz transformation in the limit of zero acceleration, $\beta \to \beta_o$. One obtains two types of simple generalizations of the Lorentz transformation from inertial frames to constant-linear-acceleration frames, which are characterized by the metric tensor of the form (2). [1, 4]

Let us consider a class of transformations for an inertial frame $F_I(w_I, x_I, y_I, z_I)$ and a constant-linear-acceleration frame $F(w, x, y, z)$, accelerated in the x-direction, [5] including constant spacetime translations:

$$w_I = \gamma\beta\left(x + \frac{1}{\alpha_o\gamma_o{}^2}\right) - \frac{\beta_o}{\alpha_o\gamma_o} + w_o, \qquad x_I = \gamma\left(x + \frac{1}{\alpha_o\gamma_o{}^2}\right) - \frac{1}{\alpha_o\gamma_o} + x_o, \tag{3}$$

$$y_I = y + y_o, \qquad z_I = z + z_o; \qquad \beta = \beta(w) = \text{ arbitrary function},$$

$$\gamma = \frac{1}{\sqrt{1-\beta^2}}, \qquad \gamma_o = \frac{1}{\sqrt{1-\beta_o{}^2}},$$

where $x^\mu_o = (w_o, x_o, y_o, z_o)$ are constants and the velocity function $\beta(w)$ will be specified and discussed below. This general transformation of spacetime is interesting because it includes the Wu transformation and the Møller transformation for accelerated frames as special cases. Even though $\beta(w)$ may be an arbitrary function of time w in F, it is actually a specific and unique expression when we express it in terms of spacetime coordinates of the inertial frame F_I:

$$\beta(w) = \frac{w_I + \beta_o/(\alpha_o\gamma_o) - w_o}{x_I + 1/(\alpha_o\gamma_o) - x_o}. \tag{4}$$

Since the right-hand-side of (4) does not involve any arbitrary function, this implies that, from the viewpoint of observers in the inertial frame F_I, the frame F always moves the same way no matter what is the function given to $\beta(w)$. This may be interpreted as a flexibility of time w of the accelerated frame F, as we shall see below.

When the acceleration in (3) approaches zero, i.e., $\beta \to \beta_o$, the linear-acceleration transformations (3) must reduce to the Poincaré transformations (provided $w = ct$ and $w_I = ct_I$),

$$w_I = \gamma_o(w + \beta_o x) + w_o, \quad x_I = \gamma_o(x + \beta_o w) + x_o, \quad y_I = y + y_o, \quad z_I = z + z_o. \tag{5}$$

$$\beta(w) \to \beta_o + \alpha_o w, \qquad (\text{for small } \alpha_o). \tag{6}$$

We stress that as long as the arbitrary function $\beta(w)$ satisfies this limiting property as $\alpha_o \to 0$, equation (3) will reduce to the Poincaré transformation (for $w = ct$ and $w_I = ct_I$) or the Lorentz transformation (when $x_o^\mu = 0$) in the limit of zero acceleration. If a spacetime transformation for non-inertial frames does not have this limiting 4-dimensional symmetry, it is incomplete and probably not viable.

In the following discussions, we shall set $x_o^\mu = 0$ for simplicity, and concentrate on the physical effects on spacetime of reference frames due to accelerations.

3 Wu transformations, singular wall and horizons

Making the minimal departure from the Lorentz transformation and easiest non-constant choice for $\beta(w)$ in (3), we assume $\beta(w) = \beta_o + \alpha_o w$. The spacetime transformations (3) take the following form:

$$w_I = \gamma(\beta_o + \alpha_o w)\left(x + \frac{1}{\alpha_o {\gamma_o}^2}\right) - \frac{\beta_o}{\alpha_o \gamma_o}, \qquad x_I = \gamma\left(x + \frac{1}{\alpha_o {\gamma_o}^2}\right) - \frac{1}{\alpha_o \gamma_o}, \tag{7}$$

$$y_I = y, \qquad z_I = z; \qquad \gamma = \frac{1}{\sqrt{1-\beta^2}}, \qquad \beta = \beta_o + \alpha_o w.$$

This is called the Wu transformations, which was the first generalization of the Lorentz transformation obtained on the basis on limiting 4-dimensional symmetry. [2] We also have the following transformation for differential 4-vector $dx^\mu = (dw, dx, dy, dz)$ and the explicit expression for the invariant interval ds^2:

$$dw_I = \gamma(Wdw + \beta dx), \quad dx_I = \gamma(dx + \beta W dw), \quad dy_I = dy, \quad dz_I = dz, \tag{8}$$

$$ds^2 = {dw_I}^2 - {dx_I}^2 - {dy_I}^2 - {dz_I}^2 = W^2 dw^2 - dx^2 - dy^2 - dz^2,$$

$$W = \gamma^2(\gamma_o^{-2} + \alpha_o x) > 0.$$

Note that there is a singular wall at $x = -1/(\alpha_o {\gamma_o}^2)$, where W or the metric tensor P_{oo} vanish, which is unphysical. On the other side of the singular wall, dw_I and Wdw turn out to have a different sign in the transformation (8).

The inverse transformation of (7) is given by

$$w = \frac{1}{\alpha_o}\left(\frac{w_I + \beta_o/\alpha_o\gamma_o}{x_I + 1/\alpha_o\gamma_o} - \beta_o\right) \tag{9}$$

$$x = \sqrt{(x_I + 1/\alpha_o\gamma_o)^2 - (w_I + \beta_o/\alpha_o\gamma_o)^2} - \frac{1}{\alpha_o\gamma_o^2},$$

$$y = y_I, \qquad z = z_I.$$

Contributions of singular terms in the transformations (3) and (9) cancel in the limit of zero acceleration. Thus, the inertial limit $\alpha_o \to 0$ is well-defined for the Wu transformations.

We are used to picture physical world from the viewpoint of observers in an inertial frame. Let us use the inverse Wu transformations (9) to show the change and the physical properties of space and time axes (w,x) due to velocity and acceleration. We observe that for constant w and $x_I \neq -1/(\alpha_o\gamma_o)$, the time transformation in (9) leads to straight lines. All lines correspond to constant w pass through the same point $(w_I, x_I) = (-\beta_o/(\alpha\gamma_o), -1/(\alpha_o\gamma_o))$. In contrast, for constant x, the inverse Wu transformation (9) leads to two sets of hypobolic lines, which satisfy the condition $|x_I + 1/\alpha_o\gamma_o| > |w_I + \beta_o/\alpha_o\gamma_o|$.

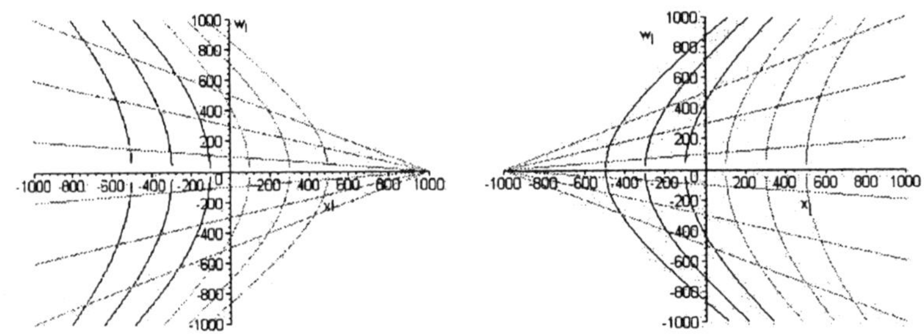

(a) Acceleration $\alpha_o = -0.001$ (b) Acceleration $\alpha_o = +0.001$

Figure 1: Spacetime graphs with velocity $\beta_o = 0$

Figure 1 shows lines of constant x and constant w in the (w_I, x_I) plane for the inverse Wu transformations (9) with zero inertial velocity, $\beta_o = 0$, and different values for constant acceleration, $\alpha = -0.001, +0.001$.

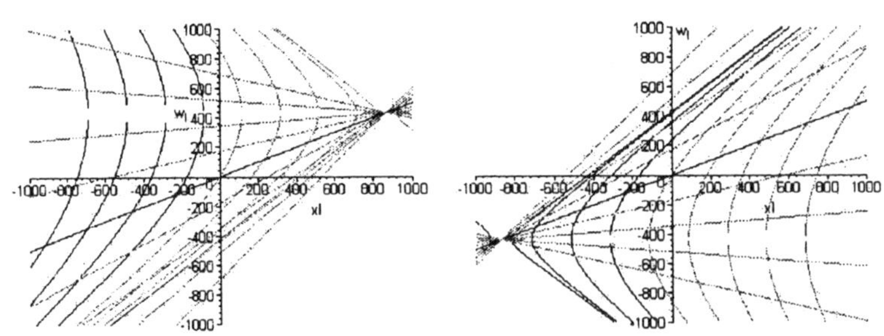

(a) Acceleration $\alpha_o = -0.001$ (b) Acceleration $\alpha_o = +0.001$

Figure 2: Spacetime graph with velocity $\beta_o = 0.5$

Figure 2 shows the same spacetime picture with the initial velocity, $\beta_o = 0.5$, and two values of accelerations $\alpha_o = -0.001, +0.001$.

Clearly, figure 3 has zero acceleration, so that all lines are straight, corresponding to an inertial frames with the velocity $\beta_o = 0.5$, as we expect from the Lorentz transformation.

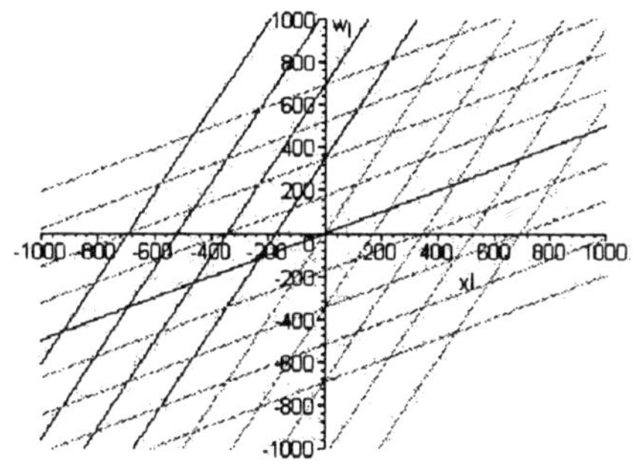

Figure 3: Spacetime graph with velocity $\beta_o = 0.5$ accelerations $\alpha_o \to 0$

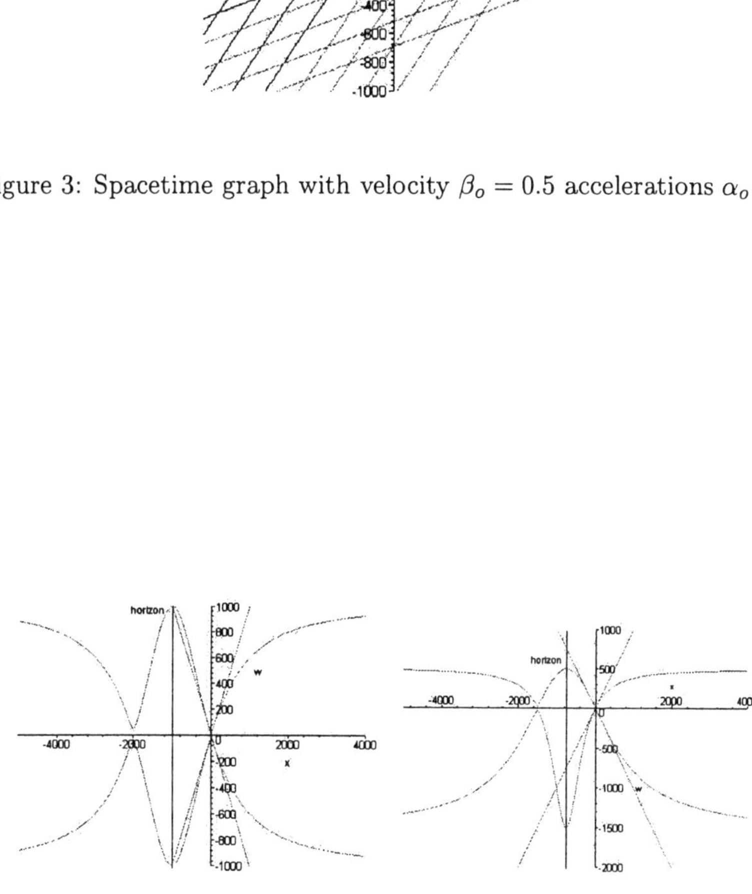

(a) Velocity $\beta_o = 0$ acceleration $\alpha_o = 0.001$

(b) Velocity $\beta_o = 0$ acceleration $\alpha_o = 0.001$

Figure 4: Lightcone graphs

Figure 4 shows the changes of lightcone given by the relation $s^2 = {w_I}^2 - {x_I}^2 = 0$, where ${w_I}^2$ and ${x_I}^2$ can be expressed in terms of w and x by using the Wu transformation (7). The straight lines represent the lightcone in an inertial reference frame for comparison.

It is interesting to observe that the region of spacetime consisting of all points $[x + 1/(\alpha\gamma^2)] > 0$ and all w from $(-1 - \beta_o)/\alpha_o$ to $(1 - \beta_o)/\alpha_o$ corresponds to the sector of the (w_I, x_I) plane between $(w_I + \beta_o/\alpha_o\gamma_o) = +(x_I + 1/\alpha_o\gamma_o)$ and $(w_I + \beta_o/\alpha_o\gamma_o) = -(x_I + 1/\alpha_o\gamma_o)$. As $\alpha_o \to 0$, this region becomes larger and larger, and eventually cover the whole plane (w_I, x_I), as one can see in Figures 1, 2 and 3. Thus, this region of spacetime is identified with the physical world of the CLA frame F. Similarly, the region of spacetime consisting of all $[x + 1/(\alpha\gamma^2)] < 0$ is represented by the opposite sector with negative $(x_I + 1/\alpha_o\gamma_o)$. This opposite region of spacetime may be called the 'mirror spacetime' of the CLA frame. The rest of the (w_I, x_I) plane cannot be reached for any real w and x for $\alpha_o \neq 0$. The time w is physical only within the range, $(-1 - \beta_o)/\alpha_o < w < (1 - \beta_o)/\alpha_o$. Outside this range, the CLA frame ceases to exist because the velocity $\beta(w)$ is greater than 1, which is unphysical. The limits of time, $w \to (\pm 1 - \beta_o)/\alpha_o$, in the CLA frame implies $w_I \to \pm\infty$ in inertial frame, (everywhere in physical space except at the singular wall). This property is interlocked with the assumption $\beta(w) = \beta_o + \alpha_o w$. If one makes a different assumption for the arbitrary function $\beta(w)$ in (3), one has a different time for a CLA frame. [1]

The Wu transformations suggest that accelerations distort spacetime and form a horizon, which corresponds to a 'wall singularity' $x = -1/(\alpha_o{\gamma_o}^2)$ (with arbitrary y and z coordinates) of the coordinates for the CLA frame F or $(x_I + 1/\alpha_o\gamma_o)^2 = (w_I + \beta_o/\alpha_o\gamma_o)^2$. This 'singular wall' separates physical spacetime from the 'mirror spacetime'. The location of the wall singularity depends on the sign of the acceleration α_o and the magnitude $|\alpha_o{\gamma_o}^2|$. The 'mirror spacetime' emerges from the quadratic equation associated with the second equation in (9):

$$\left(x + \frac{1}{\alpha_o\gamma_o^2}\right)^2 = \left(x_I + \frac{1}{\alpha_o\gamma_o}\right)^2 - \left(w_I + \frac{\beta_o}{\alpha_o\gamma_o}\right)^2$$

Note that we also have a quadratic equation for the relativistic energy-momentum relation, which leads to 'negative energy' solution. In some sense, the mirror spacetime resembles to the negative energy solution of a particle, so one may ask whether the mirror spacetime exists in some physical sense. Such a question probably cannot be answered because even if it exists there is no possibility of communication between these two sectors of spacetime separated by the singular wall.

The rate of ticking of a clock and the observable speed of light in the CLA frame F nearby this singular wall have very peculiar properties. Namely, as $x \to -1/(\alpha_o{\gamma_o}^2)$ in (9) (i.e., $(x_I + 1/\alpha_o\gamma_o) \to (w_I + \beta_o/(\alpha_o\gamma_o))$, the Wu transformation leads to the following results:

(a) The clock stands still:

$$\left(\frac{dw}{dw_I}\right)_{dx=0} = \frac{1}{\gamma W} = \frac{(x_I + 1/\alpha_o\gamma_o)^2 - (w_I + \beta_o/\alpha_o\gamma_o)^2}{\alpha_o(x_I + 1/\alpha_o\gamma_o)^3} \to 0. \tag{10}$$

(b) The speed of light in the CLA frame F increases indefinitely

$$\left(\frac{dx}{dw}\right)_{ds=0} = \pm|W| = \pm\frac{\alpha_o(x_I + 1/\alpha_o\gamma_o)^2}{\sqrt{(x_I + 1/\alpha_o\gamma_o)^2 - (w_I + \beta_o/\alpha_o\gamma_o)^2}} \to \pm\infty, \tag{11}$$

The result (a) shows that the rate of ticking of a clock at rest in F at the positon $(x, 0, 0)$, i.e., $dx = 0$, slows down in comparison with a clock at $(x_I, 0, 0)$ in the inertial frame. The result (b) is due to the fact that the law for the propagation of light is given by $ds^2 = 0$. The two properties in (10) and (11) are intimately related. We may remark that it is natural for the ratio (10) to be positive. This is consistent with $(x_I + 1/(\alpha_o\gamma_o) > 0$ and $(x_I + 1/\alpha_o\gamma_o)^2 > (w_I + \beta_o/(\alpha_o\gamma_o)^2$, where the last relation is consistent with $\beta^2 < 1$.

Note that the general transformations (3) with an arbitrary velocity function $\beta(w)$ corresponds to constant-linear-acceleration in the following sense: Suppose a particle is at rest in the CLA frame F at the position $(x, 0, 0)$ = constant, we can derive the following relation,

$$\left(\frac{dp_I^0}{dx_I}\right)_{Wu} = \left(\frac{dp_I^1}{dw_I}\right)_{Wu} = \frac{md\gamma}{\gamma\beta W dw} = \frac{\alpha_o m}{\alpha_o x + {\gamma_o}^{-2}} = \text{constant}, \tag{12}$$

This result is consistent with the constant acceleration of a charged particle in a high-energy linear accelerator, which has a constant potential drop per unit length. Thus, a charged particle gains a constant kinetic energy per unit length. Note that the usual definition of constant acceleration $d^2x_I/dw_I^2 = constant$ in classical mechanics is only an approximation for small velocities and is inconsistent with high-energy experiment.

4 Lifetime Dilation with Accelerations

Let us consider lifetime dilation experiments. [6] Suppose the particles such as pions are at rest at the origin $x' = y' = z' = 0$ in an inertial frame $F'(w', x', y', z')$ and their lifetime is measured in another inertial frame $F_I(x_I^\mu)$. The number of particles at time w', $N(w')$, is given by the exponential decay law in the inertial frame F',

$$N(w') = N_o e^{-w'/w_r}, \tag{13}$$

where N_o is the number of particles at time $w' = 0$ and w_r is the constant lifetime of the particles at rest. The numerical value of w_r is the same as the decay-length D= τ_oc, where τ_o is the usual lifetime at rest. Since the experiment is performed in the inertial frame F_I, we express the law (13) in terms of observable variables in $F_I(x_I^\mu)$ by using the Lorentz transformations between F' and F_I, i.e., the coordinates $x^\mu = (w, x, y, z)$ and x_o^μ in (5) are replaced by $x'^\mu = (w', x', y', z')$ and zero respectively. Using $x' = y' = z' = 0$, we have

$$N = N_o e^{-w_I/(\gamma_o w_r)} = N_o e^{-x_I/(\gamma_o\beta_o w_r)}. \tag{14}$$

It shows that the particle lifetime is dilated by a γ_o factor. This result of lifetime dilation has been confirmed with a very large value of $\gamma_o \sim 700$, in high energy laboratories. [7]

We stress that lifetime dilation of a particle due to constant-acceleration is a pure kinematic property and is independent of detailed interactions of particles. To discuss the lifetime dilation with constant-linear-accelerations for particles at rest in the CLA frame $F(w, x, y, z)$, we assume that the form of the exponential decay law (13) still holds to a good approximation: [8]

$$N = N_o e^{-w/w_r}, \tag{15}$$

in a linearly accelerated frame. Now instead of using the Lorentz transformations, we use their generalization, i.e., the Wu transformation (3) with ${x_o}^\mu = 0$, and express w in terms of the spacetime variables in the inertial laboratory frame $F_I(x_I^\mu)$:

$$w = \frac{1}{\alpha_o}\left(\frac{\left[w_I + \frac{\beta_o}{\alpha_o\gamma_o}\right]}{\sqrt{\left(x + \frac{1}{\alpha_o\gamma_o^2}\right)^2 + \left(w_I + \frac{\beta_o}{\alpha_o\gamma_o}\right)^2}} - \beta_o\right)$$

$$= \frac{1}{\alpha_o}\left(\left[1 - \frac{(x + 1/\alpha_o{\gamma_o}^2)^2}{(x_I + 1/\alpha_o\gamma_o)^2}\right]^{1/2} - \beta_o\right), \quad x = \text{constant} \tag{16}$$

where $(x, 0, 0)$ denotes the particle's fixed position in the CLA frame $F(x^\mu)$.

In the experiment of particle decay in flight, one measures the number of particle decay as a function of position x_I rather than time w_I in an inertial laboratory. From (15) and (16), we obtain the generalized decay law for unstable particles moving with constant-linear-accelerations:

$$N = N_o \exp\left[-\frac{1}{\alpha_o w_r}\left(\left[1 - \frac{(x + 1/\alpha_o{\gamma_o}^2)^2}{(x_I + 1/\alpha_o\gamma_o)^2}\right]^{1/2} - \beta_o\right)\right]. \tag{17}$$

In the derivation of (16) and (17), we have used the inverse Wu transformation (9).

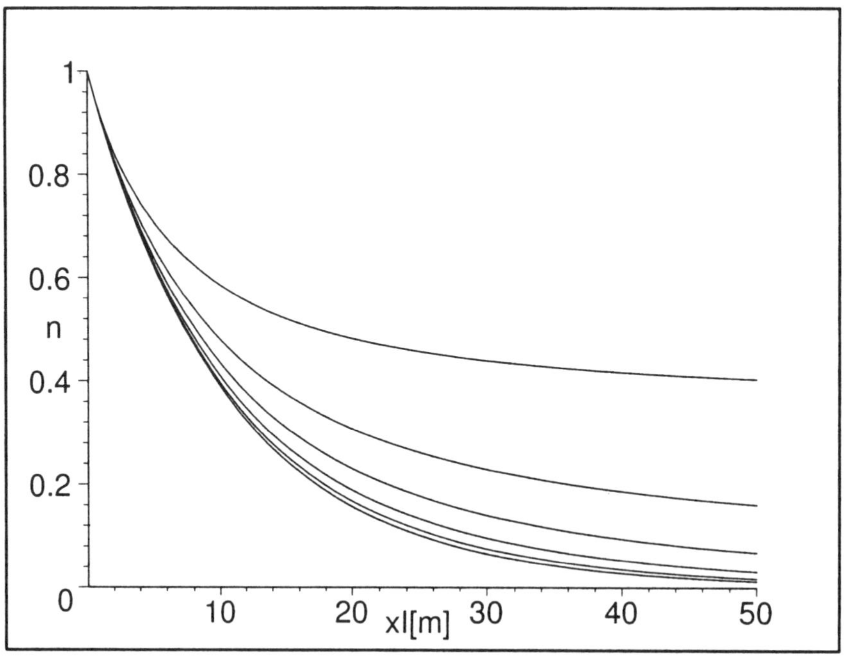

Figure 5: **Particle Decay in Flight with Different Accelerations.** This figure shows the different decays of accelerated pions (mass $m_o = 140MeV$, 'decay length' $D = 7.8m = w_r$, and $x = 0$) with $\beta_0 = 0.8$. The symbol xI[m] denotes the distance x_I measured in meters, n denotes N. From the upper to the lower graph, the constant 'acceleration' F/m_o is respectively given by $1/14m$, $1/35m$, $1/70m$, $1/140m$, $1/280m$, and $1/560m$.

For small α_o, the second equation in (3) with $x_o = 0$ and the approximation $\gamma =$

$\gamma_o(1 + \alpha_o\beta_o{\gamma_o}^2 w)$ lead to the relation for w:

$$w = \frac{1}{\beta_o}\left(\frac{x_I}{\gamma_o} - x - \alpha_o{\gamma_o}^2 x_I x + \alpha_o{\gamma_o}^2 x^2\right).$$

This relation and (9) lead to the following approximate decay formula for small acceleration α_o,

$$N = N_o \exp\left[-\frac{1}{\beta_o w_r}\left(\frac{x_I}{\gamma_o} - x - \alpha_o{\gamma_o}^2 x_I x + \alpha_o{\gamma_o}^2 x^2\right)\right], \qquad \beta_o > 0. \tag{18}$$

Clearly, when $\alpha_o \to 0$ it reduces to the lifetime dilation (15) of special relativity, if the particle is at rest at the origin of the moving frame, i.e., $x = 0$. However, when the location of particles at rest in F is at $x = a \neq 0$, one obtains a new effect of lifetime dilation due to acceleration α_o. Based on (19), one can verify that $N(a)/N(0) \neq N(0)/N(-a)$. This result in a CLA frame suggests that we no longer have the usual translational invariance of inertial frames. This new non-translational invariance in non-inertial frames, suggested by the lifetime dilation for particle decay in flight with acceleration, could be tested by experiments in the future. Of course, the measurements of $N(a), N(0)$ and $N(-a)$ are more difficult than the usual lifetime dilation. We note that the generalized law for the lifetime dilation as shown in the decay law (17) or (19) holds for both particles moving with constant velocity and constant-linear-acceleration.

To see the behavior of the particle decay at different accelerations, let us plot the curves with different values of α. In linear accelerators, the constant acceleration α is expressed in terms of a constant potential drop per unit length,

$$|\Delta\phi/\Delta x| = F_o,$$

we have [9]

$$\alpha_o = \frac{1}{{\gamma_o}^2(m_o/F_o - x)}. \tag{19}$$

Let us use $m_o = 140 MeV/c^2$ and the rest decay length D=7.8 meters for pions at rest in F moving with a constant velocity $\beta_o = 0.8$ for a demonstration. The exponential law for particle decay in flight with different potential drops per meter is shown in figures 5 and 6.

It is important to determine experimentally the correct physical time in accelerated frames before one formulates physical theories in accelerated frames. Therefore, this type of experiment to test physical time in accelerated frames is crucial to free our understanding of physics from the bondage of inertial frames. The experimental results will also motivate physicists to formulate theories in both inertial and non-inertial frames and to extend our view and understanding of the physical world.

The work was supported in part by the Jing Shin Research Fund of the UMassD Foundation and the Potz Science Fund.

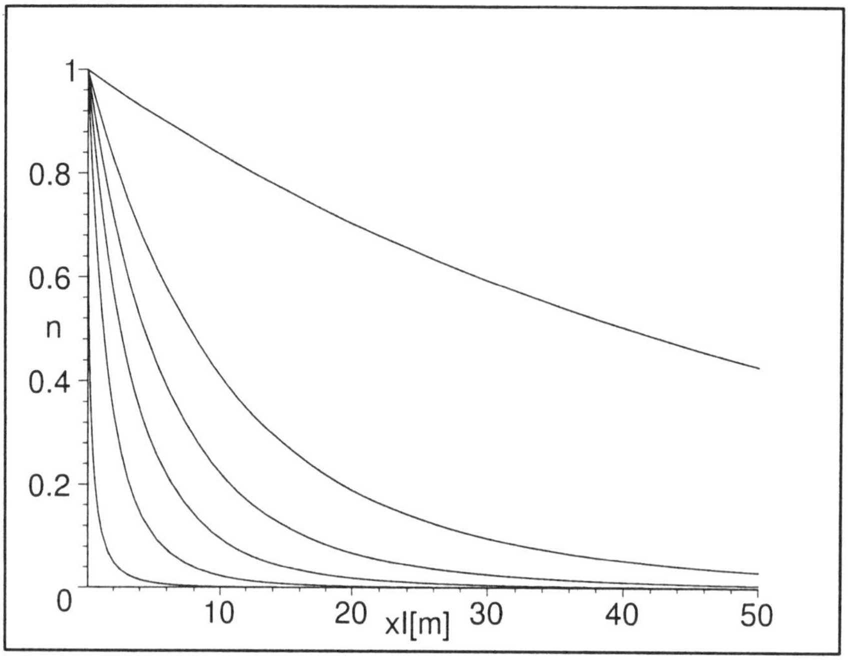

Figure 6: **Particle Decay in Flight with Different Initial Velocities.** This figure shows the different decays of accelerated pions (mass $m_o = 140MeV$, 'decay length' $D = 7.8m = w_r$, and $x = 0$) with the force $F = 1MeV/m$ acting on them. The initial velocity β_0 is, from the upper to the lower graph, given by 0.99, 0.5, 0.4, 0.3, 0.2, and 0.1 respectively.

References

[1] Daniel T. Schmitt and Jong-Ping Hsu, Intern. J. Modern Phys. A (2005, to be published).

[2] Jong-Ping Hsu and Leonardo Hsu, Nuovo Cimento B, **112**, 575 (1997) and Chin. J. Phys. **35**, 407 (1997).

[3] C. Møller, Danske Vid. Sel. Mat.-Fys. **20**, No. 19 (1943); see also *The Theory of Relativity*, (Oxford university press, 1952), Chapter VII; Ta-You Wu and Y. C. Lee, Intern. J. Theoretical Phys. 5, 307 (1972). Ta-You Wu, *Theoretical Physics, vol.4, Theory of Relativity* (Lian Jing Publishing Co., Taipei, 1978) pp. 172-175. T. Fulton and F. Rohrlich, Ann. Phys. **9**, 499 (1960); T. Fulton, F. Rohrlich and L. Witten, Nuovo Cimento XXVI, 652 (1962); E. A. Desloge and R. J. Philpott, Am. J. Phys. **55**, 252 (1987).

[4] Jong-Ping Hsu, *Einstein's Relativity and Beyond - New Symmetry Approaches*, (World Scientific, Singapore, 2000), Chapters 21-23. See also ref. 1.

[5] Jong-Ping Hsu, in *Symposium on the Frontiers of physics at Millennium* (Edited by Yue-Liang Wu and Jong-Ping Hsu, World Scientific, Singapore, 2001) pp. 321-328.

[6] Tobias Kleinschmidt and Jong-Ping Hsu, UMassD preprint (2003).

[7] N. Grossman, K. Heller, *et al.*, Phys. Rev. Lett. **59** (1987) 18.

[8] Explicit calculation of particle's decay at rest based on quantum field theory formulated in non-inertial frames appears to be non-trivial. So far, there is no satisfactory formulation of quantum field theory in non-inertial frames. The assumption of a constant lifetime of a particle at rest in CLA frames can be justified for small accelerations. Furthermore, this constant value of lifetime is not crucial for experimental tests discussed here. The position-dependence of N(x) in equation (17) or (18) is crucial for experimental tests.

[9] Their relation can be obtained by calculating $(dp_I/dx_I)_x$ with x fixed, where $p_I = m_o\gamma_o$. One has $(dp_{I0}/dx_I)_x = |(dp_{I1}/dw_I)_x| = m_o\alpha_o/({\gamma_o}^{-2} + \alpha_o x) = F_o$. See also equation (12).

Chapter 6

Quantum Gravity and 'Ghosts'*

---

*R. P. Feynman, B. S. DeWitt, L. D. Faddeev, V. N. Popov, S. Mendelstam, E. S. Fradkin, I. V. Tyutin, F. J. Dyson

Vol. XXIV (1963) ACTA PHYSICA POLONICA Fasc. 6 (12)

QUANTUM THEORY OF GRAVITATION*

By R. P. Feynman

(Received July 3, 1963)

My subject is the quantum theory of gravitation. My interest in it is primarily in the relation of one part of nature to another. There's a certain irrationality to any work in gravitation, so it's hard to explain why you do any of it; for example, as far as quantum effects are concerned let us consider the effect of the gravitational attraction between an electron and a proton in a hydrogen atom; it changes the energy a little bit. Changing the energy of a quantum system means that the phase of the wave function is slowly shifted relative to what it would have been were no perturbation present. The effect of gravitation on the hydrogen atom is to shift the phase by 43 seconds of phase in every hundred times the lifetime of the universe! An atom made purely by gravitation, let us say two neutrons held together by gravitation, has a Bohr orbit of 10^8 light years. The energy of this system is 10^{-70} rydbergs. I wish to discuss here the possibility of calculating the Lamb correction to this thing, an energy, of the order 10^{-120}. This irrationality is shown also in the strange gadgets of Prof. Weber, in the absurd creations of Prof. Wheeler and other such things, because the dimensions are so peculiar. It is therefore clear that the problem we are working on is not the correct problem; the correct problem is what determines the size of gravitation? But since I am among equally irrational men I won't be criticized I hope for the fact that there is no possible, practical reason for making these calculations.

I am limiting myself to not discussing the questions of quantum geometry nor what happens when the fields are of very short wave length. I am not trying to discuss any problems which we don't already have in present quantum field theory of other fields, not that I believe that gravitation is incapable of solving the problems that we have in the present theory, but because I wish to limit my subject. I suppose that no wave lengths are shorter than one-millionth of the Compton wave length of a proton, and therefore it is legitimate to analyze everything in perturbation approximation; and I will carry out the perturbation approximation as far as I can in every direction, so that we can have as many terms as we want, which means that we can go to ten to the minus two-hundred and something rydbergs.

I am investigating this subject despite the real difficulty that there are no experiments. Therefore there is so real challenge to compute true, physical situations. And so I made

* Based on a tape-recording of Professor Feynman's lecture at the *Conference on Relativistic Theories of Gravitation*, Jabłonna, July, 1962. — Ed.

believe that there were experiments; I imagined that there were a lot of experiments and that the gravitational constant was more like the electrical constant and that they were coming up with data on the various gravitating atoms, and so forth; and that it was a challenge to calculate whether the theory agreed with the data. So that in each case I gave myself a specific physical problem; not a question, what happens in a quantized geometry, how do you define an energy tensor *etc.*, unless that question was necessary to the solution of the physical problem, so please appreciate that the plan of the attack is a succession of increasingly complex physical problems; if I could do one, then I was finished, and I went to a harder one imagining the experimenters were getting into more and more complicated situations. Also I decided not to investigate what I would call familiar difficulties. The quantum electrodynamics diverges; if this theory diverges, it's not something to be investigated unless it produces any specific difficulties associated with gravitation. In short, I was looking entirely for unfamiliar (that is, unfamiliar to meson physics) difficulties. For example, it's immediately remarked that the theory is non-linear. This is not at all an unfamiliar difficulty; the theory, for example, of the spin 1/2 particles interacting with the electromagnetic field has a coupling term $\bar{\psi}A\psi$ which involves three fields and is therefore non-linear; that's not a new thing at all. Now, I thought that this would be very easy and I'd just go ahead and do it, and here's what I planned. I started with the Lagrangian of Einstein for the interacting field of gravity and I had to make some definition for the matter since I'm dealing with real bodies and make up my mind what the matter was made of; and then later I would check whether the results that I have depend on the specific choice or they are more powerful. I can only do one example at a time; I took spin zero matter; then, since I'm going to make a perturbation theory, just as we do in quantum electrodynamics, where it is allowed (it is especially more allowed in gravity where the coupling constant is smaller), $g_{\mu\nu}$ is written as flat space as if there were no gravity plus $\varkappa$ times $h_{\mu\nu}$, where $\varkappa$ is the square root of the gravitational constant. Then, if this is substituted in the Lagrangian, one gets a big mess, which is outlined here.

$$\mathcal{L} = \frac{1}{\varkappa^2}\int R\sqrt{g}\,d\tau + \frac{1}{2}\int \left(\sqrt{g}\,g^{\mu\nu}\varphi_{,\mu}\varphi_{,\nu} - m^2\sqrt{g}\,\varphi^2\right)d\tau \tag{1}$$

$$g_{\mu\nu} = \delta_{\mu\nu} + \varkappa h_{\mu\nu}.$$

Substituting and expanding, and simplifying the results by a notation (a bar over a tensor means

$$\bar{x}_{\mu\nu} \equiv \frac{1}{2}\left(x_{\mu\nu} + x_{\nu\mu} - \delta_{\mu\nu}x_{\sigma\sigma}\right);$$

notice that if $x_{\mu\nu}$ is symmetric, $\bar{\bar{x}}_{\mu\nu} = x_{\mu\nu}$) we get

$$\mathcal{L} = \int \left(h_{\mu\nu,\sigma}\bar{h}_{\mu\nu,\sigma} - 2\bar{h}_{\mu\sigma,\sigma}\bar{h}_{\mu\sigma,\sigma}\right) + \frac{1}{2}\int \left(\varphi_{,\mu}^2 - m^2\varphi^2\right)d\tau +$$

$$+ \varkappa\int\left(\bar{h}_{\mu\nu}\varphi_{,\mu}\varphi_{,\nu} - m^2\frac{1}{2}h_{\sigma\sigma}\varphi^2\right) + \varkappa\int \text{“}hhh\text{”} + \varkappa^2\int \text{“}hh\varphi\varphi\text{”} + \text{etc.} \tag{2}$$

First, there are terms which are quadratic in h; then there are terms which are quadratic in φ, the spin zero meson field variable; then there are terms which are more complicated than quadratic; for example, here is a term with two φ's and one h, which I will write $h\varphi\varphi$ (I have written that one out, in particular); there are terms with three h's; then there are terms which involve two h's and two φ's; and so on and so on with more and more complicated terms. The first two terms are considered as the free Lagrangian of the gravitational field and of the matter.

Now we look first at what we would want to solve problem classically, we take the variation of this with respect to h, from the first term we produce a certain combination of second derivatives, and on the other side a mess involving higher orders than first. And the same with the φ, of course.

$$h_{\mu\nu,\sigma\sigma}-\bar{h}_{\sigma\nu,\sigma\mu}-\bar{h}_{\sigma\mu,\sigma\nu} = \bar{S}_{\mu\nu}(h,\varphi) \tag{3}$$

$$\varphi_{,\sigma\sigma}-m^2\varphi = \chi(\varphi, h). \tag{4}$$

We will speak in the following way: (3) is a wave equation, of which $S_{\mu\nu}$ is the source, just like (4) is the wave equation of which χ is the source. The problem is to solve those equations in succession, and to use the usual methods of calculation of the quantum theory. Inasmuch as I wanted to get into the minimum of difficulties, I just took a guess that I use the same plan as I do in electricity; and the plan in electricity leads to the following suggestion here: that if you have a source, you divide by the operator on the left side of (3) in momentum space to get the propagator field. So I have to solve this equation (3). But as you all know it is singular; the entire Lagrangian in the beginning was invariant under a complicated transformation of g, which in the form of h is the following; if you add to h a gradient plus more, the entire system is invariant:

$$h'_{\mu\nu} = h_{\mu\nu}+2\xi_{\mu,\nu}+2h_{\mu\sigma}\,\xi_{\sigma,\nu}+\xi_\sigma\,h_{\mu\nu,\sigma}, \tag{5}$$

where ξ_μ is arbitrary, and μ and ν should be made symmetric in all these equations. As a consequence of this same invariance in the complet Lagrangian one can show that the source $S_{\mu\nu}$ must have zero divergence $S_{\mu\nu,\nu} = 0$. In fact equations (3) would not be consistent without this condition as can be seen by barring both sides and taking the divergence — the left side vanishes identically. Now, because of the invariance of the equations, in the same way that the Maxwell equations cannot be solved to get a unique vector potential — so these can't be solved and we can't get a unique propagator. But because of the invariance under the transformation some arbitrary choice of a condition on $h_{\mu\nu}$ can be made, analogous to the Lorentz condition $A_{\mu,\mu}=0$ in quantum electrodynamics. Making the simplest choice which I know, I make choice $\bar{h}_{\mu\sigma,\sigma} = 0$. This is four conditions and I have free the four variables ξ_μ that I can adjust to make the condition satisfied by $h'_{\mu\nu}$. Then this equation (3) is very simple, because two terms in (3) fall away and all we have is that the d'Alemberian of h is equal to S. Therefore the generating field from a source $S_{\mu\nu}$ will equal the $\bar{S}_{\mu\nu}$ times $1/k^2$ in Fourier series, where k^2 is the square of the frequency, wave vector; the time part might be called the frequency ω, the space part $\boldsymbol{k}$. This is the analogue of the equation in electricity that says that the field is $1/k^2$ times the current. In the method of quantum field theory, you have a source which generates something, and that may interact later with something else; the iteraction, of course, is $S_{\mu\nu}\, h_{\mu\nu}$; so that, I say, one source may create a potential which acts on another source. So, to take the very simplest example of two interacting systems, let's say S and S', the result would be the following: h would be generated by $S_{\mu\nu}$, and then it would interact with $S'_{\mu\nu}$, so we would get for the interaction of two systems, of two particles, the fundamental interaction that we investigate

$$\varkappa^2 \bar{S}_{\mu\nu} \frac{1}{k^2} S_{\mu\nu}. \tag{6}$$

This represents the law of gravitational interaction expressed by means of an interchange of a virtual graviton. To understand the theory better and to see how far we already arrived we expand it out in components. Let index 4 represent the time, and 3 the direction of **k**, so that 1 and 2 are transverse. The condition $\mathrm{k}_\mu S_{\mu\nu} = 0$ becomes $\omega S_{4\nu} = \mathrm{k} S_{3\nu}$ where k is the magnitude of **k**. Using this, many of the terms involving number 3 component of S can be replaced by terms in number 4 components. After some rearranging there results

$$-2\bar{S}_{\mu\nu} \frac{1}{k^2} S_{\mu\nu} = \frac{1}{\mathrm{k}^2}\,[S_{44} S'_{44}] + \frac{1}{\mathrm{k}^2}\,[S_{44}(S'_{11} + S'_{22}) + S'_{44}(S_{11} + S_{22}) +$$

$$+ S_{43} S'_{43} - 4 S_{41} S'_{41} - 4 S_{42} S'_{42}] + \frac{1}{\mathrm{k}^2 - \omega^2 + i\varepsilon}\,[(S_{11} - S_{22})\,(S'_{11} - S'_{22}) + 4 S_{12} S'_{12}]. \tag{7}$$

There is a singular point in the last term when $\omega = \mathrm{k}$, and to be precise we put in the $+\,i\varepsilon$ as is well-known from electrodynamics. You note that in the first two terms instead of one over a four-dimensional $\omega^2 - \mathbf{k}^2$ we have here just $1/\mathbf{k}^2$, the momentum itself. S_{44} is the energy density, so this first term represents the two energy densities interacting with no ω dependence which means, in the Fourier transform an interaction instantaneous in time; and $1/\mathbf{k}^2$ means $1/r$ in space, so there's an instantaneous $1/r$ interaction between masses, Newton's law. In the next term there's another instantaneous term which says that Newton's mass law should be corrected by some other components analogous to a kind of magnetic interaction (not quite analogous because the magnetic interaction in electricity already involves a $\mathrm{k}^2 - \omega^2 + i\varepsilon$ propagator rather than just k^2. But the $\mathrm{k}^2 - \omega^2 + it$ in gravitation comes even later and is a much smallei term which involves velocities to the fourth). So if we really wanted to do problems with atoms that were held together gravitationally it would be very easy; we would take the first term, and possibly even the second as the interaction. Being instantaneous, it can be put directly into a Schrödinger equation, analogous to the e^2/r term for electrical interaction. And that take care of gravitation to a very high accuracy, without a quantized field theory at all. However, for still higher accuracy we have to do the radiative corrections, which come from the last term.

Radiation of free gravitons corresponds to the situation that there is a pole in the propagator. There is a pole in the last term when $\omega = \mathrm{k}$, of course, which means that the wave number and the frequency are related as for a mass zero particle. The residue of the pole, we see, is the product of two terms; which means that there are two kinds of waves, one generated by $S_{11} - S_{22}$ and the other generated by S_{12}, and so we have two kinds of transverse polarized waves, that is there are two polarization states for the graviton. The linear combination $S_{11}-S_{22}\pm 2iS_{12}$ vary with angle Θ of rotation in the 1—2 plane as $e^{\pm 2i\theta}$ so the gravitaton has spin 2, component ± 2 along direction of polarization. Everything is clear directly from the expression (7); I just wanted to illustrate that the propagator (6) of quantum mechanics and all that we know about the classical situation are in evident coincidence.

In order to proceed to make specific calculations by means of diagrams, beside the propagator we need to know just what the junctions are, in other words just what the S's are for a particular problem; and I shall just illustrate how that's done in one example. It is done by looking at the non-quadratic terms in the Lagrangian I've written one out completely. This one has an h and two φ's in the Lagrangian (2). The rules of the quantum mechanics for writting this thing are to look at the h and two φ's: one φ each refers to the in and out particle, and the one h corresponds to the graviton; so we immediately see in that term a two particle interaction through a graviton (see Fig. 1). And we can immediately

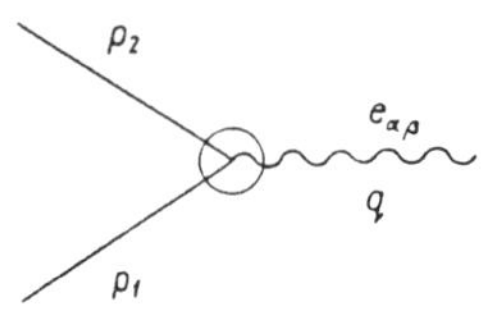

Fig. 1

read off the answer for the interaction this way: if the p_1 and p_2 are the momenta of the particles and q the momentum of the graviton; and $e_{\alpha\beta}$ is the polarization tensor of the plane wave representing the graviton, that is $h_{\alpha\beta}=e_{\alpha\beta}\, e^{iq\cdot x}$, the Fourier expansion of this term gives the amplitude for the coupling of two particles to a graviton

$$p^1_\mu p^2_\nu \bar{e}_{\mu\nu}-\frac{1}{2}m^2 e_{\sigma\sigma}\,. \tag{8}$$

So this is a coupling of matter to gravity; it is first order, and then there are higher terms; but the point I'm trying to make is that there is no mystery about what to write down — everything is perfectly clear, from the Lagrangian. We have the propagator, we have the couplings, we can write everything. A term like hhh implies a definite formula for the interaction of three gravitons; it is very complicated, and I won't write it down, but you can read it right off directly by substituting momenta for the gradients. That such a term exists is, of course, natural, because gravity interacts with any kind of energy, including its own, so if it interacts with an object-particles it will interact with gravitons; so this is the scattering of a graviton in a gravitational field, which must exist. So that everything is directly readable and all we have to do is proceed to find out if we get a sensible physics. I've already indicated that the physics of direct interactions is sensible; and I go ahead now to compute a number of other things.

To take just one example, we compute the Compton effect, or the analogue rather, of the Compton effect, in which a graviton comes in and out on a particle. The amplitude for this is a sum of terms corresponding to the diagrams of Fig. 2. The amplidute for the first diagram of Fig. 2 is the coupling (8) times the propagator for the intermediate meson which reads $(p^2-m^2)^{-1}$, which is the Fourier transform of the equation (4) which is the propagation of the spin zero particle. Then there is another coupling of the same form as (8). We multiply these together, to get the amplitude for that diagram

$$\left(p^2_\mu p_\nu \bar{e}^b_{\mu\nu} - \frac{1}{2} e_{\mu\mu} m^2\right) \frac{1}{p^2-m^2} \left(p_\sigma p^1_\tau e^a_{\sigma\tau} - \frac{1}{2} \bar{e}_{\sigma\sigma} m^2\right),$$

where we should substitute $p = p^2+q^b = p^1+q^a$. Then you must add similar contributions from the other diagrams.

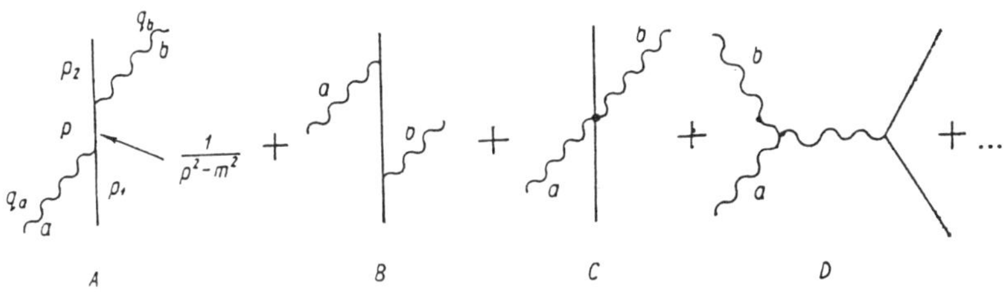

Fig. 2

The third one comes in because there are terms with two h's and two φ's in the Lagrangian. One adds the four diagrams together and gets an answer for the Compton effect. It is rather simple, and quite interesting; that it is simple is what is interesting, because the labour is fantastic in all these things.

But the thing I would like to emphasize is this; in this problem we used a certain wave $e^a_{\alpha\beta}$ for the incoming graviton number "a" say; the question is could we use a different one? According to the theory, it should really be invariant under coordinate transformations and so on, but what it corresponds to here is the analogue of gauge invariance, that you can add to the potential a gradient (see (5)). And therefore it should be that if I changed $e_{\alpha\beta}$ of a particular graviton to $e_{\alpha\beta}+q_\alpha\xi_\beta$ where ξ is arbitrary, and q_α is the momentum of the graviton, there should be no change in the physics. In short, the amplitude should be unchanged; and it is. The amplitude for this particular process is what I call gauge-invariant, or coordinate-transforming invariant. At first sight this is somewhat puzzling, because you would have expected that the invariance law of the whole thing is more complicated, including the last two terms in (5), which I seem to have omitted. But those terms have been included; you see asymptotically all you have to do is worry about the second term, the last two in h's times ξ's are in fact generated by the last diagram, Fig. 2D; when I put a gradient in here for this one, what this means is if I put for the incoming wave a pure gradient, I should get zero. If I put the gradient $q_\alpha\xi_\beta$ in for $e^a_{\alpha\beta}$ on this term D, I get a coupling between ξ and the other field $e^a_{\alpha\beta}$ because of the three graviton coupling. The result, as far as the matter line is concerned is that it is acted on in first order by a resultant field $e^b_{\mu\sigma}\xi_\sigma q^a_\nu + \frac{1}{2} q^b_\sigma e_{\mu\nu}\xi_\sigma$ which is just the last two terms in (5). The rule is that the field which acts on the matter itself must be invariant the way described by (5); but here in Fig. 2 I've already calculated all the corrections, the generator and all the necessary non-linear modifications if I take all the diagrams into account. In short, asymptotically far away if I include all kinds of diagrams such as D, the invariance need be checked only for a pure gradient added to an incoming wave. It takes care of the non-linearities by calculating them through the interaction.

I would like, now, to emphasize one more point that is very important for our later discussion. If I add a gradient, I said, the result was zero. Let's call a the one graviton coming in and b the other one in every diagram. The result is zero if I use a gradient for a, only if b is a free graviton with no source; that is if it is either really an honest graviton with $(q^b)^2 = 0$, or a pure potential, which is a solution of the free wave equation. That is unlike electrodynamics, where the field b could have been any potential at all and adding a gradient to a would have made no difference. But in gravity, it must be that b is a pure wave; the reason is very simple. There is no way to avoid this by changing any propagators; this is not a disease — there is a physical reason. The reason can be seen as follows: If this b had a source let me modify my diagrams to show the source of b, suppose some other matter particle made the b, so we add onto each b line a matter line at the end, like Fig. 3a. (*E.g.* Fig. 2a becomes Fig. 3b *etc.*)

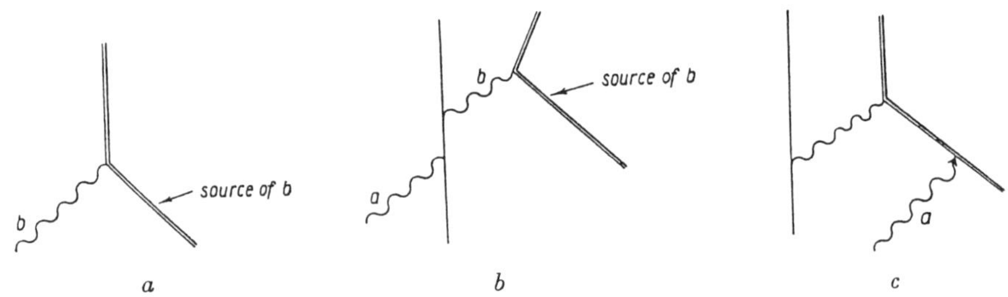

Fig. 3

Now, if b isn't a free wave, but it had a source, the situation is this. If this "a" field is taken as a gradient field which operates everywhere on everything in the diagram it should give zero. But we forgot something; there's another type of diagram, if the "a" is supposed to act on everything, one of which looks like Fig. 3c, in which the "a" itself acts on the source of b and then b comes over to interact with the original matter. In other words, among all the diagrams where there is a source, there's also these of type 3c. The sum of all diagrams is zero; but the sum of those like Fig. 2 without those of type 3c is not zero, and therefore if I were to just calculate the diagrams of Fig. 2 and forget about the source of b and then put a gradient in for "a" the result cannot be zero, but must be getting ready to cancel the terms from the likes of 3c when I do it right. That will turn out to be an important point to emphasize. I have done a lot of problems like this, without closed loops but I won't bore you with all the problems and answers; there's nothing new, I mean nothing interesting, in the sense that no apparent difficulties arise.

However, the next step is to take situations in which we have what we call closed loops, or rings, or circuits, in which not all momenta of the problem are defined. Let me just mention something. I've analyzed this method both by doing a number of problems, and by a mathematical high-class elegant technique — I can do high class mathematics too, but I don't believe in it, that's the difference. I have to check it in a problem. I can prove that no matter how complicated the problem is, if you take it in the order in which there are no rings, in which every momentum is determined, the invariance is satisfied, the system is independent of what choice I made of gauge and of the propagator I made in the beginning; and everything is all right, there are no difficulties. I emphasize that this contains all the classical cases, and so I 'm really saying there are no difficulties in the classical gravitation theory. This is not meant as a grand discovery, because after all, you've been worrying about all these difficulties that I say don't exist, but only for you to get an idea of the calibration — what I mean by difficulties! If we take the next case, let's say the interaction of two particles in a higher order, then you get diagrams of which I'll only begin to write a few of them. One that looks like this in which two gravitons are exchanged,

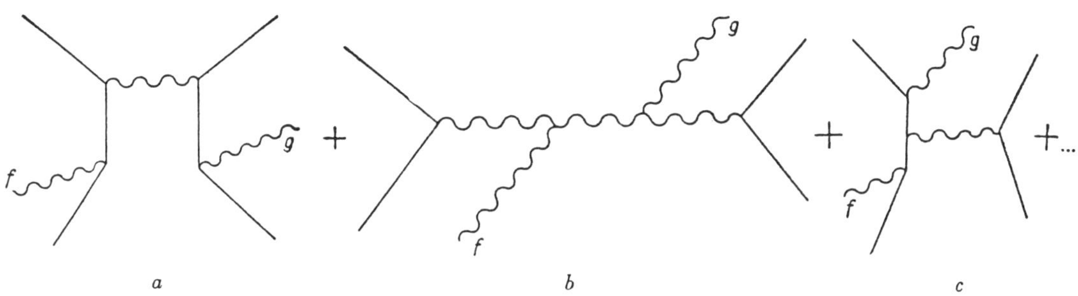

Fig. 4

or, for instance, a graviton gets split into two gravitons and then come back — these are only the beginning of a whole series of frightening-looking pictures, which correspond to the problem of calculating the Lamb shift, or the radiative corrections to the hydrogen atom. When I tried to do this, I did it in a straightforward way, following all the rules, putting in the propagator $1/k^2$, and so on. I had some difficulties, the thing didn't look gauge invariant but that had to do with the way I was making the cutoffs, because the stuff is infinite. Shortage of time does'nt permit me to explain the way I got around all those things, because in spite of getting around all those things the result is nevertheless definitely incorrect. It's gauge-invariant, it's perfectly O.K. looking, but it is definitely incorrect. The reason I knew it was incorrect is the following. In order to get it gauge-invariant, I had to do a lot of pushing and pulling, and I got the feeling that the thing might not be unique. I figured that maybe somebody else could do it another way or something, and I was rather suspicious, so I tried to get more tests for it; and a student of mine, by the name of Yura, tested to see if it was unitary; and what that means is the following: Let me take instead of this scattering problem, a problem of Fig. 4 in which time runs vertically, a problem which gives the same diagrams but in which time is running horizontally, which is the annihilation of a pair, to produce another pair, and we are calculating second order corrections to that problem. Let's suppose for simplicity that in the final state the pair is in the same state as before.

Then, adding all these diagrams gives the amplitude that if you have a pair, particle and antiparticle, they annihilate and recreate themselves; in other words it's the amplitude that the pair is still in the same state as a function of time. The amplitude to remain in the same state for a time T in general is of the form

$$e^{-i\left(E_0-i\frac{\gamma}{2}\right)T}$$

you see that the imaginary part of the phase goes as $e^{-\frac{\gamma}{2}T}$; which means that the probability of being in a state must decrease with time. Why does the probability decrease in time? Because there's another possibility, namely, these two objects could come together, annihilate, and produce a real pair of gravitons. Therefore, it is necessary that this decay rate of the closed loop diagrams in Fig. 4 that I obtain by directly finding the imaginary part of the sum agrees with another thing I can calculate independently, without looking at the closed loop diagrams. Namely, what is the rate at which a particle and antiparticle annihilate into two gravitons? And this is very easy to calculate (same set of diagrams as Fig. 2, only turned on its side). I calculated this rate from Fig. 2, checked whether this rate agrees with the rate at which the probability of the two particles staying the same decreases (imaginary part of Fig. 4), and it does not check. Somethin'gs the matter.

This made me investigate the entire subject in great detail to find out what the trouble is. I discovered in the process two things. First, I discovered a number of theorems, which as far as I know are new, which relate closed loop diagrams and diagrams without closed loop diagrams (I shall call the latter diagrams "trees"). The unitarity relation which I have just been describing, is one connection between a closed loop diagram and a tree; but I found a whole lot of other ones, and this gives me more tests on my machinery. So let me just tell you a little bit about this theorem, which gives other rules. It is rather interesting. As a matter of fact, I proved that if you have a diagram with rings in it there are enough theorems altogether, so that you can express any diagram with circuits completely in terms of diagrams with trees and with all momenta for tree diagrams in physically attainable regions and on the mass shell. The demonstration is remarkably easy. There are several ways of demonstrating it; I'll only chose one. Things propagate from one place to another, as I said, with amplitude $1/k^2$. When translated into space, that's a certain propagation function which you might call $K_+(1, 2)$, a function of two positions, 1, 2, in space-time. It represents, in the past, incoming waves and in the future, it represents outgoing waves; so you have waves come in and out; and that's the conventional propagator, with the $i\varepsilon$ and so on, as usually represented. However, this is only a solution of the propagators's equation, the wave equation I mean; it is a special solution, as you all know. There are other solutions; for instance there is a solution which is purely retarded, which I'll call K_{ret} and which exists only inside the future light-cone. Now, if you have two Green's functions for the same equation they must differ by some solution of the homogeneous equation, say K_x. That means K_x is a solution of the free wave equation and $K_+ = K_{\text{ret}} + K_x$. In a ring like Fig. 4a we have a whole product of these K_+'s. For example, for four points 1, 2, 3, 4 in a ring we have a product like this: $K_+(1, 2)K_+(2, 3)K_+(3, 4)K_+(4, 1)$ (all K's are not the same, some of them belong to the gravitons and some are propagators for the particles and so on).

706

But now let us see what happens if we were to replace one (or more) of these K_+ by K_x, say $K_+(1,2)$ is $K_x(1,2)$? Then between 1, 2 we have just free particles, you've broken the ring; you've got an open diagram, because K_x is free wave solution, and this means it's an integral over all real momenta of free particles, on the mass shell and perfectly honest. Therefore if we replace one of K_+ by K_x then that particular line is opened; and the process is changed to one in which there is a forward scattering of an extra particle; there's a fake particle that belongs to this propagator that has to be integrated over, but it's a free diagram — it is now a tree, and therefore perfectly definite and unique to calculate. But I said that I could open every diagram; the reason is this. First I note that if I put $K_{\rm ret}$ for every K in a ring, I get zero

$$K_{\rm ret}(1,2)K_{\rm ret}(2,3)K_{\rm ret}(3,4)K_{\rm ret}(4,1) = 0 \tag{9}$$

for to be non zero t_1 must be greater than t_2, $t_2 > t_3$, $t_3 > t_4$ and $t_4 > t_1$ which is impossible. Now make the substitution $K_{\rm ret} = K_+ - K_x$ in (9). You get either all K_+ in each factor, which is the closed loop we want; or at least one K_x, which are represented by tree diagrams. Since the sum is zero, closed loops can be represented as integrals over tree diagrams. I was surprised I had never noticed this thing before.

Well, then I checked whether these diagrams of Fig. 4 when opened into trees agreed with the theorem. I mean I hoped that the theorem proved for other meson theories would agree in principle for the gravity case, such that on opening a virtual graviton line the tree would correspond to forward scattering of free graviton waves. And it does not work in the gravity case. But, you say, how could it fail, after you just demonstrated that it ought to work? The reason it fails is the following: This argument has to do with the position of the poles in the propagators; a typical propagator is a factor $1/(k^2-m^2+i\varepsilon)$, the $+i\varepsilon$ due to the poles, and all I'm doing here is changing the rule about the poles and picking up an extra delta function $\delta(k^2-m^2)$ as a consequence, which is the free wave coming in and out. What I want these free waves to represent in the gravity case are physical gravitons and not something wrong. They do represent waves of $q^2 = 0$ of course, but, as it turns out, not with the correct polarization to be free gravitons. I'd like to show it. It has to do with the numerator, not the denominator. You see the propagator that I wrote before, which was $S_{\mu\nu}$ times $1/(k^2+i\varepsilon)$ times $\overline{S'_{\mu\nu}}$, is being replaced by $S_{\mu\nu}\,\delta(q^2)\overline{S'_{\mu\nu}}$. Now when I make $q^2 = 0$ I have a free wave instead of arbitiary momentum. This should be a real graviton or else there's going to be physical trouble. It ins't; although it is of zero momentum, it is not transverse. It does not make any difference in understanding the point so forget one index in $S_{\mu\nu}$ — it's a lot of extra work to carry the other index so just imagine there's one index: $S_\mu S_\mu\,\delta(q^2)$. This combination $S_\mu S'_\mu$, is $S_4S'_4-S_3S'_3-S_1S'_1-S_2S'_2$, where 4 is the time and 3 is the direction, say, of momentum of the four-vector q. Then 1 and 2 are transverse, and those are the only two we want. (Please appreciate I iemoved one index — I can make it more elaborate, but it is the same idea.) That is we want only $-S_1S'_1-S_2S'_2$ instead of the sum over four. Now what about this extra term $S_4S'_4-S_3S'_3$? Well, it is S_4-S_3 times $S'_4+S'_3$ plus S_4+S_3 times $S'_4-S'_3$. But S_4-S_3 is proprtional to $q_\mu S_\mu$ (suppressing one index) because q_4 in this notation is the frequency and equals q_3, if we assume the 3-direction is the direction of the momentum. So S_4-S_3 is the response of the system to a gradient potential, which we proved was zero in our invariance discussion. Therefore, we have shown $(S_4-S_3)/(S_4'+S_3')=0$ and this should be accounted for by purely transverse wave contributions. But it ins't, and it isn't because *the proof that the response to a gradient potential is zero required that the other particle that was interacting was an honest free graviton.* And four plus three in $S_4'+S_3'$ is not honest — it's not transverse, it is not a correct kind of graviton. You see, the only way you can get a polarization 4+5 going in the 4—3 direction is to have what I call longitudinal response; it's not a transverse wave. Such a wave could only be generated by an artificial source here of some silly kind; it is not a free wave. When there's an artificial source for one graviton, even the another is a pure gradient, the sum of all the diagrams does not give zero. If the beam is not exactly that of a free wave, perfectly transverse and everything, the argument that the gradient has to be zero must fail, for the reason outlined previously.

Although this gradient for S_4-S_3 is what I want and I hoped it was going to be zero I forgot that the other end of it — $S_4'+S_3'$ is a funny wave which is not a gradient, and which is not a free wave — and therefore you do not get zero and should not get zero, and something is fundamentally wrong.

Incidentally I investigated further and discovered another very interesting point. There is another theory, more well-known to meson physicists, called the Yang-Mills theory, and I take the one with zero mass; it is a special theory that has never been investigated in great detail. It is very analogous to gravitation; instead of the coordinate transformation group being the source of everything, it's the isotopic spin rotation group that's the source of everything. It is a non-linear theory, that's like the gravitation theory, and so forth. At the suggestion of Gell-Mann I looked at the theory of Yang-Mills with zero mass, which has a kind of gauge group and everything the same; and found exactly the same difficulty. And therefore in meson theory it was not strictly unknown difficulty, because it should have been noticed by meson physicists who had been fooling around the Yang-Mills theory. They had not noticed it because they're practical, and the Yang-Mills theory with zero mass obviously does not exist, because a zero mass field would be obvious; it would come out of nuclei right away. So they didn't take the case of zero mass and investigate it carefully. But this disease which I discovered here is a disease which exist in other theories. So at least there is one good thing: gravity isn't alone in this difficulty. This observation that Yang-Mills was also in trouble was of very great advantage to me; it made everything much easier in trying to straighten out the troubles of the preceding paragraph, for several reasons. The main reason is if you have two examples of the same disease, then there are many things you don't worry about. You see, if there is something different in the two theories it is not caused by that. For example, for gravity, in front of the second derivatives of $g_{\mu\nu}$ in the Lagrangian there are other g's, the field itself. I kept worrying something was going to happen from that. In the Yang-Mills theory this is not so, that's not the cause of the trouble, and so on. That's one advantage — it limits the number of possibilities. And the second great advantage was that the Yang-Mills theory is enormously easier to compute with than the gravity theory, and therefore I continued most of my investigations on the Yang-Mills theory, with the idea, if I ever cure that one, I'll turn around and cure the other. Because I can demonstrate one thing; line for line it's a translation like music transcribed to a different score; everything has its analogue precisely, so it is a very good example to work with. Incidentally, to give you some idea of the difference in order to calculate this diagram Fig. 4b the Yang-Mills case took me about a day; to calculate the diagram in the case of gravitation I tried again and again and was never able to do it; and it was finally put on a computing machine—I don't mean the arithmetic, I mean the algebra of all the terms coming in, just the algebra; I did the integrals myself later, but the algebra of the thing was done on a machine by John Matthews, so I couldn't have done it by hand. In fact, I think it's historically interesting that it's the first problem in algebra that I know of that was done on a machine that has not been done by hand.

Well, what then, now you have the difficulty; how do you cure it? Well I tried the following idea: I assumed the tree theorem to be true, and used it in reverse. If every closed ring diagram can be expressed as trees, and if trees produce no trouble and can be computed, then all you have to do is to say that the closed loop diagram is the sum of the corresponding tree diagrams, that it should be. Finally in each tree diagram for which a graviton line has been opened, take only real transverse graviton to represent that term. This then serves as the definition of how to calculate closed-loop diagrams; the old rules, involving a propagator $1/k^2+i\varepsilon$ *etc.* being superseded. The advantage of this is, first, that it will be gauge invariant, second, it will be unitary, because unitarity is a relation between a closed diagram and an open one, and is one of the class of relations I was talking about, so there's no difficulty. And third, it's completely unique as to what the answer is; there's no arbitrary fiddling around with different gauges and so forth, in the inside ring as there was before. So that's the plan.

Now, the plan requires, however, one more point. It's true that we proved here that every ring diagram can be broken up into a whole lot of trees; but, a given tree is not gauge invariant. For instance the tree diagram of Fig. 2A is not. Each one of the four diagrams of Fig. 2 is not gauge-invariant, nor is any combination of them except the sum of all four. So the thing is the following. Suppose I take all the processes, all of them that belong together in a given order; for example, all the diagrams of fourth order, of which Fig. 4 illustrates three; I break the whole mess into trees, lots of trees. Then I must gather

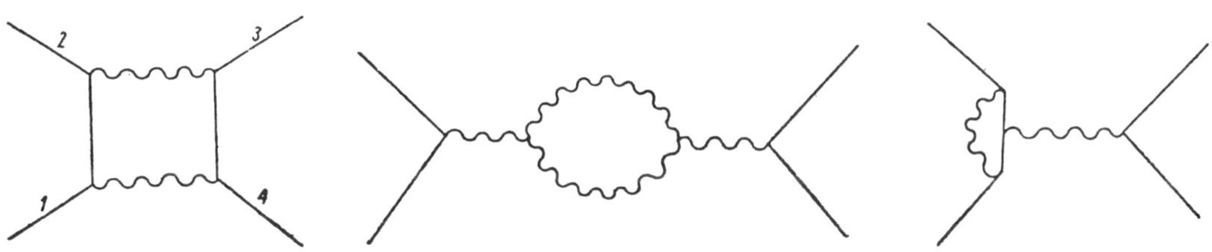

Fig. 5

the trees into baskets again, so that each basket contains the total of all of the diagrams of some specific process (for example the four diagrams of Fig. 2), you see, not just some particular tree diagram but the complet set for some process. The business of gathering the tree diagrams together in bunches representing all diagrams for complet processes is important, for only such a complet set is gauge invariant. The question is: Will any odd tree diagrams be left out or can they all be gathered into processes? The question is: Can we express the closed ring diagrams for some process into a sum over various other processes of tree diagrams for these processes?

Well, in the case with one ring only, I am sure it can be done, I proved it can be done and I have done it and it's all fine. And therefore the problem with one ring is fundamentally solved; because we say, you express it in terms of open parts, you find the processes that they correspond to, compute each process and add them together.

You might be interested in what the rule is for one ring; it's the sum of several pieces: first it is the sum of all the processes which you get in the lower order, in which you scatter one extra particle from the system. For instance, in Fig. 4 we have the rings for two particles scattering. There is no external graviton but there are two internal ones; now we compute in the same order a new problem in which there are two particles scattering, but while that's happening another particle, for example a graviton scatters forward. Some of the diagrams for this are illustrated in Fig. 5. State f the same state as g; so another graviton comes in and is scattered forward. In other words we do the forward scattering of an extra graviton. In addition, from breaking matter lines we have terms for the forward scattering of an extra positron, plus the forward scattering of an extra electron, and so on; one adds the forward scattering of every possible extra particle together. That is the first contribution. But when you break up the trees, you also sometimes break two lines, and then you get diagrams like Fig. 6 with two extra particles scattering (here a graviton and electron) so it turns out you must now subtract all the diagrams with two extra particles of all kinds scattering. Then add all diagrams with 3 extra particles scattering and so on. It's a nice rule, its's quite beautiful; it took me quite a while to find; I have other proofs for orther cases that are easy to understand.

Now, the next thing that anybody would ask which is a natural, interesting thing to ask, is this. Is it possible to go back and to find the rule by which you could have integrated the closed rings directly? In other words, change the rule for integrating the closed rings, so that when you integrate them in a more natural fashion, with the new method, it will

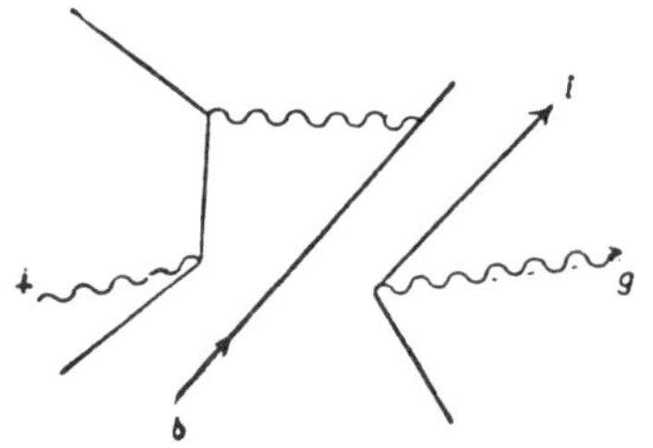

Fig. 6

give the same answer as this unique, absolute, definite thing of the trees. It's not necessary to do this, because, of course, I've defined everything; but it's of great interest to do this, because maybe I'll understand what I did wrong before. So I investigated that in detail. It turns out there are two changes that have to be made — it's a little hard to explain in

710

terms of the gravitation of which I'll only tell about one. Well, I'll try to explain the other, but it might cause some confusion. Because I have to explain in general what I'm doing when I do a ring. Most what it corresponds to is this: first you subtract from the Lagrangian this

$$\int \sqrt{g}\, \bar{H}^{\mu\nu}{}_{;\nu} H^{\sigma}_{\mu;\sigma}\, d\tau .$$

In that way the equation of motion that results is non-singular any more. Let me write what it really is so that there's no trouble. You say to me what is this, there's a g in it and an H in it? Yes. In doing a ring, there's a field variation over which you're integrating, which I call H; and there's a g — which is the representative of all the outside disturbances which can be summarized as being an effective external field g. And so you add to the complicated Lagrangian that you get in the ordinary way an extra term, which makes it no longer singular. That's the first thing; I found it out by trial and error before, when I made it gauge invariant. But then secondly, you must subtract from the answer, the result that you get by imagining that in the ring which involves only a graviton going around, instead you calculate with a different particle going around, an artificial, dopey particle is coupled to it. It's a vector particle, artificially coupled to the external field, so designed as to correct the error in this one. The forms are evidently invariant, as far as your g-space is concerned; these are like tensors in the g world; and therefore it's clear that my answers are gauge invariant or coordinate transformable, and all that's necessary. But are also quantum-mechanically satisfactory in the sense that they are unitary.

Now, the next question is, what happens when there are two or more loops? Since I only got this completely straightened out a week before I came here, I haven't had time to inwestigate the case of 2 or more loops to my own satisfaction. The preliminary investigations that I have made do not indicate that it's going to be possible so easily gather the things into the right barrels. It's surprising, I can't understand it; when you gather the trees into processes, there seems to be some loose trees, extra trees. I don't understand them at the moment, and I therefore do not claim that this method of quantization can be obviously and evidently carried on to the next order. In short, therefore, we are still not sure, of the radiative corrections to the radiative corrections to the Lamb shift, the uncertainty lies in energies of the order of magnitude of 10^{-255} rydbergs. I can therefore relax from the problem, and say: for all practical purposes everything is all right. In the meantime, unfortunately, although I could retire from the field and leave you experts who are used to working in gravitation to worry about this matter, I can't retire on the claim that the number is so small and that the thing is now really irrational, if it was not irrational before. Because, unfortunately, I also discovered in the process that the trouble is present in the Yang-Mills theory; and secondly I have incidentally discovered a tree-ring connection which is of very great interest and importance in the meson theories and so on. And so I'm stuck to have to continue this investigation, and of course you all appreciate that this is the secret reason for doing any work, no matter how absurd and irrational and academic it looks; we all realize that no matter how small a thing is, if it has physical interest and is thought about carefully enough, you're bound to think of something that's good for something else.

DISCUSSION

Møller: May I, as a non-expert, ask you a very simple and perhaps foolish question. Is this theory really Einstein's theory of gravitation in the sense that if you would have here many gravitons the equations would go over into the usual field equations of Einstein?

Feynman: Absolutely.

Møller: You are quite sure about it?

Feynman: Yes, in fact when I work out the fields and I don't say in what order I'm working, I have to do it in an abstract manner which includes any number of gravitons; and then the formulas are definitely related to the general theory's formulas; and the invariance is the same; things like this that you see labelled as loops are very typical quantum-mechanical things; but even here you see a tendency to write things with the right derivatives, gauge invariant and everything. No, there's no question that the thing is the Einsteinian theory. The classical limit of this theory that I'm working on now is a non-linear theory exactly the same as the Einsteinian equations. One thing is to prove it by equations; the other is to check it by calculations. I have mathematically proven to myself so many things that aren't true. I'm lousy at proving things — I always make a mistake. I don't notice when I'm doing a path integral over an infinite number of variables that the Lagrangian does not depend upon one of them, the integral is infinite and I've got a ratio of two infinities and I could get a different answer. And I don't notice in the morass of things that something, a little limit or sign, goes wrong. So I always have to check with calculations; and I'm very poor at calculations — I always get the wrong answer. So it's a lot of work in these things. But I've done two things. I checked it by the mathmatics, that the forms of the mathematical equations are the same; and then I checked it by doing a considerable number of problems in quantum mechanics, such as the rate of radiation from a double star held together by quantum-mechanical force, in several orders and so on, and it gives the same answer in the limit as the corresponding classical problem. Or the gravitational radiation when two stars — excuse me, two particles — go by each other, to any order you want (not for stars, then they have to be particles of specified properties; because obviously the rate of radiation of the gravity depends on the give of the starstides are produced). If you do a real problem with real physical things in in then I'm sure we have the right method that belongs to the gravity theory. There's no question about that. It can't take care of the cosmological problem, in which you have matter out to infinity, or that the space is curved at infinity. It could be done I'm sure, but I haven't investigated it. I used as a background a flat one way out at infinity.

Møller: But you say you are not sure it is renormalizable.

Feynman: I'm not sure, no.

Møller: In the limit of large number of gravitons this would not matter?

Feynman: Well, no; you see, there is still a classical electrodynamics; and it's not got to do with the renormalizability of quantum electrodynamics. The infinities come in different places. It's not a related problem.

Rosen: I'm not sure of this, not being one of the experts; but I have the impression that because of the non-linearity of the Einstein equations there exists a difficulty of the

712

following kind. If the linear equations have a solution in the form of an infinite plane monochromatic wave, there does not seem to correspond to that a more exact solution; because you get piling up of energies in space and the solution then diverges at infinity. Could that have any bearing on the accuracy of this kind of calculation?

Feynman: No, I take that into account by a series of corrections. A single graviton is not the same thing as an infinite gravitational wave, because there's a limited energy in it. There's only one $\hbar\omega$.

Rosen: But you're using a momentum expansion which involves infinite waves.

Feynman: Yes, there are corrections. You see what happens if one calculates the corrections. If you have here a graviton coming in this way, then there are corrections for such a ring as this and so on. And these produce first, a divergence as usual; but second, a term in the logarithm of q^2; which means that if this thing is absolutely a free plane

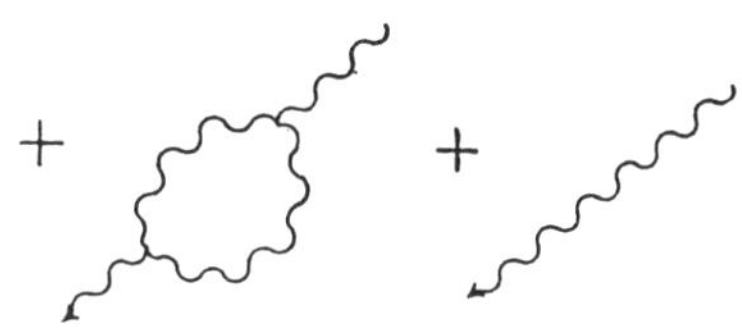

wave, there's no meaning to the correction. So it must be understood in this way, that the thing was emitted some time far in the past, and is going to be absorbed some time in the future; and has not absolutely been going on forever. Then there's a very small coefficient in front of the logarithm and then for any reasonable q^2, like the diameter of the universe or something, I can still get a sensible answer; this is the shadow of the phenomenon you're talking about, that the corrections to the propagation of a graviton, dependent on the logarithm of the momentum squared carried by the graviton and which would be infinite if it were really a zero momentum graviton exactly. And so a free graviton just like that does not quite exist. And this is the correction for that. Strictly we would have to work with wave packets, but they can be of very large extent compared to the wave length of the gravitons.

Anderson: I'd like to ask if you get the same difficulty in the electromagnetic case that you did in the Yang-Mills and gravitational cases?

Feynman: No, sir, you do not. Gauge invariance of diagrams such as Fig. 2 (there is no $2D$) is satisfied whether b is a free wave or not. That is because photons are not the source of photons; they are uncharged.

Anderson: The other thing I would like to suggest is that in putting of things into baskets, you might be able to get easily by always only starting out with vacuum diagrams and opening those successively.

Feynman: I tried that and it didn't go successfully.

Ivanenko: If I understood you correctly, you had used in the initial presentation the transmutation of two particles into gravitons. Yes?

Feynman: It was one of the examples.

Ivanenko: Yes. This process was considered, perhaps in a preliminary manner, by ourselves and by Prof. Weber and Brill. I ask you two questions. Do you possess the effective cross-section? Can you indicate the effects for which high-energy processes play an important role?

Feynman: I never went to energies more than one billion-billion BeV. And then the cross-sections of any of these processes are infinitesimal.

Ivanenko: They increase very, very sharply with energy. Yes, because the radiation is quadrupole, so it increases sharply in contrast to the electromagnetic transmutation of an electron-positron pair.

Feynman: It increases very sharply indeed. On the other hand, it starts out so low that one has to go pretty far to get anywhere. And the distance that you have to go is involved in this thing — the thing that's the analogue of $e^2/\hbar c$ in electricity, which is 1/137 is non--existent in gravitation; it depends on the problem; this is so because of the dimensions of G. So if E is the energy of some process, then if you take $GE^2/\hbar c$ you get an equivalent to this $e^2/\hbar c$. It may be less than that, but at least it can't be any bigger than this. So in order to make this thing to be of the order of 1%, in which case the rate is similar to the rate of photon annihilation, at ordinary energies, we need the GE^2 to be of the order of hc, and as has been pointed out many times, that's an energy of the order 10^{-5} grams, which is 10^{18} BeV. You can figure out the answer right away; just take the energy that you are interested in, square, multiply by G and divide by hc; if that becomes something, then you're getting somewhere. You still might not get somewhere, because the cross-section might not go up that fast, but at least it can't get up any worse than that. So I think that in order to get an appreciable effect, you've got to go to ridiculous energies. So you either have a ridiculously small effect or a ridiculous energy.

Weber: I have a cross-section which may be a partial answer to Ivanenko's question. Could I write it on the board? We have carried out a canonical quantization, which is not as fancy as the one you have just heard about; but considering the interaction of photons and gravitons; and it turns out that even in the linear approximation that one has the possibility of the graviton production by scattering of photons in a Coulomb field. And the scattering cross-section for this case turns out to be $8\pi^2$ times the constant of gravitation times the energy of the scatterer times the thickness of the scatterer in the direction of propagation of the photon through it divided by c^4. This assumes that all of the dimensions of the scatterer are large in comparison with the wave length of the photon. We obtained this result by quantization, and noticed that it didn't have Planck's constant in it, so we turned around and calculated it classically. Now, if one puts numbers in this, one finds that the scattering cross-section of a galaxy due to a uniform magnetic field through it is 10^{28} cm^2, a much larger number than the object that you talked about. This represents a conversion of photons into gravitons of about 1 part in 10^{16}. This is of course too small to measure. Also, we considered the possibility of using this cross-section for a laboratory experiment in which one had a scatterer consisting, say of a million gauss magnetic field over something like a cubic meter. This turned out to be entirely impossible, a result in total contradiction to what has appeared in the Russian literature. In fact, the theory of fluctuations shows that for a laboratory experiment involving the production of gravitons

714

by scattering of photons in a Coulomb field, the scattered power has to be greater than twice the square root of kT times the photon power divided by the averaging time of the experiment. I believe that the incorrect results that have appeared in the literature have been due to the statement that ΔP has to be greater than kT over τ; dimensionally these things are the same, but order of magnitude-wise this kind of experiment for the scatterer of which I spoke requires something like 10^{50} watts. Maybe I can say something about this afternoon; I don't want to take any more time.

De Witt: I should like to ask Prof. Feynman the following questions. First, to give us a careful statement of the tree theorem; and then outline, if he can to a brief extent, the nature of the proof of the theorem for the one-loop case, whish I understand does work. And then, to also show in a little bit more detail the structure and nature of the fictitious particle needed if you want to renormalize everything directly with the loops. And if you like, do it for the Yang-Mills, if things are prettier that way.

Feynman: I usually don't find that to go into the mathematical details of proofs in a large company is a very effective way to do anything; so, although that's the question that you asked me — I'd be glad to do it — I could instead of that give a more physical explanation of why there is such a theorem; how I thought of the theorem in the first place, and things of this nature; although I do have a proof — I'm not trying to cover up.

De Witt: May we have a statement of the theorem first?

Feynman: That I do not have. I only have it for one loop, and for one loop the careful statement of the theorem is... — look, let me do it my way. First — let me tell you how I thought of this crazy thing. I was invited to Brussels to give a talk on electrodynamics — the 50th anniversary of the 1911 Solvay Conference on radiation. And I said I'd make believe I'm coming back, and I'm telling an imaginary audience of Einstein, Lorentz and so on what the answer was. In other words, there are going to be intelligent guys, and I'll tell them the answer. So I tried to explain quantum electrodynamics in a very elementary way, and started out to explain the self-energy, like the hydrogen Lamb shift. How can you explain the hydrogen Lamb shift easily? It turns out you can't at all — they didn't even know there was an atomic nucleus. But, never mind. I thought of the following. I would explain to Lorentz that his idea that he mentioned in the conference, that classically the electromagnetic field could be represented by a lot of oscillators was correct. And that Planck's idea that the oscillators are quantized was correct, and that Lorentz's suggestion, which is also in that thing, that Planck should quantize the oscillators that the field is equivalent to, was right. And it was really amusing to discover that all that was in 1911. And that the paper in which Planck concludes that the energy of each oscillator was not $n\hbar\omega$ but $(n+1/2)\hbar\omega$ which was also in that, was also right; and that this produced a difficulty, because each of the harmonic oscillators of Lorentz in each of the modes had a frequency of $\hbar\omega/2$ which is an infinite amount of energy, because there are an infinite number of modes. And that that's a serious problem in quantum electrodynamics and the first one we have to remove. And the method we use to remove it is to simply redefine the energy so that we start from a different zero, because, of course, absolute energy doesn't mean anything. (In this gravitational context, absolute energy does mean something, but it's one of the technical points I can't discuss, which did require a certain skill to get rid of, in making a gravity theory; but never mind.) Now look — I make a little hole in the box and I let in a little bit of hydrogen gas from a reservoir; such a small amount of hydrogen gas, that the density is low enough that the index of refraction in space differs from one by an amount proportional to A, the number of atoms. With the index being somewhat changed, the frequency of all the normal modes is altered. Each normal mode has the same wavelenght as before, because it must fit into the box; but the frequencies are all altered. And therefore the $\hbar\omega$'s should all be shifted a trifle, because of the shift of index, and therefore there's a slight shift of the energy. Although we subtract $\hbar\omega/2$ for the vacuum, there's a correction when we put the gas in; and this correction is proportional to the number of atoms, and can be associated with an energy for each atom. If you say, yes, but you had that energy already when you had the gas in back in the reservoir, I say, but let us only compare the difference in energy between the $2S$ and $2P$ state. When we change the excitation of the hydrogen gas from $2S$ to $2P$ then it changes its index without removing anything; and the energy difference that is needed to change the energy from $2S$ to the $2P$ for all these atoms is not only the energy that you calculate with disregard of the zero point energy; but the fact is that the zero point energy is changed very slightly. And this very slight difference should be the Lamb effect. So I thought, it's a nice argument; the only question is, is it true. In the first place it's interesting, because as you well know the index differs from one by an amount which is proportional to the forward scattering for γ rays of momentum k and therefore that shift in energy is essentially the sum over all momentum states of the forward scattering for γ rays of momentum k. So I looked at the forward scattering and compared it with the right formula for the Lamb shift, and it was not true, of course; it's too simple an argument. But then I said, wait, I forgot something. Dirac, explained to us that there are negative energy states for the electron but that the whole sea of negative energy states is filled. And, of course, if I put the hydrogen atoms in here all those electrons in negative energy states are also ascattering off the hydrogen atoms; and therefore their states are all shifted; and therefore the energy levels of all those are shifted a tiny bit. And therefore there's shift in the energy due to those. And so there must be an additional term which is the forward scattering of positrons, which is the same as scattering of negative energy electrons. Actually, for the symmetry of things it is better to take half the case where you make the positrons the holes and the other half where you make the electrons the holes; so it should be 1/2 forward scattering by electrons, 1/2 scattering by positrons and scattering by γ rays — the sum of all those forward scattering amplitudes ought to equal the self--energy of the hydrogen atom. And thats' right. And it's simple, and it's very peculiar. The reason it's peculiar is that these forward scatterings are real processes. At last I had discovered a formula I had always wanted, which is a formula for energy differences (which are defined in terms of virtual fields) in terms of actual measurable quantities, no matter how difficult the experiment may be — I mean I have to be able to scatter these things. Many times in studying the energy difference due to electricity (I suppose) between the proton and the neutron, I had hoped for a theorem which would go something like this — this energy difference between proton and neutron must be equal to the following sum of a bunch of cross-sections for a number of processes, but all real physical processes, I don't care how hard they are to measure. So this is the beginning of such a formula. It's rather surprising.

It's not the same as the usual formula — it's equal to it but it's not the same. I have no formulation of the laws of quantum gravidynamics; I have a proposal on how to make the calculations. When I make the proposal on how to do the closed loops, the obvious proposal does not work; it gives non-unitarity and stuff like that. So the obvious proposal is no good; it works O.K. for trees; so how am I going to define the answer for would correspond to a ring? The one I happen to have chosen is the following: I take the ring in general for any meson theory, one closed ring can be written as equivalent to a whole lot of processes each one of which is trees. I then define, as my belief as to what the ring ought to be in the grand theory, that it's going to be also equal to the corresponding physical set of trees. When I said this is equal to this. I didn't worry about gauge or anything else; what I means was, if these weren't gravitons but photons or any other neutral object — it doesn't make any difference what they are — this theorem is right. So I suppose it's right also for real gravitons, and I suppose also that what's being scattered is only transverse and is only a real free graviton with $q^2 = 0$. Therefore, I say let this ring equal this set of trees. Every one of these terms can be completely computed — it's a tree. And it's gauge invariant; that is, if I added an extra potential on the whole thing, another outside disturbance of a type which is nothing but a coordinate transformation — in short a pure gradient wave — to the whole diagram then it comes on to all of these processes; but it makes no effect on any of them, and therefore makes no effect on the sum; and therefore I know my definition of this ring is gauge-invariant. Second, unitarity is a property of the breaking of this diagram; the imaginary part of this equals something; if you take the imaginary part of this side, it's already broken up, in fact, and you can prove immediately that it's the correct unitarity rule. Therefore it's going to be unitarity and so on and so on. And so I therefore define gravity with one ring in this way. Now what prevents me from doing it with two rings? The lack of a complet statement of what two rings is equal to in terms of processes; that is I can open the ring all right; but I can't put the pieces — the broken diagrams — back together again into complete sets that each one is a complete physical process. In other words some of them correspond to the scattering of a graviton, but leaving out some diagrams. But the scattering of a graviton leaving out diagrams is no longer gauge invariant, I mean, not evidently gauge invariant, and so the power of the whole thing collapses. I don't know what to do with it. So that's the situation; that's why it is crucial to the particular plan. There's always, of course, another way out. And that's the following (and that's what I tried to describe at the end of the talk — maybe I talked too fast): After all now I've defined what this results is equal to — by definition not that you should do a loop some way and get this, but that a loop is equal to this by defition, and I'm not going to do a loop any other way. But, of course, from a practical point of view or from the point of view purely of interest, the question is, can you come back now and calculate the ring directly by some particular mathematical shenanigans, and get the same answer as you get by adding the trees. And I found the way to do that. I have another way, in other words, to do the ring integral directly. I have to subtract something from a vector particle going around the instead of a graviton to get the answer right. So I known the rule, and I know why the rule is, and I have a proof of the rule for one loop. I have two ways of extending. I can either break this two loop diagram open and get it back into the processes, like I did with the one ring — where so far I'm stuck. Or,

I can take the rule which I found here and try to guess the generalization for any number of rings. Also stuck. But I've only had a week, gentlemen; I've only been able to straighten out the difficulty of a single ring a week ago when I got everything cleaned up. It's more than a week — I had to take a lot of time checking and checking; but I was only finished checking to make sure of everything for this conference. And of course you're always asking me about the thing I haven't had time to make sure about yet, and I'm sorry; I worked hard to be sure of something, and now you ask me about those things I haven't had time. I hoped that I would be able to get it. I still have a few irons to try; I'm not completely stuck—maybe.

DeWitt: Because of the interest of the tricky extra particle that you mentioned at the end, and its possible connection, perhaps, with some work of Dr Białynicki-Birula, have you got far enough on that so that you could repeat it with just a little more detail? The structure of it and what sort of an equation it satisfies, and what is its propagator? These are technical points, but they have an interest.

Feynman: Give me ten minutes. And let me show how the analysis of these tree diagrams, loop diagrams and all this other stuff is done mathematical way. Now I will show you that I too can write equations that nobody can understand. Before I do that I should like to say that there are a few properties that this result has that are interesting. First of all in the Yang-Mills case there also exists a theory which violates the original idea of symmetry of the isotopic spin (from which was originally invented) by the simple assumption that the particle has a mass. That means to add to the Lagrangian a term $-\mu^2 a_\mu a^\mu$ where a_μ is an isotopic vector. You add this to the Lagrangian. This destroys the gauge invariance of the theory — it's just like electrodynamics with a mass, it's no longer gauge-invariant, it's just a dirty theory. Knowing that there is no such field with zero mass people say: „let's put the mass term on". Now when you put a mass term on it is no longer gauge invariant. But then it is also no longer singular. The Lagrangian is no longer singular for the same reason that it is not invariant. And therefore everything can be solved precisely. The propagator instead of being $\delta_{\mu\nu}$ between two currents is

$$\frac{\delta_{\mu\nu}-q_\mu q_\nu/\mu^2}{q^2-\mu^2}, \tag{10}$$

where q_μ is the momentum of propagating particle. The factor $1/(q^2-\mu^2)$ is typical for mass μ but the part $-q_\mu q_\nu/\mu^2$ is an important term which can be taken to be zero in electrodynamics but it is not obvious whether it can be taken to be zero in the case of Yang-Mills theory. In fact it has been proved it cannot be taken to be zero; this propagator is used between two currents. I am using the Yang-Mills example instead of the gravity example. I really want only the case $\mu^2 = 0$, and am asking whether I can get there by first calculating finite μ^2, then taking the limit $\mu^2 = 0$.

Now, with $\mu^2 \neq 0$ this is a definite propagator and there are no ambiguities at the closed rings, the closed loops. I have no freedom, I must compute this propagator. I mean there is no reason for trouble, and there is no trouble. There is no gauge invariance either.

And of course I checked. I broke the rings and I computed by the broken ring theorem method a closed loop problem of fair complexity (which in fact was the interaction of two electrons). I computed it by the open ring method and by the closed ring method, and of course it agreed, there is no reason that it shouldn't. It turned out that for tree diagrams you don't have to worry about this $q_\mu q_\nu/\mu^2$ term, you can drop it — but not for the closed ring — only for tree. Therefore the tree diagrams have the definite limit as μ^2 goes to zero. And yet I have the closed ring diagram which is equal to the tree diagram when the mass is anything but zero, and therefore it ought to be true that the limit as μ^2 goes to zero of the ring is equal to the case when $\mu = 0$. It sounds like a great idea why don't you define the desired $\mu^2 = 0$ theory that way? Answer: You can't put μ^2 equal zero in the form (10). You can't do it because of the $q_\mu q_\nu/\mu^2$. So it was necessary next to see if there is a way to re-express the ring diagrams, for the case with $\mu^2 \neq 0$, in a new form with a propagator different from (10), that didn't have a μ^2 in it, in such a form that you can take the limits as μ^2 goes to zero. Then that would be a new way to do the μ equal zero case; and that's the way I found the formula. I'll try to explain how to find that theory.

We start with a definite theory, the Yang-Mills theory with a mass (the reason I do that is that there's no ambiguity about what I am trying to do) and later on I take the mass to zero, then the theory works something like this. You have the Lagrangian $\mathcal{L}(A, \varphi)$ which involves the vector potential of this field and the fields φ representing the matter with which this object is interacting for zero mass, to which, for finite mass we add the term $\mu^2 A_\mu A_\mu$. This is the Lagrangian that has to be integrated and the idea is that you integrate this over all fields A and φ; and that is the answer for the amplitude of the problem

$$X = \int e^{\int \mathcal{L}(A,\varphi)d\tau + \mu^2 A_\mu A_\mu d\tau}\, DAD\varphi. \tag{11}$$

But wait, what about the initial and final conditions? You have certain particles coming in and going out. To simplify things (this is not essential) I'll just study the case that corresponds only to gravitons in and out. I'll call them gravitons and mesons even though they are vector particles. The question is first, what is the right answer if you have gravitons represented by plane waves, A_1, A_2, A_3 ... going in (positive frequency in A_1) or out (negative frequency). You make the following field up. Let A_{asym} be defined as α times the wave function A_1 that represents the first graviton coming in a plane wave, plus β times A_2 plus γ times A_3 and so on.

$$A_{\text{asym}} = \alpha A_1 + \beta A_2 + \gamma^{A_3}\ldots, \quad A \to A_{\text{asym}}. \tag{12}$$

Then you calculate this integral (11) subject to the condition that A approaches A_{asym} at infinity. The result of this is of course a function of α, β, γ ... and so on. Then what you want for X is just the term first order in α, β, γ ... That means just one of each these gravitons coming in and out. That's the right formula for a regular theory, for meson theory, You calculate the integral subject to the asymptotic condition, when you imagine all these waves, but you take the first order perturbation with respect to each one of the incoming waves. You never let the same photon operate twice; a photon operating twice is not a photon, it is a classical wave. So you take the derivative of this with respect to α, β, γ and so on, then setting them all equal to zero. That's problem. (In general there's φ asymptotic too.)

Now the way I happened to do this is the following: Let us call A_0 the A which satisfies the classical eqatiuons of motion, which in this particular case will be

$$\left.\frac{\partial \mathcal{L}}{\partial A}\right|_{A^\circ} + \mu^2 A^\circ = 0 \tag{13}$$

I solve this subject to the condition that A_0 equals A_{asym}. In other words, I find what is the maximum or minimum — whatever it is — of the action in (11), subject to the asymptotic condition. That's the beginning of analysing this.

The next thing is to make the simple substitution $A = A_0 + B$ and put it back in equation (11). Then if you take $\mathcal{L}$ od A_0+B (if B is negligible you get $\mathcal{L}$ of A_0 and so forth) so you get something like this

$$e^{i[\mathcal{L}(A_0)+\mu^2 A_0 A_0]} \int e^{[\mathcal{L}(A^0+B)-\mathcal{L}(A^0)]+\mu^2 BB+2\mu^2 AB}\, DB. \tag{14}$$

The integral is over all B, and B must go to zero asymptotically. This business can be expanded in powers of B.

$$\mathcal{L}(A+B)-\mathcal{L}(A)+\mu^2 BB+2\mu^2 AB = \text{Quad}\,(B)+\text{Cubic}\,(B)+\ldots+\mu^2 BB. \tag{15}$$

The zeroth power B is evidently zero. The first power of B is also zero because A_0 minimized the original thing. So this starts out quadratic in B plus cubic in B plus *etc.*, that's what this is here. These quadratic forms Quad (B) and so on of course depend on A_0, the cubic form involves A_0 in some complex, maybe very complicated, locked-up mess, but as far as B is concerned it is second power and higher powers.

Now I would like to point something out. First — it turns out if you analyze it, that the contribution of the first factor here alone (if you had forgotten the intergal and called it one) is exactly the contribution of all trees to the problem. So that's like the classical theories related to trees. Next, if you drop the term cubic in B in the exponent completely and just integrated the result over DB, that corresponds to the contribution from one ring, or from two isolated rings, or three isolated rings, but not interlocked rings. If you start to include the cubic term is has to come in a second power to do anything, because of the evenness and oddness of function. And as soon as it comes in second power, the cubic term, having three of these things come together twice, makes a terrible thing like ∞ which is a double ring. So you don't get to a double ring until you bring a cubic term down to the second order. So if I disregard that and just work with this second order term Quad (B) $+\mu^2 BB$, I'm studying the contribution from one ring. If I study this I am working from the trees. And now you see I have in my hands an expression for the contribution of a ring correct in all orders no matter how many lines come in. I also have expressions for the contributions from trees and so on. I can compare them in different mathematical circumstances, and it's on this basis that I have been able to prove everything I have been able to prove relating one ring to trees.

Now, let me explain how the theorem was obtained that takes the case for the mass and for a ring. Now we have to discuss a ring, which is a formula like this

$$X = \int e^{(\text{Quadr}(B)+\mu^2 B^2)}\, DB. \tag{15}$$

The quadratic form involves A_0 so the answer depends on A_0 — it's some complicated functional of A_0. Anyway I won't say that all the time, I'll just remember that. We have to integrate over all B. And the difficulty is — not difficulty, but the point is — that this quadratic form in B is singular, because it came from the piece of the action that has an invariance and this invariance keeps chasing us along. And there are certain transformations of B which leave this Quard B part unchanged in first order. That transformation in the Yang-Mills theory is

$$\vec{B}'_\mu = \vec{B}_\mu + \vec{\nabla}\alpha + (\vec{\alpha}\times\vec{A}) = \vec{B}_\mu + \alpha_{;\mu}\,. \tag{16}$$

where the vectors are in isotopic spin space and α is considered as first order. This transformation leaves the quadratic form invariant so the Quad (B) thing by itself is singular. But it doesn't make any difference, because of the addition of the $\mu^2 BB$. If $\mu^2 \neq 0$, there is no problem, but if $\mu^2 \to 0$, I'd be in trouble.

I discovered that if I make this change (16) in the actual Lagrangian and carry everything up to second order it is exact, in fact because it's only second order. If I do it with the exact change, the thing isn't invariant, it is only invariant to first order in α. But if I make the substitution exactly, then I get a certain addition to the Lagrangian, in other words the Lagrangian of B' (this includes the μ^2, the Lagrangian plus the μ^2 term in B) is the Lagrangian plus the μ^2 term in B plus something like this

$$\mu^2 B_\mu \cdot a_{,\mu} + \frac{1}{2}\,\mu^2 a_{,\mu}\, a_{;\mu}$$

I have to explain that the semicolon is analogous to the semicolon in gravity. The semicolon derivative $X_{;\mu}$ means the ordinary derivative of X minus A cross X and that's the analogue of the Christoffel symbols. Anyway, I find out what happens to L when I make this transformation. Now comes the idea, the trick, the nonsense: you start with the following thing; you, say, suppose instead of writing the original terms down, instead of writing the original Lagrangian I were to write the following:

$$\int e^{\mathcal{L}(B) + \frac{1}{2}(B_{\mu,\mu} - \alpha_{;\mu\mu} + \mu^2\alpha)^2}\, \mathcal{D}a\mathcal{D}B.$$

Now I say that the integral over α is some constant or other. So all I have done is to multiply my original integral by $\mathcal{L}$ of B (by $\mathcal{L}$ of B I mean the whole thing, I mean this whole thing is going to be $\mathcal{L}$ of B). If I can claim that when I integrate α I get something which is independent of B, which is not self-evident. If I integrate over all α it does not look as if it is independent of B — but after a moment's consideration you see that it is. Because if I can solve a certain equation, which is $\alpha^{\mu}_{;\mu} - \mu^2\alpha = B^{\mu}_{\mu}$, I can shift the value of α by that amount, and then this term would disappear. In other words if I can solve this, and call this solution α_0 and change α to α_0, then the B would cancel and it would only be α' here. I did it a little abstractly which is a little easier to explain, therefore, this term that I've added can be thought of as an integral of the following nature: Integral of some B, plus an operator acting on α (this complicated operator is the second derivative and so on) squared $\mathcal{D}\alpha$. And then by that substitution I've just mentioned, this becomes equal to 1/2 the operator on A

times α' squared $\mathcal{D}\alpha$, which is equal to the integral e to the one half of α times A, the operator A, times the operator A times α integrated over primed α. Now when you integrate a quadratic form, which is a quadratic with an operator like this you get one over the square root of the determinant of the operator. So this thing is one over the square root of the determinant of the operator AA. The determinant of the operator A times A is square of the determinant of A. So this is one over the determinant of the operator A, or better it is one over square root of the determinant of the operator A squared, you'll see in a minute why I like to write it in this way. In other words, when I've written this thing down I've written the answer that I want. Let's call X the unknown answer that I want. Then this is equal to X divided by this determinant's square root squared. Now comes the trick — I now make the change from B to B'. We notice that B changed to B' is simply... oh!, this is wrong, that's what's wrong, it should be just this. Now I've got it. The change from B' to B is to add something to B. Therefore to the differential of B it adds nothing, it's just shifting the B to a new value. So I make the transformation from B to B' everywhere. So then I have $d\alpha$ and dB, and now I have a new thing up here where I make use of the formula for $\mathcal{L}$ of B':

$$\mathcal{L}(B') = \mathcal{L}(B) + \mu^2 B_\mu \alpha_\mu + \frac{1}{2}\, \mu^2 \alpha_{,\mu} \alpha_{;\mu}$$

You see there is a certain cross term generated here and another cross term coming from expanding this out and the net result, with a little algebra here, is that becomes $\mathcal{L}$ of B, but the quadratic term doesn't cancel out and is left; there's one half of $B_{\mu,\mu}$ squared; that's from this term; the cross term here cancels the cross term in there; and then we have only the quadratic — I mean the α terms

$$\int e^{\mathcal{L}(B)+\frac{1}{2}(B_{\mu,\mu})^2}\, \mathcal{D}B\, e^{\frac{\mu^2}{2}(\alpha_{,\mu}\alpha_{;\mu}+\mu^2\alpha^2)}\, \mathcal{D}\alpha.$$

And the problem is now to do this integral on α; well, another miraculous thing happens. I have the operator A, but that this down thing is $\alpha A \alpha$, and therefore its result is just determinant once; or the square of this integral is equal to this determinant, or something like that. Therefore, when you get all the factors right, X, the unknown, is equal to

$$X = \Big[\int e^{\mathcal{L}(B)+\frac{1}{2}(B_{\mu,\mu})^2}\, \mathcal{D}B\Big] : \Big[\int e^{\frac{\mu^2}{2}(\alpha_{,\mu}\alpha_{;\mu}+\mu^2\alpha^2)}\, \mathcal{D}\alpha\Big].$$

Sachs: I want to ask a question about long-range hopes. Perhaps for irrational reasons people are particularly interested in those parts of the theory where is a possibility of real qualitative differences: what do the coordinates or topology mean in a quantized theory, and this kind of junk. Now I wonder if you think that this perturbation theory can eventually be jazzed up to cover also this kind of questions?

Feynman: The present theory is not a theory as it is incomplete. I do not give a rule on how to do all problems. I expect of course that if I spend more time on figuring out how to untangle the pretzels I shall be able to make it into such a theory. So let's suppose I did. Now you can ask the question would the completed job, assuming it exists, be of any interest to esoteric question about the quantization of gravity. Of course it would be, because it

722

would be the expression of the quantum theory; there is today no expression of the quantum theory which is consistent. You say: but it's perturbation theory. But it isn't. I worked on the thing analyzing it in the series of increasing accuracy, but that's only, obviously, when I am doing problems and checking, or doing things like I just did. But even there I haven't said how many times the vector potential A_0 is attacking the diagram, there is no limit to what order of external lines are involved in the calculation of A_0, for example. And so if I get my general theorem for all orders, I'll have some kind of a formulation. The fact is, that in such things as electrodynamics and other theories, it has not been possible to figure out the consequences of the quantum field theory in the case of strong interactions, because of technical difficulties which are not technical difficulties just of the gravitation theory, but exist all over the quantum field theory. I do not expect that the gravitational problems will be any easier in that region than they are in any other field theory, so I can say very little there. But at least one should certainly formulate the theory that you're trying to calculate first, and then find out what the consequences are, before trying to do it the other way round. So I think that you'll be frustrated by the difficulties that do appear whenever any theory diverges. On other hand, if you ask about the physical significance of the quantization of geometry, in other words about the philosophy behind it; what happens to the metric, and all such questions, those I believe will be answerable, yes. I think you would be able to figure out the physics of it afterwards, but I won't to think about that until I have it completely formulated, I don't want to start to work out the anser to something unless I know what the equation is I am trying to analyze. But I dont' have the doubt that you will be able to do something, because after all you are describing the phenomena that you would expect, and if you describe the phenomena then you expect you can then find some kind of framework in which to talk to help to understand the phenomena.

PHYSICAL REVIEW VOLUME 162, NUMBER 5 25 OCTOBER 1967

Quantum Theory of Gravity. II. The Manifestly Covariant Theory*

BRYCE S. DEWITT
Institute for Advanced Study, Princeton, New Jersey
and
Department of Physics, University of North Carolina, Chapel Hill, North Carolina†

(Received 25 July 1966; revised manuscript received 9 January 1967)

Contrary to the situation which holds for the canonical theory described in the first paper of this series, there exists at present no tractable pure operator language on which to base a manifestly covariant quantum theory of gravity. One must construct the theory by analogy with conventional S-matrix theory, using the c-number language of Feynman amplitudes when nothing else is available. The present paper undertakes this construction. It begins at an elementary level with a treatment of the propagation of small disturbances on a classical background. The classical background plays a fundamental role throughout, both as a technical instrument for probing the vacuum (i.e., analyzing virtual processes) and as an arbitrary fiducial point for the quantum fluctuations. The problem of the quantized light cone is discussed in a preliminary way, and the formal structure of the invariance group is displayed. A condensed notation is adopted which permits the Yang-Mills field to be studied simultaneously with the gravitational field. Generally covariant Green's functions are introduced through the imposition of covariant supplementary conditions on small disturbances. The transition from the classical to the quantum theory is made via the Poisson bracket of Peierls. Commutation relations for the asymptotic fields are obtained and used to define the incoming and outgoing states. Because of the non-Abelian character of the coordinate transformation group, the separation of propagated disturbances into physical and nonphysical components requires much greater care than in electrodynamics. With the aid of a canonical form for the commutator function, two distinct Feynman propagators relative to an arbitrary background are defined. One of these is manifestly covariant, but propagates nonphysical as well as physical quanta; the other propagates physical quanta only, but lacks manifest covariance. The latter is used to define external-line wave functions and non-radiatively-corrected amplitudes for scattering, pair production, and pair annihilation by the background field. The group invariance of these amplitudes is proved. A fully covariant generalization of the complete S matrix is next proposed, and Feynman's *tree theorem* on the group invariance of non-radiatively-corrected n-particle amplitudes is derived. The big problem of radiative corrections is then confronted. The resolution of this problem is carried out in steps. The single-loop contribution to the vacuum-to-vacuum amplitude is first computed with the aid of the formal theory of continuous determinants. This contribution is then functionally differentiated to obtain the lowest-order radiative corrections to the n-quantum amplitudes. These amplitudes split automatically into *Feynman baskets*, i.e., sums over tree amplitudes (bare scattering amplitudes) in which all external lines are on the mass shell. This guarantees their group invariance. The invariance can be made partially manifest by converting from the noncovariant Feynman propagator to the covariant one, and this leads to the formal appearance of *fictitious quanta* which compensate the nonphysical modes carried by the covariant propagator. Although avoidable in principle, these quanta necessarily appear whenever manifestly covariant expressions are employed, e.g., in renormalization theory. The fictitious quanta, however, appear only in closed loops and are coupled to real quanta through vertices which vanish when the invariance group is Abelian. The vertices are nonsymmetric and always occur with a uniform orientation around any fictitious quantum loop. The problem of splitting radiative corrections into Feynman baskets becomes more difficult in higher orders, when overlapping loops occur. This problem is approached with the aid of the Feynman functional integral. It is shown that the "measure" or "volume element" for the functional integration plays a fundamental role in the decomposition into Feynman baskets and in guaranteeing the invariance of radiative corrections under arbitrary changes in the choice of basic field variables. The "measure" has two effects. Firstly, it removes from all closed loops the *noncausal chains* of cyclically connected advanced (or retarded) Green's functions, thereby breaking them open and ensuring that at least one segment of every loop is on the mass shell. Secondly it adds certain nonlocal corrections to the operator field equations, which vanish in the classical limit $\hbar \to 0$. The question arises why these removals and corrections are always neglected in conventional field theory without apparent harm. It is argued that the usual procedures of renormalization theory automatically take care of them. In practice the criteria of locality and unitarity are replaced by analyticity statements and Cutkosky rules. It is virtually certain that the "measure" may be similarly ignored (set equal to unity) in gravity theory, and that attention may therefore be confined to *primary diagrams*, i.e., diagrams which contain Feynman propagators only, with no noncausal chains removed. A general algorithm is given for obtaining the primary diagrams of arbitrarily high order, including all fictitious quantum loops, and the group invariance of the amplitudes thereby defined is proved. Essential to all these derivations is the use of a background field satisfying the classical "free" field equations. It is never necessary to employ external sources, and hence the well-known difficulties arising with sources in a non-Abelian context are avoided.

1. INTRODUCTION

IN the first paper of this series[1] an attempt was made to show what happens when canonical Hamiltonian quantization methods are applied to the gravitational field. Attention was focused on some of the bizarre features of the resulting formalism which arise in the case of finite worlds, and which are of possible cosmological and even metaphysical significance. Such

* This research was supported in part by the Air Force Office of Scientific Research under Grant AFOSR-153-64, and in part by the National Science Foundation under Grant GP7437.

† Permanent address.

[1] B. S. DeWitt, Phys. Rev. 160, 1113 (1967). This paper will be referred to as I.

prosaic questions as the scattering, production, absorption, and decay of individual quanta were left untouched. The main reason for this was that the canonical theory does not lend itself easily to the study of these questions when physical conditions are such that the effects of vacuum processes must be taken into account. A manifestly covariant formalism is needed instead. It is the task of the present paper to provide such a formalism.

We must begin by making clear precisely what is meant by "manifest covariance." In conventional S-matrix theory (whether based on a conventional field theory or not) "manifest covariance" means "manifest Lorentz covariance." In the context of a theory of gravity the question arises whether it should mean more than this, since the classical theory from which one starts has "manifest *general* covariance." Here one must be careful. There is an important difference between general covariance and ordinary Lorentz covariance, and neither one implies the other. Lorentz covariance is the expression of a *geometrical* symmetry possessed by a system. In gravity theory it has relevance at most to the asymptotic state of the field. As has been emphasized by Fock,[2] the word "relativity" in the name "general relativity" has connotations of symmetry which are misleading. Far from being more relativistic than special relativity, general relativity is in fact less relativistic. For as soon as space-time acquires bumps (i.e., curvature) it becomes absolute in the sense that one may be able to specify position or velocity with respect to these bumps, provided they are sufficiently pronounced and distinguishable from one another. Only when the bumps coalesce into regions of uniform curvature does space-time regain its relativistic properties. It never becomes *more* relativistic than flat space-time, which is characterized by the 10-parameter Poincaré group.

The technical method of distinguishing between the Poincaré group and the general coordinate transformation group is to confine the operations of the latter group to a finite (but arbitrary) region of space-time. The asymptotic coordinates are then left undisturbed by general coordinate transformations, and only the operations of the Poincaré group (if that is indeed the asymptotic symmetry group of the problem) are allowed to change them. The general coordinate transformation group thus becomes a *gauge* group which, although historically an offspring of the Poincaré group and the equivalence principle, plays technically the rather obscure role of providing the analytic means by which the Einstein equations can be obtained from a variational principle and their essential locality displayed.[3]

This, however, is not the whole story, for the general coordinate transformation group still has, even as a gauge group, profound physical implications. Some of these we have already encountered in I, and some we shall encounter in the present paper. Others will appear in the final paper of this series, which is to be devoted to applications of the covariant theory. If it were not for these implications there would be little interest in pushing our investigations further, for there is no likelihood that such "prosaic" processes as graviton-graviton scattering or curvature induced vacuum polarization will ever be experimentally observed.[4] The real reason for studying the quantum theory of gravity is that by uniting quantum theory and general relativity one may discover, at no cost in the way of new axioms of physics, some previously unknown consequences of general coordinate invariance, which suggest new interesting things that can be done with quantum field theory as a whole.

Our problem will be to develop a formalism which makes manifest the extent to which general covariance permeates the theory. This will be accomplished by introducing, instead of a flat background, an adjustable c-number background metric. Use of such a metric has the following fundamental technical advantages: (1) It facilitates the introduction of particle propagators which are generally covariant rather than merely Lorentz-covariant. (2) It reduces the study of radiative corrections to the study of the vacuum. (3) It makes possible the generally covariant isolation of divergences, which is essential to any renormalization program. (4) It renders theorems analogous to the Ward identity almost trivial. (5) It makes possible, in principle, the extension of the theory of radiative corrections to worlds for which space-time is not asymptotically flat and which may even be closed and finite. These advantages are typical of what we shall mean by the phrase "manifest covariance." Use of the phrase, however, is not to be understood as implying that the simple trick of introducing a variable background metric makes *everything* obvious. The generally covariant propagators will not be unique but will be choosable in various ways, analogous to the gauge choices in quantum electrodynamics, and we shall have to undertake a separate investigation, just as in quantum electrodynamics, to verify that the choice is irrelevant. This investigation turns out to be much more complicated than in the case of quantum electrodynamics.

Of the five advantages listed above as stemming from the use of a variable background metric only the first two will appear in the present paper. The third

[2] V. Fock, *The Theory of Space-Time and Gravitation* (Pergamon Press, New York, 1959).

[3] The content of the Einstein equations can be expressed in an intrinsic coordinate-independent form only at the cost of introducing nonlocal structures. (See, for example, Ref. 32). It can be argued [see S. Weinberg, Phys. Rev. 138, B988 (1965)] that the general coordinate transformation group is simply a consequence of the zero rest mass of the gravitational field and its long-range character.

[4] Although one might hope for some very indirect cosmological evidence for such processes.

and fourth will be demonstrated in the following paper of this series, while the fifth remains a program for the future. It is not out of place here, however, to speculate briefly on this ultimate program. As long as the conventional S matrix is our chief concern it is appropriate to choose a background metric which is asymptotically flat. We shall see that Lorentz invariance of the S matrix then follows almost trivially from the formalism, in the limit in which the background metric becomes everywhere Minkowskian. Now it is obvious that scattering processes are also possible in an infinite world which is not asymptotically flat. In such a world it should be possible to construct a generalized S matrix in which the conventional plane-wave momentum eigenfunctions are replaced by wave functions appropriate to the altered asymptotic geometry. The asymptotic geometry itself would be fixed by choosing the background metric appropriately.

In a closed world no rigorous S matrix exists. The continuum of scattering states is replaced by a regime of discrete quantization, and, as we have seen in I, the wave function of the universe may even be unique. It may be conjectured that the formalism most appropriate to this case is obtained by choosing the background metric to be *not* a c number but rather an operator depending on a small number (e.g., *one*) of quantum variables similar to the operator R representing the radius of the Friedmann universe studied in I. These variables would be quantized by the canonical method, while the full q-number metric would continue to be treated by manifestly covariant methods. (Conditions of constraint would, of course, have to be imposed on the latter metric to take into account the fact that some of its degrees of freedom have been transferred to the background metric.) The resulting simultaneous use of both the canonical and covariant theories might help to reveal the relationship between them.

As has been remarked in I, no rigorous mathematical link has thus far been established between the canonical and covariant theories. In the case of infinite worlds it is believed that the two theories are merely two versions of the same theory, expressed in different languages, but no one knows for sure. The analysis of radiative corrections has turned out to be of such intricacy that the covariant theory has had to be developed completely within its own framework and independently of the canonical theory. Although the structure of the covariant theory is suggested by the formalism of field operators, and hence maintains a few points of contact with conventional field theory, the language of operators is dropped at a certain key stage and c-number criteria are thenceforth exclusively employed to maintain internal consistency. It turns out that the language of operators is a peculiarly unwieldy one in which to discuss questions of consistency when the invariance group of the theory is non-Abelian. The language of graphs and the S matrix is much more direct.

The latter language, embracing as it does many different particle theories at once, is also much less dependent on the detailed Lagrangian structure of the field theory on which it is based. It assumes that virtual processes may be described by an infinite set of basic diagrams, the combinatorial properties of which are the same for all field theories. In working out the details of how this language is to be extended to the non-Abelian case, we have attempted to develop it within as broad a framework as possible. Every theorem in this paper will therefore apply not only to the gravitational field but also to the Yang-Mills field[5] which, like the gravitational field, possesses a non-Abelian invariance group.[6]

Section 2 begins with the introduction of a notation which is sufficiently general to embrace all boson field theories and at the same time condensed enough to reduce the highly complex analysis of subsequent sections to manageable proportions. A table is included to facilitate comparison of the condensed notation with the detailed forms which the various symbols take in the case of the Yang-Mills and gravitational fields. The notation is particularly useful in dealing with the second functional derivative of the action, which plays the role of the differential operator governing the propagation of infinitesimal disturbances on an arbitrary background field. It is also useful in dealing with the higher functional derivatives, which are the bare vertex functions of the theory. The problem of the quantized light cone is discussed in a preliminary way in Sec. 3, and its relationship to the "nonrenormalizability" of the theory is noted. Attention is called to the various roles of the background metric, one of which is to define the concepts of "past" and "future." Green's theorem for an arbitrary differential operator is then derived.

Section 4 introduces a notation for the basic structures governing the action of the invariance group on the field variables. The relationship between manifest covariance and linearity of the group transformation laws is emphasized. In Sec. 5 it is pointed out that the infinitesimal disturbances themselves are determined only modulo an Abelian transformation group. This group, which is the tangent group of the full group, affects only the field variables but not physical observables. The latter are necessarily group-invariant. Infinitesimal disturbances satisfying retarded or ad-

[5] C. N. Yang and R. L. Mills, Phys. Rev. **96**, 191 (1954).

[6] The term "invariance group," as used in this paper, will always refer to the infinite dimensional "gauge" group of the theory, and not to the finite dimensional (≤ 10) asymptotic isometry group, which is undetermined *a priori*. It is not hard to show that the Yang-Mills field and its "gauge" group can be given a metrical interpretation which suggests a physical kinship between the Yang-Mills and gravitational fields which is closer than the formal mathematical similarities between them alone indicate. [See B. S. DeWitt, *Dynamical Theory of Groups and Fields* (Gordon and Breach Science Publishers, Inc., New York, 1965), problem 77, p. 139.]

vanced boundary conditions can be computed with the aid of corresponding Green's functions provided supplementary conditions are imposed. For convenience these supplementary conditions are chosen in a manifestly covariant way, but their essential arbitrariness is emphasized.

Use of the covariant Green's functions in connection with Cauchy data for infinitesimal disturbances is discussed in Sec. 6, and the fundamental reciprocity relations of propagator theory are established. Transition from the classical to the quantum theory is made via the Poisson bracket of Peierls (see Ref. 20), which is determined solely by the behavior of infinitesimal disturbances. The reciprocity relations are used to show that Peierls' Poisson bracket satisfies all the usual identities. Section 7 introduces the important concept of the asymptotic fields, which obey the field equations of the linearized theory. From the asymptotic fields one can construct asymptotic invariants, which may be used to characterize completely the physical state of the field. The asymptotic invariants are conditional invariants, i.e., invariants *modulo* the field equations. It is emphasized that their commutators (i.e., Poisson brackets) are nonetheless well defined. A direct proof is given that the asymptotic invariants satisfy the commutation relations of the linearized theory, a result which is nontrivial when a group is present. This result is used in Sec. 8 to construct the creation and annihilation operators for real (i.e., physical) quanta in the remote past and future. The detailed structures of the asymptotic Yang-Mills and gravitational fields must be investigated separately, but a condensed notation (for the asymptotic wave functions) is again introduced, which embraces both fields at once and emphasizes their similarities. A table is included to facilitate the comparison. The quanta of both fields are transverse and differ only in spin. States are labeled by helicity, which is readily shown to be Lorentz-invariant.

Continuing the uniform treatment of the two fields, Sec. 9 shows that the asymptotic commutator functions of both can be expressed in a standard canonical form. A special notation is introduced for the projection of the canonical form into the physical subspace. With the aid of this projection two distinct Feynman propagators are defined relative to an arbitrary background field. Both serve to describe the propagation of field quanta in nonasymptotic regions as well as at infinity. One is manifestly covariant but propagates nonphysical as well as physical quanta; the other propagates physical quanta only but lacks manifest covariance. The latter is used in Sec. 10 to define the external line wave functions which enter into the ultimate definition of the S matrix. These functions serve to generalize the asymptotic wave functions to the case in which an arbitrary background field is present. They satisfy a number of important relations following from a fundamental lemma which is proved in this section. The lemma is used again in Sec. 11 to prove that the non-radiatively-corrected amplitudes for scattering, pair production and pair annihilation by the background field are group-invariant. "Group invariance" here implies invariance under group transformations of the background field, under gauge changes of the propagators, and under radiation gauge changes in the asymptotic wave functions. The amplitudes are also shown to satisfy a set of relations which are the relativistic generalizations of the well known optical theorem for nonrelativistic scattering.

Construction of the full S matrix of the theory is begun in Sec. 12. The field operators are separated into two parts, a classical background satisfying the classical field equations, and a quantum remainder. Vacuum states associated with the remote past and future are defined *relative* to the background field. Vacuum matrix elements of chronological products are constructed by varying the vacuum-to-vacuum amplitude with respect to the background field. It turns out that all physical amplitudes can be obtained in this way despite the fact that the variations in the background field are subject to the constraint that the classical field equations never be violated. The well-known difficulties arising with the use of external sources in a non-Abelian context are thus avoided. When no invariance group is present the vacuum matrix elements of chronological products are expressible in terms of functions having the combinatorial structure of tree diagrams. Use of these functions constitutes an essential part of the program for constructing the S matrix as given in this paper. Since these functions are initially defined only in the absence of an invariance group, however, we are at this point forced to abandon the strict operator formalism. Section 13 displays the structure of the S matrix and its unitarity conditions when no invariance group is present. Section 14 then begins the long and intricate task of generalizing this structure to the case in which a group is present. Aside from an invariance lemma which is used to suggest the desired generalization, the important proof of this section is the *tree theorem*. The tree theorem says that the lowest-order (i.e., non-radiatively corrected) contributions to any scattering process can always be calculated by elementary methods, using any choice of gauge for the propagators of the internal lines and any choice of gauge for the external-line wave functions. The result will be independent of the gauge choices provided all the tree diagrams contributing to the given process are summed together.

There remains only the question of the vacuum-to-vacuum amplitude itself. Since all radiative corrections can be obtained by functionally differentiating this amplitude with respect to the background field, a proof of its group invariance would complete the proof of the invariance of the entire S matrix. The real problem, however, is to *construct* the amplitude, and the invariance criterion must therefore be used as a guide rather than as an *a posteriori* consistency check. Section 15 pauses briefly to review the question of Lorentz invariance, to point out that the theory should also be invariant under changes in the specific variables with which one works, and to comment upon the utility of using c-number language exclusively. Section 16 then plunges into the main problem. The single-loop contribution to the vacuum-to-vacuum amplitude is computed with the aid of the formal theory of continuous determinants, and various alternative forms for it are given. There is no ambiguity about this contribution, and its group invariance is readily demonstrated. This contribution is functionally differentiated in Sec. 17 to yield the lowest-order contribution to single quantum production by the background field. The latter splits into two parts, one involving the covariant propagator for normal quanta and the other involving the covariant propagator for a set of *fictitious quanta* which compensate the nonphysical quanta that the first propagator also carries. The fictitious quanta are coupled to real quanta through asymmetric vertices which vanish when the invariance group is Abelian. With the aid of the fundamental lemma of Sec. 10 and a collection of new identities it is shown that the fictitious quanta can be formally avoided by replacing the covariant propagator by the noncovariant one which carries physical quanta only. The covariant propagators, however, are needed for the practical implementation of any renormalization program.

The lowest-order radiative corrections to the n-quantum amplitudes are analyzed in Sec. 18. These amplitudes split automatically into *Feynman baskets*, i.e., sums over tree amplitudes (lowest-order scattering amplitudes) in which all external lines are on the mass shell. The tree theorem then guarantees their group invariance. This invariance can be made partially manifest by converting from the noncovariant propagator to the covariant one, and the fictitious quanta again make their appearance.

The problem of splitting the radiative corrections into Feynman baskets becomes more difficult in higher orders, when overlapping loops occur. This problem is approached in Sec. 19 with the aid of the Feynman functional integral. When no invariance group is present it is shown that the "measure" or "volume element" for the functional integration plays a fundamental role in the decomposition into Feynman baskets and in guaranteeing the invariance of the vacuum-to-vacuum amplitude under arbitrary changes in the choice of basic field variables. The "measure" has two effects. Firstly, it removes from all closed loops the *noncausal chains* of cyclically connected advanced (or retarded) Green's functions, thereby breaking them open and insuring that at least one segment of every loop is on the mass shell. Secondly, it adds certain nonlocal corrections to the operator field equations, which vanish in the classical limit $\hbar \rightarrow 0$. The question arises why these removals and corrections are always neglected in conventional field theory without apparent harm. It is argued that the usual procedures of renormalization theory automatically take care of them and that in practice the criteria of locality and unitarity are replaced by analyticity statements and Cutkosky rules (see Ref. 52). A detailed investigation of these corrections when a group is present is undertaken in Sec. 20. The two-loop Feynman-basket decomposition of the preceding section is appropriately generalized and the result is reexpressed in terms of covariant propagators, including the fictitious quanta. It turns out that the total two-loop amplitude is obtainable from a set of covariant *primary diagrams* (containing Feynman propagators only, and hence off-mass-shell contributions in all lines) by a process of removing noncausal chains and adding nonlocal corrections, which is completely analogous to that of the no-group case. Moreover, the primary diagrams, taken together, are group-invariant as they stand, *independently of the tree theorem.* This suggests that even when a group is present the noncausal chains and nonlocal corrections may be neglected as in conventional field theory. The problem therefore becomes one of finding a general algorithm for obtaining the primary diagrams of arbitrarily high order, including all fictitious quantum loops. The remainder of Sec. 20 is devoted to the construction of such an algorithm. The generator for the algorithm is a Feynman functional integral for the vacuum-to-vacuum amplitude, which includes fields representing the fictitious quanta. The group invariance of this integral is explicitly demonstrated, and the fictitious quanta are shown formally to obey Fermi statistics despite their integral spin. No physical criteria are violated, however, since the fictitious quanta never occur outside of closed loops. Finally, the rules for inserting external lines into the primary vacuum diagrams are given, and the asymmetric vertices contained in the fictitious quantum loops are shown to have a uniform orientation around each loop.

2. NOTATION. INFINITESIMAL DISTURBANCES. BARE VERTEX FUNCTIONS

A quantum field theory begins with the selection of an action functional S. If the theory is *local* this functional is expressible in the form

$$S \equiv \int \mathcal{L} dx, \quad dx \equiv dx^0 dx^1 dx^2 dx^3, \tag{2.1}$$

where $\mathcal{L}$—the Lagrangian (density)—is a function of the dynamical variables and a finite number of their space-time derivatives at a single point. Various criteria such as covariance, self-consistency of the field equations, the existence of the vacuum as a state of lowest energy, and positive definiteness of the quantum-mechanical Hilbert space in practice drastically limit the possible choices for $\mathfrak{L}$. However, many different choices exist for the Lagrangian of a given field. Thus it is always possible to add a trivial divergence to the Lagrangian without changing the field equations at all. Moreover, the field variables may be replaced by arbitrary functions of themselves; this replaces the field equations by linear combinations of themselves. Finally, even the number of field variables is not unique; for example, alternative Lagrangians may be found leading to field equations which express some of the variables in terms of derivatives of others. What is important is that the choice of Lagrangian is basically irrelevant to the development of the theory of a given field and should be determined only by convenience. The quantum theory of a given field must be constructed in such a way that it is invariant under changes in the mode of description of the field.

It will prove convenient in what follows to adopt a highly condensed notation. The field variables (assumed here to be real) will be denoted by φ^i,[7] and commas followed by indices from the middle of the Greek alphabet will be used to denote differentiation with respect to the space-time coordinates. The first part of the Greek alphabet will be reserved for *group* indices, to be introduced presently. Primes will be used to distinguish different points of space-time; they will also appear on associated indices, or on field symbols themselves, when it is desired to avoid cumbersome explicit appearances of the x's. In most cases, however, the primes will be simply omitted. This corresponds to making the indices i, j, etc. do double duty as discrete labels for field components and as continuous labels over the points of space-time. That is, an index such as i will really stand for the quintuple (i, x^0, x^1, x^2, x^3) and the summation convention for repeated indices will be extended to include integrations over the x's. The significance of the indices thus becomes almost purely combinatorial. When this notation is employed it is necessary to remember that expressions such as M_{ij} are really elements of continuous matrices and that the symbol $\delta^i{}_j$ involves a 4-dimensional δ function.

For most purposes the form of the field equations is more important than the value of the action functional. Therefore, the domain of integration in (2.1) is unimportant; when otherwise unspecified it is to be understood as being large enough to embrace all points at which it may be desired to perform functional differentiations. Functional differentiation with respect to the field variables will be denoted by a comma followed by one or more Latin indices. Thus the field equations will be expressed in the symbolic form

$$S_{,i}=0. \tag{2.2}$$

[7] In this paper no restriction is imposed on the range of Latin indices. Other conventions, to the extent they overlap, are the same as in I.

Suppose the form of the action functional suffers the following change:

$$S\rightarrow S+\epsilon A\,, \tag{2.3}$$

where ϵ is an infinitesimal constant. Such a change may be thought of as being brought about by weak coupling to some external agent. The coupling produces an infinitesimal disturbance $\delta\varphi^i$ in the field, which satisfies the linear inhomogeneous equation

$$S_{,ij}\delta\varphi^j=-\epsilon A_{,i}. \tag{2.4}$$

That is, $\varphi^i+\delta\varphi^i$ satisfies the field equations of the system $S+\epsilon A$ if φ^i satisfies those of the system S. The undisturbed field φ^i may be regarded as a *background field* upon which the disturbance $\delta\varphi^i$ propagates. The concept of the background field proves to be a useful one in the covariant theory, and will occur repeatedly in what follows.

For local theories the quantity $S_{,ij}$ has the form of a linear combination of δ functions and derivatives of δ functions, with functions of the field variables and their derivatives as coefficients. In Eq. (2.4) $S_{,ij}$ therefore plays the role of a linear differential operator with variable coefficients. The reader will find it useful to consult Table I, which lists the explicit forms which this and various other abstract symbols of the general formalism take in the cases of the Yang-Mills field and the gravitational field, respectively.

In the case of linear theories $S_{,ij}$ corresponds to a linear differential operator with constant coefficients, and the higher functional derivatives $S_{,ijk}$, etc., vanish. In nonlinear theories the higher functional derivatives are known as *bare vertex functions*. They describe the basic interactions between *finite* disturbances, the propagation of which, as will be seen later, provides a direct classical model for the quantum S matrix.

It is frequently convenient to introduce a further condensation of notation, namely to make the replacement

$$S_{,i_1\cdots i_n}\rightarrow S_n \tag{2.5}$$

and to drop the indices altogether. Equations (2.2) and (2.4) are then replaced by

$$S_1=0 \tag{2.6}$$

and

$$S_2\delta\varphi=-\epsilon A_1\,, \tag{2.7}$$

respectively. If the basic field variables are properly chosen the number of nonvanishing bare vertex functions is finite in the case of both the Yang-Mills and gravitational fields. Thus, for the Yang-Mills field we have $S_n=0$ for $n>4$ when the field variables are chosen as in Table I, while for the gravitational field we have $S_n=0$ for $n>9$ if the quantities $\varphi^{\mu\nu}\equiv g^{5/18}g^{\mu\nu}-\eta^{\mu\nu}$ are

TABLE I. Expressions for the Yang-Mills and gravitational fields corresponding to quantities appearing in the abstract formalism.

Abstract symbol or equation	Corresponding expression for the Yang-Mills field	Corresponding expression for the gravitational field
φ^i	$A^\alpha{}_\mu;\ \mu=0,1,2,3;\ \alpha=1\cdots n$	$\varphi_{\mu\nu}\equiv g_{\mu\nu}-\eta_{\mu\nu};\ \mu,\nu=0,1,2,3;\ \varphi_{\mu\nu}\equiv\varphi_{\nu\mu}$
S	$S\equiv -\frac{1}{4}\int F_{\alpha\mu\nu}F^{\alpha\mu\nu}dx,$ $F^\alpha{}_{\mu\nu}\equiv A^\alpha{}_{\nu,\mu}-A^\alpha{}_{\mu,\nu}+c^\alpha{}_{\beta\gamma}A^\beta{}_\mu A^\gamma{}_\nu.$ The indices μ, ν are raised and lowered by means of the Minkowski metric $\eta_{\mu\nu}\equiv\mathrm{diag}(-1,1,1,1)$ and its inverse $\eta^{\mu\nu}$. The indices α, β, γ are raised and lowered by means of the Cartan metric, $\gamma_{\alpha\beta}\equiv -c^{-2}c^\gamma{}_{\alpha\delta}c^\delta{}_{\beta\gamma}$ and its inverse $\gamma^{\alpha\beta}$. The c's are the structure constants of a compact n-dimensional semi-simple Lie group, and the constant c^2 is chosen so that $\det(\gamma_{\alpha\beta})=1$.	$S\equiv\int g^{1/2}\,{}^{(4)}R dx,$ $g\equiv-\det(g_{\mu\nu}),\quad {}^{(4)}R\equiv R_\mu{}^\mu,\quad R_{\mu\nu}\equiv R_{\sigma\mu\nu}{}^\sigma,$ $R_{\mu\nu\sigma}{}^\tau\equiv\Gamma_{\nu\sigma}{}^\tau{}_{,\mu}-\Gamma_{\mu\sigma}{}^\tau{}_{,\nu}+\Gamma_{\nu\sigma}{}^\rho\Gamma_{\mu\rho}{}^\tau-\Gamma_{\mu\sigma}{}^\rho\Gamma_{\nu\rho}{}^\tau,$ $\Gamma_{\mu\nu}{}^\sigma\equiv\frac{1}{2}g^{\sigma\tau}(g_{\mu\tau,\nu}+g_{\nu\tau,\mu}-g_{\mu\nu,\tau}).$ The indices μ, ν, ρ, σ, τ are raised and lowered by means of the metric tensor $g_{\mu\nu}$ and its inverse $g^{\mu\nu}$. In the remaining entries of this table the symbol ${}^{(4)}R$ is replaced simply by R.
$0=S_{,i}$	$0=\delta S/\delta A^\alpha{}_\mu\equiv -F_\alpha{}^{\mu\nu}{}_{;\nu}.$	$0=\delta S/\delta\varphi_{\mu\nu}\equiv -g^{1/2}(R^{\mu\nu}-\frac{1}{2}g^{\mu\nu}R).$
$\delta^i{}_j$	$\delta^\alpha{}_{\mu\beta'}{}^{\nu'}\equiv\delta^\alpha{}_\beta\delta_\mu{}^\nu\delta(x,x')$	$\delta_{\mu\nu}{}^{\sigma'\tau'}\equiv\frac{1}{2}(\delta_\mu{}^\sigma\delta_\nu{}^\tau+\delta_\nu{}^\sigma\delta_\mu{}^\tau)\delta(x,x').$
$S_{,ij}$	$\delta^2S/\delta A^\alpha{}_\mu\delta A^{\beta'}{}_{\nu'}\equiv\delta_\alpha{}^\mu{}_{\beta'}{}^{\nu'}{}_{;\sigma}{}^\sigma-\delta_\alpha{}^\sigma{}_{\beta'}{}^{\nu'}{}_{;}{}^\mu{}_\sigma+c^\epsilon{}_{\alpha\gamma}F^{\gamma\mu}{}_\sigma\delta_\epsilon{}^\sigma{}_{\beta'}{}^{\nu'}$	$\delta^2S/\delta\varphi_{\mu\nu}\delta\varphi_{\sigma'\tau'}\equiv\frac{1}{4}g^{1/2}(g^{\mu\rho}g^{\nu\lambda}+g^{\mu\lambda}g^{\nu\rho}-g^{\mu\nu}g^{\rho\lambda})g^{\iota\kappa}$ $\times(\delta_{\rho\lambda}{}^{\sigma'\tau'}{}_{;\iota\kappa}+\delta_{\iota\kappa}{}^{\sigma'\tau'}{}_{;\rho\lambda}-\delta_{\rho\iota}{}^{\sigma'\tau'}{}_{;\lambda\kappa}-\delta_{\lambda\kappa}{}^{\sigma'\tau'}{}_{;\rho\iota})$ $-\frac{1}{2}g^{1/2}(R^{\mu\nu}-\frac{1}{2}g^{\mu\nu}R)\delta_\rho{}^{\rho\sigma'\tau'}$ $+g^{1/2}(R^\mu{}_\rho\delta^{\rho\nu\sigma'\tau'}+R^\nu{}_\rho\delta^{\rho\mu\sigma'\tau'}-\frac{1}{2}R\delta^{\mu\nu\sigma'\tau'}$ $-\frac{1}{2}g^{\mu\nu}R^{\rho\lambda}\delta_{\rho\lambda}{}^{\sigma'\tau'})$
$\delta\xi^\alpha$	The infinitesimal group parameters are functions $\delta\xi^\alpha(x)$ which assign to each point x a corresponding infinitesimal transformation of the generating Lie group. Under inner automorphisms they transform according to the adjoint representation of the full group.	The infinitesimal group parameters are the functions $\delta\xi^\mu(x)$ appearing in the infinitesimal coordinate transformation $\bar{x}^\mu=x^\mu+\delta\xi^\mu$. Under inner automorphisms they transform as contravariant vectors. Note that group and coordinate indices coincide in the case of the general coordinate transformation group.
$R^i{}_\alpha$	$R^\alpha{}_{\mu\beta'}\equiv-\delta^\alpha{}_{\beta';\mu},\quad \delta^\alpha{}_{\beta'}\equiv\delta^\alpha{}_\beta\delta(x,x')$	$R_{\mu\nu\sigma'}\equiv-\delta_{\mu\sigma';\nu}-\delta_{\nu\sigma';\mu},\quad \delta_{\mu\sigma'}\equiv g_{\mu\sigma}\delta(x,x')$
$\delta\varphi^i=R^i{}_\alpha\delta\xi^\alpha$	$\delta A^\alpha{}_\mu=-\delta\xi^\alpha{}_{;\mu}=-\delta\xi^\alpha{}_{,\mu}-c^\alpha{}_{\gamma\beta}A^\gamma{}_\mu\delta\xi^\beta$ Semicolons denote invariant differentiation. A field	$\delta\varphi_{\mu\nu}=-\delta\xi_{\mu;\nu}-\delta\xi_{\nu;\mu}=-g_{\mu\nu,\sigma}\delta\xi^\sigma-g_{\sigma\nu}\delta\xi^\sigma{}_{,\mu}-g_{\mu\sigma}\delta\xi^\sigma{}_{,\nu}$ Semicolons denote covariant differentiation. A field

	quantity φ which has the group transformation law $\delta\varphi = G_\alpha \varphi \delta\xi^\alpha,$ where the G_α are the generators of a matrix representation of the generating Lie group, is defined to have the invariant derivative $\varphi_{;\mu} \equiv \varphi_{,\mu} + G_\alpha A^\alpha{}_\mu \varphi.$ Invariant differentiation leaves transformation properties intact. It has the commutation law $\varphi_{;\mu\nu} - \varphi_{;\nu\mu} = -G_\alpha F^\alpha{}_{\mu\nu}\varphi.$	quantity φ which has the group transformation law $\delta\varphi = -\varphi_{,\mu}\delta\xi^\mu + G^\nu{}_\mu \varphi \delta\xi^\mu{}_{,\nu} = -\varphi_{;\mu}\delta\xi^\mu + G^\nu{}_\mu\varphi\delta\xi^\mu{}_{;\nu}$ where the $G^\nu{}_\mu$ are the generators of a matrix representation of the linear group, is defined to have the covariant derivative $\varphi_{;\mu} \equiv \varphi_{,\mu} + G^\nu{}_\sigma \Gamma_{\mu\nu}{}^\sigma \varphi.$ Covariant differentiation adds one covariant index. It has the commutation law $\varphi_{;\mu\nu} - \varphi_{;\nu\mu} = -G^\sigma{}_\tau R_{\mu\nu\sigma}{}^\tau \varphi.$
$S_{,i}R^i{}_\alpha \equiv 0$	$F^{\alpha\mu\nu}{}_{;\mu\nu} \equiv 0$ This identity is a consequence of the antisymmetry of $F^\alpha{}_{\mu\nu}$ and of the structure constants $c^\gamma{}_{\alpha\beta}$. $F^\alpha{}_{\mu\nu}$ transforms according to the adjoint representation of the group and also satisfies the cyclic identity $F^\alpha{}_{\mu\nu;\sigma} + F^\alpha{}_{\nu\sigma;\mu} + F^\alpha{}_{\sigma\mu;\nu} \equiv 0.$	$-2[g^{1/2}(R^{\mu\nu} - \frac{1}{2}g^{\mu\nu}R)]_{\cdot\nu} \equiv 0.$ This identity results from contracting the *Bianchi identity* $R_{\mu\nu\sigma}{}^\tau{}_{;\rho} + R_{\nu\rho\sigma}{}^\tau{}_{;\mu} + R_{\rho\mu\sigma}{}^\tau{}_{;\nu} \equiv 0$ which can be verified by straightforward computation using the fact that the *Riemann tensor* $R_{\mu\nu\sigma}{}^\tau$ transforms as a mixed tensor of the fourth rank.
$R^i{}_{\alpha,j}$	$\delta R^\alpha{}_{\mu\beta'}/\delta A^{\gamma''}{}_{\nu''} \equiv c^\alpha{}_{\beta\gamma}\delta_\mu{}^\nu \delta(x,x')\delta(x,x'') \equiv c^\alpha{}_{\delta\epsilon}\delta^\delta{}_{\beta'}\delta^\epsilon{}_{\mu\gamma''}{}^{\nu''}$	$\delta R_{\mu\nu\sigma'}/\delta\varphi_{\rho''\tau''} \equiv -\delta_{\mu\nu}{}^{\rho''\tau''}{}_{;\lambda}\delta^\lambda{}_{\sigma'} - \delta_{\lambda\nu}{}^{\rho''\tau''}\delta^\lambda{}_{\sigma';\mu} - \delta_{\mu\lambda}{}^{\rho''\tau''}\delta^\lambda{}_{\sigma';\nu}$
$c^\alpha{}_{\beta\gamma}$	$c^\alpha{}_{\beta'\gamma''} \equiv c^\alpha{}_{\beta\gamma}\delta(x,x')\delta(x,x'')$	$c^\mu{}_{\nu'\sigma''} \equiv \delta^\mu{}_\nu \delta_{,\sigma}(x,x')\delta(x,x'') - \delta^\mu{}_\sigma \delta_{,\nu}(x,x'')\delta(x,x')$ $= \delta^\mu{}_{\nu';\tau}\delta^\tau{}_{\sigma''} - \delta^\mu{}_{\sigma'';\tau}\delta^\tau{}_{\nu'}$
γ_{ij}	$\gamma_\alpha{}^\mu{}_{\beta'}{}^{\nu'} \equiv -\delta_\alpha{}^\mu{}_{\beta'}{}^{\nu'}$	$\gamma^{\mu\nu\sigma'\tau'} \equiv -\frac{1}{2}g^{1/2}(\delta^{\mu\nu\sigma'\tau'} - \frac{1}{2}g^{\mu\nu}\delta_\rho{}^{\rho\sigma'\tau'})$
$\hat{F}_{\alpha\beta}$	$\hat{F}_{\alpha\beta'} \equiv \delta_{\alpha\beta';\sigma}{}^\sigma$	$\hat{F}_{\mu\nu'} \equiv g^{1/2}(\delta_{\mu\nu';\sigma}{}^\sigma + R_\mu{}^\sigma\delta_{\sigma\nu'})$
$\tilde{\gamma}_{\alpha\beta}$	$\tilde{\gamma}_{\alpha\beta'} \equiv -\delta_{\alpha\beta'}$	$\tilde{\gamma}_{\mu\nu'} \equiv -g^{1/2}\delta_{\mu\nu'}$
$\tilde{\gamma}^{-1\alpha\beta}$	$\tilde{\gamma}^{-1\alpha\beta'} \equiv -\delta^{\alpha\beta'}$	$\tilde{\gamma}^{-1\mu\nu'} \equiv -\delta^{\mu\nu'}g'^{-1/2}$
F_{ij}	$F_\alpha{}^\mu{}_{\beta'}{}^{\nu'} \equiv \delta_\alpha{}^\mu{}_{\beta'}{}^{\nu'}{}_{;\sigma}{}^\sigma + 2c^\epsilon{}_{\alpha\gamma}F^{\gamma\mu}{}_\sigma \delta_\epsilon{}^\sigma{}_{\beta'}{}^{\nu'}$	$F^{\mu\nu\sigma'\tau'} \equiv \frac{1}{4}g^{1/2}(g^{\mu\rho}g^{\nu\lambda} + g^{\mu\lambda}g^{\nu\rho} - g^{\mu\nu}g^{\rho\lambda})$ $\times(\delta_{\rho\lambda}{}^{\sigma'\tau'}{}_{;\kappa}{}^\kappa - 2R_{\rho\iota\lambda\kappa}\delta^{\iota\kappa\sigma'\tau'} - R_{\rho\lambda}\delta_\kappa{}^{\kappa\sigma'\tau'}$ $+ 2R_\rho{}^\kappa\delta_{\kappa\lambda}{}^{\sigma'\tau'} - g_{\rho\lambda}R_{\iota\kappa}\delta^{\iota\kappa\sigma'\tau'} - R\delta_{\rho\lambda}{}^{\sigma'\tau'}$ $+ \frac{1}{2}g_{\rho\lambda}R\delta_\kappa{}^{\kappa\sigma'\tau'})$ The last five terms inside the parentheses may be omitted when the field equations are satisfied.

chosen as the basic field variables.[8] With the conventional choice of Table I the number of nonvanishing gravitational vertex functions is infinite.

For a local theory a typical term in S_n involves the product of $n-1$ δ functions or derivatives of δ functions. In momentum space with a constant (e.g., flat) background field these reduce to a single δ function, which expresses the conservation of momentum of the n field quanta taking part in the elementary process described by the vertex in question. The calculation of specific processes is usually most conveniently performed in momentum space; the development of the general theory, however—in particular, the demonstration of the covariance of renormalization procedures—is best done in coordinate space.

Because of the commutativity of functional differentiation the bare vertex functions $S_{,ijk\cdots}$ are completely symmetric in their indices, and $S_{,ij}$ corresponds to a self-adjoint linear operator. When employing the notation (2.5) we may regard the symbol S_2 as actually representing this operator. Note: The abstract notation must be used with a measure of caution because the associative law of matrix multiplication does not always hold. If Φ^i and Ψ^i are two functions which do not vanish rapidly at infinity, the value of the expression $\Phi^i S_{,ij} \Psi^j$ may depend on which implicit integration is performed first. This ambiguity may be removed by using arrows to distinguish the two possibilities: $\Phi^{\sim}\overrightarrow{S}_2\Psi$ and $\Phi^{\sim}\overleftarrow{S}_2\Psi$.[9]

The present discussion will be limited to boson fields. For the extension of the formalism to the case of fermion fields, which involves anticommutative differentiation and antisymmetric vertex functions, the reader may consult the reference given in Ref. 6. This reference contains detailed proofs of some of the important theorems to be stated in what follows. We shall therefore restrict ourselves here to sketch-proofs or simple statements of these theorems but will take the occasion to improve their presentation.

PHYSICAL REVIEW VOLUME 162, NUMBER 5 25 OCTOBER 1967

Quantum Theory of Gravity. III. Applications of the Covariant Theory*

BRYCE S. DEWITT
Institute for Advanced Study, Princeton, New Jersey
and
Department of Physics, University of North Carolina, Chapel Hill, North Carolina†

(Received 25 July 1966; revised manuscript received 9 January 1967)

The basic momentum-space propagators and vertices (including those for the fictitious quanta) are given for both the Yang-Mills and gravitational fields. These propagators are used to obtain the cross sections for gravitational scattering of two scalar particles, scattering of gravitons by scalar particles, graviton-graviton scattering, two-graviton annihilation of scalar-particle pairs, and graviton bremsstrahlung. Special features of these cross sections are noted. Problems arising in renormalization theory and the role of the Planck length are discussed. The gravitational Ward identity is derived, and the structure of the radiatively corrected 1-graviton vertex for a scalar particle is displayed. The Ward identity is only one of an infinity of identities relating the many-graviton vertex functions of the theory. The need for such identities may be eliminated in principle by computing radiative corrections directly in coordinate space, using the theory of manifestly covariant Green's functions. As an example of such a calculation, the contribution of conformal metric fluctuations to the vacuum-to-vacuum amplitude is summed to all orders. The physical significance of the renormalization terms is discussed. Finally, Weinberg's treatment of the infrared problem is examined. It is not difficult to show that the fictitious quanta contribute negligibly to infrared amplitudes, and hence that Weinberg's use of the DeDonder gauge is justified. His proof that the infrared problem in gravidynamics can be handled just as in electrodynamics is thereby made rigorous.

1. INTRODUCTION

IN the first two papers of this series[1] two distinct mathematical approaches to the quantum theory of gravity were developed, one based on the so-called canonical or Hamiltonian theory and the other on the manifestly covariant theory of propagators and diagrams. So far no rigorous mathematical link between the two has been established. In part this is due to the kinds of questions each asks. The canonical theory leads almost unavoidably to speculations about the meaning of "amplitudes for different 3-geometries" or "the wave function of the universe." The covariant theory, on the other hand, concerns itself with "microprocesses" such as scattering, vacuum polarization, etc. Some of the questions raised by the canonical theory were explored in I. In this third and final paper of the series we examine some of the consequences of the covariant theory.

* This research was supported in part by the Air Force Office of Scientific Research under Grant AFOSR-153-64 and in part by the National Science Foundation under Grant GP7437.

† Permanent address.

[1] B. S. DeWitt, Phys. Rev. **160**, 1113 (1967); preceding paper, *ibid*. **162**, 1195 (1967). These papers will be referred to as I and II, respectively. The notation of the present paper is the same as that of II, which should be consulted for the definition of unfamiliar symbols, e.g., S_n for the n-pronged bare vertex and $V_{(\alpha i)\beta}$ for the asymmetric vertex coupling real and fictitious quanta.

Armed with the formalism constructed in II one can in principle carry out the calculation of any "microprocess" to any order of perturbation theory in a manner which is completely invariant and unambiguous except for the arbitrary high-energy cutoff which must be introduced to render divergent integrals finite. A few of these calculations have actually been performed, and the only thing which prevents more of them from being done is the extreme tediousness of the algebra involved and the lack of any experimental motivation for them. It is a pity that Nature displays such indifference to so intriguing and beautiful a subject, for the calculations themselves are of considerable intrinsic interest. The present paper contains several examples. They are by no means exhaustive but have been selected as useful landmarks in a still largely unexplored territory. Not all of these were originally carried out by the author, but it is hoped that their unified presentation here will make their results more accessible than hitherto.

Section 2 begins with the rules of calculation in momentum space. The basic structural elements of the theory, namely the propagators for real and fictitious quanta, the vertices S_3, S_4, $V_{(\alpha i)\beta}$, and the coupling with matter fields, are given for both the Yang-Mills and gravitational fields. The standard Feynman rules are summarized. The results of a few lowest-order scattering calculations based on these rules are given and discussed in Sec. 3. Included are the cross section for gravitational scattering of two scalar particles, the cross section for scattering of gravitons by scalar particles, the corresponding annihilation cross section, and the graviton-graviton cross section. Section 4 is devoted to the problem of gravitational bremsstrahlung. The role of the energy quadrupole moment tensor and the absence of the forward peak at high energies, characteristic of photon bremsstrahlung, are noted.

Section 5 discusses some of the problems which arise in renormalization theory. Although the Yang-Mills theory looks as if it may be renormalizable (provided its infrared difficulties can be disposed of), quantum gravidynamics is definitely not renormalizable in the usual sense. Tentative proposals for dealing with this situation are briefly described, as is also the evidence that gravity contains its own cutoff—at the Planck length. Illustration of the actual details of the renormalization program, by explicit calculation of a radiative correction, is postponed to Sec. 7.

The gravitational Ward identity and its implications for gravitational form factors are derived in Sec. 6. The general structure of the radiatively corrected 1-graviton vertex is displayed in the case of a scalar particle. It is emphasized that the gravitational Ward identity is only one of an infinity of identities relating the many-graviton vertex functions of the theory. The need for Ward identities can be eliminated by computing radiative corrections directly in coordinate space rather than in momentum space. An example of such a calculation is given in Sec. 7, where the contribution of conformal metric fluctuations to the vacuum-to-vacuum amplitude is summed to all orders. The calculation, which is manifestly covariant throughout, makes use of an integral representation for the amplitude. A resumé is given of that part of the mathematical theory of covariant Green's functions which is needed.

Section 8 concludes the paper with a review of Weinberg's treatment of the infrared problem (see Ref. 37). If Yang-Mills quanta are assumed to be massless then, since they can act as their own sources, they give rise to the special infrared divergences which plague massless electrodynamics. Weinberg showed that gravity miraculously escapes these difficulties; its infrared divergences can be handled by the standard methods familiar in ordinary quantum electrodynamics. His proofs, however, were incomplete, since he did not have available a fully elaborated quantum theory. In particular he used the DeDonder gauge without taking into account the fictitious quanta. It is not difficult to show that the fictitious quanta contribute negligibly in the infrared limit. Weinberg's results are therefore rigorous.

2. RULES OF CALCULATION IN MOMENTUM SPACE

We begin with the vertex functions for the Yang-Mills field interacting with itself. We have seen in II that when the standard field variables are used only S_3 and S_4 are nonvanishing for this case. In momentum space these become (apart from a δ function expressing conservation of momentum)

$$\frac{\delta^3 S}{\delta A^{\alpha}{}_{\mu}\delta A^{\beta'}{}_{\nu'}\delta A^{\gamma''}{}_{\sigma''}} \to i c_{\alpha\beta\gamma}[(p'^{\mu}-p''^{\mu})\eta^{\nu\sigma} + (p''^{\nu}-p^{\nu})\eta^{\sigma\mu} + (p^{\sigma}-p'^{\sigma})\eta^{\mu\nu}], \tag{2.1}$$

$$\begin{aligned}\frac{\delta^4 S}{\delta A^{\alpha}{}_{\mu}\delta A^{\beta'}{}_{\nu'}\delta A^{\gamma''}{}_{\sigma''}\delta A^{\delta'''}{}_{\tau'''}} \to & -c_{\alpha\beta\epsilon}c^{\epsilon}{}_{\gamma\delta}(\eta^{\mu\sigma}\eta^{\nu\tau}-\eta^{\mu\tau}\eta^{\nu\sigma}) \\ & -c_{\alpha\gamma\epsilon}c^{\epsilon}{}_{\beta\delta}(\eta^{\mu\nu}\eta^{\sigma\tau}-\eta^{\mu\tau}\eta^{\sigma\nu}) \\ & -c_{\alpha\delta\epsilon}c^{\epsilon}{}_{\gamma\beta}(\eta^{\mu\sigma}\eta^{\tau\nu}-\eta^{\mu\nu}\eta^{\tau\sigma}).\end{aligned} \tag{2.2}$$

The correspondence of momenta with indices is $p\alpha\mu$, $p'\beta'\nu'$, $p''\gamma''\sigma''$, $p'''\delta'''\tau'''$. All momenta are incoming (to the vertex), and momentum conservation implies $p+p'+p''=0$ for S_3 and $p+p'+p''+p'''=0$ for S_4. Indices on the structure constants are raised and lowered by means of the Cartan metric $\gamma_{\alpha\beta}$. When all indices are in the lower position the structure constants are completely antisymmetric.

In addition to the above vertices, the fictitious vertex $V_{(\alpha i)\beta}$ is needed for the calculation of radiative corrections. For the Yang-Mills field it takes the form

$$V_{(\alpha\gamma''{}^{\sigma''})\beta'} \to -ic_{\alpha\beta\gamma}p'^{\sigma} = -ic_{\alpha\gamma\beta}(p^{\sigma}+p''^{\sigma}). \quad (2.3)$$

The propagators for the normal and fictitious quanta are, respectively,

$$G \to \gamma^{\alpha\beta}\eta_{\mu\nu}/p^2, \quad (2.4)$$

$$\hat{G} \to \gamma^{\alpha\beta}/p^2, \quad (2.5)$$

with p^2 being understood to have the usual small negative imaginary part.

The corresponding quantities for the gravitational field are much more complicated. In this case we shall employ the momentum-index combinations $p\mu\nu$, $p'\sigma'\tau'$, $p''\rho''\lambda''$, $p'''\iota'''\kappa'''$. The vertices must not only be symmetric in each index pair but must also remain unchanged under arbitrary permutations of the momentum-index triplets. At least 171 separate terms are required in the complete expression for S_3 in order to exhibit this full symmetry, and for S_4 the number is 2850. However, these numbers can be greatly reduced by counting only the combinatorially distinct terms[2] and leaving it understood that the appropriate symmetrizations are to be carried out. In this way S_3 is reduced to 11 terms and S_4 to 28 terms, as follows:

$$\frac{\delta^3 S}{\delta\varphi_{\mu\nu}\delta\varphi_{\sigma'\tau'}\delta\varphi_{\rho''\lambda''}} \to$$

$$\begin{aligned}\mathrm{Sym}[&-\tfrac{1}{4}P_3(p\cdot p'\eta^{\mu\nu}\eta^{\sigma\tau}\eta^{\rho\lambda})-\tfrac{1}{4}P_6(p^{\sigma}p^{\tau}\eta^{\mu\nu}\eta^{\rho\lambda})+\tfrac{1}{4}P_3(p\cdot p'\eta^{\mu\sigma}\eta^{\nu\tau}\eta^{\rho\lambda})+\tfrac{1}{2}P_6(p\cdot p'\eta^{\mu\nu}\eta^{\sigma\rho}\eta^{\tau\lambda})+P_3(p^{\sigma}p^{\lambda}\eta^{\mu\nu}\eta^{\tau\rho})\\&-\tfrac{1}{2}P_3(p^{\tau}p'^{\mu}\eta^{\nu\sigma}\eta^{\rho\lambda})+\tfrac{1}{2}P_3(p^{\rho}p'^{\lambda}\eta^{\mu\sigma}\eta^{\nu\tau})+\tfrac{1}{2}P_6(p^{\rho}p^{\lambda}\eta^{\mu\sigma}\eta^{\nu\tau})+P_6(p^{\sigma}p'^{\lambda}\eta^{\tau\mu}\eta^{\nu\rho})+P_3(p^{\sigma}p'^{\mu}\eta^{\tau\rho}\eta^{\lambda\nu})\\&-P_3(p\cdot p'\eta^{\nu\sigma}\eta^{\tau\rho}\eta^{\lambda\mu})],\end{aligned} \quad (2.6)$$

$$\frac{\delta^4 S}{\delta\varphi_{\mu\nu}\delta\varphi_{\sigma'\tau'}\delta\varphi_{\rho''\lambda''}\delta\varphi_{\iota'''\kappa'''}} \to$$

$$\begin{aligned}\mathrm{Sym}[&-\tfrac{1}{8}P_6(p\cdot p'\eta^{\mu\nu}\eta^{\sigma\tau}\eta^{\rho\lambda}\eta^{\iota\kappa})-\tfrac{1}{8}P_{12}(p^{\sigma}p^{\tau}\eta^{\mu\nu}\eta^{\rho\lambda}\eta^{\iota\kappa})-\tfrac{1}{4}P_6(p^{\sigma}p'^{\mu}\eta^{\nu\tau}\eta^{\rho\lambda}\eta^{\iota\kappa})+\tfrac{1}{8}P_6(p\cdot p'\eta^{\mu\sigma}\eta^{\nu\tau}\eta^{\rho\lambda}\eta^{\iota\kappa})\\&+\tfrac{1}{4}P_6(p\cdot p'\eta^{\mu\nu}\eta^{\sigma\tau}\eta^{\rho\iota}\eta^{\lambda\kappa})+\tfrac{1}{4}P_{12}(p^{\sigma}p^{\tau}\eta^{\mu\nu}\eta^{\rho\iota}\eta^{\lambda\kappa})+\tfrac{1}{2}P_6(p^{\sigma}p'^{\mu}\eta^{\nu\tau}\eta^{\rho\iota}\eta^{\lambda\kappa})-\tfrac{1}{4}P_6(p\cdot p'\eta^{\mu\sigma}\eta^{\nu\tau}\eta^{\rho\iota}\eta^{\lambda\kappa})\\&+\tfrac{1}{4}P_{24}(p\cdot p'\eta^{\mu\nu}\eta^{\sigma\rho}\eta^{\tau\lambda}\eta^{\iota\kappa})+\tfrac{1}{4}P_{24}(p^{\sigma}p^{\tau}\eta^{\mu\rho}\eta^{\nu\lambda}\eta^{\iota\kappa})+\tfrac{1}{4}P_{12}(p^{\rho}p'^{\lambda}\eta^{\mu\sigma}\eta^{\nu\tau}\eta^{\iota\kappa})+\tfrac{1}{2}P_{24}(p^{\sigma}p'^{\rho}\eta^{\tau\mu}\eta^{\nu\lambda}\eta^{\iota\kappa})\\&-\tfrac{1}{2}P_{12}(p\cdot p'\eta^{\nu\sigma}\eta^{\tau\rho}\eta^{\lambda\mu}\eta^{\iota\kappa})-\tfrac{1}{2}P_{12}(p^{\sigma}p'^{\mu}\eta^{\tau\rho}\eta^{\lambda\nu}\eta^{\iota\kappa})+\tfrac{1}{2}P_{12}(p^{\sigma}p^{\rho}\eta^{\tau\lambda}\eta^{\mu\nu}\eta^{\iota\kappa})-\tfrac{1}{2}P_{24}(p\cdot p'\eta^{\mu\nu}\eta^{\tau\rho}\eta^{\lambda\iota}\eta^{\kappa\sigma})\\&-P_{12}(p^{\sigma}p^{\tau}\eta^{\nu\rho}\eta^{\lambda\iota}\eta^{\kappa\mu})-P_{12}(p^{\rho}p'^{\lambda}\eta^{\nu\iota}\eta^{\kappa\sigma}\eta^{\tau\mu})-P_{24}(p_{\sigma}p'^{\rho}\eta^{\tau\iota}\eta^{\kappa\mu}\eta^{\nu\lambda})-P_{12}(p^{\rho}p'^{\iota}\eta^{\lambda\sigma}\eta^{\tau\mu}\eta^{\nu\kappa})\\&+P_6(p\cdot p'\eta^{\nu\rho}\eta^{\lambda\sigma}\eta^{\tau\iota}\eta^{\kappa\mu})-P_{12}(p^{\sigma}p^{\rho}\eta^{\mu\nu}\eta^{\tau\iota}\eta^{\kappa\lambda})-\tfrac{1}{2}P_{12}(p\cdot p'\eta^{\mu\rho}\eta^{\nu\lambda}\eta^{\sigma\iota}\eta^{\tau\kappa})-P_{12}(p^{\sigma}p^{\rho}\eta^{\tau\lambda}\eta^{\mu\iota}\eta^{\nu\kappa})\\&-P_6(p^{\rho}p'^{\iota}\eta^{\lambda\kappa}\eta^{\mu\sigma}\eta^{\nu\tau})-P_{24}(p^{\sigma}p'^{\rho}\eta^{\tau\mu}\eta^{\nu\iota}\eta^{\kappa\lambda})-P_{12}(p^{\sigma}p'^{\mu}\eta^{\tau\rho}\eta^{\lambda\iota}\eta^{\kappa\nu})+2P_6(p\cdot p'\eta^{\nu\sigma}\eta^{\tau\rho}\eta^{\lambda\iota}\eta^{\kappa\mu})].\end{aligned} \quad (2.7)$$

The "Sym" standing in front of these expressions indicates that a symmetrization is to be performed on each index pair $\mu\nu$, $\sigma\tau$, etc. The symbol P indicates that a summation is to be carried out over all distinct permutations of the momentum-index triplets, and the subscript gives the number of permutations required in each case.

Expressions (2.6) and (2.7) can be obtained in a straightforward manner by repeated functional differentiation of the Einstein action. This procedure, however, is exceedingly laborious. A more efficient (but still lengthy) method is to make use of the hierarchy of identities (II, 17.31). It is a remarkable fact that once $S_2{}^0$ is known all the higher vertex functions, and hence the complete action functional itself, are determined by the general coordinate invariance of the theory. It is convenient, in the actual computation of the vertices via (II, 17.31), to invent diagrammatic schemes for displaying the combinatorics of indices. Since each reader will devise the scheme which suits him best we shall not shackle him by describing one here. We also make no attempt to display S_5 or any higher vertices.

The vertex $V_{(\alpha i)\beta}$ has the following form for the gravitational field:

$$\begin{aligned}V_{(\mu^{\sigma''\tau''})\nu'} \to\ &\tfrac{1}{2}\mathrm{Sym}[2p''_{\mu}p'^{\sigma}\delta_{\nu}{}^{\tau}-p''_{\mu}p'_{\nu}\eta^{\sigma\tau}\\&+(p_{\nu}p'^{\sigma}-p'_{\nu}p^{\sigma})\delta_{\mu}{}^{\tau}+p\cdot p'\delta_{\mu}{}^{\sigma}\delta_{\nu}{}^{\tau}],\end{aligned} \quad (2.8)$$

where the momentum-index combinations are $p\mu$, $p'\nu'$, $p''\sigma''\tau''$, and the symmetrization is to be performed on the index pair $\sigma\tau$. The propagators for the normal and fictitious quanta are given by

$$G \to (\eta_{\mu\sigma}\eta_{\nu\tau}+\eta_{\mu\tau}\eta_{\nu\sigma}-\eta_{\mu\nu}\eta_{\sigma\tau})/p^2, \quad (2.9)$$

$$\hat{G} \to \eta^{\mu\nu}/p^2. \quad (2.10)$$

[2] The choice of terms is not completely unique since momentum conservation may be used to replace a given term by other terms. We give here what we believe (but have not proved) to be the expressions containing the smallest number of terms.

If one wishes to calculate processes involving the interaction of the Yang-Mills and/or gravitational field with matter, additional vertices describing this interaction must be included. As prototypes of such vertices, we shall display those which arise from interactions with scalar (or pseudoscalar) particles. The latter particles contribute to the total action functional an expression of the form

$$S_\varphi = -\frac{1}{2}\int g^{1/2}(\bar{\varphi}_{;\mu}\varphi_{;}{}^{\mu} + m^2\bar{\varphi}\varphi)dx\,, \tag{2.11}$$

where the covariant derivative is defined in Table I of II and where

$$\bar{\varphi} \equiv \varphi^{\sim}\gamma\,, \quad \gamma G_\alpha \gamma^{-1} = -G_\alpha{}^{\sim}\,, \tag{2.12}$$

γ being the matrix which connects the two forms of a self-contragredient representation (of the Yang-Mills Lie group) generated by the matrices G_α and $-G_\alpha{}^{\sim}$, respectively. We find

$$\frac{\delta^3 S_\varphi}{\delta\bar{\varphi}'\delta\varphi\delta A^{\alpha''}{}_{\mu''}} \to -i(p'^\mu - p^\mu)G_\alpha\,, \tag{2.13}$$

$$\frac{\delta^4 S_\varphi}{\delta\bar{\varphi}'\delta\varphi\delta A^{\alpha''}{}_{\mu''}\delta A^{\beta'''}{}_{\nu'''}} \to (G_\alpha G_\beta + G_\beta G_\alpha)\eta^{\mu\nu}\,, \tag{2.14}$$

$$\frac{\delta^3 S_\varphi}{\delta\bar{\varphi}'\delta\varphi\delta\varphi_{\mu''\nu''}} \to -\tfrac{1}{2}[\eta^{\mu\nu}(m^2 - p\cdot p') + p^\mu p'^\nu + p^\nu p'^\mu]\,, \tag{2.15}$$

$$\begin{aligned}\frac{\delta^4 S_\varphi}{\delta\bar{\varphi}'\delta\varphi\delta\varphi_{\mu''\nu''}\delta\varphi_{\sigma'''\tau'''}} \to\ & \tfrac{1}{4}[(\eta^{\mu\sigma}\eta^{\nu\tau} + \eta^{\mu\tau}\eta^{\nu\sigma} - \eta^{\mu\nu}\eta^{\sigma\tau})(m^2 - p\cdot p') \\ & + (p^\mu p'^\sigma + p^\sigma p'^\mu)\eta^{\nu\tau} + (p^\nu p'^\tau + p^\tau p'^\nu)\eta^{\mu\sigma} \\ & + (p^\mu p'^\tau + p^\tau p'^\mu)\eta^{\nu\sigma} + (p^\nu p'^\sigma + p^\sigma p'^\nu)\eta^{\mu\tau} \\ & - (p^\mu p'^\nu + p^\nu p'^\mu)\eta^{\sigma\tau} - (p^\sigma p'^\tau + p^\tau p'^\sigma)\eta^{\mu\nu}.\end{aligned} \tag{2.16}$$

The corresponding vertices which describe the interaction of the gravitational and/or Yang-Mills fields with particles having spin are obtained by straightforward computation from the pertinent action functional. The latter is obtained in each case via the "principle of minimal coupling" (which, in the case of gravity, is nothing but the "strong equivalence principle") from the corresponding action functional in the absence of gravitational and Yang-Mills fields, by replacing ordinary derivatives by covariant derivatives, the Minkowski metric $\eta_{\mu\nu}$ by $g_{\mu\nu}$, and the volume element dx by $g^{1/2}\,dx$. We do not give here the results of such calculations for particles with spin but merely point out (what is more useful for the reader) that the three-pronged vertices, when sandwiched between normalized wave functions, always reduce, in the limit of zero momentum transfer, with particle momenta on the mass shell, to precisely the forms (2.13) and (2.15) regardless of the magnitude of the particle spin. This may be proved in each instance as a straightforward consequence of the gauge invariance of the theory and, when extended to the radiatively corrected vertices, constitutes a boundary condition on the Yang-Mills and gravitational form factors.[3] [See also Sec. 6.]

It is to be emphasized that the inclusion of additional fields in no way affects the formal theoretical structure developed in II. The topology and invariance properties of diagrams remain completely unchanged. One simply permits the field indices i, j, etc., to extend over a greater range of values in order to accommodate the components of the new fields which have been added. The only differences are differences of detail such as, for example, the sign modifications due to statistics which appear when some of the added components are those of fermion fields, or changes in the structure of the invariance group which arise from having both the Yang-Mills and gravitational fields simultaneously present and interacting with each other.[4]

The rules for combining vertices and propagators into transition amplitudes are completely standard. With the notational conventions of the present paper they may be summarized as follows: (1) An expression such as (2.1), (2.2), (2.3), (2.6), etc., for each vertex; (2) an expression such as (2.4), (2.5), (2.9), (2.10), etc. for each propagator; (3) a factor $(-i)/(2\pi)^4$ for each independent closed loop; (4) an additional factor (-1) for each closed fermion or fictitious-quantum loop, or when necessary to assure antisymmetry of fermion amplitudes; (5) an over-all factor $i(2\pi)^4$ times a δ function assuring total energy-momentum conservation; (6) a wave function $u^i{}_A$ (see Table II of II) or its complex conjugate evaluated at $x=0$ for each external line; (7) integration over all the independent momenta.

Gauge invariance may be invoked as a useful consistency check in all calculations. However, it must be applied to the entire amplitude for a given process and not merely to a single diagram. It is therefore algebraically more laborious than corresponding checks in electrodynamics. It is no longer possible to exploit charge conservation by following individual lines

[3] These are analogs of the electromagnetic form factors. The gravitational form factors are also sometimes referred to as stress-energy, mass, or mechanical form factors.

[4] These are, in fact, the most important differences. It is worth mentioning that when fermion fields are included it is usually convenient to replace the metric field $g_{\mu\nu}$ by a *vierbein* field. Otherwise the group transformation laws are no longer linear. [See B. S. DeWitt and C. M. DeWitt, Phys. Rev. **87**, 116 (1952).] We also mention that the combined *vierbein*-general-coordinate-transformation group has the structure of a semi-direct product based on the automorphisms of the *vierbein* group under general coordinate transformations. In the combined group only the *vierbein* group is an invariant subgroup. The coordinate transformation group is its factor group. Similar statements apply to the combined Yang-Mills-general coordinate-transformation group. The analysis of these cases is therefore correspondingly complicated.

through diagrams, for now the conserved quantity—Yang-Mills charge, energy-momentum—leaks all over every diagram. Moreover, when Yang-Mills quanta or gravitons interact with themselves, the closed loops form traffic jams of spurious charge which can be unsnarled only by calling the fictitious quanta to the rescue.

3. SCATTERING CROSS SECTIONS

We now display some of the lowest-order amplitudes and scattering cross sections which the covariant theory yields. One of the simplest is the amplitude for the scattering of two identical scalar particles by exchange of a single Yang-Mills quantum. This has the form

$$i(2\pi)^{-2}\delta(p_1'+p_2'-p_1-p_2)j_\alpha(1)\cdot j^\alpha(2)/q^2$$
$$+\text{exchange and virtual annihilation terms}, \quad (3.1)$$

where

$$q=p_1'-p_1=p_2-p_2', \quad (3.2)$$

$$j_{\alpha\mu}=\tfrac{1}{2}i(E'E)^{-1/2}\chi'^\dagger\gamma G_\alpha\chi(p'_\mu+p_\mu), \quad (3.3)$$

the χ's being the internal (Yang-Mills group) states of the particles and the remaining notation being conventional. The same form (3.1) also holds for particles with spin, but the expression for the current $j_{\alpha\mu}$ is then more complicated.

Since the initial and final momenta are on the mass shell we have the conservation laws

$$j_\alpha(1)\cdot q=j_\alpha(2)\cdot q=0, \quad (3.4)$$

which permit the scattering amplitude to be reexpressed in the form

$$-\frac{j_{\alpha 0}(1)j^\alpha{}_0(2)}{\mathbf{q}^2}+\frac{j_{\alpha 1}(1)j^\alpha{}_1(2)+j_{\alpha 2}(1)j^\alpha{}_2(2)}{q^2},$$
$$+\text{exchange and virtual annihilation terms}, \quad (3.5)$$

where a factor $i(2\pi)^{-2}\delta(p_1'+p_2'-p_1-p_2)$ has been removed, and the 3-axis has been chosen in the direction of the spatial part $\mathbf{q}$ of the space-like 4-vector q. The first term of (3.5) represents the instantaneous "Coulomb" interaction of the particles; the second represents a "delayed" interaction propagated by transverse quanta, the factors $j_{\alpha 1}$ and $j_{\alpha 2}$ being separately coupled to the two states of linear polarization of these quanta.

The corresponding amplitude arising from exchange of a graviton is

$$i(4\pi)^{-2}\delta(p_1'+p_2'-p_1-p_2)T_{\mu\nu}(1)$$
$$\times(\eta^{\mu\sigma}\eta^{\nu\tau}+\eta^{\mu\tau}\eta^{\nu\sigma}-\eta^{\mu\nu}\eta^{\sigma\tau})T_{\sigma\tau}(2)/q^2$$
$$+\text{exchange and virtual annihilation terms}, \quad (3.6)$$

where

$$T_{\mu\nu}=\tfrac{1}{2}(E'E)^{-1/2}[p_\mu p'_\nu+p_\nu p'_\mu-\eta_{\mu\nu}(p\cdot p'+m^2)]. \quad (3.7)$$

Again we have conservation laws

$$T_{\mu\nu}(1)q^\nu=T_{\mu\nu}(2)q^\nu=0, \quad (3.8)$$

which permit the amplitude to be recast in the form

$$\tfrac{1}{4}\{T_{00}(1)T_{00}(2)-4T_{01}(1)T_{01}(2)-4T_{02}(1)T_{02}(2)$$
$$-T_{03}(1)T_{03}(2)-T_{00}(1)[T_{11}(2)+T_{22}(2)]$$
$$-[T_{11}(1)+T_{22}(1)]T_{00}(2)\}/\mathbf{q}^2+\tfrac{1}{4}\{[T_{11}(1)-T_{22}(1)]$$
$$\times[T_{11}(2)-T_{22}(2)]+4T_{12}(1)T_{12}(2)\}/q^2$$
$$+\text{exchange and virtual annihilation terms}. \quad (3.9)$$

The first term yields an instantaneous "Newtonian" interaction, while the second gives rise to a "delayed" interaction propagated by transverse gravitons. In this case the factors which couple separately to the two states of linear polarization are $T_{11}-T_{22}$ and $2T_{12}$, respectively.

From (3.6) it is straightforward to compute the differential cross section for gravitational scattering of identical scalar particles in the center-of-mass frame. One finds[5]

$$\frac{d\sigma}{d\Omega}=\frac{G^2E^2}{16}\left[\frac{(1+3v^2)(1-v^2)+4v^2(1+v^2)\cos^2(\theta/2)}{v^2\sin^2(\theta/2)}+\frac{(1+3v^2)(1-v^2)+4v^2(1+v^2)\sin^2(\theta/2)}{v^2\cos^2(\theta/2)}+(3-v^2)(1+v^2)+2v^4\sin^2\theta\right]^2, \quad (3.10)$$

where $v=|\mathbf{p}|/E$ and the gravitation constant has been reintroduced through the units convention $16\pi G=1$. The nonrelativistic and extreme relativistic limits of this cross section are, respectively,

$$\left(\frac{d\sigma}{d\Omega}\right)_{\rm NR}=\frac{G^2m^2}{16}\left[\frac{1}{v^2\sin^2\frac{1}{2}\theta}+\frac{1}{v^2\cos^2\frac{1}{2}\theta}+3\right]^2, \quad (3.11)$$

$$\left(\frac{d\sigma}{d\Omega}\right)_{\rm ER}=4G^2E^2[\cot^2\tfrac{1}{2}\theta+\tan^2\tfrac{1}{2}\theta+\tfrac{1}{2}+\tfrac{1}{4}\sin^2\theta]^2. \quad (3.12)$$

In a similar manner one may compute the cross section for scattering of gravitons by scalar particles. The relevant diagrams are shown in Fig. 1, the heavy lines denoting particles and the light lines gravitons. Diagrams (a) and (b) vanish in the rest frame of the target particle, and one finds for unpolarized gravitons[5]

$$\frac{d\sigma}{d\Omega}=\frac{G^2m^2}{[1+2\epsilon\sin^2\frac{1}{2}\theta]^2\sin^4\frac{1}{2}\theta}[\cos^8\tfrac{1}{2}\theta+\sin^8\tfrac{1}{2}\theta$$
$$+2\epsilon(\cos^6\tfrac{1}{2}\theta+\sin^6\tfrac{1}{2}\theta)\sin^2\theta$$
$$+\epsilon^2(\cos^4\tfrac{1}{2}\theta+\sin^4\tfrac{1}{2}\theta)\sin^4\theta], \quad (3.13)$$

$$\left(\frac{d\sigma}{d\Omega}\right)_{\rm NR}=\frac{G^2m^2}{\sin^4\frac{1}{2}\theta}(\cos^8\tfrac{1}{2}\theta+\sin^8\tfrac{1}{2}\theta), \quad (3.14)$$

$$\left(\frac{d\sigma}{d\Omega}\right)_{\rm ER}=4G^2m^2(\cos^4\tfrac{1}{2}\theta+\sin^4\tfrac{1}{2}\theta)\cot^4\tfrac{1}{2}\theta, \quad (3.15)$$

[5] C. F. Cooke, Ph.D. thesis, University of North Carolina, 1964 (unpublished).

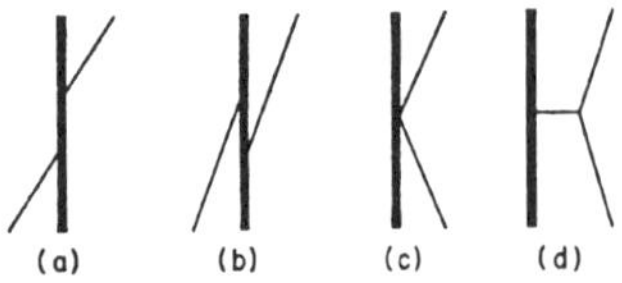

FIG. 1. Lowest-order diagrams for scattering of a graviton by a material particle. The heavy line denotes the particle and the light lines denote gravitons.

where ϵ is the energy of the graviton measured in units of m. It will be noted that these cross sections have no resemblance to those for Compton scattering, but on the contrary, continue to display the sharp forward "Rutherford peak" characteristic of long-range interactions. This feature is due to diagram (d) of Fig. 1 whose presence, as may be readily checked, is essential for the gauge invariance of the scattering amplitude. Owing to the equivalence principle gravitons, like photons, are deflected by a gravitational field (in particular by the long-range static field of any material particle), and the above cross sections are dominated by this effect.

By the well-known "substitution rule" the diagrams of Fig. 1 yield also the amplitude for annihilation of a pair of scalar particles into gravitons. We record here only the low- and high-energy limits of the total annihilation cross section in the center-of-mass frame:

$$\sigma_{\mathrm{NR}} = 2\pi G^2 m^2 / v\,, \tag{3.16}$$

$$\sigma_{\mathrm{ER}} = (38\pi/3) G^2 E^2\,. \tag{3.17}$$

The cross section for the inverse process, namely, the production of a scalar pair by colliding gravitons (again in the center-of-mass frame) is identical with (3.17) at high energies. Near threshold, on the other hand, it is given by

$$\sigma = 2\pi G^2 m^2 (\epsilon^2 - 1)^{1/2}/\epsilon\,, \quad \epsilon \geq 1\,. \tag{3.18}$$

The only elastic process which remains to be considered is the scattering of one graviton by another. This process has some unusual features. It turns out that the helicity of the colliding gravitons is *individually* conserved. That is, there is no spin flip, in spite of the presence of derivative coupling. If both gravitons are right (left) handed before collision then both are right (left) handed after the collision. If one is right handed and the other left handed then they maintain this relationship also, through the collision.

The helicity of extremely relativistic particles, and of massless quanta in particular, is notoriously rigid. In the classical theory, for example, the spin of such a particle suffers no precession under geodetic motion in an external gravitational field but remains always pointing parallel or antiparallel to the trajectory. However, no general principle has yet been discovered which implies that helicity conservation must hold to all orders of perturbation theory. It has so far been demonstrated only in lowest order, by carrying out a brute-force computation of the relevant amplitudes.

The tediousness of the algebra involved in obtaining the graviton-graviton cross section may be inferred from the complexity of the vertex functions (2.6) and (2.7) which are involved in the diagrams which represent the amplitude (Fig. 1 with the heavy lines replaced by graviton lines). Fortunately, the presence of the polarization tensors in the external-line wave functions, and the momentum condition $p^2=0$ for free quanta, eliminate many of the terms from these expressions. Nevertheless, a large amount of cancellation between terms still has to be dug out of the algebra, and this, combined with the fact that the final results are ridiculously simple, leads one to believe that there must be an easier way. The cross sections which one finds are

$$\frac{d\sigma_{++}}{d\Omega} = \frac{d\sigma_{--}}{d\Omega} = 4G^2E^2 \left[\frac{\cos^6\frac{1}{2}\theta}{\sin^2\frac{1}{2}\theta} + \frac{\sin^6\frac{1}{2}\theta}{\cos^2\frac{1}{2}\theta} + 4 - \tfrac{1}{2}\sin^2\theta \right]^2, \tag{3.19}$$

$$\frac{d\sigma_{+-}}{d\Omega} = 4G^2E^2 \frac{\cos^{12}\frac{1}{2}\theta}{\sin^4\frac{1}{2}\theta}\,, \tag{3.20}$$

showing again the forward Rutherford peaking.

We shall not record here the corresponding cross sections involving Yang-Mills quanta, since these depend, in their finer details, on which Lie group is chosen as generator of the Yang-Mills group and on which representations are chosen for the material particles. There is also a serious difficulty with the Yang-Mills field in regard to the infrared catastrophe, which will be discussed in Sec. 8. Since our primary interest in this article is the gravitational field, we refer the reader with a special interest in Yang-Mills cross sections to the dissertations of Remler[6] and Dotson.[7] It is, however, perhaps worth remarking that in the case of the scattering of one Yang-Mills quantum by another the phenomenon of helicity conservation is again found to hold, with or without the inclusion of graviton exchange forces in the total amplitude, and regardless of the choice of the Lie group. Moreover, an extension of the helicity conservation rule to processes involving real gravitons in interaction with Yang-Mills quanta apparently exists. Thus individual helicities remain unchanged when a graviton and a Yang-Mills quantum collide elastically. If the diagrams contributing to this process are turned on their sides so as to yield the amplitudes for annihilation of two Yang-Mills quanta into a pair of gravitons (or the reverse process) further selection rules emerge. One

[6] E. A. Remler, Ph.D. thesis, University of North Carolina, 1964 (unpublished).

[7] A. C. Dotson, Ph.D. thesis, University of North Carolina, 1964 (unpublished).

finds that it is impossible to produce two gravitons having opposite helicities by annihilation of Yang-Mills quanta, or conversely to produce a Yang-Mills pair having opposite helicities by the reverse process. The quanta in both the initial and final states must have identical helicities if the amplitude is to be nonvanishing. Helicity selection rules exist even for the process in which two Yang-Mills quanta coalesce to produce a single Yang-Mills quantum and a graviton. If both initial quanta have the same helicity the final quanta must have this helicity too; if the initial helicities are opposite the final helicities must be opposite. The same obviously holds for the reverse process.

4. GRAVITATIONAL BREMSSTRAHLUNG

Since the problem of gravitational radiation from accelerating masses has bedeviled classical relativists for years it is a pleasant surprise to discover that its treatment within the quantum framework is quite simple.[8] Consider a scattering diagram in which one of the lines represents a scalar particle (real or virtual) of momentum p. Let the diagram be modified by the emission of a graviton of momentum q from this line. If the momenta of all lines subsequent to the inserted graviton vertex are held fixed while those prior to the vertex are adjusted in such a way as to conserve momentum and keep external lines on the mass shell, then the only additional effect of the graviton emission is to introduce into the corresponding amplitude, a factor

$$\frac{e_\pm{}^{\mu*}e_\pm{}^{\nu*}}{(2\pi)^{3/2}}\frac{1}{\sqrt{q^0}}\times\frac{p_\mu(p_\nu+q_\nu)+p_\nu(p_\mu+q_\mu)-\eta_{\mu\nu}[m^2+p\cdot(p+q)]}{(p+q)^2+m^2-i0},\tag{4.1}$$

which follows from Eq. (2.15) and Table II of II. Alternatively, if the momenta *prior* to the vertex are held fixed we get a factor which differs from (4.1) by the replacement $q\to -q$.

If the graviton is emitted from an external line these factors reduce to

$$\frac{e_\pm{}^{\mu*}e_\pm{}^{\nu*}}{(2\pi)^{3/2}}\frac{1}{\sqrt{q^0}}\frac{p_\mu p_\nu+\frac{1}{2}\eta(p_\mu q_\nu+p_\nu q_\mu-\eta_{\mu\nu}p\cdot q)}{q^2+2\eta p\cdot q-i0},\tag{4.2}$$

where $\eta=+1$ or -1 according as the external line is outgoing or incoming, and p is held fixed on the mass shell. In the long-wavelength limit $q\to 0$ (4.2) itself reduces to

$$\frac{\eta}{2(2\pi)^{3/2}\sqrt{q^0}}\frac{(e_\pm{}^*\cdot p)^2}{p\cdot q-i\eta 0},\tag{4.3}$$

and simply multiplies the original amplitude. This limiting form actually holds for *all* external lines, regardless of the spin character of their associated particles. It even holds when the external line is a graviton line, provided the emission vertex is inserted not merely into a single diagram but into the sum of all diagrams contributing to the original amplitude. This may be verified in a straightforward manner by plugging in the 3-pronged graviton vertex (2.6) and eliminating the terms involving q. Of the remaining terms only those survive which yield a net contribution of the form (4.3); the rest disappear in virtue of the gauge invariance of the total original amplitude.[9]

The multiplicative factor (4.3) exhibits the well-known infrared divergence and can be obtained from a purely classical model. We note that the infrared divergence shows up only when the emission takes place from lines on the mass shell; it does not occur when the emission is from internal lines of a scattering diagram. The external lines therefore dominate the soft graviton emission. This means that the precise details of the scattering process have little relevance in the limit $q\to 0$, and that the long-wavelength end of the emission spectrum is determined primarily by the asymptotic trajectories of the incoming and outgoing particles, just as in the case of photon bremsstrahlung. For wavelengths large compared to the space-time region in which the collision takes place (the size of this region is determined by the magnitudes of typical energies exchanged in the collision) the effective graviton source is a stress tensor of the form

$$T^{\mu\nu}(x)=\sum_n \eta_n m_n V_n{}^\mu V_n{}^\nu\int_0^{\eta_n\infty}\delta(x-V_n\tau)d\tau,\tag{4.4}$$

which idealizes the particles to classical points colliding at the coordinate origin. Here m_n and V_n are, respectively, the mass and 4-velocity of the nth particle, and the sign factor η_n tells whether the particle is incoming or outgoing. The summation is over all the external lines, and the velocities are subject to the energy-momentum conservation law:

$$\sum_n \eta_n m_n V_n=0.\tag{4.5}$$

The classical emission spectrum is obtained by projecting (4.4) onto the graviton wave functions $u_{\mu\nu\pm}(x,\mathbf{q})$ (see Table II of II). The corresponding quantum amplitude is

$$\begin{aligned}\tfrac{1}{2}i\int u_{\mu\nu\pm}{}^*(x,\mathbf{q})T^{\mu\nu}(x)dx &= i\sum_n\frac{\eta_n m_n(e_\pm{}^*\cdot V_n)^2}{2(2\pi)^{3/2}\sqrt{q^0}}\int dx\int_0^{\eta_n\infty}d\tau\, e^{-iq\cdot x}\delta(x-V_n\tau)\\ &=\sum_n\frac{\eta_n m_n}{2(2\pi)^{3/2}\sqrt{q^0}}\frac{(e_\pm{}^*\cdot V_n)^2}{V_n\cdot q-i\eta_n 0},\end{aligned}\tag{4.6}$$

[8] The comparison is actually unfair. The questions which classical relativists ask are usually quite detailed—e.g., the precise damped motion of radiating sources, or the nonlinear properties of coherent waves of large amplitude—and are inevitably much more difficult.

[9] The gauge invariance holds in every order of perturbation theory.

which, in view of the relation $p_n = m_n V_n$, is just (4.3) summed over all the external lines.

When the collision is nonrelativistic (4.6) reduces to

$$\tfrac{1}{2}(2\pi q^0)^{-3/2}\mathbf{e}_\pm{}^* \cdot \Delta\mathfrak{T} \cdot \mathbf{e}_\pm{}^*, \tag{4.7}$$

where the graviton gauge is chosen so that the components $e_\pm{}^0$ of the polarization vectors $e_\pm$ vanish, and $\Delta\mathfrak{T}$ is the change in the spatial integral of the total 3-stress dyadic as a result of the collision:

$$\Delta\mathfrak{T} = \Delta \int \mathbf{T} d\mathbf{x} = \sum_n \eta_n \mathbf{p}_n \mathbf{v}_n, \tag{4.8}$$

$$\mathbf{T} = \sum_n \theta(\eta_n x^0)\mathbf{p}_n \mathbf{v}_n \delta(\mathbf{x} - \mathbf{v}_n x^0), \tag{4.9}$$

$$\mathbf{v}_n \equiv \mathbf{V}_n / V_n{}^0 = \mathbf{p}_n / E_n \approx \mathbf{p}_n / m_n. \tag{4.10}$$

Now it is well known[10] that energy-momentum conservation permits the integral of the 3-stress dyadic to be reexpressed as one half the second time derivative of the second moment $\int \mathbf{x}\mathbf{x} T_{00} d\mathbf{x}$ of the energy density. Moreover, since $\mathbf{e}_\pm{}^* \cdot \mathbf{e}_\pm{}^* = 0$, the trace of $\Delta\mathfrak{T}$ may be removed from (4.7).[11] Therefore the emission amplitude may be written in the alternative form:

$$\tfrac{1}{4}(2\pi q^0)^{-3/2}\mathbf{e}_\pm{}^* \cdot \Delta(d^2\mathbf{Q}/dt^2) \cdot \mathbf{e}_\pm{}^*, \tag{4.11}$$

where $\Delta(d^2\mathbf{Q}/dt^2)$ denotes the change in the second time derivative of the energy quadrupole moment tensor

$$\mathbf{Q} = \int (\mathbf{x}\mathbf{x} - \tfrac{1}{3}\mathbf{1}\mathbf{x}^2) T_{00} d\mathbf{x}, \tag{4.12}$$

showing that soft gravitons are emitted predominantly in the quadrupole mode.

It is of interest to examine the angular distribution of the emitted radiation. From (4.6) one sees that each external line makes a contribution to the emission amplitude, which has an angular distribution of the form

$$\frac{\sin^2\theta}{1 - v\cos\theta} e^{2i\varphi}, \tag{4.13}$$

where θ is the angle between $\mathbf{v}$ and $\mathbf{q}$, and φ is a helicity phase angle. In the case of photon bremsstrahlung the $\sin\theta$ appears linearly instead of quadratically in the numerator, with the consequence that for relativistic collisions ($v \approx 1$) the emission is concentrated sharply in the forward directions of all the particles (initial as well as final). This peaking may be attributed to the individual Lorentz-contracted Coulomb fields, which resemble bundles of plane waves having momenta confined to narrow cones. These bundles (particularly their outer regions) have difficulty readjusting to the altered particle trajectories arising from the collision and hence partly escape as radiation.

In the gravitational case the sharp forward emission is absent.[12] In fact for an extremely relativistic collision ($|\mathbf{p}_n| \approx E_n$) which is confined to a plane (e.g., 2-particle scattering) it is easy to verify that the total sum (4.6) yields an amplitude which vanishes for emission *in* the plane.[13] This implies that, unlike photon emission, graviton emission is a cooperative phenomenon which cannot be traced to the individual particle fields. Indeed the *real* gravitational field of a particle, namely the Riemann tensor, falls off as the inverse cube rather than the inverse square of the distance, and hence its outer regions contribute negligibly to the emission. This has obvious implications for investigations of classical 2-body radiation as well as for attempts to introduce Weizsäcker-Williams approximation schemes into quantum calculations.

5. RENORMALIZATION AND THE PLANCK LENGTH

In lowest-order perturbation theory the formal rules of the manifestly covariant theory yield results which agree with the classical theory in the correspondence principle limit. In higher orders, divergences appear, just as they do for other field theories, and almost nothing is known about how to extract finite and physically meaningful radiative corrections from the results. In the case of quantum gravidynamics the severity of the divergences is such that the theory is not, by standard criteria, renormalizable. This is due to the quadratic momentum dependence of the vertices $S_n (n \geqslant 3)$, which in turn may be traced to the dependence of the light cone on the background field, i.e., to the field dependence of the coefficients of the second time derivatives appearing in S_2. Thus by counting momentum powers one finds for the superficial degree of divergence of any diagram

$$D = -2L_i + 2\sum_n V_n + 4K, \tag{5.1}$$

where L_i denotes the number of internal lines, V_n the number of n-pronged vertices, and K the number of independent momentum integrations. Now it is not difficult to show that

$$K = L_i - \sum_n V_n + 1. \tag{5.2}$$

---

[10] See, for example, L. D. Landau and E. M. Lifshitz, *The Classical Theory of Fields*, translated by M. Hammermesh (Addison-Wesley Publishing Company, Inc., Reading, Massachusetts, 1962), rev. 2nd Ed.

[11] In view of the nonrelativistic energy conservation law $\sum_n \eta_n (m_n + \frac{1}{2}\mathbf{p}_n \cdot \mathbf{v}_n) = 0$, this trace is just twice the rest mass lost in the collision and already vanishes for elastic collisions.

[12] This was first pointed out by R. P. Feynman in a mimeographed letter to V. F. Weisskopf dated January 4 to February 11, 1961 (unpublished).

[13] Introducing unit vectors Ω and Ω_n in the directions of $\mathbf{q}$ and $\mathbf{p}_n$, respectively, one may write the amplitude in this case in the form

$$\text{const} \times \sum_n \eta_n E_n \frac{(\Omega \times \Omega_n)^2}{1 - \Omega \cdot \Omega_n} = \text{const} \times \sum_n \eta_n E_n (1 + \Omega \cdot \Omega_n),$$

which vanishes by energy-momentum conservation.

Therefore

$$D=2(K+1), \quad K\geqslant 1, \tag{5.3}$$

which increases without limit as the order of the diagram increases.

In the case of the Yang-Mills field the situation is better. Here we have

$$D=-2L_i+V_3+4K, \tag{5.4}$$

which, in combination with the readily verified combinatorial law

$$L_e+2L_i=\sum_n nV_n=3V_3+4V_4, \tag{5.5}$$

yields

$$D=4-L_e, \tag{5.6}$$

where L_e is the number of external lines. In order to compensate the divergences one may introduce counter terms into the original Lagrangian in the conventional manner. The most divergent counter term is always the simplest. It is necessarily of the form const$\times F_{\alpha\mu\nu}F^{\alpha\mu\nu}$, provided the divergence has been handled in a manifestly group-invariant manner. But such a counter term can be detected by as many as four external lines. Hence D is never actually greater than zero. This is quite analogous to the situation which occurs with vacuum polarization diagrams in quantum electrodynamics: Although $L_e=2$, D is reduced from 2 to 0 by gauge invariance. One therefore expects that, with proper handling of the overlapping divergences, a careful analysis will show that none of the ultraviolet divergences of the Yang-Mills theory is worse that logarithmic, and that only a single counter term, of the above mentioned type, is needed for each group-invariant set of diagrams. If that is true it is then not difficult to show that renormalization merely rescales the structure constants $c^\alpha{}_{\beta\gamma}$. Mathematically, this is equivalent to a rescaling of coordinates in the group manifold; physically, it corresponds to a change in the strength of the coupling of the Yang-Mills field to itself and to other Yang-Mills-charge bearing fields.

It should be remarked that although the above results [Eqs. (5.3) and (5.6)] have been stated for the case in which each field interacts only with itself, they hold also when other fields bearing stress-energy or Yang-Mills charge are present, provided the spins of the added fields are not greater than $\frac{1}{2}$ and their other mutual interactions are renormalizable.[14] Unfortunately, in the case of the Yang-Mills field the ultraviolet divergences are not the whole story; infrared difficulties of a special type also make their appearance. These will be discussed in Sec. 8.

In the case of gravity there are no infrared problems beyond those which can be handled by conventional methods. Equation (5.3), however, casts a rather dismal light on the ultraviolet problem. Faced with its brutal consequences there are several paths one may try to follow to make life bearable. One of these is to soften the degree of divergence by abandoning S-matrix unitarity (i.e., positive definiteness of Hilbert space) through the introduction of field equations of the fourth differential order. This may be accomplished by adding terms of the form $\int g^{1/2\,(4)}R^2dx$ and $\int g^{1/2}R_{\mu\nu}R^{\mu\nu}dx$ to the Einstein action, which changes Eqs. (5.1) and (5.3) to

$$D=-4L_i+4\sum_n V_n+4K=4 \tag{5.7}$$

for all diagrams of order greater than zero. Such a procedure in effect introduces a separate unit of mass (or length) into the theory, and if this mass is chosen sufficiently big, the S matrix will be *nearly* unitary, significant departures from unitarity occurring only under extreme conditions, when collision energies approach that of the unit of mass.

Nevertheless, it would be nice if any breakdown in conventional ideas which may be necessary were to emerge from the quantized Einstein theory in its unmutilated form. There is already a unit of mass in the theory: the absolute unit $(\hbar c/16\pi G)^{1/2}=3.07\times10^{-6}$g $\sim10^{18}$ BeV, and one is loath to introduce another. One might try to use the extra mass merely as a regulator, in a spirit similar to that of the ξ-limiting proposal of Lee and Yang[15] for the charged vector boson. Equation (5.7) suggests that the regulated theory may be renormalizable, requiring only three counter terms, respectively, quartically, quadratically, and logarithmically divergent. If this is so one might attempt to let the regulator mass become infinite *after* the renormalization has been performed. However, there is no guarantee that the renormalized amplitudes will themselves remain finite in the limit, nor that unitarity, which is violated by the regulator, will then be restored. If unitarity stays violated one is not sure whether this represents a fundamental feature of the quantized Einstein theory or is merely a consequence of the regulator approach; one would be inclined to suspect the latter. Halpern[16] has criticized unorthodox uses of regulators in handling nonrenormalizable field theories, and has shown that they often lead to illegal modifications of analytic properties.

If regulators are to be excluded then perturbation theory cannot be used except in a formal way. One must necessarily sum infinite classes of diagrams and hope that the increasingly strong divergence of the successive terms of the series, as expressed by Eq. (5.3), will lead to high-energy damping and a finite result for the total amplitude. The author has shown[17] that this hope is actually fulfilled for at least one

[14] In the case of gravity additional fields of spin 1 are allowed if they are massless.

[15] T. D. Lee and C. N. Yang, Phys. Rev. **128**, 885 (1962); **128**, 899 (1962).

[16] M. B. Halpern, Phys. Rev. **140**, B1570 (1965).

[17] B. S. DeWitt, Phys. Rev. Letters **13**, 114 (1964).

simple class of diagrams, namely, those which represent two scalar particles exchanging gravitons in the ladder approximation. It turns out that the "leading terms" (i.e., the most divergent) of the Bethe-Salpeter amplitude can be summed exactly, and, owing to certain remarkable cancellations, the sum of the ladder-type contributions to the gravitational self-energy can be expanded in a power series in the bare mass, with no approximations whatever. The method can also be extended to the case of charged scalar particles, with one or more of the graviton ladder rungs replaced by photons, and a simple expression can be obtained for the lowest-order electromagnetic self-energy. The self-energies and renormalization constants found in this way are all *finite*.

The finiteness of these quantities may be traced to the behavior of the particle-particle scattering amplitude. In the limit of very high momentum transfer the singularity of the gravitational interaction kernel is displaced off the light cone in coordinate space and onto a hyperboloid lying at a distance $\lambda=(4\hbar G/\pi c^3)^{1/2}=1.82\times10^{-33}$ cm in spacelike directions. This is roughly equivalent to endowing the scalar particles with the properties of hard spheres of diameter λ, and may be regarded as a manifestation of the smearing out of the light cone due to quantum fluctuations.

Similar results have been found for spin-$\frac{1}{2}$ electrons by Khriplovich,[18] and there seems to be no reason why, with enough labor, they may not also be extended to particles of higher spin, including the graviton in interaction with itself. Thus gravity may indeed prove to be the universal regulator which renders all field theories finite.

It should be remarked that the self-energy functions which are obtained by summing ladder graphs appear to correspond to "good" spectral functions, which do a minimum of violence to unitarity. This suggests that no illegal analytic operations have inadvertently crept into the summation procedure. An improved calculational method, which insures analytic legality in general, has been developed by Halpern.[16] He sums first the absorptive parts of any amplitude and then obtains the full amplitude by a dispersion integral. The technique is applicable to gravity theory as well as to other nonrenormalizable theories, and is amenable to N/D approximation schemes. It is probably the safest method currently available, but it is very complicated to apply.

Although the finite results which have been obtained thus far are very suggestive, one must remember that they derive from restricted classes of diagrams. They are therefore not γ-invariant but depend on the particular gauge chosen for the internal graviton lines. So far calculations have been restricted to those gauges which avoid "dangerous" singularities in the resulting integral equations, or otherwise simplify the computational labor. It is clear that the results can give at best only a qualitative insight into the true analytic structure of the theory.

[18] I. B. Khriplovich, report, Siberian Section, Academy of Science, USSR, Novosibirsk, 1965 (unpublished).

6. THE GRAVITATIONAL WARD IDENTITY

Although the computational difficulties involved in extracting physical information from quantum gravidynamics are formidable, the theory has a redeeming feature in its general covariance, which serves as a cross check on the consistency of various calculations and imposes constraints on the permissible forms of various amplitudes. One of these constraints has recently been discussed by Brout and Englert.[19] These authors derive a generalized Ward identity relating the gravitational vertex function of a scalar particle to the self-energy function arising from all its interactions. Their derivation is easily generalized to the case of a particle of arbitrary spin.

Denote the field of the particle by φ^A. In addition to the functions $R^i{}_\alpha$ (or, in expanded notation, $R_{\mu\nu\sigma'}$) characterizing the coordinate transformation behavior of the gravitational field (see II) we now have corresponding functions $R^A{}_\alpha$ for φ^A. The explicit structure of these functions may be inferred from Table I of II:

$$R^A{}_{\mu'}=-\varphi^A{}_{,\mu}\delta(x,x')+G^\nu{}_\mu{}^A{}_B\varphi_B\delta_{,\nu}(x,x'). \qquad (6.1)$$

We note that $R^A{}_{\mu'}$ vanishes in the limit $\varphi^A\to0$, and that its functional derivative has the momentum-space form

$$R^A{}_{\mu',B''}\to-i\delta^A{}_Bp''{}_\mu+iG^\nu{}_\mu{}^A{}_Bp'{}_\nu, \qquad (6.2)$$

in which the association of momenta with indices is pA, $p'\mu'$, $p''B''$ $(p+p'+p''=0)$.

Let us denote the full (radiatively corrected) propagator for the particle by S^{AB}. It is the sum of the bare propagator G^{AB} and a function obtained by applying the operator $G^{AB}\delta/\delta\varphi^B$ twice to the vacuum-to-vacuum amplitude. Since the vacuum-to-vacuum amplitude is an invariant the propagator S^{AB}, like G^{AB}, transforms in the manner indicated by the position of its indices.[20] Its inverse must transform contragrediently:

$$S^{-1}{}_{AB,i}R^i{}_\alpha+S^{-1}{}_{AB,C}R^C{}_\alpha = -S^{-1}{}_{CB}R^B{}_{\alpha,A}-S^{-1}{}_{AC}R^C{}_{\alpha,B}. \qquad (6.3)$$

Equation (6.3) is the gravitational Ward identity. To get it into more familiar form one must reexpress it in momentum space, with all the background fields set equal to zero. In this limit $S^{-1}{}_{AB,i}$ becomes the negative of the gravitational vertex function, which is conven-

[19] R. Brout and F. Englert, Phys. Rev. 141, 1231 (1966). See also K. Just and K. Rossberg, Nuovo Cimento 40, 1077 (1965).

[20] This will be true even if φ^A possesses a gauge group of its own, provided the gauge conditions which determine G^{AB} are covariant. Note that the "background field" now includes φ^A in addition to the metric field.

tionally denoted by $\Gamma^{\mu\nu}$, the particle indices being suppressed and the index i being replaced by the more explicit $\mu\nu$. Making use of (6.2) and the momentum space form of $R^i{}_\alpha$, which is given in Table II of II, one readily finds

$$2\Gamma_\mu{}^\nu(p',p)q_\nu=S^{-1}(p')p_\mu-S^{-1}(p)p'_\mu -[S^{-1}(p')G^\nu{}_\mu-G^\nu{}_\mu{}^\sim S^{-1}(p)]q_\nu, \quad (6.4)$$

where p and p' are, respectively, the incoming and outgoing particle momenta and $q=p'-p$ is the incoming graviton momentum. This, with the spin terms involving $G^\nu{}_\mu$ omitted is the equation given by Brout and Englert. It holds, as a simple consequence of general covariance, no matter how many other fields are coupled to the field φ^A and involved in the structure of the vertex function.

Now introduce the vertex and wave-function renormalization constants Z_1 and Z_2. They are defined by

$$u^\dagger(p)\Gamma_{\mu\nu}(p,p)u(p)=Z_1^{-1}u^\dagger(p)\gamma_{\mu\nu}(p,p)u(p), \quad p^2=-m^2, \quad (6.5)$$

$$S^{-1}(p)=Z_2^{-1}[G^{-1}(p)+\bar{\Sigma}(p)], \quad [\partial\bar{\Sigma}(p)/\partial p^\mu]_{p^2=-m^2}=0, \quad (6.6)$$

where $\gamma_{\mu\nu}$ and G are the bare vertex and propagation functions, respectively, m is the particle rest mass, and $u(p)$ is a particle wave function satisfying

$$S^{-1}(p)u(p)=G^{-1}(p)u(p)=0 \quad (6.7)$$

on the mass shell. From (6.6) we may infer

$$[\partial S^{-1}(p)/\partial p^\mu]_{p^2=-m^2}=Z_2^{-1}[\partial G^{-1}(p)/\partial p^\mu]_{p^2=-m^2}. \quad (6.8)$$

On the other hand, (6.4) yields, in the limit $p'\to p$,

$$2\Gamma_{\mu\nu}(p,p)=p_\mu\partial S^{-1}(p)/\partial p^\nu-\eta_{\mu\nu}S^{-1}(p) -S^{-1}(p)G_{\nu\mu}-G_{\nu\mu}{}^\sim S^{-1}(p), \quad (6.9)$$

whence, in virtue of (6.7),

$$2u^\dagger(p)\Gamma_{\mu\nu}(p,p)u(p)=p_\mu u^\dagger(p)[\partial S^{-1}(p)/\partial p^\nu]u(p), \quad p^2=-m^2. \quad (6.10)$$

Now, since (6.4) is a consequence simply of general covariance, it holds also if $\Gamma_{\mu\nu}$ and S^{-1} are replaced by $\gamma_{\mu\nu}$ and G^{-1}, respectively. Therefore we have

$$2u^\dagger(p)\gamma_{\mu\nu}(p,p)u(p)=p_\mu u^\dagger(p)[\partial G^{-1}(p)/\partial p^\nu]u(p), \quad p^2=-m^2. \quad (6.11)$$

From (6.5), (6.8), and (6.11) it follows that

$$Z_1=Z_2. \quad (6.12)$$

When both vertex and wave-function radiative corrections are taken into account the two renormalizations cancel, and there remains only the graviton renormalization Z_3 arising from vacuum polarization,[21] which has the effect of modifying the gravitation constant.

The cancellation of *divergences* which is implied by (6.12) applies only to the leading term of the vertex function, in the limit $p'\to p$, and only on the mass shell. In order that no divergences occur in the remaining terms, or off the mass shell, the interactions which the field φ^A experiences with other fields must be of the renormalizable type (or else summable to finite values). The example of the scalar particle provides an adequate illustration of the conditions which must be satisfied. In this case we have[22]

$$G^{-1}(p)=p^2+m^2, \quad (6.13)$$

$$\gamma_{\mu\nu}(p',p)=\tfrac{1}{2}[p_\mu p'_\nu+p_\nu p'_\mu-\eta_{\mu\nu}(p\cdot p'+m^2)] \quad (6.14a)$$

$$=\gamma^0{}_{\mu\nu}(p',p)-\tfrac{1}{2}\eta_{\mu\nu}\delta m^2, \quad (6.14b)$$

where the index 0 refers to the bare mass, and we may write

$$S^{-1}(p)=p^2+m_0{}^2+\Sigma(p^2), \quad \delta m^2=m^2-m_0{}^2=\Sigma(-m^2), \quad (6.15)$$

$$\Gamma_{\mu\nu}(p',p)=\gamma^0{}_{\mu\nu}(p',p)+\Lambda_{\mu\nu}(p',p). \quad (6.16)$$

The functions Σ and Λ are related by the Ward identity as follows:

$$2\Lambda_{\mu\nu}(p',p)q^\nu=\Sigma(p'^2)p_\mu-\Sigma(p^2)p'_\mu. \quad (6.17)$$

It is not hard to show that the general solution of (6.17) is

$$\Lambda_{\mu\nu}(p',p)=\frac{1}{2}\frac{\Sigma(p'^2)-\Sigma(p^2)}{p'^2-p^2}(p_\mu p'_\nu+p_\nu p'_\mu+\tfrac{1}{2}q_\mu q_\nu) -\tfrac{1}{4}[\Sigma(p'^2)+\Sigma(p^2)]\eta_{\mu\nu} +F(p'^2,p^2,q^2)(q^2\eta_{\mu\nu}-q_\mu q_\nu), \quad (6.18)$$

where F is an arbitrary function. Therefore the graviton vertex of a scalar particle is characterized on the mass shell, by a single function of q^2. This is the gravitational form factor.

Now introduce the renormalized self-energy function $\bar{\Sigma}$, defined by

$$\Sigma(p^2)=\delta m^2+(Z_2{}^{-1}-1)(p^2+m^2)+Z_2{}^{-1}\bar{\Sigma}(p^2), \quad \bar{\Sigma}(-m^2)=0, \quad [d\bar{\Sigma}(p^2)/dp^2]_{p^2=-m^2}=0. \quad (6.19)$$

In terms of this function Eq. (6.18) takes the form

$$\Lambda_{\mu\nu}(p',p)=(Z_2{}^{-1}-1)\gamma_{\mu\nu}(p',p)-\tfrac{1}{2}\delta m^2\eta_{\mu\nu} +\tfrac{1}{2}Z_2{}^{-1}\frac{\bar{\Sigma}(p'^2)-\bar{\Sigma}(p^2)}{p'^2-p^2}(p_\mu p'_\nu+p_\nu p'_\mu+\tfrac{1}{2}q_\mu q_\nu) -\tfrac{1}{4}Z_2{}^{-1}[\bar{\Sigma}(p'^2)+\bar{\Sigma}(p^2)]\eta_{\mu\nu} +[F(p'^2,p^2,q^2)-\tfrac{1}{4}(Z_2{}^{-1}-1)] \times(q^2\eta_{\mu\nu}-q_\mu q_\nu), \quad (6.20)$$

[21] The polarization of the vacuum by a gravitational field is of the quadrupole type. Examples of renormalization terms to which it leads are given in Sec. 7.

[22] Equation (6.14a) is obtained from (2.15) by making the replacement $p'\to -p'$, since p' is here an outgoing and not an incoming momentum.

which suggests that we also introduce a renormalized form factor $\bar{F}$, defined by

$$F(p'^2,p^2,q^2)=\tfrac{1}{4}(Z_2^{-1}-1)+Z_2^{-1}\bar{F}(p'^2,p^2,q^2). \quad (6.21)$$

Combining (6.14b), (6.16), and (6.20) we then get

$$\bar{\Gamma}_{\mu\nu}(p',p)\equiv Z_2\Gamma_{\mu\nu}(p',p) \quad (6.22a)$$

$$=\gamma_{\mu\nu}(p',p)+\frac{1}{2}\frac{\bar{\Sigma}(p'^2)-\bar{\Sigma}(p^2)}{p'^2-p^2}(p_\mu p'_\nu+p_\nu p'_\mu+\tfrac{1}{2}q_\mu q_\nu)$$
$$-\tfrac{1}{4}[\bar{\Sigma}(p'^2)+\bar{\Sigma}(p^2)]\eta_{\mu\nu}++\bar{F}(p'^2,p^2,q^2)$$
$$\times(q^2\eta_{\mu\nu}-q_\mu q_\nu), \quad (6.22b)$$

which reduces, on the mass shell, to

$$\bar{\Gamma}_{\mu\nu}(p',p)=\gamma_{\mu\nu}(p',p)+\bar{F}(-m^2,\ -m^2,\ q^2)(q^2\eta_{\mu\nu}-q_\mu q_\nu),$$
$$p'^2=p^2=-m^2. \quad (6.23)$$

The Z_2 factor in (6.22a) takes into account the wave-function renormalization arising from self-energy insertions in the external lines.

If the scalar particle is coupled to other fields through nonrenormalizable interactions then the functions $\bar{\Sigma}$ and $\bar{F}$ will diverge in perturbation theory. In particular, they will diverge if virtual gravitons are permitted to contribute to the vertex function. Thus unless an arbitrary cutoff is used, or someone discovers a way to sum gravitational interactions to all orders, the gravitational field must be allowed to act only through the external graviton line. Although the identity (6.12) continues to hold formally in the nonrenormalizable case, it is then of no utility. Because of the divergence which remains in $\bar{F}$, Eq. (6.23) will yield an infinite cross section for the scattering of the particle in an external gravitational field.

In the renormalizable case $\bar{\Sigma}$ and $\bar{F}$ are finite, and expression (6.23) has a well-defined limit as $q\rightarrow 0$, namely,

$$\bar{\Gamma}_{\mu\nu}(p,p)=p_\mu p_\nu,\quad p^2=-m^2. \quad (6.24)$$

More generally, with particles of arbitrary spin one finds

$$u^\dagger(p)\bar{\Gamma}_{\mu\nu}(p,p)u(p)=(2\pi)^{-3}p_\mu p_\nu/2E,\quad p^2=-m^2, \quad (6.25)$$

when the wave functions $u(p)$ are chosen to correspond to δ-function normalization with respect to 3-momentum. As Brout and Englert point out,[19] the universality of (6.25) implies that the equivalence principle relating gravitational and inertial mass holds in the quantum theory as well as the classical theory. In particular the motion of a nonrelativistic particle in a slowly varying gravitational field is independent of its mass.

If a high-energy cutoff is permitted then the Ward identity may be applied to gravity itself, i.e., to the three-graviton vertex. In this case the wave function renormalization constants Z_2 and Z_3 coincide, and Eq. (6.12) tells us that $Z_1=Z_2=Z_3$. This leaves only the magnitude of the gravitation constant G, in terms of arbitrarily chosen (e.g., international mks) mass standards, to be determined by experiment.[23]

It is clear that the gravitational Ward identity is only one of an infinity of identities, derivable from Eq. (17.31) of II, which relate vertex functions involving n gravitons to those involving n-1 gravitons. Such identities become superfluous if calculations are performed in coordinate space rather than in momentum space, for then the general covariance of the theory can be kept constantly manifest. That such calculations are actually feasible will be demonstrated in the next section.

7. RENORMALIZATION IN COORDINATE SPACE. CONFORMAL VACUUM FLUCTUATIONS

The chief tool for studying quantum gravidynamics directly in space-time is the theory of Green's functions in hyperbolic Riemannian manifolds developed by Hadamard.[24] The basic structural element of this theory is the *geodetic interval*, denoted by σ,[25] which is defined as one half the square of the distance along the geodesic between any two space-time points x and x'. The geodetic interval is a symmetric function of x and x' which transforms as a *biscalar*, i.e., as a scalar separately at x and x'. It satisfies the differential equation[26]

$$\sigma=\tfrac{1}{2}\sigma_{;\mu}\sigma_{;}{}^{\mu}=\tfrac{1}{2}\sigma_{;\mu'}\sigma_{;}{}^{\mu'}, \quad (7.1)$$

and the boundary condition

$$\lim_{x'\rightarrow x}\sigma_{;\mu\nu}=-\lim_{x'\rightarrow x}\sigma_{;\mu\nu'}=g_{\mu\nu}. \quad (7.2)$$

In a general Riemannian manifold σ is not single-valued, except when x and x' are sufficiently close to one another.[27] The geodesics emanating from a given point will often, beyond a certain distance, begin to cross over one another. The locus of points at which the onset of overlap occurs forms an envelope of the

---

[23] The necessity of measuring G disappears if absolute units are adopted, with $\hbar=c=16\pi G=1$. However, the *masses* of the elementary particles must then be measured in absolute units, which is operationally the same thing as measuring both G and the masses in mks units.

[24] J. Hadamard, *Lectures on Cauchy's Problem in Linear Partial Differential Equations* (Yale University Press, New Haven, Connecticut, 1923).

[25] B. S. DeWitt and R. W. Brehme, Ann. Phys. (N. Y.) **9**, 220 (1960). See also J. L. Synge, *Relativity: The General Theory* (North-Holland Publishing Company, Amsterdam, 1960). Synge calls this function the *world function* and denotes it by the symbol Ω.

[26] The semicolons denote covariant differentiation. For a scalar this is the same as ordinary differentiation. $\sigma_{;\mu}$ is a vector of length equal to the distance along the geodesic between x and x', tangent to geodesic at x, and oriented in the direction $x'\rightarrow x$. $\sigma_{;\mu'}$ is a vector of equal length, tangent to the geodesic at x', and oriented in the opposite direction.

[27] In some manifolds (e.g., some compact manifolds) every pair of points may be linked by more than one geodesic. It is always possible, however, to define a single-valued function σ in the neighborhood of x by starting at x and following each geodesic emanating from x until it hits a caustic.

family of geodesics, known as a *caustic surface*. The equation for the caustic surface relative to a given point is $D^{-1}=0$, where

$$D\equiv-\det(-\sigma_{;\mu\nu'})\,. \quad (7.3)$$

D is a *bidensity* of unit weight at both x and x', which satisfies the boundary condition

$$\lim_{x'\to x} D=g\,. \quad (7.4)$$

It is convenient to replace D by the biscalar

$$\Delta\equiv g^{-1/2}Dg'^{-1/2}\,,\quad \lim_{x'\to x}\Delta=1\,, \quad (7.5)$$

whose values at given points are independent of the choice of coordinate system. By covariantly differentiating Eq. (7.1), one can derive the differential equation[25]

$$\Delta^{-1}(\Delta\sigma_{;}{}^{\mu})_{;\mu}=4 \quad\text{or}\quad \sigma_{;}{}^{\mu}{}_{\mu}=4-\sigma_{;}{}^{\mu}(\ln\Delta)_{;\mu}\,, \quad (7.6)$$

which shows that Δ increases or decreases along each geodesic from x' according as the rate of divergence of the neighboring geodesics from x', which is measured by $\sigma_{;}{}^{\mu}{}_{\mu}$, is less than or greater than 4, the rate in flat space-time. If the divergence rate becomes negatively infinite a caustic surface develops and Δ blows up.

We shall illustrate the use of σ and Δ in the theory of Green's functions by considering the Feynman propagator of the simplest of all fields: the massless scalar field. The defining equation is[28]

$$g^{1/2}G_{;\mu}{}^{\mu}(x,x')=-\delta(x,x')\,, \quad (7.7)$$

together with appropriate boundary conditions. The introduction of boundary conditions is most easily accomplished in the abstract formalism which replaces (7.7) by

$$FG=-1\,, \quad (7.8)$$

where

$$F=-p_\mu g^{1/2}g^{\mu\nu}p_\nu\,, \quad (7.9)$$

$$G(x,x')=\langle x|G|x'\rangle\,, \quad (7.10)$$

the $|x'\rangle$ being eigenvectors of a commuting set of Hermitian operators x^μ in a fictitious Hilbert space,[29] and the p's being Hermitian "momenta" "canonically conjugate" to the x's. The formal solution of (7.8) which incorporates the Feynman boundary conditions is

$$g^{1/4}Gg^{1/4}=\frac{-1}{g^{-1/4}Fg^{-1/4}+i0}$$

$$=i\int_0^\infty \exp(ig^{-1/4}Fg^{-1/4}s)ds\,, \quad (7.11)$$

the factors $g^{\pm1/4}$ being inserted to insure the covariance of operator functions.[30]

Taking matrix elements of (7.11) one obtains

$$g^{1/4}G(x,x')g'^{1/4}=i\int_0^\infty \langle x,s|x',0\rangle ds\,, \quad (7.12)$$

where

$$\langle x,s|x',0\rangle\equiv\langle x|\exp(ig^{-1/4}Fg^{-1/4}s)|x'\rangle\,, \quad (7.13)$$

satisfying

$$-i\frac{\partial}{\partial s}\langle x,s|x',0\rangle=\langle x,s|x',0\rangle_{;\mu}{}^{\mu} \quad (7.14)$$

and the boundary condition

$$\langle x,0|x'0\rangle=\langle x|x'\rangle=\delta(x,x')\,. \quad (7.15)$$

The "Schrödinger equation" (7.14) is solved by the ansatz

$$\langle x,s|x',0\rangle=-i(4\pi)^{-2}D^{1/2}s^{-2}e^{(i/2s)\sigma}\sum_{n=0}^{\infty}a_n(is)^n\,, \quad (7.16)$$

$$a_0=1\,, \quad (7.17)$$

which is suggested by its known solution in flat space-time.[29]

Inserting (7.16) into (7.14) and making use of (7.1) and (7.6), one finds

$$\sigma_{;}{}^{\mu}a_{n;\mu}+na_n=\Delta^{-1/2}(\Delta^{1/2}a_{n-1})_{;\mu}{}^{\mu}\,,\qquad n=1,2,3\cdots. \quad (7.18)$$

These recursion relations may be solved by successive quadratures along each geodesic emanating from x'. Hadamard[24] and Riesz[31] have shown that the solutions as well as the series (7.16) converge up to the first caustic. Formally we may write

$$a_n=\prod_{m=1}^{n}\left[(i\sigma_{;}{}^{\mu}p_\mu+m)^{-1}\Delta^{-1/2}g^{-1/4}Fg^{-1/4}\Delta^{1/2}\right]\cdot 1\,, \quad (7.19)$$

where ip_μ and $g^{-1/4}Fg^{-1/4}$ are now to be understood as the gradient and Laplace-Beltrami operators, respectively. Setting

$$(i\sigma_{;}{}^{\mu}p_\mu+m)^{-1}=\int_{-\infty}^{0}\left[\exp(i\sigma_{;}{}^{\mu}p_\mu+m)t'_m\right]dt'_m\,, \quad (7.20)$$

and making the variable transformation

$$\begin{array}{ll} t'_1=t_1\,, & t_1=t'_1\,,\\ t'_2=t_2-t_1\,, & t_2=t'_1+t'_2\,,\\ \cdots & \cdots\\ t'_n=t_n-t_{n-1}\,, & t_n=t'_1+t'_2+\cdots+t'_n\,, \end{array} \quad (7.21)$$

---

[28] In Eq. (7.7) $G(x,x')$ is to be understood as a biscalar and the δ function $\delta(x,x')$ as a bidensity of unit weight at x and zero weight at x'.

[29] Cf., J. Schwinger, Phys. Rev. **82**, 664 (1951).

[30] For example

$$\langle x|(g^{1/4}Gg^{1/4})^2|x'\rangle=g^{1/4}g'^{1/4}\int G(x,x'')g''^{1/2}G(x'',x')dx''.$$

[31] M. Riesz, Acta Math. **81** (1949).

we may recast (7.19) into the form

$$a_n=\int_{-\infty}^{0}dt_1\int_{-\infty}^{t_1}dt_2\cdots\int_{-\infty}^{t_{n-1}}dt_n\mathfrak{Q}(t_1)\cdots\mathfrak{Q}(t_n)\cdot 1\,, \quad (7.22)$$

where

$$\mathfrak{Q}(t)\equiv\exp[(i\sigma_{;}{}^{\mu}p_\mu+1)t]\,\Delta^{-1/2}g^{-1/4}Fg^{-1/4}\Delta^{1/2}\times\exp(-i\sigma_{;}{}^{\mu}p_\mu t)\,. \quad (7.23)$$

Substitution into (7.16) then yields

$$\langle x,s|x',0\rangle=-i(4\pi)^{-2}D^{1/2}s^{-2}T[\exp i(\mathfrak{R}s+\sigma/2s)]\cdot 1\,, \quad (7.24)$$

where

$$\mathfrak{R}\equiv\int_{-\infty}^{0}\mathfrak{Q}(t)dt\,, \quad (7.25)$$

and where the symbol T indicates that the operators $\mathfrak{Q}(t)$ appearing in the exponential, via (7.25), are to be "chronologically ordered" with respect to the parameters t.

Now the chronological ordering operation commutes with differentiation or integration with respect to the parameter s. Hence Eq. (7.24) may be inserted directly into (7.12). The result is a formal generalization of a well-known expression

$$G(x,x')=-\frac{\Delta^{1/2}}{8\pi}T\left[\frac{H_1^{(2)}((2\mathfrak{R}\sigma)^{1/2})}{(2\mathfrak{R}\sigma)^{1/2}}\right]\cdot 1\,, \quad (7.26)$$

$H_1^{(2)}$ being the Hankel function of the second kind of order 1. This formula has the series expansion

$$G(x,x')=\frac{i\Delta^{1/2}}{8\pi^2}T\left\{\frac{1}{\sigma+i0}-\mathfrak{R}[2\gamma-\ln 2+\ln(-\mathfrak{R}-i0)+\ln(\sigma+i0)]\left[\tfrac{1}{2}-\frac{2\mathfrak{R}\sigma}{2^2\cdot 4}+\frac{(2\mathfrak{R}\sigma)^2}{2^2\cdot 4^2\cdot 6}-\cdots\right]\right.$$
$$\left.-2\mathfrak{R}\left[\tfrac{1}{4}-\frac{2\mathfrak{R}\sigma}{2^2\cdot 4}(1+\tfrac{1}{4})+\frac{(2\mathfrak{R}\sigma)^2}{2^2\cdot 4^2\cdot 6}(1+\tfrac{1}{2}+\tfrac{1}{6})-\cdots\right]\right\}\cdot 1\,, \quad (7.27)$$

$$\gamma=0.5772\cdots\,, \quad (7.28)$$

where the instructlions "$\sigma+i0$" and "$-\mathfrak{R}-i0$" indicate what is evident from (7.24), namely, that $G(x,x')$ is the boundary va ue of a function of σ and $\mathfrak{R}$ which is analytic in the upper-half σ plane and the upper-half $\mathfrak{R}$ plane. The singularity structure in σ reflects the usual behavior of the Feynman propagator on the light cone ($\sigma=0$). The remaining singularity structure symbolized by the logarithm of $-\mathfrak{R}-i0$, on the other hand, is far from simple owing to the presence of the chronological ordering operation.

In the perturbative approach to quantum gravidynamics we must deal not with the scalar propagator (7.27) but with the vector and tensor propagators $\hat{G}^{\alpha\beta}$ and G^{ij}. However, the latter have structures closely similar to (7.27); the only difference is that the operators $\mathfrak{Q}(t)$ out of which $\mathfrak{R}$ is built are slightly more complicated, and the "1" standing on the right of Eqs. (7.19), (7.22), (7.24), (7.26), and (7.27) is replaced by the *geodetic parallel displacement function*.[25] Therefore we can gain a qualitative understanding of the renormalization program in coordinate space already by studying the scalar propagator. Moreover, there is an interesting nonperturbative treatment of the vacuum-to-vacuum amplitude in which the scalar propagator itself directly enters:

Consider the Feynman functional integral, Eq. (20.33) of II, which may be rewritten in the form

$$\exp i\bar{w}[\varphi]=Z\int\exp i(S[\varphi+\phi]-S[\varphi]-S_{,i}[\varphi]\phi^i)d\phi\,, \quad (7.29)$$

where S is the Einstein action and $\varphi_{\mu\nu}=g_{\mu\nu}-\eta_{\mu\nu}$ (see Table I of II). Because of the coordinate invariance of the theory the functional integration is redundant and ambiguous, and since no one has yet discovered an analytically accessible nonredundant subspace for the integration, we are forced to accept Eq. (20.12) of II as the effective definition of the integral. However, there is an *incomplete* nonredundant subspace which is easily accessible, namely, the subspace of all conformally equivalent geometries. One may simply set

$$\phi_{\mu\nu}=\chi g_{\mu\nu}\,,\quad \varphi_{\mu\nu}+\phi_{\mu\nu}=\bar{g}_{\mu\nu}-\eta_{\mu\nu}\,,\quad \bar{g}_{\mu\nu}=(1+\chi)g_{\mu\nu}\,, \quad (7.30)$$

and integrate over χ, to obtain the partial contribution to $\langle 0,\infty|0,-\infty\rangle$ arising from conformal fluctuations in the vacuum geometry. The special interest of this integration is that it can be performed exactly, giving the conformal contribution to all orders of perturbation theory. The only "fly in the ointment" is that this is the *one* contribution for which high-energy damping cannot be expected to produce a finite cutoff. There is no smearing out of the light cone, because conformal metric fluctuations leave the light cone invariant.

It is easy to show that

$$\bar{g}^{1/2}\,{}^{(4)}\bar{R}=(1+\chi)g^{1/2}\,{}^{(4)}R-3g^{1/2}(1+\chi)^{-1}\chi_{;\mu}\chi_{;}{}^{\mu}-3g^{1/2}\chi_{;\mu}{}^{\mu}\,, \quad (7.31)$$

and hence

$$S[\varphi+\phi]-S[\varphi]-S_{,i}[\varphi]\phi^i=\int[\bar{g}^{1/2}\,{}^{(4)}\bar{R}-g^{1/2}\,{}^{(4)}R+g^{1/2}(R^{\mu\nu}-\tfrac{1}{2}g^{\mu\nu}\,{}^{(4)}R)\phi_{\mu\nu}]dx$$
$$=-3\int g^{1/2}(1+\chi)^{-1}\chi_{;\mu}\chi_{;}{}^{\mu}\,dx\,. \quad (7.32)$$

The following change of variables then suggests itself:

$$\chi=\phi+\tfrac{1}{4}\phi^2,\quad 1+\chi=(1+\tfrac{1}{2}\phi)^2. \qquad (7.33)$$

This change not only simplifies expression (7.32) but at the same time guarantees the integrity of the signature of space-time. We may allow ϕ to range from $-\infty$ to ∞ without danger of encountering unphysical geometries and at the trivial cost of counting each distinct geometry twice at every point instead of only once. Thus we write

$$\exp(i\bar{w}_{\text{conformal}}) = Z\int \exp\left(-3i\int g^{1/2}\phi_{;\mu}\phi_{;}{}^{\mu}dx\right)d\phi, \qquad (7.34)$$

from which we immediately obtain

$$\bar{W}_{\text{conformal}}\equiv \bar{w}_{\text{conformal}}[\varphi]-\bar{w}_{\text{conformal}}[0] = -\tfrac{1}{2}i\ln\frac{\det(g^{1/4}Gg^{1/4})}{\det G_0}. \qquad (7.35)$$

The formal determinant may be evaluated by a variational technique with the aid of Eq. (7.11). Under a change in the background field (7.35) suffers the variation

$$\begin{aligned}\delta\bar{W}_{\text{conformal}}&=-\tfrac{1}{2}i\operatorname{tr}[g^{1/4}Gg^{1/4}\delta(g^{-1/4}Fg^{-1/4})]\\ &=\tfrac{1}{2}\operatorname{tr}\int_0^\infty \exp(ig^{-1/4}Fg^{-1/4}s)\delta(g^{-1/4}Fg^{-1/4})\,ds\\ &=-\tfrac{1}{2}i\delta\operatorname{tr}\int_0^\infty s^{-1}\exp(ig^{-1/4}Fg^{-1/4}s)\,ds, \qquad (7.36)\end{aligned}$$

which may be immediately integrated to yield

$$\bar{W}_{\text{conformal}}=-\tfrac{1}{2}i\operatorname{tr}\int_0^\infty s^{-1}\exp(ig^{-1/4}Fg^{-1/4}s)ds + \text{constant}. \qquad (7.37)$$

The trace symbol here means "integrate the diagonal matrix element over space-time." Hence, making use of (7.12), (7.13), (7.16) and (7.27), we find

$$\bar{W}_{\text{conformal}}=\int \mathcal{L}_{\text{conformal}}dx+\text{constant}, \qquad (7.38)$$

where

$$\begin{aligned}\mathcal{L}_{\text{conformal}}&=-\tfrac{1}{2}i\int_0^\infty s^{-1}\langle x,s|x,0\rangle ds = g^{1/2}[i\partial G(x,x')/\partial\sigma]_{x'=x}\\ &=\frac{g^{1/2}}{8\pi^2}\Bigg[T\bigg\{\frac{1}{(\sigma+i0)^2}+\frac{1}{2(\sigma+i0)}\Re\\ &\quad+[2\gamma-\ln 2-\tfrac{5}{16}+\ln(-\Re-i0)\\ &\quad+\ln(\sigma+i0)]\Re^2\bigg\}\cdot 1\Bigg]_{x'=x}. \qquad (7.39)\end{aligned}$$

Several comments are now in order. First we remark that although the final result is divergent, *the degree of divergence is bounded*. The singularity at $x'=x$ is therefore not an essential one as one might have expected on the basis of Eq. (5.3). As a matter of fact (7.39) is identical in structure with the contributions which the propagators G^{ij} and $\hat{G}^{\alpha\beta}$ of the full theory make in lowest perturbation order (i.e., the single closed loops of $\bar{W}_{(1)}$).[32]

It may be conjectured that inclusion of the nonconformal vacuum fluctuations will eliminate the divergences altogether, and that a rough approximation to the exact vacuum-to-vacuum amplitude can be obtained simply by making the replacement $\sigma(x,x)\to\frac{1}{2}\Lambda^{-2}$ in (7.39), where Λ is a high-energy cutoff of the order of unity in absolute units. The "$i0$" attached to each σ in (7.39) reflects the presence of unremoved noncausal chains. In passing from $\bar{W}$ to W these imaginary infinitesimals should be discarded. We obtain, therefore, the estimate

$$W=\int \mathcal{L}'dx+\text{constant}, \qquad (7.40)$$

$$\mathcal{L}'\sim\frac{g^{1/2}}{8\pi^2}\bigg\{T\bigg[4\Lambda^4+\Lambda^2\Re+\bigg(\ln\bigg(\frac{-\Re-i0}{\Lambda^2}\bigg)-0.545\cdots\bigg)\Re^2\bigg]\cdot 1\bigg\}_{x'=x}. \qquad (7.41)$$

Still cruder estimates of W can be obtained by finding approximations to the complicated quantity $\Re$. By repeated covariant differentiation of Eqs. (7.1) and (7.6), and use of the commutation laws for the indices thereby induced, one can show that

$$\lim_{x'\to x}\Delta^{-1/2}g^{-1/4}Fg^{-1/4}\Delta^{1/2}\cdot 1 = \lim_{x'\to x}\Delta^{-1/2}\Delta^{-1/2}{}_{;\mu}{}^{\mu}=\tfrac{1}{6}\,{}^{(4)}R. \qquad (7.42)$$

This quantity raised to the nth power can be extracted from expression (7.19) or (7.22) for a_n. Moreover, it is clear that the operator $\mathfrak{Q}(t)$ has the dimensions of the curvature scalar and in the limit $x'\to x$, is a kind of nonlocal, or mean curvature averaged over a certain neighborhood of x. If we represent the purely nonlocal part schematically by $\Delta\Re$, we may write

$$\mathfrak{Q}(t)\underset{x'\to x}{\sim} e^{t}(\tfrac{1}{6}\,{}^{(4)}R+\Delta\Re), \qquad (7.43)$$

$$\Re\underset{x'\to x}{\sim}\tfrac{1}{6}\,{}^{(4)}R+\Delta\Re. \qquad (7.44)$$

The crudest approximation to $\mathcal{L}'$ is then obtained

[32] The manifestly covariant occurrence of three distinct types of divergences: quartic, quadratic, and logarithmic, already in lowest order, implies that the conjecture of Brout and Englert (Ref. 19) that quantum gravidynamics is conventionally renormalizable is unfounded.

simply by omitting $\Delta\mathfrak{R}$ altogether:

$$\mathcal{L}'\sim\frac{\Lambda^4}{2\pi^2}g^{1/2}+\frac{\Lambda^2}{48\pi^2}g^{1/2}\,{}^{(4)}R+\frac{1}{288\pi^2}g^{1/2}\,{}^{(4)}R^2\times\left[\ln\left(\frac{-{}^{(4)}R-i0}{6\Lambda^2}\right)-0.545\cdots\right]. \quad (7.45)$$

Expression (7.45) is prototypical of the contributions which *all* fields make to the geometrical part of the vacuum-to-vacuum amplitude. (The only deviations from it occur with massive fields, for which $\frac{1}{6}\,{}^{(4)}R$ gets replaced by $-m^2+\frac{1}{6}\,{}^{(4)}R$, and with fermion fields, for which the sign of each term is reversed.) These contributions originate in the vacuum polarization which the background geometry induces, and give rise to nonobservable renormalizations as well as physically real radiative corrections.

The first term in (7.45) is a "cosmological" term representing the zero-point vacuum energy which every field, including the gravitational field itself, possesses. It is eliminated by redefining the zero point.

The second term in (7.45) renormalizes the gravitational interaction strength. The relation between the renormalized and "bare" gravitation constants (G and G_0, respectively) is

$$G=ZG_0, \quad (7.46)$$

$$Z\sim(1+\Lambda^2/48\pi^2)^{-1}. \quad (7.47)$$

In the theory of the pure gravitational field Z is the only renormalization constant which occurs (provided, of course, the exact theory is really finite.) Because of the manifest covariance of (7.45) it is clear that the same renormalization applies to all vertex functions no matter how many graviton prongs they possess. No Ward identity is needed.

The third term in (7.45) is the only one having observable physical consequences.[33] In the classical limit of long wavelengths and large coherent amplitudes it may be regarded as a correction to the Einstein Lagrangian. Hill[34] has applied such a correction term to the problem of gravitational collapse of the Friedmann universe, with encouraging results. He finds that if the sign of the coefficient in front is negative, as would be the case for the contribution from a fermion field, this term succeeds in turning the collapse cycle around before infinite curvature is reached.[35] It may be objected that in applying the correction to the Friedmann model one violates the boundary conditions of asymptotic flatness which were assumed to get it in the first place. However, vacuum polarization is basically a local phenomenon, and global conditions should have little relevance here.

8. THE INFRARED PROBLEM

The most important contributors to the gravitational polarization of the vacuum, and to the modifications in Einstein's equations which this polarization produces, are the massless fields, including gravity itself. These are also the fields which most readily yield real quanta. The effect of real quantum production on the vacuum-to-vacuum amplitude is taken into account by the "$-i0$" attached to $-\mathfrak{R}$ in the logarithm of (7.41). Owing to the complexity of $\mathfrak{R}$, however, the branch-point behavior of the logarithm is very involved, and it is not easy to investigate directly in coordinate space whether or not serious infrared difficulties lie hidden in this expression.

In a closed finite world such difficulties cannot arise since there is a natural low-energy cutoff; troubles occur only in infinite worlds. Let us for simplicity confine our attention to flat backgrounds.[36] It is then appropriate to revert to momentum space to study the problem. The analysis for this case is straightforward and has been carried out by Weinberg;[37] we shall summarize his results.

The amplitude for a single soft graviton to be produced in a given process has already been derived [Eq. (4.6)]. The corresponding amplitude for the emission of N soft gravitons in all possible ways from a given diagram is just the product of N single-graviton amplitudes. The form of these amplitudes is such that an infrared divergence arises in the computation of the rate at which any physical process takes place when arbitrary numbers of soft gravitons having total energy less than E are simultaneously emitted. This divergence disappears if the contributions from virtual soft gravitons are also included. Weinberg shows that the correct total rate is given by an expression of the form

$$\Gamma(E)=\Gamma_0(E/\Lambda)^B b(B), \quad (8.1)$$

$$b(B)\equiv\frac{1}{\pi}\int_{-\infty}^{\infty}\frac{\sin\sigma}{\sigma}\exp\left(B\int_0^1\frac{e^{i\omega\sigma}-1}{\omega}d\omega\right)d\sigma \quad (8.2)$$

$$=1-\tfrac{1}{12}\pi^2B^2+\cdots, \quad (8.3)$$

where Γ_0 is the rate without graviton emission and Λ is a parameter marking the dividing line between "soft" and "hard" virtual gravitons. If Λ is chosen to be of the order of the typical energies involved in the physical process, Eq. (8.1) gives a fair estimate of the rigorous value which would be obtained for $\Gamma(E)$ if the contributions from ultraviolet virtual gravitons were also included and appropriate renormalizations performed. The soft gravitons make appreciable contributions only if attached to the external lines of the Γ_0

[33] The nonlocal part of the Λ^2 term, which has been omitted from (7.45), also has observable consequences.

[34] T. W. Hill, Ph.D. thesis, University of North Carolina, 1965 (unpublished).

[35] Not, however, until a density of the order of unity in absolute units is reached. At this density all the matter in the visible universe has been compressed to a region the size of a nucleon.

[36] Conclusions reached for this case are presumably valid also for infinite worlds having other background geometries.

[37] S. Weinberg, Phys. Rev. 140, B516 (1965).

diagrams, and hence the only radiative corrections which should be included in Γ_0 are those which involve internal lines and vertices. The quantity B is given by

$$B=\frac{G}{2\pi}\sum_{m,n}\frac{1+v_{nm}^2}{(1-v_{nm}^2)^{1/2}}\frac{\eta_n\eta_m m_n m_m}{v_{nm}}\ln\frac{1+v_{nm}}{1-v_{nm}}, \quad (8.4)$$

$$v_{nm}\equiv[1-(m_n m_m/p_n\cdot p_m)^2]^{1/2}; \quad (8.5)$$

it depends only on the parameters of the external lines.

These results are completely standard. Except for the detailed form of the quantity B they are identical with the corresponding results in quantum electrodynamics. The question which now must be asked is: What happens when emission takes place from particles which are themselves massless? In quantum electrodynamics such emission is known to give rise to a new and more serious kind of infrared divergence which cannot be removed in any simple or completely natural way. This circumstance has been invoked as the "reason" why massless charged particles do not occur in Nature. In the case of the Yang-Mills and gravitational fields the difficulty presents itself in a peculiarly acute form since these fields are themselves both massless and "charged." Moreover, although there is no experimental evidence for the existence of the Yang-Mills field, gravity is an established fact, as is also its interaction with photons.

Since the Yang-Mills field is a vector field its divergence difficulties are similar to those of massless electrodynamics and hence are difficult if not impossible to remove. Because of the noncommutativity of the emission vertices for Yang-Mills quanta it is not possible to sum the effect of arbitrary numbers of real and virtual quanta into a closed expression like (8.1). Moreover, the situation is further complicated by the fact that there is a non-negligible amplitude for the soft quanta themselves to emit soft quanta. However, there is no evidence whatever that the situation would improve if one could find a way to take all these extra complications rigorously into account.

In the case of the gravitational field, on the other hand, the difficulties miraculously disappear. This happy state of affairs is a consequence of the detailed structure of expression (8.4), which in turn derives from the special form of the graviton emission vertex: $\Gamma_{0\mu\nu}\to p_\mu p_\nu$ as $q\to 0$. We shall now show how it comes about.

Let us use indices from the first part of the alphabet to distinguish the massless particles from the others. We shall continue to use the symbols m_a, m_b, etc. but with the understanding that these masses ultimately tend to zero. We may then write

$$\frac{1+v_{ab}^2}{(1-v_{ab}^2)^{1/2}}\frac{\eta_a\eta_b m_a m_b}{v_{ab}}\ln\frac{1+v_{ab}}{1-v_{ab}}\to -4\eta_a\eta_b p_a\cdot p_b\ln\left(-\frac{2p_a\cdot p_b}{m_a m_b}\right), \quad (8.6)$$

which permits (8.4) to be decomposed as follows:

$$\begin{aligned}B=&-(2G/\pi)\sum_{a,b}\eta_a\eta_b p_a\cdot p_b\ln(-2p_a\cdot p_b)\\&-(4G/\pi)\sum_{a,m}\eta_a\eta_m p_a\cdot p_m\ln(-2p_a\cdot p_m/m_m)\\&+\frac{G}{2\pi}\sum_{m,n}\frac{1+v_{nm}^2}{(1-v_{nm}^2)^{1/2}}\frac{\eta_n\eta_m m_n m_m}{v_{nm}}\ln\frac{1+v_{nm}}{1-v_{nm}}\\&+(2G/\pi)\sum_{a,b}\eta_a\eta_b p_a\cdot p_b\ln(m_a m_b)\\&+(4G/\pi)\sum_{a,m}\eta_a\eta_m p_a\cdot p_m\ln m_a. \quad (8.7)\end{aligned}$$

With the aid of the energy-momentum conservation law

$$\sum_a\eta_a p_a+\sum_m\eta_m p_m=0, \quad (8.8)$$

the last two terms of (8.7) may be combined into

$$(2G/\pi)\sum\eta_a\eta_b p_a\cdot p_b\,(\ln m_b-\ln m_a),$$

which vanishes by symmetry. The masses m_a, m_b thus disappear from (8.7), and since $p_a\cdot p_b\ln(-2p_a\cdot p_b)$ vanishes when either $a=b$ or $\mathbf{p}_a$ is parallel to $\mathbf{p}_b$, it is evident that B *is completely free of divergences*.

The only uncertainty which remains in Weinberg's analysis, and which he himself points out, concerns his use of the DeDonder gauge for the virtual gravitons. Except when the stress-energy tensor is conserved at each virtual graviton vertex it is not easy to see that the choice of gauge is immaterial. But stress-energy conservation of this simple type holds only when the particle lines on both sides of the vertex are on the mass shell. [See Eq. (3.8)]. Since only the external lines satisfy this condition Weinberg must appeal to the fact that the other lines are only *slightly off* the mass shell and hence violate the conservation conditions only minimally.[38]

Weinberg's act of faith on this question can be rigorously justified within the framework of the complete theory developed in II. We known from this theory that the choice of gauge for internal lines is irrelevant provided: (a) it is applied consistently and (b) *all* diagrams contributing to a given process are included. Now Weinberg omits the diagrams which involve infrared *fictitious* quanta. But it is not hard to show that the contributions of these diagrams all vanish as the infrared momenta go to zero, and hence may be neglected. This is a consequence of the fact that the fictitious quanta always occur in closed loops containing uniformly oriented vertices $V_{(\alpha i)\beta}$. Because of the special

[38] In electrodynamics it is not difficult to show that gauge invariance holds when every vertex along each charged particle line is taken into account. In gravidynamics, however, every line is "charged," and the "charge" splits up or recombines at every vertex.

form (2.8) which these vertices possess, the uniform orientation guarantees that at least one of the vertices in each infrared loop is proportional to an infrared momentum.

We conclude this section by repeating Weinberg's calculation of B in the nonrelativistic limit and correcting a minor mistake in his result. The quantity v_{nm}^2 is first expanded in the form

$$v_{nm}^2=(\mathbf{v}_n-\mathbf{v}_m)^2-\mathbf{v}_n^2\mathbf{v}_m^2+2(\mathbf{v}_n^2+\mathbf{v}_m^2)\mathbf{v}_n\cdot\mathbf{v}_m-3(\mathbf{v}_n\cdot\mathbf{v}_m)^2+\cdots, \quad (8.9)$$

where $\mathbf{v}_n=\mathbf{p}_n/E_n$. This expansion is then inserted into

$$\frac{1+v_{nm}^2}{(1-v_{nm}^2)^{1/2}}\frac{\eta_n\eta_m m_n m_m}{v_{nm}}\ln\frac{1+v_{nm}}{1-v_{nm}}=2\eta_n\eta_m m_n m_m\left(1+\frac{11}{6}v_{nm}^2+\frac{63}{40}v_{nm}^4+\cdots\right) \quad (8.10)$$

to obtain a lengthy expression for B correct to the fourth order in the velocities. This expression can be greatly simplified with the aid of the energy-momentum conservation laws

$$\sum_n \eta_n m_n(1+\tfrac{1}{2}\mathbf{v}_n^2+\tfrac{3}{8}\mathbf{v}_n^4+\cdots)=0,$$

$$\sum_n \eta_n m_n \mathbf{v}_n(1+\tfrac{1}{2}\mathbf{v}_n^2+\cdots)=0,$$

and one finally obtains the compact formula

$$B=(4G/5\pi)\,\mathrm{tr}(\Delta d^2\mathbf{Q}/dt^2)^2, \quad (8.11)$$

where $\Delta d^2\mathbf{Q}/dt^2$ is the dyadic previously defined by Eqs. (4.11) and (4.12), having the explicit traceless form[39]

$$\Delta d^2\mathbf{Q}/dt^2=\sum_n \eta_n m_n(\mathbf{v}_n\mathbf{v}_n-\tfrac{1}{3}\mathbf{1}\mathbf{v}_n^2). \quad (8.12)$$

[39] By inadvertently dropping a term Weinberg obtains a dyadic which is not traceless.

Bryce DeWitt (Photo courtesy of Cecile DeWitt-Morette)

FEYNMAN DIAGRAMS FOR THE YANG-MILLS FIELD

L. D. FADDEEV and V. N. POPOV
Mathematical Institute, Leningrad, USSR

Received 1 June 1967

Feynman and De Witt showed, that the rules must be changed for the calculation of contributions from diagrams with closed loops in the theory of gauge invariant fields. They suggested also a specific recipe for the case of one loop. In this letter we propose a simple method for calculation of the contribution from arbitrary diagrams. The method of Feynman functional integration is used.

It is known, that one can associate the field of the Yang-Mills type with an arbitrary simple group G [1-3]. It is appropriate to describe this field by means of the matricies $B_\mu(x)$ with values in the Lie algebra of this group.

The gauge group consits of the transformations

$$B_\mu \to B_\mu^\Omega = \Omega B_\mu \Omega^{-1} + \epsilon^{-1}\partial_\mu \Omega\Omega^{-1}$$

where $\Omega(x)$ is an arbitrary function with values in the group G.

The Lagrange function

$$\mathcal{L}(x) = -\tfrac{1}{4}\mathrm{Sp}G_{\mu\nu}G^{\mu\nu};\; G_{\mu\nu} = \partial_\nu B_\mu - \partial_\mu B_\nu + \epsilon[B_\mu, B_\nu]$$

is invariant with respect to these transformations. It is clear, that

$$\mathcal{L} = \mathcal{L}_0 + \mathcal{L}_1$$

where $\mathcal{L}_0$ is a quadratic form, and $\mathcal{L}_1$ is the sum of trilinear and quartic forms in B. In the quantization of the Feynman type $\mathcal{L}_1$ generates vertices with three and four external lines and $\mathcal{L}_0$ is to define the propagator function. However the form $\mathcal{L}_0$ is singular and the longitudinal part of the propagator can not be found unambiguously. This ambiguity does not influence the physical results in quantum electrodynamics. It seems that Feynman [4] was the first to show that the matter is not so simple in the cases of Yang-Mills and gravitational fields. Namely the contribution of the closed loop diagrams depends essentially on the longitudinal part of the propagator and spoils the transversality and unitarity properties of scattering amplitudes. Feynman himself described the necessary change of rules for calculation the contribution from diagrams with one closed loop. A more detailed derivation of the new rules was given by De Witt [5]. However it seems that nobody gave the generalization of these rules for arbitrary diagrams.

The formal considerations below are to give a simple explanation of the described difficulties and a quite workable recipe to circumvent them.

We know from Feynman [6] that every element of the S-matrix can be written down as the functional integral

$$\langle \mathrm{in}|\mathrm{out}\rangle \sim \int \exp\{iS[B]\}\prod_x \mathrm{d}B(x)$$

up to an (infinite) normalizing factor. Here $S[B] = \int \mathcal{L}(x)\mathrm{d}x$ is the action functional and one is to integrate over all fields $B(x)$ with the asymptote at $t = x_0 \to \pm\infty$ prescribed by in-and out-states. The diagrams appear naturally in the perturbative calculation of this integral.

In the case of gauge invariant theory it is necessary to transform this integral a little. In fact, we can say, using the natural geometrical language, that the integrand is constant on the "orbits" $B_\mu \to B_\mu^\Omega$ of the gauge group in the manifold of all fields $B_\mu(x)$. It follows that the integral itself is proportional to the volume of this orbit which can be expressed as the integral $\int \prod_x \mathrm{d}\Omega(x)$ over all matrices $\Omega(x)$. This integral should be factorized before using the perturbation theory.

There exist several methods for this purpose. The idea of one of them is to integrate over the orbits and some transversal surface. It is appropriate to choose for the latter the "plane" $\partial_\mu B^\mu = 0$. Then the integral reduces to the following

$$\int \exp\{\mathrm{i}S[B]\}\ \Delta[B] \prod_x \delta\left(\partial_\mu B^\mu(x)\right) \mathrm{d}B(x) \int \prod_x \mathrm{d}\Omega(x)$$

where the factor $\prod_x \delta\left(\partial_\mu B^\mu\right)$ symbolizes that we integrate over transverse fields and $\Delta[B]$ is to be chosen such that the condition

$$\Delta[B] \int \prod_x \delta\left(\partial_\mu (B^\mu)^\Omega\right) \mathrm{d}\Omega = \mathrm{const}$$

holds. It is the nontriviality of $\Delta[B]$ which distinguishes the theories of Yang-Mills and gravitational fields from quantum electrodynamics.

We must know $\Delta[B]$ only for transverse fields and in this case all contribution to the last integral is given by the neighbourhood of the unit element of the group. After appropriate linearization we come to the condition

$$\Delta[B] \int \prod_x \delta\left(\Box u - \epsilon[B^\mu, \partial_\mu u]\right) \mathrm{d}u(x) = \mathrm{const}$$

where $\Box$ is the D'Alembert operator and $u(x)$ are functions with values in the Lie algebra of the group G.

Formally $\Delta[B]$ is equal to the determinant of the operator

$$Au = \Box u - \epsilon[B^\mu, \partial_\mu u] \equiv A_0 u - \epsilon V(B) u$$

After extracting the trivial infinite factor $\det A_0$ we obtain the following expression for $\ln \Delta[B]$

$$\ln \Delta[B] = \ln(\det A / \det A_0) = \mathrm{Sp} \ln(1 - \epsilon A_0^{-1} V(B))$$

Developing the right hand side in a power series in ϵ we have the following expressions for the coefficients

$$\int \mathrm{d}x_1 \ldots \int \mathrm{d}x_n\ \mathrm{Sp}\left(B^{\mu_1}(x_1) \ldots B^{\mu_n}(x_n)\right) \times$$
$$\times\ \partial_{\mu_1} G(x_1 - x_2) \ldots \partial_{\mu_n} G(x_n - x_1)$$

where $G(x)$ is a Green function of the D'Alembert operator. This expression corresponds to the closed loop with the scalar particle propagating along it and interacting with the transverse vector particles according to the law $\sim \epsilon\ \mathrm{Sp}(\varphi[B^\mu \partial_\mu \varphi])$.

There results the diagram technique with the following features:

1. The pure transversal Green function is to be used as a propagator for the vector particles (Landau gauge).

2. It is necessary to take into account the new vertex with two scalar and one vector external line in addition to the ordinary vertices with three and four lines.

Concrete calculations with these changes in the rules give transverse and unitary expressions for the scattering amplitudes.

It must be stressed that the Landau gauge is essential for the new rules. It is connected with the chosen method of extracting the fact or $\int \prod_x \mathrm{d}\Omega(x)$.

An other method leads to the expression

$$\int \exp\{\mathrm{i}S[B] - \tfrac{1}{2}\mathrm{i} \int \mathrm{Sp}(\partial_\mu B^\mu)^2\, \mathrm{d}x\}\ \varphi[B] \prod_x \mathrm{d}B$$

where the factor $\varphi[B]$ must be found from the condition

$$\varphi[B] \int \exp\{-\tfrac{1}{2}\mathrm{i} \int \mathrm{Sp}\left(\partial_\mu (B^\mu)^\Omega\right)^2 \mathrm{d}x\} \prod_x \mathrm{d}\Omega = \mathrm{const}$$

This integral gives the perturbation series with Feynman propagator, but the calculation of $\varphi[B]$ is more cumbersome than that of $\Delta[B]$.

We conclude with the comment that one can proceed in an analogous way with gravitation theory. The analog for $\Delta[B]$ is the determinant of the Beltrami - Laplace operator in a harmonic coordinate system.

References

1. C.N.Yang and R.L.Mills, Phys. Rev. 96 (1954) 191.
2. R.Utiyama, Phys. Rev. 101 (1956) 1597.
3. S.L.Glashow and M.Gell-Mann, Ann. of Phys. 15 (1961) 437.
4. R.P.Feynman, Acta Physica Polonica, 24 (1963) 697.
5. B.S.De Witt, Relativity, groups and topology. (Blackie and Son Ltd 1964) pp 587-820.
6. R.P.Feynman, Phys. Rev. 80 (1950) 440.

PHYSICAL REVIEW VOLUME 175, NUMBER 5 25 NOVEMBER 1968

Feynman Rules for Electromagnetic and Yang-Mills Fields from the Gauge-Independent Field-Theoretic Formalism*

STANLEY MANDELSTAM
Department of Physics, University of California, Berkeley, California 94720
(Received 17 June 1968)

The Feynman rules for the Yang-Mills field, originally derived by Feynman and DeWitt from S-matrix theory and the tree theorem, are here derived as a consequence of field theory. Our starting point is the gauge-independent, path-dependent formalism which we proposed earlier. The path-dependent Green's functions in this theory are expressed in terms of auxiliary, path-independent Green's functions in such a way that the path-dependence equation is automatically satisfied. The formula relating the path-dependent to the auxiliary Green's functions is similar to the classical formula relating the path-dependent field variables to the potentials. By using a notation similar but not identical to Schwinger's functional notation, the infinite set of equations satisfied by the Green's function can be replaced by a single equation. When the equation for the auxiliary Green's functions of electromagnetism is solved in a perturbation series, the usual Feynman rules result. For the Yang-Mills field, however, one obtains extra terms; such terms correspond precisely to the closed loops of fictitious scalar particles introduced by Feynman, DeWitt, and Faddeev and Popov.

1. INTRODUCTION

THE discovery of the Feynman rules for the Yang-Mills and gravitational fields by Feynman himself[1] has solved a long-standing problem in relativistic quantum mechanics. Feynman only derived his procedure for diagrams with a single closed loop, but DeWitt[2] has recently extended the procedure to diagrams of arbitrary complexity. Another general proof of the prescription for the Yang-Mills field has been given by Faddeev and Popov,[3] who used a functional integration procedure which is probably equivalent to that of DeWitt.

Feynman and DeWitt obtained their prescription by a somewhat indirect method. From the Feynman rules for nongauge particles they obtained the "tree theorem," which relates the contribution to the S matrix from a closed-loop diagram to the contribution from a diagram where the loop is opened at one point. They then assumed that the tree theorem was valid in theories with gauge particles; they were thus able to derive the Feynman rules for the S matrix. The validity of the tree theorem guarantees that the S matrix is unitarity, and their results can almost certainly be derived from an analyticity-unitarity calculation in perturbation theory.

The question arises whether one can obtain the Feynman rules within the framework of a field theory of the Yang-Mills field or of gravity, and it is the purpose of the present paper to attempt to do so. We shall take as our basis the path-dependent theory of gauge fields which we suggested earlier.[4] The theory was originally formulated for electromagnetism and for gravity, but it can be applied to any gauge field. The path-dependent theory of the Yang-Mills field has been treated by Bialynicki-Birula.[5]

In the present paper we shall rederive the Feynman rules for the electromagnetic field from the path-dependent formalism, and we shall then derive the more complicated Feynman rules for the Yang-Mills field. We shall derive the Feynman rules for the gravitational field in the following paper.

The fundamental principle of the path-dependent formalism was to avoid the introduction of non-gauge-invariant quantities. Thus the electromagnetic potentials were not introduced, but were replaced by the electromagnetic field variables $F_{\mu\nu}$. Similarly, the charged field variables $\phi(x)$ were replaced by the path-dependent but gauge-invariant variables $\Phi(x,P)$. For practical purposes one would like to introduce the potentials as auxiliary variables, as one does in classical field theory. By doing so one would be able to calculate in terms of path-independent variables; one would transfer to the path-dependent variables at the end of the calculation. It is well known, however, that one cannot introduce covariant potentials without enlarging the Hilbert space and employing an indefinite metric. For electromagnetism one can use noncovariant potentials such as those of the Coulomb gauge. One can then derive the Feynman rules after a certain amount of algebraic calculation. It is possible to formulate the Yang-Mills theory in terms of noncovariant gauges, such as Schwinger's modification of the Coulomb gauge, or the Arnowitt-Fickler gauge.[6] However, the method which was used in electromagnetism for deriving the covariant Feynman rules from such gauges is not applicable here, at any rate without essential modification. To our knowledge no such consistent formalism has been given for the gravitational field.

* Research supported in part by the Air Force Office of Scientific Research, Office of Aerospace Research under Grant No. AF-AFOSR-68-1471.

[1] R. P. Feynman, Acta Phys. Polon. **24**, 697 (1963).
[2] B. S. DeWitt, Phys. Rev. **162**, 1195 (1967); **162**, 1239 (1967).
[3] L. D. Faddeev and V. N. Popov, Phys. Letters **25B**, 29 (1967).
[4] S. Mandelstam, Ann. Phys. (N. Y.) **19**, 1, 25 (1962).
[5] I. Bialynicki-Birula, Bull. Acad. Polon. Sci. **11**, 135 (1963).
[6] J. Schwinger, Phys. Rev. **127**, 324 (1962); R. L. Arnowitt and S. I. Fickler, *ibid.* **127**, 1821 (1962).

In the present treatment we shall avoid noncovariant quantities and we shall therefore not introduce potentials as quantum-mechanical operators. Instead, we shall introduce auxiliary Green's functions. In formalisms of quantum electrodynamics which employ potentials, whether in the Coulomb or Lorentz gauges, one can define Green's functions

$$G_\mu(x_1,\cdots;y_1,\cdots;z_1,\cdots)\\=\langle 0|T\{\phi(x_1)\cdots\phi^*(y_1)\cdots A_\mu(z_1)\cdots|0\rangle.$$

One can also define path-dependent but gauge-invariant Green's functions

$$G_{\mu\nu}(x_1,P_1,\cdots;y_1,P_1',\cdots;z_1,\cdots)\\=\langle 0|T\{\Phi(x_1,P_1)\cdots\Phi^*(y_1,P_1')\cdots F_{\mu\nu}(z_1)\cdots\}|0\rangle.$$

The latter Green's functions can be expressed in terms of the former. In our present approach, the path-independent Green's functions will be introduced, not as vacuum-expectation values of time-ordered products, but as auxiliary functions in their own right. The physical, path-dependent Green's functions of our theory will then be expressed in terms of the auxiliary Green's functions by using the same formulas as in theories with potentials. The connection between the path-dependent and path-independent Green's functions will guarantee that the path-dependence equations are satisfied, as we shall verify explicitly. We then have to find the equations which the auxiliary Green's functions must satisfy in order that the path-dependent Green's functions satisfy the correct equations.

For electrodynamics, such an approach has already been carried out by Sarker.[7] He found that the equations satisfied by the auxiliary Green's functions are similar, but not identical, to the equations satisfied by the Green's functions of the Lorentz-gauge theory. The difference is due to the fact that he started with the Maxwell equations $\partial F_{\mu\nu}(x)/(\partial x_\mu)+j_\nu=0$, whereas the Lorentz-gauge theory starts with the equations $\Box^2 A_\nu(x)+j_\nu=0$. Nevertheless, he showed that the Green's functions calculated by the usual Feynman rules do satisfy the correct equations. The Feynman rules were thus derived from a procedure which was covariant throughout and which did not make use of an enlarged Hilbert space.

When we carry out a similar treatment for the Yang-Mills field, we shall again find that the equations satisfied by our auxiliary Green's functions are slightly different from the corresponding equations in the (incorrect) Lorentz-gauge theory. As with electromagnetism, the difference is due to the dropping of a term $-\partial^2 A_\nu/\partial x_\mu\partial x_\nu$ in the Lorentz gauge. In this case, however, we shall find that the difference is important, and that the solution to our equations contains terms besides those given by the Lorentz-gauge Feynman rules.

[7] A. Q. Sarker, Ann. Phys. (N. Y.) 24, 19 (1963).

Our results will be the same as those found by Feynman, DeWitt, and Faddeev and Popov. They showed that the correct prescription was to take all Feynman diagrams of the Lorentz-gauge theory, together with Feynman diagrams containing closed loops of fictitious scalar particles. In our treatment we shall find that integrals corresponding to closed loops of scalar particles appear directly in the solution of the Green's-function equations. We may associate such integrals with closed loops of scalar particles if we wish, but this is purely a mnemonic device. The fictitious particles never occur in external lines, nor do they appear in the intermediate states of the unitarity condition.

In our present formulation of the theory, the Feynman rules are thus rules for calculating auxiliary Green's functions. We can then proceed to calculate the gauge-invariant, path-dependent Green's functions, since we shall already have expressed them in terms of the auxiliary Green's functions. By using the reduction formulas we can then calculate the S matrix. The fundamental reduction formulas of the theory involve the path-dependent Green's functions. However, one can use these reduction formulas to derive further reduction formulas involving the auxiliary Green's functions. Thus, from the Feynman rules for the auxiliary Green's functions, one can derive Feynman rules for the S matrix by the usual reinterpretation of the external lines.

The equations for the Green's functions are coupled integral equations between an infinite number of such functions. Moreover, when expressing path-dependent Green's functions in terms of auxiliary Green's functions, one finds that a single path-dependent Green's function is equal to the sum of an infinite number of auxiliary Green's functions. It would be clumsy, if in principle possible, to carry out manipulations with such infinite systems of equations. We require a shorthand for expressing the infinite sets of equations as single equations. The Schwinger functional notation provides us with such a shorthand; Schwinger's functional differential equation is equivalent to the complete set of equations for the Green's functions. Unfortunately it does not appear to be an easy matter to express the equations for path-dependent Green's functions in Schwinger's notation. We shall therefore use another notation in which our fundamental quantity corresponds to Schwinger's $\delta/\delta\eta$ rather than to η. We shall indicate the connection between our notation and Schwinger's but we shall not assume knowledge of his notation.

In the following section we shall illustrate some of our methods by using the $\lambda\phi^3$ theory. We shall find the differential equations for the Green's functions and shall use them to construct the perturbation expansion. We shall then develop our notation for simplifying the writing of the differential equations. Essentially what we shall do is to form a linear space of all Green's functions and to write the differential equations as equations for vectors in this space. In Sec. 3 we shall treat the

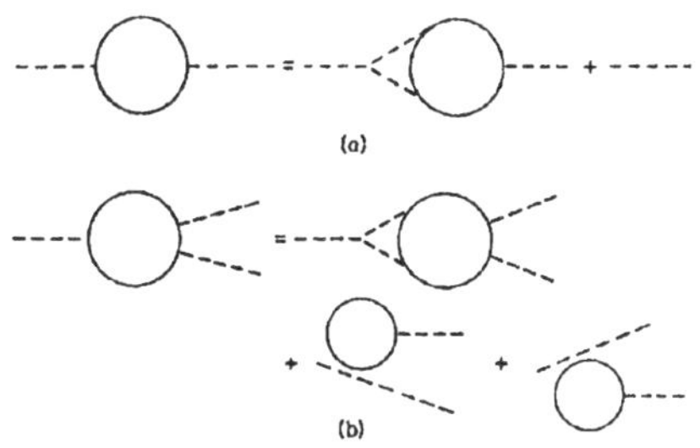

FIG. 1. Diagrammatic representation of Eqs. (2.5).

electromagnetic field. We shall write down the equations for the path-dependent Green's functions and shall reexpress them in our shorthand notation. Working within this notation, we shall then express our path-dependent Green's functions in terms of new, path-independent, auxiliary Green's functions. We shall determine the equations which the auxiliary Green's functions should satisfy in order that the path-dependent Green's functions satisfy the required equations. On solving them, we shall find that they lead to the ordinary Feynman rules. In Sec. 4 we shall treat the Yang-Mills field in a similar way. Here, however, we shall find that the perturbation expansion contains terms besides those given by naive Feynman rules.

2. DIFFERENTIAL EQUATIONS FOR GREEN'S FUNCTIONS

In this section we shall summarize the method of determining Green's functions by solving differential equations, and shall also develop our shorthand notation. The method is certainly not new but, as far as we are aware, there is no easily available reference in which it is described, and we therefore felt it worthwhile to describe its application to non-gauge fields before passing on to the gauge fields in which we are interested.

We shall treat the simple case of a neutral scalar field with $\lambda\phi^3$ coupling. The field equations will be

$$(\Box^2-\mu^2)\phi(x)-\tfrac{1}{2}\lambda\{\phi(x)\}^2=0, \qquad (2.1)$$

and the ϕ's will satisfy the commutation relations

$$[\phi(\mathbf{x},t)\phi(\mathbf{y},t)]=[\dot\phi(\mathbf{x},t)\dot\phi(\mathbf{y},t)]=0, \qquad (2.2a)$$

$$[\dot\phi(x,t),\phi(y,t)]=-i\delta^3(\mathbf{x}-\mathbf{y}). \qquad (2.2b)$$

We can now define Green's functions

$$G_2(x_1,x_2)=\langle 0|T\{\phi(x_1),\phi(x_2)\}|0\rangle, \qquad (2.3a)$$

$$G_3(x_1,x_2,x_3)=\langle 0|T\{\phi(x_1),\phi(x_2),\phi(x_3)\}|0\rangle, \qquad (2.3b)$$

$$G_4(x_1,x_2,x_3,x_4)=\langle 0|T\{\phi(x_1),\phi(x_2),\phi(x_3),\phi(x_4)\}|0\rangle, \quad \text{etc.} \qquad (2.3c)$$

For formal purposes we may also define the Green's functions

$$G_0=\langle 0|0\rangle=1, \qquad (2.3d)$$

$$G_1(x)=\langle 0|\phi(x)|0\rangle. \qquad (2.3e)$$

One method of obtaining the perturbation series for the Green's functions is to use the differential equations satisfied by them. This is the method we shall use in the following sections when treating gauge fields. Thus, G_2 will satisfy the equation

$$(\Box_1^2-\mu^2)G_2(x_1,x_2)=\tfrac{1}{2}\lambda G_3(x_1,x_1,x_2)+i\delta^4(x_1-x_2). \qquad (2.4a)$$

Equation (2.4a) is obtained by applying the differential equation (2.1) to the factor $\phi(x_1)$ of (2.3a). The first term on the right of (2.4a) arises from the interaction term in (2.1), while the second term is obtained by applying the differential operator $-\partial^2/\partial x_0^2$ to the time ordering itself. In deriving this term it is of course necessary to use the commutation relation (2.2b).

The higher Green's functions will satisfy similar equations. Thus G_3 will satisfy the equation

$$(\Box_1^2-\mu^2)G_3(x_1,x_2,x_3)=\tfrac{1}{2}\lambda G_4(x_1,x_1,x_2,x_3)+i\delta^4(x_1-x_2)G_1(x_3)+i\delta^4(x_1-x_3)G_1(x_2). \qquad (2.4b)$$

Equations (2.4a) and (2.4b) can be integrated to yield the formulas

$$G_2(x_1,x_2)=-\tfrac{1}{2}i\lambda\int dx_4\tfrac{1}{2}\Delta_F(x_1-x_4)G_3(x_4,x_4,x_2)+\tfrac{1}{2}\Delta_F(x_1-x_2), \qquad (2.5a)$$

$$G_3(x_1,x_2,x_3)=-\tfrac{1}{2}i\lambda\int dx_4\tfrac{1}{2}\Delta_F(x_1-x_4)G_4(x_4,x_4,x_2,x_3)+\tfrac{1}{2}\Delta_F(x_1-x_2)G_1(x_3)+\tfrac{1}{2}\Delta_F(x_1-x_3)G_1(x_2). \qquad (2.5b)$$

Equations (2.5) are illustrated diagramatically in Fig. 1.

If we are working in perturbation theory, the first Green's function on the right of (2.5a) or (2.5b) will be required to one order lower than that on the left, since it contains an explicit factor λ. The second term on the right of (2.5a) is known explicitly, while that on the right of (2.5b) only involves G_1. Hence, if we construct the perturbation series order by order and, within each order, construct the functions G_1, G_2, $\cdots$ successively, the right-hand side of (2.5) will be known in terms of previously calculated functions. We can therefore construct the entire perturbation series in this manner, and it is not difficult to see that we obtain the usual prescription for Feynman diagrams.

In a field theory with a simple Lagrangian, such as the $\lambda\phi^3$ theory, it is sufficient to write down the first few equations (2.4) and (2.5); the form of the subsequent equations is then fairly obvious. When writing down equations for gauge fields and performing manipulations with them, however, it would be somewhat cumbersome to proceed in this manner. We require a notation in which the whole series of equations (2.4) can be simply displayed. In the remainder of the section we shall develop such a notation. We emphasize that we are doing nothing more than constructing a shorthand for expressing the equations satisfied by the Green's functions.

We shall work with the linear space of the totality of functions $C_0, C_1(x_1), C_2(x_1,x_2), \cdots$. A typical vector in the linear space may be written

$$\begin{matrix} C_0 \\ C_1(x_1) \\ C_2(x_1x_2) \\ \cdots \end{matrix} \tag{2.6}$$

The linear space is thus the sum of a series of subspaces

$$\begin{matrix} C_0 & 0 & 0 & \\ 0 & C_1(x_1) & 0 & \cdots \\ 0 & 0 & C_2(x_1,x_2) & \\ \cdots & \cdots & \cdots & . \end{matrix}$$

The first subspace consists of a single vector, but each of the other subspaces is itself an infinite-dimensional subspace. Thus, the second subspace is the space of all functions of a single variable x_1, the third subspace is the space of all symmetric functions of two variables x_1 and x_2, and so on. We shall denote the vector (2.6) by the symbol $|C)$. We are interested in the particular vector

$$\begin{matrix} G_0 \\ G_1(x_1) \\ G_2(x_1,x_2) \\ \cdots \end{matrix} \tag{2.7}$$

where $G_0=\langle 0|0\rangle=1$, and the remaining G's are the Green's functions of our theory. We shall denote this vector by the symbol $|G)$.

We now define an operator $\tilde{\phi}(x)$ acting in this space as follows:

$$\tilde{\phi}(x)\begin{Bmatrix} C_0 \\ C_1(x_1) \\ C_2(x_1,x_2) \\ \cdots \end{Bmatrix}=\begin{Bmatrix} C_1(x) \\ C_2(xx_1) \\ C_3(xx_1x_2) \\ \cdots \end{Bmatrix}. \tag{2.8}$$

The variable x in the vector on the right is regarded as a fixed parameter, while $x_1, x_2, \cdots$ are the variables corresponding to our linear space.

To avoid misunderstanding we must emphasize that the operator $\tilde{\phi}(x)$ *corresponds* to the quantum-mechanical operator $\phi(x)$, but that it is a totally different type of operator acting on a totally different type of space. Once we have expressed the field theory in terms of Green's functions, we need no longer consider the quantum-mechanical Hilbert space; the whole theory has been expressed in terms of the c-number equations (2.4). In order to express these c-number equations in a simple way, we have defined a new linear space on which the operators $\tilde{\phi}(x)$ act. The $\tilde{\phi}$'s do not satisfy the quantum-mechanical commutation rules. In fact, all the $\tilde{\phi}$'s and their space and time derivatives commute with one another. The linear space of vectors $|\ )$ is thus a totally different space from the quantum-mechanical Hilbert space of vectors $|\ \rangle$.

It would be inconvenient if we had to display equations such as (2.8) whenever we used them, and, in fact, there is a standard notation for expressing (2.8) in a more compact form. This notation is expressed in terms of vectors in the *dual space*. A vector in the dual space is written in the form $(H|$, and is *defined* by means of its scalar products $(H|C)$ with all vectors in our space $|C)$. The scalar products must depend linearly on $|C)$. We define the vectors $(H_0|$, $(H_1(x_1)|$, $(H_2(x_1,x_2)|, \cdots$ in the dual space by the equation.

$$\begin{aligned} (H_0|C)&=C_0, \\ (H_1(x_1)|C)&=C_1(x_1), \\ (H_2(x_1,x_2)|C)&=C_2(x_1,x_2), \end{aligned} \tag{2.9}$$

C being the general vector (2.6). There is a single vector $(H_0|$, a vector $(H_1(x_1)|$ for each value of x_1, a vector $(H_2(x_1,x_2)|$ for each combination of variables x_1, x_2, and so on. A vector in the original space is uniquely defined by its scalar products with all vectors $(H_0|$, $(H_1(x_1)|$, $(H_2(x_1,x_2)|, \cdots$ in the dual space.

We next construct the dual space of the space of vectors $(H|$. If $|C')$ is a vector in this space, it is defined by its scalar product will all vectors $(H|$:

$$(H_n(x_1,\cdots x_n)|C')=C_n'(x_1,\cdots x_n).$$

However, the totality of functions C_0', $C_1'(x)$, $C_2'(x_1,x_2)$, $\cdots$ define a vector in the original space with which we started. There is thus a one-one correspondence between our original space $|C)$ and our new space $|C')$, which is such that corresponding vectors have identical scalar products with all vectors $(H|$. The space of vectors $|C')$, which is the dual of the space of vectors $(H|$, may therefore be regarded as identical to our original space of vectors $|C)$.

Given any operator O acting in the dual space, we can define an operator O acting in the original space as follows:

$$(H|\{O|C)\}=\{(H|O\}|C) \tag{2.10}$$

for all vectors H in the dual space. We can therefore use the notation $(H|O|C)$ to express the scalar product. Equation (2.10) can also be used to define an operator in the dual space once the operator in the original space is known.

It is now easy to express $\tilde{\phi}$ as an operator in the dual space. By (2.9) and (2.8)

$$\begin{aligned} (H_0|\{\tilde{\phi}(x)|C)\}&=C_1(x), \\ (H_1(x_1)|\{\tilde{\phi}(x)|C)\}&=C_2(x,x_1), \\ (H_2(x_1,x_2)|\{\tilde{\phi}(x)|C)\}&=C_3(x,x_1,x_2). \end{aligned} \tag{2.11}$$

Now, by (2.9),

$$(H_1(x)|C)=C_1(x),$$
$$(H_2(x,x_1)|C)=C_2(x,x_1), \qquad (2.12)$$
$$(H_3(x,x_1,x_2)|C)=C_3(x,x_1,x_2).$$

From (2.11) and (2.12), we conclude that

$$(H_0|\bar{\phi}(x)=(H_1(x)|,$$
$$(H_1(x_1)|\bar{\phi}(x)=(H_2(x,x_1)|, \qquad (2.13)$$
$$(\rho_2(x_1,x_2)|\bar{\phi}(x)=(H_3(x,x_1,x_2)|,$$

since $|C)$ is an arbitrary vector. For the vector $(H_n(x_1\cdots x_n)|$, Eq. (2.13) states that

$$(H_n(x_1\cdots x_n)|\bar{\phi}(x)=(H_{n+1}(x,x_1\cdots x_n)|. \qquad (2.14)$$

The notation of the dual space therefore allows us to express (2.8) in the compact form (2.14), and we shall use this notation in the remainder of the paper. We should emphasize that the use of the dual space does not involve any deep mathematics, but is purely a concession to the printer. It would be perfectly possible to rewrite every subsequent equation in this paper without using the dual space. Every equation of the form (2.14) would then be replaced by an equation of the form (2.8).

We now define a new operator η, which we shall require when writing the right-hand side of (2.4). It is defined as follows:

$$(H_n(x_1,x_2,\cdots,x_n)|\eta(x)$$
$$=\sum_{r=1}^{n}(H_{n-1}(x_1,x_2,\cdots,[x_r],\cdots x_n)|\delta(x-x_r). \qquad (2.15)$$

The vector $(H_{n-1}(x_1,x_2,\cdots,[x_r],\cdots,x_n)|$ denotes the vector which is obtained from $(H_n(x_1,x_2,\cdots,x_r,\cdots,x_n)|$ by removing the variable x_r. From (2.14) and (2.15) it is easily seen that η and ϕ obey the commutation relations:

$$[\eta(x_1),\bar{\phi}(x_2)]=-\delta^4(x_1-x_2). \qquad (2.16)$$

Furthermore, from (2.15),

$$(H_0|\eta(x)=0. \qquad (2.17)$$

Equations (2.16) and (2.17) are sufficient to determine η, since $(H_{n+1}|\eta$ can be found from (2.14) and (2.16) once $(H_n|\eta|$ is known. We may therefore regard (2.16) and (2.17) as the definitions of η; we can then easily obtain (2.15).

Having defined the operators $\bar{\phi}$ and η, we can express the Green's functions equations in a compact form. Our Eq. (2.9), which defines the vectors $(H_n|$, shows that

$$(H_n(x_1\cdots x_n)|G)=G(x_1\cdots x_n). \qquad (2.18)$$

Hence Eq. (2.4a) may be written as follows:

$$(\Box_1^2-\mu^2)(H_2(x_1,x_2)|G)=\tfrac{1}{2}\lambda(H_3(x_1,x_1,x_2)|G)$$
$$+i\delta_4(x_1-x_2)(H_0|G). \qquad (2.19)$$

We now use (2.14) and (2.15) to express the vectors $(H_2(x_1,x_2)|$, $(H_3(x_1,x_2,x_2)|$, and $(H_0|\delta(x_1-x_2)$ in terms of the single vector $(H_1(x_2)|$:

$$(H_2(x_1,x_2)|=(H_1(x_2)|\bar{\phi}(x_1), \qquad (2.20a)$$
$$(H_3(x_1,x_1,x_2)|=(H_1(x_2)|\bar{\phi}^2(x_1), \qquad (2.20b)$$
$$(H_0|\delta(x_1-x_2)=(H_1(x_2)|\eta(x_1). \qquad (2.20c)$$

Equation (2.19) therefore becomes

$$(\Box_1^2-\mu^2)(H_1(x_2)|\bar{\phi}(x_1)|G)=\tfrac{1}{2}\lambda(H_1(x_2)|\bar{\phi}^2(x_1)|G)$$
$$+i(H_1(x_2)|\eta(x_1)|G). \quad (2.21)$$

Since the operator $(\Box_1^2-\mu^2)$ in the first term of (2.21) acts only on the variable x_1, it may be taken inside the scalar product. Thus

$$(H_1(x_2)|(\Box_1^2-\mu^2)\bar{\phi}(x_1)|G)=\tfrac{1}{2}\lambda(H_1(x_2)|\bar{\phi}^2(x_1)|G)$$
$$+i(H_1(x_2)|\eta(x_1)|G). \qquad (2.22a)$$

Similarly, Eq. (2.4b) may be written in the form

$$(H_2(x_2x_3)|(\Box_1^2-\mu^2)\bar{\phi}(x_1)|G)=\tfrac{1}{2}\lambda(H_2(x_2,x_3)|\bar{\phi}^2(x_1)|G)$$
$$+i(H_2(x_2,x_3)|\eta(x_1)|G). \quad (2.22b)$$

The general Eq. (2.4) can be obtained by replacing the vectors $(H_1|$ and $(H_2|$ in (2.23) by the general vector $(H_n|$:

$$(H_n(x_2\cdots x_{n+1})|(\Box_1^2-\mu^2)\bar{\phi}(x_1)|G)$$
$$=\tfrac{1}{2}\lambda(H_n(x_2\cdots x_{n+1})|\bar{\phi}^2(x_1)|G)$$
$$+i(H_n(x_2\cdots x_{n+1})|\eta(x_1)|G). \quad (2.22c)$$

Since the vector $(H_n(x_2\cdots x_{n+1})|$ in (2.22c) is arbitrary, we may rewrite the equation as an equation for the vector G:

$$(\Box_1^2-\mu^2)\bar{\phi}(x_1)|G)=\tfrac{1}{2}\lambda\bar{\phi}^2(x_1)|G)+i\eta(x_1)|G). \quad (2.23)$$

We have thus replaced the infinite set of equations for the Green's functions by a single equation for the vector (G).

We can easily integrate (2.23) to give the result:

$$\left(\bar{\phi}(x)+\tfrac{1}{2}i\lambda\int dx'\tfrac{1}{2}\Delta_F(x-x')\bar{\phi}^2(x')\right.$$
$$\left.-\int dx'\tfrac{1}{2}\Delta_F(x-x')\eta(x')\right)|G)=0. \quad (2.24)$$

Equation (2.24) is of course equivalent to the series of Eqs. (2.5) in our new notation.

It is worthwhile noticing that Eq. (2.23) has the same form as the field Eq. (2.1), except for the term $i\eta(x)$. This last term corresponds to the δ-function terms in the equations for the Green's functions.

We can relate our notation to Schwinger's functional notation by making the correspondence

$$\bar{\phi}(x)\to\delta/\delta\eta(x), \quad \eta(x)\to\eta(x),$$

and regarding the operators as acting on the Schwinger functional. We have chosen to define ϕ as our fundamental quantity and to define η in terms of it, whereas Schwinger proceeds in the reverse direction. The reason why we have set up a new notation is that quantities similar to ϕ may easily be defined in the path-dependent formalism, whereas the quantities analogous to η are not quite so simple. Our notation is therefore more easily applicable to the theories treated in this paper.

4. YANG-MILLS FIELD

The massless Yang-Mills field appears to possess all the essential complications of the gravitational field while lacking some of the algebraic complications. It is therefore instructive to consider this field before going on to the gravitational field. We shall treat a self-interacting Yang-Mills field, since interaction with other fields does not introduce any new features.

The path-dependent formalism for the Yang-Mills field has been examined by Bialynicki-Birula.[5] The procedure followed is analogous to that used for the electromagnetic field, with the difference that the Yang-Mills field plays the dual role of the gauge field and the charged field. The field equations are simpler *in appearance* than the Maxwell equations of electrodynamics, since there is no additional current term. They take the form

$$\frac{\partial F_{\mu\nu}{}^{\alpha}(x,P)}{\partial x_{\mu}}=0. \tag{4.1a}$$

We shall always use indices from the beginning of the Greek alphabet to denote components in isotopic space, indices from the middle of the Greek alphabet to denote components in ordinary space. Note that F, being a charged field, is path-dependent.

The path-dependence equation is a straightforward generalization of (3.5):

$$\delta_z F_{\mu\nu}{}^{\alpha}(x,P)=g\epsilon_{\alpha\beta\gamma}F_{\mu\nu}{}^{\gamma}(x,P)F_{\rho\sigma}{}^{\beta}(z,P')\sigma_{\rho\sigma}. \tag{4.2}$$

As usual, $\delta_z F_{\mu\nu}{}^{\alpha}$ is the change of the variable $F_{\mu\nu}{}^{\alpha}$ caused by a change in the path by an infinitesimal area $\sigma_{\rho\sigma}$ at the point z. The path P' is that portion of P leading to z. Equation (4.2) requires the following consistency condition, which is analogous to the homogeneous Maxwell equations:

$$\epsilon_{\mu\nu\rho\sigma}\frac{\partial F_{\mu\nu}{}^{\alpha}(x,P)}{\partial x_{\rho}}=0. \tag{4.1b}$$

The equal-time commutators between the F's will contain terms analogous to (3.3) and (3.4). Thus

$$[F_{ij}{}^{\alpha}(\mathbf{x},t,P),F_{i'j'}{}^{\beta}(\mathbf{y},t,P')]=0, \tag{4.3a}$$

$$\begin{aligned}[F_{0i}{}^{\alpha}(\mathbf{x},t,P),F_{jk}{}^{\beta}(\mathbf{y},t,P')] &= -i\delta_{\alpha\beta}\left(\delta_{ik}\frac{\partial}{\partial y_j}-\delta_{ij}\frac{\partial}{\partial y_k}\right)\delta^3(\mathbf{x}-\mathbf{y}) \\ &\quad +i\epsilon_{\alpha\beta\gamma}\int^{y} d\xi_i\delta^3(\mathbf{x}-\boldsymbol{\xi})F_{jk}{}^{\gamma}(\mathbf{y},t),\end{aligned} \tag{4.3b}$$

$$\begin{aligned}[F_{0i}{}^{\alpha}(\mathbf{x},t,P),F_{0j}{}^{\beta}(\mathbf{y},t,P')] &= i\epsilon_{\alpha\beta\gamma}\int_{P'}^{y} d\xi_i\delta^3(\mathbf{x}-\boldsymbol{\xi})F_{0j}{}^{\gamma}(\mathbf{y},t) \\ &\quad +i\epsilon_{\alpha\beta\gamma}\int_{P}^{x} d\xi_j\delta^3(\mathbf{y}-\boldsymbol{\xi})F_{0i}{}^{\gamma}(\mathbf{x},t).\end{aligned} \tag{4.3c}$$

It is not difficult to check that Eqs. (4.1)-(4.3) are consistent with one another and with Lorentz transformations. In fact, the equations of motion and commutation relations may be derived from the Lagrangian

$$\mathcal{L}=-\tfrac{1}{4}[F_{\mu\nu}{}^{\alpha}(x,P)]^2.$$

One may define path-dependent Green's functions in the usual way. As in the electromagnetic case, it is necessary to include δ-function terms if the Green's functions are to be covariant. The definitions are therefore as follows:

$$\hat{G}^{\alpha\beta}{}_{\mu\nu,ij}(x_1,P_1,x_2,P_2) = \langle 0|T\{F_{\mu\nu}{}^{\alpha}(x_1,P_1),F_{ij}{}^{\beta}(x_2,P_2)\}|0\rangle, \tag{4.4a}$$

$$\hat{G}^{\alpha\beta}{}_{0i,0j}(x_1,P_1,x_2,P_2) = \langle 0|T\{F_{0i}{}^{\alpha}(x_1,P_1),F_{0j}{}^{\beta}(x_2,P_2)\}|0\rangle +i\delta_{\alpha\beta}\delta_{ij}\delta^4(x_1-x_2). \tag{4.4b}$$

[10] B. Zumino, J. Math. Phys. 1, 1 (1960).

Higher Green's functions may be defined in a similar way.

The field equations (4.1a) may be rewritten as equations for the Green's function, analogous to Eqs. (3.8) for the electromagnetic Green's functions. Thus the two-point Green's function satisfies the equation

$$\frac{\partial}{\partial x_{1\mu}}\hat{G}^{\alpha\beta}{}_{\mu\nu,\rho\sigma}(x_1,P_1,x_2,P_2) = i\left(\frac{\partial}{\partial x_{2\rho}}\delta_{\sigma\nu}-\frac{\partial}{\partial x_{2\sigma}}\delta_{\rho\nu}\right)\delta_{\alpha\beta}\delta^4(x_1-x_2). \quad (4.5a)$$

The four-point Green's function satisfies the equation

$$\frac{\partial}{\partial x_{1\mu}}\hat{G}^{\alpha\beta\gamma\delta}{}_{\mu\nu,\rho\sigma,\kappa\lambda,\tau\omega}(x_1P_1,x_2P_2,x_3P_3,x_4P_4) = i\left(\frac{\partial}{\partial x_{2\rho}}\delta_{\sigma\nu}-\frac{\partial}{\partial x_{2\sigma}}\delta_{\rho\nu}\right)\delta_{\alpha\beta}\delta^4(x_1-x_2)\hat{G}^{\gamma\delta}{}_{\kappa\lambda,\tau\omega}(x_3P_3,x_4P_4)$$
$$+\text{two similar terms with } 2\leftrightarrow 3,\ 2\leftrightarrow 4$$
$$-ig\epsilon_{\alpha\beta\epsilon}\int_{P_2}^{x_2} d\xi_\nu\delta^4(x_1-\xi)G^{\epsilon\gamma\delta}{}_{\rho\sigma,\kappa\lambda,\tau\omega}(x_2P_2,x_3P_3,x_4P_4)$$
$$+\text{two similar terms with } 2\leftrightarrow 3,\ 2\leftrightarrow 4. \quad (4.5b)$$

The right-hand side of (4.5b) is obtained by differentiating the time ordering and applying the commutation relations (4.3). Higher Green's functions will satisfy equations similar to (4.5b).

Equation (4.1b) implies that the Green's functions satisfy equations such as

$$\epsilon_{\mu\nu\rho\sigma}\frac{\partial}{\partial x_{1\rho}}\hat{G}^{\alpha\beta}{}_{\mu\nu,\kappa\lambda}(x_1,P_1,x_2,P_2)=0. \quad (4.5c)$$

The path-dependence equations satisfied by the Green's function can be obtained from (4.2), by following reasoning identical to that used for the electromagnetic field. Again the term arising from the variation of the time-ordering cancels against the term obtained from the δ function in the Green's function [the second term on the right of (4.4b)], and the final result is

$$\delta_{x_3}\hat{G}^{\alpha\beta}{}_{\mu\nu,\rho\sigma}(x_1,P_1,x_2,P_2)=g\epsilon_{\alpha\gamma\epsilon}\int d\sigma_{\kappa\lambda}(x_3)\hat{G}^{\epsilon\beta\gamma}{}_{\mu\nu,\rho\sigma,\kappa\lambda}\times(x_1,P_1,x_2,P_2,x_3,P_1'). \quad (4.6)$$

In this equation, δ_{x_3} is the variation of G caused by a variation of the path P_1 at the point x_3, and P_1' is the portion of P_1 leading to x_3. The higher Green's functions satisfy similar path-dependence equations.

Condensed Notation

The condensed notation which we shall use is very similar to that used in the two previous sections, and we need not explain it in detail again. We construct the linear space of the totality of all functions $\hat{C}_{\mu\nu}\ldots^{\alpha\cdots}(x_1,P_1,\cdots)$. We then construct the dual space and define the vector $(\hat{H}_{\mu\nu}\ldots^{\alpha\cdots}(x_1,P_1,\cdots)|$ in the usual way. We next define the operators $\tilde{F}^\alpha{}_{\mu\nu}(x,P)$ by the equations

$$(\hat{H}_{\rho\sigma}\ldots^{\beta\cdots}(x_1,P_1,\cdots)|\tilde{F}^\alpha{}_{\mu\nu}(x,P) =(\hat{H}_{\rho\sigma,\mu\nu}\ldots^{\beta\alpha\cdots}(x_1,P_1,x,P,\cdots)|. \quad (4.7)$$

We also define an operator $U(x,P)$ which corresponds to the right-hand side of (4.5):

$$(\hat{H}_{\kappa\lambda},\ldots^{\beta\cdots}(x_1,P_1),\cdots|U_\nu{}^\alpha(x,P) =\sum_r\left(\frac{\partial}{\partial x_{r\rho}}\delta_{\sigma\nu}-\frac{\partial}{\partial x_{r\sigma}}\delta_{\rho\nu}\right)\times(\hat{H}_{\kappa\lambda},\ldots_{[\rho\sigma]}\ldots^{\beta\cdots[\gamma]\cdots}(x_1,P_1,\cdots[x_r,P_r]\cdots)|$$
$$\times\delta_{\alpha\gamma}\delta^4(x-x_r)-g\sum_r\epsilon_{\alpha\gamma\delta}\int_{P_r}^{x_r}d\xi_\nu \times(\hat{H}_{\kappa\lambda},\ldots^{\beta\cdots[\gamma]\delta\cdots}(x_1,P_1,\cdots)|\delta^4(x_r-\xi), \quad (4.8)$$

where γ and $\rho\sigma$ are the indices corresponding to the coordinates x_r, and the superscript $[\gamma]\delta$ in (4.8) indicates that γ is to be replaced by δ. The operator U may be defined by its commutation relations

$$[U_\nu{}^\alpha(x_1,P_1),\tilde{F}_{\rho\sigma}{}^\beta(x_2,P_2)] =-\left(\frac{\partial}{\partial x_{2\rho}}\delta_{\nu\sigma}-\frac{\partial}{\partial x_{2\sigma}}\delta_{\nu\rho}\right)\delta_{\alpha\beta}\delta^4(x_1-x_2) +g\epsilon_{\alpha\beta\gamma}\int_{P_2}^{x_2}d\xi_\nu\delta^4(x_1-\xi)F_{\rho\sigma}{}^\gamma(x_2,P_2). \quad (4.9)$$

Note that the right-hand side of (4.9) has terms corresponding to the right-hand sides of both (3.19a) and (3.19b), this is because the variable F in the Yang-Mills field plays the roles of the gauge field and the charged field. Equation (4.9) must be supplemented by the equation

$$(H_0|U_\nu{}^\alpha(x_1,P_1)=0, \quad (4.10)$$

to complete the definition of U.

In our present notation, the field equations (4.5) become

$$\left(\frac{\partial}{\partial x_\mu}\tilde{F}_{\mu\nu}{}^\alpha(x,P)-iU_\nu{}^\alpha(x,P)\right)|\hat{G})=0. \quad (4.11)$$

The path-dependence equation (4.6) is

$$\delta_z\tilde{F}_{\mu\nu}{}^\alpha(x,P)=g\epsilon_{\alpha\beta\gamma}\tilde{F}_{\kappa\lambda}{}^\beta(x,P')\tilde{F}_{\mu\nu}{}^\gamma(x,P)\sigma_{\kappa\lambda}, \quad (4.12)$$

where P' represents the portion of P leading to the point z.

Auxiliary Variables

Following the procedure used in electrodynamics, we shall attempt to express our path-dependent quantities $\tilde{F}$ in terms of auxiliary path-independent quantities $\tilde{A}$. The formulas relating the $\tilde{F}$'s to the $\tilde{A}$'s will be the same as the formulas relating the field variables F to the potentials A. Thus, following the results of Ref. 5, we make the connection as follows:

$$\tilde{F}_{\mu\nu}{}^{\alpha}(x,P)=V_{\alpha\gamma}(x,P)\tilde{f}_{\mu\nu}{}^{\gamma}(x)\,, \tag{4.13a}$$

where

$$\tilde{f}_{\mu\nu}{}^{\gamma}(x)=\frac{\partial\tilde{A}_{\nu}{}^{\gamma}(x)}{\partial x_{\mu}}-\frac{\partial\tilde{A}_{\mu}{}^{\gamma}(x)}{\partial x_{\nu}}+g\epsilon_{\gamma\delta\epsilon}\tilde{A}_{\mu}{}^{\delta}(x)\tilde{A}_{\nu}{}^{\epsilon}(x)\,, \tag{4.13b}$$

$$\begin{aligned}V_{\alpha\gamma}(x,P)=\delta_{\alpha\gamma}&+g\epsilon_{\alpha\beta\gamma}\int_P^x d\xi_\rho\tilde{A}_\rho{}^\beta(\xi)\\&+g^2\epsilon_{\alpha\beta\epsilon}\epsilon_{\epsilon\delta\gamma}\int_P^x d\xi_\rho\int_P^\xi d\xi_\sigma{}'\tilde{A}_\sigma{}^\beta(\xi')\tilde{A}_\rho{}^\delta(\xi)\\&+g^3\epsilon_{\alpha\beta\epsilon}\epsilon_{\epsilon\delta\eta}\epsilon_{\eta\zeta\gamma}\int_P^x d\xi_\rho\int_P^\xi d\xi_\sigma\int_P^{\xi'} d\xi_\tau{}''\tilde{A}_\tau{}^\beta(\xi'')\\&\times\tilde{A}_\sigma{}^\zeta(\xi')\tilde{A}_\rho{}^\zeta(\xi)+\cdots.\end{aligned} \tag{4.13c}$$

If we compare (4.13) with (3.24) and (3.34), we observe that we have to take the curl of A *and* multiply it by the function V in order to obtain the path-dependent function $\tilde{F}$. Again, this is because the Yang-Mills field is both a gauge field and a charged field.

Before going further it will be useful to obtain two identities satisfied by the function V. For this purpose it is convenient to obtain an alternative definition of V, due to Bialynicki-Birula.[6] We introduce the following unitary matrix in the Pauli spin space:

$$W(x,P)=L\exp\left\{-ig\int_P^x d\xi_\rho A_\rho{}^\alpha(\xi)\tau_\alpha\right\}, \tag{4.14}$$

where the matrices τ_α are the Pauli matrices. The symbol L indicates that the τ's are to be ordered from the beginning to the end of the path when expanding the exponential. It is then not difficult to see that

$$W^{\dagger}(x,P)\tau_\gamma W(x,P)=V_{\alpha\gamma}(x,P)\tau_\alpha. \tag{4.15}$$

From (4.15) we can derive the identities

$$V_{\alpha\gamma}V_{\alpha\gamma'}=V_{\gamma\alpha}V_{\gamma'\alpha}=\delta_{\gamma\gamma'}\,, \tag{4.16a}$$

$$\epsilon_{\zeta\delta\epsilon}V_{\delta\alpha}V_{\epsilon\beta}=\epsilon_{\alpha\beta\gamma}V_{\zeta\gamma}\,, \tag{4.16b}$$

$$\epsilon_{\zeta\delta\epsilon}V_{\alpha\delta}V_{\beta\epsilon}=\epsilon_{\alpha\beta\gamma}V_{\gamma\zeta}. \tag{4.16c}$$

For future reference we shall add the following trivial identity involving the ϵ's;

$$\epsilon_{\alpha\beta\gamma}\epsilon_{\alpha\delta\epsilon}+\epsilon_{\alpha\beta\delta}\epsilon_{\alpha\epsilon\gamma}+\epsilon_{\alpha\beta\epsilon}\epsilon_{\alpha\gamma\delta}=0. \tag{4.16d}$$

Another symbol which it will be convenient to introduce is the following:

$$\begin{aligned}V_{\alpha\gamma}(x',x,P)=\delta_{\alpha\gamma}&+g\epsilon_{\alpha\beta\gamma}\int_{x',P}^x d\xi_\rho\tilde{A}_\rho{}^\beta(\xi)+g^2\epsilon_{\alpha\beta\epsilon}\epsilon_{\epsilon\delta\gamma}\\&\times\int_{x',P}^x d\xi_\rho\int_{x',P}^x d\xi_\sigma{}'\tilde{A}_\sigma{}^\beta(\xi')A_\rho{}^\delta(\xi)+\cdots,\end{aligned} \tag{4.17}$$

where x' is a point on the path P. In other words, $V_{\alpha\gamma}(x',x,P)$ is defined in a similar way to $V_{\alpha\gamma}(x,P)$ except that all integrals are taken from x' to x instead of from $-\infty$ to x. The following relation may be verified directly:

$$V_{\alpha\gamma}(x,P)=V_{\alpha\delta}(x',P')V_{\delta\gamma}(x',x,P)\,, \tag{4.18}$$

where as usual P' is the portion of P leading to x'.

If we differentiate (4.13c) with respect to the endpoint of the path of integration, we obtain the equation

$$\frac{\partial}{\partial x_\mu}V_{\alpha\gamma}(x,P)=g\epsilon_{\gamma\delta\epsilon}\tilde{A}_\mu{}^\epsilon(x)V_{\epsilon\delta}(x,P). \tag{4.19}$$

As we shall see below, this equation will enable us to express derivatives of path-dependent functions in terms of derivatives of auxiliary functions. Equation (4.19) can be written in integral form,

$$V_{\alpha\gamma}(x,P)=\delta_{\alpha\gamma}+g\epsilon_{\gamma\delta\epsilon}\int_P^x d\xi_\mu\tilde{A}_\mu{}^\epsilon(\xi)V_{\alpha\delta}(\xi,P)\,, \tag{4.20}$$

where we have used the boundary condition $V_{\alpha\gamma}=\delta_{\alpha\gamma}$ when $g=0$ or when $x\to\infty$. By expanding (4.20) in a power series in g, we recover the definition (4.13c). Equation (4.19) or (4.20) may therefore be taken as a definition of V in place of (4.13c).

We can now show that the definitions (4.13) do lead to the correct path-dependence equations (4.12) for F, and we have carried out the algebra in Appendix A. The formulas (4.13) can therefore be used to define path-dependent quantities in terms of auxiliary quantities.

The operator η will be defined in exactly the same way as the analogous operators were defined in the two preceding sections:

$$[\eta_\mu{}^\alpha(x_1),\tilde{A}_\nu{}^\beta(x_2)]=-\delta_{\alpha\beta}\delta_{\mu\nu}\delta^4(x_1-x_2)\,, \tag{4.21}$$

$$(H_0|\eta_\mu{}^\alpha(x)=0. \tag{4.22}$$

We require an expression for the corresponding path-dependent quantity $U_\nu{}^\alpha(x,P)$, defined by (4.9) and (4.10), in terms of the auxiliary variables. In electrodynamics the quantity $U(x,P)$ was equal to $V(x,P)\times\eta(x)$, and this suggests that the Yang-Mills operator $U_\nu{}^\alpha(x,P)$ might be given by a similar equation:

$$U_\nu{}^\alpha(x,P)\approx V_{\alpha\gamma}(x,P)\eta_\nu{}^\gamma(x). \tag{4.23}$$

We shall verify (4.23) in Appendix A. The equality between the two sides of (4.23) is to be interpreted in the sense that they both have the same commutation relations (4.9) with the operators $F_{\rho\sigma}{}^{\beta}(x',P')$.

Once we have defined the operators $\tilde{A}_{\mu}{}^{\alpha}(x)$, we can construct our enlarged dual space of vectors $(H_{\mu\ldots}{}^{\alpha\cdots}(x_1,\cdots)|$. We can then construct the linear space of vectors $|G)$ and can define auxiliary, path-independent Green's functions G. The path-dependent Green's functions $\hat{G}$ can be expressed in terms of the G's by formulas analogous to (3.29). We shall not give the details, which are the exact analog of the corresponding details in electrodynamics. We can also write equations similar to (3.32) for the operators η.

Gauge Transformations

The gauge transformations are given by the equation

$$\tilde{A}_{\mu}{}^{\alpha}(x)\to\tilde{A}_{\mu}{}^{\alpha}(x)+\lambda\left(\frac{\partial\chi_{\alpha}(x)}{\partial x_{\mu}}+g\epsilon_{\alpha\beta\gamma}\chi_{\gamma}(x)\tilde{A}_{\mu}{}^{\beta}(x)\right). \quad (4.24)$$

If we compare (4.24) with (3.36) we notice that terms corresponding to the right-hand side of (3.36a) and (3.36c) appear on the right of (4.24). This is once more due to the fact that the Yang-Mills field plays the dual role of gauge field and charged field.

Let us first investigate how the operators defined in (4.13) transform when $\tilde{A}$ undergoes the transformation (4.24). From (4.13b) one can verify directly that

$$\tilde{f}_{\mu\nu}{}^{\alpha}(x)\to\tilde{f}_{\mu\nu}{}^{\alpha}(x)+\lambda g\epsilon_{\alpha\beta\gamma}\chi_{\gamma}(x)\tilde{A}_{\mu}{}^{\beta}(x). \quad (4.25)$$

The function V transforms in a similar way:

$$V_{\alpha\gamma}(x,P)\to V_{\alpha\gamma}(x,P)+\lambda g\epsilon_{\gamma\delta\epsilon}\chi_{\epsilon}(x)V_{\alpha\delta}(x,P). \quad (4.26)$$

The easiest way of verifying (4.26) is to show that (4.19), which may be taken as the defining equation for V, remains invariant when $\tilde{A}$ undergoes the transformation (4.24) and V the transformation (4.26). Under such a transformation, the two sides of (4.19) transform as follows:

$$\frac{\partial}{\partial x_{\mu}}V_{\alpha\gamma}(x,P)\to\frac{\partial}{\partial x_{\mu}}V_{\alpha\gamma}(x,P)+\lambda g\epsilon_{\gamma\delta\epsilon}\left(\frac{\partial\chi_{\epsilon}(x,P)}{\partial x_{\mu}}V_{\alpha\delta}(x,P)+\chi_{\epsilon}\frac{\partial}{\partial x_{\mu}}V_{\alpha\delta}(x,P)\right)=\frac{\partial}{\partial x_{\mu}}V_{\alpha\gamma}(x,P)+\lambda g\epsilon_{\gamma\delta\epsilon}$$
$$\times\frac{\partial\chi_{\epsilon}(x,P)}{\partial x_{\mu}}V_{\alpha\delta}(x,P)+\lambda g^2\epsilon_{\gamma\delta\epsilon}\epsilon_{\delta\zeta\eta}\chi_{\epsilon}(x)\tilde{A}_{\mu}{}^{\eta}(x)V_{\alpha\zeta}(x,P), \quad (4.27)$$

from (4.19),

$$g\epsilon_{\gamma\delta\epsilon}\tilde{A}_{\mu}{}^{\epsilon}(x)V_{\alpha\delta}(x,P)\to g\epsilon_{\gamma\delta\epsilon}\tilde{A}_{\mu}{}^{\epsilon}(x)V_{\alpha\delta}(x,P)+\lambda g\epsilon_{\gamma\delta\epsilon}\frac{\partial\chi_{\epsilon}(x)}{\partial x_{\mu}}V_{\alpha\delta}(x,P)$$
$$+\lambda g^2\epsilon_{\gamma\delta\epsilon}\epsilon_{\epsilon\zeta\eta}\chi_{\eta}(x)\tilde{A}_{\mu}{}^{\zeta}(x)V_{\alpha\delta}(x,P)+\lambda g^2\epsilon_{\gamma\delta\epsilon}\epsilon_{\delta\zeta\eta}\tilde{A}_{\mu}{}^{\epsilon}(x)\chi_{\eta}(x)V_{\alpha\zeta}(x,P)$$
$$=g\epsilon_{\gamma\delta\epsilon}\tilde{A}_{\mu}{}^{\epsilon}(x)V_{\alpha\delta}(x,P)+\lambda g\epsilon_{\gamma\delta\epsilon}\frac{\partial\chi_{\epsilon}(x)}{\partial x_{\mu}}V_{\alpha\delta}(x,P)+\lambda g^2\epsilon_{\gamma\epsilon\eta}\epsilon_{\epsilon\delta\zeta}\chi_{\eta}(x)\tilde{A}_{\mu}{}^{\zeta}(x)V_{\alpha\delta}(x,P), \quad (4.28)$$

from (4.16d). Comparing (4.27) and (4.28), we observe that the changes in the two sides of Eq. (4.19) are the same, so that (4.19) is invariant under the transformation (4.24), (4.26). Thus the change in V is given by (4.26).

We can now find the change in $\tilde{F}$ as defined by (4.13):

$$\tilde{F}_{\mu\nu}{}^{\alpha}(x,P)\to\tilde{F}_{\mu\nu}{}^{\alpha}(x,P)+\lambda g\epsilon_{\gamma\delta\epsilon}\chi_{\epsilon}(x)V_{\alpha\delta}(x,P)\tilde{f}_{\mu\nu}{}^{\gamma}(x)+\lambda g\epsilon_{\gamma\delta\epsilon}V_{\alpha\gamma}(x,P)\chi_{\epsilon}(x)\tilde{f}_{\mu\nu}{}^{\delta}(x) \quad \text{[from (4.25) and (4.26)]}$$
$$=\tilde{F}_{\mu\nu}{}^{\alpha}(x,P).$$

The variable $\tilde{F}_{\mu\nu}{}^{\alpha}(x,P)$ is therefore invariant under the transformation (4.24), and we are justified in calling it a gauge transformation.

To define a generator of the gauge transformation, we construct the operator

$$Y_{\beta}(y)=\frac{\partial\eta_{\mu}{}^{\beta}(y)}{\partial y_{\mu}}+\epsilon_{\beta\epsilon\delta}\eta_{\mu}{}^{\delta}(y)\tilde{A}_{\mu}{}^{\epsilon}(y)$$
$$=\left(\delta_{\beta\delta}\frac{\partial}{\partial y_{\mu}}+\epsilon_{\beta\epsilon\delta}\tilde{A}_{\mu}{}^{\epsilon}(y)\right)\eta_{\mu}{}^{\delta}(y). \quad (4.29)$$

The integral $\lambda\int dy Y_{\beta}(y)\chi_{\beta}(y)$, when commuted with $\tilde{A}_{\mu}{}^{\alpha}(x)$, does give (4.24). We thus conclude that the integral $\int dy Y_{\beta}(y)\chi_{\beta}(y)$ commutes with all our path-dependent variables $\tilde{F}_{\mu\nu}{}^{\alpha}(x,P)$. Furthermore, since V undergoes the transformation (4.26), we conclude that

$$\left[\int dy\, Y_{\beta}(y)\chi_{\beta}(y),V_{\alpha\gamma}(x,P)\right]=-g\epsilon_{\gamma\delta\epsilon}\chi_{\delta}(x)V_{\alpha\epsilon}(x,P). \quad (4.30)$$

Field Equations

Our aim is now to express the field equations (4.11) as equations for the auxiliary variables. The first term of (4.11) is easily transformed:

$$\frac{\partial}{\partial x_\mu}\tilde{F}_{\mu\nu}{}^{\alpha}(x,P)$$

$$=\frac{\partial}{\partial x_\mu}V_{\alpha\gamma}(x,P)\tilde{f}_{\mu\nu}{}^{\gamma}(x)$$

$$=V_{\alpha\gamma}(x,P)\frac{\partial}{\partial x_\mu}\tilde{f}_{\mu\nu}{}^{\gamma}(x)+\epsilon_{\gamma\delta\epsilon}V_{\alpha\delta}(x)\tilde{A}_\mu{}^{\epsilon}(x)\tilde{f}_{\mu\nu}{}^{\gamma}(x)$$

[from (4.19)]

$$=V_{\alpha\gamma}(x,P)\left(\frac{\partial}{\partial x_\mu}\tilde{f}_{\mu\nu}{}^{\gamma}(x)+\epsilon_{\gamma\delta\epsilon}\tilde{A}_\mu{}^{\delta}(x)\tilde{f}_{\mu\nu}{}^{\epsilon}(x)\right). \quad (4.31)$$

We have seen that the second term $-iU_\nu{}^{\alpha}(x,P)$ of (4.11) is equivalent to the operator $-iV_{\alpha\gamma}(x,P)\eta_\nu{}^{\gamma}(x)$ in the sense that they both have the same commutation relations with the operator $F_{\rho\sigma}{}^{\beta}(x',P')$. We may therefore be tempted to rewrite the field equations (4.11) as follows:

$$V_{\alpha\gamma}(x,P)\left(\frac{\partial}{\partial x_\mu}\tilde{f}_{\mu\nu}{}^{\gamma}(x)+\epsilon_{\gamma\delta\epsilon}\tilde{A}_\mu{}^{\delta}(x)\times\tilde{f}_{\mu\nu}{}^{\epsilon}(x)-i\eta_\nu{}^{\gamma}(x)\right)|G\rangle=0. \quad (4.32)$$

However, we shall show below that the consistency of Eq. (4.32) requires that its last term satisfy a divergence condition similar to the corresponding condition in the Maxwell equation of electrodynamics, and we shall have to generalize it if the condition is to be satisfied.

We shall follow the procedure used in electrodynamics and shall make use of the fact that the commutation relations (4.9), which define the operator $U_\nu{}^{\alpha}(x,P)$ uniquely in the original linear space, do not define it uniquely in the enlarged linear space. We employ this freedom to find a definition of $U_\nu{}^{\alpha}(x,P)$ which gives consistent field equations. We begin by writing

$$U_\nu{}^{\alpha}(x,P)=V_{\alpha\gamma}(x,P)\eta_\nu{}^{\gamma}(x)+\int dy\, Y_\beta(y)\chi_{\alpha\beta}'(x,y) \quad (4.33)$$

where χ' is arbitrary. Since the second term commutes with every gauge-invariant operator, the right-hand side of (4.33) maintains the correct commutation relations (4.9). All the terms in the equation of motion (4.32) have a factor $V_{\alpha\gamma}(x,P)$ in front of them, and it will be convenient for us if the last term in (4.33) also such a factor. We therefore define

$$\chi_{\alpha\beta}'(x,y)=V_{\alpha\gamma}(x,P)\chi_{\gamma\beta}(x).$$

Thus,

$$U_\nu{}^{\alpha}(x,P)=V_{\alpha\gamma}(x,P)\eta_\nu{}^{\gamma}(x)+\int dy\, Y_\beta(y)V_{\alpha\gamma}(x,P)\chi_{\gamma\beta}(x,y). \quad (4.34)$$

The factor $V_{\alpha\gamma}$ in the second term of (4.34) is still not in front of the other factors, but we can move it into this position by using the commutation relation (4.30) The equation then becomes

$$U_\nu{}^{\alpha}(x,P)=V_{\alpha\gamma}(x,P)\eta_\nu{}^{\gamma}(x)+\int dy V_{\alpha\gamma}(x,P)Y_\beta(y)\chi_{\gamma\beta}(x,y)-g\epsilon_{\gamma\delta\epsilon}V_{\alpha\epsilon}(x,x)\chi_{\gamma\delta}(x,x)=V_{\alpha\gamma}(x,P)\theta_\nu{}^{\gamma}(x), \quad (4.35)$$

where

$$\theta_\nu{}^{\gamma}(x)=\eta_\nu{}^{\gamma}(x)+\int dy\, Y_\beta(y)\chi_{\gamma\beta}(x,y)+g\epsilon_{\gamma\delta\epsilon}\chi_{\epsilon\delta}(x,x)$$

$$=\eta_\nu{}^{\gamma}(x)+\int dy\left(\frac{\partial\eta_\mu{}^{\beta}(y)}{\partial y_\mu}+g\epsilon_{\beta\epsilon\delta}\eta_\mu{}^{\delta}(y)\tilde{A}_\mu{}^{\epsilon}(y)\right)\chi_{\gamma\beta}(x,y)+g\epsilon_{\gamma\delta\epsilon}\chi_{\epsilon\delta}(x,x), \quad (4.36)$$

from (4.29).

We can thus generalize the field equations (4.32) to read

$$V_{\alpha\gamma}(x,P)\left(\frac{\partial}{\partial x_\mu}\tilde{f}_{\mu\nu}{}^{\gamma}(x)+\epsilon_{\gamma\delta\epsilon}\tilde{A}_\mu{}^{\delta}(x)\tilde{f}_{\mu\nu}{}^{\epsilon}(x)-i\theta_\nu{}^{\gamma}(x)\right)|G\rangle=0, \quad (4.37)$$

with θ given by (4.36). If we multiply (4.37) by $V_{\alpha\gamma'}$, sum over α, and apply (4.16a), we obtain the equation

$$\left(\frac{\partial}{\partial x_\mu}\tilde{f}_{\mu\nu}{}^{\gamma}(x)+\epsilon_{\gamma\delta\epsilon}\tilde{A}_\mu{}^{\delta}(x)\tilde{f}_{\mu\nu}{}^{\epsilon}(x)-i\theta_\nu{}^{\gamma}(x)\right)|G\rangle=0. \quad (4.38)$$

The path dependence has been removed from (4.38), and we shall adopt it as our field equation. By taking the gauge-invariant derivative

$$\delta_{\alpha\gamma}\frac{\partial}{\partial x_\nu}+g\epsilon_{\alpha\beta\gamma}\tilde{A}_\nu{}^{\beta}$$

of the factor within the parentheses we can easily show that the last term must satisfy the consistency condition

$$\left(\delta_{\alpha\gamma}\frac{\partial}{\partial x_\nu}+g\epsilon_{\alpha\beta\gamma}\tilde{A}_\nu{}^{\beta}(x)\right)\theta_\nu{}^{\gamma}(x)=0. \quad (4.39)$$

We have to choose the function χ in the definition (4.36) of θ so that (4.39) is satisfied.

In order to orient ourselves we shall first find a function $\theta_\nu{}^{\alpha}$, of the form $\eta_\nu{}^{\alpha}+\theta_\nu{}'^{\alpha}$, which satisfies (4.39). The term $\theta_\nu'^\alpha$ will not have precisely the form of the second term of (4.36), but we shall then be able to modify it so that all conditions are satisfied. The following function clearly satisfies (4.39):

$$\theta_\nu^{(1)\alpha}(x)=\eta_\nu^\alpha(x)-\frac{\partial}{\partial x_\nu}\left[\left(I\frac{\partial}{\partial x_\lambda}+gE_\beta\tilde{A}_\lambda^\beta(x)\right)\frac{\partial}{\partial x_\lambda}\right]^{-1}_{\alpha\gamma}\times\left[\delta_{\gamma\delta}\frac{\partial}{\partial x_\mu}+g\epsilon_{\gamma\epsilon\delta}\tilde{A}_\mu^\epsilon(x)\right]\eta_\mu^\delta(x),\quad(4.40)$$

where the matrices I and E_β represent the symbols $\delta_{\alpha\gamma}$ and $\epsilon_{\alpha\beta\gamma}$, considered as matrices in α and γ. The superscript (1) on θ indicates that it is not our final definition of this operator. Equation (4.40) can be rewritten without using reciprocals of operators as follows:

$$\theta_\nu^{(1)\alpha}(x)=\eta_\nu^\alpha(x)-\frac{\partial}{\partial x_\nu}\int dy\, O_{\alpha\gamma}(x,y)\times\left(\delta_{\gamma\delta}\frac{\partial}{\partial y_\mu}+g\epsilon_{\gamma\epsilon\delta}\tilde{A}_\mu^\epsilon(y)\right)\eta_\mu^\delta(y),\quad(4.41)$$

where

$$\left(\frac{\partial}{\partial x_\lambda}\delta_{\alpha\gamma}+g\epsilon_{\alpha\beta\gamma}\tilde{A}_\lambda^\beta(x)\right)\frac{\partial}{\partial x_\lambda}O_{\gamma\delta}(x,y)=\delta_{\alpha\delta}\delta^4(x-y).\quad(4.42)$$

The right-hand side of (4.41) resembles that of (4.36). The differences are, first that the operator η in the second term of (4.41) is ordered to the right of the other operators whereas it should be ordered to the left, and second that the last term of (4.36) is missing. Let us therefore change (4.42) to bring it into the correct form (4.36):

$$\theta_\nu^\alpha(x)=\eta_\nu^\alpha(x)-\int dy\left(\frac{\partial\eta_\mu^\gamma(y)}{\partial y_\mu}+g\epsilon_{\gamma\epsilon\delta}\eta_\mu^\delta(y)\tilde{A}_\mu^\epsilon(y)\right)\times\frac{\partial}{\partial x_\nu}O_{\alpha\gamma}(x,y)-g\epsilon_{\alpha\beta\gamma}\frac{\partial}{\partial x_\nu}O_{\beta\gamma}(x,y)\,|_{x=y},\quad(4.43)$$

with the function O still defined by (4.42). In Appendix B we shall calculate the value of the expression

$$\left(\delta_{\alpha\gamma}\frac{\partial}{\partial x_\nu}+g_{\alpha\beta\gamma}\tilde{A}_\nu^\beta(x)\right)\theta_\nu^\gamma(x)\quad(4.44)$$

and shall show that it is zero, so that (4.43) is a permissible, consistent choice for the function θ.

Rules for Feynman Diagrams

We can rewrite Eq. (4.38) in the form

$$\left[\frac{\partial}{\partial x_\mu}\left(\frac{\partial\tilde{A}_\nu^\alpha(x)}{\partial x_\mu}-\frac{\partial\tilde{A}_\mu^\alpha(x)}{\partial x_\nu}\right)+g\tilde{J}_\nu^\alpha(x)-i\theta_\nu(x)\right]|G)=0,\quad(4.45)$$

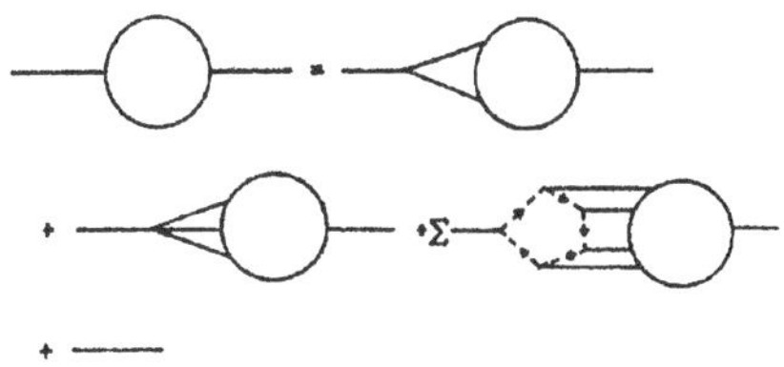

FIG. 3. Diagrammatic representation of the equation for the two-point Green's function in the Yang-Mills theory.

where

$$\tilde{J}_\nu^\alpha(x)=\epsilon_{\alpha\beta\gamma}\left(-\tilde{A}_\nu^\beta(x)\frac{\partial A_\mu^\gamma(x)}{\partial x_\mu}+2\tilde{A}_\mu^\beta(x)\frac{\partial A_\nu^\gamma(x)}{\partial x_\mu}-A_\mu^\beta(x)\frac{\partial A_\mu^\gamma(x)}{\partial x_\nu}\right)+g\epsilon_{\alpha\beta\gamma}\epsilon_{\gamma\delta\epsilon}A_\mu^\beta(x)A_\mu^\delta(x)A_\nu^\epsilon(x).\quad(4.46)$$

By integrating (4.45) in the usual way and using (4.43) for θ, we obtain the result

$$\left(A_\nu^\alpha(x)-ig\int dx'\tfrac{1}{2}\Delta_F(x-x')\tilde{J}_\nu^\alpha(x')-\int dx'\tfrac{1}{2}\Delta_F(x-x')\eta_\nu^\alpha(x')-g\epsilon_{\alpha\beta\gamma}\int dx'\tfrac{1}{2}\Delta_F(x-x')\times\frac{\partial}{\partial x_\nu'}O_{\beta\gamma}(x',y)\,|_{x'=y}\right)|G)=0,\quad(4.47)$$

where O is defined by (4.42). We have omitted the middle term of (4.43), as it is a pure divergence. If we wish we may generalize (4.47) by replacing the propagator $\delta_{\nu\rho}\frac{1}{2}\Delta(x-x')$ by $[\delta_{\nu\rho}-c(\partial^2/\partial x_\nu\partial x_\rho)\Box^{-2}]\frac{1}{2}\Delta_F\times(x-x')$; we then obtain other gauges such as the Landau gauge.

If the last term of (4.47) had been absent, we would have obtained Feynman rules similar to those for electrodynamics. The equation for the Green's function could then be represented graphically as in Fig. 3, without the second-last diagram. For simplicity we have exhibited the equation for the two-point Green's function; equations for higher Green's functions can be similarly represented. The three- and four-point vertices have the following factors associated with them:

$$v_3(p_1,\alpha,\mu;\,p_2,\beta,\nu;\,p_3,\gamma,\rho)=i(2\pi)^4\epsilon_{\alpha\beta\gamma}[(p_2-p_3)_\mu\delta_{\nu\rho}+(p_3-p_1)_\nu\delta_{\mu\rho}+(p_1-p_2)_\rho\delta_{\mu\nu}]\quad(4.48)$$

$$v_4(p_1,\alpha,\mu;\,p_2,\beta,\nu;\,p_3,\gamma,\rho;\,p_4,\delta,\sigma)=-(2\pi)^4\epsilon_{\alpha\beta\epsilon}\epsilon_{\epsilon\gamma\delta}(\delta_{\mu\rho}\delta_{\nu\sigma}-\delta_{\mu\sigma}\delta_{\nu\rho})-(2\pi)^4\epsilon_{\alpha\gamma\epsilon}\epsilon_{\epsilon\beta\delta}(\delta_{\mu\nu}\delta_{\rho\sigma}-\delta_{\mu\sigma}\delta_{\nu\rho})-(2\pi)^4\epsilon_{\alpha\delta\epsilon}\epsilon_{\epsilon\beta\gamma}(\delta_{\mu\nu}\delta_{\rho\sigma}-\delta_{\mu\rho}\delta_{\nu\sigma}).\quad(4.49)$$

One could then construct Feynman diagrams by arranging the vertices (4.48) and (4.49) in all possible ways. In fact, Eq. (4.47) without the last term is identical to the equation we would have obtained by starting from the Lagrangian

$$\mathcal{L}=\frac{1}{2}\left(\frac{\partial A_\nu{}^\alpha}{\partial x_\mu}\right)^2+\tfrac{1}{2}\epsilon_{\alpha\beta\gamma}\left(\frac{\partial A_\nu{}^\alpha}{\partial x_\mu}-\frac{\partial A_\mu{}^\alpha}{\partial x_\nu}\right)A_\mu{}^\beta A_\nu{}^\gamma$$
$$+\tfrac{1}{4}\epsilon_{\alpha\beta\gamma}\epsilon_{\alpha\delta\epsilon}A_\mu{}^\beta A_\nu{}^\gamma A_\mu{}^\delta A_\nu{}^\epsilon \quad (4.50)$$

and writing down "naive" Feynman rules in the usual way.

The presence of the last term in (4.47) shows that the naive Feynman rules are not correct and that there are additional terms in the perturbation expansion. From (4.42), we can expand O as a perturbation series in g as follows:

$$O_{\beta\gamma}(x,y)=-i\sum_{n=0}^{\infty}\int dx_1\cdots dx_r\tfrac{1}{2}\Delta_F(x-x_1)ig\epsilon_{\beta\delta\epsilon}A_\lambda{}^\delta(x_1)$$
$$\times\frac{\partial}{\partial x_{1\lambda}}\tfrac{1}{2}\Delta_F(x_1-x_2)ig\epsilon_{\epsilon\zeta\eta}A_\mu{}^\zeta(x_2)\cdots\epsilon_{\theta\iota\gamma}A_\sigma{}^\iota(x_r)$$
$$\times\frac{\partial}{\partial x_{r\sigma}}\frac{1}{2}\Delta_F(x_n-y). \quad (4.51)$$

When (4.51) is substituted in the last term of (4.47), we obtain the result

$$\sum_{n=0}^{\infty}\int dx'dx_1\cdots dx_r\tfrac{1}{2}\Delta_F(x-x')ig\epsilon_{\gamma\alpha\beta}\frac{\partial}{\partial x_\nu'}\frac{1}{2}\Delta_F(x'-x_1)$$
$$\times ig\epsilon_{\beta\delta\epsilon}A_\lambda{}^\delta(x_1)\frac{\partial}{\partial x_{1\lambda}}\frac{1}{2}\Delta_F(x_1-x_2)$$
$$\times ig\epsilon_{\epsilon\zeta\eta}A_\mu{}^\zeta(x_2)\cdots\epsilon_{\theta\iota\gamma}A_\sigma{}^\iota(x_r)\frac{\partial}{\partial x_{r\sigma}}\frac{1}{2}\Delta_F(x_n-x'). \quad (4.52)$$

The expression (4.52) has the form of an integral which occurs in Feynman diagrams, and the contribution (4.52) to (4.47) has been represented by the second-last diagram of Fig. 3. The summation sign represents the sum over polygons with any number of dashed lines, and corresponds to the summation over n in (4.52). The dashed lines and vertices are associated respectively with the factors $\frac{1}{2}\Delta_F(x_r-x_{r+1})$ and $ig\epsilon(\partial/\partial x_r)$ in (4.52). Thus, in momentum space, the following factors are associated with the dashed lines and the vertices at which they end:

dashed lines: $$\frac{i}{(2\pi)^4}\frac{\delta_{\alpha\beta}}{-p^2+i\epsilon}; \quad (4.53a)$$

vertices: $$w(p_1\alpha,p_2\beta\sigma,p_3\gamma)=-(2\pi)^4 g\epsilon_{\alpha\beta\gamma}p_{3\sigma}; \quad (4.53b)$$

an over-all factor -1. (4.53c)

In (4.53c), the quantities p_1, α and p_3, γ refer to the dashed lines, the quantities p_2, $\beta\sigma$ to the solid lines representing the Yang-Mills quanta. We notice that the vertex factor is not symmetric in the two dashed lines; it involves a factor $p_{3\sigma}$ but no factor $p_{1\sigma}$. It is for this reason that we have drawn arrows on the dashed lines in Fig. 3. The factor $p_{3\sigma}$ in (4.53b) is associated with the line directed away from the vertex.

The presecription for constructing Feynman diagrams is therefore to draw three-particle and four-particle vertices with factors (4.48) and (4.49), and also polygons with any number of dashed lines and with factors (4.53) associated with them. The three- and four-point vertices, as well as the vertices of the polygons, are then to be joined by solid Yang-Mills lines in all possible ways.

The Feynman rules for our theory are the same as those for a theory with ficititious scalar particles as well as the Yang-Mills particles. The Feynman diagrams contain three- and four-point vertices involving the Yang-Mills lines alone. The factors (4.48) and (4.49) are associated with these vertices. In addition, the diagrams contain vertices involving two scalar lines and one Yang-Mills line. Associated with such vertices are the factors (4.53b). There is a further factor -1 associated with each closed loop of scalar particles. The scalar lines only occur as internal lines and only in closed loops.

Note added in manuscript. Faddeev and Popov (unpublished) have shown that their functional-integration prescription[8] can be related to Schwinger's formulation of the Yang-Mills theory.[6] This therefore provides an alternative derivation of the Feynman rules from a quantized field theory. Faddeev and Popov have restricted themselves to Landau gauge.

PHYSICAL REVIEW D VOLUME 2, NUMBER 12 15 DECEMBER 1970

S Matrix for Yang-Mills and Gravitational Fields

E. S. FRADKIN AND I. V. TYUTIN
Physical Lebedev Institute, Academy of Sciences, Moscow, U.S.S.R.
(Received 19 January 1970)

A method is suggested (and applied to the Yang-Mills and gravitational fields) for the construction of the generating functional (S matrix) for fields possessing an invariance group. The unitarity and gauge independence of the S matrix on the mass shell are seen explicitly.

I. INTRODUCTION

THERE has lately been considerable intensification in the study of theories partially or completely invariant under non-Abelian groups of transformations. This is in connection with the discovery of vector mesons and their classification into multiplets, with the use of vector mesons to account for the form factors of particles, and with the current-algebra approach. Intermediate vector bosons are introduced in many schemes of weak interaction. An important example of a theory with a non-Abelian group of invariance is that of the gravitational field.

In the present paper a procedure for constructing the Feynman rules is proposed for theories possessing a gauge group, such as the theories of the massless Yang-Mills and gravitation fields. It is known that some additional (gauge) condition must be imposed on the dynamical variables in order that a consistent quantum field theory may be formulated on the basis of a Lagrangian density invariant under a local transformation group. In covariant gauges this can conveniently be done by the use of Lagrange multipliers. The basic idea of the method proposed is to choose the Lagrange multiplier in such a way that one is led to free equations of motion for the additional field. This fact guarantees the unitarity of the S matrix in physical space. The Feynman rules obtained coincide with those proposed in Refs. 1–5. The difference of the method under consideration from that of Refs. 2–5 is that we have succeeded in obtaining a set of consistent dynamical equations completely describing the theory. On the one hand, these equations make it possible to elucidate the reason for the additional diagrams to appear, and, on the other hand, guarantee the unitarity of the physical S matrix. Section II is devoted to the construction of the S matrix for the massless Yang-Mills field in arbitrary gauge and to the proof of the gauge invariance of the S matrix. In Sec. III constructing the Feynman rules in the Coulomb and axial gauges is considered on the basis of the canonical quantization procedure. The S matrix obtained coincides with that found in Sec. II, which fact reaffirms its unitarity. It is shown, furthermore, that taking the additional conditions consistently into account makes it possible to obtain self-consistent equations for the massless Yang-Mills field in the presence of an external source.

In Sec. IV the Feynman rules for the gravitational field are constructed in covariant gauges. These rules coincide with those suggested in Refs. 2, 3, and 5. In the framework of our approach we also obtain the S matrix for a noncovariant (Dirac) gauge, for which the Feynman rules have been obtained by Popov and Faddeev[6] using a method closely connected with the canonical formulation of the gravitational field. In addition, by our method, the equivalence of the S matrix in covariant and noncovariant gauges is proved.

We use the following notation. Greek μ, ν, λ, ... and the Latin i, j, k indices take the values 0, 1, 2, 3 and 1, 2, 3, respectively. In Secs. II and III, $g_{\mu\nu}$ means the Minkowski tensor $(+, ---)$ and δ_{ik} means the unit tensor. By the summation over repeated indices is everywhere meant $a_\mu b_\mu \equiv a_0 b_0 - a_k b_k$; $\partial_\mu \equiv \partial/\partial x^\mu$; $\Box = \partial_\mu\partial_\mu$; $\nabla \equiv \partial_k \partial_k$. In Sec. IV, $g_{\mu\nu}$ means the metric tensor, and the Minkowski tensor is designated as $\delta_{\mu\nu}$. The usual summation over repeated indices means $a_\mu b^\mu \equiv \sum_{\mu=0}^{3} a_\mu b^\mu$. We use the system of units $\hbar = c = 1$ in Secs. II and III and $c^3/16\pi k = 1$ in Sec. IV (where k is the gravitational constant).

[1] R. P. Feynman, Acta Phys. Polon. **24**, 697 (1963).
[2] B. S. DeWitt, Phys. Rev. **162**, 1195 (1967); **162**, 1239 (1967).
[3] L. D. Faddeev and V. N. Popov, Phys. Letters **25B**, 30 (1967); V. N. Popov and L. D. Faddeev, ITP report, Kiev, 1967 (unpublished).
[4] S. Mandelstam, Phys. Rev. **175**, 1580 (1968).
[5] S. Mandelstam, Phys. Rev. **175**, 1604 (1968).

IV. CONSTRUCTION OF FEYNMAN RULES FOR GRAVITATIONAL FIELD

In this section we shall obtain the general rules for construction of the S matrix for the gravitation field, prove the gauge invariance of the S matrix, and in more detail consider two covariant gauges (the harmonic condition and its linearized form) and the Dirac[17] noncovariant gauge. The Feynman rules found by us coincide with those of Refs. 2, 3, and 5.

[6] V. N. Popov and L. D. Faddeev (to be published).
[17] P. A. M. Dirac, Phys. Rev. **114**, 924 (1959).

The classical gravitational field is described by the action[18]

$$W_0=\int dx\, L_0(x)\,, \tag{4.1}$$

$$L_0=\tfrac{1}{2}(\sqrt{-g})g^{\mu\nu}R_{\mu\nu}\,. \tag{4.2}$$

Here $g_{\mu\nu}$ is the metric tensor, $g\equiv\det g_{\mu\nu}$, $g^{\mu\nu}g_{\nu\lambda}=\delta^\mu{}_\lambda$, and $R_{\mu\nu}$ is the curvature tensor of second rank (Ricci tensor)

$$R_{\mu\nu}=\partial_\nu\Gamma^\sigma{}_{\mu\sigma}-\partial_\sigma\Gamma^\sigma{}_{\mu\nu}+\Gamma^\rho{}_{\mu\sigma}\Gamma^\sigma{}_{\nu\rho}-\Gamma^\rho{}_{\mu\nu}\Gamma^\sigma{}_{\rho\sigma}\,. \tag{4.3}$$

$\Gamma^\mu{}_{\nu\lambda}$ is the Christoffel symbol

$$\Gamma^\mu{}_{\nu\lambda}=\tfrac{1}{2}g^{\mu\sigma}(\partial_\nu g_{\sigma\lambda}+\partial_\lambda g_{\sigma\nu}-\partial_\sigma g_{\nu\lambda})\,. \tag{4.4}$$

As we shall show later, the most convenient choice of the concrete form of the dynamical variables depends on the gauge condition.

The variables we shall use belong to the following class:

$$g^{(\beta)}{}_{\mu\nu}\equiv g^\beta g_{\mu\nu} \quad\text{and}\quad g_{(\beta)}{}^{\mu\nu}\equiv g^\beta g^{\mu\nu}\,. \tag{4.5}$$

In the general case we shall designate the variables $g_{(\beta)}{}^{\mu\nu}$ and $g^{(\beta)}{}_{\mu\nu}$ as $\kappa_{\mu\nu}$. The Einstein equations for the gravitational field can be obtained from (4.1) by two methods:

In the first-order formalism the expression (4.1) should be varied with respect to $\kappa_{\mu\nu}$ and $\Gamma^\mu{}_{\nu\lambda}$, considered as independent variables. From the equations obtained by varying (4.1) with respect to $\Gamma^\mu{}_{\nu\lambda}$, the expression (4.4) for $\Gamma^\mu{}_{\nu\lambda}$ can be deduced.

In the second-order formalism the variation should be made only with respect to $\kappa_{\mu\nu}$. It is assumed in this case that (4.1) is expressed only in terms of $\kappa_{\mu\nu}$ with the help of (4.4).

Both methods lead, of course, to the same field equations.

As is well known, (4.1) and (4.2) are invariant under the gauge transformations of $\kappa_{\mu\nu}$ and $\Gamma^\mu{}_{\nu\lambda}$, the infinitesimal form of which is

$$g_{(\beta)}{}^{\mu\nu}\to g_{(\beta)}{}^{S\mu\nu}\equiv g_{(\beta)}{}^{\mu\nu}-\xi^\gamma\partial_\gamma g_{(\beta)}{}^{\mu\nu}+g_{(\beta)}{}^{\mu\gamma}\partial_\gamma\xi^\nu +g_{(\beta)}{}^{\nu\gamma}\partial_\gamma\xi^\mu-2\beta g_{(\beta)}{}^{\mu\nu}\partial_\gamma\xi^\gamma\,, \tag{4.6}$$

$$g^{(\beta)}{}_{\mu\nu}\to g^{(\beta)S}{}_{\mu\nu}\equiv g^{(\beta)}{}_{\mu\nu}-\xi^\gamma\partial_\gamma g^{(\beta)}{}_{\mu\nu}-g^{(\beta)}{}_{\mu\gamma}\partial_\nu\xi^\gamma -g^{(\beta)}{}_{\nu\gamma}\partial_\mu\xi^\gamma-2\beta g^{(\beta)}{}_{\mu\nu}\partial_\gamma\xi^\gamma\,, \tag{4.6$'$}$$

$$\Gamma^\mu{}_{\nu\lambda}\to\Gamma^{S\mu}{}_{\nu\lambda}\equiv\Gamma^\mu{}_{\nu\lambda}-\xi^\gamma\partial_\gamma\Gamma^\mu{}_{\nu\lambda}+\Gamma^\sigma{}_{\nu\lambda}\partial_\sigma\xi^\mu -\Gamma^\mu{}_{\nu\sigma}\partial_\lambda\xi^\sigma-\Gamma^\mu{}_{\lambda\sigma}\partial_\nu\xi^\sigma-\partial_\nu\partial_\lambda\xi^\mu\,. \tag{4.7}$$

$\xi^\mu(x)$ are arbitrary infinitesimal functions of x^μ. In the first-order formalism, gauge transformations of both $\kappa_{\mu\nu}$ and $\Gamma^\mu{}_{\nu\lambda}$ should be made; in the second-order formalism, only gauge transformations of $\kappa_{\mu\nu}$ should be made. The gauge variation of (4.1) has the form

$$\delta W_0=\int dx\, \xi^\mu(x)[\mathbf{R}_{\mu\nu\lambda}G^{\nu\lambda}(x)+\mathbf{R}_{\mu\nu\lambda}{}^\sigma G_\sigma{}^{\nu\lambda}(x)]\,. \tag{4.8}$$

[18] L. D. Landau and E. M. Lifshitz, *The Classical Theory of Fields* (Pergamon, London, 1962), 2nd ed.

In the case $\kappa_{\mu\nu}=g_{(\beta)}{}^{\mu\nu}$ the first term in (4.8) should be written as

$$\int dx\, \xi^\mu(x)\mathbf{R}_\mu{}^{\nu\lambda}G_{\nu\lambda}(x)\,. \tag{4.9}$$

The second term in (4.8) is absent in the second-order formalism. We use the following designations:

$$G^{\mu\nu}(x)\equiv G_{(\beta)}{}^{\mu\nu}(x)\equiv\delta W_0/\delta g^{(\beta)}{}_{\mu\nu}(x)\,, \tag{4.10}$$

$$G_{\mu\nu}(x)\equiv G^{(\beta)}{}_{\mu\nu}(x)\equiv\delta W_0/\delta g_{(\beta)}{}^{\mu\nu}(x)\,, \tag{4.11}$$

$$G_\mu{}^{\nu\lambda}(x)\equiv\delta W_0/\delta\Gamma^\mu{}_{\nu\lambda}(x)\,. \tag{4.12}$$

Equation (4.12) exists in the second-order formalism only. We find the differential operators R with the help of (4.6) and (4.7):

$$\mathbf{R}_{\mu\nu\lambda}\equiv\mathbf{R}^{(\beta)}{}_{\mu\nu\lambda}\equiv(2\beta-1)\partial_\mu g^{(\beta)}{}_{\nu\lambda} +2\beta g^{(\beta)}{}_{\nu\lambda}\partial_\mu+2\partial_\nu g^{(\beta)}{}_{\mu\lambda}+2g^{(\beta)}{}_{\mu\lambda}\partial_\nu \tag{4.13}$$

for the case $\kappa_{\mu\nu}=g^{(\beta)}{}_{\mu\nu}$;

$$\mathbf{R}_\mu{}^{\nu\lambda}\equiv\mathbf{R}^{(\beta)}{}_\mu{}^{\nu\lambda}\equiv(2\beta-1)\partial_\mu g_{(\beta)}{}^{\nu\lambda}+2\beta g_{(\beta)}{}^{\nu\lambda}\partial_\mu -2\delta_\mu{}^\nu\partial_\gamma g_{(\beta)}{}^{\gamma\lambda}-2\delta_\mu{}^\nu g_{(\beta)}{}^{\gamma\lambda}\partial_\gamma \tag{4.14}$$

for the case $\kappa_{\mu\nu}=g_{(\beta)}{}^{\mu\nu}$;

$$\mathbf{R}_{\mu\nu\lambda}{}^\sigma\equiv-\partial_\mu\Gamma^\sigma{}_{\nu\lambda}-\delta^\sigma{}_\mu\partial_\gamma\Gamma^\gamma{}_{\nu\lambda}-\delta^\sigma{}_\mu\Gamma^\gamma{}_{\nu\lambda}\partial_\gamma +2\partial_\lambda\Gamma^\sigma{}_{\mu\nu}+2\Gamma^\sigma{}_{\mu\nu}\partial_\lambda-\delta^\sigma{}_\mu\partial_\nu\partial_\lambda\,. \tag{4.15}$$

Since the $\xi^\mu(x)$ are arbitrary, we obtain an important identity from (4.8):

$$\mathbf{R}_{\mu\nu\lambda}G^{\nu\lambda}(x)+\mathbf{R}_{\mu\nu\lambda}{}^\sigma G_\sigma{}^{\nu\lambda}(x)\equiv0\,. \tag{4.16}$$

Note that (4.16) is satisfied by arbitrary $\kappa_{\mu\nu}$ and $\Gamma^\mu{}_{\nu\lambda}$. According to (4.16), four identities exist among the Einstein equations. As noted in Sec. II, this means that four additional (gauge) conditions should be imposed on $\kappa_{\mu\nu}$ and $\Gamma^\mu{}_{\nu\lambda}$ for consistent construction of the quantum theory.

As in Sec. II, we shall use the method of Lagrange multipliers. Consider the class of gauges determined by the function

$$\psi_\mu\equiv\psi_\mu(x;\kappa,\Gamma)\,. \tag{4.17}$$

Three concrete forms of the gauge functions will be considered later. The Lagrangian is

$$L=L_0+\psi_\mu\bar{B}_\psi{}^\mu+\tfrac{1}{2}\alpha\bar{B}_\psi{}^\mu\delta_{\mu\nu}\bar{B}_\psi{}^\nu\,, \tag{4.18}$$

where $\delta_{\mu\nu}$ is the Minkowski tensor. The case $\alpha\neq0$ will be considered only for these gauge functions (4.17) which depend linearly on $\kappa_{\mu\nu}$ and are independent of $\Gamma^\mu{}_{\nu\lambda}$.

$$\bar{B}_\psi{}^\mu(x)=\int dy\,\mathbf{D}_{\psi\nu}{}^\mu(x,y;\kappa,\Gamma)B^\nu(y)\,. \tag{4.19}$$

Varying (4.18) with respect to $\kappa_{\mu\nu}$, $\Gamma^\mu{}_{\nu\lambda}$, and B^μ, we obtain the field equations

$$\psi_\mu=-\alpha\delta_{\mu\nu}\bar{B}_\psi{}^\nu\,,\quad G^{\mu\nu}+\frac{\delta\psi_\sigma}{\delta\kappa_{\mu\nu}}\bar{B}^\sigma=0\,, \tag{4.20}$$

in the case of the second-order formalism. In the first-order formalism the following equation should be added:

$$G_\mu{}^{\nu\lambda}+\frac{\delta\psi_\sigma}{\delta\Gamma^\mu{}_{\nu\lambda}}\bar{B}^\sigma=0. \quad (4.21)$$

With the help of the identity (4.16), we obtain the following consequences of the field equations:

$$(Q^\psi{}_{\mu\nu}\bar{B}^\nu)(x)=0, \quad (4.22)$$

$$Q^\psi{}_{\mu\nu}\equiv Q^\psi{}_{\mu\nu}(x,y;\kappa,\Gamma)$$

$$=\mathbf{R}_{\mu\lambda\gamma}\frac{\partial\psi_\nu(y;\kappa,\Gamma)}{\partial\kappa_{\lambda\gamma}(x)}+\mathbf{R}_{\mu\lambda\gamma}{}^\sigma\frac{\partial\psi_\nu(y;\kappa,\Gamma)}{\partial\Gamma^\sigma{}_{\lambda\gamma}(x)}. \quad (4.23)$$

We impose the restriction on ψ_μ that in the limit

$$\kappa_{\mu\nu}=\delta_{\mu\nu}, \quad \Gamma^\mu{}_{\nu\lambda}=0,$$

the operator $Q^\psi{}_{\mu\nu}$ should be a nonsingular differential operator $Q^{(0)}{}_{\psi\mu\nu}$. Choose the $\mathbf{D}_\psi$ function in the form

$$\mathbf{D}_\psi{}^\mu{}_\nu=[Q_\psi{}^{-1}]^{\mu\nu}\overleftarrow{Q}_{\psi\lambda\nu}{}^{(0)}. \quad (4.24)$$

Then the B field satisfies the free equation

$$Q_\psi{}^{(0)}{}_{\mu\nu}B^\nu=0. \quad (4.25)$$

Therefore, the S matrix is unitary, and the Einstein equations are valid in the physical subspace.

The generating functional is equal to

$$Z_\alpha{}^\psi=\int d(\kappa,\Gamma)dB^\mu\exp\Big[i\int dx(L_0+\psi_\mu\bar{B}_\psi{}^\mu+\tfrac{1}{2}\alpha\bar{B}^\mu\delta_{\mu\nu}\bar{B}^\nu+\kappa_{\mu\nu}J^{\mu\nu})\Big]$$

$$=\int d(\kappa,\Gamma)\exp\Big[i\int dx\Big(L_0-\frac{1}{2\alpha}\psi_\mu\delta^{\mu\nu}\psi_\nu+\kappa_{\mu\nu}J^{\mu\nu}\Big)+\operatorname{Tr}\ln Q_\psi(\overleftarrow{Q}_\psi{}^{(0)})^{-1}\Big]. \quad (4.26)$$

In the gauge

$$\psi_\mu=0,$$

i.e., for $\alpha=0$, we have

$$Z_0{}^\psi=\int d(\kappa,\Gamma)\delta\{\psi_\mu(x;\kappa,\Gamma)\}\times\exp\Big\{i\int dx(L_0+\kappa_{\mu\nu}J^{\mu\nu})+\operatorname{Tr}\ln Q_\psi(\overleftarrow{Q}_\psi{}^{(0)})^{-1}\Big\}. \quad (4.27)$$

In (4.26) and (4.27), $d(\kappa,\Gamma)$ is equal to

$$\prod_x\prod_{\mu\le\nu}d\kappa_{\mu\nu}(x) \quad (4.28)$$

in the second-order formalism, and to

$$\prod_x\prod_{\mu\le\nu}d\kappa_{\mu\nu}(x)\prod_{\mu;\nu\le\lambda}d\Gamma_{\nu\lambda}{}^\mu(x) \quad (4.29)$$

in the first-order formalism.

Note that the generating functionals (4.26), written in terms of different variables $g^{(\beta_1)}{}_{\mu\nu}$ or $g^{(\beta_2)}{}_{\mu\nu}$, do not coincide in general. In this connection there arises the question of a choice of "the true variables" ("the true measure") in terms of which the generating functional can be written as the functional integral of $\exp(iL)$ over "the true variables." The proposed method does not permit the value of the β to be determined. It could be found in principle with the help of the correctly formulated canonical formalism for the gravitational field. However, we shall not investigate this problem in the present work.

We now give some arguments which show that the S matrix corresponding to the generating functional (4.26) and (4.27) is independent of the choice of the variables of the functional integration belonging to class (4.5). Suppose we integrate over $\kappa^1{}_{\mu\nu}$ in (4.26). The corresponding Jacobian has in general the form

$$d(\kappa^1,\Gamma)/d(\kappa,\Gamma)=\operatorname{Det}^{4\gamma}\kappa(x)\delta(x-y), \quad (4.30)$$

where γ is some number and

$$\kappa(x)\equiv\det\kappa_{\mu\nu}(x). \quad (4.31)$$

Then we have

$$\tilde{Z}_\alpha{}^\psi\equiv\int d(\kappa^1,\Gamma)\exp\Big[i\int dx\Big(L_0-\frac{1}{2\alpha}\psi_\mu\delta^{\mu\nu}\psi_\nu+\kappa_{\mu\nu}J^{\mu\nu}\Big)+\operatorname{Tr}\ln Q_\psi(\overleftarrow{Q}_\psi{}^{(0)})^{-1}\Big]$$

$$=\int d(\kappa,\Gamma)\exp\Big[i\int dx\Big(L_0-\frac{1}{2\alpha}\psi_\mu\delta^{\mu\nu}\psi_\nu+\kappa_{\mu\nu}J^{\mu\nu}\Big)+\operatorname{Tr}\ln Q_\psi(-\kappa)^\gamma(\overleftarrow{Q}_\psi{}^{(0)})^{-1}\Big]$$

$$=\int d(\kappa,\Gamma)dB^\mu\exp\Big[i\int dx(L_0+\psi_\mu\tilde{B}_\psi{}^\mu+\tfrac{1}{2}\alpha\tilde{B}_\psi{}^\mu\delta_{\mu\nu}\tilde{B}_\psi{}^\nu+\kappa_{\mu\nu}J^{\mu\nu})\Big], \quad (4.26')$$

$$\tilde{B}_\psi{}^\mu(x)=(Q_\psi{}^{-1\mu\nu}(-\kappa)^{-\gamma}\overleftarrow{Q}_{\psi\nu\lambda}{}^{(0)}B^\lambda)(x). \quad (4.19')$$

We can see from (4.26′) that $\tilde{Z}_\alpha{}^\psi$ corresponds to the field equations for $\kappa_{\mu\nu}$, $\Gamma^\mu{}_{\nu\lambda}$, and B^μ which are obtained from (4.20)–(4.22) by the substitution $\bar{B}_\psi{}^\mu\to\tilde{B}_\psi{}^\mu$.

$$\psi_\mu=-\alpha\delta_{\mu\nu}\tilde{B}_\psi{}^\nu, \quad G^{\mu\nu}+\frac{\delta\psi_\sigma}{\delta\kappa_{\mu\nu}}\tilde{B}_\psi{}^\sigma=0, \quad (4.20')$$

$$G_\mu{}^{\nu\lambda}+\frac{\delta\psi_\sigma}{\delta\Gamma^\mu{}_{\nu\lambda}}\tilde{B}_\psi{}^\sigma=0\,, \qquad (4.21')$$

$$(Q_{\mu\nu}{}^\psi\tilde{B}_\psi{}^\nu)(x)=0\,. \qquad (4.22')$$

It is seen from (4.19′) and (4.22′) that B^ν satisfies the free equation

$$Q_\psi{}^{(0)}{}_{\mu\nu}B^\nu=0\,. \qquad (4.25')$$

Let us write $[-\kappa(x)]^{-\gamma}$ in the form

$$[-\kappa(x)]^{-\gamma}=1+a(x)\,. \qquad (4.32)$$

The expression (4.19′) then acquires the form

$$\tilde{B}_\psi{}^\mu(x)=\bar{B}_\psi{}^\mu(x)+(Q_\psi{}^{-1\mu\lambda}a\overleftarrow{Q}_{\psi\lambda\nu}{}^{(0)}B^\nu)(x)\,. \quad (4.33)$$

It is clear that in the second term of (4.33), integrating by parts can change the direction of the $Q_\psi{}^{(0)}$ operation (at least in the perturbation theory) and one can use (4.25′).

Finally we observe that the field equations (4.20′)-(4.22′) coincide with (4.20)-(4.22).

Thus the generating functional (4.26′) leads to the same field equations for $\kappa_{\mu\nu}$, $\Gamma_{\nu\lambda}{}^\mu$, B^μ, and consequently to the same S matrix, as the generating functional (4.26) does.

Note also that all the Heisenberg operators belonging to class (4.5) must lead to the same S matrix according to the Borchers theorem. (We ignore the question of the meaning of $\kappa^1{}_{\mu\nu}$ as an operator function of $\kappa_{\mu\nu}$. See also the analogous statement for the case of nonlinear chiral Lagrangians in Ref. 19.)

Now we pass to the proof of gauge invariance. We first prove that the S matrix is independent of the type of gauge condition

$$\psi_\mu(x;\kappa,\Gamma)=0\,, \qquad (4.34)$$

i.e., that the S matrix corresponding to the generating functional (4.27) is independent of the form of the function ψ_μ.

Define the function $\Delta_\psi(\kappa,\Gamma)$ by the relation

$$\phi_\psi(\kappa,\Gamma)\equiv\Delta_\psi(\kappa,\Gamma)\int d\mu(S)\;\delta\{\psi_\mu(x;\kappa^S,\Gamma^S)\}=1\,. \quad (4.35)$$

Here S is an element of the coordinate gauge transformation group, and $d\mu(S)$ is the measure of group integration. (For more details on the coordinate gauge transformation group see Ref. 12.)

Let us explain some peculiarities of the coordinate group transformations. Under the transformation

$$x^\mu\rightarrow\bar{x}^\mu(x)\,, \qquad (4.36)$$

the corresponding transformation of the metric tensor is

$$g_{\mu\nu}{}^{(\beta)}(x)\rightarrow g_{\mu\nu}{}^{(\beta)\prime}(x)=J_z{}^{2\beta}\frac{\partial z^\alpha(x)}{\partial x^\mu}\frac{\partial z^\rho(x)}{\partial x^\nu}g_{\alpha\rho}{}^{(\beta)}(z)\,, \quad (4.37)$$

$$z^\mu(x)=\bar{x}^{-1\mu}(x)\,,\quad J_z(x)=\det\partial z^\mu(x)/\partial x^\nu\,, \quad (4.38)$$

and the Jacobian of the transformation of the integration measure $d(\kappa)$ is

$$D_\beta\equiv d(g_{\mu\nu}{}^{(\beta)\prime})/d(g_{\mu\nu}{}^{(\beta)})$$
$$=\mathrm{Det}J_z{}^{5+20\beta}(x)\delta^{10}(z(x)-y)\,. \quad (4.39)$$

In order to calculate the D_β, we note that the matrix inverse to the $\delta(z(x)-y)$ is

$$\delta(x-z(y))J_z(y)\,. \qquad (4.40)$$

If we formally use the rule of the calculation of the determinant of the product of the matrices, then

$$\mathrm{Det}\delta(z(x)-y)=\mathrm{Det}^{-1/2}J_z(x)\delta(x-y)\,. \quad (4.41)$$

If

$$\mathrm{Det}J_z(x)\delta(x-y)\neq1\,, \qquad (4.42)$$

then the invariant gravitation measure is

$$\prod_x\prod_{\mu\leq\nu}dg_{\mu\nu}(x)\,. \qquad (4.43)$$

This result does not agree with the form of the invariant measure

$$\prod_x\prod_{\mu\leq\nu}g^{(-5/2)}(x)dg_{\mu\nu}(x)=\prod_x\prod_{\mu\leq\nu}dg_{\mu\nu}{}^{(-5/20)}(x)\,, \quad (4.44)$$

which is proposed by a number of authors.[20]

Note that the integration measure over the group of the gauge coordinate transformations has a form[12] which is analogous to (4.39):

$$d\mu_L=\mathrm{Det}^4J_{\bar{x}}(x)\delta(\bar{x}(x)-y)\prod_{x,\mu}d\bar{x}^\mu(x)\,, \quad (4.45)$$

$$d\mu_R=\mathrm{Det}^{-1}J_{\bar{x}}(x)\delta(x-y)\prod_{x,\mu}d\bar{x}^\mu(x)\,. \quad (4.46)$$

If (4.42) is true, then the left and the right measures are different. When proving the gauge independence of the S matrix, this fact should be taken into account [in particular, in (4.35) one should use the $d\mu_L$].

The formal proof of the invariance of the S matrix can be made in the general case $\mathrm{Det}J(x)\delta(x-y)\neq1$ (then it is necessary to assume the β-independence of the S matrix). However, taking into account that the arbitrary functions $z^\mu(x)$ are the coordinates themselves, we can expect that

$$\mathrm{Det}J(x)\delta(x-y)=1\,. \qquad (4.47)$$

Indeed, the $\delta(z(x)-y)$ can be considered as the matrix

[19] S. Coleman, J. Wess, and B. Zumino, Phys. Rev. 177, 2239 (1969).

[20] C. W. Misner, Rev. Mod. Phys. 29, 497 (1957); J. R. Klauder, Nuovo Cimento 19, 1059 (1961); B. Laurent, Arkiv Fysik 16, 279 (1959); B. S. DeWitt, J. Math. Phys. 3, 1073 (1962).

of the permutation of the points. The corresponding finite-dimensional matrix has determinant 1. Therefore, the $\delta(z(x)-y)$ has determinant 1 if it is calculated by the use of the finite-dimensional approximations. In this case (4.47) follows from (4.41) and

$$d\mu_L = d\mu_R = d\mu = \prod_{x,\mu} d\bar{x}^\mu(x). \tag{4.48}$$

Furthermore, $d\mu$ has the property (2.44), and the measures $d(\kappa)$ and $d(\kappa,\Gamma)$ for arbitrary β's are invariant under the gauge coordinate transformations. Below we assume that (4.47) is true.

As in Sec. II the property (2.44) enables one to prove the gauge invariance of $\Delta_\psi(\kappa,\Gamma)$. We must know the function $\Delta_\psi(\kappa,\Gamma)$ only for $\kappa_{\mu\nu}$ and $\Gamma^\mu{}_{\nu\lambda}$ satisfying the condition (4.34). In this case the group integral is concentrated in the neighborhood of the unit element. The gauge transformations have the form (4.6), (4.7), and

$$d\mu(S) \approx \prod_{x,\mu} d\xi^\mu(x). \tag{4.49}$$

As in Sec. II we obtain

$$\Delta_\psi^{-1}(\kappa,\Gamma)|_{\psi_\mu=0} = \int d\xi^\mu(x)\, \delta\{(Q_{\mu\nu}{}^{\psi T}\xi^\nu)(x)\} = \mathrm{Det}^{-1}Q_\psi. \tag{4.50}$$

Keeping (4.50) in mind, the expression (4.27) can be rewritten in the form

$$Z_0{}^\psi = \int d(\kappa,\Gamma)\delta\{\psi_\mu(x;\kappa,\Gamma)\}\Delta_\psi(\kappa,\Gamma) \times \exp\left[i\int dx(L_0+\kappa_{\mu\nu}J^{\mu\nu})\right]. \tag{4.51}$$

Consider another gauge condition:

$$\psi_\mu{}^1(\kappa,\Gamma)=0. \tag{4.52}$$

Multiply (4.51) by $\phi_{\psi^1}(\kappa,\Gamma)$ and perform the gauge transformation

$$\kappa_{\mu\nu} \to \kappa_{\mu\nu}{}^{S^{-1}}, \quad \Gamma^\mu{}_{\nu\lambda} \to \Gamma^{S^{-1}\mu}{}_{\nu\lambda}.$$

The quantities L_0, Δ_ψ, and Δ_{ψ^1} are invariant. Furthermore, the following substitution can be made on the mass shell:

$$\kappa^{S^{-1}}{}_{\mu\nu}J^{\mu\nu} \to \kappa_{\mu\nu}J^{\mu\nu} \tag{4.53}$$

as discussed in Sec. II.

Then we obtain

$$Z_0{}^\psi|_{\mathrm{m.s.}} = \int d(\kappa,\Gamma)\int d\mu(S)\, \delta\{\psi_\mu(x;\kappa,\Gamma)\}\Delta_\psi(\kappa,\Gamma)\delta\{\psi_\mu{}^1\} \times \Delta_{\psi^1}\exp\left[i\int dx(L_0+\kappa_{\mu\nu}J^{\mu\nu})\right] = Z_0{}^{\psi^1}|_{\mathrm{m.s.}}. \tag{4.54}$$

Thus we prove that the S matrix of the gravitation field is independent of the type of gauge condition (4.34).

The proof that the S matrix is independent of the α's for the case when the gauge function is fixed can be made similarly to Sec. II. For this purpose substitution of (4.6) or (4.6′) should be made in expression (4.26) with

$$\xi^\mu(x) = -(\delta\alpha/2\alpha)([Q_\psi{}^{T-1}]^{\mu\nu}\psi_\nu)(x). \tag{4.55}$$

Then

$$L_0 - \frac{1}{2\alpha}\psi_\mu\delta^{\mu\nu}\psi_\nu + \kappa_{\mu\nu}J^{\mu\nu} \to L_0 - \frac{1}{2(\alpha+\delta\alpha)}\psi_\mu\delta^{\mu\nu}\psi_\nu + \kappa_{\mu\nu}J^{\mu\nu}, \tag{4.56}$$

and the variation of the term Tr ln in (4.26) is compensated by the resulting Jacobian. We omit the corresponding cumbersome calculations.[2]

Let us pass to the consideration of some particular gauges.

A. Harmonic Condition

Consider the class of gauges determined by the function

$$\psi_1{}^\nu(x) \equiv \partial_\mu \hat{g}^{\mu\nu}(x), \quad \hat{g}^{\mu\nu} = (\sqrt{-g})g^{\mu\nu}. \tag{4.57}$$

The harmonic condition corresponds to

$$\psi_1{}^\mu = 0. \tag{4.58}$$

We shall use $\hat{g}^{\mu\nu}$ as independent variables. By means of (4.14) (with $\beta=\frac{1}{2}$), (4.23), and (4.24), we find

$$Q_1{}^\mu{}_\nu = \delta^\mu{}_\nu(\hat{g}^{\lambda\sigma}\partial_\lambda\partial_\sigma + \partial_\lambda\hat{g}^{\lambda\sigma}\partial_\sigma) + \partial_\lambda\hat{g}^{\lambda\mu}\partial_\nu. \tag{4.59}$$

The generating functional is equal to[21]

$$Z_\alpha{}^1 = \int d(\hat{g})\exp\left[i\int dx\left(L_0 - \frac{1}{2\alpha}\psi_1{}^\mu\delta_{\mu\nu}\psi_1{}^\nu + \hat{g}^{\mu\nu}J_{\mu\nu}\right) + \mathrm{Tr}\ln Q_1\overleftarrow{\Box}^{-1}\right]. \tag{4.60}$$

In transverse gauge ($\alpha=0$), we have

$$Z_0{}^1 = \int d(\hat{g})\exp\left[i\int dx(L_0+\hat{g}^{\mu\nu}J_{\mu\nu}) + 4\,\mathrm{Tr}\ln\hat{g}^{\mu\nu}\partial_\mu\partial_\nu\overleftarrow{\Box}^{-1}\right]\delta\{\partial_\mu\hat{g}^{\mu\nu}(x)\}. \tag{4.61}$$

The Feynman rules for calculation of the generating functional (4.60) in powers of

$$\hat{h}^{\mu\nu} \equiv \hat{g}^{\mu\nu} - \delta^{\mu\nu} \tag{4.62}$$

[21] E. S. Fradkin and I. V. Tyutin, CNR Laboratorio di Cibernetica report, Napoli, 1969 (unpublished).

are

$$\text{(a) } L^1{}_{\text{int}}=L_0-(1/2\alpha)\partial_\mu\hat{h}^{\mu\nu}\delta_{\nu\lambda}\partial_\sigma\hat{h}^{\lambda\sigma}-L_{(0)}{}^1 -i\,\text{Tr}\ln\{\delta_\nu{}^\mu+[(\hat{h}^{\lambda\sigma}\partial_\lambda\partial_\sigma+\partial_\lambda\hat{h}^{\lambda\sigma}\partial_\sigma)\delta_\nu{}^\mu+\partial_\lambda\hat{h}^{\lambda\mu}\partial_\nu]\Box\}. \quad (4.63)$$

These are taken as the interaction Lagrangian.

(b) $L_{(0)}{}^1$ is the Lagrangian of the linearized theory

$$L_{(0)}{}^1=\tfrac{1}{4}\partial^\mu\hat{h}^{\nu\lambda}\partial_\mu\hat{h}_{\nu\lambda}-[(\alpha+1)/2\alpha]\partial_\mu\hat{h}^{\mu\nu}\partial^\lambda\hat{h}_{\lambda\nu}-\tfrac{1}{8}\partial_\mu\hat{h}\partial^\mu\hat{h},\quad \hat{h}\equiv\delta_{\mu\nu}\hat{h}^{\mu\nu}. \quad (4.64)$$

The lowering and raising of indices in (4.64) are done by the Minkowski tensors $\delta_{\mu\nu}$ and $\delta^{\mu\nu}$.

(c) The free propagator of $\hat{h}^{\mu\nu}$ is calculated from (4.64) to be

$$D_1{}^{\mu\nu,\lambda\sigma}(p)=-i\bigg[(2+\alpha)\delta^{\mu\nu}\delta^{\lambda\sigma}-\delta^{\mu\lambda}\delta^{\nu\sigma}-\delta^{\mu\sigma}\delta^{\nu\lambda}+\frac{2(\alpha+1)}{p^2}(\delta^{\mu\nu}p^\lambda p^\sigma+\delta^{\lambda\sigma}p^\mu p^\nu)+\frac{\alpha+1}{p^2}(\delta^{\mu\lambda}p^\nu p^\sigma+\delta^{\mu\sigma}p^\nu p^\lambda+\delta^{\nu\lambda}p^\mu p^\sigma+\delta^{\nu\sigma}p^\mu p^\lambda)\bigg]\frac{1}{p^2}. \quad (4.65)$$

The Feynman rules for the gravitational field in the gauge $\partial_\mu\hat{g}^{\mu\nu}=0$ were also obtained by Fadeev and Popov.[3]

B. Linearized Form of Harmonic Condition

Consider the class of gauges described by the function

$$\psi_\mu{}^2\equiv\delta^{\sigma\lambda}(\partial_\sigma g_{\lambda\mu}-\tfrac{1}{2}\partial_\mu g_{\lambda\sigma}). \quad (4.66)$$

We choose $g_{\mu\nu}$ as independent variables. We then obtain with the help of (4.13) (with $\beta=0$), (4.32), and (4.24)

$$Q_{\mu\nu}{}^2=g_{\mu\nu}\Box+(\partial_\alpha g_{\mu\beta}-\tfrac{1}{2}\partial_\mu g_{\alpha\beta})\times(\delta^{\sigma\alpha}\partial_\sigma\delta_\nu{}^\beta+\delta^{\sigma\beta}\partial_\sigma\delta_\nu{}^\alpha-\delta^{\alpha\beta}\partial_\nu). \quad (4.67)$$

The generating functional is[21]

$$Z_\alpha{}^2=\int d(g)\exp\bigg[i\int dx\Big(L_0-\frac{1}{2\alpha}\psi_\mu{}^2\delta^{\mu\nu}\psi_\nu{}^2+g_{\mu\nu}J^{\mu\nu}\Big)+\text{Tr}\ln Q^2\overleftarrow{\Box}^{-1}\bigg]. \quad (4.68)$$

The Feynman rules for the perturbation calculation of (4.68) in powers of

$$h_{\mu\nu}\equiv g_{\mu\nu}-\delta_{\mu\nu} \quad (4.69)$$

are

$$\text{(a) } L_{\text{int}}{}^2=L_0-(1/2\alpha)\psi_\mu{}^2\delta^{\mu\nu}\psi_\nu{}^2-L_{(0)}{}^2 -i\,\text{Tr}\ln\{\delta_{\mu\nu}+[h_{\mu\nu}\Box+(\partial_\alpha h_{\beta\mu}-\tfrac{1}{2}\partial_\mu h_{\beta\alpha})\times(\delta^{\sigma\alpha}\partial_\sigma\delta_\nu{}^\beta+\delta^{\sigma\beta}\partial_\sigma\delta_\nu{}^\alpha-\delta^{\alpha\beta}\partial_\nu)]\Box\}. \quad (4.70)$$

These should be taken as the interaction Lagrangian.

(b) $L_{(0)}{}^2$ is the Lagrangian of the linearized theory

$$L_{(0)}{}^2=\tfrac{1}{4}\partial_\mu h_{\nu\lambda}\partial^\mu h^{\nu\lambda}-\frac{\alpha+1}{2\alpha}\partial_\mu h^{\mu\nu}\partial^\lambda h_{\nu\lambda}+\frac{\alpha+1}{2\alpha}\partial_\mu h\partial_\nu h^{\mu\nu}-\frac{2\alpha+1}{8\alpha}\partial_\mu h\partial^\mu h,\quad h\equiv\delta^{\mu\nu}h_{\mu\nu}. \quad (4.71)$$

The raising of indices in (4.71) is accomplished by the Minkowski tensor $\delta^{\mu\nu}$.

(c) The free propagator of $h_{\mu\nu}$ is calculated from (4.71) to be

$$D_{\mu\nu,\lambda\sigma}{}^2(p)=-i[\delta_{\mu\nu}\delta_{\lambda\sigma}-\delta_{\mu\lambda}\delta_{\nu\sigma}-\delta_{\mu\sigma}\delta_{\nu\lambda}+[(\alpha+1)/p^2](\delta_{\mu\lambda}p_\nu p_\sigma+\delta_{\mu\sigma}p_\nu p_\lambda+\delta_{\nu\lambda}p_\mu p_\sigma+\delta_{\nu\sigma}p_\mu p_\lambda)]p^{-2}. \quad (4.72)$$

The Feynman rules for $\alpha=-1$ were also given by Mandelstam.[5]

C. Dirac Gauge

We give the arguments which show that the S matrix obtained by Popov and Faddeev[6] in the Dirac gauge[17] coincides with the S matrix in the covariant gauges. Consider the following set of gauge conditions:

$$\psi_3{}^0\equiv(\sqrt{-g})e^{ik}\Gamma_{ik}{}^0=0,\qquad \psi_3{}^k\equiv\partial_i[(-g^{(3)})^{1/3}e^{ik}]=0. \quad (4.73)$$

Here

$$g^{(3)}\equiv\det g_{ik},\quad g=(1/g^{00})g^{(3)},\quad e^{ik}g_{kj}=\delta_j{}^i. \quad (4.74)$$

In gauge (4.73) it is natural to use the first-order formalism. We choose $g^{\mu\nu}$ and $\Gamma^\mu{}_{\nu\lambda}$ as independent variables. With the help of (4.14) (with $\beta=0$), (4.15), and (4.23) one finds

$$Q^0{}_{30}=[-2\overleftarrow{\partial}_i e^{ik}\Gamma_{0k}{}^0-2\overleftarrow{\partial}_i(g^{0k}/g^{00})e^{il}\Gamma_{lk}{}^0+\overleftarrow{\partial}_i e^{lk}\Gamma_{lk}{}^i-\overleftarrow{\partial}_i\overleftarrow{\partial}_k e^{ik}]\sqrt{-g}, \quad (4.75)$$

$$Q_3{}^0{}_j=0, \quad (4.76)$$

$$Q^j{}_{30}=-2\overleftarrow{\partial}_i T^{ij}-\overleftarrow{e}^{il}(-g^{(3)})^{1/3}(\tfrac{1}{3}g_{il}e^{kj}\partial_k-\delta_i{}^j\partial_l), \quad (4.77)$$

$$T^{ij}=(1/g^{00})g^{0m}e^{il}(-g^{(3)})^{1/3}\times(\tfrac{1}{3}g_{lm}e^{kj}\partial_k-\tfrac{1}{2}\delta_m{}^j\partial_l-\tfrac{1}{2}\delta_l{}^j\partial_m), \quad (4.78)$$

$$Q^i{}_{3j}=-\tfrac{1}{3}\overleftarrow{\partial}_j(-g^{(3)})^{1/3}e^{li}\partial_l+\delta_j{}^i(-g^{(3)})^{1/3}e^{lm}\partial_l\partial_m. \quad (4.79)$$

From (4.75)-(4.79), we obtain

$$Q_3{}^{(0)}{}_0{}^0=\nabla,\quad Q_3{}^{(0)}{}_j{}^i=-\delta_j{}^i\nabla-\tfrac{1}{3}\delta^{il}\partial_l\partial_j,\quad Q_3{}^{(0)}{}_i{}^0=Q_3{}^{(0)}{}_0{}^i=0. \quad (4.80)$$

The generating functional is

$$Z^3=\int d(g^{\mu\nu},\Gamma)\delta\{\psi_3{}^\mu\}\exp\Big[i\int dx(L_0+g^{\mu\nu}J_{\mu\nu})+\mathrm{Tr}\ln Q_{30}{}^0\overleftarrow{\nabla}^{-1}+\mathrm{Tr}\ln B_j{}^i(Q_3{}^{(0)-1})_k{}^j\Big],\quad(4.81)$$

$$B_j{}^i\equiv Q_3{}^T{}_j{}^i=(-g^{(3)})^{1/3}(\delta_j{}^ie^{lm}\partial_l\partial_m+\tfrac{1}{3}e^{li}\partial_l\partial_j).$$

According to the general arguments given in this section, the generating functional (4.81) on the mass shell is equivalent to (4.60), (4.61), and (4.68).

Now we transform expression (4.75). Using (4.25) and (4.80) the field equations for B can be obtained:

$$\nabla B_0=0,\quad(\delta_j{}^i\nabla+\tfrac{1}{3}\delta^{il}\partial_l\partial_j)B_i=0.\quad(4.82)$$

The only physical solution of (4.82) is $B_\mu\equiv0$. Thus $g^{\mu\nu}$ and $\Gamma_{\nu\lambda}{}^\mu$ satisfy (4.20) and (4.21) with $\bar{B}_\mu{}^{(3)}\equiv0$; i.e., relations (4.4) for $\Gamma_{\nu\lambda}{}^\mu$ are valid, and $g^{\mu\nu}$ satisfies the usual Einstein equations.

Substituting (4.4) into (4.75), one obtains

$$Q_{30}{}^0=(\sqrt{-g^{(3)}})e^{ik}\nabla_i\partial_k(\sqrt{\alpha})-(\sqrt{\alpha})(\sqrt{-g^{(3)}})e^{ik}\nabla_i\partial_k,\quad(4.83)$$

$$\nabla_i\partial_k\equiv\partial_i\partial_k-\gamma^l{}_{ik}\partial_l.\quad(4.84)$$

Here $\gamma_{ik}{}^l$ is the three-dimensional Christoffel symbol, and $\alpha=(g^{00})^{-1}$. Let us find the expression for $\Gamma^0{}_{ik}$ with the help of the Einstein equations for $g^{\mu\nu}$, and substitute it into the relation

$$e^{ik}\dot{\Gamma}^0{}_{ik}+\dot{e}^{ik}\Gamma^0{}_{ik}=\dot{\psi}_3{}^0=0,\quad(4.85)$$

which must be true according to (4.73). From (4.85) one obtains

$$(\sqrt{-g^{(3)}})e^{ik}\nabla_i\partial_k(\sqrt{\alpha})+(\sqrt{\alpha})(\sqrt{-g^{(3)}})R^{(3)}=0.\quad(4.86)$$

Here $R^{(3)}=e^{ik}R^{(3)}{}_{ik}$, and $R^{(3)}{}_{ik}$ is the three-dimensional curvature tensor of second rank.

Finally, the expression for $Q_{30}{}^0$ takes the form

$$Q_{30}{}^0=-(\sqrt{\alpha})A,\quad(4.87)$$

$$A=(\sqrt{-g^{(3)}})R^{(3)}+(\sqrt{-g^{(3)}})e^{ik}\nabla_i\partial_k.\quad(4.88)$$

Thus we can see that the expression

$$Z'=\int d(g^{\mu\nu},\Gamma)\delta\{\psi_3{}^\mu\}\exp\Big[i\int dx(L_0+g^{\mu\nu}J_{\mu\nu})+\mathrm{Tr}\ln A\sqrt{\alpha}+\mathrm{Tr}\ln B_j{}^i-\mathrm{Tr}\ln Q_3{}^{(0)}\Big]\quad(4.89)$$

for the generating functional with gauge condition (4.73) can be used instead of (4.81). The generating functional (4.89) leads to the same field equations for $g^{\mu\nu}$ and $\Gamma_{\nu\lambda}{}^\mu$ and consequently to the same S matrix as (4.81) does.

Now let us integrate in (4.89) over all the $\Gamma^\mu{}_{\nu\lambda}$ except $\Gamma^0{}_{ik}$. One can show[6] that L_0 takes the form

$$L_0\to\pi^{ik}\dot{g}_{ik}-H(\pi,g).\quad(4.90)$$

Here $H(\pi,g)$ is the Hamiltonian of the gravitation field, the explicit form of which we do not need. π^{ik} are the canonical momenta for g_{ik}:

$$\pi^{ik}=[\sqrt{(-g^{(3)})}/\sqrt{g^{00}}](e^{ik}e^{lm}-e^{il}e^{km})\Gamma^0{}_{lm}.\quad(4.91)$$

With the help of (4.91) the gauge condition (4.73) can be rewritten exactly in the form given by Dirac[17]:

$$\psi_3{}^0\equiv g_{ik}\pi^{ik}=0,\quad\psi_3{}^i\equiv\partial_i[(-g^{(3)})^{1/3}e^{ik}]=0.\quad(4.92)$$

Let us pass from the integration over $\Gamma^0{}_{ik}$ and $g^{\mu\nu}$ in (4.89) to the integration over π^{ik} and $g_{\mu\nu}$. The resulting Jacobian can be omitted. The proof of this fact is analogous to that of the possibility of arbitrary choice of the functional integration variables belonging to class (4.5).

The final expression for the generating functional in gauge (4.73) or (4.92) is

$$Z''=\int d(g,\pi)\delta\{\psi_3{}^\mu\}\times\exp\Big[i\int dx(\pi^{ik}\dot{g}_{ik}-H(\pi,g)+g_{ik}J^{ik})+\mathrm{Tr}\ln A\overleftarrow{\nabla}^{-1}+\mathrm{Tr}\ln B_j{}^i(Q_3{}^{(0)-1})_k{}^j\Big].\quad(4.93)$$

Expression (4.93) has been obtained by Popov and Faddeev[6] with the help of another method closely connected with the canonical quantization procedure.

Note once more that the S matrix corresponding to (4.93) is equal to that in covariant gauges.

V. CONCLUSION

The present paper has been devoted to constructing the S matrix in theories invariant under gauge groups. Though the only cases considered were those of the Yang-Mills field and gravitation, the method developed can in principle be applied to arbitrary theories (the theories of the Yang-Mills and the gravitational fields are apparently the only gauge theories of physical interest[12]), particularly in the cases where no connection with the canonical scheme can be traced. Furthermore, the method suggested proves to be convenient for constructing the perturbation expansion of the S matrix in theories partially invariant under a gauge group, the power of divergence in the S matrix being considerably reduced.

In this paper no attention was paid to possible interactions with other particles. The latter would not affect our considerations, however.

We would like to discuss briefly the problems which have not yet been solved.

(1) Owing to divergences, there is an important problem of introducing a regularization which will not affect the group properties of the theory. Recall that non-gauge-invariant regularization in electrodynamics creates the photon mass. From the more recent view, the resulting photon mass is due to Schwinger terms or, in the end, to the singular character of products of field operators at coincident points. In nonlinear theories this problem becomes even more complicated. The Schwinger terms affect even the renormalization constant, as for instance in the case of the Yang-Mills field.

(2) There is an interesting question whether the gravitation field is renormalizable in the framework of perturbation theory. (We mean here the usual perturbation expansion with respect to a coupling constant and not the method of Fradkin and Efimov.[22]) It is convenient to treat this problem using the variables $\hat{h}^{\mu\nu}$ and $\Gamma^{\mu}{}_{\nu\lambda}$ in the first-order formalism where there are two vertices: a vertex $\Gamma\Gamma\hat{h}$ and the vertex responsible for the interaction of $\hat{h}^{\mu\nu}$ with the fictitious B field. The formal estimate of degrees of growth leads to the conclusion that the theory is of unrenormalizable type.

ACKNOWLEDGMENTS

The authors are grateful to the participants in the theoretical seminar at the P. N. Lebedev Institute for useful discussions. One of the authors (E. F.) thanks Dr. Popov and Dr. Faddeev for being so kind as to acquaint him with Ref. 6 prior to publication, and Professor E. R. Caianiello for his kind hospitality. He is also grateful to Professor Abdus Salam for hospitality at the International Centre for Theoretical Physics, Trieste.

[22] E. S. Fradkin, Nucl. Phys. **49**, 624 (1963); **76**, 588 (1966); G. V. Efimov, Zh. Eksperim. i Teor. Fiz. **44**, 2107 (1963) [Soviet Phys. JETP **17**, 1417 (1963)]; Nuovo Cimento **32**, 1046 (1964).

BULLETIN OF THE
AMERICAN MATHEMATICAL SOCIETY
Volume 78, Number 5, September 1972

MISSED OPPORTUNITIES[1]

BY FREEMAN J. DYSON

> It is important for him who wants to discover not to confine himself to one chapter of science, but to keep in touch with various others.
>
> JACQUES HADAMARD

1. **Introduction.** The purpose of the Gibbs lectures is officially defined as "to enable the public and the academic community to become aware of the contribution that mathematics is making to present-day thinking and to modern civilization." This puts me in a difficult position. I happen to be a physicist who started life as a mathematician. As a working physicist, I am acutely aware of the fact that the marriage between mathematics and physics, which was so enormously fruitful in past centuries, has recently ended in divorce. Discussing this divorce, the physicist Res Jost remarked the other day, "As usual in such affairs, one of the two parties has clearly got the worst of it." During the last twenty years we have seen mathematics rushing ahead in a golden age of luxuriant growth, while theoretical physics left on its own has become a little shabby and peevish. So I am forced to give this lecture an emphasis different from that intended by the founders. Instead of talking about "the contribution that mathematics is making to present-day thinking" in my field, I shall talk about the contribution that mathematics ought to have made but did not. I shall examine in detail some examples of missed opportunities, occasions on which mathematicians and physicists lost chances of making discoveries by neglecting to talk to each other. My purpose in calling attention to such incidents is not to blame the mathematicians or to excuse the physicists for our failure in the last twenty years to equal the great achievements of the past. My purpose is not to lament the past but to mould the future.

It is obviously absurd for me to imagine that I can mould the future with a one-hour lecture. The fact that Hilbert in 1900 [1] and Minkowski in 1908 [2] succeeded in doing it does not give me any confidence that I can do it too. But at least I have learned from Hilbert and Minkowski that one does not influence people by talking in generalities. Hilbert and Minkowski gave specific suggestions of things that mathematicians and physicists could profitably think about. I shall try to follow their

[1] Josiah Willard Gibbs Lecture, given under the auspices of the American Mathematical Society, January 17, 1972; received by the editors January 17, 1972.

style. I shall try to convince you by examining actual cases that the progress of both mathematics and physics has in the past been seriously retarded by our unwillingness to listen to one another. And I will end with an attempt to identify some areas in which opportunities for future discoveries are now being missed.

6. **General coordinate invariance.** Up to now, my examples of missed opportunities have been mathematical discoveries which actually occurred, although they could have occurred a long time earlier. In such cases one can be sure that an opportunity existed, but it existed only in the past. I now come to the more difficult task of identifying missed opportunities that are still open. Here one can no longer be sure that the opportunity is real, but if it is real then it has the virtue of existing in the present.

The past opportunities which I discussed have one important feature in common. In every case there was an empirical finding that two disparate or incompatible mathematical concepts were juxtaposed in the description of a single situation. Taking the four examples in turn, the pairs of disparate concepts were respectively: modular functions and Lie algebras, field equations and particle dynamics, Lorentz invariance and Galilean invariance, quaternion algebra and Grassmann algebra. In each case the opportunity offered to the pure mathematician was to create a wider conceptual framework within which the pair of disparate elements would find a harmonious coexistence. I take this to be my methodological principle in looking for opportunities that are still open. I look for situations in which the juxtaposition of a pair of incompatible concepts is acknowledged but unexplained.

The most glaring incompatibility of concepts in contemporary physics is that between Einstein's principle of general coordinate invariance and all the modern schemes for a quantum-mechanical description of nature. Einstein based his theory of general relativity [**28**] on the principle that God did not attach any preferred labels to the points of space-time. This principle requires that the laws of physics should be invariant under the Einstein group E, which consists of all one-to-one and twice-differentiable transformations of the coordinates. By making full use of the invariance under E, Einstein was able to deduce the precise form of his law of gravitation from general requirements of mathematical simplicity without any arbitrariness. He was also able to reformulate the whole of classical physics (electromagnetism and hydrodynamics) in E-invariant fashion, and so determine unambiguously the mutual interactions of matter, radiation and gravitation within the classical domain. There is no part of physics more coherent mathematically and more satisfying aesthetically than this classical theory of Einstein based upon E-invariance.

On the other hand, all the currently viable formalisms for describing nature quantum-mechanically use a much smaller invariance group. The analysis of Bacry and Lévy-Leblond [**21**] indicates the extreme range of quantum-mechanical kinematical groups that have been contemplated. In practice all serious quantum-mechanical theories are based either on the Poincaré group P or the Galilei group G. This means that a class of preferred inertial coordinate-systems is postulated a priori, in flat contradiction to Einstein's principle. The contradiction is particularly uncomfortable, because Einstein's principle of general coordinate invariance has such an attractive quality of absoluteness. A physicist's intuition tells him that, if Einstein's principle is valid at all, it ought to be valid for the whole of physics, quantum-mechanical as well as classical. If the principle were not universally valid, it is difficult to understand why Einstein achieved such deeply coherent insights into nature by assuming it to be so.

To make the mathematical incompatibility more definite, I will focus attention on one of the competing schemes for describing a quantum-mechanical universe. I choose the scheme which is most carefully based on rigorous mathematical definitions and which is also general enough to encompass a wide variety of physical systems. This scheme is the "Algebra of Local Observables" of Haag and Kastler [**29**]. The six axioms of Haag and Kastler are the following.

(1) *Existence of local observables.* To every region B (i.e. open set with compact closure) in 4-dimensional space-time there corresponds an abstract C^*-algebra $\mathscr{A}(B)$.

(2) *Isotony.* If $B_1 \supset B_2$ then $\mathscr{A}(B_1) \supset \mathscr{A}(B_2)$, and $\mathscr{A}(B_1)$ and $\mathscr{A}(B_2)$ have a common unit element.

(3) *Existence of quasilocal observables.* The union of all $\mathscr{A}(B)$ is a normed $*$-algebra, the completion of which is a C^*-algebra $\mathscr{A}$, the algebra of quasilocal observables. It is supposed that all physically measurable quantities are elements of $\mathscr{A}$.

(4) *Poincaré invariance.* To every element L of the Poincaré group there corresponds an automorphism α_L of $\mathscr{A}$, such that $\alpha_L(\mathscr{A}(B)) = \mathscr{A}(LB)$ for every region B.

(5) *Local commutativity.* If two regions B_1 and B_2 are completely spacelike with respect to each other (i.e. if B_2 lies outside every light-cone with vertex in B_1), then $\mathscr{A}(B_1)$ and $\mathscr{A}(B_2)$ commute.

(6) *Primitivity.* $\mathscr{A}$ admits a faithful algebraically irreducible representation.

These axioms, taken together with the axioms defining a C^*-algebra [**30**], are a distillation into abstract mathematical language of all the general truths that we have learned about the physics of microscopic systems during the last 50 years. They describe a mathematical structure of great elegance whose properties correspond in many respects to the facts of experimental physics. In some sense, the axioms represent the most serious attempt that has yet been made to define precisely what physicists mean by the words "observability, causality, locality, relativistic invariance," which they are constantly using or abusing in their everyday speech.

If we look at the axioms in detail, we see that (1), (2), (3) and (6) are consistent with Einstein's general coordinate invariance, but (4) and (5) are inconsistent with it. Axioms (4) and (5), the axioms of Poincaré invariance and local commutativity, require the Poincaré group to be built into the structure of space-time. If we try to replace the Poincaré group P by the Einstein group E, we have no way to define a space-like relationship between two regions, and axiom (5) becomes meaningless. I therefore propose as an outstanding opportunity still open to the pure mathematicians, *to create a mathematical structure preserving the main features of the Haag-Kastler axioms but possessing E-invariance instead of P-invariance.*

I had better warn any mathematician who intends to respond to my challenge that his task will not be easy. No merely formal rearrangement of the Haag-Kastler axioms can possibly be sufficient. For we know that Einstein could construct his E-invariant classical theory of 1916 only by bringing in the full resources of Riemannian differential geometry. He needed a metric tensor to give his space-time a structure independent of coordinate-systems. Therefore an E-invariant axiom of local commutativity to replace axiom (5) will require at least some quantum-mechanical analog of Riemannian geometry. Some analog of a metric tensor must be introduced in order to give a meaning to space-like separation. The answer to my challenge will necessarily involve a delicate weaving together of concepts from differential geometry, functional analysis, and abstract algebra. With these words of warning I leave the problem to you.

References

1. D. Hilbert, *Mathematische Probleme*, Lecture to the Second Internat. Congress of Math. (Paris, 1900), Arch. Math. und Phys. (3) **1** (1901), 44–63; 213–237; English transl., Bull. Amer. Math. Soc. **8** (1902), 437–479.

2. H. Minkowski, *Raum und Zeit*, Lecture to the 80th Assembly of Natural Scientists (Köln, 1908), Phys. Z. **10** (1909), 104–111. English transl., *The principle of Relativity*, Aberdeen Univ. Press, Aberdeen, 1923.

21. H. Bacry and J.-M. Lévy-Leblond, *Possible kinematics*, J. Mathematical Phys. **9** (1968), 1605–1614. MR **38** # 6821. To save time I have slightly misstated their conclusion; each of the groups D, P' and N can occur in two alternative forms, so that the number of possibilities is strictly speaking 11 rather than 8.

28. A. Einstein, *Die Grundlage der allgemeinen Relativitätstheorie*, Ann. Phys. **49** (1916), 769–822.

29. R. Haag and D. Kastler, *An algebraic approach to quantum field theory*, J. Mathematical Phys. **5** (1964), 848–861. MR **29** #3144.

30. J. Dixmier, *Les C*-algèbres et leurs représentations*, Cahiers Scientifiques, fasc. 29, Gauthier-Villars, Paris, 1964. MR **30** #1404.

A Brief Remark for "Missed Opportunities"

Freeman Dyson

Looking back on this lecture thirty years later, I have the impression that things have improved. Mathematicians and physicists are listening more to one another now than they were then. Ideas and methods are spreading more easily between the two disciplines. On the other hand, there is now a widening gulf of incomprehension between two groups of physicists, one group doing string theory and the other group doing other kinds of physics. The string theorists and the mathematicians understand each other, but the gap that used to separate mathematics from physics now separates string theory from the rest of physics.

Chapter 7

Gauge Theories of Gravity*

---

*K. Hayashi, T. Nakano, R. Utiyama and T. Fukuyama, C. N. Yang, W. T. Ni, Y. M. Cho, J. P. Hsu

Progress of Theoretical Physics, Vol. 38, No. 2, August 1967

Extended Translation Invariance and Associated Gauge Fields

Kenji HAYASHI and Tadao NAKANO*

Department of Physics, Kyoto University, Kyoto
**Department of Physics, Osaka City University, Osaka*

(Received April 1, 1967)

Gauge fields together with non-linear field equations to govern them are introduced by requiring that the Lagrangian should be invariant under an extended translation in space-time, i. e. a translation in which four parameters are replaced by four arbitrary coordinate-dependent functions. A prescription is given to convert a non-invariant canonical ("pseudo") energy-momentum tensor into an invariant one.

The symmetric part of these field equations is examined for the two cases: (1) under linear and non-relativistic approximation, it reduces to the classical gravitational-field equation, (2) for static and spherically symmetric field, its solution is shown to correspond to Schwarz-schild's solution. The antisymmetric part has no classical analogues, for there are no sources of skew-symmetric energy-momentum tensors in the classical experiments. A reasonable method is proposed to eliminate this redundant field.

§ 1. Introduction

Since it was suggested that the electromagnetic interaction is best understood in terms of a principle of gauge invariance, under a gauge transformation with a coordinate-dependent function, there have been a number of attempts to deduce the existence of gauge fields coupled to conserved currents, starting with the idea of extended transformations.[1]

It was shown that the invariance under the n-parameter Lie group of trans-formation referred to space-time and/or fields leads to the conservation of n generators. Further, invariance requirement under an extended transformation, i.e. a transformation whose n parameters are replaced by n space-time dependent functions, necessitates the introduction of n (generally) non-commuting vector fields together with field equations which they must obey.[2),3)]

The purpose of this paper is to deduce the existence of a gravitational field from the translational invariance in an extended sense just mentioned above. In order to construct the gravitational interaction, Utiyama has proposed to introduce 24 new field variables by postulating the invariance under an extended four-dimensional rotation which is specified by six skew-symmetric arbitrary functions $\omega_{ij}(x)$.[3)] However, the self-inconsistency of his scheme was pointed out by Kibble who has claimed that it is necessary to consider the extension of full 10-parameter inhomogeneous Lorentz group in place of the restricted six-parameter group.[4)] Then, our method is different from both of them and will be shown to be one of the simplest ways of discussing the gravitational interaction within the Lagrangian formalism of the unquantized fields in that we need the minimal transformation group (translation group) and its extension necessary and sufficient to deduce it.*)

In the following section, a general formulation of the extended translation is given within the classical Lagrangian framework and a prescription is presented to convert a "pseudo" energy-momentum tensor into an invariant energy-momentum tensor. In § 3, we apply it to the system consisting of the spinor field and the new fields.

We identify the symmetric part of these new fields with the classical gravitational field by means of the linear approximation to the non-linear field variables. In § 4, for the purposes of comparison we shall consider the static and spherically symmetric field in which the exact solution of Einstein's equation of gravitation has been well known and verified by the observations. In § 5, elimination of the antisymmetric part of the new fields will be attempted and the final section is devoted to a discussion of the results.

§ 2. General formulation

We start with the Lagrangian density**),***)

$$L_0 = L_0(q^A, q^A_{,k}, x_k), \qquad q^A_{,k} = \partial q^A/\partial x_k\,,$$

where q^A are a set of fields, $(A=1, 2, \cdots, N)$. The action integral referred to an arbitrary four-dimensional domain Σ,

$$I(\Sigma) = \int_\Sigma L_0(x)\, d^4x\,, \tag{2·1}$$

is invariant under the following infinitesimal transformation:

$$\left.\begin{aligned} &q^A(x) \to q'^A(x') = q^A(x) + \delta q^A(x)\,, \\ &x_k \to x_k' = x_k + \delta x_k\,, \end{aligned}\right\} \tag{2·2}$$

if the following identity holds true at any world points (independent of the behavior of q^A and its derivatives):

$$L_0 + L_0 \delta x_{k,k} = \delta * L_0 + (L_0 \delta x_k)_{,k} \equiv 0\,, \tag{2·3}$$

where $\delta * L_0 = \delta L_0 - L_{0,k}\delta x_k$ is often called a substantial variation in the sense that a variation caused by the coordinate transformation is subtracted.****) Up to the variation of first order, it follows that

---

*) Einstein's theory of general relativity has been based on the general covariance under the extended translation within the classical mechanics.

**) It is assumed that the Lagrangian to be considered hereinafter contains the first derivatives of the field variable at most.

***) We use the imaginary fourth coordinate $x_4 = ict$.

****) A summation convention for dummy indices is used throughout this paper.

$$I(\Sigma)\longrightarrow I'(\Sigma)=\int_{\Sigma} L_0'(x)\,(1+\delta x_{k'k})\,d^4x\,, \tag{2·1'}$$

which is used in deriving the above identity. Consider a translation

$$\delta q^A=0\,,\quad \delta x_k=\epsilon_k\,,\quad (\epsilon_k:\text{ infinitesimal parameter}), \tag{2·4}$$

then the following identity must obtain in order to preserve the invariance under this transformation:

$$\frac{\partial L_0}{\partial x_k}=\frac{dL_0}{dx_k}-\frac{\partial L}{\partial q^A}\frac{\partial q^A}{\partial x_k}-\frac{\partial L}{\partial q^A_{\cdot l}}\frac{\partial q^A_{\cdot l}}{\partial x_k}\equiv 0\,, \tag{2·5}$$

which obviously implies that the invariant Lagrangian under the translation has no explicit x-dependence, hence we shall consider exclusively the Lagrangian

$$L_0=L_0(q^A,\, q^A_{\cdot k})\,. \tag{2·6}$$

Equation (2·3) is rewritten as

$$[L_0]_{q^A}\delta{*}q^A+S_{k'k}\equiv 0\,, \tag{2·7}$$

and the equation of motion is abbreviated:

$$[L_0]_{q^A}=\frac{\partial L_0}{\partial q^A}-\left(\frac{\partial L_0}{\partial q^A_{\cdot k}}\right)_{\cdot k}=0\,,^{*)} \tag{2·8}$$

where

$$\begin{aligned} &S_k=L_{0'q^A_{,k}}\delta q^A-T_{lk}\delta x_l\,, \qquad \partial L_0/\partial q^A_{,k}=L_{0,q^A_{,k}}\,,\\ &T_{lk}=L_{0,q^A_{,k}}q^A_{,l}-\delta_{lk}L_0\,. \end{aligned} \tag{2·9}$$

If the action integral is invariant under the translation (2·4), the conservation law of the energy-momentum tensor defined above follows

$$T_{kl'l}=0 \tag{2·10}$$

on account of the field equation (2·8).

Next we consider the extended translation**)

$$\delta q^A=0\,,\qquad \delta x^\mu=\varepsilon^\mu(x), \tag{2·11}$$

$$(\varepsilon^\mu(x);\ \text{infinitesimal arbitrary function}).$$

The invariance property of L_0 under the translation (2·4) breaks down in this case; the variation of the derivative does not vanish,

$$\delta q^A_{\cdot\mu}=(\delta q^A)_{\cdot\mu}-q^A_{\cdot\nu}\delta x^\nu_{\cdot\mu}=-\varepsilon^\nu_{\cdot\mu}q^A_{\cdot\nu}\,. \tag{2·12}$$

We shall further require the invariance of the action integral under the extended

*) This is called the Euler equation and derived by postulating $\delta{*}I=0$ under the condition that $\delta{*}q^A$ should vanish on the boundary surface of the integration domain.

**) In this case the Greek indices are used for conveniences.

translation, by defining the covariant derivative through which the new field $a_k^{\ \mu}(x)$ is introduced so as to satisfy our postulate:

$$D_k q^A = \{\delta_k^{\ \mu} + a_k^{\ \mu}(x)\} q^A_{,\mu} = b_k^{\ \mu}(x) q^A_{,\mu}\,, \tag{2·13}$$

$$\delta D_k q^A = 0\,. \tag{2·14}$$

In order to satisfy Eq. (2·14), it follows immediately,

$$\delta b_k^{\ \mu} = \varepsilon^\mu_{,\nu} b_k^{\ \nu}. \tag{2·15}$$

Therefore we recover the invariance of the action integral even under the extended translation (i) by simply replacing $q^A_{,k}$ in the original Lagrangian by the covariant derivative $D_k q^A$ defined above;

$$L_0(q^A, q^A_{,k}) \longrightarrow L'(q^A, q^A_{,k}, b_k^{\ \mu}) = L''(q^A, D_k q^A)$$
$$= L_0(q^A, q^A_{,k} \longrightarrow D_k q^A)\,, \tag{2·16}$$

hence its variation associated with (2·11) vanishes identically

$$\delta L' = 0\,, \tag{2·16'}$$

and further (ii) by multiplying L' by a certain function $b(x)$ so as to satisfy the required identity (2·3):

$$\delta \mathbf{L} + \mathbf{L}\varepsilon^\mu_{,\mu} = 0\,,$$
$$\mathbf{L} = bL'. \tag{2·17}$$

Accordingly, the transformation property of b has to be

$$\delta b = -\varepsilon^\mu_{,\mu} b\,. \tag{2·18}$$

Next tasks are then to construct such a function $b(x)$ and the invariant field strength from $b_k^{\ \mu}$ and its first derivatives. For these purposes, it is necessary to define the field $b_{k\mu}$ inverse to $b_k^{\ \mu}$ from the following orthogonal relations:

$$b_{k\mu} b_l^{\ \mu} = \delta_{kl}\,,$$
$$b_{k\mu} b_k^{\ \nu} = \delta_\mu^{\ \nu}. \tag{2·19}$$

Consequently, it follows

$$\delta b_{k\mu} = -\varepsilon^\nu_{,\mu} b_{k\nu}\,,$$

hence we choose

$$b = \det(b_{k\mu})\,,$$

because it has the desired property (2·18). In other words, the invariant volume element becomes bd^4x instead of d^4x. Suppose that we obtain a free Lagrangian L^G for the new field, the action integral turns out

$$I(\Sigma) = \int_\Sigma d^4x \mathbf{L}\,,$$

where

$$\boldsymbol{L}=\boldsymbol{L}'+\boldsymbol{L}^G=b(L'+L^G),^{*)} \tag{2·20}$$

and L^G consists of the invariant field strength. We shall write for short

$$(Q^\alpha)=(q^A,\ b_{k\mu}). \tag{2·21}$$

The invariance of the action integral follows from the following identity analogous to (2·3):

$$\delta*\boldsymbol{L}+(\boldsymbol{L}\varepsilon^\mu)_{,\mu}=0\,,$$

which is just shown above to hold by means of our prescription. The above identity is rewritten as before,

$$[\boldsymbol{L}]_{Q^\alpha}\delta*Q^\alpha+\boldsymbol{S}^\mu_{;\mu}=0\,, \tag{2·22}$$

where

$$\left.\begin{aligned}\boldsymbol{S}^\mu&=\boldsymbol{L}_{,Q^\alpha_{;\mu}}\delta Q^\alpha-\boldsymbol{T}_\nu{}^\mu\varepsilon^\nu,\\ \boldsymbol{T}_\nu{}^\mu&=\boldsymbol{L}_{,Q^\alpha_{;\mu}}Q^\alpha_{;\nu}-\delta_\nu{}^\mu\boldsymbol{L}={}^m\boldsymbol{T}_\nu{}^\mu+\tilde{\boldsymbol{t}}_\nu{}^\mu,\end{aligned}\right\}^{**)} \tag{2·23}$$

$$\left.\begin{aligned}{}^m\boldsymbol{T}_\nu{}^\mu&=\boldsymbol{L}'_{,q^A_{;\mu}}q^A_{;\nu}-\delta_\nu{}^\mu\boldsymbol{L}',\\ \tilde{\boldsymbol{t}}_\nu{}^\mu&=\boldsymbol{L}^G_{,b_{k\lambda,\mu}}b_{k\lambda;\nu}-\delta_\nu{}^\mu\boldsymbol{L}^G.\end{aligned}\right\} \tag{2·24}$$

From (2·22), we obtain

$$\begin{aligned}&(-\varepsilon^\mu)\{[\boldsymbol{L}']_{q^A}q^A_{;\mu}+[\boldsymbol{L}]_{b_{k\nu}}b_{k\nu;\mu}-([\boldsymbol{L}]_{b_{k\nu}}b_{k\mu})_{;\nu}\}\\ &-\{[\boldsymbol{L}]_{b_{k\mu}}b_{k\nu}\varepsilon^\nu+\boldsymbol{L}^G_{,b_{k\lambda;\mu}}b_{k\nu}\varepsilon^\nu_{,\lambda}+({}^m\boldsymbol{T}_\nu{}^\mu+\tilde{\boldsymbol{t}}_\nu{}^\mu)\varepsilon^\nu\}_{;\mu}\equiv 0\,.\end{aligned} \tag{2·25}$$

As ε^ν, $\varepsilon^\nu_{;\mu}$ and $\varepsilon^\nu_{;\mu\lambda}$ are chosen arbitrarily inside the integration domain Σ, the second term of Eq. (2·25) resolves itself into the three identities

$$\varepsilon^\nu;\qquad ([\boldsymbol{L}]_{b_{k\mu}}b_{k\nu})_{;\mu}+({}^m\boldsymbol{T}_\nu{}^\mu+\tilde{\boldsymbol{t}}_\nu{}^\mu)_{;\mu}\equiv 0\,, \tag{2·26}$$

$$\varepsilon^\nu_{;\mu};\qquad [\boldsymbol{L}]_{b_{k\mu}}b_{k\nu}+({}^m\boldsymbol{T}_\nu{}^\mu+\tilde{\boldsymbol{t}}_\nu{}^\mu)+(\boldsymbol{L}^G_{,b_{k\mu;\lambda}}b_{k\nu})_{;\lambda}\equiv 0\,, \tag{2·27}$$

$$\varepsilon^\nu_{;\mu\lambda}\boldsymbol{L}^G_{,b_{k\lambda;\mu}}b_{k\nu}\equiv 0\,, \tag{2·28}$$

among which there are only two independent identities, for the last identity implies

$$\boldsymbol{L}^G_{,b_{k\lambda;\mu}}+\boldsymbol{L}^G_{,b_{k\mu;\lambda}}=0 \tag{2·28'}$$

and the differentiation of the second identity yields the first one, by making use of (2·28′). Furthermore it suggests that the invariant field strength must contain an antisymmetric combination of $b_{k\lambda;\mu}$ with respect to the Greek subscripts and

*) In the standard terminology of tensor algebra, $\boldsymbol{L}$ is named the "tensor density".

**) ${}^m\boldsymbol{T}_\nu{}^\mu$, $\tilde{\boldsymbol{t}}_\nu{}^\mu$ are the "canonical" energy-momentum tensors.

then the contraction of these indices has to be performed; finally we have*)

$$c_{klm}=2b_{k[\mu,\nu]}b_l{}^{\mu}b_m{}^{\nu},$$
$$=2b_{k\nu}b_{[l}{}^{\mu}b_{m]}{}^{\nu}{}_{,\mu}$$

with

$$\delta c_{klm}=0\ .$$

It should be noticed that there exist infinitely many conservation laws for generalized energy-momentum tensors in addition to the one implied directly by (2·26), ($f^{(\nu)}$: an arbitrary function),

$$\left.\begin{aligned}&\boldsymbol{T}^{(f)\mu}=f^{(\nu)}({}^m\boldsymbol{T}_\nu{}^\mu+\tilde{\boldsymbol{t}}_\nu{}^\mu)-f^{(\nu)}_{,\lambda}(\boldsymbol{L}^G_{,b_{k\mu,\lambda}}b_{k\nu}),\\&\boldsymbol{T}^{(f)\mu}_{,\mu}=0\ .^{**)}\end{aligned}\right\}\tag{2·29}$$

Under the extended translation only the Greek indices are associated with the transformation properties: Then there arises a question, "What role does the Latin index play?" To see it, we consider the four-dimensional rotation of the field variables only,

$$\left.\begin{aligned}&\delta x^\mu=0\ ,\\&\delta q^A=Tq^A,\end{aligned}\right\}\tag{2·30}$$

under the assumption that L' is kept invariant under (2·30) and L_0 under the Lorentz transformation specified by

$$\delta x_k=\omega_{kl}x_l\ ,\qquad (\omega_{(kl)}=0)\ ,$$
$$\delta q^A=Tq^A.$$

The assumed invariance properties respectively yield the following identities:

$$L''_{,q^A}Tq^A+L''_{,D_kq^A}TD_kq^A+L'_{,b_k{}^\mu}\delta b_k{}^\mu=0\ ,\tag{2·31}$$
$$L_{0,q^A}Tq^A+L_{0,q^A_{,k}}Tq^A_{,k}-L_{0,q^A_{,k}}q^A_{,l}\omega_{lk}=0\ ,$$

and (2·31) passes into

$$(\delta b_k{}^\mu-\omega_{kl}b_l{}^\mu)L''_{,D_kq^A}q^A_{,\mu}=0\ ,$$

by making use of the relation (2·16). Hence the transformation property is established,

$$\delta b_k{}^\mu=\omega_{kl}b_l{}^\mu,$$

that is, the Latin index is related to the four-dimensional rotation and $b_k{}^\mu$ transforms as a four vector under it (the same is true for $b_{k\mu}$ too).

The quantity $b_k{}^\mu$ is the contravariant vector as it transforms contragradiently

---

*) $A_{(\mu\nu)}=(1/2)(A_{\mu\nu}+A_{\nu\mu})$, $A_{[\mu\nu]}=(1/2)(A_{\mu\nu}-A_{\nu\mu})$.

**) The conserved quantities will be explicitly given later in an invariant manner.

to $\delta q^A_{,\mu}$, and $b_{k\mu}$ the covariant vector as it transforms cogradiently to (2·12). $b_k{}^\mu$ is to be referred to as a " vierbein " system because of its dual character under the extended translation and the four-dimensional rotation specified by (2·11) and (2·30), respectively. These situations are made clear and are summed up by the following statement. Under the combination of these two independent transformations,

$$\delta x^\mu = \varepsilon^\mu(x), \tag{2·30'}$$

$$\delta q^A = Tq^A, \tag{2·30''}$$

L' stays invariant if $b_k{}^\mu$ (or equivalently $b_{k\mu}$) transforms as follows:

$$\begin{aligned} \delta b_k{}^\mu &= \varepsilon^\mu_{,\nu} b_k{}^\nu + \omega_{kl} b_l{}^\mu, \\ \delta b_{k\mu} &= -\varepsilon^\nu_{,\mu} b_{k\nu} + \omega_{kl} b_{l\mu} . \end{aligned} \tag{2·30'''}$$

The field strength c_{klm} is reducible under the four-dimensional rotation; the irreducible parts of it are calculated by means of the standard method:

i) an irreducible tensor of rank 3,

$$c^T_{klm} = c_{(kl)m} - (1/3)(\delta_{kl} c_m{}^V - \delta_{m(k} c_{l)}{}^V),$$

ii) a vector,

$$c_k{}^V = c_{mmk} = (b b_k{}^\mu)_{,\mu}/b ,$$

iii) an axial vector,

$$c_k{}^A = i\varepsilon_{klmn} c_{lmn}/6 .$$

The tensor c_{klm} is represented in terms of these irreducible tensors,

$$c_{klm} = (4/3) c^T_{k[lm]} + (2/3)\delta_{k[l} c^V_{m]} + i\varepsilon_{klmn} c_n{}^A .$$

We shall require that L^G should be of the quadratic form in the first derivative of $b_k{}^\mu$. Thus, we choose

$$\boldsymbol{L}^G = b(\alpha c^T_{klm} c^T_{klm} + \beta c_k{}^V c_k{}^V + \gamma c_k{}^A c_k{}^A + \delta)^{*)} \tag{2·32}$$

with the arbitrary constants α, β, γ, δ. Inserting the above Lagrangian into (2·27), we obtain after some algebra

$$\boldsymbol{B}_{kl} = {}^m\boldsymbol{T}_{kl} , \tag{2·33}$$

$$\boldsymbol{B}_{kl} = -b_m{}^\mu b \boldsymbol{F}_{klm,\mu} + c_m{}^V \boldsymbol{F}_{kml} + (1/2) c_{lmn} \boldsymbol{F}_{kmn} - c_{mnk} \boldsymbol{F}_{mnl} + \delta_{kl} \boldsymbol{L}^G, \tag{2·34}$$

$$\boldsymbol{F}_{klm} = 4b\{\alpha c^T_{k[lm]} + \beta \delta_{k[l} c^V_{m]} - (1/6) i\gamma \varepsilon_{klmn} c_n{}^A\}, \tag{2·35}$$

$$\left.\begin{aligned} {}^m\boldsymbol{T}_{kl} &= b_k{}^\nu b_{l\mu} {}^m\boldsymbol{T}_\nu{}^\mu = b\,{}^mT_{kl} \\ &= \boldsymbol{L}''_{,D_l q^A} D_k q^A - \delta_{kl} \boldsymbol{L}'',^{**)} \end{aligned}\right\} \tag{2·36}$$

---

*) We shall assume invariance under space reflection. If we do not assume it, we can add a linear sum of the following two terms: $bc_k{}^V c_k{}^A$, $i\varepsilon_{klmn} bc^T{}_{jkl} c^T{}_{jmn}$.

**) It is also rewritten as $-\boldsymbol{L}'_{,b_{k\mu}} b_{l\mu}$.

where all the tensors are converted into the local tensors in order to preserve the invariance under the extended translation.

We shall manage to write down the equation of motion (2·33) in a simpler form analogous to the divergent form (2·27),

$$-(b_l{}^\mu b_m{}^\nu \boldsymbol{F}_{klm})_{,\nu} = b_l{}^\mu ({}^m\boldsymbol{T}_{kl} + \boldsymbol{t}_{kl}), \tag{2·37}$$

where

$$\begin{aligned}\boldsymbol{t}_{kl} &= b_k{}^\nu b_{l\mu} \boldsymbol{t}_\nu{}^\mu = \boldsymbol{L}^G_{,b_l{}^\mu} b_k{}^\mu \\ &= c_{mnk} \boldsymbol{F}_{mnl} - \delta_{kl} \boldsymbol{L}^G,\end{aligned} \tag{2·38}$$

$$\begin{aligned}\tilde{\boldsymbol{t}}_{kl} &= b_k{}^\nu b_{l\mu} \tilde{\boldsymbol{t}}_\nu{}^\mu \\ &= \boldsymbol{t}_{kl} + b_k{}^\mu b_m{}^\nu b_{n\mu,\nu} \boldsymbol{F}_{nml}\,.\end{aligned} \tag{2·39}$$

It should be emphasized that $\boldsymbol{t}_{kl}$ remains invariant under the extended translation while $\tilde{\boldsymbol{t}}_{kl}$ is not invariant, as is easily shown (hence a tilde has been attached to the canonical energy-momentum tensor density $\tilde{\boldsymbol{t}}_\nu{}^\mu$).*) From (2·37), we obtain the conservation law

$$\{b_l{}^\mu ({}^m\boldsymbol{T}_{kl} + \boldsymbol{t}_{kl})\}_{,\mu} = 0\,, \tag{2·40}$$

which is essentially the same as (2·26), although $\boldsymbol{t}_\nu{}^\mu$ is preferred to $\tilde{\boldsymbol{t}}_\nu{}^\mu$. In a manner similar to that stated above, (2·29) turns out

$$\left.\begin{aligned}&(b_l{}^\mu \boldsymbol{T}_l{}^{(f)})_{,\mu} = 0\,, \\ &\boldsymbol{T}_l{}^{(f)} = f_{(k)} ({}^m\boldsymbol{T}_{kl} + \boldsymbol{t}_{kl}) - f_{(k),\nu} b_m{}^\nu \boldsymbol{F}_{klm}\,, \\ &(f_{(k)}(x) \text{ is an arbitrary function}).\end{aligned}\right\} \tag{2·41}$$

Hence there exist indenumerable conserved quantities

$$\boldsymbol{P}^{(f)} = \int d^3x b_l{}^0 \boldsymbol{T}_l{}^{(f)}, \tag{2·42}$$

where of course the conserved vector corresponding to (2·40) is included by a particular choice in $f_{(k)}(x)$:

$$\begin{aligned}&(f_{(k)}) = (f^l_{(k)}) = (\delta_k{}^l), \\ &\boldsymbol{P}_k = \int d^3x b_l{}^0 ({}^m\boldsymbol{T}_{kl} + \boldsymbol{t}_{kl})\,.\end{aligned} \tag{2·43}$$

A function $f_{(k)}$ needs not to be a vector under the four-dimensional rotation.

Before closing this section, we shall resolve (2·33) into the symmetric and skew-symmetric parts**)

*) $\tilde{\boldsymbol{t}}_{kl}$ corresponds to the famed "pseudo" energy-momentum tensor of the gravitational field which does not transform as a tensor under the general coordinate transformation.

**) Just recall that the symmetrized (not the canonical) energy-momentum tensor of the matter source is of physical significance.

$$\boldsymbol{B}_{(kl)} = {}^m\boldsymbol{T}_{(kl)}, \tag{2·44}$$

$$\boldsymbol{B}_{[kl]} = {}^m\boldsymbol{T}_{[kl]}, \tag{2·45}$$

and further attempt to convert them into the divergent forms similar to (2·37),

$$\left.\begin{aligned} &-\{bb_l{}^\mu b_m{}^\nu(3\alpha(c_{klm}-(3/2)i\varepsilon_{klmn}c_n{}^A)-2(\alpha-2\beta)\delta_{k[l}c^V_{m]})\}_{,\nu} \\ &\qquad = b_l{}^\mu({}^m\boldsymbol{T}_{(kl)}+\boldsymbol{t}^1_{kl})+2(\alpha+\beta)bb_l{}^\mu b^\nu_{[k}c^V_{l]},_\nu , \end{aligned}\right\} \tag{2·46}$$

$$\left.\begin{aligned} &-\{bb_l{}^\mu b_m{}^\nu(3/2)i(\alpha-(4/9)\gamma)\varepsilon_{klmn}c_n{}^A\}_{,\nu} \\ &\qquad = b_l{}^\mu({}^m\boldsymbol{T}_{[kl]}+\boldsymbol{t}^2_{kl})-2(\alpha+\beta)bb_l{}^\mu b^\nu_{[k}c^V_{l]},_\nu , \end{aligned}\right\} \tag{2·47}$$

with

$$\left.\begin{aligned} \boldsymbol{t}^2_{kl} &= \boldsymbol{t}_{kl}-\boldsymbol{t}^1_{kl} \\ &= b(3/4)i\{\alpha-(4/9)\gamma\}\varepsilon_{kmnj}c_{lmn}c_j{}^A, \end{aligned}\right\} \tag{2·48}$$

where $\boldsymbol{t}^2_{kl}$ does not contain an arbitrary constant β; it can be rewritten in terms of the irreducible tensor components,

$$\boldsymbol{t}^2_{kl} = (1/2)ib\{\alpha-(4/9)\gamma\}\{2\varepsilon_{kmnj}c^T_{lmn}c_j{}^A+\varepsilon_{klmj}c_m{}^Vc_j{}^A+3i(\delta_{kl}c_m{}^Ac_m{}^A-c_k{}^Ac_l{}^A)\}. \tag{2·48′}$$

In deriving (2·46) and (2·47) a special care is taken in order to eliminate the second derivate of the field variables from the definitions of $\boldsymbol{t}^1_{kl}$ and $\boldsymbol{t}^2_{kl}$, by dint of the useful identity

$$b_k{}^\mu c_{klm},_\mu + c_k{}^V c_{klm} = 2b^\mu_{[l}c^V_{m]},_\mu . \tag{2·49}$$

If we put

$$\alpha+\beta=0, \tag{2·50}$$

the above two equations (2·46) and (2·47) yield after the differentiation

$$\{b_l{}^\mu({}^m\boldsymbol{T}_{(kl)}+\boldsymbol{t}^1_{kl})\}_{,\mu}=0, \tag{2·51}$$

$$\{b_l{}^\mu({}^m\boldsymbol{T}_{[kl]}+\boldsymbol{t}^2_{kl})\}_{,\mu}=0. \tag{2·52}$$

§ 3. Linear approximation in spinor-vierbein interaction

The field equations proposed in § 2 are non-linear with respect to the field variables $b_k{}^\mu$. We know that a linear theory (Newton's theory) accounts, with a considerable degree of accuracy, for the motion of bodies under the gravitational forces. We shall discuss the interaction between the spinor field and the vierbein field $b_k{}^\mu$ by assuming both the difference

$$b_{k\mu}-\delta_{k\mu}=a_{k\mu}$$

and its first derivatives are so small as compared to unity that the quadratic terms in $a_k{}^\mu$ and/or its derivatives lead only to secondary effects and are hereafter neglected. In this linear approximation, all the Greek indices are replaced by the Latin indices as there remain no differences between them. From the orthogonal relations (2·19), it follows that

$$a_k{}^l = -a_{lk}\,,$$

and the various non-linear quantities pass into the linearized ones,

$$\left.\begin{aligned} c_{klm} &= 2a_{k[l'm]}\,,\\ c^T_{klm} &= a_{(kl)'m} - a_{(km'l)} - (1/3)\,(\delta_{kl}c_m{}^V - \delta_{m(k}c^V_{l)})\,,\\ c_k{}^V &= 2a_{l[l'k]}\,,\\ c_k{}^A &= (1/3)\,i\varepsilon_{klmn}a_{lm'n}\,. \end{aligned}\right\} \tag{3·1}$$

The field equations (2·44) and (2·45) become

$$\begin{aligned} &-3\alpha\Box a_{(kl)} + (2\alpha-\beta)\,(a_{(km)'ml} + a_{(lm)'mk})\\ &\quad + (\alpha-2\beta)\,\{\delta_{kl}(\Box a - a_{(mn)'mn}) - a_{\cdot kl}\} + (\alpha+\beta)\,(a_{[km]'ml} + a_{[lm]'mk})\\ &\quad + \delta\cdot\delta_{kl} = {}^mT_{(kl)}\,, \end{aligned} \tag{3·2}$$

$$\begin{aligned} &-\{\alpha-(4/9)\gamma\}\Box a_{[kl]} + \{\beta+(4/9)\gamma\}\,(a_{[lm]'mk} - a_{[km]'ml})\\ &\quad - (\alpha+\beta)\,(a_{(lm)'mk} - a_{(km)'ml}) = {}^mT_{[kl]} \end{aligned} \tag{3·3}$$

with

$$a = a_{mm}\,, \qquad \Box = \partial_m\partial_m\,.$$

Provided that the relation (2·50) holds, these two equations are completely decomposed into the symmetric and anti-symmetric parts,

$$-3\alpha\{\Box a_{(kl)} + a_{\cdot kl} - a_{(km)'ml} - a_{(lm)'mk} - \delta_{kl}(\Box a - a_{(mn)'mn})\} + \delta\cdot\delta_{kl} = {}^mT_{(kl)}\,, \tag{3·4}$$

$$-\{\alpha-(4/9)\gamma\}\,(\Box a_{[kl]} + a_{[lm]'mk} - a_{[km]'ml}) = {}^mT_{[kl]}\,. \tag{3·5}$$

With the help of the convenient notations

$$\left.\begin{aligned} &\alpha = (3\kappa^2)^{-1}, \qquad \alpha-(4/9)\gamma = \lambda^{-2},\\ &a_{(kl)} - (1/2)\,\delta_{kl}a = \kappa S_{kl}\,,\\ &a_{[kl]} = \lambda A_{kl}\,, \end{aligned}\right\} \tag{3·6}$$

we shall be able to simplify the form of these equations to some extent,

$$\Box S_{kl} - 2S_{(km'ml)} + \delta_{kl}(S_{mn'mn} - \kappa\delta) = -\kappa\,{}^mT_{(kl)}\,,^{*)} \tag{3·7}$$

$$\Box A_{kl} - 2A_{[km'ml]} = -\kappa\,{}^mT_{[kl]}\,. \tag{3·8}$$

Further by imposing the generalized Lorentz conditions on S_{kl} and A_{kl}

*) $\kappa\delta$ corresponds to the cosmological constant, hence it will be neglected hereinafter.

$$\left.\begin{aligned} S_{km'm}&=0\,, \\ A_{km'm}&=0\,, \end{aligned}\right\} \tag{3·9}$$

we obtain

$$\Box S_{kl} = -\kappa^{m}T_{(kl)}\,, \tag{3·10}$$

$$\Box A_{kl} = -\lambda^{m}T_{[kl]}\,. \tag{3·11}$$

On the other hand, we choose the Lagrangian of a spinor field, say an electron, as a matter field,

$$L_0 = (1/2)\,(\bar{\psi}\gamma_k\psi_{,k} - \bar{\psi}_{,k}\gamma_k\psi) + m\bar{\psi}\psi\,, \tag{3·12}$$

hence we obtain the invariant substitute of it by the prescription stated in § 2,

$$L' = (1/2)\,b_k{}^{\mu}(\bar{\psi}\gamma_k\psi_{,\mu} - \bar{\psi}_{,\mu}\gamma_k\psi) + m\bar{\psi}\psi\,. \tag{3·13}$$

The equation of motion derived from $\boldsymbol{L}' = bL'$ is

$$b_k{}^{\mu}\gamma_k\psi_{,\mu} + (1/2)\,c_k{}^{V}\gamma_k\psi + m\psi = 0\,, \tag{3·14}$$

which is reduced by the linear approximation (3·1) and the condition of divergence-free (3·9) to

$$\gamma_k\{\partial_k - \kappa(S_{kl} - (1/2)\,\delta_{kl}S)\,\partial_l + \lambda A_{kl}\partial_l - (1/4)\,\kappa S_{,k}\}\,\psi + m\psi = 0\,, \tag{3·14′}$$

$$(S_{mm} = S)\,.$$

On multiplying the dual operator, we obtain the differential equation of second order,

$$\begin{aligned} &\{(1+\kappa S)\Box - 2\kappa S_{kl}\partial_k\partial_l - (1/4)\,\kappa(\Box S)\}\,\psi \\ &\quad + i\,\{\kappa(S_{km',l}\partial_m - (1/2)\,S_{,l}\partial_k) + \lambda A_{lm',k}\partial_m\}\,\sigma_{kl}\psi - m^2\psi = 0\,, \end{aligned} \tag{3·15}$$

where

$$\gamma_k\gamma_l = \delta_{kl} + i\sigma_{kl}\,.$$

Now let us proceed to the non-relativistic limit of (3·10), (3·11) and (3·15);

$$\left.\begin{aligned} &\Box S_{00} = -\kappa^{m}T_{00} = -\kappa\rho\,, \\ &\Box S_{a0} = 0 = \Box S_{ab}\,, \qquad (a, b = 1, 2, 3)\,, \\ &\Box A_{kl} = 0\,, \end{aligned}\right\} \tag{3·16}$$

$$E\psi = \{p^2/(2m) - (\kappa m/2)\,S_{00} + (\kappa^2/8m)^{m}T_{00}\}\,\psi, \tag{3·17}$$

where the well-known relations

$$i\partial_t\psi = (m+E)\,\psi\,, \qquad -i\partial_a\psi = p_a\psi\,, \qquad (a = 1, 2, 3)$$

---

*) One is permitted to set $S_{kl} = 0 = A_{kl}$ except for S_{00} only, as there are no components of sources for them.

are used and ρ denotes the density of matter sources.

Equation (3·17) turns out,

$$E\psi = \{p^2/(2m) - (\kappa m/2) S_{00}\} \psi \,, \tag{3·18}$$

where we neglected the last term containing κ^2.

Upon comparing the potential term in this Schroedinger equation with the Newtonian potential of gravitation φ, we find that

$$\varphi = -(\kappa/2) S_{00} = a_{44} \,. \tag{3·19}$$

If the field does not change quickly with time, i.e. S_{00} is almost static, it follows from (3·16) and (3·19)

$$\Delta\varphi = -(\kappa/2)\Delta S_{00} = (\kappa^2/2)\rho \,,$$

with which the Newtonian equation

$$\Delta\varphi = 4\pi k\rho$$

is to be compared. In this manner we are able to determine the coupling constant of the symmetric field,

$$\left.\begin{array}{l} \kappa = \sqrt{8\pi k} \,, \\ (k = 1.06 \times 10^{-9} g^{-2} = 5.2 \times 10^{-67}\ \mathrm{cm}^2 \quad \text{in the natural units}). \end{array}\right\} \tag{3·20}$$

We close this section by remarking that the coupling constant of the anti-symmetric field A_{kl} cannot be determined as there seems no source of skew-symmetric energy-momentum tensors in the classical experiments.

§ 4. Comparison with Einstein's theory

In this section we shall compare the symmetric part of our equation (2·44) with Einstein's equation of gravitation, by defining the symmetric metric tensor by

$$g^{\mu\nu} = b_k{}^\mu b_k{}^\nu \,, \qquad g_{\mu\nu} = b_{k\mu} b_{k\nu} \,. \tag{4·1}$$

The Christoffel three-index symbol of the second kind is given by

$$\Gamma^\lambda_{\mu\nu} = b_l{}^\lambda \{ b_{k(\mu} b_{m\nu)} c_{klm} + b_{l(\mu,\nu)} \} \,. \tag{4·2}$$

The Einstein equation takes the form

$$G_{\mu\nu} = R_{\mu\nu} - (1/2) R g_{\mu\nu} + \kappa^2 g_{\mu\nu} \delta = \kappa^{2m} T_{(\mu\nu)} \,,$$

where

$$R_{\mu\nu} = 2\{\Gamma^\lambda_{\mu[\nu,\lambda]} - \Gamma^\rho_{\mu[\lambda} \Gamma^\lambda_{\nu]\rho}\} \,,$$

$$R = g^{\mu\nu} R_{\mu\nu} \,.$$

Transition from the Greek indices into the Latin ones (as already shown in (2·36), for example) yields

$$G_{kl} = b_k{}^{\mu} b_l{}^{\nu} G_{\mu\nu} = \kappa^{2m} T_{(kl)} \,, \tag{4·3}$$

with which our symmetric equation (2·44) (devided by b)

$$\kappa^2 B_{(kl)} = \kappa^{2m} T_{(kl)} \tag{4·4}$$

should be compared. By the use of the previous notation (3·7) provided that the relation (2·50) holds, the difference of these two equations is represented by

$$\kappa^2 B_{(kl)} - G_{kl} = -(4\lambda^2)^{-1}(-8i\varepsilon_{(kmni} c_{l)mn} c_i{}^A + 6c_k{}^A c_k{}^A + 3\delta_{kl} c_m{}^A c_m{}^A) \,. \tag{4·5}$$

It should be noticed that each term of the difference contains an axial vector $c_k{}^A$. Now we have to check whether it vanishes or not.

Apart from the linear approximation where the above difference vanishes exactly, there has been the well-known solution of Schwarzschild in the spherically symmetric field. In particular, we consider its static case: the field variables do not depend on time and a mass point is situated at the origin under the influence of the spherically symmetric forces. In this situation, the components of $b_{k\mu}$ transform according to the laws

$$\left.\begin{aligned} &\delta b_{a\alpha} = \omega_{ab} b_{b\alpha} + \omega_{\alpha\beta} b_{a\beta} \,, \\ &\delta b_{a4} = \omega_{ab} b_{b4} \,, \\ &\delta b_{4\alpha} = \omega_{\alpha\beta} b_{4\beta} \,, \\ &\delta b_{44} = 0 \,, \quad (a, b, \alpha, \beta = 1, 2, 3) \,, \end{aligned}\right\} \tag{4·6}$$

under the three-dimensional rotation

$$\left.\begin{aligned} &\delta x^{\alpha} = \omega_{\alpha\beta} x^{\beta}, \quad (\omega_{(\alpha\beta)} = 0) \,, \\ &\delta x^4 = 0 \,. \end{aligned}\right. \tag{4·7}$$

It is easy to construct the general form of $b_{k\mu}$ such that it has the required transformation properties mentioned above.

$$\left.\begin{aligned} &b_{a\alpha} = \delta_{a\alpha}(1 + A) + BX_a X_\alpha \,, \\ &b_{a4} = iCX_a \,, \\ &b_{4\alpha} = iDX_\alpha \,, \\ &b_{44} = 1 + E \,, \end{aligned}\right\} \tag{4·8}$$

where

$$X_a = x^a / r \,, \qquad r^2 = x^2 \,,$$

and A, B, $\cdots$, E are functions of r only.

A simple calculation yields by making use of these general forms,

$$c_k{}^A = 0 \,. \tag{4·9}$$

Consequently, the equivalence between the solution of Schwarzschild and ours is established.

§ 5. Elimination of the antisymmetric field

We shall discuss various choices in the arbitray constants introduced in the free Lagrangian L^G for the $b_k{}^\mu$ field (2·32), except for a trivial case, $\alpha=\beta=\gamma=0$. In particular we lay emphasis on the possible vanishing of the skew-symmetric energy-momentum tensor generated by a matter field about which no definite statements have been given in the previous sections. The well-known procedure of symmetrizing energy-momentum tensors by means of adding the canonical spin angular-momentum to it cannot be applied to the present case.

To be specific, we shall base our arguments on the spinor Lagrangian (3·12). As it is invariant under the translation in the usual sense (see (2·4)), there are the conservation laws of the symmetric and antisymmetric energy-momentum tensors separately,

$$T_{(kl)\cdot l}=0\,, \tag{5·1}$$

$$T_{[kl]\cdot l}=0\,, \tag{5·2}$$

where

$$T_{lk}=(1/2)\,(\bar{\psi}\gamma_k\psi_{\cdot l}-\bar{\psi}_{\cdot l}\gamma_k\psi)\,.$$

As stated at the end of § 2, we could obtain the separate conservation laws (2·51) and (2·52) closely similar to the above ones, provided that α and β satisfy the relation given by (2·50). Now we shall assume it, although we cannot find any *a priori* reason to require such separate conservation laws in an extended sense; however, the result obtained in § 4 seems to support our assumption. Various cases are investigated in order.

Case 1: $\alpha+\beta=0$.

The field equations for the $b_k{}^\mu$ field become

$$-\{b_l{}^\mu b_m{}^\nu b3\alpha[c_{klm}-(3/2)\,i\varepsilon_{klmn}c_n{}^A-2\delta_{k[l}c^V_{m]}]\}_{\cdot\nu}=b_l{}^\mu({}^m\boldsymbol{T}_{(kl)}+\boldsymbol{t}^1_{kl})\,, \tag{5·3}$$

$$-\{b_l{}^\mu b_m{}^\nu b(3/2)\,i[\alpha-(4/9)\gamma]\varepsilon_{klmn}c_n{}^A\}_{\cdot\nu}=b_l{}^\mu({}^m\boldsymbol{T}_{[kl]}+\boldsymbol{t}^2_{kl})\,. \tag{5·4}$$

As is easily shown, it is impossible to make both sides of Eq. (5·4) vanish identically without further conditions.

Case 2: $\alpha+\beta=0$ and $\alpha-(4/9)\gamma=0$.

We find

$$\boldsymbol{t}^2_{kl}=0\,, \qquad \boldsymbol{t}^1_{kl}=\boldsymbol{t}_{kl} \tag{5·5}$$

and

$$\boldsymbol{B}_{[kl]}=0\,. \tag{5·6}$$

However, the skew-symmetric part of the energy-momentum tensor ${}^m\boldsymbol{T}_{[kl]}$ cannot vanish identically so long as the matter field exists. It is interesting to check whether its derivatives should vanish or not. By making use of the equation of motion (3·14) and the generalized Klein-Gordon equation which is derived from (3·14) by multiplying a proper dual operator,

$$(\square^G - m^2)\psi = \{i\sigma_{mn}D_mD_n + c_m{}^VD_m + (1/2)\,b_m{}^\mu c^V_{m,\mu} + (i/2)\,\sigma_{mn}b_m{}^\mu c^V_{n,\mu} + (1/4)\,c_m{}^Vc_m{}^V\}\psi,^{*)}$$

$$\square^G = D_mD_m\,, \tag{5·7}$$

we observe the connection between the energy-momentum tensor and its derivative as follows

$${}^m\boldsymbol{T}_{lk} = (1/2)\,bb_l{}^\mu(\bar\psi\gamma_k\psi_{,\mu} - \bar\psi_{,\mu}\gamma_k\psi) = b^mT_{lk}\,, \tag{5·8}$$

$$(b_l{}^{\mu m}\boldsymbol{T}_{kl})_{,\mu} = -c_{mkl}{}^m\boldsymbol{T}_{ml}\,, \tag{5·8'}$$

$$(b_l{}^{\mu m}\boldsymbol{T}_{lk})_{,\mu} = -b(1/4)\,\varepsilon_{klmn}c_{jmn}(\bar\psi\gamma_5\gamma_l\psi_{,j} - \bar\psi_{,j}\gamma_5\gamma_l\psi) + (1/2)\,(bb_m{}^\mu c_{mkl}\bar\psi\gamma_l\psi)_{,\mu}\,, \tag{5·9}$$

$$(b_l{}^\mu \boldsymbol{t}_{kl})_{,\mu} = -c_{mkl}\boldsymbol{t}_{ml} + 2b_k{}^\nu b_{n[\mu,\nu]}(b_l{}^\mu b_m{}^\lambda \boldsymbol{F}_{nlm})_{,\lambda}\,. \tag{5·10}$$

In fact, the sum of (5·8) and (5·10) vanishes if the field equation for the $b_k{}^\mu$ field is employed. Thus, the derivative of the skew-symmetric energy-momentum tensor cannot be made zero with any choices in the free parameters.

Case 3: $\alpha+\beta=0$, $\alpha-(4/9)\gamma=0$ and we add an axial-vector interaction to the matter field.

First, we consider the local homogeneous Lorentz transformation (compare with (2·30)),

$$\left.\begin{aligned} &\delta x^\mu = 0\,,\\ &\delta\psi = (i/4)\,\omega_{kl}(x)\,\sigma_{kl}\psi\,, \quad (\omega_{(kl)}(x)=0),\\ &\delta\bar\psi = -(i/4)\,\omega_{kl}(x)\,\bar\psi\sigma_{kl}\,. \end{aligned}\right\} \tag{5·11 **)}$$

Under this transformation, the modified Lagrangian L' (3·13) is no longer invariant. According to our recipe we should introduce new fields in order to make the theory invariant.***) Instead of introducing new fields, however, we shall here add some interaction terms consisting of the gravitational field strength. There are the vector, the axial-vector and the tensor couplings constructed in terms of $c_k{}^V$, $c_k{}^A$ and c^T_{klm} in the form of tri-linear interactions. On examining the respective transformation properties, we find a promising one, that is, an

---

*) Its linearized form is given by (3·15). It should be noticed that the covariant derivative does not commute each other, yielding the invariant field strength, $[D_k, D_l]=c_{mkl}D_m$.

**) $\omega_{kl}(x)$ and its first derivative are assumed to vanish on the boundary surface of the integration domain.

***) Detailed discussion of it will be made in a forthcoming paper.

axial-vector coupling,

$$L^A = -(3i/4)c_k{}^A\bar{\psi}\gamma_5\gamma_k\psi\,, \tag{5·12}$$

and the modified Lagrangian

$$L' + L^A \tag{5·13}$$

remains invariant under the extended four-dimensional rotation (and of course under the extended translation). From (5·13), the equation of motion is replaced by

$$b_k{}^\mu\gamma_k\psi_{\cdot\mu} + (1/2)c_k{}^V\gamma_k\psi - (3i/4)c_k{}^A\gamma_5\gamma_k\psi + m\psi = 0\,. \tag{5·14}$$

It should be noticed, on the other hand, that the action integral

$$I(\Sigma) = \int_\Sigma d^4x\boldsymbol{L}^G$$

is kept invariant even under (5·11), because of the particular choice in the arbitrary coefficient (see (5·6)),

$$\delta I = \int_\Sigma \boldsymbol{B}_{kl}\omega_{kl}d^4x = \int_\Sigma \boldsymbol{B}_{(kl)}\omega_{kl}d^4x = 0\,.$$

Now the total Lagrangian density becomes

$$\boldsymbol{L}' + \boldsymbol{L}^A + \boldsymbol{L}^G.$$

Let us calculate the contribution from L^A to the energy-momentum tensor ${}^mT_{kl}$ (5·8); it is denoted by T^A_{kl},

$$\begin{aligned}T^A_{kl} = (1/4)\,[\,(1/2)\,\varepsilon_{kijm}c_{lij} - \varepsilon_{lijm}c_{ijk} - \varepsilon_{kljm}c_j{}^V]\,\bar{\psi}\gamma_5\gamma_m\psi \\ - (1/4)\,\varepsilon_{kljm}b_j{}^\nu[\bar{\psi}_{\cdot\nu}\gamma_5\gamma_m\psi + \bar{\psi}\gamma_5\gamma_m\psi_{\cdot\nu}]\,.\end{aligned} \tag{5·15}$$

The sum of the anti-symmetric parts

$$\begin{aligned}{}^mT_{[kl]} + T^A_{[kl]} = {} & (1/2)\,b^\mu_{[k}(\bar{\psi}\gamma_{l]}\psi_{\cdot\mu} - \bar{\psi}_{\cdot\mu}\gamma_{l]}\psi) - (3i/4)\,c^A_{[k}\bar{\psi}\gamma_5\gamma_{l]}\psi \\ & - (1/4)\,\{c_m{}^V\bar{\psi}\gamma_k\gamma_l\gamma_m\psi + b_m{}^\mu(\bar{\psi}\gamma_k\gamma_l\gamma_m\psi_{\cdot\mu} + \bar{\psi}_{\cdot\mu}\gamma_k\gamma_l\gamma_m\psi) \\ & - \delta_{kl}[c_m{}^V\bar{\psi}\gamma_m\psi + b_m{}^\mu(\bar{\psi}\gamma_m\psi_{\cdot\mu} + \bar{\psi}_{\cdot\mu}\gamma_m\psi)\,]\} + (1/2)\,c^V_{[k}\bar{\psi}\gamma_{l]}\psi \\ & + (1/2)\,b^\mu_{[k}(\bar{\psi}\gamma_{l]}\psi_{\cdot\mu} + \bar{\psi}_{\cdot\mu}\gamma_{l]}\psi)\end{aligned}$$

is shown to vanish after some algebra with the aid of the equation of motion (5·14) and the similar one for $\bar{\psi}$. Consequently, we have managed to symmetrize the energy-momentum tensor of the matter field; in this case the additional interaction Lagrangian of an axial-vector type plays the similar role to that of the canonical spin angular momentum played in a conventional manner for a flat-space. The additional interaction gives rise to a spin interaction with a spinor field, which is explicitly observed in the non-relativistic limit.

§ 6. Discussion

We have discussed the extension of the translation group and introduced the new field including the gravitational field in a manner quite analogous to the electromagnetic case. It should be stressed that the geometrical interpretation in terms of a Riemannian space may be given if desired and if necessary.

What differs from the electromagnetic field lies in the fact that our field strength is decomposed into the three irreducible parts. The free Lagrangian for the new fields is constructed with the four arbitrary constants.*)

In performing the linear approximation to the complicated non-linear field equations and making comparison between the Einstein's equation of gravitation and ours, we have assumed one relation among the arbitrary constants. This assumption is equivalent to the requirement that there should exist the conservation laws of the energy-momentum tensors quite similar to those implied by the translational invariance in a narrow sense for a free matter field. The relation assumed above has enabled us 1) to decompose the linearized field equations into one for the symmetric field variable and the other for the antisymmetric field variable, respectively, and 2) to identify the symmetric solution of the non-linear field equations with Schwarzschild's solution in the spherical symmetry.

We have been incapable of making any definite and conclusive statement concerning the antisymmetric part of the field equations, except for the proposed prescription to eliminate it as a redundant field by adding the tri-linear interaction Lagrangian of the axial-vector type besides the particular choice in the free parameters. In this case the total Lagrangian keeps invariant even under the extended four-dimensional rotation.

In a forthcoming paper, the extension of the homogeneous Lorentz group and its detailed consequences will be reported, where it is emphatically aimed to introduce a massive gauge field in an invariant-theoretic way.

References

1) For example, C. N. Yang and R. L. Mills, Phys. Rev. **96** (1954), 191.
2) See, for example, Kōdi Husimi, Lecture on Variational Principle in Field Theory in 1943 at the Meeting of Japan Physical Society (edited by the Japan Ministry of Education, 1944). Earlier references are listed in detail therein.
3) R. Utiyama, Phys. Rev. **101** (1956), 1597.
4) T. W. B. Kibble, J. Math. Phys. **2** (1961), 212.
D. W. Sciama, *Festschrift for Infeld* (Pergamon Press, New York, 1961), p. 415. We are indebted to Dr. Kei Semba for informing us of this paper.

*) The field strength of the electromagnetic field is irreducible for itself and there remains only one free parameter in producing the free electromagnetic Lagrangian.

612

Progress of Theoretical Physics, Vol. 45, No. 2, February 1971

Gravitational Field as a Generalized Gauge Field

Ryoyu UTIYAMA and Takeshi FUKUYAMA

Department of Physics, Osaka University, Osaka

(Received August 17, 1970)

It is shown that a symmetric tensor field of the second rank $A_{\mu\nu}(x)$ should be introduced in order to retain the invariance of the action-integral under a generalized translation $x^\mu \to x^\mu + \xi^\mu(x)$, provided that the original action-integral is invariant under inhomogeneous Lorentz transformations. It is further proved that the generalized gauge field $A_{\mu\nu}$ should appear in the Lagrangian in exactly the same fashion as the metric tensor $g_{\mu\nu}$ does in Einstein's theory of gravitation.

Some general feature is also discussed with respect to a law of conservation of some physical quantity which becomes no longer valid when the interaction with the generalized gauge field takes place, provided that the associated group is non-Abelian.

§ 1. Introduction

A gravitational field was first interpreted as a kind of generalized gauge fields by one of the present authors[1] by introducing a system of tetrads $h_a{}^\mu(x)$ and extending the Lorentz transformation of the tetrads at each world point to a larger group depending upon six arbitrary functions of x instead of six parameters. Besides this article, some authors[2),3)] tried to introduce a gravitational field by extending the translation group to a general transformation of coordinates

$$x^\mu \to x^\mu + \xi^\mu(x),$$

but their arguments seem rather unsatisfactory and complicated.

Many groups of transformations depending on parameters have been found in connection with the different kinds of conservation laws. Among these groups it is well known that the group of phase-transformations of complex fields was extended to the gauge transformation depending on an arbitrary scalar function $\lambda(x)$ connected with the existence of an electromagnetic field. The invariance under rotations in the iso-spin space was extended to the invariance under a generalized rotation group by an adjoined introduction of the Yang-Mills field. The most well-known group, namely the translation group, has been conjectured to be related with the gravitational field because the gravitational field is, following Einstein's equation, produced by the energy-momentum tensor of material fields, the conservation of which holds owing to the invariance of the material system under a translation of coordinates. In spite of such a conjecture, however, there has not been any convincing article which shows the gravitational field being derivable from the postulate that the action integral of a matrial system is invariant under a group of generalized translations depending upon four arbitrary functions of x.

The aim of the present paper is to show that a tensor field of *the second rank* should be introduced in order to retain the invariance of the action-integral and that this tensor field should appear in the original Lagrangian of the material field in exactly the same way as the metric tensor $g_{\mu\nu}$ does in Einstein's theory of gravitation. This conclusion is derived from the assumption that the original Lagrangian is invariant under inhomogeneous Lorentz transformations but the invariance under rotations of tetrads has not been assumed.

In addition to the derivation of the gravitational field, some general feature is discussed with respect to the laws of conservation. It is shown that a physical quantity owned by some field, say $\phi_A(x)$, which is conserved owing to the invariance of the action-integral of ϕ_A under some parameter-group of transformations, becomes unable to satisfy the law of conservation when the original ϕ-field begins to interact with a generalized gauge field associated with the group mentioned above, provided that this group of transformations is non-Abelian. The conservation is recovered only when the quantity carried by the generalized gauge field is taken into account together with that possessed by the field ϕ_A.

The present procedure of introducing the interaction of a gravitational field with a material system might include its application in a derivation of an S-matrix for a material system interacting with a gravitational field if the Lorentz-invariant S-matrix is known for this matrial system without the gravitational interaction.

§ 2. Fundamental postulate

Consider a field $\phi_A(x)$ $(A=1, 2\cdots N)$ with a Lagrangian density

$$L(\phi_A, \phi_{A,\mu}), \qquad \phi_{A,\mu}=\partial\phi_A/\partial x^\mu.$$

Let us assume that the action-integral

$$I=\int L d^4x$$

is invariant under the following groups of transformations:

i) translation

$$x^\mu \to x^{\mu\prime}=x^\mu+a^\mu, \qquad (a^\mu=\text{constant parameter})$$

$$\delta\phi_A=\phi_A'(x')-\phi_A(x)=0,$$

ii) Lorentz transformation

$$x^\mu \to x^{\mu\prime}=x^\mu+\delta x^\mu=x^\mu+\epsilon^\mu{}_\nu\cdot x^\nu,$$

$$\delta\phi_A=\phi_B\cdot C^{B\mu}_{A\nu}\cdot\frac{\partial\delta x^\nu}{\partial x^\mu}=\phi_B\cdot C^{B\mu}_{A\nu}\cdot\epsilon^\nu{}_\mu,$$

where $\epsilon^\mu{}_\nu$ is an infinitesimal parameter and is restricted by the condition

$$\epsilon_{\mu\nu}=\eta_{\mu\rho}\cdot\epsilon^{\rho}{}_{\nu}=-\epsilon_{\nu\mu}\,,$$

$$\eta_{\mu\nu}=\eta^{\mu\nu}=\begin{cases}1 & \mu=\nu=1,2,3\,,\\ 0 & \mu\neq\nu\,,\\ -1 & \mu=\nu=0\,.\end{cases}$$

Here it has been assumed that the field ϕ_A is a kind of tensor and the transformation-coefficient $C^{B\mu}_{A\nu}$ is an appropriate sum of products of Kronecker's δ.

Our fundamental postulate is that the action integral should be invariant under the generalized translations which is a generalization of (i) and (ii) depending upon four infinitesimal arbitrary functions of x, in place of four parameters a^{μ}, i.e.

$$x^{\mu}\rightarrow x^{\mu\prime}=x^{\mu}+\xi^{\mu}(x). \tag{2·1}$$

In order to realize this postulate, the original arguments of the Lagrangian, for example $\partial_{\lambda}\phi_A$, should be replaced with an appropriately defined "covariant derivative" $\nabla_{\lambda}\phi_A$ by introducing a new field $A_K(x)$.

Let a Lagrangian

$$L_1(\phi_A,\phi_{A,\lambda},A_K,A_{K,\lambda})$$

be a substitute for the original one. Since the parameter a^{μ} in (i) behaves as a vector under Lorentz transformations, it is plausible to assume that the new field A_K is a covariant tensor of the r-th rank, following the conclusion of the theory of generalized gauge fields.[4] Thus the field $A_{\mu_1\cdots\mu_r}(x)$ has to be transformed under the transformation (2·1) in the following way:

$$\delta A_{\mu_1\cdots\mu_r}(x)=A_{\nu_1\cdots\nu_r}\cdot D\binom{\nu_1\cdots\nu_r}{\mu_1\cdots\mu_r}{}^{\nu}_{\mu}\cdot\xi^{\mu}{}_{,\nu}\,, \tag{2·2}$$

where the transformation coefficient D is

$$D\binom{\nu_1\cdots\nu_r}{\mu_1\cdots\mu_r}{}^{\nu}_{\mu}=-\sum_{i=1}^{r}\delta^{\nu_i}_{\mu}\cdot\delta^{\nu}_{\mu_i}\{\delta^{\nu_1}_{\mu_1}\cdots\delta^{\nu_{i-1}}_{\mu_{i-1}}\cdot\delta^{\nu_{i+1}}_{\mu_{i+1}}\cdots\delta^{\nu_r}_{\mu_r}\}. \tag{2·3}$$

The expression of $\delta\phi_A$ for the Lorentz transformation has the form

$$\delta\phi_A=\phi_B\cdot C^{B\,\nu}_{A\mu}\cdot\xi^{\mu}{}_{,\nu}$$

for the variation of ϕ_A under the transformation (2·1). This expression of $\delta\phi_A$ gives rise to not only terms being proportional to $\partial\xi/\partial x$ but also terms having $\partial^2\xi/\partial x\cdot\partial x$ in the variation of the action-integral. In order to cancel these terms, it is necessary (a) to let the Lagrangian L_1 depend upon $\partial A_K/\partial x$ in addition to A_K if A_K is a tensor as we have assumed, or (b) to change the definition (2·2) of δA_K in such a way that a term having $\partial^2\xi/\partial x\cdot\partial x$ appears in the definition of δA_K, provided that $\partial A_K/\partial x$ should not appear in L_1. The approach (b) means an abandonment of the tensor character of the A-field. In the present paper, however, we assume the A-field to be a tensor, and consequently L_1 depends on both A_K and $\partial A_K/\partial x$,

The postulate that the new action-integral

$$I_1=\int L_1 d^4x$$

should be invariant under (2·1) leads to the following various identities (see Appendix A):

$$\partial_\nu\{[L_1]^A\cdot C^{B\nu}_{A\rho}\cdot\phi_B+[L_1]^{a_1\cdots a_r}\cdot D\binom{b_1\cdots b_r}{a_1\cdots a_r}^\nu_\rho\cdot A_{b_1\cdots b_r}\}$$
$$+[L_1]^A\cdot\phi_{A,\rho}+[L_1]^{a_1\cdots a_r}\cdot A_{a_1\cdots a_r,\rho}\equiv 0\,, \tag{2·4}$$

$$\partial_\lambda\{[L_1]^A\cdot C^{B\lambda}_{A\rho}\cdot\phi_B+[L_1]^{a_1\cdots a_r}\cdot D\binom{b_1\cdots b_r}{a_1\cdots a_r}^\lambda_\rho\cdot A_{b_1\cdots b_r}-T^\lambda_{(1)\rho}\}\equiv 0\,, \tag{2·5}$$

$$[L_1]^A\cdot C^{B\mu}_{A\rho}\cdot\phi_B+[L_1]^{a_1\cdots a_r}\cdot D\binom{b_1\cdots b_r}{a_1\cdots a_r}^\mu_\rho\cdot A_{b_1\cdots b_r}$$
$$+\partial_\lambda\left\{\frac{\partial L_1}{\partial\phi_{A,\lambda}}\cdot C^{B\mu}_{A\rho}\cdot\phi_B+\frac{\partial L_1}{\partial A_{a_1\cdots a_r,\lambda}}\cdot D\binom{b_1\cdots b_r}{a_1\cdots a_r}^\mu_\rho\cdot A_{b_1\cdots b_r}\right\}\equiv T^\mu_{(1)\rho}\,, \tag{2·6}$$

$$\left\{\frac{\partial L_1}{\partial\phi_{A,\nu}}\cdot C^{B\mu}_{A\rho}\cdot\phi_B+\frac{\partial L_1}{\partial A_{a_1\cdots a_r,\nu}}\cdot D\binom{b_1\cdots b_r}{a_1\cdots a_r}^\mu_\rho\cdot A_{b_1\cdots b_r}\right\}$$
$$+\{\mu\text{ and }\nu\text{ interchanged}\}\equiv 0\,, \tag{2·7}$$

where the following abbreviations have been used:

$$[L_1]^A=\frac{\partial L_1}{\partial\phi_A}-\partial_\lambda\left(\frac{\partial L_1}{\partial\phi_{A,\lambda}}\right),$$

$$[L_1]^{a_1\cdots a_r}=\frac{\partial L_1}{\partial A_{a_1\cdots a_r}}-\partial_\lambda\left(\frac{\partial L_1}{\partial A_{a_1\cdots a_r,\lambda}}\right),$$

$$T^\mu_{(1)\rho}=\frac{\partial L_1}{\partial\phi_{A,\mu}}\phi_{A,\rho}+\frac{\partial L_1}{\partial A_{a_1\cdots a_r,\mu}}\cdot A_{a_1\cdots a_r,\rho}-\delta^\mu_\rho L_1\,.$$

§ 3. Determination of the type of gauge field $A_{(a_1\cdots a_r)}$

The identity (2·7) shows that $\phi_{A,\nu}$ and $A_{a_1\cdots a_r,\lambda}$ should be included in L_1 only through a particular linear combination $\nabla_\lambda\phi_A$ of the following type:

$$\nabla_\lambda\phi_A=\partial_\lambda\phi_A+\phi_B\cdot M^{Ba_1\cdots a_r\nu}_{A\lambda}(x)\cdot A_{a_1\cdots a_r,\nu}\,, \tag{3·1}$$

where the coefficient M is to be determined later and probably depends on x. $\nabla_\lambda\phi_A$ is called in what follows a "covariant derivative" of ϕ_A.

The Lagrangian L_1 can be rewritten in terms of $\nabla_\lambda\phi_A$ as

$$L_1(\phi_A,\phi_{A,\lambda},A,\partial A/\partial x)=L_2(\phi_A,\nabla_\lambda\phi_A,A)\,. \tag{3·2}$$

Making use of (3·1) and rewriting (2·7) in terms of L_2, we have

$$\frac{\partial L_2}{\partial\nabla_\lambda\phi_A}\cdot\phi_B\cdot[\{\delta^\nu_\lambda\cdot C^{B\mu}_{A\rho}+M^{Ba_1\cdots a_r\nu}_{A\lambda}\cdot D\binom{b_1\cdots b_r}{a_1\cdots a_r}^\mu_\rho\cdot A_{b_1\cdots b_r}\}$$
$$+\{\nu\text{ and }\mu\text{ interchanged}\}]\equiv 0\,.$$

This expression suggests that the coefficient M should be a linear combination of $C^{B\mu}_{A\rho}$:

$$M^{B a_1\cdots a_r\nu}_{A\lambda}(x)=C^{B\beta}_{A\alpha}\cdot Y^{a_1\cdots a_r\nu\alpha}_{\lambda\beta}(x)\,. \tag{3·3}$$

By substituting (3·3) for M's in the above identity, (2·7) gives

$$\begin{aligned}-2\delta^\alpha_\rho\cdot\delta^{(\nu\mu)}_{(\lambda\beta)}&\equiv-\delta^\alpha_\rho(\delta^\nu_\lambda\delta^\mu_\beta+\delta^\nu_\beta\delta^\mu_\lambda)\\&\equiv A_{b_1\cdots b_r}(x)\cdot[\{Y^{a_1\cdots a_r\nu\alpha}_{\lambda\beta}\cdot D\left(^{b_1\cdots b_r}_{a_1\cdots a_r}\right)^\mu_\rho\}+\{\mu\text{ and }\nu\text{ interchanged}\}]\,.\end{aligned} \tag{3·4}$$

Since the left-hand side of (3·4) is symmetric with respect to the subscripts λ and β, the same should also be true for the right-hand side. Thus we find*)

$$Y^{a_1\cdots a_r\nu\alpha}_{\lambda\beta}=Y^{a_1\cdots a_r\nu\alpha}_{\beta\lambda}=Y^{a_1\cdots a_r\nu\alpha}_{(\beta\lambda)}\,.$$

In what follows, for the sake of simplicity, let the discussion be restricted to the case that the field A is an irreducible tensor of the r-th rank with a full symmetry:

$$A_{(a_1\cdots a_r)}(x)\,.$$

The double contraction of (3·4) by putting $\nu=\lambda$ and $\mu=\beta$ and making use of the definition (2·3) gives a relation

$$20\delta^\alpha_\rho=2r\cdot Y^{(a_1\cdots a_{r-1}\cdot\mu)\nu\alpha}_{(\nu\mu)}(x)\cdot A_{(a_1\cdots a_{r-1}\rho)}(x)\,.$$

If we define $A^{(a_1\cdots a_r)}(x)$ by

$$A^{(a_1\cdots a_r)}(x)\propto Y^{(a_1\cdots a_{r-1}\mu)\nu a_r}_{(\mu\nu)}\,, \tag{3·5}$$

the above relation is written as

$$A^{(a_1\cdots a_{r-1}\alpha)}\cdot A_{(a_1\cdots a_{r-1}\rho)}\propto\delta^\alpha_\rho\,. \tag{3·6}$$

This result shows that r should be >1 otherwise (3·6) leads to a contradiction.

The relation (3·5) allows us to represent Y in terms of $A^{(a_1\cdots)}$ as

$$Y^{(a_1\cdots a_r)\nu\alpha}_{(\lambda\beta)}=A^{(b_1\cdots b_r)}(x)\cdot Z^{(a_1\cdots a_r)\nu\alpha}_{(b_1\cdots b_r)(\lambda\beta)}\,, \tag{3·7}$$

where the coefficient Z on the right-hand side is an appropriate sum of products of Kronecker's δ.

Substituting (3·7) for Y in (3·4) and making use of the definition (2·3) of D, we have an important relation

$$2\delta^\alpha_\rho\delta^{(\mu\nu)}_{(\lambda\beta)}\equiv r\cdot A^{(b_1\cdots b_r)}\cdot A_{(a_1\cdots a_{r-1}\rho)}\{Z^{(a_1\cdots a_{r-1}\mu)\nu\alpha}_{(b_1\cdots b_r)(\lambda\beta)}+Z^{(a_1\cdots a_{r-1}\nu)\mu\alpha}_{(b_1\cdots b_r)(\lambda\beta)}\}\,. \tag{3·8}$$

From our assumption that $A_{(a_1\cdots a_r)}$ and $A^{(b_1\cdots b_r)}$ are both fully symmetric with respect to their suffices, it is plausible to assume that the undetermined coefficient Z is also symmetric with respect to both superscripts $(a_1\cdots a_r)$ and subscripts $(b_1\cdots b_r)$, in addition to the symmetric pair of subscripts $(\lambda\beta)$. Since we have no information about the symmetry of Z with respect to the extra superscripts

*) $(\beta\lambda)$ means that Y is symmetric with respect to suffices inside the parentheses.

ν and α, Z having the above mentioned symmetry with respect to the suffices can be expressed in terms of the products of Kronecker's δ in the following way:

$$Z^{(a_1\cdots a_r)\nu\alpha}_{(b_1\cdots b_r)(\lambda\beta)}=a\cdot\delta^{(a_1\cdots a_r)}_{(b_1\cdots b_r)}\delta^{(\nu\alpha)}_{(\lambda\beta)}+b\sum_{i=1}^{r}[\delta^{(a_1\cdots a_r)}_{(b_1\cdots\lambda\cdots b_r)}\cdot\delta^{\nu}_{b_i}\delta^{\alpha}_{\beta}+\delta^{(a_1\cdots a_r)}_{(b_1\cdots\beta\cdots b_r)}\delta^{\nu}_{b_i}\delta^{\alpha}_{\lambda}]$$

$$+c\sum_{i=1}^{r}[\delta^{(a_1\cdots a_r)}_{(b_1\cdots\lambda\cdots b_r)}\cdot\delta^{\alpha}_{b_i}\delta^{\nu}_{\beta}+\delta^{(a_1\cdots a_r)}_{(b_1\cdots\beta\cdots b_r)}\delta^{\alpha}_{b_i}\delta^{\nu}_{\mu}]+d\sum_{i,j(i\neq j)}^{r}\delta^{(a_1\cdots a_r)}_{(b_1\cdots\lambda\cdots\beta\cdots b_r)}\delta^{(\nu\alpha)}_{(b_ib_j)}, \tag{3·9}$$

where a, b, c and d are undetermined constants. $\delta^{(\cdots)}_{(\cdots)}$ in (3·9) means

$$\delta^{(a_1\cdots a_r)}_{(b_1\cdots b_r)}=\frac{1}{r!}\sum_{(\mathrm{perm})}\delta^{a_1}_{b_i}\delta^{a_2}_{b_j}\cdots\delta^{a_r}_{b_k},$$

where the summation should be taken over all the permutations of $(b_1\cdots b_r)$.

Inserting the expression (3·9) into (3·8), and taking contractions of (3·8) with respect to many different pairs of suffices, we arrive at the result

$$r=2, \qquad a=b=0, \qquad c=-d=\tfrac{1}{4} \tag{3·10}$$

with the normalization

$$A^{(\mu\rho)}(x)\cdot A_{(\rho\nu)}(x)=\delta^{\mu}_{\nu}. \tag{3·11}$$

The details of the derivation of this result are given in Appendix B. (3·10) determines Z, Y and $\nabla_{\lambda}\phi_{A}$ as follows:

$$Z^{(a_1a_2)\nu\alpha}_{(b_1b_2)(\lambda\beta)}=\tfrac{1}{4}[\delta^{(a_1a_2)}_{(b_1\lambda)}\delta^{\alpha}_{b_2}\delta^{\nu}_{\beta}+\delta^{(a_1a_2)}_{(b_2\lambda)}\delta^{\alpha}_{b_1}\delta^{\nu}_{\beta}$$

$$+\delta^{(a_1a_2)}_{(b_1\beta)}\delta^{\alpha}_{b_2}\delta^{\nu}_{\lambda}+\delta^{(a_1a_2)}_{(b_2\beta)}\delta^{\alpha}_{b_1}\delta^{\nu}_{\lambda}-2\delta^{(a_1a_2)}_{(\lambda\beta)}\delta^{(\nu\alpha)}_{(b_1b_2)}], \tag{3·12}$$

$$Y^{(a_1a_2)\nu\alpha}_{(\lambda\beta)}(x)\cdot A_{(a_1a_2),\nu}=\tfrac{1}{2}A^{\alpha a}\{A_{a\lambda,\beta}+A_{\beta a,\lambda}-A_{\lambda\beta,a}\}=\Delta^{\alpha}_{\beta\lambda}(x), \tag{3·13}$$

$$\nabla_{\lambda}\phi_{A}=\partial_{\lambda}\phi_{A}+\phi_{B}\cdot C^{B\beta}_{A\alpha}\cdot\Delta^{\alpha}_{\beta\lambda}. \tag{3·14}$$

$\Delta^{\alpha}_{\beta\lambda}$ in (3·13) is nothing but the Christoffel's $\Gamma^{\alpha}_{\beta\lambda}$ provided that our $A_{\mu\nu}$ is identified with the metric tensor $g_{\mu\nu}$.

§ 4. Derivation of Lagrangian

Before beginning a discussion about (2·6), let us consider a little extension of the Lorentz-invariance of the original Lagrangian L.

The Lorentz-invariance of the original Lagrangian can be made manifest by writing explicitly the metric tensor $\eta_{\mu\nu}$. It is easily seen, however, that this manifestly invariant expression of the action-integral can also be invariant under an affine transformation*)

$$x^{\mu}\to x^{\mu\prime}=a^{\mu}{}_{\nu}\cdot x^{\nu}+a^{\mu}, \qquad \det(a^{\mu}{}_{\nu})\neq 0, \tag{4·1}$$

*) In the case of an affine transformation, since the coefficient $a^{\mu}{}_{\nu}$ has no such restriction as $\eta_{\mu\nu}a^{\mu}{}_{\alpha}\cdot a^{\nu}{}_{\beta}=\eta_{\alpha\beta}$, there cannot exist such a covariant affine tensor as $\eta_{\mu\nu}$ whose components are kept unchanged under the transformation (4·1).

if the metric tensor $\eta_{\mu\nu}$ in L is replaced with a constant covariant affine tensor $\zeta_{\mu\nu}$ and at the same time, if L is replaced with

$$\boldsymbol{L}(\phi_A, \phi_{A,\lambda}, \zeta_{\mu\nu}) = \sqrt{-\zeta}\cdot L(\phi_A, \phi_{A,\lambda}, \zeta_{\mu\nu}),$$

$$\zeta = \det(\zeta_{\mu\nu}). \tag{4·2}$$

The invariance of $I = \int \boldsymbol{L} d^4x$ under an infinitesimal affine transformation

$$x^\mu \to x^{\mu\prime} = x^\mu + \Delta^\mu{}_\nu \cdot x^\nu + a^\mu$$

gives the following identities:

$$[\boldsymbol{L}]^A \cdot \phi_{A,\rho} = -\partial_\lambda \boldsymbol{T}^\lambda{}_\rho \tag{4·3}$$

and

$$[\boldsymbol{L}]^A \cdot \phi_B \cdot C^{B\mu}_{A\nu} - [\boldsymbol{L}]^A \cdot \phi_{A,\nu} \cdot x^\mu$$
$$-2\frac{\partial \boldsymbol{L}}{\partial \zeta_{\mu\rho}} \cdot \zeta_{\nu\rho} + \partial_\lambda \left[\frac{\partial \boldsymbol{L}}{\partial \phi_{A,\lambda}} \cdot \phi_B \cdot C^{B\mu}_{A\nu} - \boldsymbol{T}^\lambda{}_\nu \cdot x^\mu \right] \equiv 0, \tag{4·4}$$

where

$$\boldsymbol{T}^\lambda{}_\nu = \frac{\partial \boldsymbol{L}}{\partial \phi_{A,\lambda}} \cdot \phi_{A,\nu} - \delta^\lambda_\nu \cdot \boldsymbol{L}.$$

Making use of (4·3), we can rewrite (4·4) as

$$\frac{\partial \boldsymbol{L}}{\partial \phi_A} \cdot C^{B\mu}_{A\nu} \cdot \phi_B + \frac{\partial \boldsymbol{L}}{\partial \phi_{A,\lambda}} \phi_{B,\lambda} \cdot C^{B\mu}_{A\nu} - 2\frac{\partial \boldsymbol{L}}{\partial \zeta_{\mu\rho}} \cdot \zeta_{\nu\rho} \equiv \boldsymbol{T}^\mu{}_\nu. \tag{4·5}$$

Now let us return to a discussion about (2·6) which is rewritten in terms of L_2 defined by (3·2) as follows:

$$\frac{\partial L_2}{\partial \phi_A} \cdot \phi_B \cdot C^{B\mu}_{A\rho} - 2\frac{\partial L_2}{\partial A_{\mu\nu}} A_{\rho\nu} + \frac{\partial L_2}{\partial \nabla_\lambda \phi_A} \cdot C^{B\mu}_{A\rho} \cdot \nabla_\lambda \phi_B$$
$$-\left(\frac{\partial L_2}{\partial \nabla_\mu \phi_A} \cdot \nabla_\rho \phi_A - \delta^\mu_\rho L_2 \right) \equiv \frac{\partial L_2}{\partial (\nabla_\lambda \phi_A)} \cdot \{\cdots\}, \tag{4·6}$$

where the following relations have been employed:

$$\frac{\partial L_1}{\partial \phi_A} = \frac{\partial L_2}{\partial \phi_A} + \frac{\partial L_2}{\partial \nabla_\lambda \phi_A} \cdot C^{A\alpha}_{B\beta} \cdot \Delta^\beta_{\alpha\lambda},$$

$$\frac{\partial L_1}{\partial A_{ab}} = \frac{\partial L_2}{\partial A_{ab}} - \frac{\partial L_2}{\partial \nabla_\lambda \phi_A} \cdot \phi_B \cdot C^{B\alpha}_{A\beta} \cdot A^{ma} A^{nb} \cdot Z^{(rs)\nu\beta}_{(mn)(\lambda\alpha)} \cdot A_{rs,\nu}.$$

The expression $\{\cdots\}$ on the right-hand side of (4·6) vanishes owing to the commutation relation

$$[\boldsymbol{C}^\alpha_\beta, \boldsymbol{C}^\mu_\rho] = \delta^\mu_\beta \cdot \boldsymbol{C}^\alpha_\rho - \delta^\alpha_\rho \cdot \boldsymbol{C}^\mu_\beta, \tag{4·7}$$

where $\boldsymbol{C}^\alpha_\beta$ is a $N \times N$ matrix whose (A, B) element is $(A|\boldsymbol{C}^\alpha_\beta|B) \equiv C^{A\alpha}_{B\beta}$. The commutation relation (4·7) holds owing to the fact that the matrix $\boldsymbol{C}^\alpha_\beta$ is a representation of the N-th order of generators of the group of affine transformations and, in fact, we can easily derive the relation (4·7) by considering a simple example.

Comparing (4·5) with (4·6) and remembering that the right-hand side of (4·6) identically vanishes, we are led to the conclusion that L_2 should have the same functional form as $\boldsymbol{L}$, namely

$$L_2 \equiv \sqrt{|A|}\cdot L(\phi_A, \nabla_\lambda \phi_A, A_{\mu\nu})$$
$$= \boldsymbol{L}(\phi_A, \nabla_\lambda \phi_A, A_{\mu\nu}),$$

where

$$A = \det(A_{\mu\nu}).$$

The discussion so far developed neither compelles us to interpret $A_{\mu\nu}$ as the gravitational field nor gives any information on the signature of $A_{\mu\nu}$, but in order to let the field equation of ϕ_A be hyperbolic, $A_{\mu\nu}$ should have the signature $-$, $+$, $+$, $+$. In place of our $A_{\mu\nu}$, one can consider a particular tensor field $B_{\mu\nu}$ which can be derived by introducing a system of curvilinear coordinates u^μ into the Minkowskian space, that is,

$$ds^2 = \eta_{\alpha\beta}\cdot dx^\alpha \cdot dx^\beta = B_{\mu\nu}(u)\cdot du^\mu \cdot du^\nu,$$

$$B_{\mu\nu}(u) = \frac{\partial x^\alpha}{\partial u^\mu}\cdot\frac{\partial x^\beta}{\partial u^\nu}\cdot \eta_{\alpha\beta}\,.$$

The question whether our $A_{\mu\nu}$ is identical with the fictitious gravitational field $B_{\mu\nu}$ or is an entity being completely different from $B_{\mu\nu}$, giving a non-vanishing "curvature tensor",*) is to be answered by the field equation of $A_{\mu\nu}$. Thus if Einstein's equation is taken as a field equation, our $A_{\mu\nu}$ describes a permanent gravitational field produced by the material field ϕ_A.

§ 5. Law of conservation derived from identities (2·4) and (2·5)

The equation of the field $\phi_A(x)$

$$[L_1]^A = 0 \tag{5·1}$$

gives rise to an interesting relation when it is inserted into the identities (2·4) and (2·5).

By recalling the definition (2·3), the identity (2·5) becomes

$$\frac{\partial T^\lambda_{(1)\rho}}{\partial x^\lambda} = \frac{\partial}{\partial x^\lambda}\boldsymbol{S}^\lambda{}_\rho\,, \tag{5·2}$$

while (2·4) reads

*) The term "curvature tensor" means the Riemann-Christoffel's curvature tensor when $A_{\mu\nu}$ is substituted for the metric tensor $g_{\mu\nu}$ in the ordinary definition of the curvature tensor.

$$\frac{\partial \boldsymbol{S}^{\nu}{}_{\rho}}{\partial x^{\nu}}=\frac{1}{2}\boldsymbol{S}^{ab}\cdot A_{ab,\rho}\,, \tag{5·3}$$

where the following notations have been employed:

$$\boldsymbol{S}^{\mu\nu}=-2[L_1]^{\mu\nu}=-2\left\{\frac{\partial L_1}{\partial A_{\mu\nu}}-\partial_\lambda\left(\frac{\partial L_1}{\partial A_{\mu\nu,\lambda}}\right)\right\},$$

$$\boldsymbol{S}^{\mu}{}_{\nu}=\boldsymbol{S}^{\mu\rho}\cdot A_{\rho\nu}\,. \tag{5·4}$$

The relationship between $\boldsymbol{S}^{\mu}{}_{\nu}$ and $T^{\mu}_{(1)\nu}$ is given by (2·6) with the aid of (5·1):

$$\boldsymbol{S}^{\mu}{}_{\rho}=T^{\mu}_{(1)\rho}-\partial_\lambda F^{[\lambda\mu]}_{\rho}, \tag{5·5}$$

where

$$F^{[\lambda\mu]}_{\rho}\equiv-F^{[\mu\lambda]}_{\rho}=\frac{\partial L_1}{\partial\phi_{A,\lambda}}\cdot C^{B\mu}_{A\rho}\cdot\phi_B-2\frac{\partial L_1}{\partial A_{\mu a,\lambda}}A_{\rho a}\,.$$

The antisymmetry of F with respect to the superscripts $[\lambda\mu]$ is due to the identity (2·7).

The relations (5·5) and (5·2) and the fact that $\boldsymbol{S}^{\mu\nu}$ is a symmetric tensor density, as easily seen from the definition (5·4), show that $\boldsymbol{S}^{\mu\nu}$ should be regard ed as a energy-momentum tensor density of the field ϕ_A interacting with the field $A_{\mu\nu}$.

Taking into account the transformation property of $\boldsymbol{S}^{\mu\nu}$ under (2·1), we can rewrite (5·3) in a covariant form

$$\nabla_\mu S^{\nu}{}_{\rho}=0\,, \tag{5·3$'$}$$

where

$$S^{\nu}{}_{\rho}=\boldsymbol{S}^{\nu}{}_{\rho}/\sqrt{|A|}\,,$$

and the covariant derivative of $S^{\nu}{}_{\rho}$ is

$$\nabla_\lambda S^{\nu}{}_{\rho}=\partial_\lambda S^{\nu}{}_{\rho}+\Delta^{\nu}_{\lambda\sigma}\cdot S^{\sigma}{}_{\rho}-\Delta^{\sigma}_{\lambda\rho}\cdot S^{\nu}{}_{\sigma}\,.$$

The existence of the non-vanishing right-hand side of (5·3) means that the energy-momentum of ϕ_A is not conserved owing to the interaction of ϕ_A with A. It is well known that (5·3) can be transformed into the expression

$$\partial_\nu\{\boldsymbol{S}^{\nu}{}_{\rho}+\boldsymbol{t}^{\nu}{}_{\rho}\}=0 \tag{5·6}$$

by the aid of the field equation of A, where $\boldsymbol{t}^{\nu}{}_{\rho}$ represents a pseudo energy-momentum tensor density of the field A.

This result that the energy-momentum of ϕ_A can no longer be conserved when the interaction of ϕ_A with A takes place, is a consequence of the general theory of the non-Abelian gauge fields on which a brief explanation will be given in the next section.

§ 6. General remarks on the law of conservation*)

Consider a field $\phi_A(x)$, $(A=1, 2, \cdots N)$, the field equation of which is derived from the action-integral

$$I=\int L_{(\phi)}(\phi_A, \phi_{A,\mu})\, d^4x\,.$$

Let us assume that I is invariant under a group of transformations depending on parameters ϵ_a, $(a=1, 2\cdots r)$:**)

$$\delta\phi_A(x)=\phi_A'(x)-\phi_A(x)=\phi_B\cdot \overset{a}{C}{}^B_A\cdot\epsilon_a\,.$$

For simplicity, the transformation of coordinates is excluded from our discussion.

The invariance of I leads to a set of identities

$$[L_{(\phi)}]^A\cdot\phi_B\cdot\overset{a}{C}{}^B_A+\partial_\lambda\left(\frac{\partial L_{(\phi)}}{\partial\phi_{A,\lambda}}\cdot\phi_B\cdot\overset{a}{C}{}^B_A\right)\equiv 0\,. \qquad (6\cdot 1)$$

The postulate that I should be invariant even under an extended group which depends upon arbitrary function $\lambda_a(x)$'s instead of ϵ_a's, necessitates an introduction of a generalized gauge field $A_{a_\mu}(x)$ with a transformation property

$$\delta A_{a_\mu}=A_{b_\mu}\cdot M^b{}_a{}^c\cdot\lambda_c(x)+\lambda_{a,\mu}\,. \qquad (6\cdot 2)$$

This postulate of invariance gives rise to the following identities, if one follows a similar line of argument to that given in Appendix A:

$$[L_1]^A\cdot\overset{a}{C}{}^B_A\cdot\phi_B+\frac{\partial L_1}{\partial A_{b_\mu}}\cdot M^c{}_b{}^a\cdot A_{c_\mu}-\partial_\mu\left(\frac{\partial L_1}{\partial A_{a_\mu}}\right)\equiv 0\,, \qquad (6\cdot 3)$$

$$\partial_\lambda\left[\frac{\partial L_1}{\partial\phi_{A,\lambda}}\cdot\overset{a}{C}{}^B_A\cdot\phi_B+\frac{\partial L_1}{\partial A_{a\lambda}}\right]\equiv 0\,, \qquad (6\cdot 4)$$

$$\frac{\partial L_1}{\partial\phi_{A,\lambda}}\cdot\overset{a}{C}{}^B_A\cdot\phi_B+\frac{\partial L_1}{\partial A_{a\lambda}}\equiv 0\,, \qquad (6\cdot 5)$$

where the new Lagrangian L_1 is

$$L_1(\phi_A, \phi_{A,\lambda}, A_{a_\mu})\,.$$

The identity (6·5) implies that L_1 should depend upon $\phi_{A,\lambda}$ and $A_{a\lambda}$ only through an "invariant derivative" $\nabla_\lambda\phi_A$ defined by

$$\nabla_\lambda\phi_A=\partial_\lambda\phi_A-A_{a\lambda}\cdot\overset{a}{C}{}^B_A\cdot\phi_B\,. \qquad (6\cdot 6)$$

Thus L_1 can be written as

$$L_1(\phi_A, \phi_{A,\lambda}, A_{a\lambda})=L_2(\phi_A, \nabla_\lambda\phi_A)\,. \qquad (6\cdot 7)$$

*) The first half of the content of this section is a review of the paper I.I.I., but our definition of $j^{(a)}$ is different from that given by (1.27) on page 1601 of I.I.I.

**) For brevity, coordinate transformations are not considered in this section.

The identity (6·3) is transformed into the following expression with the aid of (6·6) and (6·7):

$$\frac{\partial L_2}{\partial \phi_A}\cdot \overset{a}{C}{}^B_A\cdot \phi_B+\frac{\partial L_2}{\partial \nabla_\lambda \phi_A}\cdot \overset{a}{C}{}^B_A\cdot \nabla_\lambda \phi_B\equiv 0\,, \qquad (6\cdot3)'$$

where Lie's commutation relations*)

$$[\overset{a}{C}, \overset{b}{C}]=f^{a\ b}_{\ c}\cdot \overset{c}{C}$$

have been used together with the reasonable assumption

$$M^{a\ b}_{\ c}=f^{a\ b}_{\ c}.$$

A comparison of (6·3)′ with (6·1) suggests that L_2 should be chosen to have the same functional form as $L_{(\phi)}$, namely,

$$L_1(\phi_A, \phi_{A,\lambda}, A_{a\lambda})=L_{(\phi)}(\phi_A, \nabla_\lambda \phi_A).$$

Let us define the "(a)-current" $j^{(a)\lambda}_{(\phi)}$ by

$$j^{(a)\lambda}_{(\phi)}=\frac{\partial L_{(\phi)}}{\partial \nabla_\lambda \phi_A}\cdot \overset{a}{C}{}^B_A\cdot \phi_B\equiv -\frac{\partial L_1}{\partial A_{a\lambda}}\,. \qquad (6\cdot8)$$

It is easily seen from the definition (6·8) that $j_{(\phi)}$ has a transformation property

$$\delta j^{(a)\lambda}_{(\phi)}=j^{(b)\lambda}_{(\phi)}\cdot f^{c\ a}_{\ b}\cdot \lambda_c(x),$$

which leads to the definition of the "invariant derivative" of $j_{(\phi)}$:

$$\nabla_\lambda j^{(a)\mu}_{(\phi)}=\partial_\lambda j^{(a)\mu}_{(\phi)}-A_{c\lambda}\cdot f^{c\ a}_{\ b}\cdot j^{(b)\mu}_{(\phi)}. \qquad (6\cdot9)$$

Making use of (6·9) and assuming the field equation of ϕ_A

$$[L_1]^A=0\,,$$

we can derive the equations of continuity for $j_{(\phi)}$ from the identity (6·3):

$$\nabla_\mu j^{(a)\mu}_{(\phi)}=0\,. \qquad (6\cdot10)$$

(6·10) shows that the "(a)-charge" of the ϕ_A-field defined by

$$Q^{(a)}=\int_{-\infty}^{\infty} j^{(a)0}_{(\phi)}d^3x$$

is no longer conserved except for the Abelian case $f^{a\ b}_{\ c}\equiv 0$.

Let us assume that a Lagrangian density $L_{(A)}(A_{a\lambda}, A_{a\lambda,\mu})$ is chosen in such a way as that the action-integral $I_A=\int L_{(A)}d^4x$ is invariant under the transformation (6·2). Then in a completely similar way we can derive many identities, among which the following identity corresponds to (6·4):

---

*) The (B, A) element of the $N\times N$ matrix $\overset{a}{C}$ is defined by

$$(B|\overset{a}{C}|A)=\overset{a}{C}{}^B_A.$$

$$\partial_\lambda\left[\frac{\partial L_{(\phi)}}{\partial \nabla_\lambda \phi_A}\cdot \overset{a}{C}{}^B_A\cdot\phi_B+\frac{\partial L_{(A)}}{\partial A_{b\mu,\lambda}}\cdot f^c{}_b{}^a\cdot A_{c\mu}+[L_{\rm tot}]^{a\lambda}\right]\equiv 0\,, \tag{6·11}$$

where

$$L_{\rm tot}=L_{(\phi)}(\phi_A, \nabla_\lambda\phi_A)+L_{(A)}(A_{a\mu}, A_{a\mu,\lambda})\,.$$

If the field equation of $A_{a\mu}$

$$[L_{\rm tot}]^{a\mu}=\frac{\partial L_{(A)}}{\partial A_{a\mu}}-\partial_\lambda\left(\frac{\partial L_{(A)}}{\partial A_{a\mu,\lambda}}\right)-j^{(a)\mu}_{(\phi)}=0 \tag{6·12}$$

is employed, (6·11) reads

$$\partial_\lambda[j^{(a)\lambda}_{(\phi)}+j^{(a)\lambda}_{(A)}]=0\,, \tag{6·13}$$

where the second term defined by

$$j^{(a)\lambda}_{(A)}=\frac{\partial L_{(A)}}{\partial A_{b\mu,\lambda}}\cdot f^c{}_b{}^a\cdot A_{c\mu} \tag{6·14}$$

is interpreted as the "(a)-charge" carried by the A-field. The relation (5·6) in § 5 is a particular example of (6·13). In the former case, the energy and momentum stand for the "(a)-charge" of the present section.

The field equation (6·12) shows that the A-field emerges from a current density of the "(a)-charge". This interpretation (6·12) together with the fact that the A-field possesses the "(a)-charge" as is shown by (6·13), leads to the conclusion that the A-field can be produced by itself. This is the reason why a generalized gauge field associated with a non-Abelian group should obey a set of non-linear field equations.

Appendix

A. *Derivation of identities (2·4)∼(2·7)*

Consider a variation of

$$I_1=\int_\Omega L_1(\phi_A, \phi_{A,\lambda}, A_{(a_1\cdots)}, A_{(a_1\cdots),\lambda})\,d^4x$$

due to the variations of ϕ_A and $A_{(a_1\cdots)}$

$$\begin{aligned}&\phi_A'(x')-\phi_A(x)=\delta\phi_A=\phi_B\cdot C^{B\nu}_{A\mu}\cdot\xi^\mu{}_{,\nu}\,,\\&A'_{(a_1\cdots)}(x')-A_{(a_1\cdots)}(x)=\delta A_{(a_1\cdots)}=A_{(b_1\cdots)}\cdot D\left(\begin{smallmatrix}b_1\cdots\\a_1\cdots\end{smallmatrix}\right)^\nu_\mu\cdot\xi^\mu{}_{,\nu}\end{aligned} \tag{A·1}$$

which are caused by a transformation of coordinates

$$x^\mu\to x^{\mu\prime}=x^\mu+\delta x^\mu=x^\mu+\xi^\mu(x)\,.$$

δI_1 of the first order with respect to ξ^μ is given by

$$\delta I_1=\int_\Omega L_1\{\phi_A+\delta\phi_A, \phi_{A,\lambda}+\delta\phi_{A,\lambda}, \cdots\}\frac{\partial(x^{0\prime},\cdots x^{3\prime})}{\partial(x^0,\cdots x^3)}\cdot d^4x-I_1$$

$$= \int_\Omega \left\{ \frac{\partial L_1}{\partial \phi_A} \delta\phi_A + \frac{\partial L_1}{\partial \phi_{A,\lambda}} \delta\phi_{A,\lambda} + \frac{\partial L_1}{\partial A_{(a_1\cdots)}} \cdot \delta A_{(a_1\cdots)} + \frac{\partial L_1}{\partial A_{(a_1\cdots),\lambda}} \delta A_{(a_1\cdots),\lambda} + \lambda_1 \cdot \xi^\mu_{,\mu} \right\} d^4x \,.$$

Let us introduce another kind of variation defined by

$$\bar{\delta}\phi_A(x) \equiv \phi_A'(x) - \phi_A(x) \equiv \delta\phi_A(x) - \phi_{A,\lambda}\cdot\delta x^\lambda, \tag{A·2}$$

which has a convenient property

$$\frac{\partial}{\partial x^\lambda}\bar{\delta}\phi_A = \bar{\delta}\left(\frac{\partial \phi_A}{\partial x^\lambda}\right).$$

In terms of this variation, δI_1 is written as

$$\delta I_1 = \int_\Omega \Big[[L_1]^A \cdot \bar{\delta}\phi_A + [L_1]^{(a_1\cdots a_r)} \cdot \bar{\delta} A_{(a_1\cdots a_r)} + \partial_\lambda \left\{ \frac{\partial L_1}{\partial \phi_{A,\lambda}} \delta\phi_A + \frac{\partial L_1}{\delta A_{(a_1\cdots),\lambda}} \delta A_{(a_1\cdots)} - T^\lambda{}_\rho \cdot \delta x^\rho \right\} \Big] d^4x \,. \tag{A·3}$$

Substituting the definitions (A·1) for $\delta\phi_A$ and $\delta A_{(a_1\cdots)}$ in the above expression and putting $\delta I_1 \equiv 0$, we are led to the identities. Especially when $\xi^\mu(x) = a^\mu + \Delta^\mu{}_\nu \cdot x^\nu$ and $A_{(a_1\cdots)}(x)$ is replaced with a constant affine tensor $\zeta_{\mu\nu}$ (and consequently $A_{(a_1\cdots),\lambda} = 0$), the coefficient of each parameter a^μ or $\Delta^\mu{}_\nu$ in δI_1, should identically vanish because the domain of integration can be arbitrarily chosen. The relations thus obtained are the identities (4·3) and (4·4) where $\boldsymbol{L}$ takes the place of the present L_1.

On the contrary if $\xi^\mu(x)$ is an arbitrary function of x, the first and the second terms of the right-hand side in (A·3) can be transformed by a partial-integration to the following form:

$$\begin{aligned}\delta I_1 = &- \int_\Omega \xi^\rho \cdot \big[\partial_\nu \{ [L_1]^A \cdot C^{B\nu}_{A\rho} \cdot \phi_B + [L_1]^{(a_1\cdots a_r)} \cdot D\big(^{b_1\cdots b_r}_{a_1\cdots a_r}\big)^\nu_\rho \cdot A_{(b_1\cdots b_r)} \} + [L_1]^A \cdot \phi_{A,\rho} + [L_1]^{(a_1\cdots a_r)} \cdot A_{(a_1\cdots a_r),\rho} \big] d^4x \\ &+ \int_\Omega \partial_\lambda \Big[[L_1]^A \cdot C^{B\lambda}_{A\rho} \cdot \phi_B \cdot \xi^\rho + [L_1]^{(a_1\cdots a_r)} \cdot D\big(^{b_1\cdots b_r}_{a_1\cdots a_r}\big)^\lambda_\rho A_{(b_1\cdots)} \cdot \xi^\rho \\ &+ \frac{\partial L_1}{\partial \varphi_{A,\lambda}} C^{B\mu}_{A\rho} \cdot \varphi_B \cdot \xi^\rho{}_{,\mu} + \frac{\partial L_1}{\partial A_{(a_1\cdots a_r),\lambda}} \cdot D\big(^{b_1\cdots b_r}_{a_1\cdots a_r}\big)^\mu_\rho \cdot A_{(b_1\cdots b_r)} \cdot \xi^\rho{}_{,\mu} - T^\lambda{}_\rho \cdot \xi^\rho \Big] d^4x = 0 \,.\end{aligned} \tag{A·4}$$

If ξ^ρ and $\xi^\rho{}_{,\mu}$ are chosen to vanish on the boundary surface of Ω, the second integral in (A·4) vanishes and we have

$$\delta I_1 = - \int_\Omega \xi^\rho [\cdots]_\rho \cdot d^4x \equiv 0 \,, \tag{A·5}$$

where ξ^ρ can take any value inside Ω. Thus we have the identity

$$[\cdots]_\rho = [\partial_\nu\{[L_1]^A \cdot C^{B\nu}_{A\rho} \cdot \phi_B + [L_1]^{(a_1\cdots)} \cdot D(^{b_1\cdots}_{a_1\cdots})^\nu_\rho \cdot A_{(b_1,\cdots)}\} \\ + [L_1]^A \cdot \phi_{A,\rho} + [L_1]^{(a_1\cdots)} \cdot A_{(a_1\cdots),\rho}] \equiv 0 \tag{A·5'}$$

which is nothing but the identity (2·4). Inserting (A·5)′ into (A·4), we have

$$\delta I_1 = \int_\Omega \partial_\lambda [\cdots]^\lambda \cdot d^4x \equiv 0$$

for arbitrary ξ^μ's. By putting equal to zero the coefficients of ξ^ρ, and its derivatives in the above identity, (2·5) ~ (2·7) can be derived.

B. *Determination of* $A_{(a_1\cdots a_r)}$

Let us begin with the substitution of (3·9) for Z's in the identity (3·8). The latter is written as

$$2\delta^\alpha_\rho \delta^{(\mu\nu)}_{(\lambda\beta)} \equiv r\cdot a[A^\mu_\rho \cdot \delta^{(\nu\alpha)}_{(\lambda\beta)} + A^\nu_\rho \cdot \delta^{(\mu\alpha)}_{(\lambda\beta)}] \\ + r\cdot b[2A^\nu_\rho \cdot \delta^{(\alpha\mu)}_{(\lambda\beta)} + 2A^\mu_\rho \cdot \delta^{(\alpha\nu)}_{(\lambda\beta)} + 2(r-1)\{\delta^\alpha_\beta A^{\mu\nu}_{\lambda\rho} + \delta^\alpha_\lambda A^{\mu\nu}_{\beta\rho}\}] \\ + r\cdot c[4A^\alpha_\rho \delta^{(\mu\nu)}_{(\lambda\beta)} + (r-1)\{\delta^\nu_\beta A^{\mu\alpha}_{\lambda\rho} + \delta^\nu_\lambda A^{\mu\alpha}_{\beta\rho} + \delta^\mu_\beta A^{\nu\alpha}_{\lambda\rho} + \delta^\mu_\lambda A^{\nu\alpha}_{\beta\rho}\}] \\ + r(r-1)\cdot d\cdot[2(r-2)\cdot A^{\mu\nu\alpha}_{\lambda\beta\rho} + \delta^\mu_\beta A^{\nu\alpha}_{\lambda\rho} + \delta^\mu_\lambda A^{\nu\alpha}_{\beta\rho} + \delta^\nu_\beta A^{\mu\alpha}_{\lambda\rho} + \delta^\nu_\lambda A^{\mu\alpha}_{\beta\rho}], \tag{B·1}$$

where the following abbreviations have been employed:

$$A^\alpha_\rho \equiv A^{(\alpha a_2\cdots a_r)} A_{(\rho a_2\cdots a_r)},$$

$$A^{\alpha\nu}_{\beta\rho} \equiv A^{(\alpha\nu a_3\cdots a_r)} A_{(\beta\rho a_3\cdots a_r)},$$

etc.

The double contraction of (B·1) by putting $\mu=\lambda$ and $\nu=\beta$ leads to

$$A^\alpha_\rho\{5ra + 2(2r+3)\cdot r\cdot b + 10\cdot r(r+3)\cdot c + 2r(r-1)(r+3)\cdot d\} \equiv 20\delta^\alpha_\rho, \tag{B·2}$$

which allows to put

$$A^\alpha_\rho \equiv A^{(\alpha a_2\cdots a_r)} \cdot A_{(\rho a_2\cdots a_r)} = \delta^\alpha_\rho \tag{B·3}$$

and consequently

$$A^{(a_1\cdots a_r)} \cdot A_{(a_1\cdots a_r)} = 4\,. \tag{B·3'}$$

Inserting (B·3) into (B·2), we have

$$5ra + 2(2r+3)rb + 10r(r+3)c + 2r(r-1)(r+3)d = 20\,. \tag{B·2'}$$

In a similar way, the contraction of (B·1) by putting $\alpha=\rho$ and $\mu=\lambda$ leads to

$$6ra + 2r(2r+3)b + 10r(r+3)\cdot c + 2r(r-1)(r+3)d = 20\,, \tag{B·4}$$

where the normalization (B·3) and (B·3)′ have been employed. The third type of contraction $\alpha=\lambda$ and $\mu=\rho$ gives a relation

$$15ra + 10r(2r+3)b + 2r(7r+3)c + 2r(2r+3)(r-1)d = 10\,. \tag{B·5}$$

A subtraction of (B·2)′ from (B·4) gives

$$ra=0$$

or

$$a=0\,. \tag{B·6}$$

If (B·1) is contracted by putting $\mu=\lambda$, we obtain a relation

$$\{2(r-1)\cdot r\cdot b+6(r-1)\cdot r\cdot c+2r(r-1)(r+1)d\}A^{\alpha\nu}_{\beta\rho}$$
$$=\{5-rb-r(r+a)c-r(r-1)\cdot d\}\delta^{\alpha}_{\rho}\delta^{\nu}_{\beta}-2r(r+2)b\cdot\delta^{\alpha}_{\beta}\delta^{\nu}_{\rho}. \tag{B·7}$$

Since the left-hand side of (B·7) is symmetric with respect to the pairs of suffices (α,ν) and (β,ρ) the same should be true for the right-hand side. Thus we have

$$5-rb-r(r+9)c-r(r-1)\cdot d=-2r(r+2)\cdot b\,, \tag{B·8}$$

and consequently (B·7) becomes

$$\{2r(r-1)b+6(r-1)\cdot r\cdot c+2r(r-1)(r+1)\cdot d\}A^{\alpha\nu}_{\beta\rho}=-4r(r+2)\cdot b\cdot\delta^{(\alpha\nu)}_{(\beta\rho)}\,. \tag{B·7)′}$$

Similarly, contractions $\mu=\rho$ and $\alpha=\lambda$ lead to

$$\{cr+r(r-1)^2d\}A^{\alpha\nu}_{\lambda\beta}=\{1-(2r+3)rb-r(r+1)c-r(r-1)d\}\delta^{(\alpha\nu)}_{(\lambda\beta)} \tag{B·9}$$

and

$$\{5r(r-1)b+r(r-1)c+r(r-1)^2d\}A^{\alpha\nu}_{\lambda\beta}=\{1-5rb-r(r+1)c-r(r-1)d\}\delta^{(\alpha\nu)}_{(\lambda\beta)} \tag{B·10}$$

respectively.

In (B·7)′, (B·9) and (B·10), if the coefficients of $A^{::}_{::}$ do not vanish, these relations show that $A^{::}_{::}$ should be proportional to $\delta^{(::)}_{(::)}$. Recalling the normalization (B·3), we have to put

$$A^{\alpha\nu}_{\beta\rho}=\tfrac{2}{5}\delta^{(\alpha\nu)}_{(\beta\rho)}\,. \tag{B·11}$$

Thus (B·9) and (B·10) give

$$5(2r+3)rb+(5r+7)r\cdot c+(2r+3)(r-1)\cdot r\cdot d=5 \tag{B·9)′}$$

and

$$5(2r+3)rb+(7r+3)rc+(2r+3)(r-1)rd=5 \tag{B·10)′}$$

respectively.

A subtraction of (B·9)′ from (B·10)′ gives

$$(2r-4)\cdot r\cdot c=0\,. \tag{B·12}$$

If we choose $r=2$, then (B·11) becomes unsolvable relations

$$A^{\alpha\nu}(x)\cdot A_{\beta\rho}(x)=\tfrac{2}{5}\delta^{(\alpha\nu)}_{(\beta\rho)}\,.$$

On the contrary if $c=0$ is chosen, we have three relations:

$$(\mathrm{B}\cdot 8)\to(2r+3)\cdot r\cdot b=r(r-1)d-5\,,$$

$$(\mathrm{B}\cdot 9)',\ (\mathrm{B}\cdot 10)'\to 5(2r+3)\cdot r\cdot b+(2r+3)(r-1)\cdot r\cdot d=5\,,$$

$$(\mathrm{B}\cdot 7)'\to 3(2r+3)\cdot b+(r-1)(r+1)d=0\,.$$

These three equations are unsolvable with respect to b and d. Therefore (B·12) cannot hold.

Thus we arrive at a conclusion that all the coefficients of both sides of relations (B·7)′, (B·9) and (B·10) should vanish. Vanishing of the right-hand side of (B·7)′ gives

$$b=0 \tag{B·13}$$

while the left-hand side of (B·7)′ gives a relation when it vanishes:

$$3c+(r+1)d=0\,. \tag{B·14}$$

Similarly, (B·9) gives a couple of relations

$$c+(r-1)^2 d=0\,, \tag{B·15}$$

$$1=r(r+1)c+r(r-1)d\,. \tag{B·16}$$

Finally from (B·10) we obtain relations

$$c+(r-1)d=0\,, \tag{B·17}$$

$$1=r(r+1)c+r(r-1)d\,. \tag{B·18}$$

A comparison of (B·15) with (B·17) gives

$$(r-2)\cdot d=0\,.$$

Since the case $d=0$ leads to a contradiction as easily seen, we obtain

$$r=2\,.$$

It is easily verified that the solution of our problem is

$$a=b=0\,,\qquad c=-d=\tfrac{1}{4}\,.$$

References

1) R. Utiyama, Phys. Rev. **101** (1956), 1597.
This paper is referred to as I.I.I.
2) T. W. B. Kibble, J. Math. Phys. **2** (1961), 212.
3) T. Nakano & K. Hayashi, Prog. Theor. Phys. 38 (1967), 491.
4) cf. I.I.I.

VOLUME 33, NUMBER 7 PHYSICAL REVIEW LETTERS 12 AUGUST 1974

Integral Formalism for Gauge Fields

C. N. Yang
Institute of Theoretical Physics, University of Wrocław, Wrocław, Poland, and
Institute of Theoretical Physics, State University of New York, Stony Brook, New York 11990
(Received 10 June 1974)

A new integral formalism for gauge fields is described. Further developments are presented, including gravitation equations related to, but not identical with, Einstein's equations.

It was pointed out by Weyl many years ago that the electromagnetic field can be formulated in terms of an Abelian gauge transformation. This idea was extended[1] in 1954 to the concept of gauge fields for non-Abelian groups. That formulation, like the Weyl formulation for electromagnetism, was based on the replacement of ∂_μ by $\partial_\mu - ieB_\mu$. One might call such formulations differential formulations. It is the purpose of the present paper to reformulate the concept of gauge fields in an *integral formalism*. The new formalism is conceptually superior to the differential formalism and allows for natural developments of additional concepts. It further allows a mathematical and physical discussion of the gravitational field *as a gauge field*, resulting in equations related, but not identical, to Einstein's.

The basic point is the fact that *electromagnetism is a nonintegrable phase factor*, a fact discussed many years ago by Dirac, Peierls, and others, and more recently by many authors.[2] This fact is now generalized as follows:

Definition of a gauge field.—Consider a manifold with points on it labeled by x^μ ($\mu = 1, 2, \ldots, n$) and consider a gauge G which is a Lie group with generators X_k ($k = 1, 2, \ldots, m$). [For $G = U(1)$ we have electromagnetism; for G non-Abelian we have non-Abelian gauge fields.] Define a path-dependent (i.e., nonintegrable) phase factor φ_{AB} as an element of the group G associated with path AB between two points A and B on the manifold. The association is to have the group property: $\varphi_{ABC} = \varphi_{AB}\varphi_{BC}$, where the paths AB and BC are segments of ABC. Furthermore for an infinitiesimal path A to $A + dx^\mu$ the phase factor is close to the identity I of G, so that[3]

$$\varphi_{A(A+dx)} = I + b_\mu{}^k(x) X_k \, dx^\mu . \tag{1}$$

The function $b_\mu{}^k(x)$ defined on the manifold will be called a *gauge potential;* φ_{AB} will be called a *gauge phase factor.*

With this definition additional concepts and theorems are naturally developed. We summarize some of these below. Details will be published elsewhere.

Gauge field strength.—Consider a path $ABCDA$ forming the border of an infinitesimal parallelogram with sides dx and dx'. φ_{ABCDA} can be computed by multiplying four phase factors like (1) together, resulting in

$$\varphi_{ABCDA} = I + f_{\mu\nu}{}^k X_k \, dx^\mu dx^{\nu\prime} , \tag{2}$$

where

$$f_{\mu\nu}{}^k = \frac{\partial b_\mu{}^k}{\partial x^\nu} - \frac{\partial b_\nu{}^k}{\partial x^\mu} - b_\mu{}^i b_\nu{}^j C_{ij}{}^k = -f_{\nu\mu}{}^k \tag{3}$$

in which $C_{ij}{}^k$ is the structure constant of G:

$$X_k X_j - X_j X_k = C_{kj}{}^i X_i . \tag{4}$$

$f_{\mu\nu}{}^k$ will be called a *gauge field,* or gauge field strength. They are the Faraday-Maxwell fields when $G = U(1)$.

Gauge transformation.—A gauge transformation in the integral formalism is defined by a transformation

$$\varphi_{AB} \to \varphi_{AB}' = \xi_A \varphi_{AB} \xi_B{}^{-1} , \tag{5}$$

where ξ_A is an element of G which depends on the point A. It is clear that under (5)

$$\varphi_{ABCDA} \to \varphi_{ABCDA}' = \xi_A \varphi_{ABCDA} \xi_A{}^{-1} . \tag{6}$$

Thus

$$f_{\mu\nu}{}^{k\prime} = \langle k | R_{\mathrm{adj}} | j \rangle f_{\mu\nu}{}^j , \tag{7}$$

where R_{adj} is the adjoint representation for the element ξ_A. The simple transformation property (7) is the definition for the concept that $f_{\mu\nu}{}^k$ is *gauge covariant.* Generalization to other representations R of G for a gauge-covariant quantity $\psi_{\alpha\beta\gamma}{}^K$ is immediate[3]:

$$\psi_{\alpha\beta\gamma}{}^{K\prime} = \langle K | R(\xi_A) | J \rangle \psi_{\alpha\beta\gamma}{}^J . \tag{8}$$

$b_\mu{}^k$ is not gauge covariant; $f_{\mu\nu}{}^k$ is.

Gauge-covariant differentiation.—To *retain*

gauge covariance in differentiation we define

$$\psi^K|_\mu = \frac{\partial \psi^K}{\partial x^\mu} + b_\mu{}^k \langle K|Z_k|J\rangle \psi^J, \tag{9}$$

where Z_k is the matrix representation of X_k. Generalization to other cases is obvious. An interesting theorem is that

$$f_{\mu\nu}|_\lambda{}^k + f_{\nu\lambda}|_\mu{}^k + f_{\lambda\mu}|_\nu{}^k = 0, \tag{10}$$

which is the gauge-Bianchi identity.

Introduction of a Riemannian metric.—So far we need no metric for the manifold. Now we introduce a metric for it and discuss arbitrary coordinate transformations. We come then naturally to *Riemannian covariant* quantities and *doubly covariant derivatives.* $b_\mu{}^k$ is Riemannian covariant, since φ_{AB} is coordinate-system independent. $f_{\mu\nu}{}^k$ is doubly covariant. We have

$$\begin{aligned}
\psi^K{}_{\|\mu} &= \psi^K|_\mu, \\
\psi^{K\nu}{}_{\|\mu} &= \psi^{K\nu}|_\mu + \begin{Bmatrix}\nu\\ \mu\alpha\end{Bmatrix}\psi^{K\alpha}, \\
f_{\mu\nu\|\lambda}{}^k &= f_{\mu\nu}|_\lambda{}^k - \begin{Bmatrix}\alpha\\ \mu\lambda\end{Bmatrix} f_{\alpha\nu}{}^k - \begin{Bmatrix}\alpha\\ \nu\lambda\end{Bmatrix} f_{\mu\alpha}{}^k,
\end{aligned} \tag{11}$$

etc. It is easily shown that

$$f_{\mu\nu\|\lambda}{}^k + f_{\nu\lambda\|\mu}{}^k + f_{\lambda\mu\|\nu}{}^k = 0 \tag{12}$$

which is satisfied by *all* gauge fields on *all* Riemannian manifolds.

Source of gauge fields.— We *define,* in analogy with electromagnetism, a source four-vector $J_\mu{}^k$ for a gauge field:

$$J_\mu{}^k = g^{\nu\lambda} f_{\mu\nu\|\lambda}{}^k = f_{\mu\nu}{}^{k\|\nu}. \tag{13}$$

After some computation one derives a theorem:

$$g^{\mu\lambda} J_{\mu\|\lambda}{}^k = 0 \quad \text{(conserved current)}, \tag{14}$$

which in electromagnetism states charge conservation. In Ref. 1 this was Eq. (14). One can also generalize Eqs. (15) and (16) of Ref. 1, leading to the concept of "total charge."

Parallel-displacement gauge field.—For any Riemannian manifold, the important concept of parallel displacement defines, along any path AB, a *linear* relationship between any vector V_A at A and its parallel vector V_B at B. Thus parallel displacement is defined by an $n\times n$ matrix M_{AB} which gives this linear relationship. M_{AB} is a representation of an element of GL(n). Thus we have the following:

Theorem.—Parallel displacement defines a gauge field with G being GL(n). The index k has n^2 values and we write $k=(\alpha\beta)$. The gauge potential and gauge fields are respectively

$$b_\mu{}^{(\alpha\beta)} = \begin{Bmatrix}\alpha\\ \beta\mu\end{Bmatrix}, \quad f^{(\alpha\beta)}{}_{\mu\nu} = -R^\alpha{}_{\beta\mu\nu}. \tag{15}$$

It is important to recognize that in this definition we have chosen a fixed coordinate system. A coordinate transformation would generate a linear transformation in the vector spaces V_A and V_B. In other words $M_{AB} \to N_A M_{AB} N_B{}^{-1}$. Comparison with (5) shows thus that a coordinate transformation generates a simultaneous gauge transformation of the parallel-displacement gauge potential. In fact, the usual nonlinear term in the transformation of $\left\{{\alpha \atop \alpha\gamma}\right\}$ is precisely the nonlinear term needed in the gauge transformation of the gauge noncovariant quantity $b_\mu{}^{(\alpha\beta)}$. In this connection we observe that for GL(n),

$$C^{(\alpha\beta)}_{(\lambda\mu)(\eta\zeta)} = \delta_{\mu\eta}\delta_{\alpha\lambda}\delta_{\beta\zeta} - \delta_{\lambda\zeta}\delta_{\alpha\eta}\delta_{\beta\mu}. \tag{16}$$

Thus by definitions (9) and (11)

$$\begin{aligned}
\psi^{(\alpha\beta)}{}_{\|\mu} &= \frac{\partial \psi^{(\alpha\beta)}}{\partial x^\mu} + b_\mu{}^{(\lambda\nu)} C^{(\alpha\beta)}_{(\lambda\nu)(\eta\zeta)} \psi^{(\eta\zeta)} \\
&= \psi^\alpha{}_{\beta;\mu},
\end{aligned} \tag{17}$$

where the semicolon represents the usual Riemannian covariant differentiation with α and β treated as usual contravariant and covariant indices. The rule works also in general. E.g.,

$$f^{(\alpha\beta)}{}_{\mu\nu\|\lambda} = -R^\alpha{}_{\beta\mu\nu;\lambda}. \tag{18}$$

Nontrivial sourceless gauge fields.—Gauge fields for which $f_{\mu\nu}{}^k \neq 0$ and $J_\mu{}^k = 0$ are of physical interest. So far only nonanalytic examples are known.[4]

We now can construct two general types of general types of examples.

(a) Consider the natural Riemannian geometry of a semisimple Lie group. Its parallel-displacement gauge field is sourceless and analytic.

(b) Consider the same Riemannian manifold of a group G as above in (a). Define φ_{AB} as that for an infinitesimal path AB, $\varphi_{AB} = (A^{-1}B)^{1/2}$. This gauge phase factor which is itself an element of G gives a gauge field which is analytic and sourceless.

Pure spaces.—A Riemannian manifold for which the parallel-displacement gauge field is sourceless will be called a pure space. A necessary and sufficient condition for a pure space is

$$R_{\mu\alpha;\beta} = R_{\mu\beta;\alpha} \tag{19}$$

A four-dimensional Einstein space, i.e., one for

which $R_{\alpha\beta}=0$, is a pure space.

Gravitational field as a gauge field.—The electromagnetic field and the usual gauge fields are special cases of gauge fields, satisfying (12) and (13). A natural question is whether one should identify these *same equations* for the parallel-displacement gauge field as the equations for the gravitational field. There are advantages in this identification and we shall come back to this topic in a later communication. If one adopts this identification then gravitational equations are third-order differential equations[5] for $g_{\mu\nu}$. A pure gravitational field is then described by a pure space as defined above.

Variational principles.—Equation (13) with $J_\mu^k=0$ follows from a variational principle $\delta\int\sqrt{-g}\,d^n x=0$, where

$$L=f_{\mu\nu}{}^k f_{\alpha\beta}{}^j g^{\mu\alpha}g^{\nu\beta}C_{ka}{}^b C_{jb}{}^a. \tag{20}$$

In the variation $g_{\mu\nu}$ is kept fixed and $b_\mu{}^k$ is varied, and $f_{\mu\nu}{}^k$ is given by (3); $C_{ka}{}^b$ are not varied. One could also find a variational principle which is satisfied by a pure space (19). Choose $C_{ka}{}^b$ to be the structure constants for GL(n), given by (16). Write the L of (20) as a functional of $b_\mu{}^{(\alpha\beta)}$ and $g^{\lambda\nu}$:

$$L=L(b_\mu{}^{(\alpha\beta)},g^{\lambda\nu}), \tag{21}$$

which of course also contains derivatives of $b_\mu{}^k$ and $g^{\lambda\nu}$. Now form the variation

$$\delta\int\left[L(b_\mu{}^{(\alpha\beta)},g^{\lambda\nu})-L\left(\left\{{\alpha \atop \beta\mu}\right\},g^{\lambda\nu}\right)\right]\times\sqrt{-g}\,d^n x=0, \tag{22}$$

in which $b_\mu{}^{(\alpha\beta)}$ and $g^{\lambda\nu}$ are independently varied. The resultant equations are satisfied by (15) and (19).

It is a great pleasure to acknowledge the warm hospitality extended to me during my visit to the Institute of Theoretical Physics at Wrocław where this paper was written.

[1]C. N. Yang and R. L. Mills, Phys. Rev. 96, 191 (1954).

[2]S. Mandelstam, Ann. Phys. (New York) 19 1, 25 (1962); I. Białynicki-Birula, Bull. Acad. Pol. Sci., Ser. Sci. Math. Astron. Phys. 11, 135 (1963).

[3]We use the summation convention for repeated indices. Greek indices run from 1 to n. Lower case Latin indices run from 1 to m. m of course is also the dimension of the adjoint representation of G. Upper case Latin indices run from 1 to M, where M is the dimension of a representation of G.

[4]T. T. Wu and C. N. Yang, in *Properties of Matter under Unusual Conditions*, edited by H. Mark and S. Fernbach (Wiley, New York, 1969), p. 349.

[5]R. Utiyama, Phys. Rev. 101, 1957 (1956), had concluded that Einstein's equations are gauge-field equations. We believe that was an unnatural interpretation of gauge fields.

Yang's Comments Concerning Gravitational and Gauge Fields

C. N. Yang. Talking about Physics Research and Teaching

The 5th talk at the Graduate School of the
Chinese University of Science and Technology,
Beijing, China, (1986/5/27 - 1986/6/12).

XX: What is your view concerning the unification of gravitational field and gauge field?

Yang: Comparing their formulas, one finds that they are very similar. There is no question that they have an intimate relation. But as to what kind of relation after all, it is still a controversial problem.

Both $F_{\mu\nu}$ in gauge field and $R_{\mu\nu}$ in gravitational field are curvatures in geometry. $R_{\mu\nu}$ is the second order derivative of $g_{\mu\nu}$. Therefore, Einstein's gravitational field equation is a second order differential equation of $g_{\mu\nu}$, while the equation of motion $\partial^{\mu} F_{\mu\nu} = ...$ for gauge field is a first order differential equation of curvature. This is what happens in electromagnetism. The concept that both $F_{\mu\nu}$ and $R_{\mu\nu}$ are curvatures is absolutely primary. I think that this is absolutely unchangeable. Since $\partial^{\mu} F_{\mu\nu} = ...$ is the first order differential equation of curvature, the gravitational field should be the third order differential equation of $g_{\mu\nu}$. This is a clue that Einstein's theory of gravity needs to be modified. In 1974, when the geometric structure was clarified, I proposed a new gravitational equation. However, I knew how to write down the left-hand-side of the equation and could not write down the right-hand-side. At present, this problem has not been solved.

VOLUME 35, NUMBER 5 PHYSICAL REVIEW LETTERS 4 AUGUST 1975

Yang's Gravitational Field Equations*

Wei-Tou Ni
Institute of Physics, National Tsing Hua University, Hsinchu, Taiwan, Republic of China
(Received 27 March 1975)

Yang's gravitational field equations *in vacuo* can be regarded as "derivative" equations of both Einstein's equations and Nordström's equations, and embrace all their solutions. Yang's equations admit monopole gravitational radiations; therefore no analog of the Birkhoff theorem can be valid. The most general static spherical-symmetric solution contains four arbitrary parameters. In particular, $ds^2=-dt^2+(1+c_1/r+c_2r^2)^{-1}dr^2+r^2(d\theta^2+\sin^2\theta \times d\varphi^2)$ is a two-parameter exact solution. This metric possesses no gravitational red shifts.

Recently by using an integral formalism for gauge fields, Yang[1] has proposed the following gravitational field equations for pure spaces:

$$R_{ij;k}-R_{ik;j}=0. \tag{1}$$

It is interesting to notice that (1) are satisfied by all vacuum solutions of Einstein's general relativity

$$G_{ij}+\Lambda g_{ij}=0 \tag{2}$$

and Nordström's second theory[2-4] (in Einstein-Fokker form)

$$R=0,\quad C_{ijkl}=0, \tag{3}$$

where C_{ijkl} is the Weyl tensor. In fact, from Bianchi identities and the definition of the Weyl tensor, a simple calculation shows that (1) are equivalent to

$$G_{ij;k}-G_{ik;j}=0, \tag{4}$$

or

$$R_{,i}=0,\quad C_{ijkl;}{}^{l}=0. \tag{5}$$

Equations (4) are differentiated equations[5] (antisymmetrized in j and k) of (2); while (5) are differentiated equations (summed over l) of (3). Hence Yang's theory is a derivative theory of both Einstein's theory and Nordström's theory. It is amusing and intriguing that the two eminent and structurally different theories of gravity could be embraced in a single set of equations.

Since Nordström's theory admits monopole radiations, Yang's theory admits them too. Therefore no analog of the Birkhoff theorem of general relativity could be valid for Yang's theory. More specifically, (1) have the following time-dependent spherical-symmetric solution (the time dependence cannot be transformed away by coordinate changes):

$$ds^2=\left(\frac{c_0\,tf(r-t)}{r}+\frac{g(r+t)}{r}\right)^2(-dt^2+dr^2+r^2d\Omega^2), \tag{6}$$

where c_0 is an arbitrary constant, and f and g are two arbitrary functions. Hence spherical symmetry does not imply time independence.[6]

In the static spherical-symmetric case, the metric can be put in the form

$$ds^2=-e^{2\Phi(r)}dt^2+e^{2\Lambda(r)}dr^2+r^2(dr^2+\sin^2\theta\,d\psi^2). \tag{7}$$

For this metric (1) reduce to

$$\Lambda''-2\Lambda'^2+\Lambda'\Phi'+\Phi'^2+r^{-2}(1-e^{2\Lambda})=0 \tag{8}$$

and

$$\Phi'''+2\Phi''\Phi'-3\Phi''\Lambda'-2\Lambda'\Phi'^2-\Lambda''\Phi'+2\Lambda'^2\Phi'+2r^{-1}[\Phi''-2\Phi'\Lambda'-\Lambda''+2\Lambda'^2]-2r^{-2}\Phi' -2r^{-3}(1-e^{2\Lambda})=0. \tag{9}$$

VOLUME 35, NUMBER 5 PHYSICAL REVIEW LETTERS 4 AUGUST 1975

An integration of (9) gives

$$\Phi'' + \Phi'^2 - \Phi'\Lambda' + 2r^{-2}(\Phi' - \Lambda') + r^{-2}(1 - e^{2\Lambda}) + \tfrac{1}{2}R_0 e^{2\Lambda} = 0, \qquad (10)$$

where R_0 is the constant scalar curvature. Both (8) and (10) are second order in Λ and Φ'; hence the general solution of (8) and (10) has four arbitrary parameters. (Because of nonlinearity there might be a discrete number of such sets of four parameters.) The additive constant in Φ can be removed by a change of time scale. Therefore the general static spherical-symmetric solution has four arbitrary parameters. This demonstrates that the solution of (1) is much richer than that of (2) (two parameters) or (3) (one parameter). As a matter of fact

$$ds^2 = -dt^2 + (1 + c_1/r + c_2 r^2)^{-1} dr^2 + r^2(d\theta^2 + \sin^2\theta\, d\varphi^2) \qquad (11)$$

is a solution of (1) but not of (2) or (3); (11) possesses no gravitational red shifts.[7] The problems of boundary conditions and sources for (1) deserve extensive studies to clear up this richness of solutions. In view of the present success of the renormalization of gauge theories, these studies could contribute to the solution of the long-standing problems of quantum gravity.

I am grateful to K.S. Cheng, J. C. Shaw, and E. Yen for their helpful discussions on the background of Yang's paper.

---

*Work supported in part by the National Science Council, Republic of China.

[1]C. N. Yang, Phys. Rev. Lett. 33, 445 (1974).

[2]G. Nordström, Ann. Phys. (Leipzig) 42, 533 (1913), and 43, 1101 (1914).

[3]A. Einstein and A. D. Fokker, Ann. Phys. (Leipzig) 44, 321 (1914).

[4]W.-T. Ni, Astrophys. J. 176, 769 (1972).

[5]The second term in (2) is a constant term and, hence, does not contribute to (4).

[6]In view of this result, the recent claim of Thompson [Phys. Rev. Lett. 34, 507 (1975)] that the Birkhoff theorem of general relativity generalizes to (1) is invalid.

[7]W.-T. Ni, report presented at the annual meeting of the Chinese Physical Society, Chung-Li, Taiwan, 26 January 1975 (unpublished).

Einstein Lagrangian as the translational Yang-Mills Lagrangian*

Y. M. Cho†

The Enrico Fermi Institute, and The Department of Physics, University of Chicago, Chicago, Illinois 60637

(Received 1 December 1975; revised manuscript received 6 May 1976)

The gauge theory of translation with a Yang-Mills–type Lagrangian quadratic in the field strengths is shown to be precisely Einstein's theory of gravitation and the corresponding gauge transformation is identified as the general coordinate transformation. The gauge potentials of the translation group are interpreted as the nontrivial part of the vierbein fields and the gauge field strengths are given in terms of the anholonomity of the local orthonormal basis one starts with.

I. INTRODUCTION

Gauge theories of the Yang-Mills type and Einstein's theory of gravitation have a common feature: the self-interaction of the fields. Then one is led to ask whether Einstein's theory itself is a gauge theory. Of course this is an old question and many people[1-3] have suggested that Einstein's theory can be viewed as the gauge theory of the four-dimensional translation group T_4. Unfortunately certain features seem not to have been fully clarified so far, and it is precisely these features that bear out the complete relationship between the Yang-Mills and Einstein theories.

In this paper we show that if one only applies the gauge principle (this includes a Yang-Mills—type Lagrangian quadratic in the field strengths) for the group of translation T_4 of space-time, the gauge theory that one obtains is unique and becomes precisely Einstein's theory of gravitation. In this T_4 gauge formalism of Einstein's theory the translational gauge potentials are identified as the nontrivial part of the vierbein fields and the gauge field strengths are given in terms of the commutator coefficients (i.e., the anholonomity) of the local orthonormal basis one starts with.

To prove that the unique gauge theory of the translation is Einstein's theory, it is important to observe that although the gauge group T_4 is Abelian, it is not an internal-symmetry group and acts on space-time itself. Fortunately the geometric meaning of gauge theories has been well understood by now in terms of the bundle picture.[4-6] The power of this bundle picture has been appreciated by Cho and Freund[5,6] in unifying gauge theories with gravitation and also recently by Wu and Yang.[7] In the following we will first prove our claim in a formal way constructing the eight-dimensional bundle of the translation group T_4 over space-time and then will give a precise physical meaning to this translational bundle. For the details about the bundle formalism of gauge theories we refer the reader to Ref. 5.

II. THE GAUGE THEORY OF THE TRANSLATION GROUP

Let us assume that the structural group G of our bundle P is T_4 with four commuting generators ξ_α $(\alpha=1,2,3,4)$,

$$[\xi_\alpha, \xi_\beta]=0, \tag{1}$$

and that the base manifold M is the four-dimensional space-time with an orthonormal basis at each point, i.e., four orthonormal vector fields e_i $(i=1,2,3,4)$ with the commutation relations

$$[e_i, e_j] = T_{ij}{}^k e_k. \tag{2}$$

Of course the basis independence of a theory is one of the basic principles in physics, and one can choose any other basis if one wants to, but for obvious reasons the local orthonormal basis (2) is the natural one to start with in our problem. Notice that this orthonormal basis is not in general a coordinate basis since the basis vectors do not commute. If we introduce a coordinate basis ∂_μ $(\mu=1,2,3,4)$ with

$$[\partial_\mu, \partial_\nu]=0,$$

then e_i can be written in terms of the vierbein fields h_i^μ,

$$e_i = h_i^\mu \partial_\mu,$$

and correspondingly we have

$$T_{ij}{}^k = (\partial_i h_j^\mu - \partial_j h_i^\mu) h_\mu^k. \tag{3}$$

Here $\partial_i = h_i^\mu \partial_\mu$ is the directional derivative in the direction of e_i and h_μ^k are the inverse vierbein fields

$$h_\mu^i h_k^\mu = \delta_k^i, \quad h_i^\mu h_\nu^i = \delta_\nu^\mu.$$

Observe that due to the commutation relation (2) of the basis e_i, the directional derivatives ∂_i do not commute either.

At this point we would like to emphasize that all the above expressions are just a matter of a formalism and we have *not* assumed that our space-time is curved. Eventually we will create a curvature by introducing gauge fields associated with T_4 symmetry, but we start with a *flat* space-time and so far our h_i^μ remain trivial.

Now, given a connection form[8] $\omega = \omega^\alpha \xi_\alpha$ in the bundle P, the gauge potentials B_i^α are as usual given by the connection coefficients of $\bar{e}_i$, the lift of e_i into a four-dimensional gauge-defining submanifold σ, i.e., a cross section of P:

$$\omega^\alpha(\bar{e}_i) = \kappa B_i^\alpha, \tag{4}$$

where we have introduced a dimensional constant κ (of dimension of a length) to give the canonical dimension to the gauge potentials B_i^α. This κ will serve as the coupling constant for the gauge group T_4 and will be related to the gravitational constant later on.

With $\hat{e}_i$ $(i = 1, 2, 3, 4)$ as the horizontal lift of e_i and ξ_α^* $(\alpha = 1, 2, 3, 4)$ as the fundamental vector fields which are vertical, we clearly have

$$\begin{aligned} &[\xi_\alpha^*, \xi_\beta^*] = 0, \\ &[\xi_\alpha^*, \hat{e}_k] = 0, \\ &[\hat{e}_i, \hat{e}_j] = T_{ij}{}^k \hat{e}_k - \kappa G_{ij}{}^\alpha \xi_\alpha^*, \end{aligned} \tag{5}$$

where G_{ij}^α are the vertical components of the commutator coefficients of the horizontal lift vector fields $\hat{e}_i$. The first two equations come from the definition and the third is due to the fact that the projection of $[\hat{e}_i, \hat{e}_j]$ down to the base manifold is the same as $[e_i, e_j]$. Notice that because of the Abelian character of the ξ_α^*'s the group action on the bundle space is really a translation and there is no "rotation" whatsoever. Mathematically this means that the holonomy group of P is T_4 and not the Lorentz group.

Let us recall that in the bundle picture the gauge field strengths are given by the vertical commutator coefficients of two horizontal vector fields, i.e., by $G_{ij}{}^\alpha$. To find the gauge field strengths in terms of potentials B_i^α, notice that from Eq. (4) and from the definition of ω^α,

$$\begin{aligned} &\omega^\alpha(\hat{e}_i) = 0, \\ &\omega^\alpha(\xi_\beta^*) = \delta_\beta^\alpha, \end{aligned}$$

it follows that

$$\hat{e}_i = \bar{e}_i - \kappa B_i^\alpha \xi_\alpha^*,$$

and

$$\begin{aligned} [\hat{e}_i, \hat{e}_j] &= [\bar{e}_i - \kappa B_i^\alpha \xi_\alpha^*, \bar{e}_j - \kappa B_j^\beta \xi_\beta^*] \\ &= T_{ij}{}^k \bar{e}_k - \kappa(\partial_i B_j^\alpha - \partial_j B_i^\alpha)\xi_\alpha^* \\ &= T_{ij}{}^k \hat{e}_k - \kappa[(\partial_i B_j^\alpha - \partial_j B_i^\alpha) - T_{ij}{}^k B_k^\alpha]\xi_\alpha^*. \end{aligned} \tag{5'}$$

Thus one has

$$G_{ij}{}^\alpha = (\partial_i B_j^\alpha - \partial_j B_i^\alpha) - T_{ij}{}^k B_k^\alpha . \tag{6}$$

Now following Kibble's suggestion[2] we interpret the gauge potentials B_i^μ as the nontrivial part of the vierbein fields

$$h_i^\mu = \delta_i^\mu + \kappa B_i^\mu . \tag{7}$$

This means that *we now have created the curvature of space-time by introducing the gauge potentials B_i^μ for T_4 and making the vierbein fields h_i^μ nontrivial.* Notice that the decomposition of h_i^μ into δ_i^μ and B_i^μ is basis-dependent since δ_i^μ is not invariant under a rotation of the local orthonormal basis e_i.

From Eqs. (3), (6), and (7) one finds that

$$\begin{aligned} G_{ij}{}^\alpha &= (\partial_i B_j^\alpha - \partial_j B_i^\alpha) - T_{ij}{}^k B_k^\alpha \\ &= \frac{1}{\kappa} T_{ij}{}^k (h_k^\alpha - \kappa B_k^\alpha) \\ &= \frac{1}{\kappa} T_{ij}{}^k \delta_k^\alpha . \end{aligned} \tag{6'}$$

The gauge field strengths G_{ij}^α are thus determined by the commutation coefficients $T_{ij}{}^k$ of the orthonormal basis vectors that one starts with. We would like to emphasize here that once the connection ω (i.e., the gauge potentials in physical terms) is given, the gauge field strengths G_{ij}^α are uniquely determined from the geometrical structure of the bundle and are not something that one can define otherwise as sometimes suggested.[3]

Gauge transformations in this picture are changes of bundle cross sections.[4,5] If we change the cross section σ to σ' by a four-translation $\theta^\alpha(x)$ $(\alpha = 1, 2, 3, 4)$ in the four-dimensional fiber space [geometrically $\theta^\alpha(x)$ are simply a set of transition functions that relate σ' to σ], we clearly have

$$\bar{e}_i' = \bar{e}_i + (\partial_i \theta^\alpha)\xi_\alpha^*$$

and

$$B'{}_i^\alpha = \frac{1}{\kappa}\omega^\alpha(\bar{e}_i') = B_i^\alpha + \frac{1}{\kappa}\partial_i\theta^\alpha . \tag{8}$$

From Eq. (7) this means that under the gauge transformation we have

$$\begin{aligned} h_i^\mu \to h'{}_i^\mu &= h_i^\mu + \partial_i \theta^\mu \\ &= h_i^\alpha(\delta_\alpha^\mu + \partial_\alpha \theta^\mu) \\ &= h_i^\alpha X_\alpha^\mu , \end{aligned}$$

where

$$X_\alpha^\mu = \delta_\alpha^\mu + \partial_\alpha \theta^\mu ,$$

and now *the gauge transformation is unambiguously identified as a general coordinate transformation in the coordinate basis* ∂_μ:

$$\begin{aligned} \partial_\mu \to \partial_\mu' &= (\delta_\mu^\alpha + \partial_\mu \theta^\alpha)\partial_\alpha \\ &= X_\mu^\alpha \partial_\alpha . \end{aligned} \tag{9}$$

Notice that gauge transformations (or equivalently general coordinate transformations) do not change the orthonormal basis e_i: They change the coordinate basis ∂_μ. Also notice that under a local gauge transformation one has

$$G_{ij}'^{\alpha} = (\partial_i B_j'^{\alpha} - \partial_j B_i'^{\alpha}) - T_{ij}{}^k B'^{\alpha}_k$$
$$= G_{ij}{}^{\alpha} + \frac{1}{\kappa}(\partial_i \partial_j - \partial_j \partial_i)\theta^{\alpha} - \frac{1}{\kappa} T_{ij}{}^k \partial_k \theta^{\alpha}$$
$$= G_{ij}{}^{\alpha}, \tag{10}$$

i.e., $G_{ij}{}^{\alpha}$ are invariant under gauge transformations as expected for an Abelian gauge group.

Once we have the gauge field strengths $G_{ij}{}^{\alpha}$, we can write in the manner of Yang and Mills the most general Lagrangian quadratic in these $G_{ij}{}^{\alpha}$. Using Eq. (6′) we have

$$\mathcal{L} = \sqrt{-g}\, G_{ij}{}^{\alpha} G_{kl}{}^{\beta}(a\eta^{ik}\eta^{jl}\eta_{\alpha\beta} + b\eta^{ik}\delta^l_\alpha\delta^j_\beta + c\eta^{ik}\delta^j_\alpha\delta^l_\beta)$$
$$= \frac{1}{\kappa^2}\sqrt{-g}\,(aT_{ijk}T_{ijk} + bT_{ijk}T_{ikj} + cT_{ijj}T_{ikk}),$$

where

$$\sqrt{-g} = \det(h^i_\mu),$$

a, b, and c are for the time being arbitrary constants, and we have used the *flat* metric $\eta_{\alpha\beta}$ for the fiber space. Notice that in our formalism we do *not* need a Riemannian metric *a priori*. The crucial question now is whether the constants a, b, and c are really arbitrary. To answer this question let us point out that the above Lagrangian is basis dependent as it is obtained using Eq. (6′). Now, if the theory is going to have any meaning at all, it should not depend upon which orthonormal frame one starts with. This means that if one chooses a different set of e_i's, the Lagrangian should differ only by a total divergence. We now show that this consistency requirement removes all the arbitrariness in a, b, and c.

Notice that under an infinitesimal change of orthonormal frame, one has

$$h^\mu_i(x) \to h'^\mu_i(x) = h^\mu_i(x) + \omega_{ik} h^\mu_k(x), \tag{11}$$

where $\omega_{ik}(x) = -\omega_{ki}(x)$ are six infinitesimal functions so that

$$\delta\mathcal{L} = \sqrt{-g}\,(2aT_{ijk}\delta T_{ijk} + 2bT_{ijk}\delta T_{ikj} + 2cT_{ijj}\delta T_{ikk})$$
$$= \sqrt{-g}\,[2aT_{ijk}(\partial_i\omega_{jk} - \partial_j\omega_{ik}) + 2bT_{ijk}(\partial_i\omega_{kj} - \partial_k\omega_{ij}) + 2cT_{ijj}\partial_k\omega_{ki}]$$
$$= \sqrt{-g}\,[(4a-2b)T_{ijk}\partial_i\omega_{jk} - 2bT_{ijk}\partial_k\omega_{ij} + 2cT_{ijj}\partial_k\omega_{ki}]$$
$$= \sqrt{-g}\,[(4a-2b)T_{ijk}\partial_i\omega_{jk} - (2b+c)T_{ijk}\partial_k\omega_{ij}] - 2c\partial_\mu(h^\mu_i\sqrt{-g}\,\partial_k\omega_{ki}). \tag{12}$$

Here the last equality comes from the following identity:

$$\partial_\mu(h^\mu_i\sqrt{-g}\,\partial_k\omega_{ki}) = -\sqrt{-g}(\tfrac{1}{2}T_{ijk}\partial_k\omega_{ij} + T_{ijj}\partial_k\omega_{ki}).$$

Now in the last line of Eq. (12) the third term is explicitly a total divergence. But each of the first two terms cannot be made into a total divergence and one is forced to choose $a:b:c = 1:2:-4$ to satisfy the consistency requirement. So the Lagrangian should have the form

$$\mathcal{L}_{T_4} = \frac{1}{\kappa^2}\sqrt{-g}\,(\tfrac{1}{4}T_{ijk}T_{ijk} + \tfrac{1}{2}T_{ijk}T_{ikj} - T_{ijj}T_{ikk}). \tag{13}$$

We would like to emphasize that *any other linear combination in $\mathcal{L}$ does not yield a meaningful theory.*

Now it is readily seen that $\mathcal{L}_{T_4}$ is (again up to a divergence) precisely Einstein's Lagrangian. This completes our argument that the four-dimensional translational gauge theory with Lagrangian quadratic in the field strengths is precisely Einstein's theory of gravitation. The fact that the Lagrangian (13) is equivalent to Einstein's one is of course well known. But the geometrical meaning of this Lagrangian does not seem to have been fully understood. *We now understand it as the translational gauge formalism of Einstein's theory of gravitation.*

At this point one may wonder whether we have required a Lorentz gauge invariance by imposing the independence of the theory under the local Lorentz transformation (11) of the orthonormal basis e_i. Even so, however, one does not need to introduce Lorentz gauge fields in one's theory if one has only scalars and internal gauge fields as one's source fields.[9] This is so because scalars are singlets under the Lorentz transformation and also the internal gauge fields do not couple directly to the gauge fields of the Lorentz group owing to the gauge invariance of the internal symmetry. In any event it should be made clear that the independence of the theory under the local Lorentz transformation (11) is a consistency condition that one has to require for one's theory.[10] In the presence of spinor source fields, of course, this consistency condition naturally leads us to introduce the gauge fields of the Lorentz group to the theory and one obtains the Einstein-Cartan[11] theory of gravitation as has been argued by Kibble.[2,12] In this case the translational gauge group is replaced by the Poincaré group.[9]

III. PHYSICAL INTERPRETATION

We now wish to make clear how in the presence of source fields the translation group T_4 acts on them. By doing so we will give a precise physical meaning to our bundle of translation group. Remember that we have treated our bundle just like a principal fiber bundle of an internal-symmetry group except for one crucial difference, i.e., the identification of Eq. (7). This equation interlocks the fiber space of the translation group T_4 with space-time and allows us to speak of our T_4 as a space-time symmetry rather than an internal symmetry. We will first justify this basic equation and clarify the meaning of the translation group T_4.

Let us consider a scalar field $\phi(x)$ as the source field for simplicity and start with an action integral written in the usual coordinate basis of a global Minkowski frame:

$$I = \int (\tfrac{1}{2}\eta^{\mu\nu}\partial_\mu\phi\partial_\nu\phi - \tfrac{1}{2}m^2\phi^2)d^4x. \qquad (14)$$

Clearly under a global translation of the coordinate

$$x^\mu \to x'^\mu = x^\mu + \epsilon^\mu, \qquad (15)$$

where ϵ^μ are four infinitesimal parameters, one has

$$\delta'\phi \equiv \phi(x') - \phi(x) = \epsilon^\mu\partial_\mu\phi(x). \qquad (16)$$

This suggests that one may view the generators ξ_μ of T_4 as the coordinate derivatives ∂_μ. For the moment let us take this point of view. Now under a local translation one has to introduce a covariant derivative to keep the action integral (14) invariant. But since the covariant basis is the one in which the components of the metric remain flat, it is quite natural to identify ∂_i as the covariant derivative of ∂_μ. Then one is led to the following equation:

$$\begin{aligned}\partial_i &= h_i^\mu\partial_\mu \\ &= (\delta_i^\mu + \kappa B_i^\mu)\partial_\mu.\end{aligned} \qquad (17)$$

In short, interpreting ξ_μ as ∂_μ and identifying the local orthonormal basis as the covariant basis one easily obtains Eq. (7). In fact this rather intuitive interpretation has been given by Kibble.[2]

Now one can easily write down the action integral which is invariant under a local translation and independent of a choice of a local orthonormal basis. Including the kinetic term of the translational gauge fields one has[13]

$$I = \int \sqrt{-g}\Big[\tfrac{1}{2}\eta^{ik}\partial_i\phi\partial_k\phi - \tfrac{1}{2}m^2\phi^2 + \frac{1}{\kappa^2}(\tfrac{1}{4}T_{ijk}T_{ijk} + \tfrac{1}{2}T_{ijk}T_{ikj} - T_{ijj}T_{ikk})\Big]d^4x. \qquad (18)$$

Clearly the Lagrangian (18) describes Einstein's theory of gravitation in the presence of the scalar source field $\phi(x)$ provided

$$\kappa^2 = 16\pi G, \qquad (19)$$

where G is the gravitational constant. Thus the coupling parameter κ of the group T_4 is indeed related to the gravitational constant.

Now we will give another interpretation, i.e., the one for our bundle formalism of the translational gauge theory, which allows us to keep the complete parallelism between a gauge theory of an internal symmetry and that of the space-time symmetry T_4. Notice that in the above interpretation we started from the usual coordinate basis of a global Minkowski frame. But clearly one should be able to start with any other basis as well. Indeed it may be more desirable to construct the theory in a basis-independent way. So let us start from the beginning with a local orthonormal frame e_i and write down the action integral (14) as

$$I = \int (\tfrac{1}{2}\eta^{ik}\partial_i\phi\partial_k\phi - \tfrac{1}{2}m^2\phi^2)\sqrt{-g}\,d^4x. \qquad (14')$$

In this case one is led to have the following covariant derivative D_i for the bundle of T_4:

$$D_i = \partial_i + B_i^\mu\xi_\mu. \qquad (20)$$

This is dictated by the geometry of the bundle[5] since the covariant derivative of ∂_i in the bundle formalism is given by the horizontal lift $\hat{\partial}_i$ of ∂_i. Also in this picture the group actions are interpreted to transform the field components along the fiber space keeping the *physical* space-time points invariant. This means that under the translation (15) one should have

$$\begin{aligned}\delta\phi &= \delta\theta^\mu\xi_\mu^*\phi = \delta\theta^\mu\frac{\partial}{\partial\theta^\mu}\phi \\ &= \phi'(x') - \phi(x) = 0,\end{aligned} \qquad (21)$$

where θ^μ ($\mu = 1, 2, 3, 4$) are the fiber-space coordinate variables as before and we have identified the generators ξ_μ^* as $\partial/\partial\theta^\mu$, which is again dictated by the geometry of the bundle.[5] Notice the difference between this equation and Eq. (16). Thus in this bundle picture ξ_μ^* annihilate the source fields as the fields remain invariant at each physical space-time point and do not depend upon the fiber-space variables. Observe here that this fiber-space dependence of the source fields has been derived, not assumed, from what one means by the translational invariance. The way the gauge fields of T_4 couple to source fields is then given by Eq. (7):

$$\begin{aligned}D_i\phi &= (\partial_i - \kappa B_i^\mu\xi_\mu^*)\phi \\ &= \partial_i\phi = (\delta_i^\mu + \kappa B_i^\mu)\partial_\mu\phi.\end{aligned} \qquad (22)$$

Thus the gauge fields of T_4 couple to source fields *as if* the generators of T_4 were the coordinate derivatives ∂_μ and one arrives at the action integral (18) as before due to Eq. (7).

One can choose either of the interpretations above. The first one emphasizes too much the usual global Minkowski coordinate frame but gives an intuitively clear meaning to the translational symmetry, whereas the other has the merit of treating the theory in a basis-independent way and allows us to keep the parallelism between the gauge theory of the space-time symmetry T_4 and that of an internal symmetry, with the identification of Eq. (7).

IV. CONCLUSIONS

We have shown that the gauge theory of the four-dimensional translation group is unique and becomes precisely the vierbein formalism of Einstein's theory of gravitation as far as one chooses the Lagrangian to be the lowest possible combinations, quadratic in field strengths. The gauge potentials of the translation group are interpreted as the nontrivial part of the vierbein fields and the corresponding gauge transformations are shown to be the general coordinate transformations.

Yang[14] has recently proposed a GL(4) gauge theory of gravitation of a Yang-Mills quadratic type which differs from Einstein's theory and may conflict with observation.[15] A Yang-Mills–type gauge theory of gravitation which gives Einstein's theory is, as we have seen, the gauge theory of the translation group T_4.

ACKNOWLEDGMENTS

It is a great pleasure to thank Professor P. G. O. Freund, Professor R. Geroch, and Professor Y. Nambu for many illuminating discussions, especially about the consistency condition, and for careful readings of the manuscript. I am also grateful to Dr. J. Friedman and Dr. R. Wald for many discussions.

---

*Work supported in part by the National Science Foundation under Contract No. PHY74-08833 A01.

†Present address: Department of Physics, New York University, New York, New York 10003.

[1]See, e.g., S. L. Glashow and M. Gell-Mann, Ann. Phys. (N.Y.) 15, 437 (1961); R. P. Feynman, *Lectures on Gravitation* (Caltech, Pasadena, Calif., 1963).

[2]T. W. B. Kibble, J. Math. Phys. 2, 212 (1961).

[3]See, e.g., K. Hayashi and T. Nakano, Prog. Theor. Phys. 38, 491 (1967); G. D. Kerlick, thesis, Princeton Univ., 1975 (unpublished); J. M. Nester, report, 1974 (unpublished). See also, F. A. Kaempffer, Phys. Rev. 165, 1420 (1968).

[4]A. Trautman, Rep. Math. Phys. 1, 29 (1970).

[5]Y. M. Cho, J. Math. Phys. 16, 2029 (1975).

[6]Y. M. Cho and P. G. O. Freund, Phys. Rev. D 12, 1711 (1975).

[7]T. T. Wu and C. N. Yang, Phys. Rev. D 12, 3845 (1975). See also L. N. Chang, K. I. Macrae, and F. Mansouri, *ibid.* 13, 235 (1976).

[8]Our notation here is the same as the one in Ref. 5.

[9]Y. M. Cho, Phys. Rev. D (to be published).

[10]There appear different opinions in the literature with which we do not agree. See, e.g., K. Hayashi, Nuovo Cimento 16, 639 (1973); Gen. Relativ. Gravit. 4, 1 (1974).

[11]É. Cartan, C. R. Acad. Sci. 174, 593 (1922); Ann. École Normale 40, 325 (1923); 41, 1 (1924).

[12]Kibble's work followed after the pioneering work of Utiyama. See R. Utiyama, Phys. Rev. 101, 1597 (1956). See also D. W. Sciama, *Recent Developments in General Relativity* (Pergamon, New York, 1962).

[13]In this paper our signature of the metric is $(+,-,-,-)$.

[14]C. N. Yang, Phys. Rev. Lett. 33, 445 (1974).

[15]R. Pavelle, Phys. Rev. Lett. 34, 1114 (1975).

Volume 119B, number 4,5,6 PHYSICS LETTERS 23/30 December 1982

DE SITTER AND POINCARÉ GAUGE-INVARIANT FERMION LAGRANGIANS AND GRAVITY ☆

J.P. HSU
Physics Department, Southeastern Massachusetts University, North Dartmouth, MA 02747, USA

Received 20 July 1982

We present a new fermion lagrangian which possesses exact symmetry under the local de Sitter group. The lagrangian involves new "scale gauge fields" related to the newtonian force and the usual Yang–Mills "phase gauge fields" related to a new "gravitational spin force" between two fermions. Generalization of the usual gauge theory for *external* symmetry groups is also discussed.

It has been suggested that gravity is related to gauge fields of four-dimensional symmetry such as the de Sitter group [1,2]. The idea is quite interesting because the de Sitter group possesses the maximum four-dimensional symmetry [3] and is the unique generalization of the Poincaré group. It also suggests the existence of a new "gravitational spin force" between objects with nonzero net spin densities. The de Sitter group is a rotational group in de Sitter space, which is the hypersurface of a four-dimensional sphere of a hyperbolic character in one direction, embedded in a five-dimensional space. The radius of the sphere is denoted by L. The de Sitter group reduces to the Poincaré group in the flat space limit $L \to \infty$.

One important ingredient in a realistic gauge theory of gravity is the fermion field – a source of the gravitational field. But in previous discussions [1,2] one either ignored the fermion field or discussed a fermion lagrangian which has only *approximate* symmetry under local de Sitter gauge transformations. It appears that one cannot get a fermion lagrangian with *exact* external gauge symmetry if one just employs the usual Yang–Mills fields, i.e. "phase gauge fields" [4].

In this paper, we present a new fermion lagrangian, which has exact symmetry under the local de Sitter group. It is necessary that the lagrangian involves new "scale gauge fields" in addition to the usual Yang–Mills "phase gauge fields". They have different transformation property and, therefore, must be treated as different and independent fields.

Let us consider the generalization of $\gamma^\mu \partial_\mu \psi$ in the form for a non-abelian external symmetry group:

$$\Gamma^\mu D_\mu \psi , \tag{1}$$

where Γ^μ involves both the Dirac matrices and scale gauge fields e^μ_A, and the gauge-covariant derivative D_μ contains phase gauge fields $b^A_\mu = (b^i_\mu, b^a_\mu)$:

$$\Gamma^\mu \equiv e^\mu_A \gamma^A = e^\mu_k \gamma^k + e^\mu_{jk}\, \mathrm{i}(\gamma^j\gamma^k - \gamma^k\gamma^j)/4L$$
$$\equiv E^\mu_A Z^A , \tag{2}$$

$$D_\mu \equiv \partial_\mu - \mathrm{i} g b_\mu , \qquad b_\mu \equiv b^A_\mu Z_A , \tag{3}$$

$$Z_A = (Z_i, Z_a) = (\gamma_i/2L, \mathrm{i}(\gamma_j\gamma_k - \gamma_k\gamma_j)/4) , \quad a \equiv jk ,$$

$$\{\gamma_j, \gamma_k\} = 2\eta_{jk} , \quad \eta_{jk} = (1, -1, -1, -1) ,$$

$$E^\mu_A = (2Le^\mu_i, e^\mu_{jk}/L) .$$

The quantity Z_A is the matrix representation of the SO(3,2) de Sitter group generators:

$$[Z_B, Z_C] = \mathrm{i} f^A_{BC} Z_A , \quad A = i, jk; \text{etc.} \tag{4}$$

The local de Sitter gauge transformations are given by

$$\psi \to \psi' = \xi_\mathrm{d} \psi ,$$
$$b_\mu \to b'_\mu = \xi_\mathrm{d} b_\mu \xi_\mathrm{d}^{-1} + (\partial_\mu \xi_\mathrm{d})\, \xi_\mathrm{d}^{-1}/(\mathrm{i}g) , \tag{5}$$

☆ The work is supported in part by Southeastern Massachusetts University.

Volume 119B, number 4,5,6 PHYSICS LETTERS 23/30 December 1982

$$\Gamma^\mu \to \Gamma'^\mu = \xi_d \Gamma^\mu \xi_d^{-1}, \qquad (5\text{ con'd})$$

where

$$\xi_d = \exp[i\omega^A(x) Z_A] .$$

The gauge functions $\omega^A(x) = (\omega^i(x), \omega^a(x))$ are real and arbitrary.

It can be shown that $\bar\psi(\Gamma^\mu D_\mu + m)\psi$ is invariant under the local de Sitter gauge transformations (5):

$$\bar\psi'(\Gamma'^\mu D'_\mu + m)\psi' = \bar\psi(\Gamma^\mu D_\mu + m)\psi . \qquad (6)$$

We stress that this symmetry property holds if and only if both e_i^μ and e_{jk}^μ are introduced.

Note that the field e_A^μ is dimensionless and is related to a change in the scale rather than a change in the phase [4], so that e_A^μ may be termed a "scale gauge field". In view of the presence of this new scale gauge field, the present gauge theory is a generalization of the Yang–Mills theory. Such a generalization appears to be necessary because the de Sitter group is an *external* symmetry group, in which the generator Z_A does not commute with γ_k, in contrast to the case of an internal symmetry group.

The phase field strength $F_{\mu\nu}^A$ is given by

$$(D_\mu D_\nu - D_\nu D_\mu)\psi \equiv igF_{\mu\nu}^A Z_A \psi , \qquad (7)$$

i.e.

$$F_{\mu\nu}^A = \partial_\mu b_\nu^A - \partial_\nu b_\mu^A + g f_{BC}^A b_\mu^B b_\nu^C . \qquad (8)$$

One can verify that $F_{\mu\nu} \equiv F_{\mu\nu}^A Z_A$ is gauge covariant and transforms as follows:

$$F_{\mu\nu} \to F'_{\mu\nu} = \xi_d F_{\mu\nu} \xi_d^{-1} . \qquad (9)$$

Thus, $\mathrm{Tr}(F_{\mu\nu}F_{\alpha\beta})$ is a gauge-invariant quantity:

$$\mathrm{Tr}(F'_{\mu\nu}F'_{\alpha\beta}) = \mathrm{Tr}(F_{\mu\nu}F_{\alpha\beta}) , \qquad (10)$$

which is usually used as the lagrangian for the phase field b_μ^A.

We observe that $\mathrm{Tr}(\Gamma^\mu\Gamma^\nu)$ is gauge invariant:

$$\mathrm{Tr}(\Gamma'^\mu\Gamma'^\nu) = \mathrm{Tr}(\Gamma^\mu\Gamma^\nu) . \qquad (11)$$

Thus, we expect that e_A^μ enters the lagrangian for the scale field through the combination

$$\begin{aligned}\bar g^{\mu\nu} &\equiv \mathrm{Tr}(\Gamma^\mu\Gamma^\nu/4) \\ &= \eta^{ik} e_i^\mu e_k^\nu + e_{ij}^\mu e_{mn}^\nu \eta^{im}\eta^{jn}/2L^2 \\ &\to g^{\mu\nu} \quad \text{as } L \to \infty , \qquad (12)\end{aligned}$$

$$g^{\mu\nu} \equiv e_i^\mu e_k^\nu \eta^{ik} . \qquad (12\text{ con'd})$$

For large L, $\bar g^{\mu\nu}$ is approximately the same as $g^{\mu\nu}$. In the limit $L \to \infty$, it is natural to interpret e_i^μ as the vierbein component and $g^{\mu\nu}$ as the spacetime metric. Thus we can interpret $\bar g^{\mu\nu}$ as the spacetime metric in the present theory with the de Sitter gauge group. We are able to define the affine connection $\bar\Gamma_{\mu\nu}^\alpha$ and the Riemann curvature tensor $\bar R^\alpha{}_{\lambda\mu\nu}$ in terms of the new gauge-invariant metric $\bar g^{\mu\nu}$ by the usual relations:

$$\bar\Gamma_{\mu\nu}^\alpha = \tfrac12 \bar g^{\lambda\alpha}(\partial_\nu \bar g_{\mu\lambda} + \partial_\mu \bar g_{\lambda\nu} - \partial_\lambda \bar g_{\nu\mu}) , \quad \bar g^{\mu\lambda}\bar g_{\lambda\nu} = \delta_\nu^\mu ,$$

$$\bar R^\alpha{}_{\lambda\mu\nu} = \partial_\nu \bar\Gamma_{\lambda\mu}^\alpha - \partial_\mu \bar\Gamma_{\lambda\nu}^\alpha + \bar\Gamma_{\lambda\mu}^\beta \bar\Gamma_{\beta\nu}^\alpha - \bar\Gamma_{\lambda\nu}^\beta \bar\Gamma_{\beta\mu}^\alpha . \qquad (13)$$

In this way, $\bar\Gamma_{\mu\nu}^\alpha$, $\bar R^\alpha{}_{\lambda\mu\nu}$ and $\bar g^{\mu\nu}$ are all invariant under the local de Sitter gauge transformations (5). The invariant lagrangian for these fields is uniquely determined by the principle of gauge invariance and the principle of general covariance:

$$\begin{aligned}&\int d^4x (-\det \bar g_{\mu\nu})^{1/2} [\tfrac12 (i\bar\psi\Gamma^\mu D_\mu\psi + \text{h.c.}) \\ &\quad - m\bar\psi\psi - \tfrac18 \mathrm{Tr}\, F_{\mu\nu}F^{\mu\nu} + (8\pi G)^{-1}\bar R] , \qquad (14)\end{aligned}$$

where G is a constant and $\bar R = \bar R^\alpha{}_{\mu\nu\alpha}\bar g^{\mu\nu}$. Field equations for e_A^μ, b_μ^A and ψ can be determined from the lagrangian (14).

When we take the limit $L \to \infty$, some components of gauge fields, i.e. b_μ^i and e_{jk}^μ, disappear from the theory and we obtain the following lagrangian:

$$\begin{aligned}\mathcal{L} &= (-\det g_{\mu\nu})^{1/2} \\ &\quad \times [\mathcal{L}_\psi - \tfrac18 \mathrm{Tr}(F_{\mu\nu}^a Z_a F_{\alpha\beta}^b Z_b) g^{\mu\alpha}g^{\nu\beta} + (8\pi G)^{-1} R] ,\end{aligned}$$

$$\begin{aligned}R &\equiv R^\alpha{}_{\mu\nu\alpha} g^{\mu\nu} \\ &= (\partial_\alpha \Gamma_{\mu\nu}^\alpha - \partial_\nu \Gamma_{\mu\alpha}^\alpha + \Gamma_{\mu\nu}^\lambda \Gamma_{\lambda\alpha}^\alpha - \Gamma_{\mu\alpha}^\lambda \Gamma_{\lambda\nu}^\alpha) g^{\mu\nu} ,\end{aligned}$$

$$\Gamma_{\mu\nu}^\alpha = \tfrac12 g^{\lambda\alpha}(\partial_\nu g_{\mu\lambda} + \partial_\mu g_{\lambda\nu} - \partial g_{\nu\mu}) ,$$

$$g^{\mu\nu} = e_i^\mu e_k^\nu \eta^{ik} , \qquad g_{\mu\nu} = e_\mu^i e_\nu^k \eta_{ik} ,$$

$$e_\mu^i e_i^\nu = \delta_\mu^\nu , \qquad e_\mu^k e_i^\mu = \delta_i^k ,$$

$$\mathrm{Tr}(F_{\mu\nu}^a Z_a F_{\alpha\beta}^b Z_b) = 2F_{\mu\nu}^{ik} F_{\alpha\beta}^{mn} \eta_{im}\eta_{kn} ,$$

$$a = ik , \quad b = mn , \qquad (15)$$

Volume 119B, number 4,5,6 PHYSICS LETTERS 23/30 December 1982

$$\mathcal{L}_\psi = \tfrac{1}{2}\mathrm{i}\bar{\psi}\Gamma^\mu(\partial_\mu - \mathrm{i}gb_\mu)\,\psi - \tfrac{1}{2}\mathrm{i}\bar{\psi}(\overleftarrow{\partial} + \mathrm{i}gb_\mu)\,\Gamma^\mu\psi - m\bar{\psi}\psi\,, \qquad (15\text{ con'd})$$

where the last term in (15) is identical with Einstein's lagrangian. The lagrangian (15) is invariant under local Poincaré gauge transformations (i.e. the transformations (5) with $L \to \infty$) and general coordinate transformations (with the spacetime metric $g^{\mu\nu}$). The Poincaré gauge-invariant lagrangian (15) differs from that discussed by Kibble [5]. From the viewpoint of symmetry, the lagrangian (15) is more satisfactory than Kibble's lagrangian involving fermions.

Physically, the radius L of the de Sitter space is probably very large. In this case, physical effects of b^i_μ and e^μ_{jk} are negligible. Experimentally, the difference between the de Sitter gauge-invariant lagrangian (14) and the Poincaré gauge-invariant lagrangian (15) cannot be distinguished in the near future. So the *important fields* are b^a_μ and e^μ_i, which are generated by the spin density and mass density, respectively. This can be seen from the field equations derived from (15). For example, the source of b^a_μ in the approximation of static and weak fields is

$$\nabla^2 b^{ik}_\mu(r) = g\epsilon^{ikj}\bar{U}(r)\,\sigma_j U(r)\,, \quad \mu = 0\,, \qquad = 0\,, \quad \mu \neq 0\,, \qquad (16)$$

where $U(r)$ and σ_j are respectively the positive-energy Pauli spinor and the Pauli matrices. Of course, the result (16) can also be derived from the de Sitter gauge-invariant lagrangian (14) with the approximation of very large L.

These gauge fields are interpreted as follows: *Gravity is related to scale gauge fields rather than the usual Yang–Mills gauge fields* because the scale gauge field is generated by the mass density, according to gauge-invariant lagrangians. The massless Yang–Mills field b_μ is generated by the spin density of fermions and corresponds to a new long-range force between two bodies with nonzero spin densities. The strength of this new force is determined by a new dimensionless coupling constant g^2, which is independent of the newtonian gravitational constant G. These hold for both finite L (i.e. the de Sitter group) and infinite L (i.e. the Poincaré group).

Our interpretation of gauge fields for external symmetry groups differs from previous interpretations by Kibble and others [5–7] (see also refs. [2,8]). We may remark that a "contact spin force" between fermions has been discussed by Kibble based on a non-gauge-invariant fermion lagrangian. Since the external four-dimensional symmetry group is a fundamental symmetry of nature, the prediction of the new long-range gravitational spin force should be taken seriously.

We conclude that gauge field theory based on external four-dimensional symmetry groups dictates the presence of a new "scale gauge field", which differs from the Yang–Mills" phase gauge field" [9]. Furthermore, the theory predicts a new long-range gravitational spin force between fermions. It appears that the quantization of these fields cannot be accomplished by a straightforward application of the usual quantization procedure for Yang–Mills fields. This needs further investigation.

References

[1] P.C. West, Phys. Lett. 76B (1978) 569;
P.K. Townsend, Phys. Rev. D15 (1977) 2795;
Wu Yung-shi, Lee Ken-dao and Kuo Han-ying, Kexue Tongbao 19 (1974) 509.

[2] J.P. Hsu, Phys. Rev. Lett. 42 (1979) 934, 1920(E); Nuovo Cimento 61B (1981) 249;
see also: S. Fujita, ed., The Ta-You Wu Festschrift: Science of matter (Gordon and Breach, 1978) pp. 65–73;
L.L. Smalley, preprint.

[3] F.J. Dyson, Bull. Am. Math. Soc. 78 (1972) 635.

[4] C.N. Yang and R.L. Mills, Phys. Rev. 96 (1954) 191;
C.N. Yang, Ann. NY Acad. Sci. 294 (1977) 86; Phys. Today (June 1980) 42.

[5] T.W.B. Kibble, J. Math. Phys. 2 (1961) 212.

[6] See, for example, R. Utiyama, Phys. Rev. 101 (1956) 1597;
C.N. Yang, Phys. Rev. Lett. 33 (1974) 445.

[7] P.C. West, Phys. Lett. 76B (1978) 569;
W. Drechsler, Phys. Lett. 107B (1981) 415.

[8] F.W. Hehl, J. Nitsch and P. von der Heyde, Gravitation and Poincaré gauge field theory with quadratic lagrangian, Univ. of Cologne preprint (1980).

[9] J.P. Hsu, Generalized theory with external gauge symmetry, SMU preprint (1982).

Chapter 8

Alternate Approaches to Gravity: Roads Less Traveled By*

*P. A. M. Dirac, K. Hayashi, T. Shirafuji, A. A. Logunov, M. A. Mestvirishvili, J. P. Hsu

PHYSICAL REVIEW VOLUME 114, NUMBER 3 MAY 1, 1959

Fixation of Coordinates in the Hamiltonian Theory of Gravitation

P. A. M. DIRAC*
Institute for Advanced Study, Princeton, New Jersey
(Received December 10, 1958)

The theory of gravitation is usually expressed in terms of an arbitrary system of coordinates. This results in the appearance of weak equations connecting the Hamiltonian dynamical variables that describe a state at a certain time, leading to supplementary conditions on the wave function after quantization. It is then difficult to specify the initial state in any practical problem.

To remove the difficulty one must eliminate the weak equations by fixing the coordinate system. The general procedure for this elimination is here described. A particular way of fixing the coordinate system is then proposed and its effect on the Poisson bracket relations is worked out.

INTRODUCTION AND NOTATION

THE problem of putting Einstein's equations for the gravitational field into the Hamiltonian form, as a preliminary to quantization, has recently received a good deal of attention, because of the development of mathematical methods sufficiently powerful to make it tractable.

The Hamiltonian form involves the concept of a physical state "at a certain time," which means in a relativistic theory a state on a certain three-dimensional space-like surface in space-time. At first people[1,2] chose the space-like surface independent of the coordinates x^μ, which enabled them to preserve the four-dimensional symmetry of the equations. Later it was realized[3,4] that one could effect a substantial simplification, at the expense of giving up four-dimensional symmetry, by choosing a system of coordinates such that the three-dimensional surfaces $x^0=$constant are all space-like and dealing with the physical states on these surfaces.

The main features of the Hamiltonian formalism will be recapitulated here. The notation will be that used by the author,[4] with the exception that the sign of the $g_{\mu\nu}$ will be changed throughout, to make g_{00} negative. Greek suffixes take on the values 0, 1, 2, 3, lower-case Roman suffixes take on the values 1, 2, 3, the determinant of the $g_{\mu\nu}$ is $-J^2$, the determinant of the g_{rs} is K^2, and the reciprocal matrix to g_{rs} is e^{rs}. A lower suffix added to a field quantity denotes an ordinary derivative, while $_{|\mu}$ added to it denotes the covariant derivative.

We shall deal with the gravitational field in interaction with other fields, or possibly particles. Spinor fields are excluded, as they require a special treatment.

* The author's stay at the Institute for Advanced Study was supported by the National Science Foundation.

[1] F. A. E. Pirani and A. Schild, Phys. Rev. 79, 986 (1950).

[2] Bergmann, Penfield, Schiller, and Zatzkis, Phys. Rev. 80, 81 (1950).

[3] Pirani, Schild, and Skinner, Phys. Rev. 87, 452 (1952).

[4] P. A. M. Dirac, Proc. Roy. Soc. (London) A246, 333 (1958).

We have an action density of the form

$$\mathcal{L}=\mathcal{L}_G+\mathcal{L}_M,$$

where $\mathcal{L}_G$ is the action density of the gravitational field alone, involving the $g_{\mu\nu}$ and their first derivatives, and $\mathcal{L}_M$ is the action density of the other fields, involving the other field quantities, q_M say, and their first derivatives and involving also the $g_{\mu\nu}$, but not derivatives of the $g_{\mu\nu}$.

The gravitational action density is

$$\mathcal{L}_G=(16\pi\gamma)^{-1}Jg^{\mu\nu}(\Gamma_{\mu\rho}{}^{\sigma}\Gamma_{\nu\sigma}{}^{\rho}-\Gamma_{\mu\nu}{}^{\rho}\Gamma_{\rho\sigma}{}^{\sigma}), \tag{1}$$

where γ is the gravitational constant, occurring in the numerator of Newton's law of force. To save writing, we shall take

$$16\pi\gamma=1. \tag{2}$$

HAMILTONIAN FORM OF GRAVITATIONAL THEORY

We shall deal with the physical state on the surface $x^0=t$ and shall set up Hamiltonian equations of motion to determine how the state varies as t varies. The Hamiltonian is, by the usual definition

$$H=\int\left(g_{\mu\nu 0}\frac{\partial\mathcal{L}_G}{\partial g_{\mu\nu 0}}+\sum q_{M0}\frac{\partial\mathcal{L}_M}{\partial q_{M0}}-\mathcal{L}\right)d^3x, \tag{3}$$

where the sum is over all the nongravitational dynamical coordinates q_M.

It is evident that there must be a good deal of arbitrariness in the equations of motion on account of the arbitrariness in the system of coordinates x^μ. In the first place we see that the $g_{\mu 0}$ can vary with t in an arbitrary way. To describe the geometry of the surface $x^0=t$ and also the system of coordinates x^r in it, we need the g_{rs} at all points on the surface, but we do not need the $g_{\mu 0}$, which refer only to intervals going outside the surface. Different values for the $g_{\mu 0}$ correspond to different choices of a neighboring surface $x^0=t+\epsilon$ and to different systems of coordinates x^r in the neighboring surface, and these are completely arbitrary with a given initial surface $x^0=t$.

We get the simplest form for the equations of motion if we describe the physical state on the surface $x^0=t$ entirely in terms of dynamical variables that are independent of the $g_{\mu 0}$. Let us consider the kind of quantities that can enter into such a description.

Suppose there is a vector field A_μ. The three covariant components A_r on the surface remain invariant under a change of coordinates which leaves the coordinates of each point on the surface invariant. So these A_r will enter into the description. We cannot have A_0, but we have instead the normal component of A, namely $A_\mu l^\mu$, where l^μ is the unit normal. Similarly for a tensor $B_{\mu\nu}$, which may be the covariant derivative $A_{\mu|\nu}$ of A_μ, we have the quantities B_{rs}, $B_{r\nu}l^\nu$, $B_{\mu s}l^\mu$, $B_{\mu\nu}l^\mu l^\nu$. Each of these quantities is unaffected by a change of coordinates which leaves the points on the surface invariant and is thus independent of the $g_{\mu 0}$.

It should be noted that, for a vector A_μ, the ordinary and covariant derivatives A_{rs} and $A_{r|s}$ are both independent of the $g_{\mu 0}$. Their difference, namely $\Gamma_{rs}{}^{\mu}A_\mu$, is thus independent of the $g_{\mu 0}$. We may take A_μ here to be the unit normal, namely

$$l_\mu=g_\mu{}^0/(-g^{00})^{\frac{1}{2}}, \tag{4}$$

and we find that the quantity

$$\Gamma_{rs}{}^0/(-g^{00})^{\frac{1}{2}} \tag{5}$$

is independent of the $g_{\mu 0}$. This quantity may be called the "invariant velocity" of g_{rs}, as it consists of the ordinary velocity g_{rs0} multiplied by a certain factor and with certain terms added on, so as to produce a quantity independent of the choice of coordinate system outside the surface $x^0=t$.

With the physical state described in this way, one easily finds[4] that for a dynamical variable η not involving the $g_{\mu 0}$, $d\eta/dx^0$ is of the form

$$d\eta/dx^0=\int\{(-g^{00})^{-\frac{1}{2}}\xi_L+g_{r0}e^{rs}\xi_s\}d^3x, \tag{6}$$

with ξ_L and ξ_s independent of the $g_{\mu 0}$. We need equations of motion to determine ξ_L, ξ_s for any η. The coefficients $(-g^{00})^{-\frac{1}{2}}$, g_{r0} in (6) are arbitrary and not restricted by the equations of motion.

One gets equations of motion of the form (6) from a Hamiltonian of the form

$$H=\int\{(-g^{00})^{-\frac{1}{2}}\mathcal{H}_L+g_{r0}e^{rs}\mathcal{H}_s\}d^3x, \tag{7}$$

with $\mathcal{H}_L$ and $\mathcal{H}_s$ independent of the $g_{\mu 0}$ and vanishing weakly. It has been shown[4] that the Hamiltonian (3) takes the form (7) provided the dynamical coordinates describing the nongravitational fields are chosen to be independent of the $g_{\mu 0}$, in the way discussed above, and provided one takes for $\mathcal{L}_G$, instead of (1), an expression which differs from (1) by a perfect differential and which does not contain the velocities $g_{\mu 00}$, namely

$$\mathcal{L}_G=Jg^{\mu\nu}(\Gamma_{\mu\rho}{}^{\sigma}\Gamma_{\nu\sigma}{}^{\rho}-\Gamma_{\mu\nu}{}^{\rho}\Gamma_{\rho\sigma}{}^{\sigma}) +(Jg^{00})_0(g^{r0}/g^{00})_r-(Jg^{00})_r(g^{r0}/g^{00})_0. \tag{8}$$

With this $\mathcal{L}_G$, the momenta $p^{\mu 0}$ conjugate to $g_{\mu 0}$ vanish weakly, which results in the degrees of freedom described by $g_{\mu 0}$, $p^{\mu 0}$ dropping out from the Hamiltonian formalism. The weak equations $p^{\mu 0}\approx 0$ give, when one passes to the quantum theory, the conditions $p^{\mu 0}\psi=0$, which show that the wave function ψ does not involve the $g_{\mu 0}$.

The surviving gravitational momenta are

$$p^{rs}=K(e^{ra}e^{sb}-e^{rs}e^{ab})\Gamma_{ab}{}^0/(-g^{00})^{\frac{1}{2}}. \tag{9}$$

They are built up from the invariant velocities (5). The fundamental Poisson bracket (P.b.) relations for them are

$$[g_{ab},p'^{rs}]=\tfrac{1}{2}(g_a{}^rg_b{}^s+g_b{}^rg_a{}^s)\delta(x-x'). \tag{10}$$

The expressions for $\mathcal{H}_L$ and $\mathcal{H}_s$ in (7) are found to be

$$\mathcal{H}_L = K^{-1}(p^{rs}p_{rs} - \tfrac{1}{2}p_r{}^r p_s{}^s) + B + \{K^{-1}(K^2 e^{rs})_{,r}\}_{,s} + \mathcal{H}_{ML}, \quad (11)$$

$$\mathcal{H}_s = p^{ab}g_{abs} - 2(p^{ab}g_{as})_b + \mathcal{H}_{Ms}, \quad (12)$$

where

$$B = \tfrac{1}{4}K g_{rsu} g_{abv}\{(e^{ra}e^{sb} - e^{rs}e^{ab})e^{uv} + 2(e^{ru}e^{ab} - e^{ra}e^{bu})e^{sv}\}, \quad (13)$$

and $\mathcal{H}_{ML}$, $\mathcal{H}_{Ms}$ are the contributions arising from the nongravitational fields. It should be noted that the terms $B + \{K^{-1}(K^2e^{rs})_r\}_s$ are equal to the density of the three-dimensional scalar curvature of the surface $x^0 = t$.

We have the weak equations

$$\mathcal{H}_L \approx 0, \quad \mathcal{H}_s \approx 0. \quad (14)$$

They are χ equations or secondary constraints. To see where they come from, we note that Einstein's field equations are

$$R_\mu{}^\nu - \tfrac{1}{2}g_\mu{}^\nu R_\sigma{}^\sigma = \tfrac{1}{2}T_\mu{}^\nu, \quad (15)$$

where $T_\mu{}^\nu$ is the stress tensor produced by the nongravitational fields. The left-hand side of (15) contains second derivatives of the $g_{\alpha\beta}$ and thus in general contains accelerations $g_{\alpha\beta00}$. The right-hand side of (15) contains no derivatives of the $g_{\alpha\beta}$. Now the well-known identities

$$(R_\mu{}^\nu - \tfrac{1}{2}g_\mu{}^\nu R_\sigma{}^\sigma)_{|\nu} \equiv 0$$

may be written

$$(R_\mu{}^0 - \tfrac{1}{2}g_\mu{}^0 R_\sigma{}^\sigma)_0 \equiv -(R_\mu{}^r - \tfrac{1}{2}g_\mu{}^r R_\sigma{}^\sigma)_r +, \quad (16)$$

where the $+$ at the end indicates that some further terms, not involving third derivatives of the $g_{\mu\nu}$, must be added on. The right-hand side of (16) evidently does not contain any third time derivatives $g_{\alpha\beta000}$. Thus the left-hand side cannot involve third time derivatives, so $R_\mu{}^0 - \tfrac{1}{2}g_\mu{}^0 R_\sigma{}^\sigma$ cannot involve accelerations $g_{\alpha\beta00}$. Thus if we take $\nu = 0$ in (15), we get equations involving only dynamical coordinates and velocities. By substituting for the velocities here in terms of the momenta, we get four equations between dynamical coordinates and momenta only, which yield (14).

The main part of the Hamiltonian is obtained by putting $g_{\mu0} = -\delta_{\mu0}$ in (7) and is thus

$$H_{\text{main}} = \int \mathcal{H}_L d^3x = \int \{K^{-1}(p^{rs}p_{rs} - \tfrac{1}{2}p_r{}^r p_s{}^s) + B + \mathcal{H}_{ML}\} d^3x, \quad (17)$$

after removal of a surface integral at infinity. The removal of this surface integral does not disturb the validity of H_{main} for giving equations of motion, but it results in H_{main} not vanishing weakly.

We can write the total Hamiltonian (7) in the form

$$H = H_{\text{main}} + \int \{(-g^{00})^{-\frac{1}{2}} - 1\}\mathcal{H}_L d^3x + \int g_{r0}e^{rs}\mathcal{H}_s d^3x, \quad (18)$$

when it appears as H_{main} with arbitrary linear combinations of $\mathcal{H}_L$ and of $\mathcal{H}_s$, for various values of x^r, added on. These additional terms in the Hamiltonian produce terms in the equations of motion in addition to those produced by H_{main}, corresponding to the surface $x^0 = t$ undergoing arbitrary deformations and having arbitrary changes of its coordinate system x^r as t varies.

NEED FOR FIXATION OF THE COORDINATES

To specify a physical state at a particular time in the classical theory, we must choose numerical values for all the dynamical coordinates and momenta so as to satisfy the constraints (14). This involves solving some differential equations, so it is not such a straightforward matter as specifying a state in particle dynamics.

In the quantum theory the situation is more complicated. The constraints (14) go over into the conditions on the wave function

$$\mathcal{H}_L\psi = 0, \quad (19)$$

$$\mathcal{H}_s\psi = 0. \quad (20)$$

To specify a state at a particular time involves obtaining a solution of Eqs. (19), (20), which are functional equations.

Equation (20) expresses merely that ψ must be invariant under changes of the coordinate system x^r in the surface $x^0 = t$. To get ψ to satisfy this equation is thus not difficult. Equation (19) expresses the requirement that the state shall be specified in a way that is independent of deformations of the surface $x^0 = t$. The treatment of such deformations is essentially as complicated as the treatment of the passage from the surface $x^0 = t$ to a neighboring surface $x^0 = t + \epsilon$, so to get ψ to satisfy (19) is essentially as complicated as solving the equations of motion. Thus we have the situation that we cannot specify the initial state for a problem without solving the equations of motion. The formalism is thus not suitable for dealing with practical problems.

The difficulty does not arise in the weak-field approximation, because then many of the terms in (19) get neglected and the remaining ones, if expressed in terms of Fourier components, are easy to handle.

To obtain a practical formalism of greater accuracy than the weak-field approximation, it is necessary to introduce into the theory some new constraint that fixes the surface $x^0 = t$, so that we no longer have the possibility of making arbitrary deformations in it. Then the supplementary condition (19) gets eliminated. We may also introduce some further constraints that fix the coordinate system x^r in the surface. While not essential for getting a practical formalism, such further constraints serve to simplify the formalism by eliminating the conditions (20), and so making the task of specifying the initial state a trivial one.

The fixation of coordinates is advantageous also in the weak-field approximation, because it leads to some degrees of freedom dropping out from the formalism, the procedure being similar to the elimination of the longitudinal waves in electrodynamics.

When dealing with gravitational waves, people usually restrict the coordinate system by introducing the harmonic conditions

$$(Jg^{\mu\nu})_{\nu}=0.$$

These conditions would be quite unsuitable in the present formalism because they involve the $g_{\mu 0}$, which the present formalism allows to be completely arbitrary. Any restriction imposed on the $g_{\mu 0}$ would not help one in dealing with Eqs. (14) or (19) and (20). We need some restrictions which affect only the variables involved in (14), namely g_{rs} and p^{rs}, and possibly also the nongravitational variables.

GENERAL METHOD

Let us examine the general principles which come into play when we introduce some new restrictions or constraints on the dynamical variables in a Hamiltonian theory. Suppose we have a number of weak equations $\chi_n\approx 0$ $(n=1, 2,\cdots N)$, which may be either primary or secondary constraints. We are taking N to be finite for definiteness, but the same principles apply with N infinite. Suppose further that these weak equations are all first-class, so that

$$[\chi_n,\chi_{n'}]=0.$$

Now introduce some new restrictions, say the M independent equations

$$Y_m\approx 0, \quad m=1, 2,\cdots, M$$

with $M\leqslant N$. They are, of course, weak equations. Suppose that none of them (and no linear combination of them) has zero P.b. with all the χ's, so that they are all second-class constraints. They will cause M of the χ's to become second-class, while $N-M$ of the χ's (or linear combination of them) remain first-class.

Suppose χ_1, $\chi_2,\cdots\chi_M$ become second-class, while $\chi_{M+1},\cdots, \chi_N$ remain first-class. We now have the $2M$ second-class constraints $\chi_m\approx 0$, $Y_m\approx 0$ $(m=1, 2,\cdots M)$. Let us write $\chi_m=Y_{M+m}$, so that the $2M$ second-class constraints become $Y_s\approx 0$ $(s=1, 2,\cdots, 2M)$.

There is no place for second-class weak equations in the quantum theory, so we have to transform them in some way. We shall see that we can change them into strong equations (holding as equations between operators in the quantum theory) provided we adopt a new definition of P.b.'s, which corresponds to the number of effective degrees of freedom being reduced by M.

In simple cases we can pick out directly the degrees of freedom that have to be dropped and those that survive. Let us take the special case when M of the equation $Y_s\approx 0$ are

$$p_m\approx 0, \quad m=1, 2,\cdots, M. \tag{21}$$

The remaining M of them must then contain all the variables q_m independently, (otherwise the p_m would not all be second-class) and so it must be possible to solve them for the q_m and write them as

$$q_m\approx f_m(q_{M+1}, q_{M+2}\cdots p_{M+1}, p_{M+2}\cdots). \tag{22}$$

We now see that the degrees of freedom associated with q_m, p_m $(m=1, 2,\cdots M)$ cease to play an effective role in the dynamics. We can use Eqs. (21) and (22) to eliminate the variables p_m and q_m from the theory, which implies using these equations as definitions or as strong equations. We then work with P.b.'s that refer only to the other degrees of freedom.

In the general case one retains all the dynamical variables and merely changes their P.b.'s to correspond to the reduction in the number of degrees of freedom. To do this one first sets up the matrix of all the P.b.'s $[Y_s,Y_{s'}]$. It can be shown[5] that this matrix has a nonvanishing determinant, provided there is no linear combination of the Y_s that is first-class. One must then obtain the reciprocal matrix $C_{ss'}$, satisfying

$$C_{ss'}[Y_{s'},Y_{s''}]=\delta_{ss''}. \tag{23}$$

Note that $C_{ss'}$ is a skew matrix, like $[Y_s,Y_{s'}]$. One then defines new P.b.'s by the formula

$$[\xi,\eta]^*=[\xi,\eta]-[\xi,Y_s]C_{ss'}[Y_{s'},\eta]. \tag{24}$$

It can be checked[5] that the new P.b.'s satisfy all the fundamental relations that P.b.'s ought to satisfy.

From (23) and (24) we see at once that $[\xi,Y_s]^*=0$ for any ξ. Thus the Y_s now have zero P.b. with everything, so that we can consider the equations $Y_s=0$ as strong equations and use them before working out P.b.'s.

In applying this method to the gravitational case we desire, of course, that the change in the P.b.'s shall not be too complicated. In particular, we would like to have no change at all in the P.b. of two quantities, neither of which involves the gravitational variables g_{rs}, p^{rs}. This result is ensured provided the two conditions hold: (i) The Y_m $(m=1, 2,\cdots M)$ involve only the gravitational variables; (ii) The P.b.'s $[Y_m,Y_{m'}]$ all vanish. The proof is as follows.

We have already $(\chi_m,\chi_{m'}]\approx 0$ from the assumption that the χ's were originally first-class. With the further condition $[Y_m,Y_{m'}]\approx 0$ we have $[Y_s,Y_{s'}]\approx 0$ except when $1\leqslant s\leqslant M$ and $M+1\leqslant s'\leqslant 2M$ or vice versa. This

[5] P. A. M. Dirac, Can. J. Math. 2, 129 (1950).

leads to $C_{ss'}\approx 0$ except when $1\leqslant s\leqslant M$ and $M+1\leqslant s' \leqslant 2M$ or vice versa. The surviving elements of C are thus $C_{m,\ M+m'}=-C_{M+m',\ m}$. The elements $C_{m,\ M+m'}$ form a matrix of M rows and columns, which is the reciprocal of the matrix $[\chi_{m'},Y_m]$.

The formula (24) now reduces to

$$[\xi,\eta]^*-[\xi,\eta]=-C_{m,\ M+m'}\{[\xi,Y_m][\chi_{m'},\eta] -[\xi,\chi_{m'}][Y_m,\eta]\}. \qquad (25)$$

If ξ and η do not involve the gravitational variables, the condition (i) above leads to $[\xi,Y_m]=0$ and $[Y_m,\eta]=0$, so the right-hand side of (25) vanishes.

The introduction of the new constraints into the theory, when combined with the appropriate change in the P.b.'s, leaves the Hamiltonian first-class. It follows that the Hamiltonian equations of motion preserve all the constraints.

FIXATION OF THE SURFACE

To fix the surface $x^0=t$, the natural conditions to take are

$$p_r{}^r=g_{rs}p^{rs}\approx 0. \qquad (26)$$

This involves bringing into the theory one Y equation for each point of the surface.

One easily checks that

$$[\mathcal{H}_s,g'_{uv}p'^{uv}]=g_{uv}p^{uv}\delta_s(x-x')\approx 0,$$

so the conditions (26) do not disturb the first-class character of the equations $\mathcal{H}_s\approx 0$. This means that the conditions (26) do not restrict the coordinate system x^r in the surface, a result which is evident from the tensor character of (26). The conditions (26) mean geometrically that the surface shall have a maximum three-dimensional "area." The equations (26) and $\mathcal{H}_L\approx 0$ are now second-class and we can use them to eliminate one degree of freedom at each point of space.

We have

$$[g_{rs},p'_u{}^u]=g_{rs}\delta(x-x').$$

It follows that the ratios of the g_{rs} at any point have zero P.b.'s with $p_u{}^u$ at all points of the surface. Let us put

$$K=\kappa^3,\quad \tilde{g}_{rs}=g_{rs}\kappa^{-2},\quad \tilde{e}^{rs}=e^{rs}\kappa^2. \qquad (27)$$

Then $\tilde{g}_{rs}$ involves only such ratios and has zero P.b. with $p_u{}^u$ at all points. There are five independent $\tilde{g}_{rs}$, as their determinant is unity. The $\tilde{e}^{rs}$ form the reciprocal matrix to the matrix $\tilde{g}_{rs}$, and also have the determinant unity.

We have

$$[K^2,p'_u{}^u]=3K^2\delta(x-x'),$$

and so

$$[\ln\kappa,p'_u{}^u]=\tfrac{1}{2}\delta(x-x'). \qquad (28)$$

Put

$$\tilde{p}^{rs}=(p^{rs}-\tfrac{1}{3}e^{rs}g_{ab}p^{ab})\kappa^2, \qquad (29)$$
$$\tilde{p}_{rs}=\tilde{g}_{ra}\tilde{g}_{sb}\tilde{p}^{ab}.$$

We find that $\tilde{p}^{rs}$ and $\tilde{p}_{rs}$ have zero P.b. with $p_u{}^u$ and κ at all points.

Let us change our basic dynamical coordinates from the six g_{rs} to the five independent $\tilde{g}_{rs}$ and $\ln\kappa$. The momentum conjugate to $\ln\kappa$ is now, from (28), just $2p_u{}^u$, and the momenta conjugate to the $\tilde{g}_{rs}$ are certain functions of the $\tilde{p}^{rs}$ and $\tilde{g}_{rs}$.

The conditions (26) now take the form (21) and we have the equations $\mathcal{H}_L\approx 0$ playing the role of (22). To put them into the form of (22) we must solve them, with the help of (26), to get κ expressed in terms of quantities having zero P.b. with $p_u{}^u$ and κ. Such quantities are the $\tilde{g}_{rs}$, $\tilde{e}^{rs}$, $\tilde{p}^{rs}$, $\tilde{p}_{rs}$, and the nongravitational variables.

From (11), the equation $\mathcal{H}_L\approx 0$ gives,

$$-\{\kappa^{-3}(\kappa^4\tilde{e}^{rs})_r\}_s\approx\kappa^{-3}\tilde{p}^{rs}\tilde{p}_{rs}+B+\mathcal{H}_{ML}, \qquad (30)$$

in which we look upon the g_{rs} in B and $\mathcal{H}_{ML}$ as expressed in terms of the $\tilde{g}_{rs}$ and κ. This is a difficult equation to solve generally for κ. However, for gravitational fields that are not too strong, the important terms are those that involve second derivatives of κ, i.e., those on the left-hand side. We can therefore obtain the solution by a method of successive approximation, first putting $\kappa=1$ on the right and solving the resulting simplified equation, then substituting the first approximation for κ on the right and solving to get the second approximation, and so on. We shall consider this equation further in the next section, with reference to a particular system of coordinates, and for the present we shall assume that the solution has been obtained.

Following the method of the preceding section for dealing with the second-class equations (21) and (22), we express H_{main} and $\mathcal{H}_s$ in terms of the variables $\tilde{g}_{rs}$, $\tilde{e}^{rs}$, $\tilde{p}^{rs}$, $\tilde{p}_{rs}$, $p_u{}^u$, and κ, and then eliminate $p_u{}^u$ and κ from them by means of (26) and the solution of (30), which we may now use as strong equations. The elimination from $\mathcal{H}_s$ is trivial, as we get from (12), using (26),

$$\mathcal{H}_s=\tilde{p}^{ab}\tilde{g}_{abs}-2(\tilde{p}^{ab}\tilde{g}_{as})_b+\mathcal{H}_{Ms}. \qquad (31)$$

If the nongravitational field variables are suitably chosen, $\mathcal{H}_{Ms}$ will not contain κ. The elimination from H_{main} leads to an expression

$$H^*_{\text{main}}=\int(\kappa^{-3}\tilde{p}^{rs}\tilde{p}_{rs}+B+\mathcal{H}_{ML})d^3x, \qquad (32)$$

in which κ is understood to have the appropriate value. The integrand here may be considered as the energy density or mass density. The complete Hamiltonian is now

$$H^*_{\text{main}}+\int g_{r0}e^{rs}\mathcal{H}_sd^3x. \qquad (33)$$

The term corresponding to the freedom of deformation of the surface, i.e., the middle term of (18), has disappeared.

We now have a Hamiltonian formalism in which the degree of freedom described by $p_u{}^u$ and $\ln\kappa$ has dropped out. The Hamiltonians (32) and (33) are first-class even with the condition (26), so they lead to equations of motion that preserve (26). The procedure of substituting for κ in the derivation of $H^*{}_{\text{main}}$ caused the introduction of the right amount of $\mathcal{H}_L$ into the Hamiltonian to ensure the preservation of (26).

FIXATION OF COORDINATES IN THE SURFACE

To get the theory into a more convenient form, one must also fix the coordinate system x^r in the surface. The most natural conditions to take for this purpose, from the geometrical point of view, are the harmonic conditions in three dimensions:

$$(Ke^{rs})_{,s}\approx 0. \tag{34}$$

However, (34) does not have zero P.b. with (26), so if we adopt (34) together with (26) we must change the P.b. relationships between the nongravitational variables. To avoid this inconvenience, it is better to replace (34) by

$$\tilde{e}^{rs}{}_{,s}\approx 0, \tag{35}$$

which does have zero P.b. with (26).

With the coordinates fixed by (35), Eq. (30) reduces to

$$-4\nabla^2\kappa=\kappa^{-3}\tilde{p}^{rs}\tilde{p}_{rs}+B+\mathcal{H}_{ML}. \tag{36}$$

where ∇^2 denotes the Laplacian operator with respect to the metric $\tilde{g}_{rs}$, namely

$$\nabla^2=\tilde{e}^{rs}\partial^2/\partial x^r\partial x^s. \tag{37}$$

The right-hand side in (36) equals the integrand in (32) and is the mass density. To interpret (36), let us restore the gravitational constant into the theory in accordance with (2). It then becomes

$$-(4\pi\gamma)^{-1}\nabla^2\kappa=16\pi\gamma\kappa^{-3}\tilde{p}^{rs}\tilde{p}_{rs}+(16\pi\gamma)^{-1}B+\mathcal{H}_{ML}. \tag{38}$$

We now see that $\kappa-1$ is the Newtonian potential generated by the mass density in a space with the metric $\tilde{g}_{rs}$. The fact that κ occurs in the right-hand side of (38) can be understood as due to the Newtonian potential itself having some influence on the mass density which generates it.

Let us examine the term with B in (38). The expression (13) for B, written in terms of the new variables, is

$$B=\tfrac{1}{4}\kappa^{-1}(\kappa\tilde{g}_{rsu}+2\kappa_u\tilde{g}_{rs})(\kappa\tilde{g}_{abv}+2\kappa_v\tilde{g}_{ab})\times\{(\tilde{e}^{ra}\tilde{e}^{sb}-\tilde{e}^{rs}\tilde{e}^{ab})\tilde{e}^{uv}+2(\tilde{e}^{ru}\tilde{e}^{ab}-\tilde{e}^{ra}\tilde{e}^{bu})\tilde{e}^{sv}\}.$$

With the help of the equation

$$\tilde{g}_{rsu}\tilde{e}^{rs}=0,$$

which follows from the determinant of the $\tilde{g}_{rs}$ being unity, and of the equation

$$\tilde{g}_{rsu}\tilde{e}^{ru}=0,$$

which follows from (35), this reduces to

$$B=\tfrac{1}{4}\kappa\tilde{g}_{rsu}\tilde{g}_{abv}(\tilde{e}^{ra}\tilde{e}^{sb}\tilde{e}^{uv}-2\tilde{e}^{ra}\tilde{e}^{bu}\tilde{e}^{sv})-2\kappa^{-1}\kappa_u\kappa_v\tilde{e}^{uv}. \tag{39}$$

The last term here, divided by $16\pi\gamma$, can be interpreted as the mass density (or energy density) of the Newtonian field with the potential $\kappa-1$. It is negative definite, corresponding to the Newtonian force being attractive. The remaining terms of B, together with the first term on the right-hand side of (38), give the energy density of the gravitational waves.

THE NEW POISSON BRACKETS

With the coordinates fixed by (35), the P.b.'s of the gravitational variables with one another and with the nongravitational variables will be altered. The new P.b.'s are given by formula (25) with Y_m replaced by $\tilde{e}^{ru}{}_u$ and $\chi_{m'}$ replaced by $\mathcal{H}'_s$. It thus reads

$$[\xi,\eta]^*-[\xi,\eta]=-\int\int C_r{}^s(x,x')\{[\xi,\tilde{e}^{ru}{}_u][\mathcal{H}'_s,\eta]-[\xi,\mathcal{H}'_s][\tilde{e}^{ru}{}_u,\eta]\}d^3xd^3x'. \tag{40}$$

The coefficient $C_r{}^s(x,x')$ is the reciprocal of the matrix $[\mathcal{H}'_s,\tilde{e}^{ru}{}_u]$ and thus satisfies

$$\int C_v{}^s(x'',x')[\mathcal{H}'_s,\tilde{e}^{ru}{}_u]d^3x'=g_v{}^r\delta(x-x''). \tag{41}$$

Evaluating the P.b. here, we get

$$\int C_v{}^s(x'',x')\{g_s{}^r\tilde{e}^{ab}\delta_{ab}(x-x')+\tfrac{1}{3}\tilde{e}^{ra}\delta_{sa}(x-x')\}d^3x'=g_v{}^r\delta(x-x''),$$

which reduces to

$$\nabla^2C_v{}^r(x',x)+\tfrac{1}{3}\tilde{e}^{ra}C_v{}^s(x',x)_{sa}=g^r{}_v\delta(x-x'), \tag{42}$$

with ∇^2 defined by (37).

This equation may be considered for fixed x', when it is a differential equation for the unknown functions $C_v{}^r(x',x)$ in the variables x. The important domain for x is now the neighborhood of x', since when x is far from x' the functions $C_v{}^r(x',x)$ are small. We can therefore get an approximate solution by considering the space as flat in this domain, so that the $\tilde{e}^{ab}$ are constants. With this approximation we get, on differentiating (42) with respect to x^r,

$$\nabla^2C_v{}^s{}_s=\tfrac{3}{4}\delta_v(x-x'). \tag{43}$$

The solution of this equation is

$$C_v{}^s{}_s=-\frac{1}{4\pi}\times\frac{3}{4}\left(\frac{1}{|x-x'|}\right)_v,$$

where $|x-x'|$ denotes the distance from x to x' with respect to the metric $\tilde{g}_{rs}$,

$$|x-x'|=\{\tilde{g}_{rs}(x^r-x'^r)(x^s-x'^s)\}^{\frac{1}{2}}. \tag{44}$$

Equation (42) now becomes

$$\nabla^2 C_v{}^r = g^r{}_v \delta(x-x') + \frac{1}{16\pi}\tilde{e}^{ra}\left(\frac{1}{|x-x'|}\right)_{va},$$

whose solution is

$$C_v{}^r(x',x) = -\frac{1}{4\pi}g^r{}_v\frac{1}{|x-x'|} + \frac{1}{32\pi}\tilde{e}^{ra}|x-x'|_{va}. \quad (45)$$

One could get the solution of (42) to a higher accuracy by substituting for the $\tilde{e}^{ab}$ in the left-hand side of (42), (remembering that $\tilde{e}^{ab}$ occurs also in the operator ∇^2,) their Taylor expansions in powers of $x-x'$ and using the first approximation for $C_v{}^r$ in those terms in which it occurs with a factor $x^r - x'^r$. By a process of successive approximation one could get the solution to any desired accuracy.

With the coefficients $C_r{}^s(x,x')$ in (40) determined, the new P.b.'s are determined. It should be noted that the new P.b. of any nongravitational variable with $\tilde{g}_{rs}$ or $\tilde{e}^{rs}$ vanishes. However, its new P.b. with $\tilde{p}^{rs}$ does not vanish.

QUANTIZATION

To pass over to the quantum theory, we must make all our dynamical variables into operators satisfying commutation relations corresponding to the new P.b.'s. We must then pick out a complete set of commuting observables. We may take these to consist of the $\tilde{e}^{rs}$ at all points x^r, together with a complete set of commuting nongravitational observables, ζ say. We can then set up the wave function as a function of these variables,

$$\psi(\tilde{e}^{rs},\zeta).$$

The effective domain of ψ is that for which the $\tilde{e}^{rs}$ are restricted to have the determinant unity and to satisfy $\tilde{e}^{rs}{}_s = 0$. ψ may be considered as undefined outside this domain. When we operate on ψ with $\tilde{p}^{ab}$ or with any dynamical variable in the theory, we get another wave function defined in the same domain, on account of $\tilde{p}^{ab}$ commuting with the determinant of the $\tilde{e}^{rs}$ and with $\tilde{e}^{rs}{}_s$.

There are no supplementary conditions to be imposed on ψ. We can choose it arbitrarily to correspond to the initial state in any problem. There is just one equation for ψ, the Schrödinger equation

$$i\hbar(d\psi/dx^0) = H^*{}_{\text{main}}\psi,$$

which fixes the state at later times.

For the theory to be self-consistent it is necessary that the space-like surface on which the state is defined shall always remain space-like. The condition for this is that K^2, the determinant of the g_{rs}, shall remain always positive. In the present formalism this means $\kappa^6 > 0$, with κ determined by (36). If the mass density is always positive, (36) shows that $\kappa > 1$ and there is no trouble. Difficulties arise only where there is a large negative density. This occurs very close to a point particle, on account of the last term in (39). The gravitational treatment of point particles thus brings in one further difficulty, in addition to the usual ones in the quantum theory.

PHYSICAL REVIEW D VOLUME 19, NUMBER 12 15 JUNE 1979

New general relativity

Kenji Hayashi
Institute of Physics, University of Tokyo (Komaba) Tokyo, Japan

Takeshi Shirafuji
Physics Department, Saitama University, Saitama, Japan
(Received 6 February 1979)

A gravitational theory is formulated on the Weitzenböck space-time, characterized by the vanishing curvature tensor (absolute parallelism) and by the torsion tensor formed of four parallel vector fields. This theory is called new general relativity, since Einstein in 1928 first gave its original form. New general relativity has three parameters c_1, c_2, and λ, besides the Einstein constant κ. In this paper we choose $c_1 = 0 = c_2$, leaving open λ. We prove, among other things, that (i) a static, spherically symmetric gravitational field is given by the Schwarzschild metric, that (ii) in the weak-field approximation an antisymmetric field of zero mass and zero spin exists, besides gravitons, and that (iii) new general relativity agrees with all the experiments so far carried out.

I. INTRODUCTION

In 1928 Einstein introduced the notion of absolute parallelism and tried to unify gravitation and electromagnetism, using tetrads with 16 degrees of freedom.[1] His attempt failed because there was no Schwarzschild solution in his simplified field equation.[2] Later, in 1961 Møller revived Einstein's idea,[3] and Pellegrini and Plebanski found a Lagrangian formulation for absolute parallelism.[4] Recently this formalism was reconsidered by Møller.[5]

In 1967, quite independently, Hayashi and Nakano started to formulate the gauge theory of the space-time translation group[6]: This theory was of no geometrical construction, but it was shown that, for a static, isotropic gravitational field, a symmetric part of their field equations is identical with the Einstein field equation in general relativity, and that, in the weak-field approximation, an antisymmetric part describes the propagation of an antisymmetric field, whose source is related to the intrinsic spin of spin-$\frac{1}{2}$ fundamental particles.[6] Miyamoto and Nakano estimated effects of exchanging this field in the microscopic system.[7] In later years Hayashi further developed the gauge theory into a more elaborate framework[8] and fixed the final form in 1973.[9] Quite recently Hayashi pointed out the connection between the gauge theory of the space-time translation group and absolute parallelism.[10]

Now we wish to unify these two developments mentioned above, following the geometry of underlying space-time structure. The Riemann-Cartan space-time U_4 is a paracompact, Hausdorff, connected C^∞ four-dimensional manifold endowed with a locally Lorentzian metric $\underline{g}$ and a linear affine connection $\underline{\Gamma}$ which is metric,

$$D_\lambda g_{\mu\nu} = \partial_\lambda g_{\mu\nu} - \Gamma^\rho_{\mu\lambda} g_{\rho\nu} - \Gamma^\rho_{\nu\lambda} g_{\mu\rho} = 0 . \tag{1.1}$$

From this equation we get

$$\Gamma^\lambda_{\mu\nu} = \left\{ {\lambda \atop \mu\nu} \right\} + K^\lambda_{\cdot\mu\nu} , \tag{1.2}$$

where the first term denotes the Levi-Civita connection,

$$\left\{ {\lambda \atop \mu\nu} \right\} = \tfrac{1}{2} g^{\lambda\rho} (\partial_\mu g_{\nu\rho} + \partial_\nu g_{\mu\rho} - \partial_\rho g_{\mu\nu}) , \tag{1.3}$$

and the second stands for the contortion tensor,

$$K^\lambda_{\cdot\mu\nu} = \tfrac{1}{2} (T^\lambda_{\cdot\mu\nu} - T_{\mu\cdot\nu}^{\cdot\lambda} - T_{\nu\cdot\mu}^{\cdot\lambda}) \tag{1.4}$$

with the torsion tensor

$$T^\lambda_{\cdot\mu\nu}(\Gamma) = \Gamma^\lambda_{\mu\nu} - \Gamma^\lambda_{\nu\mu} . \tag{1.5}$$

In terms of the affine connection the curvature tensor is given by

$$R^\rho{}_{\sigma\mu\nu}(\Gamma) = \partial_\mu \Gamma^\rho_{\sigma\nu} - \partial_\nu \Gamma^\rho_{\sigma\mu} + \Gamma^\rho_{\lambda\mu} \Gamma^\lambda_{\sigma\nu} - \Gamma^\rho_{\lambda\nu} \Gamma^\lambda_{\sigma\mu} . \tag{1.6}$$

The Riemann-Cartan space-time has both the curvature tensor and the torsion tensor. From this space-time follow two very interesting models of space-time. One is the well-known Riemann space-time V_4, which is obtained from the U_4 by setting the torsion tensor to be identically vanishing. From (1.2) follows the Levi-Civita connection. It is well known that general relativity is the theory of gravitation on this space-time, and that it ascribes gravitation to the Riemann-Christoffel curvature tensor formed of the Levi-Civita connection.

Another interesting model is the Weitzenböck space-time A_4,[11] which is obtained from the U_4 by setting the curvature tensor to be identically vanishing,

$$R^\rho{}_{\sigma\mu\nu}(\Gamma^*) = 0 . \tag{1.7}$$

Or, to put it equivalently,[12] the Weitzenböck space-time is obtained by requiring the U_4 to admit absolute parallelism, i.e., to have a quadruplet (specified by $k=0,1,2,3$) of linearly independent *parallel vector fields*, $\underline{b}=\{\underline{b}_k\}=\{b_k{}^\mu\}$, which is defined by

$$D^*_\nu b_k{}^\lambda = \partial_\nu b_k{}^\lambda + \Gamma^{*\lambda}_{\mu\nu} b_k{}^\mu = 0\,. \tag{1.8}$$

Solving this equation we find the nonsymmetric affine connection,

$$\Gamma^{*\lambda}_{\mu\nu} = b_k{}^\lambda \partial_\nu b^k{}_\mu\,, \tag{1.9}$$

and the torsion tensor,

$$T^\lambda{}_{\cdot\mu\nu}(\Gamma^*) = b_k{}^\lambda(\partial_\nu b^k{}_\mu - \partial_\mu b^k{}_\nu)\,. \tag{1.10}$$

Here $\underline{b}^*=\{\underline{b}^k\}=\{b^k{}_\mu\}$ is also a quadruplet of parallel vector fields, which is inverse to $\underline{b}$. It is straightforward to see that the curvature tensor indeed vanishes identically [see (1.7)]. See Fig. 1 for reduction of the Riemann-Cartan space-time.

We will give the name, new general relativity, to the theory of gravitation on the Weitzenböck space-time, since Einstein in 1928, after inventing general relativity, considered absolute parallelism for the first time, and the main consequences of the present theory will be analogous to those of general relativity so far as macroscopic phenomena are concerned. New general relativity attributes gravitation to the torsion tensor formed of the parallel vector fields.

As is well known, general relativity is formulated by the following fundamental assumptions, which we will compare with those of new general relativity: (A) Underlying space-time is the Riemann space-time, which has the metric tensor as the basic structure. All physical laws are expressed by equations that are covariant or form invariant under the group of general coordinate transformations. (B) The equivalence principle. (C) Gravitational field equations are derivable from the action principle. (D) The field equations are partial differential equations in the field variables of not higher than the second order. (E) The gravitational field is exhaustively described by the metric tensor alone.

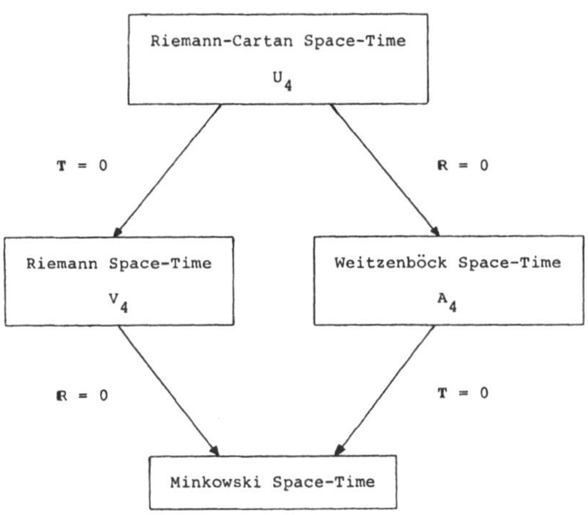

FIG. 1. The reduction of space-time is made in two particular cases: One is the Riemann space-time V_4 with a curvature tensor only ($\underline{R}$), and the other the Weitzenböck space-time A_4 with a torsion tensor alone ($\underline{T}$).

In new general relativity the fundamental assumptions are as follows: (A′) Underlying space-time is the Weitzenböck space-time, which has a quadruplet of the parallel vector fields as the fundamental structure. These parallel vector fields give rise to the metric tensor as a by-product. All physical laws are expressed by equations that are covariant or form invariant under the group of general coordinate transformations. (B′) The equivalence principle is valid only in classical physics. (C′) and (D′) are the same as (C) and (D), but at this time we start from the microscopic action principle. (E′) The gravitational field is exclusively described by a quadruplet of the parallel vector fields. As is closely related to (E′), we need to assume: (F′) All physical laws are expressed by equations that are covariant or form invariant under the group of *global* Lorentz transformations. When general relativity is extended to the domain of microscopic system, one must use tetrads and has to assume: (F) All physical laws are expressed by equations that are covariant or form invariant under the group of *local* Lorentz transformations.

We shall formulate new general relativity in the following manner: In Sec. II geometry of the Weitzenböck space-time is described in some detail, with emphasis on spinor wave functions defined in this space-time. In Sec. III microscopic matter Lagrangians are considered, such as of the electromagnetic field, of spin-$\frac{1}{2}$ fundamental particles and so forth. Their *equations of motion* are derived and then approximated by the WKB method to yield, in the classical limit, the *geodesics* of the metric $\underline{g}$,[13] along which point particles and light rays are defined to move. In Sec. IV a gravitational Lagrangian is constructed by the requirement of invariance under (1) the group of general coordinate transformations, (2) the group of global Lorentz transformations (3) the parity operation, and by the demand that (4) the Lagrangian be quadratic in the torsion tensor. Gravitational field equations are derived, with three unknown parameters, c_1, c_2, and c_3. In Sec. V a static, spherically symmetric field outside a massive neutral body is determined, with two parameters, c_1 and c_2; in this case a term proportional to c_3 is vanishing identically. In Sec. VI comparison with all the experiments so far carried out is made; firstly, we clarify how the equivalence principle is violated in microscopic systems only, and secondly, upper bounds for the parameters, c_1 and c_2, are obtained. In Sec. VII the free parameters, c_1 and c_2, are classified into two classes with c_3 arbitrary; $c_1=0=c_2$ and $c_1\neq 0\neq c_2$. (Other cases are forbidden.) The rest of the present paper concerns with the former choice of free parameters, and hence the static, isotropic field is given by the Schwarzschild solution. In this case new general relativity has only one free parameter, $\lambda=9/(4c_3)$, besides the Einstein gravitational constant κ. In Sec. VIII the group of local Lorentz transformations, which we do not assume, is seen to emerge as the dynamical symmetry group for a static, isotropic field. This new situation demands the extension of absolute parallelism. In Sec. IX, as microscopic applications, the weak-field approximation to gravitational field equations is performed. In Sec. X the coupling of an antisymmetric field to matter is discussed; it propagates in vacuum, mediating a long-range, spin-spin force among spin-$\frac{1}{2}$ fundamental particles with a coupling strength $\sqrt{\lambda}$, which is estimated by precise experimental values of quantum electrodynamics. In Sec. XI the Birkhoff theorem, that a spherically symmetric gravitational field in empty space must be static, with a metric given by the Schwarzschild solution, is proved in new general relativity. In Sec. XII we draw conclusions.

In our conventions the middle part of the Greek alphabet, $\mu,\nu,\lambda,\ldots$, refers to 0, 1, 2, and 3, while the initial part, $\alpha,\beta,\gamma,\ldots$, denotes 1, 2, and 3. In a similar way the middle part of the Latin alphabet, $i,j,k,\ldots$, means 0, 1, 2, and 3, while the initial part, $a,b,c,\ldots$, denotes 1, 2, and 3.

II. GEOMETRY OF THE WEITZENBÖCK SPACE-TIME

The space-time M is assumed to be a paracompact, Hausdorff, connected C^∞ four-dimensional manifold with a locally Lorentzian metric $\underline{g}$. Let U be a local coordinate neighborhood of $p\in M$ with local coordinates $x=\{x^\mu\}$, then we can introduce the coordinate basis $\underline{E}=\{\underline{E}_\mu\}=\{(\partial/\partial x^\mu)_p\}$ with $\mu=0$, 1, 2, and 3, and the dual basis $\underline{E}^*=\{\underline{E}^\mu\}=\{(dx^\mu)_p\}$. Every vector $\underline{V}$ at p can be written as $\underline{V}=V^\mu\underline{E}_\mu$. In particular the metric tensor $\underline{g}$ is written as

$$\underline{g}=g_{\mu\nu}\underline{E}^\mu\otimes\underline{E}^\nu\,, \tag{2.1}$$

where the metric components are simply the inner product of the coordinate basis vectors,

$$g_{\mu\nu}=g(\underline{E}_\mu,\underline{E}_\nu)=g(\underline{E}_\nu,\underline{E}_\mu)\,. \tag{2.2}$$

These components are used to raise and lower Greek indices.

By definition there exists a global system of four orthonormal vector fields $\underline{b}(p)=\{\underline{b}_i(p)\}$, such that

$$g(\underline{b}_i,\underline{b}_j)=g(\underline{b}_j,\underline{b}_i)=\eta_{ij}\,, \tag{2.3}$$

where $\eta=(\eta_{ij})=\mathrm{diag}(-1,+1,+1,+1)$. Thus the vector fields, $\underline{b}(p)=\{\underline{b}_i(p)\}$, are expressed in the old basis by

$$\underline{b}_i=b_i{}^\mu\underline{E}_\mu\,; \tag{2.4a}$$

equivalently, a global system of four orthonormal vector fields $\underline{b}^*(p)=\{\underline{b}^i(p)\}$, which are dual to $\underline{b}(p)=\{\underline{b}_i(p)\}$, is written in the old basis by

$$\underline{b}^i=b^i{}_\mu\underline{E}^\mu\,. \tag{2.4b}$$

Conversely, it follows that

$$\underline{E}_\mu=b^i{}_\mu\underline{b}_i\,, \tag{2.4c}$$

$$\underline{E}^\mu=b_i{}^\mu\underline{b}^i\,. \tag{2.4d}$$

Here the coefficients, $\{b_i{}^\mu\}$ or $\{b^i{}_\mu\}$, are 16 functions, and must staisfy

$$b_i{}^\mu b^i{}_\nu=\delta^\mu{}_\nu\,,\quad b_i{}^\mu b^j{}_\mu=\delta^j{}_i\,, \tag{2.5a}$$

$$b^i{}_\mu\eta_{ij}b^j{}_\nu=g_{\mu\nu}\,,\quad b_i{}^\mu g_{\mu\nu}b_j{}^\nu=\eta_{ij}\,. \tag{2.5b}$$

It should be remarked that Latin indices of $\underline{b}=\{\underline{b}_i\}$ and $\underline{b}^*=\{\underline{b}^i\}$ mean that they are Lorentz vectors; $\underline{b}$ is the covariant vector and $\underline{b}^*$ is the contravariant vector. From (2.4) it follows that for any vector $\underline{V}=V^\mu\underline{E}_\mu=V^i\underline{b}_i$ the components satisfy

$$V^\mu=b_i{}^\mu V^i\,,\quad V^i=b^i{}_\mu V^\mu\,. \tag{2.6}$$

This rule of converting Greek to Latin indices and vice versa is applied to any tensor of higher rank.

Now, in the Weitzenböck space-time the covariant derivative, denoted by $\underline{D}^*$, defines *absolute parallelism* with respect to the global system of the four orthonormal vector fields $\underline{b}$. By definition $\underline{D}^*$ satisfies

$$\underline{D}^*\underline{b}_i=0\,, \tag{2.7a}$$

or in the coefficient form,

$$D^*_\nu b_i{}^\lambda\equiv\partial_\nu b_i{}^\lambda+\Gamma^{*\lambda}_{\mu\nu}b_i{}^\mu=0\,. \tag{2.7b}$$

From (2.4), (2.5), and (2.7) we find

$$D^*_\nu\underline{E}_\mu=\Gamma^{*\lambda}_{\mu\nu}\underline{E}_\lambda \tag{2.8}$$

with the affine connection of absolute parallelism,

$$\Gamma^{*\lambda}_{\mu\nu}=b_i{}^\lambda\partial_\nu b^i{}_\mu=-b^i{}_\mu\partial_\nu b_i{}^\lambda\,. \tag{2.9}$$

Here the global system of the four orthonormal vector fields is called *a quadruplet of the parallel vector fields*, or simply the *parallel vector fields*. For a vector field $\underline{V}(x)=V^i(x)\underline{b}_i=V^\mu(x)\underline{E}_\mu$, the covariant derivative is given by

$$\underline{D}^*\underline{V} = (D^*_\nu V^i)\underline{E}^\nu \otimes \underline{b}_i = (D^*_\nu V^\mu)\underline{E}^\nu \otimes \underline{E}_\mu \,, \qquad (2.10)$$

where

$$D^*_\nu V^i = \partial_\nu V^i \,, \qquad (2.11)$$

$$D^*_\nu V^\mu = \partial_\nu V^\mu + \Gamma^{*\mu}_{\lambda\nu} V^\lambda \,. \qquad (2.12)$$

Thus, for the components V^i with respect to a quadruplet of the parallel vector fields, the covariant derivative coincides with the usual derivative.

In the Weitzenböck space-time absolute parallelism of vectors at different points of M is defined in the following way: Consider a vector $\underline{V}(p) = V^i \underline{b}_i(p)$ at p and a vector $\underline{W}(q) = W^i \underline{b}_i(q)$ at q, where the point q can be arbitrarily separated from p. The parallelism of $\underline{V}$ and $\underline{W}$ is manifest: If their components are equal with each other,

$$V^i = W^i \,, \qquad (2.13)$$

then the two vectors, $\underline{V}(p)$ and $\underline{W}(q)$, are parallel with each other and of equal length.

In passing we make the remark that Latin indices are used to denote components with respect to a quadruplet of the parallel vector fields, and are raised and lowered by the Minkowski metric tensor, $\{\eta_{ij}\}$ or $\{\eta^{ij}\}$.

The affine connection, $\Gamma^* = \{\Gamma^{*\lambda}_{\mu\nu}\}$, is not symmetric with respect to the exchange of lower two indices. The torsion tensor is given by

$$T^\lambda{}_{\cdot\mu\nu} \equiv \Gamma^{*\lambda}_{\mu\nu} - \Gamma^{*\lambda}_{\nu\mu} = b_i{}^\lambda (\partial_\nu b^i{}_\mu - \partial_\mu b^i{}_\nu) \,. \qquad (2.14)$$

The curvature tensor formed of Γ^* identically vanishes [see (1.7)], since parallel transfer of a vector is path independent owing to absolute parallelism. Thus the Weitzenböck space-time is characterized by the torsion tensor alone, and reduces to the Minkowski space-time provided the torsion tensor vanishes globally. See Fig. 1 for reduction of the Riemann-Cartan space-time. *In the Minkowski space-time the parallel vector fields, which define absolute parallelism, coincide with the coordinate basis of a Cartesian coordinate system.*

When a quadruplet of the parallel vector fields $\underline{b}$ is subject to a global, proper, orthochronous Lorentz transformation,

$$\underline{b}_i = A^j{}_i \underline{b}'_j \,, \qquad (2.15a)$$

$$A^j{}_i \eta_{jm} A^m{}_n = \eta_{in} \,, \quad \det A = 1 \,,$$

$$A^0{}_0 \geq 1 \,, \quad \partial_\nu A^j{}_i = 0 \,, \qquad (2.15b)$$

new absolute parallelism defined by new parallel vector fields $\underline{b}'$ is equivalent to the original one. So geometry of the Weitzenböck space-time is invariant under the global, proper, orthochronous Lorentz group, $L^\dagger_+ \equiv \{A = (A^j{}_i) \in \mathrm{GL}(4,R),\ A^j{}_i \eta_{jm} A^m{}_n = \eta_{in},\ \det A = 1,\ A^0{}_0 \geq 1,\ \partial_\nu A^j{}_i = 0\}$. In conformity with this invariance of underlying geometry, we demand that *physical laws should be invariant under the action of* $L^\dagger_+$. We call this the global Lorentz invariance.

In the Weitzenböck space-time, spinors are introduced as quantities which transform like two-valued representations of the proper, orthochronous Lorentz group $L^\dagger_+$.[14] Most elementary spinors are four kinds of two-component spinors, i.e., contravariant spinor $\{\xi^A\}$, covariant spinor $\{\chi_A\}$, dotted contravariant spinor $\{\xi^{\dot A}\}$, and dotted covariant spinor $\{\chi_{\dot A}\}$ for $A = 1$ and 2. Dotted spinors transform like the complex conjugate of undotted spinors. A spinor of higher rank is a quantity which transforms like a direct product of two-component spinors. A vector $\underline{V} = V^i \underline{b}_i$ is identified with a mixed spinor of second rank with components $V^{\dot A B}$,

$$V^{\dot A B} = \Sigma_i{}^{\dot A B} V^i \,, \qquad (2.16)$$

where $\{\Sigma_i{}^{\dot A B}\}$ is a set of Hermitian 2×2 matrices satisfying

$$\Sigma_i{}^{\dot A B} \epsilon_{\dot A \dot C} \Sigma_j{}^{\dot C D} + \Sigma_j{}^{\dot A B} \epsilon_{\dot A \dot C} \Sigma_i{}^{\dot C D} = -\eta_{ij} \epsilon^{BD} \,, \qquad (2.17)$$

where

$$(\epsilon_{\dot A \dot C}) = \begin{pmatrix} 0 & 1 \\ -1 & 0 \end{pmatrix} = (\epsilon^{AC}) \,. \qquad (2.18)$$

One of the simplest choices, which we take in this paper, is

$$(\Sigma_0{}^{\dot A B}) = -I/\sqrt{2} \,, \quad (\Sigma_a{}^{\dot A B}) = \sigma_a/\sqrt{2} \,, \qquad (2.19)$$

where $\{\sigma_1, \sigma_2, \sigma_3\}$ is a set of the Pauli matrices.

The four-component Dirac spinor ψ is defined by a direct sum of a covariant spinor and a dotted contravariant spinor, and is written as a single column matrix

$$\psi = \begin{pmatrix} \chi_1 \\ \chi_2 \\ \xi^{\dot 1} \\ \xi^{\dot 2} \end{pmatrix} \equiv \begin{pmatrix} \chi \\ \xi \end{pmatrix} . \qquad (2.20)$$

The conjugate Dirac spinor $\bar\psi$ is obtained from ψ by

$$\bar\psi = \psi^\dagger \begin{pmatrix} 0 & I \\ I & 0 \end{pmatrix} = (\xi^1, \xi^2, \chi_{\dot 1}, \chi_{\dot 2}) \,. \qquad (2.21)$$

Now we extend the definition of absolute parallelism to include spinors. Consider a spinor at p, say a contravariant two-component spinor $\{\xi^A(p)\}$, and another spinor at q of the same type, say, $\{\xi^A(q)\}$. If components of these spinors are equal, i.e., $\xi^A(p) = \xi^A(q)$, then two spinors are defined to be parallel and of the same magnitude. From (2.16) it follows that absolute parallelism of spinors implies absolute parallelism of vectors and tensors: In fact, for two vectors $\underline{V}$ at point p and $\underline{W}$ at another point q, equality of spinor components, $V^{\dot{A}B}(p)=W^{\dot{A}B}(q)$, implies $V^i(p)=W^i(q)$, because $\{\Sigma_i{}^{\dot{A}B}\}$ is independent of space-time position.

When a spinor at point p is parallel transferred to another point q, its components are kept unchanged owing to absolute parallelism. Therefore, *the covariant derivative* D^*_μ *of spinors coincides with the usual derivative* ∂_μ.

Finally, we make the following important remark: The parallel vector fields $\underline{b}$ are different from the so-called tetrad fields $\underline{e}$ by an arbitrary, position-dependent Lorentz transformation, which is called a local Lorentz transformation.

III. MATTER LAGRANGIAN AND EQUATIONS OF MOTION FOR TEST PARTICLES

A. Matter Lagrangian

In new general relativity we do not identify the six extra degrees of freedom of the parallel vector fields with the electromagnetic field strength, since we now know that such an attempt failed.[2] Instead, we take the electromagnetic potential $\underline{A}=\{A_\mu\}$ as the dynamical variable independent of the parallel vector fields. The matter part of the action is then represented as a sum of the action of fundamental particles and fields, i.e., of the electromagnetic field and several kinds of spin-$\frac{1}{2}$ fundamental particles;

$$I_M=\int d^4x\sqrt{-g}\,L_M$$
$$=\int d^4x\sqrt{-g}\left(L_{\text{em}}+\sum_i L_D^{(i)}+L_{\text{int}}\right), \qquad (3.1)$$

where g is

$$g=\det(g_{\mu\nu})=-[\det(b^i{}_\mu)]^2<0\,, \qquad (3.2)$$

and L_{int} represents nongravitational interactions among fundamental particles and fields. Here the index i in the second term labels spin-$\frac{1}{2}$ fundamental particles such as quarks, electrons, muons, electron-neutrinos, muon-neutrinos, etc., all of which can be described in fairly good approximation by spinor wave functions obeying the Dirac equation. If there exist other fundamental fields besides the electromagnetic field, their action must be added to (3.1): Gauge fields for internal symmetry of fundamental particles, if they exist, can be included in (3.1) in a similar manner to the electromagnetic field.

The gauge invariance of the electromagnetic interaction shall be assumed to hold in new general relativity, because this invariance plays the fundamental role in quantum electrodynamics. The electromagnetic Lagrangian density L_{em} is then given by

$$L_{\text{em}}=-\tfrac{1}{4}g^{\mu\rho}g^{\nu\sigma}F_{\mu\nu}F_{\rho\sigma}\,, \qquad (3.3)$$

with

$$F_{\mu\nu}=\partial_\mu A_\nu-\partial_\nu A_\mu\,, \qquad (3.4)$$

which is of the same form as the electromagnetic Lagrangian density used in general relativity.

Absolute parallelism is applied to spinor wave functions of fundamental spin-$\frac{1}{2}$ particles, and the Dirac Lagrangian density L_D is given by[15]

$$L_D=\tfrac{1}{2}i\hbar b_k{}^\mu[\bar\psi\gamma^k D^*_\mu\psi-(D^*_\mu\bar\psi)\gamma^k\psi]-m\bar\psi\psi\,. \qquad (3.5a)$$

In this paper we use the unit, $\hbar=c=1$, but throughout this section we write $\hbar$ explicity for convenience of taking the semiclassical limit. For spinors the covariant derivative D^*_μ coincides with the usual derivative

$$D^*_\mu\psi=\partial_\mu\psi\,. \qquad (3.6a)$$

If we use the covariant derivative ∇_μ of general relativity,

$$\nabla_\mu\psi=(\partial_\mu+\tfrac{1}{2}i\Delta_{ij\mu}S^{ij})\psi\,, \qquad (3.6b)$$

with respect to the Ricci rotation coefficients $\{\Delta_{ij\mu}\}$,

$$\Delta_{ijk}=b_k{}^\mu\Delta_{ij\mu}=-\tfrac{1}{2}(T_{ijk}-T_{jik}-T_{kij})\,, \qquad (3.7)$$

then L_D of (3.5a) can be rewritten as

$$L_D=\tfrac{1}{2}i\hbar b_k{}^\mu[\bar\psi\gamma^k\nabla_\mu\psi-(\nabla_\mu\bar\psi)\gamma^k\psi]$$
$$-\tfrac{3}{4}\hbar a_k\bar\psi\gamma^5\gamma^k\psi-m\bar\psi\psi\,, \qquad (3.5b)$$

where $\{a^\mu\}$ is the axial-vector part of the torsion tensor,

$$a^\mu=b_k{}^\mu a^k=\tfrac{1}{6}\epsilon^{\mu\nu\rho\sigma}T_{\nu\rho\sigma}\,. \qquad (3.8)$$

This covariant derivative ∇_μ, when applied to tensors whose indices are written in Greek, becomes the usual covariant derivative with respect to the Levi-Civita connection (1.3). Here the completely antisymmetric tensors, $\epsilon=\{\epsilon^{\mu\nu\rho\sigma}\}$ and $\epsilon^*=\{\epsilon_{\mu\nu\rho\sigma}\}$, with respect to the coordinate basis are defined by[16]

$$\epsilon^{\mu\nu\rho\sigma}=(1/\sqrt{-g})\,\delta^{\mu\nu\rho\sigma}\,,$$
$$\epsilon_{\mu\nu\rho\sigma}=\sqrt{-g}\,\delta_{\mu\nu\rho\sigma}\,, \qquad (3.9)$$

where $\delta=\{\delta^{\mu\nu\rho\sigma}\}$ and $\delta^*=\{\delta_{\mu\nu\rho\sigma}\}$ are the completely antisymmetric tensor densities of weight -1 and $+1$, respectively, with normalizations $\delta^{0123}=+1$ and $\delta_{0123}=-1$. So the completely antisymmetric tensors with respect to the parallel vector fields are defined by

$$\epsilon^{ijmn} = b^i{}_\mu b^j{}_\nu b^m{}_\rho b^n{}_\sigma \epsilon^{\mu\nu\rho\sigma}, \qquad (3.10)$$

$$\epsilon_{ijmn} = b_i{}^\mu b_j{}^\nu b_m{}^\rho b_n{}^\sigma \epsilon_{\mu\nu\rho\sigma},$$

where $\epsilon^{(0)(1)(2)(3)} = +1$ and $\epsilon_{(0)(1)(2)(3)} = -1$ with Lorentz (Latin) indices in parentheses.

Variation of L_M with respect to A_μ gives

$$\partial_\nu(\sqrt{-g}F^{\mu\nu}) = \sqrt{-g}\,j^\mu, \qquad (3.11)$$

where the electromagnetic current is defined by

$$j^\mu = \frac{\delta}{\delta A_\mu} L_{\text{int}}. \qquad (3.12)$$

Equation (3.11) is just the Maxwell equation in general relativity, and hence the law of electromagnetism is entirely free from the influence of absolute parallelism. In space-time with a given background metric $\underline{g}$, electromagnetic waves propagate in the same manner as in general relativity: In the short-wavelength limit, in particular, light rays propagate along the null geodesics of the metric $\underline{g}$.

The Dirac equation is derived from L_D by taking variation with respect to $\bar{\psi}$,

$$[i\hbar b_k{}^\mu \gamma^k(D^*_\mu + \tfrac{1}{2}v_\mu) - m]\psi = 0, \qquad (3.13a)$$

or equivalently,

$$(i\hbar b_k{}^\mu \gamma^k \nabla_\mu - \tfrac{3}{4}\hbar a_k \gamma^5 \gamma^k - m)\psi = 0, \qquad (3.13b)$$

where $\{v_\mu\}$ is the vector part of the torsion tensor,

$$v_\mu = T^\lambda{}_{\cdot\lambda\mu}, \qquad (3.14)$$

and only the gravitational interaction is included.

B. Equations of motion for massive Dirac particles

We shall derive *two equations of motion* for a freely falling Dirac particle, i.e., *the equation of orbit* and *the equation of spin precession*,[17] by applying the WKB approximation method[18] to the Dirac equation (3.13).

The particle of spin $\frac{1}{2}$ is usually represented by a four-component spinor wave function obeying the first-order Dirac equation. However, it is well known that it can equally well be described by a two-component spinor wave function obeying a second-order wave equation.[19] So there are two equivalent ways to take the classical limit for the particle of spin $\frac{1}{2}$, in accordance with which a wave equation is considered; a first-order wave equation or a second-order one.

For our present purpose of deriving the spin equation in addition to the orbit equation, it is much more convenient to start from a second-order wave equation rather than from the Dirac equation (3.13). We thus introduce a two-component spinor wave function ϕ by[20]

$$\phi = \tfrac{1}{2}(1+\gamma^5)\psi. \qquad (3.15)$$

Then, by means of the Dirac equation (3.13), this can be "solved" with respect to ψ,

$$\psi = [1 + (i/m)\hbar b_k{}^\mu \gamma^k(D^*_\mu + \tfrac{1}{2}v_\mu)]\phi, \qquad (3.16)$$

and so we find that ϕ satisfies the second-order wave equation,

$$[i\hbar b_j{}^\mu \gamma^j(D^*_\mu + \tfrac{1}{2}v_\mu) - m] \times [i\hbar b_k{}^\nu \gamma^k(D^*_\nu + \tfrac{1}{2}v_\nu) + m]\phi = 0, \qquad (3.17)$$

to which we apply the WKB approximation method.

We seek a semiclassical solution of (3.17) with the following form:

$$\phi = \exp\left(\frac{i}{\hbar}S\right)\phi_0, \qquad (3.18)$$

by assuming that $\hbar$ is very small compared to S. Using (3.18) in (3.17) and then putting each order of $(\hbar/i)$ to zero, we find (up to the first order)

$$(\hbar/i)^0:\ g^{\mu\nu}(\partial_\mu S)(\partial_\nu S) + m^2 = 0, \qquad (3.19)$$

$$(\hbar/i)^1:\ \{2g^{\mu\nu}(\partial_\mu S)(D^*_\nu + \tfrac{1}{2}v_\nu) - b_j{}^\mu b_k{}^\nu[D^*_\mu(\partial_\nu S)]\gamma^j\gamma^k\}\phi_0 = 0. \qquad (3.20a)$$

The last equation is rewritten as

$$\{2g^{\mu\nu}(\partial_\mu S)\nabla_\nu + g^{\mu\nu}[\nabla_\mu(\partial_\nu S)] + \tfrac{3}{2}i\epsilon_{ijmn}b^{m\mu}(\partial_\mu S)a^n S^{ij}\}\phi_0 = 0 \qquad (3.20b)$$

with help of the relation between D^*_ν and ∇_ν,

$$D^*_\nu\phi_0 = (\nabla_\nu + \tfrac{1}{2}iK_{ij\nu}S^{ij})\phi_0,$$
$$D^*_\nu(\partial_\mu S) = \nabla_\nu(\partial_\mu S) - K^\lambda{}_{\cdot\mu\nu}\partial_\lambda S, \qquad (3.21)$$

with $\{K_{ij\nu}\}$ and $\{K_{\lambda\mu\nu}\}$ being the contortion tensor defined by

$$K_{ij\nu} = b_i{}^\lambda b_j{}^\mu K_{\lambda\mu\nu} = \tfrac{1}{2}b_i{}^\lambda b_j{}^\mu(T_{\lambda\mu\nu} - T_{\mu\lambda\nu} - T_{\nu\lambda\mu}) = -\Delta_{ij\nu}. \qquad (3.22)$$

The applicability condition of the semiclassical solution (3.18) is that when it is used in (3.17) the terms of order $(\hbar/i)^0$ are much larger than those of order $(\hbar/i)^1$. Estimating $|\partial_\mu S|/\hbar \sim 1/(\text{wavelength}) \equiv 1/\lambda$, $D^*_\nu\phi_0 \sim \phi_0/w$ with w the width of the wave packet, and $|D^*_\nu(\partial_\mu S)| \sim \hbar/\lambda L$ with L being the distance over which the parallel vector fields $\{b^k{}_\mu\}$ vary considerably, we obtain the following inequality:

$$L \gg \lambda, \quad w \gg \lambda. \qquad (3.23)$$

Equation (3.19) is the Hamilton-Jacobi equation which describes the motion of freely falling particles in general relativity.[21] The complete solution $S(x; \alpha_1, \alpha_2, \alpha_3)$ with three free parameters, α_1, α_2, and α_3, determines the classical orbit by

$$\frac{\partial S}{\partial \alpha_a} = \beta_a(=\text{const}), \quad (a = 1, 2, 3). \qquad (3.24)$$

When the trajectory $x^\mu(\tau)$ defined by (3.24) is parametrized by the proper time τ, it satisfies the geodesic equation,

$$\frac{dx^\mu}{d\tau}=\frac{1}{m}g^{\mu\nu}\partial_\nu S\equiv U^\mu \text{ (four-velocity)}, \quad (3.25)$$

$$\frac{d^2x^\mu}{d\tau^2}+\left\{{\mu \atop \lambda\nu}\right\}\frac{dx^\lambda}{d\tau}\frac{dx^\nu}{d\tau}=0. \quad (3.26)$$

Given the solution $S(x;\alpha_1,\alpha_2,\alpha_3)$ of (3.19), Eq. (3.20) can be solved to define the spinor wave function ϕ_0 in terms of S. By virtue of (3.16), the semiclassical expression for the Dirac spinor wave function in terms of S and ϕ_0 is given by

$$\psi=\exp\left(\frac{i}{\hbar}S\right)\psi_0, \quad (3.27)$$

$$\psi_0=\left(1-\frac{\partial_\mu S}{m}b_k{}^\mu\gamma^k\right)\phi_0$$
$$=(1-b_k{}^\mu U_\mu\gamma^k)\phi_0. \quad (3.28)$$

The probability current, $j^\mu=b_k{}^\mu\bar\psi\gamma^k\psi$, then takes the following form in the semiclassical approximation:

$$j^\mu=\rho U^\mu, \quad (3.29)$$

where ρ is defined by

$$\rho=-2b_k{}^\mu U_\mu\bar\phi_0\gamma^k\phi_0. \quad (3.30)$$

Equation (3.20b) of ϕ_0 ensures that j^μ satisfies the continuity equation,

$$\nabla_\mu j^\mu=\frac{1}{\sqrt{-g}}\partial_\mu(\sqrt{-g}\,j^\mu)=0. \quad (3.31)$$

The expression (3.29) for the probability current shows that, in the semiclassical approximation, the probability may be regarded as following along the classical trajectory.

We can form a wave packet by superposing the solutions of (3.27) with different values of parameters, $\alpha=(\alpha_1,\alpha_2,\alpha_3)$:

$$\psi(x)=\int\rho(\alpha)d^3\alpha\exp\left[\frac{i}{\hbar}S(x;\alpha)\right]\psi_0(x;\alpha), \quad (3.32a)$$

where $\rho(\alpha)$ is a weight function with a "sharp peak" of width $\Delta\alpha$ at $\alpha=\bar\alpha\equiv(\bar\alpha_1,\bar\alpha_2,\bar\alpha_3)$. Here a sharp peak means that the following conditions are satisfied: (1) The ratio $\hbar/\Delta\alpha$ is negligibly small compared to the macroscopic scale, and (2) the ratio $\hbar/(\Delta\alpha)^2$ satisfies the inequality

$$\frac{\hbar}{(\Delta\alpha)^2}\gg\frac{\partial^2 S}{\partial\alpha_a\partial\alpha_b}. \quad (3.32b)$$

Since $\psi_0(x;\alpha)$ is not highly oscillating with respect to α, it follows from the inequality (3.32b) that Eq. (3.32a) can be rewritten as

$$\psi(x)=\exp\left[\frac{i}{\hbar}S(x;\bar\alpha)\right]\psi_0(x;\bar\alpha)$$
$$\times\int\rho(\alpha)d^3\alpha\exp\left[\frac{i}{\hbar}\sum_a(\alpha_a-\bar\alpha_a)\frac{\partial S}{\partial\bar\alpha_a}\right]. \quad (3.32c)$$

In the integral of (3.32c) compensation takes place almost everywhere except in the space-time region satisfying

$$\frac{\partial S}{\partial\bar\alpha_a}\lesssim\frac{\hbar}{\Delta\alpha}. \quad (3.33a)$$

According to the condition (1) stated above, the right-hand side of (3.33a) is negligibly small compared to the macroscopic scale. Therefore, in the macroscopic scale, the wave packet (3.32c) has nonvanishing amplitude only along a world line $x^\mu(\tau)$ defined by

$$\frac{\partial S}{\partial\bar\alpha_a}=0. \quad (3.33b)$$

Here τ is the proper time along the world line. *The wave packet thus propagates along a classical trajectory $x^\mu(\tau)$ satisfying the geodesic equation (3.26).*

Now we turn our consideration to the motion of spin for a spin-$\frac{1}{2}$ particle described by the spinor wave function (3.32c). The spin polarization is described by the spinor wave function $\psi_0(x(\tau);\bar\alpha)$, since other two factors of (3.32c) are scalar functions which have nothing to do with intrinsic spin polarization. We introduce a new spinor wave function $\psi_0'(x(\tau);\bar\alpha)$ by

$$\psi_0'=\frac{1}{\sqrt\rho}\psi_0. \quad (3.34a)$$

Then it is normalized like

$$b_k{}^\mu\bar\psi_0'\gamma^k\psi_0'=U^\mu \text{ (four-velocity)}, \quad (3.34b)$$

and accordingly it can be taken as the normalized spinor wave function which describes spin polarization. From (3.20a)–(3.20b) and (3.28)–(3.31), it follows after a little algebra that the normalized spinor wave function ψ_0' satisfies

$$\left(\frac{D^*}{d\tau}+iU^\mu K_{\mu ij}S^{ij}\right)\psi_0'=0, \quad (3.35a)$$

or equivalently,

$$\left(\frac{\nabla}{d\tau}+\frac{3i}{4}\epsilon_{ijmn}b^m{}_\mu U^\mu a^n S^{ij}\right)\psi_0'=0, \quad (3.35b)$$

where $D^*/d\tau$ and $\nabla/d\tau$ mean covariant differentiation along a classical trajectory $x^\mu(\tau)$, $D^*/d\tau=U^\mu D^*_\mu$ and $\nabla/d\tau=U^\mu\nabla_\mu$, respectively. Equations (3.35) describe the temporal change of the spin state as a spin-$\frac{1}{2}$ particle moves along a classical trajectory $x^\mu(\tau)$. The second term of (3.35b) represents the effect of absolute parallelism on the spin precession in new general relativity.

We define the spin vector $\{S^\mu\}$ by

$$S^\mu = -\tfrac{1}{2} b_k{}^\mu \overline{\psi}_0' \gamma^5 \gamma^k \psi_0', \tag{3.36a}$$

which has only three independent components, because (3.28) and (3.34a) give

$$U^\mu S_\mu = 0 . \tag{3.36b}$$

It follows from (3.35) that $\{S^\mu\}$ satisfies

$$\frac{D^*}{d\tau} S^\mu = -(K^{\lambda\mu\nu} - K^{\lambda\nu\mu}) U_\lambda S_\nu , \tag{3.37a}$$

or equivalently,

$$\frac{\nabla}{d\tau} S^\mu = -\tfrac{3}{2} \epsilon^{\mu\nu\rho\sigma} U_\nu a_\rho S_\sigma . \tag{3.37b}$$

These are the classical equations of spin precession in new general relativity. The right-hand side of (3.37b) represents the effect of absolute parallelism. When the axial-vector part of the torsion tensor vanishes, (3.37b) reduces to the equation of spin precession in general relativity. For a nonrelativistic particle in a weak gravitational field,[22] (3.37b) becomes

$$\frac{d\vec{S}}{dt} = -\tfrac{3}{2} \vec{a} \times \vec{S}, \tag{3.38}$$

where $\vec{S}$ and $\vec{a}$ are the space components of $\{S^\mu\}$ and $\{a^\mu\}$, respectively.

C. Equations of motion for neutrinos and antineutrinos

Neutrinos and antineutrinos are described by two-component spinor wave functions which are obtained from (four-component) Dirac spinor wave functions of (2.20) by putting $\chi = 0$ for neutrinos and $\xi = 0$ for antineutrinos, respectively. We shall consider only antineutrinos, since neutrinos can be treated in a similar manner. For antineutrinos, the Dirac equation (3.13a) becomes the Weyl equation for a right-handed massless particle,[23]

$$i\hbar b_k{}^\mu \sigma^k (D_\mu^* + \tfrac{1}{2} v_\mu) \chi = 0 . \tag{3.39}$$

The semiclassical solution, $\chi = \exp(iS/\hbar)\chi_0$, must satisfy

$$b_k{}^\mu (\partial_\mu S) \sigma^k \chi_0 = 0 , \tag{3.40}$$

and hence we get

$$\det(b_k{}^\mu \partial_\mu S \sigma^k) = -g^{\mu\nu} (\partial_\mu S)(\partial_\nu S) = 0 , \tag{3.41}$$

which is just the Hamilton-Jacobi equation for a massless particle in general relativity. The classical trajectory $x^\mu(\sigma)$ defined by (3.24) satisfies the following equations:

$$\frac{dx^\mu}{d\sigma} = g^{\mu\nu} \partial_\nu S \equiv p^\mu \text{ (four-momentum)}, \tag{3.42}$$

$$\frac{d^2 x^\mu}{d\sigma^2} + \left\{ {\mu \atop \lambda\nu} \right\} \frac{dx^\lambda}{d\sigma} \frac{dx^\nu}{d\sigma} = 0 , \tag{3.43}$$

where σ is the affine parameter along the trajectory: The normalization of σ is fixed by (3.42).

The four momentum $\{p^\mu\}$ is null, due to (3.41). The current $\{j^\mu = b_k{}^\mu \chi^\dagger \sigma^k \chi\}$ is also null, because χ is a two-component spinor. Since these two null vectors, $\{p^\mu\}$ and $\{j^\mu\}$, are orthogonal with each other by virtue of (3.40), they must be proportional to one another,

$$j^\mu = \rho p^\mu , \tag{3.44}$$

where ρ is a positive-definite scalar function. We can apply the same argument as for the massive case, to show that *antineutrinos move along the classical trajectory satisfying (3.43) in the short-wavelength limit.*

We take $\chi_0/\sqrt{\rho}$ as the normalized spinor wave function for antineutrinos. The spin vector (3.36a) then becomes

$$S^\mu = \frac{1}{2\rho} b_k{}^\mu \chi_0^\dagger \sigma^k \chi_0 = \frac{1}{2\rho} j^\mu = \tfrac{1}{2} p^\mu , \tag{3.45}$$

showing that antineutrinos are of helicity $+\frac{1}{2}$ as they should. Therefore, *the classical equations of motion for neutrinos and antineutrinos in new general relativity are the same as those in general relativity.*

IV. GRAVITATIONAL LAGRANGIAN IN VACUUM

We shall construct a gravitational Lagrangian density in vacuum,

$$I_G = \int d^4x \sqrt{-g} L_G . \tag{4.1}$$

For this purpose we enumerate the basic postulates which the above action must obey:

(1) Invariance under the group of general coordinate transformations; for arbitrary change of coordinates the parallel vector fields transform like

$$b'_k{}^\mu(x') = (\partial x'^\mu / \partial x^\nu) b_k{}^\nu(x) . \tag{4.2}$$

(2) Invariance under the group of global, proper, orthochronous Lorentz transformations $L_+^\dagger$; for its element $A = (A^j{}_k)$ with $A^t \eta A = \eta$, $\det A = 1$, $A^0{}_0 \geq 1$, and $\partial_\mu A = 0$, the parallel vector fields change like

$$b_k{}^\mu(x) = A^j{}_k b'_j{}^\mu(x) . \tag{4.3}$$

(3) L_G be invariant under the parity operation; by parity operation we mean the Lorentz transformation, $\underline{b}_{(0)} \to \underline{b}_{(0)}$ and $\underline{b}_{(a)} \to -\underline{b}_{(a)}$, where Lorentz

indices are enclosed by parentheses.

(4) L_G be quadratic terms in the torsion tensor, besides a cosmological term.

Now, the torsion tensor is given by

$$T^{\lambda}_{\cdot\mu\nu}=b_k{}^{\lambda}(\partial_\nu b^k{}_\mu-\partial_\mu b^k{}_\nu), \tag{4.4}$$

which is *reducible* with respect to the group of global Lorentz transformations. It is convenient to perform an *irreducible* decomposition, from which we can construct a gravitational Lagrangian density. The torsion tensor is decomposed into three irreducible parts under this group[24a]:

$$t_{\lambda\mu\nu}=\tfrac{1}{2}(T_{\lambda\mu\nu}+T_{\mu\lambda\nu})+\tfrac{1}{6}(g_{\nu\lambda}v_\mu+g_{\nu\mu}v_\lambda)-\tfrac{1}{3}g_{\lambda\mu}v_\nu, \tag{4.5}$$

$$v_\mu=T^{\lambda}_{\cdot\lambda\mu}, \tag{4.6}$$

$$a_\mu=\tfrac{1}{6}\epsilon_{\mu\nu\rho\sigma}T^{\nu\rho\sigma}, \tag{4.7}$$

where $\epsilon^*=\{\epsilon_{\mu\nu\rho\sigma}\}$ is the completely antisymmetric tensor, introduced in (3.9). In fact, let $\rho(m,n)$ be an irreducible representation of the proper orthochronous Lorentz group, where $2m$ and $2n$ take non-negative integer numbers.[24b] Then the tensor $\{t_{\lambda\mu\nu}\}$ transforms according to $\rho(\frac{3}{2},\frac{1}{2})\oplus\rho(\frac{1}{2},\frac{3}{2})$ of 16 dimensions (the Young table [21] minus traces), the vector $\{v_\mu\}$ according to $\rho(\frac{1}{2},\frac{1}{2})$ of 4 dimensions, and finally the axial-vector $\{a_\mu\}$ according to $\rho(\frac{1}{2},\frac{1}{2})$ of 4 dimensions (the Young table [111]). The torsion tensor is conversely written in terms of the three irreducibles,

$$T_{\lambda\mu\nu}=\tfrac{2}{3}(t_{\lambda\mu\nu}-t_{\lambda\nu\mu})+\tfrac{1}{3}(g_{\lambda\mu}v_\nu-g_{\lambda\nu}v_\mu)+\epsilon_{\lambda\mu\nu\rho}a^\rho. \tag{4.8}$$

The tensor $\{t_{\lambda\mu\nu}\}$ has the following properties derived from the defining equation (4.5):

$$t_{\lambda\mu\nu}=t_{\mu\lambda\nu}, \tag{4.9}$$

$$g^{\mu\nu}t_{\lambda\mu\nu}=0=g^{\lambda\mu}t_{\lambda\mu\nu}, \tag{4.10}$$

$$t_{\lambda\mu\nu}+t_{\mu\nu\lambda}+t_{\nu\lambda\mu}=0. \tag{4.11}$$

The above postulates of (1) to (4) require that the most general Lagrangian density be of the form

$$L_G=a_1(t^{\lambda\mu\nu}t_{\lambda\mu\nu})+a_2(v^\mu v_\mu)+a_3(a^\mu a_\mu)+a_0, \tag{4.12}$$

where a_1, a_2, and a_3 are free parameters, while a_0 is a cosmological term.

In Appendix A we will treat a case of lifting up the postulate of (3), by adding to (4.12) parity-violating terms like $(v^\mu a_\mu)$ and $(\epsilon_{\mu\nu\rho\sigma}t^{\lambda\mu\nu}t_\lambda{}^{\rho\sigma})$. In the rest of this paper we shall neglect the cosmological term, so we have the gravitational action of

$$I_G=\int d^4x\sqrt{-g}\,L_G=\int d^4x\sqrt{-g}\,[a_1(t^{\lambda\mu\nu}t_{\lambda\mu\nu})+a_2(v^\mu v_\mu)+a_3(a^\mu a_\mu)]. \tag{4.13}$$

Next we observe the identity

$$\int d^4x\sqrt{-g}\,R(\{\ \})=\int d^4x\sqrt{-g}\,[-\tfrac{2}{3}(t^{\lambda\mu\nu}t_{\lambda\mu\nu})+\tfrac{2}{3}(v^\mu v_\mu)-\tfrac{3}{2}(a^\mu a_\mu)], \tag{4.14}$$

where $R(\{\ \})$ denotes the Riemann scalar curvature. It is given by the contraction of the Ricci tensor, which is again the contraction of the Riemann-Christoffel curvature tensor:

$$R^{\rho}_{\cdot\sigma\mu\nu}(\{\ \})=\partial_\mu\{^{\rho}_{\sigma\nu}\}-\partial_\nu\{^{\rho}_{\sigma\mu}\}+\{^{\rho}_{\lambda\mu}\}\{^{\lambda}_{\sigma\nu}\}-\{^{\rho}_{\lambda\nu}\}\{^{\lambda}_{\sigma\mu}\}, \tag{4.15}$$

$$R_{\mu\nu}(\{\ \})=R^{\lambda}_{\cdot\mu\lambda\nu}(\{\ \}), \tag{4.16}$$

$$R(\{\ \})=g^{\mu\nu}R_{\mu\nu}(\{\ \}). \tag{4.17}$$

Here the symbol $\{^{\rho}_{\sigma\nu}\}$ denotes the Levi-Civita connection (1.3) of the Riemann space-time. Since the Weitzenböck space-time has the vanishing of the curvature tensor [see (1.7)], the identity of (4.14) should be taken as purely mathematical. Using $\kappa=8\pi G/c^4=8\pi G$ with G the Newton gravitational constant, we finally rewrite the gravitational action in the following form:

$$I_G=\int d^4x\sqrt{-g}\left(\frac{1}{2\kappa}R(\{\ \})+c_1(t^{\lambda\mu\nu}t_{\lambda\mu\nu})+c_2(v^\mu v_\mu)+c_3(a^\mu a_\mu)\right). \tag{4.18}$$

Comparing (4.13) with (4.18), we find that the parameters are effectively (under integration symbol) related to each other by

$$c_1=a_1+\frac{1}{3\kappa},\quad c_2=a_2-\frac{1}{3\kappa},\quad c_3=a_3+\frac{3}{4\kappa}. \tag{4.19}$$

It should be mentioned that one of the free parameters, c_1, c_2, and c_3, must be nonzero; otherwise, the left-hand side of a gravitational field equation would become symmetric, while the right-hand side, the energy-momentum tensor of spin-$\frac{1}{2}$ fundamental particles, would become nonsymmetric. This is a contradiction.

It is easy to derive a gravitational field equation. For the sake of completeness we write down a gravitational field equation when matter is present, by adding to the vacuum gravitational action a matter action I_M, which satisfies the postulates of (1) to (4),

$$I = I_G + I_M , \tag{4.20}$$

$$I_M = \int d^4x \sqrt{-g}\, L_M . \tag{4.21}$$

From this action follows the following field equation, by taking variation with respect to the parallel vector fields $b^k{}_\nu$ and then multiplying with $\eta^{kj} b_j{}^\mu$:

$$G^{\mu\nu}(\{\,\}) + 2\kappa D^*_\lambda F^{\mu\nu\lambda} + 2\kappa v_\lambda F^{\mu\nu\lambda} + 2\kappa H^{\mu\nu} - \kappa g^{\mu\nu} L' = \kappa T^{\mu\nu} . \tag{4.22}$$

Here the first term denotes the Einstein tensor of general relativity,

$$G^{\mu\nu}(\{\,\}) \equiv R^{\mu\nu}(\{\,\}) - \tfrac{1}{2} g^{\mu\nu} R(\{\,\}) , \tag{4.23}$$

and the tensor $\{F^{\mu\nu\lambda}\}$ stands for

$$\begin{aligned} F^{\mu\nu\lambda} &= c_1(t^{\mu\nu\lambda} - t^{\mu\lambda\nu}) + c_2(g^{\mu\nu} v^\lambda - g^{\mu\lambda} v^\nu) \\ &\quad - \tfrac{1}{3} c_3 \epsilon^{\mu\nu\lambda\rho} a_\rho \\ &= -F^{\mu\lambda\nu} . \end{aligned} \tag{4.24}$$

The fourth term $\{H^{\mu\nu}\}$ is defined by

$$H^{\mu\nu} = T^{\rho\sigma\mu} F_{\rho\sigma}{}^{\cdot\,\cdot\,\nu} - \tfrac{1}{2} T^{\nu\rho\sigma} F^{\mu}{}_{\cdot\,\rho\sigma} = H^{\nu\mu} , \tag{4.25}$$

which is shown to be symmetric upon inserting the irreducible decomposition (4.8) of the torsion tensor. Finally, L' is given by

$$L' = c_1(t^{\lambda\mu\nu} t_{\lambda\mu\nu}) + c_2(v^\mu v_\mu) + c_3(a^\mu a_\mu) . \tag{4.26}$$

A source term is, as usual, defined by

$$\begin{aligned} \delta \int d^4x \sqrt{-g}\, L_M &= \int d^4x \sqrt{-g}\, T_k{}^{\cdot\mu} \delta b^k{}_\mu \\ &= - \int d^4x \sqrt{-g}\, T_\mu{}^{\cdot k} \delta b_k{}^\mu . \end{aligned} \tag{4.27}$$

Therefore, an energy-momentum tensor is given by

$$\sqrt{-g}\, T_{\mu\nu} = -\eta_{kj} b^j{}_\nu \delta \sqrt{-g}\, L_M / \delta b_k{}^\mu , \tag{4.28a}$$

or equivalently,

$$\sqrt{-g}\, T^{\mu\nu} = \eta^{kj} b_j{}^\mu \delta \sqrt{-g}\, L_M / \delta b^k{}_\nu . \tag{4.28b}$$

For instance, the energy-momentum tensor of the electromagnetic field and the Dirac field is calculated from the above formulas with (3.3) and (3.5) to be

$$T_{\mu\nu} = F_{\mu\rho} F_{\nu\sigma} g^{\rho\sigma} + g_{\mu\nu} L_{em} , \tag{4.29}$$

$$\begin{aligned} T_{\mu\nu} &= -\tfrac{1}{2} i b^k{}_\nu [\bar\psi \gamma_k \partial_\mu \psi - (\partial_\mu \bar\psi) \gamma_k \psi] + g_{\mu\nu} L_D \\ &= -\tfrac{1}{2} i b^k{}_\nu [\bar\psi \gamma_k \nabla_\mu \psi - (\nabla_\mu \bar\psi) \gamma_k \psi] \\ &\quad + \tfrac{1}{4} \epsilon_{\lambda\rho\sigma\nu} b_k{}^\lambda K^{\rho\sigma}{}_{\cdot\,\cdot\,\mu} \bar\psi \gamma^5 \gamma^k \psi \\ &\quad + g_{\mu\nu}(L_D^{GR} - \tfrac{3}{4} a_k \bar\psi \gamma^5 \gamma^k \psi) , \end{aligned} \tag{4.30a}$$

with the Dirac Lagrangian L_D^{GR} used in general relativity

$$L_D^{GR} = \tfrac{1}{2} i b_k{}^\mu [\bar\psi \gamma^k \nabla_\mu \psi - (\nabla_\mu \bar\psi) \gamma^k \psi] - m \bar\psi \psi . \tag{4.30b}$$

It is useful to split the gravitational field equation into the symmetric and antisymmetric parts:

$$G^{\mu\nu}(\{\,\}) + 2\kappa D^*_\lambda F^{(\mu\nu)\lambda} + 2\kappa v_\lambda F^{(\mu\nu)\lambda} + 2\kappa H^{\mu\nu} - \kappa g^{\mu\nu} L' = \kappa T^{(\mu\nu)} , \tag{4.31}$$

$$2 D^*_\lambda F^{[\mu\nu]\lambda} + 2 v_\lambda F^{[\mu\nu]\lambda} = T^{[\mu\nu]} , \tag{4.32}$$

where

$$\begin{aligned} F^{(\mu\nu)\lambda} &= \tfrac{1}{2}(F^{\mu\nu\lambda} + F^{\nu\mu\lambda}) , \\ F^{[\mu\nu]\lambda} &= \tfrac{1}{2}(F^{\mu\nu\lambda} - F^{\nu\mu\lambda}) , \\ T^{(\mu\nu)} &= \tfrac{1}{2}(T^{\mu\nu} + T^{\nu\mu}) , \\ T^{[\mu\nu]} &= \tfrac{1}{2}(T^{\mu\nu} - T^{\nu\mu}) . \end{aligned} \tag{4.33}$$

Furthermore, it is often useful to rewrite the field equation in Latin indices:

$$G^{ij}(\{\,\}) + \frac{2\kappa}{\sqrt{-g}} \partial_\lambda(\sqrt{-g}\, F^{ij\lambda}) + 2\kappa H^{ij} - \kappa \eta^{ij} L' = \kappa T^{ij} , \tag{4.34}$$

where

$$\begin{aligned} G^{ij}(\{\,\}) &= b^i{}_\mu b^j{}_\nu G^{\mu\nu}(\{\,\}) , \\ F^{ij\lambda} &= b^i{}_\mu b^j{}_\nu F^{\mu\nu\lambda} , \\ H^{ij} &= b^i{}_\mu b^j{}_\nu H^{\mu\nu} , \quad T^{ij} = b^i{}_\mu b^j{}_\nu T^{\mu\nu} . \end{aligned} \tag{4.35}$$

Finally the symmetric and antisymmetric parts of the gravitational field equation are derived as follows:

$$G^{ij}(\{\,\}) + \frac{2\kappa}{\sqrt{-g}} \partial_\lambda(\sqrt{-g}\, F^{(ij)\lambda}) + 2\kappa H^{ij} - \kappa \eta^{ij} L' = \kappa T^{(ij)} , \tag{4.36}$$

$$2 \partial_\lambda(\sqrt{-g}\, F^{[ij]\lambda}) = \sqrt{-g}\, T^{[ij]} . \tag{4.37}$$

V. THE STATIC, ISOTROPIC GRAVITATIONAL FIELD IN VACUUM

Let us consider a static, isotropic gravitational field produced by a static, spherical body, assuming that the spin of constituent particles of a body, if it exists, can be completely neglected. The state of a central body then does not change under space inversion, besides time reversal and space rotation. Therefore, it is possible to find a set of coordinates, $x^0 = t$, x^1, x^2, and x^3, such that the parallel vector fields $\underline{b} = \{\underline{b}_k\} = \{b_k{}^\mu\}$ are *form invariant* under time reversal, space inversion, and space rotation,

$$t \to -t , \quad \underline{b}_{(0)} \to -\underline{b}_{(0)} \quad \text{(time reversal)} , \tag{5.1a}$$

$$x^\alpha \to -x^\alpha, \quad \underline{b}_{(a)} \to -\underline{b}_{(a)}$$
$$\text{(space inversion)}, \tag{5.1b}$$
$$x^\alpha \to R_{\alpha\beta} x^\beta, \quad \underline{b}_{(a)} \to R_{ac} \underline{b}_{(c)}$$
$$\text{(space rotation)}, \tag{5.1c}$$

where, to avoid confusion, Latin indices in b_k are enclosed in parentheses, and $R = (R_{ac}) = (R_{\alpha\beta})$ is a position-independent 3×3 orthogonal matrix

$$RR^t = R^t R = I, \quad \det R = 1. \tag{5.1d}$$

Then, as is shown in Appendix B, by using the freedom to redefine the radius

$$x'^\alpha = \Phi(r) x^\alpha, \quad r \equiv (x^\alpha x^\alpha)^{1/2},$$

we can assume, without loss of generality, that the parallel vector fields $\underline{b} = \{\underline{b}_k\} = \{b_k{}^\mu(x)\}$ have a diagonal form,

$$b_{(0)}{}^0 = C(r), \quad b_{(0)}{}^\alpha = 0 = b_{(a)}{}^0,$$
$$b_{(a)}{}^\alpha = D(r)\delta_a^\alpha, \tag{5.2}$$

with two unknown functions of r, C, and D. The invariant distance ds^2 is then expressed in the isotropic form,

$$ds^2 = g_{\mu\nu} dx^\mu dx^\nu$$
$$= \eta_{ij} b^i{}_\mu b^j{}_\nu dx^\mu dx^\nu$$
$$= -A(r)dt^2 + B(r) dx^\alpha dx^\alpha, \tag{5.3a}$$

where the metric coefficients, A and B, are related to C and D by

$$A = 1/C^2, \quad B = 1/D^2. \tag{5.3b}$$

The set of coordinates $\{x^\mu\}$ is therefore the isotropic coordinate system.

We write the gravitational field equation (4.22) as

$$I^{\mu\nu} = \kappa T^{\mu\nu}, \tag{5.4a}$$

where $\{I^{\mu\nu}\}$ is defined by

$$I^{\mu\nu} \equiv G^{\mu\nu}(\{\,\}) + 2\kappa D^*_\lambda F^{\mu\nu\lambda} + 2\kappa v_\lambda F^{\mu\nu\lambda}$$
$$+ 2\kappa H^{\mu\nu} - \kappa g^{\mu\nu} L'. \tag{5.4b}$$

For a static, isotropic gravitational field with (5.2), $\{I^{\mu\nu}\}$ is given by

$$I^{00} = -\frac{1+\kappa(c_1+4c_2)}{AB}\left[\epsilon\left(\frac{A'}{A}\right)' + (1-2\epsilon)\left(\frac{B'}{B}\right)' + \frac{2}{r}\left(\epsilon\frac{A'}{A} + (1-2\epsilon)\frac{B'}{B}\right) + \frac{\epsilon}{4}\left(\frac{A'}{A}\right)^2 + \frac{\epsilon}{2}\left(\frac{A'}{A}\right)\left(\frac{B'}{B}\right) + \tfrac{1}{4}(1-4\epsilon)\left(\frac{B'}{B}\right)^2\right], \tag{5.5a}$$

$$I^{0\alpha} = 0 = I^{\alpha 0}, \tag{5.5b}$$

$$\begin{aligned} I^{\alpha\beta} = {} & \frac{\delta^{\alpha\beta}}{2B^2}[1+\kappa(c_1+4c_2)]\left[(1-2\epsilon)\left(\frac{A'}{A}\right)' + \left(\frac{B'}{B}\right)' + \frac{1}{r}\left((1-2\epsilon)\frac{A'}{A} + \frac{B'}{B}\right) + \tfrac{1}{2}(1-3\epsilon)\left(\frac{A'}{A}\right)^2 + \epsilon\left(\frac{A'}{A}\right)\left(\frac{B'}{B}\right)\right] \\ & - \frac{1}{2B^2}\frac{x^\alpha x^\beta}{r^2}[1+\kappa(c_1+4c_2)]\left[(1-2\epsilon)\left(\frac{A'}{A}\right)' + \left(\frac{B'}{B}\right)' - \frac{1}{r}\left((1-2\epsilon)\frac{A'}{A} + \frac{B'}{B}\right) + \tfrac{1}{2}(1-4\epsilon)\left(\frac{A'}{A}\right)^2\right. \\ & \left. - (1-3\epsilon)\left(\frac{A'}{A}\right)\left(\frac{B'}{B}\right) - \frac{1}{2}\left(\frac{B'}{B}\right)^2\right], \end{aligned} \tag{5.5c}$$

where the parameter ϵ is a constant defined by

$$\epsilon \equiv \frac{\kappa(c_1+c_2)}{1+\kappa(c_1+4c_2)}, \tag{5.6}$$

and a prime means differentiation with respect to r. It is shown in Appendix C that the constant $[1+\kappa(c_1+4c_2)]$ is a nonzero number.

There is no appearance of the parameter c_3, but only the parameters, c_1 and c_2, owing to a static, isotropic gravitational field. In other words, we can say nothing about the parameter c_3 in this case. Now we proceed to study a solution of the field equation (5.4) with (5.5) for the following three cases: (A) the Newtonian limit, (B) the post-Newtonian approximation, and (C) an exact solution in vacuum.

A. The Newtonian limit

We assume that a central gravitating body is a nonrelativistic system with all the components of $T^{\alpha\beta}$ being negligibly small compared to T^{00}; $T^{00} \gg |T^{\alpha\beta}| \simeq 0$. Then the gravitational field is weak; the metric coefficients, A and B, are nearly unity, $A \simeq 1 \simeq B$, and terms quadratic in A' and B' can be ignored in the field equation (5.4) with (5.5). We then find that the gravitational field equation in the Newtonian limit is given by

$$-[1+\kappa(c_1+4c_2)]\left\{\epsilon A'' + (1-2\epsilon)B'' + \frac{2}{r}[\epsilon A' + (1-2\epsilon)B']\right\} = \kappa T^{00}, \tag{5.7a}$$

$$(1-2\epsilon)A'+B'=0. \tag{5.7b}$$

The external solution satisfying the boundary condition,

$$\lim_{r\to\infty} A(r)=\lim_{r\to\infty} B(r)=1, \tag{5.8}$$

is

$$A(r)=1-\frac{2}{(1-\epsilon)(1-4\epsilon)[1+\kappa(c_1+4c_2)]}\frac{Gm}{r}, \tag{5.9a}$$

$$B(r)=1+\frac{2(1-2\epsilon)}{(1-\epsilon)(1-4\epsilon)[1+\kappa(c_1+4c_2)]}\frac{Gm}{r}, \tag{5.9b}$$

with m the total mass of the source,

$$m=\int T^{00}(x)d^3x=4\pi\int r^2T^{00}(r)dr. \tag{5.10}$$

It was found in Sec. III that the trajectory of a test particle is determined by the geodesic equation (3.26), which reduces for a nonrelativistic particle to

$$\frac{d^2x^\alpha}{dt^2}=\frac{1}{2}\frac{\partial}{\partial x^\alpha}g_{00}(x)$$

$$=-\frac{1}{(1-\epsilon)(1-4\epsilon)[1+\kappa(c_1+4c_2)]}\frac{\partial}{\partial x^\alpha}\left(-\frac{Gm}{r}\right). \tag{5.11a}$$

Here the solution (5.9a) is used in the final step. We demand that the trajectory of a nonrelativistic test particle, specified by $x^\alpha(t)$, obeys the Newton equation of motion

$$\frac{d^2x^\alpha}{dt^2}=-\frac{\partial}{\partial x^\alpha}\phi, \tag{5.11b}$$

where ϕ is a gravitational potential, which takes the form

$$\phi=-Gm/r \tag{5.11c}$$

for a gravitational field around a spherical body with mass m. Accordingly, the parameters, c_1 and c_2, must satisfy the condition

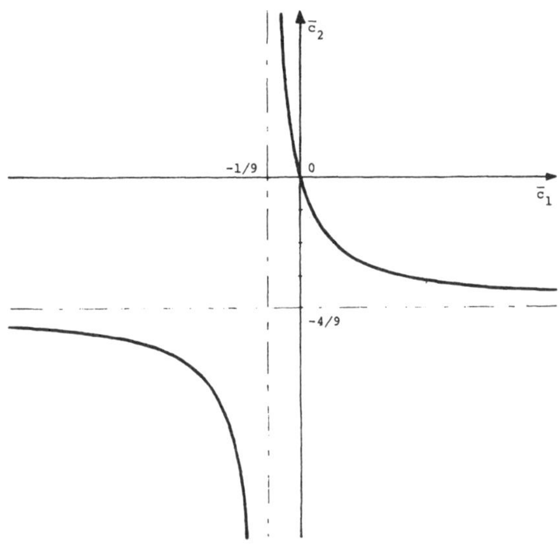

FIG. 2. The curve of (5.12b): $\bar{c}_1=\kappa c_1$ and $\bar{c}_2=\kappa c_2$. $\bar{c}_2=-4\bar{c}_1/(1+9\bar{c}_1)$.

$$(1-\epsilon)(1-4\epsilon)[1+\kappa(c_1+4c_2)]=1, \tag{5.12a}$$

which we shall assume hereafter. This condition is called the Newton approximation condition. In terms of $\bar{c}_1=\kappa c_1$ and $\bar{c}_2=\kappa c_2$, the Newton approximation condition reads as

$$4\bar{c}_1+\bar{c}_2+9\bar{c}_1\bar{c}_2=0. \tag{5.12b}$$

From this follow the two cases, $c_1=0=c_2$ and $c_1\neq 0\neq c_2$. See Fig. 2 for the curve specified by (5.12b). Now, combining (5.6) and (5.12b), we find

$$\bar{c}_1=-\frac{\epsilon}{3(1-\epsilon)},\quad \bar{c}_2=\frac{4\epsilon}{3(1-4\epsilon)}. \tag{5.12c}$$

Since ϵ is observable in solar-system experiments, as will be shown in Sec. VI, we draw the curves of (5.12c) versus ϵ in Fig. 3.

B. Vacuum solution in the post-Newtonian approximation

The field equation (5.4) with (5.5) can be rewritten in vacuum as follows,

$$\epsilon\left(\frac{A'}{A}\right)'+(1-2\epsilon)\left(\frac{B'}{B}\right)'+\frac{2}{r}\left(\epsilon\frac{A'}{A}+(1-2\epsilon)\frac{B'}{B}\right)+\frac{\epsilon}{4}\left(\frac{A'}{A}\right)^2+\frac{\epsilon}{2}\left(\frac{A'}{A}\right)\left(\frac{B'}{B}\right)+\tfrac{1}{4}(1-4\epsilon)\left(\frac{B'}{B}\right)^2=0, \tag{5.13a}$$

$$(1-2\epsilon)\left(\frac{A'}{A}\right)'+\left(\frac{B'}{B}\right)'+\frac{1}{r}\left((1-2\epsilon)\frac{A'}{A}+\frac{B'}{B}\right)+\tfrac{1}{2}(1-3\epsilon)\left(\frac{A'}{A}\right)^2+\epsilon\left(\frac{A'}{A}\right)\left(\frac{B'}{B}\right)=0, \tag{5.13b}$$

$$(1-2\epsilon)\left(\frac{A'}{A}\right)'+\left(\frac{B'}{B}\right)'-\frac{1}{r}\left((1-2\epsilon)\frac{A'}{A}+\frac{B'}{B}\right)+\tfrac{1}{2}(1-4\epsilon)\left(\frac{A'}{A}\right)^2-(1-3\epsilon)\left(\frac{A'}{A}\right)\left(\frac{B'}{B}\right)-\frac{1}{2}\left(\frac{B'}{B}\right)^2=0. \tag{5.13c}$$

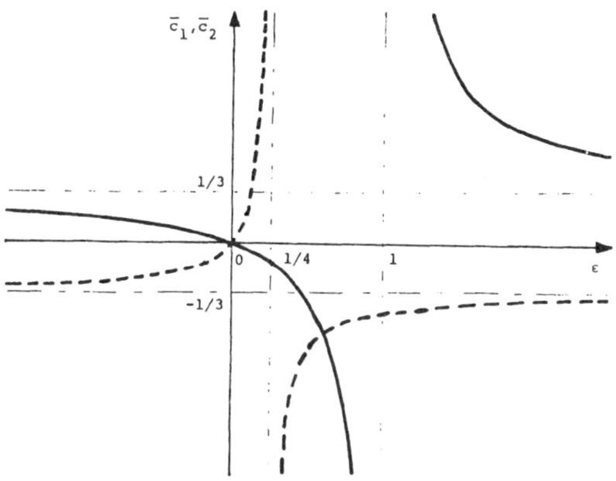

FIG. 3. The curves of (5.12c): The solid curve is for $\bar{c}_1$, and the dashed curve is for $\bar{c}_2$.

In the spatial region far outside the Schwarzschild radius, i.e., $r \gg GM$, the metric coefficients, $A(r)$ and $B(r)$, can be expanded in a small parameter (GM/r),

$$A(r) = 1 - 2\frac{GM}{r} + 2\beta\left(\frac{GM}{r}\right)^2 + \cdots, \tag{5.14a}$$

$$B(r) = 1 + 2\gamma\frac{GM}{r} + 2\delta\left(\frac{GM}{r}\right)^2 + \cdots, \tag{5.14b}$$

where M is the gravitational mass of a central gravitating body, and β, γ, and δ are expansion parameters to be determined by the field equation. Using (5.14) in (5.13), and putting each order of (GM/r) equal to zero, we find that the parameters, β, γ, and δ, are given by

$$\beta = 1 - \epsilon/2, \quad \gamma = 1 - 2\epsilon, \qquad \delta = \tfrac{3}{4}(1 - 3\epsilon + \tfrac{8}{3}\epsilon^2). \tag{5.15}$$

It is to be noticed that the Newton approximation condition (5.12a) is not used to derive (5.15), although the above results are consistent with (5.12a).

C. Exact vacuum solution

The gravitational field studied in the previous two subsections is weak. Now we derive an *exact* solution of the vacuum field equation (5.13), which allows us to study a strong gravitational field in new general relativity.

After slight modification of $[2\times(5.13\text{b}) - (5.13\text{c})]$ we obtain

$$\frac{d}{dr}\left[r^3\left((1-2\epsilon)\frac{A'}{A} + \frac{B'}{B}\right)\right] + \frac{r^3}{2}\left(\frac{A'}{A} + \frac{B'}{B}\right)\left((1-2\epsilon)\frac{A'}{A} + \frac{B'}{B}\right) = 0. \tag{5.16}$$

This equation can be integrated to give

$$(1-2\epsilon)\frac{A'}{A} + \frac{B'}{B} = (AB)^{-1/2}\frac{f_1}{r^3}, \tag{5.17a}$$

with f_1 an integration constant, which can be fixed by using (5.14) with (5.15) in (5.17a):

$$f_1 = (1-\epsilon)(1-4\epsilon)(GM)^2. \tag{5.17b}$$

In the same way we get from $[3\times(5.13\text{b}) - (5.13\text{c}) - 2\times(5.13\text{a})]$

$$(1-3\epsilon)\frac{A'}{A} + 2\epsilon\frac{B'}{B} = (AB)^{-1/2}\frac{f_2}{r^2}, \tag{5.18a}$$

with f_2 an integration constant given by

$$f_2 = 2(1-\epsilon)(1-4\epsilon)(GM). \tag{5.18b}$$

From the combination $[(1-5\epsilon)\times(5.17\text{a}) + 2\epsilon\times(5.18\text{a})]$ it follows that

$$\frac{d}{dr}(AB)^{1/2} = 2\epsilon\frac{GM}{r^2} + \frac{1-5\epsilon}{2}\frac{(GM)^2}{r^3}, \tag{5.19}$$

and, therefore, remembering that the boundary condition for A and B, denoted by (5.8), is expressed by

$$\lim_{r\to\infty} AB = 1, \tag{5.20}$$

we obtain

$$(AB)^{1/2} = 1 - 2\epsilon\frac{GM}{r} - \frac{1-5\epsilon}{4}\left(\frac{GM}{r}\right)^2 = \left(1 - \frac{GM}{pr}\right)\left(1 + \frac{GM}{qr}\right), \tag{5.21}$$

where two constants, p and q, are defined by

$$p \equiv \frac{2}{1-5\epsilon}\{[(1-\epsilon)(1-4\epsilon)]^{1/2} - 2\epsilon\} = 2 + \epsilon + O(\epsilon^2),$$
$$q \equiv \frac{2}{1-5\epsilon}\{[(1-\epsilon)(1-4\epsilon)]^{1/2} + 2\epsilon\} = 2 + 9\epsilon + O(\epsilon^2). \tag{5.22a}$$

Here ϵ is assumed to be

$$\epsilon \leq \tfrac{1}{4}, \tag{5.22b}$$

which covers the important case of $\epsilon = 0$; for $\frac{1}{4} < \epsilon < 1$, p and q become complex values. Substitution of (5.21) into (5.17a) and (5.18a) finally gives

$$A(r) = \left(1 - \frac{GM}{pr}\right)^p\left(1 + \frac{GM}{qr}\right)^{-q}, \qquad B(r) = \left(1 - \frac{GM}{pr}\right)^{2-p}\left(1 + \frac{GM}{qr}\right)^{2+q}. \tag{5.23}$$

It can be shown by direct calculations that this solution indeed satisfies the field equation (5.13).

The parallel vector fields of (5.2) are thus given by

$$\begin{aligned} b_{(0)}{}^{0} &= \frac{1}{\sqrt{A}} = \left(1-\frac{GM}{pr}\right)^{-p/2}\left(1+\frac{GM}{qr}\right)^{q/2}, \\ b_{(0)}{}^{\alpha} &= 0 = b_{(a)}{}^{0}, \\ b_{(a)}{}^{\alpha} &= \frac{\delta_a^\alpha}{\sqrt{B}} = \left(1-\frac{GM}{pr}\right)^{-1+p/2}\left(1+\frac{GM}{qr}\right)^{-1-q/2}\delta_a^\alpha, \end{aligned} \tag{5.24}$$

in a static, isotropic gravitational field. The invariant distance ds^2 of (5.3a) becomes

$$\begin{aligned} ds^2 &= -\left(1-\frac{GM}{pr}\right)^{p}\left(1+\frac{GM}{qr}\right)^{-q}dt^2+\left(1-\frac{GM}{pr}\right)^{2-p}\left(1+\frac{GM}{qr}\right)^{2+q}dx^\alpha dx^\alpha \\ &= -\left(1-\frac{GM}{pr}\right)^{p}\left(1+\frac{GM}{qr}\right)^{-q}dt^2+\left(1-\frac{GM}{pr}\right)^{2-p}\left(1+\frac{GM}{qr}\right)^{2+q}[dr^2+r^2(d\theta^2+\sin^2\theta\, d\phi^2)], \end{aligned} \tag{5.25}$$

where we have introduced the spherical polar coordinates by

$$x^1 = r\sin\theta\cos\phi,\quad x^2 = r\sin\theta\sin\phi, \quad x^3 = r\cos\theta. \tag{5.26}$$

If the parameter ϵ of (5.6) is *exactly zero*, then two constants, p and q, are exactly equal to 2, and hence this metric coincides with the Schwarzschild metric written in the isotropic coordinates[25]:

$$\begin{aligned} ds^2 &= -\frac{(1-GM/2r)^2}{(1+GM/2r)^2}dt^2+\left(1+\frac{GM}{2r}\right)^4 dx^\alpha dx^\alpha \\ &= -\frac{(1-GM/2r)^2}{(1+GM/2r)^2}dt^2 \\ &\quad +\left(1+\frac{GM}{2r}\right)^4[dr^2+r^2(d\theta^2+\sin^2\theta\, d\phi^2)]. \end{aligned} \tag{5.27}$$

VI. COMPARISON WITH EXPERIMENTS

A. The equivalence principle

It has been verified experimentally to very high accuracy[26] that the world line of a freely falling test body is independent of its composition and structure. The equivalence principle implies that the unique world line of a test body coincides with the geodesics of the metric $\underline{g}$. It was shown in Sec. III that by taking the short-wavelength limit of the Maxwell and Dirac equations the photon and Dirac particles in the classical limit are to travel along the geodesics of the metric $\underline{g}$. Thus, new general relativity is compatible with the equivalence principle in this limit.

In general relativity implications of the equivalence principle are concisely expressed by the conservation law,

$$\nabla_\nu T_{\rm GR}^{\mu\nu} = 0, \tag{6.1}$$

where $\{T_{\rm GR}^{\mu\nu}\}$ is the matter energy-momentum tensor appearing on the right-hand side of the Einstein field equation. It follows from the conservation law that the world line of a freely falling test body is the geodesics of the metric $\underline{g}$. The characteristic feature of general relativity is that the conservation law of (6.1) is a consequence of the Einstein gravitational field equation, and hence that mechanical equations of motion for matter are consequences of the same gravitational field equation.

Now we shall show that almost the same property holds also in new general relativity based on the Weitzenböck space-time. From the invariance of the gravitational action under the group of general coordinate transformations follows the identity[27]

$$\sqrt{-g}\, B^{k\nu}\partial_\mu b_{k\nu} - \partial_\nu(\sqrt{-g}\, B_\mu^{\cdot\nu}) \equiv 0, \tag{6.2}$$

where

$$\sqrt{-g}\, B^{k\nu} = \sqrt{-g}\, b^k{}_\mu B^{\mu\nu} \equiv -\delta\sqrt{-g}\, L_G/\delta b_{k\nu}. \tag{6.3}$$

After slight modification, this identity can be rewritten as

$$\nabla_\nu B^{\mu\nu} - K^{\nu\lambda\mu}B_{\nu\lambda} \equiv 0, \tag{6.4}$$

where $\{K^{\nu\lambda\mu}\}$ is the contortion tensor given by (3.22). From the definition of (6.3) it follows that the gravitational field equation takes the form

$$B^{\mu\nu} = T^{\mu\nu}, \tag{6.5}$$

with the matter energy-momentum tensor $\{T^{\mu\nu}\}$ defined by (4.28). Using (6.5) in the identity (6.4), we get the response equation to gravitation,

$$\nabla_\nu T^{\mu\nu} - K^{\nu\lambda\mu}T_{\nu\lambda} = 0. \tag{6.6}$$

This is the conservation law of new general relativity, corresponding to the conservation law (6.1) of general relativity. The energy-momentum tensor $\{T^{\mu\nu}\}$ is not symmetric in new general relativity. However, an antisymmetric part $\{T^{[\mu\nu]}\}$ is due to the contribution from the intrinsic spin of spin-$\frac{1}{2}$ fundamental particles. For macroscopic bodies such as a test body employed in terrestrial experiments and astrophysical objects such as planets and stars, effects due to the intrinsic spin of spin-$\frac{1}{2}$ fundamental particles can be ignored, and hence their energy-momentum tensor can be supposed to be symmetric and of the same form as that of general relativity. Therefore, an energy-momentum tensor of macroscopic bodies satisfies the conservation law

$$\nabla_\nu T^{\mu\nu} = 0 \tag{6.7}$$

owing to the antisymmetric property of the contortion tensor $\{K^{\nu\lambda\mu}\}$ with respect to ν and λ. The only exception seems to be compact stellar objects such as neutron stars and black holes: The spin direction of neutrons may happen to be aligned over the macroscopic scale inside neutron stars. If this is indeed the case, the gravitational response of neutron matter should be described by Eq. (6.6) instead of by the conservation law (6.1)

The equivalence principle is thus satisfied for macroscopic bodies in new general relativity, and the world line of a test body coincides with the geodesics of the metric $\underline{g}$, although the metric $\underline{g}$ itself may be different from that of general relativity.

In the microscopic scale new general relativity violates the equivalence principle, since effects due to the intrinsic spin of spin-$\frac{1}{2}$ fundamental particles cannot be ignored there, and an antisymmetric part of the energy-momentum tensor should be seriously taken into account. The motion of the intrinsic spin of a freely falling spin-$\frac{1}{2}$ fundamental particle, for example, does not satisfy the equivalence principle. As was shown in (3.37b), the spin vector $\{S^\mu\}$ obeys the equation of motion

$$\nabla S^\mu/d\tau = -\tfrac{3}{2}\epsilon^{\mu\nu\rho\sigma}\, U_\nu a_\rho S_\sigma \tag{6.8a}$$

with

$$U^\mu \equiv \frac{dx^\mu}{d\tau}, \quad \frac{\nabla S^\mu}{d\tau} \equiv \frac{dS^\mu}{d\tau} + \left\{{}^{\mu}_{\lambda\nu}\right\} U^\lambda S^\nu , \tag{6.8b}$$

where τ is the proper time. In order for this equation of motion for the spin vector to meet with the equivalence principle, the right-hand side should vanish. Therefore, unless the axial-vector part $\{a^\mu\}$ of the torsion tensor happens to vanish identically, the equation of motion for the spin vector violates the equivalence principle.[28] Another important implication of new general relativity for microscopic phenomena is the prediction of universal spin-spin interaction, caused by an antisymmetric part of the energy-momentum tensor. This interaction, if it exists, contributes to the hyperfine splitting of the atomic energy levels, and it shall be discussed in Sec. X.

B. Comparison with solar-system experiments

Since the invariant distance ds^2 of (5.3a) is written in the isotropic coordinates, the post-Newtonian parameters of the expansion (5.14a)–(5.14b), β and γ, are the Eddington-Robertson parameters.[29] Thus, by virtue of (5.15), the Eddington-Robertson parameters of new general relativity are given by

$$\beta = 1 - \epsilon/2, \quad \gamma = 1 - 2\epsilon . \tag{6.9}$$

The values of β and γ have been measured by the solar-system experiments:

$$\gamma = \begin{cases} 1.00 \pm 0.06 \,(\text{retardation of radio waves}^{30}), & (6.10a) \\ 1.014 \pm 0.018 \;\; (\text{solar deflection}^{31}), & (6.10b) \end{cases}$$

$$\tfrac{1}{3}(2 + 2\gamma - \beta) = 1.003 \pm 0.005$$
$$(\text{perihelion advances}^{32}), \tag{6.11}$$

$$\eta \equiv 4\beta - \gamma - 3 = -0.001 \pm 0.015$$
$$(\text{lunar laser ranging}^{33}). \tag{6.12}$$

From (6.9) it follows that the Nordtvedt parameter η is vanishing in new general relativity;

$$\eta = 0 . \tag{6.13}$$

For the sake of safety, we here adopt the value (6.10b) for γ. Using (6.9) in (6.10b) and (6.11), we get

$$\epsilon = \begin{cases} -0.007 \pm 0.009 & \text{from (6.10b)}, \\ -0.003 \pm 0.004 & \text{from (6.11)}. \end{cases} \tag{6.14a}$$

Combining these two values for ϵ as if they were independent, we are led to

$$\epsilon = -0.004 \pm 0.004 . \tag{6.14b}$$

This value of ϵ satisfies our assumption of $\epsilon \leqslant \frac{1}{4}$; see (5.22b).

By virtue of (5.12c), two dimensionless constants, κc_1 and κc_2, can be expressed as

$$\begin{aligned} \kappa c_1 &= -\frac{\epsilon}{3(1-\epsilon)} = -\frac{\epsilon}{3} + O(\epsilon^2), \\ \kappa c_2 &= \frac{4\epsilon}{3(1-4\epsilon)} = \frac{4\epsilon}{3} + O(\epsilon^2). \end{aligned} \tag{6.15}$$

Use of (6.14b) then gives

$$\kappa c_1 = 0.001 \pm 0.001, \quad \kappa c_2 = -0.005 \pm 0.005 . \tag{6.16}$$

Rewriting the gravitational Lagrangian density L_G of (4.18) as

$$L_G = \frac{1}{2\kappa}[R(\{\}) + 2\kappa c_1(t^{\lambda\mu\nu} t_{\lambda\mu\nu}) + 2\kappa c_2(v^\mu v_\mu) + 2\kappa c_3(a^\mu a_\mu)], \tag{6.17}$$

we find that the strength of the c_1 and c_2 terms are severely restricted by the solar-system experiments.

VII. THE CASE OF $c_1=0=c_2$

We have seen in the last section that the c_1 and c_2 terms of L_G are, if they exist, very severely restricted by the solar-system experiments. Therefore, we shall henceforce ignore these two terms and assume that $c_1 = 0 = c_2$. The case of $c_1 \neq 0 \neq c_2$ shall be discussed in a separate paper.

The gravitational Lagrangian density L_G then becomes

$$L_G = \frac{1}{2\kappa} R(\{\}) + c_3(a^\mu a_\mu), \tag{7.1}$$

and the gravitational field equation, (4.31) and (4.32), can be expressed as

$$G^{\mu\nu}(\{\}) + K^{\mu\nu} = \kappa T^{(\mu\nu)}, \tag{7.2}$$

$$b_i{}^\mu b_j{}^\nu \partial_\rho(\sqrt{-g}\, J^{ij\rho}) = \lambda\sqrt{-g}\; T^{[\mu\nu]}, \tag{7.3}$$

where we have introduced a new parameter λ by

$$\lambda = \frac{9}{4c_3}, \tag{7.4}$$

and $\{K^{\mu\nu}\}$ and $\{J^{\mu\nu\rho}\}$ are defined by

$$K^{\mu\nu} = \frac{\kappa}{\lambda}\{\tfrac{1}{2}[\epsilon^{\mu\rho\sigma\lambda}(T^{\nu}_{\cdot\rho\sigma} - T_{\rho\sigma}^{\cdot\cdot\nu}) + \epsilon^{\nu\rho\sigma\lambda}(T^{\mu}_{\cdot\rho\sigma} - T_{\rho\sigma}^{\cdot\cdot\mu})]a_\lambda - \tfrac{3}{2}a^\mu a^\nu - \tfrac{3}{4}g^{\mu\nu}a^\rho a_\rho\}, \tag{7.5}$$

$$J^{\mu\nu\rho} = b_i{}^\mu b_j{}^\nu J^{ij\rho} = -\tfrac{3}{2}\epsilon^{\mu\nu\rho\sigma} a_\sigma. \tag{7.6}$$

Taking the combination of $[(7.2) + (\kappa/\lambda) \times (7.3)]$, the gravitational field equation is rewritten as

$$G^{\mu\nu}(\{\}) + L^{\mu\nu} = \kappa T^{\mu\nu}, \tag{7.7}$$

with $\{L^{\mu\nu}\}$ defined by

$$L^{\mu\nu} = \frac{\kappa}{2\lambda}\{a_\lambda[\epsilon^{\mu\rho\sigma\lambda}(T^{\nu}_{\cdot\rho\sigma} - T_{\rho\sigma}^{\cdot\cdot\nu}) + \epsilon^{\nu\rho\sigma\lambda}(T^{\mu}_{\cdot\rho\sigma} - T_{\rho\sigma}^{\cdot\cdot\mu})] - 3a^\mu a^\nu - \tfrac{3}{2}g^{\mu\nu}a^\rho a_\rho + 3\epsilon^{\mu\nu\rho\sigma}(b^i{}_\rho\, \partial_\sigma a_i + a_\rho v_\sigma)\}, \tag{7.8}$$

where $\{a_i = b_i{}^\mu a_\mu\}$ is a scalar with respect to general coordinate transformations.

As is evident by the definition of the torsion tensor (4.4) and its irreducible components of (4.5)–(4.7), the second term $\{L^{\mu\nu}\}$ of (7.7) does not transform like a tensor under a *local* Lorentz transformation

$$\begin{aligned} \underline{b}_k(x) &= A^j{}_k(x)\underline{b}'_j(x), \\ A^j{}_k(x)\eta_{jm}A^m{}_n(x) &= \eta_{kn}. \end{aligned} \tag{7.9}$$

The energy-momentum tensor of the electromagnetic field depends on the parallel vector fields $\underline{b}$ only through the metric tensor, $g_{\mu\nu} = b^i{}_\mu \eta_{ij} b^j{}_\nu$, and hence it is locally Lorentz invariant. The energy-momentum tensor of spin-$\frac{1}{2}$ fundamental particles, however, is not locally Lorentz invariant, due to the second term of the second line of (4.30a), i.e., $\frac{1}{4}\epsilon_{\lambda\rho\sigma\nu} b_k{}^\lambda K^{\rho\sigma}_{\cdot\cdot\mu}\bar{\psi}\gamma^5\gamma^k\psi$. Thus, the energy-momentum tensor of matter is not locally Lorentz invariant, unless effects due to the intrinsic spin of spin-$\frac{1}{2}$ fundamental particles can be neglected. Therefore, *the gravitational field equation of* (7.7) *is not invariant under a local Lorentz transformation.*

The gravitational field equation is considerably simplified in the particular case which satisfies the following two conditions: (1) The axial-vector part of the torsion tensor vanishes identically,

$$a^\mu = \tfrac{1}{6}\,\epsilon^{\mu\nu\rho\sigma} b^k{}_\nu(\partial_\sigma b_{k\rho} - \partial_\rho b_{k\sigma}) = 0, \tag{7.10}$$

and (2) effects due to the intrinsic spin of spin-$\frac{1}{2}$ fundamental particles can be neglected. The first condition implies that the left-hand side of (7.7) becomes the Einstein tensor $G^{\mu\nu}(\{\})$. The second condition, on the other hand, allows us to treat spin-$\frac{1}{2}$ fundamental particles as if they were spinless; the energy-momentum tensor $\{T^{\mu\nu}\}$ on the right-hand side of (7.7) can then be identified with the energy-momentum tensor $\{T^{\mu\nu}_{\rm GR}\}$ used in general relativity. Thus, in this particular case the gravitational field equation (7.7) is identical with the Einstein field equation,

$$G^{\mu\nu}(\{\}) = \kappa T^{\mu\nu}_{\rm GR}. \tag{7.11}$$

For example, suppose that the metric in the invariant distance,

$$ds^2 = -A(x)(dx^0)^2 + B(x)(dx^1)^2 + C(x)(dx^2)^2 + D(x)(dx^3)^2, \tag{7.12}$$

is an exact solution of the Einstein field equation (7.11), where $A(x)$, $B(x)$, $C(x)$, and $D(x)$ are functions of x. Define the parallel vector fields $\underline{b} = \{\underline{b}_k\}$ by

$$\underline{b}_{(0)} = \frac{1}{\sqrt{A}} \underline{E}_0, \quad \underline{b}_{(1)} = \frac{1}{\sqrt{B}} \underline{E}_1, \quad \underline{b}_{(2)} = \frac{1}{\sqrt{C}} \underline{E}_2, \quad \underline{b}_{(3)} = \frac{1}{\sqrt{D}} \underline{E}_3, \tag{7.13}$$

with $\underline{E}_\mu = \partial/\partial x^\mu$ and Latin indices in $\underline{b}_k$ enclosed in parentheses. Then they form a system of four orthonormal vectors with their contravariant components given by

$$b_{(0)}{}^0 = 1/\sqrt{A}\,, \quad b_{(1)}{}^1 = 1/\sqrt{B}\,, \quad b_{(2)}{}^2 = 1/\sqrt{C}\,, \quad b_{(3)}{}^3 = 1/\sqrt{D}\,, \quad b_k{}^\mu = 0 \text{ otherwise}, \tag{7.14a}$$

and their covariant components given by

$$b^{(0)}{}_0 = \sqrt{A}\,, \quad b^{(1)}{}_1 = \sqrt{B}\,, \quad b^{(2)}{}_2 = \sqrt{C}\,, \quad b^{(3)}{}_3 = \sqrt{D}\,, \quad b^k{}_\mu = 0 \text{ otherwise}. \tag{7.14b}$$

In this case the axial-vector part of the torsion tensor, formed of $\underline{b}$, vanishes identically;

$$a^\mu = 0\,. \tag{7.15}$$

Therefore, the parallel vector fields (7.13) are an exact solution of the gravitational field equation (7.7) with the source term, $T^{\mu\nu} = T^{\mu\nu}_{\mathrm{GR}}$. The metric of the form (7.12) covers, among others, a number of static vacuum solutions with high symmetry of the Einstein field equation,[34] such as the Schwarzschild solution, the Reissner-Nordström solution and the Weyl solution, and the Friedmann model[35] in cosmology.

VIII. GEOMETRY OF THE EXTENDED WEITZENBÖCK SPACE-TIME

In this section we shall consider the particular case discussed in the last section: Namely, we shall assume that the parallel vector fields, $\underline{b} = \{\underline{b}_k\}$, satisfy both the condition (7.10) and the condition that spin-$\frac{1}{2}$ fundamental particles can be treated as if they were spinless. The gravitational field equation (7.7) is then apparently of the same form as the Einstein field equation, but the geometrical background of these two equations are quite different: In general relativity the Einstein equation defines the Riemann space-time, while in new general relativity the gravitational field equation is to define the parallel vector fields of the Weitzenböck space-time.

Let the parallel vector fields, $\underline{b} = \{\underline{b}_k\}$, be a solution of the gravitational field equation (7.7): Namely, we suppose that $\underline{b} = \{\underline{b}_k\}$ simultaneously satisfies both the condition (7.10) and the Einstein field equation (7.11). New parallel vector fields, $\underline{b}' = \{\underline{b}'_k\}$, obtained from $\underline{b}$ by a *local* Lorentz transformation (7.9), also satisfy the Einstein field equation by virtue of the local Lorentz invariance of the Einstein field equation. The condition (7.10), on the other hand, is not fulfilled by $\underline{b}'$ in general, because the axial-vector part of the torsion tensor, $\{a^\mu\}$, transforms like

$$a'^\mu(x) = a^\mu(x) - \tfrac{1}{3}\epsilon^{\mu\nu\rho\sigma} b^j{}_\nu b^k{}_\rho A^m{}_j(x) A_{mk}(x)_{,\sigma} \tag{8.1}$$

under a local Lorentz transformation (7.9). Here A_{mk} is defined by $A_{mk} = \eta_{mj} A^j{}_k$.

The new parallel vector fields $\underline{b}'$ thus satisfy the gravitational field equation (7.7), if and only if the transformation matrix $[A^j{}_k(x)]$ obeys the condition,

$$\epsilon^{\mu\nu\rho\sigma} b^j{}_\nu b^k{}_\rho A^m{}_j(x) A_{mk,\sigma} = 0\,, \tag{8.2}$$

which ensures the condition (7.10) for $\underline{b}'$. In the present particular case, therefore, *the gravitational field equation* (7.7) *is invariant under those local Lorentz transformations which satisfy the condition* (8.2), *and the parallel vector fields are defined by the gravitational field equation with ambiguity of making those local Lorentz transformations*. This ambiguity does not lead to any observable effects, because the Maxwell and Dirac equations, (3.11) and (3.13), respectively, are also invariant under those transformations.

In the Weitzenböck space-time the parallel vector fields should be defined only with arbitrariness of making a *global* Lorentz transformation, and there is no room for making any local Lorentz transformations. In the present particular case, however, the new parallel vector fields $\underline{b}'$ connected with $\underline{b}$ by a local Lorentz transformation satisfying (8.2) should be regarded as equivalent to b', because the Maxwell, Dirac, and gravitational field equations are all invariant under the transformation from $\underline{b}$ to $\underline{b}'$. We are thus forced to generalize the concept of absolute parallelism in the following manner: *Absolute parallelism defined by* $\underline{b}'$ *shall be regarded as equivalent to that defined by* $\underline{b}$, *provided that* $\underline{b}$ *and* $\underline{b}'$ *are connected with each other by a local Lorentz transformation subjecting to* (8.2). We shall refer to this new parallelism as *extended absolute parallelism*, and space-time endowed with extended absolute parallelism shall be called the *extended Weitzenböck space-time*. The geometry of the extended Weitzenböck space-time then is invariant under those local Lorentz transformations which fulfil the condition (8.2).

For given parallel vector fields $\underline{b}$, we denote by $\Lambda(\underline{b})$ the set of those local Lorentz transformations which fulfil the condition (8.2). The set $\Lambda(\underline{b})$ does not form a Lie group: Namely, for two elements of $\Lambda(\underline{b})$, A and A', the inverse A^{-1} and the product $A'A$ do not belong to $\Lambda(\underline{b})$ in general. However, for an infinitesimal local Lorentz transformation,

$$A^j{}_k(x) = \delta^j{}_k + \omega^j{}_k(x), \quad \omega_{jk} + \omega_{kj} = 0, \quad |\omega_{jk}| \ll 1, \tag{8.3}$$

the condition (8.2) becomes

$$\epsilon^{\mu\nu\rho\sigma} b^j{}_\nu b^k{}_\rho \omega_{jk}(x)_{,\sigma} = 0, \tag{8.4}$$

by neglecting the second- and higher-order terms of ω_{jk}. Since the condition (8.4) is linear in ω_{jk}, the infinitesimal neighborhood of the unit element in $\Lambda(\underline{b})$ has some of a Lie-algebra property: The inverse, $(A^{-1})^j{}_k = \delta^j{}_k - \omega^j{}_k$, and the product, $(A'A)^j{}_k = \delta^j{}_k + \omega'{}_{jk} + \omega_{jk}$, satisfy (8.4) for any two infinitesimal local Lorentz transformations, A and A', belonging to $\Lambda(\underline{b})$.

As an example of the extended Weitzenböck space-time, consider the static isotropic space-time, which has the Schwarzschild metric for the present case of $c_1 = 0 = c_2$. Written in the isotropic coordinates used in Sec. V, the Schwarzschild metric is expressed in the isotropic form,

$$ds^2 = -A(r)dt^2 + B(r)dx^\alpha dx^\alpha, \tag{8.5}$$

with

$$A(r) = \frac{(1 - GM/2r)^2}{(1 + GM/2r)^2}, \quad B(r) = \left(1 + \frac{GM}{2r}\right)^4, \tag{8.6}$$

and the parallel vector fields $\underline{b}$ defined by (5.24) are

$$\underline{b}_{(0)} = \frac{1}{\sqrt{A}} \underline{E}_0, \quad \underline{b}_{(a)} = \frac{1}{\sqrt{B}} \delta^\alpha_a \underline{E}_\alpha, \tag{8.7}$$

with $\{\underline{E}_\mu\}$ the coordinate basis; $\underline{E}_\mu = \partial/\partial x^\mu$. The axial-vector part of the torsion tensor, formed of $\underline{b}$, thus vanishes identically, and so the static, isotropic space-time is the extended Weitzenböck space-time. Besides the parallel vector fields $\underline{b}$ of (8.7), there exists an infinitely large number of parallel vector fields, which are all equivalent to $\underline{b}$ and with each other: All parallel vector fields are related to $\underline{b}$ of (8.7) by local Lorentz transformations satisfying the condition (8.2). The condition (8.4) for an infinitesimal local Lorentz transformation specified by (8.3) can easily be solved in this case: The solution which leaves $\underline{b}$ static is

$$\omega_{(a)(b)} = \delta^\alpha_a \delta^\beta_b (H_{\alpha,\beta} - H_{\beta,\alpha}), \quad \omega_{(0)(a)} = -\omega_{(a)(0)} = \delta^\alpha_a H_{0,\alpha}, \tag{8.8}$$

with H_μ arbitrary small functions independent of t, where Latin (Lorentz) indices in ω_{jk} are enclosed by parentheses.

For a finite local Lorentz transformation it is not easy to solve the Eq. (8.2). However, we can find a special kind of the parallel vector fields in the static, isotropic space-time by looking for such a set of coordinates $\{x^\mu\}$ that the Schwarzschild metric is expressed in the diagonal form of (7.12). The parallel vector fields, defined by (7.13), in such coordinates are equivalent to the parallel vector fields of (8.7) defined in the isotropic coordinates. An example is given by the spherical polar coordinates (t, r, θ, ϕ) introduced by (5.26): The Schwarzschild metric reads as

$$ds^2 = -A(r)dt^2 + B(r)dr^2 + C(r)d\theta^2 + D(r,\theta)d\phi^2 \tag{8.9}$$

with A and B still given by (8.6), and

$$C(r) = r^2\left(1 + \frac{GM}{2r}\right)^4, \quad D(r,\theta) = r^2\left(1 + \frac{GM}{2r}\right)^4 \sin^2\theta. \tag{8.10}$$

Thus, the system of four orthonormal vectors, $\underline{b}'$,

$$\underline{b}'_{(0)} = \frac{1}{\sqrt{A}} \underline{E}_t, \quad \underline{b}'_{(1)} = \frac{1}{\sqrt{B}} \underline{E}_r, \quad \underline{b}'_{(2)} = \frac{1}{\sqrt{C}} \underline{E}_\theta, \quad \underline{b}'_{(3)} = \frac{1}{\sqrt{D}} \underline{E}_\phi, \tag{8.11}$$

with

$$\underline{E}_t = \partial/\partial t, \quad \underline{E}_r = \partial/\partial r, \quad \underline{E}_\theta = \partial/\partial\theta, \quad \underline{E}_\phi = \partial/\partial\phi, \tag{8.12}$$

is also a solution of the gravitational field equation in vacuum, and can be taken as the parallel vector fields, which are related to $\underline{b}$ of (8.7) by a local space rotation

$$\underline{b}'_{(0)} = \underline{b}_{(0)}, \quad \underline{b}'_{(a)} = R_{ac} \underline{b}_{(c)}, \tag{8.13}$$

with

$$(R_{ac}) = \begin{pmatrix} \sin\theta\cos\phi & \sin\theta\sin\phi & \cos\theta \\ \cos\theta\cos\phi & \cos\theta\sin\phi & -\sin\theta \\ -\sin\phi & \cos\phi & 0 \end{pmatrix}. \tag{8.14}$$

The parallel vector fields $\underline{b}'$ are usually used as tetrad fields in quantum field theory for the Schwarzschild space-time.[36] The Schwarzschild metric is of the form (7.12) also in the Kruskal-Szekeres coordinates,[37] and so we can use this coordinate system to form parallel vector fields $\underline{b}''$ by (7.13).

IX. THE WEAK-FIELD APPROXIMATION

Further insights into new general relativity can be gained by applying the set of the gravitational field equation, (7.2) and (7.3), to weak-field situations

$$b^k{}_\mu(x) = \delta^k{}_\mu + a^k{}_\mu(x), \quad |a^k{}_\mu| \ll 1, \quad b_k{}^\mu(x) = \delta_k{}^\mu + c_k{}^\mu(x), \quad |c_k{}^\mu| \ll 1, \tag{9.1a}$$

since in this case the particle spectrum of the new general relativity can be clarified by the use of the unitary, irreducible representations of the Poincaré group. In this situation we can expand the field equation in $a^k{}_\mu$ and $c_k{}^\mu$ and can keep only lowest terms. Thus we need not distinguish Latin indices from Greek indices, which are now raised and lowered by the Minkowski metric tensor, $\{\eta^{\mu\nu}\}$ or $\{\eta_{\mu\nu}\}$: We shall use Greek indices throughout this section. From (2.5a) follows

$$a^{\nu}_{\cdot\mu} + c^{\cdot\nu}_{\mu} = 0, \tag{9.1b}$$

and hence we take $\{a_{\mu\nu}\}$ as the basic field variable. We shall decompose the weak field $\{a_{\mu\nu}\}$ into its symmetric and antisymmetric parts,

$$a_{\mu\nu} = \tfrac{1}{2} h_{\mu\nu} + A_{\mu\nu}, \tag{9.2}$$

with $h_{\mu\nu} = h_{\nu\mu}$ and $A_{\mu\nu} = -A_{\nu\mu}$. The components of the metric tensor are then written as

$$g_{\mu\nu} = \eta_{\mu\nu} + h_{\mu\nu}. \tag{9.3}$$

The antisymmetric field makes no contribution to the space-time metric in this approximation, implying that it is associated with the intrinsic spin of spin-$\frac{1}{2}$ fundamental particles.

The Einstein tensor becomes to lowest order in $h_{\mu\nu}$

$$G^{(1)}_{\mu\nu}(\{\ \}) = -\tfrac{1}{2}[\Box \bar{h}_{\mu\nu} - \partial^\lambda(\partial_\mu \bar{h}_{\nu\lambda} + \partial_\nu \bar{h}_{\mu\lambda}) + \eta_{\mu\nu}\partial_\rho\partial_\sigma \bar{h}^{\rho\sigma}], \tag{9.4}$$

where $\Box = \partial^\mu\partial_\mu$ and $\bar{h}_{\mu\nu}$ are defined as usual by

$$\bar{h}_{\mu\nu} = h_{\mu\nu} - \tfrac{1}{2}\eta_{\mu\nu} h, \quad h = \eta^{\mu\nu} h_{\mu\nu}. \tag{9.5}$$

The second term of (7.2), i.e., $\{K^{\mu\nu}\}$, is of second order in $h_{\mu\nu}$ and $A_{\mu\nu}$, and hence can be ignored. The left-hand side of (7.3), when indices are lowered, becomes in the weak-field approximation

$$b_{i\mu} b_{j\nu}\partial_\rho(\sqrt{-g}\, J^{ij\rho}) = -[\Box A_{\mu\nu} - \partial^\lambda(\partial_\mu A_{\nu\lambda} - \partial_\nu A_{\mu\lambda})], \tag{9.6a}$$

because the axial-vector part $\{a^\mu\}$ of the torsion tensor is given by

$$a^\mu = \tfrac{1}{3}\epsilon^{\mu\nu\rho\sigma}\partial_\nu A_{\rho\sigma}. \tag{9.6b}$$

Thus in the weak-field approximation the symmetric and antisymmetric parts of the gravitational field equation are given by

$$\Box \bar{h}_{\mu\nu} - \partial^\lambda(\partial_\mu \bar{h}_{\nu\lambda} + \partial_\nu \bar{h}_{\mu\lambda}) + \eta_{\mu\nu}\partial_\rho\partial_\sigma \bar{h}^{\rho\sigma} = -2\kappa T_{(\mu\nu)}, \tag{9.7}$$

$$\Box A_{\mu\nu} - \partial^\lambda(\partial_\mu A_{\nu\lambda} - \partial_\nu A_{\mu\lambda}) = -\lambda T_{[\mu\nu]}. \tag{9.8}$$

It follows from these equations that the symmetric field $\{h_{\mu\nu}\}$ and the antisymmetric field $\{A_{\mu\nu}\}$ are completely decoupled from each other. The $\{h_{\mu\nu}\}$ obeys the linearized Einstein field equation. The nonsymmetric energy-momentum tensor $\{T_{\mu\nu}\}$ is taken to lowest order in the weak fields; namely, it is independent of $\{h_{\mu\nu}\}$ and $\{A_{\mu\nu}\}$, and satisfies the ordinary conservation law in special relativity,

$$\partial_\nu T^{\mu\nu} = 0, \tag{9.9}$$

by virtue of the response equation (6.6).

Multiplying ∂^ν on both sides of (9.7) and (9.8), we find that both the symmetric and antisymmetric parts of $\{T_{\mu\nu}\}$ satisfy the conservation law,

$$\partial_\nu T^{(\mu\nu)} = 0, \tag{9.10a}$$

$$\partial_\nu T^{[\mu\nu]} = 0. \tag{9.10b}$$

By virtue of (9.9), these two equations are not independent of each other. The conservation law (9.10b) imposes a severe restriction on the form of spin tensor of matter. In fact, due to the Tetrode formula in special relativity,[39]

$$2T^{[\mu\nu]} = \partial_\rho S^{\mu\nu\rho}, \tag{9.11}$$

Eq. (9.10b) is automatically satisfied if and only if a spin tensor, $\{S^{\mu\nu\rho}\}$, is totally antisymmetric with respect to its three indices. Thus, the gravitational field equation demands that a spin tensor be expressed as

$$S^{\mu\nu\rho} = \epsilon^{\mu\nu\rho\sigma} J_{5\sigma} \tag{9.12}$$

by an axial-vector current, $\{J_{5\sigma}\}$. For Dirac particles, $\{J_{5\sigma}\}$ is given by

$$J_{5\sigma} = -\tfrac{1}{2}\bar{\psi}\gamma^5\gamma_\sigma\psi. \tag{9.13}$$

As can be checked by direct calculation, the linearized field equations of (9.7)–(9.8) are invariant under gauge transformations,

$$h'_{\mu\nu} = h_{\mu\nu} - \partial_\mu J_\nu - \partial_\nu J_\mu, \tag{9.14a}$$

$$A'_{\mu\nu} = A_{\mu\nu} + \partial_\mu H_\nu - \partial_\nu H_\mu, \tag{9.14b}$$

with J_μ and H_μ arbitrary small functions which leave the fields weak. In a particular case, $J_\mu = 2H_\mu = \Lambda_\mu$, these gauge transformations give rise to an infinitesimal coordinate transformation, $x^\mu \to x'^\mu = x^\mu + \Lambda^\mu(x)$. By means of these gauge freedoms, we can put the gauge conditions,

$$\partial_\nu \bar{h}^{\mu\nu} = 0, \tag{9.15}$$

$$\partial_\nu A^{\mu\nu} = 0, \tag{9.16}$$

which we shall assume henceforth. Then the field equations of (9.7)–(9.8) become

$$\Box \bar{h}_{\mu\nu} = -2\kappa T_{(\mu\nu)}, \tag{9.17}$$

$$\Box A_{\mu\nu} = -\lambda T_{[\mu\nu]}. \tag{9.18}$$

We shall restrict our discussions to the antisymmetric field $\{A_{\mu\nu}\}$, because the physics of the $\{h_{\mu\nu}\}$ field is well known.[39] The retarded solution of (9.18) is given by

$$A_{\mu\nu}(\vec{x}, t) = \frac{\lambda}{4\pi} \int d^3x' \frac{T_{[\mu\nu]}(\vec{x}', t - |\vec{x} - \vec{x}'|)}{|\vec{x} - \vec{x}'|}. \tag{9.19}$$

Suppose that we observe the gravitational field in the space region far outisde a source; therefore, we can calculate the solution of (9.19) to lowest order in $1/r \equiv 1/|\vec{x}|$, using the expansion

$$|\vec{x} - \vec{x}'| = r - \frac{\vec{x}\cdot\vec{x}'}{r} + O\left(\frac{1}{r}\right),$$
$$|\vec{x} - \vec{x}'|^{-1} = \frac{1}{r} + \frac{\vec{x}\cdot\vec{x}'}{r^3} + O\left(\frac{1}{r^3}\right). \tag{9.20}$$

We assume that the energy-momentum tensor can be expressed as Fourier integral or as a sum of Fourier components; suppose we calculate for a single Fourier component,

$$T_{[\mu\nu]}(\vec{x}, t) = T_{[\mu\nu]}(\vec{x}, \omega)\, e^{-i\omega t} + \overline{T}_{[\mu\nu]}(\vec{x}, \omega)\, e^{+i\omega t}. \tag{9.21}$$

where a bar means complex conjugation. In the wave zone the solution (9.19) then becomes just like a plane wave

$$A_{\mu\nu}(\vec{x}, t) = d_{\mu\nu}(\vec{x}, \omega)\, e^{ikx} + \bar{d}_{\mu\nu}(\vec{x}, \omega)\, e^{-ikx}, \tag{9.22}$$

with the wave vector,

$$\vec{k} \equiv \omega\hat{x}, \quad k^0 \equiv \omega, \quad (\hat{x} = \vec{x}/r), \tag{9.23}$$

and the polarization tensor

$$d_{\mu\nu}(\vec{x}, \omega) = \frac{\lambda}{4\pi r} \int d^3x'\, T_{[\mu\nu]}(\vec{x}', \omega)\, e^{-i\vec{k}\cdot\vec{x}'}. \tag{9.24}$$

The wave vector $\{k^\mu\}$ is a null vector, and the polarization tensor satisfies the conditions,

$$k^\nu d_{\mu\nu}(\vec{x}, \omega) = 0, \quad k^\nu \bar{d}_{\mu\nu}(\vec{x}, \omega) = 0 \tag{9.25}$$

by virtue of the conservation law (9.10b). Since r is very large, the $\vec{x}$ dependence of $d_{\mu\nu}(\vec{x}, \omega)$ can be neglected, and so the plane wave (9.22) satisfies the d'Alembert equation, $\Box A_{\mu\nu}(\vec{x}, t) = 0$, and the gauge condition (9.16).

The energy-momentum tensor $\{t^{\mu\nu}\}$ of the $\{A_{\mu\nu}\}$ field is given by

$$t^{\mu\nu} = -[\partial L_A/\partial(\partial_\nu A_{\rho\sigma})]\partial^\mu A_{\rho\sigma} + \eta^{\mu\nu} L_A$$
$$= -\left(\frac{3}{2\lambda}\right) a_\lambda \epsilon^{\lambda\rho\sigma\nu}\partial^\mu A_{\rho\sigma} + \left(\frac{9}{4\lambda}\right)\eta^{\mu\nu}(a_\lambda a^\lambda), \tag{9.26}$$

where L_A is the linearized Lagrangian density of the $\{A_{\mu\nu}\}$ field,

$$L_A = c_3(a_\mu a^\mu) = \left(\frac{9}{4\lambda}\right)(a_\mu a^\mu). \tag{9.27}$$

We use the plane wave solution (9.22) in (9.26), and average $t^{\mu\nu}$ over a space-time region much larger than $|\vec{k}|^{-1}$. The average kills all terms proportional to $\exp(\pm 2ikx)$, and we are left with only the $\vec{x}$-independent terms,

$$\langle t^{\mu\nu}\rangle = (2/\lambda)\,\mathrm{Re}(k^\mu k^\nu d^{\rho\sigma}\bar{d}_{\rho\sigma} - 2k^\mu k^\rho d^{\nu\sigma}\bar{d}_{\rho\sigma}) + (2/\lambda)\eta^{\mu\nu}k_\rho k^\sigma d^{\rho\lambda}\bar{d}_{\sigma\lambda}. \tag{9.28}$$

Using the condition (9.25), we then find

$$\langle t^{\mu\nu}\rangle = (4/\lambda)\,|d_{12}|^2 k^\mu k^\nu, \tag{9.29}$$

where we have chosen the direction of $\vec{k}$ as the third axis. Therefore, *only the* (12) *component is physically significant, and the energy density* t^{00} *is positive definite if the constant* λ *is positive.*

The (12) component, d_{12}, does not change at all under a rotation around the third axis; in fact, for such a rotation the rotation matrix (R_{ab}) satisfies $R_{13} = 0 = R_{23}$, and hence the d_{12} transforms like a scalar,

$$d'_{12} = R_{1a}R_{2b}d_{ab} = (R_{11}R_{12} - R_{12}R_{21})d_{12}$$
$$= \det(R_{ab})d_{12} = d_{12}. \tag{9.30}$$

The physically significant component d_{12} is thus of helicity zero. In the terminology of elementary particle physics *the* $\{A_{\mu\nu}\}$ *field is a massless field of spin* 0.

In the above discussion the $\{A_{\mu\nu}\}$ field is assumed to be a classical field. The quantization of the $\{A_{\mu\nu}\}$ field can be performed consistently, and the resulting quantized theory does not involve ghost states.[40]

The space components of the solution (9.22), which decrease as $1/r$, contribute to the energy-momentum tensor, but the (0α) components do not, and hence they are of no physical significance. This fact suggests that the next terms of $A_{0\alpha}(\vec{x}, t)$, which are proportional to $1/r^2$, are important. In order to eliminate the $(1/r)$ term from $A_{0\alpha}(\vec{x}, t)$, we rewrite the (0α) components of (9.19) as follows:

$$A_{0\alpha}(\vec{x}, t) = \frac{\lambda}{4\pi} \int d^3x'\, T_{[0\alpha]}(\vec{x}', t - r) \times \left[\frac{1}{r} + \frac{\vec{x}\cdot\vec{x}'}{r^3} + O\left(\frac{1}{r^3}\right)\right] + \frac{\partial}{\partial t}\xi_\alpha, \tag{9.31}$$

with ξ_α given by

$$\xi_\alpha = \frac{\lambda}{4\pi}\sum_{n=0}^{\infty}\frac{1}{(n+1)!}\frac{\partial^n}{\partial t^n}\int d^3x'\, T_{[0\alpha]}(\vec{x}', t - r) \times \frac{(r - |\vec{x} - \vec{x}'|)^{n+1}}{|\vec{x} - \vec{x}'|}. \tag{9.32}$$

The integral in (9.31) can be rewritten by using the relations, (9.11) and (9.12):

$$\int d^3x'\, T_{[0\alpha]}(\vec{x}', t - r) = \epsilon_{0\alpha\beta\gamma}\int d^3x'\, J_{5\gamma',\beta}(\vec{x}', t - r) = 0,$$

$$\int d^3x' T_{[0\alpha]}(\vec{x}', t-r)(\vec{x}\cdot\vec{x}')$$
$$= \tfrac{1}{2}\epsilon_{0\alpha\beta\gamma}\int d^3x' J_{5\gamma',\beta}(\vec{x}', t-r)(\vec{x}\cdot\vec{x}')$$
$$= \tfrac{1}{2}\epsilon_{\alpha\beta\gamma}\int d^3x' J_{5\gamma}(\vec{x}', t-r)x^\beta$$
$$= \tfrac{1}{2}\epsilon_{\alpha\beta\gamma}x^\beta S^\gamma(t-r),$$

where $\vec{S}=\{S^\alpha\}$ is a *total intrinsic spin* of the source,

$$S_\alpha(t)=\int d^3x J_{5\alpha}(\vec{x},t)=\tfrac{1}{2}\epsilon_{\alpha\beta\gamma}\int d^3x S^{\beta\gamma0}(\vec{x},t). \tag{9.33}$$

We now perform a gauge transformation (9.14b) with H_μ given by

$$H_\alpha=\xi_\alpha,\quad H_0=0. \tag{9.34}$$

Then, dropping a prime on $A_{0\alpha}$, we finally get

$$A_{0\alpha}(\vec{x},t)=\frac{\lambda}{8\pi}\epsilon_{\alpha\beta\gamma}\frac{x^\beta S^\gamma(t-r)}{r^3}+O\left(\frac{1}{r^3}\right). \tag{9.35}$$

Since ξ_α decrease as $1/r$, the change of the space components, $\delta A_{\alpha\beta}=\partial_\alpha\xi_\beta-\partial_\beta\xi_\alpha$, decreases as $1/r^2$, and hence the $(1/r)$ terms of $A_{\alpha\beta}$ do not change under this gauge transformation. The expression (9.35) is to be compared with the asymptotic expression for $h_{0\alpha}(\vec{x},t)$,[41]

$$h_{0\alpha}(\vec{x},t)=\frac{\kappa}{4\pi}\epsilon_{\alpha\beta\gamma}\frac{x^\beta M^\gamma(t-r)}{r^3}, \tag{9.36}$$

where $\vec{M}=\{M^\alpha\}$ is a *total angular momentum* of the source,

$$M_\alpha(t)=\epsilon_{\alpha\beta\gamma}\int d^3x\, x^\beta T^{(\gamma0)}(\vec{x},t). \tag{9.37}$$

See Table I for an illustration of $\{h_{\mu\nu}\}$ and $\{A_{\mu\nu}\}$ in the asymptotic region.

TABLE I. Asymptotic expressions for $h_{\mu\nu}$ and $A_{\mu\nu}$ far from a weakly gravitating system. The result for $h_{\mu\nu}$ is well known (Ref. 41), but we list it here for comparison's sake.

$\mu\nu$	$h_{\mu\nu}$	$A_{\mu\nu}$
00	$-\left[1-\frac{2GM}{r}+O\left(\frac{1}{r^2}\right)\right]$	0
0α	$\frac{\kappa}{4\pi}\epsilon_{\alpha\beta\gamma}\frac{x^\beta M^\gamma}{r^3}+O\left(\frac{1}{r^3}\right)$	$\frac{\lambda}{8\pi}\epsilon_{\alpha\beta\gamma}\frac{x^\beta S^\gamma}{r^3}+O\left(\frac{1}{r^3}\right)$
$\alpha\beta$	$\left(1+\frac{2GM}{r}\right)\delta_{\alpha\beta}$ + [gravitational radiation terms that die out as $O(1/r)$]	[gravitational radiation terms that die out as $O(1/r)$]

X. COUPLING OF AN ANTISYMMETRIC FIELD

As is wellknown, the symmetric field $\{h_{\mu\nu}\}$ can be neglected in atomic phenomena. So we shall study the coupling of an antisymmetric field, assuming that the metric tensor is the Minkowski metric tensor,

$$g_{\mu\nu}=\eta_{\mu\nu}. \tag{10.1}$$

It is then convenient to employ a Cartesian coordinate system $\{x^\mu\}$: The tetrad fields associated with it, which we denote by $\underline{e}=\{\underline{e}_k\}=\{e_k{}^\mu=\delta_k{}^\mu\}$, are related to the parallel vector fields $\underline{b}$ by a local Lorentz transformation,

$$\underline{e}_k=\Lambda^j{}_k(x)\underline{b}_j,$$
$$\Lambda^j{}_k(x)=\delta^j{}_k-\delta_k{}^\mu\delta^{j\nu}A_{\mu\nu}(x). \tag{10.2}$$

Here we assume that an antisymmetric field $\{A_{\mu\nu}\}$ is so weak that we can neglect the second- and higher-order terms of $\{A_{\mu\nu}\}$.

In Sec. II, the Dirac spinor wave function was introduced by referring to the parallel vector fields $\underline{b}$; we denote it here by ψ_b. The Dirac spinor wave function ψ_e, which is defined by referring to the tetrad fields $\underline{e}$, is related to ψ_b by the local Lorentz transformation (10.2);

$$\psi_e=U(\Lambda)\psi_b,\quad U(\Lambda)=1-\tfrac{1}{2}iA_{\mu\nu}S^{\mu\nu}. \tag{10.3}$$

It should be remarked here that the spinor wave function ψ_e is usually used in atomic physics to describe the electron.

Suppose that ψ_b satisfies the Dirac equation (3.13b), then Eq. (10.3) implies that ψ_e satisfies

$$(i\gamma^\mu\partial_\mu-\tfrac{3}{4}a_\mu\gamma^5\gamma^\mu-m)\psi_e=0 \tag{10.4}$$

by virtue of the following property of the covariant derivative ∇_μ:

$$U(\Lambda)\nabla_\mu^{(b)}\psi_b=\nabla_\mu^{(e)}\psi_e=\partial_\mu\psi_e. \tag{10.5}$$

Here $\nabla_\mu^{(b)}$ and $\nabla_\mu^{(e)}$ mean the covariant derivative defined by the Ricci rotation coefficients formed of $\underline{b}$ and $\underline{e}$, respectively: $\nabla_\mu^{(e)}$ coincides with the usual derivative ∂_μ, since $e_k{}^\mu=\delta_k{}^\mu$.

Now we apply the Dirac equation (10.4) to the electron in the hydrogen atom, including the electromagnetic interaction between the electron and the proton by the minimal principle

$$\partial_\mu\to\partial_\mu+ieA_\mu, \tag{10.6}$$

where $(-e)$ is the electric charge of the electron, and the electromagnetic potential $\{A^\mu\}=(A^0,\vec{A})$ is given by

$$A^0=-e^2/r=-A_0,$$
$$\vec{A}=(eg_p/2M_p)(\vec{\nabla}\times\vec{S}_p)\frac{1}{r}. \tag{10.7}$$

Here the vector potential $\vec{A}$ is due to the magnetic

moment of the proton; M_p, $\vec{S}_p$, and g_p are the mass, the spin, and the gyromagnetic ratio of the proton, respectively. The Dirac equation (10.4) then becomes

$$[i\gamma^\mu(\partial_\mu + ieA_\mu) - \tfrac{3}{4}a_\mu\gamma^5\gamma^\mu - m]\psi_e = 0\ , \qquad (10.8)$$

for the electron in hydrogen atom.

For the proton at rest at the origin, the axial-vector current of (9.13) is given by

$$J_{50} = 0,\quad \vec{J}_5 = \vec{S}_p\,\delta^3(\vec{x})\ . \qquad (10.9)$$

Use of this in (9.11)–(9.12) shows that space-space components of the antisymmetric part of $T_{\mu\nu}$, $T_{[\alpha\beta]}$, vanish identically; therefore, we find that

$$A_{\alpha\beta} = 0 \qquad (10.10a)$$

around the proton. On the other hand, the (0α) components of an antisymmetric field are given by (9.35):

$$(A_{0\alpha}) = -\frac{\lambda}{8\pi}(\vec{\nabla}\times\vec{S}_p)\frac{1}{r}\ . \qquad (10.10b)$$

Using (10.10a)–(10.10b) in (9.6b), we obtain the axial-vector part of the torsion tensor around the proton at rest,

$$a^0 = 0,\quad \vec{a} = -\frac{\lambda}{12\pi}\vec{\nabla}\times(\vec{\nabla}\times\vec{S}_p)\frac{1}{r}\ . \qquad (10.11)$$

In order to evaluate the effects due to an antisymmetric field, we rewrite the Dirac equation (10.8) into two-component wave equations,

$$\left(i\frac{\partial}{\partial t} + \frac{e^2}{r} + \frac{3}{4}\vec{a}\cdot\vec{\sigma}\right)\phi = \vec{\sigma}\cdot(\vec{p} + e\vec{A})\chi \qquad (10.12a)$$

$$\left(i\frac{\partial}{\partial t} + 2m + \frac{e^2}{r} + \frac{3}{4}\vec{a}\cdot\vec{\sigma}\right)\chi = \vec{\sigma}\cdot(\vec{p} + e\vec{A})\phi\ , \qquad (10.12b)$$

where we put

$$\psi_e = \begin{pmatrix}\phi\\ \chi\end{pmatrix}e^{-imt}\ , \qquad (10.13)$$

and used the standard representation of the γ matrices.[42] Here $\vec{p}$ denotes the momentum operator; $p^\alpha = -i\partial/\partial x^\alpha$. In the Pauli approximation, in which (10.12b) may be approximated to

$$\chi = \frac{1}{2m}(\vec{p} + e\vec{A})\cdot\vec{\sigma}\phi\ , \qquad (10.12b')$$

we get

$$i\frac{\partial\phi}{\partial t} = H\phi\ , \qquad (10.14)$$

$$H = \frac{1}{2m}(\vec{p} + e\vec{A})^2 - \frac{e^2}{r} + \left(\frac{e^2 g_p}{4mM_p} + \frac{\lambda}{16\pi}\right) \times\vec{\sigma}\cdot\{\vec{\nabla}\times(\vec{\nabla}\times\vec{S}_p)\}\frac{1}{r}\ , \qquad (10.15)$$

The last term of (10.15), which consists of two parts, describes the spin-spin interaction of the electron and the proton: One is due to the magnetic moment of the proton, and the other due to an antisymmetric field.

The spin-spin coupling due to an antisymmetric field is not restricted to the case of the electron and the proton, but quite universal. For any two spin-$\frac{1}{2}$ particles, A and B, separated by r, we can show in the similar way that *in the nonrelativistic approximation the coupling with an antisymmetric field leads to universal spin-spin coupling,*

$$H_{\text{spin-spin}} = \frac{\lambda}{8\pi}\vec{S}_A\cdot[\vec{\nabla}\times(\vec{\nabla}\times\vec{S}_B)]\frac{1}{r} = \frac{\lambda}{8\pi}\left[\frac{8\pi}{3}(\vec{S}_A\cdot\vec{S}_B)\delta^3(\vec{x}) - \frac{1}{r^3}\left((\vec{S}_A\cdot\vec{S}_B) - \frac{3}{r^2}(\vec{S}_A\cdot\vec{x})(\vec{S}_B\cdot\vec{x})\right)\right], \qquad (10.16)$$

where $\vec{S}_A$ and $\vec{S}_B$ are the spin vectors of the spin-$\frac{1}{2}$ particles, A and B, respectively. This spin-spin coupling makes a contribution to the hyperfine splitting of energy levels in atoms and muonium (the bound state of an electron and a positive muon).

Let us first consider the hyperfine structure interval $\Delta\nu(H)$ of the ground state of the hydrogen atom. We denote by $\Delta\nu_{\text{QED}}(H)$ the theoretical value which is based on conventional quantum electrodynamics and on the assumption that the proton is a Dirac particle without internal structure. Adding possible corrections to $\Delta\nu_{\text{QED}}(H)$, we express $\Delta\nu(H)$ as

$$\Delta\nu(H) = \Delta\nu_{\text{QED}}(H)[1 + \delta_N^{(2)} + \delta_A(H)]\ . \qquad (10.17)$$

Here $\delta_N^{(2)}$ is the correction due to internal structure of the proton: The precise value of $\delta_N^{(2)}$ is not known at present, but it is estimated to be 1–2 ppm.[43] The last term $\delta_A(H)$ is a possible correction which arises from universal spin-spin coupling of (10.16): From the expression (10.15) for the Hamiltonian, we obtain

$$\delta_A(H) = \frac{\lambda/16\pi}{e^2 g_p/4mM_p} = 0.012\times\frac{\lambda}{4\pi}\,(\text{GeV})^2\ . \qquad (10.18)$$

The theoretical value $\Delta\nu_{\text{QED}}(H)$ is in good agreement with the experimental value[44];

$$\frac{\Delta\nu_{\text{exp}}(H) - \Delta\nu_{\text{QED}}(H)}{\Delta\nu_{\text{exp}}(H)} = (2.5 \pm 4.0)\times 10^{-6}\ . \qquad (10.19)$$

Since the correction $\delta_N^{(2)}$ is of the order of 1 ppm, we estimate the upper limit on $\delta_A(H)$ as

$$\delta_A(H) \lesssim 5\times10^{-6} . \tag{10.20}$$

Because of ambiguity in the correction $\delta_H^{(2)}$, it seems difficult to estimate the value of $\delta_A(H)$ with higher precision than (10.20). Combining (10.18) and (10.20), we get

$$\frac{\lambda}{4\pi} \lesssim 4\times10^{-4}\ (\mathrm{GeV})^{-2} . \tag{10.21}$$

Next we shall consider the hyperfine structure interval of the ground state of muonium. The theoretical value $\Delta\nu_{\mathrm{QED}}(e\mu)$ based on conventional quantum electrodynamics agrees well with the experimental value[45];

$$\frac{\Delta\nu_{\mathrm{exp}}(e\mu)-\Delta\nu_{\mathrm{QED}}(e\mu)}{\Delta\nu_{\mathrm{exp}}(e\mu)} = -(0.45\pm1.5)\times10^{-6} . \tag{10.22}$$

The value of $\Delta\nu_{\mathrm{exp}}(e\mu)$ is known with much higher precision than $\Delta\nu_{\mathrm{QED}}(e\mu)$, due to uncertainty in our knowledge of the fundamental constants μ_μ/μ_p and α; here μ_μ and μ_p are the magnetic moments of the muon and the proton, respectively, and α is the fine-structure constant. Possible fractional correction to $\Delta\nu_{\mathrm{QED}}(e\mu)$, $\delta_A(e\mu)$, which arises from universal spin-spin coupling of (10.16), is obtained from (10.18) by replacing the proton parameters, M_p and g_p, with the muon parameters, M_μ and g_μ;

$$\delta_A(e\mu)=\frac{\lambda/16\pi}{e^2g_\mu/4mM_\mu}=0.0037\times\frac{\lambda}{4\pi}\ (\mathrm{GeV})^2 . \tag{10.23}$$

Here M_μ is the muon mass and g_μ is the gyromagnetic ratio of the muon. From (10.22) we estimate the upper limit on $\delta_A(e\mu)$ as

$$\delta_A(e\mu) \lesssim 10^{-6} . \tag{10.24}$$

This upper limit can be improved, provided that the fundamental constants μ_μ/μ_p and α would be known with higher precision. Using (10.23) in (10.24), we obtain

$$\frac{\lambda}{4\pi} \lesssim 3\times10^{-4}\ (\mathrm{GeV})^{-2} . \tag{10.25}$$

Summing up, we conclude from (10.21) and (10.25) that *the square of the coupling strength of an antisymmetric field is bounded by* $\lambda/4\pi \lesssim 3\times10^{-4}\ (\mathrm{GeV})^{-2}$. This result is in agreement with the quantum-field-theoretical estimation of Miyamoto and Nakano.[7]

XI. TIME-DEPENDENT SPHERICALLY SYMMETRIC FIELDS

We now turn to a spherically symmetric, but *not necessarily static* gravitational field in vacuum. When we considered the static, isotropic gravitational field in Sec. V, we assumed that the state of a central gravitating spherical body does not change under space inversion, besides it is invariant under time reversal and space rotation. This is the case either if constituent particles of a spherical body are spinless, or if the spin of constituent particles is randomly distributed and can be ignored. If the spin of constituent particles of a spherical body happens to be polarized to outward (or inward) radial direction, however, the spin state of the gravitating body changes under space inversion: In fact, if the spin of constituent particles is polarized to *outward* radial direction, then after space inversion the spin is polarized to *inward* radial direction. Therefore, we assume here that *the parallel vector fields* $\underline{b}=\{\underline{b}_k\}=\{b_k{}^\mu\}$ *and their dual* $\underline{b}^*=\{\underline{b}^k\}=\{b^k{}_\mu\}$ *are form invariant under space rotation* (5.1c), *but not necessarily form invariant under time reversal* (5.1a) *and space inversion* (5.1b). It is shown in Appendix B that we can then take the following expression for $\underline{b}^*=\{\underline{b}^k\}=\{b^k{}_\mu\}$:

$$(b^k{}_\mu)=\begin{pmatrix} C & 0 \\ Hx^a & D\delta^a_\alpha+F\epsilon_{a\alpha\beta}x^\beta \end{pmatrix}, \tag{11.1a}$$

where C, D, F, and H are unknown functions of t and $r=(x^\alpha x^\alpha)^{1/2}$. The parallel vector fields $\underline{b}=\{\underline{b}_k\}=\{b_k{}^\mu\}$ are then represented as

$$(b_k{}^\mu)=\begin{pmatrix} 1/C & -\dfrac{H}{CD}x^\alpha \\ 0 & \left(D\delta^\alpha_a+\dfrac{F^2}{D}x^a x^\alpha+F\epsilon_{a\alpha\beta}x^\beta\right)\Big/(D^2+r^2F^2) \end{pmatrix} \tag{11.1b}$$

and the invariant distance ds^2 is expressed in a rotationally invariant form,

$$ds^2=-(C^2-r^2H^2)dt^2+2DH\,dt(x^\alpha dx^\alpha) + (D^2+r^2F^2)dx^\alpha dx^\alpha - F^2(x^\alpha dx^\alpha)^2 . \tag{11.2}$$

In empty space the antisymmetric part (7.3) of the gravitational field equation reads

$$\partial_\rho(\sqrt{-g}\ \epsilon^{ijmn}b_m{}^\rho b_n{}^\sigma a_\sigma)=0 . \tag{11.3}$$

The axial-vector part $\{a^\mu\}$ of the torsion tensor is expressed in terms of the unknown functions, C, D, F, and H, as

$$\sqrt{-g}\ a^\mu=\begin{cases} P, & \text{for } \mu=0 , \\ Qx^\alpha, & \text{for } \mu=\alpha , \end{cases} \tag{11.4}$$

with P and Q defined by

$$P = 2DF + \tfrac{2}{3}r(DF' - D'F)\,,\qquad Q = -\tfrac{4}{3}HF + \tfrac{2}{3}(\dot{D}F - D\dot{F})\,, \tag{11.5}$$

where a dot and a prime denote $\partial/\partial t$ and $\partial/\partial r$, respectively. Using (11.4) in (11.3), we get

$$2F\frac{HP+DQ}{D^2+r^2F^2}x^a = 0 \tag{11.6}$$

for $(i,j) = (0,a)$, and

$$\partial_0\left(\frac{1}{C}(HP+DQ)\epsilon_{abc}x^c\right) - \partial_\alpha\left(\frac{H}{CD}(HP+DQ)\epsilon_{abc}x^c x^\alpha\right) + \frac{\epsilon_{abc}x^c}{r}\left[\left(\frac{C}{D}P\right)' + 2r\frac{F^2}{D^2+r^2F^2}\left(\frac{C}{D}P\right)\right] = 0 \tag{11.7}$$

for $(i,j) = (a,b)$. Equation (11.6) gives

$$F = 0\,, \tag{11.8a}$$

or

$$HP + DQ = 0\,. \tag{11.8b}$$

It follows from (11.5) that if $F = 0$, then P and Q vanish identically. On the other hand, if (11.8b) is satisfied, (11.7) gives

$$\left(\frac{C}{D}P\right)' + 2r\frac{F^2}{D^2+r^2F^2}\frac{C}{D}P = 0\,, \tag{11.9}$$

which can be readily integrated to give

$$\frac{C}{D}P = f(t)\exp\left(-\int_{r_0}^{r}\frac{2rF^2}{D^2+r^2F^2}dr\right)\,, \tag{11.10}$$

with $f(t)$ an unknown function of t. We impose the boundary condition at spatial infinity as

$$\lim_{r\to\infty} b^k{}_\mu = \delta^k{}_\mu\,,\quad \lim_{r\to\infty} b_k{}^\mu = \delta_k{}^\mu\,. \tag{11.11}$$

Then unknown functions, C, D, F and H, satisfy

$$\lim_{r\to\infty} C(t,r) = \lim_{r\to\infty} D(t,r) = 1\,, \tag{11.12}$$

$$\lim_{r\to\infty} rH(t,r) = \lim_{r\to\infty} rF(t,r) = 0\,, \tag{11.13}$$

and hence from (11.5) if follows that

$$\lim_{r\to\infty} rP(t,r) = \lim_{r\to\infty} rQ(t,r) = 0\,. \tag{11.14}$$

Because of the boundary condition (11.13) for F, the integral in the exponent of (11.10) converges for $r\to\infty$, and so the exponential factor of (11.10) approaches a finite positive value for $r\to\infty$. Therefore, in order to satisfy the boundary condition (11.14), the unknown function $f(t)$ must vanish, and hence we get

$$P(t,r) = 0,\quad Q(t,r) = 0\,, \tag{11.15}$$

by virtue of (11.8b) and (11.10). It then follows from (11.4) that *the axial-vector part of the torsion tensor must vanish,*

$$a^\mu = 0 \tag{11.16}$$

for a spherically symmetric gravitational field in vacuum. The symmetric part (7.2) of the gravitational field equation now becomes the Einstein equation in vacuum,

$$G_{\mu\nu}(\{\;\}) = 0\,. \tag{11.17}$$

According to the Birkhoff theorem[45] in general relativity, a spherically symmetric solution of (11.17) must be static and is given by the Schwarzschild solution.

We have thus shown that *a spherically symmetric solution of the gravitational field equations* (7.2)–(7.3) *with source terms absent must coincides with the static, isotropic field in vacuum studied in Secs.* V *and* VII, i.e., the Schwarzschild solution. This is just *the Birkhoff theorem* of new general relativity.

XII. CONCLUSION

We have formulated new general relativity and proved the following:

(1) The equations of motion for spin-$\frac{1}{2}$ fundamental particles and photons are approximated by the WKB approximation method to yield, in the classical limit, the geodesics of the metric $\underline{g}$, the extremal curve. This is the "corresponding principle" in new general relativity.

(2) In the case of $c_1 = 0 = c_2$ the gravitational action is of the form

$$I = \int d^4x\sqrt{-g}\left(\frac{1}{2\kappa}R(\{\;\}) + \frac{9}{4\lambda}(a_\mu a^\mu) + L_M\right)\,. \tag{12.1}$$

Here κ is the Einstein gravitational constant, $\kappa = 8\pi G/c^4 = 8\pi G$, and λ is a new parameter, bounded by $\lambda/4\pi < 10^{-4}\,\hbar c/(\mathrm{GeV})^2$ from precise experiments in quantum electrodynamics. (We leave open the possibility that λ would be equal to κ, i.e., $\lambda = \kappa$.) What differs from general relativity is the second term, which consists of the axial-vector part $\{a^\mu\}$ of the torsion tensor. From this action follows the gravitational field equation,

$$G^{\mu\nu}(\{\;\}) + L^{\mu\nu} = \kappa T^{\mu\nu}\,, \tag{12.2}$$

where

$$L^{\mu\nu} = \frac{\kappa}{2\lambda}\{a_\lambda[\epsilon^{\mu\rho\sigma\lambda}(T^{\nu}_{\cdot\rho\sigma} - T_{\rho\sigma}^{\cdot\cdot\nu}) + \epsilon^{\nu\rho\sigma\lambda}(T^{\mu}_{\cdot\rho\sigma} - T_{\rho\sigma}^{\cdot\cdot\mu})] - 3a^\mu a^\nu - \tfrac{3}{2}g^{\mu\nu}a_\rho a^\rho + 3\epsilon^{\mu\nu\rho\sigma}(b^i{}_\rho a_{i,\sigma} + a_\rho v_\sigma)\}\,. \tag{12.3}$$

Here $a_i = b_i{}^\mu a_\mu$ is a vector with respect to global Lorentz transformations, but a scalar with respect to general coordinate transformations. Obviously, the field equation is invariant under global Lorentz transformations but violates the local Lorentz invariance in general.

(3) In the static, isotropic gravitational field the axial-vector part of the torsion tensor is identically vanishing, and the solution is given by the Schwarzschild solution.

(4) New general relativity agrees with all the experiments which have so far been carried out, as general relativity does.

(5) In the weak-field approximation to the gravitational field equation, it splits into two separate equations; one is for the symmetric field $\{\bar{h}_{\mu\nu}\}$, and the other is for the antisymmetric field $\{A_{\mu\nu}\}$,

$$\Box\bar{h}_{\mu\nu} = -2\kappa T_{(\mu\nu)}\,, \tag{12.4}$$

$$\Box A_{\mu\nu} = -\lambda T_{[\mu\nu]}\,, \tag{12.5}$$

with the conditions, $\partial_\nu\bar{h}^{\mu\nu}=0$ and $\partial_\nu A^{\mu\nu}=0$. The first equation describes the propagation of a graviton with zero mass and spin 2, and the second means the propagation of a zero-mass and zero-spin particle, which exerts spin-dependent force among spin-$\frac{1}{2}$ fundamental particles.

(6) In microscopic processes the equivalence principle is violated by means of the antisymmetric field described by (12.5), which is coupled to spin-$\frac{1}{2}$ fundamental particles. However, in the macroscopic scale the equivalence principle is recovered.

(7) In new general relativity the Birkhoff theorem, that a spherically symmetric gravitational field in empty space must be static, with a metric given by the Schwarzschild solution, is proved.

At this point we summarize several important features of new general relativity in comparison with general relativity; see Table II.

Finally, we emphasize that new general relativity, originally due to Einstein in 1928, is a gravitational theory that is acceptable on the experimental and theoretical grounds.

At present it seems impossible to detect the differences between general relativity and new general relativity. Among other things, it is highly expected to see what are Kerr-like solutions, i.e., stationary and axially symmetric solutions, in new general relativity, since the Kerr solution in general relativity has the *total angular momentum*, to which the axial-vector part $\{a^\mu\}$ of the torsion tensor in new general relativity may contribute.

ACKNOWLEDGMENT

One of the authors (K.H.) wishes to thank members of the Max Planck Institut für Physik und Astrophysik for discussions and suggestions during his stay there in 1976–1978, when the present work began.

APPENDIX A: PARITY-VIOLATING TERMS, $(v^\mu a_\mu)$ AND $(\epsilon_{\mu\nu\rho\sigma}t^{\lambda\mu\nu}t_\lambda{}^{\cdot\rho\sigma})$

In Sec. IV we constructed the gravitational Lagrangian density L_G of (4.12) by postulating the basic principles (1)–(4). We now lift up the postulate of (3), then we can add to L_G of (4.12) parity-violating terms like $(v^\mu a_\mu)$ and $(\epsilon_{\mu\nu\rho\sigma}t^{\lambda\mu\nu}t_\lambda{}^{\cdot\rho\sigma})$:

$$L_G = a_1(t^{\lambda\mu\nu}t_{\lambda\mu\nu}) + a_2(v^\mu v_\mu) + a_3(a^\mu a_\mu) + a_4(v^\mu a_\mu) + a_5(\epsilon_{\mu\nu\rho\sigma}t^{\lambda\mu\nu}t_\lambda{}^{\cdot\rho\sigma})\,, \tag{A1}$$

where a cosmological term is neglected. Because of the identity,

$$\sqrt{-g}\,\epsilon_{\mu\nu\rho\sigma}t^{\lambda\mu\nu}t_\lambda{}^{\cdot\rho\sigma} = \tfrac{9}{2}\sqrt{-g}\,v^\mu a_\mu - \tfrac{27}{4}(\sqrt{-g}\,a^\mu)_{,\mu}\,, \tag{A2}$$

we can drop the a_5 term, absorbing it into the a_4 term. Accordingly, the gravitational action of (4.18) involves one additional parity violating term;

$$I_G = \int d^4x\sqrt{-g}\,L_G = \int d^4x\sqrt{-g}\left(\frac{1}{2\kappa}R(\{\ \}) + c_1(t^{\lambda\mu\nu}t_{\lambda\mu\nu}) + c_2(v^\mu v_\mu) + c_3(a^\mu a_\mu) + c_4(v^\mu a_\mu)\right), \tag{A3}$$

where the parameters, c_1, c_2, and c_3, are given by (4.19) and c_4 is given by

$$c_4 = a_4\,. \tag{A4}$$

The gravitational field equations are then given by (4.22) with the tensor $\{F_{\mu\nu\lambda}\}$ of (4.24) involving an additional term,

$$F^{\mu\nu\lambda} = c_1(t^{\mu\nu\lambda} - t^{\mu\lambda\nu}) + c_2(g^{\mu\nu}v^\lambda - g^{\mu\lambda}v^\nu) - \frac{c_3}{3}\epsilon^{\mu\nu\lambda\rho}a_\rho + \frac{c_4}{2}(g^{\mu\nu}a^\lambda - g^{\mu\lambda}a^\nu - \tfrac{1}{3}\epsilon^{\mu\nu\lambda\rho}v_\rho) = -F^{\mu\lambda\nu}\,. \tag{A5}$$

The tensor $\{H^{\mu\nu}\}$ is still defined by (4.25), and L' is

$$L' = c_1(t^{\lambda\mu\nu}t_{\lambda\mu\nu}) + c_2(v^\mu v_\mu) + c_3(a^\mu a_\mu) + c_4(v^\mu a_\mu)\,. \tag{A6}$$

It is to be noticed that the choice of parameters in Ref. 4 corresponds to the case of $c_3 = 0 \neq c_4$.

In a static, isotropic gravitational field, for which the parallel vector fields $\underline{b}$ are of a diagonal form (5.2), there is no appearance of the parameters, c_3 and c_4, in the gravitational field equation. Therefore, all the results of Sec. V still hold true independently of the parity-violating

TABLE II. Comparison of new general relativity with general relativity.

	General relativity	New general relativity
Space-time	Riemann space-time	Weitzenböck space-time -absolute parallelism-
Connection	Levi-Civita connection $\Gamma^{\lambda}_{\mu\nu}=\{{}^{\lambda}_{\mu\nu}\}$	Nonsymmetric affine connection $\Gamma^{*\lambda}_{\mu\nu}=b_k{}^{\lambda}\partial_\nu b^k{}_\mu$
Basic structure	Metric tensor $\underline{g}=\{g_{\mu\nu}\}$	Parallel vector fields $\underline{b}=\{b_k{}^{\mu}\}\to\underline{g}=\{g_{\mu\nu}\}$
Gravitation	Riemann-Christoffel curvature tensor	Torsion tensor $T^{\lambda}{}_{\bullet\mu\nu}=b_k{}^{\lambda}(\partial_\nu b^k{}_\mu-\partial_\mu b^k{}_\nu)$
Transformation group	General coordinate transformation group (Local Lorentz group)	General coordinate transformation group Global Lorentz group
The Birkhoff theorem	Yes	Yes
Static, isotropic gravitational field	The Schwarzschild solution	The Schwarzschild solution
Static, axially symmetric gravitational field	The Kerr solution	Not yet found
Very strong, static, isotropic field	Black holes	Black holes
Newtonian approximation	Yes	Yes
Weak-field approximation	Symmetric field $\{\bar{h}_{\mu\nu}\}$; $\Box\bar{h}_{\mu\nu}=-2\kappa T_{\mu\nu}$ with $\partial_\nu\bar{h}^{\mu\nu}=0$	Symmetric field $\{\bar{h}_{\mu\nu}\}$; $\Box\bar{h}_{\mu\nu}=-2\kappa T_{(\mu\nu)}$ with $\partial_\nu\bar{h}^{\mu\nu}=0$ Antisymmetric field $\{A_{\mu\nu}\}$; $\Box A_{\mu\nu}=-\lambda T_{[\mu\nu]}$ with $\partial_\nu A^{\mu\nu}=0$
Quantum	Graviton; spin 2 and massless	Graviton; spin 2 and massless Scalar particle; positive energy, spinless and massless
Theory	Macroscopic	Microscopic
Equivalence principle	Yes	Yes, for macroscopic phenomena No, for microscopic phenomena

c_4 term of L_G. In particular, the values of the parameters, c_1 and c_2, are severely restricted by the solar-system experiments, as is shown in (6.16).

We shall thus assume henceforth in this appendix that $c_1=0=c_2$. Furthermore, in order to elucidate effects of the parity-violating c_4 term of L_G, we here apply the gravitational field equation to weak-field situations, where (9.1a) is satisfied. Using the notation introduced in Sec. IX, we find that the symmetric and antisymmetric parts of the gravitational field equation are given by

$$G^{(1)}_{\mu\nu}(\{\ \})-\tfrac{1}{3}\kappa c_4\partial^\lambda(\partial_\mu\bar{A}_{\nu\lambda}+\partial_\nu\bar{A}_{\mu\lambda})=\kappa T_{(\mu\nu)}\,, \quad \text{(A7)}$$

$$-\tfrac{4}{9}c_3[\Box A_{\mu\nu}+\partial^\lambda(\partial_\mu A_{\nu\lambda}-\partial_\nu A_{\mu\lambda})] + c_4[-\tfrac{1}{3}\partial^\lambda(\partial_\mu\bar{A}_{\nu\lambda}-\partial_\nu\bar{A}_{\mu\lambda}) +\tfrac{1}{3}\epsilon_{\mu\nu\lambda\rho}\partial^\lambda\partial_\sigma(\tfrac{1}{2}\bar{h}^{\rho\sigma}-A^{\rho\sigma})]=T_{[\mu\nu]}\,, \quad \text{(A8)}$$

where $\{\bar{A}_{\mu\nu}\}$ is the dual of $\{A_{\mu\nu}\}$,

$$\bar{A}_{\mu\nu}=\tfrac{1}{2}\epsilon_{\mu\nu\rho\sigma}A^{\rho\sigma}\,. \quad \text{(A9)}$$

The nonsymmetric energy-momentum tensor $\{T_{\mu\nu}\}$ satisfies the ordinary conservation law (9.9).

Corresponding to the invariance of the gravitational field equation (4.22) under general coordinate transformations, the linearized field equation (A7)–(A8) is invariant under gauge transformation,

$$h'_{\mu\nu}=h_{\mu\nu}-(\partial_\mu\Lambda_\nu+\partial_\nu\Lambda_\mu),$$
$$A'_{\mu\nu}=A_{\mu\nu}+\tfrac{1}{2}(\partial_\mu\Lambda_\nu-\partial_\nu\Lambda_\mu), \tag{A10}$$

where Λ_μ are four small but otherwise arbitrary functions that leave $\{h_{\mu\nu}\}$ and $\{A_{\mu\nu}\}$ weak. For the symmetric field $\{h_{\mu\nu}\}$, the most convenient choice of gauge is to put the harmonic condition (9.15), which we shall assume henceforth.

It is to be noticed that the field equation (A7)–(A8) is not invariant under space inversion and time reversal, when c_4 does not vanish. This is, of course, a direct consequence of the fact that the gravitational Lagrangian density of (A3) involves the term $c_4(v^\mu a_\mu)$, which changes sign under space inversion and time reversal. This apparent parity violation, however, does not lead to any observable effects in the weak-field approximation, as will be shown below.

Multiplying ∂^ν on (A7) [or (A8)], we get

$$-\tfrac{2}{3}c_4\Box\partial^\nu\overline{A}_{\mu\nu}=\partial^\nu T_{\nu\mu}, \tag{A.11}$$

of which the retarded solution is given by

$$\partial^\nu\overline{A}_{\mu\nu}(\vec{x},t)=\frac{3}{8\pi c_4}\int d^3x'\,\frac{\partial^\nu T_{\nu\mu}(\vec{x}',t-|\vec{x}-\vec{x}'|)}{|\vec{x}-\vec{x}'|}, \tag{A.12}$$

since c_4 is assumed to be nonvanishing. Using (9.4), (9.15), and (A12) in (A7), the field equation of $h_{\mu\nu}$ reads

$$\Box\overline{h}_{\mu\nu}=-2\kappa T_{(\mu\nu)}-\frac{\kappa}{2\pi}\int d^3x'\,\frac{\partial^\lambda T_{\lambda(\mu,\nu)}(\vec{x}',t-|\vec{x}-\vec{x}'|)}{|\vec{x}-\vec{x}'|}, \tag{A13}$$

where

$$T_{\lambda(\mu,\nu)}\equiv\tfrac{1}{2}(\partial_\mu T_{\lambda\nu}+\partial_\nu T_{\lambda\mu}). \tag{A14}$$

The retarded solution of (A13) is

$$\Box\overline{h}_{\mu\nu}(\vec{x},t)=\frac{\kappa}{2\pi}\int d^3x'\,\frac{T_{(\mu\nu)}(\vec{x}',t-|\vec{x}-\vec{x}'|)}{|\vec{x}-\vec{x}'|}+\frac{\kappa}{8\pi^2}\int\int d^3x'd^3x''\,\frac{\partial^\lambda T_{\lambda(\mu,\nu)}(\vec{x}'',t-|\vec{x}-\vec{x}'|-|\vec{x}'-\vec{x}''|)}{|\vec{x}-\vec{x}'|\,|\vec{x}'-\vec{x}''|}, \tag{A15}$$

which is interpreted as the gravitational radiation produced by the source $\{T_{\mu\nu}\}$. Inspection of (A15) shows that if $\{\partial_\lambda T^{\lambda\mu}\}$ does not identically vanish, the field $\{h_{\mu\nu}\}$ propagates *inside* the light cone as if it is massive.

It seems natural, however, to restrict the theoretical framework of gravitation by requiring that gravitational radiation should propagate *on* the light cone with the speed of light. In view of this criterion, the case of $c_4\neq 0$ should be disregarded unless the energy-momentum tensor satisfies

$$\partial_\lambda T^{\lambda\mu}=0, \tag{A.16}$$

in addition to the ordinary energy-momentum conservation law (9.9). Therefore, *we shall assume* (A16) *hereafter*. Then the spin tensor $\{S^{\lambda\mu\nu}\}$ is totally antisymmetric with respect to its three indices, and is represented as (9.12).

It follows from (A13) and (A16) that the symmetric field $\{h_{\mu\nu}\}$ satisfies the field equation

$$\Box\overline{h}_{\mu\nu}=-2\kappa T_{(\mu\nu)}, \tag{A.17}$$

which is nothing but the field equation (9.17) in the case of $c_4=0$. Consequently, we find that *the symmetric field $\{h_{\mu\nu}\}$ is not influenced at all by the parity-violating c_4 term of L_G.*

From (A12) and (A16) it follows that

$$\partial_\nu\overline{A}^{\mu\nu}=0, \tag{A.18a}$$

which, in view of (9.6b), is equivalent to the vanishing of the axial-vector part of the torsion tensor;

$$a^\mu=0. \tag{A.18b}$$

Therefore, $\{A_{\mu\nu}\}$ can be represented as curl of a vector field $\{B_\mu\}$,

$$A_{\mu\nu}=\partial_\mu B_\nu-\partial_\nu B_\mu. \tag{A.19}$$

Using (9.11)–(9.12), (A18a), and (A19) in (A8), we find that the field equation of $\{A_{\mu\nu}\}$ is rewritten as

$$\Box A_{\mu\nu}=\frac{3}{2c_4}(\partial_\mu J_{5\nu}-\partial_\nu J_{5\mu}), \tag{A.20}$$

the retarded solution of which is given by (A19), with B_μ defined by

$$B_\mu=-\frac{3}{8\pi c_4}\int d^3x'\,\frac{J_{5\mu}(\vec{x}',t-|\vec{x}-\vec{x}'|)}{|\vec{x}-\vec{x}'|}. \tag{A21}$$

It follows from (A18b) that an antisymmetric field does not couple with spin-$\frac{1}{2}$ fundamental particles [see the Dirac equation (3.13b)]. On the other hand, the electromagnetic field is decoupled from an antisymmetric field, since the former interacts with the gravitational field through the metric tensor $\{g_{\mu\nu}\}$. Consequently, an antisymmetric field (A19) does not interact with fundamental particles and fields, and so *it is entirely devoid of physical reality.*

The present case of $c_4 \neq 0$ is invariant under the gauge transformation (A10). Using this gauge freedom, we have put the harmonic condition (9.15), which is necessary to eliminate unphysical components of the symmetric field $\{h_{\mu\nu}\}$. We are still left with the freedom to perform a gauge transformation (A10) with Λ_μ satisfying the d'Alembert equation,

$$\Box \Lambda_\mu = 0 . \tag{A22}$$

It follows from (A21), however, that $\{B_\mu\}$ satisfies the *inhomogeneous* d'Alembert equation,

$$\Box B_\mu = \frac{3}{2c_4} J_{5\mu} , \tag{A23}$$

if matter exists. Accordingly, a gauge transformation (A10) with (A22) is *insufficient* to make an antisymmetric field (A19) vanishing in that space-time region where there exists nonvanishing source, $\{J_{5\mu}\} \neq 0$.

Therefore, an antisymmetric field (A19), although it is unphysical, cannot be eliminated by a symmetry transformation of the present case of $c_4 \neq 0$. This situation is to be contrasted with that of the electromagnetic field, in which unphysical components of the electromagnetic potential $\{A_\mu\}$ can be eliminated by choosing an appropriate gauge. It is unreasonable to accept a theory involving such unphysical degrees of freedom that cannot be removed by a symmetry transformation of a theory. Consequently, we should disregard the case of $c_4 \neq 0$.

APPENDIX B: SPHERICALLY SYMMETRIC PARALLEL VECTOR FIELDS

Consider the parallel vector fields for a spherically symmetric (but time-dependent in general) system. We mean by "spherically symmetric" that it is possible to choose a "quasi-Minkowskian" coordinates, $x^1, x^2, x^3, x^0 = t$, such that the parallel vector fields $\underline{b}^* = \{\underline{b}^k\} = \{b^k{}_\mu\}$ are *form invariant* under space rotation

$$x^\alpha \to R_{\alpha\beta} x^\beta , \quad \underline{b}^{(a)} \to R_{ac} \underline{b}^{(c)} , \tag{B1}$$

where $R = (R_{ac}) = (R_{\alpha\beta})$ is an orthogonal 3×3 matrix

$$RR^t = R^t R = I , \quad \det R = 1 .$$

The most general expression of $\{b^k{}_\mu\}$ can then be given by

$$(b^k{}_\mu) = \begin{pmatrix} C & G x^\alpha \\ H x^a & D\delta_{a\alpha} + E x^a x^\alpha + F \epsilon_{a\alpha\beta} x^\beta \end{pmatrix} , \tag{B2}$$

(rows labelled by k, columns by μ), where C, D, E, F, G, and H are unknown functions of t and $r = (x^\alpha x^\alpha)^{1/2}$. We are, however, still free to redefine the time coordinate and the radius by

$$t' = \phi(t, r) , \quad x'^\alpha = \psi(t, r) x^\alpha , \tag{B3}$$

with ϕ and ψ arbitrary functions of t and r.

Under arbitrary coordinate transformation $x^\mu \to x'^\mu$, the parallel vector fields $\{b^k{}_\mu\}$ transform like covariant vectors

$$b'^k{}_\mu(x') = (\partial x^\nu / \partial x'^\mu) b^k{}_\nu(x) . \tag{B4}$$

For a redefinition (B3) of t and r, the transformation coefficients $(\partial x^\nu / \partial x'^\mu)$ are given by

$$\begin{aligned} &\partial t / \partial t' = (\psi + r\psi')/\Delta , \\ &\partial x^\alpha / \partial t' = -(\dot\psi/\Delta) x^\alpha , \\ &\partial t / \partial x'^\alpha = -(\phi'/r\Delta) x^\alpha , \\ &\partial x^\alpha / \partial x'^\beta = \frac{1}{\psi}\left(\delta^\alpha_\beta + \frac{1}{\Delta}(\phi'\dot\psi - \dot\phi\psi')\frac{x^\alpha x^\beta}{r}\right) , \end{aligned} \tag{B5}$$

where

$$\begin{aligned} &\dot\phi = \partial\phi/\partial t , \quad \phi' = \partial\phi/\partial r , \\ &\dot\psi = \partial\psi/\partial t , \quad \psi' = \partial\psi/\partial r , \\ &\Delta \equiv \frac{\partial(t', r')}{\partial(t, r)} = \psi\dot\phi + r(\dot\phi\psi' - \phi'\dot\psi) . \end{aligned} \tag{B6}$$

Using (B5) in (B4), we obtain

$$b'^{(0)}{}_0 = \frac{1}{\Delta}[(\psi + r\psi')C - r^2\dot\psi G] , \tag{B7a}$$

$$b'^{(0)}{}_\alpha = \frac{1}{\Delta}\left(-\frac{\phi'}{r} C + \dot\phi G\right) x^\alpha , \tag{B7b}$$

$$b'^{(a)}{}_0 = \frac{1}{\Delta}[(\psi + r\psi')H - \dot\psi(D + r^2 E)] x^a , \tag{B7c}$$

$$b'^{(a)}{}_\alpha = \frac{D}{\psi}\delta^a{}_\alpha + \frac{1}{\Delta}\left(\dot\phi E - \frac{\phi'}{r} H - \frac{\dot\phi\psi' - \phi'\dot\psi}{r\psi} D\right) x^a x^\alpha + \frac{F}{\psi}\epsilon_{a\alpha\beta} x^\beta . \tag{B7d}$$

Inspection of (B7b) shows that the (0α) components, $b^{(0)}{}_\alpha$, can be eliminated by setting

$$\frac{C}{r}\phi' - G\dot\phi = 0 , \quad \psi = 1 . \tag{B8}$$

In particular, if C, D, E, F, G, and H are all time independent, ϕ can be taken as

$$\phi = t + \bar\phi(t) , \quad \frac{C}{r}\frac{d\bar\phi}{dr} - G = 0 . \tag{B9}$$

where $\bar\phi(r)$ is a function of r. Now we assume G to be zero, then the E term in $b^{(a)}{}_\alpha$ can be eliminated by putting

$$\phi = t , \quad E - \frac{D}{r}\frac{\psi'}{\psi} = 0 . \tag{B10}$$

The parallel vector fields (B2) then take the following form:

$$(b^k{}_\mu)=\begin{pmatrix} C & 0 \\ Hx^a & D\delta_{a\alpha}+F\epsilon_{a\alpha\beta}x^\beta \end{pmatrix}. \tag{B11}$$

Further reduction of $\{b^k{}_\mu\}$ is impossible: Any of C, D, F, and H cannot be put to zero in addition to E and G. This is evident from (B7a) and (B7d) for C, D, and F. To prove this for H, we assume that the $(a0)$ components, $b^{(a)}{}_0$, were eliminated from (B11) by a suitable redefinition of t and r, then Eqs. (B7b)–(B7d) show that the function ψ must satisfy

$$\psi'=0, \quad \psi H-\psi D=0 .$$

However, these two conditions of ψ are not compatible with each other, since D and H are, in general, functions of r.

Now we turn to a static, spherically symmetric system. We assume furthermore that the spin of constituent particles of the system, if it exists, can be completely neglected: This means that there is no physical distinction between the left- and right-handed coordinate system to describe it. Then the parallel vector fields are *form invariant* under time reversal and space inversion,

$$t\to -t, \quad \underline{b}_{(0)}\to -\underline{b}_{(0)} \text{ (time reversal)} \tag{B12}$$

$$x^\alpha\to -x^\alpha, \quad \underline{b}_{(a)}\to -\underline{b}_{(a)} \text{ (space inversion)} \tag{B13}$$

in addition to space rotation (B1). Hence the $(a0)$ components, $b^{(a)}{}_0$, and the F term in the $(a\alpha)$ components, $b^{(a)}{}_\alpha$, must vanish. The parallel vector fields (B11) then become

$$(b^k{}_\mu)=\begin{pmatrix} C & 0 \\ 0 & D\delta_{a\alpha} \end{pmatrix}, \tag{B14}$$

where C and D are unknown functions of r alone.

APPENDIX C: PROOF OF $1+(c_1+4c_2)\neq 0$

The field equations for the static isotropic gravitational field become

$$-\kappa(c_1+c_2)A''-[1-\kappa(c_1-2c_2)]B'' - \frac{2}{r}\{\kappa(c_1+c_2)A'+[1-\kappa(c_1-2c_2)]B'\}=T^{00}, \tag{C1a}$$

$$[1-\kappa(c_1-2c_2)]A'+[1+\kappa(c_1+4c_2)]B'=0 \tag{C1b}$$

in the Newtonian limit. (Dividing (C1) by $[1+\kappa(c_1+4c_2)]$ gives (5.7).) Assume that $1+\kappa(c_1+4c_2)=0$, then (C1b) gives $[1-\kappa(c_1-2c_2)]A'=0$. If $A'=0$, the equation of motion (5.11a) for a nonrelativistic test particle becomes

$$\frac{d^2x^\alpha}{dt^2}=\frac{1}{2}\frac{\partial}{\partial x^\alpha}g_{00}(x)=\frac{1}{2}\frac{\partial}{\partial x^\alpha}A=0,$$

which contradicts the Newton equation of motion. If $1-\kappa(c_1-2c_2)=0$, on the other hand, c_1 and c_2 are given by $c_1=1/3\kappa$ and $c_2=-1/3\kappa$, respectively. Using these values of c_1 and c_2 in (C1a) gives

$$0=\kappa T^{00},$$

which is a contradiction. Therefore, $[1+\kappa(c_1+4c_2)]$ should not be zero.

---

[1] A. Einstein, (a) Sitzungsber. Preuss. Akad. Wiss. 217 (1928); (b) 224 (1928); (c) 2 (1929); (d) 156 (1929); (e) 18 (1930); (f) 401 (1930).

[2] A. Einstein and W. Mayer, Sitzungsber. Preuss. Akad. Wiss. 110 (1930).

[3] C. Møller, K. Dan. Vidensk. Selsk. Mat. Fys. Skr. 1, No. 10 (1961).

[4] C. Pellegrini and J. Plebanski, K. Dan. Vidensk. Selsk. Mat. Fys. Skr. 2, No. 4 (1962).

[5] C. Møller, K. Dan. Vidensk. Selsk. Mat. Fys. Skr. 89, No. 13 (1978).

[6] K. Hayashi and T. Nakano, Prog. Theor. Phys. 38, 491 (1967).

[7] S. Miyamoto and T. Nakano, Prog. Theor. Phys. 45, 295 (1971).

[8] K. Hayashi, (a) Gen. Relativ. Gravit. 4, 1 (1973); (b) Lett. Nuovo Cimento 5, 529 (1972); (c) 5, 739 (1972); (d) 5, 883 (1972); (e) Phys. Lett. 43B, 497 (1973); (f) 44B, 497 (1973).

[9] K. Hayashi, Nuovo Cimento 16A, 639 (1973).

[10] K. Hayashi, Phys. Lett. 69B, 441 (1977).

[11] R. Weitzenböck, *Invariantentheorie* (Noordhoff, Groningen, 1923); Chap. XIII, Sec. 7.

[12] See, e.g., E. T. Davies and K. Yano, in *Convegno Internazionale Celebrativo del Centenario della Nascita di Tulli Levi-Civita, Atti dei Convegni Lincei* (Academia Nazionale dei Lincei, Roma, 1975); p. 53.

[13] Throughout this paper we mean by "geodesics" the shortest (or longest) possible path between two points, "length" being measured by the metric g.

[14] See, for example, K. Hayashi and T. Shirafuji, Prog. Theor. Phys. 57, 302 (1977).

[15] Our convention of the γ matrices is as follows:

$$\{\gamma^i, \gamma^j\}=-2\eta^{ij}, \quad S^{ij}=(i/4)[\gamma^i, \gamma^j],$$

$$\gamma^5=i\gamma^0\gamma^1\gamma^2\gamma^3.$$

In the spinor representation (2.20) of ψ, the γ matrices are

$$\gamma^0=\begin{pmatrix} 0 & I \\ I & 0 \end{pmatrix}, \quad \gamma^a=\begin{pmatrix} 0 & -\sigma^a \\ \sigma^a & 0 \end{pmatrix}, \quad \gamma^5=\begin{pmatrix} I & 0 \\ 0 & -I \end{pmatrix}.$$

[16]C. Møller, *The Theory of Relativity* (Clarendon, Oxford, 1952).

[17]Although the magnitude of spin vanishes in the classical limit $\hbar \to 0$, the spin polarization has the meaningful classical limit. The classical equation of spin precession in a homogeneous electromagnetic field is now well established and employed in the experimental study of the anomalous magnetic moment of muons and electrons. See, for example, V. Bargmann, L. Michel, and V. L. Telegdi, Phys. Rev. Lett. 2, 435 (1959).

[18]The WKB approximation method was first applied to the Dirac equation in the electromagnetic field by W. Pauli, Helv. Phys. Acta 5, 179 (1932). The classical equation of spin precession in the homogeneous magnetic field was later derived by this method in S. I. Rubinow and J. B. Keller, Phys. Rev. 131, 2789 (1963); K. Rafanelli and R. Shiller, *ibid.* 135, B279 (1964).

[19]See, for example, R. P. Feynman and M. Gell-Mann, Phys. Rev. 109, 193 (1958).

[20]In the spinor representation, ϕ is indeed a two-component spinor. We can as well use $\phi = \frac{1}{2}(1-\gamma^5)\psi$ instead of (3.15) without any change in the result of the classical limit.

[21]The Hamilton-Jacobi equation in classical mechanics is treated in, for example, H. Goldstein, *Classical Mechanics* (Addison-Wesley, Reading, Mass., 1950). Application to particle motion in general relativity can be found in C. W. Misner, K. S. Thorne, and J. A. Wheeler, *Gravitation* (Freeman, San Francisco, 1973).

[22]Then we can put $U^\mu \approx (1,0,0,0)$, $b_k{}^\mu \approx \delta_k{}^\mu$, and so $S^\mu \approx (0,\vec{S})$ by virtue of (3.28) and (3.30).

[23]The σ matrix (σ^k) is defined by $(\sigma^k) = (I, \sigma^1, \sigma^2, \sigma^3)$.

[24a]K. Hayashi and A. Bregman, Ann. Phys. (N.Y.) 75, 562 (1973); p. 597.

[24b]I. M. Gel'fand, R. A. Minlos, and Z. Ya. Shapiro, *Representations of the Rotation and Lorentz Groups and Their Applications* (Pergamon, Oxford, 1963).

[25]R. C. Tolman, *Relativity, Thermodynamics and Cosmology* (Oxford Univ. Press, Oxford, England, 1934), Eq. (82.14).

[26]P. J. Roll, R. Krotkov, and R. H. Dicke, Ann. Phys. (N.Y.) 26, 442 (1964); V. B. Braginsky and V. I. Panov, Zh. Eksp. Teor. Fiz. 61, 873 (1971) [Sov. Phys.-JETP 34, 464 (1971)].

[27]K. Hayashi, Lett. Nuovo Cimento 5, 529 (1972).

[28]For a spinning macroscopic test body such as a torque-free gyroscope, the situation is different, and the equation of the spin precession can be derived from the conservation law (6.7) by applying the method developed by Papapetrou in general relativity; A. Papapetrou, Proc. R. Soc. London A209, 248 (1951); E. Corinaldesi and A. Papapetrou, *ibid.* A209, 259 (1951). See also L. Schiff, Proc. Natl. Acad. Sci. U. S. A. 46, 871 (1960); Phys. Rev. Lett. 4, 215 (1960).

[29]A. S. Eddington, *The Mathematical Theory of Relativity* (Cambridge Univ. Press, Cambridge, England, 1924), 2nd edition, p. 105; H. P. Robertson, in *Space Age Astronomy*, edited by A. J. Deutsch and W. B. Klempler (Academic, New York, 1962), p. 228.

[30]J. D. Anderson *et al.*, Astrophys. J. 200, 221 (1975).

[31]E. B. Fomalont and R. A. Sramek, Phys. Rev. Lett. 36, 1475 (1976).

[32]Footnote 27 of I. I. Shapiro *et al.*, Phys. Rev. Lett. 36, 555 (1976).

[33]J. G. Williams *et al.*, Phys. Rev. Lett. 36, 551 (1976); I. I. Shapiro *et al.*, *ibid.* 36, 555 (1976).

[34]See, for example, J. Ehlers and W. Kundt, in *Gravitation*, edited by L. Witten (Wiley, New York, 1962); and W. Kinnersley, in *General Relativity and Gravitation*, edited by G. Shaviv and N. Rosen (Wiley, New York, 1975).

[35]A. Friedmann, Z. Phys. 10, 377 (1922); 21, 326 (1924).

[36]D. G. Boulware, Phys. Rev. D 11, 1404 (1975); 12, 350 (1975).

[37]M. D. Kruskal, Phys. Rev. 119, 1743 (1960); G. Szekeres, Publ. Mat. Debrecen 7, 285 (1960).

[38]The spin tensor $\{S^{\lambda\mu\nu}\}$ is taken to lowert order in the weak field, and so it is independent of the weak field. The Tetrode formula (9.11) is equivalent to the total angular momentum conservation law,

$$\partial_\nu M^{\lambda\mu\nu} = 0,$$

with $M^{\lambda\mu\nu}$ defined by

$$M^{\lambda\mu\nu} = x^\lambda T^{\mu\nu} - x^\mu T^{\lambda\nu} + S^{\lambda\mu\nu}.$$

[39]See, for example, S. Weinberg, *Gravitation and Cosmology* (Wiley, New York, 1972), Chap. 10; or C. W. Misner, K. S. Thorne, and J. A. Wheeler, *Gravitation* (Freeman, San Francisco, 1973), Chap. 18.

[40]For quantization of the $\{A_{\mu\nu}\}$ field, see K. Hayashi, Phys. Lett. 44B, 497 (1973).

[41]See the second reference of Ref. 39, p. 449.

[42]Here we use the standard representation of the γ matrices:

$$\gamma^0 = \begin{pmatrix} I & 0 \\ 0 & -I \end{pmatrix}, \quad \gamma^a = \begin{pmatrix} 0 & \sigma_a \\ -\sigma^a & 0 \end{pmatrix}, \quad \gamma^5 = \begin{pmatrix} 0 & I \\ I & 0 \end{pmatrix}.$$

[43]S. D. Drell and J. D. Sullivan, Phys. Rev. 154, 1477 (1967).

[44]B. N. Taylor, W. H. Parker, and D. N. Langenberg, Rev. Mod. Phys. 41, 375 (1969).

[45]The hyperfine structure of muonium is reviewed both theoretically and experimentally by V. M. Hughes and T. Kinoshita, in *Muon Physics*, edited by V. W. Hughes and C. S. Wu (Academic, New York, 1977), Vol. I, Chap. II.

[46]G. Birkhoff, *Relativity and Modern Physics* (Harvard Univ. Press, Cambridge, Mass., 1923), p. 253.

PHYSICAL REVIEW D VOLUME 24, NUMBER 12 15 DECEMBER 1981

Brief Reports

Brief Reports are short papers which report on completed research which, while meeting the usual **Physical Review** *standards of scientific quality, does not warrant a regular article. (Addenda to papers previously published in the* **Physical Review** *by the same authors are included in Brief Reports.) A Brief Report may be no longer than 3½ printed pages and must be accompanied by an abstract. The same publication schedule as for regular articles is followed, and page proofs are sent to authors.*

Addendum to "New general relativity"

Kenji Hayashi
Institute of Physics, Tokyo University (Komaba), Tokyo 153, Japan

Takeshi Shirafuji
Physics Department, Saitama University, Urawa, Saitama 338, Japan

(Received 28 July 1981)

We make a short comment on our new general relativity formulated on the Weitzenböck space-time. The new general relativity considered here has one free parameter besides the Einstein constant κ. The total action is invariant under a class of local Lorentz transformations, besides being invariant under general coordinate and global Lorentz transformations. The consequences of this "restricted local Lorentz invariance" are studied.

In a previous paper[1] we studied a gravitational theory based on the Weitzenböck space-time with absolute parallelism. This theory attributes gravity to the torsion of space-time, defined by[2]

$$T_{\lambda\mu\nu}=b^{i}{}_{\lambda}(\partial_\nu b_{i\mu}-\partial_\mu b_{i\nu}) \tag{1}$$

with $\underline{b}=\{\underline{b}_i\}=\{b_i{}^\mu\}$ a quartet of the parallel vector fields (or simply the parallel vector fields). Postulating that the gravitational Lagrangian density should be quadratic in the torsion tensor and conserve parity, we found that it is represented by

$$L_G=a_1(t^{\lambda\mu\nu}t_{\lambda\mu\nu})+a_2(v^\mu v_\mu)+a_3(a^\mu a_\mu)\,, \tag{2a}$$

where a_1, a_2, and a_3 are free parameters. Or equivalently, it is rewritten by

$$L_G=(1/2\kappa)R(\{\;\})+c_1(t^{\lambda\mu\nu}t_{\lambda\mu\nu})+c_2(v^\mu v_\mu)+c_3(a^\mu a_\mu)+\text{a total derivative} \tag{2b}$$

with $\kappa=8\pi G$, $c_1=a_1+(1/3\kappa)$, $c_2=a_2-(1/3\kappa)$, and $c_3=a_3+(3/4\kappa)$, where G denotes the Newtonian gravitational constant. Here $t^{\lambda\mu\nu}$, v^μ, and a^μ are three irreducible building blocks of the torsion tensor, while $R(\{\;\})$ is the ordinary scalar curvature defined by the Christoffel symbol. We have compared the theory with solar-system experiments, and found that the parameters c_1 and c_2 are severely restricted by the currently available experimental data; $\kappa c_1=0.001\pm0.001$ and $\kappa c_2=-0.005\pm0.005$. As for the parameter c_3, solar-system experiments do not give any restriction at all.

In view of this severe restriction on κc_1 and κc_2, we have proposed a particular model for which the parameters c_1 and c_2 are exactly vanishing; $c_1=c_2=0$. It is the purpose of this paper to make a brief comment on the internal consistency of this model; in particular, we put forward an argument against a recent statement of internal inconsistency.[3]

The Lagrangian density of (2b) now becomes

$$L_G=(1/2\kappa)R(\{\;\})+c_3(a^\mu a_\mu)\,, \tag{3}$$

with a total-derivative term neglected, and it has the following invariance property besides the invariance under general coordinate and global Lorentz transformations: namely, the L_G of (3) is invariant under those *local* Lorentz transformations,

$$b_i{}^\mu=b'_j{}^\mu A^j{}_i(x)\,,$$

or simply

$$b=b'A(x)\,, \tag{4}$$

with $A(x)=\{A^j{}_i(x)\}$ and $A^T\eta A=\eta$, which leave the axial-vector part of the torsion tensor a^μ invariant. Using the definition of a^μ, $a^\mu=\frac{1}{6}\epsilon^{\mu\nu\rho\sigma}T_{\nu\rho\sigma}$, we see that the constraint imposed on $A(x)$ is given by

$$\epsilon^{ijmn}b_i{}^\mu b_n{}^\nu A_{kj}(x)\partial_\nu A^k{}_m(x)=0\,, \tag{5}$$

where $A_{kj}(x)=\eta_{km}A^m{}_j(x)$. This property of L_G, which we shall hereafter refer to as the "restricted local Lorentz invariance," is a characteristic feature of the present model with $c_1=c_2=0$, and it has some consequences which we shall now discuss.

Consider the variation of the action under an infinitesimal local Lorentz transformation (4) constrained by (5), i.e., $A^j{}_i(x)=\delta^j{}_i+\omega^j{}_i(x)$ with

$$\epsilon^{ijmn}b_i{}^\mu b_n{}^\nu\partial_\nu\omega_{jm}(x)=0\,, \tag{6}$$

where $\omega_{ij}(x)=\eta_{im}\omega^m{}_j(x)$, $\omega_{ij}(x)+\omega_{ji}(x)=0$, and $|\omega_{ij}(x)| \ll 1$. The matter part of the action changes like

$$\delta\int d^4x\sqrt{-g}\,L_M$$
$$=\int d^4x\sqrt{-g}\left(T^{[ij]}+\frac{1}{\sqrt{-g}}\frac{\delta}{\delta q}(\sqrt{-g}\,L_M)\frac{i}{2}S^{ij}q\right)\omega_{ij}(x)\,, \tag{7}$$

where T^{ij} is the energy-momentum tensor of matter,

$$\sqrt{-g}\,T^{ij}=b^j{}_\mu\delta(\sqrt{-g}\,L_M)/\delta b_{i\mu}\,, \tag{8}$$

and S^{ij} is the infinitesimal Lorentz generator for the matter fields which we denote collectively by q. The gravitational part of the action, on the other hand, does not change under this transformation:

$$\delta\int d^4x\sqrt{-g}\,L_M=-\int d^4x\sqrt{-g}B^{[ij]}\omega_{ij}(x)=0\,, \tag{9}$$

where B^{ij} is defined by

$$\sqrt{-g}B^{ij}=-b^j{}_\mu\delta(\sqrt{-g}\,L_G)/\delta b_{i\mu}\,. \tag{10}$$

Therefore, the gravitational field equation, which now reads as

$$B^{ij}=T^{ij}\,, \tag{11}$$

requires that the $T^{[ij]}$ should also satisfy

$$\int d^4x\sqrt{-g}T^{[ij]}\omega_{ij}(x)=0 \tag{12}$$

when the matter fields obey their field equations $\delta(\sqrt{-g}L_M)/\delta q=0$. According to (7), this condition of $T^{[ij]}$ is automatically guaranteed if and only if the matter part of the action is invariant under infinitesimal local Lorentz transformations constrained by (6). Namely, it is required by consistency of the gravitational field equation that *the matter part of the Lagrangian density L_M should have the same restricted local Lorentz invariance as the gravitational Lagrangian density of (3).*

The Lagrangian density of the fundamental particles of spin $\frac{1}{2}$, which is derived from the special-relativistic one by the minimal prescription, is given by

$$\begin{aligned}L_D&=(i/2)b_k{}^\mu(\bar{q}\gamma^k\partial_\mu q-\partial_\mu\bar{q}\gamma^k q)-m\bar{q}q\\&=[(i/2)b_k{}^\mu(\bar{q}\gamma^k\nabla_\mu q-\nabla_\mu\bar{q}\gamma^k q)-m\bar{q}q]\\&\quad+\tfrac{3}{4}(\bar{q}\gamma_k\gamma^5 q)a^k\,,\end{aligned} \tag{13}$$

where ∇_μ denotes the covariant derivative with respect to the Ricci rotation coefficients formed of $\{b_i{}^\mu\}$: the terms in the square brackets are invariant under any local Lorentz transformations. Therefore, the Lagrangian density L_D of (13) has the required restricted local Lorentz invariance. For the Rarita-Schwinger field of spin $\frac{3}{2}$, however, the *minimal* Lagrangian density does not possess this invariance property, and one must add some nonminimal coupling terms in an *ad hoc* manner to ensure the restricted local Lorentz invariance.

For the gauge fields of internal symmetrv of the fundamental particles, such as photons, $W^\pm$ mesons, Z mesons, and gluons, the Lagrangian density is constructed by the usual Yang-Mills procedure, and hence it is described in terms of the metric tensor alone, namely, the Lagrangian density for the gauge fields of internal symmetry is invariant under any local Lorentz transformations.

The equations of motion for matter fields are derived from L_M by the action principle, and so they are covariant under the local Lorentz transformations constrained by (5). That is, we have

$$\delta(\sqrt{-g}\,L_M)/\delta q=[\delta(\sqrt{-q}\,L_M)/\delta q']U(A)\,, \tag{14}$$

where we denote the Lorentz transformation rule of q by $q'=U(A)q$.

It is thus impossible to distinguish experimentally (i.e., by observing the motion of test bodies) two parallel vector fields b and b' from one another, if these two parallel vector fields are related to each other by a local Lorentz transformation satisfying (5). In other words, these two parallel vector fields should be interpreted as physically equivalent with each other. We can therefore divide the set of parallel vector fields into equivalence classes. Observing the motion of test bodies, one cannot unambiguously specify a single quartet of parallel vector fields but only an equivalence class of parallel vector fields. Consequently, the underlying space-time of the present model with $c_1=c_2=0$ is not a Weitzenböck space-time but a new class of space-time which may be classified somewhere between the Riemann and the Weitzenböck space-times. We shall call this new class of space-time the *extended Weitzenböck space-time.*

The gravitational field equation symbolically denoted by (11) reads

$$G^{\mu\nu}(\{\ \})+(2\kappa c_3/9)\Big\{a_\lambda[\epsilon^{\mu\rho\sigma\lambda}(T^\nu{}_{\rho\sigma}-T_{\rho\sigma}{}^\nu)+(\mu\leftrightarrows\nu)]-3a^\mu a^\nu-\tfrac{3}{2}g^{\mu\nu}a^\rho a_\rho+3\epsilon^{\mu\nu\rho\sigma}(b^i{}_\rho\partial_\sigma a_i+a_\rho v_\sigma)\Big\}=\kappa T^{\mu\nu}\,, \tag{15}$$

where $G^{\mu\nu}(\{\,\})$ is the ordinary Einstein tensor formed of the metric tensor. This equation is *not form invariant* under the local Lorentz transformations constrained by (5), because it involves terms like $T_{\lambda\mu\nu}$, $b^i{}_\rho \partial_\sigma a_i$, $\partial_\rho v_\sigma$, and $T^{\mu\nu}$, which are not covariant under those local Lorentz transformations. This fact poses a serious question about internal consistency of the present model: namely, does the gravitational field equation indeed define only an equivalence class of parallel vector fields? If the present model is internally consistent, the answer should be affirmative. When a quartet of parallel vector fields $\underline{b}$ is a solution of the gravitational field equation, then any other member $\underline{b}'$ of the same equivalence class should also be a solution at the same time.

At present we do not have a definite answer to the above question. However, we shall give a formal argument which suggests that the answer may indeed be affirmative.

Consider two parallel vector fields, $\underline{b}=\{b_i{}^\mu\}$ and $\underline{b}'=\{b'_i{}^\mu\}$, which are related to each other by a local Lorentz transformation (4) constrained by (5). Since the constraint equation (5) involves the parallel vector fields $\underline{b}$ explicitly, its solution [i.e., the transformation matrix $A(x)$] depends on $\underline{b}$. This is the reason why the gravitational field equation of (15) is not form invariant under the local Lorentz transformations constrained by (5).

We shall assume that the solution $A(x)$ of (5) changes smoothly as $\underline{b}$ is varied. Then, when $\underline{b}$ is varied to $\underline{b}+\delta\underline{b}$, the $A(x)$ changes to $A(x)[1+\Omega(x)] = \{A^j{}_k(x)[\delta^k{}_i+\Omega^k{}_i(x)]\}$, where $\Omega_{ij}(x)+\Omega_{ji}(x)=0$ and $|\Omega_{ij}(x)| \ll 1$ with $\Omega_{ij}(x)=\eta_{ik}\Omega^k{}_j(x)$. According to the constraint of (5) with $\underline{b}$ replaced by $\underline{b}+\delta\underline{b}$, the $\Omega^j{}_i(x)$ must satisfy

$$\epsilon^{ijmn}\{b_i{}^\mu b_n{}^\nu[\partial_\nu\Omega_{jm}+2A^k{}_j(\partial_\nu A_{kp})\Omega^p{}_m] + (b_i{}^\mu \delta b_n{}^\nu + b_n{}^\nu \delta b_i{}^\mu)A^k{}_j(\partial_\nu A_{km})]\}=0\,, \tag{16}$$

for which we assume that the solution depending linearly on δb may be represented as an integral form:

$$\Omega_{ij}(x)=\int d^4x'[-g(x')]^{1/2}X_{ijk\mu}(x,x')\delta b^{k\mu}(x')\,. \tag{17}$$

From the restricted local Lorentz invariance of the total Lagrangian density $L=L_G+L_M$, we have

$$\begin{aligned} L(\underline{b},q) &= L(\underline{b}',U(A)q=q')\,,\\ L(\underline{b}+\delta\underline{b},q) &= L(\underline{b}'+\delta\underline{b}',U(A(1+\Omega))q)\,, \end{aligned} \tag{18}$$

where $\delta\underline{b}'$ is defined by $\underline{b}+\delta\underline{b}=(\underline{b}'+\delta\underline{b}')A(1+\Omega)$. Taking the difference of these two equations integrated over whole space-time, and then using (17), we obtain the following transformation rule for the Euler derivative $\delta(\sqrt{-g}\,L)/\delta b^{i\mu}$:

$$\frac{\delta}{\delta b^{i\mu}}(\sqrt{-g}\,L)=\frac{\delta}{\delta b'^{j\mu}}(\sqrt{-g}\,L)A^j{}_i(x)+\sqrt{-g}\int d^4x' X_{mni\mu}(x',x)\left(\frac{\delta}{\delta b'^{k\lambda}}(\sqrt{-g}\,L)A^{km}b^{n\lambda}+\frac{\delta}{\delta q'}(\sqrt{-g}\,L)U(A)\frac{i}{2}S^{mn}q\right)\,, \tag{19}$$

where $A^{km}(x)=\eta^{mj}A^k{}_j(x)$.

The second term under the integral sign in (19) arises due to the fact that the local Lorentz transformation matrix $A(x)$ does depend on the parallel vector fields, and it represents the peculiar feature of the transformation law of the gravitational field equation in the present model with $c_1=c_2=0$. Because of this second term in (19), the gravitational field equation *is not covariant* under the local Lorentz transformations constrained by (5). Nevertheless, it does follow from (14) and (19) that if $\underline{b}'$ and q' obey their field equations

$$\delta(\sqrt{-q}\,L)/\delta b'^{i\mu}=0\,,\quad \delta(\sqrt{-g}\,L)/\delta q'=0\,, \tag{20}$$

then $\underline{b}$ and q also satisfy their field equations at the same time. This result indicates that the parallel vector fields are determined by the gravitational field equation up to a freedom of making the local Lorentz transformations constrained by (5).

We can thus conclude that the present model with $c_1=c_2=0$ is internally consistent if the Lorentz transformation matrix $A(x)$ constrained by (5) depends smoothly on $\underline{b}$. More specifically, we have assumed the integral representation (17) for $\Omega_{ij}(x)$, which we have not yet succeeded in proving.

[1]K. Hayashi and T. Shirafuji, Phys. Rev. D 19, 3524 (1979).

[2]We use the same notations and conventions as in Ref. 1.

[3]W. Kopczyński, University of Cologne report (unpublished).

Progress of Theoretical Physics, Vol. 74, No. 1, July 1985

Relativistic Theory of Gravitation

A. A. LOGUNOV and M. A. MESTVIRISHVILI

USSR State Committee for Utilization of Atomic Energy
Institute for High Energy Physics, Serpukhov, Moscow

(Received October 22, 1984)

In the present paper a relativistic theory of gravity (RTG) is unambiguously constructed on the basis of the special relativity and geometrization principle. In this, a gravitational field is treated as the Faraday-Maxwell spin-2 and spin-0 physical field possessing energy and momentum. The source of a gravitational field is the total conserved energy-momentum tensor of matter and of a gravitational field in Minkowski space. In the RTG, the conservation laws are strictly fulfilled for the energy-momentum and for the angular momentum of matter and a gravitational field. The theory explains the whole available set of experiments on gravity. In virtue of the geometrization principle, the Riemannian space in our theory is of field origin, since it appears as an effective force space due to the action of a gravitational field on matter. The RTG leads to an exceptionally strong prediction: The Universe is not closed but just "flat". This suggests that in the Universe a "missing mass" should exist in a form of matter.

§ 1. Introduction

In this paper the relativistic theory of gravitation (RTG) is constructed on the basis of the special relativity, and the ideas by Poincaré, Minkowski, Einstein and Hilbert get their further development. Also the investigations of the authors[1),2)] are reflected and developed here.

First the principle of relativity was applied to mechanical phenomena only. But then Henri Poincaré formulated it as the universal principle for all physical phenomena:[3)] "The laws of physical phenomena should be the same both for an observer at rest and for one who is in the state of a uniformly translational motion. So, we do not and cannot have any means to distinct whether we are in such a motion or not". Even by now they used to think that the essence of the principle of relativity is restricted by the existence of only one class of coordinate systems, the so-called inertial reference frames within which physical processes take place in the same way. However, as shown in Ref. 4), the pseudo-Euclidean space-time geometry discovered by Minkowski allows to formulate the generalized principle of relativity, valid both for the class of inertial and that of noninertial frames. The generalized principle of relativity was formulated in Ref. 4) : Whichever physical reference system is chosen, inertial or noninertial, one can always find an infinite set of other frames, where physical phenomena are simultaneous with those in the initial reference frame. Thus, we do not and cannot have any experimental means to distinguish in what particular reference frame out of this infinite set we are.

The discovery of the pseudo-Euclidean space-time geometry allows to formulate physical laws both in inertial and noninertial reference frames, and thus to disprove the erroneous statements[5)] on inapplicability of the special theory of relativity to accelerated reference frames. This means that when describing physical phenomena in Minkowski space subject to a physical problem we may choose any reference frame adequate for the given problem and, hence, set a corresponding metric tensor γ^{ik} of Minkowski space. According to the ideology of the general relativity (GR), the special principle of relativity cannot be applied for gravitational phenomena. It was that a very central point in which almost seventy years ago Einstein and Hilbert turned away from a special theory of relativity when costructing GR. This resulted in giving up the conservation laws for the energy-momentum and angular momentum, as well as in the development of unphysical concepts on the nonlocalizalility of gravitational energy, and of many other things, which have nothing to do with gravity. These two eminent scientists left the surprisingly simple Minkowski space with the maximal, ten-parameter, group of space motion and entered the maze of the Riemannian geometry, which entangled the following generations of physicists engaged in gravity. Some authors even consider giving up the energy-momentum conservation laws in GR to be the most important principal step of this theory which overthrew the concept of energy. But it would be too thoughtless if we renounced the most important law of nature, i. e., the conservation law of energy-momentum and the angular momentum of a closed system, without sound experimental grounds. It was shown in Refs. 1) that, since the GR does not and cannot have conservation laws for the energy-momentum of both matter and a gravitational field, then the inert mass defined in Einstein theory has no physical sense, the gravitational radiation flux, as it is defined in the GR, can always be annihilated by the correponding choice of the admissible reference frame, and hence, the Einstein quadrupole formula for the gravitational field radiation does not follow from GR. The general relativity does not basically suggest that a binary system looses energy because of gravitational radiation. The GR does not have the classical Newtonian limit and, consequently, does not satisfy the most fundamental principle of physics, i. e., the correspondence principle. This is what the absence of energy-momentum conservation laws leads to, should one reject dogmatism, think seriously over the heart of the problem and perform almost an elementary analysis. All of it testifies to the fact that the GR is not a satisfactory physical theory. Therefore, the problem of constructing a classical theory of gravity which would satisfy all the requirements imposed on a physical theory, is quite vital.

As opposed to the GR, our theory is based on the special principle of relativity which we, following Poincaré, consider universal and, consequently, applicable to gravitational phenomena as well. Thus, in our approach the conservation laws for the energy-momentum and angular momentum are fulfilled strictly and have a covariant character. Therefore, our theory contains no pseudotensors and as a consequence no unphysical concepts of the gravitational energy nonlocalizability arises. Figuratively speaking, our overriding problem is to construct without leaving Minkowski space an effective field Riemannian space with the help of a tensor gravitation field and the geometrization principle, with the conservation laws for matter being strictly fulfilled. This will allow us to use, if necessary, Riemannian space already inspired with the conservation laws for matter. Note, that Riemannian space constructed in such a way is, literally, of a field origin since the effective force space is generated by a gravitation field of Faraday-Maxwell type. Thus, in the present paper we shall carry out this program developing the ideas of Refs. 6). In this we manage to preserve with necessity the Hilbert-Einstein equations supplementing them with four new field equations. According to new equations, a gravitation field has, in the general case, only 2 and 0 spins. This theory changes the conventional concepts of space-time influenced by the GR, takes us out of the maze of the Riemannian geometry and is in spirit of the modern theories in elementary particle physics. Everything turns out surprisingly simple and natural. The only thing to wonder is that the way to this simplicity and lucidity took 70 years. It follows from our theory, that the Einstein general principle of relativity has neither physical sense nor any physical content.[7)]

The theory developed in this paper is based on the concept of a gravitation field being a physical field of Faraday-Maxwell type and possessing energy-momentum. Thus, a gravitation field, as well as all other physical fields, is characterized by the energy-momentum tensor of the system. We consider a gravitation field as a spin 2 and spin 0 physical field, and a free gravitation field to have spin 2. The space-time geometry for all physical fields is pseudo-Euclidean (Minkowski space). Thus, the conservation laws for the energy-momentum and angular momentum of a closed system are rigorously fulfilled. This is the principal distinction between our theory and the Einstein GR. Another important problem arising in the costruction of a theory of gravity, is that concerning the interaction between a gravitation field and matter. We think a gravitation field to be universal, and to act on all forms of matter identically. We construct our theory on the basis of the geometrization principle,[1)] which says that the equations of motion of matter under the action of the tensor gravitation field ϕ^{ik} in Minkowski space with the metric tensor γ^{ik}, may be identically represented as the equations of motion of matter in the effective Riemannian space-time with the metric tensor g^{ik} depending on the gravitational field ϕ^{ik} and the metric tensor γ^{ik}. In this way we introduce the concept of an effective Riemannian space of field nature. Proceeding from Minkowski space and the geometrization principle, the Lagrangian density has a general form

$$L = L_g(\tilde{\gamma}^{ik}, \tilde{\phi}^{ik}) + L_M(\tilde{g}^{ik}, \Phi_A)\,, \tag{A}$$

where $\tilde{\phi}^{ik} = \sqrt{-\gamma}\,\phi^{ik}$ is the density of the tensor of a field variable in the gravitational field ϕ^{ik}. $\tilde{g}^{ik} = \sqrt{-g}\,g^{ik}$ is the density of the metric tensor of the Riemannian space g^{ik}. $\tilde{\gamma}^{ik} = \sqrt{-\gamma}\,\gamma^{ik}$ is the density of the metric tensor of Minkowski space. Φ_A are the fields of matter.

In this theory, the Lagrangian density of the gravitational field depends on the metric tensor γ^{ik} and the gravitational field ϕ^{ik}. That is why it crucially differs from the GR, where the Lagrangian density depends only on the metric tensor of the Riemannian space g^{ik}. Thus, in our theory, contrary to the GR, the geometrization of the Lagrangian density of the gravitational field is not complete.

§ 2. Geometrization principle and general relations in the relativistic theory of a tensor field

Without any loss of generality, we shall assume the tensor density $\tilde{g}^{ik}$ of the metric tensor of Riemannian space-time to be a local function depending on the tensor density $\tilde{\gamma}^{ik}$ of the metric tensor of Minkowski space and the tensor density $\tilde{\phi}^{ik}$ of the gravitation field.

Let the Lagrangian density L_M depends only on the fields Φ_A, on their first-order covariant derivatives, and also on the tensor density $\tilde{g}^{ik}$ in virtue of the geometrization principle. The Lagrangian density of the gravitational field in taken to depend on the tensor density $\tilde{\gamma}^{ik}$, their first-order partial derivatives, as well as on the gravitation field density $\tilde{\phi}^{ik}$ and its first-order covariant derivatives with respect to the Minkowski metric. To obtain the coservation laws, we use the invariance of the action in the infinitesimal covariant shift. Indeed, since the action is a scalar, the variations of the actions of matter, δJ_M, and of the gravitational field, δJ_g, will be zeros under an arbitrary infinitesimal coordinate transformation. Calculate first the variation of the action of matter under the transformation

$$x'^i = x^i + \xi^i(x)\,, \tag{1·1}$$

ξ^i being the infinitesimal shift four-vector

$$\delta J_M = \int d^4x \left[\frac{\delta L_M}{\delta \tilde{g}^{\mu\nu}} \delta_L \tilde{g}^{\mu\nu} + \frac{\delta L_M}{\delta \Phi_A} \delta_L \Phi_A + \mathrm{div} \right] = 0\,, \tag{1·2}$$

where div are divergence terms inessential for our consideration.

The Euler variation is defined as usual:

$$\frac{\delta L}{\delta \phi} = \frac{\partial L}{\partial \phi} - \partial_n \left(\frac{\partial L}{\partial(\partial_n \phi)} \right).$$

Under coordinate transformation(1·1) the variations $\delta_L \tilde{g}^{\mu\nu}$, $\delta_L \Phi_A$ are easy to calculate, provided one uses their transformation law:

$$\delta_L \tilde{g}^{\mu\nu} = \tilde{g}^{\alpha\nu} D_\alpha \xi^\mu + \tilde{g}^{\alpha\mu} D_\alpha \xi^\nu - D_\alpha(\xi^\alpha \tilde{g}^{\mu\nu})\,, \tag{1·3}$$

$$\delta_L \Phi_A = -\xi^\alpha D_\alpha \Phi_A + F_{A;k}^{B;n} \Phi_B D_n \xi^k\,. \tag{1·4}$$

Here and in what follows D_ν is the covariant derivative with respect to the Minkowski metric. Putting these expressions into (1·2) and integrating them by parts, one obtains

$$\delta J_M = \int d^4x \left\{ -\xi^\mu \left[D_\alpha \left(2\frac{\delta L_M}{\delta \tilde{g}^{\mu\nu}} \tilde{g}^{\alpha\nu} \right) - D_\mu \left(\frac{\delta L_M}{\delta \tilde{g}^{\alpha\beta}} \right) \tilde{g}^{\alpha\beta} \right. \right.$$

$$\left. \left. + D_\nu \left(\frac{\delta L_M}{\delta \Phi_A} F_{A;\mu}^{B;\nu} \Phi_B \right) + \frac{\delta L_M}{\delta \Phi_A} D_\mu \Phi_A \right] + \mathrm{div} \right\} = 0\,.$$

In virtue of arbitrariness of the vector ξ^μ, one finds from the condition $\delta J_M = 0$ a strong identity

$$D_\alpha \left(2\frac{\delta L_M}{\delta \tilde{g}^{\mu\nu}} \tilde{g}^{\alpha\nu} \right) - D_\mu \left(\frac{\delta L_M}{\delta \tilde{g}^{\alpha\beta}} \right) \tilde{g}^{\alpha\beta} = -D_\nu \left(\frac{\delta L_M}{\delta \Phi_A} F_{A;\mu}^{B;\nu} \Phi_B \right) - \frac{\delta L_M}{\delta \Phi_A} D_\mu \Phi_A\,. \tag{1·5}$$

Introduce the notations:

$$\tilde{T}_{\mu\nu} = 2\frac{\delta L_M}{\delta \tilde{g}^{\mu\nu}}\,, \qquad T_{\mu\nu} = 2\frac{\delta L_M}{\delta g^{\mu\nu}}\,. \tag{1·6}$$

Then the lhs of relation (1·5) may be represented as:

$$D_\alpha(\tilde{T}_{\mu\nu} \tilde{g}^{\alpha\nu}) - \frac{1}{2} \tilde{g}^{\alpha\beta} D_\mu \tilde{T}_{\alpha\beta} = \partial_\alpha(\tilde{T}_{\mu\nu} \tilde{g}^{\alpha\nu}) - \frac{1}{2} \tilde{g}^{\alpha\beta} \partial_\mu \tilde{T}_{\alpha\beta}\,.$$

The rhs of this equality may easily be brought to the form

$$\partial_\alpha(\tilde{T}_{\mu\nu} \tilde{g}^{\alpha\nu}) - \frac{1}{2} \tilde{g}^{\alpha\beta} \partial_\mu \tilde{T}_{\alpha\beta} = \tilde{g}_{\mu\nu} \nabla_\alpha \left(\tilde{T}^{\alpha\nu} - \frac{1}{2} \tilde{g}^{\alpha\nu} \tilde{T} \right). \tag{1·7}$$

Proceeding from this, strong identity (1·5) may be written in the form

$$\tilde{g}_{\mu\nu}\nabla_\alpha\left(\tilde{T}^{\alpha\nu}-\frac{1}{2}\tilde{g}^{\alpha\nu}\tilde{T}\right)=-D_\nu\left(\frac{\delta L_M}{\delta\Phi_A}F^{B;\nu}_{A;\mu}\Phi_B\right)-\frac{\delta L_M}{\delta\Phi_A}D_\mu\Phi_A\,. \tag{1·8}$$

Due to the least action principle, the equations for matter fields have the form

$$\frac{\delta L_M}{\delta\Phi_A}=0\,. \tag{1·9}$$

Taking into account the above equation, one may find from strong identity (1·8) the weak identity:

$$\nabla_\mu\left(\tilde{T}^{\mu\nu}-\frac{1}{2}\tilde{g}^{\mu\nu}\tilde{T}\right)=0\,. \tag{1·10}$$

Note, that the density of the energy-momentum tensor of matter in Riemannian space $T^{\mu\nu}$ is expressed via $\tilde{T}^{\mu\nu}$ as follows:

$$\sqrt{-g}\,T^{\mu\nu}=\tilde{T}^{\mu\nu}-\frac{1}{2}\tilde{g}^{\mu\nu}\tilde{T}\,. \tag{1·11}$$

Thus, expression (1·10) entails the covariant equation for matter conservation in Riemannian space:

$$\nabla_\mu T^{\mu\nu}=0\,. \tag{1·12}$$

If the number of equations for matter is four, then instead of equations for matter (1·9) one can always use equivalent equations (1·12).

The variation of the action integral may be written in the equivalent form

$$\delta J_M=\int d^4x\left[\frac{\delta L_M}{\delta\tilde{\phi}^{\mu\nu}}\delta_L\tilde{\phi}^{\mu\nu}+\frac{\delta L_M}{\delta\tilde{\gamma}^{\mu\nu}}\delta_L\tilde{\gamma}^{\mu\nu}+\frac{\delta L_M}{\delta\Phi_A}\delta_L\Phi_A+\mathrm{div}\right]=0\,. \tag{1·13}$$

Here the variations $\delta_L\tilde{\phi}^{\mu\nu}$, $\delta_L\tilde{\gamma}^{\mu\nu}$ under coordinate transformation (1·1) equal

$$\delta_L\tilde{\phi}^{\mu\nu}=\tilde{\phi}^{\alpha\nu}D_\alpha\xi^\mu+\tilde{\phi}^{\alpha\mu}D_\alpha\xi^\nu-D_\alpha(\xi^\alpha\tilde{\phi}^{\mu\nu})\,, \tag{1·14}$$

$$\delta_L\tilde{\gamma}^{\mu\nu}=\tilde{\gamma}^{\alpha\nu}D_\alpha\xi^\mu+\tilde{\gamma}^{\alpha\mu}D_\alpha\xi^\nu-\tilde{\gamma}^{\mu\nu}D_\alpha\xi^\alpha\,. \tag{1·15}$$

Substituting expressions for the variations $\delta_L\tilde{\phi}^{\mu\nu}$, $\delta_L\tilde{\gamma}^{\mu\nu}$, $\delta_L\Phi_A$ into (1·13) and integrating by parts in virtue of arbitrariness of ξ^μ, one comes to the strong identity

$$D_\alpha\left(2\frac{\delta L_M}{\delta\tilde{\phi}^{\mu\nu}}\tilde{\phi}^{\alpha\nu}\right)-D_\mu\left(\frac{\delta L_M}{\delta\tilde{\phi}^{\alpha\beta}}\right)\tilde{\phi}^{\alpha\beta}+D_\alpha\left(2\frac{\delta L_M}{\delta\tilde{\gamma}^{\mu\nu}}\tilde{\gamma}^{\alpha\nu}\right)-D_\mu\left(\frac{\delta L_M}{\delta\tilde{\gamma}^{\alpha\beta}}\tilde{\gamma}^{\alpha\beta}\right)$$

$$=-D_\nu\left(\frac{\delta L_M}{\delta\Phi_A}F^{B;\nu}_{A;\mu}\Phi_B\right)-\frac{\delta L_M}{\delta\Phi_A}D_\mu\Phi_A\,. \tag{1·16}$$

It should be noted that this identity is valid irrespective of the fulfillment of the equations of motion for matter and for gravitation field.

Let us introduce the notations:

$$\tilde{t}^{\mu\nu}_M=-2\frac{\delta L_M}{\delta\tilde{\gamma}_{\mu\nu}}\,. \tag{1·17}$$

Comparing identities (1·8) and (1·16), we find

$$\tilde{g}_{\mu\nu}\nabla_\alpha\left(\tilde{T}^{\alpha\nu}-\frac{1}{2}\tilde{g}^{\alpha\nu}\tilde{T}\right)=\tilde{\gamma}_{\mu\nu}D_\alpha\left(\tilde{t}^{\alpha\nu}_M-\frac{1}{2}\tilde{\gamma}^{\alpha\nu}\tilde{t}_M\right)+D_\alpha\left(2\frac{\delta L_M}{\delta\tilde{\phi}^{\mu\nu}}\tilde{\phi}^{\alpha\nu}\right)-D_\mu\left(\frac{\delta L_M}{\delta\tilde{\phi}^{\alpha\beta}}\right)\tilde{\phi}^{\alpha\beta}. \tag{1·18}$$

Analogously, it follows from the invariance of the action of the gravitation field under coordinate transformations (1·1) that:

$$\tilde{\gamma}_{\mu\nu}D_\alpha\left[\tilde{t}^{\alpha\nu}_g-\frac{1}{2}\tilde{\gamma}^{\alpha\nu}\tilde{t}_g\right]+D_\alpha\left(2\frac{\delta L_g}{\delta\tilde{\phi}^{\mu\nu}}\tilde{\phi}^{\alpha\nu}\right)-D_\mu\left(\frac{\delta L_g}{\delta\tilde{\phi}^{\alpha\beta}}\right)\tilde{\phi}^{\alpha\beta}=0. \tag{1·19}$$

Here

$$\tilde{t}^{\mu\nu}_g=-2\frac{\delta L_g}{\delta\tilde{\gamma}_{\mu\nu}}. \tag{1·20}$$

Adding expressions (1·18) and (1·19), we get

$$\tilde{g}_{\mu\nu}\nabla_\alpha\left(\tilde{T}^{\alpha\nu}-\frac{1}{2}\tilde{g}^{\alpha\nu}\tilde{T}\right)=\tilde{\gamma}_{\mu\nu}D_\alpha\left(\tilde{t}^{\alpha\nu}-\frac{1}{2}\tilde{\gamma}^{\alpha\nu}\tilde{t}\right)+D_\alpha\left(2\frac{\delta L}{\delta\tilde{\phi}^{\mu\nu}}\tilde{\phi}^{\alpha\nu}\right)-D_\mu\left(\frac{\delta L}{\delta\tilde{\phi}^{\alpha\beta}}\right)\tilde{\phi}^{\alpha\beta}. \tag{1·21}$$

Here

$$\tilde{t}^{\mu\nu}=\tilde{t}^{\mu\nu}_g+\tilde{t}^{\mu\nu}_M. \tag{1·22}$$

In virtue of the least action principle, the equations of the gravitation field have the form

$$\frac{\delta L}{\delta\tilde{\phi}^{\mu\nu}}=\frac{\delta L_g}{\delta\tilde{\phi}^{\mu\nu}}+\frac{\delta L_M}{\delta\tilde{\phi}^{\mu\nu}}=0. \tag{1·23}$$

If we take these equations into account from (1·21), we obtain an extremely important identity:

$$\tilde{g}_{\mu\nu}\nabla_\alpha\left(\tilde{T}^{\alpha\nu}-\frac{1}{2}\tilde{g}^{\alpha\nu}\tilde{T}\right)=\tilde{\gamma}_{\mu\nu}D_\alpha\left(\tilde{t}^{\alpha\nu}-\frac{1}{2}\tilde{\gamma}^{\alpha\nu}\tilde{t}\right). \tag{1·24}$$

As may easily be verified, the density of the energy-momentum tensor of matter in Riemannian space is equal to

$$\sqrt{-g}\,T^{\mu\nu}=\tilde{T}^{\mu\nu}-\frac{1}{2}\tilde{g}^{\mu\nu}\tilde{T}. \tag{1·25}$$

Similarly, the density of the total energy-momentum tensor in Minkowski space equals

$$\sqrt{-\gamma}\,t^{\mu\nu}=\tilde{t}^{\mu\nu}-\frac{1}{2}\tilde{\gamma}^{\mu\nu}\tilde{t}. \tag{1·26}$$

On applying these expressions, relation (1·24) takes the form

$$\nabla_\mu T_\nu^\mu=D_\mu t_\nu^\mu. \tag{1·27}$$

This equality reflects the geometrization principle.

In a pseudo-Euclidean space, the covariant divergence of the sum of densities of the energy-momentum tensors of matter and gravitation field is exactly equal to that in an effective Riemannian space, but only of the density of the energy-momentum tensor of matter. Provided the equations of motion for matter are satisfied, one has

$$D_\mu t_\nu^\mu = \nabla_\mu T_\nu^\mu = 0 . \qquad (1\cdot28)$$

From the covariant equation for matter conservation in Riemannian space, it is not clear, what is conserved while from the conservation law for the total energy-momentum tensor t_ν^μ in Minkowski space it is clear that both the energy-momentum of matter and gravitation field are conserved. Thus, in this theory Riemannian space appears as a result of the action of the gravitation field on all forms of matter. That is why it is an effective Riemannian space of field origin. Minkowski space finds its exact physical reflection in the conservation laws for the energy-momentum tensor and the angular momentum of matter and gravitation field taken together.

Since there are ten Killing vectors in a flat space, there are, consequently, ten conservable integral quantities for a closed system of fields. If the number of equations of motion for matter is four, then instead of them we may use the equations expressing the total energy-momentum tensor conservation in Minkowski space:

$$D_\mu (t^\mu_{g\nu} + t^\mu_{M\nu}) = 0 . \qquad (1\cdot29)$$

This equation, alongside of those for a gravitation field, defines all the unknown characteristics of matter and gravitation field. It is worth noting that both matter and gravitation field in our theory are characterized by energy momentum tensors. As a result, in our theory, contrary to the GR, no pseudotensors arise and, hence, there are no unphysical concepts of the nonlocalizability of gravitation energy.

If we, following Hilbert and Einstein, choose the Lagrangian density of a gravitation field in a completely geometrized form, i. e., depending on the metric tensor g^{ik} of Riemannian space and its derivatives only, e. g.,

$$L_g = \sqrt{-g}\, R ,$$

where R is the scalar curvature of Riemannian space, then in virtue of the field equations the energy-momentum tensor density of the free gravitation field in Minkowski space would always be equal to zero:

$$\frac{\delta L_g}{\delta \gamma^{\mu\nu}} = \frac{\delta L_g}{\delta g^{\sigma\tau}} \frac{\partial g^{\sigma\tau}}{\partial \gamma^{\mu\nu}} = 0 . \qquad (1\cdot30)$$

Thus, it is impossible, in principle, to construct a completely geometrized Lagrangian of the tensor physical field, possessing energy and momentum. Therefore, the theory constructed on the basis of the completely geometrized Lagrangian cannot describe the physical gravitation field of Faraday-Maxwell type in Minkowski space. As was stated in literature (see, e. g., Ref. 8)), in Minkowski space one can unambiguously find with the help of the spin-2 tensor field the GR gravitation field Lagrangian which equals the scalar curvature R. However, these papers have no physical sense, as long as the energy-momentum tensor is zero, as is seen from (1·30). Therefore, all these papers are meaningless from the viewpoint of physics, and their results are inaccurate.

§ 3. Relation between canonical energy-momentum tensor and Hilbert tensor

The Lagrangian density of the gravitational field depends on the metric tensor density $\tilde{\gamma}^{\mu\nu}$, the density $\tilde{\phi}^{\mu\nu}$ of the tensor gravitation field and on their first-order derivatives.

Under coordinate transformation (1·1), the action variation δJ_g is zero and, hence,

$$\delta J_g=\int_\Omega d^4x\Big[D_\lambda J^\lambda+\frac{\delta L_g}{\delta\tilde{\phi}^{\mu\nu}}\delta_L\tilde{\phi}^{\mu\nu}+\frac{\delta L_g}{\delta\tilde{\gamma}^{\mu\nu}}\delta_L\tilde{\gamma}^{\mu\nu}\Big]=0\,. \tag{2·1}$$

Here

$$J^\lambda=-\xi^\alpha\tau_\alpha^\lambda+K_\mu^{\alpha\lambda}D_\alpha\xi^\mu\,, \tag{2·2}$$

where the density of the canonical tensor τ_α^λ is equal to

$$\tau_\alpha^\lambda=-\delta_\alpha^\lambda L_g+D_\alpha\tilde{g}^{\mu\nu}\frac{\partial L_g}{\partial(\partial_\lambda\tilde{\phi}^{\mu\nu})}=-\delta_\alpha^\lambda L_g+D_\alpha g^{\mu\nu}\frac{\partial L_g}{\partial(\partial_\lambda g^{\mu\nu})}\,, \tag{2·3}$$

and the density of the tensor of rank three $K_\mu^{\alpha\lambda}$ is

$$K_\mu^{\alpha\lambda}=2\frac{\partial L_g}{\partial(\partial_\lambda\tilde{\phi}^{\mu\nu})}\tilde{\phi}^{\alpha\nu}-\delta_\mu^\alpha\frac{\partial L_g}{\partial(\partial_\lambda\tilde{\phi}^{\sigma\tau})}\tilde{\phi}^{\sigma\tau}+2\frac{\partial L_g}{\partial(\partial_\lambda\tilde{\gamma}^{\mu\nu})}\tilde{\gamma}^{\alpha\nu}-\delta_\mu^\alpha\frac{\partial L_g}{\partial(\partial_\lambda\tilde{\gamma}^{\sigma\tau})}\tilde{\gamma}^{\sigma\tau}\,. \tag{2·4}$$

Putting into (2·1) formulae (1·14) and (1·15) for the variations $\delta_L\tilde{\phi}^{\mu\nu}$, $\delta_L\tilde{\gamma}^{\mu\nu}$, we shall obtain in virtue of arbitrariness of the volume Ω the following strong identity:

$$\begin{aligned}0=\xi^\alpha\Big[D_\alpha\tilde{\phi}^{\mu\nu}\frac{\delta L_g}{\delta\tilde{\phi}^{\mu\nu}}+D_\lambda\tau_\alpha^\lambda\Big]-K_\mu^{\alpha\lambda}D_\alpha D_\lambda\xi^\mu+D_\alpha\xi^\mu\Big[\tau_\mu^\alpha-D_\lambda K_\mu^{\alpha\lambda}\\ -2\frac{\delta L_g}{\delta\tilde{\phi}^{\mu\nu}}\tilde{\phi}^{\alpha\nu}+\delta_\mu^\alpha\frac{\delta L_g}{\delta\tilde{\phi}^{\sigma\tau}}\tilde{\phi}^{\sigma\tau}-2\frac{\delta L_g}{\delta\tilde{\gamma}^{\mu\nu}}\tilde{\gamma}^{\alpha\nu}+\delta_\mu^\alpha\frac{\delta L_g}{\delta\tilde{\gamma}^{\sigma\tau}}\tilde{\gamma}^{\sigma\tau}\Big]\,.\end{aligned} \tag{2·5}$$

Since the shift vector ξ^α is arbitrary, the last expression leads to the following identities:

$$D_\lambda\tau_\alpha^\lambda=-\frac{\delta L_g}{\delta\tilde{\phi}^{\mu\nu}}D_\alpha\tilde{\phi}^{\mu\nu}\,, \tag{2·6}$$

$$\tau_\mu^\alpha-D_\lambda K_\mu^{\alpha\lambda}=2\frac{\delta L_g}{\delta\tilde{\phi}^{\mu\nu}}\tilde{\phi}^{\alpha\nu}-\delta_\mu^\alpha\frac{\delta L_g}{\delta\tilde{\phi}^{\sigma\tau}}\tilde{\phi}^{\sigma\tau}+2\frac{\delta L_g}{\delta\tilde{\gamma}^{\mu\nu}}\tilde{\gamma}^{\alpha\nu}-\delta_\mu^\alpha\frac{\delta L_g}{\delta\tilde{\gamma}^{\sigma\tau}}\tilde{\gamma}^{\sigma\tau}\,, \tag{2·7}$$

$$K_\mu^{\alpha\lambda}=-K_\mu^{\lambda\alpha}\,. \tag{2·8}$$

Henceforth our theory is based on the linear relation between the metric tensor density $\tilde{g}^{\mu\nu}$ of the effective Riemannian space and the density $\tilde{\phi}^{\mu\nu}$ of the tensor gravitation field

$$\tilde{g}^{\mu\nu}=\tilde{\gamma}^{\mu\nu}+\tilde{\phi}^{\mu\nu}\,. \tag{2·9}$$

In this case we shall obtain the equalities:

$$\frac{\delta L_g}{\delta\tilde{\phi}^{\mu\nu}}=\frac{\delta L_g}{\delta\tilde{g}^{\mu\nu}}\,,\quad\frac{\partial L_g}{\partial(\partial_\sigma\tilde{\phi}^{\mu\nu})}=\frac{\partial L_g}{\partial(D_\sigma\tilde{g}^{\mu\nu})}\,.$$

Using these equalities we find

$$\frac{\partial L_g}{\partial(\partial_\lambda\tilde{\gamma}^{\mu\nu})}=\frac{\partial L_g}{\partial(D_\lambda\tilde{g}^{\mu\nu})}-\tilde{g}^{\alpha\beta}\frac{\partial L_g}{\partial(D_\tau\tilde{g}^{\alpha\beta})}\frac{\partial\Gamma_{\sigma\tau}^\sigma}{\partial(\partial_\lambda\tilde{\gamma}^{\mu\nu})}+\tilde{g}^{\sigma\beta}\frac{\partial L_g}{\partial(D_\tau\tilde{g}^{\alpha\beta})}\frac{\partial\Gamma_{\sigma\tau}^\alpha}{\partial(\partial_\lambda\tilde{\gamma}^{\mu\nu})}\,.$$

Here $\Gamma_{\mu\nu}^\sigma$ are Christoffel symbols of Minkowski space

$$\Gamma_{\mu\nu}^\sigma=\frac{1}{2}\gamma^{\sigma\tau}[\partial_\mu\gamma_{\tau\nu}+\partial_\nu\gamma_{\tau\mu}-\partial_\tau\gamma_{\mu\nu}]\,.$$

As a result of elementary calculations, one obtains for $K_\mu^{\alpha\lambda}$ the following expression:

$$K_\mu^{\alpha\lambda}=\frac{\partial L_g}{\partial(D_\lambda \tilde{g}^{\mu\nu})}\tilde{g}^{\alpha\nu}-\frac{\partial L_g}{\partial(D_\alpha \tilde{g}^{\mu\nu})}\tilde{g}^{\lambda\nu}+\tilde{g}^{\sigma\tau}\tilde{\gamma}_{\sigma\mu}\left[\frac{\partial L_g}{\partial(D_\alpha \tilde{g}^{\tau\nu})}\tilde{\gamma}^{\lambda\tau}-\frac{\partial L_g}{\partial(D_\lambda \tilde{g}^{\tau\nu})}\tilde{\gamma}^{\sigma\tau}\right]$$

$$+\frac{\partial L_g}{\partial(D_\tau \tilde{g}^{\sigma\nu})}\tilde{\gamma}_{\tau\mu}(\tilde{g}^{\alpha\nu}\tilde{\gamma}^{\lambda\sigma}-\tilde{g}^{\lambda\nu}\tilde{\gamma}^{\alpha\sigma})\,. \tag{2·10}$$

Since the density of the energy-momentum tensor of the gravitation field is

$$t_{g\mu}^{\alpha}=2\frac{\delta L_g}{\delta\tilde{\gamma}^{\mu\nu}}\tilde{\gamma}^{\alpha\nu}-\delta_\mu{}^\alpha\frac{\delta L_g}{\delta\tilde{\gamma}^{\sigma\tau}}\tilde{\gamma}^{\sigma\tau}\,, \tag{2·11}$$

identity (2·7) may be written in the form

$$t_{g\mu}^{\alpha}=\tau_\mu{}^\alpha-D_\lambda K_\mu^{\alpha\lambda}-2\frac{\delta L_g}{\delta\tilde{\phi}^{\mu\nu}}\tilde{\phi}^{\alpha\nu}+\delta_\mu{}^\alpha\frac{\delta L_g}{\delta\tilde{\phi}^{\sigma\tau}}\tilde{\phi}^{\sigma\tau}\,. \tag{2·12}$$

It is just the identity which establishes the relation between the Hilbert tensor density in Minkowski space and that of the canonical energy momentum tensor.

For further use it is convenient to introduce the following quantity as a characteristic of the gravitation field:

$$\overset{\circ}{t}{}_{g\mu}^{\alpha}=\tau_\mu{}^\alpha-D_\lambda K_\mu^{\alpha\lambda}\,. \tag{2·13}$$

In virtue of identity (2·12), this quantity coincides exactly with the density of the Hilbert energy-momentum tensor in the case of a free gravitation field.

§ 4. The basic identity

As was shown in Ref. 9), the symmetric tensor ϕ_{ik} of rank two may be represented as the sum of the irreducible representations: one with spin 2, one with spin 1 and two with spin 0:

$$\phi_{ik}=[P_2+P_1+P_0+P_{0'}]_{ik}^{lm}\Phi_{lm}\,. \tag{3·1}$$

The quantities $\boldsymbol{P}_s$ are convenient to write in the momentum representation. Let us introduce the auxiliary operators:

$$X_{ik}=\frac{1}{\sqrt{3}}\left[\gamma_{ik}-\frac{q_iq_k}{q^2}\right],\quad Y_{ik}=\frac{q_iq_k}{q^2} \tag{3·2}$$

with the help of which $\boldsymbol{P}_s$ may be represented in the form

$$P_0=X_{ni}X^{ml}\,,\qquad P_{0'}=Y_{ni}Y^{ml}\,, \tag{3·3}$$

$$P_1=\frac{\sqrt{3}}{2}[X_i{}^lY_n{}^m+X_n{}^mY_i{}^l+X_i{}^mY_n{}^l+X_n{}^lY_i{}^m]\,, \tag{3·4}$$

$$P_2=\frac{3}{2}[X_i{}^lX_n{}^m+X_i{}^mX_n{}^l]-X_{ni}X^{ml}\,. \tag{3·5}$$

In the x-representation the projection operators $\boldsymbol{P}_s$ are nonlocal integrodifferential ones:

$$P_{ni}^{ml}\phi_{ml}=\int d^4yP_{ni}^{ml}(x,y)\phi_{ml}(y).$$

With the help of expressions (3·3)~(3·5) one may easily make sure that only operators $\boldsymbol{P}_2$ and $\boldsymbol{P}_0$ are conserved:

$$q_lP_{2ni}^{ml}=q_lP_{0ni}^{ml}=0, \qquad q_mP_{2ni}^{ml}=q_mP_{0ni}^{ml}=0. \tag{3·6}$$

As may be easily verified, the tensor field has the only local operator of a lower order which is linear in field. It equals

$$f_{ik}=\Box[(\boldsymbol{P}_2-2\boldsymbol{P}_0)\Phi]_{ik} \tag{3·7}$$

and its divergence is identically equal to zero.

$$\partial^i f_{ik}=0. \tag{3·8}$$

The field f_{ik} describes only spins 2 and 0, i.e., in a more detailed form

$$f_{ik}=\Box\theta_{ik}-\partial_i\partial^m\theta_{mk}-\partial_k\partial^m\theta_{mi}+\gamma_{ik}\partial^m\partial^l\theta_{ml}, \tag{3·9}$$

$$\theta_{ik}=\phi_{ik}-\frac{1}{2}\gamma_{ik}\phi. \tag{3·10}$$

In the covariant derivatives this operator becomes

$$J^{mn}=D_\mu D_\sigma(\tilde{g}^{\sigma n}\gamma^{\mu m}+\tilde{g}^{\sigma m}\gamma^{\mu n}-\tilde{g}^{mn}\gamma^{\mu\sigma}-\tilde{g}^{\mu\sigma}\gamma^{mn}). \tag{3·11}$$

Or, in another form

$$J^{mn}=-\gamma^{\mu\sigma}D_\mu D_\sigma\tilde{g}^{mn}+D^mD_\sigma\tilde{g}^{\sigma n}+D^nD_\sigma\tilde{g}^{\sigma m}-\gamma^{mn}D_\mu D_\sigma\tilde{g}^{\mu\sigma}.$$

One may easily get convinced that the following identity takes place:

$$D_mJ^{mn}\equiv 0. \tag{3·12}$$

We have called it basic, since it is of fundamental importance for the construction of the relativistic theory of gravity.

§ 5. Equations of the relativistic theory of gravitation

In the present section we construct relativistic equations for matter and for gravitation field in the framework of the special relativity and geometrization principle.

The relation between the effective metric of a field Riemannian space and gravitation field may be chosen in the simplest form:

$$\tilde{g}^{ik}=\sqrt{-g}\,g^{ik}=\sqrt{-\gamma}\,\gamma^{ik}+\sqrt{-\gamma}\,\phi^{ik}. \tag{4·1}$$

In our theory, the field variable of the gravitation field is the tensor ϕ^{ik}. Let us assume that in the general case the gravitation field has only spins 2 and 0 and that a free gravitation field has spin 2. Such physical requirements, as is seen in § 4, lead in Galilean coordinates to the following four equations for the gravitational field:

$$\partial_i\phi^{ik}=\partial_i\tilde{g}^{ik}=0. \tag{4·2}$$

Similar conditions were sometimes used earlier[2),7),10)] in the GR as a specific class of harmonic coordinate conditions to solve "island-type" problems. It was Fock[7] who paid special attention to the importance of harmonic coordinate conditions in solving "island" problems. He wrote as follows: "The above remarks concerning the privileged character of the harmonic system of coordinates should not be understood, in any case, as some kind of prohibition of the use of other coordinate systems. Nothing is more alien to our point of view than such an interpretation". He went on: "Likewise, in the case of the Theory of Gravitation, the existence of harmonic coordinates, defined apart from a Lorentz transformation, though a fact of primary theoretical and practical importance, does not in any way preclude the use of other, non-harmonic, coordinate systems." From the point of view of our theory, when solving "island" problems, Fock was unconsciously working with ordinary Galilean coordinates in an inertial reference frame, which are, as known from the special relativity, definitely singled out coordinates. Therefore, in Fock's calculations for "island" systems the harmonic conditions were not the coordinate ones as he thought them to be but, as will be seen from our theory, field equations in Galilean coordinates of an inertial reference frame. It was due to this very fact that they played such an important role in his specific calculations, which neither Fock nor others even suspected.

Thus, Fock considered harmonic conditions no more than as priviledged coordinate conditions applicable for "island"-type problems only. This is quite natural, since he and all his eminent predecessors were captured by the Riemannian geometry, which basically gave no possibility to make a deeper insight into the problem. In order to make step ahead and impose these conditions as universal covariant field equations, it was necessary to give up the ideology of the GR, leave the maze of the Riemannian geometry and apply a special principle of relativity in defiance of the GR, as well as to introduce the concepts of a gravitation field as a Faraday-Maxwell physical field, possessing some energy and momentum. All of it was translated into reality in our theory, with the choice of coordinates being arbitrary and set only by the metric tensor γ^{ik} of Minkowski space, as is generally accepted in elementary particle theory. As for Eq. (4·2), in our theory they are comprehensive and universal because of being the gravitation field equations. They have nothing to do with the choice of coordinates. In Minkowski space these equations are written in the covariant form

$$\sqrt{-\gamma}\, D_i \phi^{ik} = D_i \tilde{g}^{ik} = 0\,. \tag{4·3}$$

Judging by § 3, we conclude that these field equations exclude automatically spins 1 and 0′ from a gravitational tensor field. Thus, we have already costructed four covariant equations (4·3) for the fourteen unknown variables of a gravitation field and of matter. To construct other ten equations, we draw a simple, but farreaching analogy with an electromagnetic field. As is known, Maxwell electrodynamic equations may be written in the covariant form as follows:

$$\gamma^{ik} D_i D_k A^\nu - D^\nu D_k A^k = \frac{4\pi}{c} j^\nu\,. \tag{4·4}$$

Due to conservation of electromagnetic current one has

$$D_\nu j^\nu = 0\,. \tag{4·5}$$

The lhs of Maxwell equations (4·4) is constructed in such a way that its divergence is identically zero. It follows that spin 0 is excluded from the vector field with spins 1 and 0. By analogy with electrodynamics, we shall construct equations for the tensor gravitation field. The only tensor of the second rank which conserves is the energy-momentum tensor of matter and a gravitation field in Minkowski space.

$$D_\mu(t_g^{\mu\nu}+t_M^{\mu\nu})=0\,. \qquad (4\cdot6)$$

Therefore, it would be natural to choose it as a complete source of the gravitation field. We have already obtained four equations (4·3) describing a gravitation field. Therefore, in order to achieve a complete description of the unknown ten field variables and of four variables of matter, it is necessary to write ten more equations. As proved in § 4, the only identically conserved tensor linear operator is $J^{\mu\nu}$. Hence, by analogy with electrodynamics, for the remaining ten equations one should necessarily take the following ones:

$$D_\sigma D_\tau[\tilde{g}^{\sigma\nu}\gamma^{\tau\mu}+\tilde{g}^{\sigma\mu}\gamma^{\tau\nu}-\tilde{g}^{\mu\nu}\gamma^{\sigma\tau}-\tilde{g}^{\tau\sigma}\gamma^{\mu\nu}]=\lambda(t_g^{\mu\nu}+t_M^{\mu\nu})\,. \qquad (4\cdot7)$$

This type of equations guarantees automatically that the conservation law for the energy-momentum tensor of matter and a gravitational field will be fulfilled in Minkowski space:

$$D_\mu(t_g^{\mu\nu}+t_M^{\mu\nu})=0\,. \qquad (4\cdot8)$$

Besides, as a consequence, the covariant equation for matter conservation in Riemannian space is also satisfied:

$$\nabla_\mu T^{\mu\nu}=0\,. \qquad (4\cdot9)$$

Since the divergence of the lhs of Eq.(4·7) is identically zero, four components corresponding to spins 1 and 0′ are excluded from the tensor field ϕ^{ik} containing representations with spins 2, 1, 0, 0′. Systems of Eqs. (4·3) and (4·7) determine completely the unknown variables of matter and a gravitation field. The principal problem is to find out if there is a Lagrangian density for the gravitation field with spins 2 and 0 which would automatically lead to equations (4·7) due to the least action principle. The most general Lagrangian density of the gravitation field ϕ^{ik} describing spins 2 and 0 quadratic in the first-order derivatives has the form

$$L_g=a\tilde{g}_{km}\tilde{g}_{nq}\tilde{g}^{lp}D_l\tilde{g}^{kq}D_p\tilde{g}^{mn}+b\tilde{g}_{kq}D_m\tilde{g}^{pq}D_p\tilde{g}^{mk}+c\tilde{g}_{mk}\tilde{g}_{nq}\tilde{g}^{lp}D_l\tilde{g}^{mk}D_p\tilde{g}^{nq}\,. \qquad (4\cdot10)$$

The convolution of covariant derivatives taken with respect to Minkowski metric is realized with the help of the effective metric tensor g^{ik} of Rimannian space. In this way we guarantee the action of the gravitation field on itself which is similar to the one on matter. The constants a, b and c in the Lagrangian are arbitrary so far.

Because of the geometrization principle, the Lagrangian density of matter is

$$L_M=L_M(\tilde{g}^{ik},\,\Phi_A)\,. \qquad (4\cdot11)$$

In virtue of the least action principle, the system of equations for a gravitation field is as follows:

$$\frac{\delta L_g}{\delta\phi^{ik}}+\frac{\delta L_M}{\delta\phi^{ik}}=\frac{\delta L_g}{\delta\tilde{g}^{ik}}+\frac{\delta L_M}{\delta\tilde{g}^{ik}}=0\,. \qquad (4\cdot12)$$

Here relation(4·1) was taken into account. For this system of equations to be presented in the form of(4·7), it is necessary to choose the constants a, b and c for the Lagrangian density in a definite and unique way.

This suggests that the Lagrangian of the gravitation field with spins 2 and 0 in Minkowski space is determined unambiguously. In order to make such a choice of the coefficients a, b and c let us calculate the density of the energy momentum tensor for matter and a gravitational field.

Let us introduce the notations

$$\tilde{t}^{mn}_M=-2\frac{\delta L_M}{\delta\tilde{\gamma}_{mn}}, \qquad \tilde{t}_M=-2\tilde{\gamma}_{mn}\frac{\delta L_M}{\delta\tilde{\gamma}_{mn}}. \tag{4·13}$$

Taking into account the definitions of $\tilde{\gamma}^{mn}$ we find

$$\frac{\partial\tilde{\gamma}^{mn}}{\partial\gamma^{pq}}=\sqrt{-\gamma}\left(\delta^{mn}_{pq}-\frac{1}{2}\gamma^{mn}\gamma_{pq}\right). \tag{4·14}$$

Similarly

$$\frac{\partial\tilde{g}^{mn}}{\partial g^{pq}}=\sqrt{-g}\left(\delta^{mn}_{pq}-\frac{1}{2}g^{mn}g_{pq}\right). \tag{4·15}$$

Using equality (4·14) we obtain

$$t^{mn}_M=\frac{1}{\sqrt{-\gamma}}\left[\tilde{t}^{mn}_M-\frac{1}{2}\tilde{\gamma}^{mn}\tilde{t}_M\right]. \tag{4·16}$$

Calculating the variation of the total Lagrangian over $\tilde{\gamma}_{mn}$ and with regard for the field equations

$$\frac{\delta L_g}{\delta\tilde{g}^{mn}}+\frac{\delta L_M}{\delta\tilde{g}^{mn}}=0, \tag{4·17}$$

one finds

$$t^{mn}=2bJ^{mn}+D_\sigma\{(2a+b)[H^{\sigma n}_\nu\gamma^{\nu m}+H^{\sigma m}_\nu\gamma^{\nu n}-H^{mn}_\nu\gamma^{\nu\sigma}]-2(a+2c)\tilde{g}^{\nu\sigma}\tilde{g}_{\lambda k}\gamma^{mn}D_\nu\tilde{g}^{\lambda k}\}, \tag{4·18}$$

where

$$H^{\sigma n}_\nu=(\tilde{g}^{\sigma l}D_l\tilde{g}^{kn}+\tilde{g}^{nl}D_l\tilde{g}^{k\sigma})\tilde{g}_{k\nu}. \tag{4·19}$$

In order that no new equations on the field ϕ^{ik} would arise from the equality

$$D_m t^{mn}=0,$$

which would otherwise lead to a re-definition of the system of equations, it is necessary and sufficient for the coefficients a, b and c to satisfy the following conditions:

$$a=-b/2, \qquad c=b/4. \tag{4·20}$$

Thus, with such a choice of the constants one comes to the identity

$$D_m t^{mn}\equiv 0$$

which was enclosed in Eq. (4·7). With regard to the choice of the coefficients as in (4·20), expression (4·18) takes the form

$$D_\sigma D_\nu(\tilde{g}^{\sigma n}\gamma^{\nu m}+\tilde{g}^{\sigma m}\gamma^{\nu n}-\tilde{g}^{mn}\gamma^{\nu\sigma}-\tilde{g}^{\nu\sigma}\gamma^{mn})=\frac{1}{2b}(t_g^{mn}+t_M^{mn}), \tag{4·21}$$

which coincides with Eq.(4·17) obtained earlier by analogy with electrodynamics, provided one puts

$$2b=1/\lambda .$$

So, the only Lagrangian density of the form

$$L_g=-\frac{1}{32\pi}[\tilde{g}_{kq}D_m\tilde{g}^{pq}D_p\tilde{g}^{km}-\frac{1}{2}\tilde{g}_{km}\tilde{g}_{nq}\tilde{g}^{lp}D_l\tilde{g}^{kq}D_p\tilde{g}^{mn}+\frac{1}{4}\tilde{g}_{km}\tilde{g}_{nq}\tilde{g}^{lp}D_l\tilde{g}^{km}D_p g^{nq}] \tag{4·22}$$

leads to the field equations in the form of (4·21). According to the correspondence principle, the constant λ is chosen to be equal

$$\lambda=-16\pi . \tag{4·23}$$

This Lagrangian density can be presented in the form

$$L_g=\frac{1}{32\pi}[\tilde{G}^l_{mn}D_l\tilde{g}^{mn}-\tilde{g}^{mn}\tilde{G}^k_{mk}\tilde{G}^l_{nl}], \tag{4·24}$$

where the tensor of rank three, $\tilde{G}^k_{ml}$, is defined by the formula

$$\tilde{G}^k_{ml}=\frac{1}{2}\tilde{g}^{pk}(D_m\tilde{g}_{lp}+D_l\tilde{g}_{mp}-D_p\tilde{g}_{lm}) . \tag{4·25}$$

It may also be presented in the form

$$L_g=-\frac{1}{16\pi}\sqrt{-g}\,g^{mn}[G^k_{ml}G^l_{nk}-G^l_{mn}G^k_{lk}] . \tag{4·26}$$

Such a Lagrangian was considered for the first time by Rosen in Ref. 2). In (4·26) the tensor of the third rank, G^k_{ml}, is equal to

$$G^k_{ml}=\frac{1}{2}g^{pk}(D_m g_{pl}+D_l g_{pm}-D_p g_{lm}) . \tag{4·27}$$

With consideration for Eq. (4·3), the complete system of equations for matter and gravitational field will be[6)]

$$\gamma^{\mu\sigma}D_\mu D_\sigma\tilde{g}^{mn}=16\pi(t_g^{mn}+t_M^{mn}) , \tag{4·28}$$

$$D_m\tilde{g}^{mn}=0 , \tag{4·29}$$

or in the Galilean coordinate system

$$\Box\tilde{g}^{mn}=16\pi(t_g^{mn}+t_M^{mn}) , \qquad \partial_m\tilde{g}^{mn}=0 .$$

Should we confine ourselves only to the first system of equations, (4·28), then the separation of the Riemannian space metric into the Minkowski space metric and tensor gravitational field would be of a conditional character and would not have any physical meaning. The second system of four field equations, (4·29), separates decisively all that relates to the inertia forces from all that is connected with gravitational field. The two systems of Eqs. (4·28) and (4·29), are generally covariant. Corresponding physical conditions are imposed, as usual, within a given, for example, Galilean coordinate system on the behaviour of gravitational field. In the framework of GR, one cannot formulate the conditions for the metric g^{ik} remaining in Riemannian space since the asymptotics of the metric always depends on the choice of the three-dimensional coordinate system. It should also be noted that the equations of matter motion are contained in the given system of equations. The density for the energy-momentum tensor of gravitational field in Minkowski space is equal to

$$t_g^{mn}=-\frac{1}{16\pi}J^{mn}-\frac{\sqrt{-\gamma}}{8\pi}\gamma^{mp}\gamma^{nq}\Big[R_{pq}-\frac{1}{2}\sqrt{\frac{g}{\gamma}}\gamma_{pq}R\Big] \tag{4·30}$$

for Lagrangian density (4·22). Here, as we can see, there automatically appears the second-rank curvature tensor R_{pq} in Riemannian space. Similarly, the tensor density of matter energy-momentum in Minkowski space is equal to

$$t_M^{mn}=\sqrt{\frac{\gamma}{g}}\gamma^{mp}\gamma^{nq}\Big[T_{pq}-\frac{1}{2}g_{pq}T\Big]+\frac{1}{2}\gamma^{mn}T \tag{4·31}$$

for Lagrangian density (4·11).

In deriving expressions (4·30) and (4·31) we used the identity

$$\frac{\delta L_M}{\delta\gamma_{pq}}=\frac{\delta L_M}{\delta g^{mn}}\frac{\partial g^{mn}}{\partial\gamma_{pq}}, \tag{4·32}$$

and the equality

$$\frac{\partial g^{mn}}{\partial\gamma_{pq}}=-\sqrt{\frac{\gamma}{g}}\gamma^{pi}\gamma^{qk}\Big[\delta_{ik}^{mn}-\frac{1}{2}g_{ik}g^{mn}\Big]-\frac{1}{2}\gamma^{pq}g^{mn}, \tag{4·33}$$

which directly follows from the expression for coupling (4·1). Putting the expressions for the energy-momentum tensors of matter and gravitational field in field equations (4·7), we transform them to the form of Hilbert-Einstein equations

$$\sqrt{-g}\,R_{\mu\nu}=8\pi\Big(T_{\mu\nu}-\frac{1}{2}g_{\mu\nu}T\Big). \tag{4·34}$$

Thus, system of Eq. (4·7) is equivalent to the system of Hilbert-Einstein equations. The complete system of equations for matter and gravitational field, (4·28) and (4·29), is equivalent to the system of equations

$$\sqrt{-g}\,R_{\mu\nu}=8\pi\Big(T_{\mu\nu}-\frac{1}{2}g_{\mu\nu}T\Big), \tag{4·35}$$

$$D_\mu\tilde{g}^{\mu\nu}=0. \tag{4·36}$$

It is worth mentioning that Eq. (4·36) is general and universal since these are field equations describing a gravitational field with spins 2 and 0. The choice of a reference frame (or coordinate system) is determined by the metric tensor $\gamma^{\mu\nu}$ of Minkowski space. Hence, Eq. (4·36) do not impose any restrictions on the choice of a coordinate system. Consequently, the system of Eq. (4·36) excludes spins 1 and 0′ in the density of tensor field ϕ^{ik}, leaving only spins 2 and 0. The required six components of gravitational field, corresponding to these spins, and four components of matter are defined from field equations (4·28) or from their equivalent Hilbert-Einstein equations (4·35). The system of equations for gravitational field, (4·28) and (4·29), may be expressed in a somewhat different form through the Hilbert energy-momentum tensor density in Riemannian space. However, for this purpose we will have to obtain some relations if use is made of a specific expression for the Lagrangian density of gravitational field obtained by us earlier, (4·24):

$$L_g=\frac{1}{32\pi}[\tilde{G}^l_{mn}D_l\tilde{g}^{mn}-\tilde{g}^{mn}\tilde{G}^k_{mk}\tilde{G}^l_{nl}],$$

where

$$\tilde{G}^\lambda_{\mu\nu}=\frac{1}{2}\tilde{g}^{\lambda\sigma}(D_\mu\tilde{g}_{\sigma\nu}+D_\nu\tilde{g}_{\sigma\mu}-D_\sigma\tilde{g}_{\mu\nu}).$$

Using these expressions, we calculate then the density of the tensor of the third rank, $K^{\alpha\lambda}_\mu$, with formula (2·10).

With account for the equality

$$\frac{\partial L_g}{\partial(D_\lambda\tilde{g}^{\mu\nu})}=\frac{1}{16\pi}[\tilde{G}^\lambda_{\mu\nu}+\frac{1}{2}\tilde{g}^{\lambda\sigma}\tilde{g}_{\mu\nu}\tilde{G}^\tau_{\tau\sigma}],$$

we shall have

$$16\pi K^{\alpha\lambda}_\mu=[\tilde{g}^{\alpha\nu}\tilde{G}^\lambda_{\mu\nu}-\tilde{g}^{\lambda\nu}\tilde{G}^\alpha_{\mu\nu}]+\tilde{g}^{\sigma\tau}\tilde{\gamma}_{\sigma\mu}[\tilde{\gamma}^{\lambda\tau}\tilde{G}^\alpha_{\tau\nu}-\tilde{\gamma}^{\alpha\tau}\tilde{G}^\lambda_{\tau\nu}]+\tilde{\gamma}_{\mu\tau}\tilde{G}^\tau_{\sigma\nu}[\tilde{g}^{\alpha\nu}\tilde{\gamma}^{\lambda\sigma}-\tilde{g}^{\lambda\nu}\tilde{\gamma}^{\alpha\sigma}].$$

Putting the value for $\tilde{G}^\lambda_{\mu\nu}$ into this equality, we shall obtain

$$16\pi K^{\alpha\lambda}_\mu=\tilde{g}_{\mu\nu}D_\sigma(\tilde{g}^{\lambda\sigma}\tilde{g}^{\alpha\nu}-\tilde{g}^{\alpha\sigma}\tilde{g}^{\lambda\nu})-\gamma_{\mu\nu}D_\beta[\tilde{g}^{\lambda\beta}\gamma^{\alpha\nu}+\tilde{g}^{\alpha\nu}\gamma^{\lambda\beta}-\tilde{g}^{\alpha\beta}\gamma^{\lambda\nu}-\tilde{g}^{\lambda\nu}\gamma^{\alpha\beta}]. \quad (4·37)$$

Using this expression and definition (2·13), we shall have for $\overset{\circ}{t}{}^\alpha_{g\mu}$ the following expression:

$$\overset{\circ}{t}{}^\alpha_{g\mu}=\tau_\mu{}^\alpha-\frac{1}{16\pi}D_\lambda\sigma^{\alpha\lambda}_\mu-\frac{1}{16\pi}\gamma_{\mu\nu}J^{\alpha\nu}, \quad (4·38)$$

where the density of the asymmetric tensor $\sigma^{\alpha\lambda}_\mu$ is equal to

$$\sigma^{\alpha\lambda}_\mu=\tilde{g}_{\mu\nu}D_\sigma[\tilde{g}^{\lambda\sigma}\tilde{g}^{\alpha\nu}-\tilde{g}^{\alpha\sigma}\tilde{g}^{\lambda\nu}], \quad (4·39)$$

and the well-known expression (3·11) is denoted through $J^{\alpha\nu}$.

Expression (4·38) is right the one we will need in what follows. In preparing for further calculations we derive now an identity often used in literature. In Galilean coordinates, the Lagrangian density of gravitational field, (4·24), takes the form

$$L_g=\frac{1}{32\pi}[\tilde{G}^l_{mn}\partial_l\tilde{g}^{mn}-\tilde{g}^{mn}\tilde{G}^k_{mk}\tilde{G}^l_{nl}],$$

where in this case

$$\tilde{G}^\lambda_{\mu\nu}=\frac{1}{2}\tilde{g}^{\lambda\sigma}(\partial_\mu\tilde{g}_{\sigma\nu}+\partial_\nu\tilde{g}_{\sigma\mu}-\partial_\sigma\tilde{g}_{\mu\nu}).$$

The quantities $\tilde{G}^\lambda_{\mu\nu}$ are the tensors of the third rank with respect to the linear transformation of the coordinates. Therefore L_g will be a scalar density with respect to the same transformations. From the invariance of action with respect to the linear transformations, we have

$$\delta J_g = \int_\Omega d^4x \left[\partial_\lambda J^\lambda + \frac{\delta L_g}{\delta \tilde{g}^{\mu\nu}} \delta_L \tilde{g}^{\mu\nu} \right] = 0 . \qquad (4\cdot40)$$

Here

$$J^\lambda = -\xi^\alpha \tau_\alpha^\lambda + \tilde{K}_\mu^{\alpha\lambda} \partial_\alpha \xi^\mu , \qquad (4\cdot41)$$

where the density of the canonical tensor τ_α^λ is equal to

$$\tau_\alpha^\lambda = -\delta_\alpha^\lambda L_g + \partial_\alpha \tilde{g}^{\mu\nu} \frac{\partial L_g}{\partial(\partial_\lambda \tilde{g}^{\mu\nu})} , \qquad (4\cdot42)$$

and the density of the tensor of the third rank, $\tilde{K}_\mu^{\alpha\lambda}$, is in this case

$$\tilde{K}_\mu^{\alpha\lambda} = 2\frac{\partial L_g}{\partial(\partial_\lambda \tilde{g}^{\mu\nu})} \tilde{g}^{\alpha\nu} - \delta_\mu^\alpha \frac{\partial L_g}{\partial(\partial_\lambda \tilde{g}^{\sigma\tau})} \tilde{g}^{\sigma\tau} . \qquad (4\cdot43)$$

Putting into (4·40), the formula for the variation of $\delta_L \tilde{g}^{\mu\nu}$ with respect to linear transformations in virtue of the arbitrariness of the volume Ω, we obtain the identity

$$\xi^\alpha \left[\partial_\lambda \tau_\alpha^\lambda + \frac{\delta L_g}{\delta \tilde{g}^{\mu\nu}} \partial_\alpha \tilde{g}^{\mu\nu} \right] + \partial_\alpha \xi^\mu \left[\tau_\mu^\alpha - \partial_\lambda \tilde{K}_\mu^{\alpha\lambda} - 2\frac{\delta L_g}{\delta \tilde{g}^{\mu\nu}} \tilde{g}^{\alpha\nu} + \delta_\mu^\alpha \frac{\delta L_g}{\delta \tilde{g}^{\sigma\tau}} \tilde{g}^{\sigma\tau} \right] = 0 . \qquad (4\cdot44)$$

Whereof there straightforwardly follow the identities

$$\partial_\lambda \tau_\alpha^\lambda = -\frac{\delta L_g}{\delta \tilde{g}^{\mu\nu}} \partial_\alpha \tilde{g}^{\mu\nu} , \qquad (4\cdot45)$$

$$\tau_\mu^\alpha - \partial_\lambda \tilde{K}_\mu^{\alpha\lambda} = 2\frac{\delta L_g}{\delta \tilde{g}^{\mu\nu}} \tilde{g}^{\alpha\nu} - \delta_\mu^\alpha \frac{\delta L_g}{\delta \tilde{g}^{\sigma\tau}} \tilde{g}^{\sigma\tau} . \qquad (4\cdot46)$$

Since

$$\frac{\delta L_g}{\delta \tilde{g}^{\mu\nu}} = -\frac{1}{16\pi} R_{\mu\nu}, \qquad (4\cdot47)$$

then from identity (4·46) we have

$$\tau_\mu^\alpha - \partial_\lambda \tilde{K}_\mu^{\alpha\lambda} = -\frac{\sqrt{-g}}{8\pi} \left[R_\mu^\alpha - \frac{1}{2} \delta_\mu^\alpha R \right] . \qquad (4\cdot48)$$

Considering the equality

$$\frac{\partial L_g}{\partial(\partial_\lambda \tilde{g}^{\mu\nu})} = \frac{1}{16\pi} \left[\tilde{G}_{\mu\nu}^\lambda + \frac{1}{2} \tilde{g}^{\lambda\tau} \tilde{g}_{\mu\nu} \tilde{G}_{\tau\sigma}^\sigma \right]$$

and the expressions for $\tilde{G}_{\mu\nu}^\lambda$ in Galilean coordinates we will come to

$$16\pi \tilde{K}_\mu^{\alpha\lambda} = \partial_\nu (\delta_\mu^\lambda \tilde{g}^{\alpha\nu} - \delta_\mu^\nu \tilde{g}^{\alpha\lambda}) + \sigma_\mu^{\alpha\lambda} , \qquad (4\cdot49)$$

where $\sigma_\mu^{\alpha\nu}$ is the density of the antisymmetric tensor,

$$\sigma_\mu^{\alpha\lambda} = -\sigma_\mu^{\lambda\alpha} = \tilde{g}_{\mu\nu} \partial_\sigma (\tilde{g}^{\lambda\sigma} \tilde{g}^{\alpha\nu} - \tilde{g}^{\alpha\sigma} \tilde{g}^{\lambda\nu}) . \qquad (4\cdot50)$$

Putting the expression for $\tilde{K}_\mu^{\alpha\lambda}$ into (4·48), we shall obtain the identity

$$\tau_\mu{}^\alpha - \frac{1}{16\pi}\partial_\lambda \sigma_\mu^{\alpha\lambda} = -\frac{\sqrt{-g}}{8\pi}\left[R_\mu{}^\alpha - \frac{1}{2}\delta_\mu{}^\alpha R\right]. \tag{4·51}$$

In the curvature tensor, one can always identically replace, leaving it unchanged, the conventional derivatives by the covariant ones in the Minkowski metric, therefore expression (4·51) may be presented in the covariant form:

$$\tau_\mu{}^\alpha - \frac{1}{16\pi}D_\lambda \sigma_\mu^{\alpha\lambda} = -\frac{\sqrt{-g}}{8\pi}\left[R_\mu{}^\alpha - \frac{1}{2}\delta_\mu{}^\alpha R\right]. \tag{4·52}$$

In this case the canonical tensor density in (4·52) will be equal to expression (2·3):

$$\tau_\mu{}^\alpha = -\delta_\mu{}^\alpha L_g + \frac{\partial L_g}{\partial(\partial_\lambda g^{\mu\nu})}D_\alpha g^{\mu\nu},$$

where the Lagrangian density L_g is already presented in terms of the derivatives covariant in the Minkowski metric, (4·24).

Using identity (4·52) we may present the expression for $\overset{\circ}{t}{}^\alpha_{g\mu}$ (4·38) in the form

$$\overset{\circ}{t}{}^\alpha_{g\mu} = -\frac{\sqrt{-g}}{8\pi}\left[R_\mu{}^\alpha - \frac{1}{2}\delta_\mu{}^\alpha R\right] - \frac{1}{16\pi}\gamma_{\mu\nu}J^{\alpha\nu}. \tag{4·53}$$

As we have already established, the system of equations of matter and gravitational field, (4·28) and (4·29), is equivalent to the system of Eqs. (4·35) and (4·36). With the help of expression (4·53) the system of equations of matter and gravitational field may also be rewritten in another equivalent form:[6)]

$$\gamma_{\nu\sigma}\gamma^{\alpha\beta}D_\alpha D_\beta \tilde{g}^{\mu\sigma} = 16\pi[T_\nu{}^\mu + \overset{\circ}{t}{}^\mu_{g\nu}], \tag{4·54}$$

$$D_\mu \tilde{g}^{\mu\nu} = 0. \tag{4·55}$$

Here $T_\nu{}^\mu$ is the Hilbert energy-momentum tensor density (1·6) for the matter in Riemannian space. It is quite obvious that in virtue of (4·54) and (4·55) the conservation law for energy-momentum tensor of matter and gravitational field has the form

$$D_\mu(T_\nu{}^\mu + \overset{\circ}{t}{}^\mu_{g\nu}) = 0. \tag{4·56}$$

The covariant matter conservation law in Riemannian space may identically be presented in the form

$$\nabla_\mu T_\nu{}^\mu = \partial_\mu T_\nu{}^\mu - \frac{1}{2}T^{\mu\sigma}\partial_\nu g_{\mu\sigma} \equiv D_\mu T_\nu{}^\mu - G^\lambda_{\mu\nu}T_\lambda{}^\mu = 0. \tag{4·57}$$

From comparison of (4·56) and (4·57), we have

$$G^\lambda_{\mu\nu}T_\lambda{}^\mu = -D_\mu \overset{\circ}{t}{}^\mu_{g\nu}. \tag{4·58}$$

As seen from this expression, the matter acquires energy and momentum right from gravitational field, the total energy-momentum tensor of matter and gravitational field being always conserved rigorously. The construction of RTG on the basis of Minkowski space and geometrization principle allowed us to deal only with covariant quantities at every stage of our reasonings. Here we give briefly some of our results following from this theory. The post-Newtonian parameters in it are equal to

$$\gamma=\beta=1\,,\quad \alpha_1=\alpha_2=\alpha_3=\xi_1=\xi_2=\xi_3=\xi_4=\xi_w=0\,.$$

Therefore the theory describes the whole set of gravitation experiments available at present. It strictly satisfies the correspondence principle and predicts the existence of gravitational waves in the Faraday-Maxwell spirit, which possess both energy and momentum. In virtue of the geometrization principle, the curvature of the effective field Riemannian space appears as a result of gravitational field action on matter. Gravitational waves in this theory propagate as electromagnetic ones. Since our theory is based on the special relativity principle, the inertial mass of the system is well defined

$$m=\int d^3x\,[t_g^{00}+t_M^{00}]$$

and is a scalar with respect to the three-dimensional transformations of coordinates. The quantity

$$p^\nu=\int d^3x\,[t_g^{0\nu}+t_M^{0\nu}]$$

is an energy-momentum four-vector with respect to any coordinate transformations; similarly, the angular momentum is also a tensor with respect to any coordinate transformations in four-dimensional Minkowski space. It may also be shown that for any island static system, the inertial mass is exactly equal to its active gravitating mass. The given theory provides a prediction of an extraordinary force, it leads to a strictly definite development of the Universe.[6] According to it, the Universe is not closed, it is "flat" in virtue of Eq (4·29),

$$ds^2=c^2d\tau^2-V(\tau)(dx^2+dy^2+dz^2)\,.$$

The Universe expansion is defined by the function $V(\tau)$ which is easily calculated from the field equations. Equations (4·28) and (4·29) make us easily convinced that the total energy density of matter and gravitational field is always zero at any instant of time of the Universe development. The nowaday density of all the forms of matter, ρ_o, should be equal to its critical density ρ_c,

$$\rho_o=\rho_c=3H^2/8\pi G\,,$$

where H is the Hubble constant.

The Universe expands infinitely, the deceleration parameter q being equal to

$$q=1/2\,.$$

The Universe age T is defined by the formula

$$T=2/3H\,.$$

The density of the observed usual matter is much smaller than the critical density ρ_c. Since the given theory is based on fundamental general physical principles and is constructed unambiguously, its predictions on the character of the Universe development are so much general that they necessarily demand an obligatory existence of a "missing" mass is some form of matter. Hence, in the Universe there must exist a "missing" mass so that the total density of matter be equal to the critical value of ρ_c.

Acknowledgements

The authors express their deep gratitude to V. A. Ambartsumyan, N. N. Bogolubov, A. A. Vlasov, S. S. Gershtein and A. N. Tavkhelidze for valuable discussions.

References

1) A. A. Logunov et al., Theor. Math. Phys. **40** (1979), 291.
V. I. Denisov and A. A. Logunov, Theor. Math. Phys. **50** (1982), 3; *Elementary Particle and Atomic Nucleus Physics* (1982), v. 13, part 3, p. 757; *Modern Problems of Mathematics* (Moscow, VINITI of Acad. of Sciences, USSR, 1982), v. 21.
2) N. Rosen, Phys. Rev. **57** (1940), 147; Ann. of Phys. **22** (1963), 1.
A. Papapetrou, Proc. Roy. Irish Acad. **A52** (1948), 11.
S. Gupta, Proc. Roy. Soc. **A65** (1952), 608.
W. Thirring, Ann. of Phys. **16** (1961), 69.
3) H. Poincaré, *Present and Future of Mathematical Physics*(Bulletin des Sciences Mathematiques, December, 1904), v. 28, ser. 2, p. 302; *The Monist*(January, 1905), v. XV, No. 1; *Relativity Principle*, ed. A. A. Tyapkin (Atomizdat, Moscow, 1973).
4) A. A. Logunov, *Lectures on the Theory of Relativity*(Moscow Univ. 1984), (In Russian).
5) A. Einstein, *Collected works* (Nauka, Moscow, 1965), v. 1.
W. Pauli, *Theory of Relativity* (Pergamon Press, 1965).
C. Møller, *The Theory of Relativity* (Clarendon Press, Oxford, 1972).
L. I. Mandelstam, *Lectures on Optics, Relativity Theory and Quantum Mechanics* (Nauka, Moscow, 1972), p. 218.
6) A. A. Logunov and A. A. Vlasov, *Minkowski Space as the Basis of the Physical Gravitation Theory* (Moscow Univ., Moscow, 1984); *TMF* (1984), v. 60, p. 319; *Spherically Symmetric Solution in Gravitation Theory Based on Minkowski Space* (Moscow Univ., Moscow, 1984); *TMF* (1984), v. 60, p. 163.
A. A. Vlasov, A. A. Logunov and M. A. Mestvirishvili, IHEP Preprint 84-156 (Serpukhov, 1984).
7) V. A. Fock, *Theory of Space, Time and Gravitation* (Pergamon Press, London, 1959).
8) V. I. Ogievetsky and I. V. Polubarinov, Ann. of Phys. **35** (1965), 167; JINR Preprint P-2106 (1965).
9) C. Fronsdal, Sup. Nuovo Cim. **9** (1958), 416.
K. J. Barnes, J. Math. Phys. **6** (1965), 788.
10) De-Donder, *La Gravifique Einsteinienne.* (Paris, 1921); *Theorie des Champs Gravifiques* (Paris, 1926)
V. A. Fock, J. of Phys. **1** (1939), 81; Rev. Mod. Phys. **29** (1957), 235.

Authors' comments:
'.....I think it appropriate to say that in the paper "Relativistic Theory of Gravitation" some aspects that are of great importance and appeared in the later works had not been taken into consideration. Therefore, the book "The Theory of Gravity" Moscow NAUKA, 2001 (also see gr-qc/0210005 V2 21 Oct 2002), is more useful and complete.'

Yang-Mills Gravity: A Union of the Einstein-Grossmann Metric with Yang-Mills Tensor Fields in Flat Spacetime with Translation Symmetry

Jong-Ping Hsu
Department of Physics,
University of Massachusetts Dartmouth
North Dartmouth, MA 02747-2300, USA

Based on spacetime translation gauge symmetry, we discuss a new Yang-Mills gravity for interacting spinor and tensor fields in flat spacetime. The theory possesses both a Yang-Mills action with quadratic gauge curvature of spin-2 fields and a particle action involving an effective Einstein-Grossmann metric for the motions of matter. The effective Einstein-Grossmann metric is naturally and completely determined by the translation gauge symmetry in flat spacetime. Such a Yang-Mills gravity holds in both inertial and non-inertial frames. Its interaction Lagrangian of the spin-2 gauge field has the maximum coupling of 4-vertex and is much simpler than that in the Hilbert-Einstein action of metric fields. The results of the post-Newtonian approximation are consistent with gravitational experiments. Yang-Mills gravity in flat spacetime can shed light on renormalizable quantum gravity.

In general relativity, the structure of couplings for $g_{\mu\nu}$ is very complicated and also has non-trivial difficulties in both technical and conceptual aspects from the viewpoint of quantum field theory. However, as a classical field theory, one of the strengths of general relativity lies in its successful equation of motions for objects and light rays, based on the Einstein-Grossmann metric $g_{\mu\nu}dx^\mu dx^\nu$. [1] As for renormalizable quantum field theory, Yang-Mills fields have the best track record in theory and experiment, provided the underlying spacetime is flat. In this paper, we show that Yang-Mills fields with spacetime translation gauge symmetry have special features which provide a natural union of the Einstein-Grossmann metric and the gravitational Yang-Mills field. The union implies: (i) The framework is applicable to all general frames of reference (both inertial and non-inertial) in which the spacetime is characterized by the vanishing Riemann-Christoffel curvature tensor. (ii) The 'effective Einstein-Grossmann metric' originates physically from a spin-2 Yang-Mills field in flat spacetime.

The formulations for electromagnetic and Yang-Mills fields associated with internal gauge groups have been developed extensively. They are based on the replacement $\partial_\mu \to \partial_\mu + igB_\mu^a\tau^a$, where τ^a is the constant matrix representations of the gauge groups which have little to do with external spacetime. For external gauge groups related to spacetime, e.g., the de Sitter group or the Poincaré group, the gauge invariant Lagrangian involving fermions turns out to be richer in content. [2]

In this paper, we concentrate on a specific simple external gauge group of translations T(4) in flat spacetime. The translation group T(4) is the Abelian subgroup of the Poincaré group. It is particularly interesting because it is the minimal gauge group related to the conserved energy-momentum tensor which couples to a spin-2 field $\phi_{\mu\nu}$. However, the generators of the translational group are the displacement operators, $p_\mu = i\partial_\mu$ (c=$\hbar$=1) in inertial frames. In a general frame (inertial or non-inertial) with a metric tensor $P_{\mu\nu}$, we replace ∂_μ by D_μ, i.e., the partial covariant derivative with respect to the Levi-Civita connection (or the metric tensor $P_{\mu\nu}$) in flat spacetime. Thus, the replacement in Yang-Mills gravity takes a different form: $D_\mu \to D_\mu - ig\phi_{\mu\nu}p^\nu \equiv J_{\mu\nu}D^\nu$, where $D_\mu = P_{\mu\nu}D^\nu$, in such a gauge theory. The generators of this non-compact translation group do not have a constant matrix representation. It is precisely this unique property that leads naturally to an 'effective Einstein-Grossmann metric' in flat spacetime. Such an effective metric emerges from the Lagrangian of matter fields (as shown in equations (5) and (6) below). Furthermore, the displacement operator of the translation gauge group dictates that the coupling constant g in $J_{\mu\nu}$ must have the dimension of length and that the interaction cannot have both attractive and repulsive forces, in sharp contrast to the dimensionless real coupling constants in electrodynamics and other Yang-Mills theories associated with internal gauge groups.

The new formalism of external gauge symmetry for translations in flat spacetime leads to a gauge-invariant action involving fermions. It suggests (a) the massless Yang-Mills spin-2 field in flat spacetime as the gravitational gauge field [3], (b) a new gravitational gauge equation in both inertial and non-inertial frames and (c) an effective metric $G_{\mu\nu}dx^\mu dx^\nu$ for the motion of classical objects. In the post-Newtonian approximation, the present gauge field equation is consistent with classical tests such as the perihelion shift of the Mercury and the time delay of radar echoes.

Let us consider the local spacetime translation with an arbitrary infinitesimal vector gauge-function $\Lambda^\mu(x)$,

$$x^\mu \to x'^\mu = x^\mu + \Lambda^\mu(x), \qquad x \equiv x^\lambda = (w, x, y, z). \tag{1}$$

The basic point is that this transformation has a dual interpretation: (i) a shift of the spacetime coordinates by an infinitesimal vector gauge-function $\Lambda^\mu(x)$, and (ii) an arbitrary infinitesimal transformation. These two mathematical implications of the transformation (1) dictates the following gauge transformation of spacetime translations for physical quantities in the Lagrangian of fields (e.g., $L_{\psi\phi}\sqrt{-P}$ in (12) below):

$$Q^{\mu_1...\mu_m}_{\alpha_1...\alpha_n}(x) \to Q^{\$\mu_1...\mu_m}_{\alpha_1...\alpha_n}(x) = \left(Q^{\nu_1...\nu_m}_{\beta_1...\beta_.}(x) - \Lambda^\lambda(x)\partial_\lambda Q^{\nu_1...\nu_m}_{\beta_1...\beta_n}(x)\right) \tag{2}$$

$$\times \frac{\partial x'^{\mu_1}}{\partial x^{\nu_1}} \cdots \frac{\partial x'^{\mu_m}}{\partial x^{\nu_m}} \frac{\partial x^{\beta_1}}{\partial x'^{\alpha_1}} \cdots \frac{\partial x^{\beta_n}}{\partial x'^{\alpha_n}},$$

where $\mu_1, \nu_1, \alpha_1, \beta_1$, etc. are spacetime indices.

The spinor $\psi(x)$ and $D_\mu\psi$ transform, by definition, as a scalar function $Q(x)$ and a covariant vector $Q_\mu(x)$ respectively under the translation gauge transformation. We use D_μ to denote the partial covariant derivative with respect to the metric tensor $P_{\mu\nu}(x)$ in a general reference frame. The translation gauge transformation (2) is formally similar to the Lie variations in the coordinate transformation.

For an example of $P_{\mu\nu}(x)$ in the flat spacetime of a general reference frame, let us consider the accelerated Wu transformation between an inertial frame $F_I(w_I, x_I, y_I, z_I)$ and a non-inertial frame $F(w, x, y, z)$. Suppose F_I is at rest and the frame F moves in the x-direction with an initial velocity β_o and a constant linear acceleration α_o [4]:

$$w_I = \gamma u(x + \frac{1}{\alpha_o\gamma_o^2}) - \frac{\beta_o}{\alpha_o\gamma_o}, \qquad x_I = \gamma(x + \frac{1}{\alpha_o\gamma_o^2}) - \frac{1}{\alpha_o\gamma_o}, \quad y_I = y, \quad z_I = z, \tag{3}$$

where $u = \beta_o + \alpha_o w$, $\gamma = 1/\sqrt{1-(\beta_o+\alpha_o w)^2}$ and $\gamma_o = 1/\sqrt{1-\beta_o^2}$. In the zero acceleration limit, $\alpha_o \to 0$, the Wu transformations (3) reduce to the usual 4-dimensional transformations, $w_I = \gamma_o(w + \beta_o x)$, $x_I = \gamma_o(x + \beta_o w)$, $y_I = y$, $z_I = z$. In the special case $\beta_o = 0$, the Wu transformations (3) are the same as the accelerated Møller transformations, [5] provided one makes a change of the time variable, $w \to w^* = (1/\alpha_o)tanh^{-1}(\alpha_o w)$. The Wu transformation preserves the spacetime interval ds^2:

$$ds^2 = dw_I{}^2 - dr_I{}^2 = W^2dw^2 - dr^2 = P_{\mu\nu}dx^\mu dx^\nu, \quad W^2 = \gamma^4\left(\frac{1}{\gamma_o^2} + \alpha_o x\right)^2. \tag{4}$$

All constant-linear-acceleration frames of reference have the metric tensor of the form $P_{\mu\nu} = (W^2, -1, -1, -1)$. [6] The existence of the finite Wu transformations (3) implies that the spacetime of the constant-linear-acceleration frames is flat, i.e., having zero Riemann-Christoffel curvature tensor. The metric tensor $P_{\mu\nu}$ for a general frame of reference with zero Riemann-Christoffel curvature tensor may be called the Poincaré metric tensor. In the limit of zero acceleration, $\alpha_o \to 0$, $P_{\mu\nu}$ in (4) reduces to the Minkowski metric tensor $\eta_{\mu\nu}$ of inertial frames.

In order to see the connection of such a spin-2 field and the gravitational field, let us consider the kinetic term in the Lagrangian of a scalar field Φ: $(1/2)P^{\mu\nu}D_\mu\Phi D_\nu\Phi$. In the presence of the spin-2 field $\phi_{\mu\nu}$, the translation gauge symmetry dictates the replacement,

$$D_\mu\Phi \to (P_{\mu\alpha} + g\phi_{\mu\alpha})D^\alpha\Phi \equiv J_{\mu\alpha}D^\alpha\Phi.$$

Thus, we have

$$\frac{1}{2}P^{\mu\nu}(D_\mu\Phi)(D_\nu\Phi) \to \frac{1}{2}P^{\mu\nu}J_{\mu\alpha}(D^\alpha\Phi)J_{\nu\beta}(D^\beta\Phi) = \frac{1}{2}G_{\alpha\beta}(D^\alpha\Phi)(D^\beta\Phi), \tag{5}$$

$$G_{\alpha\beta} = P^{\mu\nu}J_{\mu\alpha}J_{\nu\beta} = P_{\alpha\beta} + 2g\phi_{\alpha\beta} + g^2\phi_{\alpha\lambda}\phi_{\beta\sigma}P^{\lambda\sigma}, \qquad D_\mu\Phi = \partial_\mu\Phi,$$

where $G_{\alpha\beta}$ formally resembles a metric tensor. One can see from (5) that it appears as if the geometry of the spacetime is changed from an Euclidean spacetime to a non-Euclidean spacetime due to the presence of the spin-2 field $\phi_{\mu\nu}$. As suggested by the action for a quantum field involving the kinetic term in (5), the action S_p for the motion of classical objects effectively is assumed to take the form,

$$S_p = -\int m ds_{ei}, \qquad ds_{ei}^2 = G_{\mu\nu}dx^\mu dx^\nu, \tag{6}$$

where $G_{\mu\nu}dx^\mu dx^\nu$ denotes the effective Einstein-Grossmann metric for motions of classical objects.

This action (6) for particles suggests a simple and natural union of Einstein-Grossmann metric for motions of classical objects and Yang-Mills fields for gravity with flat-spacetime translation gauge group: Namely, the spin-2 gauge field and its interaction with fermion matter actually takes place in flat spacetime; only the equation of motion of classical objects is derived from the classical action S_p which happens to have a form similar to the Einstein-Grossmann metric $g_{\mu\nu}dx^\mu dx^\nu$. We stress that the effective metric tensor $G_{\mu\nu}$ in (5) and (6) is completely determined by the Yang-Mills action with translation gauge symmetry.

The present theory of Yang-Mills gravity is formulated on the basis of the translation gauge symmetry and the postulate of the effective metric tensor in (6) for the motion of a classical particle in such a spin-2 fields. However, from the field-theoretic viewpoint, the real physical spacetime is still flat and the fundamental metric tensor is still $P_{\mu\nu}$ in general frames of reference. Thus, $G_{\mu\nu}$ in (6) is treated as merely an "effective metric tensor" for the motion of a classical object in the presence of the spin-2 gauge field, in the sense that the 4-dimensional effective interval is $ds_{ei}^2 = G_{\mu\nu}dx^\mu dx^\nu$ in the action S_p for classical objects such as planets, stars or light rays. We note that the Poincaré metric tensor $P_{\mu\nu}$ is a purly geometrical property of spacetime and does not contain physical field $\phi_{\mu\nu}$. This property is important because it enables the Yang-Mills gravity to have a very simple coupling, namely, the maximum coupling is a 4-vertex (in Feymann diagrams).

The translational gauge invariance requires that a symmetric spin-2 field, $\phi_{\mu\nu} = \phi_{\nu\mu}$, must couple to the fermion field ψ via the energy-momentum tensor. We postulate the following gauge-invariant fermion action S_ψ in a general frame:

$$S_\psi = \int L_\psi\sqrt{-P}d^4x, \quad L_\psi = \left(i\overline{\psi}\Gamma^\mu\Delta_\mu\psi - m\overline{\psi}\psi\right), \tag{7}$$

$$[\Gamma^\mu, \Gamma^\nu] = 2P^{\mu\nu}(x), \quad \Gamma^\mu = \gamma^a e^\mu_a, \quad P = det(P_{\mu\nu}),, \tag{8}$$

$$\Delta_\mu \psi = J_{\mu\nu} D^\nu \psi, \quad D^\nu \psi = \partial^\nu \psi, \quad J_{\mu\nu} = P_{\mu\nu} + g\phi_{\mu\nu}, \tag{9}$$

where γ^a and e^μ_a are respectively the constant Dirac matrices and the tetrads, while Δ_μ is the translational gauge-covariant derivative. If one considers $e^\mu{}_b J_{\mu\nu}$ in the action (7) as an 'effective tetrad', $E_{a\nu} = e^\mu{}_a J_{\mu\nu}$, then the corresponding "effective metric tensor" is $E_{a\alpha} E_{b\beta} \eta^{ab} = e^\mu{}_a J_{\mu\alpha} e^\nu{}_b J_{\nu\beta} \eta^{ab} = P^{\mu\nu} J_{\mu\alpha} J_{\nu\beta} = G_{\alpha\beta}$, which is the same as that in (5). We may remark that the Lagrangian L_ψ in (7) can be symmetrized.

The translational gauge transformation (2) leads to $L_\psi \to L_\psi - \Lambda^\lambda \partial_\lambda L_\psi$. Therefore, the fermion Lagrangian $\sqrt{-P} L_\psi$ only changes by a divergence,

$$\sqrt{-P} L_\psi \to \sqrt{-P} L_\psi - \partial_\lambda (\Lambda^\lambda L_\psi \sqrt{-P}), \tag{10}$$

where we have used the gauge transformation,

$$\sqrt{-P} \to \sqrt{-P^\$} = [(1 - \Lambda^\sigma \partial_\sigma)\sqrt{-P}](1 - \partial_\lambda \Lambda^\lambda).$$

This gauge transformation can be obtained by using

$$P_{\mu\nu} - \Lambda^\lambda \partial_\lambda P_{\mu\nu} - P_{\mu\beta} \partial_\nu \Lambda^\beta - P_{\alpha\nu} \partial_\mu \Lambda^\alpha = [(1 - \Lambda^\sigma \partial_\sigma) P_{\alpha\beta}](\delta^\alpha_\mu - \partial_\mu \Lambda^\alpha)(\delta^\beta_\nu - \partial_\nu \Lambda^\beta)$$

for an infinitesimal vector function $\Lambda^\mu(x)$. The last term in (10) does not contribute to the action (7) because of Gauss' theorem. Thus, the action S_ψ is translational gauge invariant.

As usual, the gauge-curvature (or the gauge-field strength) is given by the commutator of two gauge covariant derivatives $\Delta_\mu = J_{\mu\nu} D^\nu$:

$$\left[J_{\mu\nu} D^\nu, J_{\alpha\beta} D^\beta\right] = \left(J_{\mu\nu}(D^\nu J_{\alpha\beta}) - J_{\alpha\nu}(D^\nu J_{\mu\beta})\right) D^\beta \equiv C_{\mu\alpha\beta} D^\beta. \tag{11}$$

The gauge-curvature $C_{\mu\alpha\beta}$ satisfies the algebraic identities, (i) cyclicity, $C_{\mu\alpha\beta} + C_{\alpha\beta\mu} + C_{\beta\mu\alpha} = 0$ and (ii) antisymmetry, $C_{\mu\alpha\beta} = -C_{\alpha\mu\beta}$, which can be directly verified. By examining all possible terms of the scalar quadratic gauge-curvature such as $C_{\mu\beta\alpha} C^{\beta\alpha\mu}, C_{\alpha\mu}{}^\alpha C^{\beta\mu}{}_\beta$, etc. by using the anti-symmetric property $C_{\mu\alpha\beta} = -C_{\alpha\mu\beta}$ and by interchanging dummy indices, we find that there are only two independent quadratic terms, $C_{\mu\alpha\beta} C^{\mu\beta\alpha}$ and $C_{\mu\alpha}{}^\alpha C^{\mu\beta}{}_\beta$, for the T(4) gauge curvature. All other terms can be expressed in terms of them, e.g., $C_{\mu\alpha\beta} C^{\mu\alpha\beta} = 2C_{\mu\alpha\beta} C^{\mu\beta\alpha}$. Thus, the T(4) gauge invariant Lagrangian density is naturally assumed to be a linear combination of these two independent terms, $C_{\mu\alpha\beta} C^{\mu\beta\alpha} + f C_{\mu\alpha}{}^\alpha C^{\mu\beta}{}_\beta$. For simplicity, the constant f is chosen to be $f = -1$. That is, one has the simplest linearized equation when a gauge condition is imposed. In this case, the linearized gauge field equation turns out to be the same as the linearlized Einstein's equation in general relativity. Thus, we postulate the action for the spinor-tensor fields to be

$$S_{\psi\phi} = \int L_{\psi\phi} \sqrt{-P} d^4x, \quad L_{\psi\phi} = -\frac{1}{2g^2}\left(C_{\mu\alpha\beta} C^{\mu\beta\alpha} - C_{\mu\alpha}{}^\alpha C^{\mu\beta}{}_\beta\right) + L_\psi, \tag{12}$$

where $L_{\psi\phi}\sqrt{-P}$ only changes by a divergence under the gauge transformation (2). Note that the quadratic gauge-curvature in (12) can be written as $(-1/2g^2)(\frac{1}{2} C_{\mu\alpha\beta} C^{\mu\alpha\beta}$ - $C_{\mu\alpha}{}^\alpha C^{\mu\beta}{}_\beta)$.

For simplicity, let us choose inertial frames rather than general non-inertial frames in the following discussions of experimental implications. One of the ways to do this is to consider the case,

$$P_{\mu\nu} = \eta_{\mu\nu}, \qquad J_{\mu\nu} = \eta_{\mu\nu} + g\phi_{\mu\nu}, \tag{13}$$

and all tensors indices are raised and lowered by $\eta_{\mu\nu}$. For weak fields, the linearized gauge equation for spin-2 field can be derived from (12):

$$[\partial_\lambda \partial^\lambda \phi_{\mu\nu} - \partial^\lambda \partial_\mu \phi_{\lambda\nu} - \eta_{\mu\nu}\partial_\lambda(\partial^\lambda \phi^\alpha_\alpha - \partial^\alpha \phi^\lambda_\alpha) + \partial_\mu \partial_\nu \phi^\alpha_\alpha - \partial_\mu \partial^\beta \phi_{\nu\beta}] = -gT_{\mu\nu}, \tag{14}$$

where μ and ν are understood to be symmetrized, and $T_{\mu\nu} = \overline{\psi} i\gamma_\mu \partial_\nu \psi$ is the energy-momentum tensor of a free fermion. In the calculations of the effective metric tensor and the gravitational quadrupole radiation, it is not necessary to symmetrized μ and ν explicitly in the gauge-field equation. The gauge symmetry allows us to impose the usual gauge condition for the massless spin-2 field in (14),

$$\partial^\mu \phi_{\mu\nu} = \frac{1}{2}\partial_\nu \phi^\lambda_\lambda. \tag{15}$$

The gauge equation (14) takes a simple form:

$$\partial_\lambda \partial^\lambda \phi_{\mu\nu} - \frac{1}{2}\eta_{\mu\nu}\partial_\lambda \partial^\lambda \phi^\alpha_\alpha = -gT_{\mu\nu}, \tag{16}$$

which turns out to be formally the same as Einstein's equation for weak gravitational field.

The classical particle action $S_p = -\int m ds_{ei}$, with the effective interval $(ds_{ei})^2 = G_{\mu\nu}dx^\mu dx^\nu$, leads to the variation

$$\delta S_p = -mG_{\mu\nu}(dx^\mu/ds_{ei})\delta x^\nu, \tag{17}$$

if one considers only the actual path with one of its end point variable. [8] Thus we have the Hamilton-Jacobi equation for a particle with mass m,

$$I^{\mu\nu}(\partial_\mu S)(\partial_\nu S) - m^2 = 0, \qquad I^{\mu\nu}G_{\nu\lambda} = \delta^\mu_\lambda. \tag{18}$$

In the Newtonian limit, (16) with the only non-vanishing component, $T_{00} = m\delta^3(\mathbf{r})$, leads to $g\phi_{00} = -g^2 m/(8\pi r)$. Also, I^{00} in the Hamilton-Jacobi equation (18) has the usual result $I^{00} = 1 + 2Gm/r$ in this limit, where G is the gravitational constant. Based on these results, together with $I^{\mu\nu}$ in (18) and $G_{\mu\nu}$ in (6), we obtain the relation

$$g = \sqrt{8\pi G}, \qquad \left(g\phi_{00} = g\phi_{11} = -\frac{Gm}{r}, etc.\right) \tag{19}$$

by solving (16) with the spherical coordinate, $x^\mu = (w, r, \theta, \phi)$ to the first order approximation.

In order to show that the theory is viable beyond the first-order approximation (19), let us compare the result (18) with the perihelion shift of the Mercury, [7, 8] which is sensitive to the coefficient appearing in the second-order term in I^{00} or G_{00}. We solve the non-linear gauge field equations by the method of successive approximation and carry out the related post-Newtonian approximation to the second order. To accomplish this calculation and to have a well-defined gauge fields, we include the usual gauge condition (15), so that the total Lagrangian takes the form in a general frame:

$$L_t\sqrt{-P} = \{L_{\psi\phi} - (\xi/(2g^2)[D_\mu J^{\mu\alpha} - \frac{1}{2}D^\alpha J^\mu_\mu][D^\nu J_{\nu\alpha} - \frac{1}{2}D_\alpha J^\nu_\nu]\}\sqrt{-P}. \tag{20}$$

We derive the following exact gauge field equations in a general frame:

$$H^{\mu\nu} + \xi A^{\mu\nu} = -g^2 T^{\mu\nu}, \tag{21}$$

$$H^{\mu\nu} \equiv D_\lambda(J^\lambda_\rho C^{\rho\mu\nu} - J^\lambda_\alpha C^{\alpha\beta}{}_\beta P^{\mu\nu} + C^{\mu\beta}{}_\beta J^{\nu\lambda})$$
$$-C^{\mu\alpha\beta} D^\nu J_{\alpha\beta} + C^{\mu\beta}{}_\beta D^\nu J^\alpha_\alpha - C^{\lambda\beta}{}_\beta D^\nu J^\mu_\lambda,$$

$$A^{\mu\nu} = D^\mu\left(D^\lambda J_\lambda{}^\nu - \frac{1}{2}D^\nu J^\lambda_\lambda\right) - \frac{1}{2}P^{\mu\nu}\left(D^\alpha D^\lambda J_{\lambda\alpha} - \frac{1}{2}D^\alpha D_\alpha J^\lambda_\lambda\right),$$

where μ and ν are understood to be symmetrized.

It suffices to consider an inertial frame and a static and spherically symmetric system, in which gauge fields are produced by an object at rest with mass m. Based on symmetry considerations, [9] the non-vanishing components of the exterior solutions $\phi_{\mu\nu}(r)$ are $\phi_{00}(r), \phi_{11}(r), \phi_{22}(r)$ and $\phi_{33}(r) = \phi_{22} sin^2\theta$, where $x^\mu = (w, r, \theta, \phi)$.

To solve the static gauge field, let us write

$$J_{00} = J^0_0 = S, \quad -J_{11} = J^1_1 = R, \quad -\frac{J_{22}}{r^2} = J^2_2 = -\frac{J_{33}}{r^2 sin^2\theta} = J^3_3 = T. \tag{22}$$

For $(\mu,\nu) = (0,0), (1,1), (2,2), (3,3)$, the gauge-field equation (21) can be written respectively as

$$\frac{d}{dr}\left(R^2\frac{dS}{dr}\right) + \frac{2R^2}{r}\frac{dS}{dr} + \left(R\frac{d}{dr} + \frac{dR}{dr} + \frac{2}{r}R\right)\left(-R(\frac{dS}{dr} + 2\frac{dT}{dr}) - \frac{2}{r}T^2 + \frac{2}{r}TR\right) \tag{23}$$
$$+ \ \xi\left[\frac{1}{4}\frac{d^2}{dr^2}(S - R + 2T) + \frac{1}{2r}\frac{d}{dr}(S - 3R + 4T) - \frac{1}{r^2}(R - T)\right] = 0,$$

$$R(\frac{dS}{dr})^2 + 2r^3\frac{d}{dr}\frac{T}{r}\left[\frac{R}{r}\frac{d}{dr}\frac{T}{r} + \frac{T^2}{r^3}\right] + \left(\frac{dS}{dr} + 2\frac{dT}{dr} - \frac{2T}{r}\right) \times \tag{24}$$
$$\left(-R(\frac{dS}{dr} + 2\frac{dT}{dr}) - \frac{2}{r}T^2 + \frac{2}{r}TR\right)$$
$$+ \ \xi\left[\frac{1}{4}\frac{d^2}{dr^2}(S - R + 2T) - \frac{1}{2r}\frac{d}{dr}(S + R) + \frac{3}{r^2}(R - T)\right] = 0,$$

$$\left(R\frac{d}{dr} + \frac{dR}{dr} + \frac{5R}{r} - \frac{2T}{r}\right)\left[\frac{R}{r}\frac{d(T/r)}{dr} + \frac{T^2}{r^3}\right] + \tag{25}$$
$$+ \left[\frac{1}{r^2}(R\frac{d}{dr} + \frac{dR}{dr}) + \frac{3R}{r^3} - \frac{2T}{r^3}\right]\left[-R\left(\frac{dS}{dr} + 2\frac{dT}{dr}\right) - \frac{2}{r}T^2 + \frac{2}{r}TR\right]$$

$$+ \ \xi\left[\frac{1}{4r^2}\frac{d^2}{dr^2}(S-R+2T)-\frac{1}{r^3}\frac{d}{dr}(R-T)+\frac{1}{r^4}(R-T)\right]=0,$$

after tedious but straightforward calculations. The equation for the component $(\mu,\nu)=(3,3)$ is the same as equation (25).

The second-order approximation of the fields in (23)-(25) are solved and, hence, (22) leads to the following results:

$$S=1+\frac{a_o}{r}+\frac{a_o^2}{2r^2},\qquad R=1-\frac{a_o}{r}-\frac{a_o^2}{r^2}\left(\frac{1}{2}+\frac{2}{\xi}\right),\tag{26}$$

$$T=1-\frac{a_o}{r}+\frac{2a_o^2}{r^2}\left(\frac{1}{\xi}-1\right).$$

The parameter a_o can be determined by the Newtonian approximation of $g\phi_{00}$

$$S=1+g\phi_{00}=1-\frac{Gm}{r},\qquad a_o=-Gm.\tag{27}$$

Equations (22) and (26) lead to the second order approximation of gauge field $g\phi_{00}$:

$$g\phi_{00}=-\frac{Gm}{r}+\frac{G^2m^2}{2r^2},\qquad g\phi_{11}=-\frac{Gm}{r}+\frac{G^2m^2}{r^2}\left(\frac{1}{2}+\frac{2}{\xi}\right),\tag{28}$$

$$g\phi_{22}=-r^2\left[\frac{Gm}{r}+\frac{2G^2m^2}{r^2}\left(\frac{1}{\xi}-1\right)\right],\qquad g\phi_{33}=g\phi_{22}sin^2\theta;$$

which is consistent with the ξ-independent first order solutions in (19).

Based on (21) and (6) with $P_{\mu\nu}=(1,-1,-r^2,-r^2sin^2\theta)$, we obtain the effective metric tensor,

$$G_{00}(r)=1-\frac{2Gm}{r}+\frac{2G^2m^2}{r^2},\qquad G_{11}(r)=-\left[1+\frac{2Gm}{r}-\frac{4G^2m^2}{r^2\xi}\right],\tag{29}$$

$$G_{22}(r)=-r^2\left[1+\frac{2Gm}{r}+\frac{G^2m^2}{r^2}\left(\frac{4}{\xi}-3\right)\right],\qquad G_{33}(r)=G_{22}(r)sin^2\theta;$$

which are ξ-dependent, just like the gauge field $\phi_{\mu\nu}$. In order to see the ξ-independent physical results and their agreement with gravitational experiments, let us carry out the expansion to the second order in all components G_{00}, G_{11}, G_{22} and G_{33} in the usual spherical coordinate. For any given value of ξ, we can make a change of variable: $\rho^2=r^2(1+2Gm/r+[G^2m^2/r^2][4/\xi-3])$, where the zeroth and the first orders terms are independent of ξ. We obtain the following effective metric tensors $G_{\mu\nu}(\rho)$:

$$G_{00}(\rho)=1-\frac{2Gm}{\rho}+\frac{A_0}{\rho^2},\qquad G_{11}(\rho)=-\left(1+\frac{2Gm}{\rho}-\frac{2G^2m^2}{\rho^2}\right),\tag{30}$$

$$G_{22}(\rho)=-\rho^2,\qquad G_{33}(\rho)=-\rho^2sin^2\theta,\qquad A_0=0,$$

which are ξ-independent. This is interesting because, although Yang-Mills gravity is formulated in flat spacetime, the spacetime translation gauge symmetry leads to an effective metric tensor $G_{\mu\nu}(\rho)$ in (5), (6) and (30) to the second order. Note that only the second order term in $G_{11}(\rho)$, i.e., $2G^2m^2/\rho^2$, differs from the corresponding term in Einstein's theory. As we shall see below, this difference is too small to be detected with the available apparatus. [9] We have shown that the effective metric tensor $G_{\mu\nu}(\rho)$ in the spherical coordinates are independent of the gauge parameter ξ in the post-Newtonian approximation.

The inverse of non-vanishing components of $G_{\mu\nu}$ are given by

$$
\begin{gathered}
I^{00}(\rho) = \frac{1}{G_{00}(\rho)} = 1 + \frac{2Gm}{\rho} + \frac{4G^2m^2}{\rho^2}, \\
I^{11}(\rho) = \frac{1}{G_{11}(\rho)} = -\left[1 - \frac{2Gm}{\rho} + \frac{6G^2m^2}{\rho^2}\right], \\
I^{22}(\rho) = \frac{1}{G_{22}(\rho)} = \frac{-1}{\rho^2}, \qquad I^{33}(\rho) = \frac{1}{G_{33}(\rho)} = \frac{-1}{\rho^2 sin^2\theta}.
\end{gathered}
\tag{31}
$$

Let us choose $\theta = \pi/2$, the Hamilton-Jacobi equation for a planet with mass m_p has the following form:

$$
I^{00}(\rho)\left(\frac{\partial S}{\partial w}\right)^2 + I^{11}(\rho)\left(\frac{\partial S}{\partial \rho}\right)^2 + I^{22}(\rho)\left(\frac{\partial S}{\partial \phi}\right)^2 - m_p^2 = 0. \tag{32}
$$

By the general procedure for solving this equation, we look for an action S in the form $S = -E_o w + M\phi + f(\rho)$. The action S is found to be

$$
S = -E_o w + M\phi + \int \frac{1}{\sqrt{|I^{11}|(\rho)}}\sqrt{E_o^2 I^{00}(\rho) - m_p^2 - \frac{M^2}{\rho^2}}d\rho, \tag{33}
$$

where E_o and M are respectively constant energy and angular momentum. As usual, the trajectory is determined by $\partial S/\partial M = constant$:

$$
\phi = \int \frac{(M/\rho^2)d\rho}{\sqrt{E_o^2 I^{00}|I^{11}| - m_p^2|I^{11}| - M^2|I^{11}|/\rho^2}}, \tag{34}
$$

$$
I^{00}|I^{11}| = 1 + \frac{6G^2m^2}{\rho^2},
$$

where I^{00} and I^{11} are given by (31). This term $I^{00}|I^{11}|$ differs from the corresponding term in Einstein's theory, in which $g^{00}|g^{11}| = 1$. For the approximate trajectory, we write (34) as a differential equation with $\sigma = 1/\rho$, We have

$$
M^2\left(\frac{d\sigma}{d\phi}\right)^2 = E_o^2 I^{00}|I^{11}| - m_p^2|I^{11}| - M^2\sigma^2|I^{11}|. \tag{35}
$$

We differentiate equation (35) with respect to ϕ and obtain

$$
\frac{d^2\sigma}{d\phi^2} = \frac{1}{P} - \sigma(1 - Q) + 3Gm\sigma^2, \tag{36}
$$

$$P = \frac{M^2}{m_p^2 Gm}, \qquad Q = \frac{6Gm}{P}\left(\frac{E_o^2 - m_p^2}{m_p^2}\right).$$

Einstein's theory does not have this type of correction term Q because it has the relation $g^{00}|g^{11}| = 1$. The new correction term Q is of the order of $(Gm/P)\beta^2$ which is the result of the relation $I^{00}|I^{11}| > 1$ in (34). This correction is extremely small and undetectable because of the velocity β of the planet is very small in comparison with the speed of light.

To see the effect of this correction Q to the perihelion shift, we solve (36) by a change of variable $\overline{\sigma} = \sigma(1-Q)$. We can write equation (36) as

$$\frac{d^2\overline{\sigma}}{d\overline{\phi}^2} = \frac{1}{P} - \overline{\sigma} + \frac{3Gm}{(1-Q)^2}\overline{\sigma}^2, \qquad \overline{\phi} = \phi\sqrt{1-Q}). \tag{37}$$

By the usual successive approximation, [7] we obtain the solution

$$\overline{\sigma} = \frac{1}{P}\left[1 + e\ cos\left(\overline{\phi}(1 - \frac{3Gm}{P(1-Q)^2})\right)\right]. \tag{38}$$

It can be written as

$$\sigma = \frac{1}{P(1-Q)}\left[1 + e\ cos\left(\phi(1 - \frac{\alpha}{P})\right)\right], \tag{39}$$

$$\alpha = 3Gm\left(1 + \frac{E_o^2 - m_p^2}{m_p^2}\right),$$

where e is the eccentricity. The major semiaxis a can be expressed in terms of e and P: $a = P/(1-e^2)$. The perihelion shift for one revolution of the planet is give by

$$\delta\phi = \frac{2\pi\alpha}{P} = \frac{6\pi Gm}{P}\left[1 + \frac{E_o^2 - m_p^2}{m_p^2}\right], \tag{40}$$

where the second term shows the difference between the present Yang-Mills gravity and Einstein's theory. This result shows that the observable perihelion shift is independent of the gauge parameter ξ which appears in the second order approximation of the solution of the spin-2 gauge field $g\phi$. Since the observational accuracy of the perihelion shift of the Mercury is about 1 %, the prediction (40) of Yang-Mills gravity can be tested only when $(E_o^2 - m_p^2)/m_p^2 \approx \beta^2 \approx 0.01$. Thus, it is not possible to test the small correction in (40) of Yang-Mills gravity in the solar system.

For the bending of light, the eikonal equation (47) with $I^{\mu\nu}$ given by (31), one obtains the trajectory of the ray which is the same as (36) with $m_p \to 0$ and E_0 replaced by $\omega_o = -\partial\psi/\partial w$ (c=1). Following the usual procedure, we obtain

$$\Delta\phi \approx \frac{4Gm\omega_o}{M}\left(1 + \frac{3\pi Gm\omega_o}{4M}\right), \tag{41}$$

where the second term in the bracket is negligible. So there is no observable difference between Yang-Mills gravity and Einstein's theory in the known experiments.

If one compares the fermion equation, $(i\Gamma^{\mu}(P_{\mu\nu}+g\phi_{\mu\nu})D^{\nu}-m)\psi=0$, derived from (12) with the Dirac equation in quantum electrodynamics [i.e., $(i\gamma^{\mu}\partial_{\mu}-e\gamma^{\mu}A_{\mu}-m)\psi=0$], one can see a distinct difference: Namely, the kinematic term $i\gamma^{\mu}\partial_{\mu}$ and the electromagnetic coupling term $e\gamma^{\mu}A_{\mu}$ have a different relative sign, if one takes the complex conjugate of the Dirac equations. This implies the presence of both repulsive and attractive forces between two charges. However, there is no change in the relative sign of the kinematical term and the spin-2 coupling term in our fermion equation when one takes the complex conjugate. This provides a natural explanation that the gravitational force is only attractive between two fermions, in contrast to that in electrodynamics.

The Yang-Mills gravity with the translational symmetry [10] has a well-defined conservation law for the energy-momentum tensor, just as that in ordinary field theory. It is believed that such a spacetime gauge theory can shed light on quantum gravity because (i) the maximum interaction vertex is 4-vertex, just as that in the usual Yang-Mills theory with internal Lie groups, and (ii) it is based on gauge symmetry, which could minimize the ultraviolet divergences. In light of these discussions, it is possible to understand gravity based on the spacetime gauge theory with the translational symmetry in general frames of reference.

The author would like to thank Zhenhua Ning for his help. The work is supported in part by the Potz Science Fund and the Jing Shin Research Fund of the UMass Dartmouth Foundation.

Note added. The energy-momentum tensor of gravitation $t_{\mu\nu}$ is defined by the field equation (21) (with $\xi=0$) written in the following form, $D^{\lambda}D_{\lambda}\phi^{\mu\nu}=-g(T^{\mu\nu}+t^{\mu\nu})$, in a general frame. Using the usual approximations and gauge condition (15), we can calculate the average energy-momentum of a gravitational plane wave and the power. For example, the power P_o emitted per unit solid angle in the direction $\mathbf{x}/|\mathbf{x}|$ can be written as

$$\frac{dP_o}{d\Omega}=\frac{G\omega^2}{\pi}\left(T^{\lambda\rho}(\mathbf{k},\omega)T^{*}_{\lambda\rho}(\mathbf{k},\omega)-\frac{1}{2}T(\mathbf{k},\omega)T^{*}(\mathbf{k},\omega)\right),$$

where $T(\mathbf{k},\omega)$ is defined as follows: [9] Suppose one observes this radiation in the wave zone, one can write the polarization tensor in terms of the Fourier transform of $T_{\mu\nu}$:

$$e_{\mu\nu}(\mathbf{x},\omega)=\frac{-g}{4\pi r}[T_{\mu\nu}(\mathbf{k},\omega)-\frac{1}{2}\eta_{\mu\nu}T(\mathbf{k},\omega)],\qquad T=T^{\lambda}_{\lambda},$$

$$T_{\mu\nu}(\mathbf{k},\omega)\equiv\int d^3\mathbf{x}'T_{\mu\nu}(\mathbf{x}',\omega)]exp(-i\mathbf{k}\cdot\mathbf{x}'),$$

where the polarization tensor $e_{\mu\nu}(\mathbf{x},\omega)$ is defined by the relation: $\phi_{\mu\nu}(\mathbf{x},\mathbf{t})\approx[e_{\mu\nu}(\mathbf{x},\omega)\times exp(-ik_{\lambda}x^{\lambda})+c.c.]$. The approximate result for the power emitted per solid angle in Yang-Mills gravity turns out to be the same as that obtained in Einstein's theory. [9]

References

[1] A. Einstein and M. Grossmann, Z. Math. Physik, **62** 225 (1913). See also F. J. Dyson, Bull. Am. Math. Soc. 78 (1972) 635, and S. Weinberg, *Gravitation and Cosmology* (John Wiley and Sons, 1972) pp. 285-289.

[2] Jong-Ping Hsu, Phys. Lett., 119B (1982) 328. Such a gauge theory predicts a new gravitational spin force produced by fermion spin densities, in addition to the usual 'gravitational force' produced by the mass density. The usual Yang-Mills-type formalism (with Faddeev-Popov's gauge compensation terms) for internal gauge groups is more difficult to be applied to this case.

[3] The idea of an 'effective Riemannian spacetime' due to the presence of a symmetric spin-2 field in flat spacetime was extensively discussed by Logunov and others. Their theory has a different gauge transformation and a completely different Lagrangian. See A. A. Logunov, *The Theory of Gravity* (Trans. by G. Pontecorvo, Moscow, Nauka, 2001) and references therein. For a discussion of the spin-2 field, see also H. van Dam and M. Veltman, Nucl. Phys. B22 (1970) 397, and S. Weinberg, *The Quantum Theory of Fields. vol. I. Foundations* (Cambridge Univ. Press, 1995) pp. 246-255.

[4] Jong-Ping Hsu and Leonardo Hsu, Nuovo Cimento B, 112 (1997) 575 and Chin. J. Phys. 35 (1997) 407. Jong-Ping Hsu, *Einstein's Relativity and Beyond - New Symmetry Approaches*, (World Scientific, Singapore, 2000), Chapters 21-23. Daniel Schmidt and Jong-Ping Hsu, Intern. J. Modern Phys. A (2005, to be published).

[5] C. Møller, Danske Vid. Sel. Mat.-Fys. 20 (1943) No. 19; Ta-You Wu and Y. C. Lee, Intern. J. Theoretical Phys. 5 (1972) 307.

[6] For arbitrary linear accelerations with limiting 4-dimensional symmetry, one has $P_{\mu\nu}dx^\mu dx^\nu = W^2 dw^2 + 2U dw dx - dr^2$. See J. P. Hsu, Chin. J. Phys. 40 (2002) 265.

[7] L. Landau and E. Lifshitz, *The Classical Theory of Fields* (trans. by M. Hamermesh, Addison-Wesley, 1951) p. 58 and pp. 312-316. See also S. Weinberg, ref. 1, pp. 185-201.

[8] Wei-Tou Ni, in *The Proceedings of the Fourth International Workshop on Gravitation and Astrophysics* (Ed. L. Liu, J. Luo, X. Z. Li and J. P. Hsu, World Scientific, 2000) pp.1-19.

[9] S. Weinberg, *Gravitation and Cosmology* (John Wiley and Sons, 1972) p. 178 and pp. 259-273.

[10] Within the framework of curved spacetime, gauge gravity with translation gauge symmetry was discussed by Y. M. Cho, Phys. Rev. **14**, 2515 (1976). His formulation and results are very much different from ours. For example, Cho made additional assumption $\xi_\mu \phi = 0$ for the gauge covariant derivative, so that one has $\Delta_i \psi = (\partial_i + B_i^\mu \xi_\mu)\psi = \partial_i \psi$, where ξ_μ are the translation group generators. Furthermore, he assumed $\partial_i \psi = h_i^\mu \partial_\mu \psi$, where $h_i^\mu = \delta_i^\mu + f B_i^\mu$. This assumption is equivalent to assuming the curved spacetime. Yang-Mills' formulation of gauge symmetry and gauge fields do not make these additional assumptions.

Chapter 9

Experimental Tests of Gravitational Theories*

*W. T. Ni, J. H. Taylor, Jr.

EMPIRICAL FOUNDATIONS OF THE RELATIVISTIC GRAVITY

WEI-TOU NI
Center for Gravitation and Cosmology, Solar-System Division,
Purple Mountain Observatory, Chinese Academy of Sciences,
No. 2, Beijing W. Rd., Nanjing, China 210008 wtni@pmo.ac.cn

Received 16 February 2005

In 1859, Le Verrier discovered the mercury perihelion advance anomaly. This anomaly turned out to be the first relativistic-gravity effect observed. During the 141 years to 2000, the precision of laboratory and space experiments, and astrophysical and cosmological observations on relativistic gravity have improved by 3 orders of magnitude. In 1999, we envisaged a 3-6 order improvement in the next 30 years in all directions of tests of relativistic gravity. In 2000, the interferometric gravitational wave detectors began their runs to accumulate data. In 2003, the measurement of relativistic Shapiro time-delay of the Cassini spacecraft determined the relativistic-gravity parameter γ to be 1.000021 ± 0.000023 of general relativity --- a 1.5-order improvement. In October 2004, Ciufolini and Pavlis reported a measurement of the Lense-Thirring effect on the LAGEOS and LAGEOS2 satellites to be 0.99 ± 0.10 of the value predicted by general relativity. In April 2004, Gravity Probe B (Stanford relativity gyroscope experiment to measure the Lense-Thirring effect to 1 %) was launched and has been accumulate science data for more than 170 days now. μSCOPE (MICROSCOPE: MICRO-Satellite à trainée Compensée pour l'Observation du Principle d'Équivalence) is on its way for a 2007 launch to test Galileo equivalence principle to 10^{-15}. LISA Pathfinder (SMART2), the technological demonstrator for the LISA (Laser Interferometer Space Antenna) mission is well on its way for a 2008 launch. STEP (Satellite Test of Equivalence Principle), and ASTROD (Astrodynamical Space Test of Relativity using Optical Devices) are in the good planning stage. Various astrophysical tests and cosmological tests of relativistic gravity will reach precision and ultra-precision stages. Clock tests and atomic interferometry tests of relativistic gravity will reach an ever-increasing precision. These will give revived interest and development both in experimental and theoretical aspects of gravity, and may lead to answers to some profound questions of gravity and the cosmos.

1. Introduction

A dimensionless parameter $\xi(\mathbf{x}, t)$ characterizing the strength of gravity at a spacetime point $\wp$ [with coordinates $(\mathbf{x}, t)$] due to a gravitating source is the ratio of the negative of the potential energy, mU (due to this source), to the inertial mass-energy mc^2 of a test body at $\wp$, i.e.,

$$\xi(\mathbf{x}, t) = U(\mathbf{x}, t) / c^2. \tag{1}$$

Here $U(\mathbf{x}, t)$ is the gravitational potential.

For a point source with mass M in Newtonian gravity,

$$\xi(\mathbf{x}, t) = GM / Rc^2, \qquad (2)$$

where R is the distance to the source. For a nearly Newtonian system, we can use Newtonian potential for U. The strength of gravity for various configurations is tabulated in Table 1.

Table 1. The strength of gravity for various configurations.

Source	Field Position	Strength of Gravity ξ
Sun	Solar Surface	2.1×10^{-6}
Sun	Mercury Orbit	2.5×10^{-7}
Sun	Earth Orbit	1.0×10^{-8}
Sun	Jupiter Orbit	1.9×10^{-9}
Earth	Earth Surface	0.7×10^{-9}
Earth	Moon's Orbi	1.2×10^{-11}
Galaxy	Solar System	10^{-5} - 10^{-6}
Significant Part of Observed Universe	Our Galaxy	1 - 10^{-2}

The development of gravity theory stems from experiments. Newton's theory of gravity [1] is empirically based on Kepler's laws [2] (which are based on Brahe's observations) and Galileo's law of free-falls [3] (which is based on Galileo's experiment of motions on inclined planes and with free-all trajectories). Toward the middle of nineteenth century, astronomical observations accumulated a precision which enabled Le Verrier [4] to discover the mercury perihelion advance anomaly in 1859. This anomaly is the first relativistic-gravity effect observed. Michelson-Morley experiment [5], via various developments [6], prompted the final establishment of the special relativity theory in 1905 [7, 8]. Motivation for putting electromagnetism and gravity into the same theoretical framework [7], the precision of Eötvös experiment [9] on the equivalence, the formulation of Einstein equivalence principle [10] together with the perihelion advance anomaly led to the road for general relativity theory [11] in 1915.

In 1919, the observation of gravitational deflection of light passing near the Sun during a solar eclipse [12] made general relativity famous and popular. The perihelion advance of Mercury, the gravitational deflection of light passing near the Sun together with the establishment of the gravitational redshift constitute three classical tests of general relativity. With the development of technology and advent of space era, Shapiro [13] proposed a fourth test --- time delay of radar echoes in gravitational field. This, together with the more precise test of the equivalence principle [14] and Pound-Rebka redshift experiment [15] in 1960's, marked the beginning of a new era for testing relativistic gravity. Since the beginning of this era, we have seen 3-4 order improvements for old tests together with many new tests. The technological development is ripe that we are now in a position to discern 3-5 order further improvements in testing relativistic gravity in the coming 25 years (2005-2030). This will enable us to test the second order relativistic-gravity effects. A road map of experimental progress in gravity together with its theoretical implication is shown in Table 2.

Table 2. A road map (Highlights) for gravity. D denotes dynamical effect; EP denotes equivalence principle effect.

Experiment	Precision †	Theory
Brahe's observations (1584-1600) (2'-5' accuracy)	10^{-3} (D)	Kepler's laws
Galileo's experiment on inclined planes (1592)	5×10^{-3} (EP) 5×10^{-3} (D)	(i) Galieleo's EP (ii) The motion with constant force has constant acceleration
Newton's pendulum experiment (1687 [1])	10^{-3} (EP)	§
Observation of celestial-body motions of the solar system (~1687)	10^{-3} (EP) 10^{-3} - 10^{-4} (D)	Newton's inverse square law
Anomalous advance of Mercury's perihelion based on 397 meridian and 14 transit observations (1859 [4])	10^{-8} (D)	*
Anomalous advance of Mercury's perihelion based on transit observations from 1677 to 1881 (1882 [16])	10^{-9} (D)	*
Michelson-Morley Experiment (1887 [5])	10^{-9}	Special relativity *
Eötvös experiment (1889 [9])	10^{-8} (EP)	§, *
Light deflection experiment (1919 [12])	5×10^{-7} (D)	#
Roll-Krotkov-Dicke experiment (1964 [14])	10^{-11} (EP)	§
Binary pulsar observation (1979 [17])	10^{-13} (D)	¶
Solar system test (1970-1999)	10^{-10} - 10^{-11} (D)	‖
Supernova Cosmology Experiment (1998-1999 [18-21])	10^{-2} (D)	General relativity with cosmological constant
Cassini time-delay experiment [56]	10^{-11} - 10^{-12} (D)	#
LAGEOS gravity experiment [67]	10^{-11} (D)	%, #
GP-B Experiment [68]	10^{-12} (D)	%
μSCOPE Experiment [109]	10^{-15} (EP)	¦
STEP (2010-2030)	10^{-17} - 10^{-19} (EP)	¦
Bepi-Colombo, GAIA, ASTROD (2010-2030)	10^{-14} - 10^{-17} (D)	‡

† In terms of dominant observable effects

§ Confirmation of Galileo's equivalence principle

* Leading to general relativity

Confirming the prediction of general relativity

¶ Confirming the quadrupole radiation formula of general relativity

‖ In terms of relativistic parameters, the precision is 10^{-3}

%Testing frame dragging of general relativity

¦Testing Galileo's equivalence principle

‡ In terms of relativistic parameters, the precision will be 10^{-5} - 10^{-9}

This review is a five-year update from a previous review article (W.-T. Ni, "Empirical tests of the relativistic gravity: the past, the present and the future", in *Recent advances and cross-century outlooks in physics: interplay between theory and experiment: proceedings of the Conference held on March 18-20, 1999 in Atlanta, Georgia,* editors, Pisin Chen, and Cheuk-Yin Wong, [Singapore: World Scientific, 2000] ; and pp. 1-19 in *Gravitation and Astrophysics,* editors, Liu L, Luo J, Li X-Z and Hsu J-P [Singapore: World Scientific, 2000]).

In section 2, we review the Mercury's perihelion advance and events leading to general relativity. In section 3, we discuss the classical tests. In section 4, we review precision measurement tests and the foundations of relativistic gravity. In section 5, we review solar system tests since the revival (1960). In sections 6 and 7, we discuss astrophysical tests and cosmological tests respectively. In section 8, we discuss gravitational-wave observations in relation to testing relativistic gravity. In section 9, we discuss next generation experiments in progress, planned and proposed. In section 10, we give an outlook. In the appendix, we discuss empirical tests associated with Edditington-Robertson formalism.

2. Mercury's Perihelion Advance and Events Leading to General Relativity

In 1781, Herschel discovered the planet Uranus. Over years, Uranus persistently wandered away from its expected Newtonian path. In 1834, Hussey suggested that the deviation is due to perturbation of an undiscovered planet. In 1846, Le Verrier predicted the position of this new planet. On September 25, 1846, Galle and d'Arrest found the new planet, Neptune, within one degree of arc of Le Verrier's calculation. This symbolized the great achievement of Newton's theory. [22]

With the discovery of Neptune, Newton's theory of gravitation was at its peak. As the orbit determination of Mercury reached 10^{-8}, relativistic effect of gravity showed up. In 1859, Le Verrier discovered the anomalous perihelion advance of Mercury [4].

In 1840, Arago suggested to Le Verrier to work on the subject of Mercury's motion. Le Verrier published a provisional theory in 1843. It was tested at the 1848 transit of Mercury and there was not close agreement. As to the cause, Le Verrier [23] wrote "Unfortunately, the consequences of the principle of gravitation have not be deduced in many particulars with a sufficient rigour: we will not be able to decide, when faced with a disagreement between observation and theory, whether this results completely from analytical errors or whether it is due in part to the imperfection of our knowledge of celestial physics." [24]

In 1859, Le Verrier [4] published a more sophisticated theory of Mercury's motion. This theory was sufficently rigorous for any disagreement with observation to be taken quite confidently as indicating a new scientific fact. In this paper, he used two sets of observations --- a series of 397 meridian observations of Mercury taken at the Paris Observatory between 1801 and 1842, and a set of observations of 14 transits of Mercury. The transit data are more precise and the uncertainty is of the order of 1". The calculated planetary perturbations of Mercury is listed in Table 3. In addition to these perturbations, there is a 5025"/century general precession in the observational data due to the precession of equinox. The fit of observational data with theoretical calculations has discrepancies. These discrepancies turned out to be due to relativistic-gravity effects. Le Verrier attributed these discrepancies to an additional 38" per century anomalous advance in the perihelion of Mercury.

Newcomb [16] in 1882, with improved calculations and data set, obtained 42".95 per century anomalous perihelion advance of Mercury. The value more recently was (42".98 ± 0.04)/century [25].

In the last half of the 19th century, efforts to account for the anomalous perihelion advance of Mercury went into two general directions: (i) searching for the planet Vulcan, intra-Mercurial matter and the like; (ii) modification of the gravitation law. Both kinds of efforts were not successful. For modification of the gravitational law, Clairaut's hypothesis, Hall's hypothesis and velocity-dependent force laws were considered. The successful solution awaited the development of general relativity.

In 1887, the result of Michelson-Morley experiment posed a serious problem to Newtonian mechanics. A series of developments [6] led to Poincaré's adding to the five classical principles of Physics the Principle of Relativity [26, 27] in 1904 --- "*The laws of physical phenomena must be the same for a fixed observer and for an observer in rectilinear and uniform motion so that we have no possibility of perceiving whether or not we are dragged in such a motion*", and to the seminal works of Poincaré [7] and Einstein [8] in 1905. In [7], Poincaré attempted to develop a relativistic theory of gravity and mentioned gravitational-wave propagating with the speed of light based on Lorentz invariance.

A crucial milestone toward a viable relativistic theory of gravity was established when Einstein proposed his equivalence principle in 1907 [10]. This principle had a firm empirical basis due to the precision experiment of Eötvös [9]. In this same paper, Einstein predicted gravitational redshift. With Special Relativity and the Einstein Equivalence Principle, geometrization of classical physics in the large came naturally in the setting of pseudo-Riemannian manifold. In the years up to 1915 were full of debates and arguments between a number of physicists (Abraham, Einstein, Nordstrom,....) concerned with developing a new relativistic gravitational theory [28]. In 1915, Einstein's General Relativity was proposed [11, 29] and the anomalous perihelion advance of Mercury was explained.

Table 3. Planetary perturbations of the perihelion of Mercury [4]

Venus	280".6/century
Earth	83".6/century
Mars	2".6/century
Jupiter	152".6/century
Saturn	7".2/century
Uranus	0".1/century
Total	526".7/century

3. Classical Tests

The perihelion advance anomaly of Mercury, the deflection of light passing the limb of the Sun and the gravitational redshift are the three classical tests of relativistic gravity. Using EEP, Einstein [30] derived the deflection of light passing the limb of the Sun in 1911. This agrees with the deflection of light derived by using particle model of light in the late 18th century. Before 1915, observations on light deflection were not successful due to war and weather. Einstein's general relativity doubled the prediction of the deflection of light (1".75). The 1919 British solar eclipse expeditions reported reasonably good agreement with the prediction of Einstein's relativity. Before 1960, there were several such observations. The accuracy of these observations was not better than 10 - 20%.

After Einstein [10] proposed the gravitational redshift, Freundlich [31] started the long effort to disentangle the gravitational redshift of solar and other stellar spectral lines from other causes. Over the next five decades, astronomers did not agree on whether there is gravitational redshift empirically [32]. This question is finally settled and gravitational redshift confirmed by Pound and Rebka [15] using Mössbauer effect. The improved result of Pound and Snider [15] confirmed the redshift prediction to 1 % accuracy.

4. Precision Measurement and Foundations of Relativistic Gravity

The foundation of relativistic gravity at present rests on the equivalence of local physics everywhere in spacetime. This equivalence is called the Einstein equivalence principle (EEP) [10]. Its validity guarantees the universal implementation of metrology and standards. Precision metrology or measurement, in turn, test its validity. Possible violations will give clues to the origin of gravity.

The most tested part of equivalence is the Galileo equivalence principle (the universality of free-all). In the study of the theoretical relations between the Galileo equivalence principle and the Einstein equivalence principle, we [33, 34] proposed the χ-g framework summarized in the following interaction Lagrangian density

$$L_I = - (1/(16\pi))\chi^{ijkl} F_{ij} F_{kl} - A_k j^k (-g)^{(1/2)} - \Sigma_I m_I (ds_I)/(dt)\, \delta(\mathbf{x}-\mathbf{x}_I), \tag{3}$$

where $\chi^{ijkl} = \chi^{klij} = -\chi^{klji}$ is a tensor density of the gravitational fields (e.g., g_{ij}, φ, etc.) and j^k, $F_{ij} \equiv A_{j,i} - A_{i,j}$ have the usual meaning. The gravitation constitutive tensor density χ^{ijkl} dictates the behavior of electromagnetism in a gravitational field and has 21 independent components in general. For a metric theory (when EEP holds), χ^{ijkl} is determined completely by the metric g^{ij} and equals $(-g)^{1/2} [(1/2) g^{ik} g^{jl} - (1/2) g^{il} g^{kj}]$. Here we use this framework to look into the foundation of relativistic gravity empirically.

The condition for no birefringence (no splitting, no retardation) for electromagnetic wave propagation in all directions in the weak field limit gives ten constraints on the χ's. With these ten constraints, χ can be written in the following form

$$\chi^{ijkl}=(-H)^{1/2}[(1/2)H^{ik} H^{jl}-(1/2)H^{il} H^{kj}]\psi + \varphi e^{ijkl}, \tag{4}$$

where $H = \det(H_{ij})$ is a metric which generates the light cone for electromagnetic propagation, and e^{ijkl} is the completely antisymmetric symbol with $e^{0123} = 1$ [35-37]. Recently, Lämmerzahl and Hehl have shown that this non-birefringence guarantees, without approximation, Riemannian light cone, i.e., Eq. (4) [38].

Eq. (4) is verified empirically to high accuracy from pulsar observations and from polarization measurements of extragalactic radio sources and will be discussed in §6 on the astrophysical tests. Let us now look into the empirical constraints for H^{ij} and φ. In Eq. (3), ds is the line element determined from the metric g_{ij}. From Eq. (4), the gravitational coupling to electromagnetism is determined by the metric H_{ij} and two scalar fields φ and ψ. If H_{ij} is not proportional to g_{ij}, then the hyperfine levels of the lithium atom, the beryllium atom, the mercury atom and other atoms will have additional shifts. But this is not observed to high accuracy in Hughes-Drever experiments [39]. Therefore H_{ij} is proportional to g_{ij} to a certain accuracy. Since a change of H^{ik} to λH^{ij} does not affect χ^{ijkl} in Eq. (4), we can define $H_{11} = g_{11}$ to remove this scale freedom. [35, 40]

In Hughes-Drever experiments [39] $\Delta m/m \leq 0.5 \times 10^{-28}$ or $\Delta m/m_{e.m.} \leq 0.3 \times 10^{-24}$ where $m_{e.m.}$ is the electromagnetic binding energy. Using Eq. (4) in Eq. (3), we have three kinds of contributions to $\Delta m/m_{e.m.}$. These three kinds are of the order of (i) $(H_{\mu\nu} - g_{\mu\nu})$, (ii) $(H_{0\mu} - g_{0\mu})v$, and (iii) $(H_{00} - g_{00})v^2$ respectively [35, 40]. Here the Greek indices μ, ν denote space indices. Considering the motion of laboratories from earth rotation, in the solar system and in our galaxy, we can set limits on various components of $(H_{ij} - g_{ij})$ from Hughes-Drever experiments as follows:

$$\begin{aligned} &| H_{\mu\nu} - g_{\mu\nu} | / U \leq 10^{-18} \\ &| H_{0\mu} - g_{0\mu} | / U \leq 10^{-13} - 10^{-14}, \\ &| H_{00} - g_{00} | / U \leq 10^{-10}. \end{aligned} \tag{5}$$

where U (~ 10^{-6}) is the galactical gravitational potential.

Eötvös-Dicke experiments [9, 14, 41-43] are performed on unpolarized test bodies; the latest such experiments [43] reach a precision of 3×10^{-13}. In essence, these experiments show that unpolarized electric and magnetic energies follow the same trajectories as other forms of energy to certain accuracy. The constraints on Eq. (4) are

$$| 1-\psi | / U < 10^{-10} \tag{6}$$

and

$$| H_{00} - g_{00} | / U < 10^{-6} \tag{7}$$

where U is the solar gravitational potential at the earth.

In 1976, Vessot and Levine [44] used an atomic hydrogen maser clock in a space probe to test and confirm the metric gravitational redshift to an accuracy of 1.4×10^{-4} [45]. The space probe attained an altitude of 10,000 km above the earth's surface. With Eq. (6), the constraint on Eq. (4) is

$$| H_{00} - g_{00} | / U \leq 1.4 \times 10^{-4}. \tag{8}$$

Thus, we see that for the constraint on $| H_{00} - g_{00} | / U$, Hughes-Drever experiments give the most stringent limit. However, STEP mission concept [46] proposes to improve the WEP experiment by five orders of magnitude. This will again lead in precision in determining H_{00}.

The theory (3) with χ^{ijkl} given by

$$\chi^{ijk} = (-g)^{1/2} [(1/2) g^{ik} g^{jl} - (1/2) g^{il} g^{kj} + \varphi e^{ijkl}], \tag{9}$$

where φ is a scalar or pseudoscalar function of the gravitational field and $e^{ijkl} = (-g)^{-1/2} e^{ijkl}$ is studied in [47] and [48]. In (3), particles considered have charges but no spin. To include spin-1/2 particles, we can add the Lagrangian for Dirac particles. Experimental tests of the equivalence principle for polarized-bodies are reviewed in [49].

To include QCD and other gauge interactions, we have generalized the χ-g framework [50]. Now we are working on a more comprehensive generalization to include a framework to test special relativity, and a framework to test the gravitational interactions of scalar particles and particles with spins together with gauge fields.

5. Solar System Tests

For last forty years, we have seen great advances in the dynamical testing of relativistic gravity. This is largely due to interplanetary radio ranging and lunar laser ranging. Interplanetary radio ranging and tracking provided more stimuli and progresses at first. However with improved accuracy of 2 cm from 20-30 cm and long-accumulation of observation data, lunar laser ranging reaches similar accuracy in determining relativistic parameters as compared to interplanetary radio ranging. Table 4 gives such a comparison.

Table 4. Relativity-parameter determination from interplanetary radio ranging and from lunar laser ranging.

Parameter	Meaning	Value from Solar System Determinations	Value from Lunar Laser Ranging
β	PPN [51] Nonlinear Gravity	1.000±0.003 [25] (perihelion shift with J_2 (Sun)=10^{-7} assumed) 0.9990±0.0012 [52] (Solar-System Tests with J_2 (Sun)=(2.3±5.2)x10^{-7} fitted) 1.0000±0.0001 [53] (EPM2004 fitting)	1.003±0.005 [54] 1.00012±0.0011 [55, 56]
γ	PPN Space Curvature	1.000±0.002 [25] (Viking ranging time delay) 0.9985±0.0021 [52] (Solar-System Tests) 1.000021±0.000023 [56](Cassini S/C Ranging) 0.9999±0.0001 [53] (EPM2004 fitting)	1.000±0.005 [54]
K_{gp}	Geodetic Precession		0.997±0.007 [54] 0.9981±0.0064 [55]
E	Strong Equivalence Principle		(3.2±4.6)x10^{-13} [54] (-2.0±2.0)x10^{-13} [55, 43]
$\dot{G}/G$	Temporal Change in G	(2±4)x10^{-12}/yr [57] (Viking Lander Ranging) ±10x10^{-12}/yr [58] (Viking Lander Ranging) ±2.0x10^{-12}/yr [59](Mercury & Venus Ranging) ±(1.1-1.8)x10^{-12}/yr [60] (Solar-System Tests)	(1±8) x 10^{-12}/yr [54] (0.4±0.9)x 10^{-12}/yr [55]

In the last column of Table 4, the values come from two references [54] and [55]. In [55], Williams *et al.* used a total of 15 553 LLR normal-point data in the period of March 1970 to April 2004 from Observatoire de la Côte d'Azur, McDonald Observatory and Haleakala Observatory in their determination. Each normal point comprises from 3 to about 100 photons. The weighted rms scatter after their fits for the last ten years of ranges is about 2 cm (about 5×10^{-11} of range).

In 2003, Bertotti, Iess and Tortora [56] reported a measurement of the frequency shift of radio photons due to relativistic Shapiro time-delay effect from the Cassini spacecraft as they passed near the Sun during the June 2002 solar conjunction. From this measurement, they determined γ to be 1.000021 ± 0.000023.

With the advent of VLBI (Very Long Baseline Interferometry) at radio wavelengths, the gravitational deflection of radio waves by the Sun from astrophysical radio sources has been observed and accuracy of observation had been improved to 1.7×10^{-3} for γ [61, 62] in 1995. Recent analysis using VLBI data from 1979-1999 improved this result by about four times to 0.99983 ± 0.00045 [63]. Fomalont and Kopeikin [64] measured the effect of retardation of gravity by the field of moving Jupiter via VLBI observation of light bending from a quasar.

The solar-system measurements have made possible the creation of high-accuracy planetary and lunar ephemerides. Two most complete series of ephemerides are the numerical DE ephemerides of JPL [65] and the EPM ephemerides of the Institute of Applied Astromomy [53]. They are of the same level of accuracy and can be used to fit experiments/observations and to determine astronomical constants. Krasinsky and Brumberg [66] used these two series of ephemerides to analyze the major planet motions and the AU (Astronomical Unit); Pitjeva [53] have recently used the EPM framework to determine the AU and obtain 1 AU = 149 597 870 696.0 m. The JPL DE410 determination of this number is 1 AU = 149 597 870 697.4 m. The difference of 1.4 m represents the realistic error in the determination of the AU. Pitjeva's [53] determination of β and γ is obtained simultaneously with estimations for the solar oblateness and the possible variability of the gravitational constant.

In 1918, Lense and Thirring predicted that the rotation of a body like Earth will drag the local inertial frames of reference around it in general relativity. In 2004, Ciufolini and Pavlis [67] reported a measurement of this Lense-Thirring effect on the two Earth satellites, LAGEOS and LAGEOS2; it is 0.99 ± 0.10 of the value predicted by general relativity. In the same year, Gravity Probe B (a space mission to test general relativity using cryogenic gyroscopes in orbit) was launched in April and aims at measurement of Lense-Thirring effect to about 1 % [68].

With the Hipparcos mission, very accurate measurements of star positions at various elongations from the Sun were accumulated. Most of the measurements were at elongations greater than 47° from the Sun. At these angles, the relativistic light deflections are typically a few mas; it is 4.07 mas according to general relativity at right angles to the solar direction for an observer at 1 AU from the Sun. In the Hipparcos measurements, each abscissa on a reference great-circle has a typical precision of 3 mas for a star with 8-9 mag. There are about 3.5 million abscissae generated, and the precision in angle or similar parameter determination is in the range. Frœschlé, Mignard and Arenou [69] analyzed these Hipparcos data and determined the light deflection parameter γ to be 0.997 ± 0.003. This result demonstrated the power of precision optical astrometry.

6. Astrophysical Tests

In the early days, astronomical observations of the solar system provided the basis for developing gravitation theories. With increasing precise observations, astrophysics and cosmology are increasingly more important for such developments. Precise timing of pulsars provides:

(i) confirmation of quadrupole radiation formula for gravitational radiation [17],
(ii) additional testing ground for Post-Newtonian Parameters [17],
(iii) test of nonbirefringence of propagation of electromagnetic wave in a gravitational field, and
(iv) upper limit of background gravitational-wave radiation [70-73].

We refer (i), (ii) and (iv) to references cited. Here, we discuss (iii).

With the null-birefringence observations of pulsar pulses and micropulses before 1980, the relations (4) for testing EEP are empirically verified to $10^{-14} - 10^{-16}$ [35-37]. With the present pulsar observations, these limits would be improved; a detailed such analysis is in [74]. Analyzing the data from polarization measurements of extragalactic radio sources, Haugan and Kauffmann [75] inferred that the resolution for null-birefringence is 0.02 cycle at 5 GHz. This corresponds to a time resolution of 4×10^{-12} s and gives much better constraints. With a detailed analysis and more extragalactic radio observations, (4) would be tested down to 10^{-28}-10^{-29} at cosmological distances. In 2002, Kostelecky and Mews [76] used polarization measurements of light from cosmologically distant astrophysical sources to yield stringent constraints down to 2×10^{-32}. The electromagnetic propagation in Moffat's nonsymmetric gravitational theory fits the χ-g framework. Krisher [77], and Haugan and Kauffmann [75] have used the pulsar data and extragalactic radio observations to constrain it.

7. Cosmological Tests

In an attempt to find a static cosmology, Einstein add a cosmological constant Λ to his equation. The term containing Λ can be interpreted as a modification of Einstein's equation or it can be just interpreted as vacuum stress-energy.

Although Einstein considered the proposal of this term his biggest blunder in his life, the value of Λ needs to be determined using cosmological observations.

Recent evidence suggests that Type Ia supernovae (SNeIa) can be used as precise cosmological distance indicators [78]. Early results with these SNeIa observations imply that there is not enough gravitating matter to close the universe [18, 19] and that currently the expansion of the Universe is accelerating [20, 21], indicating Λ-density (cosmological term, dark energy or quintessence) is larger than the ordinary-matter density. More supernovae observations together with more precise cosmic background anisotropy measurements will be important in testing and determining the gravitational equation in the cosmological context.

In section 4, we mentioned a nonmetric theory [33, 34, 47] in discussing the foundations of relativistic gravity. Theories with spontaneous direction [79] and axion theories also have such an electromagnetic interaction. The effect of φ [in (9)] in this theory is to change the phase of two different circular polarizations in gravitation field and gives polarization rotation for linearly polarized light [47, 79, 80]. Using polarization observations of radio galaxies, Carroll, Field and Jackiw [79, 80] put a limit of 0.1 on $\Delta\varphi$ over cosmological distances. Using a different analysis of polarization observation of radio galaxies, Nodland and Ralston [81] found indication of anisotropy in electromagnetic propagation over cosmological distances with a birefringence scale of order 10^{25} m (i.e., about 0.1 - 0.2 Hubble distance). This gave $\Delta\varphi \sim 5$ - 10 over Hubble distance). Later analyses [82-86] did not confirm this result and put a limit of $\Delta\varphi \leq 1$ over cosmological distance scale.

The natural coupling strength φ is of order 1. However, the isotropy of our observable universe to 10^{-5} may leads to a change $\Delta\varphi$ of φ over cosmological distance scale 10^{-5} smaller [87]. Hence, observations to test and measure $\Delta\varphi$ to 10^{-6} are significant and they are promising. In 2002, DASY microwave interferometer observed the polarization of the cosmic background. With the axial interaction (9), the polarization anisotropy is shifted relative to the temperature anisotropy. In 2003, WMAP (Wilkenson Microwave Anisotropy Probe) [88] found that the polarization and temperature are correlated. This gives a constraint of 10^{-1} of $\Delta\varphi$ [89]. Planck Surveyor [90] will be launched in 2007 with better polarization-temperature measurement and will give a sensitivity to $\Delta\varphi$ of 10^{-2}-10^{-3}. A dedicated future experiment on cosmic microwave background radiation will reach 10^{-5}-10^{-6} $\Delta\varphi$-sensitivity. This is very significant as a positive result may indicate that our patch of inflationary universe has a ‘spontaneous polarization’ in fundamental law of electromagnetic propagation influenced by neighboring patches and we can 'observe' neighboring patches through a determination of this fundamental physical law; if a negative result turns out at this level, it may give a good constraint on superstring theories as axions are natural to superstring theories.

In section 4, we mentioned Hughes-Drever experiments [39] to test the spatial anisotropy. Here we mention test of cosmic spatial anisotropy using polarized electrons. Following Phillips' pioneer work [91], we [92] and Berglund *et al.* [93] improved the sensitivity. In 2000, we used a rotatable torsion balance carrying a transversely spin-polarized ferrimagnetic Dy_6Fe_{23} mass to test the cosmic spatial anisotropy, and have achieved an order-of-magnitude improvement [94, 95] over previous experiments; the anomalous transverse energy splitting of spin states of electron due to spatial anisotropy is constrained to be less than 6×10^{-29} Gev. Eöt-Wash group also improved on their experiment and gave a constraint of 1.2×10^{-28} Gev [96].

8. Gravitational-Wave Observations

The importance of gravitational-wave detection is twofold: (i) as probes to fundamental physics and cosmology, especially black hole physics and early cosmology, and (ii) as tools in astronomy and astrophysics to study compact objects and to count them. We follow [97] to extend the conventional classification of gravitational-wave frequency bands [98] into the ranges:

(i) High-frequency band (1-10 kHz): This is the frequency band that ground gravitational-wave detectors are most sensitive to.

(ii) Low-frequency band (100 nHz - 1 Hz): This is the frequency band that space gravitational-wave experiments are most sensitive to.

(iii) Very-low-frequency band (300 pHz-100 nHz): This is the frequency band that the pulsar timing experiments are most sensitive to.

(iv) Extremely-low-frequency band (1 aHz - 10 fHz): This is the frequency band that the cosmic microwave anisotropy and polarization experiments are most sensitive to.

The cryogenic resonant bar detectors have already reached a strain sensitivity of $(10^{-21})/(\mathrm{Hz})^{1/2}$ in the kHz region. Five such detectors --- ALLEGRO, AURIGA, EXPLORER, NAUTILUS and NIOBE --- have been on the air, forming a network with their bar axis quasi-parallel in a continuous search for bursts. TAMA (300 m armlength) interferometer started accumulating data in 2000. GEO, and kilometer size laser-interferometric gravitational-wave detectors --- LIGO and VIRGO --- took runs and started to accumulate data also with strain sensitivity goal aimed at $10^{-23}/(\mathrm{Hz})^{1/2}$ in the frequency around 100 Hz. Various limits on the gravitation-wave strains for different sources become significant. For example, analysis of data collected during the second LIGO science run set strain upper limits as low as a few times 10^{-24} for some pulsar sources; these translate into limits on the equatorial ellipticities of the pulsars, which are smaller than 10^{-5} for the four closest pulsars [99].

Space interferometer (LISA [100, 101], ASTROD [102, 103]) for gravitational-wave detection hold the most promise. LISA (Laser Interferometer Space Antenna) [100] is aimed at detection of low-frequency (10^{-4} to 1 Hz) gravitational waves with a strain sensitivity of $4 \times 10^{-21}/(\mathrm{Hz})^{1/2}$ at 1 mHz. There are abundant sources for LISA: galactic binaries (neutron stars, white dwarfs, etc.). Extra-galactic targets include supermassive black hole binaries, supermassive black hole formation, and cosmic background gravitational waves. A date of launch is hoped for 2013.

For the very-low-frequency band and for the extremely-low-frequency band, it is more convenient to express the sensitivity in terms of energy density per logarithmic frequency interval divided by the cosmic closure density ρ_c for a cosmic background of gravitational waves, i.e., $\Omega_g(f)(=(f/\rho_c)d\rho_g(f)/df)$.

The upper limits from pulsar timing observations on a gravitational wave background are about $\Omega_g \leq 10^{-7}$ in the frequency range 4-40 nHz [70], and $\Omega_g \leq 4 \times 10^{-9}$ at 6×10^{-8} Hz [72]. More pulsar observations with extended periods of time will improve the limits by two orders of magnitude in the lifetime of present ground and space gravitational-wave-detector projects. The COBE microwave-background quadrupole anisotropy measurement [104, 105] gives a limit Ω_g (1 aHz) ~ 10^{-9} on the extremely-low-frequency gravitational-wave background [106, 107]. Ground and balloon experiments probe smaller-angle anisotropies and, hence, higher-frequency background. WMAP [108] and Planck Surveyor [90] space missions could probe anisotropies with l up to 2000 and with higher sensitivity.

9. The Next 25 Years (2005-2030)

During last 146 years, the precision of laboratory and space experiments, and astrophysical and cosmological observations on relativistic gravity have improved by 3-4 orders of magnitude. In 2000, the ground long interferometric gravitational wave detectors began their runs to accumulate data. In April 2004, Gravity Probe B (Stanford relativity gyroscope experiment to measure the Lense-Thirring effect to 1 %) [68] was launched and has been accumulate science data for more than 170 days now. μSCOPE (MICROSCOPE: MICRO-Satellite à trainée Compensée pour l'Observation du Principle d'Équivalence) [109] is on its way for a 2007 launch to test Galileo equivalence principle to 10^{-15}. LISA Pathfinder (SMART2), the technological demonstrator for the LISA (Laser Interferometer Space Antenna) mission is well on its way for a 2008 launch. STEP (Satellite Test of Equivalence Principle) [46], and ASTROD (Astrodynamical Space Test of Relativity using Optical Devices) [102, 103] are in the good planning stage. Various astrophysical tests and cosmological tests of relativistic gravity will reach precision and ultra-precision stages. Clock tests and atomic interferometry tests of relativistic gravity will reach an ever-increasing precision. In the next 25 years, we envisage a 3-5 order improvement in all directions of tests of relativistic gravity. These will give revived interest and development both in experimental and theoretical aspects of gravity, and may lead to answers to some profound questions of gravity and the cosmos.

In this section, we illustrate this expectation by looking into various ongoing / proposed experiments related to the determination of the PPN space curvature parameter γ. Some motivations for determining γ precisely to 10^{-5} – 10^{-9} are given in [110, 111]. Table 5 lists the aimed accuracy of such experiments.

GP-B is an ongoing experiment [68] using quartz gyro at low-temperature to measure the Lense-Thirring precession and the geodetic precession. The geodetic precession gives a measure of γ.

Bepi-Colombo [112] is planned for a launch in 2013 to Mercury. A simulation predicts that the determination of γ can reach 2×10^{-6} [113].

GAIA (Global Astrometric Interferometer for Astrophysics) [114] is an astrometric mission concept aiming at the broadest possible astrophysical exploitation of optical interferometry using a modest baseline length (~3m). GAIA is planned to be launched in

2013. At the present study, GAIA aims at limit magnitude 21, with survey completeness to visual magnitude 19^{-20}, and proposes to measure the angular positions of 35 million objects (to visual magnitude V=15) to 10 μas accuracy and those of 1.3 billion objects (to V=20) to 0.2 mas accuracy. The observing accuracy of V=10 objects is aimed at 4 μas. To increase the weight of measuring the relativistic light deflection parameter γ, GAIA is planned to do measurements at elongations greater than 35° (as compared to essentially 47° for Hipparcos) from the Sun. With all these, a simulation shows that GAIA could measure γ to $1\times10^{-5} - 2\times10^{-7}$ accuracy [115].

Table 5. Aimed accuracy of PPN space parameter γ for various ongoing / proposed experiments. The types of experiments (deflection, retardation or geodetic precession) are given in the parentheses.

Ongoing / Proposed experiment	Aimed accuracy of γ
GP-B [68] (geodetic precession)	1×10^{-5}
Bepi-Colombo [113] (retardation)	2×10^{-6}
GAIA [115] (deflection)	$1\times10^{-5} - 2\times10^{-7}$
ASTROD I [116] (retardation)	1×10^{-7}
LATOR [117]	1×10^{-8}
ASTROD [118]	1×10^{-9}

In the ranging experiments, the retardations (Shapiro time delays) of the electromagnetic waves are measured to give γ. In the astrometric experiments, the deflections of the electromagnetic waves are measured to give γ. These two kinds of experiments complement each other in determining γ. The ASTROD I (Single Spacecraft Astrodynamical Space Test of Relativity using Optical Devices) mission concept [116] is to use a drag-free spacecraft orbiting around the Sun using 2-way laser pulse ranging and laser-interferometric ranging between Earth and spacecraft to measure γ and other relativistic parameters precisely. The γ parameter can be separated from the study of the Shapiro delay variation. The uncertainty on the Shapiro delay measurement depends on the uncertainties introduced by the atmosphere, timing systems of the ground and space segments, and the drag-free noise. A simulation shows that an uncertainty of 10^{-7} on the determination of γ is achievable.

LATOR (Laser Astrometric Test Of Relativity) [117] proposed to use laser interferometry between two micro-spacecraft in solar orbit, and a 100 m baseline multi-channel stellar optical interferometer placed on the ISS (International Space Station) to do spacecraft astrometry for a precise measurement of γ.

For ASTROD (Astrodynamical Space Test of Relativity) [102, 103, 118], 3 spacecraft, advanced drag-free systems, and mature laser interferometric ranging will be used and the resolution is subwavelength. The accuracy of measuring γ and other parameters will depend on the stability of the lasers and/or clocks. An uncertainty of 1×10^{-9} on the determination of γ is achievable in the time frame of 2015-2025.

10. Outlook

Physics is an empirical science, so is gravitation. The road map for gravitation is clearly empirical. As precision is increased by orders of magnitude, we are in a position to explore deeper into the origin of gravitation. The current and coming generations are holding such promises.

Acknowledgements

I would like to thank the National Natural Science Foundation (Grant No. 10475114), and the Foundation of Minor Planets of Purple Mountain Observatory for supporting this work.

Appendix

Since many readers are more familiar with the parametrization given by Eddington (A. S. Eddington, *The Mathematical Theory of Relativity* [2nd ed., Cambridge University Press, 1924]) and Robertson (H. P. Robertson, p. 228 in *Space Age Astronomy*, ed. by A. J. Deutsch and W. H. Klemperer [Academic Press, New York, 1962]) in testing relativistic gravity, with the recommendation of J. P. Hsu (one of the editor of this book), I add this explanatory appendix.

The Eddington-Robertson parametrization of metric is

$$ds^2 = (1-2\alpha(GM/r)+2\beta(GM/r)^2+\ldots)dt^2 - (1+2\gamma(GM/r)+\ldots)\,(dr^2+r^2d\theta^2+r^2\sin^2\theta d\varphi^2),$$

where α, β, γ are the Eddington-Robertson relativistic parameters. For matter, α can be absorbed into G; the Newtonian limit then requires G to be the Newtonian gravitational constant. To test the Einstein Equivalence Principle, for electromagnetism, we could set α, β and γ to α_{em}, β_{em} and γ_{em}, and the metric corresponds to H_{ij} in section 4; for matter, we then have α, β and γ equal to $\alpha_{matter}(=1)$, β_{matter} and γ_{matter}, and the metric corresponds to g_{ij}. The constraints (5, 7, 8) become

$$2|\gamma_{em} - \gamma_{matter}| / U \leq 10^{-18},$$
$$2|\alpha_{em} - 1| / U \leq 10^{-10}, \tag{5'}$$

$$2|\alpha_{em} - 1| / U < 10^{-6}, \tag{7'}$$

$$2|\alpha_{em} - 1| / U \leq 1.4 \times 10^{-4}. \tag{8'}$$

For dynamical tests of β and γ, Table 4 and Table 5 and the associated discussions in section 5 and section 9 apply.

References

1. I. Newton, *Philosophiae Naturalis Principia Mathematica* (London, 1687).
2. J. Kepler, *Astronomia nova de motibus stellae Martis* (Prague, 1609); *Harmonice mundi* (Linz, 1619).

3. G. Galilei, *Discorsi e dimostriazioni matematiche intorno a due nuove scienze* (Elzevir, Leiden, 1638)
4. U. J. J. Le Verrier, Theoric du mouvement de Mercure, *Ann. Observ. imp. Paris (Mém.)* **5**, 1-196 (1859).
5. A. A. Michelson and E. W. Morley, *Am. J. Sci.* **34**, 333 (1887).
6. H. A. Lorentz, *Kon. Neder. Akad. Wet. Amsterdam. Versl. Gewone Vergad. Wisen Natuurkd. Afd.* **6**, 809 (1904), and references therein.
7. H. Poincaré, em *C. R. Acad. Sci.* **140**, 1504 (1905), and references therein.
8. A. Einstein, *Ann. Phys.* **17**, 891 (1905).
9. R. V. Eötvös, *Math. Naturwiss. Ber. Ungarn* **8**, 65 (1889).
10. A. Einstein, *Jahrb. Radioakt. Elektronik* **4**, 411 (1907); Corrections by Einstein in *Jahrb. Radioakt. Elektronik* **5**, 98 (1908); English translations by H. M. Schwartz in *Am. J. Phys.* **45**, 512, 811, 899 (1977).
11. A. Einstein, *Preuss. Akad. Wiss. Berlin, Sitzber.* 778, 799, 831, 844 (1915).
12. F. Dyson, A. Eddington and C. Davidson, *Phil. Trans. Roy. Soc.* **220A**, 291 (1920); F. Dyson, A. Eddington and C. Davidson, *Mem. Roy. Ast. Soc.* **62**, 291 (1920).
13. I. I. Shapiro, *Phys. Rev. Lett.* **13**, 789 (1964).
14. P. G. Roll, R. Krotkov and R. H. Dicke, *Ann. Phys.* (U. S. A.) **26**, 442 (1964).
15. R. V. Pound and G. A. Rebka, *Phys. Rev. Lett.* **4**, 337 (1960); R. V. Pound and J. L. Snider, *Phys. Rev. Lett.* **13**, 539 (1964).
16. S. Newcomb, "Discussion and results of observations on transits of Mercury from 1677 to 1881", *Astr. Pap. am. Ephem. naut. Alm.*, **1**, 367-487 (U.S. Govt. Printing Office, Washington, D.C., 1882).
17. J. H. Taylor, *Rev. Mod. Phys.* **66**, 711 (1994); and references therein.
18. P. M. Granavich, *et al.*, *Astrophys. J.* **493**, 53 (1998).
19. S. Perlmutter, *et al.*, *Nature* **391**, 51 (1998).
20. A. G. Riess, *et al.*, *Astrophys. J.* **116**, 1009 (1998).
21. S. Perlmutter, *et al.*, *Astrophys. J.*, **517**(2), 565-586 (1999).
22. P. Moore, *The Story of Astronomy*, 5th revised edition (New York, Grosset & Dunlap Publishing, 1977).
23. U. J. J. Le Verrier, *C. R. Acad. Sci. Paris* **29**, 1 (1849); the English translation is from [24].
24. N. T. Roseveare, *Mercury's Perihelion from Le Verrier to Einstein* (Oxford, Clarendon Press, 1982); the reader is referred to this book for a thorough study of the history related to the Mercury's perihelion advance.
25. I. I. Shapiro, in *General Relativity and Gravitation*, ed. N. Ashby, D. F. Bartlett and W. Wyss, p. 313 (Cambridge, Cambridge University Press, 1990).
26. H. Poincaré, L'état actuel et l'avenir de la physique mathématique, *Bulletin des Sciences Mathématiques*, Tome **28**, 2e série (reorganized 39-1), 306 (1904); the English translation is from [27].
27. C. Marchal, *Sciences* **97-2** (April, 1997) and English translation provided by the author.
28. J. Stachel, p. 249 in *Twentieth Century Physics, Vol. 1*, ed. L. M. Brown, A. Pais and B. Pippard (New York, AIP Press, 1995).
29. D. Hilbert, *Konigl. Gesell. d. Wiss. Göttingen, Nachr., Math.-Phys. Kl.*, 395 (1915).
30. A. Einstein, *Ann. Phys. (Germany)* **35**, 898 (1911).
31. K. Hentschel, Erwin Finlay-Freundlich and testing Einstein's theory of relativity, *Arch. Hist. Exact Sci.* **47**, 143 (1994); and references therein.
32. E. G. Forbes, *Ann. Sci.* **17**, 143 (1961).
33. W.-T. Ni, *Bull. Am. Phys. Soc* **19**, 655 (1974).
34. W.-T. Ni, *Phys. Rev. Lett.* **38**, 301 (1977).
35. W.-T. Ni, "Equivalence Principles and Precision Experiments" pp. 647-651, in Precision Measurement and Fundamental Constants II, ed. by B. N. Taylor and W. D. Phillips, Natl. Bur. Stand. (U.S.), Spec. Publ. **617** (1984).
36. W.-T. Ni, "Timing Observations of the Pulsar Propagations in the Galactic Gravitational Field as Precision Tests of the Einstein Equivalence Principle", pp. 441-448 in Proceedings of the Second Asian-Pacific Regional Meeting of the International Astronomical Union, ed. by B.

Hidayat and M. W. Feast (Published by Tira Pustaka, Jakarta, Indonesia, 1984).
37. W.-T. Ni, "Equivalence Principles, Their Empirical Foundations, and the Role of Precision Experiments to Test Them", pp. 491-517 in Proceedings of the 1983 International School and Symposium on Precision Measurement and Gravity Experiment, Taipei, Republic of China, January 24-February 2, 1983, ed. by W.-T. Ni (Published by National Tsing Hua University, Hsinchu, Taiwan, Republic of China, June, 1983).
38. C. Lämmerzahl and F.W. Hehl, *Phys. Rev.* D **70**, 105022 (7) (2004).
39. V. W. Hughes, H. G. Robinson, and V. Beltran-Lopez, *Phys. Rev. Lett.* **4**, 342 (1960); V. Beltran-Lopez, H. G. Robinson, and V. W. Hughes, *Bull. Am. Phys. Soc.* **6**, 424 (1961); R. W. P. Drever, Phil. Mag. **6**, 683 (1962); J. F. Ellena, W.-T. Ni and T.-S. Ueng, *IEEE Transactions on Instrumentation and Measurement* **IM-36**, 175 (1987); T. E. Chupp, R. J. Hoara, R. A. Loveman, E. R. Oteiza, J. M. Richardson, M. E. Wagshul, *Phys. Rev. Lett.* **63**, 1541 (1989).
40. W.-T. Ni, "Implications of Hughes-Drever Experiments", pp. 519-529 in Proceedings of the 1983 International School and Symposium on Precision Measurement and Gravity Experiment, Taipei, Republic of China, January 24-February 2, 1983, ed. by W.-T. Ni (Published by National Tsing Hua University, Hsinchu, Taiwan, Republic of China, June, 1983).
41. R. V. Eötvös, D. Pekar and E. Fekete, *Ann. Phys. (Leipzig)* **68**, 11 (1922).
42. V. B. Braginsky and V. I. Panov, *Zh. Eksp. Teor. Fiz.* **61**, 873 (1971) [Sov. Phys. JETP **34**, 463 (1972)].
43. Y. Su, *et al.*, *Phys. Rev.* **D 50**, 3614 (1994); S. Baessler, *et al.*, *Phys. Rev. Lett.* **83**, 3585 (1999); E. G. Adelberger, *Class. Quantum Grav.* **18**, 2397 (2001).
44. R. F. C. Vessot, and M. W. Levine, *Gen. Rel. Grav.* **10**, 181 (1979).
45. R. F. C. Vessot, *et al.*, *Phys. Rev. Lett.* **45**, 2081 (1980).
46. ESA SCI(93)4, STEP (Satellite test of the equivalence principle) report on the phase A study (1993).
47. W.-T. Ni, A Nonmetric Theory of Gravity, preprint, Montana State University, Bozeman, Montana, USA (1973). The paper is available via http://gravity5.phys.nthu.edu.tw/webpage/article4/index.html.
48. W.-T. Ni, "Spin, Torsion and Polarized Test-Body Experiments", pp. 531-540 in Proceedings of the 1983 International School and Symposium on Precision Measurement and Gravity Experiment, Taipei, Republic of China, January 24-February 2, 1983, ed. by W.-T. Ni (Published by National Tsing Hua University, Hsinchu, Taiwan, Republic of China, June, 1983).
49. W.-T. Ni, Searches for the role of polarization and spin in gravitation, review article in preparation for Reports on Progress in Physics (2005).
50. W.-T. Ni, *Phys. Lett.* **A 120**, 174 (1987).
51. C. M. Will, and K. Nordtvedt, Jr., *Astrophys. J.*, **177**, 757 (1972); and references therein.
52. J. D. Anderson, E. L. Lau, S. Turyshev, J. G. Williams, and M. M. Nieto, *Bulletin of the American Astronomical Society* **34**, 660 (2002).
53. E. Pitjeva "Precise determination of the motion of planets and some astronomical constants from modern observations", 12 p, to be published in IAU Coll. N 196 / Transit of Venus: new views of the solar system and galaxy (ed. D.W. Kurtz), Cambridge: Cambridge University Press, 2005.
54. J. G. Williams, X. X. Newhall, and J. O. Dickey, *Phys. Rev.* **D 53**, 6730 (1996).
55. J. G. Williams, S. G. Turyshev, and D. H. Boggs, *Phys. Rev. Lett.* **93**, 261101(4) (2004).
56. B. Bertotti, L. Iess, and P. Tortora, *Nature* **425**, 374-376 (2003).
57. R. W. Hellings, P. J. Adams, J. D. Anderson, M. S. Keesey, E. L. Lau, E. M. Standish, V. M. Canuto and I. Goldman, *Phys. Rev. Lett.* **51**, 1609 (1983).
58. R. D. Reasenberg, *Philos. Trans. R. Soc. London. Sec.* **A 310**, 227 (1983); J. F. Chandler, R. D. Reasenberg, and I. I. Shapiro, *Bull. Amer. Astron. Soc.* **25**, 1233 (1993).
59. J. D. Anderson, J. K. Campbell, R. F. Jurgens, E. L. Lau, X. X. Newhall, M. A. Slade III, and E. M. Standish Jr., "Recent Developments in Solar-System Tests of General Relativity", in *Proceedings of the 6th Marcel Grossmann Meeting on General Relativity*, Ed. H. Sato and T. Nakamura, p. 353 (Singapore, World Scientific, 1992).
60. J. G. Williams, J. D. Anderson, D. H. Boggs, E. L. Lau, and J. O. Dickey, *Bulletin of the*

American Astronomical Society **33**, p. 836 (2001).
61. D. S. Robertson, W. E. Carter, and W. H. Dillinger, *Nature* **349**, 768 (1991).
62. E. Lebach, *et al., Phys. Rev. Lett.* **75,** 1439 (1995).
63. S. S. Shapiro, J. L. Davis, D. E. Lebach, and J. S. Gregory, *Phys. Rev. Lett.* **92**, 121101(4) (2004).
64. E. B. Fomalont, and S. M. Kopeikin, *Astrophys. J.*, **598**, 704 (2003).
65. E. M. Standish, "JPL Interoffice Memorandum", 312.N-03-009, 16 p (2003).
66. G. A. Krasinsky and V.A. Brumberg, *Celest. Mech. & Dyn. Astron.* **90**, 267--288 (2004).
67. I. Ciufonlini, and E. C. Pavlis, *Nature* **431**, 958-960 (2004); N. Ashby, *Nature* **431**, 918-919 (2004).
68. GP-B (Gravity Probe B) http://einstein.stanford.edu/; C. W. F. Everitt, S. Buchman, D. B. DeBra, G. M. Keiser, J. M. Lockhart, B. Muhlfelder, B. W. Parkinson, J. P. Turneaure, and other members of the Gravity Probe B team, "Gravity Probe B: Countdown to Launch", pp. 52-82 in Gyros Clocks, Interferometers ...: Testing Relativistic Gravity in Space, eds. C. Lämmerzahl, C. W. F. Everitt, F. W. Hehl (Berlin, Springer-Verlag, 2001).
69. M. Frœschlé, F. Mignard and F. Arenou, p. 49 in *Proceedings of the ESA Symposium "Hipparcos-Venice'97"*, 13-16 May, Venice, Italy, ESA SP-402 (July 1997).
70. M. P. McHugh, G. Zalamansky, F. Vernotte, and E. Lantz, Phys. Rev. D 54, 5993 (1996), and references therein.
71. G. Zalamansky, C. Robert, F. Vernotte *et al.*, *Mon. Not. Roy. Astron. Soc.* **288** (2), 533-537 (1997).
72. A. N. Lommen, Proceedings of the 270 WE-Heraeus Seminar on: "*Neutron Stars, Pulsars and Supernova remnants*", Physikzentrum, Bad Honnef, Germany, Jan. 21-25, 2002, eds. W. Becker, H. Lesch and J. Trümper, MPE Report 278, pp.114-125; astro-ph/0208572.
73. V. A. Potapov, Y. P. Ilyasov, V. V. Oreshko, and A. E. Rodin, Astronomy Letters **29** (4), 241-245 (2003); S. M. Kopeikin, V. A. Potapov, *Mon. Not. Roy. Astron. Soc.* **355** (2), 395-412 (2004), and references therein.
74. H.-W. Huang, "Pulsar Timing and Equivalence Principle Tests", Master thesis, National Tsing Hua University (Hsinchu, 2002).
75. M. P. Haugan and T. F. Kauffmann, *Phys. Rev.* **D 52**, 3168 (1995).
76. V. A. Kostelecky and M. Mewes, *Phys. Rev.* **D 66**, 056005(24) (2002).
77. T. P. Krisher, *Phys. Rev.* **D 44**, R2211 (1991).
78. A. G. Riess, W. H. Press and R. P. Kirshner, *Astrophys. J.*, **473**, 588 (1996), and references therein.
79. S. M. Carroll, G. B. Field, R. Jackiw, *Phys. Rev.* **D 41**, 1231 (1990).
80. S. M. Carroll and G. B. Field, *Phys. Rev.* **D 43**, 3789 (1991).
81. B. Nodland and J. P. Ralston, *Phys. Rev. Lett.* **78**, 3043 (1997).
82. J. F. C. Wardle, R. A. Perley, and M. H. Cohen, *Phys. Rev. Lett.* **79**, 1801 (1997).
83. D. J. Eisenstein and E. F. Bunn, *Phys. Rev. Lett.* **79**, 1957 (1997).
84. S. M. Carroll and G. B. Field, *Phys. Rev. Lett.* **79**, 2394 (1997).
85. T. J. Loredo, E. A. Flanagan, and I. M. Wasserman, *Phys, Rev.* **D 56**, 7507 (1997).
86. S. M. Carroll, *Phys. Rev. Lett.* **81**, 3067 (1998).
87. W.-T. Ni, "Spin in Gravity", pp. 439-456 in Gyros, Clocks, Interferometers...: Testing Relativistic Gravity in Space, eds. C. Lämmerzahl, C. W. F. Everitt, F. W. Hehl (Berlin, Springer-Verlag, 2001).
88. A. Kogut, D. N. Spergel, C. Barnes, *et al.*, *Astophys. J. Suppl.* **148** (1), 161-173 (2003); A. Kogut, *New Astron. Rev.*, **47**(11-12), 977-986 (2003).
89. W.-T. Ni, *Chin. Phys. Lett.* **22**(1), 33-35 (2005).
90. www.rssd.esa.int/index.php?project=PLANCK
91. R. Phillips, *Phys. Rev. Lett.* **59**, 1784 (1987).
92. S.-L. Wang, W.-T. Ni and S.-s. Pan, *Modern Phys. Lett.* **A 8**, 3715 (1993).
93. C. J. Berglund *et al., Phys. Rev. Lett.* **75**, 1879 (1995).
94. L.-S. Hou and W.-T. Ni, "Test of Spatial Anisotropy for Polarized Electrons Using a Rotatable Torsion Balance", p. 143 in *Proceedings of the International Workshop on Gravitation and Astrophysics, Tokyo, 1997*, edited by K. Kuroda (ICRR, Tokyo, 1998).

95. L.-S. Hou, W.-T. Ni and Y.-C. M. Li, *Phys. Rev. Lett.* **90** (20), 201101 (4) (2003); *Preprint* physics/0009012 (2000).
96. B. R. Heckel *et al.*, in *CPT and Lorentx Symmetry II*, V. A. Kostelecky, ed. (World Scientific, Singapore, 2002).
97. K. S. Thorne, *Gravitational Waves*, p. 160 in *Paticle and Nuclear Astrophysics and Cosmology in the Next Millennium}*, ed. E. W. Kolb and R. D. Peccei (World Scientific, Singapore, 1995).
98. W.-T. Ni, "ASTROD and gravitational waves", pp. 117-129 in Gravitational Wave Detection, edited by K. Tsubono, M.-K. Fujimoto and K. Kuroda (Universal Academy Press, Tokyo, Japan, 1997).
99. B. Abbott *et al.*, "Limits on gravitational wave emission from selected pulsars using LIGO data, arXiv: gr-qc/0410007 v2 19 Jan., 2005 (2005).
100. LISA, Pre-Phase A Report, second edition, July (1998).
101. LISA, Laser Interferometer Space Antenna: A Cornerstone Mission for the Observation of Gravitational Waves, ESA System and Technology Study Report, **ESA-SCI 11** (2000).
102. A. Bec-Borsenberger, J. Christensen-Dalsgaard, M. Cruise, A. Di Virgilio, D. Gough, M. Keiser, A. Kosovichev, C. Laemmerzahl, J. Luo, W.-T. Ni, A. Peters, E. Samain, P. H. Scherrer, J.-T. Shy, P. Touboul, K. Tsubono, A.-M. Wu, and H.-C. Yeh, "Astrodynamical Space Test of Relativity using Optical Devices ASTROD---A Proposal Submitted to ESA in Response to Call for Mission Proposals for Two Flexi-Mission F2/F3", January 31, 2000.
103. W.-T. Ni, *Int. J. Mod. Phys.* **D 11** (7): 947-962 (2002).
104. G. F. Smoot *et al.*, *Astrophys. J.* **396**, L1 (1992).
105. C. L. Bennett *et al.*, *Astrophys. J.* **464**, L1 (1996).
106. L. M. Krauss and M. White, *Phys. Rev. Lett.* **69**, 969 (1992).
107. R. L. Davis, H. M. Hodges, G. F. Smoot, P. J. Steinhardt, and M. S. Turner, *Phys. Rev. Lett.* **69**, 1856 (1992).
108. C. L. Bennett, M. Halpern, G. Hinshaw, *et al.*, *Astrophys. J. Suppl.* **148** (1), 1-27 (2003).
109. P. Touboul, M. Rodrigues, G. Métris and B. Tatry, MICROSCOPE, testing the equivalence principle in space, *C. R. Acad. Sci. Ser.* IV **2**(9), 1271–1286 (2001).
110 T. Damour and K. Nordtvedt, Jr., *Phys. Rev. Lett.* **70**, 2217 (1993).
111. T. Damour, F. Piazza, and G. Veneziano, *Phys. Rev.* **D 66**, 046007 (2002); *Preprint* hep-th/0205111 (2002).
112. http://www.esa.int/esaSC/120391_index_0_m.html
113. A. Milani, D. Vokrouhlicky, D. Villani, C. Bonanno, and A. Rossi, *Phys. Rev.* **D 66** (8), 082001(21) (2004).
114. http://www.esa.int/esaSC/120377_index_0_m.html
115. A. Vecchiato, M. G. Lattanzi, B. Bucciarelli, *et al.*, *Astron. Astrophys.* **399** (1), 337-342, (2003).
116. W.-T. Ni, G. Bao, Y. Bao, *et al.*, *J. Korean Phys. Soc.* **45**: S118-S123 (2004).
117. S. G. Turyshev, M. Shao, and K. Nordtvedt, *Class. Quantum Grav.* **21**, 2773-2799 (2004).
118. W.-T. Ni, S. Shiomi and A.-C. Liao, *Class. Quantum Grav.* **21,** S641 (2004).

Binary pulsars and relativistic gravity*

Joseph H. Taylor, Jr.

Princeton University, Princeton, New Jersey 08544

I. SEARCH AND DISCOVERY

Work leading to the discovery of the first pulsar in a binary system began more than twenty years ago, so it seems reasonable to begin with a bit of history. Pulsars burst onto the scene (Hewish *et al.*, 1968) in February 1968, about a month after I completed my Ph.D. at Harvard University. Having accepted an offer to remain there on a post-doctoral fellowship, I was looking for an interesting new project in radio astronomy. When *Nature* announced the discovery of a strange new rapidly pulsating radio source, I immediately drafted a proposal, together with Harvard colleagues, to observe it with the 92 m radio telescope of the National Radio Astronomy Observatory. By late spring we had detected and studied all four of the pulsars which by then had been discovered by the Cambridge group, and I began thinking about how to find further examples of these fascinating objects, which were already thought likely to be neutron stars. Pulsar signals are generally quite weak, but have some unique characteristics that suggest effective search strategies. Their otherwise noise-like signals are modulated by periodic, impulsive waveforms; as a consequence, dispersive propagation through the interstellar medium makes the narrow pulses appear to sweep rapidly downward in frequency. I devised a computer algorithm for recognizing such periodic, dispersed signals in the inevitable background noise, and in June 1968 we used it to discover the fifth known pulsar (Huguenin *et al.*, 1968).

Since pulsar emissions exhibited a wide variety of new and unexpected phenomena, we observers put considerable effort into recording and studying their details and peculiarities. A pulsar model based on strongly magnetized, rapidly spinning neutron stars was soon established as consistent with most of the known facts (Gold, 1968). The model was strongly supported by the discovery of pulsars inside the glowing, gaseous remnants of two supernova explosions, where neutron stars should be created (Large, *et al.*, 1968; Staelin and Reifenstein, 1968), and also by an observed gradual lengthening of pulsar periods (Richards and Comella, 1969) and polarization measurements that clearly suggested a rotating source (Radhakrishnan and Cooke, 1969). The electrodynamical properties of a spinning, magnetized neutron star were studied theoretically (Goldreich and Julian, 1969) and shown to be plausibly capable of generating broadband ratio noise detectable over interstellar distances. However, the rich diversity of the observed radio pulses suggested magnetospheric complexities far beyond those readily incorporated in theoretical models. Many of us suspected that detailed understanding of the pulsar emission mechanism might be a long time coming—and that, in any case, the details might not turn out to be fundamentally illuminating.

In September 1969 I joined the faculty at the University of Massachusetts, where a small group of us planned to build a large, cheap radio telescope especially for observing pulsars. Our telescope took several years to build, and during this time it became clear that whatever the significance of their magnetospheric physics, pulsars were interesting and potentially important to study for quite different reasons. As the collapsed remnants of supernova explosions, they could provide unique experimental data on the final stages of stellar evolution, as well as an opportunity to study the properties of nuclear matter in bulk. Moreover, many pulsars had been shown to be remarkably stable natural clocks (Manchester and Peters, 1972), thus providing an alluring challenge to the experimenter, with consequences and applications about which we could only speculate at the time. For such reasons as these, by the summer of 1972 I was devoting a large portion of my research time to the pursuit of accurate timing measurements of known pulsars, using our new telescope in western Massachusetts, and to planning a large-scale pulsar search that would use bigger telescopes at the national facilities.

I suspect it is not unusual for an experiment's motivation to depend, at least in part, on private thoughts quite unrelated to avowed scientific goals. The challenge of a good intellectual puzzle, and the quiet satisfaction of finding a clever solution, must certainly rank highly among my own incentives and rewards. If an experiment seems difficult to do, but plausibly has interesting consequences, one feels compelled to give it a try. Pulsar searching is the perfect example: it's clear that there must be lots of pulsars out there, and, once identified, they are not so very hard to observe. But finding each one for the first time is a formidable task, one that can become a sort of detective game. To play the game you invent an efficient way of gathering clues, sorting, and assessing them, hoping to discover the identities and celestial locations of all the guilty parties.

Most of the several dozen pulsars known in early 1972 were discovered by examination of strip-chart records, without benefit of further signal processing. Nevertheless, it was clear that digital computer techniques would

*Nobel Lecture, presented to the Royal Swedish Academy of Sciences on 8 December 1993.

be essential parts of more sensitive surveys. Detecting new pulsars is necessarily a multidimensional process; in addition to the usual variables of two spatial coordinates, one must also search thoroughly over wide ranges of period and dispersion measure. Our first pulsar survey, in 1968, sought evidence of pulsar signals by computing the discrete Fourier transforms of long sequences of intensity samples, allowing for the expected narrow pulse shapes by summing the amplitudes of a dozen or more harmonically related frequency components. I first described this basic algorithm (Burns and Clark, 1969) as part of a discussion of pulsar search techniques, in 1969. An efficient dispersion-compensating algorithm was conceived and implemented soon afterward (Manchester *et al.*, 1972; Taylor, 1974), permitting extension of the method to two dimensions. Computerized searches over period and dispersion measure, using these basic algorithms, have by now accounted for discovery of the vast majority of nearly 600 known pulsars, including forty in binary systems (Taylor *et al.*, 1993; Camilo, 1994).

In addition to private stimuli related to "the thrill of the chase," my outwardly expressed scientific motivation for planning an extensive pulsar survey in 1972 was a desire to double or triple the number of known pulsars. I had in mind the need for a more solid statistical basis for drawing conclusions about the total number of pulsars in the Galaxy, their spatial distribution, how they fit into the scheme of stellar evolution, and so on. I also realized (Taylor, 1972) that it would be highly desirable "... to find even *one* example of a pulsar in a binary system, for measurement of its parameters could yield the pulsar mass, an extremely important number." Little did I suspect that just such a discovery would be made, or that it would have much greater significance that anyone had foreseen! In addition to its own importance, the binary pulsar PSR 1913+16 is now recognized as the harbinger of a new class of unusually short-period pulsars with numerous important applications.

An up-to-date map of known pulsars on the celestial sphere is shown in Fig. 1. The binary pulsar PSR 1913+16 is found in a clump of objects close to the Galactic plane around longitude 50°, a part of the sky that passes directly overhead at the latitude of the Arecibo Observatory in Puerto Rico. Forty of these pulsars, including PSR 1913+16, were discovered in the survey that Russell Hulse and I carried out with the 305 m Arecibo telescope (Hulse and Taylor, 1974, 1975a, 1975b). Figure 2 illustrates the periods and spin-down rates of known pulsars, with those in binary systems marked by larger circles around the dots. All radio pulsars slow down gradually in their own rest frames, but the slow-down rates vary over nine orders of magnitude. Figure 2 makes it clear that binary pulsars are special in this regard. With few exceptions, they have unusually small values of both period and period derivative—an important factor which helps to make them especially suitable for high-precision timing measurements.

Much of the detailed implementation and execution of our 1973–74 Arecibo survey was carried out by Russell Hulse. He describes that work, and particularly the discovery of PSR 1913+16, in his accompanying lecture (Hulse, 1994). The significant consequences of our discovery have required accurate timing measurements extending over many years, and since 1974–76 I have pursued these with a number of other collaborators. I shall now turn to a description of these observations.

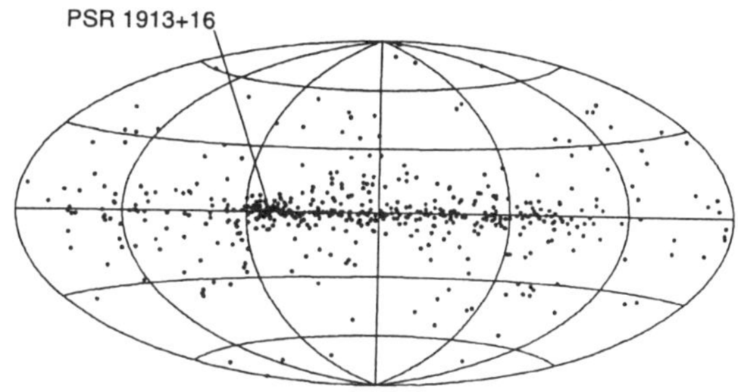

FIG. 1. Distribution of 558 pulsars in Galactic coordinates. The Galactic center is in the middle, and longitude increases to the left.

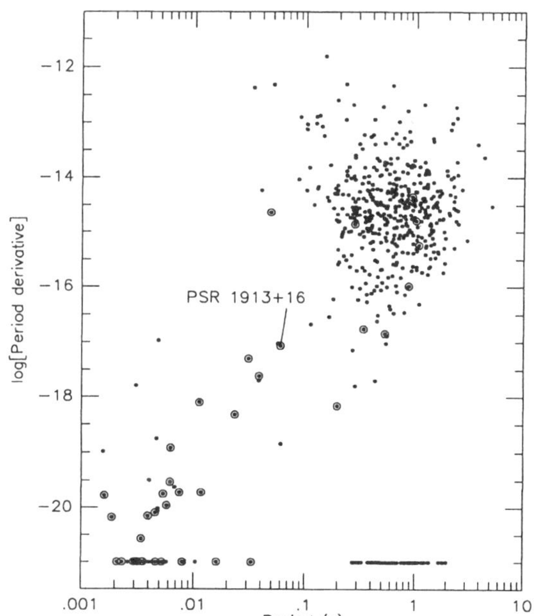

FIG. 2. Periods and period derivatives of known pulsars Binary pulsars, denoted by larger circles around the dots, generally have short periods and small derivatives. Symbols aligned near the bottom represent pulsars for which the slow-down rate has not yet been measured.

II. CLOCK-COMPARISON EXPERIMENTS

Pulsar timing experiments are straightforward in concept: one measures pulse times of arrival (TOAs) at the telescope, and compares them with time kept by a stable reference clock. A remarkable wealth of information about a pulsar's spin, location in space, and orbital motion can be obtained from such simple measurements. For binary pulsars, especially, the task of analyzing a sequence of TOAs often assumes the guise of another intricate detective game. Principal clues in this game are the recorded TOAs. The first and most difficult objective is the assignment of unambiguous pulse numbers to each TOA, despite the fact that some of the observations may be separated by months or even years from their nearest neighbors. During such inevitable gaps in the data, a pulsar may have rotated through as many as 10^7-10^{10} turns, and in order to extract the maximum information content from the data, these integers must be recovered *exactly*. Fortunately, the correct sequence of pulse numbers is easily recognized, once attained, so you can tell when the game has been "won."

A block diagram of equipment used for recent pulsar timing observations (Taylor, 1991) at Arecibo is shown in Fig. 3. Incoming radio-frequency signals from the antenna are amplified, converted to intermediate frequency, and passed through a multichannel spectrometer equipped with square-law detectors. A bank of digital signal averagers accumulates estimates of a pulsar's periodic wave form in each spectral channel, using a precomputed digital ephemeris and circuitry synchronized with the observatory's master clock. A programmable synthesizer, its output frequency adjusted once a second in a phase-continuous manner, compensates for changing Doppler shifts caused by accelerations of the pulsar and the telescope. Average profiles are recorded once every few minutes, together with appropriate time tags. A log is kept of small measured offsets (typically of order 1 μs) between the observatory clock and the best available standards at national time-keeping laboratories, with time transfer accomplished via satellites in the Global Positioning System.

An example of pulse profiles recorded during timing observations of PSR 1913+16 is presented in Fig. 4, which shows intensity profiles for 32 spectral channels spanning the frequency range 1383–1423 MHz, followed by a "de-dispersed" profile at the bottom. In a five-minute observation such as this, the signal-to-noise ratio is just high enough for the double-peaked pulse shape of PSR 1913+16 to be evident in the individual channels. Pulse arrival times are determined by measuring the phase offset between each observed profile and a long-term average with much higher signal-to-noise ratio. Differential dispersive delays are removed, the adjusted offsets are averaged over all channels, and the resulting mean value is added to the time tag to obtain an equivalent TOA. Nearly 5000 such five-minute measurements have been obtained for PSR 1913+16 since 1974, suing essentially this technique. Through a number of improvements in the data-taking systems (Taylor *et al.*, 1976; McCulloch *et al.*, 1979; Taylor *et al.*, 1979; Taylor and Weisberg, 1982, 1989; Stinebring *et al.*, 1992), the typical uncertainties have been reduced from around 300 μs in 1974 to 15–20 μs since 1981.

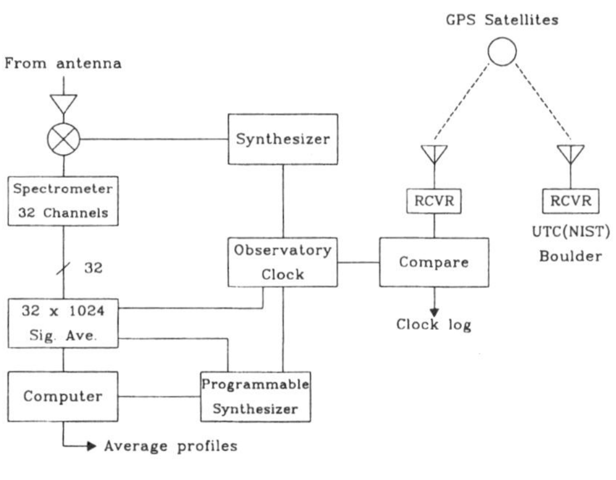

FIG. 3. Simplified block diagram of equipment using for timing pulsars at Arecibo.

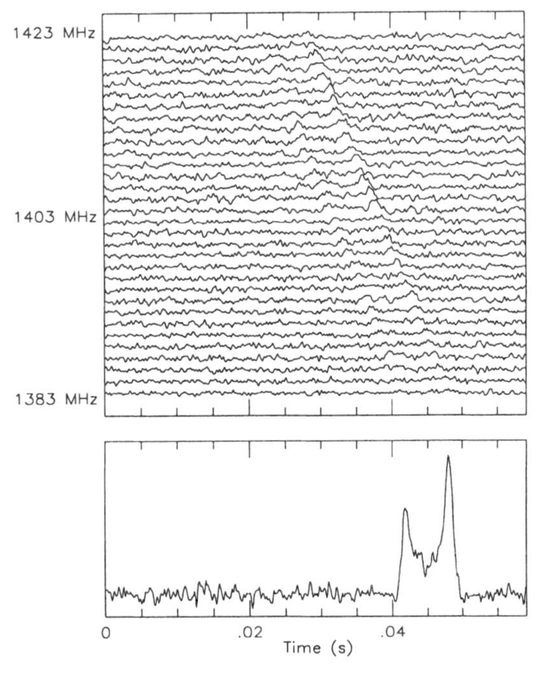

FIG. 4. Pulse profiles obtained on April 24, 1992 during a five-minute observation of PSR 1913+16. The characteristic double-peaked shape, clearly seen in the de-dispersed profile at the bottom, is also discernible in the 32 individual spectral channels.

III. MODEL FITTING

In the process of data analysis, each measured topocentric TOA, say t_{obs}, must be transformed to a corresponding proper time of emission T in the pulsar frame. Under the assumption of a deterministic spin-down law, the rotational phase of the pulsar is given by

$$\phi(T)=\nu T+\frac{1}{2}\dot{\nu}T^2 , \qquad (1)$$

where ϕ is measured in cycles, $\nu \equiv 1/P$ is the rotation frequency, P the period, and $\dot{\nu}$ the slowdown rate. Since a topocentric TOA is a relativistic space-time event, it must be transformed as a four-vector. The telescope's location at the time of a measurement is obtained from a numerically integrated solar-system model, together with published data on the Earth's unpredictable rotational variations. As a first step one normally transforms to the solar-system barycenter, using the weak-field, slow-motion limit of general relativity. The necessary equations include terms depending on the positions, velocities, and masses of all significant solar-system bodies. Next, one accounts for propagation effects in the interstellar medium; and finally, for the orbital motion of the pulsar itself.

With presently achievable accuracies, all significant terms in the relativistic transformation can be summarized in the single equation

$$\begin{aligned} T=t_{obs}-t_0+\Delta_C-D/f^2+\Delta_{R\odot}(\alpha,\delta,\mu_\alpha,\mu_\delta,\pi) \\ +\Delta_{E\odot}-\Delta_{S\odot}(\alpha,\delta) \\ -\Delta_R(x,e,P_b,T_0,\omega,\dot{\omega},\dot{P}_b)-\Delta_E(\gamma)-\Delta_S(r,s) . \qquad (2) \end{aligned}$$

Here t_0 is a nominal equivalent TOA at the solar-system barycenter; Δ_C represents measured clock offsets; D/f^2 is the dispersive delay for propagation at frequency f through the interstellar medium; $\Delta_{R\odot}$, $\Delta_{E\odot}$, and $\Delta_{S\odot}$ are propagation delays and relativistic time adjustments within the solar system; and Δ_R, Δ_E, and Δ_S are similar terms for effects within a binary pulsar's orbit. Subscripts on the various Δ's indicate the nature of the time-dependent delays, which include "Römer," "Einstein," and "Shapiro" delays in the solar system and in the pulsar orbit. The Römer terms have amplitudes comparable to the orbital periods times $v/2\pi c$, where v is the orbital velocity and c the speed of light. The Einstein terms, representing the integrated effects of gravitational redshift and time dilation, are smaller by another factor ev/c, where e is the orbital eccentricity. The Shapiro time delay is a result of reduced velocities that accompany the well-known bending of light rays propagating close to a massive object. The delay amounts to about 120 μs for one-way lines of sight grazing the Sun, and the magnitude depends logarithmically on the angular impact parameter. The corresponding delay within a binary pulsar orbit depends on the companion star's mass, the orbital phase, and the inclination i between the orbital angular momentum and the line of sight.

Figure 5 illustrates the combined orbital delay $\Delta_R+\Delta_E+\Delta_S$ for PSR 1913+16, plotted as a function of orbital phase. Despite the fact that the Einstein and Shapiro effects are orders of magnitude smaller than the Römer delay, they can still be measured separately if the precision of available TOAs is high enough. In fact, the available precision is very high indeed, as one can see from the lone data point shown in Fig. 5 with 50 000σ error bars.

Equations (1) and (2) have been written to show explicitly the most significant dependences of pulsar phase on as many as nineteen *a priori* unknowns. In addition to the rotational frequency ν and spin-down rate $\dot{\nu}$, these phenomenological parameters include a reference arrival time t_0, the dispersion constant D, celestial coordinates α and δ, proper-motion terms μ_α and μ_δ, and annual parallax π. For binary pulsars the terms on the third line of Eq. (2), with as many as ten significant orbital parameters, are also required. The additional parameters include five that would be necessary even in a purely Keplerian analysis of orbital motion: the projected semimajor axis $x\equiv a_1 \sin i/c$, eccentricity e, binary period P_b, longitude of periastron ω, and time of periastron T_0. If the experimental precision is high enough, relativistic effects can yield the values of five further "post-Keplerian" parameters: the secular derivatives $\dot{\omega}$ and $\dot{P}_b$, the Einstein parameter γ, and the range and shape of the orbital Shapiro delay, r and $s\equiv \sin i$. Several earlier versions of this formalism for treating timing measurements of binary pulsars exist (Blandford and Teukolsky, 1976; Epstein, 1977; Haugan, 1985), and have been historically important to our progress with the PSR 1913+16 experiment. The elegant framework outlined here was derived during 1985–86 by Damour and Deruelle (1985, 1986).

Model parameters are extracted from a set of TOAs by calculating the pulsar phases $\phi(T)$ from Eq. (1) and minimizing the weighted sum of squared residuals,

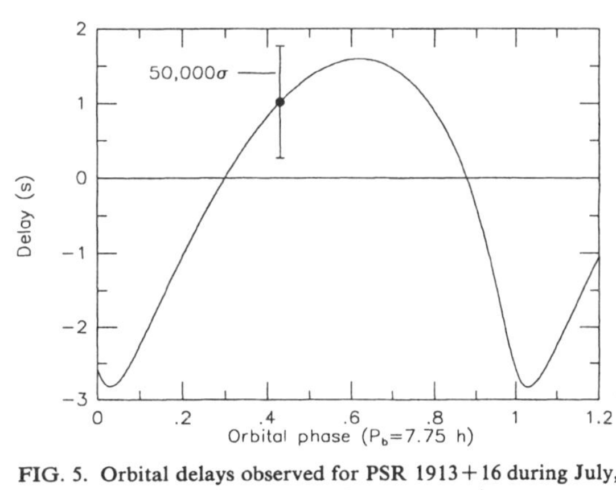

FIG. 5. Orbital delays observed for PSR 1913+16 during July, 1988. The uncertainty of an individual five-minute measurement is typically 50 000 times smaller than the error bar shown.

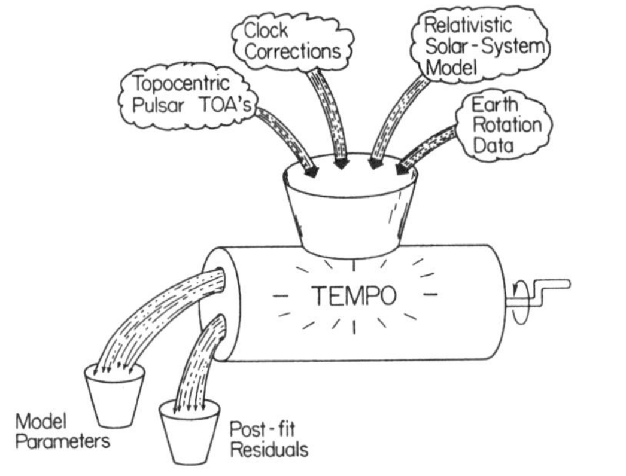

FIG. 6. Schematic diagram of the analysis of pulsar timing measurements carried out by the computer program TEMPO. The essential functions are all described in the text.

$$\chi^2=\sum_{i=1}^{N}\left[\frac{\phi(T_i)-n_i}{\sigma_i/P}\right]^2, \tag{3}$$

with respect to each parameter to be determined. In this equation, n_i is the closest integer to $\phi(T_i)$, and σ_i is the estimated uncertainty of the ith TOA. In a valid and reliable solution the value of χ^2 will be close to the number of degrees of freedom, i.e., the number of measurements N minus the number of adjustable parameters. Parameter errors so large that the closest integer to $\phi(T_i)$ may not be the correct pulse number are invariably accompanied by huge increases in χ^2; this is the reason for my earlier statement that correct pulse numbering is easily recognizable, once attained. In addition to providing a list of fitted parameter values and their estimated uncertainties, the least-squares solution produces a set of post-fit residuals, or differences between measured TOAs and those predicted by the model (see Fig. 6). The post-fit residuals are carefully examined for evidence of systematic trends that might suggest experimental errors, or some inadequacy in the astrophysical model, or perhaps deep physical truths about the nature of gravity.

Necessarily some model parameters will be easier to measure than others. When many TOAs are available, spaced over many months or years, it generally follows that at least the pulsar's celestial coordinates, spin parameters, and Keplerian orbital elements will be measurable with high precision, often as many as 6–14 significant digits. As we will see, the relativistic parameters of binary pulsar orbits are generally much more difficult to measure—but the potential rewards for doing so are substantial.

IV. THE NEWTONIAN LIMIT

Thirty-five binary pulsar systems have now been studied well enough to determine their basic parameters, including the Keplerian orbital elements, with good accuracy. For each system the orbital period P_b and projected semimajor axis x can be combined to give the mass function.

$$f_1(m_1,m_2,s)=\frac{(m_2 s)^3}{(m_1+m_2)^2}=\frac{x^3}{T_\odot(P_b/2\pi)^2}\ . \tag{4}$$

Here m_1 and m_2 are the masses of the pulsar and companion in units of the Sun's mass, $M_\odot$; I use the shorthand notations $s\equiv\sin i$, $T_\odot\equiv GM_\odot/c^3=4.925\,490\,947\times10^{-6}$ s, where G is the Newtonian constant of gravity. In the absence of other information, the mass function cannot provide unique solutions for m_1, m_2, or s. Nevertheless, likely values of m_2 can be estimated by assuming a pulsar mass close to $1.4M_\odot$ (the Chandrasekhar limit for white dwarfs) and the median value $\cos i=0.5$, which implies $s=0.87$. With this approach one can distinguish three categories of binary pulsars, which I shall discuss by reference to Fig. 7: a plot of binary pulsar companion masses versus orbital eccentricities.

Twenty-eight of the binary systems in Fig. 7 have orbital eccentricities $e<0.25$ and low-mass companions likely to be degenerate dwarfs. Most of these have nearly circular orbits; indeed, the only ones with eccentricities more than a few percent are located in globular clusters, and their orbits have probably been perturbed by near collisions with other stars. Five of the binaries have much larger eccentricities and likely companion masses of $0.8M_\odot$ or more; these systems are thought to be pairs of neutron stars, one of which is the detectable pulsar. The large orbital eccentricities are almost certainly the result of rapid ejection of mass in the supernova explosion creating the second neutron star. Finally, at the upper right of Fig. 7 we find two binary pulsars that

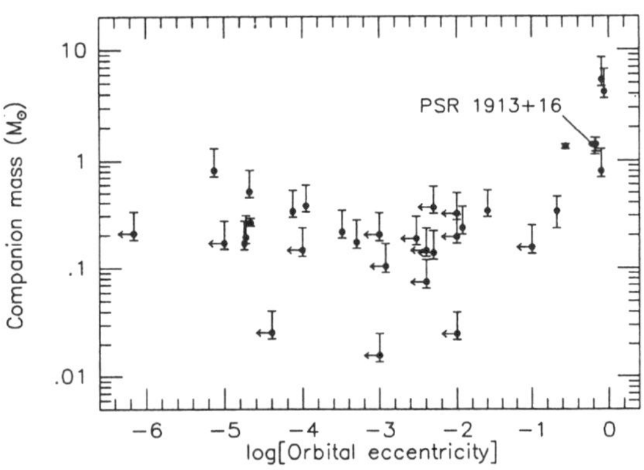

FIG. 7. Masses of the companions of binary pulsars, plotted as a function or orbital eccentricity. Near the marked location of PSR 1913+16, three distinct symbols have merged into one; these three binary systems, as well as their two nearest neighbors in the graph, are thought to be pairs of neutron stars. The two pulsars at the upper right are accompanied by high-mass main-sequence stars, while the remainder are believed to have white-dwarf companions.

move in eccentric orbits around high-mass main-sequence stars. These systems have not yet evolved to the stage of a second supernova explosion. Unlike the binary pulsars with compact companions, these two systems have orbits that could be significantly modified by complications such as tidal forces or mass loss.

V. GENERAL RELATIVITY AS A TOOL

As Russell Hulse and I suggested in the discovery paper for PSR 1913+16 (Hulse and Taylor, 1975a), it should be possible to combine measurements of relativistic orbital parameters with the mass function, thereby determining masses of both stars and the orbital inclination. In the post-Keplerian (PK) framework outlined above, each measured PK parameter defines a unique curve in the (m_1,m_2) plane, valid within a specified theory of gravity. Experimental values for any two PK parameters (say, $\dot{\omega}$ and γ, or perhaps r and s) establish the values of m_1, m_2, and s unambiguously. In general relativity the equations for the five most significant PK parameters are as follows (Damour and Deruelle, 1986; Taylor and Weisberg, 1989; Damour and Taylor, 1992):

$$\dot{\omega}=3\left[\frac{P_b}{2\pi}\right]^{-5/3}(T_{\odot}M)^{2/3}(1-e^2)^{-1}\ , \tag{5}$$

$$\gamma=e\left[\frac{P_b}{2\pi}\right]^{1/3}T_{\odot}^{2/3}M^{-4/3}m_2(m_1+2m_2)\ , \tag{6}$$

$$\dot{P}_b=-\frac{192\pi}{5}\left[\frac{P_b}{2\pi}\right]^{-5/3}\left[1+\frac{73}{24}e^2+\frac{37}{96}e^4\right]\times(1-e^2)^{-7/2}T_{\odot}^{5/3}m_1m_2M^{-1/3}\ , \tag{7}$$

$$r=T_{\odot}m_2\ , \tag{8}$$

$$s=x\left[\frac{P_b}{2\pi}\right]^{-2/3}T_{\odot}^{-1/3}M^{2/3}m_2^{-1}\ . \tag{9}$$

Again the masses m_1, m_2, and $M\equiv m_1+m_2$ are expressed in solar units. I emphasize that the left-hand sides of Eqs. (5) through (9) represent directly measurable quantities, at least in principle. Any two such measurements, together with the well-determined values of e and P_b, will yield solutions for m_1 and m_2, as well as explicit predictions for the remaining PK parameters.

The binary systems most likely to yield measurable PK parameters are those with large masses and high eccentricities and which are astrophysically "clean," so that their orbits are overwhelmingly dominated by the gravitational interactions between two compact masses. The five pulsars clustered near PSR 1913+16 in Fig. 7 would seem to be especially good candidates, and this has been borne out in practice. In the most favorable circumstances, even binary pulsars with low-mass companions and nearly circular orbitals can yield significant post-Keplerian measurements. The best present example is PSR 1855+09: its orbital plane is nearly parallel to

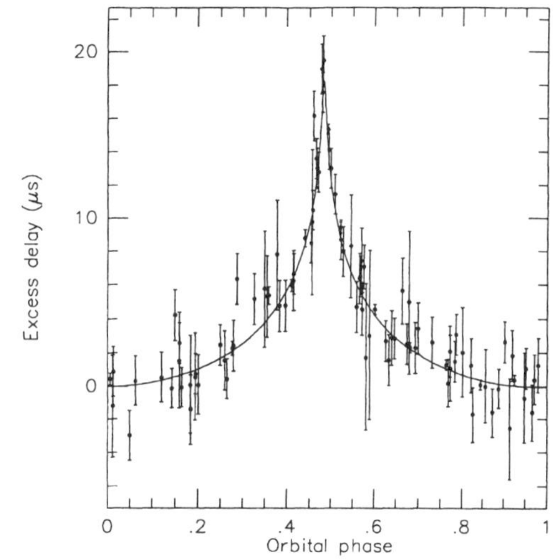

FIG. 8. Measurements of the Shapiro time delay in the PSR 1855+09 system. The theoretical curve corresponds to Eq. (10), and the fitted values of r and s can be used to determine the masses of the pulsar and companion star.

the line of sight, greatly magnifying the orbital Shapiro delay. The relevant measurements (Rawley *et al.*, 1988; Ryba and Taylor, 1991; Kaspi *et al.*, 1994) are illustrated in Fig. 8, together with the fitted function $\Delta_S(r,s)$, in this case closely approximated by

$$\Delta_S=-2r\log(1-s\cos[2\pi(\phi-\phi_0)])\ , \tag{10}$$

where ϕ is the orbital phase in cycles and $\phi_0=0.4823$ the phase of superior conjunction. The fitted values of r and s yield the masses $m_1=1.50^{+0.26}_{-0.14}$, $m_2=0.258^{+0.028}_{-0.016}$. In a

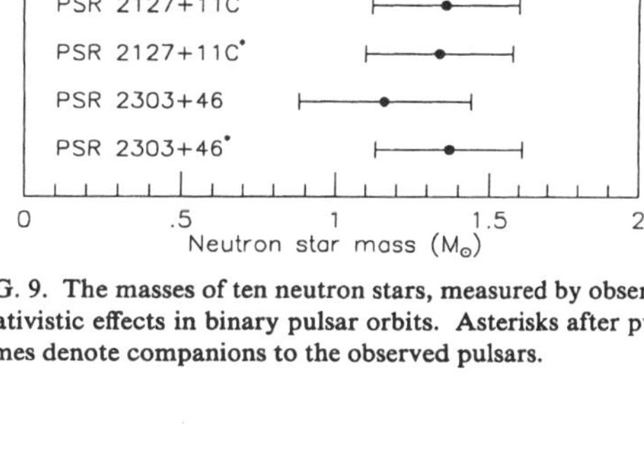

FIG. 9. The masses of ten neutron stars, measured by observing relativistic effects in binary pulsar orbits. Asterisks after pulsar names denote companions to the observed pulsars.

similar way, all binary pulsars with two measurable PK parameters yield solutions for their component masses. At present, most of the experimental data on the masses of neutron stars (see Fig. 9) come from such timing analyses of binary pulsar systems (Taylor and Dewey, 1988; Thorsett *et al.*, 1993, and references therein).

VI. TESTING FOR GRAVITATIONAL WAVES

If three or more post-Keplerian parameters can be measured for a particular pulsar, the system becomes over-determined, and the extra experimental degrees of freedom transform it into a calibrated laboratory for testing relativistic gravity. Each measurable PK parameter beyond the first two provides an explicit, quantitative test. Because the velocities and gravitational energies in a high-mass binary pulsar system can be significantly relativistic, strong-field and radiative effects come into play. Two binary pulsars, PSRs 1913+16 and 1534+12, have now been timed well enough and long enough to yield three or more PK parameters. Each one provides significant tests of gravitation beyond the weak-field, slow-motion limit (Damour and Taylor, 1982; Taylor *et al.*, 1992).

PSR 1913+16 has an orbital period $P_b \approx 7.8$ h, eccentricity $e \approx 0.62$, and mass function $f_1 \approx 0.13 M_\odot$. With the available data quality and time span, the Keplerian orbital parameters are actually determined with fractional accuracies of a few parts per million, or better. In addition, the PK parameters $\dot{\omega}$, γ, and $\dot{P}_b$ are determined with fractional accuracies better than 3×10^{-6}, 5×10^{-4}, and 4×10^{-3}, respectively (Taylor and Weisberg, 1989; Taylor, 1993). Within any viable relativistic theory of gravity, the values of $\dot{\omega}$ and γ yield the values of m_1 and m_2 and a corresponding prediction for $\dot{P}_b$ arising from the damping effects of gravitational radiation. At present levels of accuracy, a small kinematic correction (approximately 0.5% of the observed $\dot{P}_b$) must be included to account for accelerations of the solar system and the binary pulsar system in the Galactic gravitational field (Damour and Taylor, 1991). After doing so, we find that Einstein's theory passes this extraordinarily stringent test with a fractional accuracy better than 0.4% (see Figs. 10 and 11). The clock-comparison experiment for PSR 1913+16 thus provides direct experimental proof that changes in gravity propagate at the speed of light, thereby creating a dissipative mechanism in an orbiting system. It necessarily follows that gravitational radiation exists and has a quadrupolar nature.

PSR 1534+12 was discovered just three years ago, in a survey by Aleksander Wolszczan (1991) that again used the huge Arecibo telescope to good advantage. This pulsar promises eventually to surpass the results now available from PSR 1913+16. It has orbital period $P_b \approx 10.1$ h, eccentricity $e \approx 0.27$, and mass function $f_1 \approx 0.31 M_\odot$. Moreover, with a stronger signal and narrower pulse than PSR 1913+16, its TOAs have considerably smaller measurement uncertainties, around 3 μs for five-minute

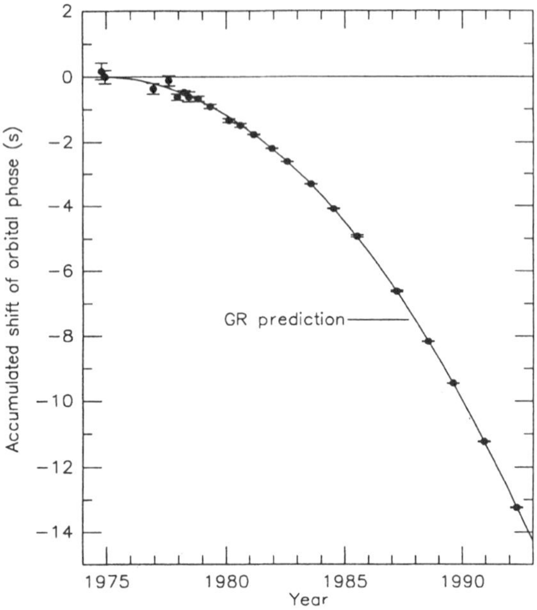

FIG. 10. Accumulated shift of the times of periastron in the PSR 1913+16 system, relative to an assumed orbit with constant period. The parabolic curve represents the general relativistic prediction for energy losses from gravitational radiation.

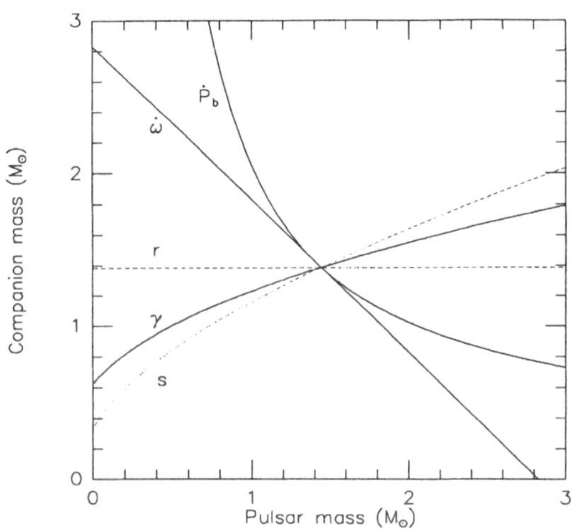

FIG. 11. Solid curves correspond to Eqs. (5)–(7) together with the measured values of $\dot{\omega}$, γ, and $\dot{P}_b$. Their intersection at a single point (within the experimental uncertainty of about 0.35% in $\dot{P}_b$), establishes the existence of gravitational waves. Dashed curves correspond to the *predicted* values of parameters r and s; these quantities should become measurable with a modest improvement in data quality.

observations. Results based on 15 months of data (Taylor *et al.*, 1992) have already produced significant measurements of four PK parameters: $\dot{\omega}$, γ, r, and s. In recent work not yet published, Wolszczan and I have measured the orbital decay rate, $\dot{P}_b$, and found it to be in accord with general relativity at about 20% level. In fact, *all* measured parameters of the PSR 1534+12 system are consistent within general relativity, and it appears that when the full experimental analysis is complete, Einstein's theory will have passed three more very stringent tests under strong-field and radiative conditions.

I do not believe that general relativity necessarily contains the last valid words to be written about the nature of gravity. The theory is not, of course, a quantum theory, and at its most fundamental level the universe appears to obey quantum-mechanical rules. Nevertheless, our experiments with binary pulsars show that, whatever the precise directions of future theoretical work may be, the correct theory of gravity must make predictions that are asymptotically close to those of general relativity over a vast range of classical circumstances.

ACKNOWLEDGMENTS

Russell Hulse and I have many individuals to thank for their important work, both experimental and theoretical, without which our discovery of PSR 1913+16 could not have borne fruit so quickly or so fully. Most notable among these are Roger Blandford, Thibault Damour, Lee Fowler, Peter McCulloch, Joel Weisberg, and the skilled and dedicated technical staff of the Arecibo Observatory.

REFERENCES

Blandford, R., and S. A. Teukolsky, 1976, "Arrival-time analysis for a pulsar in a binary system," Astrophys. J. **205**, 580–591.

Burns, W. R. and B. G. Clark, 1969, "Pulsar search techniques," Astron. Astrophys. **2**, 280–287.

Camilo, F., 1994, "Millisecond pulsar searches," in *Lives of the Neutron Stars, NATO ASI Series*, edited by A. Alpar (Kluwer, Dordrecht).

Damour, T., and N. Deruelle, 1985, "General relativistic celestial mechanics of binary systems. I. The post-Newtonian motion," Ann. Inst. Henri Poincaré Phys. Théor. **43**, 107–132.

Damour, T., and N. Deruelle, 1986, "General relativistic celestial mechanics of binary systems. II. The post-Newtonian timing formula," Ann. Inst. Henri Poincaré Phys. Théor. **44**, 263–292.

Damour, T., and J. H. Taylor, 1991, "On the orbital period change of the binary pulsar PSR 1913+16," Astrophys. J. **366**, 501–511.

Damour, T., and J. H. Taylor, 1992, "Strong-field tests of relativistic gravity and binary pulsars," Phys. Rev. D **45**, 1840–1868.

Epstein, R., 1977, "The binary pulsar: Post Newtonian timing effects," Astrophys. J. **216**, 92–100.

Gold, T., 1968, "Rotating neutron stars as the origin of the pulsating radio sources," Nature **218**, 731–732.

Goldreich, P., and W. H. Julian, 1969, "Pulsar electrodynamics," Astrophys. J. **157**, 869–880.

Haugan, M. P., 1985, "Post-Newtonian arrival-time analysis for a pulsar in a binary system," Astrophys. J. **296**, 1–12.

Hewish, A., S. J. Bell, J. D. H. Pilkington, P. F. Scott, and R. A. Collins, 1968, "Observation of a rapidly pulsating radio source," Nature **217**, 709–713.

Huguenin, G. R., J. H. Taylor, L. E. Goad, A. Hartai, G. S. F. Orsten, and A. K. Rodman, 1968, "New pulsating radio source," Nature **219**, 576.

Hulse, R. A., 1994, "The discovery of the binary pulsar," in Les Prix Nobel (The Nobel Foundation).

Hulse, R. A., and J. H. Taylor, 1974, "A high sensitivity pulsar survey," Astrophys. J. **191**, L59–L61.

Hulse, R. A. and J. H. Taylor, 1975a, "Discovery of a pulsar in a binary system," Astrophys. J. **195**, L51–L53.

Hulse, R. A., and J. H. Taylor, 1975b, "A deep sample of new pulsars and their spatial extent in the galaxy," Astrophys. J. **201**, L55-L59.

Kaspi, V. M., J. H. Taylor, and M. Ryba, 1994, "High-precision timing of millisecond pulsars. III. Long-term monitoring of PSRs B1855+09 and B1937+21," Astrophys. J. (in press).

Large M. I., A. E. Vaughan, and B. Y. Mills, 1968, "A pulsar supernova association," Nature **220**, 340–341.

Manchester, R. N., and W. L. Peters, 1972, "Pulsar parameters from timing observations," Astrophys. J. **173**, 221–226.

Manchester, R. N., J. H. Taylor, and G. R. Huguenin, 1972, "New and improved parameters for twenty-two pulsars," Nature Phys. Sci. **240**, 74.

McCulloch, P. M., J. H. Taylor, and J. M. Weisberg, 1979, "Tests of a new dispersion-removing radiometer on binary pulsar PSR 1913+16, "Astrophys. J. **227**, L133-L137.

Radhakrishnan, V., and D. J. Cooke, 1969, "Magnetic poles and the polarization structure of pulsar radiation," Astrophys. Lett. **3**, 225–229.

Rawley, L. A., J. H. Taylor, and M. M. Davis, 1988, "Fundamental astrometry and millisecond pulsars," Astrophys. J. **326**, 947–953.

Richards, D. W., and J. M. Comella, 1969, "The period of pulsar NP 0532," Nature **222**, 551–552.

Ryba, M. F., and J. H. Taylor, 1991, "High precision timing of millisecond pulsars. I. Astrometry and masses of the PSR 1855+09 system," Astrophys. J. **371**, 739–748.

Staelin, D. H., and E. C. Reifenstein, III, 1968, "Pulsating radio sources near the Crab Nebula," Science **162**, 1481–1483.

Stinebring, D. R., V. M. Kaspi, D. J. Nice, M. F. Ryba, J. H. Taylor, S. E. Thorsett, and T. H. Hankins, 1992, "A flexible data acquisition system for timing pulsars," Rev. Sci. Instrum. **63**, 3551–3555.

Taylor, J. H., 1972, "A high sensitivity survey to detect new pulsars," research proposal submitted to the US National Science Foundation, September, 1972.

Taylor, J. H., 1974, "A sensitive method for detecting dispersed radio emission," Astron. Astrophys. Suppl. Ser. **15**, 367.

Taylor, J. H., 1991, "Millisecond pulsars: Nature's most stable clocks," Proc. IEEE **79**, 1054–1062.

Taylor, J. H., 1993, "Testing relativistic gravity with binary and millisecond pulsars," in *General Relativity and Gravitation 1992*, edited by R. J. Gleiser, C. N. Kozameh, and O. M. Moreschi (Institute of Physics, Bristol), pp. 287–294.

Taylor, J. H., and R. J. Dewey, 1988, "Improved parameters for four binary pulsars," Astrophys. J. **332**, 770–776.

Taylor, J. H., L. A. Fowler, and P. M. McCulloch, 1979, "Measurements of general relativistic effects in the binary pulsar PSR 1913+16," Nature **277**, 437.

Taylor, J. H., R. A. Hulse, L. A. Fowler, G. W. Gullahorn, and J. M. Rankin, 1976, "Further observations of the binary pulsar PSR 1913+16," Astrophys. J. **206**, L53–L58.

Taylor, J. H., R. N. Manchester, and A. G. Lyne, 1993, "Catalog of 558 pulsars," Astrophys. J. Suppl. Ser. **88**, 529–568.

Taylor, J. H., and J. M. Weisberg, 1982, "A new test of general relativity: Gravitational radiation and the binary pulsar PSR 1913+16," Astrophys. J. **253**, 908–920.

Taylor, J. H., and J. M. Weiberg, 1989, "Further experimental tests of relativistic gravity using the binary pulsar PSR 1913+16," Astrophys. J. **345**, 434–450.

Taylor, J. H., A. Wolszczan, T. Damour, and J. M. Weisberg, 1992, "Experimental constraints on strong-field relativistic gravity," Nature **355**, 132–136.

Thorsett, S. E., Z. Arzoumanian, M. M. McKinnon, and J. H. Taylor, 1993, "The masses of two binary neutron star systems," Astrophys. J. **405**, L29–L32.

A. Wolszczan, 1991, "A nearby 37.9 ms radio pulsar in a relativistic binary system," Nature **350**, 688–690.

Chapter 10

Other Perspectives*

*T. T. Wu, C. N. Yang, R. L. Mills, H. L. O'Raifeartaigh, N. Straumann, P. M. Ho, S. Weinberg, P. J. E. Peebles and B. Ratra,

PHYSICAL REVIEW D VOLUME 12, NUMBER 12 15 DECEMBER 1975

Concept of nonintegrable phase factors and global formulation of gauge fields

Tai Tsun Wu*
Gordon McKay Laboratory, Harvard University, Cambridge, Massachusetts 02138

Chen Ning Yang†
Institute for Theoretical Physics, State University of New York, Stony Brook, New York 11794
(Received 8 September 1975)

Through an examination of the Bohm-Aharonov experiment an intrinsic and complete description of electromagnetism in a space-time region is formulated in terms of a nonintegrable phase factor. This concept, in its global ramifications, is studied through an examination of Dirac's magnetic monopole field. Generalizations to non-Abelian groups are carried out, and result in identification with the mathematical concept of connections on principal fiber bundles.

I. MOTIVATION AND INTRODUCTION

The concept of the electromagnetic field was conceived by Faraday and Maxwell to describe electromagnetic effects in a space-time region. According to this concept, the field strenght $f_{\mu\nu}$ describes electromagnetism. It was later realized,[1] however, that $f_{\mu\nu}$ by itself does not, in quantum theory, completely describe all electromagnetic effects on the wave function of the electron. The famous Bohm-Aharonov experiment, first beautifully performed by Chambers,[2] showed that in a multiply connected region where $f_{\mu\nu}=0$ everywhere there are physical experiments for which the outcome depends on the loop integral

$$\frac{e}{\hbar c}\oint A_\mu dx^\mu \tag{1}$$

around an unshrinkable loop. This raises the question of what constitutes *an intrinsic and complete description* of electromagnetism. In the present paper we wish to discuss this question and also its generalization to non-Abelian gauge fields.

An examination of the Bohm-Aharonov experiment indicates that in fact only *the phase factor*

$$\exp\left(\frac{ie}{\hbar c}\oint A_\mu dx^\mu\right), \tag{2}$$

and *not the phase* (1), is physically meaningful. In other words, the phase (1) contains more information than the phase factor (2). But the additional information is not measurable. This simple point, probably implicitly recognized by many authors, is discussed in Sec. II. It leads to the concept of nonintegrable (i.e., path-dependent) phase factor as the basis of a description of electromagnetism.

This concept has been taken[3] as the basis of the definition of a gauge field. The discussions in Ref. 3, however, centered only on the local properties of gauge fields. To extend the concept to global problems we analyze in Sec. III the field produced by a magnetic monopole. We demonstrate how the quantization of the pole strength, a striking result due to Dirac,[4] is understood in this concept of electromagnetism. The demonstration is closely related to that in the original Dirac paper. Dirac discussed the phase factor of the wave function of an electron (which, among other things, depends on the electron energy). Our emphasis is on the nonintegrable electromagnetic phase factor (which does not depend on such quantities as the energy of the electron).

The monopole discussion leads to the recognition that in general the phase factor (and indeed the vector potential A_μ) can only be properly defined in each of many overlapping regions of space-time. In the overlap of any two regions there exists a gauge transformation relating the phase factors defined for the two regions. This discussion is made more precise in Sec. IV. It leads to the definition of global gauges and global gauge transformations.

In Sec. V generalizations to non-Abelian gauge groups are made. The special cases of SU_2 and SO_3 gauge fields are discussed in Secs. VI and VII. A surprising result is that the monopole types are quite different for SU_2 and SO_3 gauge fields and for electromagnetism.

The mathematics of these results is in fact well known to the mathematicians in *fiber bundle theory*. An identification table of terminologies is given in Sec. V. We should emphasize that our interest in this paper does not lie in the beautiful, deep, and general mathematical development in fiber bundle theory. Rather we are concerned with the necessary *concepts to describe the physics of gauge theories*. It is remarkable that these concepts have already been intensively studied as mathematical constructs.

Section VII discusses a "*gedanken*" generalized Bohm-Aharonov experiment for SU_2 gauge fields.

Unfortunately, the experiment is not feasible unless the mass of the gauge particle vanishes. In the last section we make several remarks.

II. DESCRIPTION OF ELECTROMAGNETISM

The Bohm-Aharonov experiment explores the electromagnetic effect on an electron beam (Fig. 1) in a doubly connected region where the electromagnetic field is zero. As predicted[1] by Aharonov and Bohm, the fringe shift is dependent on the phase factor (2), which is equal to

$$\exp\left(\frac{-ie}{\hbar c}\Omega\right),$$

where Ω is the magnetic flux in the cylinder. Thus two cases a and b for which

$$\Omega_a - \Omega_b = \text{integer} \times (hc/e) \tag{3}$$

give the same interference fringes in the experiment. This we shall state and prove as follows.

Theorem 1: If (3) is satisfied, no experiment outside of the cylinder can differentiate between cases a and b.

Consider first an electron outside of the cylinder. We look for a gauge transformation on the electron wave function ψ_a and the vector potential $(A_\mu)_a$ for case a, which changes them into the corresponding quantities for case b, i.e. we try to find $S=e^{-i\alpha}$ such that

$$S = S_{ab} = (S_{ba})^{-1},$$

$$\psi_b = S^{-1}\psi_a, \quad \text{or } \psi_b = e^{i\alpha}\psi_a, \tag{4}$$

$$(A_\mu)_b = (A_\mu)_a - \frac{i\hbar c}{e} S\frac{\partial S^{-1}}{\partial x^\mu}, \quad \text{or } (A_\mu)_b = (A_\mu)_a + \frac{\hbar c}{e}\frac{\partial \alpha}{\partial x^\mu}. \tag{5}$$

For this gauge transformation to be definable, S must be *single-valued*, but α itself need not be. Now $(A_\mu)_b - (A_\mu)_a$ is curlless; hence (5) can always be solved for α. But it is multiple-valued with an increment of

$$\Delta\alpha = \frac{e}{\hbar c}\oint [(A_\mu)_b - (A_\mu)_a]\, dx^\mu$$
$$= \frac{e}{\hbar c}(\Omega_b - \Omega_a) \tag{6}$$

every time one goes around the cylinder. If (3) is satisfied, $\Delta\alpha = 2\pi \times$ integer and S is single-valued. Case a and case b outside of the cylinder are then gauge-transformable into each other, and no physically observable effects would differentiate them. The same argument obviously holds if one studies the wave function of an interacting system of particles provided the charges of the particles are all integral multiples of e. Thus we have shown the validity of Theorem 1.

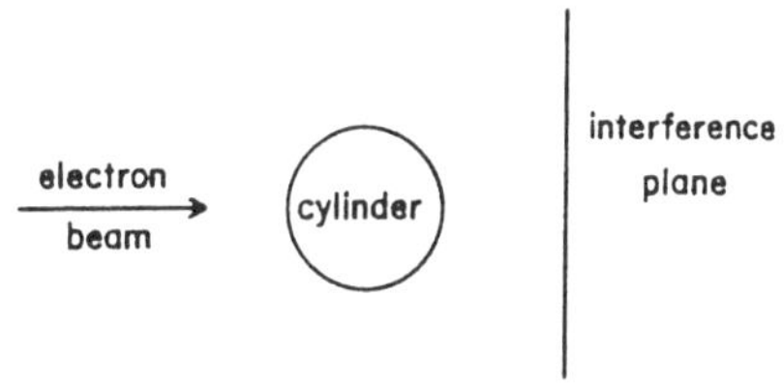

FIG. 1. Bohm-Aharonov experiment (Refs. 1, 2). A magnetic flux is in the cylinder. Outside of the cylinder the field strength $f_{\mu\nu}=0$.

We conclude: (a) The field strength $f_{\mu\nu}$ underdescribes electromagnetism, i.e., different physical situations in a region may have the same $f_{\mu\nu}$. (b) The phase (1) overdescribes electromagnetism, i.e., different phases in a region may describe the same physical situation. What provides a complete description that is neither too much nor too little is the phase factor (2).

Expression (2) is less easy to use (especially when one makes generalizations to non-Abelian groups) as a fundamental concept than the concept of a phase factor for any path from P to Q

$$\Phi_{QP} = \exp\left(\frac{ie}{\hbar c}\int_P^Q A_\mu dx^\mu\right) \tag{7}$$

provided that an arbitrary gauge transformation

$$\exp\left(\frac{ie}{\hbar c}\int_P^Q A_\mu dx^\mu\right)$$
$$\to \exp\left(\frac{ie}{\hbar c}a(Q)\right)\exp\left(\frac{ie}{\hbar c}\int_P^Q A_\mu dx^\mu\right)\exp\left(\frac{-ie}{\hbar c}a(P)\right) \tag{8}$$

does not change the prediction of the outcome of any physical measurements. Following Ref. 3, we shall call the phase factor (7) a nonintegrable (i.e., path-dependent) phase factor.

Electromagnetism is thus the gauge-invariant manifestation of a nonintegrable phase factor. We shall develop this theme further in the next section.

III. FIELD DUE TO A MAGNETIC MONOPOLE

The definition of a nonintegrable phase factor (7) in a general case may present problems. To illustrate the problem, let us study the magnetic monopole field of Dirac.[4] Consider a static magnetic monopole of strength $g \neq 0$ at the origin $\vec{r}=0$ and take the region R of space-time under consideration to be all space-time minus the origin $\vec{r}=0$. We shall now show the following:

Theorem 2: There does not exist a singularity-free A_μ over all R.

If a singularity-free A_μ does exist throughout R, consider the loop integral $\oint A_\mu dx^\mu$ for time $t=0$ around a circle at fixed spherical coordinates r and θ with azimuthal angle $\phi = 0 - 2\pi$. This integral, denoted by $\Omega(r,\theta)$ for $r>0$, is equal to the magnetic flux through a cap bounded by the loop, or more explicitly $\Omega(r,\theta) = 2\pi g(1-\cos\theta)$. At $\theta=0$, $\Omega(r,0)=0$. Increasing θ leads to a continuous increase in Ω till one approaches $\theta=\pi$, at which

$$\Omega(r,\pi) = 4\pi g\,. \tag{9}$$

But at $\theta=\pi$ the loop shrinks to a point. Therefore $\Omega(r,\pi)=0$ since A_μ has no singularity. We have thus reached a contradiction and Theorem 2 is proved.

With an A_μ which has singularities, the nonintegrable phase factor becomes undefined if the path goes through a singularity. This difficulty *must* be resolved in order to use a nonintegrable phase factor as a fundamental concept to describe electromagnetism. It can be resolved in the following way. Let us seek to divide R into two overlapping regions R_a and R_b and to define $(A_\mu)_a$ and $(A_\mu)_b$, each singularity-free in their respective regions, so that (i) their curls are equal to the magnetic field and (ii) in the overlapping region $(A_\mu)_a$ and $(A_\mu)_b$ are related by a gauge transformation. One possible choice is to take the regions to be

$$\begin{aligned} R_a&:\ 0\le\theta<\pi/2+\delta\quad 0<r,\quad 0\le\phi<2\pi\,,\ \text{all } t\\ R_b&:\ \pi/2-\delta<\theta\le\pi\quad 0<r,\quad 0\le\phi<2\pi\,,\ \text{all } t \end{aligned} \tag{10}$$

with an overlap extending throughout $\pi/2-\delta<\theta<\pi/2+\delta$. (We assume $0<\delta\le\pi/2$.) Take

$$\begin{aligned} (A_t)_a &= (A_r)_a = (A_\theta)_a = 0\,,\quad (A_\phi)_a = \frac{g}{r\sin\theta}(1-\cos\theta)\,,\\ (A_t)_b &= (A_r)_b = (A_\theta)_b = 0\,,\quad (A_\phi)_b = \frac{-g}{r\sin\theta}(1+\cos\theta)\,. \end{aligned} \tag{11}$$

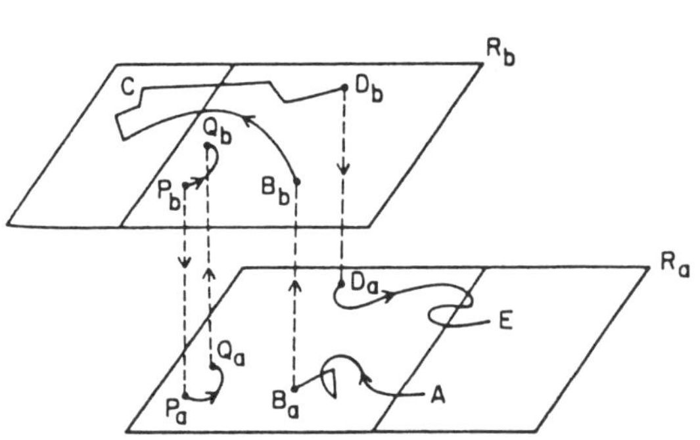

FIG. 2. Schematic diagram illustrating the relationship between R_a and R_b.

The gauge transformation in the overlap of the two regions is

$$S = S_{ab} = \exp(-i\alpha) = \exp\left(\frac{2ige}{\hbar c}\phi\right)\,. \tag{12}$$

This is an allowed gauge transformation if and only if S is single-valued, i.e.,

$$\frac{2ge}{\hbar c} = \text{integer} = D\,, \tag{13}$$

which is Dirac's quantization. With (13) we have

$$S_{ab} = \exp(iD\phi)\,. \tag{12'}$$

To define the phase factor for a path we refer to Fig. 2, where a point in the overlapping region, such as point P, is regarded as two points P_a and P_b. If a path is entirely within region a or b, we define Φ along the path by (7) with $(A_\mu)_a$ or $(A_\mu)_b$ in the integrand in the exponent. If the path $Q-P$ is entirely within the overlapping region we have then two possible phase factors $\Phi_{Q_aP_a}$ and $\Phi_{Q_bP_b}$. It is easy to prove that

$$\Phi_{Q_bP_b} = S^{-1}(Q)\Phi_{Q_aP_a}S(P)\,, \tag{14}$$

i.e.,

$$\Phi_{Q_aP_a}S(P) = S(Q)\Phi_{Q_bP_b}\,, \tag{14'}$$

which merely states that $(A_\mu)_a$ and $(A_\mu)_b$ are related by a gauge transformation with the transformation factor (12).

For a path that crisscrosses in and out of the overlapping region, such as $A-B-C-D-E$ in Fig. 2, the definition of Φ is

$$\Phi_{EDCBA} = \Phi_{ED_a}S_{ab}(D)\Phi_{D_bCB_b}S_{ba}(B)\Phi_{B_aA}\,. \tag{15}$$

Notice that fixing the path but sliding the points B and D along it does not change Φ_{EDCBA} [because of formulas like (14')] so long as B and D remain in the overlapping region.

The phase factor so defined satisfies the group property, e.g.,

$$\begin{aligned} \Phi_{EDCBA} &= \Phi_{ED_a}\Phi_{D_aCBA}\\ &= \Phi_{ED_b}\Phi_{D_bCBA}\\ &= \Phi_{EDC}\Phi_{CBA}\,,\ \text{etc.} \end{aligned} \tag{16}$$

The relationship between the electromagnetic field and the phase factor around a loop is the same as usual. One only has to be careful that if the starting and terminating point A is in the overlapping region, the phase factor is taken to be $\Phi_{A_aBA_a} = \Phi_{A_bBA_b}$, and not $\Phi_{A_aBA_b}$ or $\Phi_{A_bBA_a}$. The phase factor around the loop is then equal to

$$\exp\left(\frac{ie}{\hbar c}\right)\Omega\,,$$

where Ω is the magnetic flux through a cap bordered by the loop. Notice that because of Dirac's quantization condition, the phase factor is the same whichever way one chooses the cap provided it does not pass through the point $\vec{r}=0$ (any t).

We have satisfactorily resolved the difficulty mentioned at the beginning of this section, provided Dirac's quantization condition (13) is satisfied. We shall now prove the following.

Theorem 3: If (13) is not satisfied (the above method of resolving the difficulty would not work since) there exists no division of R into overlapping regions $R_a, R_b, R_c, \ldots$ so that condition (i) and (ii) stated above, properly generalized to the case of more than two regions, would hold.

To prove this statement, observe that if such a division is possible, one could generalize (15) and arrive at a satisfactory definition of the phase factor. The phase factor around a loop is then a continuous function of the loop. Take the loop to be a parallel on the sphere r fixed, $t=0$, θ fixed, $\phi=0-2\pi$. The phase factor defined by the generalization of (15) is equal to

$$\exp\left[\frac{ie}{\hbar c}\Omega(r,\theta)\right]=\exp\left[\frac{ie}{\hbar c}2\pi g(1-\cos\theta)\right]. \qquad (17)$$

This is not equal to unity when $\theta=\pi$, since (13) is assumed to be invalid. Thus we have a contradiction.

Theorem 3 shows that if Dirac's quantization condition (13) is not satisfied, then the field of a magnetic monopole of strength g cannot be taken as a realizable physical situation in R. (Of course, if one excludes the half-line $x=y=0$, $z<0$, or any half-line starting from $\vec{r}=0$ leading to infinity, then it is possible to have any value for g.) This conclusion is the same as Dirac's, but viewed from a somewhat different point of emphasis.

IV. GENERAL DEFINITION OF GAUGE AND GLOBAL GAUGE TRANSFORMATION

Assuming that (13) holds, to round out our concept of a nonintegrable phase factor the question of the flexibility in the choice of the overlapping regions and the flexibility in the choice of A_μ in the regions must be faced. Both of these questions are related to gauge transformations.

Consider a gauge transformation ξ in R_b (ξ will be assumed to be many times differentiable, but not necessarily analytic), resulting in a new potential $(A_\mu)_b'$. We shall illustrate schematically the transformation by "elevating" the region b in Figure 3(a).

One could extend the region b. One could also contract it, provided the whole R remain covered.

One could create a new region by considering a subregion of b as an additional region R_c [Figure 3(b)], and define the gauge transformation connecting them as the identity transformation so that $(A_\mu)_c=(A_\mu)_b$. One can then "elevate" R_c and contract R_b, which results in Fig. 3(c).

Through operations of the kind mentioned in the last three paragraphs, which we shall call *distortions*, we arrive at a large number of possibilities, each with a particular choice of overlapping regions and with a particular choice of gauge transformation from the original $(A_\mu)_a$ or $(A_\mu)_b$ to the new A_μ in each region. Each of such possibilities will be called a *gauge* (or *global gauge*). This definition is a natural generalization of the usual concept, extended to deal with the intricacies of the field of a magnetic monopole.

For each choice of gauge there is a definition of a nonintegrable phase factor for every path. The group condition $\Phi_{C_cBA_a}=\Phi_{C_cB_b}\Phi_{B_bA_a}$ is always satisfied.

Notice that the original gauge we started with was characterized by (a) specifying [in (10)] the regions [R_a and R_b] and (b) specifying the gauge transformation factor (12′) in the overlap (between R_a and R_b). *It does not refer to any specific* A_μ. [A distortion may of course lead to no changes in characterizations (a) and (b). Thus two different gauges may share the same characterizations (a) and (b).] In the case of the monopole field, we had chosen the vector potential to be given by (11). But, in fact, we can attach to this gauge any $(A_\mu)_a$ and $(A_\mu)_b$ provided they are gauge-transformed into each other by (12′) in the region of overlap. (The resultant $f_{\mu\nu}$ is, of course, not a monopole field in general.) *Thus a gauge is a concept not tied to any specific vector potential.* We shall call the process of distortion leading from one gauge to another a *global gauge transformation.* It is also a concept not tied to any specific vector potential. It is a natural generalization of the usual gauge transformation.

The collection of gauges that can be globally gauge-transformed into each other will be said to

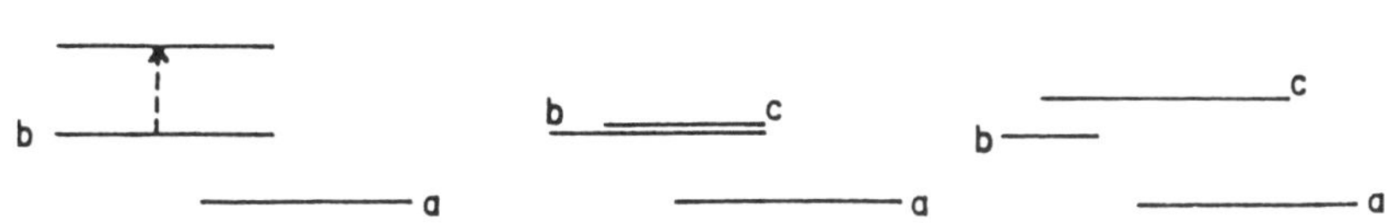

FIG. 3. Distortions allowed in gauge transformation.

belong to the same *gauge type*.

The phase factor around a loop starts and ends at the same point in the same region. Thus it does not change under any global gauge transformation, i.e. we have, for Abelian gauge fields, the following.

Theorem 4a: The phase factor around any loop is invariant under a global gauge transformation.

It follows trivially from this, by taking an infinitesimal loop, that

Theorem 5a: The field strength $f_{\mu\nu}$ is invariant under a global gauge transformation.

For a given value of D, the gauge defined by (10) and (12) will be denoted by $\mathcal{G}_D$. For $D \neq D'$, the relationship, or rather the lack of relationship, between $\mathcal{G}_D$ and $\mathcal{G}_{D'}$ is shown by Theorem 6.

Theorem 6: For $D \neq D'$, $\mathcal{G}_D$ and $\mathcal{G}_{D'}$ are not related by a global gauge transformation, i.e., they are not of the same gauge type.

To prove this theorem we use Theorem 7.

Theorem 7: Between two gauge fields defined on the same gauge there exists a continuous interpolating gauge field defined on the same gauge.

To prove Theorem 7, we simply make a linear interpolation between the two original gauge fields which we shall denote by $(A_\mu)^{(\alpha)}$ and $(A_\mu)^{(\beta)}$:

$$A^{(\gamma)} = t(A_\mu)^{(\alpha)} + (1-t)(A_\mu)^{(\beta)}, \quad 0 \le t \le 1. \qquad (18)$$

In an overlap between regions a and b this interpolating vector potential assumes values $(A_\mu)_a^{(\gamma)}$ and $(A_\mu)_b^{(\gamma)}$ which are related by the proper gauge transformation belonging to this overlap. Thus we have proved Theorem 7.

Now go back to Theorem 6 and assume it to be invalid. Then we can gauge-transform the vector potential belonging to the monopole of strength $D'\hbar c/2e$ to the gauge $\mathcal{G}_D$. For this gauge we have then two monopole fields of different pole strengths. Using Theorem 7 we interpolate between them and obtain unquantized magnetic monopoles, which contradict Theorem 3.

Notice that although in this proof of Theorem 6 we have used two specific gauge fields, the theorem itself does not refer to any specific gauge fields at all.

By the same argument as used in the proof of Theorem 7, any gauge field defined on $\mathcal{G}_D$ must have a magnetic monopole of strength $D\hbar c/2e$ at the excluded point $\vec{r}=0$, in addition to possible fields produced by electric charges and currents. Thus the total magnetic flux around the origin $\vec{r}=0$ is equal to $(2\pi\hbar c/e)D$ for any gauge field defined on $\mathcal{G}_D$. We shall state this as a theorem and give another proof of it.

Theorem 8: Consider gauge $\mathcal{G}_D$ and define any gauge field on it. The total magnetic flux through a sphere around the origin $\vec{r}=0$ is *independent of the gauge field* and only depends on the gauge:

$$\oiint f_{\mu\nu}\,dx^\mu dx^\nu = \frac{-i\hbar c}{e} \oint \frac{\partial}{\partial x^\mu}(\ln S_{ab})\,dx^\mu, \qquad (19)$$

where S is the gauge transformation defined by (12) for the gauge $\mathcal{G}_D$ in question, and the integral is taken around any loop around the origin $\vec{r}=0$ in the overlap between R_a and R_b, such as the equator on a sphere $r=1$.

To prove this theorem we observe that the flux through the upper half of the sphere $r=1$ is equal to the following integral around the equator:

$$\oint (A_\mu)_a\,dx^\mu. \qquad (20a)$$

The flux through the lower half is equal to a similar integral around the equator:

$$-\oint (A_\mu)_b\,dx^\mu. \qquad (20b)$$

Hence

$$\text{total flux} = \oint [(A_\mu)_a - (A_\mu)_b]\,dx^\mu$$

$$= \frac{-i\hbar c}{e} \oint \frac{\partial}{\partial x^\mu}(\ln S_{ab})\,dx^\mu, \qquad (21)$$

which completes the proof. Using (13) and (12), the right-hand side of (21) is equal to $4\pi g$, as expected.

If one starts with any gauge which is of the same gauge type as $\mathcal{G}_D$, and makes a global gauge transformation on it, the total flux is not changed by Theorem 5a. Thus (19), which depends only on the gauge, is in fact the same for all gauges of the same type. Notice that if there are more regions in a gauge than two, (19) should be replaced by a sum of line integrals along paths that are in the various overlaps between the regions. For a case of three regions there are three paths, which are illustrated in Fig. 4. Along each path the integral is of the form (19) with S denoting the gauge transformation factor, such as (12), between the two regions containing the path. To prove Theorem 8 in this case one need only add three loop integrals to-

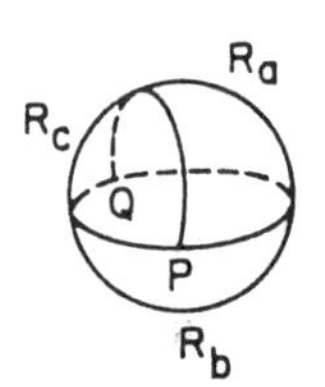

FIG. 4. Case of three regions for Theorem 8. The three paths from P to Q are in the three overlapping regions between (R_a, R_b), (R_b, R_c), and (R_c, R_a).

gether, each of the form of (20a) and (20b), and notice that along each path the integrand is always the difference of the vector potential A_μ between two regions, very much as in (21).

The first proof we gave above of Theorem 8 is easy and is "obvious" to a physicist. The second proof is more involved but is more intrinsic. The theorem is a special case of the Chern-Weil theorem which evolved from the famous Gauss-Bonnet-Allendoerfer-Weil-Chern theorem, a seminal development in contemporary mathematics.[5] We want to emphasize two consequences of the theorem. (i) The right-hand side of (19) is independent of the gauge field, and only depends on the gauge type. (ii) The right-hand side of (19) has as integrand the gradient of $\ln S$. Since S is single-valued, the integral must be equal to an *integral multiple* of a constant (in this case $2\pi i$). A remarkable fact is that these consequences remain valid in the general mathematical theorem, which is very deep.

V. GENERALIZATION TO NON-ABELIAN GAUGE FIELD

So far we have only considered electromagnetism and described it in terms of an Abelian gauge field that corresponds to the group U_1, or equivalently SO_2. On the basis of the discussions in the preceding section, the generalization to the non-Abelian case can be carried out without much difficulty. For a local region this has been done in Ref. 3. Extension to global considerations is our present focus of interest.

A gauge is defined by (a) a particular choice of overlapping regions and (b) a particular choice of *single-valued* gauge transformations S_{ab} in the overlapping regions. The choice of gauge transformations clearly must satisfy the following two conditions.

(1) In the overlapping region $R_a \cap R_b$, the gauge transformations S_{ba} from a to b and S_{ab} from b to a are related by

$$S_{ab} S_{ba} = 1 ,$$

where 1 is the identity element of the gauge group.

(2) If three regions R_a, R_b, and R_c overlap, then there are gauge transformations $S_{ab}, S_{ba}, S_{ac}, S_{ca}, S_{bc}, S_{cb}$ so that

$$S_{ab} S_{bc} S_{ca} = 1 , \text{ etc.}$$

in $R_a \cap R_b \cap R_c$.

As in the case of electromagnetism, both the concept of a gauge and the concept of a global gauge transformation are not tied to any specific gauge potentials, denoted in general by b^k_μ.

The *nonintegrable phase factor* for a given path is now an element of the gauge group. We shall still call it a phase factor. Since these phase factors do not in general commute with each other, Theorems 4a and 5a for the Abelian case need to be modified as follows.

Theorem 4: Under a global gauge transformation, the phase factor around any loop remains in the same class. The class does not depend on which point is taken as the starting point around the loop.

Theorem 5: The field strength $f^k_{\mu\nu}$ is covariant under a global gauge transformation.

Only theorem 4 is not immediately transparent. For a loop $ABCA$, under a gauge transformation[3]

$$\Phi_{ABCA} \to \Phi'_{ABCA} = \xi(A) \Phi_{ABCA} \xi^{-1}(A) .$$

Thus Φ'_{ABCA} and Φ_{ABCA} are in the same class. Also around the same loop if we change the starting point from A to C,

$$\Phi_{CABC} = \Phi_{CA} \Phi_{ABCA} \Phi_{AC} .$$

Hence changing the starting point does not change the class.

Theorem 4 defines *the class of a loop*. This concept is the generalization of the phase factor for electromagnetism around a loop with the magnetic flux as the exponent. It is a gauge-invariant concept.

These concepts have been extensively studied by the mathematicians in the framework of more general[6] mathematical constructs. A translation of terminology is given in Table I.

TABLE I. Translation of terminology.

Gauge field terminology	Bundle terminology
gauge (or global gauge)	principal coordinate bundle
gauge type	principal fiber bundle
gauge potential b_μ^k	connection on a principal fiber bundle
S_{ba} (see Sec. V)	transition function
phase factor Φ_{QP}	parallel displacement
field strength $f_{\mu\nu}^k$	curvature
source [a] J_μ^K	?
electromagnetism	connection on a $U_1(1)$ bundle
isotopic spin gauge field	connection on a SU_2 bundle
Dirac's monopole quantization	classification of $U_1(1)$ bundle according to first Chern class
electromagnetism without monopole	connection on a trivial $U_1(1)$ bundle
electromagnetism with monopole	connection on a nontrivial $U_1(1)$ bundle

[a] I.e., electric source. This is the generalization (see Ref. 3) of the concept of electric charges and currents.

Chen Ning Yang and Tai Tsun Wu in Leiden (1984)
(Photo courtesy of Judy Wong)

Gauge fields

Robert Mills
Physics Department, The Ohio State University, Columbus, Ohio 43210

(Received 15 January 1987; accepted for publication 6 February 1989)

This article is a survey of the history and ideas of gauge theory. Described here are the gradual emergence of symmetry as a driving force in the shaping of physical theory; the elevation of Noether's theorem, relating symmetries to conservation laws, to a fundamental principle of nature; and the force of the idea ("the gauge principle") that the symmetries of nature, like the interactions themselves, should be *local* in character. The fundamental role of gauge fields in mediating the interactions of physics springs from Noether's theorem and the gauge principle in a remarkably clean and elegant way, leaving, however, some tantalizing loose ends that might prove to be the clue to a future deeper level of understanding. The example of the electromagnetic field as the prototype gauge theory is discussed in some detail and serves as the basis for examining the similarities and differences that emerge in generalizing to non-Abelian gauge theories. The article concludes with a brief examination of the dream of total unification—all the forces of nature in a single unified gauge theory, with the differences among the forces due to the specific way in which the fundamental symmetries are broken in the local environment.

FOREWORD

This article appeared first in Chinese translation, in Ziran Zazhi (Nature Journal, published in Shanghai), August 1987 and, subsequently, in a shortened English version in the proceedings of a conference on the History of Modern Gauge Theories (July 1987, Utah State University, Logan, UT), edited by Arnold Rosenblum and published by Plenum Press. Both Ziran Zazhi and Plenum Press have kindly given their permission for its use in the *American Journal of Physics*.

I hope to be forgiven the absence of a bibliography. The discursive and descriptive nature of the article has not seemed to require references and it would be difficult to do justice to the voluminous literature on the subject without major changes in the character of the article.

I. INTRODUCTION: THE SWEEP OF CHANGE

The 20th century has been a time of tremendous ferment and change in our understanding of the world around us. It has not been simply a matter of discovering new phenomena or adding new features to existing theories though indeed those things have happened to a remarkable extent: It goes much deeper than that. The laws of physics, as we try to state them, have a basic character that reflects our understanding of the nature of reality at the deepest philosophical level; it is this most basic character that has been repeatedly and violently upset just within the lifetime of many of us.

No one figure stands larger in 20th-century physics, or symbolizes more fully this revolution or, rather, this series of revolutions, in our understanding of nature than Albert Einstein (1879–1955). In devising the special and general theories of relativity, Einstein singlehandedly revised, radically and permanently, our conception of space and time. He also played a key role in the even deeper revolution represented by quantum theory, which shattered our deterministic view of physical laws (though Einstein himself could not accept this), and gave the *observer* a basic role in fixing physical reality whose significance we have still, in my view, not fully grasped. The welding of these two conceptual revolutions, of relativity and quantum theory, into the view of nature known as quantum field theory leads to yet another radical departure, not implied by either relativity or quantum theory alone, from our conventional understanding, involving in this case the very nature of elementary particles. These are no longer seen as primordial indestructible matter, but as "field quanta," that is, the quantum excitations of different fundamental fields, in exact analogy to the way in which photons, the particlelike units of electromagnetic energy, are understood as quantum excitations of the electromagnetic field.

Finally, I want to draw attention to Einstein's dream of a unified *geometrical* picture of all the fundamental forces of nature, generalizing his most fruitful and most beautiful creation, the general theory of relativity. This theory sees gravity as the manifestation of a curvature of space and time by which the motion of all objects—whether planets or ping-pong balls—is seen not as the result of actual gravitational "forces" acting on the object, but as the object's attempt to follow the straightest possible path (the "geodesic") among the bumps and hollows of space-time itself. Einstein spent most of his later years in a herculean, but vain, effort to extend this geometrical approach to include electromagnetic forces. It is ironic, therefore, that the last 30 years, starting, in fact, the year before Einstein died, have seen the development of a point of view, the principle of *gauge invariance*, that generalizes in an entirely unexpected way the geometrical character of Einstein's general theory of relativity and succeeds in providing a satisfactory description of all the forces of nature. It is not yet a *unified* theory—there are three independent gauge theories in the currently accepted picture—but most physicists feel that full unification is only a matter of time.

The gauge principle, which might also be described as a principle of *local symmetry*, is a statement about the invariance properties of physical laws. It requires that *every continuous symmetry be a local symmetry*, a phrase that is explained and discussed in detail later. This requirement involves a far greater degree of symmetry in nature than is normally conceived and the idea would probably never have been proposed if there were not familiar examples of local symmetry already known, namely, general relativity and electromagnetic theory. It is so severe a constraint that there is only a very limited class of theories that can meet it and extremely little arbitrariness in the forms of interaction that are allowed.

It is the purpose of this article to explore this idea of gauge invariance—to tell something of the history of the idea, to give a survey of some of the physics needed to understand the principle, and to describe the logic of the gauge invariance principle and how it gives rise to very particular forms of physical theories. Three key concepts run through the discussion—distinct, but tightly interrelated: symmetry, conservation laws, and gauge fields. We shall look first at how these interrelationships appear in the case of classical electromagnetic theory, which is the simplest example of a gauge field theory, and then examine how this provides the stimulus and basis for the generalization to what are now called non-Abelian gauge theories. I shall then give a sketch of the developments in more recent years—the resolution of a variety of formidable difficulties and the formulation of the "Standard Model," the current widely accepted picture of the elementary particles and the gauge fields by which they interact.

While our present understanding of physics involves a unification of *type* in the sense that all the force fields of nature are seen to be of the same character, namely, *gauge fields*, I shall discuss, finally, the hope that most physicists share of a more complete unification—the hope of showing that all the forces are associated with a *single* gauge theory, with a single multicomponent gauge field whose different components are related to each other in some completely symmetrical way. This dream has already been partially realized in the Standard Model through the unification of the electromagnetic and weak interactions, but there still are major obstacles in the way of its final realization.

II. THE BEGINNINGS OF THE GAUGE IDEA

The key ideas leading up to the introduction of generalized gauge fields came from Noether, Weyl, and London. The underlying trend, of which gauge symmetry is a particular manifestation, is the growing realization in this century of the importance of symmetry to our basic understanding of the universe, to the point where it is now felt that it is the underlying symmetry of physical laws that *drives* the system—that determines the structure of the laws and the number and character of the elementary particles. This is a characteristically 20th-century development. Prior to this, symmetries were seen as accidental and if the theories of physics showed certain symmetrical structures, that was nice, but not of fundamental importance.

A. Noether

Emmy Noether (1882–1935) is regarded in mathematical circles as one of the most important mathematicians of this century, though not in fact for the work that has made her known to physicists. The theorem that bears her name, which has been the keystone in the development of symmetry as a guiding force in physics and for which (and for which alone) she is known in physics, is hardly known to mathematicians, who honor her for her work on commutative rings and algebraic number theory. In each case, though, she is noted for probing the underlying concepts upon which mathematical disciplines are based.

Noether's theorem, proved in 1918 while she was still (as a woman) striving for recognition at the University of Göttingen, relates in particular to variational principles as they apply to physics. The substance of the theorem, for our purposes, is that *for every symmetry of nature there is a corresponding conservation law and for every conservation law there is a symmetry*. In the Lagrangian formulation of a physical theory, the symmetry in question is a symmetry of the Lagrangian; since the form of the Lagrangian determines the equations of motion of the system being described, this means that it is symmetry of those equations of motion, that is, of the physical theory itself.

Physics is characterized by a number of deeply significant conservation laws such as energy, linear and angular momentum, and electric charge, together with a number of others that have emerged in more recent years. Noether's theorem shows explicitly how these are related to the very structure—the symmetry, in fact—of physical laws. All theorems are proved on the basis of given hypotheses, and in the case of Noether's theorem, the most important assumption is that the equations of motion of physics are derivable from a *variational principle*, known as Hamilton's principle. Hamilton's principle states that for the true trajectory of a system—its history as a function of time—the time integral of the Lagrangian is stationary with respect to small changes of that trajectory away from its true shape. There is no obvious reason *why* the equations of motion should have this character. It is easy to devise a universe whose equations of motion do not satisfy any such variational principle, but in our universe they always do. Hamilton's principle was first discovered in connection with mechanical systems, where the Lagrangian turns out to be the difference between the kinetic and potential energies, but the principle is easily extended to include velocity-dependent forces of certain types, including the magnetic force on a moving charged particle. (Hamilton's principle cannot be extended to include dissipative forces, but elementary forces are never dissipative.) Finally, it has turned out that systems of a completely nonmechanical nature, such as the electromagnetic field, also can be described in this way: Maxwell's equations, the equations of motion of the electromagnetic field, can also be derived from Hamilton's principle, though with a Lagrangian that seems to have nothing to do with such things as kinetic and potential energy. We have become so used to this state of affairs that when we are trying to devise a new theory we invariably look for the correct Lagrangian, assuming almost without question that the new theory must also obey Hamilton's principle. It always seems to work and, in consequence, we always have Noether's theorem also: the relation between symmetries and conservation laws.

When we make the transition to quantum theory, Noether's theorem is still true, although in most ways of doing quantum theory the proof looks very different. One such proof is outlined in Sec. III. It seems to me quite possible that Noether's theorem is the more fundamental fact—that the physical theories that we devise to describe the universe about us have the structure they do *because* of this fundamental relationship between symmetries and conservation laws. If this is so, then Noether's theorem becomes a *principle* rather than a theorem, like the principles of equivalence in special and general relativity; we should say then that classical physical laws take the Lagrangian form and quantum theory takes its characteristic Hamiltonian form as a consequence of Noether's principle.

B. Weyl and London

The idea of gauge invariance also had its inception in 1918. Hermann Weyl (1885–1955), a friend of Noether's at Göttingen, had been deeply influenced by Einstein and shared his vision of seeing electromagnetism as a manifestation of some kind of local symmetry, similar to the local symmetry that characterizes the general theory of relativity. In the case of general relativity, the symmetry in question is an invariance of the form of the basic equations under arbitrary curvilinear coordinate transformations, corresponding to the physical requirement that physical laws appear the same to all observers regardless of the state of motion—accelerating, rotating, or whatever—of their reference frames.

The invariance that Weyl hoped to exploit was an invariance with respect to change of *scale*: the requirement that physical laws be the same if the scale of all length measurements is changed by the same overall factor. Weyl wanted to require a *local* gauge invariance in which the scale changes are allowed to be different at different points in space and time, analogous to the curvilinear coordinate transformations of general relativity. The associated conservation law, by Noether's theorem, was to be the conservation of electric charge.

Einstein pointed out serious flaws in the idea and it lay dormant until 1927, by which time Schrödinger had introduced his wave equation for quantum theory (in 1926) and complex wavefunctions were seen to play a role in physics. In 1927, Fritz London (1900–1954) pointed out that the symmetry associated with electric charge conservation was not a scale invariance, but a *phase* invariance, i.e., the invariance of quantum theory under an arbitrary change in the complex phase of the wavefunction (explained in detail in Sec. II). The invariance uner a "global" phase change—multiplication of the wavefunction by a constant phase factor $e^{i\theta}$—was trivial in fact; the nontrivial fact was that the existence of the electromagnetic field allows a much broader kind of invariance, invariance under a *local* phase change, in which the phase factor varies arbitrarily from one point to another in space-time. That is, θ becomes an arbitrary function of x, y, z, and t, the coordinates of space and time. How this works is explained in Sec. III.

Weyl also played a part in this modification of his idea and continued to use the name "gauge symmetry" to describe it, although it was now a misnomer, since the word "gauge" historically refers to a choice of length scale, rather than to the assignment of complex phases.

C. Yang and Mills

For some 25 years the idea of gauge invariance (almost always thought of in terms of *local* gauge invariance) was seen as a specific characteristic of electromagnetic theory, useful in various ways (e.g., as a check on the validity of calculation procedures), but not of more fundamental significance. Local gauge invariance was felt to have additional implications within electromagnetic theory, such as zero mass for the photon (although it was hard to make a rigorous proof of this) and the constancy, among the different elementary particles, of the elementary unit of electric charge.

The idea that local gauge invariance might have a more universal significance in physics began to be considered in the early 1950s, particularly by C. N. Yang (Yang Chen Ning) (1922–), who was then at the Institute for Advanced Study in Princeton, NJ. For some years prior to this, since the time when Yang was a graduate student at the Southwest Associated University in Kunming, China, he had been much impressed with the relationship between charge conservation and gauge invariance and, in particular, by the fact that the whole structure of electromagnetic theory would be uniquely determined by the sole requirement of gauge invariance. After coming to the United States in 1945, as a graduate student at the University of Chicago, Yang began the attempt to generalize the gauge invariance argument to other conservation laws, in particular the conservation of isospin. Many conservation laws of various sorts have appeared since then, but at that time the only conservation law that bore a useful similarity to electric charge was the conservation of isospin (usually referred to then as "isotopic spin"). Isospin was an imperfect conservation law, violated by electromagnetic and weak interactions, but apparently strictly true for strong interactions. One could easily imagine a world with only the strong interaction, where the conservation of isospin and the associated symmetries would be exactly valid. If the gauge invariance idea could be generalized, the result should be a complete theory of the strong interaction, with isospin as the "charge" responsible for the interactions and the newly invented gauge field as the "glue" playing the same role as the electromagnetic field in electrodynamics.

During the academic year 1953–1954, Yang was a visitor to Brookhaven National Laboratory, about 80 km east of New York City on Long Island. Here the Cosmotron, then the biggest particle accelerator in the world (accelerating protons of energies of 2–3 GeV) was just beginning to produce the abundance of new and unfamiliar particles that have transformed the face of physics in the years since. I was at Brookhaven also, on a postdoctoral appointment, and was assigned to the same office as Yang. (I was still belatedly writing my dissertation, the study of a possible contribution to the fourth-order Lamb shift, under the guidance of Normal Kroll at Columbia University in New York.) Yang, who has demonstrated on a number of occasions his generosity to young physicists beginning their careers, told me about his idea of generalizing gauge invariance and we discussed it at some length. Having some background in quantum electrodynamics, I was able to contribute something to the discussions, especially with regard to the quantization procedures, and to a small degree in working out the formalism; however, the key ideas were Yang's. The predicted quanta would have a spin of 1, like the photon, but would also have isospin of 1, like the pi meson: This means that they would form a charge triplet, with positive, negative, and neutral states, just like the pion. The question of renormalizability was far beyond us, as was the question of the mass of the gauge field quantum. These questions were not to be resolved for another 10 –15 years, by which time there had been an order-of-magnitude increase in the sophistication of physicists' understanding of quantum field theory.

At about the same time (also in 1954), Ronald Shaw, a student of Abdus Salam at Cambridge University in England, was also thinking deeply about possible generalizations of the idea of gauge invariance, influenced in particular by lecture notes of Schwinger. Shaw's unpublished doctoral dissertation (1954) on *The Problem of Particle Types and Other Contributions to the Theory of Elementary*

Particles includes a section ("Invariance under General Isotopic Spin Transformations") that closely and independently parallels the argument of the 1954 paper of Yang and myself and duplicates the basic equations of non-Abelian gauge theory. There seems to be no question that the time was ripe for this development.

III. THE GAUGE PHILOSOPHY: LOCAL SYMMETRY

The idea at the core of gauge theory, as mentioned earlier, is the local symmetry principle: *Every continuous symmetry of nature is a local symmetry.* Let me now explain more fully what these words mean. First, what is a "symmetry of nature"? We say that an *object* is symmetrical if its appearance is unchanged by some transformation—we can say that its properties are "invariant" under the transformation. Look at some examples: An equilateral triangle looks the same if it is rotated by 120° or if it is reflected with respect to any of its altitudes, while a sphere has a much richer class of symmetries, being invariant under a rotation of *any* magnitude about *any* of its diameters or a reflection in *any* of its median planes. Another type of symmetry, called a "translational" symmetry, is illustrated by an infinitely long cylinder, which is invariant under any displacement parallel to its axis. The point, then, is that a symmetry necessarily involves some invariance property.

Now the world as we see it is not particularly symmetrical because the objects in it are irregular in shape and location. The *rules of behavior* of the physical world have a great deal of symmetry, however; this is what I mean by *symmetries of nature.* The fact that experiments work the same in China and in the US reflects an invariance of the laws of physics under spatial displacements and rotations, and hence a symmetry. The fact that the same experimental results are obtained at different times reveals the *time displacement symmetry* of those laws, etc.

The transformations we normally consider are *global*—a single rotation, for example, on the entire universe. Invariance under such a transformation is referred to as a "global symmetry." The meaning of a *local* symmetry, on the other hand, would be that the objects or physical laws in question are invariant under a "local transformation," — which is in fact a large number of separate transformations with a different one at every point in space and time. As an example, consider a long circular cylinder, which is clearly invariant under rotations about its axis. Now imagine that the cylinder is sliced into a large number N, say, of very thin disks. The system is now invariant under a transformation in which every ring is rotated through a different angle and the resulting symmetry, a local symmetry, is a much richer symmetry than the original one. The original global symmetry is an invariance under a set of transformations described by a single parameter, the angle of rotation, while in the local symmetry case, we have invariance under a much larger set of transformations described by N different angles of rotation.

We now apply this idea, not to objects, but to the laws of physics, and ask that somehow they manage to have this much richer type of symmetry that we saw in the case of the thinly sliced cylinder. Why would one think it possible that the laws of physics should have such an extended symmetry? It would seem extremely unlikely if it were not for the fact that we already know of two examples in nature: electromagnetic theory and general relativity, both of which, as we shall see, show just this kind of local symmetry.

In general relativity, the relevant local symmetry is an invariance under arbitrary curvilinear coordinate transformations—as if one were making a different space-time rotation at every point. This is a sort of four-dimensional analog of the sliced cylinder described above, with the difference that we are now talking about the invariance, not of an object, but of physical laws. It is well known that insisting on this general invariance leads inevitably to a complete theory of the gravitational force.

In the electromagnetic case, which I talk about in more detail in Sec. IV, the symmetry in question is an invariance under changes in phase of complex fields or wavefunctions,

$$\psi \rightarrow \psi e^{i\theta}, \tag{1}$$

and it becomes a local symmetry if θ is taken to be an arbitrary function of the space-time coordinates (x,y,z,t). This may not seem to have anything to do with electromagnetism but, in fact, as we shall see, this local symmetry can be realized only if we introduce an additional field with all the familiar properties of the electromagnetic field.

The electromagnetic case was in fact the inspiration for the further development of the gauge field idea. The fact that examples of local symmetry were already known to exist in nature strongly suggested that this might be a general principle and that we should examine other observed symmetries of nature to see if the same thing happens in every case. In Sec. IV, I will show how the gauge principle works in detail, first for the electromagnetic case and then for the non-Abelian generalization.

IV. CONSERVED QUANTITIES, SYMMETRIES, AND GAUGE FIELDS

We now need to find out how the assumption of local symmetry can lead to a physical theory and how it can determine the character of that theory. We find that in every case there is a characteristic logical pattern that emerges, represented graphically in Fig. 1, connecting conserved quantities, symmetries of nature, and gauge fields. First, as we have seen, there is Noether's theorem, which states that for every conservation law there is an associated symmetry and vice versa; second, there is the fact, mentioned above, that the requirement of local symmetry leads to a gauge field theory of a particular well-determined character; and third, we find that the gauge field theory determined in this way necessarily includes interactions between the gauge field and the conserved quantity with which we started. Thus we have the astonishing fact that

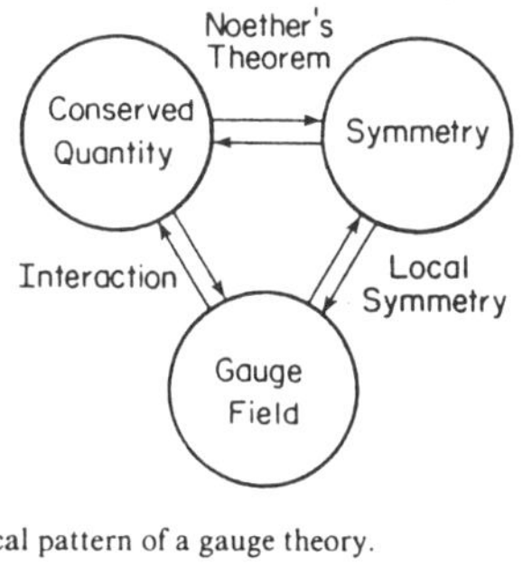

Fig. 1. The logical pattern of a gauge theory.

for every true conservation law there is a complete theory of a gauge field for which the given conserved quantity is the source. The only restriction is that the conservation law be associated with a continuous symmetry (this would exclude, for example, parity, which is associated with reflection symmetry). The resulting theory has just one free parameter, the interaction strength. We shall now discuss in detail the different links in this logical pattern.

A. Noether's theorem

Noether's theorem, connecting conservation laws with symmetry, was originally developed in 1918 as a theorem in the calculus of variations, with an immediate application to classical Lagrangian mechanics. This theorem takes a particularly elegant and general form, however, in relation to Hamiltonian mechanics, whether of the classical or quantum mechanical type. I shall present Noether's theorem in the quantum context, but it will easily be seen by those who know about such things that the entire discussion can be brought back to the classical domain by simply replacing commutators with Poisson brackets throughout.

The first thing to observe is that the Hermitian linear operators of quantum theory play a double role: Each operator represents, of course, one of the dynamical variables of the theory; it also serves, however, as the generator of a class of transformations. Suppose, for example, that $\hat{A}$ is the linear operator associated with the observable A, with the usual relationship that if the system is in the state represented by the state vector Ψ, then the expectation value of A is given by

$$\langle A \rangle = \Psi^* \hat{A} \Psi, \tag{2}$$

which illustrates the relation of the operator $\hat{A}$ to the corresponding physical observable A. [Here, the caret indicates an operator and the asterisk indicates conjugation. Note that the left side of Eq. (2) represents a *physical* quantity, the mean value of a physical variable, while the right side is a *mathematical* expression.] On the other hand, the same operator can be used to generate a unitary transformation on the state vectors of the system:

$$\Psi \to \Psi' = e^{-i\lambda \hat{A}} \Psi, \tag{3}$$

where λ is an arbitrary real parameter. Such a transformation on the state vector may represent a transformation on the physical state itself, so that, for example, if $\hat{A}$ is taken as $\hat{P}_z$, the z component of the momentum operator, then Ψ' represents the state that differs from Ψ by a displacement λ in the z direction. Every continuous family of transformations is generated in this way and is thus associated uniquely with one of the physical variables of the theory.

Now, one of the well-known features of these linear operators is that they do not in general commute with each other (that is, multiplication of operators is not commutative) and whether or not two operators commute has physical significance. This significance takes different forms depending on which role each of the two operators is playing. Thus if two operators $\hat{A}$ and $\hat{B}$ commute, we can make four different statements about the variables and transformations associated with these operators:

(1) The transformations generated by $\hat{A}$ and $\hat{B}$ commute with each other. Thus *rotations* about different axes do not commute, while *displacements* in different directions do commute, that is, you get a different net rotation depending on the sequence in which the rotations are performed. This is directly associated with the fact that the different components of the angular momentum operator do not commute with each other, while the different components of the linear momentum operator do commute with each other.

(2) The dynamical variable A is invariant under the transformations generated by $\hat{B}$. For example, the x component of position of a particle is invariant under displacements in the y direction (the operator $\hat{x}$ commutes with the operator $\hat{P}_y$), but is not invariant under displacements in the x direction (the operator $\hat{x}$ does not commute with the operator $\hat{P}_x$). In mathematical form, if $\hat{A}$ represents the variable A and $\hat{B}$ generates the transformation

$$\Psi' = e^{-i\lambda \hat{B}} \Psi, \tag{4}$$

then the expectation value of A is unaltered by the transformation

$$\Psi'^* \hat{A} \Psi' = \Psi^* \hat{A} \Psi. \tag{5}$$

(3) The dynamical variable $\hat{B}$ is invariant under the transformations generated by $\hat{A}$ [as in (2)].

(4) The dynamical variables A and B are simultaneously measurable with arbitrary precision—there is no uncertainty principle relating them. (Another way of saying this is that they have a complete set of simultaneous eigenstates, i.e., states for which both A and B have exactly predictable values.)

Any one of these four statements is true *only* if the corresponding operators commute, so we can say that they are equivalent—each implies the others. In particular, the equivalence of (2) and (3) says that the dynamical variable A is invariant under the transformations generated by $\hat{B}$ if and only if the dynamical variable B is invariant under the transformations generated by $\hat{A}$.

Now, our interest is in the case that $\hat{B}$, say, is taken to be the Hamiltonian operator $\hat{H}$, which is the operator associated with the total energy of the system. The transformations generated by $\hat{H}$ are time displacements, which is why $\hat{H}$ is the operator that appears in the Schrödinger equation,

$$i\hbar \frac{d\Psi}{dt} = \hat{H}\Psi, \tag{6}$$

whose integrated form,

$$\Psi(t) = e^{-i\hat{H}t}\Psi(0), \tag{7}$$

corresponds to the general form (3). Thus we see why it is the *energy* operator that governs the dynamics of the system.

To say that a dynamical variable A is invariant under the transformations generated by $\hat{H}$ is to say that A is a constant of the motion, that is, its expectation value in any state will be invariant under any time displacement. On the other hand, the statement that H is invariant under the transformations generated by $\hat{A}$ is to say that $\hat{A}$ represents a symmetry of the dynamical laws because it is $\hat{H}$ that determines those dynamical laws. We now see (and this is our modified version of Noether's theorem) that these two statements are equivalent since each is equivalent to the statement that the operators $\hat{A}$ and $\hat{H}$ commute:

The dynamical variable A is conserved if and only if the dynamical laws are invariant under the transformations generated by $\hat{A}$.

B. Local symmetry and gauge fields: The electromagnetic case

We now come to the next link in the diagram shown in Fig. 1, where we relate the assumption of local symmetry to the existence of gauge fields. Suppose we consider some particular conservation law such as the electric charge Q, and identify the associated symmetry transformations generated by the Hermitian operator $\hat{Q}$. The action of this transformation on a complex field $\psi(x)$ associated with particles of charge ne is to change its phase by an amount proportional to n:

$$\psi(x) \to \psi'(x) = e^{-ine\theta}\psi(x). \tag{8}$$

In what follows it is convenient to work with infinitesimal transformations for which Eq. (8) reduces, by Taylor expansion of the exponential, to

$$\psi(x) = (1 - ine\theta)\psi(x). \tag{9}$$

Here, x represents the space-time coordinates and θ is the arbitrary infinitesimal parameter of the transformation. You can think of $\psi(x)$ as essentially equivalent to the Schrödinger or Dirac wavefunction. The invariance comes from the fact that any quantity that is physically observable involves the factors ψ^* and ψ, so that the phase factors cancel. If several different fields are involved, as in a production or decay process, then the conservation of charge in the process is exactly related to the cancellation of these phase factors. The cancellation works the same for derivatives of ψ as long as θ is taken as a constant with respect to x.

Now, however, I want to introduce the idea of *local* symmetry, as discussed in Sec. III, that is, I want to make a different transformation at each different point in space-time, which corresponds exactly to allowing θ to be an arbitrary function $\theta(x)$:

$$\psi'(x) = e^{-ine\theta(x)}\psi(x) \tag{10}$$

$$\to [1 - ine\theta(x)]\psi(x) \tag{11}$$

or

$$\delta\psi(x) = \psi'(x) - \psi(x) \tag{12}$$

$$= -ine\theta(x)\psi(x). \tag{13}$$

The invariance of expressions such as $\psi^*\psi$ is easy to see, but the invariance is lost in expressions that involve derivatives of ψ since derivatives of θ appear as well. Thus (letting ∂_μ represent the partial derivative with respect to x^μ, $x^0 = ct$, $x^1 = x$, $x^2 = y$, $x^3 = z$) we find that $\partial_\mu\psi(x)$ transforms as

$$\delta[\partial_\mu\psi(x)] = -ine\theta(x)\partial_\mu\psi(x) - ine[\partial_\mu\theta(x)]\psi(x), \tag{14}$$

with the first term of standard form [Eq. (13)] and an awkward second term involving the derivative of $\theta(x)$. In order to achieve local symmetry, we have to get rid of the second term in some way. To do this, we must replace the conventional derivative $\partial_\mu\psi$ by a "covariant" derivative $D_\mu\psi$ whose transformation law is *required* to be of the standard form,

$$\delta[D_\mu\psi(x)] = -ine\theta(x)D_\mu\psi(x). \tag{15}$$

This will make expressions such as $\psi^*D_\mu\psi$, for instance, invariant under the transformation. How can we accomplish this? We simply introduce the electromagnetic potentials (as you may have anticipated if you are familiar with the standard form of the Schrödinger equation in the presence of the electromagnetic field), that is, we combine the conventional derivative ∂_μ with a vector field A_μ,

$$D_\mu = \partial_\mu + ineA_\mu, \tag{16}$$

with the transformation law for A_μ chosen in such a way that Eq. (15) is satisfied. If you substitute expression (16) into Eq. (15), and use Eq. (14) for the variation of $\partial_\mu\psi(x)$, then the terms that are left, after removing the terms that cancel, give

$$-ine(\partial_\mu\theta)\psi + ine\,\delta A_\mu\psi = 0, \tag{17}$$

which is satisfied if A satisfies the transformation law

$$\delta A_\mu(x) = \partial_\mu\,\theta(x), \tag{18}$$

the familiar gauge transformation for electromagnetic potentials.

These expressions can be rewritten in a more familiar notation: $A_0 = \phi$, the scalar potential, and $-A_i$ $(i = 1,2,3)$ are the components of the vector potential **A**, so that Eq. (18) is equivalent to

$$\delta\varphi = \frac{\dot\theta}{c}, \quad \left(\dot\theta = \frac{\partial\theta}{\partial t}\right), \tag{19}$$

$$\delta\mathbf{A} = -\nabla\theta. \tag{20}$$

What have we proved? We started with a conservation law, the conservation of electric charge with its associated symmetry, and then (as if we had never heard of the electromagnetic field) we showed that the requirement that the symmetry be *local* forces us to introduce a gauge field, which turns out to be nothing but the familiar electromagnetic field. Before we turn to the more general kinds of gauge fields, let us continue with the example of the electromagnetic field and see how the gauge invariance condition fixes the form of the theory.

The standard procedure for setting up a physical theory is to start with a classical Lagrangian formulation and then go through a well-defined "quantization" procedure to generate the appropriate quantum mechanical theory. The classical equations of motion are generated from the Lagrangian by a variational principle (Hamilton's principle), which is discussed in Sec. II and explained in standard mechanics texts. Then the quantization procedure (a prescription for constructing the canonical momentum operators, the Hamiltonian operator, and the quantum mechanical equations of motion) guarantees that the resulting theory satisfies the important Correspondence Principle, which is the requirement that the theory be entirely consistent with the classical theory in the macroscopic domain, where the classical theory is known to be valid: We need only discuss the Lagrangian since the form of the theory is determined once that is known.

In the case of a field theory, the important quantity is the Lagrangian density L (the Lagrangian itself is the space integral of L), which depends on the various fields and their partial derivatives and which must be a relativistic invariant in order that the resulting theory be Lorentz invariant. Indeed, the Lagrangian density must display *all* the symmetries that are required of the theory that it is to generate.

Thus we see that if we want to construct a dynamical theory of our new gauge field, we need to construct an appropriate Lagrangian density L which is both Lorentz and gauge invariant. If we add some fairly standard requirements about the physical reasonableness of the field equations, we find that L is uniquely determined apart from trivial factors. If our gauge field interacts with just a single charged particle field ψ, then we expect to get two field equations, one for ψ and one for the gauge field A_μ. Now suppose that we already know the Lagrangian density for ψ in the absence of the gauge field; It certainly involves derivatives of ψ and to achieve gauge invariance we have found [Eq. (16)] that the ordinary partial derivative must be replaced by the gauge covariant derivative D_μ. Thus L now automatically includes an interaction term involving A_μ and ψ that shows exactly how the charged particles behave in the presence of a given gauge field A_μ. If ψ is a Dirac field (for a relativistic spin-1/2 fermion), for example, then the appropriate terms in L take the form

$$L_D = \bar{\psi}(i\gamma^\mu D_\mu - m)\psi, \tag{21}$$

where $\bar{\psi}$ is related to ψ^*, the conjugate of ψ. (The speed of light c has been taken as equal to 1 and the repeated index μ is summed over by a standard "summation convention.") In the classical limit, it can easily be shown that the interaction term in (21),

$$L_{\rm int} = -ne\bar{\psi}\gamma^\mu A_\mu \psi, \tag{22}$$

corresponds exactly to the Lorentz force (the familiar electrostatic and magnetic forces) on the charged particle.

To get the dynamical equations for the gauge field, we need to add a term to L that is again both Lorentz and gauge invariant and involves derivatives of A_μ, so that the resulting equations for A_μ will have the character of field equations. We first look for *gauge invariant* expressions involving A_μ and find that the only one we can make is the combination

$$f_{\mu\nu} = \partial_\nu A_\mu - \partial_\mu A_\nu. \tag{23}$$

Since the variation of A_μ under a gauge transformation is simply the space-time gradient of $\theta(x)$ [Eq. (18)], the cross derivatives of $\theta(x)$ that appear in Eq. (23) cancel, leaving $f_{\mu\nu}$ invariant. The gauge invariant combination $f_{\mu\nu}$ is the familiar tensor representation of the electric and magnetic fields **E** and **H**; the elements f_{i0} ($i = 1,2,3$) are the components of **E** and the elements f_{ij} give the components of **H**. Next, we need to construct a relativistic invariant (Lorentz scalar) out of $f_{\mu\nu}$ to serve as the Lagrangian density for the field; the only way to do this that produces physically reasonable field equations is to take the scalar product

$$L_{\rm em} = -\tfrac{1}{4} f_{\mu\nu} f^{\mu\nu}, \tag{24}$$

where μ and ν are summed over by the summation convention mentioned above and the superscripts ("contravariant" indices) indicate an additional factor of -1 when either μ or ν has the value 1, 2, or 3. The factor $\frac{1}{4}$ merely serves to set the scale of the electromagnetic field vectors. When the standard procedure is followed for generating the classical field equations from the Lagrangian density (24), with the interaction term (22) the result is simply Maxwell's equations. Interaction term generates a source term for Maxwell's equations, which is found to be just the charge-current density

$$j^\mu = ne\bar{\psi}\gamma^\mu\psi. \tag{25}$$

In the covariant notation that we have been using, the field equations, which as I said are just Maxwell's equations, take the form

$$\partial_\nu f^{\mu\nu} = j^\mu. \tag{26}$$

One might ask whether there should be some kind of gauge-covariant derivative, such as D_μ, in Eq. (26); the answer is that there is no difference in this case since the extra term in D_μ [Eq. (16)] is proportional to the charge of the relevant particle and the charge of the photon is zero.

We have now completed the pattern shown in Fig. 1. The requirement of local symmetry has not only generated a gauge field of uniquely determined structure, but has dictated almost uniquely the form of the interaction—the precise form of the forces on the charged particle and the precise way in which the electric charge-current density serves as the source for the gauge field.

C. Local symmetry and gauge fields: The general case

Now we are ready to look at the more general case, in which we start with an arbitrary conservation law and, by the same logic that we used above, develop the theory of a gauge field whose relationship to the conserved quantity will be the same as that of the electromagnetic field to the electric charge.

The most obvious candidate for the conserved quantity might seem to be energy or momentum, but these in fact turn out to be much more complicated. The associated local symmetry is the invariance under local coordinate transformations and the associated gauge field, as mentioned in Sec. III, is the gravitational field; however, there are a number of sources of confusion that make this a poor case to present here. I discuss this case in more detail in Sec. VI.

The more typical case of a non-Abelian gauge field is obtained when the symmetry involved is an *internal* symmetry, i.e., one that is not associated with any kind of coordinate transformation. The phase invariance associated with charge conservation is one example of an internal symmetry; there can be others of this type associated with other conserved particle numbers such as baryon or lepton number. These examples are exactly equivalent in form to the electromagnetic case, so that the result of the logic would be a gauge field identical in structure to the electromagnetic field, but coupled to the appropriate conserved particle number. This is not what happens.

The other kind of internal symmetry is associated with families of identical particles, such as isospin multiplets (if one can treat them as identical) or multiplets of colored quarks. The conserved quantities in this case are associated with the quantum numbers that label the members of the multiplets, together with certain operators that induce transitions from one member of a multiplet to another. There is thus a family of operators that on one hand correspond to the conserved dynamical variables, e.g., isospin or color and, on the other hand, correspond to a group of transformations, the "symmetry group" of the multiplets. The fact that these operators do not in general commute with each other is precisely what makes this case so very different from the electromagnetic case, as we shall see in what follows. In each case it is the mathematical structure of the symmetry group that determines the structure of the gauge field and the form of the interaction. The symmetry groups have names according to their structure, so that one speaks of U(1), the simple phase symmetry group of electromagnetic theory or SU(2), the isospin symmetry group. [SU(2) is also the symmetry group of spatial rotations in quantum theory, which is why the language of isospin is so similar to the language of angular momentum, even though they have nothing to do with each other physically.]

I want to describe in qualitative terms what happens when we apply the gauge philosophy to one of these noncommuting ("non-Abelian") symmetry groups. The historical example was the isospin case and indeed this is the simplest case to consider, though it should be emphasized that the same mathematical forms appear for any non-Abelian symmetry group, with only minor changes. I shall follow the example of the electromagnetic field step by step and try to clarify the differences.

First, the conserved electric charge operator $\hat{Q}$ is replaced by the family of operators $\hat{T}_\alpha$, which we take as the three components of isospin. The "z component" $\hat{T}_3$ (which has nothing to do with the z direction in real space) labels, by its eigenvalues, the members of an isospin multiplet, while the x and y components $\hat{T}_1$ and $\hat{T}_2$, which do not commute with $\hat{T}_3$, generate transformations that mix the different values of T_3. [In the case of color, the symmetry group is called SU (3) and there are eight operators $\hat{T}_\alpha$, two of which commute with each other so that their eigenvalues can serve to label the members of a color multiplet, a two-dimensional array.]

It is characteristic of all these symmetry groups that the set of operators $\hat{T}_\alpha$ is closed under the commutation relation, that is, the commutator is always just a linear combination of the same operators $\hat{T}_\alpha$. Thus we write

$$[\hat{T}_\alpha,\hat{T}_\beta] = ic_{\alpha\beta\gamma}\hat{T}_\gamma, \tag{27}$$

where the coefficients $c_{\alpha\beta\gamma}$, called the "structure constants," are a set of numerical constants that completely characterize the local structure of the symmetry group; again, γ is summed over. In the case of isospin, the coefficients take the values 0, ± 1 in such a way as to give the usual angular-momentum-type commutation relations

$$[\hat{T}_1,\hat{T}_2] = i\hat{T}_3, \text{ etc.} \tag{28}$$

The complex field ψ now has several components, depending on the size of the multiplet we want to describe. For the isospin case we can consider as an example just the "nucleon" doublet, consisting of the proton and neutron, so that

$$\psi = \begin{pmatrix}\psi_p\\ \psi_n\end{pmatrix}. \tag{29}$$

Equation (29) is the wavefunction for a single nucleon, which may be in the "proton state," with amplitude ψ_p, or in the "neutron state," with amplitude ψ_n. [The basic multiplet for color SU(3) would be a color *triplet* of quarks described by a three-component wavefunction ψ.]

The general transformation on the nucleon wavefunction ψ, generalizing Eq. (9), is

$$\psi' = (1 - ig\theta_\alpha T_\alpha)\psi, \tag{30}$$

$$= (1 - ig\boldsymbol{\theta}\cdot\mathbf{T})\psi, \tag{31}$$

where ψ' represents a different mixture of the proton and neutron states than does ψ. The *change* in ψ is given by

$$\delta\psi = -ig\boldsymbol{\theta}\cdot\mathbf{T}\psi. \tag{32}$$

There are now *three* arbitrary infinitesimal parameters θ_α describing the transformation and the symbols T_α here represent 2×2 matrices which govern the mixing of the components ψ_p and ψ_n in the transformation. These matrices (proportional to the Pauli spin matrices) have the same commutation relations (27) as the operators $\hat{T}_\alpha$ and provide what is called a "representation" of the symmetry group. $\boldsymbol{\theta}$ and $\mathbf{T}$ are the three-component vectors whose components are the parameters θ_α and the matrices T_α.

Now, in exact parallel to the electromagnetic case, we try to impose the condition of *local* symmetry, which means that we allow $\boldsymbol{\theta}$ to be an arbitrary vector-valued function of x, so that

$$\delta\psi(x) = -ig\boldsymbol{\theta}(x)\cdot\mathbf{T}\psi(x), \tag{33}$$

in close analogy to Eq. (13). Again, we find that while terms not involving derivatives cause no problems, the variation of the *derivative* of ψ has an additional term involving the derivative of $\boldsymbol{\theta}$, as in Eq. (14). To get rid of these we again need to replace the conventional derivative ∂_μ with a covariant derivative D_μ, defined in terms of a new field whose transformation rule can have a term proportional to $\partial_\mu\boldsymbol{\theta}(x)$. The correct form for $D_\mu\psi$ is given by

$$D_\mu\psi = (\partial_\mu + ig\mathbf{B}_\mu\cdot\mathbf{T})\psi, \tag{34}$$

with a new field $\mathbf{B}_\mu(x)$ that is a vector with respect to Lorentz transformations and also a vector with respect to isospin rotations; It therefore has 12 components in all. When the condition of covariance is imposed on $D_\mu\psi$, namely,

$$\delta(D_\mu\psi) = -g\boldsymbol{\theta}\cdot\mathbf{T}(D_\mu\psi) \tag{35}$$

[like ψ itself, Eq. (33)], then we find that the transformation rule for $\mathbf{B}_\mu$ is completely determined. Because of the fact that the matrices T_α do not commute, we find that the transformation rule is more complicated than that for the electromagnetic potentials A_μ:

$$\delta\mathbf{B}_\mu = \partial_\mu\boldsymbol{\theta} - g\boldsymbol{\theta}\times\mathbf{B}_\mu. \tag{36}$$

The second term on the right side of (36), which is absent in the electromagnetic case [Eq. (18)], is extremely important and is associated in one way or another with all the interesting and novel features of non-Abelian gauge fields (i.e., fields associated with non-Abelian symmetry groups): This term reflects the isovector character of our new field (it is the normal expression for the change in a vector under rotation) and the fact that the quanta of this field must carry isospin. This is in contrast to the case of photons, which do not carry electric charge. The new quanta interact with each other in a nonlinear way, just as photons would if they were charged.

It may be helpful to think about this analogy a little more carefully. If photons were electrically charged, then they would interact with each other by way of the electromagnetic field, i.e., in the language of quantum field theory, by the exchange of virtual photons! Even if no other charged particles were present, the photons would not be free. Another way to visualize this state of affairs is in terms of the classical electromagnetic field equations: There would be nonzero electric charge and current densities wherever the electromagnetic field strengths were nonzero and the dynamical behavior of the fields would be influenced by these charge and current densities in a complicated nonlinear way. Many of the usual properties of the electromagnetic field, in particular the linear superposition of wave solutions, would be lost.

Photons are not charged, of course, but the B quanta *are* in the sense that they carry isospin and isospin is the source—the analog of charge—for the B field. The description of the last paragraph may not apply to the electromagnetic case, but it applies almost exactly to the B field. Except for this self-interaction, the field equations for each isospin component of the B field are identical to those for the electromagnetic field.

Let us now complete the description of the B field, still following the logical sequence of the electromagnetic case. Recall first that the components $\mathbf{B}_\mu(x)$ are analogous to the electromagnetic potentials $A_\mu(x)$, so that we need to construct the electric and magnetic field analogs. These analogs take the covariant form $\mathbf{f}_{\mu\nu}$, which now has the character of an isovector, like $\mathbf{B}_\mu$, and is given [cp. Eq. (23)] by

$$\mathbf{f}_{\mu\nu} = \partial_\nu \mathbf{B}_\mu - \partial_\mu \mathbf{B}_\nu - g\mathbf{B}_\mu \times \mathbf{B}_\nu. \tag{37}$$

The "electric" field, whose spatial components $\mathbf{E}_i$ are given by $\mathbf{f}_{i0}$, has the character of an isovector also, so that it has nine components in all instead of the familiar three components. The spatial components $\mathbf{H}_i$ of the "magnetic" field, given by the elements $\mathbf{f}_{jk}$, have the same character.

The final steps, as with the electromagnetic field, are to construct the Lagrangian density and field equations for the B field. Because it must be Lorentz and gauge invariant, the Lagrangian density can only be

$$L_B = -\tfrac{1}{4}\mathbf{f}_{\mu\nu}\cdot\mathbf{f}^{\mu\nu} \tag{38}$$

[cp. Eq. (24)], while the field equations derived from (38) have the form

$$D_\nu \mathbf{f}^{\mu\nu} = \mathbf{j}^\mu, \tag{39}$$

like Eq. (26). Here $\mathbf{j}^\mu$ is the isospin current density associated with the other particles present—in our example, the neutron–proton field,

$$\mathbf{j}^\mu = g\bar{\psi}\gamma^\mu \mathbf{T}\psi. \tag{40}$$

The covariant derivative D_μ of the field $\mathbf{f}$ is no longer equivalent to the regular derivative ∂_μ since $\mathbf{f}$ is an isovector. The rule is

$$D_\nu \mathbf{f}^{\mu\nu} = \partial_\nu \mathbf{f}^{\mu\nu} + g\mathbf{B}_\nu \times \mathbf{f}^{\mu\nu}, \tag{41}$$

in close analogy with Eqs. (36) and (37).

Now, the form of Eq. (39) does not tell us that $\mathbf{j}^\mu$ is conserved, and in view of our discussion we should not expect it to be since the B field carries isospin as well. However, if the second term on the right side of (41) is taken to the right side of Eq. (39), then Eq. (39) takes the form

$$\partial_\nu \mathbf{f}^{\mu\nu} = \mathbf{j}^\mu - g\mathbf{B}_\nu \times \mathbf{f}^{\mu\nu}, \tag{42}$$

and since the four-dimensional divergence of the left side of (39) is now automatically zero, we see that it is the right side of (42) that represents the full conserved isospin current, including the contribution of the B field.

It is important to mention that the structures we have developed are seen by mathematicians as having a deeply geometrical character, which is realized in the mathematical theory of fiber bundles. The space in which this geometrical character manifests itself is a kind of product space (the "associated fiber bundle," to be more precise) in which a local vector space, with elements $\psi(x)$, is attached to each point x of space-time. The term "fiber bundle" refers to the analogy with a collection of thin threads or fibers bound together into a much thicker bundle. Each thread corresponds to one of the local vector spaces at a point x; the collection of these into a product space is the bundle. (There is also a "principal fiber bundle" consisting of the collection of parameter spaces, each attached to a point x, associated with the local transformations discussed above.)

The process of moving from one point x to a neighboring point requires us to define a "connection" between the two associated local vector spaces, giving rise to the covariant derivatives introduced in Eqs. (16) and (34), and which bear a very close relationship to the covariant derivatives of general relativity theory. The failure of the covariant derivatives to commute is an indication of a kind of "curvature" in this extended space, with the quantities $\mathbf{f}_{\mu\nu}$ giving a precise characterization of that curvature, again in close analogy to the curvature tensor of general relativity. It seems very clear that this geometrical structure represents a wonderful and unexpected vindication of Einstein's vision of a unified geometrical picture of all the forces of nature.

This completes our outline of the non-Abelian gauge field logic, using the example of isospin as the conserved quantity that serves as the source for the gauge field $\mathbf{B}$. At the time the theory was invented, isospin was the only reasonable candidate for this role and the hope was that the resulting gauge field might serve as the carrier of the strong interaction. There was a general feeling at the time that the idea, while beautiful, was beset by too many difficulties to be acceptable. A number of years had to pass and a number of sophisticated developments had to occur before it was discovered how the difficulties could be resolved and thus the non-Abelian gauge fields seen as an acceptable description of the fundamental interactions of nature.

V. RESOLVING DIFFICULTIES: THE GAUGE THEORY OF THE ELECTROWEAK INTERACTION

As mentioned briefly in Sec. II, the mass of the gauge quantum and the renormalizability of the theory were seen from the beginning as major problems. The resolution of these problems occurred over the course of time as part of the long and intricate process by which the modern gauge theories of the electroweak and strong interactions were developed. A proper account of these developments is beyond the scope of this article, but I shall give a brief description of how these particular difficulties were finally resolved.

The problem with mass is essentially that the forces to be described are short-range forces and thus require massive interaction quanta (as explained below), while the introduction of such a mass into a gauge theory destroys the gauge invariance, as well as making it virtually impossible that the theory be renormalizable.

The question of renormalizability has to do with the handling of certain infinite expressions that seem to arise inevitably when doing any kind of relativistic quantum field theory. For certain kinds of theory, including quantum electrodynamics, there is a logical procedure (discussed below) for dealing with these infinities that permits us to make realistic finite calculations for real physical effects. The procedures are not mathematically rigorous, but they work; in the case of quantum electrodynamics (QED) they work with astonishing precision. Theories for which these procedures work are called " renormalizable"; nonrenormalizable theories are useless in practice and apparently unphysical.

A. Mass of gauge quanta

Our experience with gauge fields led us to suppose that the quanta would have to have zero mass in order for the theory to satisfy gauge invariance. Now, the range of an interaction is inversely proportional to the mass of the mediating quanta and zero-mass quanta imply long-range interactions, like the electromagnetic and gravitational forces. The strong and weak interactions, however, are of extremely short range and fall off exponentially at distances of the order of the nuclear size. (This is why the nuclei of neighboring atoms, for example, interact with each other only by electromagnetic forces, even though the nuclear force is much stronger.) Thus we seem to run into a contradiction and we ask the following question: Is it possible for the gauge quanta to acquire a mass in some way without violating gauge invariance?

In what we now believe is a reliable picture of fundamental particles and interactions at our present level of understanding, there are three separate gauge theories: the Glashow–Weinberg–Salam theory for electromagnetic and weak interactions, the color gauge theory for strong interactions, and general relativity for the gravitational interaction. The first two of these theories, together with the spectrum of elementary particles associated with them, make up what is now referred to as the "Standard Model." In each of these three gauge theories the question of the mass of the quanta takes a different form. In gravitation theory, the mass is simply zero and the forces are long range. In quantum chromodynamics (QCD), the theory of the color gauge field that mediates the strong interaction, the mass is zero, but the confinement of color charge (discussed below) prevents its manifestation as a long-range force. Finally, in the electroweak theory, which unifies the electromagnetic and weak interactions, all the gauge quanta except the photon acquire a mass through a mechanism known as "spontaneous symmetry breaking," which does not in fact spoil either the gauge invariance of the theory or its renormalizability.

The principle of spontaneous symmetry breaking is that the actual symmetry of a system may be less than the symmetry of the underlying physical laws. This is obvious in the world around us, which is full of objects that are neither translation nor rotation invariant, even though the physical laws of which they are a manifestation *are* both translation and rotation invariant. The new discovery is that even the vacuum state—the supposedly blank canvas on which the universe is painted—can fail to exhibit the full symmetry of the laws of physics. If the vacuum is unique, that is, if there is only one state of lowest energy, then it must indeed have the full symmetry of the laws, but if it is not unique, i.e., if the vacuum is a degenerate state, then this is no longer the case and for each unsymmetrical vacuum state there will be others of the same minimal energy related to the first by the various symmetry transformations under which the physical laws are invariant. A famous example is the ferromagnet, which, when cooled to absolute zero and thus brought to its state of lowest energy, must come to a magnetized state with some definite orientation even though any orientation is allowed. This ground state of the ferromagnet is the analog of the vacuum state for our universe.

The symmetry of the underlying theory, then, is spontaneously broken in the sense that the universe we see, existing within the framework of such an unsymmetrical vacuum state, will exhibit this broken symmetry even in the way that the physical laws appear to operate. Now, the broken symmetry that we want to look at is the gauge symmetry of the electromagnetic and weak interactions, viewed together as associated with a single four-component gauge field. There is no evidence that the vacuum state for such a theory *is* degenerate, so the authors of the theory forced the vacuum to be degenerate by means of a device due to Higgs, which consists of the artificial introduction of an additional field, called the "Higgs field," with properties chosen so as to *make* the vacuum state degenerate. In the example worked out in Sec. IV, where the symmetry was the isospin symmetry SU(2), the Higgs field might be taken to be an isospin vector field, with a quartic form for the field energy chosen in such a way that the field configuration of minimum energy is a uniform nonzero field with a definite but arbitrary orientation in the isospin vector space. Like the magnetization of the ferromagnet, this oriented Higgs field gives the observed universe a preferred orientation, now in isospin space, and provides an elegant model for the observed broken symmetry.

In the standard model, however, isospin symmetry is treated as accidental and is not associated with a gauge field [although efforts to extend the standard model to a larger symmetry group that will unify the strong and electroweak interactions ("Grand Unified Theories," or "GUT's") invariably incorporate the symmetry of "flavor," of which isospin is one facet]. In the gauge theory of electroweak interactions, the symmetry group is taken to be SU(2)×U(1), where U(1) is the phase symmetry discussed in Sec. IV in connection with the electromagnetic field [Eq. (8)] and SU(2) is the same symmetry group used to describe isospin symmetry, but now applied to a somewhat different symmetry referred to as "weak isospin." Weak isospin symmetry, unlike ordinary isospin, relates the weakly interacting particles, so that the electron and its neutrino form a doublet (with extra complications associated with left- and right-handed spin asymmetries that I will not go into here), as well as the muon and tau with their respective neutrinos. The isospin multiplets of hadrons (the strongly interacting particles) are modified so that, for example, the proton is paired, not with the neutron alone, but with a quantum mixture ("linear superposition") of the neutron, the Λ° and the Σ°.

This theory predicts the existence of four gauge quanta: a neutral photonlike object, sometimes called the X°, associated with the U(1) symmetry and a weak-isospin triplet associated with the SU(2) symmetry, whose members are referred to as W°, $W^\pm$. The Higgs symmetry-breaking mechanism now has several consequences: The $W^\pm$ particles acquire a mass and the X° and the W° are mixed, so that the neutral particles you see in nature are two different linear combinations of these two particles. One of these, christened the Z°, has a mass and the other, the familiar photon, is massless. The masses of the $W^\pm$ and the Z° are governed by the structure of the uniform Higgs field background and do not affect the basic gauge invariance of the theory. The interaction strength associated with all four particles is essentially the same and the observed weakness of the weak interaction, mediated by the $W^{\pm}$ and the Z°, is understood as a consequence of their masses, which are taken to be large enough (90–100 proton masses) to agree with what we see.

B. Renormalizability

Renormalizability was the other problem, reflecting one of the deep questions in quantum field theory, a reminder of the fact that the theory has no rigorous mathematical foundation and indeed has some serious flaws if viewed from a strictly mathematical viewpoint. Let us review just enough of the history of this problem to see how it relates to the new non-Abelian gauge theories.

The big problem is what are called "divergences," or "infinities." Since the early days of electromagnetic theory and on into the era of QED, certain calculations of what should be physically reasonable effects have given infinite answers. The first is the infinite electrostatic self-energy of the classical electron (or of any charged point particle). The energy density of the electric field is proportional to $|\mathbf{E}|^2$ and the field at a distance r from a particle of charge e is proportional to e/r^2; thus one ought to be able to find the total energy in the electric field of the particle by integrating.

If the particle has radius a, say, the result is an integral proportional to $e^2\int r^{-2}\,dr$, where the integration is from a to ∞, giving e^2/a. If the electron is a point particle, then $a = 0$ and the energy is infinite. Is this physically reasonable? One might argue that this is an unobservable energy and that, therefore, its value is physically meaningless. Difficulties arise on account of relativity theory, however, which tells us that energy is equivalent to mass and that such a point particle should therefore have infinite inertia, which would certainly be an observable effect. A *finite* radius a, on the other hand, also produces contradictions with the requirements of relativistic invariance and we are left with a dilemma that is still with us, though in a more abstract form, 80 years later.

Since the advent of QED, the difficulty has shown up in the form of divergent integrals in the calculation of all sorts of physical effects. The historic example was the "Lamb shift," referring to the measurement by Lamb and Retherford in 1947 of the extremely small shift of the spectral lines of hydrogen due to quantum effects in the electromagnetic field. The experiment revealed, and the theory *in principle* predicted, a contribution smaller by a factor α (the "fine-structure constant," approximately equal to $\frac{1}{137}$) than the smallest previously identified contribution. The actual calculation of this contribution was straightforward to set up using standard principles of quantum mechanical "perturbation theory," as it is called, and gave the proper factor α, multiplied, however, by an expression involving divergent integrals, that is, with a value equal to infinity. This highly unphysical result shows that there was something seriously wrong with the basic ideas of QED and caused considerable anxiety among physicists at the time.

The problem was solved, after a fashion, when it was realized that the fundamental parameters involved, the mass and electric charge of the electron, needed to be *renormalized*. These two parameters, m and e, appear in the theory from the beginning and represent the mass and charge of the *noninteracting* electron, usually referred to as the "bare mass" and the "bare charge." What was realized, then, was that the electromagnetic interaction, in addition to shifting atomic energy levels, would also alter the observed mass and charge. To obtain a prediction of the observed values, one needs to do another calculation; the resulting expressions are equal to the "bare" values plus corrections of order α, α^2, etc. These corrections are called the "renormalized" mass and charge. I called this a *prediction*, but that is not correct because in fact we do not know what the bare values are and we do know what are the observed, or renormalized, values. What we must do, then, is use these expressions to deduce the *bare* parameters from the *observed* values. This process is called "renormalization" and would be necessary and indeed quite straightforward even if there were no divergences in the theory. Predictions of experimental quantities such as the Lamb shift must be reexpressed in terms of the *renormalized* mass and charge since these are the observed values.

What happens, though, is that in the calculation of the renormalized mass and charge the same sort of divergent integrals appear as were found in the original Lamb shift calculation. What does this mean? First, it does *not* mean that the renormalized values are infinite (since they are the observed values), but that the *bare* values differ by some infinite factor from the observed values. The process of renormalization, then, requires us to manipulate these infinite quantities, which is clearly not allowed mathematically, but seems to be necessary to arrive at any kind of answer.

What we do in practice is find some artificial way of making the integrals converge so that the expressions we need to manipulate are finite. There are a number of ways of doing this, closely analogous to keeping a finite radius a for the electron in the calculation of the electron self-energy discussed above (and indeed the self-energy *is* the mass). The divergent expressions are now finite for finite a and become infinite only when a is set equal to zero. The renormalization procedure can now be carried through and the expressions for the bare charge and mass in terms of the observed values substituted into the expression for the Lamb shift (or any other experimentally observable quantity). The astonishing result is that the troublesome integrals cancel out and disappear completely and when the cutoff parameter—a in our example—is set equal to zero the result is still finite. The theory could now be compared with experiment (remember that the effect predicted is *extremely tiny*) and the agreement was spectacular. The initial agreement was at the level of a few parts per million and has steadily improved over the years.

Clearly, we have in some sense a correct theory, even though the procedures are mathematically unsound. In the case of QED, the results are to a large degree independent of what cutoff scheme is used; there are a number of different ways of replacing the divergent integrals with finite expressions and the end result, when the divergent parts have canceled, seems to be always the same. A theory for which this procedure works is called "renormalizable." It is easy to find otherwise reasonable theories that are *not* renormalizable since there are a number of things that can go wrong; thus a very important fact about QED is that it *is* renormalizable and can therefore be used to make highly precise predictions about the tiny effects such as the Lamb shift that validate its claim to be true.

It is evident, now, that if non-Abelian gauge theories are to be taken seriously, their renormalizability becomes a question of great importance. The nonlinear character of these theories, which is related directly, as discussed in Sec. IV, to their self-interaction, makes this question very difficult indeed, even if the quanta are massless. To make matters worse, the first such theory to show signs of really being true was the Glashow–Weinberg–Salam theory of the electroweak interaction and as we have seen the gauge quanta (apart from the photon) are massive. If the masses are put in directly, in violation of gauge invariance, the theory is found to be nonrenormalizable, but what if the masses arise from spontaneous symmetry breaking? This remained an open question for some time and glimpses of the truth were beginning to be seen by several physicists when, in 1971, Gerard 't Hooft, a young graduate student from the Netherlands, solved the problem completely. In a truly remarkable piece of work, 't Hooft confirmed the renormalizability of massless gauge theories, already shown implicitly by others, and went on to renormalize the massive case provided the masses were generated by spontaneous symmetry breaking, the Higgs mechanism.

This major breakthrough cleared the way for wide acceptance of the Glashow–Weinberg–Salam theory of the electroweak interaction. The authors of the idea had intended to produce an *example* of a theory with broken symmetry using the Higgs mechanism in an *ad hoc* way and hardly expected that the result would correspond to reality. In fact, however, the Higgs mechanism not only provided the quanta with masses, but allowed for complete renormalizability and resulted, surprisingly, in a practical theory that has enjoyed a remarkable degree of experimental confirmation.

VI. QCD AND THE STANDARD MODEL

We turn now to a brief description of the second of the two gauge theories that make up the Standard Model, namely, quantum chromodynamics (QCD), or color gauge theory, the gauge theory of the strong interactions. Again, there is a long and complicated history leading up to the point where such a theory could even be proposed since the major ingredients of the theory are impossible to observe directly. The color gauge field is unobservable; the conserved color charge is unobservable; and the quarks, which are the particles that carry the color charge, are unobservable. Actually, the last comment is not quite fair because although free quarks have never been seen and are probably impossible to produce, one can nevertheless say that the quarks inside the neutron or proton have been observed indirectly through the deep inelastic scattering of high-energy probes, in a manner very similar to the observation of the atomic nucleus by Rutherford and his coworkers in 1909, through the scattering of alpha particles.

The first stages leading to the development of color gauge theory involved the realization that the hadrons (the proton, neutron, and other strongly interacting particles, both mesons and baryons) could not be elementary particles. For one thing, there was no logical reason for identifying some of the hadrons as more fundamental than the rest, while schemes for treating all of them as equally fundamental proved unfruitful. For another thing, the hadrons had more structure than seemed appropriate for elementary particles, even taking into account the big effect that virtual quantum processes would have as a result of their large interaction strength. Finally, the observed symmetry patterns among the hadrons, described by quantum numbers known collectively as "flavor" (which is like a multidimensional extension of isospin) strongly suggested a composite structure, since it would take only a small number of subnuclear particles, with the correct fundamental symmetry characteristics, to permit the construction of the large variety of observed particles.

These subnuclear particles, christened "quarks" by Murray Gell-Mann, who was responsible for the present form of this hypothesis, were at first three in number. Two of the quarks, called the u and d (for "up" and "down," constituted an isospin doublet and generated all the observed isospin multiplets, while the third, the s (for "strange") quark, was responsible for the additional dimension of flavor known as "strangeness."

Each of the known baryons (strongly interacting fermions) could then be modeled as a bound state of three quarks and each of the mesons (strongly interacting bosons) could be modeled as a bound state of a quark and an antiquark. It was also found to be essential (after some distress with the idea and a number of unsuccessful efforts to avoid it) to give the quarks fractional electric charge, so that the u quark has charge $+\frac{2}{3}$ and the d and s quarks each have charge $-\frac{1}{3}$, in units of e, the proton charge.

In more recent years, two additional flavors of quark have been identified, the c (for "charm") and b (for "bottom," or sometimes "beauty"), while a sixth flavor, to be known as t (for "top," or "truth") is widely felt to be needed to complete the picture. This would then match the six varieties of lepton in some deeper flavor symmetry that may some day unify the strong and electroweak interactions. In the standard model, though, flavor symmetry is not included as a true symmetry and is not associated with a gauge field. The many efforts to model the strong interactions in such a way were unsuccessful and the next level of truth was found to involve a deeper degree of complexity, as we shall now see.

The next chapter in this story came with the realization that the different flavors did not provide enough quarks to explain what we observed. In the first place, there was a difficulty with the Pauli exclusion principle. In order to explain several of the observed baryons, one had to suppose that three quarks of the same variety were bound in what was essentially the same orbital state. The quarks, though, had to be spin-$\frac{1}{2}$ fermions, just like electrons, which meant that there could be no more than two in the same orbital state, corresponding to the two possible spin states. (The idea that quarks might be neither bosons nor fermions, but obey some more complicated kind of "parastatistics," was also explored, particularly by O. W. Greenberg in 1964, but has not proved fruitful.) The other difficulty was in finding a force that would bind the quarks together so strongly that they could never (or only rarely) escape and yet would not show up as a comparably strong force among the observed hadrons, seen as bound configurations of quarks. In 1965, Han and Nambu pointed the way to a solution of these difficulties by suggesting (in connection with a model of quarks with integer rather than fractional charge) that there might be additional quantum numbers needed to describe quarks, so that *each* flavor of quark could come in three varieties, later to be called "colors," and linked by a new symmetry, with the structure known as SU(3). This dealt nicely with the exclusion principle and the "color charge," as one might call it, could provide the basis for a binding force for the quarks.

It was not until 1972, though, when gauge theories had become more popular, that the idea of binding the quarks with a gauge field was put forward. In that year, Gell-Mann, Fritzsch, and Bardeen introduced the term "color" to describe these additional degrees of freedom and proposed a gauge theory based on the color SU(3) symmetry group. This also followed the observation by Adler that one can calculate the rate of decay of the pi meson into two photons and that the result depends strongly on the number of *types* of constituent particle. Adler found that the decay rate is off by a factor of about 3 if one used only the flavored quarks, but gives very reasonable agreement if one uses the color triplets.

One of the strong motivations for the new color SU(3) model was the need to explain why quarks were never observed. The hope was that there might be what is called "color confinement," which would require that the force between color charges is strong enough at large distances to prevent the charges' ever being separated. It is as if the Coulomb potential between electric charges, instead of falling off as $1/r$, were logarithmic, or even linear, at large distances. Then it would take an infinite amount of work to separate two opposite charges to an infinite distance from each other and any unbalanced charges would inevitably find each other and recombine into neutral composites. As it is, in fact, with the long-range Coulomb force is such that free electric charges have a strong tendency to recombine and most matter at the macroscopic scale is very close to being electrically neutral.

In the case of quarks, the color charge is seen as a vector in a two-dimensional plane and the three different colors correspond to vectors at angles of 120° from each other, as shown in Fig. 2(a). In order to obtain a color neutral combination, it is necessary to add the color vectors in multiples of three. Thus, to get vectors to add to zero, three vectors, one of each type, can be added, as in Fig. 2(b), or a vector can be added to its negative (corresponding to an antiquark), as in Fig. 2(c). The first possibility corresponds to the baryons, consisting of three quarks and the second possibility corresponds to the mesons, consisting of a quark–antiquark pair. The use of the term "color" for this conserved vector quantity was motivated by the analogy to the three primary colors, usually taken as red, blue, and green or red, blue, and yellow. In order to achieve white, the neutral color, you must combine all three primary colors, or else combine a color with its complementary color, in a nice analogy with the two ways of adding the three color vectors as described above. In contrast to ordinary color, though, the three types of color charge are totally indistinguishable from each other, reflecting the symmetry upon which the gauge theory is based.

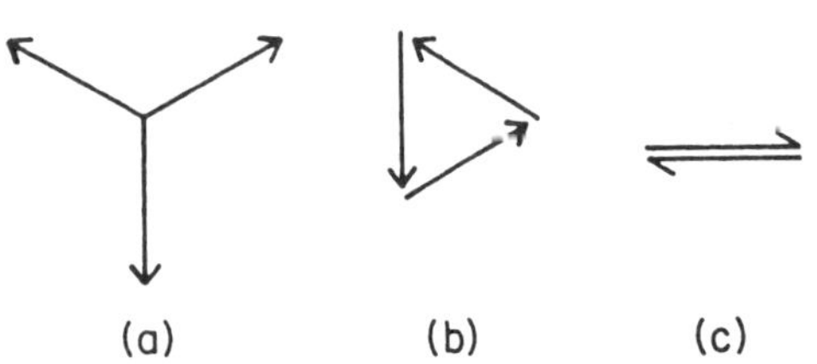

Fig. 2. Vector addition for color charge.

(A complication needs to be mentioned here: The conserved color charge is in actuality a vector in an eight-dimensional space, in analogy to the three-dimensional isospin space. Just as the isospin components fail to commute, so that the eigenvalues of just one of them, referred to as I_3, label the members of each isospin multiplet, in the same way just two of the eight components of color can be taken to be commuting, so that the quark multiplets are labeled by those two parameters and give rise to the two-dimensional vectors discussed above.)

The gauge quanta have come to be referred to as "gluons," in reference to the long-standing puzzlement as to the nature of the "glue" that holds nuclear matter together; the color gauge field is often referred to as the "gluon field." The quantum field theory of the color gauge field is called "quantum chromodynamics" (QCD) in analogy to quantum electrodynamics (QED), the quantum theory of the electromagnetic field. ("Chromo-" is a prefix, from a Greek root, meaning "color.") The gluons are massless (there are no broken symmetries in this theory), but because they inevitably carry color charge (see Sec. IV), they are also confined, just as are the quarks. This, together with the color neutrality of all the physical hadrons, prevents the gluon field from generating a long-range force. Many people have speculated that there might be color neutral bound configurations of gluons, referred to as "glueballs," but they would necessarily be massive.

Whether the gluon field does in fact produce confinement is still an unanswered question, but there are strong indications that it does, as a direct consequence of the nonlinear character of such a gauge field. In contrast to confinement, which concerns the behavior of the field at large distances from the source, there is also a very important question as to what happens at small distances. As mentioned at the beginning of this section, high-energy probes have pointed to the presence of quarks in the interior of baryons and furthermore have indicated that the quarks move very freely at these small distances, as if the forces between them were much weaker than the theory seems to demand. This weakness of the interaction at small distances has come to be called "asymptotic freedom," in reference to the fact that very small distances correspond to very large momentum transfers (related to the Heisenberg uncertainty principle). It was very important, then, that any proposed theory show this sort of behavior; it has been one of the triumphs of QCD that Gross and Wilczek and, independently, Politzer, were able to show in 1972 that this was indeed the case.

With this settling of the problem of asymptotic freedom, all the major difficulties in formulating gauge theories of the strong, electromagnetic, and weak interactions were essentially solved, giving rise to what we have referred to as the Standard Model, the picture of the world as consisting of quarks and leptons, together with the two basic gauge fields associated with the strong and electroweak interactions.

I have not yet discussed the gravitational field, which seems to me to be the mystery at the center of the puzzle. It is clear that Einstein's general theory of relativity can be understood in many ways as the gauge theory associated with the symmetries of space and time, but it does not fit easily into the pattern I have tried to describe. The local symmetry involved is the invariance of the theory under arbitrary curvilinear coordinate transformations, which can be seen as a kind of local Poincaré invariance (the complete symmetry of special relativity). The conserved quantities associated with Poincaré invariance, namely, energy, momentum, and angular momentum, are to some extent the sources for the gravitational field, as the gauge philosophy requires, but there are important differences. In the first place, the local transformations associated with the coordinate transformations of general relativity are general affine transformations—a much broader class than the Poincaré group and too broad in fact to describe the symmetries we see in our local environment. Closely associated with this is the fact that the quantities in general relativity that play the role of the gauge potentials, namely, the Cristoffel symbols (or the "affine connection" in mathematical terms), are not dynamically independent gauge potentials at all, but are derived from the metric tensor, which in some way then acts the part of the gauge potential.

In the second place, the sources of the gravitational field in conventional general relativity are just the energy and momentum distributions, associated through Noether's theorem not with the whole Poincaré group, but just with a subgroup of it, namely, the translation symmetry of space and time, as if general relativity were the gauge theory of the translation symmetries alone. Many people, then, have tried to extend the gauge symmetry to the full Poincaré group—theories with "torsion," in the usual terminology—which would predict forces associated with angular momentum distributions. This remains very speculative.

Finally, gravitation theory has proved extremely difficult to quantize, with the problems I have described appearing in their worst possible form, together with other problems not present in the other kinds of gauge theory.

Nonetheless, these three basic interaction theories seem to give a remarkably coherent picture of the universe as we know it. Classical gravitation theory, which is general relativity as Einstein developed it, has survived many experimental tests and fits beautifully with all our conceptions of stellar behavior and cosmology. The Glashow–Weinberg–Salam theory of the electroweak interaction has also had remarkable success in describing experimental results and, while it is extremely difficult to do any but the crudest calculations with QCD, there are many indications that it does give a correct description of the strong interactions. In Sec. VII I shall try to give a summary of the ideas presented here and some account of the directions that are being explored in trying to find a single unified theory encompassing all the fundamental interactions—a "theory of everything," as some people have called it.

VII. THE FUTURE: A THEORY OF EVERYTHING?

In the preceding pages I have tried to describe some of the history of the idea of gauge symmetrics and gauge fields starting with Noether's theorem—or Noether's *principle*, perhaps (see Sec. III)—relating symmetrics and conservation laws; going on to develop the logic of gauge theory and the unique way in which the gauge philosophy generates theories associated with given natural symmetrics; and ending with a coherent, if still incomplete, picture in which all the basic interactions of nature are understood in terms of three fundamental gauge fields: the gluon field, the electroweak field, and the gravitational field.

In this presentation, I have greatly oversimplified the complex developments leading up to the Standard Model and I must do the same now with the multitudinous developments and efforts of recent years extending beyond that model. Two principal lines of thought have been directed toward the goal of *unification*. There has long been a desire on the part of scientists, stimulated especially by Einstein, to see coherence in the laws of physics, a feeling that all the laws should be part of a single pattern, consequences in a unified way of a single fundamental principle governing all of nature. The history of physics has seen many unifications and our present picture, consisting of the Standard Model plus general relativity, represents spectacular progress toward this end in that all the diverse areas of 19th- and 20th-century physics are drawn together into a very compact set of elementary constituents and interaction laws; furthermore, the interactions are all associated with gauge theories that follow the same basic principles.

It is clear, though, that the job is not finished. The dream is to unify the basic interaction fields into a single theory, presumably a gauge theory associated with some great and natural symmetry group that comprises all the symmetries of our current picture and breaks down through some kind of spontaneous symmetry breaking, like that associated with the electroweak gauge theory, into the different subsymmetries. Gravity will undoubtedly be the most difficult to incorporate into such a unified theory; thus the first step was expected to be a unification of the strong and electroweak theories into a GUT. In such a scheme it is evident that the six (presumed) flavors of quarks will be symmetrically related to the six different leptons and inevitably, on account of the symmetry breaking, will be able very weakly to transform into each other. If quarks can turn into leptons, though, it means that an otherwise stable hadron (the hadrons are the baryons and mesons, the strongly interacting particles) would ultimately decay. In fact, the only stable hadron is the proton (though the neutron can also be stable when bound in a nucleus), so the first characteristic of a GUT would be proton decay. The lifetime would have to exceed around 10^{31} years, which is greater than the age of our universe by a factor of around 10^{21}, so you might expect it to be impossible to observe. However, there are many protons in the matter around us, for example, about 10^{30} in a ton of water, so there is in fact a real chance of seeing them. Major experimental efforts have been made to do so in a number of different laboratories, using huge vats of liquid surrounded by electronic eyes, in deep underground mines in order to minimize cosmic ray background. Thus far the results have been negative and the limits obtained on the decay rate are sufficiently low to have forced us to abandon at least the most reasonable forms of GUT. The efforts continue, but the interest in GUT's has decreased considerably. One of the untidy elements in the search for a GUT is the "Higgs," the presumed quantum of the Higgs field, introduced in an *ad hoc* way, as you will recall, to provide a symmetry breaking mechanism for the electroweak gauge theory. Such a particle seems to have no fundamental reason of its own for existing since it seems esthetically unsatisfactory to suppose that nature invented it for that purpose alone. What many hope is that the dynamics of the fermion and gauge fields will be found to produce spontaneous symmetry breaking without the help of such a device, a mechanism referred to as "dynamical symmetry breaking."

Beyond the unification of the strong and electroweak forces there lies the dream of total unification of gravity into a single "theory of everything." This sort of work is much more speculative than in the case of GUT's because of the extra degrees of complexity in gravitation and its failure to conform to the pattern of the other gauge theories. The major reason for these differences is that gravitation is associated with what is called a "dynamical symmetry group" having to do with transformations on space and time, the medium in which dynamical events take place, while the other gauge theories are associated with "internal symmetry groups" whose transformations involve only nondynamical degrees of freedom, the flavor and color parameters labeling the different elementary particles.

While it may turn out that the role of gravity is indeed intrinsically different from that of the other forces, there are nevertheless two lines of attack on the problem of unification that are elegant enough to stand at least some chance of being true, namely, Kaluza–Klein and supersymmetry theories. The idea in supersymmetric theories is to extend the symmetry groups of nature to include transformations that mix boson and fermion fields. This can be done without violating the conservation of fermion number by allowing the cross terms in the transformation matrices, the terms associated with boson–fermion mixing, to be what are called "Grassmann elements," which are totally anticommuting analogs of ordinary *c* numbers. While the supersymmetry transformations never turn bosons completely into fermions, or vice versa, such theories do predict matching pairs of bosons and fermions, including even the gauge quanta (photons and "photinos," gluons and "gluinos," etc.). This proliferation of particles may be somewhat embarrassing, but what is startling is that a spin-2 gauge field arises very naturally in this context, with many of the right properties to be a lively candidate for the gravitational field. It is linked to a spin-3/2 "gravitino" field which might or might not have observable consequences.

Another very significant approach is the Kaluza–Klein idea, growing out of the idea of Kaluza (1921) and Klein (1926), that the basic structure of space-time might in fact have more than the usual four dimensions. If the additional dimensions are tightly curled up, so that you could not go more than 10^{-40} cm, say, in such a direction without returning to your starting point, then at the level where ordinary physics takes place only four dimensions would be observed. The basic symmetries of nature, however, would be realized in this higher dimensioned space-time and the basic quantum field theory would be a field theory in that higher number of dimensions. On one hand, dynamical symmetries involving the extra dimensions would appear at our level as internal symmetries, allowing for the possibility (not completely realized at this point) of relating the two types of symmetry, while, on the other hand, the problems with divergences and other ambiguities that beset quantum field theories would at least take a different form and might even prove to be soluble.

One exciting development that combines these ideas is known as "ten-dimensional superstring theory." The theory is a supersymmetric gauge theory in a ten-dimensional space-time, with six of the dimensions assumed to be highly compact, as suggested above, and with the additional feature that the fundamental entities out of which matter is to be built are not point particles, but "strings," tiny open or closed loops that can be thought of as dislocations in the basic structure of space and time and whose transformation symmetries are the basis for an associated gauge theory. With this precise combination of dimensions and structures it seems that many of the difficulties mentioned above might find their natural resolution, such as the canceling of divergences and anomalies in the gauge theories, the natural appearance of the gravitational field, and the natural concealment of the extra bosons and fermions generated by supersymmetry. There are many aspects of such a theory that could never be directly tested since they involve the structure of space-time and matter at a totally inaccessible level: There are those who regard the exercise as futile for this reason. One could argue, though, that if such a theory should be found to provide a consistent basis for understanding physics at the level at which we do observe the world and if there is no other way of doing this that anyone can think of, then elegance alone is a sufficient reason for treating it as true.

There seems to be little doubt now that the ultimate theory, if it is ever accurately identified, will turn out to be a gauge theory. My own feeling is that there will have to be at least one more major conceptual revolution before that final goal is achieved. While gauge theories are easy to formulate at the classical level, the process of quantizing gauge theories is quite awkward, involving either noncovariant procedures or the introduction of unphysical degrees of freedom, all of which suggests to me that we may be starting from an incorrect understanding of quantum theory itself. If the most basic theory of the universe is a quantum gauge theory, then a gauge theory should be the *most* natural thing (if not perhaps the *only* thing) that can be quantized, rather than the most awkward; indeed, you should be able to formulate a quantum gauge theory directly, without going through the intermediate stage of the classical theory. Will the ultimate theory be a gauge theory of the full group of unitary transformations on the Hilbert space of quantum theory, so that quantum states themselves will be understood *locally*, rather than globally?

MAGNETIC MONOPOLES, FIBER BUNDLES, AND GAUGE FIELDS

Chen Ning Yang

Institute for Theoretical Physics
State University of New York at Stony Brook
Stony Brook, New York 11794

The reports in this monograph have shown great enthusiasm and exuberance for the unification of various interactions through the concept of gauge fields. I would like to emphasize a point that has not yet been explicitly stated by any of the other authors: gauge fields are deeply related to some profoundly beautiful ideas of contemporary mathematics, ideas that are the driving forces of part of the mathematics of the last 40 years. Recalling the relationship between physics and mathematics in earlier periods, general relativity and Riemannian geometry, quantum mechanics and Hilbert space, it is all too obvious that physicists may again be zeroing in on a fundamental new secret of nature.

The mathematical development referred to above is the theory of fiber bundles. It may appear, a priori, that this theory is quite abstract and is unrelated to the structure of the physical world. To show that this is not true, we will start with a simple demonstration that electromagnetism and quantum mechanics together lead naturally to "nontrivial fiber bundles." We will then trace the early history of the gauge field concept and its generalization, emphasizing three related but different conceptual motivations, each of which leads to a general formulation of gauge fields.

Magnetic Monopoles and Nontrivial Bundles

The magnetic monopole is the magnetic charge. Though the idea of magnetic monopoles probably was discussed in classic physics early in the history of electricity and magnetism, modern discussions of this concept date back only to 1931, when the important paper of Dirac[1] pointed out that magnetic monopoles in quantum mechanics exhibit some extra and subtle features. In particular, with the existence of a magnetic monopole of strength g, electric charges and magnetic charges must necessarily be quantized, in quantum mechanics. We will give a new derivation of this result below.

If one wants to describe the wave function of an electron in the field of a magnetic monopole, it is necessary to find the vector potential $\mathbf{A}$ around the monopole. Dirac chose a vector potential that has a string of singularities. The necessity of such a string of singularities is obvious if we prove the following theorem[2]:

Theorem: Consider a magnetic monopole of strength $g \neq 0$ at the origin, and consider a sphere of radius R around the origin. There does not exist a vector potential $\mathbf{A}$ for the monopole magnetic field that is singularity free on the sphere.

This theorem can be proved easily in the following way. If there were a singularity-free $\mathbf{A}$, we would consider the loop integral

$$\oint A_\mu dx^\mu$$

around a parallel on the sphere, as indicated in FIGURE 1. According to Stoke's theorem, this loop integral is equal to the total magnetic flux through the cap α:

$$\oint A_\mu \mathrm{d}x^\mu = \Omega_\alpha. \tag{1}$$

Similarly, we can apply Stoke's theorem to cap β, obtaining

$$\oint A_\mu \mathrm{d}x^\mu = \Omega_\beta. \tag{2}$$

Here, Ω_α and Ω_β are the total upward magnetic fluxes through caps α and β, both of which are bordered by the parallel. Subtracting these two equations, we obtain

$$0 = \Omega_A - \Omega_B, \tag{3}$$

which is equal to the total flux *out* of the sphere, which, in turn, is equal to $4\pi g \neq 0$. We have thus reached a contradiction.

Having proved this theorem, we observe that R is arbitrary. Thus, one concludes that there must be a string(s) of singularities in the vector potential to describe the monopole field. Yet, we know that the magnetic field around the monopole is singularity free. This fact suggests that the string of singularities is not a real physical difficulty. Indeed, the situation is reminiscent of the problem that one faces when one wants to find a parametrization of the surface of the globe. The coordinate system that we usually use, latitude and longitude, is not singularity free. It has singularities at the north pole and at the south pole. Yet, the surface of the globe is evidently devoid of singularities. We deal with this situation usually in the manner illustrated in FIGURE 2. We consider a rubber sheet with nicely defined coordinates and stretch and wrap it downward onto the globe, so that it covers more than the northern hemisphere. Similarly, we consider another rubber sheet with nicely defined coordinates and stretch and wrap it upward, so it covers more than

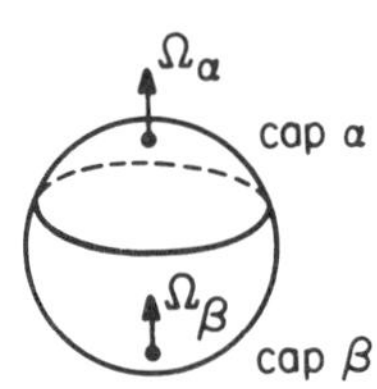

FIGURE 1. A sphere of radius R with a magnetic monopole at its center. The parallel divides the sphere into two caps α and β.

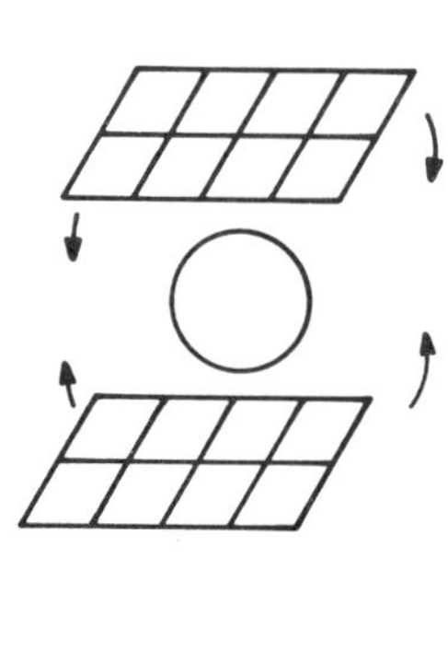

FIGURE 2. Method of parametrizing the globe.

the southern hemisphere. We now have a double system of coordinates to describe the points on the globe. The description is analytic in the domain covered by each sheet, if the globe had experienced no violence in the stretching and wrapping. In the overlapping region covered by both sheets, one has two coordinate systems that are transformable into each other by an analytic nonvanishing Jacobian. This double coordinate system is an entirely satisfactory way to parametrize the globe.

Following this idea, we will now attempt to exorcise the string of singularities in the monopole problem by dividing space into two regions. We will call the points outside of the origin, above the lower cone in FIGURE 3, region R_a. Similarly, we will call the points outside of the origin, under the upper cone, R_b. The union of these two regions gives all points outside of the origin. In R_a, we will choose a vector potential for which there is only one nonvanishing component of A, the azimuthal component:

$$(A_r)_a = (A_\theta)_a = 0, \qquad (A_\phi)_a = \frac{g}{r\sin\theta}(1-\cos\theta), \tag{4}$$

It is important to notice that this vector potential has no singularities anywhere in R_a. Similarly, in R_b, we choose the vector potential

$$(A_r)_b = (A_\theta)_a = 0 \qquad (A_\phi)_b = \frac{-g}{r\sin\theta}(1 \times \cos\theta), \tag{5}$$

which has no singularities in R_b. It is simple to prove that the curl of either of these two potentials gives correctly the magnetic field of the monopole.

In the region of overlap, because both of the two sets of vector potentials share the same curl, the difference between them must be curlless and therefore must be a gradient. Indeed, a simple calculation shows

$$(A_\mu)_a - (A_\mu)_b = \partial_\mu \alpha, \ \text{where } \alpha = 2g\phi, \tag{6}$$

where ϕ is the azimuthal angle. The Schrödinger equation for an electron in the monopole field is thus

$$\frac{1}{2m}(p - eA_a)^2\psi_a + V\psi_a = E\psi_a, \ \text{in } R_a,$$

$$\frac{1}{2m}(p - eA_b)^2\psi_b + V\psi_b = E\psi_b, \ \text{in } R_b,$$

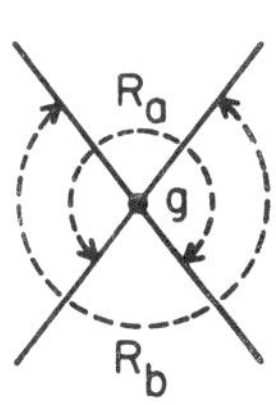

FIGURE 3. Division of space outside of monopole g into overlapping regions R_a and R_b.

where ψ_a and ψ_b are, respectively, the wave functions in the two regions. The fact that the two vector potentials in these two equations are different by a gradient tells us, by the well-known gauge principle, that ψ_a and ψ_b are related by a phase factor transformation

$$\psi_a = S\psi_b, \quad S = \exp(ie\alpha), \tag{7}$$

or

$$\psi_a = [\exp(2iq\phi)]\,\psi_b, \; q = eg. \tag{8}$$

Around the equator, which is entirely in R_a, ψ_a is single valued. Similarly, because the equator is also entirely in R_b, ψ_b is single valued around the equator. Therefore, S must return to its original value when one goes around the equator. That fact implies Dirac's quantization condition:

$$2q = \text{integer}. \tag{9}$$

Hilbert Space of Sections

Two ψs, ψ_a and ψ_b, in R_a and R_b, respectively, that satisfy the condition of transition (Equation **8**) in the overlap region are called a *section* by the mathematicians. We see that around a monopole, the electron wave function is a *section* and *not an ordinary function*. We will call these functions wave sections.

Different wave sections (which belong to different energies, for example) clearly satisfy the same condition of transition (Equation **8**) with the same q. Thus, we need to develop [3] the concept of a Hilbert space of sections. To develop this concept, we define the scalar product of two sections ξ, η (for the *same* q) by

$$(\eta, \xi) = \int \eta^* \xi \, d^3 r. \tag{10}$$

(The question of convergence at $r=0$ and $r=\infty$ is ignored here.) Notice that in the overlap

$$(\eta_a)^* \xi_a = (\eta_b)^* \xi_b, \tag{11}$$

so that Equation **10** is well defined.

It is clear that if ξ is a section, $x\xi$ is also a section, because

$$x\xi_a = S(x\xi_b).$$

Thus, x is an *operator* in the Hilbert space of sections. Similarly, we prove that the components of $(\mathbf{p} - e\mathbf{A})$ are operators, but those of $\mathbf{p}$ are not. Furthermore, $\mathbf{x}$ and $\mathbf{p} - e\mathbf{A}$ are both Hermitian.

Following Fierz,[4] we will now attempt to construct angular momentum operators. Define

$$\mathbf{L} = \mathbf{r} \times (\mathbf{p} - e\mathbf{A}) - \frac{q\mathbf{r}}{r}. \tag{12}$$

It is clear that L_x, L_y, and L_z are Hermitian operators on the Hilbert space of sections. The following commutation rules can be easily verified:

$$\begin{aligned} &[L_x, x] = 0, \quad [L_x, y] = iz, \quad [L_x, z] = -iy, \\ &[L_x, p_x - eA_x] = 0, \quad [L_x, p_y - eA_y] = i(p_z - eA_z), \\ &[L_x, p_z - eA_z] = -i(p_y - eA_y). \end{aligned} \tag{13}$$

It follows from these commutation rules that

$$[L_x, L_y] = iL_z, \text{ etc.} \tag{14}$$

Equation **13**, together with its consequence (Equation **14**), show that L_x, L_y, and L_z are the *angular momentum operators*.[4] We emphasize that neither the Hilbert space nor these operators possess any "singularities." (The singularities of A_a and A_b are not real singularities, because they occur outside of R_a and R_b, respectively.)

MONOPOLE HARMONICS $Y_{q,l,m}$

Because $[r^2, \mathbf{L}] = 0$, we can diagonalize r^2 and study operators $\mathbf{L}$ for fixed r^2. That is, we will study sections of the form

$$\delta(r^2 - r_0^2)\xi,$$

where ξ is a section dependent only on angular coordinates θ and ϕ. $\mathbf{L}$ operates, then, on "angular sections."

Equation **14** shows that $[L^2, L_z] = 0$. Simultaneous diagonalization produces the familiar multiplets with eigenvalues $l(l+1)$ and m,

$$L^2 Y_{q,l,m} = l(l+1) Y_{q,l,m}; \; L_z Y_{q,l,m} = m Y_{q,l,m}, \tag{15}$$

where $l = 0, 1/2, 1, \ldots$, and for each value of l, m ranges from $-l$ to $+l$ in integral steps of increment. $Y_{q,l,m}$ are eigensections, which are called[3] monopole harmonics. The allowed values of l and m are

$$l = |q|, \; |q|+1, \; |q|+2, \ldots, \qquad m = -l, \; -l+1, \ldots, l. \tag{16}$$

Each of these l, m combinations occurs exactly once. One can choose each Y normalized, so that

$$\int_0^\pi \sin\theta d\theta \int_0^{2\pi} |Y_{q,l,m}|^2 d\phi = 1. \tag{17}$$

Different $Y_{q,l,m}$ (for fixed q) are orthogonal, a fact one easily proves in the usual way from Equation **15**.

The explicit values of $Y_{q,l,m}$ in terms of Jacobi polynomials were given in Reference 3. They were obtained from Equation **15**, in exactly the same way one usually obtains the spherical harmonics $Y_{l,m}$. Indeed,

$$Y_{l,m} = Y_{0,l,m}.$$

The collection of $Y_{q,l,m}$ for fixed q and values of l, m given by Equation **16** form[3] a complete orthonormal set of angular sections.

Each $(Y_{q,l,m})_a$ is analytic in R_a; so is $(Y_{q,l,m})_b$ in R_b. Thus, all of the discontinuities, cusps, and singularities in **A** and in ψ are removed in a very smooth way.

Remarks: (A) It is important to realize that the above-described way of using $(A)_a$ and $(A)_b$ together to describe the magnetic field of a monopole has an additional advantage: It gives the magnetic field **H** correctly *everywhere*. In older papers, one often used a single **A** with a string of singularities. Because, by definition,

$$\nabla \cdot (\nabla \times A) = 0,$$

the magnetic field described by $\nabla \times A$ must have *continuous* flux lines. Thus, its flux lines consist of the dotted lines of FIGURE 4, plus the bundle of lines described by the solid line, so as to make the net flux at the origin zero. Thus, $\nabla \times A$ does not correctly describe the magnetic field of the monopole, a point already emphasized by Wentzel.[5]

(B) For ordinary spherical harmonics, there are many important theorems, such as the spherical harmonics addition theorem and the decomposition of products of spherical harmonics by use of Clebsch-Gordon coefficients. These theorems can be generalized to monopole harmonics.[6]

(C) In the approximately 40 years since Dirac's first paper on monopoles, the subject has been beset with difficulties due to singularities. Now that we have removed the difficulty of string singularities through the introduction of the concept of sections, it is revealed that there is yet another difficulty, which we will call the Lipkin-Weisberger-Peshkin[7] difficulty. This difficulty occurs[8] in studying the radial wave function of a Dirac electron around a monopole (TABLE 1). It can be removed through the introduction of a small extra magnetic moment for the Dirac electron.

(D) It is instructive to go back to the reasoning represented in FIGURE 1 and attempt to repeat the steps for the combined A_a, A_b description of the magnetic field. Choose the parallel to be the equator. Then,

FIGURE 4. Magnetic flux lines due to **A**. Because $\nabla \cdot (\nabla \times A) = 0$, flux lines are everywhere continuous. Therefore, there is "return flux" along the solid line.

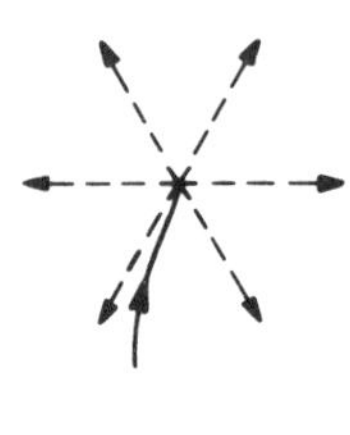

TABLE 1
DIFFICULTIES AND METHODS OF SOLUTION FOR STUDYING THE MOTION OF A DIRAC ELECTRON IN THE FIELD OF A MAGNETIC MONOPOLE

Angular Wave Function	Radial Wave Function
Difficulty of string singularity, solved by introducing sections	Lipkin-Weisberger-Peshkin difficulty, solved by introducing extra magnetic moment

$$\oint (A_\mu)_{\mathrm{a}} \mathrm{d}x^\mu = \Omega_\alpha,$$

$$\oint (A_\mu)_{\mathrm{b}} \mathrm{d}x^{\mathrm{b}} = \Omega_\beta.$$

Thus,
$$4\pi g = \Omega_\alpha - \Omega_\beta = \oint [(A_\mu)_\alpha - (A_\mu)_\beta] \mathrm{d}x^\mu,$$

which is, by Equation **6**, equal to the increment of α around the equation, that is, $2g(2\pi) = 4\pi g$.

We have arrived at an identity. I have provided this simple argument because it is exactly the gist of the proof of the famous Gauss-Bonnet-Allendoerfer-Weil-Chern theorem and the later Chern-Weil theorem, which play seminal roles in contemporary mathematics.

In fact, gauge fields, of which electromagnetism is the simplest example, are conceptually identical to some mathematical concepts in fiber bundle theory. TABLE 2 gives [2] translations for the terminologies used by physicists, on the one hand, and mathematicians, on the other. We notice that, in particular, Dirac's monopole quantization (Equation **9**) is identical to the mathematical concept of classification of U(1) bundles according to the first Chern class.

The last two entries of TABLE 2 identify electromagnetism with and without magnetic monopoles with connections to trivial and nontrivial U(1) bundles. Why is electromagnetism without monopoles "trivial"? We can gain some understanding by looking at a paper loop and a Moebius strip (FIGURE 5). If they are cut along the dotted lines, each would break into two pieces. Looking at the resultant pieces, we cannot differentiate between the two. The paper loop and the Moebius strip are different only in the way the resultant pieces are put together. For the latter, a twist of one of the resultant pieces is necessary. The difference between a trivial and a nontrivial bundle resides only in the processes of *joining:* for the nontrivial bundle, a twist is needed in the joining process. In the case of electromagnetism, the joining process is given by Equation **7** or **8**. If there is no monopole, $S=1$, and the bundle is trivial. If there is a monopole, $S\neq 1$, and the bundle is nontrivial. (We may describe the nontrivial nature by saying that a *twist of phase* is necessary.)

EARLY HISTORY OF THE CONCEPT OF GAUGE FIELDS

Einstein's discovery of the relationship between gravitation and the geometry of space-time stimulated work by many great geometers: Levi-Civita, Cartan, Weyl, and others. In his book, ***Raum, Zeit und Materie*** (space, time and matter), Weyl[9]

TABLE 2
TRANSLATION OF TERMINOLOGIES

Gauge Field Terminology	Bundle Terminology
Gauge (or global gauge)	principal coordinate bundle
Gauge type	principal fiber bundle
Gauge potential b_μ^k	connection on a principal fiber bundle
S (Equation **8**)	transition function
Phase factor Φ_{QP}	parallel displacement
Field strength $f_{\mu\nu}^k$	curvature
Source (electric) J_μ^k	?
Electromagnetism	connection on a U_1 bundle
Isotopic spin gauge field	connection on a SU_2 bundle
Dirac's monopole quantization	classification of U_1 bundle according to first Chern class
Electromagnetism without monopole	connection on a trivial U_1 bundle
Electromagnetism with monopole	connection on a nontrivial U_1 bundle

FIGURE 5. Examples of trivial (*left*) and nontrivial (or Moebius strips, *right*) fiber bundles.

attempted to unify gravity and electromagnetism through the use of the geometric concept of a space-time-dependent scale change. The basic idea is summarized below.

	$\bullet$	$\xrightarrow{dx\mu}$ $\bullet$
scale	1	$1+S_\mu dx^\mu$
f	f	$f+(\partial f/\partial x^\mu)dx^\mu$
scale change	f	$f+(\partial/\partial x^\mu+S_\mu)f dx^\mu$

In the summary above, the first line indicates how the scale changes in going from a point x^μ to a neighboring point $x^\mu+dx^\mu$ of space-time. The second line shows how a function of space-time changes as a result of the change in argument from x^μ to $x^\mu+dx^\mu$. Finally, if the scale change is applied to the function f, one obtains at $x^\mu+dx^\mu$ the product

$$(f+\partial f/\partial x^\mu dx^\mu)\quad(1+S_\mu dx^\mu).$$

Expanding to first order in the small displacement gives the last line in the summary. The increment in f is, then,

$$(\partial/\partial x^\mu+S_\mu)f dx^\mu. \tag{18}$$

Weyl tried to incorporate electromagnetism into a geometric theory by identifying the vector potential A_μ with a space-time-dependent S_μ, generating scale changes as described. This attempt proved, however, unsuccessful.

In 1925, the concepts of quantum mechanics emerged. A key concept in quantum mechanics is the replacement of the momentum p_μ in the classic Hamiltonian by an operator:

$$p_\mu \rightarrow -ih(\partial/\partial x^\mu).$$

For a charged particle, the replacement is

$$p_\mu - (e/c)A_\mu \rightarrow -ih[\partial/\partial x^\mu - i(e/hc)A_\mu]. \tag{19}$$

In 1927, Fock[10] observed that one could base quantum electrodynamics on this operator. London[11] pointed out the similarity of Fock's to Weyl's earlier work. Comparing Equations **18** and **19**, Weyl's identification would be correct if one makes the replacement

$$S_\mu \rightarrow -i(e/hc)A_\mu.$$

In other words, instead of a *scale change*

$$(1 + S_\mu \mathrm{d}x^\mu),$$

one considers a *phase change*

$$[1 - i(e/hc)A_\mu \mathrm{d}x^\mu] \simeq \exp[-i(e/hc)A_\mu \mathrm{d}x^\mu], \tag{20}$$

which can be thought of as an *imaginary scale change.* Weyl put all of these expressions together[12] in a remarkable paper (which also first discussed the two-component theory of a spin-1/2 particle) in which the transformation of the electromagnetic potential

$$A_\mu \rightarrow A'_\mu = A_\mu + \partial_\mu \alpha \text{ (second-type transformation)}, \tag{21}$$

and the associated phase transformation

$$\psi \rightarrow \psi' = \psi \exp(ie\alpha/hc) \text{ (first-type transformation)}, \tag{22}$$

of the wave function of a charged particle were explicitly discussed.[13]

Although the phase change factor (Equation **20**) is no longer a scale factor, Weyl kept the earlier terminology*† that he used in 1918–20 and called both the transformation (Equation **20**) and the associated phase change of wave functions "gauge" transformations.

Generalization: With the discovery of many new particles after World War II, physicists explored various couplings between the "elementary particles." Many possible couplings can be written down, and the desire to find *a principle to choose among the many possibilities* was one of the motivations[17,18] for an attempt to generalize Weyl's gauge principle for electromagnetism. The point here is that for electromagnetism, the gauge principle determines, all at once, the way in which *any* particle of charge qe, a *conserved* quantity, serves as a *source* of the electromagnetic field. Because the isotopic spin **I** is also conserved, a natural question was, "Does there exist a generalized gauge principle that determines the way in which **I** serves as the source of a new field?"

Another motivation for an attempt at generalization is the observation that the conservation of **I** implies that the proton and the neutron are similar. Which to call a proton or, indeed, which superposition of the two to call a proton, is a convention that one can select arbitrarily (if the electromagnetic interaction is switched off). If one requires this freedom of choice to be independent for observers at different space-time points, that is, if one requires *localized* freedom of choice, one is led to a generalization of the gauge principle.

These two motivations were, of course, intertwined and led quite naturally to the formulation[18] of non-Abelian gauge fields.

A third approach[19] to a generalized gauge principle came later and is the "integral formalism" of gauge fields. It starts from the observation that the gauge principle of Weyl deals with a phase factor (Equation **20**) between two neighboring points. Along a path from space-time point A to space time point B, the resultant phase factor is

$$\Phi_{\mathrm{BA}} = \exp[-i(e/hc)\int_{\mathrm{A}}^{\mathrm{B}} A_\mu \mathrm{d}x^\mu], \qquad \textbf{(23)}$$

which is path dependent, that is, nonintegrable. (Dirac[1] had already discussed, in 1931, "non-integrable phases for wave functions.") If one analyzes the meaning of electromagnetism in quantum mechanics, especially through a discussion of the Bohm-Aharonov experiment,[20‡] one reaches the conclusion[2] that "electromagnetism is the gauge invariant manifestation of a non-integrable phase factor."

Once this conclusion is reached, a natural generalization is to replace a

* The idea of scale invariance, discussed in Reference 9, was developed earlier, in 1918–19, in three papers by Weyl (submitted on May 2 and June 8, 1918 and on January 7, 1919). In the first two of them, he used the term *Massstab Invarianz* (see Reference 14); in the third paper, he settled on the term *Eich Invarianz.*

The English translation of *Eich Invarianz* was "calibration invariance" in Henry Brose's 1921 translation of the fourth edition of Weyl's book *Space, Time and Matter*[15] (republished by Dover). The translation "gauge invariance" was not used, I suspect, until after Weyl's 1929 article.[12] It appeared (probably not for the first time) in Dirac's article[1] of 1931.

† The transformation (Equation **21**) that leaves field strengths unchanged must have been known in the nineteenth century. It did not, however, seem to have a specific name. In the many editions of Foppl-Abraham-Becker-Sauter on electricity and magnetism, which started in 1894, *Eich* or "gauge" was not used until the 1964 English translation *Electromagnetic Fields and Interactions,*[16] in which the term "Lorentz gauge" was inserted in a footnote.

‡ The experiment was performed by Chambers.[20]

"nonintegrable phase factor" by a "nonintegrable element of a Lie group." One thus obtains naturally an integral formalism of gauge fields.

We illustrate in FIGURE 6 the three approaches to the general concept of gauge fields. The three approaches are, of course, deeply interrelated, because phases, symmetry, and conservation laws are themselves related.

It is my opinion that, conceptually, the integral formalism of gauge fields is to be preferred to the earlier differential approach. The integral formalism has more structure and more meaning. It brings to the fore problems of global topology not easily formulated in terms of the differential approach. For example, in our earlier discussion of the field around the magnetic monopole, we did not introduce the concept of nonintegrable phase factors. We did not run into any conceptual difficulties, only because we had not raised such questions as a rotation of the coordinate axes. As soon as such questions are raised, it becomes apparent that the integral formalism is more superior, because it specifies that intrinsic meaning is unrelated to the choice of coordinate axes and of regions R_a and R_b.

Differential formalism, however, is used in computing. (The relationship between differential and integral formalisms is quite similar to that between Lie algebras and Lie groups.) In fact, a gauge-Riemannian calculus has been developed.[21]

Electromagnetism is, as we have seen, a gauge field. That gravitation is a gauge field is universally accepted, although exactly how it is a gauge field is a matter still to be clarified.[19,22] Whether weak and strong interactions are also due to gauge fields is a matter that has been intensively studied in recent years,[23] together with the question of the renormalizability of non-Abelian gauge fields.[24§] If one may borrow a term used by the biologists, one would say that there is gradually forming a "dogma" that all interactions are due to gauge fields. Because of the mathematical difficulties involved in the solution of quantized gauge fields,

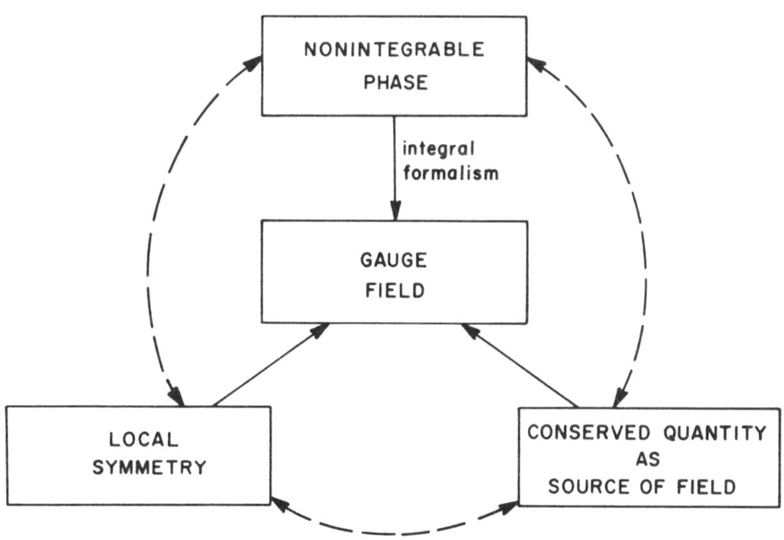

FIGURE 6. Three motivations that led to the concept of gauge fields.

§ Abers and Lee[24] also contains a review of earlier works of R. P. Feynman, L. D. Faddeev, V. N. Popov, and M. T. Veltman.

however, I believe it will be a long time before the question can be definitively answered as to exactly how strong and weak interactions are due to gauge fields.

Reflecting on how the concepts basic to gauge fields were formulated by physicists, we see that at every step, the development was tied to the problem of the conceptual description of the physical world. Firstly, Maxwell equations originated with the four fundamental experimental laws of electricity and magnetism and with Faraday's introduction of the concepts of field and flux. Maxwell's equations and the principles of quantum mechanics led to the idea of gauge invariance. Attempts to generalize this idea, motivated by physical concepts of phases, symmetry, and conservation laws, led to the theory of non-Abelian gauge fields. That non-Abelian gauge fields are conceptually identical to ideas in the beautiful theory of fiber bundles, developed by mathematicians *without reference to the physical world,* was a great marvel to me. In 1975, I discussed my feelings with Chern, and said, "This is both thrilling and puzzling, since you mathematicians dreamed up these concepts out of nowhere." He immediately protested, "No, no, these concepts were not dreamed up. They were natural and real."

References

1. Dirac, P.A.M. 1931. Proc. Roy. Soc. **A133:** 60.
2. Wu, T. T. & C. N. Yang. 1975. Phys. Rev. D **12:** 3845.
3. Wu, T. T. & C. N. Yang. 1976. Nucl. Phys. B **107:** 365.
4. Fierz, M. 1944. Helv. Phys. Acta **17:** 27.
5. Wentzel, G. 1966. Progr. Theor. Phys. Suppl. **37–38:** 163.
6. Wu, T. T. & C. N. Yang. 1977. To be published.
7. Lipkin, H. J., W. I. Weisberger & M. Peshkin. 1969. Ann. Phys. **53:** 203.
8. Kazama, Y., C. N. Yang & A. S. Goldhaber. 1977. Phys. Rev. D. In press.
9. Weyl, H. 1920. Raum, Zeit und Materie. 3rd edit. Springer Verlag. Berlin-Heidelberg. New York.
10. Fock, V. 1927. Z. Phys. **39:** 226.
11. London, F. 1927. Z. Phys. **42:** 375.
12. Weyl, H. 1929. Z. Phys. **56:** 330.
13. Pauli, W. 1933. Handbuch der Physik. 2nd edit. Vol. 24(1): 83. Geiger and Scheel.; Pauli, W. 1941. Rev. Mod. Phys. **13:** 203.
14. Weyl, H. 1918. Sitzber. Preuss Akad. Wiss.: 465; Weyl, H. 1918. Math. Z. **2:** 384; Weyl, H. 1919. Ann. Phys. **59:** 101.
15. Weyl, H. 1921. Space, Time and Matter. Dover Publications, Inc. New York, N.Y.
16. 1964. Electromagnetic Fields and Interactions. Blaisdell Publishing Co. Waltham, Mass.
17. Yang, C. N. & R. Mills. 1954. Phys. Rev. **95:** 631.
18. Yang, C. N. & R. Mills. 1954. Phys. Rev. **96:** 191.
19. Yang, C. N. 1974. Phys. Rev. Lett. **33:** 445.
20. Aharonov, Y. & D. Bohm. 1959. Phys. Rev. **115:** 485; Chambers, R. G. 1960. Phys. Rev. Lett. **5:** 3.
21. Yang, C. N. 1975. Proc. Sixth Hawaii Topical Conf. Particle Phys.
22. Utiyama, R. 1956. Phys. Rev. **101:** 1957.
23. Weinberg, S. 1967. Phys. Rev. Lett. **19:** 1264; Salam, A. 1968. *In* Elementary Particle Theory. N. Svartholm, Ed. Almquist and Forlag. Stockholm, Sweden.
24. 'tHooft, G. 1971. Nucl. Phys. B **35:** 167; Abers, E. S. & B. W. Lee. 1973. Phys. Rep. **9C:** 1.

Gauge theory: Historical origins and some modern developments

Lochlainn O'Raifeartaigh

Dublin Institute for Advanced Studies, Dublin 4, Ireland

Norbert Straumann

Institut für Theoretische Physik der Universität Zürich-Irchel, Zürich, Switzerland

One of the major developments of twentieth-century physics has been the gradual recognition that a common feature of the known fundamental interactions is their gauge structure. In this article the authors review the early history of gauge theory, from Einstein's theory of gravitation to the appearance of non-Abelian gauge theories in the fifties. The authors also review the early history of dimensional reduction, which played an important role in the development of gauge theory. A description is given of how, in recent times, the ideas of gauge theory and dimensional reduction have emerged naturally in the context of string theory and noncommutative geometry.

CONTENTS

I. INTRODUCTION

It took decades until physicists understood that all known fundamental interactions can be described in terms of gauge theories. Our historical account begins with Einstein's general theory of relativity, which is a non-Abelian gauge theory of a special type (see Secs. III and VII). That other gauge theories emerged, in a slow and complicated process, gradually from general relativity and their common geometrical structure—best expressed in terms of connections of fiber bundles—is now widely recognized. Thus H. Weyl was right when he wrote in the preface to the first edition of *Space, Time, Matter (Raum.· Zeit.· Materie)* early in 1918: "Wider expanses and greater depths are now exposed to the searching eye of knowledge, regions of which we had not even a presentiment. It has brought us much nearer to grasping the plan that underlies all physical happening" (Weyl, 1922).

It was Weyl himself who in 1918 made the first attempt to extend general relativity in order to describe gravitation and electromagnetism within a unifying geometrical framework (Weyl, 1918). This brilliant proposal contains the germs of all mathematical aspects of a non-Abelian gauge theory, as we shall make clear in Sec. II. The words gauge (*Eich-*) transformation and gauge invariance appeared for the first time in this paper, but in the everyday meaning of change of length or change of calibration.[1]

Einstein admired Weyl's theory as "a coup of genius of the first rate . . . ," but immediately realized that it was physically untenable: "Although your idea is so beautiful, I have to declare frankly that, in my opinion, it is impossible that the theory corresponds to Nature." This led to an intense exchange of letters between Einstein (in Berlin) and Weyl [at the Eidgenössische Technische Hochschule (ETH) in Zürich], part of which has now been published in Vol. 8 of *The Collected Papers of Albert Einstein* (1987). [The article of Straumann (1987) gives an account of this correspondence, which is preserved in the Archives of the ETH.] No agreement was reached, but Einstein's intuition proved to be right.

Although Weyl's attempt was a failure as a physical theory, it paved the way for the correct understanding of gauge invariance. Weyl himself reinterpreted his original theory after the advent of quantum theory in a seminal paper (Weyl, 1929), which we shall discuss at length in Sec. III. Parallel developments by other workers and interconnections are indicated in Fig. 1.

[1]The German word *eichen* probably comes from the Latin *aequare*, i.e., equalizing the length to a standard one.

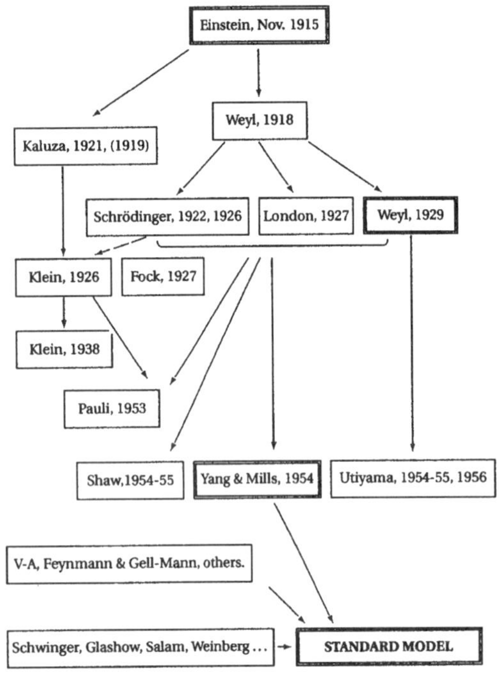

FIG. 1. Key papers in the development of gauge theories.

At the time Weyl's contributions to theoretical physics were not appreciated very much, since they did not really add new physics. The attitude of the leading theoreticians was expressed with familiar bluntness in a letter by Pauli to Weyl of July 1, 1929, after he had seen a preliminary account of Weyl's work:

> Before me lies the April edition of the *Proc. Nat. Acad. (US)*. Not only does it contain an article from you under "Physics" but shows that you are now in a "Physical Laboratory": from what I hear you have even been given a chair in "Physics" in America. I admire your courage; since the conclusion is inevitable that you wish to be judged, not for success in pure mathematics, but for your true but unhappy love for physics. (Translated from Pauli, 1979.)

Weyl's reinterpretation of his earlier speculative proposal had actually been suggested before by London and Fock, but it was Weyl who emphasized the role of gauge invariance as a *symmetry principle* from which electromagnetism can be derived. It took several decades until the importance of this symmetry principle—in its generalized form to non-Abelian gauge groups developed by Yang, Mills, and others—also became fruitful for a description of the weak and strong interactions. The mathematics of the non-Abelian generalization of Weyl's 1929 paper would have been an easy task for a mathematician of his rank, but at the time there was no motivation for this from the physics side. The known properties of the weak and strong nuclear interactions, in particular their short-range behavior, did not point to a gauge-theoretical description. We all know that the gauge symmetries of the standard model are very hidden, and it is therefore not astonishing that progress was very slow indeed.

In this paper we present only the history up to the invention of Yang-Mills theory in 1954. The independent discovery of this theory by other authors has already been described (O'Raifeartaigh, 1997). Later history covering the application of the Yang-Mills theory to the electroweak and strong interactions is beyond our scope. The main features of these applications are well known and are covered in contemporary textbooks. One modern development that we do wish to mention, however, is the emergence of both gauge theory and dimensional reduction in two fields other than traditional quantum field theory, namely, string theory and noncommutative geometry, as their emergence in these fields is a natural extension of the early history. Indeed in string theory both gauge invariance and dimensional reduction occur in such a natural way that it is probably not an exaggeration to say that, had they not been found earlier, they would have been discovered in this context. The case of noncommutative geometry is a little different, as the gauge principle is used as an input, but the change from a continuum to a discrete structure produces qualitatively new features. Amongst these is an interpretation of the Higgs field as a gauge potential and the emergence of a dimensional reduction that avoids the usual embarrassment concerning the fate of the extra dimensions.

A fuller account of the early history of gauge theory is given by O'Raifeartaigh (1997). There one can also find English translations of the most important papers of the early period, as well as Pauli's letters to Pais on non-Abelian Kaluza-Klein reductions. These works underlie the diagram in Fig. 1.

II. WEYL'S ATTEMPT TO UNIFY GRAVITATION AND ELECTROMAGNETISM

On the 1st of March 1918 Weyl writes in a letter to Einstein:

> "These days I succeeded, as I believe, to derive electricity and gravitation from a common source . . . ."

Einstein's prompt reaction by postcard indicates already a physical objection, which he explained in detail shortly afterwards. Before we come to this we have to describe Weyl's theory of 1918.

A. Weyl's generalization of Riemannian geometry

Weyl's starting point was purely mathematical. He felt a certain uneasiness about Riemannian geometry, as is clearly expressed by the following sentences early in his paper:

> But in Riemannian geometry described above there is contained a last element of geometry "at a distance" (*ferngeometrisches Element*)—with no good reason,

as far as I can see; it is due only to the accidental development of Riemannian geometry from Euclidean geometry. The metric allows the two magnitudes of two vectors to be compared, not only at the same point, but at any arbitrarily separated points. *A true infinitesimal geometry should, however, recognize only a principle for transferring the magnitude of a vector to an infinitesimally close point* and then, on transfer to an arbitrary distant point, the integrability of the magnitude of a vector is no more to be expected than the integrability of its direction.

After these remarks Weyl turns to physical speculation and continues as follows:

> On the removal of this inconsistency there appears a geometry that, surprisingly, when applied to the world, *explains not only the gravitational phenomena but also the electrical.* According to the resultant theory both spring from the same source, indeed *in general one cannot separate gravitation and electromagnetism in a unique manner.* In this theory *all physical quantities have a world geometrical meaning; the action appears from the beginning as a pure number. It leads to an essentially unique universal law; it even allows us to understand in a certain sense why the world is four dimensional.*

In brief, Weyl's geometry can be described as follows (see also Audretsch, Gähler, and Straumann, 1984). First, the space-time manifold M is equipped with a conformal structure, i.e., with a class $[g]$ of conformally equivalent Lorentz metrics g (and not a definite metric as in general relativity). This corresponds to the requirement that it should only be possible to compare lengths at one and the same world point. Second, it is assumed, as in Riemannian geometry, that there is an affine (linear) torsion-free connection which defines a covariant derivative ∇ and respects the conformal structure. Differentially this means that for any $g \in [g]$ the covariant derivative ∇g should be proportional to g:

$$\nabla g = -2A \otimes g \quad (\nabla_\lambda g_{\mu\nu} = -2A_\lambda g_{\mu\nu}), \tag{1}$$

where $A = A_\mu dx^\mu$ is a differential 1-form.

Consider now a curve $\gamma:[0,1] \to M$ and a parallel-transported vector field X along γ. If l is the length of X, measured with a representative $g \in [g]$, we obtain from Eq. (1) the following relation between $l(p)$ for the initial point $p = \gamma(0)$ and $l(q)$ for the end point $q = \gamma(1)$:

$$l(q) = \exp\left(-\int_\gamma A\right) l(p). \tag{2}$$

Thus the ratio of lengths in q and p (measured with $g \in [g]$) depends in general on the connecting path γ (see Fig. 2). The length is only independent of γ if the curl of A,

$$F = dA \quad (F_{\mu\nu} = \partial_\mu A_\nu - \partial_\nu A_\mu), \tag{3}$$

vanishes.

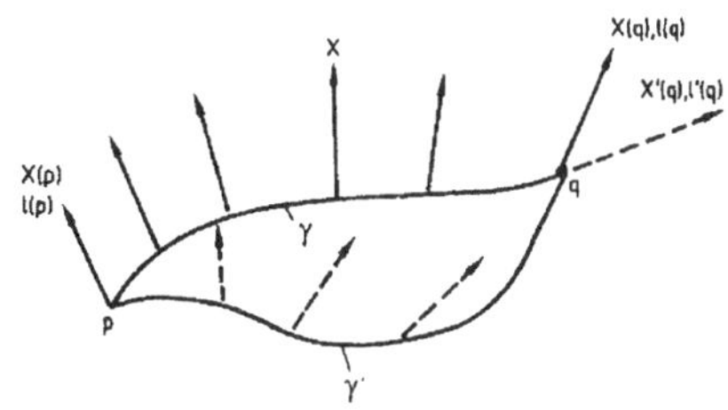

FIG. 2. Path dependence of parallel displacement and transport of length in Weyl space.

The compatibility requirement (1) leads to the following expression for the Christoffel symbols in Weyl's geometry:

$$\Gamma^\mu_{\nu\lambda} = \frac{1}{2} g^{\mu\sigma}(g_{\lambda\sigma,\nu} + g_{\sigma\nu,\lambda} - g_{\nu\lambda,\sigma}) + g^{\mu\sigma}(g_{\lambda\sigma}A_\nu + g_{\sigma\nu}A_\lambda - g_{\nu\lambda}A_\sigma). \tag{4}$$

The second A-dependent term is a characteristic new piece in Weyl's geometry, which has to be added to the Christoffel symbols of Riemannian geometry.

Until now we have chosen a fixed, but arbitrary, metric in the conformal class $[g]$. This corresponds to a choice of calibration (or gauge). Passing to another calibration with metric $\bar{g}$, related to g by

$$\bar{g} = e^{2\lambda} g, \tag{5}$$

we find that the potential A in Eq. (1) will also change to $\bar{A}$, say. Since the covariant derivative has an absolute meaning, $\bar{A}$ can easily be worked out: On the one hand we have, by definition,

$$\nabla \bar{g} = -2\bar{A} \otimes \bar{g}, \tag{6}$$

and on the other hand we find for the left side with Eq. (1)

$$\nabla \bar{g} = \nabla(e^{2\lambda} g) = 2\, d\lambda \otimes \bar{g} + e^{2\lambda} \nabla g = 2\, d\lambda \otimes \bar{g} - 2A \otimes \bar{g}. \tag{7}$$

Thus

$$\bar{A} = A - d\lambda \quad (\bar{A}_\mu = A_\mu - \partial_\mu \lambda). \tag{8}$$

This shows that a change of calibration of the metric induces a *gauge transformation* for A:

$$g \to e^{2\lambda} g, \quad A \to A - d\lambda. \tag{9}$$

Only gauge classes have an absolute meaning. [The Weyl connection is, however, gauge invariant. This is conceptually clear, but can also be verified by direct calculation from Eq. (4).]

B. Electromagnetism and gravitation

Turning to physics, Weyl assumes that his "purely infinitesimal geometry" describes the structure of space-time and consequently he requires that physical laws satisfy a double invariance: (1) They must be invariant with respect to arbitrary smooth coordinate transformations;

(2) They must be *gauge invariant*, i.e., invariant with respect to the substitutions of Eq. (9) for an arbitrary smooth function λ.

Nothing is more natural to Weyl than identifying A_μ with the vector potential and $F_{\mu\nu}$ in Eq. (3) with the field strength of electromagnetism. In the absence of electromagnetic fields ($F_{\mu\nu}=0$) the scale factor, $\exp(-\int_\gamma A)$ in Eq. (2), for length transport becomes path independent (integrable) and one can find a gauge such that A_μ vanishes for simply connected spacetime regions. In this special case, it is the same situation as in general relativity.

Weyl proceeds to find an action that is generally invariant as well as gauge invariant and that would give the coupled field equations for g and A. We do not want to enter into this, except for the following remark. In his first paper Weyl (1918) proposes what we now call the Yang-Mills action:

$$S(g,A)=-\frac{1}{4}\int \mathrm{Tr}(\Omega\wedge *\Omega). \tag{10}$$

Here Ω denotes the curvature from and $*\Omega$ its Hodge dual.[2] Note that the latter is gauge invariant, i.e., independent of the choice of $g\in[g]$. In Weyl's geometry the curvature form splits as $\Omega=\hat{\Omega}+F$, where $\hat{\Omega}$ is the metric piece (Audretsch, Gähler, and Straumann, 1984). Correspondingly, the action also splits,

$$\mathrm{Tr}(\Omega\wedge *\Omega)=\mathrm{Tr}(\hat{\Omega}\wedge *\hat{\Omega})+F\wedge *F. \tag{11}$$

The second term is just the Maxwell action. Weyl's theory thus contains formally all aspects of a non-Abelian gauge theory.

Weyl emphasizes, of course, that the Einstein-Hilbert action is not gauge invariant. Later work by Pauli (1919) and by Weyl himself (1918, 1922) soon led to the conclusion that the action of Eq. (10) could not be the correct one, and other possibilities were investigated (see the later editions of *Space, Time, Matter*).

Independent of the precise form of the action, Weyl shows that in his theory gauge invariance implies the conservation of electric charge in much the same way as general coordinate invariance leads to the conservation of energy and momentum.[3] This beautiful connection pleased him particularly: "... [it] seems to me to be the strongest general argument in favour of the present theory—insofar as it is permissible to talk of justification in the context of pure speculation." The invariance principles imply five "Bianchi-type" identities. Correspondingly, the five conservation laws follow in two independent ways from the coupled field equations and may be "termed the eliminants" of the latter. These structural connections hold also in modern gauge theories.

C. Einstein's objection and reactions of other physicists

After this sketch of Weyl's theory we come to Einstein's striking counterargument, which he first communicated to Weyl by postcard (see Fig. 3). The problem is that if the idea of a nonintegrable length connection (scale factor) is correct, then the behavior of clocks would depend on their history. Consider two identical atomic clocks in adjacent world points and bring them along different world trajectories which meet again in adjacent world points. According to Eq. (2) their frequencies would then generally differ. This is in clear contradiction with empirical evidence, in particular with the existence of stable atomic spectra. Einstein therefore concludes (see Straumann, 1987):

> ... (if) one drops the connection of the ds to the measurement of distance and time, then relativity loses all its empirical basis.

Nernst shared Einstein's objection and demanded on behalf of the Berlin Academy that it be printed in a short amendment to Weyl's article. Weyl had to accept this. One of us has described elsewhere (Straumann, 1987; see also Vol. 8 of Einstein, 1987) the intense and instructive subsequent correspondence between Weyl and Einstein. As an example, let us quote from one of the last letters of Weyl to Einstein:

> This [insistence] irritates me of course, because experience has proven that one can rely on your intuition; so unconvincing as your counterarguments seem to me, as I have to admit...
>
> By the way, you should not believe that I was driven to introduce the linear differential form in addition to the quadratic one by physical reasons. I wanted, just to the contrary, to get rid of this "methodological inconsistency (*Inkonsequenz*)" which has been a bone of contention to me already much earlier. And then, to my surprise, I realized that it looked as if it might explain electricity. You clap your hands above your head and shout: But physics is not made this way! (Weyl to Einstein 10 December 1918).

Weyl's reply to Einstein's criticism was, generally speaking, this: The real behavior of measuring rods and clocks (atoms and atomic systems) in arbitrary electromagnetic and gravitational fields can be deduced only from a dynamical theory of matter.

Not all leading physicists reacted negatively. Einstein transmitted a very positive first reaction by Planck, and Sommerfeld wrote enthusiastically to Weyl that there was "... hardly doubt, that you are on the correct path and not on the wrong one."

In his encyclopedia article on relativity Pauli (1921) gave a lucid and precise presentation of Weyl's theory, but commented on Weyl's point of view very critically. At the end he states:

[2]The integrand in Eq. (10) is indeed just the expression $R_{\alpha\beta\gamma\delta}R^{\alpha\beta\gamma\delta}\sqrt{-g}\,dx^0\wedge\cdots\wedge dx^3$ in local coordinates which is used by Weyl ($R_{\alpha\beta\gamma\delta}$=the curvature tensor of the Weyl connection).

[3]We adopt here the somewhat naive interpretation of energy-momentum conservation for generally invariant theories of the older literature.

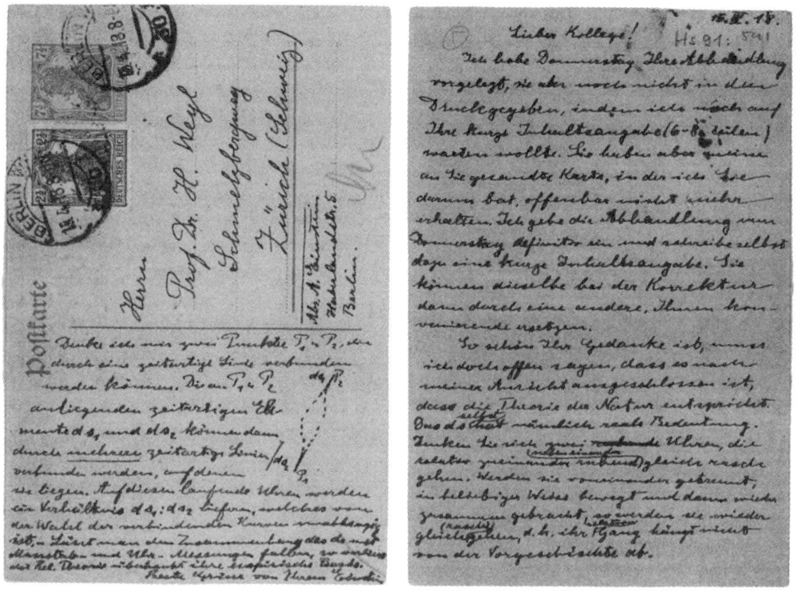

FIG. 3. Postcard from Einstein to Weyl 15 April 1918. From Archives of Eidgenössische Technische Hochschule, Zürich.

> ... In summary one may say that Weyl's theory has not yet contributed to getting closer to the solution of the problem of matter.

Eddington's reaction was at first very positive but he soon changed his mind and denied the physical relevance of Weyl's geometry.

The situation was later appropriately summarized by London (1927) as follows:

> In the face of such elementary experimental evidence, it must have been an unusually strong metaphysical conviction that prevented Weyl from abandoning the idea that Nature would have to make use of the beautiful geometrical possibility that was offered. He stuck to his conviction and evaded discussion of the above-mentioned contradictions through a rather unclear re-interpretation of the concept of "real state," which, however, robbed his theory of its immediate physical meaning and attraction.

In this remarkable paper, London suggested a reinterpretation of Weyl's principle of gauge invariance within the new quantum mechanics: The role of the metric is taken over by the wave function, and the rescaling of the metric has to be replaced by a phase change of the wave function.

In this context an astonishing early paper by Schrödinger (1922) has to be mentioned, which also used Weyl's "world geometry" and is related to Schrödinger's later invention of wave mechanics. This precursor relation was discovered by Raman and Forman (1969). [See also the discussion by C. N. Yang in Schrödinger (1987).]

Simultaneously with London, Fock (1927) arrived along a completely different line at the principle of gauge invariance in the framework of wave mechanics. His approach was similar to that of Klein, which will be discussed in detail (in Sec. IV).

The contributions of Schrödinger (1922), London (1927), and Fock (1927) are discussed in the book of O'Raifeartaigh (1997), where English translations of the original papers can also be found. Here, we concentrate on Weyl's seminal paper "Electron and Gravitation."

III. WEYL'S 1929 CLASSIC: "ELECTRON AND GRAVITATION"

Shortly before his death late in 1955, Weyl wrote for his *Selecta* (Weyl, 1956) a postscript to his early attempt in 1918 to construct a unified field theory. There he expressed his deep attachment to the gauge idea and adds (p. 192):

> Later the quantum-theory introduced the Schrödinger-Dirac potential ψ of the electron-positron field; it carried with it an experimentally-based principle of gauge-invariance which guaranteed the conservation of charge, and connected the ψ with the electromagnetic potentials ϕ_i in the same way that my speculative theory had connected the gravitational potentials g_{ik} with the ϕ_i, and measured the ϕ_i in known atomic, rather than unknown cosmological units. I have no doubt but that the correct context for the principle of gauge-invariance is here and not, as I believed in 1918, in the intertwining of electromagnetism and gravity.

This reinterpretation was developed by Weyl in one of the great papers of this century (Weyl, 1929). Weyl's classic not only gives a very clear formulation of the gauge principle, but contains, in addition, several other important concepts and results—in particular his two-component spinor theory. The richness and scope of the paper is clearly visible from the following table of contents:

> Introduction. Relationship of General Relativity to the quantum-theoretical field equations of the spinning electron: mass, gauge-invariance, distant-parallelism. Expected modifications of the Dirac theory. $-I$. Two-component theory: the wave function ψ has only two components. $-$§1. Connection between the transformation of the ψ and the transformation of a normal tetrad in four-dimensional space. Asymmetry of past and future, of left and right. $-$§2. In General Relativity the metric at a given point is determined by a normal tetrad. Components of vectors relative to the tetrad and coordinates. Covariant differentiation of ψ. $-$§3. Generally invariant form of the Dirac action, characteristic for the wave-field of matter. $-$§4. The differential conservation law of energy and momentum and the symmetry of the energy-momentum tensor as a consequence of the double-invariance (1) with respect to coordinate transformations (2) with respect to rotation of the tetrad. Momentum and moment of momentum for matter. $-$§5. Einstein's classical theory of gravitation in the new analytic formulation. Gravitational energy. $-$§6. The electromagnetic field. From the arbitrariness of the gauge-factor in ψ appears the necessity of introducing the electromagnetic potential. Gauge invariance and charge conservation. The space-integral of charge. The introduction of mass. Discussion and rejection of another possibility in which electromagnetism appears, not as an accompanying phenomenon of matter, but of gravitation.

The modern version of the gauge principle is already spelled out in the introduction:

> The Dirac field-equations for ψ together with the Maxwell equations for the four potentials f_p of the electromagnetic field have an invariance property which is formally similar to the one which I called gauge-invariance in my 1918 theory of gravitation and electromagnetism; the equations remain invariant when one makes the simultaneous replacements
>
> $$\psi \text{ by } e^{i\lambda}\psi \quad \text{and} \quad f_p \text{ by } f_p - \frac{\partial\lambda}{\partial x^p},$$
>
> where λ is understood to be an arbitrary function of position in four-space. Here the factor e/ch, where $-e$ is the charge of the electron, c is the speed of light, and $h/2\pi$ is the quantum of action, has been absorbed in f_p. The connection of this "gauge invariance" to the conservation of electric charge remains untouched. But a fundamental difference, which is important to obtain agreement with observation, is that the exponent of the factor multiplying ψ is not real but purely imaginary. ψ now plays the role that Einstein's ds played before. It seems to me that this new principle of gauge-invariance, which follows not from speculation but from experiment, tells us that the electromagnetic field is a necessary accompanying phenomenon, not of gravitation, but of the material wave-field represented by ψ. Since gauge-invariance involves an arbitrary function λ it has the character of "general" relativity and can naturally only be understood in that context.

We shall soon enter into Weyl's justification, which is, not surprisingly, strongly associated with general relativity. Before this we have to describe his incorporation of the Dirac theory into general relativity, which he achieved with the help of the tetrad formalism.

One of the reasons for adapting the Dirac theory of the spinning electron to gravitation had to do with Einstein's recent unified theory, which invoked a distant parallelism with torsion. Wigner (1929) and others had noticed a connection between this theory and the spin theory of the electron. Weyl did not like this and wanted to dispense with teleparallelism. In the introduction he says:

> I prefer not to believe in distant parallelism for a number of reasons. First my mathematical intuition objects to accepting such an artificial geometry; I find it difficult to understand the force that would keep the local tetrads at different points and in rotated positions in a rigid relationship. There are, I believe, two important physical reasons as well. The loosening of the rigid relationship between the tetrads at different points converts the gauge-factor $e^{i\lambda}$, which remains arbitrary with respect to ψ, from a constant to an arbitrary function of space-time. In other words, only through the loosening of the rigidity does the established gauge-invariance become understandable.

This thought is carried out in detail after Weyl has set up his two-component theory in special relativity, including a discussion of P and T invariance. He emphasizes thereby that the two-component theory excludes a linear implementation of parity and remarks: "It is only the fact that the left-right symmetry actually appears in Nature that forces us to introduce a second pair of ψ components." To Weyl the mass problem is thus not relevant for this.[4] Indeed he says: "Mass, however, is a gravitational effect; thus there is hope of finding a substitute in the theory of gravitation that would produce the required corrections."

[4]At the time it was thought by Weyl, and indeed by all physicists, that the two-component theory required a zero mass. In 1957, after the discovery of parity nonconservation, it was found that the two-component theory could be consistent with a finite mass. See Case (1957).

A. Tetrad formalism

In order to incorporate his two-component spinors into general relativity, Weyl was forced to make use of local tetrads (*Vierbeine*). In Sec. 2 of his paper he develops the tetrad formalism in a systematic manner. This was presumably independent work, since he does not give any reference to other authors. It was, however, mainly E. Cartan (1928) who demonstrated the usefulness of locally defined orthonormal bases—also called moving frames—for the study of Riemannian geometry.

In the tetrad formalism the metric is described by an arbitrary basis of orthonormal vector fields $\{e_\alpha(x);\ \alpha=0,1,2,3\}$. If $\{e^\alpha(x)\}$ denotes the dual basis of 1-forms, the metric is given by

$$g=\eta_{\mu\nu}e^\mu(x)\otimes e^\nu(x),\quad (\eta_{\mu\nu})=\mathrm{diag}(1,-1,-1,-1). \tag{12}$$

Weyl emphasizes, of course, that only a class of such local tetrads is determined by the metric: the metric is not changed if the tetrad fields are subject to space-time-dependent Lorentz transformations:

$$e^\alpha(x)\rightarrow\Lambda^\alpha_\beta(x)e^\beta(x). \tag{13}$$

With respect to a tetrad, the connection forms $\omega=(\omega^\alpha_\beta)$ have values in the Lie algebra of the homogeneous Lorentz group:

$$\omega_{\alpha\beta}+\omega_{\beta\alpha}=0. \tag{14}$$

(Indices are raised and lowered with $\eta^{\alpha\beta}$ and $\eta_{\alpha\beta}$, respectively.) They are determined (in terms of the tetrad) by the first structure equation of Cartan:

$$de^\alpha+\omega^\alpha_\beta\wedge e^\beta=0. \tag{15}$$

(For a textbook derivation see Straumann, 1984.) Under local Lorentz transformations [Eq. (13)] the connection forms transform in the same way as the gauge potential of a non-Abelian gauge theory:

$$\omega(x)\rightarrow\Lambda(x)\omega(x)\Lambda^{-1}(x)-d\Lambda(x)\Lambda^{-1}(x). \tag{16}$$

The curvature forms $\Omega=(\Omega^\mu_\nu)$ are obtained from ω in exactly the same way as the Yang-Mills field strength from the gauge potential:

$$\Omega=d\omega+\omega\wedge\omega \tag{17}$$

(second structure equation).

For a vector field V, with components V^α relative to $\{e_\alpha\}$, the covariant derivative DV is given by

$$DV^\alpha=dV^\alpha+\omega^\alpha_\beta V^\beta. \tag{18}$$

Weyl generalizes this in a unique manner to spinor fields ψ:

$$D\psi=d\psi+\frac{1}{4}\omega_{\alpha\beta}\sigma^{\alpha\beta}\psi. \tag{19}$$

Here, the $\sigma^{\alpha\beta}$ describe infinitesimal Lorentz transformations (in the representation of ψ). For a Dirac field these are the familiar matrices

$$\sigma^{\alpha\beta}=\frac{1}{2}[\gamma^\alpha,\gamma^\beta]. \tag{20}$$

(For two-component Weyl fields, one has similar expressions in terms of the Pauli matrices.)

With these tools the action principle for the coupled Einstein-Dirac system can be set up. In the massless case the Lagrangian is

$$\mathcal{L}=\frac{1}{16\pi G}R-i\bar{\psi}\gamma^\mu D_\mu\psi, \tag{21}$$

where the first term is just the Einstein-Hilbert Lagrangian (which is linear in Ω). Weyl discusses, of course, immediately the consequences of the following two symmetries:

(i) local Lorentz invariance,
(ii) general coordinate invariance.

B. The new form of the gauge principle

All this is a kind of a preparation for the final section of Weyl's paper, which has the title "electric field." Weyl says:

> We come now to the critical part of the theory. In my opinion the origin and necessity for the electromagnetic field is in the following. The components ψ_1,ψ_2 are, in fact, not uniquely determined by the tetrad but only to the extent that they can still be multiplied by an arbitrary "gauge-factor" $e^{i\lambda}$. The transformation of the ψ induced by a rotation of the tetrad is determined only up to such a factor. In special relativity one must regard this gauge-factor as a constant because here we have only a single point-independent tetrad. Not so in general relativity; every point has its own tetrad and hence its own arbitrary gauge-factor; because by the removal of the rigid connection between tetrads at different points the gauge-factor necessarily becomes an arbitrary function of position.

In this manner Weyl arrives at the gauge principle in its modern form and emphasizes "From the arbitrariness of the gauge factor in ψ appears the necessity of introducing the electromagnetic potential." The first term $d\psi$ in Eq. (19) now has to be replaced by the covariant gauge derivative $(d-ieA)\psi$, and the nonintegrable scale factor (2) of the old theory is now replaced by a phase factor:

$$\exp\left(-\int_\gamma A\right)\rightarrow\exp\left(-i\int_\gamma A\right),$$

which corresponds to the replacement of the original gauge group **R** by the compact group $U(1)$. Accordingly, the original Gedankenexperiment of Einstein translates now to the Aharonov-Bohm effect, as was first pointed out by Yang (1980). The close connection between gauge invariance and conservation of charge is again revealed. The current conservation follows, as in the original theory, in two independent ways: On the one hand, it is a consequence of the field equations for matter plus gauge invariance. On the other hand, however, it is also a consequence of the field equations for the electromagnetic field plus gauge invariance. This corresponds to an identity in the coupled system of field equations that has to exist as a result of gauge invariance. All this is now familiar to students of physics and does not need to be explained in more detail.

Much of Weyl's paper appeared also in his classic book *The Theory of Groups and Quantum Mechanics* (Weyl, 1981). There he mentions the transformation of his early gauge-theoretic ideas: "This principle of gauge invariance is quite analogous to that previously set up by the author, on speculative grounds, in order to arrive at a unified theory of gravitation and electricity. But I now believe that this gauge invariance does not tie together electricity and gravitation, but rather electricity and matter."

When Pauli saw the full version of Weyl's paper he became more friendly and wrote (Pauli, 1979, p. 518):

> In contrast to the nasty things I said, the essential part of my last letter has since been overtaken, particularly by your paper in *Z. f. Physik*. For this reason I have afterward even regretted that I wrote to you. After studying your paper I believe that I have really understood what you wanted to do (this was not the case in respect of the little note in the *Proc. Nat. Acad.*). First let me emphasize that side of the matter concerning which I am in full agreement with you: your incorporation of spinor theory into gravitational theory. I am as dissatisfied as you are with distant parallelism and your proposal to let the tetrads rotate independently at different space-points is a true solution.

In brackets Pauli adds:

> Here I must admit your ability in Physics. Your earlier theory with $g'_{ik}=\lambda g_{ik}$ was pure mathematics and unphysical. Einstein was justified in criticizing and scolding. Now the hour of your revenge has arrived.

Then he remarks, in connection with the mass problem,

> Your method is valid even for the massive [Dirac] case. I thereby come to the other side of the matter, namely, the unsolved difficulties of the Dirac theory (two signs of m_0) and the question of the 2-component theory. In my opinion these problems will not be solved by gravitation . . . the gravitational effects will always be much too small.

Many years later, Weyl summarized this early tortuous history of gauge theory in an instructive letter (Seelig, 1960) to the Swiss writer and Einstein biographer C. Seelig, which we reproduce in an English translation.

> The first attempt to develop a unified field theory of gravitation and electromagnetism dates to my first attempt in 1918, in which I added the principle of gauge invariance to that of coordinate invariance. I myself have long since abandoned this theory in favour of its correct interpretation: gauge invariance as a principle that connects electromagnetism not with gravitation but with the wave-field of the electron. —Einstein was against it [the original theory] from the beginning, and this led to many discussions. I thought that I could answer his concrete objections. In the end he said "Well, Weyl, let us leave it at that! In such a speculative manner, without any guiding physical principle, one cannot make Physics." Today one could say that in this respect we have exchanged our points of view. Einstein believes that in this field [Gravitation and Electromagnetism] the gap between ideas and experience is so wide that only the path of mathematical speculation, whose consequences must, of course, be developed and confronted with experiment, has a chance of success. Meanwhile my own confidence in pure speculation has diminished, and I see a need for a closer connection with quantum-physics experiments, since in my opinion it is not sufficient to unify Electromagnetism and Gravity. The wave-fields of the electron and whatever other irreducible elementary particles may appear must also be included.

Independently of Weyl, Fock (1929) also incorporated the Dirac equation into general relativity using the same method. On the other hand, Tetrode (1928), Schrödinger (1932), and Bargmann (1932) reached this goal by starting with space-time-dependent γ matrices, satisfying $\{\gamma^\mu,\gamma^\nu\}=2g^{\mu\nu}$. A somewhat later work by Infeld and van der Waerden (1932) is based on spinor analysis.

IV. THE EARLY WORK OF KALUZA AND KLEIN

Early in 1919 Einstein received a paper of Theodor Kaluza, a young mathematician (Privatdozent) and consummate linguist in Königsberg. Inspired by the work of Weyl one year earlier, he proposed another geometrical unification of gravitation and electromagnetism by extending space-time to a five-dimensional pseudo-Riemannian manifold. Einstein reacted very positively. On 21 April 1919 he writes, "The idea of achieving [a unified theory] by means of a five-dimensional cylinder world never dawned on me . . . . At first glance I like your idea enormously." A few weeks later he adds: "The formal unity of your theory is starting." For unknown reasons, Einstein submitted Kaluza's paper to the Prussian Academy after a delay of two years (Kaluza, 1921).

Kaluza was actually not the first who envisaged a five-dimensional unification. It is astonishing to note that G. Nordström had this idea already in 1914 (Nordström, 1914). We recall that Nordström had worked out in several papers (Nordström, 1912, 1913a, 1913b) a scalar theory of gravitation that was regarded by Einstein as the only serious competitor to general relativity.[5] (In collaboration with Fokker, Einstein gave this theory a generally covariant, conformally flat form.) Nordström started in his unification attempt with five-dimensional electrodynamics and imposed the "cylinder condition," that the fields should not depend on the fifth coordinate. Then the five-dimensional gauge potential ${}^{(5)}A$ splits as ${}^{(5)}A = A + \phi\, dx^5$, where A is a four-dimensional gauge potential and ϕ is a space-time scalar field. The Maxwell field splits correspondingly, ${}^{(5)}F = F + d\phi \wedge dx^5$, and hence the free Maxwell Lagrangian becomes

$$-\frac{1}{4}({}^{(5)}F|{}^{(5)}F) = -\frac{1}{4}(F|F) + \frac{1}{2}(d\phi|d\phi). \tag{22}$$

In this manner Nordström arrived at a unification of his theory of gravity and electromagnetism. [The matter source (five-current) is decomposed correspondingly.] It seems that this early attempt left, as far as we know, no traces in the literature.

We now return to Kaluza's attempt. Like Nordström he assumes the cylinder condition. Then the five-dimensional metric tensor splits into the four-dimensional fields $g_{\mu\nu}$, A_μ, and ϕ. Kaluza's identification of the electromagnetic potential is not quite the right one, because he chooses it equal to $g_{\mu 5}$ (up to a constant), instead of taking the quotient $g_{\mu 5}/g_{55}$. This does not matter in his further analysis, because he considers only the linearized approximation of the field equations. Furthermore, the matter part is only studied in a nonrelativistic approximation. In particular, the five-dimensional geodesic equation is only written in this limit. Then the scalar contribution to the four-force becomes negligible and an automatic split into the usual gravitational and electromagnetic parts is obtained.

Kaluza was aware of the limitations of his analysis, but he was confident of being on the right track, as becomes evident from the final paragraph of his paper:

> In spite of all the physical and theoretical difficulties which are encountered in the above proposal it is hard to believe that the derived relationships, which could hardly be surpassed at the formal level, represent nothing more than a malicious coincidence. Should it sometime be established that the scheme is more than an empty formalism this would signify a new triumph for Einstein's General Theory of Relativity, whose suitable extension to five dimensions is our present concern.

For good reasons the role of the scalar field was unclear to him, except in the limiting situation of his analysis, where ϕ becomes the negative of the gravitational potential. Kaluza was, however, well aware that the scalar field could play an important role, and he makes some speculative remarks in this direction.

In the classical part of his first paper, Klein (1926a) improves on Kaluza's treatment. He assumes, however, beside the condition of cylindricity, that g_{55} is a constant. Following Kaluza, we keep here the scalar field ϕ and write the Kaluza-Klein ansatz for the five-dimensional metric ${}^{(5)}g$ in the form

$${}^{(5)}g = \phi^{-1/3}(g - \phi\,\omega \otimes \omega), \tag{23}$$

where $g = g_{\mu\nu}\, dx^\mu\, dx^\nu$ is the space-time metric and ω is a differential 1-form of the type

$$\omega = dx^5 + \kappa A_\mu dx^\mu. \tag{24}$$

Like ϕ, $A = A_\mu\, dx^\mu$ is independent of x^5; κ is a coupling constant to be determined. The convenience of the conformal factor $\phi^{-1/3}$ will become clear shortly.

Klein considers the subgroup of five-dimensional coordinate transformations which respect the form (23) of the $d=5$ metric:

$$x^\mu \to x^\mu, \quad x^5 \to x^5 + f(x^\mu). \tag{25}$$

Indeed, the pull-back of ${}^{(5)}g$ is again of the form (23) with

$$g \to g, \quad \phi \to \phi, \quad A \to A + \frac{1}{\kappa} df. \tag{26}$$

Thus $A = A_\mu\, dx^\mu$ transforms like a gauge potential under the Abelian gauge group (25) and is therefore interpreted as the electromagnetic potential. This is further justified by the most remarkable result derived by Kaluza and Klein, often called the Kaluza-Klein miracle. It turns out that the five-dimensional Ricci scalar ${}^{(5)}R$ splits as follows:

$${}^{(5)}R = \phi^{1/3}\left(R + \frac{1}{4}\kappa^2 \phi F_{\mu\nu}F^{\mu\nu} - \frac{1}{6\phi^2}(\nabla\phi)^2 + \frac{1}{3}\Delta \ln \phi \right). \tag{27}$$

For $\phi \equiv 1$ this becomes the Lagrangian of the coupled Einstein-Maxwell system. In view of the gauge group (25), this split is actually no miracle, because no other gauge-invariant quantities can be formed.

For the development of gauge theory this dimensional reduction was particularly important, because it revealed a close connection between coordinate transformations in higher-dimensional spaces and gauge transformations in space-time.

With Klein we consider the $d=5$ Einstein-Hilbert action

$${}^{(5)}S = \frac{-1}{\kappa^2 L}\int {}^{(5)}R\sqrt{|{}^{(5)}g|}\, d^5x, \tag{28}$$

assuming that the higher-dimensional space is a cylinder with $0 \leq x^5 \leq L = 2\pi R_5$. Since

$$\sqrt{|{}^{(5)}g|}\, dx^5 = \sqrt{-g}\,\phi^{-1/3}\, d^4x\, dx^5 \tag{29}$$

we obtain

$${}^{(5)}S = -\int \left(\frac{1}{\kappa^2} R + \frac{1}{4}\phi F_{\mu\nu}F^{\mu\nu} - \frac{1}{6\kappa^2\phi^2}(\nabla\phi)^2 \right)\sqrt{-g}\, d^4x. \tag{30}$$

[5]For instance, Einstein extensively discussed Nordström's second version in his famous Vienna lecture "On the Foundations of the Problem of Gravitation" (23 September 1913) and made it clear that Nordström's theory was a viable alternative to his own attempt with Grossmann. [See Doc. 17 of Vol. 4 of the *Collected Papers of Albert Einstein* (Einstein, 1987)].

Our choice of the conformal factor $\phi^{-1/3}$ in Eq. (23) was made so that the gravitational part in Eq. (30) is just the Einstein-Hilbert action, if we choose

$$\kappa^2=16\pi G. \tag{31}$$

For $\phi\equiv 1$ a beautiful geometrical unification of gravitation and electromagnetism is obtained.

We pause by noting that nobody in the early history of Kaluza-Klein theory seems to have noticed the following inconsistency in putting $\phi\equiv 1$ [see, however, Lichnerowicz (1995)]: The field equations for the dimensionally reduced action (30) are just the five-dimensional equations ${}^{(5)}R_{ab}=0$ for the Kaluza-Klein ansatz (23). Among these, the ϕ equation, which is equivalent to ${}^{(5)}R_{55}=0$, becomes

$$\Box(\ln\phi)=\frac{3}{4}\kappa^2\phi F_{\mu\nu}F^{\mu\nu}. \tag{32}$$

For $\phi\equiv 1$ this implies the unphysical result $F_{\mu\nu}F^{\mu\nu}=0$. This conclusion is avoided if one proceeds in the reverse order, i.e., by putting $\phi\equiv 1$ in the action (30) and varying afterwards. However, if the extra dimension is treated as physical—a viewpoint adopted by Klein (as we shall see)—it is clearly essential that one maintain consistency with the $d=5$ field equations. This is an example of the crucial importance of scalar fields in Kaluza-Klein theories.

Kaluza and Klein both studied the $d=5$ geodesic equation. For the metric (23) this is just the Euler-Lagrange equation for the Lagrangian

$$L=\frac{1}{2}g_{\mu\nu}\dot{x}^\mu\dot{x}^\nu-\frac{1}{2}\phi(\dot{x}^5+\kappa A_\mu\dot{x}^\mu)^2. \tag{33}$$

Since x^5 is cyclic, we have the conservation law (m = mass of the particle)

$$p_5/m:=\frac{\partial L}{\partial\dot{x}^5}=\phi(\dot{x}^5+\kappa A_\mu\dot{x}^\mu)=\text{const}. \tag{34}$$

If use of this is made in the other equations, we obtain

$$\ddot{x}^\mu+\Gamma^\mu_{\alpha\beta}\dot{x}^\alpha\dot{x}^\beta=-\frac{p_5}{m}\kappa F^\mu_{\ \nu}\dot{x}^\nu-\frac{1}{2}\left(\frac{p_5}{m}\right)^2\phi^{-2}\nabla^\mu\phi. \tag{35}$$

Clearly, p_5 has to be interpreted as q/κ, where q is the charge of the particle,

$$p_5=q/\kappa. \tag{36}$$

The physical significance of the last term in Eq. (35) remained obscure. Much later, Jordan (1949, 1955) and Thiry (1948, 1951) tried to make use of the new scalar field to obtain a theory in which the gravitational constant is replaced by a dynamical field. Further work by Jordan (1949, 1955), Fierz (1956), and Brans and Dicke (1961) led to a much studied theory, which has been for many years a serious competitor of general relativity. Generalized versions (Bergmann, 1968) have recently played a role in models of inflation (see, for example, Steinhardt, 1993). The question of whether the low-energy effective theory of string theories, say, has Brans-Dicke-type interactions has lately been investigated for instance by Damour and collaborators (Damour and Polyakov, 1994).

Since the work of Fierz (published in German, Fierz, 1956) is not widely known, we briefly describe its main point. Quoting Pauli, Fierz emphasizes that, in theories containing both tensor and scalar fields, the tensor field appearing most naturally in the action of the theory can differ from the "physical" metric by some conformal factor depending on the scalar fields. In order to decide which is the "atomic-unit" metric and thus the gravitational constant, one has to look at the coupling to matter. The "physical" metric $g_{\mu\nu}$ is the one to which matter is universally coupled (in accordance with the principle of equivalence). For instance, the action for a spin-0 massive matter field ψ should take the form

$$S_\psi=\frac{1}{2}\int(g^{\mu\nu}\partial_\mu\psi\,\partial_\nu\psi-m^2\psi^2)\sqrt{-g}\,d^4x. \tag{37}$$

A unit of length is then provided by the Compton wavelength $1/m$, and test particles fall along geodesics of $g_{\mu\nu}$. Fierz specializes Jordan's theory (with two free constants) such that the Maxwell density, expressed in terms of the physical metric, is not multiplied with a spacetime-dependent function. Otherwise the vacuum would behave like a variable dielectric and this would have unwanted consequences, although the refraction is 1: The fine structure constant would become a function of spacetime, changing the spectra of galaxies over cosmological distances.

With these arguments Fierz arrives at a theory which was later called the Brans-Dicke theory. He did not, however, confront the theory with observations, because he did not believe in its physical relevance. [The intention of Fierz's publication was mainly pedagogical (Fierz, 1999, private communication).]

Equation (36) brings us to the part of Klein's first paper that is related to quantum theory. There he interprets the five-dimensional geodesic equation as the geometrical optical limit of the wave equation ${}^{(5)}\Box\Psi=0$ on the higher-dimensional space and establishes for special situations a close relation of the dimensionally reduced wave equation with Schrödinger's equation, which had been discovered in the same year. His ideas are more clearly spelled out shortly afterwards in a brief *Nature* note entitled "The Atomicity of Electricity as a Quantum Theory Law" (Klein, 1926b). There Klein says in connection with Eq. (36):

> The charge q, so far as our knowledge goes, is always a multiple of the electronic charge e, so that we may write
>
> $$p_5=n\frac{e}{\kappa}\quad[n\in\mathbf{Z}]. \tag{38}$$
>
> This formula suggests that the atomicity of electricity may be interpreted as a quantum theory law. In fact, if the five-dimensional space is assumed to be closed in the direction of x^5 with period L, and if

we apply the formalism of quantum mechanics to our geodesics, we shall expect p_5 to be governed by the following rule:

$$p_5=n\frac{h}{L}, \tag{39}$$

n being a quantum number, which may be positive or negative according to the sense of motion in the direction of the fifth dimension, and h the constant of Planck.

Comparing Eqs. (38) and (39), Klein finds the value of the period L,

$$L=\frac{h}{ec}\sqrt{16\pi G}=0.8\times 10^{-30}\ \text{cm}, \tag{40}$$

and adds:

> The small value of this length together with the periodicity in the fifth dimension may perhaps be taken as a support of the theory of Kaluza in the sense that they may explain the non-appearance of the fifth dimension in ordinary experiments as the result of averaging over the fifth dimension.

Klein concludes this note with the daring speculation that the fifth dimension might have something to do with Planck's constant:

> In a former paper the writer has shown that the differential equation underlying the new quantum mechanics of Schrödinger can be derived from a wave equation of a five-dimensional space, in which h does not appear originally, but is introduced in connection with the periodicity in x^5. Although incomplete, this result, together with the considerations given here, suggests that the origin of Planck's quantum may be sought just in this periodicity in the fifth dimension.

This was not the last time that such speculations have been put forward. The revival of (supersymmetric) Kaluza-Klein theories in the eighties (Appelquist, Chados, and Freund, 1987; Kubyshin *et al.*, 1989) led to the idea that the compact dimensions would necessarily give rise to an enormous quantum vacuum energy via the Casimir effect. There were attempts to exploit this vacuum energy in a self-consistent approach to compactification, with the hope that the size of the extra dimensions would be calculable as a pure number times the Planck length. Consequently the gauge-coupling constant would then be calculable.

Coming back to Klein we note that he would also have arrived at Eq. (39) by the dimensional reduction of his five-dimensional equation. Indeed, if the wave field $\psi(x,x^5)$ is Fourier decomposed with respect to the periodic fifth coordinate,

$$\psi(x,x^5)=\frac{1}{\sqrt{L}}\sum_{n\in Z}\psi_n(x)e^{inx^5/R_5}, \tag{41}$$

one obtains for each amplitude $\psi_n(x)$ [for the metric (23) with $\phi\equiv 1$] the following four-dimensional wave equation:

$$\left(D^\mu D_\mu-\frac{n^2}{R_5^2}\right)\psi_n=0, \tag{42}$$

where D_μ is the doubly covariant derivative (with respect to $g_{\mu\nu}$ and A_μ) with the charge

$$q_n=n\frac{\kappa}{R_5}. \tag{43}$$

This shows that the mass of the nth mode is

$$m_n=|n|\frac{1}{R_5}. \tag{44}$$

Combining Eq. (43) with $q_n=ne$, we obtain, as before, Eq. (40) or

$$R_5=\frac{2}{\sqrt{\alpha}}l_{\text{Pl}}, \tag{45}$$

where α is the fine-structure constant and l_{Pl} is the Planck length.

Equations (43) and (44) imply a serious defect of the five-dimensional theory: The (bare) masses of all charged particles ($|n|\geq 1$) are of the order of the Planck mass

$$m_n=n\frac{\sqrt{\alpha}}{2}m_{\text{Pl}}. \tag{46}$$

The pioneering papers of Kaluza and Klein were taken up by many authors. For some time the "projective" theories of Veblen (1933), Hoffmann (1933), and Pauli (1933) played a prominent role. These are, however, just equivalent formulations of Kaluza's and Klein's unification of the gravitational and the electromagnetic field (Bergmann, 1942; Ludwig, 1951).

Einstein's repeated interest in five-dimensional generalizations of general relativity has been described by Bergmann (1942) and Pais (1982) and will not be discussed here.

V. KLEIN'S 1938 THEORY

The first attempt to go beyond electromagnetism and gravitation and apply Weyl's gauge principle to the nuclear forces occurred in a remarkable paper by Oskar Klein, presented at the Kazimierz Conference on New Theories in Physics (Klein, 1938). Assuming that the mesons proposed by Yukawa were vectorial, Klein proceeded to construct a Kaluza-Klein-like theory which would incorporate them. As in the original Kaluza-Klein theory he introduced only one extra dimension but his theory differed from the original in two respects:

(i) The fields were not assumed to be completely independent of the fifth coordinate x^5 but to depend on it through a factor e^{-iex^5} where e is the electric charge.

(ii) The five-dimensional metric tensor was assumed to be of the form

$$g_{\mu\nu}(x),\quad g_{55}=1,\quad g_{\mu 5}=\beta\chi_\mu(x), \tag{47}$$

where $g_{\mu\nu}$ was the usual four-dimensional Einstein metric, β was a constant, and $\chi_\mu(x)$ was a *matrix-valued field* of the form

$$\chi_\mu(x)=\begin{pmatrix} A_\mu(x) & \tilde{B}_\mu(x) \\ B_\mu(x) & A_\mu(x) \end{pmatrix}=\sigma_3[\vec{\sigma}\cdot\vec{A}_\mu(x)], \tag{48}$$

where the σ's are the usual Pauli matrices and $\vec{A}_\mu(x)$ is what we would now call an $SU(2)$ gauge potential. This was a most remarkable ansatz considering that it implies a matrix-valued metric, and it is not clear what motivated Klein to make it. The reason that he multiplied the present-day $SU(2)$ matrix by σ_3 is that σ_3 represented the charge matrix for the fields.

Having made this ansatz, Klein proceeded in the standard Kaluza-Klein manner and obtained, instead of the Einstein-Maxwell equations, a set of equations that we would now call the Einstein-Yang-Mills equations. This is a little surprising because Klein inserted only electromagnetic gauge invariance. However, one can see how the $U(1)$ gauge invariance of electromagnetism could generalize to $SU(2)$ gauge invariance by considering the field strengths. The $SU(2)$ form of the field strengths corresponding to the $\tilde{B}_\mu$ and B_μ fields, namely,

$$F^{\tilde{B}}_{\mu\nu}=\partial_\mu\tilde{B}_\nu-\partial_\nu\tilde{B}_\mu+ie(A_\mu\tilde{B}_\nu-A_\nu\tilde{B}_\mu), \tag{49}$$

$$F^{B}_{\mu\nu}=\partial_\mu B_\nu-\partial_\nu B_\mu-ie(A_\mu B_\nu-A_\nu B_\mu), \tag{50}$$

actually follows from the electromagnetic gauge principle $\partial_\mu\to D_\mu=\partial_\mu+ie(1-\sigma_3)A_\mu$, given that the three vector fields belong to the same 2×2 matrix. The more difficult question is why the expression

$$F^{A}_{\mu\nu}=\partial_\mu A_\nu-\partial_\nu A_\mu-ie(\tilde{B}_\mu B_\nu-\tilde{B}_\nu B_\mu) \tag{51}$$

for the field strength corresponding to A_μ contained a bilinear term when most other vector-field theories, such as the Proca theory, contained only the linear term. The reason is that the geometrical nature of the dimensional reduction meant that the usual space-time derivative ∂_μ was replaced by the covariant space-time derivative $\partial_\mu+ie(1-\sigma_3)\chi_\mu/2$, with the result that the usual curl $\partial\wedge\chi$ was replaced by $\partial_\mu\chi_\nu-\partial_\nu\chi_\mu+ie/2[\chi_\mu,\chi_\nu]$, whose third component is just the expression for $F^A_{\mu\nu}$.

Being primarily interested in the application of his theory to nuclear physics, Klein immediately introduced the nucleons, treating them as an isodoublet $\psi(x)$ on which the matrix ξ_μ acted by multiplication. In this way he was led to field equations of the familiar $SU(2)$ form, namely,

$$(\gamma\cdot D+M)\psi(x)=0,\quad D_\mu=\partial_\mu+\frac{ie}{2}(1-\sigma_3)\chi_\mu. \tag{52}$$

However, although the equations of motion for the vector fields A_μ, $\tilde{B}_\mu$, and B_μ would be immediately recognized today as those of an $SU(2)$ gauge-invariant theory, this was not at all obvious at the time and Klein does not seem to have been aware of it. Indeed, he immediately proceeded to break the $SU(2)$ gauge invariance by assigning *ad hoc* mass terms to the $\tilde{B}_\mu$ and B_μ fields.

An obvious weakness of Klein's theory is that there is only one coupling constant β, which implies that the nuclear and electromagnetic forces would be of approximately the same strength, in contradiction with experiment. Furthermore, the nuclear forces would not be charge independent, as they were known to be at the time. These weaknesses were noticed by Møller, who, at the end of the talk, objected to the theory on these grounds. Klein's answer to these objections was astonishing: this problem could easily be solved he said, because the strong interactions could be made charge independent (and the electromagnetic field separated) by introducing one more vector field C_μ and generalizing the 2×2 matrix χ_μ

$$\text{from}\quad \chi_\mu=\sigma_3(\vec{\sigma}\cdot\vec{A}_\mu)\quad\text{to}\quad \sigma_3(C_\mu+\vec{\sigma}\cdot\vec{A}_\mu). \tag{53}$$

In other words, he there and then generalized what was effectively a (broken) $SU(2)$ gauge theory to a broken $SU(2)\times U(1)$ gauge theory. In this way, he anticipated the mathematical structure of the standard electroweak theory by 21 years!

Klein has certainly not forgotten his ambitious proposal of 1938, in contrast to what has been suspected by Gross (1995). In his invited lecture at the Berne Congress in 1955 (Klein, 1956) he came back to some main aspects of his early attempt and concluded with the statement:

> On the whole, the relation of the theory to the five-dimensional representation of gravitation and electromagnetism on the one hand and to symmetric meson theory on the other hand—through the appearance of the charge invariance group—may perhaps justify the confidence in its essential soundness.

VI. THE PAULI LETTERS TO PAIS

The next attempt to write down a gauge theory for the nuclear interactions was due to Pauli. During a discussion following a talk by Pais at the 1953 Lorentz Conference in Leiden (Pais, 1953), Pauli said:

> . . . I would like to ask in this connection whether the transformation group with constant phases can be amplified in way analogous to the gauge group for electromagnetic potentials in such a way that the meson-nucleon interaction is connected with the amplified group . . .

Stimulated by this discussion, Pauli worked on this problem and drafted a manuscript to Pais that begins with the heading (Pauli, 1999).

> Written down July 22–25, 1953, in order to see how it looks. Meson-Nucleon Interaction and Differential Geometry.

In this manuscript, Pauli generalizes the Kaluza-Klein theory to a six-dimensional space and arrives through dimensional reduction at the essentials of an $SU(2)$ gauge theory. The extra dimensions form a two-sphere S^2 with space-time-dependent metrics on which $SU(2)$

operates in a space-time-dependent manner. Pauli develops first in "local language" the geometry of what we now call a fiber bundle with a homogeneous space as typical fiber [in his case $S^2 \cong SU(2)/U(1)$]. Studying the curvature of the higher-dimensional space, Pauli automatically finds, for the first time, the correct expression for the non-Abelian field strength.

Since it is somewhat difficult to understand exactly what Pauli did, we give some details, using more familiar formulations and notations.

Pauli considers the six-dimensional total space $M \times S^2$, where S^2 is the two-sphere on which $SO(3)$ acts in the canonical manner. He distinguishes among the diffeomorphisms (coordinate transformations) those which leave M pointwise fixed and induce space-time-dependent rotations on S^2:

$$(x,y) \rightarrow [x, R(x)\cdot y]. \tag{54}$$

Then Pauli postulates a metric on $M \times S^2$ that is supposed to satisfy three assumptions. These lead him to what is now called the non-Abelian Kaluza-Klein ansatz: The metric $\hat{g}$ on the total space is constructed from a space-time metric g, the standard metric γ on S^2, and a Lie-algebra-valued 1-form,

$$A = A^a T_a, \quad A^a = A^a_\mu dx^\mu, \tag{55}$$

on M [T_a, $a=1,2,3$, are the standard generators of the Lie algebra of $SO(3)$] as follows: If $K^i_a \partial/\partial y^i$ are the three Killing fields on S^2, then

$$\hat{g} = g - \gamma_{ij}[dy^i + K^i_a(y)A^a] \otimes [dy^j + K^j_a(y)A^a]. \tag{56}$$

In particular, the nondiagonal metric components are

$$\hat{g}_{\mu i} = A^a_\mu(x)\gamma_{ij}K^j_a. \tag{57}$$

Pauli does not say that the coefficients of A^a_μ in Eq. (57) are the components of the three independent Killing fields. This is, however, his result, which he formulates in terms of homogeneous coordinates for S^2. He determines the transformation behavior of A^a_μ under the group (54) and finds in matrix notation what he calls "the generalization of the gauge group":

$$A_\mu \rightarrow R A_\mu R^{-1} + R^{-1}\partial_\mu R. \tag{58}$$

With the help of A_μ, he defines a covariant derivative, which is used to derive "field strengths" by applying a generalized curl to A_μ. This is exactly the field strength that was later introduced by Yang and Mills. To our knowledge, apart from Klein's 1938 paper, it appears here for the first time. Pauli says that "this is the *true* physical field, the analog of the *field strength*" and he formulates what he considers to be his "main result":

> The vanishing of the field strength is necessary and sufficient for the $A^a_\mu(x)$ in the whole space to be transformable to zero.

It is somewhat astonishing that Pauli did not work out the Ricci scalar for $\hat{g}$ as for the Kaluza-Klein theory. One reason may be connected with his remark on the Kaluza-Klein theory in Note 23 of his relativity article (Pauli, 1958) concerning the five-dimensional curvature scalar (p. 230):

> There is, however, no justification for the particular choice of the five-dimensional curvature scalar P as integrand of the action integral, from the standpoint of the restricted group of the cylindrical metric [gauge group]. The open problem of finding such a justification seems to point to an amplification of the transformation group.

In a second letter (Pauli, 1999), Pauli also studies the dimensionally reduced Dirac equation and arrives at a mass operator that is closely related to the Dirac operator in internal space (S^2, γ). The eigenvalues of the latter operator had been determined by him long before (Pauli, 1939). Pauli concludes with the statement: "So this leads to some rather unphysical 'shadow particles.'"

VII. YANG-MILLS THEORY

In his Hermann Weyl Centenary Lecture at the ETH (Yang, 1980), C. N. Yang commented on Weyl's remark "The principle of gauge-invariance has the character of general relativity since it contains an arbitrary function λ, and can certainly only be understood in terms of it" (Weyl, 1968) as follows:

> The quote above from Weyl's paper also contains something which is very revealing, namely, his strong association of gauge invariance with general relativity. That was, of course, natural since the idea had originated in the first place with Weyl's attempt in 1918 to unify electromagnetism with gravity. Twenty years later, when Mills and I worked on non-Abelian gauge fields, our motivation was completely divorced from general relativity and we did not appreciate that gauge fields and general relativity are somehow related. Only in the late 1960s did I recognize the structural similarity mathematically of non-Abelian gauge fields with general relativity and understand that they both were connections mathematically.

Later, in connection with Weyl's strong emphasis of the relation between gauge invariance and conservation of electric charge, Yang continues with the following instructive remarks:

> Weyl's reason, it turns out, was also one of the melodies of gauge theory that had very much appealed to me when as a graduate student I studied field theory by reading Pauli's articles. I made a number of unsuccessful attempts to generalize gauge theory beyond electromagnetism, leading finally in 1954 to a collaboration with Mills in which we developed a non-Abelian gauge theory. In [···] we stated our motivation as follows:

> The conservation of isotopic spin points to the existence of a fundamental invariance law similar to the conservation of electric charge. In the latter case, the electric charge serves as a source of electromagnetic field; an important concept in this case is gauge invariance, which is closely connected with (1) the equation of motion of the electro-

> magnetic field, (2) the existence of a current density, and (3) the possible interactions between a charged field and the electromagnetic field. We have tried to generalize this concept of gauge invariance to apply to isotopic spin conservation. It turns out that a very natural generalization is possible.
>
> Item (2) is the melody referred to above. The other two melodies, (1) and (3), where what had become pressing in the early 1950s when so many new particles had been discovered and physicists had to understand how they interact with each other.
>
> I had met Weyl in 1949 when I went to the Institute for Advanced Study in Princeton as a young "member." I saw him from time to time in the next years, 1949–1955. He was very approachable, but I don't remember having discussed physics or mathematics with him at any time. His continued interest in the idea of gauge fields was not known among the physicists. Neither Oppenheimer nor Pauli ever mentioned it. I suspect they also did not tell Weyl of the 1954 papers of Mills' and mine. Had they done that, or had Weyl somehow came across our paper, I imagine he would have been pleased and excited, for we had put together two things that were very close to his heart: gauge invariance and non-Abelian Lie groups.

It is indeed astonishing that during those late years neither Pauli nor Yang ever talked with Weyl about non-Abelian generalizations of gauge invariance.

With the background of Sec. VI, the following story of spring 1954 becomes more understandable. In late February, Yang was invited by Oppenheimer to return to Princeton for a few days and to give a seminar on his joint work with Mills. Here is Yang's report (Yang, 1983):

> Pauli was spending the year in Princeton, and was deeply interested in symmetries and interactions. (He had written in German a rough outline of some thoughts, which he had sent to A. Pais. Years later F. J. Dyson translated this outline into English. It started with the remark, "Written down July 22–25, 1953, in order to see how it looks," and had the title "Meson-Nucleon Interaction and Differential Geometry.") Soon after my seminar began, when I had written down on the blackboard,
>
> $(\partial_\mu - i\epsilon B_\mu)\psi,$
>
> Pauli asked, "What is the mass of this field B_μ?" I said we did not know. Then I resumed my presentation, but soon Pauli asked the same question again. I said something to the effect that that was a very complicated problem, we had worked on it and had come to no definite conclusions. I still remember his repartee: "That is not sufficient excuse." I was so taken aback that I decided, after a few moments' hesitation to sit down. There was general embarrassment. Finally Oppenheimer said, "We should let Frank proceed." I then resumed, and Pauli did not ask any more questions during the seminar.
>
> I don't remember what happened at the end of the seminar. But the next day I found the following message:
>
> February 24, Dear Yang, I regret that you made it almost impossible for me to talk with you after the seminar. All good wishes. Sincerely yours, W. Pauli.
>
> I went to talk to Pauli. He said I should look up a paper by E. Schrödinger, in which there were similar mathematics.[6] After I went back to Brookhaven, I looked for the paper and finally obtained a copy. It was a discussion of spacetime-dependent representations of the γ_μ matrices for a Dirac electron in a gravitational field. Equations in it were, on the one hand, related to equations in Riemannian geometry and, on the other, similar to the equations that Mills and I were working on. But it was many years later when I understood that these were all different cases of the mathematical theory of connections on fiber bundles.

Later Yang adds:

> I often wondered what he [Pauli] would say about the subject if he had lived into the sixties and seventies.

At another occasion (Yang, 1980) he remarked:

> I venture to say that if Weyl were to come back today, he would find that amidst the very exciting, complicated and detailed developments in both physics and mathematics, there are fundamental things that he would feel very much at home with. He had helped to create them.

Having quoted earlier letters from Pauli to Weyl, we add what Weyl said about Pauli in 1946 (Weyl, 1980):

> The mathematicians feel near to Pauli since he is distinguished among physicists by his highly developed organ for mathematics. Even so, he is a physicist; for he has to a high degree what makes the physicist; the genuine interest in the experimental facts in all their puzzling complexity. His accurate, instructive estimate of the relative weight of relevant experimental facts has been an unfailing guide for him in his theoretical investigations. Pauli combines in an exemplary way physical insight and mathematical skill.

To conclude this section, let us emphasize the main differences between general relativity and Yang-Mills theories. Mathematically, the $so(1,3)$-valued connection forms ω in Sec. III A and the Lie-algebra-valued gauge potential A are on the same footing; they are both representatives of connections in (principle) fiber bundles

[6]E. Schrödinger, Sitzungsberichte der Preussischen (Akademie der Wissenschaften, 1932), p. 105.

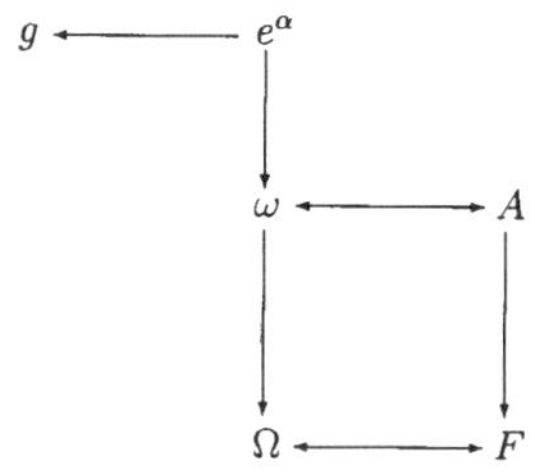

FIG. 4. General relativity vs Yang-Mills theory.

over the space-time manifold. Equation (17) translates into the formula for the Yang-Mills field strength F,

$$F = dA + A \wedge A. \tag{59}$$

In general relativity one has, however, additional geometric structure, since the connection derives from a metric, or the tetrad fields $e^{\alpha}(x)$, through the first structure equation (15). This is shown schematically in Fig. 4. [In bundle theoretical language one can express this as follows: The principle bundle of general relativity, i.e., the orthonormal frame bundle, is constructed from the base manifold and its metric, and has therefore additional structure, implying, in particular, the existence of a canonical 1-form (soldering form), whose local representatives are the tetrad fields; see, for example, Bleecker (1981).]

Another important difference is that the gravitational Lagrangian $*R = 1/2\Omega_{\alpha\beta} \wedge *(e^{\alpha} \wedge e^{\beta})$ is linear in the field strengths, whereas the Yang-Mills Lagrangian $F \wedge *F$ is quadratic.

VIII. RECENT DEVELOPMENTS

The developments after 1958 consisted in the gradual recognition that—contrary to phenomenological appearances—Yang-Mills gauge theory could describe weak and strong interactions. This important step was again very difficult, with many hurdles to overcome.

One of them was the mass problem, which was solved, probably in a preliminary way, through spontaneous symmetry breaking. Of critical significance was the recognition that spontaneously broken gauge theories are renormalizable. On the experimental side the discovery and intensive investigation of the neutral current was, of course, crucial. For the gauge description of the strong interactions, the discovery of asymptotic freedom was decisive. That the $SU(3)$ color group should be gauged was also not at all obvious. And then there was the confinement idea, which explains why quarks and gluons do not exist as free particles. All this is described in numerous modern textbooks and does not have to be repeated.

The next step in creating a more unified theory of the basic interactions will probably be much more difficult. All major theoretical developments of the last twenty years, such as grand unification, supergravity, and supersymmetric string theory, are almost completely separated from experience. There is a great danger that theoreticians may get lost in pure speculations. As in the first unification proposal of Hermann Weyl, they may create beautiful and highly relevant mathematics, which do not, however, describe Nature. In the latter case history shows, however, that such ideas can one day also become fruitful for physics. It may, therefore, be appropriate to conclude with some remarks on current attempts in string theory and noncommutative geometry.

A. Gauge theory and strings

1. Introduction

So far we have considered gravitation and gauge theory only within the context of local field theory. However, gravitation and gauge theory also occur naturally in string theory (Green, Schwarz, and Witten, 1987; Polchinski, 1998). Indeed, whereas in field theory they are optional extras that are introduced on phenomenological grounds (equality of inertial and gravitational mass, divergence-free character of the magnetic field, etc.) in string theory they occur as an intrinsic part of the structure. Thus string theory is a very natural setting for gravitation and gauge fields. One might go so far as to say that, had string theory preceded field theory historically, the gravitational and gauge fields might have emerged in a completely different manner. An interesting feature of string gauge theories is that the choice of gauge group is quite limited.

String theory is actually a natural setting not only for gravitational and gauge fields but also for the Kaluza-Klein mechanism. Historically, the most obvious difficulty with Kaluza-Klein reductions was that there was no experimental evidence and no theoretical need for any extra dimensions. String theory changes this situation dramatically. As is well known, string theory is conformally invariant only if the dimension d of the target space is $d=10$ or $d=26$, according to whether the string is supersymmetric or not. Thus, in contrast to field theory, string theory points to the existence of extra dimensions and even specifies their number.

We shall treat an important specific case of dimensional reduction within string theory, namely, the toroidal reduction from 26 to 10 dimensions, in Sec. VIII. A. 7. However, since no phenomenologically satisfactory reduction from 26 or 10 to 4 dimensions has yet emerged, and the dimensional reduction in string theory is rather similar to that in ordinary field theory (Appelquist, Chodos, and Freund, 1987; Kubyshin *et al.*, 1989), we shall not consider any other case, but refer the reader to the literature.

Instead we shall concentrate on the manner in which gauge theory and gravitation occur in the context of dimensionally unreduced string theory. We shall rely heavily on the result (Yau, 1985) that if a massless vector field theory with polarization vector ξ_{μ} and on-shell momentum p_{μ} ($p^2=0$) is invariant with respect to the transformation

$$\xi_{\mu}(p) \to \xi_{\mu}(p) + \eta(p) p_{\mu}, \tag{60}$$

where $\eta(p)$ is an arbitrary scalar, then it must be a gauge theory. Similarly, we shall rely on the result that a second-rank symmetric-tensor theory must be a gravitational theory if the polarization vector $\xi_{\mu\nu}$ satisfies

$$\xi^{\mu}_{\mu}(p)=0, \quad p^{\mu}\xi_{\mu\nu}(p)=0, \tag{61}$$

and if the theory is invariant with respect to

$$\xi_{\mu\nu}(p)\to\xi_{\mu\nu}(p)+\eta_{\mu}p_{\nu}+\eta_{\nu}p_{\mu} \tag{62}$$

for arbitrary $\eta_{\mu}(p)$, and $p^2=0$ (Weinberg, 1965; Wald, 1986, and references therein; Feynman, 1995).

2. Gauge properties of open bosonic strings

To fix our ideas we first recall the form of the path integral for the bosonic string (Green, Schwarz, and Witten, 1987; Polchinski, 1998), namely,

$$\int dh\, dX\, e^{\int d^2\sigma h^{\alpha\beta}\eta_{\mu\nu}\partial_{\alpha}X^{\mu}(\sigma)\partial_{\beta}X^{\nu}(\sigma)}, \tag{63}$$

where σ are the coordinates, $d^2\sigma$ is the diffeomorphic-invariant measure, and $h^{\alpha\beta}$ is the metric on the two-dimensional world sheet of the string, while $\eta_{\mu\nu}$ is the Lorentz metric for the 26-dimensional target space in which the string, with coordinates $X^{\mu}(\sigma)$, moves. Thus the $X^{\mu}(\sigma)$ may be regarded as fields in a two-dimensional quantum field theory. The action in Eq. (63), and hence the classical two-dimensional field theory, is conformally invariant, but, as is well known, the quantum theory is conformally invariant only if $N=26$ or $N=10$ in the supersymmetric version.

The *open* bosonic string is the one in which gauge fields occur. Indeed, one might go so far as to say that the open string is a natural nonlocal generalization of a gauge field. The ends of the open strings are usually assumed to be attached to quarks, and thus there is a certain qualitative resemblance between the open bosonic strings and the gluon flux lines that link the quarks in theories of quark confinement. We wish to make the relationship between gauge fields and open bosonic strings more precise.

As is well known (Green, *et al.* 1987; Polchinski, 1998), the vacuum state $|0\rangle$ of the open string is a scalar tachyon and the first excited states are $X^{\mu}(\tau)|0\rangle = \int d\,(\sin s)X^{\mu}_{+}(\tau,s)|0\rangle$, where τ and s are the time and space components of σ and $X^{\nu}_{+}(\tau,s)$ is the positive-frequency part of $X^{\mu}(\tau,s)$. For a suitable standard value of the normal-ordering parameter for the Noether generators of the conformal (Virasoro) group, these states are massless, i.e., $p^2=0$, where p_{μ} is the 26-dimensional momentum. Furthermore, they are the only massless states. Thus if there are gauge fields in the theory, these are the states that must be identified with them. On the other hand, since all the other (massive) states in the Fock space of the string are formed by acting on $|0\rangle$ with higher moments of $X_{-}(\sigma)$ we see that the operators $X^{\mu}_{+}(\tau)$ are the prototypes of the operators that create the whole string. It is in this sense that the string can be regarded as a nonlocal generalization of a gauge field.

The question is: how is the identification of the massless states $X_{-}(\tau)|0\rangle$ to be justified? The justification comes about through the so-called vertex operators for the emission of the on-shell massless states. The vertices are the analogs of the ordinary Feynmann diagrams in quantum field theory and take the form

$$V(\xi,p)=\int ds e^{ip\cdot X(\sigma)}\xi\cdot\partial_{s}X(\sigma), \tag{64}$$

where ξ_{μ} is the polarization vector, p_{μ} is the momentum $(p^2=0)$ of the emitted particle, and ∂_{η} is a spacelike derivative. This vertex operator is to be inserted in the functional integral (63). Although the form of this vertex is not deduced from a second-quantized theory of strings (which does not yet exist) the vertex operator (64) is generally accepted as the correct one, because it is the only vertex that is compatible with the two-dimensional structure and conformal invariance of the string, and that reduces to the usual vertex in the point-particle limit. Suppose now that we make the transformation $\xi_{\nu}\to\xi_{\nu}+\alpha(p)p_{\nu}$ in Eq. (60). Then the vertex $V(\xi,p)$ acquires the additional term

$$\eta(p)\int ds e^{ip\cdot X(\sigma)}p\cdot\partial_{s}X(\sigma)=-i\eta(p)\int ds\,\partial_{s}(e^{1p\cdot X(\sigma)}), \tag{65}$$

which is an integrable factor that attaches itself to the two ends of the string and thus displays the gauge-covariant character of $V(\xi,p)$. The important point is that this gauge covariance is not imposed from outside, but is an intrinsic property of the string. It is a consequence of the fact that the string is conformally invariant (which dictates the form of the vertex operator) and has an internal structure (manifested by the fact that it has an internal two-dimensional integration).

3. Gravitational properties of closed bosonic strings

Just as the open bosonic string is the one in which gauge fields naturally occur, the closed bosonic string is the one in which gravitational fields naturally occur. It turns out, in fact, that a gravitational field, a dilaton field, and an antisymmetric tensor field occur in the closed string in the same way that the gauge field appears in the open string. The ground state $|0\rangle$ of the closed string is again a tachyon but the new feature is that, for the standard value of the normal-ordering constant, the first excited states are massless states of the form $X^{\mu}_{+}(\sigma)X^{\nu}_{+}(\sigma)|0\rangle$ and it is the symmetric, trace, and antisymmetric parts of the two-tensor formed by the X's that are identified with the gravitational, dilaton, and antisymmetric tensor fields, respectively. The question is how the identification with the gravitational field is to be justified, and again the answer is by means of a vertex operator, this time for the emission of an on-shell graviton. The vertex operator that describes the emission of an on-shell massless spin-2 field (graviton) of polarization $\xi_{\mu\nu}$ and momentum p_{λ}, where $p^2=0$, is

$$V(\xi,p)=\int d^2\sigma\, e^{ip\cdot X(\sigma)}\sqrt{h}h^{\alpha\beta}\xi_{\mu\nu}\partial_{\alpha}X^{\mu}(\sigma)\partial_{\beta}X^{\nu}(\sigma). \tag{66}$$

Already at this stage there is a feature that does not arise in the gauge-field case: Since the vertex operator is bilinear in the field X^μ it has to be normal ordered, and it turns out that the normal ordering destroys the classical conformal invariance, unless

$$\xi^\mu_\mu = 0 \quad \text{and} \quad p^\mu \xi_{\mu\nu} = 0. \tag{67}$$

We next make the momentum-space version of an infinitesimal coordinate transformation, namely,

$$\xi_{\mu\nu} \to \xi_{\mu\nu} + \eta_\mu(p) p_\nu + \eta_\nu(p) p_\mu . \tag{68}$$

Under this transformation the vertex V_ξ picks up an additional term of the form

$$2\eta_\nu \int d^2\sigma\, h^{\alpha\beta} e^{ip\cdot X(\sigma)} p_\mu\, \partial_\alpha X^\mu(\sigma) \partial_\beta X^\nu(\sigma)$$

$$= -2i\eta_\nu \int d^2\sigma\, h^{\alpha\beta} (\partial_\alpha e^{ip\cdot X(\sigma)}) \partial_\beta X^\nu(\sigma). \tag{69}$$

In analogy with the electromagnetic case we can carry out a partial integration. However, this time the expression does not vanish completely but reduces to

$$2i\eta_\nu \int d^2\sigma e^{ip\cdot X(\sigma)} \Delta X^\nu(\sigma), \tag{70}$$

where Δ denotes the two-dimensional Laplacian. On the other hand, $\Delta X^\nu(\sigma) = 0$ is just the classical field equation for $X^\nu(\sigma)$, and it can be shown that even in the quantized version it is effectively zero (Green, Schwarz, and Witten, 1987; Polchinski, 1998). Thus, thanks to the dynamics, we have invariance with respect to the transformations (68). But Eq. (67) and invariance with respect to Eq. (68) are just the conditions (61) and (62) discussed earlier for the vertex to be a gravitational field. As in the gauge-field case, the important point is that the general coordinate invariance is not imposed, but is a consequence of the conformal invariance and internal structure of the string.

The appearance of a scalar field in this context is not too surprising, since a scalar also appeared in the Kaluza-Klein reduction. What is more surprising is the appearance of an antisymmetric tensor. From the point of view of traditional local gravitational and gauge-field theory the presence of an additional antisymmetric tensor field seems at first sight to be an embarrassment. But it turns out to play an essential role in maintaining conformal invariance (cancellation of anomalies), so its presence is to be welcomed.

4. The presence of matter

Of course, the open bosonic string is not the whole story any more than pure gauge fields are the whole story in quantum field theory. One still has to introduce quantities that correspond to fermions (and possibly scalars) at the zero-mass level. There are essentially two ways to do this. The first is the Chan-Paton mechanism (Green, Schwarz and Witten, 1987; Polchinski, 1998), which dates from the days of strong-interaction string theory. In this mechanism one simply attaches charged particles to the open ends of the string. These charged particles are not otherwise associated with any string and thus the mechanism is rather *ad hoc* and leads to a hybrid of string and field theories. But it has the merit of introducing charged particles directly and thus emphasizing the relationship between strings and gauge fields.

The Chan-Paton mechanism has the further merit of allowing a simple generalization to the non-Abelian case. This is done by replacing the charged particles by particles belonging to the fundamental representations of compact internal symmetry groups G, typically quarks $q_a(x)$ and antiquarks $\bar{q}_b(x)$. The vertex operator then generalizes to one with double labels (a,b) and represents non-Abelian gauge fields in much the same way that the simple bosonic string represents an Abelian gauge field.

An interesting restriction arises from the fact that since the string represents gauge fields, and gauge fields belong to the adjoint representation of the gauge group, the vertex function must belong to the adjoint representation. This implies that even at the tree level the tensor product of the fundamental group representation with itself must produce only the adjoint representation, and this restricts G to be one of the classical groups $SO(n)$, $Sp(2n)$, and $U(n)$. Furthermore, it is found that $U(n)$ violates unitarity at the one-loop level, which leaves only $SO(n)$ and $SP(2n)$. Finally, these groups require symmetrization and antisymmetrization in the indices a and b to produce only the adjoint representation, and this implies that the string is oriented (symmetric with respect to its end points). When all these conditions are satisfied it can be shown that the non-Abelian vertex corresponding to Eq. (64) is covariant with respect to non-Abelian gauge transformations corresponding to $\xi_\mu \to \xi_\mu + \eta(p) p_\mu$ above. But since these transformations are nonlinear the proof is more difficult than in the Abelian case.

5. Fermionic and heterotic strings: Supergravity and non-abelian gauge theory

The Chan-Paton version of gauge string has the obvious disadvantage that the charged fields (quarks) are not an intrinsic part of the theory. A second method of introducing fermions is to place them in the string itself. This is done by replacing the kinetic term $(\partial X)^2$ by a Dirac term $\bar{\Psi}\partial\!\!\!/\Psi$ in the Lagrangian density. Interesting cases are those in which the number of fermion components just matches the number of bosons, so that the Lagrangian is supersymmetric. In that case the condition for quantum conformal invariance reduces from $d=26$ to $d=10$. An interesting case from the point of view of gauge theory and dimensional reduction is the heterotic string, in which the left-handed part forms a superstring and the right-handed part forms a bosonic string in which 16 of the bosons are fermionized. For the heterotic string the Lagrangian in the bosonic-string path integral (63) is replaced by the Lagrangian

$$\sum_{\mu=1}^{\mu=10} \partial_\alpha X^\mu \partial^\alpha X_\mu - 2\sum_{\mu=1}^{\mu=10} \psi_+^\mu \partial_- \psi_{\mu+} - 2\sum_{A=1}^{A=32} \lambda_-^A \partial_+ \lambda_-^A, \tag{71}$$

where the ψ's and λ's are Majorana-Weyl fermions and the λ's belong to a representation (labeled with A) of an internal symmetry group G. It is only through the λ's that the internal symmetry group enters. The left- and right-handed parts of the theory are conformally invariant for quite different reasons. The left-handed part of the X's and the left-handed fermions ψ are conformally invariant, because together they form the left-handed part of a superstring (this is why the summation over the X's is only from 1 up to 10). The right-handed part of the X's and the right-handed fermions λ^A are conformally invariant because, from the point of view of anomalies, two Majorana-Weyl fermions are equivalent to one boson and thus the system is equivalent to the right-handed part of a 26-dimensional bosonic string. (This is why the index A runs from 1 to 32.) The fact that there are 32 fermions obviously puts strong restrictions on the choice of the internal symmetry group G.

We now examine the particle content of the theory, using the light-cone gauge, where there are no redundant fields. There are no tachyons for the left-moving fields; the first excited states are massless and take the form

$$|i\rangle_L \quad \text{and} \quad |\alpha\rangle_L, \tag{72}$$

where the $|i\rangle_L$ for $i=1\cdots 8$ are the left-handed components of a massless space-time vector in the eight transverse directions in the light-cone gauge and $|\alpha\rangle_L$ are the components of a massless fermion in one of the two fundamental spinor representations of the same space-time $SO(8)$ group. These states are all G invariant.

The first excited states for the right-moving sector are

$$|i\rangle_R \quad \text{and} \quad \lambda^A\lambda^B|0\rangle, \tag{73}$$

where the $|i\rangle_R$ are the right-handed analogs of the $|i\rangle_L$ and the $\lambda\lambda|0\rangle$ states are massless space-time scalars. The states $|i\rangle_R$ are G invariant but the states $\lambda\lambda|0\rangle$ belong to the adjoint representation of G and thus it is only through these states that the internal symmetry enters at the massless level.

The physical states are obtained by tensoring the left- and right-moving states (72) and (73). On tensoring the right-handed states with the vectors in Eq. (73) we obviously obtain states that are G invariant, and they turn out to be just the states that would occur in $N=1$ supergravity. An analysis of the vertex operators, similar to that carried out above for closed bosonic strings, confirms that these fields do indeed correspond to supergravity.

6. The internal symmetry group G

From the point of view of non-Abelian gauge theory the interesting states are those belonging to the nontrivial representations of G, and these are the ones obtained from the tensor products of Eq. (73) with the space-time scalars $\lambda\lambda|0\rangle$. At this point one must make a choice about the internal symmetry group G. The simplest choice is evidently $G=SO(32)$, and it is obtained by assigning antiperiodic boundary conditions to *all* the fermion fields λ. (Assigning periodic boundary conditions to all of them violates the masslessness condition.) Since the product states continue to belong to the adjoint representation of $SO(32)$, they are the natural candidates for states associated with non-Abelian gauge-fields, and an analysis of the vertex operators associated with these states confirms that they do indeed correspond to $SO(32)$ gauge fields.

In sum, the heterotic string produces both supergravity and non-Abelian gauge theory.

7. Dimensional reduction and the heterotic symmetry group $E_8\times E_8$

A variety of other left-handed internal symmetry groups $G\subset SO(32)$ can be obtained by assigning periodic and antiperiodic boundary conditions to the fermions λ^A of the heterotic string in a nonuniform manner. However, apart from the $SO(32)$ case just discussed, the only assignment that satisfies unitarity at the one-loop level is an equipartition of the 32 fermions into two sets of 16, with mixed boundary conditions. This would appear, at first sight, to lead to an $SO(16)\times SO(16)$ internal symmetry and gauge group, on the same grounds as $SO(32)$ above, but a closer analysis shows that it actually leads to a larger group, namely $E_8\times E_8$, which actually has the same dimension (496) as $SO(32)$. This group is quite attractive for grand unification theory, as it breaks naturally to E_6, which is one of the favorite grand unified theory groups.

Once we accept that $SO(16)\times SO(16)$ is a gauge group and that a rigid internal symmetry group $E_8\times E_8$ exists, it follows immediately that $E_8\times E_8$ must be a gauge group, because the action of the rigid generators of $E_8\times E_8$ on the $SO(16)\times SO(16)$ gauge fields produces $E_8\times E_8$ gauge fields.

This reduces the problem to the existence of a rigid $E_8\times E_8$ symmetry, but, within the context of our present methods, this is a rather convoluted process. One must introduce special representations of $SO(16)\times SO(16)$, project out some of the resulting states, and construct vertices that represent the elements of the coset $(E_8\times E_8)/[SO(16)\times SO(16)]$. Luckily there is a much more intuitive way to establish the existence of the $E_8\times E_8$ symmetry, and, as this way provides a very nice example of dimensional reduction within string theory, we shall now sketch it.

We have already remarked that, from the point of view of Virasoro anomalies, the 32 right-handed Majorana-Weyl fermions λ are equivalent to the right-handed parts of 16 bosons. This relationship can be carried farther by bosonizing the fermions according to $\lambda_\pm(\sigma)=:\exp[\pm\phi_i^R(\sigma)]:$, where $\phi^R(\sigma)$ is a right-moving bosonic field, compactified so that $0\leq\phi(\sigma)<2\pi$. In that case we may regard the right-handed part of the heterotic string as originating in the right-handed part of an ordinary 26-dimensional bosonic string, in which 16 of the 26 right-moving bosonic fields $X^R(\sigma)$ have been fermionized by letting $X_a^R(\sigma)\to\phi_a^R(\sigma)$ for $0\le\phi_a(\sigma)<2\pi$ and $a=1\cdots16$. Since the X's correspond to coordinates in the target space of the string, this is equivalent to a toroidal compactification of 16 of the target-space dimensions and thus is equivalent to a Kaluza-Klein-type dimensional reduction from 26 to 10 dimensions. It turns out that the toroidal compactification and conversion to fermions is consistent only if the lattice that defines the 16-dimensional torus is even and self-dual. But it is well known that there are only two such lattices, called D_{16}^+ and E_8+E_8, and since these have automorphism groups $SO(32)$ and $E_8\times E_8$, respectively, one sees at once where the origin of these symmetry groups lies. The further reduction from ten to four dimensions is, of course, another question. One of the more attractive proposals is that the quotient, six-dimensional space, be a Calabi-Yau space (Green, Schwarz, and Witten, 1987; Yau, 1985), but we do not wish to pursue this question further here.

B. Gauge theory and noncommutative geometry

The recent development of noncommutative geometry by Connes (1994) has permitted the generalization of gauge-theory ideas to the case in which the standard differential manifolds (Minkowskian, Euclidean, Riemannian) become mixtures of differential and discrete manifolds. The differential operators then become mixtures of ordinary differential operators and matrices. From the point of view of the fundamental physical interactions, the interest in such a generalization of gauge theory is that the Higgs field and its potential, which are normally introduced in an *ad hoc* manner, appear as part of the gauge-field structure. Indeed the Higgs field emerges as the component of the gauge potential in the "discrete direction" and the Higgs potential, like the self-interaction of the gauge field, emerges from the square of the curvature. The theory also relates to Kaluza-Klein theory because the Higgs field and potential can also be regarded as coming from a dimensional reduction in which the discrete direction in the gauge group is reduced to an internal direction.

1. Simple example

To explain the idea in its simplest form we follow Connes (1994) and use as an example the simplest nontrivial case, namely, when the continuous manifold is a four-dimensional compact Riemannian manifold with gauge group $U(1)$ and the discrete manifold consists of just two points. With respect to the new discrete (two-point) direction the zero-forms (functions) $\omega_0(x)$ are taken to be diagonal 2×2 matrices with ordinary scalar functions as entries:

$$\omega_0(x)=\begin{pmatrix} f_a(x) & 0\\ 0 & f_b(x)\end{pmatrix}\in\Omega_0, \tag{74}$$

where Ω_0 denotes the space of zero-forms. The essential new feature is the introduction of a discrete component $\mathbf{d}$ of the outer derivative d. This is defined as a self-adjoint off-diagonal matrix, i.e.,

$$\mathbf{d}=\begin{pmatrix} 0 & k\\ k & 0\end{pmatrix} \tag{75}$$

with constant entries k. (More generally one could take the off-diagonal elements in d to be complex-conjugate bounded operators, but that will not be necessary for our purpose.) The outer derivative of the zero-forms with respect to $\mathbf{d}$ is obtained by commutation,

$$\mathbf{d}\wedge\omega_0\equiv[\mathbf{d},\omega_0]=[f_b(x)-f_a(x)]\begin{pmatrix} 0 & k\\ -k & 0\end{pmatrix}. \tag{76}$$

The noncommutativity enters in the fact that $d\omega_0$ does not commute with the forms in Ω_0. The one-forms ω_1 are taken to be off-diagonal matrices,

$$\omega_1(x)=\begin{pmatrix} 0 & v_a(x)\\ v_b(x) & 0\end{pmatrix}\in\Omega_1, \tag{77}$$

where the $v(x)$'s are ordinary scalar functions and Ω_1 denotes the space of one-forms. Note that, according to Eq. (76), the discrete component of the outer derivative maps Ω_0 into Ω_1. The outer derivative of a one-form with respect to $\mathbf{d}$ is obtained by anticommutation. Thus

$$\mathbf{d}\wedge\omega_1\equiv\{\mathbf{d},\omega_1\}=[v_b(x)-v_a(x)]kI\in\Omega_0, \tag{78}$$

where I is the unit 2×2 matrix. It is easy to check that with this definition we have $\mathbf{d}\wedge\mathbf{d}\wedge=0$ on both Ω-spaces.

The $U(1)$ gauge group is a zero-form and is the direct sum of the $U(1)$ gauge groups on the two sectors of the zero-forms. Thus it has elements of the form

$$U(x)=\begin{pmatrix} e^{i\alpha(x)} & 0\\ 0 & e^{i\beta(x)}\end{pmatrix}\in U(1). \tag{79}$$

Its action on both Ω_0 and Ω_1 is by conjugation. Thus under a gauge transformation the zero-forms are invariant and the one-forms transform according to

$$\omega_1(x)\to\omega_1'(x)=U^{-1}(x)\omega_1(x)U(x). \tag{80}$$

Explicitly,

$$\omega_1'(x)=\begin{pmatrix} 0 & e^{i\lambda(x)}g_a(x)\\ e^{-i\lambda(x)}g_b(x) & 0\end{pmatrix},$$

$$\text{where } \lambda(x)=\beta(x)-\alpha(x). \tag{81}$$

The discrete component of a connection takes the form

$$V(x)=\begin{pmatrix} 0 & v(x)\\ v^*(x) & \end{pmatrix} \tag{82}$$

and thus resembles a Hermitian one-form. But, being a connection, it is assumed to transform with respect to $U(x)$ as

$$V(x)\to V_u(x)=U^{-1}(x)V(x)U(x)+e^{-1}U^{-1}(x)dU(x), \tag{83}$$

where e is a constant. The transformation law (83) is the natural extension of the conventional transformation law for connection forms.

The discrete component of the covariant outer derivative is defined to be

$$\mathbf{D}=\mathbf{d}+eV(x)=\begin{pmatrix} 0 & k+ev(x) \\ k+ev^*(x) & 0 \end{pmatrix} = e\begin{pmatrix} 0 & \phi(x) \\ \phi^*(x) & 0 \end{pmatrix}, \tag{84}$$

where

$$\phi(x)=v(x)+c, \quad c=\frac{k}{e}. \tag{85}$$

The outer derivative with $\mathbf{D}$ is formed in the same way as with $\mathbf{d}$, namely, by commutation and anticommutation on the forms Ω_0 and Ω_1, respectively. From Eq. (83) it follows in the usual way that $\mathbf{D}$ transforms covariantly with respect to the $U(1)$ gauge group, i.e.,

$$\mathbf{D}[\phi(x)]\rightarrow\mathbf{D}[\phi_\lambda(x)]=U^{-1}(x)\mathbf{D}[\phi(x)]U(x), \tag{86}$$

where

$$\phi_\lambda(x)=e^{i\lambda(x)}\phi(x). \tag{87}$$

This is consistent with the fact that $\mathbf{D}$ acts on the gauge group by commutation.

Note that, although the component $v(x)$ of the connection does not transform covariantly with respect to $U(1)$, the field $\phi(x)$ does. Since $\phi(x)$ is also a spacetime scalar, it can therefore be identified as a Higgs field. As we shall see, the fact that $\phi(x)$ rather than $v(x)$ is identified as the Higgs field is of great importance for the Higgs potential.

Having defined the covariant derivative, we can proceed to construct the curvature. In an obvious notation this can be written as

$$F_{AB}=\begin{pmatrix} F_{\mu\nu} & F_{d\mu} \\ F_{d\mu} & F_{dd} \end{pmatrix}, \tag{88}$$

where $F_{\mu\nu}$ is the conventional curvature and

$$F_{d\mu}=\partial_\mu V-d\wedge A_\mu+[A_\mu,V] = \begin{pmatrix} 0 & D_\mu\phi \\ D_\mu\phi^* & 0 \end{pmatrix}\equiv D_\mu\Phi, \tag{89}$$

where D_μ is the conventional covariant derivative. The interesting component is F_{dd}, which turns out to be

$$F_{dd}=\mathbf{d}\wedge V+eV^2. \tag{90}$$

The explicit form of Eq. (90) is easily computed to be

$$F_{dd}=(k(v+v^*)+evv^*)I=e(|\phi|^2-c^2)I. \tag{91}$$

Since it is $\phi(x)$ that must be identified as a Higgs field, the relationship between Eq. (91) and the standard $U(1)$ Higgs potential is obvious.

Before applying the above formalism to physics, however, we have to introduce fermion fields $\Psi(x)$. These are taken to be column vectors of ordinary fermions $\psi_a(x)$,

$$\Psi(x)=\begin{pmatrix} \psi_a(x) \\ \psi_b(x) \end{pmatrix}. \tag{92}$$

The action of the gauge group and the covariant derivative on them is by ordinary multiplication, i.e.,

$$U(x)\Psi(x)=\begin{pmatrix} e^{i\alpha(x)}\psi_a(x) \\ e^{i\beta(x)}\psi_b(x) \end{pmatrix} \tag{93}$$

and

$$\mathbf{D}\Psi(x)=e\begin{pmatrix} 0 & \phi(x) \\ \phi^*(x) & 0 \end{pmatrix}\begin{pmatrix} \psi_a(x) \\ \psi_b(x) \end{pmatrix} = e\begin{pmatrix} \phi(x)\psi_b(x) \\ \phi^*(x)\psi_a(x) \end{pmatrix}, \tag{94}$$

respectively. As might be expected from the fact that the fermions are $U(1)$ covariant, it is the $U(1)$-covariant field $\phi(x)$, and not the component $v(x)$ of the connection, that couples to them in Eq. (94).

2. Application to the standard model

As has already been mentioned, the immediate physical interest of the noncommutative gauge theory lies in its application to the standard model of the fundamental interactions. The new feature is that it produces the Higgs field and its potential as natural consequences of gauge theory, in contrast to ordinary field theory in which they are introduced in an *ad hoc* phenomenological manner. The mechanism by which they are produced is very like that used in Kaluza-Klein reduction so, to put the noncommutative mechanism into perspective, let us first digress a little to recall the usual Kaluza-Klein mechanism.

a. The Kaluza-Klein mechanism

Consider the gauge-fermion Lagrangian density in 4 $+n$ dimensions, namely,

$$L=\frac{1}{4}\mathrm{Tr}(F_{AB})^2+\bar{\psi}\gamma^A D_A\psi, \tag{95}$$

where $A,B=1\ldots4+n$. If we let $\mu,\nu=0\ldots3$ and $r,s=4\ldots n$ and assume that the fields do not depend on the coordinates x_r, the Dirac operator and the curvature decompose into

$$F_{AB}=\begin{pmatrix} F_{\mu\nu} & D_\mu A_r \\ -D_\mu A_r & [A_r,A_s] \end{pmatrix}$$

and

$$\gamma^A D_A=\gamma^\mu D_\mu+\gamma^r A_r, \tag{96}$$

respectively, and hence the Lagrangian (95) decomposes into

$$L=\frac{1}{4}\mathrm{Tr}(F_{\mu\nu})^2-\frac{1}{2}(D_\nu A_r)^2+\bar{\psi}\gamma^\nu D_\mu\psi +\bar{\psi}\gamma^s A_s\psi+\frac{1}{4}\mathrm{Tr}[A_s,A_r]^2. \tag{97}$$

The extra components A_r of the gauge potential are space-time scalars and may therefore be identified as Higgs fields. Thus the dimensional reduction produces a standard kinetic term, a standard Yukawa term, and a potential for the Higgs fields. The problem is that the Higgs potential is not the one required for the standard model. In particular, its minimum does not force $|A_r|$ to assume the fixed nonzero value that is necessary to produce the masses of the gauge fields and leptons.

b. The noncommutative mechanism

As we shall now see, the noncommutative mechanism is very similar to the Kaluza-Klein mechanism. But it eliminates the artificial assumption that the fields do not depend on the extra coordinates and it produces a Higgs potential that is of the same form as those used in standard models. As in the Kaluza-Klein case, the procedure is to start with the formal gauge-fermion Lagrangian (95) and expand around the conventional four-dimensional gauge and fermion fields.

From the discussion of the previous section we see that if we expand the Dirac operator and the Yang-Mills curvature in this way we obtain

$$F_{AB}=\begin{pmatrix} F_{\mu\nu} & D_\mu\Phi(x) \\ -D_\mu\Phi(x) & F_{dd} \end{pmatrix}$$

and

$$\gamma^A D_A=\gamma^\mu D_\mu+g\mathbf{D}, \tag{98}$$

respectively, where g is a constant whose value cannot be fixed as the theory does not relate the scales of D_μ and $\mathbf{D}$. The resemblance between Eq. (98) and the corresponding Kaluza-Klein expression (96) is striking.

It is clear from Eq. (98) that for the noncommutative case the formal Yang-Mills-fermion Lagrangian (95) decomposes to

$$L=\frac{1}{4}\mathrm{Tr}(F_{\mu\nu})^2-\frac{1}{2}(D_\nu\phi)^2+\bar\Psi\gamma^\nu D_\mu\Psi$$
$$+G\bar\Psi\Phi\Psi+\frac{1}{4}\mathrm{Tr}[F_{dd}(\phi)]^2, \tag{99}$$

where

$$\Phi(x)=\begin{pmatrix} 0 & \phi(x) \\ \phi^*(x) & 0 \end{pmatrix} \quad \text{and} \quad G=eg. \tag{100}$$

Since the field $\phi(x)$ is a scalar that transforms covariantly with respect to the $U(1)$ gauge group it may be interpreted as a Higgs field. Hence, in analogy with the Kaluza-Klein mechanism, the noncommutative mechanism produces a standard kinetic term, a standard Yukawa term, and a potential for the Higgs field. The difference lies in the form of the potential, which is no longer the square of a commutator. From Eq. (91) we have

$$\frac{1}{4}\mathrm{Tr}(F_{dd})^2=\frac{1}{2}e^2[|\phi(x)|^2-c^2]^2. \tag{101}$$

But this is just the renormalizable potential that is used to produce the spontaneous breakdown of $U(1)$ invariance. Putting all the new contributions together, we see that the introduction of the discrete dimension and its associated gauge potential $\phi(x)$ produces exactly the extra terms

$$-\frac{1}{2}[D_\mu\phi(x)]^2+G\bar\Psi(x)\Phi(x)\Psi(x)$$
$$+\frac{1}{2}e^2[|\phi(x)|^2-c^2]^2 \tag{102}$$

that describe the Higgs sector of the standard $U(1)$ model. Thus, when the concept of manifold is generalized in the manner dictated by noncommutative geometry, the standard Higgs sector emerges in a natural way. Note, however, that since there are three undetermined parameters in Eq. (99), the noncommutative approach does not achieve any new unification in the sense of reducing the number of parameters. However, it considerably reduces the ranges of the parameters, places strong restrictions on the matter-field representations, and even rules out the exceptional groups as gauge groups (Schucker, 1997).

Of course the above model is only a toy one, since it uses the gauge group $U(1)\times U(1)$ rather than the gauge groups $U(2)$ and $S[U(2)\times U(3)]$ of the standard electroweak and electroweak-strong models or the gauge groups of grand unified theory.

However, the general structure provided by noncommutative geometry can be applied to any gauge group. Connes himself (Connes, 1994) has applied it to the standard model. There is some difficulty in applying it to grand unified theories because of the restrictions on fermion representations, but a modified version has been applied to grand unified theories in the work of Chamseddine *et al.* (1993). As in the toy model, the noncommutative approach does not achieve any new unification in the sense of reducing the number of parameters, though, as already mentioned, it introduces some important restrictions. Most importantly, it provides a new and interesting interpretation.

ACKNOWLEDGMENTS

We are indebted to C. N. Yang for important remarks which improved the paper. A number of people gave us positive reactions and welcome comments. We thank, in particular, D. Giulini, F. Hehl, U. Lindstrom, T. Schucker, and D. Vassilevich. Special thanks go to T. Damour for some pertinent remarks which led to several improvements of the printed version.

REFERENCES

Appelquist, T., A. Chodos, and P. G. O. Freund, 1987, *Modern Kaluza-Klein Theories* (Addison-Wesley, London).

Audretsch, J., F. Gähler, and N. Straumann, 1984, Commun. Math. Phys. **95**, 41.

Bergmann, V., 1932, Sitzungsber. K. Preuss. Akad. Wiss., Phys. Math. Kl. 346.

Bergmann, P. G., 1942, *An Introduction to the Theory of Relativity* (Prentice-Hall, New York), Chaps. XVII and XVIII.

Bergmann, P. G., 1968, Int. J. Theor. Phys. **1**, 25.

Bleecker, D., 1981, *Gauge Theory and Variational Principles* (Addison-Wesley, London).

Brans, C. H., and R. H. Dicke, 1961, Phys. Rev. **124**, 925.

Cartan, E., 1928, *Leçons sur la Géométrie des Espaces de Riemann*, 2nd ed. (Gauthier-Villars, Paris).

Case, C. M., 1957, Phys. Rev. **107**, 307.

Chamseddine, A. H., G. Felder, and J. Fröhlich, 1993, Nucl. Phys. B **395**, 672.

Chandrasekharan, K., 1986, Ed., *Hermann Weyl, 1885–1985* (Springer, New York).

Connes, A., 1994, *Noncommutative Geometry* (Academic, New York).

Damour, T., and A. M. Polyakov, 1994, Nucl. Phys. B **423**, 532.

Dicke, R. H., 1962, Phys. Rev. **125**, 2163.

Einstein, A., 1987, *The Collected Papers of Albert Einstein*, edited by J. Stachel, D. C. Cassidy, and R. Schulmann (Princeton University, Princeton, NJ).

Feynman, R. P., 1995, *Feynman Lectures on Gravitation*, edited by B. Hatfield (Addison-Wesley).

Fierz, M., 1956, Helv. Phys. Acta **29**, 128.

Fierz, M., 1999, private communication.

Fock, V., 1927, Z. Phys. **39**, 226.

Fock, V., 1929, Z. Phys. **57**, 261.

Green, M., J. Schwarz, and E. Witten, 1987, *Theory of Strings and Superstrings* (Cambridge University, Cambridge, England).

Gross, D., 1995, in *The Oskar Klein Centenary Symposium*, edited by U. Lindstrom (World Scientific, Singapore), p. 94.

Hoffmann, B., 1933, Phys. Rev. **37**, 88.

Infeld, L., and B. L. van der Waerden, 1932, Sitzungsber. K. Preuss. Akad. Wiss., Phys. Math. Kl. 380 and 474.

Jordan, P., 1949, Nature (London) **164**, 637.

Jordan, P., 1955, Schwerkraft und Weltall, 2nd ed. (Vieweg, Braunschweig).

Kaluza, Th., 1921, Sitzungsber. K. Preuss. Akad. Wiss., Phys. Math. Kl., 966; for an English translation see O'Raifeartaigh, 1997.

Klein, O., 1926a, Z. Phys. **37**, 895; for an English translation see O'Raifeartaigh, 1997.

Klein, O., 1926b, Nature (London) **118**, 516.

Klein, O., 1938, *1938 Conference on New Theories in Physics, held in Kasimierz, Poland*; reproduced in O'Raifeartaigh, 1997.

Klein, O., 1956, in *Proceedings of the Berne Congress*, Helv. Phys. Acta, Suppl. IV, 58.

Kubyshin, Y. A., J. M. Mourao, G. Rudolph, and I. P. Volobujev, 1989, *Dimensional Reduction of Gauge Theories, Spontaneous Compactification and Model Building*, Springer Lecture Notes in Physics No. 349 (Springer, New York).

Lichnerowicz, A., 1955, *Théories Relativiste de la Gravitation et de l'Electromagnétisme* (Masson, Paris), Chap. 4.

London, F., 1927, Z. Phys. **42**, 375.

Ludwig, C., 1951, *Fortschritte der Projektiven Relativitatstheorie* (Vieweg, Branunschweig).

Nordström, G., 1912, Phys. Z. **13**, 1126.

Nordström, G., 1913a, Ann. Phys. (Leipzig) **40**, 856.

Nordström, G., 1913b, Ann. Phys. (Leipzig) **42**, 533.

Nordström, G., 1914, Phys. Z. **15**, 504.

O'Raifeartaigh L., 1997, *The Dawning of Gauge Theory* (Princeton University, Princeton, NJ).

Pais, A., 1953, *Conference in Honour of H. A. Lorentz, Leiden 1953*, Physica (Amsterdam) **19**, 869.

Pais, A., 1982, *Subtle is the Lord: The Science and Life of Albert Einstein* (Oxford University, New York).

Pauli, W., 1919, Phys. Z. **20**, 457.

Pauli, W., 1921, "Relativitätstheorie," *Encyklopädie der Mathematischen Wissenschaften* (Leipzig, Teubner), Vol. 5.3, p. 539.

Pauli, W., 1933, Ann. Phys. (Leipzig) **18**, 305.

Pauli, W., 1939, Helv. Phys. Acta **12**, 147.

Pauli, W., 1958, *Theory of Relativity* (Pergamon, New York).

Pauli, W., 1979, *Wissenschaftlicher Briefwechsel*, Vol. I: 1919–1929 (Springer, Berlin), p. 505. (Translation of the letter by L. O'Raifeartaigh).

Pauli, W., 1999, *Wissenschaftlicher Briefwechsel*, Vol. IV, Part II (Springer, Berlin), Letters 1614 and 1682.

Polchinski, J., 1998, *String Theory*, Vols. I, II, Cambridge Monographs on Mathematical Physics (Cambridge University, Cambridge, England).

Raman, V., and P. Forman, 1969, Hist. Stud. Phys. Sci. **1**, 291.

Schrödinger, E., 1922, Z. Phys. **12**, 13.

Schrödinger, E., 1932, Sitzungsber. K. Preuss. Akad. Wiss., Phys. Math. Kl. 105.

Schrödinger, E., 1987, *Schrödinger: Centenary Celebration of a Polymath*, edited by C. Kilmister (Cambridge University, New York/Cambridge, England).

Schucker, T., 1997, "Geometry and Forces," in *Proceedings of the 1997 EMS Summer School on Noncommutative Geometry and Applications*, Monsaraz and Lisbon, edited by P. Almeida, to appear (hep-th/9712095)

Seelig, C., 1960, *Albert Einstein* (Europa, Zürich), p. 274.

Steinhardt, P. J., 1993, Class. Quantum Grav. **10**, 33.

Straumann, N., 1984, *General Relativity and Relativistic Astrophysics*, Texts and Monographs in Physics (Springer, Berlin).

Straumann, N., 1987, Phys. Bl. (Germany) **43** (11), 414.

Tetrode, H., 1928, Z. Phys. **50**, 336.

Thiry, Y. R., 1948, C. R. Acad. Sci. **226**, p. 216, 1881.

Thiry, Y. R., 1951, These (Université de Paris).

Veblen, O., 1933, *Projektive Relativitätstheorie* (Springer, Berlin).

Wald, R. M., 1986, Phys. Rev. D **33**, 3613.

Weinberg, S., 1965, Phys. Rev. **138**, A988.

Weyl, H., 1918, "Gravitation und Elektrizität," Sitzungsber. Deutsch. Akad. Wiss. Berlin, Klossefü... pp. 465–480. See also H. Weyl, 1968, *Gesammelten Abhandlungen*, edited by K. Chadrasekharan (Springer, Berlin). An English translation is given in O'Raifeartaigh, 1997.

Weyl, H., 1922, *Space, Time, Matter* (Methuen, London, and Dover, New York). Translated from the 4th German Edition. [Raum.· Zeit.· Materie, 8. Auflage (Springer, Berlin, 1993)].

Weyl, H., 1929, "Elektron und Gravitation. I" Z. Phys. **56**, 330.

Weyl, H., 1946, "Memorabilia," in *Hermann Weyl*, edited by K. Chandrasekharan (Springer, New York), p. 85.

Weyl, H., 1956, *Selecta* (Birkhäuser, Boston).

Weyl, H., 1968, *Gesammelte Abhandlungen*, edited by K. Chandrasekharan (Springer, Berlin), Vol. III, p. 229.

Weyl, H., 1980, in *Hermann Weyl*, edited by K. Chandrasekharan (Springer, Berlin), p. 85.
Weyl, H., 1981, *Gruppentheorie und Quantenmechanik* (Wissenschaftliche Buchgesellschaft, Darmstadt), Nachdruck der 2. Aufl., Leipzig 1931. [English translation: *Group Theory and Quantum Mechanics*, Dover, New York (1950)].
Wigner, E., 1929, Z. Phys. **53**, 592.
Yang, C. N., 1980, "Hermann Weyl's Contribution to Physics," in *Hermann Weyl, 1885–1985*, edited by K. Chandrasekharan (Springer, New York), p. 7.
Yang, C. N., 1983, *Selected Papers 1945–1980 with Commentary* (Freeman, San Francisco), p. 525.
Yau, S. T., 1985, "Compact three-dimensional Kahler Manifolds with zero Ricci curvature, in *Symposium on Anomalies, Geometry, and Topology*," edited by W. Bardeen and A. White (World Scientific, Singapore), p. 395. See also references therein.

String theory as a generalization of gauge symmetry

Pei-Ming Ho

Department of Physics, National Taiwan University, Taipei, Taiwan, R.O.C.

pmho@phys.ntu.edu.tw

This is a brief review of the relation between string theory and usual gauge theories including Einstein's gravity and Yang-Mills theory. In particular, we would like to explain how string theory extends or generalizes ideas behind gauge symmetry and Einstein's general relativity. Most of the material concerning this article can be found in [1, 2], if no further reference is provided.

1 There is gravity

Let us start with some basic facts about string theory. The first and foremost to say is that string theory includes gravity. More importantly, string theory contains quantum gravity, and is the only theory of quantum gravity which admits a ultraviolet-finite and unitary perturbation theory. On the other hand, there is no consistent perturbation theory for the canonical quantization of Einstein's theory of pure gravity.

Contrary to particles, for which it is extremely hard to find a well-behaved (causal, unitary, renormalizable) interaction with the gravitons, strings are almost always accompanied by gravity. As each excited state of a string can be reinterpreted as a particle, the massless spin-two excitation of a closed string is identified with the graviton. This excitation exists in all five superstring theories (type I, type IIA, type IIB, heterotic $SO(32)$ and heterotic $E_8 \times E_8$) as well as the bosonic string theory.

A string can have (infinitely) many vibration modes. The graviton is a massless spin-2 osillation mode of a closed string. Each oscillation mode should be matched with a certain particle in spacetime. Each particle is associated with a vertex operator defined in the 1+1 dimensional quantum field theory living on the string worldsheet. Interactions among particles are determined in string theory by correlation functions of vertex operators. A Feynman diagram for a spacetime particle scattering process corresponds to a path integral of the worldsheet theory. Unlike particle theory, where interaction vertices and propagators are built in a Lagrangian, all ingredients of Feynman diagrams are fixed in string theory by the definition of a free string.

There are two complimentary approaches to see how gravitational interaction is determined in string theory. The first approach is to compute the Feynman diagram of a scattering process involving gravitons, in the flat trivial background. From the result one can construct order by order a field theory involving the metric to reproduce the scattering amplitudes.

The second approach is to consider a string propagating in a perturbative deformation of the flat background. The spacetime metric appears in the kinetic term of the string worldsheet action. The requirement of (quantum) conformal invariance then imposes a strong constraint on the metric. The constraint involves the metric and its derivatives, and can be viewed as the equation of motion of the metric. In this approach, the equation of motion for the spacetime metric is in fact a self-consistency condition. This is a remarkable feature of string theory which is not unique to gravity. Dynamics and kinematical constraints are unified in string theory.

Of course, the results of the two approaches agree with each other. They also agree with the Einstein equation at low energies for weak coupling (Newton) constant. At high energies (compared with the energy scale of the string tension), string theory modifies Einstein's theory, but with general covariance intact.

2 There is a bigger gauge symmetry

In addition to the graviton, strings have other oscillation modes including many other gauge fields. For open strings, there is a massless spin-1 oscillation mode corresponding to a gauge field in spacetime.

Non-Abelian gauge symmetry also arises in string theory in various ways. One possibility is to associate a so-called Chan-Paton factor to each endpoint of an open string. The massless spin-1 oscillation mode is then labelled by two indices, and the corresponding spacetime gauge field is promoted to a matrix. The result is that the $U(1)$ gauge symmetry is enhanced to a non-Abelian gauge symmetry. This happens for open strings in type I string theory, and the gauge group has to be $SO(32)$ for self-consistency. Another possibility is to add currents corresponding to a global symmetry on the string worldsheet, or to compactify the spacetime in a particular way so that new massless spin-1 oscillation modes appear.

Similarly, when there are multiple coincident D-branes on top of each other, an open string stretched between two D-branes naturally acquire labels specifying the two D-branes. For coincident N D-branes, there is an $U(N)$ gauge symmetry on the D-brane worldvolume.

In addition to the massless spin-1 oscillation mode, there are (infinitely) many other oscillation modes corresponding to gauge fields of higher spins. It is believed that these gauge fields are massive due to a spontaneous symmetry breaking mechanism, and the symmetry may be restored in the high energy limit. String extends general covariance and usual gauge symmetry to a much larger symmetry. There are higher spin gauge fields of arbitrarily many Lorentz indices. Furthermore, as the symmetry is so highly intertwined, it is tempting to conjecture that its self-consistency uniquely fixes the dynamics of the whole theory.

Witten's cubic string field theory [5] manifestly exihibits the gauge symmetry of bosonic open string theory. The action of the cubic string field theory is

$$S = \langle\left(\frac{1}{2}\Psi * (Q\Psi) + \frac{1}{3}\Psi * \Psi * \Psi\right)\rangle, \tag{1}$$

where $\Psi[X(\sigma)]$ is the string wave function (a state of the string worldsheet theory in the formulation using BRST quantization) and Q is the BRST charge for conformal symmetry. The product labelled by $*$ defines an associative algebra for the string states, which essentially tells us how to glue two strings together into one string. This action has the gauge symmetry

$$\Psi \rightarrow Q\Lambda + \Psi * \Lambda - \Lambda * \Psi. \tag{2}$$

We see from this formula that the string wave function is itself a gauge field for a huge gauge symmetry.

The string wave function Ψ can be represented as a generic state in the Hilbert space of the string worldsheet theory in BRST quantization. In terms of an expansion of creation operators,

$$\begin{aligned}\Psi &= \int d^dk\, [\phi(k) + iA_\mu(k)\alpha^\mu_{-1} + \alpha(k)b_{-1}c_0 + iB_\mu(k)\alpha^\mu_{-2} - B_{\mu\nu}(k)\alpha^\mu_{-1}\alpha^\nu_{-1} + \\ &\quad \beta_0(k)b_{-2}c_0 + \beta_1(k)b_{-1}c_{-1} + i\beta_\mu(k)\alpha^\mu_{-1}b_{-1}c_0 + \cdots]\, c_1|0;k\rangle, \end{aligned} \tag{3}$$

where α^μ_n are the bosonic operators representing fluctuations of the spacetime coordinate X^μ, and c_n, b_n belong to the ghost sector. The coefficients $\phi(k), A_\mu(k), \alpha(k), B_\mu(k)$, etc. are Fourier transforms of spacetime fields. Their gauge transformations are given by

$$\delta A_\mu = \partial_\mu \epsilon^0, \tag{4}$$

$$\delta\alpha = \frac{1}{2}\partial^2\epsilon^0, \tag{5}$$

$$\delta B_{\mu\nu} = -\partial_{(\mu}\epsilon^0_{\nu)} - \frac{1}{2}\eta_{\mu\nu}\epsilon^1, \tag{6}$$

$$\delta B_\mu = -\partial_\mu\epsilon^1 + \epsilon^0_\mu, \tag{7}$$

$$\delta\beta_\mu = \frac{1}{2}(\partial^2 - 2)\epsilon^0_\mu, \tag{8}$$

$$\delta\beta_0 = \frac{1}{2}(\partial^2 - 2)\epsilon^1, \tag{9}$$

$$\delta\beta_1 = -\partial^\mu\epsilon^0_\mu - 3\epsilon^1. \tag{10}$$

Obviously, A_μ is the massless spin-1 guage field mentioned above. $B_{\mu\nu}$ is a symmetric rank-2 gauge potential. At higher mass levels one finds gauge fields at higher ranks. Except A_μ, the other gauge fields are massive, signaling a spontaneous symmetry breakdown.

The physical meaning of the huge gauge symmetry of string theory has not yet been thoroughly explored. People also suspect that there is another huge hidden (global) symmetry which is spontaneously broken, and that one should study the high energy limit where the symmetry may be restored [6]. The 2 dimensional string theory is the best understood toy model of strings. It has the w_∞ symmetry and the symmetry algebra is strong enough to determine all scattering amplitudes. It has been conjectured by Gross and his collaborators [6] that the full symmetry may uniquely determine the dynamics for higher dimensional strings as well. It is possible that the huge global symmetry has the same origin as the huge gauge symmetry of string theory, or that its existence is a necessity for self-consistency due to the higher spin gauge symmetries [7].

While the general covariance of general relativity is embedded in a bigger gauge symmetry, we do not seem to fully comprehend the physical notion which generalizes the equivalence principle. What are the gendenken experiments analogous to the elevator in free fall?

3 Geometry is induced

Before the advant of string theory, spacetime is put in by hand as a stage in which particles interact and events take place. In string theory, spacetime can be a derived notion. From the viewpoint of the string worldsheet action, the spacetime coordinates of the string are scalar fields on the worldsheet. Properties of spacetime are determined by properties of these scalar fields. Of course one can also view spacetime coordinates of a point particles as a scalar field on the worldline. The crucial difference is that while string worldsheet theory determines the dynamics of spacetime geometry, the particle worldline theory does not.

As spacetime is a derived notion, the geometrical structure of spacetime is something that should be extracted from the theory. We have already mentioned above how the dynamics of spacetime metric is determined in string theory, yet Riemannian geometry is not the only geometrical structure the spacetime can possess.

Noncommutative geometry is a mathematical notion that generalizes classical geometry [8] A classical manifold is commutative, that is, the algebra of functions on the manifold is a commutative algebra. Mathematicians noticed that, although traditionally one visualizes a manifold as a collection of points (with some topology), the algebra of functions provides an alternative equivalent description to some extent. Hence it is natural to relax the definition of geometry to allow the algebra of functions to be noncommutative. Such a space no longer admits the picture of a set of points, but various geometric structure and quantities can be defined.

In a field theory, the base space can be noncommutative by imposing nontrivial commutation relations for the coordinates. In string theory, the commutativity of spacetime coordinates depends on the quantization of the scalars fields on the worldsheet. In suitable background, noncommutative geometry arises automatically.

In string theory, there are solitonic objects called D-branes. They are submanifolds of spacetime on which open strings end. Turning on a background field called NS-NS B-field, one can quantize the open string and find that the coordinates of the endpoints are noncommutative [9], with the noncommutativity depending on the B-field background. The D-brane field theory is then conveniently defined as a noncommutative field theory [10].

Another way to see the noncommutativity of a D-brane is to consider scattering amplitudes of open strings [12]. In other words, you probe the geometric property of the D-brane worldvolume using open strings. Due to the B-field background, the interactions at low energies are most conveniently described by a field theory living on a noncommutative space. With other background fields (graviphoton) turned on, the spacetime itself can also be noncommutative [13].

In addition to noncommutative geometry, there might be more exotic geometrical notions which we can learn from string theory. In general, one probes the spacetime via strings. The geometry of spacetime is an induced notion derived from decoding string interactions in a certain way. It is conceivable that there can be ambiguity in the decoding process, corresponding to the freedom of making a change of variables (field redefinition). For example, a noncommutative field theory can be reinterpreted as a commutative field theory with higher derivative interactions [11]. We will make more comments below on other ambiguities in defining spacetime properties.

Matrix model provides another way to demonstrate the idea. The BFSS model [14] and the IKKT model [15] are conjectured to be equivalent descriptions of string theory. Spacetime coordinates correspond to $N \times N$ matrices with $N \to \infty$. As matrices the coordinates are generically noncommutative. Interactions in the models would make it impossible to have large extended dimensions in spacetime if there were no supersymmetry to guarantee flat directions in the moduli space. The large scale, (roughly) commutative spacetime around us is not given to us without warrant. The choice of the effective spacetime dimension by our universe can be translated into questions about the free energy of the theory [16].

T-duality (including Mirror symmetry) is another example of the ambiguity in extracting the spacetime structure from string theory. The simplest setting of T-duality is having one spatical direction compactified on a circle. Let the radius be denoted R. It turns out that this theory is equivalent to another string theory with a dual spatial direction compactified on a circle of radius $1/R$ in string units. The two theories dual to each other are different descriptions of the same physical system. The same physical state can have totally different descriptions in the two theories. For instance, a string winding on a circle can be matched with a Kaluza-Klein mode (momentum eigenstate) in the dual theory. But a state is always matched with another state with the same energy. Physicists trying to describe a physical process can adopt either one of the two theories and its language. They may sound very different in words, but the two theories agree when it comes to the prediction of the result of a measurement in terms of pure numbers.

In general, shapes and topology of spacetime can be different between equivalent theories. Even the spacetime dimension is not independent of the choice of description/theory. Superstrings in 10 dimensions are believed to be equivalent to the M-theory in 11 dimensions. Physics is not only observer-dependent, like it already is in Einstein's theory, but is now also theory-dependent. In fact, all five superstring theories are believed to be dual to one another, and also to (a quantum version of) the 11 dimensional theory of supergravity.

4 Summary

To summarize, gauge symmetry is generalized to include gauge fields of arbitrary spins in string theory. Einstein's theory of spacetime is also extended to more general geometric structures. The notion that physics is observer-dependent, which we first learned from the theory of relativity, is magnified to the notion that physics is theory-dependent. It is also believed that Yang's inspiring saying that "Symmetry dictates interaction" is fully honoured by string theory.

An obvious difference between string theory and general relativity or Yang-Mills theory is that the discovery of the latter was motivated by beautiful theoretical notions on symmetry and geometry, but the discovery of string theory was an accident. What is the symmetry principle of string theory? This is one of the most important problems of string theory.

We have so far avoided discussions on holography and cosmological constant in this article. Holographic principle is argued to be a salient feature of quantum gravity [17]. In string theory we have seen remarkable evidences of it [18]. It is generally viewed to be of utmost importance for quantum gravity, but so far our understanding of it remains at a technical level. On the other hand, our understanding of the cosmological constant problem is not even at the technical level. It might be that the secrets of holography and cosmological constant are deeply hidden inside string theory waiting to be discovered, and their revelation will bring us a drastic conceptual breakthrough so that string theory will no longer be called string theory.

Acknowledgment

This work is supported in part by the National Science Council and National Center for Theoretical Sciences, Taiwan, R.O.C. and the Center for Theoretical Physics at National Taiwan University.

References

[1] M. B. Green, J. H. Schwarz and E. Witten, "Superstring Theory. Vol. 1: Introduction", Cambridge University Press; M. B. Green, J. H. Schwarz and E. Witten, "Superstring Theory. Vol. 2: Loop Amplitudes, Anomalies And Phenomenology", Cambridge University Press.

[2] J. Polchinski, "String theory. Vol. 1: An introduction to the bosonic string", Cambridge University Press; J. Polchinski, "String theory. Vol. 2: Superstring theory and beyond", Cambridge University Press.

[3] See, for instance, J. Isberg, U. Lindstrom, B. Sundborg and G. Theodoridis, "Classical and quantized tensionless strings," Nucl. Phys. B **411**, 122 (1994) [arXiv:hep-th/9307108]; B. Sundborg, "Stringy gravity, interacting tensionless strings and massless higher spins," Nucl. Phys. Proc. Suppl. **102**, 113 (2001) [arXiv:hep-th/0103247]; E. Sezgin and P. Sundell, "Massless higher spins and holography," Nucl. Phys. B **644**, 303 (2002) [Erratum-ibid. B **660**, 403 (2003)] [arXiv:hep-th/0205131]; C. S. Chu, P. M. Ho and F. L. Lin, "Cubic string field theory in pp-wave background and background independent Moyal structure," JHEP **0209**, 003 (2002) [arXiv:hep-th/0205218].

[4] See, for instance, C. Fronsdal, "Massless Fields With Integer Spin," Phys. Rev. D **18**, 3624 (1978); J. Fang and C. Fronsdal, "Massless Fields With Half Integral Spin," Phys. Rev. D **18**, 3630 (1978); M. A. Vasiliev, "Progress in higher spin gauge theories," *Prepared for 9th Marcel Grossmann Meeting on Recent Developments in Theoretical and Experimental General Relativity, Gravitation and Relativistic Field Theories (MG 9), Rome, Italy, 2-9 Jul 2000*; M. A. Vasiliev, "Higher spin gauge theories in various dimensions," Fortsch. Phys. **52**, 702 (2004) [arXiv:hep-th/0401177].

[5] E. Witten, "Noncommutative Geometry And String Field Theory," Nucl. Phys. B **268**, 253 (1986).

[6] D.J. Gross and P. Mende, Phys. Lett. B197,129 (1987); Nucl.Phys.B303, 407(1988); D.J. Gross, " High energy symmetry of string theory", Phys. Rev. Lett. 60,1229 (1988); Phil.Trans. R. Soc. Lond. A329,401(1989); D.J. Gross and J.L. Manes, "The high energy behavior of open string theory", Nucl.Phys.B326, 73(1989). See section 6 for details.

[7] C. T. Chan, P. M. Ho and J. C. Lee, "Ward identities and high-energy scattering amplitudes in string theory," Nucl. Phys. B **708**, 99 (2005) [arXiv:hep-th/0410194].

[8] A. Connes, "Noncommutative Geometry", Academic Press.

[9] C. S. Chu and P. M. Ho, "Noncommutative open string and D-brane," Nucl. Phys. B **550**, 151 (1999) [arXiv:hep-th/9812219].

[10] A. Connes, M. R. Douglas and A. Schwarz, "Noncommutative geometry and matrix theory: Compactification on tori," JHEP **9802**, 003 (1998) [arXiv:hep-th/9711162].

[11] N. Seiberg and E. Witten, "String theory and noncommutative geometry," JHEP **9909**, 032 (1999) [arXiv:hep-th/9908142].

[12] V. Schomerus, "D-branes and deformation quantization," JHEP **9906**, 030 (1999) [arXiv:hep-th/9903205].

[13] H. Ooguri and C. Vafa, "The C-deformation of gluino and non-planar diagrams," Adv. Theor. Math. Phys. **7**, 53 (2003) [arXiv:hep-th/0302109]; N. Seiberg, "Noncommutative superspace, N = 1/2 supersymmetry, field theory and string theory," JHEP **0306**, 010 (2003) [arXiv:hep-th/0305248].

[14] T. Banks, W. Fischler, S. H. Shenker and L. Susskind, "M theory as a matrix model: A conjecture," Phys. Rev. D **55**, 5112 (1997) [arXiv:hep-th/9610043].

[15] N. Ishibashi, H. Kawai, Y. Kitazawa and A. Tsuchiya, "A large-N reduced model as superstring," Nucl. Phys. B **498**, 467 (1997) [arXiv:hep-th/9612115].

[16] H. Aoki, S. Iso, H. Kawai, Y. Kitazawa and T. Tada, "Space-time structures from IIB matrix model," Prog. Theor. Phys. **99**, 713 (1998) [arXiv:hep-th/9802085].

[17] G. 't Hooft, "Dimensional reduction in quantum gravity," [arXiv:gr-qc/9310026]; L. Susskind, "The World as a hologram," J. Math. Phys. **36**, 6377 (1995) [arXiv:hep-th/9409089].

[18] J. M. Maldacena, "The large N limit of superconformal field theories and supergravity," Adv. Theor. Math. Phys. **2**, 231 (1998) [Int. J. Theor. Phys. **38**, 1113 (1999)] [arXiv:hep-th/9711200]; O. Aharony, S. S. Gubser, J. M. Maldacena, H. Ooguri and Y. Oz, "Large N field theories, string theory and gravity," Phys. Rept. **323**, 183 (2000) [arXiv:hep-th/9905111].

The cosmological constant problem*

Steven Weinberg

Theory Group, Department of Physics, University of Texas, Austin, Texas 78712

Astronomical observations indicate that the cosmological constant is many orders of magnitude smaller than estimated in modern theories of elementary particles. After a brief review of the history of this problem, five different approaches to its solution are described.

CONTENTS

As I was going up the stair,
I met a man who wasn't there.
He wasn't there again today,
I wish, I wish he'd stay away.

Hughes Mearns

I. INTRODUCTION

Physics thrives on crisis. We all recall the great progress made while finding a way out of various crises of the past: the failure to detect a motion of the Earth through the ether, the discovery of the continuous spectrum of beta decay, the τ-θ problem, the ultraviolet divergences in electromagnetic and then weak interactions, and so on. Unfortunately, we have run short of crises lately. The "standard model" of electroweak and strong interactions currently faces neither internal inconsistencies nor conflicts with experiment. It has plenty of loose ends; we know no reason why the quarks and leptons should have the masses they have, but then we know no reason why they should not.

Perhaps it is for want of other crises to worry about that interest is increasingly centered on one veritable crisis: theoretical expectations for the cosmological constant exceed observational limits by some 120 orders of magnitude.[1] In these lectures I will first review the history of this problem and then survey the various attempts that have been made at a solution.

II. EARLY HISTORY

After completing his formulation of general relativity in 1915–1916, Einstein (1917) attempted to apply his new theory to the whole universe. His guiding principle was that the universe is static: "The most important fact that we draw from experience is that the relative velocities of the stars are very small as compared with the velocity of light." No such static solution of his original equations could be found (any more than for Newtonian gravitation), so he modified them by adding a new term involving a free parameter λ, the cosmological constant:[2]

$$R_{\mu\nu}-\tfrac{1}{2}g_{\mu\nu}R-\lambda g_{\mu\nu}=-8\pi G T_{\mu\nu}\ . \tag{2.1}$$

Now, for $\lambda>0$, there was a static solution for a universe filled with dust of zero pressure and mass density

$$\rho=\frac{\lambda}{8\pi G}\ . \tag{2.2}$$

Its geometry was that of a sphere S_3, with proper circumference $2\pi r$, where

$$r=1/\sqrt{8\pi\rho G}\ , \tag{2.3}$$

so the mass of the universe was

$$M=2\pi^2 r^3\rho=\frac{\pi}{4}\lambda^{-1/2}G^{-1}\ . \tag{2.4}$$

In some popular history accounts, it was Hubble's discovery of the expansion of the universe that led Einstein to retract his proposal of a cosmological constant. The real story is more complicated, and more interesting.

One disappointment came almost immediately. Einstein had been pleased at the connection in his model between the mass density of the universe and its geometry, because, following Mach's lead, he expected that the mass distribution of the universe should set inertial frames. It was therefore unpleasant when his friend de Sitter, with whom Einstein remained in touch during the war, in 1917 proposed another apparently static cosmological model with no matter at all. (See de Sitter, 1917.) Its line element (using the same coordinate system as de Sitter, but in a different notation) was

$$d\tau^2=\frac{1}{\cosh^2 Hr}[dt^2-dr^2-H^{-2}\tanh^2 Hr(d\theta^2+\sin^2\theta\, d\varphi^2)]\ , \tag{2.5}$$

*Morris Loeb Lectures in Physics, Harvard University, May 2, 3, 5, and 10, 1988.

[1]For a good nonmathematical description of the cosmological constant problem, see Abbott (1988).

[2]The notation used here for metrics, curvatures, etc., is the same as in Weinberg (1972).

with H related to the cosmological constant by

$$H=\sqrt{\lambda/3} \tag{2.6}$$

and $\rho=p=0$. Clearly matter was not needed to produce inertia.

At about this time, the redshift of distant objects was being discovered by Slipher. Over the period from 1910 to the mid-1920s, Slipher (1924) observed that galaxies (or, as then known, spiral nebulae) have redshifts $z\equiv\Delta\lambda/\lambda$ ranging up to 6%, and only a few have blueshifts. Weyl pointed out in 1923 that de Sitter's model would exhibit such a redshift, increasing with distance, because although the metric in de Sitter's coordinate system is time independent, test bodies are not at rest; there is a nonvanishing component of the affine connection

$$\Gamma^r_{tt}=-H\sinh Hr\tanh Hr \tag{2.7}$$

giving a redshift proportional to distance

$$z\simeq Hr \quad \text{for } Hr\ll 1 . \tag{2.8}$$

In his influential textbook, Eddington (1924) interpreted Slipher's redshifts in terms of de Sitter's "static" universe.

But of course, although the cosmological constant *was* needed for a static universe, it was not needed for an expanding one. Already in 1922, Friedmann (1924) had described a class of cosmological models, with line element (in modern notation)

$$d\tau^2=dt^2-R^2(t)\left[\frac{dr^2}{1-kr^2}+r^2(d\theta^2+\sin^2\theta\, d\varphi^2)\right] . \tag{2.9}$$

These are comoving coordinates; the universe expands or contracts as $R(t)$ increases or decreases, but the galaxies keep fixed coordinates r,θ,φ. The motion of the cosmic scale factor is governed by an energy-conservation equation

$$\left[\frac{dR}{dt}\right]^2=-k+\tfrac{1}{3}R^2(8\pi G\rho+\lambda) . \tag{2.10}$$

The de Sitter model is just the special case with $k=0$ and $\rho=0$; in order to put the line element (2.5) in the more general form (2.9), it is necessary to introduce new coordinates,

$$\begin{aligned} t'&=t-H^{-1}\ln\cosh Hr , \\ r'&=H^{-1}\exp(-Ht)\sinh Hr , \\ \theta'&=\theta, \quad \varphi'=\varphi , \end{aligned} \tag{2.11}$$

and then drop the primes. However, we can also easily find expanding solutions with $\lambda=0$ and $\rho>0$. Pais (1982) quotes a 1923 letter of Einstein to Weyl, giving his reaction to the discovery of the expansion of the universe: "If there is no quasi-static world, then away with the cosmological term!"

III. THE PROBLEM

Unfortunately, it was not so easy simply to drop the cosmological constant, because anything that contributes to the energy density of the vacuum acts just like a cosmological constant. Lorentz invariance tells us that in the vacuum the energy-momentum tensor must take the form

$$\langle T_{\mu\nu}\rangle=-\langle\rho\rangle g_{\mu\nu} . \tag{3.1}$$

(A minus sign appears here because we use a metric which for flat space-time has $g_{00}=-1$.) Inspection of Eq. (2.1) shows that this has the same effect as adding a term $8\pi G\langle\rho\rangle$ to the effective cosmological constant

$$\lambda_{\text{eff}}=\lambda+8\pi G\langle\rho\rangle . \tag{3.2}$$

Equivalently we can say that the Einstein cosmological constant contributes a term $\lambda/8\pi G$ to the total effective vacuum energy

$$\rho_V=\langle\rho\rangle+\lambda/8\pi G=\lambda_{\text{eff}}/8\pi G . \tag{3.3}$$

A crude experimental upper bound on λ_{eff} or ρ_V is provided by measurements of cosmological redshifts as a function of distance, the program begun by Hubble in the late 1920s. The present expansion rate is today estimated as

$$\left[\frac{1}{R}\frac{dR}{dt}\right]_{\text{now}}\equiv H_0\simeq 50-100\text{ km/sec Mpc}$$
$$\simeq(\tfrac{1}{2}-1)\times 10^{-10}/\text{yr} .$$

Furthermore, we do not gross effects of the curvature of the universe, so very roughly

$$|k|/R^2_{\text{now}}\lesssim H_0^2 .$$

Finally, the ordinary nonvacuum mass density of the universe is not much greater than its critical value

$$|\rho-\langle\rho\rangle|\lesssim 3H_0^2/8\pi G .$$

Hence (2.10) shows that

$$|\lambda_{\text{eff}}|\lesssim H_0^2$$

or, in physicists' units,

$$|\rho_V|\lesssim 10^{-29}\text{ g/cm}^3\approx 10^{-47}\text{ GeV}^4 . \tag{3.4}$$

A more precise observational bound will be discussed in Sec. V, but this one will be good enough for our present purposes.

As everyone knows, the trouble with this is that the energy density $\langle\rho\rangle$ of empty space is likely to be enormously larger than 10^{-47} GeV^4. For one thing, summing the zero-point energies of all normal modes of some field of mass m up to a wave number cutoff $\Lambda\gg m$ yields a vacuum energy density (with $\hbar=c=1$)

$$\langle\rho\rangle=\int_0^{\Lambda}\frac{4\pi k^2dk}{(2\pi)^3}\frac{1}{2}\sqrt{k^2+m^2}\simeq\frac{\Lambda^4}{16\pi^2} . \tag{3.5}$$

If we believe general relativity up to the Planck scale, then we might take $\Lambda \simeq (8\pi G)^{-1/2}$, which would give

$$\langle \rho \rangle \approx 2^{-10}\pi^{-4}G^{-2} = 2\times 10^{71}\ \mathrm{GeV}^4 . \qquad (3.6)$$

But we saw that $|\langle \rho \rangle + \lambda/8\pi G|$ is less than about 10^{-47} GeV^4, so the two terms here must cancel to better than 118 decimal places. Even if we only worry about zero-point energies in quantum chromodynamics, we would expect $\langle \rho \rangle$ to be of order $\Lambda_{\mathrm{QCD}}^4/16\pi^2$, or 10^{-6} GeV^4, requiring $\lambda/8\pi G$ to cancel this term to about 41 decimal places.

Perhaps surprisingly, it was a long time before particle physicists began seriously to worry about this problem, despite the demonstration in the Casimir effect of the reality of zero-point energies.[3] Since the cosmological upper bound on $|\langle \rho \rangle + \lambda/8\pi G|$ was vastly less than any value expected from particle theory, most particle theorists simply assumed that for some unknown reason this quantity was zero. But cosmologists generally continued to keep an open mind, analyzing cosmological data in terms of models with a possibly nonvanishing cosmological constant.

In fact, as far as I know, the first published discussion of the contribution of quantum fluctuations to the effective cosmological constant was triggered by astronomical observations. In the late 1960s it seemed that an excessively large number of quasars were being observed with redshifts clustered about $z = 1.95$. Since $1+z$ is the ratio of the cosmic scale factor $R(t)$ at present to its value at the time the light now observed was emitted, this could be explained if the universe loitered for a while at a value of $R(t)$ equal to 1/2.95 times the present value. A number of authors [Petrosian, Salpeter, and Szekeres (1967); Shklovsky (1967); Rowan-Robinson (1968)] proposed that such a loitering could be accounted for in a model proposed by Lemaître (1927, 1931). In this model there is a positive cosmological constant λ_{eff} and positive curvature $k = +1$, just as in the static Einstein model, while the mass of the universe is taken close to the Einstein value (2.4). The scale factor $R(t)$ starts at $R = 0$ and then increases; however, when the mass density drops to near the Einstein value (2.2), the universe behaves for a while like a static Einstein universe, until the instability of this model takes over and the universe starts expanding again. In order for this idea to explain a preponderance of redshifts at $z = 1.95$, the vacuum energy density ρ_V would have to be $(2.95)^3$ times the present nonvacuum mass density ρ_0.

These considerations led Zeldovich (1967) to attempt to account for a nonzero vacuum energy density in terms of quantum fluctuations. As we have seen, the zero-point energies themselves gave far too large a value for $\langle \rho \rangle$, so Zeldovich assumed that these were canceled by $\lambda/8\pi G$, leaving only higher-order effects: in particular, the gravitational force between the particles in the vacuum fluctuations. (In Feynman diagram terms, this corresponds to throwing away the one-loop vacuum graphs, but keeping those with two loops.) Taking Λ^3 particles of energy Λ per unit volume gives the gravitational self-energy density of order

$$\langle \rho \rangle \approx (G\Lambda^2/\Lambda^{-1})\Lambda^3 = G\Lambda^6 . \qquad (3.7)$$

For no clear reason, Zeldovich took the cutoff Λ as 1 GeV, which yields a density $\langle \rho \rangle \approx 10^{-38}$ GeV^4, much smaller than from zero-point energies themselves, but still larger than the observational bound (3.4) on $|\langle \rho \rangle + \lambda/8\pi G|$ by some 9 orders of magnitude. Neither Zeldovich nor anyone else felt encouraged to pursue these ideas.

The real beginning of serious worry about the vacuum energy seems to date from the success of the idea of spontaneous symmetry breaking in the electroweak theory. In this theory, the scalar field potential takes the form (with $\mu^2 > 0$, $g > 0$)

$$V = V_0 - \mu^2\phi^\dagger\phi + g(\phi^\dagger\phi)^2 . \qquad (3.8)$$

At its minimum this takes the value

$$\langle \rho \rangle = V_{\min} = V_0 - \frac{\mu^4}{4g} . \qquad (3.9)$$

Apparently some theorists felt that V should vanish at $\phi = 0$, which would give $V_0 = 0$, so that $\langle \rho \rangle$ would be negative definite.[4] In the electroweak theory this would give $\langle \rho \rangle \simeq -g(300\ \mathrm{GeV})^4$, which even for g as small as α^2 would yield $|\langle \rho \rangle| \simeq 10^6$ GeV^4, larger than the bound on ρ_V by a factor 10^{53}. Of course we know of no reason why V_0 or λ must vanish, and it is entirely possible that V_0 or λ cancels the term $-\mu^4/4g$ (and higher-order corrections), but this example shows vividly how unnatural it is to get a reasonably small effective cosmological constant. Moreover, at early times the effective temperature-dependent potential has a positive coefficient for $\phi^\dagger\phi$, so the minimum then is at $\phi = 0$, where $V(\phi) = V_0$. Thus, in order to get a zero cosmological constant today, we have to put up with an enormous cosmological constant at times before the electroweak phase transition. [This is *not* in conflict with experiment; in fact, the phase transition occurs at a temperature T of order $\mu/\sqrt{g}$, so the black-body radiation present at that

[3]Casimir (1948) showed that quantum fluctuations in the space between two flat conducting plates with separation d would produce a force per unit area equal to $\hbar c\pi^2/240d^4$, or 1.30×10^{-18} dyn cm$^2/d^4$. This was measured by Sparnaay (1957), who found a force per area of $(1-4)\times 10^{-18}$ dyn cm$^2/d^4$, when d was varied between 2 and 10 μm.

[4]Veltman (1975) attributes this view to Linde (1974), himself (quoted as "to be published"), and Dreitlein (1974). However, Linde's paper does not seem to me to take this position. Dreitlein's paper proposed that Eq. (3.9) could give an acceptably small value of $\langle \rho \rangle$, with $\mu/\sqrt{g}$ fixed by the Fermi coupling constant of weak interactions, if μ is very small, of order 10^{-27} MeV. Veltman's paper gives experimental arguments against this possibility.

time has an energy density of order μ^4/g^2, larger than the vacuum energy by a factor $1/g$ (Bludman and Ruderman, 1977).] At even earlier times there were other transitions, implying an even larger early value for the effective cosmological constant. This is currently regarded as a good thing; the large early cosmological constant would drive cosmic inflation, solving several of the long-standing problems of cosmological theory (Guth, 1981; Albrecht and Steinhardt, 1982; Linde, 1982). We want to explain why the effective cosmological constant is small now, *not* why it was always small.

Before closing this section, I want to take up a peculiar aspect of the problem of the cosmological constant. The appearance of an effective cosmological constant makes it impossible to find any solutions of the Einstein field equations in which $g_{\mu\nu}$ is the constant Minkowski term $\eta_{\mu\nu}$. That is, the original symmetry of general covariance, which is always broken by the appearance of any given metric $g_{\mu\nu}$, cannot, without fine-tuning, be broken in such a way as to preserve the subgroup of space-time translations.

This situation is unusual. Usually if a theory is invariant under some group G, we would not expect to have to fine-tune the parameters of the theory in order to find vacuum solutions that preserve any given subgroup $H \subset G$. For instance, in the electroweak theory, there is a finite range of parameters in which any number of doublet scalars will get vacuum expectation values that preserve a U(1) subgroup of SU(2)$\times$U(1). So why will this not work for the translational subgroup of the group of general coordinate transformations? Suppose we look for a solution of the field equations that preserves translational invariance. With all fields constant, the field equations for matter and gravity are

$$\frac{\partial \mathcal{L}}{\partial \psi_i}=0 , \tag{3.10}$$

$$\frac{\partial \mathcal{L}}{\partial g_{\mu\nu}}=0 . \tag{3.11}$$

With N ψ's, these are $N+6$ equations for $N+6$ unknowns, so one might expect a solution without fine-tuning. The problem is that when (3.10) is satisfied, the dependence of $\mathcal{L}$ on $g_{\mu\nu}$ is too simple to allow a solution of (3.11). There is a GL(4) symmetry that survives as a vestige of general covariance even when we constrain the fields to be constants: under the GL(4) transformation

$$g_{\mu\nu}\rightarrow A^{\rho}{}_{\mu}A^{\sigma}{}_{\nu}g_{\rho\sigma} , \tag{3.12}$$

$$\psi_i \rightarrow D_{ij}(A)\psi_j ; \tag{3.13}$$

the Lagrangian transforms as a density,

$$\mathcal{L}\rightarrow \mathrm{Det}\, A \mathcal{L} . \tag{3.14}$$

When Eq. (3.10) is satisfied, this implies that $\mathcal{L}$ transforms as in (3.14) under (3.12) *alone*. This has the unique solution

$$\mathcal{L}=c(\mathrm{Det} g)^{1/2} , \tag{3.15}$$

with c independent of $g_{\mu\nu}$. With this $\mathcal{L}$, there are no solutions of Eq. (3.11), unless for some reason the coefficient c vanishes when (3.10) is satisfied.

Now that the problem has been posed, we turn to its possible solution. The next five sections will describe five directions that have been taken in trying to solve the problem of the cosmological constant.

IV. SUPERSYMMETRY, SUPERGRAVITY, SUPERSTRINGS

Shortly after the development of four-dimensional globally supersymmetric field theories, Zumino (1975) pointed out that supersymmetry in these theories would, if unbroken, imply a vanishing vacuum energy. The argument is very simple: the supersymmetry generators Q_α satisfy an anticommutation relation

$$\{Q_\alpha, Q_\beta^\dagger\}=(\sigma_\mu)_{\alpha\beta}P^\mu , \tag{4.1}$$

where α and β are two-component spin indices; σ_1, σ_2, and σ_3 are the Pauli matrices; $\sigma_0=1$; and P^μ is the energy-momentum 4-vector operator. If supersymmetry is unbroken, then the vacuum state $|0\rangle$ satisfies

$$Q_\alpha|0\rangle = Q_\alpha^\dagger|0\rangle=0 , \tag{4.2}$$

and from (4.1) and (4.2) we infer that the vacuum has vanishing energy and momentum

$$\langle 0|P^\mu|0\rangle=0 .$$

This result can also be obtained by considering the potential $V(\phi,\phi^*)$ for the chiral scalar fields ϕ^i of a globally supersymmetric theory:

$$V(\phi,\phi^*)=\sum_i \left|\frac{\partial W(\phi)}{\partial \phi^i}\right|^2 , \tag{4.3}$$

where $W(\phi)$ is the so-called superpotential. (Gauge degrees of freedom are ignored here, but they would not change the argument.) The condition for unbroken supersymmetry is that W be stationary in ϕ, which would imply that V take its minimum value,

$$\langle \rho\rangle = V_{\min}=0 . \tag{4.4}$$

Quantum effects do not change this conclusion, because with boson-fermion symmetry, the fermion loops cancel the boson ones.

The trouble with this result is that supersymmetry is broken in the real world, and in this case either (4.1) or (4.3) shows that the vacuum energy is positive-definite. *If* this vacuum energy were the sole contribution to the effective cosmological constant, then the effect of supersymmetry would be to convert the problem of the cosmological constant from a crisis into a disaster.

Fortunately this is not the whole story. It is not possible to decide the value of the effective cosmological constant unless we explicitly introduce gravitation into the theory. Any globally supersymmetric theory that involves gravity is inevitably a locally supersymmetric supergravity theory. In such a theory the effective cosmological constant *is* given by the expectation value of the potential, but the potential is now given by (Cremmer *et al.*, 1978, 1979; Barbieri *et al.*, 1982; Witten and Bagger, 1982)

$$V(\phi,\phi^*)=\exp(8\pi GK)[D_iW(\mathcal{G}^{-1})^i{}_j(D_jW)^* -24\pi G|W|^2]\ , \tag{4.5}$$

where $K(\phi,\phi^*)$ is a real function of both ϕ and ϕ^* known as the Kahler potential, D_iW is a sort of covariant derivative

$$D_iW\equiv\frac{\partial W}{\partial\phi^i}+8\pi G\frac{\partial K}{\partial\phi^i}\ , \tag{4.6}$$

and $(\mathcal{G}^{-1})^i{}_j$ is the inverse of a metric

$$\mathcal{G}^i{}_j\equiv\frac{\partial^2K}{\partial\phi^{i*}\partial\phi^j}\ . \tag{4.7}$$

The condition for unbroken supersymmetry is now $D_iW=0$. This again yields a stationary point of the potential, but now it is one at which V is generally negative. In fact, even if we fine-tuned W so that there were a supersymmetric stationary point at which $W=0$ and hence $V=0$, such a solution would not, in general, be the state of lowest energy, though it would be stable [Coleman and de Luccia (1980), Weinberg (1982)]. It should, however, be mentioned that if there is a set of field values at which $W=0$ and $D_iW=0$ for all i in lowest order of perturbation theory, then the theory has a supersymmetric equilibrium configuration with $V=0$ to all orders of perturbation theory, though not necessarily beyond perturbation theory (Grisaru, Siegel, and Rocek, 1979). The same is believed to be true in superstring perturbation theory (Dine and Seiberg, 1986; Friedan, Martinec, and Shenker, 1986; Martinec, 1986; Attick, Moore, and Sen, 1987; Morozov and Perelomov, 1987).

Without fine-tuning, we can generally find a nonsupersymmetric set of scalar field values at which $V=0$ and $D_iW\neq0$, but this would not normally be a stationary point of V. Thus in supergravity the problem of the cosmological constant is no more a disaster, but just as much a crisis, as in nonsupersymmetric theories.

On the other hand, supergravity theories offer opportunities for changing the context of the cosmological constant problem, if not yet for solving it. Cremmer *et al.* (1983) have noted that there is a class of Kahler potentials and superpotentials that, for a broad range of most parameters, automatically yield an equilibrium scalar field configuration in which $V=0$, even though supersymmetry is broken. Here is a somewhat generalized version: the Kahler potential is

$$K=-3\ln|T+T^*-h(C^a,C^{a*})|/8\pi G +\tilde{K}(S^n,S^{n*}) \tag{4.8}$$

while the superpotential is

$$W=W_1(C^a)+W_2(S^n)\ , \tag{4.9}$$

and T,C^a,S^n are all chiral scalar fields. No constraints are placed on the functions $h(C^a,C^{a*})$, $\tilde{K}(S^n,S^{n*})$, $W_1(C^a)$, or $W_2(S^n)$, except that h and $\tilde{K}$ are real, and functions all depend only on the fields indicated; in particular, the superpotential must be independent of the single chiral scalar T.

With these conditions the potential (4.5) takes the form

$$V=\exp(8\pi G\tilde{K})\left[\frac{1}{3(T+T^*+h)^3}\left[\frac{\partial W}{\partial C^a}\right](\mathcal{N}^{-1})^a{}_b \times\left[\frac{\partial W}{\partial C^b}\right]^*+(D_nW)(\mathcal{G}^{-1})^n{}_m(D_mW)^*\right]\ , \tag{4.10}$$

where $(\mathcal{N}^{-1})^a{}_b$ is the reciprocal of the matrix

$$\mathcal{N}^a{}_b=\frac{\partial^2h}{\partial C^{a*}\partial C^b}\ . \tag{4.11}$$

The matrices $\mathcal{N}^a{}_b$ and $\mathcal{G}^n{}_m$ are necessarily positive-definite, because of their role in the kinetic part of the scalar Lagrangian

$$\mathcal{L}_{\text{kin}}=-\mathcal{G}^i{}_j\frac{\partial\phi^{i*}}{\partial x^\mu}\frac{\partial\phi^j}{\partial x_\mu} =-\frac{3}{(T+T^*+h)^2}\left[\frac{\partial T}{\partial x^\mu}+\frac{\partial h}{\partial C^a}\frac{\partial C^a}{\partial x^\mu}\right]\left[\frac{\partial T}{\partial x_\mu}+\frac{\partial h}{\partial C^b}\frac{\partial C^b}{\partial x_\mu}\right]^*-\frac{3}{|T+T^*-h|}\mathcal{N}^a{}_b\frac{\partial C^{a*}}{\partial x^\mu}\frac{\partial C^b}{\partial x_\mu}-\mathcal{G}^n{}_m\frac{\partial S^{n*}}{\partial x^\mu}\frac{\partial S^m}{\partial x_\mu}\ . \tag{4.12}$$

Hence Eq. (4.10) is positive and therefore, without further fine-tuning, may be expected to have a stationary point with $V=0$, specified by the conditions

$$\frac{\partial W}{\partial C^a}=D_nW=0\ . \tag{4.13}$$

But this is not necessarily a supersymmetric configuration, because here

$$D_aW\equiv\frac{\partial W}{\partial C^a}+8\pi G\frac{\partial K}{\partial C^a}W =-\frac{3}{|T+T^*+h|}\frac{\partial h}{\partial C^a}W\ , \tag{4.14}$$

and this does not necessarily vanish. (However, to have supersymmetry broken, it is essential that the superpotential actually depend on all of the chiral scalars S^n, because otherwise the conditions $D_n W=0$ would require $W=0$ and hence $D_a W=0$.)

The superpotential W depends on C^a and S^n, but not on T, so the conditions (4.13) will generally fix the values of C^a and S^n at the minimum of V, while leaving T undetermined. The field T enters the potential only in the overall scale of the part that depends on the C^a, so such theories are called "no-scale" models. An intensive phenomenological study of these models was carried out at CERN for several years following 1983 (Ellis, Lahanas, *et al.*, 1984; Ellis, Kounnas, *et al.*, 1984; Barbieri *et al.*, 1985).

Of course, these models do not solve the cosmological constant problem, because neither Eq. (4.8) nor Eq. (4.9) is dictated by any known physical principle. In particular, in order to cancel the second term in Eq. (4.5), it is essential that the coefficient of the logarithm in the first term in (4.8) be given the apparently arbitrary value $-3/8\pi G$.

It was therefore exciting when, in some of the first work on the physical implications of superstring theory, it was found that compactification of six of the ten original dimensions yielded a four-dimensional supergravity theory with Kahler potential and superpotential of the form (4.8) and (4.9). Specifically, Witten (1985) found a Kahler potential of the form (4.8), with h quadratic in the C's and $\tilde{K}=-\ln(S+S^*)/8\pi G$, but with a superpotential that depended solely on the C's. By including nonperturbative gaugino condensation effects, Dine *et al.* (1985) were able to give the superpotential a dependence on S (though they did not treat the dependence of the Kahler potential or superpotential on the C^a fields). In this work, the S field is a complex function (now often called Y) of four-dimensional dilaton and axion fields, while the T field represents the scale of the compactified six-dimensional manifold. The factor 3 in Eq. (4.8) arises in these models because one compactifies on a complex manifold with $(10-4)/2=3$ complex dimensions (Chang *et al.*, 1988).

Intriguing as these results are, they have not been taken seriously (even by the original authors) as a solution of the cosmological constant problem. The trouble is that no one expects the simple structures (4.8) and (4.9) to survive beyond the lowest order of perturbation theory, because they are not protected by any symmetry that survives down to accessible energies.

Recently Moore (1987a, 1987b) has attempted a more specifically "stringy" attack on the problem. Early work by Rohm (1984) and Polchinski (1986) had shown that in the calculation of the vacuum energy density, the sum over zero-point energies can be converted into an integral over a complex "modular parameter" τ. (In string theories, two-dimensional conformal symmetry makes the tree-level vacuum energy vanish.) Last year Moore pointed out that for some special compactifications there is a discrete symmetry of modular space, known as Atkin-Lehner symmetry, that makes the integral over τ vanish despite the absence of space-time supersymmetry. So far, the only examples where this occurs entail a compactification to two rather than four space-time dimensions, but it does not seem unlikely that four-dimensional examples could be found. A more serious obstacle is that the Atkin-Lehner symmetry seems irretrievably tied to one-loop order.

Indeed, it is very hard to see how any property of supergravity or superstring theory could make the effective cosmological constant sufficiently small. It is not enough that the vacuum energy density cancel in lowest order, or to all finite orders of perturbative theory; even nonperturbative effects like ordinary QCD instantons would give far too large a contribution to the effective cosmological constant if not canceled by something else. According to our modern theories, properties of elementary particles, like approximate baryon and lepton conservation, are dictated by gauge symmetries of the standard model, which survive down to accessible energies. We know of no such symmetry (aside from the unrealistic example of unbroken supersymmetry) that could keep the effective cosmological constant sufficiently small. It is conceivable that in supergravity the property of having zero effective cosmological constant does survive to low energies without any symmetry to guard it, but this would run counter to all our experience in physics.

V. ANTHROPIC CONSIDERATIONS

I now turn to a very different approach to the cosmological constant, based on what Carter (1974) has called the anthropic principle.[5] Briefly stated, the anthropic principle has it that the world is the way it is, at least in part, because otherwise there would be no one to ask why it is the way it is. There are a number of different versions of this principle, ranging from those that are so weak as to be trivial to those that are so strong as to be absurd. Three of these versions seem worth distinguishing here.

(i) In one very weak version, the anthropic principle amounts simply to the use of the fact that we are here as one more experimental datum. For instance, recall M. Goldhaber's joke that "we know in our bones" that the lifetime of the proton must be greater than about 10^{16} yr, because otherwise we would not survive the ionizing particles produced by proton decay in our own bodies. No one can argue with this version, but it does not help us to explain anything, such as *why* the proton lives so long. Nor does it give very useful experimental information; certainly experimental physicists (including Goldhaber) have provided us with better limits on the proton lifetime.

[5]Recent discussions of the anthropic principle are given in the books by Davies (1982) and Barrow and Tipler (1986), and in articles by Carter (1983), Page (1987), and Rees (1987).

(ii) In one rather strong version, the anthropic principle states that the laws of nature, which are otherwise incomplete, are completed by the requirement that conditions must allow intelligent life to arise, the reason being that science (and quantum mechanics in particular) is meaningless without observers. I do not know how to reach a decision about such matters and will simply state my own view, that although science is clearly impossible without scientists, it is not clear that the universe is impossible without science.

(iii) A moderate version of the anthropic principle, sometimes known as the "weak anthropic principle," amounts to an explanation of which of the various possible eras or parts of the universe we inhabit, by calculating which eras or parts of the universe we *could* inhabit. An example is provided by what I think is the first use of anthropic arguments in modern physics, by Dicke (1961), in response to a problem posed by Dirac (1937). In effect, Dirac had noted that a combination of fundamental constants with the dimensions of a time turns out to be roughly of the order of the present age of the universe:

$$\hbar/Gcm_\pi^3=4.5\times10^{10}\text{ yr} . \qquad (5.1)$$

[There are various other ways of writing this relation, such as replacing m_π with various combinations of particle masses and introducing powers of $e^2/\hbar c$. Dirac's original "large-number" coincidence is equivalent to using Eq. (5.1) as a formula for the age of the universe, with m_π replaced by $(137m_pm_e^2)^{1/3}=183$ MeV. In fact, there are so many different possibilities that one may doubt whether there is any coincidence that needs explaining.] Dirac reasoned that if this connection were a real one, then, since the age of the universe increases (linearly) with time, some of the constants on the left side of (5.1) must change with time. He guessed that it is G that changes, like $1/t$. [Zee (1985) has applied similar arguments to the cosmological constant itself.] In response to Dirac, Dicke pointed out that the question of the age of the universe could only arise when the conditions are right for the existence of life. Specifically, the universe must be old enough so that some stars will have completed their time on the main sequence to produce the heavy elements necessary for life, and it must be young enough so that some stars would still be providing energy through nuclear reactions. Both the upper and lower bounds on the ages of the universe at which life can exist turn out to be roughly (very roughly) given by just the quantity (5.1). Hence there is no need to suppose that any of the fundamental constants vary with time to account for the rough agreement of the quantity (5.1) with the present age of the universe.

It is this "weak anthropic principle" that will be applied here. Its relevance arises from the fact that, in some modern cosmological models, the universe does have parts or eras in which the effective cosmological constant takes a wide variety of values. Here are some examples.

(1) The vacuum energy may depend on a scalar field vacuum expectation value that changes slowly as the universe expands, as in a model of Banks (1985).

(2) In a model of Linde (1986, 1987, 1988b), fluctuations in scalar fields produce exponentially expanding regions of the universe, within which further fluctuations produce further subuniverses, and so on. Since these subuniverses arise from fluctuations in the fields, they have differing values of various "constants" of nature.

(3) The universe may go through a very large number of first-order phase transitions in which bubbles of smaller vacuum energy form; within these bubbles there form further bubbles of even smaller vacuum energy, and so on. This can happen if the potential for some scalar field has a large number of small bumps, as in a model of Abbott (1985). Alternatively, the bubble walls may be elementary membranes coupled to a 3-form gauge potential $A_{\mu\nu\lambda}$, as in the work of Brown and Teitelboim (1987a, 1987b).

(4) The universe may start in a quantum state in which the cosmological constant does not have a precise value. Any "measurement" of the properties of the universe yields a variety of possible values for the cosmological constant, with *a priori* probabilities determined by the initial state (Hawking, 1987a). We will see examples of this in Secs. VII and VIII.

In models of these types, it is perfectly sensible to apply anthropic considerations to decide which era or part of the universe we could inhabit, and hence which values of the cosmological constant we could observe.

A large cosmological constant would interfere with the appearance of life in different ways, depending on the sign of λ_{eff}. For a large *positive* λ_{eff}, the universe very early enters an exponentially expanding de Sitter phase, which then lasts forever. The exponential expansion interferes with the formation of gravitational condensations, but once a clump of matter becomes gravitationally bound, its subsequent evolution is unaffected by the cosmological constant. Now, we do not know what weird forms life may take, but it is hard to imagine that it could develop at all without gravitational condensations out of an initially smooth universe. Therefore the anthropic principle makes a rather crisp prediction: λ_{eff} must be small enough to allow the formation of sufficiently large gravitational condensations (Weinberg, 1987).

This has been worked out quantitatively, but we can easily understand the main result without detailed calculations. We know that in our universe gravitational condensation had already begun at a redshift $z_c\geq4$. At this time, the energy density was greater than the present mass density ρ_{M_0} by a factor $(1+z_c)^3\geq125$. A cosmological constant has little effect as long as the nonvacuum energy density is larger than ρ_V, so one can conclude that a vacuum energy density ρ_V no larger than, say $100\rho_{M_0}$ would not be large enough to prevent gravitational condensations. [The quantitative analysis of Weinberg

(1987) shows that for $k=0$, a vacuum energy density no greater than $\pi^2(1+z_c)^3\rho_{M_0}/3$ would not prevent gravitational condensation at a redshift z_c; this is $410\rho_{M_0}$ for $z_c=4$.]

This result suggests strongly that if it is the anthropic principle that accounts for the smallness of the cosmological constant, then we would expect a vacuum energy density $\rho_V\sim(10-100)\rho_{M_0}$, because there is no anthropic reason for it to be any smaller.

Is such a large vacuum energy density observationally allowed? There are a number of different types of astronomical data that indicate differing answers to this question.

A. Mass density

If, as often assumed, the universe now has negligible spatial curvature, then

$$\Omega_V+\Omega_{M_0}=1\ , \tag{5.2}$$

where Ω_V and Ω_{M_0} are the ratios of the vacuum energy density and the present mass density to the critical density

$$\Omega_V\equiv\frac{8\pi G\rho_V}{3H_0^2}=\frac{\lambda_{\text{eff}}}{3H_0^2}\ ,\qquad \Omega_{M_0}=\frac{8\pi G\rho_{M_0}}{3H_0^2}\ . \tag{5.3}$$

The dynamics of clusters of galaxies seems to indicate that Ω_{M_0} is in the range 0.1–0.2 (Knapp and Kormendy, 1987), which with these assumptions would indicate a value for ρ_V/ρ_{M_0} in the range 4–9. If we discount the evidence from the dynamics of clusters of galaxies, then Ω_{M_0} could be as small as 0.02 (Knapp and Kormendy, 1987), corresponding to a value of $\rho_V/\rho_{M_0}\simeq 50$. [See also Bahcall *et al.* (1987).]

B. Ages

In a dust-dominated universe with $k=0$ and $\rho_V=0$, the age of the universe is $t_0=2/3H_0$. For $H_0=100$ km/sec Mpc, this is 7×10^9 yr, considerably less than the ages usually estimated for globular clusters (Renzini, 1986). On the other hand, for a dust-dominated universe with $k=0$ and $\rho_V\neq0$, the present age of an object that formed at a redshift z_c is

$$t_0(z_c)=\frac{2}{3}\left[1+\frac{\rho_{M_0}}{\rho_V}\right]^{1/2}H_0^{-1}\left\{\sinh^{-1}\left[\frac{\rho_V}{\rho_{M_0}}\right]^{1/2}-\sinh^{-1}\left[\left[\frac{\rho_V}{\rho_{M_0}}\right]^{1/2}(1+z_c)^{-3/2}\right]\right\}. \tag{5.4}$$

For instance, for $z_c=4$ and $\rho_V/\rho_{M_0}=9$ (i.e., $\Omega_{M_0}=0.1$), this gives an age $1.1H_0^{-1}$ in place of $\frac{2}{3}H_0^{-1}$. This is not in conflict with globular cluster ages even for Hubble constants near 100 km/sec Mpc.

These considerations of cosmic age and density have led a number of astronomers to suggest a fairly large positive cosmological constant, with $\rho_V>\rho_{M_0}$ [de Vaucouleurs (1982, 1983); Peebles (1984, 1987a, 1987b); Turner, Steigman, and Krauss (1984)]. However, there recently has appeared a strong argument against this view, which we shall now consider.

C. Number counts

Loh and Spillar (1986) have carried out a survey of numbers of galaxies as a function of redshift, subsequently analyzed by Loh (1986). For a uniformly distributed class of objects that are all bright enough to be detectable at redshifts $\le z_{\max}$, the number of objects observed at redshift less than $z\le z_{\max}$ in a dust-dominated universe with $k=0$ is

$$N(<z)\propto\int_{(1+z)^{-3/2}}^{1}ds\,s^{4/3}(1+\rho_V s^2/\rho_{M_0})^{-1/2}\left[\int_0^1 ds' s'^{-2/3}(1+\rho_V s'^2/\rho_{M_0})^{-1/2}\right]^2 . \tag{5.5}$$

Of course, in the real world there are always some objects too dim to be seen. Loh's analysis allowed for an unknown luminosity distribution, assuming only that its shape does not evolve with time. Under these assumptions, he found that the vacuum energy must be quite small: specifically,

$$\rho_V/\rho_{M_0}=0.1^{-0.4}_{+0.2}\ .$$

This is more than 3 orders of magnitude below the anthropic upper bound discussed earlier. If the effective cosmological constant is really this small, then we would have to conclude that the anthropic principle does *not* explain why it is so small. [However, there are reasons to be cautious in reaching this conclusion. Bahcall and Tremaine (1988) have recently reanalyzed the data of Loh and Spillar, using a plausible model of galaxy evolution in which the shape of the luminosity distribution does change with time. They considered only the case $\rho_V=0$, leaving Ω_{M_0} undetermined, and found that evolution in this model could increase or decrease the inferred value of Ω_{M_0} by as much as unity. Presumably it would also have a similarly large effect on the inferred value of ρ_V/ρ_{M_0} when $\Omega_{M_0}+\Omega_V$ is constrained to be unity. In addition, the redshifts of Loh and Spillar are photometric and therefore less certain than those obtained from shifts of individual spectral lines.]

Now let us consider a cosmological constant of the other sign, $\lambda_{\rm eff}<0$. Here the cosmological constant does not interfere with the formation of gravitational condensations. Instead (for $k=0$ or $k=+1$), the whole universe collapses to a singularity in a finite time T. The anthropic constraint here is simply that the universe last long enough for the appearance of life (Barrow and Tipler, 1986), say, $T\gtrsim 0.5H_0^{-1}$, where H_0^{-1} is the Hubble time in *our* universe. For a dust-dominated universe with $k=0$, we have

$$T=\pi(8\pi G|\rho_V|)^{1/2}\ ,\qquad H_0^{-1}=(8\pi G\rho_{M_0}/3)^{1/2}\ , \tag{5.6}$$

so the anthropic constraint here is just

$$|\rho_V|\lesssim\rho_{M_0}\ . \tag{5.7}$$

In this case the anthropic principle can explain why the cosmological constant is as small as found by Loh (1986), but not much smaller. On the other hand, a negative cosmological constant would not help with the cosmic mass and age problems.

Before closing this section, let me take up one possibility that may confront us in a few years. Suppose it really is confirmed that, as suggested by cosmic ages and densities, there is a cosmological constant with ρ_V of order ρ_{M_0}. Would we then have any alternative to an anthropic explanation for this value of ρ_V? The mass density ρ_M changes with time, so without anthropic considerations it is very hard to explain why a constant ρ_V should equal the value that ρ_M happens to have at present. But perhaps ρ_V really is not constant. For instance, Peebles and Ratra (1988) and Ratra and Peebles (1988) have considered a model in which the vacuum energy depends on a scalar field that changes as the universe expands. In order to qualify as a vacuum energy, it is only necessary for ρ_V to be accompanied with a pressure $p_V=-\rho_V$; the value of ρ_V can change if the vacuum exchanges energy with matter and radiation. The conservation of energy then relates the change of ρ_V to the change in the densities of matter (with $p_M=0$) and radiation (with $p_R=\rho_R/3$):

$$\frac{d}{dt}\rho_V+R^{-3}\frac{d}{dt}(R^3\rho_M)+R^{-4}\frac{d}{dt}(R^4\rho_R)=0\ . \tag{5.8}$$

Freese *et al.* (1987) have considered the possibility that energy is exchanged only between the vacuum and matter, or the vacuum and radiation, in such a way that either ρ_V/ρ_M or ρ_V/ρ_R remain constant, respectively (see also Reuter and Wetterich, 1987). In order for the vacuum to transfer energy to ordinary matter in such a way that ρ_V/ρ_M remains fixed, and if baryon number is conserved, then it would be necessary to create baryon-antibaryon pairs at a sufficient rate to produce a troublesome γ-ray background. Alternatively, if the vacuum transfers energy to radiation in such a way that ρ_V/ρ_R remains constant, and if ρ_V is comparable with the present mass density ρ_{M_0}, then ρ_V/ρ_R must be rather large, completely changing the results of cosmological nucleosynthesis.

One more possibility that was not considered by Freese *et al.* is that the vacuum transfers energy to radiation, avoiding the problems of baryon-antibaryon annihilation, but in such a way as to keep a fixed ratio ρ_V/ρ_M rather than ρ_V/ρ_R. However, this also does not work. With $\rho_V=c\rho_M$ and $R^3\rho_M$ constant, Eq. (5.8) yields

$$\rho_R=\rho_{R_0}\left[\frac{R_0}{R}\right]^4+3c\rho_{M_0}\left[\left[\frac{R_0}{R}\right]^3-\left[\frac{R_0}{R}\right]^4\right]$$
$$\underset{R\ll R_0}{\longrightarrow}(\rho_{R_0}-3c\rho_{M_0})\left[\frac{R_0}{R}\right]^4\ . \tag{5.9}$$

So that there is no interference with calculations of cosmological nucleosynthesis, we need

$$\rho_R\approx\rho_{R_0}\left[\frac{R_0}{R}\right]^4,$$

and therefore

$$|\rho_V|/\rho_M\equiv|c|\lesssim\frac{\rho_{R_0}}{3\rho_{M_0}}\ll 1\ . \tag{5.10}$$

Thus, even if we are willing to suppose that the vacuum energy changes with time, a vacuum energy density comparable with the present mass density seems very difficult to explain on other than anthropic grounds.

VI. ADJUSTMENT MECHANISMS

I now turn to an idea that has been tried by virtually everyone who has worried about the cosmological constant [see, e.g., Dolgov (1982); Wilczek and Zee (1983); Wilczek (1984, 1985); Peccei, Solá, and Wetterich (1987); Barr and Hochberg (1988)]. Suppose there is some scalar ϕ whose source is proportional to the trace of the energy-momentum tensor

$$\Box^2\phi\propto T^\mu{}_\mu\propto R\ . \tag{6.1}$$

(Here $T^{\mu\nu}$ is the total energy-momentum tensor that includes a possible cosmological constant term $-\lambda g^{\mu\nu}/8\pi G$.) Suppose also that $T^\mu{}_\mu$ depends on ϕ and vanishes at some field value ϕ_0. Then ϕ will evolve until it reaches an equilibrium value ϕ_0, where $T^\mu{}_\mu=0$, and the

Einstein field equations have a flat-space solution.

Of course, we do not observe such a scalar field, but for these purposes it can couple as weakly as we like; a weak coupling simply implies that the equilibrium value ϕ_0 is very large. In this respect the scalar ϕ is analogous to the axion, especially in its later "invisible" version [Kim (1979); Dine, Fischler, and Srednicki (1981)].

Even very weakly coupled, it is possible that the ϕ field could have interesting effects, because it must have very small mass. If it has any nonzero mass M_ϕ, then at energies below m_ϕ we can work with an effective Lagrangian in which ϕ has been "integrated out," and so does not appear explicitly. But massless fields like the gravitational and electromagnetic field will still appear in this effective Lagrangian, and their vacuum fluctuations will contribute to the effective cosmological constant. In order to keep $\rho_V < 10^{-48}$ GeV^4, we need the scalar field adjustment to cancel the effect of gravitational and electromagnetic field fluctuations down to frequencies 10^{-12} GeV; for this purpose we must have $m_\phi < 10^{-12}$ GeV. A field this light will have a macroscopic range: $\hbar/m_\phi c \gtrsim 0.01$ cm.

Unfortunately it seems to be impossible to construct a theory with one or more scalar fields having the assumed properties. This can be seen in very general terms. What we want is to find an equilibrium solution of the field equations in which $g_{\mu\nu}$ and all matter fields ψ_n (perhaps tensors as well as scalars) are constant in space-time. For such constant fields the Euler-Lagrange equations are simply

$$\frac{\partial \mathcal{L}}{\partial g_{\mu\nu}} = 0 , \tag{6.2}$$

$$\frac{\partial \mathcal{L}}{\partial \psi_n} = 0 . \tag{6.3}$$

As we saw in Sec. III, the problem is in satisfying the trace of the gravitational field equation. To make a solution natural, we would like this trace to be a linear combination of the ψ_n field equations; that is, we want

$$g_{\lambda\nu}\frac{\partial \mathcal{L}(g,\psi)}{\partial g_{\lambda\nu}} = \sum_n \frac{\partial \mathcal{L}(g,\psi)}{\partial \psi_n} f_n(\psi) \tag{6.4}$$

for all constant $g_{\mu\nu}$ and ψ_n. This can be restated as a symmetry condition: for constant fields the Lagrangian must be invariant under the transformation

$$\delta g_{\lambda\nu} = 2\varepsilon g_{\lambda\nu}, \quad \delta\psi_n = -\varepsilon f_n(\psi) . \tag{6.5}$$

With this condition, if we find a solution $\psi^{(0)}$ of the Euler-Lagrange equations for ψ_n,

$$\frac{\partial \mathcal{L}}{\partial \psi_n} = 0 \quad \text{at } \psi_n = \psi_n^{(0)} , \tag{6.6}$$

then the trace of the field equation for $g_{\mu\nu}$ is automatically satisfied.

The problem is that under these assumptions, it is impossible (without fine-tuning $\mathcal{L}$) to find a solution to the field equations (6.3) for the ψ_n. To see this, we replace the N fields ψ_n with $N-1$ fields σ_a (not necessarily scalars) and one scalar ϕ, in such a way that the symmetry transformation (6.5) takes the form

$$\delta g_{\lambda\nu} = 2\varepsilon g_{\lambda\nu}, \quad \delta\sigma_a = 0, \quad \delta\phi = -\varepsilon . \tag{6.7}$$

[To do this, we first define a "transverse" surface S in field space by an equation $T(\psi)=0$, where $T(\psi)$ is any function on which $\sum_n (\partial T/\partial\psi_n) f_n(\psi)$ does not vanish. We take σ_a as any set of coordinates on this $(N-1)$-dimensional surface, and define $\psi_n(\sigma;\phi)$ as the solution of the ordinary differential equation $d\psi_n/d\phi = f_n(\psi)$ subject to the condition that at $\phi=0$, ψ_n is at the point on S with coordinates σ. The condition that S be a transverse surface ensures that, at least within a finite region of field space, any point ψ_n is on just one of these trajectories.] This symmetry simply ensures that for constant fields the Lagrangian can depend on $g_{\lambda\nu}$ and ϕ only in the combination $e^{2\phi} g_{\lambda\nu}$. The general arguments of Sec. III then show that when the field equations for σ are satisfied, the Lagrangian must take the form

$$\mathcal{L} = e^{4\phi} (\mathrm{Det}\, g)^{1/2} \mathcal{L}_0(\sigma) . \tag{6.8}$$

We see that the source of ϕ is the trace of the energy-momentum tensor

$$\frac{\partial \mathcal{L}}{\partial \phi} = T^\mu{}_\mu (\mathrm{Det}\, g)^{1/2} , \tag{6.9}$$

$$T^{\mu\nu} = g^{\mu\nu} e^{4\phi} \mathcal{L}_0(\sigma) . \tag{6.10}$$

It is true that if there were a value of ϕ where $\mathcal{L}$ is stationary in ϕ, then the trace of the Einstein field equations would automatically be satisfied at this point, but clearly there is no such stationary field value (unless, of course, we fine-tune $\mathcal{L}_0$ so that it vanishes at its stationary point). To put this another way, since $\mathcal{L}$ depends only on ϕ and $g_{\mu\nu}$ only in the combination $\hat{g}_{\mu\nu} \equiv e^{2\phi} g_{\mu\nu}$ (and derivatives of ϕ and $g_{\mu\nu}$), we might as well redefine the metric as $\hat{g}_{\mu\nu}$ instead of $g_{\mu\nu}$. Then ϕ is just a scalar with only derivative couplings and clearly cannot help with our problem.[6]

As one example of many failed attempts along this line, let us consider a proposal of Peccei, Solà, and Wetterich (1987). They observed that the symmetry (6.5) or (6.7) may be broken by conformal anomalies, such as those that produce the β function of quantum chromodynamics, in such a way that the effective Lagrangian becomes[7]

$$\mathcal{L}_{\mathrm{eff}} = (\mathrm{Det}\, g)^{1/2} [e^{4\phi} \mathcal{L}_0(\sigma) + \phi \Theta^\mu{}_\mu] , \tag{6.11}$$

where $\Theta^\mu{}_\mu$ represents the effect of the conformal anoma-

[6] This remark is due to Polchinski (1987).

[7] An equation essentially equivalent to (6.11) appeared in the preprint version of the paper by Peccei, Solá, and Wetterich (1987). In the published version this equation was removed, and it was acknowledged that fine-tuning is still needed to make the cosmological constant vanish. However, this equation was quoted in the meantime in a paper by Ellis, Tsamis, and Voloshin (1987), which mostly deals with the observable consequences of the light scalar particle in this model.

ly. The source of the ϕ field is now

$$\frac{\partial\mathcal{L}}{\partial\phi}=(T^{\mu}{}_{\mu}+\Theta^{\mu}{}_{\mu})(\mathrm{Det}g)^{1/2}\ , \qquad (6.12)$$

with $T^{\mu\nu}$ the previous energy-momentum tensor (6.10). Now we can find an equilibrium solution for the ϕ field, at a value ϕ_0 such that

$$4e^{4\phi_0}\mathcal{L}_0+\Theta^{\mu}{}_{\mu}=0\ . \qquad (6.13)$$

The trouble is that this is *not* the condition for a flat-space solution; the Einstein equation for a constant metric is

$$0=\frac{\partial\mathcal{L}_{\text{eff}}}{\partial g_{\mu\nu}}\propto e^{4\phi}\mathcal{L}_0+\phi\Theta^{\mu}{}_{\mu}\ , \qquad (6.14)$$

which is not the same as (6.13). The point is that just calling the anomalous term in (6.11) $\Theta^{\mu}{}_{\mu}$ does not make it a term in the trace of the energy-momentum tensor to which $g_{\mu\nu}$ is coupled. This result is not surprising, since (6.11) does not obey the symmetry (6.7). One cannot have it both ways: either we preserve the symmetry, in which case there is no equilibrium solution for ϕ, or we break the symmetry, in which case such an equilibrium solution does not imply a solution of the field equations for a constant metric. (Also see Coughlan *et al.*, 1988; Wetterich, 1988.)

In a slightly different version of this general class of models, we can try coupling a scalar field so that it is the curvature scalar R rather than the trace of the energy-momentum tensor that directly serves as the source of the scalar field. [See, e.g., Dolgov (1982); Barr (1987); Ford (1987).] For instance, we might take the Lagrangian as

$$\mathcal{L}=\sqrt{g}\left[-\tfrac{1}{2}\partial_{\mu}\phi\partial^{\mu}\phi-\frac{\lambda}{8\pi G}-\frac{1}{16\pi G}R-U(\phi)R\right]. \qquad (6.15)$$

This has a flat-space solution with $g_{\mu\nu}=\eta_{\mu\nu}$ and $\phi=\phi_0$ (a constant), provided

$$U(\phi_0)=\infty\ . \qquad (6.16)$$

However, as the above authors observed, the effective gravitational coupling in this theory is given by

$$G_{\text{eff}}=\frac{G}{1+16\pi GU(\phi_0)}=0\ . \qquad (6.17)$$

This is not much progress; we always knew that a nonzero vacuum energy does not prevent a flat-space solution if the gravitational constant is zero.

The "no-go" theorem proved in this section should not be regarded as closing off all hope in this direction. No-go theorems have a way of relying on apparently technical assumptions[8] that later turn out to have exceptions of great physical interest. (A famous example is the Coleman-Mandula theorem.) More discouraging than any theorem is the fact that many theorists have tried to invent adjustment mechanisms to cancel the cosmological constant, but without any success so far.

VII. CHANGING GRAVITY

A number of authors have suggested changing the rules of classical general relativity in such a way that the cosmological constant appears as a constant of integration, unrelated to any parameters in the action [Van der Bij *et al.* (1982); Weinberg (1983); Wilczek and Zee (1983); Buchmüller and Dragon (1988a, 1988b)]. This does not solve the cosmological constant problem, but it does change it in a suggestive way.

I will describe one version of this idea, in which one maintains general covariance, but reinterprets the formalism so that the determinant of the metric is not a dynamical field. *Any* theory can be written in a way that is formally generally covariant, so by the usual arguments we can take the action for gravity and matter as

$$I[\psi,g]=\frac{-1}{16\pi G}\int d^4x\sqrt{g}R+I_M[\psi,g]\ , \qquad (7.1)$$

where ψ are a set of matter fields appearing in the matter action I_M. (I_M includes a possible cosmological constant term $-\lambda\int\sqrt{g}\,d^4x/8\pi G$.) The variational derivative of Eq. (7.1) with respect to the metric is

$$\frac{\delta I}{\delta g_{\mu\nu}}=\frac{1}{8\pi G}(R^{\mu\nu}-\tfrac{1}{2}g^{\mu\nu}R)+T^{\mu\nu}\ , \qquad (7.2)$$

where, as usual, $T^{\mu\nu}$ is the variational derivative of I_M with respect to $g_{\mu\nu}$. In ordinary general relativity all components of the metric are dynamical fields, so Eq. (7.2) vanishes for all μ,ν, yielding the usual Einstein field equations. However, just because we use a generally covariant formalism does not mean that we are committed to treating all components of the metric as dynamical fields. For instance, we all learn in childhood how to write the equations of Newtonian mechanics in general curvilinear spatial coordinate systems, without supposing that the 3-metric has to obey any field equations at all.

In particular, if the determinant g is not dynamical, then the action only has to be stationary with respect to variations in the metric that keep the determinant fixed,

[8]For instance, we assumed that in the solution for flat space all fields are constant, but it might be that this solution preserves only some combination of translation and gauge invariance, in which case some gauge-noninvariant fields might vary with space-time position. (This is the case for the 3-form gauge field model discussed at the end of Sec. VII and in Sec. VIII.) Furthermore, it is possible that the foliation of field space, which allows us to replace the ψ_n with σ_a and ϕ, does not work throughout the whole of field space.

i.e., for which $g^{\mu\nu}\delta g_{\mu\nu}=0$; hence only the traceless part of (7.2) needs to vanish, yielding the field equation

$$R^{\mu\nu}-\tfrac{1}{4}g^{\mu\nu}R=-8\pi G(T^{\mu\nu}-\tfrac{1}{4}g^{\mu\nu}T^{\lambda}{}_{\lambda}) . \quad (7.3)$$

This is just the traceless part of the Einstein field equations; these equations evidently contain less information than Einstein's, but as we shall see, not much less. Because the whole formalism is generally covariant, the energy-momentum tensor satisfies the usual conservation law

$$T^{\mu\nu}{}_{;\mu}=0 , \quad (7.4)$$

and of course the Bianchi identities still hold,

$$(R^{\mu\nu}-\tfrac{1}{2}g^{\mu\nu}R)_{;\mu}=0 . \quad (7.5)$$

The full Einstein field equations are automatically consistent with (7.4) and (7.5), but for the traceless part we get a nontrivial consistency condition. Taking the covariant derivative of Eq. (7.3) with respect to x^{μ} yields

$$\tfrac{1}{4}\partial_{\mu}R=8\pi G\tfrac{1}{4}\partial_{\mu}T^{\lambda}{}_{\lambda} ,$$

or, in other words, $R-8\pi GT^{\lambda}{}_{\lambda}$ is a constant, which we will call -4Λ:

$$R-8\pi GT^{\lambda}{}_{\lambda}=-4\Lambda \quad (\text{constant}) . \quad (7.6)$$

From (7.3) and (7.6), we obtain

$$R^{\mu\nu}-\tfrac{1}{2}g^{\mu\nu}R-\Lambda g^{\mu\nu}=-8\pi GT^{\mu\nu} . \quad (7.7)$$

Thus we recover the Einstein field equations, but with a cosmological constant that has nothing to do with any terms in the action or vacuum fluctuations, arising, instead, as a mere integration constant. To put this another way, Eq. (7.3) does not involve a cosmological constant; the contribution of vacuum fluctuations automatically cancel on the right-hand side of Eq. (7.3), so this equation does have flat-space solutions in the absence of matter and radiation. The remaining problem in this formulation is: why should we choose the flat-space solutions?

Before proceeding with this theory, I should pause to mention that it is closely related[9] to a proposal made long ago by Einstein (1919). After his formulation of general relativity and its application to cosmology, Einstein turned to the old problem of a field theory of matter. In a paper titled "Do Gravitational Fields Play an Essential Part in the Structure of the Elementary Particles of Matter?" he proposed to replace the original gravitational field equation with the equation

$$R_{\mu\nu}-\tfrac{1}{4}g_{\mu\nu}R=-8\pi Gt_{\mu\nu} . \quad (7.8)$$

This is consistent only if $t_{\mu\nu}$ is traceless; however, Einstein took for $t_{\mu\nu}$ not the full energy-momentum tensor of matter and radiation, but just the traceless tensor of radiation alone. This is, of course, conserved only outside matter. In such regions there is no difference between Eqs. (7.8) and (7.3), so by the same calculation as shown here, Einstein was able to recover Eq. (7.7), with Λ a constant of integration. However, inside matter, Eq. (7.8) is different from (7.3), the difference being that the right-hand side of Eq. (7.3) includes the traceless part of the energy-momentum tensor of matter. A consequence of this difference is that in charged matter R is an undetermined function, except that it is constant along world lines.

I will also take the opportunity of this pause to comment on the connection between the formulation described here and that of Zee (1985) and Buchmüller and Dragon (1988a, 1988b). These authors take as their starting point the assumption that the action is invariant not under the group of all coordinate transformations, but only under the subgroup of transformations $x^{\mu}\to x'^{\mu}$ with $\mathrm{Det}(\partial x'^{\mu}/\partial x^{\nu})=1$. This is not really in conflict with the formulation presented here; the general covariance of Eq. (7.1) is achieved at the cost of introducing a metric that is partly nondynamical (just as we can make Newtonian mechanics formally Lorentz invariant by introducing a nondynamical quantity, the velocity of the reference frame). However, in giving up general covariance, one may be led to a theory with unnecessary elements. Under transformations with $\mathrm{Det}(\partial x'/\partial x)=1$, the determinant of the metric g behaves just like any scalar field, so one can introduce arbitrary functions of g here and there in the action. There is nothing wrong with this, but it is not necessary, no different from inserting a new scalar field into the theory.

Now let us return to the theory described by the field equations (7.3). In my view, the key question in deciding whether this is a plausible classical theory of gravitation is whether it can be obtained as the classical limit of any physically satisfactory quantum theory of gravitation. To help in answering this, and also to illuminate the points raised in the previous paragraph, let us look at a simple model (Teitelboim, 1982) that shares several features with the theory of gravitation studied here.

Consider a free relativistic particle, with space-time trajectory $x^{\mu}(s)$ parametrized by a variable s. In order for the action to be invariant under arbitrary reparametrizations $s\to s'(s)$, we must introduce an "einbein" $g(s)$, with transformation rule

$$g(s)\to g'(s')=g(s)\left[\frac{ds'}{ds}\right]^{-1} . \quad (7.9)$$

The action may then be taken as

$$I[x,g]=\tfrac{1}{2}\int ds\, g^{-1}(s)\frac{dx^{\mu}(s)}{ds}\frac{dx_{\mu}(s)}{ds} -\frac{m^2}{2}\int ds\, g(s) . \quad (7.10)$$

[9]This was pointed out to me by someone in the audience of the lectures at Harvard. I thank my informant for this interesting historical reference.

The conditions that I be stationary with respect to variations in $x^\mu(s)$ and $g(s)$ are, respectively,

$$\frac{dp^\mu}{ds}=0 , \tag{7.11}$$

$$p^\mu p_\mu = -m^2 , \tag{7.12}$$

where p_μ is the canonical conjugate to x^μ:

$$p_\mu(s)=g^{-1}(s)\frac{dx_\mu(s)}{ds} . \tag{7.13}$$

However, just because we choose to write the action in a reparametrization-invariant way does not necessarily mean that we must treat the einbein $g(s)$ as a dynamical quantity. If we treat $x^\mu(s)$, but not $g(s)$, as dynamical variables, then we obtain Eq. (7.11), but not (7.12). Of course, Eq. (7.11) implies that $p_\mu p^\mu$ is a constant [just as Eq. (7.3) implies that $R-8\pi G T^\lambda{}_\lambda$ is constant]. If we like, we can call this constant $-m^2$, but this is now a mere integration constant, unrelated to anything in the original action.

Now to quantization. The Hamiltonian here is

$$H(s)=p_\mu\frac{dx^\mu}{ds}-L=\tfrac{1}{2}g(p^\mu p_\mu+m^2) , \tag{7.14}$$

so in quantum mechanics we calculate amplitudes by the functional integral

$$A=\int[dx][dp][dg]$$

$$\times\exp\left[i\int ds\left[p_\mu(s)\frac{dx^\mu(s)}{ds}-H(s)\right]\right] . \tag{7.15}$$

The einbein $g(s)$ has no canonical conjugate, and so appears here only as a Lagrange multiplier, whose integral yields a factor

$$\prod_s \delta(p^\mu p_\mu+m^2) . \tag{7.16}$$

Presumably the classical theory in which g is not dynamical would be obtained as the classical limit of a quantum theory in which we do not do a functional integral over $g(s)$, and hence do not get the factor (7.16). But then there would be nothing to keep p^μ timelike. This is such a trivial theory that it is hard to say that anything goes wrong physically; but we may anticipate that in less trivial theories, we need a field to serve as a Lagrange multiplier for every negative norm degree of freedom like p^0. This is the case, for instance, in string theories, where the integration over the world-sheet metric is needed to enforce the Virasoro conditions on physical states.

The quantum theory of gravitation can be put in similar terms. Using the Arnowitt-Deser-Misner (1962) formalism, we calculate amplitudes as functional integrals,

$$Z=\int[dh_{ij}][d\pi^{ij}][d\tilde{N}][dN^i]$$

$$\times\exp\left[i\int\left[\pi^{ij}\frac{\partial h_{ij}}{\partial t}-(\tilde{\mathcal{H}}-2\lambda)\tilde{N}-\mathcal{H}_iN^i\right]d^4x\right] . \tag{7.17}$$

Here h_{ij}, $\tilde{N}$, and N^i parametrize the 4-metric, with line element given by

$$d\tau^2=(h^{-1}\tilde{N}^2-N^iN^jh_{ij})dt^2$$

$$-2h_{ij}N^idx^jdt-h_{ij}dx^idx^j , \tag{7.18}$$

$$h\equiv\mathrm{Det}(h_{ij}) . \tag{7.19}$$

Furthermore, π^{ij} is the canonical conjugate to h_{ij}, and $\tilde{\mathcal{H}}$ and $\mathcal{H}_i$ are functions of h_{ij} and π^{ij} and their space derivatives, given by

$$\tilde{\mathcal{H}}=\tfrac{1}{2}\mathcal{G}_{ij,kl}\pi^{ij}\pi^{kl}-{}^{(3)}R , \tag{7.20}$$

$$\mathcal{H}_i=-2h_{ij}\nabla_k\pi^{jk} , \tag{7.21}$$

where ${}^{(3)}R$ is the scalar curvature and ∇_k is the covariant derivative, both calculated using the 3-metric h_{ij}, and

$$\mathcal{G}_{ij,kl}\equiv h_{ik}h_{jl}+h_{il}h_{jk}-h_{ij}h_{kl} . \tag{7.22}$$

We see that $\tilde{N}$ and N^i just act as Lagrange multipliers for $\tilde{\mathcal{H}}$ and $\mathcal{H}_i$, respectively. Moreover, from (7.18), we see that $\tilde{N}^2$ is just the quantity whose status is under question here, the determinant of the 4-metric[10]

$$\tilde{N}=(\mathrm{Det}g_{\mu\nu})^{1/2} . \tag{7.23}$$

Thus, just as the integral over the einbein $g(s)$ enforced the constraint $p^\mu p_\mu=-m^2$, the integral over Detg enforces the constraint

$$\tilde{\mathcal{H}}=2\lambda . \tag{7.24}$$

The two conditions are quite similar. Just as $\eta_{\mu\nu}$ has signature $+++-$, the quantity (7.22), viewed as a 6×6 matrix, has signature $+,+,+,+,+,-$. Hence the integration over $\mathrm{Detg}_{\mu\nu}$ has the effect of eliminating one negative norm degree of freedom for each $\mathbf{x}$, $\pi^{ij}\propto(h^{-1})^{ij}$, just as the integral over the einbein $g(s)$ allows one to eliminate the variable p^0. However, for gravity there is a "potential" term in $\tilde{\mathcal{H}}$, proportional to the 3-curvature, and it is not entirely clear to me that it really is necessary to constrain $\tilde{\mathcal{H}}$ to take a fixed value. For the present, the question of whether it is necessary to integrate over $\mathrm{Detg}_{\mu\nu}$ must be left open. [Recent work by Henneaux and Teitelboim (1988) shows that there is a sensible generally covariant quantum version of the classical theory described by Eq. (7.3).]

Before closing this section, I should note that several authors have made a rather different suggestion, which also has the effect of converting the cosmological constant from a function of parameters in the action into a constant of the motion (Aurilia *et al.*, 1980; Witten, 1983; Henneaux and Teitelboim, 1984). They proposed adding to the action a term

[10]In order to obtain this result, I have defined $\tilde{\mathcal{H}}$ and $\tilde{N}$ differently from the usual $\mathcal{H}$ and N, by moving a factor $h^{1/2}$ from N to $\mathcal{H}$.

$$I_F = -\tfrac{1}{48}\int d^4x\sqrt{g}\,F_{\mu\nu\rho\sigma}F^{\mu\nu\rho\sigma}\ , \qquad (7.25)$$

where $F_{\mu\nu\rho\sigma}$ is the exterior derivative of a 3-form gauge field $A_{\nu\rho\sigma}$,

$$F_{\mu\nu\rho\sigma} = \partial_{[\mu}A_{\nu\rho\sigma]}\ , \qquad (7.26)$$

and $g \equiv -\mathrm{Det}g_{\mu\nu}$. Since $F^{\mu\nu\rho\sigma}$ is totally antisymmetric, it can be expressed as

$$F^{\mu\nu\rho\sigma} = c\,\varepsilon^{\mu\nu\rho\sigma}/\sqrt{g}\ , \qquad (7.27)$$

where $\varepsilon^{\mu\nu\rho\sigma}$ is the Levi-Cività tensor density, with $\varepsilon^{0123} \equiv 1$, and c is a scalar field. The field equation for A is

$$F^{\mu\nu\rho\sigma}{}_{;\mu} = 0\ , \qquad (7.28)$$

so, using (7.27)

$$\frac{\partial c}{\partial x^\mu} = 0\ . \qquad (7.29)$$

But the action (7.25) then takes the form

$$I_F = +\tfrac{1}{2}c^2\int d^4x\sqrt{g}\ . \qquad (7.30)$$

In other words, whatever else contributes to the cosmological constant, there is one term that depends on the integration constant c,

$$\lambda_F = 4\pi G c^2\ . \qquad (7.31)$$

Again, this does not solve the cosmological constant problem, but it does change the way it arises.

If λ is a constant of integration, then in a quantum theory we expect the state vector of the universe to be a superposition of states with different values of λ, in which case the anthropic considerations of Sec. V would set a bound on th effective cosmological constant.

VIII. QUANTUM COSMOLOGY

The last approach to the cosmological constant problem that I shall describe here is based on the application of quantum mechanics to the whole universe. In 1984 Hawking (1984b) described how in quantum cosmology there could arise a distribution of values for the effective cosmological constant, with an enormous peak at $\lambda_{\mathrm{eff}}=0$. Very recently, this approach has been revived in an exciting paper by Coleman (1988b), using a new mechanism for producing a distribution of values for the cosmological constant (that rests in part on other work of Hawking and Coleman) and finding an even sharper peak. Related ideas have also been recently discussed by Banks (1988). Before describing the work of Coleman and Hawking, I will have to say something about quantum cosmology in general.

Most treatments of quantum cosmology are based on the "wave function of the universe," a function $\Psi[h,\phi]$ of the 3-metric and matter fields on a spacelike surface. [The 3-metric h_{ij} can be conveniently defined by adapting the space-time coordinate system so that the spacelike surface has constant t, and then decomposing the 4-metric $g_{\mu\nu}$ as in Eq. (7.19).] This wave function satisfies a sort of Schrödinger equation, known as the Wheeler-DeWitt equation [DeWitt (1967); Wheeler (1968)]:

$$\left[\frac{1}{2h^{1/2}}\frac{\delta}{\delta h_{ij}}h^{1/2}g_{ij,kl}\frac{\delta}{\delta h_{kl}} - {}^{(3)}R - 2\lambda + 8\pi G T_{00}\right]\Psi = 0\ , \qquad (8.1)$$

with notation explained in Sec. VII (except that we now include a matter energy density T_{00}, in which the canonical conjugate of a matter field Φ is replaced with $\delta/\delta\Phi$). It will be very important in what follows that we express the solution as a *Euclidean* path integral

$$\Psi \propto \int [dg][d\Phi]\exp(-S[g,\Phi])\ , \qquad (8.2)$$

where we integrate over all Euclidean-signature 4-metrics $g_{\mu\nu}$ and matter fields Φ defined on a 4-manifold M_4, that have the 3-manifold $M_3[h,\phi]$ with 3-metric h_{ij} and matter fields ϕ as a boundary. (The Wheeler-DeWitt equation is the constraint obtained from integrating the Lagrange multiplier $\tilde{N}$ as discussed in Sec. VII.) Here S is the Euclidean action[11]

$$S = \frac{1}{16\pi G}\int_{M_4}\sqrt{g}\,(R+2\lambda) + \text{matter terms} + \text{surface terms}\ . \qquad (8.3)$$

Since Eq. (8.1) is a differential equation in an infinite-dimensional space [the set of $h_{ij}(\mathbf{x})$ and $\phi(\mathbf{x})$ for all $\mathbf{x}$], it has an infinite variety of solutions, which can be specified by giving the 4-manifold in Eq. (8.2) other boundaries, besides the $M_3[h,\phi]$ on which the 3-metric and matter fields are specified. Hartle and Hawking (1983) proposed as a cosmological initial condition that the manifold M_4 should have no boundaries other than $M_3(h,\phi)$. We will see that Coleman's (1988b) approach does not depend critically on the choice of initial conditions.

There are technical problems associated with this formalism. One is an operator-ordering ambiguity: there are various ways of ordering[12] the h_{ij} fields and $\delta/\delta h_{ij}$ operators in (8.1), all of which have (8.2) as solution, but

[11]The Euclidean action S is opposite in sign to what we would get if we replaced the metric $g_{\mu\nu}$ in the action I in Eq. (7.1) with one of signature $+,+,+,+$. This sign of S is chosen so that ordinary matter makes a positive contribution to S.

[12]The insertion of factors $h^{-1/2}$ and $h^{1/2}$ in Eq. (8.1) represents one choice of operator ordering, which is made in order to allow the derivation of the conservation equation (8.8).

with different ways of calculating the measure $[dg][d\Phi]$ (Hawking and Page, 1986). Another problem, potentially more worrisome, is that for gravity the Euclidean action (8.3) is not bounded below. Gibbons, Hawking, and Perry (1978) have proposed rotating the contour of integration for the overall scale of the 4-metric so that it runs parallel to the imaginary axis. We will not need to go into these technicalities here, because it will turn out that we only need to deal with the effective action at its equilibrium point.

A problem that is more relevant to us here has to do with the probabilistic interpretation of the wave function Ψ and of Euclidean path integrals like (8.2). Hawking has proposed (1984a, 1984c) that $\exp(-S[g,\Phi])$ should be regarded as proportional to the probability of a particular metric and matter field history. It is not immediately clear what is meant by this—even supposing that we had the godlike ability to measure the gravitational and matter fields throughout space-time, it would be in a space-time of Lorentzian rather than Euclidean signature. However, since we can (sometimes) go from one signature to another by a complex coordinate transformation, it may be that a Euclidean history $g_{\mu\nu}(x)$, $\Phi(x)$ can be interpreted in terms of correlations of scalar quantities, just as if the space-time were Lorentzian. In much of Hawking's work (e.g., Hawking, 1979), these questions are avoided by using the formalism only to calculate the probability that, in the space-time history of the universe, there is a spacelike 3-surface with a given 3-metric $h_{ij}(\mathbf{x})$ and matter fields $\phi(\mathbf{x})$. For instance, with Hartle-Hawking (1983) initial conditions, we would integrate over all closed 4-manifolds that contain such a 3-surface. If this surface bisects the 4-manifold, then it can be regarded as the boundary of the two halves of the 4-manifold, and so the integral is (with some qualifications) just the square of the wave function (8.2). But questions still arise concerning the probabilistic interpretation of Ψ, particularly with regard to normalization. If $|\Psi[h,\phi]|^2$ is the probability density that there exists *some* 3-surface on which the 3-metric is $h_{ij}(\mathbf{x})$ and the matter fields are $\phi(\mathbf{x})$, then we would not simply want to set the functional integral of $|\Psi[h,\phi]|^2$ over $h_{ij}(\mathbf{x})$ and $\phi(\mathbf{x})$ equal to unity, because in this functional integral we are summing up possibilities that are not exclusive; if the universe has some $h_{ij}(\mathbf{x})$ and $\phi(\mathbf{x})$ on one 3-surface, then it may also have some other $h'_{ij}(\mathbf{x})$ and $\phi'(\mathbf{x})$ on some other 3-surface. After all, you would not expect that the probabilities that you ever in your life have flipped a coin and gotten heads, and that you ever in your life have flipped a coin and gotten tails, should add up to unity.

I would like to offer an interpretation of what is meant by treating $|\Psi[h,\Phi]|^2$ as a probability density, which seems to me implicit in Hawking's writings (and may already be stated explicitly somewhere in the literature). As everyone has recognized, the problem has to do with the role of time in quantum gravity. [See, e.g., Hartle (1987).] The problems raised here do not arise in asymptotically flat cosmologies, because in such theories there is a natural definition of time, and we generally ask for the probabilities that the fields have certain values at a definite time. However, here time is a coordinate with no objective significance, and this coordinate time is even imaginary. As Augustine (398) warned, "I must not allow my mind to insist that time is something objective."[13] Heeding this warning, suppose we choose some "time-keeping" field $a(\mathbf{x},t)$, for instance, the trace of the energy-momentum tensor, and use its value to define a local time α. Each value of α defines a 3-surface, on which the coordinate time t is a function $t(\mathbf{x},\alpha)$ defined implicitly by

$$a(\mathbf{x},t(\mathbf{x},\alpha))\equiv\alpha \ . \tag{8.4}$$

We are then interested in the probability that the tangential components of the metric and all matter fields other than $a(\mathbf{x},t)$ have specified values on this surface. Calling these quantities $b_n(\mathbf{x},t)$, we see that the probability density for the $b_n(\mathbf{x},t)$ to have the values $\beta_n(\mathbf{x})$ at local time α is

$$P_\alpha[\beta]=N\int_{M_4}[dg][d\Phi]\exp(-S[g,\Phi]) \times\prod_{n,\mathbf{x}}\delta(b_n(\mathbf{x},t(\mathbf{x},\alpha))-\beta_n(\mathbf{x}))\ , \tag{8.5}$$

with N a normalization factor, determined by the condition that the total probability of finding *any* value for the $b_n(\mathbf{x})$ at local time α should be unity:

$$1=\int P_\alpha[\beta][d\beta] =N\int_{M_4}[dg][d\Phi]\exp(-S[g,\Phi])\ . \tag{8.6}$$

[This usually makes N a function of α, because in (8.5) and (8.6) we integrate only over matter and metric histories for which Eq. (8.4) is satisfied on some 3-surface. With some boundary conditions, this condition is automatically satisfied, and then N is α independent. For instance, if M_4 has two boundaries, on which $a(x)$ is required to take values α_1 and α_2, then there are 3-surfaces on which (8.4) is satisfied for all α in the range $\alpha_1<\alpha<\alpha_2$.] Where the surface of constant a bisects the 4-space, $P_\alpha[\beta]$ can be written as proportional to the square of the wave function $\Psi[\alpha,\beta]$, but with α constant in 3-space.

[13]This quote is not merely a display of useless erudition. Book XI of Augustine's *Confessions* contains a famous discussion of the nature of time, and it seems to have become a tradition to quote from this chapter in writing about quantum cosmology. Thus Hawking (1979) quotes "What did God do before He made Heaven and Earth? I do not answer as one did merrily: He was preparing a Hell for those that ask such questions. For at no time had God not made anything, for time itself was made by God." Coleman (1988a) quotes "The past is present memory." To this, I can add one more very relevant quote: "I confess to you, Lord, that I still do not know what time is. Yet I confess too that I do know that I am saying this in time, that I have been talking about time for a long time, ..."

Coleman (1988b) short-circuits many of the problems that arise in giving a probabilistic interpretation to Euclidean path integrals by using such integrals only to calculate expectation values: the expectation value of an arbitrary scalar field $A_{g,\Phi}(x)$, which may depend on the metric and matter fields and their derivatives, is taken as

$$\langle A \rangle = \frac{\int [dg][d\Phi] A_{g,\Phi}(x) \exp(-S[g,\Phi])}{\int [dg][d\Phi] \exp(-S[g,\Phi])} . \tag{8.7}$$

The general covariance of the theory makes $\langle A \rangle$ independent of x. In fact, it should be emphasized that this sort of expectation values includes an average over the time in the history of the universe that A is measured. On the other hand, the probability $P_\alpha[\beta]$ discussed above is the expectation value of a nonlocal operator, the delta function in (8.5), and refers to a specific local time α.

(I should mention here that there is a very different and apparently unrelated approach to the problem of giving a probabilistic interpretation to the wave function Ψ. The Wheeler-DeWitt equation (8.1) is somewhat like the Klein-Gordon equation for a particle in a scalar potential and leads immediately to a somewhat similar conservation law (now given for pure gravity):

$$0 = \frac{\delta}{\delta h_{ij}(\mathbf{x})} \left[h^{1/2}(\mathbf{x}) \mathcal{G}_{ij,kl}(\mathbf{x}) \times \mathrm{Im} \left[\Psi^*[h] \frac{\delta}{\delta h_{kl}(\mathbf{x})} \Psi[h] \right] \right] . \tag{8.8}$$

Since the beginning, it was hoped that such a conservation law could be used to construct a suitable probability density (DeWitt, 1967). Usually (8.8) is stated in a minisuperspace context, where $h_{ij}(\mathbf{x})$ is constrained to depend on only a finite number of parameters. Since $\mathcal{G}_{ij,kl} h^{kl} = -h_{ij}$, it is natural to treat the overall scale of h_{ij} as a sort of global time coordinate, and take as a probability density the corresponding component of the conserved "current" in (8.8). I wish to point out here that such a construction is not limited to any particular minisuperspace formulation, but can be carried out in the general case. Take Ψ to depend on a "global time"

$$T[h] = \left[\int d^3x \, h^{1/2}(\mathbf{x}) \right]^{1/2} \tag{8.9}$$

and an arbitrary (in fact, infinite) number of other parameters $\zeta_n[h]$, all ζ_n independent of the overall scale of $h_{ij}(\mathbf{x})$:

$$0 = \int d^3x \, h^{1/2}(\mathbf{x}) h_{kl}(\mathbf{x}) \frac{\delta \zeta_n[h]}{\delta h_{kl}(\mathbf{x})} . \tag{8.10}$$

We also introduce a Jacobian $J(\zeta, T)$ and write the functional measure as

$$[dh] = J[d\zeta] dT . \tag{8.11}$$

Multiplying Eq. (8.8) with $\delta(T[h]-T)$ and doing an integral over $\mathbf{x}$ and a functional integral over $h_{ij}(\mathbf{x})$, we easily find a constancy condition

$$0 = \frac{d}{dT} \int J[d\zeta] \mathrm{Im} \left[\Psi^* \frac{\partial \Psi}{\partial T} \right] . \tag{8.12}$$

The trouble here is, of course, the same as that encountered in giving a probabilistic interpretation to the Klein-Gordon equation: the integrand in (8.12) is not, in general, positive. Banks, Fischler, and Susskind (1985) and Vilenkin (1986, 1988a), have considered minisuperspace models in which Ψ is complex, with increasing phase, for which the integrand of Eq. (8.12) is positive-definite; however, this is not the case in general, and, in particular, not for Hartle-Hawking boundary conditions. For a recent more general discussion, see Vilenkin (1988b).)

I now want to give a simplified description of Hawking's (1984b) proposed solution of the cosmological constant problem, using for this purpose parts of Coleman's (1988b) analysis. In order to make the cosmological constant into a dynamical variable, Hawking introduces a 3-form gauge field $A_{\mu\nu\lambda}$ of the sort described at the end of Sec. VII. According to the general ideas of Euclidean quantum cosmology, the probability distribution for the scalar $c(x)$ defined by Eq. (7.27) at any one point $x = x_1$ is

$$P(c) = \langle \delta(c(x_1) - c) \rangle \propto \int [dA][dg][d\Phi] \delta(c(x_1) - c) \times \exp(-S[A,g,\Phi]) . \tag{8.13}$$

It is well known that such functional integrals can be expressed as exponentials of the effective action at its stationary point.[14] In the present case, we have

$$P(c) \propto \exp(-\Gamma[A_c, g_c, \Phi_c]) , \tag{8.14}$$

where $\Gamma[A,g,\Phi]$ is the total action (the sum of one-particle irreducible graphs with external lines replaced with fields A,g,Φ) and the subscript c indicates that this quantity is to be evaluated at a point where Γ is stationary with respect to any variations in $A_{\mu\nu\lambda}(x)$, $g_{\mu\nu}(x)$, or $\Phi(x)$ that leave $c(x_1) = c$ fixed. Now, among all the possible stationary points of Γ, there is one that can be found knowing only the effective action relevant to large

[14]The usual proof, for the case without a delta function in the integrand, proceeds by adding a term $\int J\Omega$ to the action, where Ω denotes the various fields, and J is a set of corresponding currents. The path integral is then $\exp[-W(J)] \equiv \int d\Omega \exp(-S - \int J\Omega)$. The effective action is defined by the Legendre transformation $\Gamma(\Omega) = W(J_\Omega) - \int J_\Omega \Omega$, where J_Ω is the current that produces a given expectation value $\Omega = \delta W / \delta J$. The condition for zero current is that $\Gamma(\Omega)$ be stationary with respect to Ω, and at this point $\Gamma(\Omega) = W(0)$. The delta function in (8.13) can be dealt with by writing it as an integral $\int d\omega \exp\{i\omega[c(x_1) - c]\}$. One can then use the above theorem to evaluate the functional integral before integrating over ω, now with no restriction on $c(x)$, and then doing the ω integral.

4-manifolds. In this case, it is convenient to set all fields except $A_{\mu\nu\lambda}$ and $g_{\mu\nu}$ equal to their (A- and g-dependent) stationary values, in which case the effective action can be expanded in inverse powers of the size of the manifold[15]

$$\Gamma_{\rm eff}[A,g]=\frac{\lambda}{8\pi G}\int\sqrt{g}\,d^4x+\frac{1}{16\pi G}\int\sqrt{g}R\,d^4x+\tfrac{1}{48}\int d^4x\sqrt{g}F_{\mu\nu\lambda\rho}F^{\mu\nu\lambda\rho}+\cdots\,, \tag{8.15}$$

the omitted terms involving more than two derivatives of g and/or A. As we saw in Sec. VII, the condition that this be stationary in $A_{\mu\nu\lambda}$ [for variations that keep $c(x_1)$ fixed] is that $F_{\mu\nu\lambda\rho}$ have vanishing covariant divergence, from which it follows that c in Eq. (7.27) is constant; hence

$$\Gamma_{\rm eff}=\frac{\lambda(c)}{8\pi G}\int\sqrt{g}\,d^4x+\frac{1}{16\pi G}\int\sqrt{g}R\,d^4x+\cdots\,, \tag{8.16}$$

where

$$\lambda(c)=\frac{c^2}{2}+\lambda\,. \tag{8.17}$$

The condition that this be stationary in $g_{\mu\nu}$ is, of course, that $g_{\mu\nu}$ satisfy the Einstein field equations with cosmological constant $\lambda(c)$. For any such solution, $R=-4\lambda(c)$, so at the stationary point

$$\Gamma_{\rm eff}=-\frac{\lambda(c)}{8\pi G}\int\sqrt{g}\,d^4x\,. \tag{8.18}$$

With Hartle-Hawking boundary conditions, the solution of the Einstein equations for $\lambda(c)>0$ is a 4-sphere of proper circumference $2\pi r$, where

$$r=\sqrt{3/\lambda(c)}\,, \tag{8.19}$$

yielding a probability density proportional to

$$\exp(-\Gamma_{\rm eff})=\exp[3\pi/G\lambda(c)]\,. \tag{8.20}$$

On the other hand, for $\lambda(c)<0$ the solutions can be made compact by imposing periodicity conditions, but they all have $\Gamma_{\rm eff}\geq 0$. Hawking's conclusion is that the probability density has an infinite peak for $\lambda(c)\to 0+$; hence, after normalizing P,

$$P(c)=\delta(c-c_0)\,, \tag{8.21}$$

where c_0 is the value of c (assuming there is one), for which $\lambda(c)=0$.

It is important that the quantity $\lambda(c)$ is the true effective cosmological constant, previously called $\lambda_{\rm eff}$, that would be measured in gravitational phenomena at long ranges.[16] The constant λ in Eq. (8.15) includes all effects of fields other than $g_{\mu\nu}$ and $A_{\mu\nu\lambda}$, including all quantum fluctuations. Hence the result (8.21), if valid, really does solve the cosmological constant problem.

We can check that this result is not invalidated by the terms neglected in Eq. (8.16). For a large radius r, the exhibited terms in (8.16) are of order $\lambda r^4/G$ and r^2/G, respectively, while a term with $D\geq 4$ derivatives would yield a contribution to $\Gamma_{\rm eff}$ of order $(mr)^{4-D}$, where m is some combination of the Planck mass and elementary-particle masses. For $\lambda(c)\lesssim m^2$, this shifts the size of the manifold by

$$\delta r/r\approx G\lambda(c)[\lambda(c)/m^2]^{(D-4)/2}\ll 1\,.$$

The change in the stationary value of the action is then

$$\delta\Gamma_{\rm eff}\approx[\lambda(c)/m^2]^{(D-4)/2}\lesssim 1\,,$$

so these higher-derivative terms have no effect on the singularity (8.20).

Coleman (1988b) does not need to introduce a 3-form gauge field $A_{\mu\nu\lambda}$; rather, in order to make the cosmological constant into a dynamical variable, he considers the effect of topological fixtures known as wormholes.[17] An explicit example of a wormhole is provided by the metric (Hawking, 1987b, 1988)

$$ds^2=(1+b^2/x^\mu x^\mu)^2dx^\mu dx^\mu\,. \tag{8.22}$$

This appears to have a singularity at $x^\mu=0$, but the line element is invariant under the transformation

$$x^\mu\to x'^\mu=x^\mu b^2/x^\nu x^\nu\,, \tag{8.23}$$

so the region $x^\mu x^\mu<b^2$ actually has the same geometry as that with $x^\mu x^\mu>b^2$. The space described by Eq. (8.22) therefore consists of two asymptotically flat 4-spaces, joined together at the 3-surface with $x^\mu x^\mu=b^2$, a 3-sphere known as a "baby universe." This 4-metric is not a solution of the classical Einstein equations (though it does have $R=0$), but this is not very relevant; the action is

$$S=3\pi b^2/G\,, \tag{8.24}$$

so the factor $\exp(-S)$ suppresses the effects of all

[15]Such an effective action may be used as the input for calculations in which we include quantum effects only from virtual massless particles with $|q^2|$ less than some cutoff Λ^2. Such effects are, of course, finite, and their Λ dependence is to be canceled by giving the coefficients in $\Gamma_{\rm eff}$ a suitable Λ dependence. (This point of view is described by Weinberg, 1979b.) In order to prevent these quantum effects from generating an unacceptable cosmological constant, the cutoff Λ must be taken very small.

[16]This property is shared by an imaginative solution to the cosmological constant problem proposed by Linde (1988a).

[17]The importance of quantum fluctuations in space-time topology at small scales has been emphasized for many years by Wheeler (e.g., 1964), and more recently by Hawking (1978) and Strominger (1984). Such "space-time foam" was considered as a mechanism for canceling a cosmological constant by Hawking (1983).

wormholes except those of Planck dimensions or less, for which quantum effects are surely important. [A model with classical wormhole solutions, based on a 2-form axion, has been presented by Giddings and Strominger (1988a).]

If Planck-sized wormholes can connect asymptotically flat 4-spaces, then they can connect any 4-spaces that are large compared to the Planck scale. We are therefore led to consider contributions to the Euclidean path integral from large 4-spaces [like the 4-sphere in Hawking's (1984b) theory] connected to themselves and each other with Planck-sized wormholes. Each wormhole can be regarded as the creation and subsequent destruction of a baby universe [like the 3-sphere of proper circumference $4\pi b$ in Hawking's (1987b, 1988) wormhole model], and such baby universes may also appear as part of the boundary of the 4-manifold.

What are the effects of these wormholes and baby universes? At scales large compared with the scale of the baby universe, the creation or destruction of a baby universe can only show up through the insertion of a local operator in the path integral. The various types of baby universes can be classified according to the form of these local operators. The effect of creating and destroying arbitrary numbers of baby universes of all types can thus be expressed by adding a suitable term in the action

$$\tilde{S}=S+\sum_i (a_i+a_i^\dagger)\int d^4x\, O_i(x) , \qquad (8.25)$$

where a_i and $a_i^\dagger$ are the annihilation and creation operators for a baby universe of type i, and $O_i(x)$ is the corresponding local operator. [This was first stated by Hawking (1987b). Creation and annihilation operators for baby universes were earlier used by Strominger (1984). For a proof of Eq. (8.25), see Coleman (1988a) and Giddings and Strominger (1988b).] The path integral over all 4-manifolds with given boundary conditions is to be calculated as

$$\int [dg][d\Phi]e^{-S}=\int_{\text{No}}[dg][d\Phi]\langle B|e^{-\tilde{S}}|B\rangle , \qquad (8.26)$$

where No means that wormholes and baby universes are excluded, and $|B\rangle$ is a normalized baby-universe state depending on the boundary conditions. For instance, with Hartle-Hawking boundary conditions, $|B\rangle$ is the empty state

$$a_i|B\rangle=0 . \qquad (8.27)$$

These baby universes have an important effect even if none of them appear as part of the boundary of the 4-manifold, as would be the case for Hartle-Hawking boundary conditions. Hawking (1987b, 1988) has suggested that since the baby universes are unobservable, their effect is an effective loss of quantum coherence. [See also Hawking (1982); Teitelboim (1982); Strominger (1984); Lavrelashvili, Rubakov, and Tinyakov (1987, 1988); Giddings and Strominger (1988b). A contrary view was taken by Gross (1984).] Recently Coleman (1988a) has argued (convincingly, in my view) for a different interpretation [see also Giddings and Strominger (1988b)]. The state $|B\rangle$ in Eq. (8.26) may always be expanded in eigenstates of the operators $a_i+a_i^\dagger$:

$$|B\rangle=\int f_B(\alpha)\prod_i d\alpha_i|\alpha\rangle , \qquad (8.28)$$

$$(a_i+a_i^\dagger)|\alpha\rangle=\alpha_i|\alpha\rangle , \qquad (8.29)$$

$$\langle\alpha'|\alpha\rangle=\prod_i\delta(\alpha_i'-\alpha_i) , \qquad (8.30)$$

the function $f_B(\alpha)$ depending on the boundary conditions. For instance, for Hartle-Hawking conditions, $|B\rangle$ satisfies Eq. (8.27), and so

$$f_B(\alpha)=\prod_i\pi^{-1/4}\exp(-\alpha_i^2/2) . \qquad (8.31)$$

(With n baby universes on the boundary of the 4-space, this would be multiplied with a Hermite polynomial of order n.) In the state $|\alpha\rangle$, the effect of the creation and annihilation of baby universes is to change the action S to

$$S_\alpha=S+\sum_i\alpha_i\int O_i(x)d^4x . \qquad (8.32)$$

That is, the coupling constant multiplying each possible local term $\int O_i d^4x$ is changed by an amount α_i. As soon as we start to make any sort of measurements, the state of the universe breaks up into an incoherent superposition of these $|\alpha\rangle$'s, each appearing with *a priori* probability $|f_B(\alpha)|^2$; but for each term we have an ordinary wormhole-free quantum theory, with α-dependent action (8.32).

If all we want is to explain why the cosmological constant is not enormous, then our work is essentially done. The effective cosmological constant is a function of the α_i, because among the O_i there is a simple operator $O_1=\sqrt{g}$, whose coefficient contributes a term $8\pi G\alpha_1$ to λ, and also because the vacuum energy $\langle\rho\rangle$ depends on the couplings of all interactions, each of which has a term proportional to one of the α_i. Now, generic baby-universe states $|B\rangle$ will have components $|\alpha\rangle$ for which $\lambda_{\text{eff}}(\alpha)$ is very small, as well as others for which it is enormous. The anthropic considerations of Sec. VI tell us that any scientist who asks about the value of the cosmological constants can only be living in components $|\alpha\rangle$ for which λ_{eff} is quite small, for otherwise galaxies and stars could never have formed (for $\lambda_{\text{eff}}>0$), or else there would not be time for life to evolve (for $\lambda_{\text{eff}}<0$).

However, it is of great interest to ask whether the effective cosmological constant is really zero, or just small enough to satisfy anthropic bounds, in which case it should show up observationally. The probability of getting any particular value of the α_i, and hence of finding a value $\lambda_{\text{eff}}(\alpha)$, is not just given by the function $|f_B(\alpha)|^2$ arising from the boundary conditions, but is also affected by the functional integral itself.

In calculating this effect, Coleman (1988b) observed that although we are to integrate only over connected 4-manifolds, on a scale much large than the wormhole scale those manifolds that appear disconnected will really be connected by wormholes. Hence any sort of probability density or expectation value will contain as a factor a sum over disconnected manifolds consisting of arbitrary numbers of closed connected wormhole-free components.[18] Just as for Feynman diagrams, this sum is the exponential of the path integral for a single closed connected wormhole-free manifold

$$F(\alpha)=\exp\left[\int_{CC}[dg]\exp(-S_\alpha[g])\right], \tag{8.33}$$

where CC indicates that we include only closed connected wormhole-free manifolds, and $S_\alpha[g]$ is the action (8.32) with all fields other than $g_{\mu\nu}(x)$ integrated out.

The path integral in (8.33) can be evaluated by precisely the same methods as described above in connection with Hawking's (1984b) model [and used for this purpose by Coleman (1988b)]. The result is that the probability density for $\lambda_{\rm eff}$ contains a factor (for $\lambda_{\rm eff}>0$)

$$F=\exp\left[\exp\left[\frac{3\pi}{G\lambda_{\rm eff}}+O(1)\right]\right]. \tag{8.34}$$

The fact that this is now an exponential of an exponential, instead of a mere exponential, is not essential in solving the cosmological constant problem (though it is important in fixing other constants, as described at the end of this section). Either way, the probability distribution has an infinite peak at $\lambda_{\rm eff}\to 0+$, which, after normalizing so that the total probability is unity, means that $P(\alpha)$ has a factor

$$P(\alpha)\propto\delta(\lambda_{\rm eff}(\alpha)). \tag{8.35}$$

In addition, as in Hawking's case, from the way that F has been calculated it is clear that this $\lambda_{\rm eff}$ is the constant that appears in the effective action for pure gravity with all high-energy fluctuations integrated out; hence it is the cosmological constant relevant to astronomical observation.

Has the cosmological constant problem been solved? Perhaps so, but there are still some things to worry about in Coleman's approach, as also in the earlier work of Hawking. Here is a short list of qualms.

(1) Does Euclidean quantum cosmology have anything to do with the real world? It is essential to both Coleman and Hawking that the path integral be given by a stationary point of the Euclideanized action—the conclusion would be completely wiped out if in place of $\exp(3\pi/G\lambda_{\rm eff})$ we had found $\exp(3\pi i/G\lambda_{\rm eff})$. Some of the technical and conceptual difficulties of Euclidean quantum cosmology were discussed at the beginning of this section.

(2) What are the boundary conditions? It is always possible that the essential singularity in $\exp\{\exp[3\pi/G\lambda(\alpha)]\}$ is canceled by an essential zero in the *a priori* probability $|f_B(\alpha)|^2$. However, this is not the case for Hartle-Hawking boundary conditions, where $|f_B(\alpha)|^2$ is a simple Gaussian. Moreover, Coleman (1988b) has shown that in his theory such an essential zero would be destroyed by almost any perturbation of the boundary conditions; instead of its being unnatural to have zero cosmological constant, it would be highly unnatural not to. Still, the problem of boundary conditions is disturbing, because it reminds us that quantum cosmology is an incomplete theory.

(3) Are wormholes real? Coleman's calculation depends on there being a clear separation between the very large 4-manifolds, for which the long-range effective action is stationary (and large and negative), and very small wormholes, whose contribution to the action is of order unity (and generally positive). Furthermore, the wormholes have been assumed to be so well separated that we can ignore their interactions (the "dilute gas" approximation). It may be possible to construct a theory in which the wormhole scale [like b in Eq. (8.22)] is somewhat larger than the Planck scale, large enough to allow the wormhole metric to be calculated classically, but we would still have to ask whether this is actually the case. Hawking (1984b) does not need to worry about wormholes, but how do we know that the 3-form gauge field is real? A related question for both authors: even granting the existence of the stationary point of the action at which $\Gamma_{\rm eff}=-3\pi/\lambda G$, how do we know that this is the dominant stationary point?

(4) What about the other terms in the effective action? For instance, suppose we include the 6-derivative term[19]

$$\Gamma_{\rm eff}=\int d^4x\sqrt{g}\left[\frac{\lambda(\alpha)}{8\pi G(\alpha)}+\frac{R}{16\pi G(\alpha)}\right.$$
$$\left.+\zeta(\alpha)G(\alpha)R_{\mu\nu}{}^{\lambda\rho}R_{\lambda\rho}{}^{\sigma\tau}R_{\sigma\tau}{}^{\mu\nu}\right], \tag{8.36}$$

with $\zeta(\alpha)$ a dimensionless coefficient that, like λ and G, depends on the baby-universe parameters α_i. Hawking and Coleman found a stationary point of this action for which $\Gamma_{\rm eff}\to-\infty$ when $\lambda(\alpha)G(\alpha)\to 0$, but for this purpose it is essential that $\zeta(\alpha)$ remain bounded in this limit. (We recall that in our previous discussion of the higher-derivative terms in $\Gamma_{\rm eff}$, we assumed that the coefficient m^{4-D} of terms with $D\geq 4$ derivatives remains less than $\lambda^{(D-4)/2}$ as $\lambda\to 0$.) But if we can let $1/\lambda G$ go to infinity, then why not let ζ go to infinity also? In particular, why not use a dimensional factor $1/\lambda(\alpha)$ in place of $G(\alpha)$ in

[18]This sum actually includes manifolds that are truly not connected by wormholes or anything else, but their contribution is a harmless multiplicative factor, which will cancel out anyway in normalizing $P(\alpha)$.

[19]Terms involving the Ricci tensor $R_{\mu\nu}$ or its trace R are not included here, because they represent merely a redefinition of the metric; see, e.g., Weinberg (1979a). The 4-derivative term $R_{\mu\nu\lambda\rho}R^{\mu\nu\lambda\rho}$ is not included, because it can be combined with terms involving $R_{\mu\nu}$ or R to make a topological invariant and is therefore physically unmeasureable for fixed large-scale topology.

the last term of Eq. (8.36)? This would completely invalidate the analysis of the singularity in the probability density $P(\alpha)$, and could well wipe it out.

The last of these four qualms suggests some interesting possibilities. Suppose we do assume that for some reason constants like $\zeta(\alpha)$ in Eq. (8.36) are bounded. Then the effect of wormholes is not only to fix $\lambda(\alpha)$ at zero, but also to fix these other constants at their lower or upper bounds. [I think this is the correct interpretation of what Coleman (1988b) calls "the big fix."] For instance, for $\zeta(\alpha)$ bounded and $|\lambda(\alpha)G(\alpha)| \ll 1$, the action (8.36) is stationary for a sphere of proper circumference $2\pi r$, where

$$r^2 = \frac{3}{\lambda} - \frac{64\zeta\pi G^2\lambda}{3} , \tag{8.37}$$

for which the effective action takes the value

$$\Gamma_{\rm eff} = -\frac{3\pi}{G\lambda} - \frac{128\zeta G\lambda\pi^2}{3} . \tag{8.38}$$

Thus the probability distribution $\exp[\exp(-\Gamma_{\rm eff})]$ not only has an infinite peak at $\lambda(\alpha)=0$, but also contains a factor

$$\exp\left[\frac{128\zeta G\lambda\pi^2}{3}\exp\left[\frac{3\pi}{G\lambda}\right]\right] . \tag{8.39}$$

For $G\lambda \to 0$, the quantity $G\lambda \exp(3\pi/G\lambda)$ becomes infinite, so the normalized probability will have a delta function at the upper bound of $\zeta(\alpha)$. All constants in the effective action for gravitation, including terms with any numbers of derivatives, can be calculated in this way,[20] but they all have to be bounded as $\lambda(\alpha)G(\alpha) \to 0$ for any of this to make sense.

It may be that the bounds (if any) on parameters like $\zeta(\alpha)$ arise from the details of wormhole physics, in which case these remarks are not going to be useful numerically for some time. However, there is another more exciting possibility, that there are just unitarity bounds, which could be calculated working only with low-energy effective theory itself. Of course, we are not likely to be able to measure parameters like $\zeta(\alpha)$, but it would still be nice to be able to calculate them, because up to now the only really unsatisfactory feature of the quantum theory of gravitation has been the apparent arbitrariness of this infinite set of parameters.

IX. OUTLOOK

All of the five approaches to the cosmological constant problem described in Secs. IV–VIII remain interesting. At present, the fifth, based on quantum cosmology, appears the most promising. However, if wormholes (or 3-form gauge fields) do produce a distribution of values for the cosmological constant, but without an infinite peak at $\lambda_{\rm eff}=0$, then we will have to fall back on the anthropic principle to explain why $\lambda_{\rm eff}$ is not enormously larger than allowed by observation. Alternatively, it may be some change in the theory of gravity, like that described here in Sec. VII, that produces the distribution in values for $\lambda_{\rm eff}$. The approaches based on supersymmetry and adjustment mechanisms described in Secs. IV and VI seem least promising at present, but this may change.

All five approaches have one other thing in common: They show that any solution of the cosmological constant problem is likely to have a much wider impact on other areas of physics or astronomy. One does not need to explain the potential importance of supergravity and superstrings. A light scalar like that needed for adjustment mechanisms could show up macroscopically, as a "fifth force." Changing gravity by making $\mathrm{Det}g_{\mu\nu}$ not dynamical would make us rethink our quantum theories of gravitation, and wormholes might force all the constants in these theories to their outer bounds. Finally, and of greatest interest to astronomy, if it is only anthropic constraints that keep the effective cosmological constant within empirical limits, then this constant should be rather large, large enough to show up before long in astronomical observations.

Note added in proof. As might have been expected, in the time since this report was submitted for publication there have appeared a large number of preprints that follow up on various aspects of the work of Coleman (1988b) and Banks (1988). Here is a partial list: Accetta *et al.* (1988); Adler (1988); Fischler and Susskind (1988); Giddings and Strominger (1988c); Gilbert (1988); Grinstein and Wise (1988); Gupta and Wise (1988); Hosoya (1988); Klebanov, Susskind, and Banks (1988); Myers and Periwal (1988); Polchinski (1988); Rubakov (1988). I am not able to review all of these papers here. However, I do want to mention two further qualms, regarding Coleman's proposed solution of the cosmological constant problem, that are raised by some of these papers. First, Fischler and Susskind (1988), partly on the basis of conversations with V. Kaplunovsky, have pointed out that the exponential damping of large wormholes may be overcome by Coleman's double exponential. If this were the case, we would be confronted with closely packed wormholes of macroscopic as well as Planck scales. This would be a disaster for Coleman's proposed solution of

---

[20] To the extent that it will become possible to calculate functions like $\lambda(\alpha)$, $G(\alpha)$, $\zeta(\alpha)$ etc., in terms of the parameters in an underlying fundamental theory, such as a string theory, the location of the delta functions in F may allow us to infer something about the values of the α_i and of the parameters in the underlying theory. However, without such an underlying theory, it is impossible to use calculations of λG, ζ, etc., to infer anything about the observed parameters of some intermediate theory like the standard model. This is because, in addition to charges, masses, etc., the standard model implicitly also involves parameters $\lambda_0, G_0, \zeta_0, \ldots$ appearing in the effective action for gravitation. When we integrate out the quarks, leptons, and gauge and Higgs bosons, we obtain new values for λ, G, ζ, etc.; but these new values depend on an equal number of unknowns λ_0, G_0, ζ_0, etc., as well as on charges and masses.

the cosmological constant problem, and would also indicate that we do not fully understand how to use Euclidean path integrals in quantum cosmology. Next, Polchinski (1988) has found that the Euclidean path integral over closed, connected, wormhole-free manifolds inside the exponential in (8.33) has a phase that might eliminate the peak in the probability distribution at zero cosmological constant. As pointed out here in footnote 15, when we use an effective action Γ_{eff} to evaluate such path integrals, the effective action must be taken as an input to calculations in which we include quantum fluctuations in massless particle fields with momenta up to some ultraviolet cutoff Λ. This cutoff must be taken as the same as the *infrared* cutoff that was used in calculating Γ_{eff}, so that all fluctuations are taken into account. It was remarked in footnote 15 that Λ must be taken very small, to avoid reintroducing a cosmological constant, but as Polchinski now remarks, no matter how small we take Λ, the integral over fluctuations in the gravitational field with momenta less than Λ produces a phase in the integral. Since this phase appears inside the exponential in Eq. (8.33), if its real part is not positive definite there would be no exponential peak at zero cosmological constant. On the other hand, in the absence of wormholes this phase would appear as an overall factor in front of a single exponential, so it would not affect the peaking at zero cosmological constant found by Hawking (1984b).

ACKNOWLEDGMENTS

I have been greatly helped in preparing this review by conversations with many colleagues. Here is a list of a few of those to whom my thanks are especially due. Section II: G. Holton; Sec. III: E. Witten; Sec. IV: S. de Alwis, J. Polchinski, E. Witten; Sec. V: P. J. E. Peebles, P. Shapiro, E. Vishniac; Sec. VI: J. Polchinski; Sec. VII: C. Teitelboim, F. Wilczek; Sec. VIII: L. Abbott, S. Coleman, B. DeWitt, W. Fischler, S. Giddings, L. Susskind, C. Teitelboim, F. Wilczek. Of course, they take no responsibility for anything that I may have gotten wrong. Research was supported in part by the Robert A. Welch Foundation and NSF Grant No. PHY 8605978.

REFERENCES

Abbott, L., 1985, Phys. Lett. **B150**, 427.

Abbott, L., 1988, Sci. Am. **258** (No. 5), 106.

Accetta, F. S., A. Chodos, F. Cooper, and B. Shao, 1988, "Fun with the wormhole calculus," Yale University Preprint No. YCTP-P20-88.

Adler, S. L., 1988, "On the Banks-Coleman-Hawking argument for the vanishing of the cosmological constant," Institute for Advanced Study Preprint No. IASSNS-HEP-88/35.

Albrecht, A., and P. J. Steinhardt, 1982, Phys. Rev. Lett. **48**, 120.

Arnowitt, R., S. Deser, and C. W. Misner, 1962, in *Gravitation: An Introduction to Current Research,* edited by L. Witten (Wiley, New York) p. 227.

Attick, J., G. Moore, and A. Sen, 1987, Institute for Advanced Studies preprint.

Augustine, 398, *Confessions,* translated by R. S. Pine-Coffin (Penguin Books, Harmondsworth, Middlesex, 1961), Book XI.

Bahcall, J., T. Piran, and S. Weinberg, 1987, Eds., *Dark Matter in the Universe: 4th Jerusalem Winter School for Theoretical Physics* (World Scientific, Singapore).

Bahcall, S. R., and S. Tremaine, 1988, Astrophys. J. **326**, L1.

Banks, T., 1985, Nucl. Phys. B **249**, 332.

Banks, T., 1988, University of California, Santa Cruz, Preprint No. SCIPP 88/09.

Banks, T., W. Fischler, and L. Susskind, 1985, Nucl. Phys. B **262**, 159.

Barbieri, R., E. Cremmer, and S. Ferrara, 1985, Phys. Lett. B **163**, 143.

Barbieri, R., S. Ferrara, D. V. Nanopoulos, and K. S. Stelle, 1982, Phys. Lett. B **113**, 219.

Barr, S. M., 1987, Phys. Rev. D **36**, 1691.

Barr, S. M., and D. Hochberg, 1988, Phys. Lett. B **211**, 49.

Barrow, J. D., and F. J. Tipler, 1986, *The Anthropic Cosmological Principle* (Clarendon, Oxford).

Bernstein, J., and G. Feinberg, 1986, Eds., *Cosmological Constants* (Columbia University, New York).

Bludman, S. A., and M. Ruderman, 1977, Phys. Rev. Lett. **38**, 255.

Brown, J. D., and C. Teitelboim, 1987a, Phys. Lett. B **195**, 177.

Brown, J. D., and C. Teitelboim, 1987b, Nucl. Phys. B **297**, 787.

Buchmüller, W., and N. Dragon, 1988a, University of Hannover Preprint No. ITP-UH 1/88.

Buchmüller, W., and N. Dragon, 1988b, Phys. Lett. B **207**, 292.

Carter, B., 1974, in *International Astronomical Union Symposium 63: Confrontation of Cosmological Theories with Observational Data,* edited by M. S. Longair (Reidel, Dordrecht), p. 291.

Carter, B., 1983, in *The Constants of Physics, Proceedings of a Royal Society Discussion Meeting, 1983,* edited by W. H. McCrea and M. J. Rees (printed for The Royal Society, London, at the University Press, Cambridge), p. 137.

Casimir, H. B. G., 1948, Proc. K. Ned. Akad. Wet. **51**, 635.

Chang, N.-P., D.-X. Li, and J. Pérez-Mercader, 1988, Phys. Rev. Lett. **60**, 882.

Coleman, S., 1988a, Nucl. Phys. B **307**, 867.

Coleman, S., 1988b, "Why there is nothing rather than something: A theory of the cosmological constant," Harvard University Preprint No. HUTP-88/A022.

Coleman, S., and F. de Luccia, 1980, Phys. Rev. D **21**, 3305.

Coughlan, G. D., I. Kani, G. G. Ross, and G. Segré, 1988, CERN Preprint No. TH. 5014/88.

Cremmer, E., S. Ferrara, C. Kounnas, and D. V. Nanopoulos, 1983, Phys. Lett. B **133**, 61.

Cremmer, E., B. Julia, J. Scherk, S. Ferrara, L. Girardello, and P. van Nieuwenhuizen, 1978, Phys. Lett. B **79**, 231.

Cremmer, E., B. Julia, J. Scherk, S. Ferrara, L. Girardello, and P. van Nieuwenhuizen, 1979, Nucl. Phys. B **147**, 105.

Davies, P. C. W., 1982, *The Accidental Universe* (Cambridge University, Cambridge).

de Sitter, W., 1917, Mon. Not. R. Astron. Soc. **78**, 3 (reprinted in Bernstein and Feinberg, 1986).

de Vaucouleurs, G., 1982, Nature (London) **299**, 303.

de Vaucouleurs, G., 1983, Astrophys. J. **268**, 468, Appendix B.

DeWitt, B., 1967, Phys. Rev. **160**, 1113.

Dicke, R. H., 1961, Nature (London) **192**, 440.

Dine, M., W. Fischler, and M. Srednicki, 1981, Phys. Lett. B **104**, 199.

Dine, M., R. Rohm, N. Seiberg, and E. Witten, 1985, Phys. Lett. B **156**, 55.
Dine, M., and N. Seiberg, 1986, Phys. Rev. Lett. **57**, 2625.
Dirac, P. A. M., 1937, Nature (London) **139**, 323.
Dolgov, A. D., 1982, in *The Very Early Universe: Proceedings of the 1982 Nuffield Workshop at Cambridge,* edited by G. W. Gibbons, S. W. Hawking, and S. T. C. Siklos (Cambridge University, Cambridge), p. 449.
Dreitlein, J., 1974, Phys. Rev. Lett. **34**, 777.
Eddington, A. S., 1924, *The Mathematical Theory of Relativity,* 2nd Ed. (Cambridge University, London).
Einstein, A., 1917, Sitzungsber. Preuss. Akad. Wiss. Phys.-Math. Kl. 142 [English Translation in *The Principle of Relativity* (Methuen, 1923, reprinted by Dover Publications), p. 177; and in Bernstein and Feinberg, 1986].
Einstein, A., 1919, Sitzungsber. Preuss. Akad. Wiss., Phys.-Math. Kl. [English translation in *The Principle of Relativity* (Methuen, 1923, reprinted by Dover Publications), p. 191].
Ellis, J., C. Kounnas, and D. V. Nanopoulos, 1984, Nucl. Phys. B **247**, 373.
Ellis, J., A. B. Lahanas, D. V. Nanopoulos, and K. Tamvakis, 1984, Phys. Lett. B **134**, 429.
Ellis, J., N. C. Tsamis, and M. Voloshin, 1987, Phys. Lett. B **194**, A 291.
Fischler, W., and L. Susskind, 1988, "A wormhole catastrophe," Texas Preprint No. UTTG-26-88.
Ford, L. H., 1987, Phys. Rev. D **35**, 2339.
Freese, K., F. C. Adams, J. A. Frieman, and E. Mottola, 1987, Nucl. Phys. B **287**, 797.
Friedan, D., E. Martinec, and S. Shenker, 1986, Nucl. Phys. B **271**, 93.
Friedmann, A., 1924, Z. Phys. **21**, 326 [English translation in Bernstein and Feinberg, 1986, Eds., *Cosmological Constants* (Columbia University, New York)].
Gibbons, G. W., S. W. Hawking, and M. J. Perry, 1978, Nucl. Phys. B **138**, 141.
Giddings, S. B., and A. Strominger, 1988a, Nucl. Phys. B **306**, 890.
Giddings, S. B., and A. Strominger, 1988b, Nucl. Phys. B **307**, 854.
Giddings, S. B., and A. Strominger, 1988c, "Baby universes, third quantization, and the cosmological constant," Harvard Preprint No. HUTP-88/A036.
Gilbert, G., 1988, "Wormhole induced proton decay," Caltech Preprint No. CALT-68-1524.
Grinstein, B., and M. B. Wise, 1988, "Light scalars in quantum gravity," Caltech Preprint No. CALT-68-1505.
Grisaru, M. T., W. Siegel, and M. Rocek, 1979, Nucl. Phys. B **159**, 429.
Gross, D. J., 1984, Nucl. Phys. B **236**, 349.
Gupta, A. K., and M. B. Wise, 1988, "Comment on wormhole correlations," Caltech Preprint No. CALT-68-1520.
Guth, A. H., 1981, Phys. Rev. D **23**, 347.
Hartle, J. B., 1987, in *Gravitation in Astrophysics: Cargese 1986,* edited by B. Carter and J. B. Hartle (Plenum, New York), p. 329.
Hartle, J. B., and S. W. Hawking, 1983, Phys. Rev. D **28**, 2960.
Hawking, S. W., 1978, Nucl. Phys. B **144**, 349.
Hawking, S. W., 1979, in *Three Hundred Years of Gravitation,* edited by S. W. Hawking and W. Israel (Cambridge University, Cambridge).
Hawking, S. W., 1982, Commun. Math. Phys. **87**, 395.
Hawking, S. W., 1983, Philos. Trans. R. Soc. London, Ser. A **310**, 303.
Hawking, S. W., 1984a, Nucl. Phys. B **239**, 257.
Hawking, S. W., 1984b, Phys. Lett. B **134**, 403.
Hawking, S. W., 1984c, in *Relativity, Groups and Topology II,* NATO Advanced Study Institute Session XL . . . Les Houches 1983, edited by B. S. DeWitt and Raymond Stora (Elsevier, Amsterdam), p. 336.
Hawking, S. W., 1987a, remarks quoted by M. Gell-Mann, Phys. Scr. **T15**, 202 (1987).
Hawking, S. W., 1987b, Phys. Lett. B **195**, 337.
Hawking, S. W., 1988, Phys. Rev. D **37**, 904.
Hawking, S., and D. Page, 1986, Nucl. Phys. B **264**, 185.
Henneaux, M., and C. Teitelboim, 1984, Phys. Lett. B **143**, 415.
Henneaux, M., and C. Teitelboim, 1988, "The cosmological constant and general covariance," University of Texas preprint.
Hosoya, A., 1988, "A diagrammatic derivation of Coleman's vanishing cosmological constant," Hiroshima Preprint No. RRK-88-28.
Kim, J., 1979, Phys. Rev. Lett. **43**, 103.
Klebanov, I., L. Susskind, and T. Banks, 1988, "Wormholes and cosmological constant," SLAC Preprint No. SLAC-Pub.-4705.
Knapp, G. R., and J. Kormendy, 1987, Eds., *Dark Matter in the Universe: I.A.U. Symposium No. 117* (Reidel, Dordrecht).
Lavrelashvili, G. V., V. A. Rubakov, and P. G. Tinyakov, 1987, Pis'ma Zh. Eksp. Teor. Fiz. **46**, 134 [JETP Lett. **46**, 167 (1987)].
Lavrelashvili, G. V., V. A. Rubakov, and P. G. Tinyakov, 1988, Nucl. Phys. B **299**, 757.
Lemaître, G., 1927, Ann. Soc. Sci. Bruxelles, Ser. 1 **47**, 49.
Lemaître, G., 1931, Mon. Not. R. Astron. Soc. **91**, 483.
Linde, A. D., 1974, Pis'ma Zh. Eksp. Teor. Fiz. **19**, 320 [JETP Lett. **19**, 183 (1974)].
Linde, A. D., 1982, Phys. Lett. B **129**, 389.
Linde, A. D., 1986, Phys. Lett. B **175**, 395.
Linde, A. D., 1987, Phys. Scri. **T15**, 169.
Linde, A. D., 1988a, Phys. Lett. B **200**, 272.
Linde, A. D., 1988b, Phys. Lett. B **202**, 194.
Loh, E. D., 1986, Phys. Rev. Lett. **57**, 2865.
Loh, E. D., and E. J. Spillar, 1986, Astrophys. J. **303**, 154.
Martinec, E., 1986, Phys. Lett. B **171**, 189.
Moore, G., 1987a, Nucl. Phys. B **293**, 139.
Moore, G., 1987b, Institute for Advanced Study Preprint No. IASSNS-HEP-87/59, to be published in the proceedings of the Cargèse School on Nonperturbative Quantum Field Theory.
Morozov, A., and A. Perelomov, 1987, Phys. Lett. B **199**, 209.
Myers, R. C., and V. Periwal, 1988, "Constants and correlations in the Coleman calculus," Santa Barbara Preprint No. NSF-ITP-88-151.
Page, D., 1987, in *The World and I* (in press).
Pais, A., 1982, *'Subtle is the Lord . . .': The Science and the Life of Albert Einstein* (Oxford University, New York).
Peccei, R. D., J. Solà, and C. Wetterich, 1987, Phys. Lett. B **195**, 183.
Peebles, P. J. E., 1984, Astrophys. J. **284**, 439.
Peebles, P. J. E., 1987a, in *Proceedings of the Summer Study on the Physics of the Superconducting Super Collider,* edited by R. Donaldson and J. Marx (Division of Particles and Fields of the APS, New York).
Peebles, P. J. E., 1978b, Publ. Astron. Soc. Pac., in press.
Peebles, P. J. E., and B. Ratra, 1988, Astrophys. J. Lett. **325**, L17.
Petrosian, V., E. E. Salpeter, and P. Szekeres, 1967, Astrophys. J. **147**, 1222.
Polchinski, J., 1986, Commun. Math. Phys. **104**, 37.

Polchinski, J., 1987, private communication.
Polchinski, J., 1988, in preparation.
Ratra, B., and P. J. E. Peebles, 1988, Phys. Rev. D **37**, 3406.
Rees, M. J., 1987, New Sci. August 6, 1987, p. 43.
Renzini, A., 1986, in *Galaxy Distances and Deviations from Universal Expansion,* edited by B. F. Madore and R. B. Tully (Reidel, Dordrecht), p. 177.
Reuter, M., and C. Wetterich, 1987, Phys. Lett. **188**, 38.
Rohm, R., 1984, Nucl. Phys. B **237**, 553.
Rowan-Robinson, M., 1968, Mon. Not. R. Astron. Soc. **141**, 445.
Rubakov, V. A., 1988, "On the third quantization and the cosmological constant," DESY preprint.
Shklovsky, I., 1967, Astrophys. J. **150**, L1.
Slipher, V. M., 1924, table in Eddington (1924), *The Mathematical Theory of Relativity,* 2nd Ed. (Cambridge University, London), p. 162
Sparnaay, M. J., 1957, Nature (London) **180**, 334.
Strominger, A., 1984, Phys. Rev. Lett. **52**, 1733.
Teitelboim, C., 1982, Phys. Rev. D **25**, 3159.
Turner, M. S., G. Steigman, and L. M. Krauss, 1984, Phys. Rev. Lett. **52**, 2090.
Van der Bij, J. J., H. Van Dam, and Y. J. Ng, 1982, Physica **116A**, 307.
Veltman, M., 1975, Phys. Rev. Lett. **34**, 777.
Vilenkin, A., 1986, Phys. Rev. D **33**, 3560.
Vilenkin, A., 1988a, Phys. Rev. D **37**, 888.
Vilenkin, A., 1988b, Tufts Preprint No. TUTP-88-3.
Weinberg, S., 1972, *Gravitation and Cosmology* (Wiley, New York).
Weinberg, S., 1979a, in *General Relativity: An Einstein Centenary Survey,* edited by S. W. Hawking and W. Israel (Cambridge University, Cambridge), p. 800.
Weinberg, S., 1979b, Physica **96A**, 327.
Weinberg, S., 1982, Phys. Rev. Lett. **48**, 1776.
Weinberg, S., 1983, unpublished remarks at the workshop on "Problems in Unification and Supergravity," La Jolla Institute, 1983.
Weinberg, S., 1987, Phys. Rev. Lett. **59**, 2607.
Wetterich, C., 1988, Nucl. Phys. B **302**, 668.
Wheeler, J. A., 1964, in *Relativity, Groups and Topology,* Lectures Delivered at Les Houches, 1963 . . . , edited by B. DeWitt and C. DeWitt (Gordon and Breach, New York), p. 317.
Wheeler, J. A., 1968, in *Battelle Rencontres,* edited by C. DeWitt and J. A. Wheeler (Benjamin, New York).
Wilczek, F., 1984, Phys. Rep. **104**, 143.
Wilczek, F., 1985, in *How Far Are We from the Gauge Forces: Proceedings of the 1983 Erice Conference,* edited by A. Zichichi (Plenum, New York), p. 208.
Wilczek, F., and A. Zee, 1983, unpublished work quote by Zee (1985), in *High Energy Physics: Proceedings of the Annual Orbis Scientiae,* edited by S. L. Mintz and A. Perlmutter (Plenum, New York).
Witten, E., 1983, in *Proceedings of the 1983 Shelter Island Conference on Quantum Field Theory and the Fundamental Problems of Physics,* edited by R. Jackiw, N. N. Khuri, S. Weinberg, and E. Witten (MIT, Cambridge, Massachusetts), p. 273.
Witten, E., 1985, Phys. Lett. B **155**, 151.
Witten, E., and J. Bagger, 1982, Phys. Lett. B **115**, 202.
Zee, A., 1985, in *High Energy Physics: Proceedings of the 20th Annual Orbis Scientiae, 1983,* edited by S. L. Mintz and A. Perlmutter (Plenum, New York).
Zeldovich, Ya., B., 1967, Pis'ma Zh. Eksp. Teor. Fiz. **6**, 883 [JETP Lett. **6**, 316 (1967)].
Zumino, B., 1975, Nucl. Phys. B **89**, 535.

REVIEWS OF MODERN PHYSICS, VOLUME 75, APRIL 2003

The cosmological constant and dark energy

P. J. E. Peebles

Joseph Henry Laboratories, Princeton University, Princeton, New Jersey 08544

Bharat Ratra

Department of Physics, Kansas State University, Manhattan, Kansas 66506

(Published 22 April 2003)

Physics welcomes the idea that space contains energy whose gravitational effect approximates that of Einstein's cosmological constant, Λ; today the concept is termed dark energy or quintessence. Physics also suggests that dark energy could be dynamical, allowing for the arguably appealing picture of an evolving dark-energy density approaching its natural value, zero, and small now because the expanding universe is old. This would alleviate the classical problem of the curious energy scale of a millielectron volt associated with a constant Λ. Dark energy may have been detected by recent cosmological tests. These tests make a good scientific case for the context, in the relativistic Friedmann-Lemaître model, in which the gravitational inverse-square law is applied to the scales of cosmology. We have well-checked evidence that the mean mass density is not much more than one-quarter of the critical Einstein–de Sitter value. The case for detection of dark energy is not yet as convincing but still serious; we await more data, which may be derived from work in progress. Planned observations may detect the evolution of the dark-energy density; a positive result would be a considerable stimulus for attempts at understanding the microphysics of dark energy. This review presents the basic physics and astronomy of the subject, reviews the history of ideas, assesses the state of the observational evidence, and comments on recent developments in the search for a fundamental theory.

CONTENTS

I. INTRODUCTION

There is significant observational evidence for the detection of Einstein's cosmological constant, Λ, or a component of the material content of the universe that varies only slowly with time and space and so acts like Λ. We shall use the term *dark energy* for Λ or a component that acts like it. Detection of dark energy would be a new clue to an old puzzle: the gravitational effect of the zero-point energies of particles and fields. The total with other energies, that are close to homogeneous and nearly independent of time, acts as dark energy. What is puzzling is that the value of the dark-energy density has to be tiny compared to what is suggested by dimensional analysis; the startling new evidence is that it may be different from the only other natural value, zero.

The main question to consider now is whether to accept the evidence for detection of dark energy. We outline the nature of the case in this section. After reviewing the basic concepts of the relativistic world model in Sec. II, and in Sec. III reviewing the history of ideas, we present in Sec. IV a more detailed assessment of the cosmological tests and the evidence for detection of Λ or its analog in dark energy.

There is little new to report on the big issue for physics—why the dark-energy density is so small—since Weinberg's (1989) review in this journal.[1] But there have been analyses of a simpler idea: can we imagine that the present dark-energy density is evolving, perhaps approaching zero? Models are introduced in Secs. II.C and III.E, and recent work is summarized in more detail in the Appendix. Feasible advances in cosmological tests could detect evolution of the dark-energy density, and perhaps its gravitational response to large-scale fluctuations in the mass distribution. This would substantially motivate the search for a more fundamental physics model for dark energy.

A. The issues for observational cosmology

We will make two points. First, cosmology has a substantial observational and experimental basis, which supports many aspects of the standard model as almost certainly being good approximations to reality. Second, the empirical basis is not nearly as strong for cosmology as it is for the standard model of particle physics: in cosmology it is not yet a matter of measuring the parameters in a well-established theory.

To explain the second point we direct our attention to those more accustomed to experiments in the laboratory than to astronomy-related observations of astronomers' Tantalus principle: one can look at distant objects but never touch them. For example, the observations of supernovae in distant galaxies offer evidence of dark energy, under the assumption that distant and nearby supernovae are drawn from the same statistical sample (that is, that they are statistically similar enough for the purpose of this test). There is no direct way to check this, and it is easy to imagine differences between distant and nearby supernovae of the same nominal type. More distant supernovae are seen in younger galaxies, because of the travel time of light, and these younger galaxies tend to have more massive rapidly evolving stars with lower heavy-element abundances. How do we know that the properties of the supernovae are not also different? The reader is referred to Leibundgut's (2001, Sec. 4) discussion of astrophysical hazards. Astronomers have checks for this and other issues of interpretation when considering the observations used in cosmological tests. But it takes nothing away from this careful and elegant work to note that the checks are seldom convincing, because the astronomy is complicated and what can be observed is sparse. What is more, we do not know ahead of time that the physics well tested on scales ranging from the laboratory to the Solar System survives the enormous extrapolation to cosmology.

The situation is by no means hopeless. We now have significant cross-checks from the consistency of results based on independent applications of the astronomy and of the physics of the cosmological model. If the physics or astronomy was faulty we would not expect consistency from independent lines of evidence—apart from the occasional accident and the occasional tendency to stop the analysis when it approaches the "right answer." We have to demand abundant evidence of consistency, and that is starting to appear.

The case for detection of Λ or dark energy commences with the Friedmann-Lemaître cosmological model. In this model the expansion history of the universe is determined by a set of dimensionless parameters whose sum is normalized to unity,

$$\Omega_{M0}+\Omega_{R0}+\Omega_{\Lambda 0}+\Omega_{K0}=1. \quad (1)$$

The first, Ω_{M0}, is a measure of the present mean mass density in nonrelativistic matter, mainly baryons and nonbaryonic dark matter. The second, $\Omega_{R0}\sim 1\times 10^{-4}$, is a measure of the present mass in the relativistic 3-K thermal cosmic microwave background radiation, which almost homogeneously fills space, and the accompanying low-mass neutrinos. The third is a measure of Λ or the present value of the dark-energy equivalent. The fourth, Ω_{K0}, is an effect of the curvature of space. We review some details of these parameters in the next section, and of their measurements in Sec. IV.

The most direct evidence for detection of dark energy comes from observations of supernovae of a type whose intrinsic luminosities are close to uniform (after subtle astronomical corrections, a few details of which are discussed in Sec. IV.B.4). The observed brightness as a function of the wavelength shift of the radiation probes the geometry of spacetime, in what has come to be called the redshift-magnitude relation.[2] The measurements agree with the relativistic cosmological model with $\Omega_{K0}=0$, meaning no space curvature, and $\Omega_{\Lambda 0}\sim 0.7$, meaning nonzero Λ. A model with $\Omega_{\Lambda 0}=0$ is two

[1] Sahni and Starobinsky (2000); Carroll (2001); Weinberg (2001); Witten (2001); and Ellwanger (2002) present more recent reviews.

[2] The apparent magnitude is $m=-2.5\log_{10}f$ plus a constant, where f is the detected energy flux density in a chosen wavelength band. The standard measure of the wavelength shift, due to the expansion of the universe, is the redshift z defined in Eq. (7) below.

or three standard deviations off the best fit, depending on the data set and analysis technique. This is an important indication, but 2 to 3 σ is not convincing, even when we can be sure that systematic errors are under reasonable control. And we have to consider that there may be a significant systematic error from differences between distant, high-redshift, and nearby, low-redshift, supernovae.

There is a check, based on the cold-dark-matter (CDM) model[3] for structure formation. The fit of the model to the observations reviewed in Sec. IV.B yields two key constraints. First, the angular power spectrum of fluctuations in the temperature of the 3-K thermal cosmic microwave background radiation across the sky indicates that Ω_{K0} is small. Second, the power spectrum of the spatial distribution of the galaxies requires $\Omega_{M0} \sim 0.25$. Similar estimates of Ω_{M0} follow from independent lines of observational evidence. The rate of gravitational lensing prefers a somewhat larger value (if Ω_{K0} is small), and some dynamical analyses of systems of galaxies prefer lower Ω_{M0}. But the differences could all result from measurement uncertainties. Since Ω_{R0} in Eq. (1) is small, the conclusion is that $\Omega_{\Lambda 0}$ is large, in excellent agreement with the supernovae result.

Caution is in order, however, because this check depends on the CDM model for structure formation. We cannot see the dark matter, so we naturally assign it the simplest properties possible. Maybe it is significant that the model has observational problems with galaxy formation, as discussed in Sec. IV.A.2, or maybe these problems are only apparent, due to the complications of the astronomy. We are going to have to determine which is correct before we can have confidence in the role of the CDM model in cosmological tests. We will get a strong hint from current precision angular distribution measurements of the 3-K thermal cosmic microwave background radiation.[4] If the results match precisely the prediction of the relativistic model for cosmology and the CDM model for structure formation, with parameter choices that agree with the constraints from all the other cosmological tests, there will be strong evidence that we are approaching a good approximation to reality, and the completion of the great program of cosmological tests that commenced in the 1930s. But all that is in the future.

We wish to emphasize that the advances in the empirical basis for cosmology already are very real and substantial. How firm the conclusion is depends on the issue, of course. Every competent cosmologist we know accepts as established beyond reasonable doubt that the universe is expanding and cooling in a near homogeneous and isotropic way from a hotter denser state: how else could space, which is transparent now, have been filled with radiation that has relaxed to a thermal spectrum? The debate is when the expansion commenced or became a meaningful concept. Some whose opinions and research we respect question the extrapolation of the gravitational inverse-square law, in its use in estimates of masses in galaxies and systems of galaxies, and of Ω_{M0}. We agree that this law is one of the hypotheses to be tested. Our conclusion from the cosmological tests of Sec. IV is that the law passes significant, though not yet complete, tests, and that we already have a strong scientific case, resting on the abundance of cross-checks, that the matter density parameter Ω_{M0} is about one-quarter. The case for detection of $\Omega_{\Lambda 0}$ is significant too, but not yet as compelling.

For the most part the results of the cosmological tests agree wonderfully well with accepted theory. But the observational challenges to the tests are substantial: we are drawing profound conclusions from very limited information. We have to be liberal when considering ideas about what the universe is like, and conservative when accepting ideas into the established canon.

B. The opportunity for physics

Unless there is some serious and quite unexpected flaw in our understanding of the principles of physics we can be sure the zero-point energy of the electromagnetic field at laboratory wavelengths is real and measurable, as in the Casimir (1948) effect.[5] Like all energy, this zero-point energy has to contribute to the source term in Einstein's gravitational field equation. If, as seems likely, the zero-point energy of the electromagnetic field is close to homogeneous and independent of the velocity of the observer, it manifests itself as a positive contribution to Einstein's Λ, or dark energy. The zero-point energies of the fermions make a negative contribution. Other contributions, perhaps including the energy densities of fields that interact only with themselves and

[3]The model is named after the nonbaryonic cold dark matter that is assumed to dominate the masses of galaxies in the present universe. There are more assumptions in the CDM model, of course; they are discussed in Secs. III.D and IV.A.2.

[4]At the time of writing the Microwave Anisotropy Probe (MAP) satellite is collecting data; the project is described in http://map.gsfc.nasa.gov/

[5]See Bordag, Mohideen, and Mostepanenko (2001) for a recent review. The attractive Casimir force between two parallel conducting plates results from the boundary condition that suppresses the number of modes of oscillation of the electromagnetic field between the plates, thus suppressing the energy of the system. One can understand the effect at small separation without reference to the quantum behavior of the electromagnetic field, such as in the analysis of the van der Waals interaction in quantum mechanics, by taking account of the term in the particle Hamiltonian for the Coulomb potential energy between the charged particles in the two separate neutral objects. But a more complete treatment, as discussed by Cohen-Tannoudji, Dupont-Roc, and Grynberg (1992), replaces the Coulomb interaction with the coupling of the charged particles to the electromagnetic-field operator. In this picture the van der Waals interaction is mediated by the exchange of virtual photons. With either way of looking at the Casimir effect—the perturbation of the normal modes or the exchange of virtual quanta of the unperturbed modes—the effect is the same, the suppression of the energy of the system.

gravity, might have either sign. The value of the sum suggested by dimensional analysis is much larger than what is allowed by the relativistic cosmological model. The only other natural value is $\Lambda=0$. If Λ really is tiny but not zero, this introduces a most stimulating though enigmatic clue to the physics yet to be discovered.

To illustrate the problem we outline an example of a contribution to Λ. The energy density in the 3-K thermal cosmic microwave background radiation, which amounts to $\Omega_{R0}\sim 5\times 10^{-5}$ in Eq. (1) (ignoring the neutrinos), peaks at wavelength $\lambda\sim 2$ mm. At this Wien peak the photon occupation number is about one-fifteenth. The zero-point energy amounts to half the energy of a photon at the given frequency. This means the zero-point energy in the electromagnetic field at wavelengths $\lambda\sim 2$ mm amounts to a contribution $\delta\Omega_{\Lambda 0}\sim 4\times 10^{-4}$ to the density parameter in Λ or the dark energy. The sum over the modes scales as λ^{-4} [as illustrated in Eq. (37)]. Thus a naive extrapolation to visible wavelengths determines that the contribution amounts to $\delta\Omega_{\Lambda 0}\sim 5\times 10^{10}$, already a ridiculous number.

The situation can be compared to the development of the theory of weak interactions. The Fermi pointlike interaction model is strikingly successful for a considerable range of energies, but it was clear from the start that the model fails at high energy. A fix was discussed—mediate the interaction by an intermediate boson—and eventually incorporated into the even more successful electroweak theory. General relativity and quantum mechanics are extremely successful over a considerable range of length scales, provided we agree not to use the rules of quantum mechanics to count the zero-point energy density in the vacuum, even though we know we have to count the zero-point energies in all other situations. There are thoughts on improving the situation, though they seem to be less focused than was the case for the Fermi model. Perhaps a new energy component spontaneously cancels the vacuum energy density or the new component varies slowly with position and here and there happens to cancel the vacuum energy density well enough to allow observers like us to exist. Whatever the nature of the more perfect theory, it must reproduce the successes of general relativity and quantum mechanics. That includes the method of representing the material content of the observable universe—all forms of mass and energy—by the stress-energy tensor, and the relation between the stress-energy tensor and the curvature of macroscopic spacetime. One part has to be adjusted.

The numerical values of the parameters in Eq. (1) also are enigmatic, and possibly trying to tell us something. The evidence is that the parameters have the approximate values

$$\Omega_{\Lambda 0}\sim 0.7,\quad \Omega_{DM0}\sim 0.2,\quad \Omega_{B0}\sim 0.05. \tag{2}$$

We have written Ω_{M0} in two parts: Ω_{B0} measures the density of the baryons we know exist and Ω_{DM0} measures the hypothetical nonbaryonic cold dark matter we need to fit the cosmological tests. The parameters Ω_{B0} and Ω_{DM0} have similar values but represent different things—baryonic and nonbaryonic matter—and $\Omega_{\Lambda 0}$, which is thought to represent something completely different, is not much larger. Also, if the parameters were measured when the universe was one-tenth its present size the time-independent Λ parameter would contribute $\Omega_\Lambda\sim 0.003$. That is, we seem to have come on the scene just as Λ has become an important factor in the expansion rate. These curiosities surely are in part accidental, but maybe in part physically significant. In particular, one might imagine that the dark-energy density represented by Λ is rolling to its natural value, zero, but is very small now because we measure it when the universe is very old. We shall discuss efforts along this line to at least partially rationalize the situation.

C. Some explanations

We have to explain our choice of nomenclature. Basic concepts of physics say that space contains homogeneous zero-point energy, and perhaps also energy that is homogeneous or nearly so in other forms, real or effective (such as from counter terms in gravity physics, which make the net energy density cosmologically acceptable). In the literature this near homogeneous energy has been termed vacuum energy, the sum of vacuum energy and quintessence (Caldwell, Davé, and Steinhardt, 1998), and dark energy (Turner, 1999). We have adopted the last term, and we shall refer to the dark-energy density ρ_Λ that manifests itself as an effective version of Einstein's cosmological constant, but one that may vary slowly with time and position.[6]

Our subject involves two quite different traditions, in physics and astronomy. Each has familiar notation, and familiar ideas that may be "in the air" but not in recent literature. Our attempt to take account of these traditions commences with the summary in Sec. II of the basic notation with brief explanations. We expect that readers will find some of these concepts trivial and others of some use, and that the useful parts will be different for different readers.

We offer in Sec. III our reading of the history of ideas on Λ and its generalization to dark energy. This is a fascinating and we think edifying illustration of how science may advance in unexpected directions. It is relevant to an understanding of the present state of research in cosmology, because traditions inform opinions, and people have had mixed feelings about Λ ever since Einstein (1917) introduced it 85 years ago. The concept never entirely disappeared in cosmology because a series of observations hinted at its presence, and because to some cosmologists Λ fits the formalism too well to be ignored. The search for the physics of the vacuum, and its possible relation to Λ, has a long history too. Despite

[6]The dark energy should of course be distinguished from a hypothetical gas of particles with velocity dispersion large enough that the distribution is close to homogeneous.

the common and strong suspicion that Λ must be negligibly small, because any other acceptable value is absurd, all this history has made the community well prepared for the recent observational developments that argue for the detection of Λ.

Our approach in Sec. IV to the discussion of the evidence for detection of Λ, from the cosmological tests, also requires explanation. One occasionally reads that the tests will show us how the world will end. That certainly seems interesting, but it is not the main point: why should we trust an extrapolation into the indefinite future of a theory that we can at best show is a good approximation to reality?[7] As we remarked in Sec. I.A, the purpose of the tests is to check the approximation to reality, by checking the physics and astronomy of the standard relativistic cosmological model, along with any viable alternatives that may be discovered. We take our task to be the identification of the aspects of the standard theory that enter the interpretation of the measurements and thus are or may be empirically checked or measured.

II. BASIC CONCEPTS

A. The Friedmann-Lemaître model

The standard world model is close to homogeneous and isotropic on large scales, and lumpy on small scales—the effect of mass concentrations in galaxies, stars, people, etc. The length scale at the transition from nearly smooth to strongly clumpy is about 10 Mpc. We use here and throughout the standard astronomers' length unit,

$$1\ \text{Mpc}=3.1\times10^{24}\ \text{cm}=3.3\times10^{6}\ \text{light years.} \tag{3}$$

To be more definite, imagine that many spheres of radius 10 Mpc are placed at random, and the mass within each is measured. At this radius the rms fluctuation in the set of values of masses is about equal to the mean value. On smaller scales the departures from homogeneity are progressively more nonlinear; on larger scales the density fluctuations are perturbations to the homogeneous model. From now on we mention these perturbations only when relevant for the cosmological tests.

The expansion of the universe means the distance $l(t)$ between two well-separated galaxies varies with world time t as

$$l(t)\propto a(t), \tag{4}$$

where the expansion or scale factor $a(t)$ is independent of the choice of galaxies. It is an interesting exercise, for those who have not already thought to do so, to check that Eq. (4) is required to preserve homogeneity and isotropy.[8]

The rate of change of the distance in Eq. (4) is the speed

$$v=dl/dt=Hl, \quad H=\dot{a}/a, \tag{5}$$

where the overdot means the derivative with respect to world time t and H is the time-dependent Hubble parameter. When v is small compared to the speed of light this is Hubble's law. The present value of H is Hubble's constant, H_0. When needed we will use[9]

$$H_0=100h\ \ \text{km s}^{-1}\,\text{Mpc}^{-1}=67\pm7\ \ \text{km s}^{-1}\,\text{Mpc}^{-1}$$
$$=(15\pm2\ \ \text{Gyr})^{-1}, \tag{6}$$

at two standard deviations. The first equation defines the dimensionless parameter h.

Another measure of the expansion follows by considering the stretching of the wavelength of light received from a distant galaxy. The observed wavelength λ_{obs} of a feature in the spectrum that had wavelength λ_{em} at emission satisfies

$$1+z=\lambda_{\text{obs}}/\lambda_{\text{em}}=a(t_{\text{obs}})/a(t_{\text{em}}), \tag{7}$$

---

[7] Observations may now have detected Λ, at a characteristic energy scale of a millielectron volt [Eq. (47)]. We have no guarantee that an even lower-energy scale does not exist; such a scale could first become apparent through cosmological tests.

[8] We feel we have to comment on a few details about Eq. (4) to avoid contributing to debates that are more intense than seem warranted. Think of the world time t as the proper time kept by each of a dense set of observers, each moving so that all the others are isotropically moving away, and with the times synchronized to a common energy density, $\rho(t)$, in the near homogeneous expanding universe. The distance $l(t)$ is the sum of the proper distances between neighboring observers, all measured at time t, and along the shortest distance between the two observers. The rate of increase of the distance, dl/dt, may exceed the velocity of light. This is no more problematic in relativity theory than is the large speed at which the beam of a flashlight on Earth may swing across the face of the Moon (assuming an adequately tight beam). Space sections at fixed t may be noncompact, and the total mass of a homogeneous universe formally infinite. As far as is known this is not meaningful: we can only assert that the universe is close to homogeneous and isotropic over observable scales, and that what can be observed is a finite number of baryons and photons.

[9] The numerical values in Eq. (6) are determined from an analysis of all available measurements of H_0 prior to mid-1999 (Gott *et al.*, 2001). They are a very reasonable summary of the current situation. For instance, the Hubble Space Telescope Key Project summary measurement value $H_0=72\pm8\ \text{km s}^{-1}\,\text{Mpc}^{-1}$ (1σ uncertainty, Freedman *et al.*, 2001) is in very good agreement with Eq. (6), as is the recent Tammann *et al.* (2001) summary value $H_0=60\pm6\ \text{km s}^{-1}\,\text{Mpc}^{-1}$ (approximate 1σ systematic uncertainty). This is an example of the striking change in the observational situation over the previous five years: the uncertainty in H_0 has decreased by more than a factor of 3, making it one of the better-measured cosmological parameters.

where the expansion factor a is defined in Eq. (4) and z is the redshift. That is, the wavelength of freely traveling radiation stretches in proportion to the factor by which the universe expands. To understand this, imagine that a large part of the universe is enclosed in a cavity with perfectly reflecting walls. The cavity expands with the general expansion, the widths proportional to $a(t)$. Electromagnetic radiation is a sum of the normal modes that fit the cavity. At interesting wavelengths the mode frequencies are much larger than the rate of expansion of the universe, so adiabaticity says a photon in a mode stays there, and its wavelength thus must vary as $\lambda \propto a(t)$, as stated in Eq. (7). The cavity disturbs the long-wavelength part of the radiation, but the disturbance can be made exceedingly small by choosing a large cavity.

Equation (7) defines the redshift z. The redshift is a convenient label for epochs in the early universe, where z exceeds unity. A good exercise for the student is to check that when z is small Eq. (7) reduces to Hubble's law, where λz is the first-order Doppler shift in the wavelength λ, and Hubble's parameter H is given by Eq. (5). Thus Hubble's law may be written as $cz = Hl$ (where we have put in the speed of light).

These results follow from the symmetry of the cosmological model and conventional local physics; we do not need general relativity theory. When $z \gtrsim 1$ we need relativistic theory to compute the relations among the redshift and other observables. An example is the relation between redshift and apparent magnitude used in the supernova test. Other cosmological tests check consistency among these relations, and this checks the world model.

In general relativity the second time derivative of the expansion factor satisfies

$$\frac{\ddot{a}}{a} = -\frac{4}{3}\pi G(\rho + 3p). \tag{8}$$

The gravitational constant is G. Here and throughout we choose units to set the velocity of light to unity. The mean mass density, $\rho(t)$, and the pressure, $p(t)$, counting all contributions including dark energy, satisfy the local energy-conservation law

$$\dot{\rho} = -3\frac{\dot{a}}{a}(\rho + p). \tag{9}$$

The first term on the right-hand side represents the decrease of mass density due to the expansion that more broadly disperses the matter. The pdV work in the second term is a familiar local concept, and is meaningful in general relativity. But one should note that energy does not have a general global meaning in this theory.

The first integral of Eqs. (8) and (9) is the Friedmann equation

$$\dot{a}^2 = \frac{8}{3}\pi G\rho a^2 + \text{const.} \tag{10}$$

It is conventional to rewrite this as

$$\begin{aligned}\left(\frac{\dot{a}}{a}\right)^2 &= H_0^2 E(z)^2 \\ &= H_0^2[\Omega_{M0}(1+z)^3 + \Omega_{R0}(1+z)^4 \\ &\quad + \Omega_{\Lambda 0} + \Omega_{K0}(1+z)^2].\end{aligned} \tag{11}$$

The first equation defines the function $E(z)$ that is introduced for later use. The second equation assumes constant Λ; the time-dependent dark-energy case is reviewed in Secs. II.C and III.E. The first term in the last part of Eq. (11) represents nonrelativistic matter with negligibly small pressure; one sees from Eqs. (7) and (9) that the mass density in this form varies with the expansion of the universe as $\rho_M \propto a^{-3} \propto (1+z)^3$. The second term represents radiation and relativistic matter, with pressure $p_R = \rho_R/3$, whence $\rho_R \propto (1+z)^4$. The third term is the effect of Einstein's cosmological constant, or a constant dark-energy density. The last term, discussed in more detail below, is the constant of integration in Eq. (10). The four density parameters Ω_{i0} are the fractional contributions to the square of Hubble's constant, H_0^2, that is, $\Omega_{i0}(t) = 8\pi G\rho_{i0}/(3H_0^2)$. At the present epoch, $z=0$, the present value of $\dot{a}/a$ is H_0, and the Ω_{i0} sum to unity [Eq. (1)].

In this notation, Eq. (8) is

$$\frac{\ddot{a}}{a} = -H_0^2[\Omega_{M0}(1+z)^3/2 + \Omega_{R0}(1+z)^4 - \Omega_{\Lambda 0}]. \tag{12}$$

The constant of integration in Eqs. (10) and (11) is related to the geometry of spatial sections at constant world time. Recall that in general relativity events in spacetime are labeled by the four coordinates x^μ of time and space. Neighboring events at separation dx^μ have invariant separation ds defined by the line element

$$ds^2 = g_{\mu\nu}dx^\mu dx^\nu. \tag{13}$$

The repeated indices are summed, and the metric tensor $g_{\mu\nu}$ is a function of position in spacetime. If ds^2 is positive then ds is the proper (physical) time measured by an observer who moves from one event to the other; if negative, $|ds|$ is the proper distance between the events measured by an observer who is moving so the events are seen to be simultaneous.

In the flat spacetime of special relativity one can choose coordinates so the metric tensor has the Minkowskian form

$$\eta_{\mu\nu} = \begin{pmatrix} 1 & 0 & 0 & 0 \\ 0 & -1 & 0 & 0 \\ 0 & 0 & -1 & 0 \\ 0 & 0 & 0 & -1 \end{pmatrix}. \tag{14}$$

A freely falling, inertial observer can choose locally Minkowskian coordinates, such that along the path of the observer $g_{\mu\nu} = \eta_{\mu\nu}$, and the first derivatives of $g_{\mu\nu}$ vanish.

In the homogeneous world model we can choose coordinates so the metric tensor is of the form that results in the line element

$$ds^2 = dt^2 - a(t)^2\left[\frac{dr^2}{1+Kr^2} + r^2(d\theta^2 + \sin^2\theta d\phi^2)\right]$$
$$= dt^2 - K^{-1}a(t)^2[d\chi^2 + \sinh^2\chi(d\theta^2 + \sin^2\theta d\phi^2)]. \quad (15)$$

In the second expression, which assumes $K>0$, the radial coordinate is $r=K^{-1/2}\sinh\chi$. The expansion factor $a(t)$ appears in Eq. (4). If a were constant and the constant K vanished this would represent the flat spacetime of special relativity in polar coordinates. The key point for now is that Ω_{K0} in Eq. (11), which represents the constant of integration in Eq. (10), is related to the constant K:

$$\Omega_{K0} = K/(H_0 a_0)^2, \quad (16)$$

where a_0 is the present value of the expansion factor $a(t)$. Cosmological tests that are sensitive to the geometry of space constrain the value of the parameter Ω_{K0}, and Ω_{K0} and the other density parameters Ω_{i0} in Eq. (11) determine the expansion history of the universe.

It is useful for what follows to recall that the metric tensor in Eq. (15) satisfies Einstein's field equation, a differential equation we can write as

$$G_{\mu\nu} = 8\pi G T_{\mu\nu}. \quad (17)$$

The left side is a function of $g_{\mu\nu}$ and its first two derivatives; it represents the geometry of spacetime. The stress-energy tensor $T_{\mu\nu}$ represents the material contents of the universe, including particles, radiation, fields, and zero-point energies. An observer in a homogeneous and isotropic universe, moving so the universe is observed to be isotropic, would measure the stress-energy tensor to be

$$T_{\mu\nu} = \begin{pmatrix} \rho & 0 & 0 & 0 \\ 0 & p & 0 & 0 \\ 0 & 0 & p & 0 \\ 0 & 0 & 0 & p \end{pmatrix}. \quad (18)$$

This diagonal form is a consequence of the symmetry; the diagonal components define the pressure and energy density. With Eq. (18), the differential equation (17) yields the expansion-rate equations (11) and (12).

B. The cosmological constant

Special relativity is very successful in laboratory physics. Thus one might guess that any inertial observer would see the same vacuum. A freely moving inertial observer represents spacetime in the neighborhood by locally Minkowskian coordinates, with the metric tensor $\eta_{\mu\nu}$ given in Eq. (14). A Lorentz transformation to an inertial observer with another velocity does not change this Minkowski form. The same must be true of the stress-energy tensor of the vacuum, if all observers see the same vacuum, so it has to be of the form

$$T^{\Lambda}_{\mu\nu} = \rho_\Lambda g_{\mu\nu}, \quad (19)$$

where ρ_Λ is a constant, in a general coordinate labeling. When writing this contribution to the stress-energy tensor separately from the rest, we bring the field equation (17) to

$$G_{\mu\nu} = 8\pi G(T_{\mu\nu} + \rho_\Lambda g_{\mu\nu}). \quad (20)$$

This is Einstein's (1917) revision of the field equation of general relativity, where ρ_Λ is proportional to his cosmological constant Λ; his reason for writing down this equation is discussed in Sec. III.A. In many dark-energy scenarios ρ_Λ is a slowly varying function of time and its stress-energy tensor differs slightly from Eq. (19), so the observed properties of the vacuum depend on the observer's velocity.

One sees from Eqs. (14), (18), and (19) that the new component in the stress-energy tensor resembles an ideal fluid with negative pressure,

$$p_\Lambda = -\rho_\Lambda. \quad (21)$$

This fluid picture is of limited use, but the following properties are worth noting.[10]

The stress-energy tensor of an ideal fluid with four-velocity u^μ generalizes from Eq. (18) to $T^{\mu\nu}=(\rho+p)u^\mu u^\nu - p g^{\mu\nu}$. The equations of fluid dynamics follow from the vanishing of the divergence of $T^{\mu\nu}$. Let us consider the simple case of locally Minkowskian coordinates, meaning free fall, and a fluid that is close to homogeneous. By the latter we mean the fluid velocity $\vec{v}$—the space part of the four-velocity u^μ—and the density fluctuation $\delta\rho$ from homogeneity may be treated in linear perturbation theory. Then the equations of energy and momentum conservation are

$$\delta\dot{\rho} + (\langle\rho\rangle + \langle p\rangle)\nabla\cdot\vec{v} = 0, \quad (\langle\rho\rangle + \langle p\rangle)\dot{\vec{v}} + c_s^2\nabla\delta\rho = 0, \quad (22)$$

where $c_s^2 = dp/d\rho$ and the mean density and pressure are $\langle\rho\rangle$ and $\langle p\rangle$. These combine to

$$\delta\ddot{\rho} = c_s^2\nabla^2\delta\rho. \quad (23)$$

If c_s^2 is positive this is a wave equation, and c_s is the speed of sound.

The first of Eqs. (22) is the local energy-conservation law, as in Eq. (9). If $p=-\rho$, the pdV work cancels the ρdV part: the work done to increase the volume cancels the effect of the increased volume. This has to be so for a Lorentz-invariant stress-energy tensor, of course, where all inertial observers see the same vacuum. Another way to see this is to note that the energy flux density in Eqs. (22) is $(\langle\rho\rangle + \langle p\rangle)\vec{v}$. This vanishes when

[10]These arguments have been familiar, in some circles, for a long time, though in our experience, discussed more often in private than in the literature. Early statements of elements are in Lemaître (1934) and McCrea (1951); see Kragh (1999, pp. 397 and 398) for a brief historical account.

$p=-\rho$: the streaming velocity loses meaning. When c_s^2 is negative Eq. (23) shows that the fluid is unstable, in general. But when $p=-\rho$ the vanishing divergence of $T^{\mu\nu}$ becomes the condition shown in Eq. (22), that $\rho=\langle\rho\rangle+\delta\rho$ is constant.

There are two measures of gravitational interactions with a fluid: the passive gravitational mass density determines how the fluid streaming velocity is affected by an applied gravitational field, and the active gravitational mass density determines the gravitational field produced by the fluid. When the fluid velocity is nonrelativistic the expression for the former in general relativity is $\rho+p$, as one can determine by writing out the covariant divergence of $T^{\mu\nu}$. This vanishes when $p=-\rho$, consistent with the loss of meaning of the streaming velocity. The latter is $\rho+3p$, as one can see from Eq. (8). Thus a fluid with $p=-\rho/3$, if somehow kept homogeneous and static, would produce no gravitational field.[11] In the model in Eqs. (19) and (21) the active gravitational mass density is negative when ρ_Λ is positive. When this positive ρ_Λ dominates the stress-energy tensor, $\ddot{a}$ is positive: the rate of expansion of the universe increases. In the language of Eq. (20), this cosmic repulsion is a gravitational effect of the negative active gravitational mass density, not a new force law.

The homogeneous active mass represented by Λ changes the equation of relative motion of freely moving test particles in the nonrelativistic limit to

$$\frac{d^2\vec{r}}{dt^2}=\vec{g}+\Omega_{\Lambda 0}H_0^2\vec{r}, \tag{24}$$

where $\vec{g}$ is the relative gravitational acceleration produced by the distribution of ordinary matter.[12] For an illustration of the size of the last term consider its effect on our motion in a nearly circular orbit around the center of the Milky Way galaxy. The Solar System is moving at speed $v_c=220\ \mathrm{km\,s^{-1}}$ at radius $r=8$ kpc. The ratio of the acceleration g_Λ produced by Λ to the total gravitational acceleration $g=v_c^2/r$ is

$$g_\Lambda/g=\Omega_{\Lambda 0}H_0^2r^2/v_c^2\sim 10^{-5}, \tag{25}$$

a small number. Since we are near the edge of the luminous part of our galaxy, a search for the effect of Λ on the internal dynamics of galaxies such as the Milky Way does not look promising. The precision of celestial dynamics in the Solar System is much greater, but the effect of Λ is very much smaller; $g_\Lambda/g\sim 10^{-22}$ for the orbit of the Earth.

One can generalize Eq. (19) to a variable ρ_Λ, by taking p_Λ to be negative but different from $-\rho_\Lambda$. But if the dynamics were that of a fluid, with pressure a function of ρ_Λ, stability would require $c_s^2=dp_\Lambda/d\rho_\Lambda>0$, from Eq. (23), which seems quite contrived. A viable working model for a dynamical ρ_Λ is the dark energy of a scalar field with self-interaction potential chosen to make the variation of the field energy acceptably slow, as discussed next.

C. Inflation and dark energy

The negative active gravitational mass density associated with a positive cosmological constant is an early precursor of the inflation picture of the early universe; inflation in turn is one precursor of the idea that Λ might generalize into evolving dark energy.

To begin, we review some aspects of causal relations between events in spacetime. Neglecting space curvature, a light ray moves a proper distance $dl=a(t)dx=dt$ in time interval dt, so the integrated coordinate displacement is

$$x=\int dt/a(t). \tag{26}$$

If $\Omega_{\Lambda 0}=0$ this integral converges in the past—we see distant galaxies that at the time of observation cannot have seen us since the singular start of expansion at $a=0$. This "particle horizon problem" is curious: how could distant galaxies in different directions in the sky know to look so similar? The inflation idea is that in the early universe the expansion history approximates that of de Sitter's (1917) solution to Einstein's field equation for $\Lambda>0$ and $T_{\mu\nu}=0$ in Eq. (20). We can choose the coordinate labels in this de Sitter spacetime so space curvature vanishes. Then Eqs. (11) and (12) show that the expansion parameter is

$$a\propto e^{H_\Lambda t}, \tag{27}$$

where H_Λ is a constant. As one sees by working the integral in Eq. (26), here everyone can have seen everyone else in the past. The details need not concern us; for the following discussion two concepts are important. First, the early universe acts like an approximation to de Sitter's solution because it is dominated by a large effective cosmological "constant," or dark-energy density. Second, the dark energy is modeled as that of a near homogeneous field, Φ.

In this scalar field model, motivated by grand unified models of very-high-energy particle physics, the action of the real scalar field Φ (in units chosen so that Planck's constant $\hbar$ is unity) is

---

[11] Lest we contribute to a wrong problem for the student we note that a fluid with $p=-\rho/3$ held in a container would have net positive gravitational mass, from the pressure in the container walls required for support against the negative pressure of the contents. We have finessed the walls by considering a homogeneous situation. We believe Whittaker (1935) gives the first derivation of the relativistic active gravitational mass density. Whittaker also presents an example of the general proposition that the active gravitational mass of an isolated stable object is the integral of the time-time part of the stress-energy tensor in the locally Minkowskian rest frame. Misner and Putman (1959) give the general demonstration.

[12] This assumes that the particles are close enough for application of the ordinary operational definition of proper relative position. The parameters in the last term follow from Eqs. (8) and (21).

$$S=\int d^4x\sqrt{-g}\left[\frac{1}{2}g^{\mu\nu}\partial_\mu\Phi\partial_\nu\Phi-V(\Phi)\right]. \tag{28}$$

The potential-energy density V is a function of the field Φ, and g is the determinant of the metric tensor. When the field is spatially homogeneous [in the line element of Eq. (15)], and space curvature may be neglected, the field equation is

$$\ddot{\Phi}+3\frac{\dot{a}}{a}\dot{\Phi}+\frac{dV}{d\Phi}=0. \tag{29}$$

The stress-energy tensor of this homogeneous field is diagonal (in the rest frame of an observer moving so that the universe is seen to be isotropic), with time and space parts along the diagonal

$$\rho_\Phi=\frac{1}{2}\dot{\Phi}^2+V(\Phi),$$

$$p_\Phi=\frac{1}{2}\dot{\Phi}^2-V(\Phi). \tag{30}$$

If the scalar field varies slowly in time, so that $\dot{\Phi}^2\ll V$, the field energy approximates the effect of Einstein's cosmological constant, with $p_\Phi\simeq-\rho_\Phi$.

The inflation picture assumes that the near exponential expansion of Eq. (27) in the early universe lasts long enough so that every bit of the present observable universe has seen every other bit, and presumably has discovered how to relax to almost exact homogeneity. The field Φ may then start varying rapidly enough to produce the entropy of our universe, and the field or the entropy may produce the baryons, leaving ρ_Φ small or zero. But one can imagine that the late time evolution of ρ_Φ is slow. If it is slower than the evolution in the mass density in matter, there comes a time when ρ_Φ again dominates, and the universe appears to have a cosmological constant.

A model for this late time evolution assumes a potential of the form

$$V=\kappa/\Phi^\alpha, \tag{31}$$

where the constant κ has dimensions of mass raised to the power $\alpha+4$. For simplicity let us suppose the universe after inflation, but at high redshift, is dominated by matter or radiation, with mass density ρ, that drives the power-law expansion, $a\propto t^n$. Then the power-law solution to the field Eq. (29) with the potential in Eq. (31) is

$$\Phi\propto t^{2/(2+\alpha)}, \tag{32}$$

and the ratio of the mass densities in the scalar field and in matter or radiation is

$$\rho_\Phi/\rho\propto t^{4/(2+\alpha)}. \tag{33}$$

In the limit at which the parameter α approaches zero, ρ_Φ is constant, and this model is equivalent to Einstein's Λ.

When $\alpha>0$ the field Φ in this model grows arbitrarily large at large time, so $\rho_\Phi\to0$, and the universe approaches the Minkowskian spacetime of special relativity. This is within a simple model, of course. It is easy to imagine that in other models ρ_Φ approaches a constant positive value at large time, and spacetime approaches the de Sitter solution, or ρ_Φ passes through zero and becomes negative, causing spacetime to collapse to a Big Crunch.

The power-law model with $\alpha>0$ has two properties that seem desirable. First, the solution in Eq. (32) is said to be an *attractor* (Ratra and Peebles, 1988) or a *tracker* (Steinhardt, Wang, and Zlatev, 1999), meaning it is the asymptotic solution for a broad range of initial conditions at high redshift. That includes relaxation to a near homogeneous energy distribution even when gravity has collected the other matter into nonrelativistic clumps. Second, the energy density in the attractor solution decreases less rapidly than that of matter and radiation. This allows us to realize the scenario: after inflation but at high redshift the field energy density ρ_Φ is small so it does not disturb the standard model for the origin of the light elements, but eventually ρ_Φ dominates and the universe acts as if it had a cosmological constant, but one that varies slowly with position and time. We comment on details of this model in Sec III.E.

III. HISTORICAL REMARKS

These comments on what people were thinking are gleaned from the literature and supplemented by private discussions and our own recollections. More is required for a complete history of the subject, of course, but we hope we have captured the main themes and the way in which these themes have evolved into the present appreciation of the situation.

A. Einstein's thoughts

Einstein disliked the idea of an island universe in asymptotically flat spacetime, because a particle could leave the island and move arbitrarily far from all the other matter in the universe, yet preserve all its inertial properties, which he considered a violation of Mach's idea of the relativity of inertia. Einstein's (1917) cosmological model accordingly assumes that the universe is homogeneous and isotropic, on average, thus removing the possibility of arbitrarily isolated particles. Einstein had no empirical support for this assumption, yet it agrees with modern precision tests. There is no agreement as to whether this is more than a lucky guess.

Motivated by the observed low velocities of the then known stars, Einstein assumed that the large-scale structure of the universe is static. He introduced the cosmological constant to reconcile this picture with his general relativity theory. In the notation of Eq. (12), one sees that a positive value of $\Omega_{\Lambda0}$ can balance the positive values of Ω_{M0} and Ω_{R0} for consistency with $\ddot{a}=0$. The balance is unstable: a small perturbation to the mean mass density or the mass distribution causes expansion or contraction of the whole or parts of the universe. One sees this in Eq. (24): the mass distribution can be chosen so the two terms on the right-hand side cancel, but the balance can be upset by redistributing the mass.[13]

Einstein did not consider the cosmological constant to be part of the stress-energy term: his form for the field equation [in the streamlined notation of Eq. (17)] is

$$G_{\mu\nu}-8\pi G\rho_\Lambda g_{\mu\nu}=8\pi G T_{\mu\nu}. \tag{34}$$

The left-hand side contains the metric tensor and its derivatives; a new constant of nature, Λ, appears in the addition to Einstein's original field equation. One can equally place Einstein's new term on the right-hand side of the equation, as in Eq. (20), and count $\rho_\Lambda g_{\mu\nu}$ as part of the source term in the stress-energy tensor. The distinction becomes interesting when ρ_Λ takes part in the dynamics, and the field equation is properly written with ρ_Λ, or its generalization, as part of the stress-energy tensor. One would then be able to say that the differential equation of gravity physics has not changed from Einstein's original form; instead there is a new component in the content of the universe.

Having assumed that the universe is static, Einstein did not write down the differential equation for $a(t)$, and so did not see the instability. Friedmann (1922, 1924) found the evolving homogeneous solution, but had the misfortune to do so before the astronomy became suggestive. Slipher's measurements of the spectra of the spiral nebulae—galaxies of stars—showed most are shifted toward the red, and Eddington (1924, pp. 161 and 162) remarked that that might be a manifestation of the second, repulsive term in Eq. (24). Lemaître (1927) introduced the relation between Slipher's redshifts and a homogeneous matter-filled expanding relativistic world model. He may have been influenced by Hubble's work, which led to the publication (Hubble, 1929) of the linear redshift-distance relation [Eq. (5)]: as a graduate student at MIT Lemaître attended a lecture by Hubble.

In Lemaître's (1927) solution, the expanding universe traces asymptotically back to Einstein's static case. Lemaître then turned to what he called the primeval atom, which is now termed the Big Bang model. This solution expands from densities so large that they require some sort of quantum treatment, passes through a quasistatic approximation to Einstein's solution, and then continues expanding to de Sitter's (1917) empty space solution. To modern tastes, this "loitering" model requires incredibly special initial conditions, as will be discussed. Lemaître liked it because the loitering epoch allows the expansion time to be acceptably long for Hubble's (1929) estimate of H_0, which is an order-of-magnitude high.

The record shows Einstein never liked the Λ term. His view of how general relativity might fit Mach's principle was disturbed by de Sitter's (1917) solution to Eq. (34) for empty space ($T_{\mu\nu}=0$) with $\Lambda>0$.[14,15] Pais (1982, p. 288) pointed out that Einstein, in a letter to Weyl in 1923, commented on the effect of Λ in Eq. (24): "According to De Sitter, two material points that are sufficiently far apart, continue to be accelerated and move apart. If there is no quasistatic world, then away with the cosmological term." We do not know whether at this time Einstein was influenced by Slipher's redshifts or Friedmann's expanding world model.

[13]To help motivate the introduction of Λ, Einstein (1917) mentioned a modification of Newtonian gravity physics that could render the theory well defined when the mass distribution is homogeneous. In Einstein's example, similar to what was considered by Seeliger and Neumann in the mid-1890s, the modified field equation for the gravitational potential φ is $\nabla^2\varphi-\lambda\varphi=4\pi G\rho_M$. This allows the nonsingular homogeneous static solution $\varphi=-4\pi G\rho_M/\lambda$. In this example the potential for an isolated point mass is the Yukawa form, $\varphi \propto e^{-\sqrt{\lambda}r}/r$. Trautman (1965) pointed out that this is not the nonrelativistic limit of general relativity with the cosmological term. Rather, Eq. (24) follows from $\nabla^2\varphi=4\pi G(\rho_M-2\rho_\Lambda)$, where the active gravitational mass density of the Λ term is $\rho_\Lambda+3p_\Lambda=-2\rho_\Lambda$. Norton (1999) reviewed the history of ideas of this Seeliger-Neumann Yukawa-type potential in gravity physics.

[14]North (1965) reviews the confused early history of ideas on the possible astronomical significance of de Sitter's solution for an empty universe with $\Lambda>0$; we add a few comments regarding the physics that contributed to the discovery of the expanding world model. Suppose an observer in de Sitter's spacetime holds a string tied to a source of light, so the source stays at fixed physical distance $r\ll H_\Lambda^{-1}$. The source is much less massive than the observer, the gravitational frequency shift due to the observer's mass may be neglected, and the observer is moving freely. Then the observer receives light from the source shifted to the red by $\delta\lambda/\lambda=-(H_\Lambda r)^2/2$. The observed redshifts of particles moving on geodesics depend on the initial conditions. Stars in the outskirts of our galaxy are held at fixed mean distances from Earth by their motions. The mean shifts of the spectra of light from these stars include this quadratic de Sitter term as well as the much larger Doppler and ordinary gravitational shifts. The prescription for initial conditions that reproduces the linear redshift-distance relation for distant galaxies follows Weyl's (1923) principle: the world particle geodesics trace back to a near common position in the remote past, in the limiting case of the Friedmann-Lemaître model at $\Omega_{M0}\to 0$. This spatially homogeneous coordinate labeling of de Sitter's spacetime, with space sections with negative curvature, already appears in de Sitter [1917, Eq. (15)], and is repeated in Lanczos (1922). This line element is the second expression in our Eq. (15) with $a\propto\cosh H_\Lambda t$. Lemaître (1925) and Robertson (1928) present the coordinate labeling for the spatially flat case, where the line element is $ds^2=dt^2-e^{2H_\Lambda t}(dx^2+dy^2+dz^2)$ [in the choice of symbols and signature in Eqs. (15) and (27)]. Lemaître (1925) and Robertson (1928) note that particles at rest in this coordinate system present a linear redshift-distance relation, $v=H_\Lambda r$, at small v. Robertson (1928) estimated H_Λ, and Lemaître (1927) its analog for the Friedmann-Lemaître model, from published redshifts and Hubble's galaxy distances. Their estimates are not far off Hubble's (1929) published value.

[15]To the present way of thinking the lengthy debate about the singularity in de Sitter's static solution, chronicled by North (1965), seems surprising, because de Sitter (1917) and Klein (1918) had presented de Sitter's solution as a sphere embedded in 4-plus-1-dimensional flat space, with no physical singularity.

The earliest published comments we have found on Einstein's opinion of Λ within the evolving world model (Einstein, 1931; Einstein and de Sitter, 1932) make the point that, since not all the terms in the expansion-rate equation (11) are logically required, and the matter term surely is present and likely dominates over radiation at low redshift, a reasonable working model drops Ω_{K0} and $\Omega_{\Lambda 0}$ and ignores Ω_{R0}. This simplifies the expansion-rate equation to what has come to be called the Einstein–de Sitter model,

$$\frac{\dot{a}^2}{a^2}=\frac{8}{3}\pi G\rho_M, \tag{35}$$

where ρ_M is the mass density in nonrelativistic matter; here $\Omega_M=8\pi G\rho_M/(3H^2)$ is unity. The left side is a measure of the kinetic energy of expansion per unit mass, and the right-hand side a measure of the negative of the gravitational potential energy. In effect, this model universe expands with escape velocity.

Einstein and de Sitter point out that Hubble's estimate of H_0 and de Sitter's estimate of the mean mass density in galaxies are not inconsistent with Eq. (35) (and since both quantities scale with distance in the same way, this result is not affected by the error in the distance scale that affected Hubble's initial measurement of H_0). But the evidence shows now that the mass density is about one-quarter of what is predicted by this equation, as we will discuss.

Einstein and de Sitter (1932) remarked that the curvature term in Eq. (11) is "essentially determinable, and an increase in the precision of the data derived from observations will enable us in the future to fix its sign and determine its value." This is happening, 70 years later. The cosmological constant term is measurable, in principle, too, and may now have been detected. But Einstein and de Sitter said only that the theory of an expanding universe with finite mean mass density "can be reached without the introduction of" Λ.

Further to this point, in the appendix of the second edition of his book, *The Meaning of Relativity*, Einstein (1945, p. 127) states that the "introduction of the 'cosmologic member'"—Einstein's terminology for Λ—"into the equations of gravity, though possible from the point of view of relativity, is to be rejected from the point of view of logical economy," and that if "Hubble's expansion had been discovered at the time of the creation of the general theory of relativity, the cosmologic member would never have been introduced. It seems now so much less justified to introduce such a member into the field equations, since its introduction loses its sole original justification,—that of leading to a natural solution of the cosmologic problem." Einstein knew that without the cosmological constant the expansion time derived from Hubble's estimate of H_0 is uncomfortably short compared to estimates of the ages of the stars, and opined that that might be a problem with the star ages. The big error, the value of H_0, was corrected by 1960 (Sandage, 1958, 1962).

Gamow (1970, p. 44) recalls that "when I was discussing cosmological problems with Einstein, he remarked that the introduction of the cosmological term was the biggest blunder he ever made in his life." This certainly is consistent with all of Einstein's written comments that we have seen on the cosmological constant *per se*; we do not know whether Einstein was also referring to the missed chance to predict the evolution of the universe.

B. The development of ideas

1. Early indications of Λ

In the classic book, *The Classical Theory of Fields*, Landau and Lifshitz (1951, p. 338) second Einstein's opinion of the cosmological constant Λ, stating there is "no basis whatsoever" for adjustment of the theory to include this term. The empirical side of cosmology is not much mentioned in this book, however (though there is a perceptive comment on the limited empirical support for the homogeneity assumption; p. 332). In the Supplementary Notes to the English translation of his book, *Theory of Relativity*, Pauli (1958, p. 220) also endorses Einstein's position.

Discussions elsewhere in the literature on how one might find empirical constraints on the values of the cosmological parameters usually take account of Λ. The continued interest was at least in part driven by indications that Λ might be needed to reconcile theory and observations. Here are three examples.

First, the expansion time is uncomfortably short if $\Lambda=0$. Sandage's recalibration of the distance scale in the 1960s indicates $H_0\simeq 75\ \mathrm{km\,s^{-1}\,Mpc^{-1}}$. If $\Lambda=0$ this shows that the time of expansion from densities too high for stars to have existed is $<H_0^{-1}\simeq 13$ Gyr, maybe less than the ages of the oldest stars, then estimated to be greater than about 15 Gyr. Sandage (1961a) points out that the problem is removed by adding a positive Λ. The present estimates reviewed below (Sec. IV.B.3) are not far from these numbers, but still too uncertain for a significant case for Λ.

Second, counts of quasars as a function of redshift show a peak at $z\sim 2$, as would be produced by the loitering epoch in Lemaître's Λ model (Petrosian, Salpeter, and Szekeres, 1967; Shklovsky, 1967; Kardashev, 1967). The peak is now well established, centered at $z\sim 2.5$ (Croom *et al.*, 2001; Fan *et al.*, 2001). It is usually interpreted as the evolution in the rate of violent activity in the nuclei of galaxies, though in the absence of a loitering epoch the indicated sharp variation in quasar activity with time is curious (but certainly could be a consequence of astrophysics that is not well understood).

The third example is the redshift-magnitude relation. Sandage's (1961a) analysis indicates this is a promising method of distinguishing world models. The Gunn and Oke (1975) measurement of this relation for giant elliptical galaxies, with Tinsley's (1972) correction for evolution of the star population from assumed formation at high redshift, indicates curvature away from the linear relation in the direction that, as Gunn and Tinsley

(1975) discuss, could only be produced by Λ (within general relativity theory). The new application of the redshift-magnitude test, to type-Ia supernovae (Sec. IV.B.4), is not inconsistent with the Gunn-Oke measurement; we do not know whether this agreement of the measurements is significant, because Gunn and Oke were worried about galaxy evolution.[16]

2. The coincidences argument against Λ

An argument against an observationally interesting value of Λ, from our distrust of accidental coincidences, has been in the air for decades, and became very influential in the early 1980s with the introduction of the inflation scenario for the very early universe.

If the Einstein–de Sitter model in Eq. (35) were a good approximation at the present epoch, an observer measuring the mean mass density and Hubble's constant when the age of the universe was one-tenth the present value, or ten times the present age, would reach the same conclusion, that the Einstein–de Sitter model is a good approximation. That is, we would flourish at a time that is not special in the course of evolution of the universe. If, on the other hand, two or more of the terms in the expansion-rate equation (11) made substantial contributions to the present value of the expansion rate, it would mean that we are present at a special epoch, because each term in Eq. (11) varies with the expansion factor in a different way. To put this in more detail, we imagine that the physics of the very early universe, when the relativistic cosmological model became a good approximation, set the values of the cosmological parameters. The initial values of the contributions to the expansion-rate equation had to have been very different from each other, and exceedingly specially fixed, to yield two Ω_{i0}'s with comparable values. This would be a most remarkable and unlikely coincidence. The multiple coincidences required for the near vanishing of $\dot{a}$ and $\ddot{a}$ at a redshift not much larger than unity makes an even stronger case against Lemaître's loitering model, with this line of argument.

The earliest published comment we have found on this point is by Bondi (1960, p. 166), in the second edition of his book *Cosmology*. Bondi notes the "remarkable property" of the Einstein–de Sitter model: the dimensionless parameter we now call Ω_M is independent of the time at which it is computed (since it is unity). The coincidences argument follows and extends Bondi's comment. It is presented in McCrea (1971, p. 151). When Peebles was a postdoctoral research associate, in the early 1960s, in R. H. Dicke's gravity research group, the coincidences argument was discussed, but published much later (Dicke, 1970, p. 62; Dicke and Peebles, 1979). We do not know its provenance in Dicke's group, whether from Bondi, McCrea, Dicke, or someone else. We would not be surprised to learn others had similar thoughts.

The coincidences argument is sensible but not a proof, of course. The discovery of the 3-K thermal cosmic microwave background radiation gave us a term in the expansion-rate equation that is down from the dominant one by four orders of magnitude, not such a large factor by astronomical standards. This might be counted as a first step away from the argument. From the dynamics of galaxies the evidence that Ω_{M0} is less than unity is another step (Peebles, 1984, p. 442; 1986). And yet another is the development of the evidence that the Λ and dark-matter terms differ by only a factor of 3 [Eq. (2)]. This last piece is the most curious, but the community has come to accept it, for the most part. The precedent makes Lemaître's loitering model more socially acceptable.

A socially acceptable value of Λ cannot be such as to make life impossible, of course.[17] But perhaps the most productive interpretation of the coincidences argument is that it demands a search for a more fundamental underlying model. This is discussed further in Sec. III.E and the Appendix.

3. Vacuum energy and Λ

Another tradition to consider is the relation between Λ and the vacuum or dark-energy density. In one approach to the motivation for the Einstein field equation, taken by McVittie (1956) and others, Λ appears as a constant of integration (of the expression for local conservation of energy and momentum). McVittie (1956, p. 35) emphasizes that, as a constant of integration, Λ "cannot be assigned any particular value on *a priori* grounds." Interesting variants of this line of thought are still under discussion (Weinberg, 1989; Unruh, 1989; and references therein).

The notion of Λ as a constant of integration may be related to the issue of the zero point of energy. In laboratory physics one measures and computes energy differences. But the net energy matters for gravity physics, and one can imagine that Λ represents the difference between the true energy density and the sum at which one arrives by laboratory physics. Eddington (1939) and Lemaître (1934, 1949) make this point.

[16]Early measurements of the redshift-magnitude relation were meant in part to test the Steady State cosmology of Bondi and Gold (1948) and Hoyle (1948). Since Steady State cosmology assumes spacetime is independent of time its line element has to have the form of the de Sitter solution with $\Omega_{K0}=0$ and the expansion parameter in Eq. (27). The measured curvature of the redshift-magnitude relation is in the direction predicted by Steady State cosmology. But this cosmology fails other tests discussed in Sec. IV.B.

[17]If Λ were negative and the magnitude too large there would not be enough time for the emergence of life such as ours. If Λ were positive and too large the universe would expand too rapidly to allow galaxy formation. Our existence, which requires something resembling the Milky Way galaxy to contain and recycle heavy elements, thus provides an upper bound on the value of Λ. Such anthropic considerations are discussed by Weinberg (1987, 2001) and Vilenkin (2001), and references therein.

Bronstein[18] (1933) carries the idea further, allowing for transfer of energy between ordinary matter and that represented by Λ. In our notation, Bronstein expresses this picture by generalizing Eq. (9) to

$$\dot{\rho}_\Lambda = -\dot{\rho} - 3\frac{\dot{a}}{a}(\rho + p), \tag{36}$$

where ρ and p are the energy density and pressure of ordinary matter and radiation. Bronstein goes on to propose a violation of local energy conservation, a thought that no longer seems interesting. North (1965, p. 81) finds Eddington's (1939) interpretation of the zero point of energy also somewhat hard to defend. But for our purpose the important point is that the idea of Λ as a form of energy has been present, in at least some circles, for many years.

The zero-point energy of fields contributes to the dark-energy density. To make physical sense the sum over the zero-point mode energies must be cut off at a short distance or a high frequency up to which the model under consideration is valid. The integral of the zero-point energy ($k/2$) of normal modes (of wave number k) of a massless real bosonic scalar field (Φ), up to the wave-number cutoff k_c, gives the vacuum energy density the quantum-mechanical expectation value[19]

$$\rho_\Phi = \int_0^{k_c} \frac{4\pi k^2 dk}{(2\pi)^3} \frac{k}{2} = \frac{k_c^4}{16\pi^2}. \tag{37}$$

Nernst (1916) seems to have been the first to write down this equation, in connection with the idea that the zero-point energy of the electromagnetic field fills the vacuum, as a light aether, that could have physically significant properties.[20] This was before Heisenberg and Schrödinger: Nernst's hypothesis is that each degree of freedom, to which classical statistical mechanics assigns energy $kT/2$, has *Nullpunktsenergie* $h\nu/2$. This would mean that the ground-state energy of a one-dimensional harmonic oscillator is $h\nu$, twice the correct value. Nernst's expression for the energy density in the electromagnetic field thus differs from Eq. (37) by a factor of 2 (after taking account of the two polarizations), which is wonderfully close. For a numerical example, Nernst noted that if the cutoff frequency were $\nu = 10^{20}$ Hz, or ~ 0.4 MeV, the energy density of the *Lichtäther* (light aether) would be 10^{23} erg cm^{-3}, or about 100 g cm^{-3}.

By the end of the 1920s Nernst's hypothesis was replaced with the demonstration that in quantum mechanics the zero-point energy of the vacuum is as real as any other. W. Pauli, in unpublished work in the 1920s,[21] repeated Nernst's calculation, with the correct factor of 2, taking k_c to correspond to the classical electron radius. Pauli knew the value of ρ_Λ was quite unacceptable: the radius of the static Einstein universe with this value of ρ_Λ "would not even reach to the moon" (Rugh and Zinkernagel, 2002, p. 5).[22] The modern version of this "physicists' cosmological constant problem" is even

[18]Kragh (1996, p. 36) describes Bronstein's motivation and history. We discuss this model in more detail in Sec. III.E, and comment on why decay of dark energy into ordinary matter or radiation would be hard to reconcile with the thermal spectrum of the 3-K cosmic microwave background radiation. Decay into the dark sector may be interesting.

[19]Equation (37), which usually figures in discussions of the vacuum energy puzzle, gives a helpful indication of the situation: the zero-point energy of each mode is real and the sum is large. The physics is seriously incomplete, however. The elimination of spatial momenta with magnitudes $k > k_c$ only makes sense if there is a preferred reference frame in which k_c is defined. Magueijo and Smolin (2002) mention a related issue: In which reference frame is the Planck momentum of a virtual particle at the threshold for new phenomena? In both cases one may implicitly choose the rest frame for the large-scale distribution of matter and radiation. It seems strange to think that the microphysics is concerned about large-scale structure, but perhaps this happens in a sea of interacting fields. The cutoff in Eq. (37) might be applied at a fixed comoving wave number $k_c \propto a(t)^{-1}$, or at a fixed physical value of k_c. The first prescription can be described by an action written as a sum of terms $\dot{\Phi}_k^2/2 + k^2\Phi_k^2/[2a(t)^2]$ for the allowed modes. The zero-point energy of each mode scales with the expansion of the universe as $a(t)^{-1}$, and the sum over modes scales as $\rho_\Phi \propto a(t)^{-4}$, consistent with $k_c \propto a(t)^{-1}$. In the limit of exact spatial homogeneity, an equivalent approach uses the spatial average of the standard expression for the field stress-energy tensor. Indeed, DeWitt (1975) and Akhmedov (2002) show that the vacuum expectation value of the stress-energy tensor, expressed as an integral cutoff at $k = k_c$, and computed in the preferred coordinate frame, is diagonal with a space part $p_\Phi = \rho_\Phi/3$, for the massless field we are considering. That is, in this prescription the vacuum zero-point energy acts like a homogeneous sea of radiation. This defines a preferred frame of motion, where the stress-energy tensor is diagonal, which is not unexpected because we need a preferred frame to define k_c. It is unacceptable as a model for the properties of dark energy, of course. For example, if the dark-energy density were normalized to the value now under discussion, and varied as $\rho_\Lambda \propto a(t)^{-4}$, it would quite mess up the standard model for the origin of the light elements. We get a more acceptable model for the behavior of ρ_Λ from the second prescription, with the cutoff at a fixed physical momentum. If we also want to satisfy local energy conservation we must take the pressure to be $p_\Phi = -\rho_\Phi$. This does not contradict the derivation of p_Φ in the first prescription, because the second situation cannot be described by an action: the pressure must be stipulated, not derived. What is worse is that the known fields at laboratory momenta certainly do not allow this stipulation; they are well described by analogs of the action in the first prescription. This quite unsatisfactory situation illustrates how far we are from a theory of vacuum energy.

[20]A helpful discussion of Nernst's ideas on cosmology is that of Kragh (1996, pp. 151–157).

[21]This is discussed by Enz and Thellung (1960), Enz (1974), Rugh and Zinkernagel (2002, pp. 4 and 5), and Straumann (2002).

[22]In an unpublished letter in 1930, G. Gamow considered the gravitational consequences of the Dirac sea (Dolgov, 1989, p. 230). We thank A. Dolgov for helpful correspondence on this point.

more acute, because a natural value for k_c is thought to be much larger than what Nernst or Pauli used.[23]

While there was occasional discussion of this issue in the middle of the 20th century (as in the quote from N. Bohr in Rugh and Zinkernagel, 2002, p. 5), the modern era begins with the paper by Zel'dovich (1967) that convinced the community to consider the possible connection between the vacuum energy density of quantum physics and Einstein's cosmological constant.[24]

If the physics of the vacuum looks the same to any inertial observer its contribution to the stress-energy tensor is the same as Einstein's cosmological constant [Eq. (19)]. Lemaître (1934) notes this: "in order that absolute motion, i.e., motion relative to the vacuum, may not be detected, we must associate a pressure $p=-\rho c^2$ to the energy density ρc^2 of vacuum." Gliner (1965) goes further, presenting the relation between the metric tensor and the stress-energy tensor of a vacuum that is the same to any inertial observer. But it was Zel'dovich (1968) who presented the argument clearly enough and at the right time to catch the attention of the community.

With the development of the concept of broken symmetry in the now standard model for particle physics came the idea that the expansion and cooling of the universe is accompanied by a sequence of first-order phase transitions accompanying the symmetry breaking. Each first-order transition has a latent heat that appears as a contribution to an effective time-dependent $\Lambda(t)$ or dark-energy density.[25] The decrease in value of the dark-energy density at each phase transition is much larger than the acceptable present value (within relativistic cosmology); the natural presumption is that the dark energy is negligible now. This final condition seems bizarre, but the picture led to the very influential concept of inflation. We discussed the basic elements in connection with Eq. (27); we now turn to some implications.

---

[23]In terms of an energy scale ϵ_Λ defined by $\rho_\Lambda=\epsilon_\Lambda^4$, the Planck energy $G^{-1/2}$ is about 30 orders of magnitude larger than the "observed" value of ϵ_Λ. This is, of course, an extreme case, since many of the theories of interest break down well below the Planck scale. Furthermore, in addition to other contributions, one may add a counterterm to Eq. (37) to predict any value of ρ_Λ. With reference to this point, it is interesting to note that while Pauli did not publish his computation of ρ_Λ, he remarks in his famous 1933 *Handbuch der Physik* review on quantum mechanics that it is more consistent to "exclude a zero-point energy for each degree of freedom as this energy, evidently from experience, does not interact with the gravitational field" (Rugh and Zinkernagel, 2002, p. 5). Pauli was fully aware that one must take account of zero-point energies in the binding energies of molecular structure, for example (and we expect he was aware that what contributes to the energy contributes to the gravitational mass). He chose to drop the section with the above comment from the second (1958) edition of the review (Pauli, 1980, pp. iv and v). In a globally supersymmetric field theory there are equal numbers of bosonic and fermionic degrees of freedom, and the net zero-point vacuum energy density ρ_Λ vanishes (Iliopoulos and Zumino, 1974; Zumino, 1975). However, supersymmetry is not a symmetry of low-energy physics, or even at the electroweak unification scale. It must be broken at low energies, and the proper setting for a discussion of the zero-point ρ_Λ in this case is locally supersymmetric supergravity. Weinberg (1989, p. 6) notes "it is very hard to see how any property of supergravity or superstring theory could make the effective cosmological constant sufficiently small." Witten (2001) and Ellwanger (2002) review more recent developments on this issue in the superstring/M theory/branes scenario.

[24]For subsequent more detailed discussions of this issue, see Zel'dovich (1981), Weinberg (1989), Carroll, Press, and Turner (1992), Sahni and Starobinsky (2000), Carroll (2001), and Rugh and Zinkernagel (2000).

V. CONCLUDING REMARKS

We cannot demonstrate that there does not exist some other physics, applied to some other cosmology, that equally well agrees with the cosmological tests. The same applies to the whole enterprise of physical science, of course. Parts of physics are so densely checked that they are quite convincing approximations to reality. The web of tests is much less dense in cosmology, but, as we have tried to demonstrate, by no means negligible, and growing tighter.

A decade ago there was not much discussion on how to test general relativity theory on the scales of cosmology. That was in part because the theory seemed so logically compelling, and certainly in part also because there was not much evidence to work with. The empirical situation is much better now. We mentioned two tests, namely, of the relativistic active gravitational mass density and the gravitational inverse-square law. The consistency of constraints on Ω_{M0} from dynamics and the redshift-magnitude relation adds a test of the effect of space curvature on the expansion rate. These tests are developing; they will be greatly improved by work in progress. There certainly may be surprises in the gravity physics of cosmology at redshifts $z \lesssim 10^{10}$, but it is already clear that if so the surprises will be subtly hidden.

A decade ago the direction the theory of large-scale structure formation would take was not at all clear. Now the simple CDM model has had enough success that there is good reason to expect the standard model ten years from now will resemble CDM. We have listed challenges to this structure-formation model. Some may merely be a result of the complications of interpreting the theory and observations. Others may prove to be real and drive adjustments of the model. Fixes certainly will include one element from the ideas of structure formation that were under debate a decade ago: explosions that rearrange matter in ways that are difficult to compute. Fixes may also include primeval isocurvature departures from homogeneity, such as in spacetime curvature fluctuations frozen in during inflation, and perhaps in new cosmic fields. It would not be surprising if cosmic field defects, that have such a good pedigree from particle physics, also find a role in structure formation. And a central point to bear in mind is that fixes, which do not seem unlikely, could affect the cosmological tests.

A decade ago we had significant results from the cosmological tests. For example, estimates of the product $H_0 t_0$ suggested we might need positive Λ, though the precision was not quite adequate for a convincing case. That is still so; the community will be watching for further advances. We had fairly good constraints on Ω_{B0} from the theory of the origin of the light elements. The abundance measurements are improving; an important recent development is the detection of deuterium in gas clouds at redshifts $z \sim 3$. Ten years ago we had useful estimates of masses from peculiar motions on relatively small scales, but also more mixed messages from larger scales. The story seems more coherent now, but there is still room for improved consistency. We had a case for nonbaryonic dark matter, from the constraint on Ω_{B0} and from the CDM model for structure formation. The case is tighter now, most notably due to the successful fit of the CDM model prediction to the measurements of the power spectrum of the 3-K cosmic microwave background radiation anisotropy. In 1990 there were believable observations of galaxies identified as radio sources at $z \sim 3$. Now the distributions of galaxies and the intergalactic medium at $z \sim 3$ are mapped out in impressive detail, and we are seeing the development of a semiempirical picture of galaxy formation and evolution. Perhaps that will lead us back to the old dream of using galaxies as markers for cosmological tests.

Until recently it made sense to consider the constraints on one or two of the parameters of the cosmology while holding all the rest at "reasonable" values. That helps us understand what the measurements are probing; it is the path we have followed in Sec. IV.B. But the modern and very sensible trend is to consider joint fits of large numbers of parameters to the full suite of observations.[107] This includes a measure of the biasing or antibiasing of the distribution of galaxies relative to mass, rather than our qualitative argument that one usefully approximates the other. In a fully satisfactory cosmological test the parameter set will also include parametrized departures from standard physics extrapolated to the scales of cosmology.

Community responses to advances in empirical evidence are not always close to linear. The popularity of the Einstein–de Sitter model continued longer into the 1990s than seems logical to us, and the switch to the now standard ΛCDM cosmological model—with flat space sections, nonbaryonic cold dark matter, and dark energy—arguably is more abrupt than warranted by the

[107]Examples are Cole *et al.* (1997), Jenkins *et al.* (1998), Lineweaver (2001), and Percival *et al.* (2002).

advances in the evidence. Our review leads us to conclude that there is now a good scientific case arguing that the matter density parameter is $\Omega_{M0} \sim 0.25$, and a fairly good case that about three-quarters of that is not baryonic. The cases for dark energy and for the ΛCDM model are significant, too, though obscured by observational issues of whether we have an adequate picture of structure formation. But we expect that rapid advances in the observations of structure formation will soon dissipate these clouds, and, considering the record, likely reveal new clouds over the standard model for cosmology a decade from now.

A decade ago the high-energy-physics community had a well-defined challenge to show why the dark-energy density vanishes. Now there seems to be both a new challenge and clue: determine why the dark-energy density is exceedingly small but not zero. The present state of ideas can be compared with the state of research on structure formation a decade ago: in both situations there are many lines of thought but not a clear picture of which is the best direction to take. The big difference is that a decade ago we could be reasonably sure that observations in progress would guide us to a better understanding of how structure formed. Untangling the physics of dark matter and dark energy and their role in gravity physics is a much more subtle challenge, but, we hope, will be guided by advances in the exploration of the phenomenology. Perhaps in another ten years this will include detecting the evolution of the dark energy, and detecting the gravitational response of the dark-energy distribution to the large-scale mass distribution. There may be three unrelated phenomena to deal with: dark energy, dark matter, and a vanishing sum of zero-point energies and whatever goes with them. Or the phenomena may be related. Because our only evidence of dark matter and dark energy is from their gravity, it is a natural and efficient first step to suppose that their properties are as simple as allowed by the phenomenology. However, it makes sense to watch for hints of more complex physics within the dark sector.

The past eight decades have seen steady advances in the technology used for the cosmological tests, from telescopes to computers; advances in the theoretical concepts underlying the tests; and progress through the learning curves on applying the concepts and technology. We see the results: the basis for cosmology is much firmer than it was a decade ago. And the basis surely will be a lot more solid a decade from now.

Einstein's cosmological constant, and the modern variant, dark energy, have figured in a broad range of topics under discussion in physics and astronomy, in at least some circles, for much of the past eight decades. Many of these issues undoubtedly have been discovered more than once. But in our experience such ideas tend to persist for a long time at low visibility and sometimes low fidelity. Thus the community has been very well prepared for the present evidence for detection of dark energy. And for the same reason we believe that dark energy, whether constant, or rolling toward zero, or maybe even increasing, still will be an active topic of research, in at least some circles, a decade from now, whatever the outcome of the present work on the cosmological tests. Though this much is clear, we see no basis for a prediction of whether the standard cosmology a decade from now will be a straightforward elaboration of ΛCDM, or whether there will be more substantial changes of direction.

Note added in proof

The measurements of the angular distribution of the 3-K cosmic microwave background radiation by the Wilkinson Microwave Anisotropy Probe (WMAP) were released (in Bennett *et al.*, 2003; Spergel *et al.*, 2003, and references therein) after this review was completed. It is appropriate to comment on how the results of this superb experiment have changed our assessment of the cosmological tests.

The discussion in Sec. IV.C led us to conclude that the case for the detection of dark energy is not as compelling as the case for dark matter, because there are fewer cross checks. WMAP changes that. The ΛCDM model gives an excellent fit to the WMAP measurements. The parameters required by this fit, including density parameters $0.19 \leq \Omega_{M0} \leq 0.35$, $0.65 \leq \Omega_{\Lambda 0} \leq 0.81$, and $-0.02 \leq \Omega_{K0} \leq 0.06$ (all at two standard deviations), are in good agreement with other constraints (as summarized in Sec. IV). In particular, a pre-WMAP survey of the constraints on Ω_{M0} from a combination of the dynamical, baryon fraction, power spectrum, weak lensing, and cluster abundance measurements indicates $0.2 \lesssim \Omega_{M0} \lesssim 0.35$ (at two standard deviations, Chen and Ratra, 2003b), in striking accord with the WMAP estimate. The fit to the WMAP measurements and the overall consistency of parameter constraints is strong evidence that the ΛCDM model is a good approximation to reality. This evidence increases the weight of parameter estimates that depend on the ΛCDM model. And the model fit to the WMAP measurements requires the presence of dark energy, provided the Hubble constant is within acceptable distance of the astronomical measurements. This is distinct from the line of argument for $\Omega_{\Lambda 0} \sim 0.7$ described in Sec. IV. It provides the wanted cross check that makes a convincing case for the detection of dark energy.

Issues remain. The ΛCDM fit to WMAP indicates the density parameter in baryons is $0.021 \leq \Omega_{B0} h^2 \leq 0.024$ (at two standard deviations), consistent with what is indicated by the standard Big Bang nucleosynthesis model and measurements of the primeval deuterium abundance [Eq. (62) and footnote 101]. To be resolved are the somewhat different estimates of $\Omega_{B0} h^2$ from the helium and lithium abundances. The temperature anisotropy autocorrelation function is consistent with zero at angular separations greater than about 60° (consistent with the earlier but less emphatic COBE result mentioned in footnote 100), and seems not likely to be consistent with the ΛCDM prediction. Maybe this is an unlikely statistical fluctuation. Or maybe it is telling us about the physics of the dark energy.

We take this opportunity to refer to Padmanabhan's (2002) recently completed review of the cosmological constant, with particular emphasis on possible resolutions of the physicists' cosmological constant problem and the physics of the resulting dark energy models.

ACKNOWLEDGMENTS

We are indebted to Pia Mukherjee, Michael Peskin, and Larry Weaver for detailed comments on drafts of this review. We thank Uwe Thumm for help in translating and discussing papers written in German. We have also benefited from discussions with Neta Bahcall, Robert Caldwell, Gang Chen, Andrea Cimatti, Mark Dickinson, Michael Dine, Masataka Fukugita, Salman Habib, David Hogg, Avi Loeb, Stacy McGaugh, Paul Schechter, Chris Smeenk, Gary Steigman, Ed Turner, Michael Turner, Jean-Philippe Uzan, David Weinberg, and Simon White. B.R. acknowledges support from NSF CAREER Grant No. AST-9875031, and P.J.E.P. acknowledges support in part from the NSF.

REFERENCES

Abazajian, K., G. M. Fuller, and M. Patel, 2001, Phys. Rev. D **64**, 023501.
Abbott, L. F., and M. B. Wise, 1984, Nucl. Phys. B **244**, 541.
Abell, G. O., 1958, Astrophys. J., Suppl. Ser. **3**, 211.
Akhmedov, E. Kh., 2002, e-print hep-th/0204048.
Albrecht, A., and P. J. Steinhardt, 1982, Phys. Rev. Lett. **48**, 1220.
Alcock, C., and B. Paczyński, 1979, Nature (London) **281**, 358.
Allen, S. W., R. W. Schmidt, and A. C. Fabian, 2002, Mon. Not. R. Astron. Soc. **335**, 256.
Alpher, R. A., and R. Herman, 2001, *Genesis of the Big Bang* (Oxford University, Oxford).
Amendola, L., 1999, Phys. Rev. D **60**, 043501.
Amendola, L., 2000, Phys. Rev. D **62**, 043511.
Amendola, L., and D. Tocchini-Valentini, 2002, Phys. Rev. D **66**, 043528.
Aragón-Salamanca, A., C. M. Baugh, and G. Kauffmann, 1998, Mon. Not. R. Astron. Soc. **297**, 427.
Arai, K., M. Hashimoto, and T. Fukui, 1987, Astron. Astrophys. **179**, 17.
Aramendariz-Picon, C., V. Mukhanov, and P. J. Steinhardt, 2001, Phys. Rev. D **63**, 103510.
Baccigalupi, C., A. Balbi, S. Matarrese, F. Perrotta, and N. Vittorio, 2002, Phys. Rev. D **65**, 063520.
Baccigalupi, C., S. Matarrese, and F. Perrotta, 2000, Phys. Rev. D **62**, 123510.
Bacon, D. J., R. J. Massey, A. R. Refregier, and R. S. Ellis, 2002, e-print astro-ph/0203134.
Bahcall, J. N., and R. A. Wolf, 1968, Astrophys. J. **152**, 701.
Bahcall, N. A., R. Cen, R. Davé, J. P. Ostriker, and Q. Yu, 2000, Astrophys. J. **541**, 1.
Bahcall, N. A., and X. Fan, 1998, Astrophys. J. **504**, 1.
Bahcall, N. A., J. P. Ostriker, S. Perlmutter, and P. J. Steinhardt, 1999, Science **284**, 1481.
Bahcall, N. A., *et al.*, 2003, Astrophys. J. **585**, 182.
Banks, T., and W. Fischler, 2001, e-print hep-th/0102077.
Bardeen, J. M., P. J. Steinhardt, and M. S. Turner, 1983, Phys. Rev. D **28**, 679.
Barr, S. M., and D. Hochberg, 1988, Phys. Lett. B **211**, 49.
Barr, S. M., and D. Seckel, 2001, Phys. Rev. D **64**, 123513.
Barreiro, T., E. J. Copeland, and N. J. Nunes, 2000, Phys. Rev. D **61**, 127301.
Bartlett, J. G., and J. Silk, 1990, Astrophys. J. **353**, 399.
Bartolo, N., and M. Pietroni, 2000, Phys. Rev. D **61**, 023518.
Battye, R. A., M. Bucher, and D. Spergel, 1999, e-print astro-ph/9908047.
Bean, R., and A. Melchiorri, 2002, Phys. Rev. D **65**, 041302.
Bennett, C. L., *et al.*, 2003, e-print astro-ph/0302207.
Bernardeau, F., and R. Schaeffer, 1992, Astron. Astrophys. **255**, 1.
Bertolami, O., 1986, Fortschr. Phys. **34**, 829.
Bertolami, O., and P. J. Martins, 2000, Phys. Rev. D **61**, 064007.
Binétruy, P., 1999, Phys. Rev. D **60**, 063502.
Binétruy, P., and J. Silk, 2001, Phys. Rev. Lett. **87**, 031102.
Birkel, M., and S. Sarkar, 1997, Astropart. Phys. **6**, 197.
Blais-Ouellette, S., P. Amram, and C. Carignan, 2001, Astron. J. **121**, 1952.
Blanchard, A., R. Sadat, J. G. Bartlett, and M. Le Dour, 2000, Astron. Astrophys. **362**, 809.
Bludman, S. A., and M. A. Ruderman, 1977, Phys. Rev. Lett. **38**, 255.
Blumenhagen, R., B. Körs, D. Lüst, and T. Ott, 2002, Nucl. Phys. B **641**, 235.
Blumenthal, G. R., A. Dekel, and J. R. Primack, 1988, Astrophys. J. **326**, 539.
Blumenthal, G. R., S. M. Faber, J. R. Primack, and M. J. Rees, 1984, Nature (London) **311**, 517.
Bode, P., J. P. Ostriker, and N. Turok, 2001, Astrophys. J. **556**, 93.
Bond, J. R., 1988, in *The Early Universe*, edited by W. G. Unruh and G. W. Semenoff (Reidel, Dordrecht), p. 283.
Bond, J. R., R. Crittenden, R. L. Davis, G. Efstathiou, and P. J. Steinhardt, 1994, Phys. Rev. Lett. **72**, 13.
Bondi, H., 1960, *Cosmology* (Cambridge University, Cambridge).
Bondi, H., and T. Gold, 1948, Mon. Not. R. Astron. Soc. **108**, 252.
Bordag, M., U. Mohideen, and V. M. Mostepanenko, 2001, Phys. Rep. **353**, 1.
Borgani, S., P. Rosati, P. Tozzi, S. A. Stanford, P. R. Eisenhardt, C. Lidman, B. Holden, R. D. Ceca, C. Norman, and G. Squires, 2001, Astrophys. J. **561**, 13.
Boughn, S. P., and R. G. Crittenden, 2001, e-print astro-ph/0111281.
Bousso, R., 2000, J. High Energy Phys. **0011**, 038.
Branchini, E., W. Freudling, L. N. Da Costa, C. S. Frenk, R. Giovanelli, M. P. Haynes, J. J. Salzer, G. Wegner, and I. Zehavi, 2001, Mon. Not. R. Astron. Soc. **326**, 1191.
Branchini, E., I. Zehavi, M. Plionis, and A. Dekel, 2000, Mon. Not. R. Astron. Soc. **313**, 491.
Brandenberger, R. H., 2001, e-print hep-ph/0101119.
Brax, P., and J. Martin, 2000, Phys. Rev. D **61**, 103502.
Brax, P., J. Martin, and A. Riazuelo, 2000, Phys. Rev. D **62**, 103505.
Brax, P., J. Martin, and A. Riazuelo, 2001, Phys. Rev. D **64**, 083505.
Bridle, S., R. G. Crittenden, A. Melchiorri, M. P. Hobson, R. Kneissl, and A. N. Lasenby, 2002, Mon. Not. R. Astron. Soc. **335**, 1193.

Bronstein, M., 1933, Phys. Z. Sowjetunion **3**, 73.
Brown, G. S., and B. M. Tinsley, 1974, Astrophys. J. **194**, 555.
Buchalter, A., D. J. Helfand, R. H. Becker, and R. L. White, 1998, Astrophys. J. **494**, 503.
Bucher, M., and D. N. Spergel, 1999, Phys. Rev. D **60**, 043505.
Bucher, M., and N. Turok, 1995, Phys. Rev. D **52**, 5538.
Burgess, C. P., P. Martineau, F. Quevedo, G. Rajesh, and R.-J. Zhang, 2002, J. High Energy Phys. **0203**, 052.
Burles, S., K. M. Nollett, and M. S. Turner, 2001, Astrophys. J. Lett. **552**, L1.
Burstein, D., 2000, in *Cosmic Flows Workshop*, edited by S. Courteau and J. Willick, Astronomical Society of the Pacific Conference Proceedings No. 201 (Astronomical Society of the Pacific, San Francisco), p. 178.
Caldwell, R. R., 2002, Phys. Lett. B **545**, 23.
Caldwell, R. R., R. Davé, and P. J. Steinhardt, 1998, Phys. Rev. Lett. **80**, 1582.
Canuto, V., P. J. Adams, S.-H. Hsieh, and E. Tsiang, 1977, Phys. Rev. D **16**, 1643.
Canuto, V., and J. F. Lee, 1977, Phys. Lett. **72B**, 281.
Carlstrom, J. E., M. Joy, L. Grego, G. Holder, W. L. Holzapfel, S. LaRoque, J. J. Mohr, and E. D. Reese, 2001, in *Constructing the Universe with Clusters of Galaxies*, edited by F. Durret and G. Gerbal (in press), e-print astro-ph/0103480.
Carretta, E., R. G. Gratton, G. Clementini, and F. F. Pecci, 2000, Astrophys. J. **533**, 215.
Carroll, S. M., 1998, Phys. Rev. Lett. **81**, 3067.
Carroll, S. M., 2001, Living Rev. Relativ. **4**, 1.
Carroll, S. M., W. H. Press, and E. L. Turner, 1992, Annu. Rev. Astron. Astrophys. **30**, 499.
Casimir, H. B. G., 1948, Proc. K. Ned. Akad. Wet. **51**, 635.
Chaboyer, B., and L. M. Krauss, 2002, Astrophys. J. Lett. **567**, L45.
Chen, B., and F.-L. Lin, 2002, Phys. Rev. D **65**, 044007.
Chen, G., and B. Ratra, 2003a, Astrophys. J. **582**, 586.
Chen, G., and B. Ratra, 2003b, e-print astro-ph/0302002.
Chen, X., R. J. Scherrer, and G. Steigman, 2001, Phys. Rev. D **63**, 123504.
Chiba, T., 1999, Phys. Rev. D **60**, 083508.
Chiba, T., T. Okabe, and M. Yamaguchi, 2000, Phys. Rev. D **62**, 023511.
Chimento, L. P., and A. S. Jakubi, 1996, Int. J. Mod. Phys. D **5**, 71.
Choi, K., 1999, e-print hep-ph/9912218.
Cimatti, A., *et al.*, 2002, Astron. Astrophys. Lett. **391**, L1.
Coble, K., S. Dodelson, M. Dragovan, K. Ganga, L. Knox, J. Kovac, B. Ratra, and T. Souradeep, 2003, Astrophys. J. **584**, 585.
Coc, A., E. Vangioni-Flam, M. Lassé, and M. Rabibet, 2002, Phys. Rev. D **65**, 043510.
Cohen-Tannoudji, C., J. Dupont-Roc, and G. Grynberg, 1992, *Atom-Photon Interactions* (Wiley, New York), p. 121.
Colberg, J. M., *et al.*, 2000, Mon. Not. R. Astron. Soc. **319**, 209.
Cole, S., D. H. Weinberg, C. S. Frenk, and B. Ratra, 1997, Mon. Not. R. Astron. Soc. **289**, 37.
Coleman, S., and F. De Luccia, 1980, Phys. Rev. D **21**, 3305.
Colley, W. N., J. R. Gott, and C. Park, 1996, Mon. Not. R. Astron. Soc. **281**, L82.
Cooray, A. R., 1999, Astrophys. J. **524**, 504.
Copeland, E. J., N. J. Nunes, and F. Rosati, 2000, Phys. Rev. D **62**, 123503.
Corasaniti, P. S., and E. J. Copeland, 2002, Phys. Rev. D **65**, 043004.
Croom, S. M., R. J. Smith, B. J. Boyle, T. Shanks, N. S. Loaring, L. Miller, and I. J. Lewis, 2001, Mon. Not. R. Astron. Soc. **322**, L29.
Cyburt, R. H., B. D. Fields, and K. A. Olive, 2001, New Astron. **6**, 215.
Dalal, N., K. Abazajian, E. Jenkins, and A. Manohar, 2001, Phys. Rev. Lett. **87**, 141302.
Daly, R., and E. J. Guerra, 2001, e-print astro-ph/0109383.
Danese, L., G. L. Granato, L. Silva, M. Magliocchetti, and G. De Zotti, 2002, in *The Mass of Galaxies at Low and High Redshift*, edited by R. Bender and A. Renzini (Springer, Berlin), in press.
Dasgupta, K., C. Herdeiro, S. Hirano, and R. Kallosh, 2002, Phys. Rev. D **65**, 126002.
Davé, R., D. N. Spergel, P. J. Steinhardt, and B. D. Wandelt, 2001, Astrophys. J. **547**, 574.
Davis, A. C., M. Dine, and N. Seiberg, 1983, Phys. Lett. **125B**, 487.
Davis, M., G. Efstathiou, C. S. Frenk, and S. D. M. White, 1985, Astrophys. J. **292**, 371.
Davis, M., and P. J. E. Peebles, 1983a, Annu. Rev. Astron. Astrophys. **21**, 109.
Davis, M., and P. J. E. Peebles, 1983b, Astrophys. J. **267**, 465.
Davis, R. L., 1987, Phys. Rev. D **35**, 3705.
de Blok, W. J. G., and A. Bosma, 2002, Astron. Astrophys. **385**, 816.
de Blok, W. J. G., S. S. McGaugh, A. Bosma, and V. C. Rubin, 2001, Astrophys. J. Lett. **552**, L23.
Deffayet, C., G. Dvali, and G. Gabadadze, 2002, Phys. Rev. D **65**, 044023.
de la Macorra, A., and G. Piccinelli, 2000, Phys. Rev. D **61**, 123503.
de la Macorra, A., and C. Stephan-Otto, 2001, Phys. Rev. D **87**, 271301.
de Oliveira-Costa, A., M. Tegmark, L. A. Page, and S. P. Boughn, 1998, Astrophys. J. Lett. **509**, L9.
de Ritis, R., A. A. Marino, C. Rubano, and P. Scudellaro, 2000, Phys. Rev. D **62**, 043506.
de Sitter, W., 1917, Mon. Not. R. Astron. Soc. **78**, 3.
DeWitt, B. S., 1975, Phys. Rep. **19**, 295.
Dicke, R. H., 1968, Astrophys. J. **152**, 1.
Dicke, R. H., 1970, *Gravitation and the Universe* (American Philosophical Society, Philadelphia).
Dicke, R. H., and P. J. E. Peebles, 1979, in *General Relativity*, edited by S. W. Hawking and W. Israel (Cambridge University, Cambridge, England), p. 504.
Dimopoulos, K., and J. W. F. Valle, 2001, e-print astro-ph/0111417.
Dine, M., 1996, e-print hep-ph/9612389.
Di Pietro, E., and J. Demaret, 2001, Int. J. Mod. Phys. D **10**, 231.
Dirac, P. A. M., 1937, Nature (London) **139**, 323.
Dirac, P. A. M., 1938, Proc. R. Soc. London, Ser. A **165**, 199.
Dodelson, S., *et al.*, 2002, Astrophys. J. **572**, 140.
Dodelson, S., M. Kaplinghat, and E. Stewart, 2000, Phys. Rev. Lett. **85**, 5276.
Dolgov, A. D., 1983, in *The Very Early Universe*, edited by G. W. Gibbons, S. W. Hawking, and S. T. C. Siklos (Cambridge University, Cambridge, England), p. 449.

Dolgov, A. D., 1989, in *The Quest for the Fundamental Constants in Cosmology*, edited by J. Audouze and J. Tran Thanh Van (Editions Frontières, Gif-sur-Yvette), p. 227.

Doran, M., and J. Jäckel, 2002, Phys. Rev. D **66**, 043519.

Doran, M., M. Lilley, J. Schwindt, and C. Wetterich, 2001, Astrophys. J. **559**, 501.

Dreitlein, J., 1974, Phys. Rev. Lett. **33**, 1243.

Dubinski, J., and R. G. Carlberg, 1991, Astrophys. J. **378**, 496.

Durrer, R., B. Novosyadlyj, and S. Apunevych, 2003, Astrophys. J. **583**, 33.

Dvali, G., and S.-H. H. Tye, 1999, Phys. Lett. B **450**, 72.

Eddington, A. S., 1924, *The Mathematical Theory of Relativity* (Cambridge University, Cambridge, England).

Eddington, A. S., 1939, Sci. Prog. **34**, 225.

Efstathiou, G., 1999, Mon. Not. R. Astron. Soc. **310**, 842.

Efstathiou, G., 2002, Mon. Not. R. Astron. Soc. **332**, 193.

Efstathiou, G., and J. R. Bond, 1999, Mon. Not. R. Astron. Soc. **304**, 75.

Efstathiou, G., and M. J. Rees, 1988, Mon. Not. R. Astron. Soc. **230**, 5P.

Efstathiou, G., W. J. Sutherland, and S. J. Maddox, 1990, Nature (London) **348**, 705.

Einstein, A., 1917, Sitzungsber. K. Preuss. Akad. Wiss. 142 [English translation in *The Principle of Relativity* (Dover, New York, 1952), p. 177].

Einstein, A., 1931, Sitzungsber. Preuss. Akad. Wiss., Phys. Math. Kl. 235.

Einstein, A., 1945, *The Meaning of Relativity* (Princeton University, Princeton, NJ).

Einstein, A., and W. de Sitter, 1932, Proc. Natl. Acad. Sci. U.S.A. **18**, 213.

Ellwanger, U., 2002, e-print hep-ph/0203252.

Endo, M., and T. Fukui, 1977, Gen. Relativ. Gravit. **8**, 833.

Enz, C. P., 1974, in *Physical Reality and Mathematical Description*, edited by C. P. Enz and J. Mehra (Reidel, Dordrecht), p. 124.

Enz, C. P., and A. Thellung, 1960, Helv. Phys. Acta **33**, 839.

Eriksson, M., and R. Amanullah, 2002, Phys. Rev. D **66**, 023530.

Etoh, T., M. Hashimoto, K. Arai, and S. Fujimoto, 1997, Astron. Astrophys. **325**, 893.

Evrard, A. E., 1989, Astrophys. J. Lett. **341**, L71.

Faber, S. M., G. Wegner, D. Burstein, R. L. Davies, A. Dressler, D. Lynden-Bell, and R. J. Terlevich, 1989, Astrophys. J., Suppl. Ser. **69**, 763.

Falco, E. E., C. S. Kochanek, and J. A. Muñoz, 1998, Astrophys. J. **494**, 47.

Fan, X., *et al.*, 2001, Astrophys. J. **122**, 2833.

Faraoni, V., 2000, Phys. Rev. D **62**, 023504.

Feldman, H. A., J. A. Frieman, J. N. Fry, and R. Scoccimarro, 2001, Phys. Rev. Lett. **86**, 1434.

Ferrarese, L., and D. Merritt, 2000, Astrophys. J. Lett. **539**, L9.

Ferreira, P. G., and M. Joyce, 1998, Phys. Rev. D **58**, 023503.

Fields, B. D., and S. Sarkar, 2002, Phys. Rev. D **66**, 010001.

Fischler, W., A. Kashani-Poor, R. McNees, and S. Paban, 2001, J. High Energy Phys. **0107**, 003.

Fischler, W., B. Ratra, and L. Susskind, 1985, Nucl. Phys. B **259**, 730.

Fisher, K. B., C. A. Scharf, and O. Lahav, 1994, Mon. Not. R. Astron. Soc. **266**, 219.

Fixsen, D. J., E. S. Cheng, J. M. Gales, J. C. Mather, R. A. Shafer, and E. L. Wright, 1996, Astrophys. J. **473**, 576.

Flores, R. A., and J. R. Primack, 1994, Astrophys. J. Lett. **427**, L1.

Ford, L. H., 1987, Phys. Rev. D **35**, 2339.

Fosalba, P., O. Dore, and F. R. Bouchet, 2002, Phys. Rev. D **65**, 063003.

Freedman, W. L., 2002, Int. J. Mod. Phys. A **17**, 51, 58.

Freedman, W. L., *et al.*, 2001, Astrophys. J. **553**, 47.

Freese, K., F. C. Adams, J. A. Frieman, and E. Mottola, 1987, Nucl. Phys. B **287**, 797.

Frewin, L. A., and J. E. Lidsey, 1993, Int. J. Mod. Phys. D **2**, 323.

Friedland, A., H. Muruyama, and M. Perelstein, 2003, Phys. Rev. D (in press).

Friedmann, A., 1922, Z. Phys. **10**, 377 [English translation in *Cosmological Constants*, edited by J. Bernstein and G. Feinberg (Columbia University, New York, 1986), p. 49].

Friedmann, A., 1924, Z. Phys. **21**, 326 [English translation in *Cosmological Constants*, edited by J. Bernstein and G. Feinberg (Columbia University, New York, 1986), p. 59].

Frieman, J. A., C. T. Hill, A. Stebbins, and I. Waga, 1995, Phys. Rev. Lett. **75**, 2077.

Fry, J. N., 1984, Astrophys. J. **279**, 499.

Fry, J. N., 1985, Phys. Lett. **158B**, 211.

Fry, J. N., 1994, Phys. Rev. Lett. **73**, 215.

Fry, J. N., and E. Gaztañaga, 1993, Astrophys. J. **413**, 447.

Fujii, Y., 1982, Phys. Rev. D **26**, 2580.

Fujii, Y., 2000, Gravitation Cosmol. **6**, 107.

Fujii, Y., and T. Nishioka, 1990, Phys. Rev. D **42**, 361.

Fukugita, M., T. Futamase, and M. Kasai, 1990, Mon. Not. R. Astron. Soc. **246**, 24P.

Fukugita, M., C. J. Hogan, and P. J. E. Peebles, 1998, Astrophys. J. **503**, 518.

Gamow, G., 1970, *My World Line* (Viking, New York).

Ganga, K., B. Ratra, J. O. Gundersen, and N. Sugiyama, 1997, Astrophys. J. **484**, 7.

García-Bellido, J., J. Rabadán, and F. Zamora, 2002, J. High Energy Phys. **0201**, 036.

Garnavich, P. M., *et al.*, 1998, Astrophys. J. **509**, 74.

Gasperini, M., 1987, Phys. Lett. B **194**, 347.

Gasperini, M., F. Piazza, and G. Veneziano, 2002, Phys. Rev. D **65**, 023508.

Gebhardt, K., *et al.*, 2000, Astrophys. J. Lett. **539**, L13.

Geller, M. J., and P. J. E. Peebles, 1973, Astrophys. J. **184**, 329.

Georgi, H., H. R. Quinn, and S. Weinberg, 1974, Phys. Rev. Lett. **33**, 451.

Gerke, B., and G. Efstathiou, 2002, Mon. Not. R. Astron. Soc. **335**, 33.

Giovannini, M., E. Keihänen, and H. Kurki-Suonio, 2002, Phys. Rev. D **66**, 043504.

Giudice, G. F., and R. Rattazzi, 1999, Phys. Rep. **322**, 419.

Gliner, E. B., 1965, Zh. Eksp. Teor. Fiz. **49**, 542 [Sov. Phys. JETP **22**, 378 (1966)].

González-Díaz, P. F., 2000, Phys. Rev. D **62**, 023513.

Goobar, A., and S. Perlmutter, 1995, Astrophys. J. **450**, 14.

Górski, K. M., B. Ratra, R. Stompor, N. Sugiyama, and A. J. Banday, 1998, Astrophys. J., Suppl. Ser. **114**, 1.

Górski, K. M., B. Ratra, N. Sugiyama, and A. J. Banday, 1995, Astrophys. J. Lett. **444**, L65.

Górski, K. M., J. Silk, and N. Vittorio, 1992, Phys. Rev. Lett. **68**, 733.

Gott, J. R., 1982, Nature (London) **295**, 304.

Gott, J. R., 1997, in *Critical Dialogs in Cosmology*, edited by N. Turok (World Scientific, Singapore), p. 519.

Gott, J. R., M. S. Vogeley, S. Podariu, and B. Ratra, 2001, Astrophys. J. **549**, 1.
Green, A. M., and J. E. Lidsey, 2000, Phys. Rev. D **61**, 067301.
Gudmundsson, E. H., and G. Björnsson, 2002, Astrophys. J. **565**, 1.
Gunn, J. E., 1967, Astrophys. J. **147**, 61.
Gunn, J. E., and J. B. Oke, 1975, Astrophys. J. **195**, 255.
Gunn, J. E., and B. M. Tinsley, 1975, Nature (London) **257**, 454.
Gurvits, L. T., K. I. Kellermann, and S. Frey, 1999, Astron. Astrophys. **342**, 378.
Guth, A. H., 1981, Phys. Rev. D **23**, 347.
Guth, A. H., 1997, *The Inflationary Universe* (Addison-Wesley, Reading).
Guth, A. H., and S.-Y. Pi, 1982, Phys. Rev. Lett. **49**, 1110.
Guyot, M., and Ya. B. Zel'dovich, 1970, Astron. Astrophys. **9**, 227.
Halliwell, J. J., 1987, Phys. Lett. B **185**, 341.
Halpern, M., H. P. Gush, and E. H. Wishnow, 1991, in *After the First Three Minutes*, edited by S. S. Holt, C. L. Bennett, and V. Trimble (AIP, New York), p. 53.
Halverson, N. W., *et al.*, 2002, Astrophys. J. **568**, 38.
Halyo, E., 2001a, J. High Energy Phys. **0110**, 025.
Halyo, E., 2001b, e-print hep-ph/0105341.
Hamilton, A. J. S., and M. Tegmark, 2002, Mon. Not. R. Astron. Soc. **330**, 506.
Hamilton, J.-Ch., and K. Ganga, 2001, Astron. Astrophys. **368**, 760.
Harrison, E. R., 1970, Phys. Rev. D **1**, 2726.
Hawking, S. W., 1982, Phys. Lett. **115B**, 295.
He, X.-G., 2001, e-print astro-ph/0105005.
Hebecker, A., and C. Wetterich, 2001, Phys. Lett. B **497**, 281.
Helbig, P., D. Marlow, R. Quast, P. N. Wilkinson, I. W. A. Browne, and L. V. E. Koopmans, 1999, Astron. Astrophys., Suppl. Ser. **136**, 297.
Hellerman, S., N. Kaloper, and L. Susskind, 2001, J. High Energy Phys. **0106**, 003.
Hiscock, W. A., 1986, Phys. Lett. **166B**, 285.
Hivon, E., F. R. Bouchet, S. Colombi, and R. Juszkiewicz, 1995, Astron. Astrophys. **298**, 643.
Hoekstra, H., H. K. C. Yee, and M. D. Gladders, 2002, New Astron. Rev. **46**, 767.
Holden, D. J., and D. Wands, 2000, Phys. Rev. D **61**, 043506.
Holtzman, J. A., 1989, Astrophys. J., Suppl. Ser. **71**, 1.
Horvat, R., 1999, Mod. Phys. Lett. A **14**, 2245.
Hoyle, F., 1948, Mon. Not. R. Astron. Soc. **108**, 372.
Hoyle, F., 1959, in *Paris Symposium on Radio Astronomy, IAU Symposium 9*, edited by R. N. Bracewell (Stanford University, Stanford), p. 529.
Hoyle, F., and R. J. Tayler, 1964, Nature (London) **203**, 1108.
Hradecky, V., C. Jones, R. H. Donnelly, S. G. Djorgovski, R. R. Gal, and S. C. Odewahn, 2000, Astrophys. J. **543**, 521.
Hu, W., 1998, Astrophys. J. **506**, 485.
Hu, W., and P. J. E. Peebles, 2000, Astrophys. J. Lett. **528**, L61.
Hu, W., and N. Sugiyama, 1996, Astrophys. J. **471**, 542.
Huang, J.-J., 1985, Nuovo Cimento Soc. Ital. Fis., B **87B**, 148.
Hubble, E., 1929, Proc. Natl. Acad. Sci. U.S.A. **15**, 168.
Hubble, E., 1936, *Realm of the Nebulae* (Yale University, New Haven). Reprinted (Dover, New York, 1958).
Huey, G., and G. Lidsey, 2001, Phys. Lett. B **514**, 217.
Huey, G., and R. Tavakol, 2002, Phys. Rev. D **65**, 043504.
Humason, M. L., N. U. Mayall, and A. R. Sandage, 1956, Astron. J. **61**, 97.
Huterer, D., and M. S. Turner, 2001, Phys. Rev. D **64**, 123527.
Hwang, J.-c., and H. Noh, 2001, Phys. Rev. D **64**, 103509.
Ikebe, Y., T. H. Reiprich, H. Böhringer, Y. Tanaka, and T. Kitayama, 2002, Astron. Astrophys. **383**, 773.
Iliopoulos, J., and B. Zumino, 1974, Nucl. Phys. B **76**, 310.
Israel, F. P., 1998, Astron. Astrophys. Rev. **8**, 237.
Jenkins, A., C. S. Frenk, F. R. Pearce, P. A. Thomas, J. M. Colberg, S. D. M. White, H. M. P. Couchman, J. A. Peacock, G. Efstathiou, and A. H. Nelson, 1998, Astrophys. J. **499**, 20.
John, M. V., and K. B. Joseph, 2000, Phys. Rev. D **61**, 087304.
Johri, V. B., 2002, Class. Quantum Grav. **19**, 5959.
Joyce, M., and T. Prokopec, 1998, Phys. Rev. D **57**, 6022.
Juszkiewicz, R., P. G. Ferreira, H. A. Feldman, A. H. Jaffe, and M. Davis, 2000, Science **287**, 109.
Kaganovich, A. B., 2001, Phys. Rev. D **63**, 025022.
Kahn, F. D., and L. Woltjer, 1959, Astrophys. J. **130**, 705.
Kaiser, N., 1984, Astrophys. J. Lett. Ed. **284**, L9.
Kaiser, N., 1987, Mon. Not. R. Astron. Soc. **227**, 1.
Kamionkowski, M., B. Ratra, D. N. Spergel, and N. Sugiyama, 1994a, Astrophys. J. Lett. **434**, L1.
Kamionkowski, M., D. N. Spergel, and N. Sugiyama, 1994b, Astrophys. J. Lett. **426**, L57.
Kamionkowski, M., and N. Toumbas, 1996, Phys. Rev. Lett. **77**, 587.
Kardashev, N., 1967, Astrophys. J. Lett. **150**, L135.
Kay, S. T., F. R. Pearce, C. S. Frenk, and A. Jenkins, 2002, Mon. Not. R. Astron. Soc. **330**, 113.
Kazanas, D., 1980, Astrophys. J. Lett. Ed. **241**, L59.
Kim, J. E., 2000, J. High Energy Phys. **0006**, 016.
Kirzhnitz, D. A., and A. D. Linde, 1974, Zh. Eksp. Teor. Fiz. **67**, 1263 [Sov. Phys. JETP **40**, 628 (1975)].
Klein, F., 1918, Nachr. Ges. Wiss. Goettingen, Math.-Phys. Kl. December, 394.
Klypin, A., A. V. Kratsov, O. Valenzuela, and F. Prada, 1999, Astrophys. J. **522**, 82.
Knebe, A., J. E. G. Devriendt, A. Mahmood, and J. Silk, 2002, Mon. Not. R. Astron. Soc. **329**, 813.
Knox, L., and L. Page, 2000, Phys. Rev. Lett. **85**, 1366.
Kofman, L. A., and A. A. Starobinsky, 1985, Pis'ma Astron. Zh. **11**, 643 [Sov. Astron. Lett. **11**, 271 (1985)].
Kogut, A., A. J. Banday, C. L. Bennett, K. M. Górski, G. Hinshaw, G. F. Smoot, and E. L. Wright, 1996, Astrophys. J. Lett. **464**, L5.
Kolb, E. W., and S. Wolfram, 1980, Astrophys. J. **239**, 428.
Kolda, C., and W. Lahneman, 2001, e-print hep-ph/0105300.
Kolda, C., and D. H. Lyth, 1999, Phys. Lett. B **458**, 197.
Komatsu, E., B. D. Wandelt, D. N. Spergel, A. J. Banday, and K. M. Górski, 2002, Astrophys. J. **566**, 19.
Kosowsky, A., 2002, in *Modern Cosmology*, edited by S. Bonometto, V. Gorini, and U. Moschella (IOP, Bristol), p. 219.
Kragh, H., 1996, *Cosmology and Controversy* (Princeton University, Princeton, NJ).
Kragh, H., 1999, in *The Expanding Worlds of General Relativity*, edited by H. Goenner, J. Renn, J. Ritter, and T. Sauer (Birkhäuser, Boston), p. 377.
Krauss, L. M., and B. Chaboyer, 2001, e-print astro-ph/0111597.
Kruger, A. T., and J. W. Norbury, 2000, Phys. Rev. D **61**, 087303.
Kujat, J., A. M. Linn, R. J. Scherrer, and D. H. Weinberg, 2002, Astrophys. J. **572**, 1.
Kyae, B., and Q. Shafi, 2002, Phys. Lett. B **526**, 379.

Lahav, O., P. B. Lilje, J. R. Primack, and M. J. Rees, 1991, Mon. Not. R. Astron. Soc. **251**, 128.
Lanczos, K., 1922, Phys. Z. **23**, 539.
Landau, L. D., and E. M. Lifshitz, 1951, *The Classical Theory of Fields* (Pergamon, Oxford).
Landy, S. D., 2002, Astrophys. J. Lett. **567**, L1.
Larsen, F., J. P. van der Schaar, and R. G. Leigh, 2002, J. High Energy Phys. **0204**, 047.
Lau, Y.-K., 1985, Aust. J. Phys. **38**, 547.
Lazarides, G., 2002, e-print hep-ph/0204294.
Lee, A. T., *et al.*, 2001, Astrophys. J. Lett. **561**, L1.
Leibundgut, B., 2001, Annu. Rev. Astron. Astrophys. **39**, 67.
Lemaître, G., 1925, J. Math. Phys. (Cambridge, Mass.) **4**, 188.
Lemaître, G., 1927, Ann. Soc. Sci. Bruxelles, Ser. 1 **47**, 49 [Mon. Not. R. Astron. Soc. **91**, 483 (1931)].
Lemaître, G., 1934, Proc. Natl. Acad. Sci. U.S.A. **20**, 12.
Lemaître, G., 1949, in *Albert Einstein: Philosopher-Scientist*, edited by P. A. Schilpp (Library of Living Philosophers, Evanston), p. 437.
Liddle, A. R., and R. J. Scherrer, 1999, Phys. Rev. D **59**, 023509.
Lightman, A. P., and P. L. Schechter, 1990, Astrophys. J., Suppl. Ser. **74**, 831.
Lima, J. A. S., and J. S. Alcaniz, 2001, Braz. J. Phys. **31**, 583.
Lima, J. A. S., and J. S. Alcaniz, 2002, Astrophys. J. **566**, 15.
Linde, A. D., 1974, Pis'ma Zh. Eksp. Teor. Fiz. **19**, 320 [JETP Lett. **19**, 183 (1974)].
Linde, A. D., 1982, Phys. Lett. **108B**, 389.
Lineweaver, C., 2001, e-print astro-ph/0112381.
Loh, E. D., and E. J. Spillar, 1986, Astrophys. J. Lett. Ed. **307**, L1.
Lucchin, S., and S. Matarrese, 1985a, Phys. Rev. D **32**, 1316.
Lucchin, S., and S. Matarrese, 1985b, Phys. Lett. **164B**, 282.
Lynden-Bell, D., 1969, Nature (London) **223**, 690.
Lyth, D. H., and E. D. Stewart, 1990, Phys. Lett. B **252**, 336.
Magueijo, J., and L. Smolin, 2002, Phys. Rev. Lett. **88**, 190403.
Majumdar, A. S., 2001, Phys. Rev. D **64**, 083503.
Mak, M. K., J. A. Belinchón, and T. Harko, 2002, Int. J. Mod. Phys. D **11**, 1265.
Maldacena, J., and C. Nuñez, 2001, Int. J. Mod. Phys. A **16**, 822.
Maor, I., R. Brustein, J. McMahon, and P. J. Steinhardt, 2002, Phys. Rev. D **65**, 123003.
Marriage, T. A., 2002, e-print astro-ph/0203153.
Masiero, A., M. Pietroni, and F. Rosati, 2000, Phys. Rev. D **61**, 023504.
Mason, B. S., *et al.*, 2002, e-print astro-ph/0205384.
Mathis, H., and S. D. M. White, 2002, Mon. Not. R. Astron. Soc. **337**, 1193.
Matyjasek, J., 1995, Phys. Rev. D **51**, 4154.
McCrea, W. H., 1951, Proc. R. Soc. London, Ser. A **206**, 562.
McCrea, W. H., 1971, Q. J. R. Astron. Soc. **12**, 140.
McVittie, G. C., 1956, *General Relativity and Cosmology* (Chapman and Hall, London).
Medved, A. J. M., 2002, Class. Quantum Grav. **19**, 4511.
Mészáros, P., 1974, Astron. Astrophys. **37**, 225.
Milgrom, M., 1983, Astrophys. J. **270**, 365.
Miller, A. D., *et al.*, 2002a, Astrophys. J., Suppl. Ser. **140**, 115.
Miller, C. J., R. C. Nichol, C. Genovese, and L. Wasserman, 2002b, Astrophys. J. Lett. **565**, L67.
Miralda-Escudé, J., 2002, Astrophys. J. **564**, 60.
Misner, C. W., 1969, Phys. Rev. Lett. **22**, 1071.
Misner, C. W., and P. Putnam, 1959, Phys. Rev. **116**, 1045.
Mizuno, S., and K.-i. Maeda, 2001, Phys. Rev. D **64**, 123521.
Moffat, J. W., 2001, e-print hep-th/0105017.
Molaro, P., S. A. Levshakov, M. Dessauges-Zavadsky, and S. D'Odorico, 2002, Astron. Astrophys. **381**, L64.
Moore, B., 1994, Nature (London) **370**, 629.
Moore, B., S. Ghigna, F. Governato, G. Lake, T. Quinn, J. Stadel, and P. Tozzi, 1999a, Astrophys. J. Lett. **524**, L19.
Moore, B., T. Quinn, F. Governato, J. Stadel, and G. Lake, 1999b, Mon. Not. R. Astron. Soc. **310**, 1147.
Mukherjee, P., B. Dennison, B. Ratra, J. H. Simonetti, K. Ganga, and J.-Ch. Hamilton, 2002, Astrophys. J. **579**, 83.
Mukherjee, P., M. P. Hobson, and A. N. Lasenby, 2000, Mon. Not. R. Astron. Soc. **318**, 1157.
Munshi, D., and Y. Wang, 2003, Astrophys. J. **583**, 566.
Myung, Y. S., 2001, Mod. Phys. Lett. A **16**, 1963.
Narayanan, V. K., D. N. Spergel, R. Davé, and C.-P. Ma, 2000, Astrophys. J. Lett. **543**, L103.
Nernst, W., 1916, Verh. Dtsch. Phys. Ges. **18**, 83.
Netterfield, C. B., *et al.*, 2002, Astrophys. J. **571**, 604.
Newman, J. A., and M. Davis, 2000, Astrophys. J. Lett. **534**, L11.
Ng, S. C. C., N. J. Nunes, and F. Rosati, 2001, Phys. Rev. D **64**, 083510.
Ng, S. C. C., and D. L. Wiltshire, 2001, Phys. Rev. D **63**, 023503.
Nilles, H. P., 1985, in *New Trends in Particle Theory*, edited by L. Lusanna (World Scientific, Singapore), p. 119.
Nolan, L. A., J. S. Dunlop, R. Jimenez, and A. F. Heavens, 2001, e-print astro-ph/0103450.
Nomura, Y., T. Watari, and T. Yanagida, 2000, Phys. Lett. B **484**, 103.
North, J. D., 1965, *The Measure of the Universe* (Oxford University, Oxford). Reprinted (Dover, New York, 1990).
Norton, J. D., 1999, in *The Expanding Worlds of General Relativity*, edited by H. Goenner, J. Renn, J. Ritter, and T. Sauer (Birkhäuser, Boston), p. 271.
Olson, T. S., and T. F. Jordan, 1987, Phys. Rev. D **35**, 3258.
Oort, J. H., 1958, *Proceedings of the 11th Solvay Conference, Structure and Evolution of the Universe* (Stoops, Brussels), p. 163.
Ott, T., 2001, Phys. Rev. D **64**, 023518.
Oukbir, J., and A. Blanchard, 1992, Astron. Astrophys. **262**, L21.
Overduin, J. M., and F. I. Cooperstock, 1998, Phys. Rev. D **58**, 043506.
Overduin, J. M., P. S. Wesson, and S. Bowyer, 1993, Astrophys. J. **404**, 1.
Ozer, M., and M. O. Taha, 1986, Phys. Lett. B **171**, 363.
Padilla, N. D., M. E. Merchán, C. A. Valotto, D. F. Lambas, and M. A. G. Maia, 2001, Astrophys. J. **554**, 873.
Padmanabhan, T., 2002, e-print hep-th/0212290.
Pais, A., 1982, *Subtle is the Lord...* (Oxford University, New York).
Papovich, C., M. Dickinson, and H. C. Ferguson, 2002, e-print astro-ph/0201221.
Park, C.-G., C. Park, B. Ratra, and M. Tegmark, 2001, Astrophys. J. **556**, 582.
Pauli, W., 1958, *Theory of Relativity* (Pergamon, New York). Reprinted (Dover, New York, 1981).
Pauli, W., 1980, *General Principles of Quantum Mechanics* (Springer, Berlin).
Peacock, J. A., *et al.*, 2001, Nature (London) **410**, 169.
Peebles, P. J. E., 1965, Astrophys. J. **142**, 1317.

Peebles, P. J. E., 1966, Astrophys. J. **146**, 542.
Peebles, P. J. E., 1971, *Physical Cosmology* (Princeton University, Princeton, NJ).
Peebles, P. J. E., 1980a, *The Large-Scale Structure of the Universe* (Princeton University, Princeton, NJ).
Peebles, P. J. E., 1980b, Ann. N.Y. Acad. Sci. **336**, 167.
Peebles, P. J. E., 1982, Astrophys. J. Lett. Ed. **263**, L1.
Peebles, P. J. E., 1984, Astrophys. J. **284**, 439.
Peebles, P. J. E., 1986, Nature (London) **321**, 27.
Peebles, P. J. E., 1987, Astrophys. J. Lett. Ed. **315**, L73.
Peebles, P. J. E., 1989a, in *Large Scale Structure and Motions in the Universe*, edited by M. Mezzetti, G. Giuricin, and F. Mardirossian (Kluwer, Dordrecht), p. 119.
Peebles, P. J. E., 1989b, J. R. Astron. Soc. Can. **83**, 363.
Peebles, P. J. E., 1993, *Principles of Physical Cosmology* (Princeton University, Princeton).
Peebles, P. J. E., 2001, Astrophys. J. **557**, 495.
Peebles, P. J. E., 2002, e-print astro-ph/0201015.
Peebles, P. J. E., R. A. Daly, and R. Juszkiewicz, 1989, Astrophys. J. **347**, 563.
Peebles, P. J. E., S. D. Phelps, E. J. Shaya, and R. B. Tully, 2001, Astrophys. J. **554**, 104.
Peebles, P. J. E., and B. Ratra, 1988, Astrophys. J. Lett. Ed. **325**, L17.
Peebles, P. J. E., S. Seager, and W. Hu, 2000, Astrophys. J. Lett. **539**, L1.
Peebles, P. J. E., and J. Silk, 1990, Nature (London) **346**, 233.
Peebles, P. J. E., and A. Vilenkin, 1999, Phys. Rev. D **59**, 063505.
Peebles, P. J. E., and J. T. Yu, 1970, Astrophys. J. **162**, 815.
Percival, W. J., *et al.*, 2001, Mon. Not. R. Astron. Soc. **327**, 1297.
Percival, W. J., *et al.*, 2002, Mon. Not. R. Astron. Soc. **337**, 1068.
Perlmutter, S., *et al.*, 1999a, Astrophys. J. **517**, 565.
Perlmutter, S., M. S. Turner, and M. White, 1999b, Phys. Rev. Lett. **83**, 670.
Perrotta, F., C. Baccigalupi, and S. Matarrese, 2000, Phys. Rev. D **61**, 023507.
Peskin, M. E., 1997, in *Fields, Strings, and Duality*, edited by C. Efthimiou and B. Greene (World Scientific, Singapore), p. 729.
Petrosian, V., E. Salpeter, and P. Szekeres, 1967, Astrophys. J. **147**, 1222.
Phillipps, S., S. P. Driver, W. J. Couch, A. Fernandez-Soto, P. D. Bristow, S. C. Odewahn, R. A. Windhorst, and K. Lanzetta, 2000, Mon. Not. R. Astron. Soc. **319**, 807.
Phillips, M. M., 1993, Astrophys. J. Lett. **413**, L105.
Phillips, N. G., and A. Kogut, 2001, Astrophys. J. **548**, 540.
Pierpaoli, E., D. Scott, and M. White, 2001, Mon. Not. R. Astron. Soc. **325**, 77.
Plionis, M., 2002, e-print astro-ph/0205166.
Podariu, S., R. A. Daly, M. P. Mory, and B. Ratra, 2003, Astrophys. J. **584**, 577.
Podariu, S., P. Nugent, and B. Ratra, 2001, Astrophys. J. **553**, 39.
Podariu, S., and B. Ratra, 2000, Astrophys. J. **532**, 109.
Podariu, S., and B. Ratra, 2001, Astrophys. J. **563**, 28.
Podariu, S., T. Souradeep, J. R. Gott, B. Ratra, and M. S. Vogeley, 2001, Astrophys. J. **559**, 9.
Polenta, G., *et al.*, 2002, Astrophys. J. Lett. **572**, L27.
Pollock, M. D., 1980, Mon. Not. R. Astron. Soc. **193**, 825.
Premadi, P., H. Martel, R. Matzner, and T. Futamase, 2001, Astrophys. J., Suppl. Ser. **135**, 7.
Primack, J., 2002, e-print astro-ph/0205391.
Pryke, C., N. W. Halverson, E. M. Leitch, J. Kovac, J. E. Carlstrom, W. L. Holzapfel, and M. Dragovan, 2002, Astrophys. J. **568**, 46.
Pskovskii, Yu. P., 1977, Astron. Zh. **54**, 1188 [Sov. Astron. **21**, 675 (1977)].
Quevedo, F., 1996, in *Workshops of Particles and Fields and Phenomenology of Fundamental Interactions*, edited by J. C. D'Olivio, A. Fernandez, and M. A. Perez, AIP Conf. Proc. No. 359 (AIP, Woodbury, NY), p. 202.
Ratra, B., 1985, Phys. Rev. D **31**, 1931.
Ratra, B., 1989, Phys. Rev. D **40**, 3939.
Ratra, B., 1991, Phys. Rev. D **43**, 3802.
Ratra, B., 1992a, Phys. Rev. D **45**, 1913.
Ratra, B., 1992b, Astrophys. J. Lett. **391**, L1.
Ratra, B., 1994, Phys. Rev. D **50**, 5252.
Ratra, B., and P. J. E. Peebles, 1988, Phys. Rev. D **37**, 3406.
Ratra, B., and P. J. E. Peebles, 1994, Astrophys. J. Lett. **432**, L5.
Ratra, B., and P. J. E. Peebles, 1995, Phys. Rev. D **52**, 1837.
Ratra, B., and A. Quillen, 1992, Mon. Not. R. Astron. Soc. **259**, 738.
Ratra, B., R. Stompor, K. Ganga, G. Rocha, N. Sugiyama, and K. M. Górski, 1999, Astrophys. J. **517**, 549.
Ratra, B., N. Sugiyama, A. J. Banday, and K. M. Górski, 1997, Astrophys. J. **481**, 22.
Refregier, A., J. Rhodes, and E. J. Groth, 2002, Astrophys. J. Lett. **572**, L131.
Refsdal, S., 1970, Astrophys. J. **159**, 357.
Riess, A. G., *et al.*, 1998, Astrophys. J. **116**, 1009.
Rindler, W., 1956, Mon. Not. R. Astron. Soc. **116**, 662.
Robertson, H. P., 1928, Philos. Mag. **5**, 835.
Robertson, H. P., 1955, Publ. Astron. Soc. Pac. **67**, 82.
Rossi, G. C., and G. Veneziano, 1984, Phys. Lett. **138B**, 195.
Roussel, H., R. Sadat, and A. Blanchard, 2000, Astron. Astrophys. **361**, 429.
Rubakov, V. A., M. V. Sazhin, and A. V. Veryaskin, 1982, Phys. Lett. **115B**, 189.
Rubano, C., and P. Scudellaro, 2002, Gen. Relativ. Gravit. **34**, 307.
Rugh, S. E., and H. Zinkernagel, 2002, Stud. Hist. Philos. Mod. Phys. **33**, 663.
Sachs, R. K., and A. M. Wolfe, 1967, Astrophys. J. **147**, 73.
Sahni, V., M. Sami, and T. Souradeep, 2002, Phys. Rev. D **65**, 023518.
Sahni, V., and A. Starobinsky, 2000, Int. J. Mod. Phys. D **9**, 373.
Sahni, V., and L. Wang, 2000, Phys. Rev. D **62**, 103517.
Sandage, A., 1958, Astrophys. J. **127**, 513.
Sandage, A., 1961a, Astrophys. J. **133**, 355.
Sandage, A., 1961b, *The Hubble Atlas of Galaxies* (Carnegie Institution, Washington).
Sandage, A., 1962, in *Problems of Extragalactic Research*, edited by G. C. McVittie (McMillan, New York), p. 359.
Sandage, A., 1988, Annu. Rev. Astron. Astrophys. **26**, 561.
Sarkar, S., 2002, e-print hep-ph/0201140.
Sato, K., 1981a, Mon. Not. R. Astron. Soc. **195**, 467.
Sato, K., 1981b, Phys. Lett. **99B**, 66.
Sato, K., N. Terasawa, and J. Yokoyama, 1989, in *The Quest for the Fundamental Constants in Cosmology*, edited by J. Audouze and J. Tran Thanh Van (Editions Frontières, Gif-sur-Yvette), p. 193.

Schindler, S., 2001, e-print astro-ph/0107028.
Sciama, D. W., 2001, Astrophys. Space Sci. **276**, 151.
Scoccimarro, R., H. A. Feldman, J. N. Fry, and J. A. Frieman, 2001, Astrophys. J. **546**, 652.
Scott, P. F., *et al.*, 2002, e-print astro-ph/0205380.
Seljak, U., 2002, Mon. Not. R. Astron. Soc. **337**, 769.
Sellwood, J. A., and A. Kosowsky, 2001, in *Gas and Galaxy Evolution*, Astronomical Society of the Pacific No. 240, edited by J. E. Hibbard, M. Rupen, and J. H. van Gorkum (Astronomical Society of the Pacific, San Francisco), p. 311.
Sen, A. A., and S. Sethi, 2002, Phys. Lett. B **532**, 159.
Shafi, Q., and C. Wetterich, 1985, Phys. Lett. **152B**, 51.
Shandarin, S. F., H. A. Feldman, Y. Xu, and M. Tegmark, 2002, Astrophys. J., Suppl. Ser. **141**, 1.
Shiu, G., and S.-H. H. Tye, 2001, Phys. Lett. B **516**, 421.
Shklovsky, J., 1967, Astrophys. J. Lett. **150**, L1.
Shvartsman, V. F., 1969, Pis'ma Zh. Eksp. Teor. Fiz. **9**, 315 [JETP Lett. **9**, 184 (1969)].
Sigad, Y., A. Eldar, A. Dekel, M. A. Strauss, and A. Yahil, 1998, Astrophys. J. **495**, 516.
Silk, J., 1967, Nature (London) **215**, 1155.
Silk, J., 1968, Astrophys. J. **151**, 459.
Silk, J., and N. Vittorio, 1987, Astrophys. J. **317**, 564.
Singh, T. P., and T. Padmanabhan, 1988, Int. J. Mod. Phys. A **3**, 1593.
Skordis, C., and A. Albrecht, 2002, Phys. Rev. D **66**, 043523.
Smith, S., 1936, Astrophys. J. **83**, 23.
Smoot, G. F., *et al.*, 1992, Astrophys. J. Lett. **396**, L1.
Sommer-Larsen, J., and A. Dolgov, 2001, Astrophys. J. **551**, 608.
Souradeep, T., and B. Ratra, 2001, Astrophys. J. **560**, 28.
Spergel, D., and U.-L. Pen, 1997, Astrophys. J. Lett. **491**, L67.
Spergel, D. N., and P. J. Steinhardt, 2000, Phys. Rev. Lett. **84**, 3760.
Spergel, D. N., *et al.*, 2003, e-print astro-ph/0302209.
Spokoiny, B., 1993, Phys. Lett. B **315**, 40.
Starobinsky, A. A., 1982, Phys. Lett. **117B**, 175.
Starobinsky, A. A., 1998, Gravitation Cosmol. **4**, 88.
Steigman, G., 2002, private communication.
Steigman, G., D. N. Schramm, and J. E. Gunn, 1977, Phys. Lett. **66B**, 202.
Steinhardt, P. J., and N. Turok, 2002, Science **296**, 1436.
Steinhardt, P. J., L. Wang, and I. Zlatev, 1999, Phys. Rev. D **59**, 123504.
Stoehr, F., S. D. M. White, G. Tormen, and V. Springel, 2002, Mon. Not. R. Astron. Soc. **335**, L84.
Stompor, R., *et al.*, 2001, Astrophys. J. Lett. **561**, L7.
Straumann, N., 2002, e-print astro-ph/0203330.
Sugiyama, N., and N. Gouda, 1992, Prog. Theor. Phys. **88**, 803.
Sunyaev, R. A., and Ya. B. Zel'dovich, 1970, Astrophys. Space Sci. **7**, 3.
Susskind, L., 1979, Phys. Rev. D **20**, 2619.
Sutherland, W., H. Tadros, G. Efstathiou, C. S. Frenk, O. Keeble, S. Maddox, R. G. McMahon, S. Oliver, M. Rowan-Robinson, and S. D. M. White, 1999, Mon. Not. R. Astron. Soc. **308**, 289.
Tadros, H., *et al.*, 1999, Mon. Not. R. Astron. Soc. **305**, 527.
Tammann, G. A., B. Reindl, F. Thim, A. Saha, and A. Sandage, 2001, in *A New Era in Cosmology*, Astronomical Society of the Pacific Conference Proceedings, edited by T. Shanks and N. Metcalf (Astronomical Society of the Pacific, San Francisco), in press.
Tasitsiomi, A., 2002, e-print astro-ph/0205464.
Thomas, D., and G. Kauffmann, 1999, in *Spectrophotometric Dating of Stars and Galaxies*, Astronomical Society of the Pacific Conference Proceedings No. 192, edited by I. Huberg, S. Heap, and R. Cornett (Astronomical Society of the Pacific, San Francisco), p. 261.
Thuan, T. X., and Y. I. Izotov, 2002, in *Matter in the Universe*, edited by P. Jetzer, K. Pretzl, and R. Von Steiger (Kluwer, Dordrecht), in press.
Tinsley, B. M., 1972, Astrophys. J. **178**, 319.
Totani, T., Y. Yoshii, T. Maihara, F. Iwamuro, and K. Motohara, 2001, Astrophys. J. **559**, 592.
Townsend, P. K., 2001, J. High Energy Phys. **0111**, 042.
Trager, S. C., S. M. Faber, G. Worthey, and J. J. González, 2000, Astron. J. **119**, 1645.
Trautman, A., 1965, in *Lectures on General Relativity*, edited by A. Trautman, F. A. E. Pirani, and H. Bondi (Prentice-Hall, Englewood Cliffs, NJ), p. 230.
Tully, R. B., R. S. Somerville, N. Trentham, and M. A. W. Verheijen, 2002, Astrophys. J. **569**, 573.
Turner, E. L., 1990, Astrophys. J. Lett. **365**, L43.
Turner, M. S., 1999, in *The Galactic Halo*, Astronomical Society of the Pacific Conference Proceedings No. 165, edited by B. K. Gibson, T. S. Axelrod, and M. E. Putnam (Astronomical Society of the Pacific, San Francisco), p. 431.
Turner, M. S., and M. White, 1997, Phys. Rev. D **56**, R4439.
Turner, M. S., and L. Widrow, 1988, Phys. Rev. D **37**, 2743.
Unruh, W. G., 1989, Phys. Rev. D **40**, 1048.
Ureña-López, L. A., and T. Matos, 2000, Phys. Rev. D **62**, 081302.
Uson, J. M., and D. T. Wilkinson, 1984, Nature (London) **312**, 427.
Uzan, J.-P., 1999, Phys. Rev. D **59**, 123510.
Uzan, J.-P., 2003, Rev. Mod. Phys. **75**, 403.
Uzan, J.-P., and F. Bernardeau, 2001, Phys. Rev. D **64**, 083004.
Uzawa, K., and J. Soda, 2001, Mod. Phys. Lett. A **16**, 1089.
Van Waerbeke, L., Y. Mellier, R. Pelló, U.-L. Pen, H. J. McCracken, and B. Jain, 2002, Astron. Astrophys. **393**, 369.
Veltman, M., 1975, Phys. Rev. Lett. **34**, 777.
Verde, L., *et al.*, 2002, Mon. Not. R. Astron. Soc. **335**, 432.
Viana, P. T. P., R. C. Nichol, and A. R. Liddle, 2002, Astrophys. J. Lett. **569**, L75.
Vilenkin, A., 1984, Phys. Rev. Lett. **53**, 1016.
Vilenkin, A., 2001, e-print astro-ph/0106083.
Vilenkin, A., and E. P. S. Shellard, 1994, *Cosmic Strings and Other Topological Defects* (Cambridge University, Cambridge, England).
Vishwakarma, R. G., 2001, Class. Quantum Grav. **18**, 1159.
Vittorio, N., and J. Silk, 1985, Astrophys. J. Lett. Ed. **297**, L1.
Waga, I., and J. A. Frieman, 2000, Phys. Rev. D **62**, 043521.
Wang, L., and P. J. Steinhardt, 1998, Astrophys. J. **508**, 483.
Wang, X., M. Tegmark, and M. Zaldarriaga, 2002, Phys. Rev. D **65**, 123001.
Wang, Y., and G. Lovelace, 2001, Astrophys. J. Lett. **562**, L115.
Wasserman, I., 2002, Phys. Rev. D **66**, 123511.
Weinberg, S., 1987, Phys. Rev. Lett. **59**, 2607.
Weinberg, S., 1989, Rev. Mod. Phys. **61**, 1.
Weinberg, S., 2001, in *Relativistic Astrophysics*, edited by J. C. Wheeler and H. Martel, AIP Conf. Proc. No. 586 (AIP, Melville, NY), p. 893.
Weiss, N., 1987, Phys. Lett. B **197**, 42.
Weller, J., and A. Albrecht, 2002, Phys. Rev. D **65**, 103512.
Wetterich, C., 1988, Nucl. Phys. B **302**, 668.
Weyl, H., 1923, Phys. Z. **24**, 230.

White, M., and C. S. Kochanek, 2001, Astrophys. J. **560**, 539.
White, S. D. M., 1992, in *Clusters and Superclusters of Galaxies*, edited by A. C. Fabian (Kluwer, Dordrecht), p. 17.
White, S. D. M., J. F. Navarro, A. E. Evrard, and C. S. Frenk, 1993, Nature (London) **366**, 429.
Whittaker, E. T., 1935, Proc. R. Soc. London, Ser. A **149**, 384.
Wilczek, F., 1985, in *How Far Are We from the Gauge Forces*, edited by A. Zichichi (Plenum, New York), p. 157.
Wilkinson, D. T., and P. J. E. Peebles, 1990, in *The Cosmic Microwave Background: 25 Years Later*, edited by N. Mandolesi and N. Vittorio (Kluwer, Dordrecht), p. 17.
Wilson, G., N. Kaiser, and G. A. Luppino, 2001, Astrophys. J. **556**, 601.
Witten, E., 2001, in *Sources and Detection of Dark Matter and Dark Energy in the Universe*, edited by D. B. Cline (Springer, Berlin), p. 27.
Worthey, G., 1994, Astrophys. J., Suppl. Ser. **95**, 107.
Wu, J.-H. P., *et al.*, 2001a, Astrophys. J., Suppl. Ser. **132**, 1.
Wu, J.-H. P., *et al.*, 2001b, Phys. Rev. Lett. **87**, 251303.
Wu, X.-P., and F. Hammer, 1993, Mon. Not. R. Astron. Soc. **262**, 187.
Wyithe, J. S. B., and A. Loeb, 2002, Astrophys. J. **581**, 886.
Yahiro, M., G. J. Mathews, K. Ichiki, T. Kajino, and M. Orito, 2002, Phys. Rev. D **65**, 063502.
Yamamoto, K., M. Sasaki, and T. Tanaka, 1995, Astrophys. J. **455**, 412.
Zaldarriaga, M., D. N. Spergel, and U. Seljak, 1997, Astrophys. J. **288**, 1.
Zee, A., 1980, Phys. Rev. Lett. **44**, 703.
Zee, A., 1985, in *High Energy Physics*, edited by S. L. Mintz and A. Perlmutter (Plenum, New York), p. 211.
Zel'dovich, Ya. B., 1964, Astron. Zh. **41**, 19 [Sov. Astron. **8**, 13 (1964)].
Zel'dovich, Ya. B., 1967, Zh. Eksp. Teor. Fiz., Pis'ma Red. **6**, 883 [JETP Lett. **6**, 316 (1967)].
Zel'dovich, Ya. B., 1968, Usp. Fiz. Nauk **95**, 209 [Sov. Phys. Usp. **11**, 381 (1968)].
Zel'dovich, Ya. B., 1972, Mon. Not. R. Astron. Soc. **160**, 1P.
Zel'dovich, Ya. B., 1978, in *IAU Symposium 79, The Large-Scale Structure of the Universe*, edited by M. S. Longair and J. Einasto (Reidel, Dordrecht), p. 409.
Zel'dovich, Ya. B., 1981, Usp. Fiz. Nauk **133**, 479 [Sov. Phys. Usp. **24**, 216 (1981)].
Zel'dovich, Ya. B., I. Yu. Kobzarev, and L. B. Okun, 1974, Zh. Eksp. Teor. Fiz. **67**, 3 [Sov. Phys. JETP **40**, 1 (1975)].
Zimdahl, W., D. J. Schwarz, A. B. Balakin, and D. Pavón, 2001, Phys. Rev. D **64**, 063501.
Zumino, B., 1975, Nucl. Phys. B **89**, 535.
Zwicky, F., 1933, Helv. Phys. Acta **26**, 241.

Appendices*

*J. P. Hsu, D. Fine, S. L. Marateck

Marcel Grossmann

1878-1936

Marcel Grossmann studied mathematics at the Zurich Polytechnikum [later named the Eidgenössische Technische Hochschule, (ETH)] and was a staunch and helpful friend of Einstein. In 1909 Einstein dedicated his doctoral thesis to Grossmann. The collaboration of Einstein and Grossmann lead to a ground-breaking paper: "Outline of a Generalized Theory of Relativity and of a Theory of Gravitation," which was published in 1913 and was one of the two fundamental papers which established Einstein's theory of gravity. They also collaborated a second paper in 1914. Einstein had already moved to Berlin when this paper came out. Hoffmann commented: "In retrospect it is heartbreaking to see how close the collaborators came to achieving their goal. Practically all the needed mathematical ingredients were there, and, as Einstein remarked later, he and Grossmann had considered the actual field equations only to discard them for what at the time seemed compelling reasons…." [1]

Grossmann was a member of an old Swiss family from the Zurich area. His father managed a textile factory. Grossman was known for both his quick grasp of new subjects and his depth of understanding. Speaking of their time together at the Zurich Polytechnikum, Einstein described him as an excellent student of mathematics whose lecture notes of his classes were so complete and well-organized they could have been published. Grossmann appreciated the character of his friend Einstein, who concentrated on his own reading (of books by Helmholtz, Boltzmann, Maxwell and Hertz) and joyful exploration of the physical world. Grossmann told his father that Einstein had a great potential. The two met regularly, discussing their studies and much more; Einstein often visited Grossmann's home at Thalwil to play music. Since Einstein did not attend class regularly and did not pay particularly close attention to some of his mathematics classes at that time, Grossmann helped Einstein to pass the examinations by lending him the lecture notes and helping him on problems.

Marcel Grossmann told his father about Einstein's difficulties in finding a steady job more than a year after graduation in 1900. His father recommended Einstein strongly to a friend, the Director of the Swiss Patent office. As a result, Einstein found a haven in which to concentrate on physical problems, as well as the means to marry his friend and study partner Mileva Maric. Einstein considered this help to be "the greatest thing Marcel Grossmann did for me as a friend."

In 1900 Grossmann graduated from Zurich Polytechnikum and became an assistant to the geometer W. Fiedler. He continued to do research on non-Euclidean geometry and taught in high schools for the next seven years. In 1902, he earned his doctorate from the University of Zurich with the thesis "On the Metrical Properties of Collinear Structures." He also published two geometry books for high school students and three papers on non-Euclidean geometry. At a remarkably young age, Grossmann was appointed full professor of descriptive geometry at the Eidgenössische Technische

Hochschule in 1907. His doctoral thesis and his favorite subject were non-Euclidean geometry, which, as luck would have it, paved the way for his celebrated collaboration with Einstein later. Grossmann was a teacher of outstanding ability who trained many mathematicians in geometry. [2]

In the meantime, his friend Einstein was attempting to obtain a position at Berne University, so at the end of 1907 he submitted his 1905 paper on special relativity, "On the Electrodynamics of Moving Bodies," to Berne University as an inaugural thesis. The paper was rejected as being incomprehensible! [3] Einstein was bitterly disappointed and gave up his dream of becoming a university professor for a while. He became interested in a teaching position at the Technical School. Einstein wrote a letter to Grossmann for advice and explained:

"Do not imagine that I am driven to such careerist ways by megalomania or some other questionable passion; rather, I came to this hankering only because of an ardent desire to be able to continue my private scientific work under less unfavorable conditions, as you will certainly understand…."

As a professor of geometry, Grossmann organized summer courses for high school teachers. In 1910 he became one of the founders of the Swiss Mathematical Society. Within a year, he became Dean of the mathematics-physics section of Eidgenössische Technische Hochschule. As a new Dean, he made an effort to persuade Einstein to return to the ETH. As destiny would have it, Einstein agreed in 1912: "I am extraordinarily happy about the prospect of returning to Zurich." Around this time, Einstein sought to formulate mathematically his ideas on the general theory of relativity; he turned to his friend for assistance: "Grossmann, you must help me, or else I'll go crazy!" After discussions with Grossmann, Einstein was sure that he was on the right path. Grossmann introduced Einstein to the absolute differential calculus, started by E. B. Christoffel (1864) and fully developed by Ricci and Levi Civita (1901). Grossmann facilitated Einstein's unique synthesis of mathematical and theoretical physics in what is still today considered the most elegant and powerful theory of gravity: The General Theory of Relativity. [4]

Pais has an interesting, lucid and detailed analysis of the Einstein-Grossmann paper (I. Physical Part, by A. Einstein; II. Mathematical Part, by M. Grossmann):[5]

"Grossmann's concluding section starts as follows. 'The problem of the formulation of the differential equations of a gravitation field draws attention to the differential invariants.... and .... covariants of .... $ds^2 = g_{\mu\nu}dx^\mu dx^\nu$.' He then presents to Einstein the major tensor of the future theory: the 'Christoffel four-index symbol,' now better known as the Riemannian-Christoffel (curvature) tensor: $R^\lambda{}_{\mu\nu\kappa}=$....." From this tensor 'it is .... possible to derive a second-rank tensor of the second order [in the derivatives of $g_{\mu\nu}$],' the Ricci tensor: $R_{\mu\nu} = R^\lambda{}_{\mu\lambda\nu}$...."

Unfortunately, Grossmann reached an erroneous conclusion: 'It turns out, however, that in the special case of the infinitely weak, static gravitational field this tensor does not reduce to the expression $\Delta\varphi$.' Alas, the symmetry property of general covariance and the freedom of choosing a 'coordinate condition' were not properly understood by the collaborators.

A crucial change of destiny for their collaboration occurred in 1913 when Planck and Nernst came to Zurich to persuade Einstein to go to Berlin. As a result Einstein left Zurich in March 1914. During the next year, the endeavor for Einstein in Berlin was to really understand his own idea of general covariance and the key role the Ricci tensor played in his theory.

A page from the draft of Einstein's landmark paper "The Foundation of the General Theory of Relativity" was included in his collected works.[6] Einstein wrote: "...... Finally, I want to acknowledge gratefully my friend, the mathematician Grossmann, whose help not only saved me the effort of studying the pertinent mathematical literature, but who also helped me in my search for the field equations of gravitation." For some reason, Einstein did not publish this page together with his landmark paper.

Grossmann died of multiple sclerosis in 1936. Einstein wrote a letter to Grossmann's wife to convey his heartfelt appreciation of Grossmann's kindness:

"*.... Our student days together come back to me. He is a model student; I untidy and a dreamer. He on excellent terms with the teachers and grasping everything easily; I aloof and discontented, not very popular. But we were good friend and our conversations over iced coffee at the Metropol every few weeks belong among my nicest memories. Then the end of the studies.... I suddenly abandoned by everyone, facing life not knowing which way to turn. But he stood by me and through him and his father I came to Haller in the Patent Office a few years later, In a way, this saved my life; not that I would have died without it, but I would have been intellectually stunted.*"

But Einstein did not write an obituary shortly after Grossmann's death. Pais commented that "I have a sense of regret that Einstein did not do something for which he had often demonstrated a talent and sensitivity."

In the last year of Einstein's life (1955), he wrote of Grossmann, of their collaboration, and how the latter had 'checked through the literature and soon discovered that the mathematical problem had already been solved by Riemann, Ricci, and Levi-Civita.' Einstein also wrote: "The need to express at least once in my life my gratitude to Marcel Grossmann gave me the courage to write this ... autobiographical sketch." (7,8)

In 1975, the International Center for Theoretical Physics, Trieste, Italy announced the "Marcel Grossmann Meeting on the Recent Progress of the Fundamentals of General Relativity" as follows:

"Marcel Grossmann was associated with Albert Einstein in elucidating the mathematical basis of general relativity. In commemoration of his contributions, a Marcel Grossmann meeting on the Recent Progress of the Fundamentals of General Relativity will be held at the International Center for Theoretical Physics in Trieste during the period 7-12 July 1975 under the directorship of professor R. Ruffini (Institute for Advanced study, Princeton, N.J., USA) and cosponsored by the University of Trieste. The theme of the Meeting will cover recent advances in the mathematical techniques of general relativity as well as progress in the physics of relativistic field theories......"

Since this first meeting an International Conference in memory of Marcel Grossmann's contribution to theoretical physics is held regularly.

The life of Marcel Grossmann is a splendid symphony of friendship.

We would like to thank Anna Révay-Grossmann and Carlo Révay for materials regarding their grandfather Marcel Grossmann.

References

1. B. Hoffmann and H. Dukas. *Albert Einstein, Creator and Rebel* (The Viking Press, 1972) p.105.
2. L. Kollros, *Prof. Dr. Marcel Grossmann, 1878-1936* (Extrait des Actes de la Societe Helvetique des Sciences Naturelles, Geneve 1937). pp. 325-329. A. Pais, *Subtle is the Lord*...(Oxford Univ. Press, 1982), Chapter 12.
3. See ref. 1., p. 86. This is understandable: Even a thinker like Mach, whose previous criticism of absolute space and absolute motion played a major role in paving the way for Einstein, was to say harsh things about Einstein's special relativity.
4. R. Ruffini, *Marcel Grossmann,* (unpublished).
5. See paper A in Chapter 2 of this volume. A. Pais, ref. 2.
6. *The Collected Papers of Albert Einstein* (Ed. J. Stachel, Princeton Univ. Press, 1993). English translator Anna Beck. Vol. 5.
7. A. Einstein, in *Helle Zeit, Dunkle Zeit* (C. Seeling, Ed). Europa Verlag, Zurich, 1956. [A.Einstein, *Erinnerungen - Souvenirs* (100 Jahre Eidgenössische Technische Hochschule , 28 Jahrgang, 1955)].
8. A Pais, *Subtle is the Lord*... (Oxford Univ. Press, 1982), p. 225.

J. P. Hsu and D. Fine

Marcel Grossmann [photo courtesy of Anna Révay-Grossmann]

Remembering Robert L. Mills

A few months ago, I was saddened to learn that Robert L. Mills had passed away. As a fellow graduate of Columbia College, I feel a special bond with Mills and am therefore submitting this memorial note.

Mills, who shared with C. N. Yang the 1980 Rumford Premium Prize from the American Academy of Arts and Sciences for "development of a generalized gauge invariant field theory," died on 27 October 1999 from prostate cancer. His passing was a great loss to his family, friends, and the physics community.

Mills was born on 15 April 1927 in Englewood, New Jersey. He graduated from George School in Pennsylvania in early 1944 and, in March, entered Columbia College in New York. While there, he enlisted in the US Merchant Marines in the last year of World War II; he served until 1947.

On leave from the service, he attended classes at Columbia, where his father was an economics professor. In 1948, his senior year, Mills was a winner of the Putnam national college mathematics contest. The mathematical ability he displayed there was evident throughout his career as a theoretical physicist. He then studied at Cambridge University, where he received first-class honors in the mathematical tripos and a master's degree. Mills returned to Columbia and got his PhD in 1955 under Norman Kroll for a thesis on radiative corrections in quantum electrodynamics.

From 1953 to 1955, Mills was a research associate at Brookhaven National Laboratory and shared an office with Yang. During that time, they developed what is now known as the Yang–Mills theory,[1] a non-abelian local gauge invariant theory that would become one of the pivotal concepts of physics. It formed the model for non-abelian gauge theories that followed and is thus one of the bases for the standard model of elementary particles and string theory.[2] Yang–Mills also has applications to mathematics.

Robert L. Mills

From 1955 to 1956, Mills was a member of the Institute for Advanced Study in Princeton, New Jersey. He then joined the physics department of Ohio State University and became a full professor in 1962. He remained at OSU until his retirement in 1995. His research was in quantum field theory, the theory of alloys, and many-body theory. He worked with Andrew Sessler on many-body theory; later, Leon Cooper joined in the effort. That work, besides producing papers that appeared in *Physical Review* and *Physical Review Letters*, resulted in Mills's writing a book, *Propagators for Many-Particle Systems: An Elementary Treatment* (Gordon and Breach, 1969). He later wrote *Space, Time and Quanta: An Introduction to Contemporary Physics* (W. H. Freeman, 1994). For his outstanding dedication to his students, Mills received OSU's Rosalene Sedgwick Faculty Service Award. With his wife, Lee, he shared the OSU International Community Service Award. He was a visiting professor at many schools and a visiting scientist at CERN. After his retirement, he taught for a year as a Fulbright scholar at St. Patrick's College in Ireland.

According to Sessler, "Robert was even-tempered and simply a joy to work with. His coworkers enjoyed interacting with him." A memorial piece in the 2000 *OSU Physics Department Magazine* concludes with this statement: "A gentlemen of unfailing good humor and sincere and active concern for helping others, Robert Mills will be long remembered with great respect and affection."

While preparing this letter, I couldn't help but observe Mills's devotion to his friends and their devotion to him, as exemplified by his interaction with Yang and Cooper, both Nobel laureates. The following comments were obtained via private communication.

> In 1953–1954, I was visiting Brookhaven and Bob was my office mate. We discussed many things in physics, from the experimental results pouring out of the new Cosmotron, to theoretical topics like renormalization and the Ward identity. It was in that year that we found the very elegant and unique generalization of Maxwell's equation. We were pleased by the beauty of the generalization, but neither of us had anticipated its great impact on physics 20 years later.
>
> Bob spent one year, I think it was 1955–1956, at the Institute for Advanced Study in Princeton and we resumed our collaboration. One fruit of that was a paper on the overlapping divergence in the photon propagator which, however, was not written up for publication[3] until 1966, when he and his family visited us for the summer just after I had moved to [SUNY] Stony Brook.
>
> Bob was an old-fashioned man. Among all the physicists that I know, he was certainly one of the most honest and the most sincere.
>
> Bob had a brilliant mind. He was very quick at grasping new ideas. I shall treasure the memory of our intensive collaboration and of our many discussions on diverse topics ranging from accelerator theory to the theory of computability.
>
> —C. N. Yang

> Bob Mills and I, with Andy Sessler, wrote a paper[4] discussing possible superfluidity of helium-3. In it, we suggested that the electron pairing due to a phonon mediated electron–electron interaction could be duplicated among ^{3}He atoms due to the atom–atom potential. Although the solution in ^{3}He turned out to be somewhat more complicated, the basic idea was vindicated with the discovery, about ten years later, of the superfluidity of ^{3}He.
>
> Bob Mills was a talented, creative physicist. We miss him.
>
> —Leon Cooper

I would be remiss in compiling this tribute to Mills if I didn't mention the direct or indirect influence of Yang–Mills on some of the advances establishing the standard model. These advances, tours de force all, illustrate the wonderful synergy of theoretical and experimental physics and include the Glashow-Weinberg-Salam (GWS) theory, the Glashow-Iliopoulos-Maiani (GIM) model, the successful searches for neutral currents and for the gauge particles $W^{\pm}$ and Z^0, the proof of the renormalization of Yang–Mills theories, and quantum chromodynamics encompassing asymptotic freedom and quark confinement. This is a splendid legacy indeed.

Many thanks to Lee Mills for so generously giving of her time to provide me with information on her husband's career.

References

1. C. N. Yang, R. L. Mills, *Phys. Rev.* **96**, 191 (1954).
2. For details on Yang–Mills in the development of gauge theory, see L. O'Raifeartaigh, N. Strauman, *Rev. Mod. Phys.* **72**, 1 (2000); also see L. O'Raifeartaigh, *The Dawning of Gauge Theory*, Princeton U. Press, Princeton, N.J. (1997), for a reprinting of fundamental papers in gauge theory up to 1956, with commentaries.
3. R. L. Mills, C. N. Yang, *Prog. Theor. Phys. Sup.* **37**, 507 (1966).
4. L. N. Cooper, R. L. Mills, A. M. Sessler, *Phys. Rev.* **114**, 1377 (1959).

Samuel L. Marateck
(marateck@cs.nyu.edu)
Courant Institute of Mathematical Sciences
New York University
New York City